英汉 汉英 大气科学词汇

（第二版）

English-Chinese and Chinese-English Dictionary of Atmospheric Sciences

(Second Edition)

周诗健　王存忠　俞卫平　编

China Meteorological Press

内容简介

本书根据全国科学技术名词审定委员会 2009 年公布的《大气科学名词(第三版)》,以及近年来涌现出来的大气科学新词和缩略词等(尤其是气候变化研究方面),对 2007 年第一版进行了订正与增补,使该书词汇扩展到 3 万余条。书中收集的词汇涉及大气科学各分支学科以及相关的基础学科和交叉学科等学科领域。

本书可供大气科学及有关学科的科技工作者、高等院校师生和编译人员使用。

图书在版编目(CIP)数据

英汉汉英大气科学词汇/周诗健,王存忠,俞卫平编.—2版.
—北京:气象出版社,2012.9
ISBN 978-7-5029-5541-0

I. ①英… II. ①英…②王…③俞… III. ①大气科学-名词术语-英、汉 IV. ①P4-61

中国版本图书馆 CIP 数据核字(2012)第 180484 号

英汉汉英大气科学词汇(第二版)

出版发行:气象出版社	
地　　址:北京市海淀区中关村南大街 46 号	邮政编码:100081
网　　址:http://www.cmp.cma.gov.cn	E-mail:qxcbs@cma.gov.cn
总 编 室:010-68407112	发 行 部:010-68409198
责任编辑:李太宇　张锐锐	终　　审:陈云峰
印 刷 者:北京中新伟业印刷有限公司	
开　　本:889 mm×1194 mm　　1/32	印　　张:29
字　　数:950 千字	
版　　次:2012 年 11 月第 2 版	印　　次:2012 年 11 月第 2 次印刷
印　　数:6001～12000	定　　价:96.00 元

本书如存在文字不清、漏印以及缺页、倒页、脱页者,请与本社发行部联系调换。

第二版前言

自《英汉汉英大气科学词汇》第一版2007年面世以来，光阴荏苒，大气科学发展迅猛，尤其是气候变化研究已成为全球瞩目的科学问题，突破了自然科学的范畴，成为涉及国计民生、政治、经济等方方面面的重大问题。为此，值第一版售罄重印之机，我们也相应地推出第二版。

第二版根据全国科学技术名词审定委员会2009年公布的《大气科学名词（第三版）》对原有大气科学名词做了一些校正，对第一版中的差错做了订正，同时又增补了一些近期气象科技书刊中常用的英文词汇和缩略词，尤其是与气候变化、防灾减灾、低碳经济等有关的新词汇和缩略词，希望方便广大气象科技工作者查阅使用。

对于第一版中为了方便读者查阅而增列的八个附录，我们做了较大的修改：撤消了原附录4，代之以最新的关于南海热带气旋的命名方案和命名表；对其他各附录，我们也做了与时俱进的修改与增补。由于在科技书刊中，美国各州州名出现的频数日益增多，而且中译名形形色色，为了方便读者和编译者，我们增补了附录9，给出美国各州的英文名称和缩写以及它们规定的中译名，方便查阅使用和规范化。

在本书第二版编纂过程中，我们要感谢气象出版社给予的大力支持，尤其是具体负责本书运作的李太宇编审和张锐锐编辑的辛勤劳动。同时也感谢国家卫星气象中心方宗义研究员在咨询有关卫星名词定名时的有益讨论，以及中国气象科学研究院朱文琴高级工程师在汉英词汇排版和校对查错过程中的大量繁琐工作。

周诗健　王存忠　俞卫平
2012年10月15日

第一版序

准确使用规范化科学词语是进行科技交流必要的基础。现代科学技术发展日新月异，不断涌现出新的科技词语，及时规范化这些词语，编成词汇，供广大科技工作者统一使用，对推动科技交流和发展具有重要意义。

进入21世纪，大气科学愈来愈成为与社会发展与民众生活息息相关的科学领域，其内容也从传统的气象学发展成为与地理科学、地球科学、生态科学、环境科学、海洋科学、空间科学、天文科学、物理学、数学、力学、化学以及计算技术等紧密交叉的大气科学。全世界每年涉及到大气科学的国内外学术会议和学术刊物的数量十分可观。《英汉汉英大气科学词汇》的出版非常及时，适应了时代发展的需求，她将成为一本大气科学技术工作者必备而有用的工具书。

《英汉汉英大气科学词汇》是在1987年版的《英汉大气科学词汇》的基础上重新扩充编写而成的。书中共收集了3万多条词目，比1987年版增补了1.2万多条，近20年来大气科学技术新词目基本上已收入本书，并且还增加了汉英对照的大气科学词汇以及各种实用备查的附录。

编写本书的作者周诗健、王存忠、俞卫平等都是长期从事大气科学词典和词汇研究的专家，他们经验丰富，工作严谨，为编写本书付出了辛勤而平凡的劳动。

我衷心地感谢他们，也诚挚地祝贺他们，为我国大气科学事业又成功地铺垫了一块坚实的基石。

周秀骥

2007年3月

---

* 周秀骥，中国科学院院士，中国气象科学研究院名誉院长。

第一版前言

近 20 年来,随着大气科学的快速发展,其分支学科研究领域也不断拓展,出现了大量新的专业名词术语。被广泛使用的 1987 年版《英汉大气科学词汇》(气象出版社,1987)显然已不能满足广大气象科技工作者的迫切需要。为此,作为原书出版单位和全国大气科学名词审定委员会挂靠单位的气象出版社,在 2003 年下达了编写新书的任务。经过数年努力,我们编写完成了这本《英汉汉英大气科学词汇》,内容包括英汉对照和汉英对照的大气科学词汇以及各种实用备查的附录等,共收录约 3 万条词汇。

在本书的编写过程中,为了兼顾各学科以及尽可能把近 20 年来大气科学技术新词收入本书,我们以 1987 年版《英汉大气科学词汇》为蓝本,参考我社近年来出版的多种词书(如 1994 年版《中英法俄西国际气象词典》、2000 年版《世界气象组织常用缩略语词典》)、专著、译著和美国气象学会 2000 年出版的《气象学词汇》(Glossary of Meteorology),以及全国科学技术名词审定委员会审定公布的有关气象、农业、海洋、地球物理、地理、地质、天文、物理、数学、化学、电子、计算机等学科的规范术语,对本书收入的名词术语进行了规范化处理。相对于 1987 年版《英汉大气科学词汇》,我们修订和增补了 1.2 万余条与大气科学有关的专业名词术语。

为了读者使用方便,我们除给出译名外,还尽量给出一些注释说明,以便读者了解它的主要内容。对于公式、定律、理论、假说中人名的译法,我们尽量按各种规范给出,但在具体编写过程中,也发现了一些难以统一的地方。如 Boltzmann equation 玻尔兹曼方程【物】,玻耳兹曼方程【数】;Cassegrain telescope 卡塞格林望远镜【天】,Cassegrain reflector antenna 卡塞格伦反射面天线【电子】;Stefan constant 斯特藩常量【物】,Stefan problem 施特藩问题【数】,以及大气科学中常用的斯蒂芬辐射定律;van der Waals equation 范德瓦耳斯方程【物】,van der Waals radius

i

范德华半径【化】，同一人名在不同学科有不同的定名译法，我们则照原样列出，供读者择优使用。

为了满足广大气象科技工作者撰写英文论文和文摘的需要，本书比1987年版增补和加强了汉英大气科学词汇部分。参加本书编写工作的有周诗健(英汉部分A－C,汉英部分A－G,附录3,6,8)、俞卫平(英汉部分F－Q,汉英部分H－R,附录7)、王存忠(英汉部分D,E,R－Z,汉英部分S－Z,附录1,2,4,5)。纪乃晋、章澄昌、黄润恒、朱福康等参与了书稿审读工作，安红霞承担了全书的录入工作，全书最后由周诗健统编和审校。

在编写过程中，我们得到了气象出版社和全国大气科学名词审定委员会的大力支持以及业内许多专家学者的帮助，在此一并致谢。我们特别感谢中国科学院院士、中国气象科学研究院名誉院长周秀骥先生对这项工作的鼓励、支持和指导，并拨冗为本书作序。由于时间紧迫，加之编者水平所限，本书会有遗漏、欠妥甚至错误之处。欢迎读者在使用过程中提出修订、补充意见，以便今后再版时更加完善。

电子邮件地址：wangcz@cma.gov.cn

周诗健 王存忠 俞卫平
2007年3月23日

使 用 说 明

一、本书的英文词汇(不论是英文单词还是复合词),一律按英文字母顺序排列,但括号内的英文字母、词汇间的空白以及符号、希文字母和数字均不参加字序排列。如,需查 ω-equation,可直接查 equation。

二、英文词汇的换用词放在圆括号中并加"或"字;等价词放在圆括号中并加"＝"号。如:analysed chart(或 map);redundancy(＝redundance)。

三、英文缩写词放在圆括号中,如:World Meteorological Organization(WMO)。比较常用的一些英文缩写词按英文字母顺序排入正文中,可以通过英文缩写查得全文。如:WMO(World Meteorological Organization)世界气象组织。有些不常用的,则以缩写词或全文形式排一种。

四、方括号"〔〕"中的字或词表示可以省略。如:surface chart 地面〔天气〕图;supplementary〔weather〕forecast 补充〔天气〕预报。

五、一个英文词有几种不同含义的中文译名时,则以①②③…的形式分别列出,如果中文译名含义相近,则用逗号分隔。如:relief ①地形,地势②地型模型;steady rain 连续雨,绵雨。

六、中文名后的圆括号"()",表示注释。如:Lesser Heat 小暑(节气);thermal tide 热力潮(由于大气温度日变化引起的大气压力变化)。

七、中文名后的实括号"【】"标注,表示紧接标注前面的中文名词是经过全国自然科学技术名词审定委员会审定公布过的相关学科或学科分支的名词。如:polar vortex 极涡【大气】,表示该词是经过审定公布的大气科学名词;retroreflection 返射【物】,表示该词是经过审定公布的物理学名词。而如 radar horizon 雷达视线水平,雷达地平线【电子】,则表示"雷达地平线"的用法是经过审定的电子学科的中文名称,但在雷达气象学中有"雷达视线水平"的用法。标注简称所对应的学科或学科分支见表1。

八、汉英词汇中的字、词一律按汉语拼音字母和声调符号的顺序排列,但圆括号内的字不参加字(声调)序排列。一个中文词对应几个同义

i

英文词时,英文词之间用逗号分隔。

九、以英文、希文或数码字起头的中文词汇,按后面的汉字排序。

十、本书汉英大气科学词汇部分是英汉词汇部分的反查,主要是为广大气象科技工作者撰写英文论文或摘要时提供方便。它侧重于常用的气象科技名词术语,一般不常用的风名、天气现象名、单位和组织名不再列出,对于一般的多义词汇,则以英汉部分的第一释义列出。读者感到意犹未尽时可再查阅本书英汉部分或普通的汉英词典。此外,本书的八个附录也是为了方便读者使用而由编者收集编纂的。

表1 标注简称所对应的学科或学科分支

【测】	测绘学	【力】	力学
【船舶】	船舶工程	【林】	林学
【大气】	大气科学	【煤炭】	煤炭科技
【地】	地球物理学	【农】	农学
【地理】	地理学	【生】	生物
【地信】	地理信息系统	【生化】	生物化学
【地质】	地质学	【石油】	石油学
【电力】	电力学	【数】	数学
【电工】	电工学	【水产】	水产学
【电子】	电子学	【水利】	水利学
【动】	动物学	【天】	天文学
【公路】	公路交通科技	【铁道】	铁道工程
【海洋】	海洋学	【土木】	土木工程
【航海】	航海科技	【土壤】	土壤学
【航空】	航空科学技术	【物】	物理学
【化】	化学	【冶金】	冶金学
【计】	计算机科学技术	【植】	植物学
【机械】	机械工程	【自动】	自动化技术
【建筑】	建筑与园林及城市规划	【自然】	自然辩证法

目　　录

第二版前言
第一版序
第一版前言
使用说明

英汉大气科学词汇 ··· (1−476)
汉英大气科学词汇 ··· (477−888)
附录1　风力等级划分标准(蒲福风级表) ···················· (889)
附录2　云的分类 ··· (890)
附录3　二十四节气英文译名 ······································· (891)
附录4.1　世界气象组织/亚太经社理事会(WMO/ESCAP)
　　　　台风委员会关于西北太平洋和南海热带气旋的命名方案
　　　　·· (892)
附录4.2　西北太平洋和南海热带气旋命名表 ··············· (894)
附录5　热带气旋等级 ·· (895)
附录6　常见的与大气科学有关的英文版SCI期刊 ········· (897)
附录7　部分与大气科学有关的单位和机构的英文译名 ··· (903)
附录8　部分与大气科学有关的国内地名(特殊拼法)的英文拼写
　　　　·· (906)
附录9　美国各州的名称和首府以及它们的规定中译名 ··· (908)
主要参考文献 ··· (910)

英汉大气科学词汇

A

AAAS(American Association for the Advancement of Science) 美国科学促进会(出版美国著名期刊 Science 的单位)
AABW(Antarctic Bottom Water) 南极底层水
AAMW(Australasian Mediterranean Water) 大洋洲的地中海海水〔在印度尼西亚海(即大洋洲的地中海)形成的地中海水体,亦称班达海水〕
AAO(Antarctic Oscillation) 南极涛动
AASIR(Advanced Atmospheric Sounding and Imaging Radiometer) 先进的大气探测和成像辐射仪
AATSR(Advanced Along-Track Scanning Radiometer) 先进的沿轨扫描辐射仪
abatement 减退,减轻
abatement of wind 风力减弱
Abbe invariant 阿贝不变量【物】
Abbe number 阿贝数,色散系数
Abbe refractometer 阿贝折射计【物】
abbreviated ship code 简略船舶电码
ABC(Atmospheric Brown Cloud) 大气棕色云
ABC bucket ABC 吊桶(测海面水温用)
abduction 外展,误导
aberration ①像差【物】②光行差【物】
aberwind(=aperwind) 阿伯风(阿尔卑斯山春季解冻风)
abioseston 非生物悬浮物
abiotic 非生物的,无生命的
ablation 冰消作用【大气】,冰(雪)面消融,融蚀
ablation area 消冰区,冰(雪)消融区
abnormality 异常度
abnormal lapse rate 异常直减率
abnormal propagation 异常传播
abnormal refraction 异常折射
abnormal weather 异常天气
above normal 超常,正常以上
above sea level(ASL) 海拔〔高度〕
above the weather 摆脱天气影响
ab-polar current 离极气流
Abraham's tree 辐辏状卷云
abrego 阿勃列戈风(西班牙南部的西南风)
Abrolhos squalls(=abroholos) 阿布罗刘斯飑(巴西阿布罗刘斯岛附近的雷雨飑)
abrupt change of climate 气候突变【大气】
abruptness 陡度,陡峭性
abrupt slope 陡坡
abscisic acid 脱落酸(一种)制止植物生长的有机物质)
abscissa 横坐标【数】
abscission 脱落【生】
absolute acceleration 绝对加速度
absolute age determination 绝对定年法
absolute air mass 绝对大气质量
absolute altimeter 绝对高度表
absolute altitude 绝对高度
absolute angular momentum 绝对角动量【大气】
absolute annual range 绝对年较差
absolute annual range of temperature 温度绝对年较差
absolute black body 绝对黑体【大气】
absolute cavity radiometer 绝对空腔辐射计
absolute ceiling 绝对云幕高,绝对云底高度
absolute coordinate system 绝对坐标系

absolute drought　绝对干旱
absolute error　绝对误差【数】
absolute extreme　绝对极值【大气】
absolute extremum　绝对极值【数】
absolute frequency　绝对频率,频数【数】
absolute gradient current　绝对梯度流
absolute humidity　绝对湿度【大气】
absolute index of refraction　绝对折射率
absolute instability　绝对不稳定【大气】
absolute instrument　一级基准仪器,绝对准器
absolute isohypse　绝对等高线
absolute linear momentum　绝对线性动量
absolute magnitude　绝对星等【天】
absolute moisture of the soil　土壤绝对湿度
absolute momentum　绝对动量
absolute monthly maximum temperature　绝对月最高温度【大气】
absolute monthly minimum temperature　绝对月最低温度【大气】
absolute motion　绝对运动【物】
absolute parallax　绝对视差
absolute potential vorticity　绝对位涡
absolute pressure　绝对气压
absolute pyrheliometer　绝对日射表
absolute radiation scale　绝对辐射标尺
absolute reference frame　绝对参考系【物】
absolute refractive index　绝对折射率
absolute scale　绝对标尺
〔absolute〕soil moisture　土壤〔绝对〕湿度【大气】
absolute stability　绝对稳定【大气】
absolute standard barometer　绝对标准气压表【大气】
absolute temperature extremes　绝对温度极值
absolute temperature scale　绝对温标【大气】
absolute temperature　绝对温度
absolute topography　绝对形势
absolute unit　绝对单位
absolute vacuum　绝对真空

absolute vacuum gauge　绝对真空计【电子】
absolute value　绝对值【数】
absolute variability　绝对变率
absolute velocity　绝对速度
absolute vorticity　绝对涡度【大气】
absolute zero　绝对零点
absorbance　吸光度【化】,吸收度
absorbed dose　吸收剂量
absorbed solar flux　吸收太阳通量
absorbed solar radiation　吸收太阳辐射
absorbent filter　吸收过滤器
absorbent solution　吸收溶液
absorber　①吸收器②吸收体,吸收剂
absorbing agent　吸收剂
absorbing function　吸收函数
absorbing medium　吸收介质
absorbing power　吸收能力,吸收本领
absorptance　吸收比【物】
absorptiometer　①〔液体〕吸气计②吸收光度计
absorption　吸收
absorption band　吸收〔光谱〕带【大气】
absorption coefficient　吸收系数
absorption cross-section　吸收截面【大气】
absorption factor　吸收因子
absorption function　吸收函数
absorption hygrometer　吸收湿度计【大气】
absorption line　吸收〔谱〕线【大气】
absorption liquid　吸收液体
absorption loss　吸收损耗
absorption mass　吸收质量
absorption optical thickness　吸收光学厚度
absorption spectroelectrochemistry　吸收光谱电化学【化】
absorption spectrometer　吸收光谱仪
absorption spectrum　吸收谱【大气】
absorptive power　吸收本领
absorptivity(α)　①吸收率②吸光系数【化】

abstract ①摘要②抽象
abstraction ①抽象②引水③排除水分④降水损失⑤分离,提取
abstract test method 抽象测试方法【地信】
abundance 丰度
abundance of element 元素丰度【化】
abundance value 丰度值
abundant year 丰产年
abyssal environment 深海环境
abyssal plain 深海平原
abyssal zone 深海区(大陆架以外,水深2000～6000 m)
ACARS(Aircraft Communications Addressing and Reporting System) 飞机通信寻址与报告系统
ACC(antarctic circumpolar current) 绕南极洋流【大气】
accelerated erosion 加速蚀损〔作用〕
accelerated motion 加速运动
acceleration 加速度
acceleration of gravity 重力加速度
acceleration potential 加速度势
acceleration spectrum 加速度谱
accelerator ①加速器②催化器
accelerometer 加速度表
acceptance capacity 容许容量
acceptance region 接受区〔域〕
acceptor 接纳体【生化】
accessory cloud 附属云【大气】
access ①访问【计】,存取 ②接入【计】
access time 存取时间【计】
access tube 插入管(测土壤含水量用)
accidental error 偶然误差【数】
acclimation 气候适应
acclimatization 气候适应【大气】,气候驯化
accommodation 调节
accommodation coefficient 调节系数
accretion ①撞冻〔增长〕【大气】②积冰
accretion efficiency 撞冻效率

accumulated cooling 累积冷却〔量〕
accumulated precipitation 累积降水
accumulated temperature 积温【大气】
accumulated temperature curve 积温曲线
accumulation 冰积〔作用〕【大气】,累积〔作用〕
accumulation area 冰积区【大气】,累积区域(冰川的)
accumulation mode 聚积模【大气】,聚积态
accumulation raingauge 累计雨量器
accumulation zone 累积带
accumulative raingauge 累计雨量器【大气】
accumulator 累加器
accuracy 准确度【大气】【物】,精〔确〕度【数】
acdar 声〔雷〕达
ACE(aerosol characterization experiment) 气溶胶特征试验
acetaldehyde 乙醛
acetic acid 乙酸
acetone 丙酮
acetonitrile 乙腈,氰甲烷
acetylene 乙炔(C_2H_2)
ACF(autocorrelation function) 自相关函数【数】
achievements in meteorological science and technology 气象科技成果
acicular ice 屑冰,针状冰,丝状冰
acid-containing soot 含酸煤烟
acid deposition 酸沉降【大气】,酸性沉积物
acid dew 酸露【大气】
acid dew point 酸露点
acid fog 酸雾【大气】
acid frost 酸霜【大气】
acid fume 酸烟雾,酸性尘雾
acid hail 酸雹【大气】
acid haze 酸霾

acidification 酸化【化】
acidity 酸度【化】,酸性
acidity constant 酸度常数【化】
acidity profile 酸度廓线
acid mist 酸雾
acid pollution 酸污染
acid precipitation 酸性降水【大气】,酸雨【大气】
acid rain 酸性降雨【大气】,酸雨【大气】
acid snow 酸雪【大气】
acid soil 酸性土【土壤】
acknowledgement 确认【计】
aclinic line 无倾角线【大气】,磁赤道
acoustic absorption 声吸收
acoustic admittance 声导纳【物】
acoustical attenuation constant 声衰减常数
acoustical command system 声指令系统
acoustical generator 发声器,声辐射器
acoustical holography scanning technique 声全息摄影扫描法
acoustical level 声级
acoustically transparent layer 声透射层
acoustical measurement 声学测量
acoustical phase constant 声位相常数
acoustical profile 声剖面图,声廓线
acoustical reflectivity 声反射率
acoustical scintillation 声闪烁【大气】
acoustical spectrum 声谱
acoustic array 声〔学基〕阵
acoustic backscattering 声后向散射
acoustic capacitance 声容
acoustic characteristic impedance 声特征阻抗
acoustic cloud 声波反射云
acoustic compliance 声顺
acoustic conductance 声导【物】
acoustic conductivity 声导率
acoustic cutoff frequency 声截止频率
acoustic detection and ranging(ACDAR) 声学探测和测距,声〔雷〕达
acoustic dispersion 声频散
acoustic Doppler current profiler(ADCP) 声学多普勒流体廓线仪
acoustic Doppler sounder 声学多普勒探测器
acoustic Doppler system 声学多普勒系统
acoustic echo sounder 回声探测器
acoustic echo sounding 回声探测
acoustic emission 声发射
acoustic emission monitoring 声发射监测
acoustic energy 声能
acoustic environment 声环境
acoustic feedback 声反馈
acoustic field 声场
acoustic fluctuation 声〔波〕起伏
acoustic fog 声雾
acoustic frequency generator 声频发生器
acoustic gravity wave 声重力波【大气】
acoustic imaging 声成像【物】
acoustic impedance 声阻抗【物】
acoustic inertance 声质量,声狃
acoustic intensity 声强
acoustic interferometer temperature sensing 声学干涉仪测温
acoustic lens 声透镜
acoustic mass 声质量,声狃
acoustic-microwave radar 声学-微波雷达
acoustic mobility 声导纳
acoustic navigation system 声导航系统
acoustic ocean current meter 声〔学〕海流计
acoustic pollution 声污染
acoustic pressure 声压
acoustic propagation constant 声传播常数
acoustic pulse 声脉冲
acoustic radar 声〔雷〕达【大气】
acoustic raingauge 声学雨量计【大气】

acoustic reactance 声抗【物】
acoustic reflection profiling 声反射仿型
acoustic reflection 声反射
acoustic resistance 声阻【物】
acoustic resonance 声共振【物】
acoustic reverberation 声混响
acoustic scattering 声散射
acoustic sensing 声传感,声探测
acoustic signature 声特征
acoustic sounder 声探测器
acoustic sounding 声学探测【大气】
acoustic spectrum 声谱
acoustic susceptance 声纳【物】
acoustic thermometer 声学温度表【大气】,声波测温表
acoustic tomography 声学层析成像术
acoustic transducer 声换能器
acoustic transponder 声应答器
acoustic velocimeter (水内)声速计
acoustic velocity 声速
acoustic wave(AW) 声波
acoustic waveguide 声波导
acoustimeter 声级仪,噪声仪
acoustometer 声强计
acquisition ①接收,捕获②探测
acquisition range 接收范围
acre-foot 英亩-英尺〔相当于1英亩面积(4047 m^2)1英尺深(0.3048 m)的水量,约1233.5 m^3〕
acrocyanosis 冻疮
acrolein 丙烯醛
ACSYS (Arctic Climate System Study) 北极气候系统研究(计划)
actinic 光化
actinic absorption 光化吸收
actinic balance(=bolometer) 热辐射仪
actinic flux 光化通量
actinic radiation 光化辐射【物】
actinic ray 光化射线
actinogram 日射自记曲线【大气】
actinograph 日射测定计【大气】,直接辐射计
actinometer ①日射测定表【大气】②光能测定仪【化】,露光计③直接辐射表,曝光表
actinometry 日射测定学【大气】,日射测定法
actinon 锕射气【大气】
action center 活动中心
action spectral density 作用量谱密度
activated band 活动光带(极光的)
activated complex theory 活化络合物理论
activation 活化【化】,激活
activation analysis 活化分析
activation energy 活化能【大气】【化】,激活能
activation free energy 活化自由能
active accumulated temperature 活动积温【大气】
active anafront 活跃上滑锋
active aurora 活动极光
active basin area 有效流域面积
active carbon 活性炭【化】
active cavity radiometer(ACR) 有源腔体辐射仪
active center 活动中心
active cloud 活动〔积状〕云
active component 有源元件【电子】
active day 〔地磁〕扰动日
active element 有源元件
active experiment 主动实验
active file 活动文件【计】
active front 活跃锋
active glacier 活冰川
active guidance 主动制导
active katafront 主动下滑锋
active layer 活动层
active monitoring system 主动监测系统
active monitoring technique 主动监测技术
active monsoon 活跃季风【大气】,季风活跃

active network 〔色球〕活动网络
active nitrogen 活性氮
active permafrost 活动永冻土
active pollution 放射性污染【大气】
active prominence 活动日珥【天】
active reaction 活性反应
active remote sensing system 主动遥感系统
active remote sensing technique 主动遥感技术【大气】
active satellite 主动卫星,有源卫星
active scattering aerosol spectrometer(ASAS) 主动散射气溶胶粒谱仪
active scattering aerosol spectrometer probe (ASASP) 有源散射气溶胶粒谱仪探测器
active site ①活性部位②活性点
active solar region 太阳活动区
active steering 主动引导
active sun 活动太阳
active surface 活动面
active system 主动系统
active technique 主动技术
active temperature 活动温度【大气】
active volcano 活火山
activity ①活动②活性、活度
activity coefficient 活度系数
activity of a foyer of atmospherics 天电源强度
actual argument 实〔变〕元
actual density of soil 土壤实际密度
actual elevation 实际海拔〔高度〕
actual evaporation 实际蒸发
actual flying weather 实际飞行天气
actual pressure 实际气压
actual time of observation 实际观测时间【大气】
actual(或 effective) evapotranspiration 有效蒸散
acuity ①锐度②分辨能力
acute pollution 急性污染

acute triangle 锐角三角形【数】
adaptability 适应性,适应能力
adaptation brightness 适应亮度
adaptation illuminance 适应光照度,适应亮度
adaptation level 适应亮度
adaptation luminance 适应亮度,适应发光率
adaptation process 适应过程
adaptation strategies 适应性对策
adapter 适配器【计】
adaptive control 自适应控制
adaptive disease 环境适应症,水土病
adaptive filtering 自适应滤波
adaptive grid 〔自〕适应网格
adaptive neural network 自适应神经网络
adaptive observational network 〔自〕适应观测网站
adaptive observations 〔自〕适应观测
adaptive optics 自适应光学
adaptive radar 自适应雷达【电子】
adaptive regression 适应回归
adaptive telemetry 自适应遥测【电子】
A/D converter 模〔拟〕数〔字〕转换器
adding day 闰日
adding method 累加法
addition 加法【数】
addition reaction 加成反应【化】
additive constant 相加性常数
additive noise 加性噪声
address code 地址码
addressing 寻址
address marker 地址标志
ADEOS(Advanced Earth Observing Satellite) 先进地球观测卫星【大气】
adfreezing 冻附过程
adhesion efficiency 附着效率
adhesion 附着力
adhesion 粘合【化】
adhesive water 〔薄膜〕吸附水
adhesive 粘合剂【化】,胶粘剂

adiabat ①绝热〔的〕②绝热线
adiabatically enclosed system 绝热闭合系统
adiabatic approximation 绝热近似
adiabatic ascending 绝热上升【大气】
adiabatic atmosphere 绝热大气【大气】
adiabatic change 绝热变化
adiabatic chart 绝热图
adiabatic closed system 绝热闭合系统
adiabatic compression 绝热压缩
adiabatic condensation 绝热凝结
adiabatic condensation point 绝热凝结点
adiabatic condensation pressure 绝热凝结气压【大气】
adiabatic condensation temperature 绝热凝结温度【大气】
adiabatic condition 绝热条件,绝热状态（复数）
adiabatic cooling 绝热冷却【大气】
adiabatic curve 绝热曲线
adiabatic diagram 绝热图【大气】
adiabatic effect 绝热效应
adiabatic equation 绝热方程
adiabatic equilibrium 绝热平衡
adiabatic equivalent temperature 绝热相当温度【大气】
adiabatic expansion 绝热膨胀
adiabatic gradient 绝热梯度
adiabatic heating 绝热增温【大气】
adiabatic invariant 绝热不变量
adiabatic lapse rate 绝热直减率【大气】
adiabatic law 绝热定律
adiabatic liquid water content 绝热液〔态〕水含量
adiabatic meaning 绝热平均
adiabatic model 绝热模式【大气】
adiabatic motion 绝热运动
adiabatic phenomenon 绝热现象
adiabatic process 绝热过程【大气】
adiabatic psychrometer 绝热干湿表
adiabatic rate 绝热率
adiabatic region 绝热区
adiabatics 绝热曲线
adiabatic saturation point 绝热饱和点
adiabatic saturation pressure 绝热饱和气压
adiabatic saturation temperature 绝热饱和温度
adiabatic saturator 绝热饱和器（绝热干湿表用）
adiabatic sinking 绝热下沉【大气】
adiabatic state 绝热状态
adiabatic temperature change 绝热温度变化
adiabatic temperature gradient 绝热温度梯度
adiabatic trial 绝热检验【大气】,绝热试验
adiabatic warming 绝热增温
adiabatic wet-bulb temperature 绝热湿球温度
A-display A〔型〕显〔示〕
adjacent field-of-view method 相邻视场法
adjacent matrix 邻接矩阵【数】
adjoint 伴随
adjoint assimilation 伴随同化
adjoint differential equation 伴随微分方程【数】
adjoint equation 伴随方程
adjoint matrix 伴随矩阵【数】
adjoint method 伴随法
adjoint model 伴随模式
adjoint sensitivity 伴随灵敏度
adjustable cistern barometer 调槽气压表,福丁气压表
adjustable siphon mercury barometer 可调虹吸式水银气压表
adjustment 校准,调整
adjustment of circulation 环流调整【大气】
adjustment of longwave 长波调整【大气】

adjustment procedure　适应过程
adjustment process　适应过程
adjustment time　调整时间
administrative boundary　行政界线【地理】
administrative geography　行政地理【地理】
admissible concentration limit　容许浓度上限
admissible error　容许误差【数】
adret　阳坡,山阳
adsorbent　吸附剂
adsorption　吸附作用【大气】
adsorption film　吸附膜【化】
adsorption indicator　吸附指示剂【化】
adsorption isotherm　吸附等温线
adsorption layer　吸附层【化】
A/D universal weather graphic recorder　模〔拟〕数〔字〕通用天气图〔传真〕记录器,模〔拟〕数〔字〕通用天气图传真收片机
Advanced Along-Track Scanning Radiometer (AATSR)　先进的沿轨扫描辐射仪
Advanced Atmospheric Sounding and Imaging Radiometer (AASIR)　先进大气探测成像辐射仪
advanced cloud wind system　先进云风系统
Advanced Earth Observing Satellite (ADEOS)　先进地球观测卫星【大气】
Advanced Microwave Sounding Unit (AMSU)　先进微波探测装置【大气】
Advanced Moisture and Temperature Sounder (AMTS)　先进温湿探测器
Advanced Research Project Agency Network (ARPANET)　高级研究计划局网络【计】,阿帕网
Advanced Spaceborne Thermal Emission and Reflection radiometer, EOS (ASTER)　先进星载热发射和反射辐射仪(EOS)
Advanced Technology Satellite (ATS)　先进技术卫星

Advanced TIROS-N (ATN)　先进泰罗斯-N卫星【大气】
Advanced TIROS Operational Vertical Sounder (ATOVS)　先进泰罗斯业务垂直探测器
Advanced Very High Resolution Radiometer (AVHRR)　先进甚高分辨率辐射仪【大气】
Advanced Vidicon Camera System (AVCS)　先进的光导摄像系统
Advanced Weather Interactive Processing System (AWIPS)　先进天气交互处理系统
advection　平流【大气】
advection change　平流变化
advection-diffusion equation　平流扩散方程
advection effect　平流效应
advection effluent　平流流出〔量〕
advection equation　平流方程【大气】
advection fog　平流雾【大气】
advection frost　平流霜【大气】
advection inversion　平流逆温
advection jet　平流急流
advection lobe　平流瓣
advection model　平流模式
advection of effluent　流出平流
advection process　平流过程
advection propagation　平流传播
advection-radiation fog　平流辐射雾【大气】
advection scale　平流尺度
advection term　平流项
advection velocity　平流速度
advective acceleration　平流加速度
advective boundary layer　平流边界层
advective change　平流变化
advective change of temperature　温度〔的〕平流变化
advective equation　平流方程【大气】
advective flux　平流通量
advective form　平流形式

advective-gravity flow 平流重力流
advective hypothesis 平流假说
advective model 平流模式
advective pressure tendency 平流气压倾向
advective-radiation fog 平流辐射雾
advective region 平流区
advective thunderstorm 平流性雷暴【大气】
advective time scale 平流时间尺度
adverse factor 有害(不利)因子
adverse weather 不利天气,不良天气
adverse weather condition 不利气象条件
adverse wind 逆风
advice 通知,报告
advisory area 咨询〔预报〕区
advisory committee 咨询委员会
Advisory Committee for Operational Hydrology(ACOH,WMO) 业务水文学咨询委员会(WMO)
Advisory Committee of Climate Application and Data Project(ACCADP) 气候应用及资料计划咨询委员会
Advisory Committee on Marine Resources Research(ACMRR,FAO) 海洋资源研究咨询委员会(FAO)
Advisory Committee on Oceanic Meteorological Research(ACOMR,WMO) 海洋气象研究咨询委员会(WMO)
Advisory Committee on Polar Program(ACPP) 极地计划咨询委员会(美国)
Advisory Committee on Weather Control(ACWC) 天气控制咨询委员会(美国)
advisory forecast 咨询性预报
aeolian 风成的
aeolian sounds 风吹声
aeolian tones 风〔成〕音,风吹声
aeolosphere 风层
aeolotropic 各向异性的,非均质的

aeolotropism 各向异性
aeration 换气,通风
aerial ①天空〔的〕②〔架空〕天线
aerial contaminant 空气污染物
aerial contamination 空气污染
aerial detection ①空中检测②空气检测
aerial exploration 高空探测
aerial fog 气雾
aerial gain 天线增益
aerial perspective 空中透视,空间透视〔法〕
aerial photograph 航摄照片,航空摄影
aerial photography 航空摄影【测】
aerial plankton 大气浮游生物
aerial plant 气生植物【植】
aerial reconnaissance 航空侦察
aerial remote sensing 航空遥感【地理】
aerial spectrograph 航空摄谱仪【测】
aeroacoustics 航空声学
aeroagronomy 航空农〔艺〕学(飞机播种、施肥等)
aerobacteria 喜氧细菌
aerobiology 大气生物学【大气】,高空生物学
aerochemistry 高空气体化学
aeroclimatology 高空气候学【大气】
AERO code 航空天气电码
aerodrome elevation 机场海拔〔高度〕
aerodrome forecast 机场天气预报
aerodrome hazardous weather warning 机场危险天气警报【大气】
aerodrome meteorological minima(或minimums) 机场最低气象条件【大气】
aerodrome special weather report 机场特殊天气报告【大气】
aerodynamically rough surface 气体动力学粗糙面
aerodynamically smooth surface 气体动力学光滑面
aerodynamic balance 空气动力学平衡
aerodynamic coefficient 气动系数

aerodynamic contrail　空气动力凝结尾迹
aerodynamic drag　气动曳力
aerodynamic force　气动力
aerodynamic laboratory　空气动力实验室
aerodynamic method　空气动力学方法
aerodynamic resistance　气动阻力
aerodynamic roughness　气体动力〔学〕粗糙度【大气】
aerodynamic roughness length　气体动力粗糙度长度
aerodynamics　气体动力学,空气动力学【大气】
aerodynamic smoothness　气体动力平滑度
aerodynamic trail　空气动力尾迹
aeroembolism　高空病
aerogel　气凝胶
aerogram　高空图
aerograph　高空气象计
aerographical chart　高空气象图
aerography　普通气象学(罕用)
aerolite　陨石
aerologation　测高术航行〔单航向飞行最小飞行航程〕
aerological analysis　高空分析
aerological ascent　高空探测,自由大气探测
aerological days　高空观测日
aerological diagram　高空图
aerological instrument　高空仪器
aerological observatory　高空观象台
aerological sounding　高空探测
aerological station　高空〔观测〕站
aerological table　高空报表
aerological theodolite　测风经纬仪【大气】
aerology　高空气象学【大气】
aeromancy　航空天气预报
aeromedicine　航空医学
aerometeorograph　高空气象计
aerometer　气体比重计,量气计
aeronautical climate regionalization　航空气候区划【大气】
aeronautical climatography　航空气候志【大气】
aeronautical climatology　航空气候学
Aeronautical Fixed Telecommunications Network(AFTN)　航空专用电传通信网
aeronautical meteorological service　航空气象服务
aeronautical meteorological station　航空气象站
aeronautical meteorology　航空气象学【大气】
aeronautics　航空学
aeronomosphere　高层大气
aeronomy　高空大气学【大气】
aeropause　大气上界
aerophotogrammetry　航空摄影测量【测】
aerophotography　航空摄影学
aerophysics　航空物理学
aerophyte　气生植物【植】
aeroplane antenna　飞机天线
aeroplane photography　航空摄影
aerosol　气溶胶【大气】,大气微粒
aerosol analyzer　气溶胶分析仪
aerosol characterization experiment(ACE)　气溶胶特征实验
aerosol chemistry　气溶胶化学【大气】
aerosol climatic effect(ACE)　气溶胶气候效应
aerosol climatology　气溶胶气候〔学〕
aerosol composition　气溶胶成分
aerosol detector　气溶胶检测仪
aerosol distribution　气溶胶分布
aerosol electricity　气溶胶电〔学〕,大气微粒电〔学〕
aerosol layer　气溶胶层
aerosol loading　气溶胶负载,气溶胶含量
aerosol optical depth(AOD)　气溶胶光学厚度
aerosol optical thickness　气溶胶光学厚度
aerosoloscope　气溶胶仪,空气〔中〕微粒

测定器(测量表)
aerosol particle 气溶胶粒子
aerosol particle size distribution 气溶胶粒子谱【大气】
aerosol photometer 气溶胶光度计
aerosol scattering phase function 气溶胶散射相函数
aerosol size distribution 气溶胶粒径谱
aerosolsonde 气溶胶探空仪
aerosol structure 气溶胶结构
aerosol therapy 气溶胶治疗〔法〕,气雾疗法
aerosol wettability 气溶胶吸湿性
aerosonde 飞行探空仪,无人机探空
aerospace ①航空与航天②宇宙空间③大气,大气层
aerosphere 气圈,气界
aerostatic balance 空气静力平衡
aerostatics 空气静力学
aerostat ①气球②飞艇
aerostat meteorograph 气球气象仪
aero-thermo-acoustics 湍流声学,气〔动〕热声学
aerothermochemistry 空气热力化学
aerovane 风向风速仪
aestival(=estival) 夏季的,夏令
aestivation 夏眠,夏蛰
Afer(=Africino, Africo, Africuo, Africus ventus) 阿弗风(意大利的西南风)
affine transformation 仿射变换
affinity labeling 亲和标记【生化】
afforestation 造林,绿化
Afghanets 阿富汗强阵风
African easterly jet (AEJ) 非洲东风急流
African jet 非洲急流
after-exercise chill 运动后寒冷
afterglow ①余辉②晚霞
afterheat 秋老虎(罕用)
after-image 残留影像
afternoon effect 午后效应
aftershock 余震【地】

after summer 秋老虎(罕用)
AGCM (atmospheric general circulation model) 大气环流模式
age dating 计时【地质】,定年
ageostrophic acceleration 非地转加速度
ageostrophic advection 非地转平流
ageostrophic circulation 非地转环流
ageostrophic flow 非地转流
ageostrophic motion 非地转运动【大气】
ageostrophic wind 非地转风【大气】
ageostrophic wind component 非地转风分量
agglomeration ①碰撞〔增长〕②附聚,凝聚
agglutination 黏合〔作用〕,附着〔作用〕
aggregation 聚合【大气】
aggressive biometeorological index 侵占〔性〕生物气象学指数
aggressive water 侵(腐)蚀性水
aging 老化
agonic line 零值磁偏线
agricultural biometeorology programme 农业生物气象学规划
agricultural climatology(=agroclimatology) 农业气候学【大气】
agricultural drought 农业干旱
agricultural meteorological recorder 农业气象综合测录装置
agricultural meteorological station 农业气象站【大气】
agricultural meteorological station for special purpose 专业农业气象站
agricultural meteorology(=agrometeorology) 农业气象学【大气】
agricultural microclimate 农业小气候【大气】
agricultural microclimate observation 农业小气候观测
agricultural seasons 农作季节
agricultural threshold temperature 农业界限温度【大气】

agricultural weather forecast 农业天气预报
agroclimatic analogy 农业气候相似【大气】
agroclimatic analysis 农业气候分析【大气】
agroclimatic atlas 农业气候图集【大气】
agroclimatic classification 农业气候分类【大气】
agroclimatic demarcation 农业气候区划【大气】
agroclimatic division 农业气候区划【大气】
agroclimatic evaluation 农业气候评价【大气】
agroclimatic handbook 农业气候手册
agroclimatic index 农业气候指数【大气】
agroclimatic investigation 农业气候调查
agroclimate 农业气候
agroclimatic potential productivity 〔农业〕气候生产潜力【大气】
agroclimatic region 农业气候区域
agroclimatic resources 农业气候资源【大气】
agroclimatography 农业气候志【大气】
agroclimatology 农业气候学【大气】
agrometeorological code 农业气象电码
agrometeorological condition 农业气象条件
agrometeorological forecast 农业气象预报【大气】
agrometeorological hazard 农业气象灾害【大气】
agrometeorological index 农业气象指标【大气】
agrometeorological information 农业气象信息【大气】
agrometeorological model 农业气象模式【大气】
agrometeorological observation 农业气象观测【大气】
agrometeorological station 农业气象站
agrometeorological yield forecast 农业气象产量预报
agrometeorology 农业气象学【大气】
agro-met station 农〔业〕气〔象〕站
agronomy 农学,农艺学
agrotopoclimatology 农业地形气候学
AGU(American Geophysical Union) 美国地球物理联合会
agua enferma(＝aguaje) 秘鲁赤潮
agueil(＝aiguolos) 阿格伊风(法国南部的冷东风)
Agulhas Current(＝Agulhas Stream) 厄加勒斯海流
Agung 阿贡〔火山〕(印尼)
air ①空气②大气③微弱气流
air against airplane 飞机迎面气流
air and precipitation monitoring network 大气及降水监测网
air around airplane 绕机气流
air atomizer 空气雾化器
air attenuation 大气衰减
air bearing 空气轴承
airborne ①空中浮游的②机载的
airborne CCN spectrometer 机载云凝结核粒谱仪
airborne cloud collector 机载云〔含水量〕收集器
airborne dye lidar 机载染料光〔雷〕达
airborne expendable bathythermograph 〔AXBT〕机载投弃式温深仪
airborne fraction (CO_2增加的)空中承载分数
airborne gravity measurement 航空重力测量【测】
airborne laser radar 机载光〔雷〕达
airborne observation 空基观测【大气】
airborne oceanography 航测海洋学
airborne particulate 空中悬浮微粒【大气】,浮粒
airborne pollen 气传花粉
airborne radar 机载雷达
airborne radiation thermometer(ART)

机载辐射温度仪
airborne radioactivity survey　航空放射性测量【地】
airborne search radar　机载搜索雷达
airborne spectrometer　机载光谱仪
airborne weather radar　机载天气雷达【大气】
air bubble　气泡
air bump　〔空气〕颠簸
air cascade(＝air cataract)　气瀑
air change rate　换气率
air circulation　空气环流
air cleaning facility　空气净化设备
air column　气柱
air conditioning room　人工气候室
air conditioning　空调
air conductivity　大气电导率
aircraft actinometer　机载直接辐射表
aircraft bumpiness　飞机颠簸【大气】
aircraft ceiling　安全飞行最大高度
Aircraft Communications Addressing and Reporting System（ACARS）　飞机通信寻址与报告系统
aircraft dropwindsonde system（ACDWS）　机载下投式测风探空仪系统
aircraft electrification　飞机起电
aircraft emission　飞机排放
aircraft hazard　飞机事故
aircraft icing(＝aircraft ice accretion)　飞机积冰【大气】
aircraft impactor　机载撞击〔取样〕器
aircraft integration data system（AIDS）　飞机累积资料系统
aircraft measurement　机载测量〔仪器〕，航测〔记录〕
aircraft meteorological data acquisition and relay（AMDAR）　飞机气象资料接收与中继
aircraft meteorological station　飞机气象站（设在飞机内的气象站）
aircraft observation　飞机观测
aircraft reconnaissance flight　飞机〔气象〕侦察飞行
aircraft report　飞行报〔告〕
aircraft sounding　飞机探测【大气】
aircraft thermometry　飞机测温法
aircraft to satellite data relay（ASDAR）　飞机卫星资料中继
aircraft trail　飞机尾迹【大气】
aircraft turbulence　飞机湍流
aircraft wake　飞机尾流【大气】
aircraft weather reconnaissance　飞机天气侦察【大气】
air current　气流
air density　空气密度
air discharge　空中放电【大气】
air drag　空气曳力
air drainage　空气流泄
airdrome forecast　机场〔天气〕预报
airdrome pressure　场面气压【大气】
airdrome special weather report　机场特别天气报告
airdrome warning　机场预警
air-dropped（或 airborne）expendable bathythermograph（AXBT）　〔机载〕投弃式温深仪，空投温深仪
air-dry　风干
air-earth conduction current　空-地传导电流【大气】
air-earth current　空-地电流【大气】
air entrainment　空气夹卷
air exhaust　排气
airfield　机场
airfield color code　机场色标（标志能见度,最大飞行高度等）
air flow　气流
air flow multimeter　气流综合测量仪
airflow over ridge　过脊气流
airfoil section　尾翼截面（风向标的）
Air Force atmospheric model　空军大气模式,空军标准大气（美国）
air force base（A. F. B）　空军基地

Air Force Cambrige Research Laboratories (AFCRL,USAF) 〔美国〕空军剑桥研究室
Air Force Geophysics Laboratory (AFGL) 〔美国〕空军地球物理实验室
Air Force Global Weather Central (AFGWC) 〔美国〕空军全球天气中心
air fountain 气泉
airframe deicing 飞机〔机体〕除冰
airframe icing 机体积冰
air freezing index 空气冻结指数
air-fuel ratio 空气燃料比
air-gauge 压力表
airglow 气辉【大气】
airglow intensity 气辉强度
airglow photodissociation 气辉光致离解
airglow photometer 气辉光度计
airglow production 气辉形成
airglow quenching 气辉消失
airglow radiation 气辉辐射
airglow spectral line intensity 气辉谱线强度
air hoar 霜凇,树挂
air infection 空气传染
airlight 〔空〕气光【大气】(悬浮物散射光)
airlight formula 空气光公式(即 Koschmeider 公式)
air line correction 空气管路倾斜订正
air-line sounding 空气管路探测
air-line well 空气管路腔
air mass ①气团【大气】②大气质量,空气质量
air-mass analysis 气团分析【大气】
air-mass characteristic 气团特性
air-mass chart 气团图
air-mass classification 气团分类【大气】
air-mass climatology 气团气候学
air-mass fog 气团雾【大气】
air-mass frequency 气团频率
air-mass identification 气团鉴定,气团识别
air-mass meteorology 气团气象学
air-mass modification 气团变性
air-mass precipitation 气团降水
air-mass property 气团属性【大气】
air-mass shower 气团阵雨
air-mass source 气团源地【大气】
air-mass source region 气团源〔地〕区〔域〕
air-mass temperature 气团温度
air-mass thunderstorm 气团雷暴
air-mass transformation 气团变性【大气】
Air-Mass Transformation Experiment (AMTEX,Japan) 气团变性试验(日本,1974 和 1975 年两次)
air-mass transport 气团迁移
air-mass-type diagram 气团类型图
air-meter 气流表,微风表,风速表
air monitoring 大气监测
air monitoring instrument 大气〔污染〕监测仪
air monitoring network 大气监测网
air monitoring station 大气监测站
air motion 空气运动
air-ocean coupled model 海气耦合模式
air package 气块
air parcel 气块【大气】
air parcel trajectory 气块轨迹
air particle 气粒
air permeability 透气率【土壤】
airplane meteorograph 飞机气象计
airplane meteorological sounding 飞机气象探测【大气】
airplane observation 飞机观测
air plankton 大气浮游生物
air pocket 气阱,气穴
air pollution episode 空气污染事件
air poise 空气称重器,称气器
air pollutant 空气污染物
air pollutant emission 空气污染物排放【大气】

air pollutant emission standard 空气污染物排放标准【大气】
air pollution 空气污染【大气】,大气污染
air pollution alert 空气污染预警
air pollution chemistry 空气污染化学
air pollution code 空气污染法规
air pollution complaint 空气污染诉讼
air pollution control 空气污染控制
air pollution disaster 空气污染事故
air pollution forecasting 空气污染预报
air pollution index(API) 空气污染指数
air pollution law 空气污染法
air pollution legislation 空气污染立法
air pollution meteorology 空气污染气象学
air pollution model 空气污染模式【大气】
air pollution modeling 空气污染模拟【大气】
air pollution observation station 空气污染观测站
air pollution potential 空气污染潜势
air pollution potential forecast 空气污染潜势预报
air pollution regulation 空气污染条例
air pollution source 空气污染源
air pollution standard 空气污染标准
air pollution surveillance 空气污染监测
airport elevation 机场标高,机场海拔〔高度〕
airport forecast 机场预报
airport height 机场标高,机场海拔〔高度〕
airport surveillance radar 机场监视雷达【电子】
air-position indicator(API) 空中位置显示器
air pressure power spectrum 气压功率谱
air pressure 气压
air purification 空气净化
air quality 大气品位【大气】,空气质量,大气质量,大气品质

Air Quality Act 大气质量法
Air Quality Advisory Board(AQAB) 大气质量咨询委员会(美国)
air quality control region 大气质量控制区
air quality criteria 大气质量判据(标准)
air quality forecast 空气质量预报
air quality index 大气质量指数
air quality management 大气质量管理
air quality monitor 大气质量监测仪
air quality standard 空气质量标准
air quality surveillance network 大气质量监视网
air report 飞机报告
air resistance 空气阻力
air resource 空气资源
air route 航线
air route surveillance radar 航线监视雷达【电子】
Air Route Traffic Control Center(ARTCC) 航线运输指挥中心
air route〔weather〕forecast 航线天气预报【大气】
AIRS(Atmospheric Infrared Sounder) 大气红外探测器
air sampler 空气取样器
air sampling rig 空气取样装置
air-sea boundary process 海气边界过程
air-sea coupled model 海气耦合模式
air-sea exchange 海气交换【大气】
air-sea interaction(ASI) 海气相互作用【大气】,海气交互作用
air-sea interface 海气界面【大气】
air-sea temperature difference 海气温差
air self-cleaning 大气自洁作用
air self-purification 大气自净作用
airshed model 空气〔污染区〕模式
airship 飞艇,飞船
air shower 大气簇射【地】
air sickness 航空病,晕机
air sink 气汇

air-space 航空-航天的,航空-太空的
airspace 空域
airspeed ①空速,航速②气流速度
air-sphere 气界,气圈
air stagnation model 空气停滞模式
air stream 气流
air temperature 气温【大气】
air thawing index 空气解冻指数
air torrent 空气急流
air toxins 空气毒素
air traffic control 空中交通管制【电子】,空管
air traffic control radar 空管雷达【电子】
air traffic safety 航空安全
air trajectory 空气轨迹
air trap 气阱【大气】
air travel 空气轨迹
air turbulence 大气湍流
airview map 鸟瞰图【地理】
air-water interface 水-气界面
air-water interface scattering 水-气界面散射
air wave 气波
airway 航〔空〕线
airway forecasting 航线预报〔方法〕
airways forecast 航线预报
airways observation 航线观测
airways shelter 航线〔观测〕百叶箱
airway weather 航线天气
air weather group 气象勤务组
Air Weather Service(AWS) 空军天气局(美国)
airworthiness 适航性,耐航性,飞行性能
Airy disk 艾里斑【物】
Airy function 艾里函数
Airy theory of rainbow 虹的艾里理论
Airy wave 艾里波
Aitken dust counter 艾特肯计尘器【大气】,爱根计尘器【大气】
Aitken nuclei 艾特肯核【大气】,爱根核【大气】
Aitken nucleus counter 爱根计尘器
Aitken particle 爱根粒子
Akaike information criterion(AIC) AIC 准则【数】,赤池信息准则
alarm dismissal probability 漏警概率【电子】
alarm level 预警〔信度〕水平
Alaska Current 阿拉斯加海流
Alaskan high 阿拉斯加高压
Alaskan Stream 阿拉斯加〔续〕海流
albedo 反照率【大气】,反射率
albedograph 反照率计
albedometer 反照率表【大气】
albedo neutron 反照中子
albedo of the earth 地球反照率
albedo of the earth-atmosphere system 地气系统反照率【大气】
albedo of the earth planetary 地球〔行星〕反照率
albedo of underlying surface 下垫面反照率【大气】
Alberta clipper 艾伯塔切断低压
Alberta Hail Project(AHP) 艾伯塔冰雹研究计划(加拿大)
Alberta Hail Study(ALHAS) 艾伯塔冰雹研究(加拿大)
Alberta low 艾伯塔低压
albe 阿尔培风(西班牙的暖湿西南风)
alcohol-in-glass thermometer 酒精温度表
alcohol thermometer 酒精温度表
alcyone days(=halcyon days) 平安时期(冬至前后七天的风平浪静时期)
Alecto unit 亚历克托装置(一种碘化银焰弹发生器)
alee 背风
alerting zone 警报区域
Aleutian Current 阿留申海流
Aleutian low 阿留申低压【大气】
Alexander's dark band (虹霓之间的)亚历山大暗带

Alfven layer　阿尔文层【地】
Alfven wave　阿尔文波(磁流体动力波)【力】
algebraic equation　代数方程【数】
algebraic function　代数函数
algebraic integer　代数整数【数】
algebraic language　代数语言【计】
algebra　代数学【数】
Algerian Current　阿尔及利亚海流
algorithm　算法【数】
algorithmic language　算法语言【计】
alidade　照准仪
aliased spatial frequency　混淆空间频率
aliasing　混淆〔现象〕
aliasing error　混淆误差
alienation　异化
alignment　①列线②直线对准③调整④校准
alignment chart　调整列线图
alimentation　增补(冰川或雪地的)
Alisov's classification of climate　阿利索夫气候分类〔法〕
alkali fume　碱性尘雾
alkaline soil　碱性土【土壤】
alkalinity　碱度【化】,碱性
alkane　烷【化】,烷烃
alkene　烯【化】,烯烃
alkyl peroxy radicals　代烷基
alkyne　炔【化】,炔烃
Allard's law　阿拉特定律
Allen radiation belt(＝Van Allen radiation belt)　范艾伦辐射带
Allerheiligenwind　圣风(奥地利提罗耳地区晚秋后的风)
All-Hallow summer　万圣夏(11月1日万圣节前后该地反常地热)
Alliance of Small Island States(AOSIS)　小岛国联盟
allobar　变压区
allobaric　变压〔的〕
allobaric field　变压场【大气】
allobaric wind　变压风
allocation　①配置,部署②地址分配
allohypsic wind　变高风
allohypsography　变高线型,变高图法
allowable concentration　容许浓度
allowable error　容许误差
allowed transition　容许跃迁【物】
all-pass filter　全通滤波器
all-purpose rocket for collecting atmospheric soundings(ARCAS)　阿卡斯气象火箭(全能大气探测火箭)
all-round visibility　环宇能见度
All Saints' summer　万圣夏(11月1日万圣节前后该地反常地热)
all-sky camera　全天景照相机
all-sky condition　各种云天条件
all sky photometer　全天光度计【大气】
alluvial　冲积土〔层〕
alluvial aquifer　冲积含水层
alluvial deposit　冲积物【地理】
alluvial fan　冲积扇【地理】
alluvium　冲积物【地理】
all-weather　全天候〔的〕
all weather airport　全天候机场【大气】
all-weather automatic landing　全天候自动着陆【电子】
all-weather flight　全天候飞行
all-weather landing　全天候降落
all-weather wind vane and anemometer　全天候风向风速计
almanac　天文年历
Almeria-Oran Front　阿尔梅里亚-奥兰锋(西地中海锋区)
almost-equilibrium state　准平衡态
almost intransitive system　准不定转移系统,准非可递系统
almost-period　准周期
almost periodic function　殆周期函数【数】,几乎周期函数
almost periodic solution　殆周期解【数】
almucantar　地平纬圈

almwind 阿尔姆风(波兰塔特拉山脉的焚风)
aloegoe 阿洛戈风(北苏门答腊托巴湖的地方风)
aloft wind 高空风
alongshore current 沿岸流
along-slope wind system 沿坡风系
along-track scanning 沿轨道扫描
along-valley wind 沿谷风
along-valley wind system 沿谷风系
aloup de vent 阿卢普特旺特风(法国布雷维尼盆地夜间的冷风)
alpach(=aberwind) 阿伯风
alpenglow(=alpengluhen) 高山辉,染山霞
alphabetic code 字母代码
alpha ionization α射线电离
alphanumeric data 字母数字资料
alpha-particle α质点,α粒子
alpha-ray(=α-ray) α射线
alpha scale α尺度
alphascope 字母显示器
alpine climate 高山气候
alpine glacier 高山冰川
alpine glow 高山辉,染山霞
alpine permafrost 高山冻土【地理】
alpine plant 高山植物【林】
alpine tundra(=mountain tundra) 高山冻原
Alps Experiment(ALPEX) 阿尔卑斯山试验
ALSPP(Atmospheric and Land Surface Processes Project) 大气和陆面过程计划
altanus 奥坦风(法国中南部的强东南风)
altazimuth 地平经纬仪,高度方位仪
alternate airport 备降机场
alternate direction method 交替方向法
alternate forecast 备降〔机〕场天气预报
alternating current 交流电
alternating tensor 交错张量
alternating unit tensor 交错单位张量
alternative energy 替代能源
Alter shield 雨量筒防风圈
altichamber ①气压检定箱②气压测试室
alti-electrograph 高空电位计
altigraph 高度计
altimeter 高度表
altimeter correction 高度表订正
altimeter equation 高度表方程
altimeter setting 高度表拨定〔值〕【大气】,高度表拨〔正〕值
altimeter setting indicator 高度表拨正指示器
altimetry 测高术,测高学
altithermal period 冰后期高温期(约公元前5000至前2500年,此时北美西部夏季温度约比今日高1~2℃)
altitude 海拔〔高度〕【地理】,高度,高程
altitude correction 高度订正
altitude disease 高空病
altitude profile measurement 高度廓线测量
altocumulus(Ac) 高积云【大气】
altocumulus castellanus(Ac cast) 堡状高积云【大气】
altocumulus cumulogenitus(Ac cug) 积云性高积云【大气】
altocumulus duplicatus 复高积云
altocumulus floccus(Ac flo) 絮状高积云【大气】
altocumulus glomeratus 簇状高积云
altocumulus informis 无定状高积云
altocumulus lacunosus 网状高积云
altocumulus lenticularis(Ac lent) 荚状高积云【大气】
altocumulus nebulosus 薄幕状高积云
altocumulus opacus(Ac op) 蔽光高积云【大气】
altocumulus perlucidus 漏隙高积云
altocumulus radiatus 辐辏状高积云

altocumulus stratiformis 成层高积云
altocumulus translucidus(Ac tra) 透光高积云【大气】
altocumulus undulatus 波状高积云
altostratocumulus 高层积云
altostratus(As) 高层云【大气】
altostratus densus 浓密(厚)高层云
altostratus duplicatus 复高层云
altostratus fractus 碎高层云
altostratus lenticularis 荚状高层云
altostratus maculosus 斑状高层云
altostratus opacus(As op) 蔽光高层云【大气】
altostratus precipitus 降水性高层云
altostratus radiatus 辐辏状高层云
altostratus translucidus(As tra) 透光高层云【大气】
altostratus undulatus 波状高层云
alumin(i)um oxide humidity element 氧化铝湿度元件
alumin(i)um oxide hygrometer 氧化铝湿度表
alumin(i)um oxide sensor 氧化铝〔湿度〕传感器
amateur forecast 业余预报
amateur weather station 业余气象站
Amazon Basin 亚马孙平原(巴西)
ambient air 环境大气
ambient air monitoring 环境大气监测
ambient air quality 环境大气质量,环境大气品质
ambient air quality standard 环境大气质量标准
ambient air sampling 环境空气取样
ambient air sampling instrument 环境空气取样仪
ambient air standard 环境大气标准
ambient atmosphere 环境大气
ambient field 环境场
ambient liquid water content 环境液〔态〕水含量
ambient noise 环境噪声
ambient noise field 环境噪声场
ambient noise level 环境噪声级
ambient particulate concentration 环境颗粒浓度
ambient pollution burden 环境污染负荷
ambient pressure 环境气压
ambient quality standard 环境质量标准
ambient temperature 环境温度
ambiguity diagram 歧义图
ambiguity function 歧义函数
ambipolar diffusion 双极扩散
Amble diagram 安布尔图
AMDAR (aircraft meteorological data acquisition and relay) 飞机气象资料接收与下传
amendment of aviation weather forecast 航空天气订正预报【大气】
amenity 舒适(环境,气候)
American Association for the Advancement of Science(AAAS) 美国科学促进会
American Geophysical Union (AGU) 美国地球物理学联合会
American Institute of Aerological Research (AIAR) 美国高空气象研究所
American Meteorological Society (AMS) 美国气象学会
American Polar Society (APS) 美国极地研究会
American Rocket Society (ARS) 美国火箭学会
amino acid 氨基酸【化】
amino-acid dating 氨基酸定年法
AMIP(Atmospheric Models Intercomparison Project) 大气模式比较计划
ammonia 氨
ammonia cycle 氨循环
ammonification 氨化作用【生】,成氨〔作用〕
ammonium chloride 氯化铵

ammonium ion　铵离子
ammonium nitrate　硝酸铵
ammonium-nitrogen　氨态氮【土壤】
ammonium sulfate　硫酸铵
AMOC（Atlantic Moridional Overturning Circulation）　大西洋经向翻转流
amorphous　非晶〔形〕的
amorphous cloud　无定形云
amorphous frost　非晶霜
amorphous sky　碎乱天空
amorphous snow　无定形雪
amount of cloud　云量
amount of precipitation　降水量【大气】
amphidrome　无潮点
amphidromic point　无潮点
amphidromic region　无潮区
amphidromos　转风点
ample rainfall　充沛降水
amplification　放大,放大率
amplification factor　放大因子【数】
amplification matrix　放大矩阵
amplifier　放大器
amplitude　振幅
amplitude correlation function　振幅相关函数
amplitude limiter　限幅器【电子】
amplitude limiting　限幅【电子】
amplitude-modulated indicator　调幅指示器
amplitude modulation（AM）　调幅【电子】
amplitude modulator　调幅器【电子】
amplitude-period graph　振幅-周期图
amplitude response　振幅响应
amplitude spectrum　振幅谱
amplitude structure function　振幅结构函数
AMS　①（American Meteorological Society）美国气象学会②（Acta Meteorologica Sinica）（中国）气象学报的拉丁语缩写
AMSU（Advanced Microwave Sounding Unit）　先进微波探测装置【大气】

AMTEX（Air-Mass Transformation Experiment）　气团变性试验
AMTS（Advanced Moisture and Temperature Sounder）　先进温湿探测器
AN（ascending node）　升交点【大气】
An　锕射气【大气】
anabaric　正变压〔的〕,升压〔的〕
anabatic front　上滑锋【大气】
anabatic wind　上坡风
Anadyr Current　阿纳德尔海流
anaerobe　厌氧生物【动】,厌氧菌【微】
anaerobic condition　厌氧条件
anaflow　上升气流
anafront　上滑锋
anallobar　正变压线【大气】
anallobaric　正变压〔的〕
anallobaric center　正变压中心【大气】
analog bandwidth compression modem　模拟信号带宽压缩调制解调器
analog climate model　类比气候模式,相似气候模式
analog computer　模拟计算机
analog digital converter　模〔拟〕数〔字〕转换器
analogical inference　类比推理【计】
analogical learning　类比学习【计】
analog magnetometer　模拟磁强计
analog method　相似方法
analog model　相似模式
analog-to-digital conversion　模〔拟〕数〔字〕转换
analog-to-digital convert　模数转换【计】
analog-to-digital converter　模数转换器【计】
analog(ue)　①模拟设备②模拟,相似
analogue method　①模拟法②相似法
analysed chart（或 map）　分析图
analysis　分析〔学〕【数】,解析
analysis center　分析中心
analysis initialization cycle　分析初始化循环

analysis model error 分析模型误差
analysis of tropical oceanic low-levels（ATOLL） 热带海洋地区低层分析（缩写词原义"环礁"）
analysis of variance 方差分析【数】
analytical chemistry 分析化学【化】
analytical error 解析误差
analytical solution 解析解,分析解
analytical theory 解析理论
analytic continuation 解析延拓【数】
analytic function 解析函数【数】
analytic functional 解析泛函【数】
analyzer ①分析程序②分析机,分析器
anaphalanx 暖锋面
anaprop 反常传播
anathermal climatic change 增温期气候变化
anchor balloon 锚定气球
anchored platform 锚定平台
anchored station 锚定站
anchored trough 锚槽【大气】
anchor ice 锚冰【大气】,底冰
Anderson's theory 安德逊理论
andhi 对流性尘暴（常见于巴基斯坦和印度西北部）,安德海（音译）
androchore 人布植物【植】
anelastic approximate 滞弹性近似
anelastic equation 滞弹性方程
anelasticity 滞弹性【物】
anemobarometer 风速气压表
anemobiagraph 压管风速计
anemochore 风布植物【植】
anemocinemograph 电动风速计
anemoclinograph 风斜计,铅直风速仪
anemoclinometer 风斜表,铅直风速表
anemogenic curl effect 风生涡旋效应,风生涡度效应
anemogram 风力自记曲线,风速记录图
anemograph 风速计【大气】
anemology 测风学
anemometer 风速表【大气】,垂板风速表

anemometer factor 风速表系数（风速与风杯切线速度之比）
anemometer level 风速表高度
anemometer mast 测风杆
anemometer tower 测风塔
anemometer with stop-watch 停表风速表
anemometrograph 测风计,风向风速风压计
anemometry 风速测定法【大气】,测风法,测风术
anemophilae 风媒植物
anemophilous plant 风媒植物【植】
anemophily 风媒【植】
anemophobe 嫌风植物,避风植物
anemorumbograph 风向风速计【航海】
anemorumbometer 风向风速表【大气】
anemoscope 测风器
anemotachometer 风速表
anemovane 风向风速器
aneroid ①无液的,不湿的②空盒气压表
aneroid altimeter 空盒高度表
aneroid barograph 空盒气压计【大气】
aneroid barometer 空盒气压表【大气】
aneroid capsule 金属空盒,膜盒
aneroid chamber 真空膜盒,气压测量膜盒
aneroid manometer 无液压力计
aneroidogram 空盒气压曲线
aneroidograph 空盒气压计
angel 异常回波
angel echo 异常回波【大气】
angin 安金风（马来西亚的海陆风）
angin-darat 安金达拉特风（马来西亚的陆风）
angin-laut 安金劳特风（马来西亚的海风）
angle bracket 尖角括号
angle geothermometer 曲管地温表【大气】
angle in a circular segment 圆周角【数】
angle of aperture 孔径角
angle of arrival 到达角

angle of declination 偏角
angle of deflection 偏转角
angle of depression 俯角【数】
angle of deviation 偏向角【物】
angle of diffraction 衍射角【物】
angle of elevation 仰角【数】
angle of incidence 入射角
angle of inclination 倾角【数】
angle of minimum deviation 最小偏向角
angle of minimum resolution 最小分辨角【物】
angle of nutation 章动角【物】
angle of pitch 俯仰角
angle of precession 进动角【物】
angle of reflection 反射角
angle of refraction 折射角
angle of roll 滚动角,侧滚角
angle of rotation 旋转角【数】
angle of view 视角
angle of yaw 偏航角
angle stem earth thermometer 曲管地温表
angle thermometer 曲管温度表
Angola-Benguela Front 安哥拉-本格拉〔海流〕锋(海温、盐度)
Angola Current 安哥拉海流
Ångström 埃(长度单位,10^{-8} cm)
Ångström compensation pyrheliometer 埃斯特朗补偿式直接辐射表【大气】
Ångström pyrgeometer 埃斯特朗净大气辐射表
Ångström turbidity coefficient 埃斯特朗浑浊度系数
angular acceleration 角加速度【物】
angular coefficient 角系数
angular displacement 角位移【物】
angular distribution 角分布【物】
angular divergence 角散度
angular drift 碎石流
angular eclipse 环蚀
angular field-of-view 视场角
angular filter function 角滤波函数
angular frequency 角频率【物】
angular gyro-frequency 角回旋频率
angularity correction 角度校正,偏斜校正(测流断面的)
angular momentum 角动量【物】
angular momentum balance 角动量平衡【大气】
angular momentum plane 角动量平面
angular path length 角程长度
angular pattern 角型,角分布图
angular polarization 角偏振
angular position sensor 角度传感器【航海】
angular resolution 角分辨率【物】
angular spectrum 角度谱
angular spreading 角展宽【大气】
angular spreading factor 角展〔宽〕因子
angular velocity 角速度【物】
angular velocity of the earth 地球〔自转〕角速度
angular wave number 角波数
angular width 角幅
anharmonic oscillator 非谐振子【物】
anhydride 酸酐【化】
anhyetism 缺雨性
animal community 动物群落【地理】
animal ecology 动物生态学【动】
animal fog 动物雾
animal husbandry 畜牧业
animal husbandry meteorology 畜牧气象学【大气】
animal kingdom 动物界【动】
animal phenology 动物物候学
anion 负离子,阴离子
anisallobar 等正变压线
anisobaric 不等压的
anisokinetic sampling 不等动能取样法
anisotropic factor 各向异性因子
anisotropic medium 各向异性介质
anisotropic random medium 各向异性

随机介质
anisotropic scaling 各向异性尺度分析
anisotropic scattering 各向异性散射
anisotropic turbulence 各向异性湍流
anisotropy 各向异性【物】
anniversary wind 周年风【大气】,季节风【大气】
Anno Domini（A. D.） 公元（拉丁语名）
annual aberration 周年光行差【航海】
annual amount 年总量【大气】
annual anomaly 年距平
annual cycle 年循环
annual exceedance series 年超过数系列
annual flood 年最大〔洪峰〕流量,年最大洪水
annual flood series 年洪峰系列
annual maximum series 年最大〔流量〕系列
annual mean 年平均【大气】
annual minimum series 年最小〔流量〕系列
annual parallax 周年视差【天】
annual plant 一年生植物【植】
annual range 年较差【大气】
annual report 年报
annual ring 年轮【植】
annual runoff 年径流〔量〕
annual series 年系列,年极值系列（年最大最小系列）
annual storage ①年调节②年调节量,年蓄量
annual time span 年跨度
annual variation 年变〔化〕
annual wave 年波
annular solar eclipse 日环食【天】
annulus 环,环食带
annus abundans 闰年
annus deficiens 亏年
anode 阳极
anodic current 阳极电流【化】
anomalistic period 近点周期

anomalous 异常的,距平的
anomalous（cloud）line 异常云线
anomalous diffraction 反常衍射
anomalous dispersion 反常色散
anomalous extinction 反常消光
anomalous gradient wind 异常梯度风
anomalous high 反常高压
anomalous ionization 异常电离
anomalous low 反常低压
anomalous propagation 异常传播
anomaly 异常,距平
anomaly correlation 距平相关
anomaly of geopotential difference 重力位势差距平
anomaly of specific volume 比容距平
anoxia 缺氧
antarctic air（mass） 南极气团,南极空气
antarctic anticyclone 南极反气旋
Antarctic Bottom Water（AABW） 南极底层水
Antarctic Circle 南极圈
Antarctic Circumpolar Current（ACC） 绕南极洋流【大气】
Antarctic Circumpolar Water 南极绕极水,绕南极水
antarctic climate 南极气候【大气】
Antarctic Convergence Zone 南极辐合区
antarctic divergence 南极辐散〔带〕
antarctic front 南极锋【大气】
antarctic high 南极高压
Antarctic Ice Sheet 南极冰原
Antarctic Intermediate Water 南极中层水
Antarctic Oscillation（AAO） 南极涛动
Antarctic ozone distribution 南极臭氧分布
Antarctic ozone hole 南极臭氧洞【大气】
Antarctic Polar Front 南极极锋
Antarctic Polar Vortex 南极极涡
Antarctic Pole 南极
Antarctic realm 南极界【生】
antarctic sea smoke 南极海蒸气雾

antarctic stratospheric vortex　南极平流层涡旋
Antarctic Surface Water　南极表层水
Antarctic Treaty Consultative Meeting (ATCM)　南极条约协商会议
Antarctic Zone　南极区
antecedent precipitation index　前期降水指数
antecedent soil moisture　前期土壤湿度, 前期土壤水分
antenna　天线
antenna array　天线阵【电子】
antenna feed　天线馈电
antenna gain　天线增益
antenna limit　天线极限
antenna pattern　天线方向图【电子】
antenna temperature　天线温度
anthelic arc　反日弧
anthelic pillar　反日柱
anthelic point　反日点
anthelion　反日【大气】
anthesis　开花【植】
anthracometer　〔空气中〕二氧化碳测定仪
anthracometry　〔空气中〕二氧化碳测定法
anthropecology　人类生态学
anthropobiology　人类生物学
anthropocene　人类世(2000年提出的新词)
anthropochorous　人为传播的
anthropoclimatic drying power　人体干燥率
anthropoclimatology　人类气候学
anthropogenic　人为的,人类活动引起的
anthropogenic climate catastrophe　人为气候突变
anthropogenic climate change　人致气候变化【大气】,人为气候变化
anthropogenic emissions　人为排放
anthropogenic factor　人为因素
anthropogenic heat　人为热
anthropogenic impact　人类活动影响
anthropogenic sources　人类活动源
anthropogeography　人类地理学【地理】
anthroposphere　人类圈
anti-air-pollution system　防止空气污染系统
antiauxin　抗生长素
antibaric flow　反压流
anticlockwise　反时针
anticorona　对华(天象)
anticorrelation　反相关,负相关
anticrepuscular arch　反曙暮光弧
anticrepuscular ray　反曙暮辉
anticyclogenesis　反气旋生成【大气】
anticyclolysis　反气旋消散【大气】
anticyclone　反气旋【大气】
anticyclone foehn　反气旋焚风
anticyclone movement　反气旋移动
anticyclone subsidence　反气旋下沉
anticyclone tornado　反气旋龙卷
anticyclonic bora　反气旋式布拉风
anticyclonic circulation　反气旋环流【大气】
anticyclonic curvature　反气旋〔性〕曲率【大气】
anticyclonic divergence　反气旋〔性〕辐散
anticyclonic eddies　反气旋〔性〕涡旋
anticyclonic flow　反气旋气流
anticyclonic gloom　反气旋阴沉天气
anticyclonic inversion　反气旋逆温
anticyclonic phase　反气旋期
anticyclonic ridge　反气旋脊,高压脊
anticyclonic rotation　反气旋式旋转
anticyclonic shear　反气旋〔性〕切变【大气】
anticyclonic tornado　反气旋龙卷
anticyclonic vortex　反气旋〔性〕涡旋
anticyclonic vorticity　反气旋涡度【大气】
anticyclonic vorticity advection (AVA)　反气旋涡度平流
antielectron　反电子
anti El Nino　反厄尔尼诺【大气】
antifoggant　防雾剂
anti-freezing　防冻

antigravity 抗重力
antihail gun 防雹炮
antihail rocket 防雹火箭
antihelion(=anthelion) 反假日
anti-icing 防积冰
Antilles Current 安的列斯海流
anti-logarithm 反对数【数】
antilog 逆类比
antimonsoon 反季风
anti-mountain wind 反山风
antineutrino 反中微子
antineutron 反中子
antiparticle 反粒子【物】
antiphase 反相〔位〕【物】
antipleion 负偏差中心
antiproton 反质子
antiradiance 防辐射
anti-sea breeze 反海风
antiselene 反假月
anti-solar point 对日点
anti-Stokes fluorescence 反斯托克斯荧光
anti-Stokes line 反斯托克斯线
anti-symmetric matrix 反称矩阵【数】
antisymmetry(=anti-symmetry) 反对称〔性〕【物】
anti-trade 反信风【大气】
anti-transpirant 抗蒸腾的
antitriptic wind 摩擦风【大气】,减速风
anti-twilight 反曙暮光
anti-twilight arch 反曙暮光弧
anti-valley wind 反谷风
anti-wind 逆风
anvil cloud 砧状云
anvil dome 砧云丘,砧状云圆顶
AO (Arctic Oscillation) 北极涛动
AOD (aerosol optical depth) 气溶胶光学厚度
AOSIS(Alliance of Small Island States) 小岛国联盟(1991年成立)
aparktias 阿帕尔克梯风(希腊的北风)
apartment microclimate 住宅小气候

APCS (automatic weather-data processing and communication control system) 自动天气资料加工和通信控制系统
aperiodic flow 非周期流
aperiodicity 非周期性【物】
aperiodic oscillation 非周期振动
aperiodic signal 非周期信号
aperiodic solution 非周期解
aperture ①孔径②孔③光圈【测】
aperture averaging 孔径平均
aperture smooth effect 孔径修匀效应
aperwind (=alpach,aberwind) 阿伯风
aphelion 远日点【天】
Apheliotes 东风(希腊语名)
API (air pollution index) 空气污染指数
Ap index Ap指数(地磁扰动)
APO (Asian-Pacific Oscillation) 亚洲-太平洋涛动
apob 飞机仪器观测(airplane observation的缩写)
apocenter 远心点【天】
apocynthion 远月点
apodization 切趾法,衍射控象法
apogean range 远地点潮差
apogean tide 远月潮
apogean winds 阿波金风(希腊的陆风)
apogee 远地点【天】
Apollo 阿波罗
apostilb 阿普熙提(亮度单位)
APP (Asia-Pacific Partnership on Clean Development and Climate) 亚太清洁发展与气候伙伴计划(2005年制订)
APP (AVHRR Polar Pathfinder) 先进甚高分辨率辐射仪极地探索者〔计划〕
apparatus 仪器
apparent altitude 视高度【航海】
apparent brightness 视亮度
apparent diameter 视直径
apparent force 视示力【大气】
apparent form of the sky 天穹形状【大气】
apparent freezing point 视冰点

apparent gravity 视重力
apparent groundwater velocity 视地下水流速
apparent heat 视示热
apparent heat source 视热源
apparent horizon 视地平线【测】
apparent luminance 视亮度
apparent magnitude 视星等【天】
apparent moist sink 视水汽汇
apparent motion of the sun 太阳视运动
apparent noon 视午
apparent position 视位置【航海】
apparent solar day 视太阳日
apparent solar time 视太阳时
apparent stress 视应力
apparent sun 视太阳【航海】
apparent vapour source 视水汽源
apparent velocity 视速度
apparent vorticity source 视涡〔度〕源
apparent wind 视风【航海】
appendix ①附录②充气管(气球)
Appleton anomaly 阿普尔顿异常【地】
Appleton layer 阿普尔顿层
application software 应用软件【计】
Applications Technology Satellite（ATS） 应用技术卫星【大气】
applied climatology 应用气候学【大气】
applied geography 应用地理学【地理】
applied hydrology 应用水文学
applied map 专用地图【地理】
applied meteorology 应用气象学【大气】
applied optics 应用光学【物】
applied physics 应用物理〔学〕【物】
appointed aerodrome weather report 机场预约天气报告【大气】
approach channel 引水渠,引航道,引槽
approach-light contact height 目视进场高度
approach velocity 行近流速
approach visibility 倾斜能见度,进场能见度

approximate absolute temperature scale 近似绝对温标
approximate representation 近似表示【数】
approximate solution 近似解【数】
approximate value 近似值【数】
approximation 逼近【数】
approximation equation 逼近方程
approximation error 逼近误差【数】
a priori estimate 先验估计【数】
a priori probability 先验概率
a priori reason 先验理由
apsidal line 拱线
APT(automatic picture transmission) 自动图像传输【大气】
APT signal simulator 自动图像传输信号模拟器
aqual landscape 水成景观【地理】
aqua regia 王水【化】
Aqua Satellite 埃夸卫星(EOS下午轨道星)
aquasonde 含水量探空仪
aquatic plant 水生植物【林】
aqueous aerosol 湿气溶胶【大气】
aqueous vapour 水汽
aquiclude ①隔水层②弱透水层
aquifer 含水层【地质】
aquifer system 含水层系
aquifer test 含水层检验
aquifuge 不透水层【地质】,隔水层,无水层(既不储水,也不导水)
aquitard 弱透水层【地质】
AR4（IPCC Fourth Assessment Report） IPCC 第4次评估报告
aracaty 阿拉卡蒂风(巴西锡阿腊的东北风)
Arago distance 阿拉果距
Arago point 阿拉果点【大气】
Arago's neutral point 阿拉果中性点
Arakawa jacobian 荒川-雅可比近似,荒川-雅可比〔算子〕
arbor 乔木【植】
arc 弧

ARCAS-ROBIN system 阿卡斯-洛宾火箭探空系统
arc cloud 弧状云【大气】
arc cloud line 弧状云线【大气】
arc discharge 弧光放电
archaeological period 考古时期
archaeology 考古学
archaeomagnetism 古磁学
archaic 古代的
Archean Eonothem 太古宇【地质】
Archean Eon 太古宙【地质】
Archean Era 太古代
arched squall 拱状云飑
Archeozoic Era 太古代
Archimedean buoyant force 阿基米德浮力
Archimedes principle 阿基米德原理【物】
architectural meteorology 建筑气象学
archive ①档案②存档
arch twilight 曙暮光弧
arc of contact of halo 珥
arcs of Lowitz 洛维茨〔晕〕弧
arctic air mass 北极气团【大气】
arctic-alpine 寒极高山区,高寒区,高寒的
Arctic and Antarctic Scientific Research Institute(AASRI) 〔前苏联〕南北极科学研究所
arctic anticyclone 北极反气旋【大气】
arctic blackout 北极无线电衰失现象
Arctic Bottom Water 北极底层水
Arctic Circle 北极圈
arctic climate 北极气候【大气】
Arctic Climate System Study(ACSYS) 北极气候系统研究(计划)
arctic continental air(mass) 北极大陆气团,北极大陆空气
Arctic Current 北极海流
arctic desert 北极荒漠,酷寒荒原(生态)
arctic fog 北极雾
arctic front 北极锋【大气】
arctic haze 北极霾【大气】
arctic high 北极高压

Arctic Ice Dynamics Joint Experiment (AIDJEX) 北极冰动力学联合试验
Arctic Intermediate Water 北极中层水
arcticization ①北极化②北极装备,低温装备
arctic mist 北极霭,北极冰雾
Arctic Oscillation(AO) 北极涛动
arctic pack 北极浮冰群【大气】,北极陈冰(两年以上析出盐分的冰,厚度在25 m以上)
Arctic Polar Front 北极锋
Arctic Pole 北极
arctic(sea)smoke 北冰洋〔烟〕雾【大气】
arctic stratospheric vortex 北极平流层涡旋
Arctic Surface Water 北极表层水
arctic tree line 北极林木线
arctic tropopause 北极对流层顶
arctic warming 北极增温
arctic weather station 北极天气站
arctic whiteout 北极乳白天空现象,乳白辉
arctic wind 北极风
arctic zone(= north frigid zone) 北极带,北寒带
arcus(arc) 弧状(云)
ARDC model atmosphere 美国空军〔研究发展司令部的〕标准大气
ardometer 光测高温表
area array 面阵【电子】
area average 面积平均,区域平均
area covered with echoes 回波覆盖区域
area-elevation curve 面积-高程曲线
area forecast 区域预报
Area Forecast Centre(AFC) 区域预报中心
area index 面积指数
areal monitoring 区域监测
areal precipitation 面降水〔量〕【大气】,区域降水
areal reduction factor 区域衰减因子
areal velocity 面积速度,掠面速度

area mean rainfall 区域平均雨量【大气】
area of coverage ①覆盖区②观测区 ③影响区,所及范围
area of high pressure 高压区
area of influence 影响区
area precipitation 区域降水
area solidly covered with echoes 回波密实覆盖区域
area source 面源
arene 芳炔【化】
ARFOT 英制单位区域预报国际电码
ARGO(Array for Real-time Geostrophic Oceanography) 地转海洋学实时观测阵(1998年提出)
argon 氩〔气〕
argon ion laser 氩离子激光器【电子】
argon〔ion〕laser 氩〔离子〕激光器【物】
ARGOS(Automatic Relay Global Observation System) 全球观测自动中继系统
argument ①自变量②幅度,幅角
arheic(=areic, aretic, arhetic) 无流的,无河的
arid 干燥〔的〕
arid climate 干旱气候【大气】
arid cycle 干燥周期
aridity 干燥度【大气】,干燥性,干旱性
aridity coefficient 干燥系数
aridity factor 干燥因子
aridity index 干燥度指数【大气】
arid land 旱地
arid region 干〔燥〕区〔域〕
arid zone 干旱带
arid-zone hydrology 干旱带水文学
arifi 阿里法风(sirocco风的北非方言)
arithmetic address 运算地址
arithmetic mean 算术平均【数】
arithmetic series 等差级数【数】
ARMA model(autoregressive and moving average model) 自回归滑动平均模式【大气】
ARMET 公制单位区域预报国际电码

armoured thermometer 铠装温度表(测海面温度用)
armouring ①护面层②装甲的,铠装的
Army-Navy ground meteorological device (AN/GMD) 美国陆海军地面气象装备(高空探测用的地面设备)
aromatic hydrocarbon 芳(香族)烃
arouergue(=rouergue) 阿洛厄尔格风(法国中央高原南部的强西风)
around-the-clock 昼夜
arowagram 高空热力图(美国海军用)
ARPANET 高级研究计划局网络【计】,阿帕网
array ①阵列【计】②数组【计】③天线阵
array antenna 阵天线【电子】
Array for Real-time Geostrophic Oceanography(ARGO) 地转海洋学实时观测阵(1998年提出)
arrested topographic wave 截留地形波
Arrhenius equation 阿伦尼乌斯方程【化】
Arrhenius expression 阿伦尼乌斯表达式
Arrhenius theory of electrolytic dissociation 阿伦尼乌斯电离理论【化】
arrival time 到时【地】
arrival time difference 到时差【地】
arrival-time difference technique 到〔达〕时差法
arroyo 干谷、旱谷,干涸河道
artesian aquifer ①自流水〔含水层〕②承压含水量
artesian basin 自流水盆地,自流供水区域
artesian groundwater 自流地下水,承压地下水
artesian well 自流井
artificial boundary 人造边界〔条件〕
artificial boundary condition 人为边界条件
artificial climate 人造气候
artificial cloud 人造云
artificial contaminant 人为污染
artificial control 人工控制
artificial dissipation 人工消散

artificial edge   假边缘
artificial horizon   ①人工地平②人工地平仪,陀螺地平仪
artificial ice nucleus   人工冰核
artificial illumination   人工照明
artificial intelligence（AI）   人工智能【计】
artificial island   人工岛【航海】
artificial lighting   人工照明
artificial light source   人工光源
artificially initiated lightning   人工引发闪电
artificial microclimate   人工小气候【大气】
artificial neural net（=artificial neural network）   人工神经网络【电子】
artificial nucleation   人工成核作用【大气】
artificial nucleus   人工核
artificial precipitation   人工降水【大气】
artificial precipitation stimulation   人工增雨
artificial radioactivity   人工放射性【物】
artificial radio element   人造放射性元素【化】
artificial rain   人造雨
artificial recharge   人工回灌,人工地下水灌注
artificial reforestation   人工更新【林】
artificial satellite   人造卫星
artificial sodium cloud   人造钠云
artificial ventilation   人工通风
artillery meteorological condition   炮兵标定气象条件
artillery meteorological service   炮兵气象勤务
artillery meteorological standard   炮兵气象标准
ascendent   升度
ascending air   上升空气
ascending current   上升气流
ascending motion   上升运动
ascending node（AN）   升交点【天】
ascending node longitude   升交〔点〕经度
ascending node time   升交〔点〕时刻
ascending prominence   上升日珥

ascending velocity   上升速度
ascension（hydrothermal）theory   升腾〔学〕说
ascension rate of balloon   气球上升速率
ascent   上升,爬高
ascent curve   上升曲线
A scope   A 型显示器【大气】,A 示波器
A-scope display   A 型显示
A-scope indicator   A 型显示器
ASDIC（=sonar, Antisubmarine Detection Investigation Committee 的缩写） 声呐
ash   〔火山〕灰,灰分
ash air   含灰空气
ash cloud   烟灰云【大气】
ash content   灰分【航海】
ash devils   尘旋,火山灰暴
ash fall   灰尘沉降,烟尘降落
ash-grey light   灰色光
ash shower   火山灰暴
Asia and Pacific Council（ASPAC）   亚洲及太平洋理事会
Asian-Australian monsoon system   亚澳季风系统
Asian-Pacific Oscillation（APO）   亚洲-太平洋涛动
Asia-Pacific Partnership on Clean Development and Climate（APP）   亚太清洁发展与气候伙伴计划(2005 年制订)
asiderite   石陨星,陨石
asifat   阿西法阵风(阿拉伯海的热带气旋)
ASO（auxiliary ship observation）   辅助船舶观测【大气】
ASOS（automated surface observing system）   地面自动观测系统
ASPAC（Asian and Pacific Council）   亚洲及太平洋理事会
aspect   ①外貌②方向,方位③方面
aspect ratio   纵横比【物】,形态比
aspirated electrical capacitor   通风电容仪(测大气电导率用)
aspirated meteorograph   通风气象计
aspirated psychrometer   通风干湿表【大气】

aspirated quartz-crystal thermometer 通风石英晶体温度表
aspirated thermometer 通风温度表
aspiration condenser 通风电容器
aspiration meteorograph 通风气象计【大气】
aspirator 通风器,风扇抽气管
aspre 阿斯普尔风(法国中央高原的东北焚风)
assembly language 汇编语言【计】
assembly program 汇编程序
assessment 评价,评估,估定
assessment of atmospheric environment 大气环境评价【大气】
assignment statement ①指派②赋值【计】
assimilation ①同化〔作用〕②吸收
Assmann psychrometer 阿斯曼干湿表【大气】
Assmann ventilated psychrometer 阿斯曼通风干湿表
associated Legendre function 连带勒让德函数【数】
association coefficient 相伴系数,连带系数
assumption 假设
astatic 无定向〔的〕
ASTER (Advanced Spaceborne Thermal Emission and Reflection radiometer, EOS) 先进星载热发射和反射辐射仪(EOS)
asteroid 小行星
asteroid belt 小行星带
astigmatic beam 像散光束
astigmatism 像散【物】
As tra 透光高层云【大气】
astraphobia 恐雷〔电〕感
astrobiology 天体生物学【天】
astro-climatic index 天文-气候指标
astro-geodetic network 天文大地网【测】
astrogeology 天体地质学【地质】
astrolabe 等高仪【测】
astrolabe ①星盘 ②等高仪【天】

astrology 占星术【天】
astrometeorology 天文气象学
astrometry 天体测量学【天】
astronaut 宇航员
astronautics 宇宙航行,宇航学,航天学
astronavigation 天文导航【天】
astronomical constant 天文常数
astronomical coordinate 天文坐标【航海】
astronomical dating 天文学定年法
astronomical day 天文日
astronomical factor 天文因子
astronomical fix 天文船位【航海】
astronomical horizon 天文地平
astronomical latitude 天文纬度【航海】
astronomical longitude 天文经度【航海】
astronomical optics 天文光学【物】
astronomical positioning system 天文定位系统【测】
astronomical refraction 大气折射〔差〕,蒙气差
astronomical scintillation 天文闪烁
astronomical seeing 天体视宁度【天】
astronomical telescope 天文望远镜【物】
astronomical theory of climate change 气候变化的天文学理论
astronomical triangle 天文三角形【航海】
astronomical twilight 天文曙暮光
astronomical unit(AU) 天文单位【天】
astronomy 天文学【天】
astrophysical plasma 天体物理学等离〔子〕体【物】,天体等离体
astrophysics 天体物理学【天】
astrospace 宇宙空间
astrospectroscopy 天体光谱学【物】
asymmetrical top 不对称陀螺【化】
asymmetric circulation pattern 不对称环流型
asymmetric effect 东西效应(宇宙线)
asymmetric wave structure 不对称波结构
asymmetry 不对称性
asymmetry factor 不对称因子

asymmetry parameter 不对称参数
asymptote 渐近线【数】
asymptote of confluence 合流渐近线
asymptote of convergence 辐合渐近线
asymptote of divergence 辐散渐近线
asymptotic analysis 渐近分析
asymptotic expansion 渐近展开
asymptotic solution 渐近解
asymptotic theory 渐近理论
asymptotic value 渐近值【数】
asynchronous communication 异步通信
asynchronous teleconnection 非同步遥相关,非同期遥相关
asynchronous transfer mode(ATM) 异步传送模式【计】
asynoptic data 非天气图定时资料
asynoptic observation 非规定天气观测,补充观测
asynoptic time 非规定天气观测时间
ATCM (Antarctic Tready Consultative Meeting) 南极条约协商会议
athermancy 不透辐射热性
athermous 不透辐射热〔的〕
Atlantic Meridional Overturning Circulation(AMOC) 大西洋经向翻转流
Atlantic Ocean 大西洋
Atlantic Oceanographic and Meteorological Laboratories(AOML) 大西洋海洋气象实验室
Atlantic subtropical dipole 大西洋副热带偶极子
Atlantic time 大西洋时间
Atlantic Tradewind Experiment(ATEX) 大西洋信风试验
Atlantic Water 大西洋水
atlas 地图册,图集
ATM(asynchronous transfer mode) 异步传送模式【计】
atmidometer 蒸发表
atmidometry 蒸发率测定法
atmidoscope 湿度指示器

atmology 水汽学
atmometer 蒸发表
atmometry 蒸发测定法
atmoradiograph 天电强度计
atmosphere ①大气【大气】②大气圈 ③大气层
atmosphere heat flow 大气热流〔动〕
atmosphere isotopic composition 大气同位素成分
atmosphere-mixed layer ocean model 大气-海洋混合层模式
atmosphere monitoring system 大气监测系统
atmosphere noble gas 大气惰性气体
atmosphere-ocean interaction 大气-海洋交互作用〔与海气(air-sea)交互作用类似,但尺度可更大〕
atmosphere-ocean-land coupled model 大气-海洋-陆面耦合模式,气海陆耦合模式
atmospheric absorption 大气吸收【大气】
〔atmospheric〕absorptivity 〔大气〕吸收率【大气】
atmospheric accumulation 大气累积〔作用〕
atmospheric acoustics 大气声学【大气】
atmospheric aerosol 大气气溶胶
Atmospheric and Land Surface Processes Project(ALSPP) 大气和陆面过程计划
atmospheric assimilation 大气同化〔作用〕
atmospheric attenuation 大气衰减【大气】
atmospheric attenuation coefficient 大气衰减系数
atmospheric background 大气本底〔值〕【大气】
atmospheric backscattering 大气后向散射
atmospheric backscattering coefficient 大气后向散射系数
atmospheric basic equation 大气基本方程
atmospheric behaviour 大气行为,大气性状
atmospheric billow 大气波

atmospheric blocking 大气阻塞
atmospheric boil 大气闪晃
atmospheric boundary layer 大气边界层【大气】
Atmospheric Brown Cloud（ABC） 大气棕色云
atmospheric carcinogen 大气致癌物
atmospheric center of action 大气活动中心【大气】
atmospheric chemical composition 大气化学成分
atmospheric chemistry 大气化学【大气】
atmospheric circulation 大气环流【大气】
atmospheric circulation model 大气环流模式
atmospheric cleaning 大气净化【大气】
atmospheric cleaning mechanism 大气净化机制,大气清洁机制
Atmospheric Cloud Physics Laboratory（ACPL，NASA） 大气云物理实验室（NASA）
atmospheric column 气柱
atmospheric composition 大气成分【大气】
atmospheric constituents 大气组分
atmospheric contamination 大气污染
atmospheric correction 大气订正
atmospheric counter radiation 大气逆辐射【大气】
atmospheric demand 大气需水量
atmospheric density 大气密度【大气】
atmospheric diffusion 大气扩散【大气】
atmospheric diffusion equation 大气扩散方程【大气】
atmospheric dilution 大气稀释
atmospheric dispersion 大气色散
atmospheric dispersoid 大气弥散胶体
atmospheric disturbance 大气扰动【大气】
atmospheric drought 大气干旱
atmospheric duct 大气波导【大气】
atmospheric dust 大气尘埃
atmospheric dynamics 大气动力学【大气】

atmospheric dynamo 大气发电机
atmospheric electric conductivity 大气电导率【大气】
atmospheric electric field 大气电场【大气】
atmospheric effect 大气效应
atmospheric electricity ①大气电学【大气】②大气电
atmospheric electric potential chart 大气电位图
atmospheric energetics 大气能量学
atmospheric engine 大气热机
atmospheric envelope 大气层（包围地球的）
atmospheric environment 大气环境
atmospheric environment capacity 大气环境容量【大气】
Atmospheric Environment Service（AES） 大气环境局（加拿大）
atmospheric evolution 大气演化
atmospheric explosion 大气层爆炸
atmospheric extinction 大气消光【大气】
atmospheric extinction coefficient 大气消光系数
atmospheric feedback mechanism 大气反馈机制
atmospheric fluctuation 大气振荡
atmospheric forcing 大气强迫【大气】
atmospheric gauge 气压计
atmospheric general circulation model（AGCM） 大气环流模式
atmospheric geology 气界地质学
atmospheric haze 大气霾
atmospheric heat balance 大气热平衡,大气热差额
atmospheric heating source 大气加热源
atmospheric humidity 大气湿度
atmospheric hydrodynamics 大气流体动力学
atmospheric hygrometry 大气测湿法
atmospheric hypothesis 大气成分变化说
atmospheric impurity 大气杂质【大气】
Atmospheric Infrared Sounder（AIRS）

大气红外探测器
〔atmospheric〕instability 〔大气〕不稳定度【大气】
Atmospheric Instrumentation Research, Inc.(AIR) 大气探测仪器研究公司（美国）
atmospheric interference 天电,远程雷电
atmospheric inversion 大气逆温
atmospheric ion 大气离子【大气】
atmospheric ionization 大气电离〔作用〕
atmospheric layer 大气层
atmospheric long wave 大气长波【大气】
atmospheric mass 大气质量【大气】
atmospheric metamorphism 大气变性
atmospheric model 大气模式
Atmospheric Models Intercomparison Project(AMIP) 大气模式比较计划
atmospheric moisture 大气湿度
atmospheric monodispersion 大气单分散粒径分布
atmospheric noise 大气噪声【大气】,天电干扰
atmospheric obscurant 大气掩星
atmospheric opacity 大气不透明度
atmospheric optical depth 大气光学厚度【大气】
atmospheric optical mass 大气光学质量【大气】
atmospheric optical phenomena 大气光学现象【大气】
atmospheric optical spectrum 大气光谱【大气】
atmospheric optical thickness 大气光学厚度【大气】
atmospheric optics 大气光学【物】
atmospheric origin 大气起源
atmospheric oscillation 大气振荡,大气波动
atmospheric oxidant 大气氧化剂
atmospheric ozone 大气臭氧【大气】
atmospheric particle 大气粒子
atmospheric particulate content 大气微粒含量
atmospheric penetration 〔进入稠密〕大气层飞行
atmospheric phenomenon 大气现象
atmospheric photochemistry 大气光化学【大气】
atmospheric photolysis 大气光解〔作用〕【大气】
atmospheric physics 大气物理〔学〕【大气】
Atmospheric Physics and Chemistry Laboratory(APCL,NOAA) 大气物理和大气化学实验室(NOAA)
atmospheric polarization 大气偏振【大气】,大气极化
atmospheric pollutant 大气污染物【大气】
atmospheric pollutant gas 大气污染气体
atmospheric pollution 大气污染【大气】
atmospheric pollution monitoring 大气污染监测【大气】
atmospheric pollution sources 大气污染源【大气】
atmospheric polydispersion 大气多分散粒径分布
atmospheric precipitation 大气降水
atmospheric predictability 大气可预报性
atmospheric pressure 气压【大气】
atmospheric probing 大气探测
atmospheric property 大气特性
atmospheric quality standard 大气质量标准【大气】,大气品位标准【大气】
atmospheric radiation 大气辐射【大气】,长波辐射【大气】
atmospheric radiation budget 大气辐射收支
atmospheric radioactivity 大气放射性【大气】
atmospheric Rayleigh scattering cross section 大气瑞利散射截面
atmospheric refraction 大气折射【大气】
atmospheric regenerant 大气更新剂
atmospheric region 大气圈〔层〕
atmospheric remote sensing 大气遥感【大气】
atmospheric removal 大气排除
atmospheric retrieval problem 大气反演问题

atmospherics 天电【大气】
atmospheric salinity 大气盐度
atmospheric scale height 均质大气高度,大气标高【大气】
atmospheric scavenging 大气净化
atmospheric science 大气科学【大气】
atmospheric screening height 大气屏蔽高度
atmospheric seeing 大气视宁度【天】
atmospheric shell 大气圈〔层〕
atmospheric shimmer 大气闪烁
atmospheric sodium 大气钠
atmospheric sounding 大气探测
atmospheric sounding and observing 大气探测【大气】
atmospheric sounding projectile 〔高层〕大气探测火箭
atmospheric spectrum 大气光谱
〔atmospheric〕stability 〔大气〕稳定度【大气】
atmospheric stochastic noise 大气随机噪声
atmospheric stratification 大气层结【大气】,大气分层
atmospheric structure 大气结构
atmospheric subdivision 大气分层【大气】
atmospheric surface layer 大气地面层
atmospheric suspended matter 大气悬浮物【大气】
Atmospheric Technology Division (ATD, NCAR) 大气技术部(NCAR)
atmospheric temperature 大气温度
atmospheric temperature measurement 大气温度测量
atmospheric thermodynamics 大气热力学【大气】
atmospheric tidal oscillation 大气潮汐振荡
atmospheric tide 大气潮【大气】
atmospheric total ozone 大气总臭氧
atmospheric trace gas 大气痕量气体【大气】
atmospheric trace molecular spectroscopy (ATMOS) 大气痕量分子光谱仪
atmospheric transmission ①大气透射 ②大气传递
atmospheric transmission model 大气传输模式【大气】
atmospheric transmissivity 大气透射率【大气】
atmospheric transmittance 大气透射
〔atmospheric〕transparency 〔大气〕透明度【大气】
atmospheric transparency window 大气透明窗
atmospheric transport 大气输运,大气传输
atmospheric turbidity 大气浑浊度【大气】
atmospheric turbulence 大气湍流【大气】
Atmospheric Turbulence and Diffusion Laboratory (ATDL) 大气湍流和扩散实验室(美国)
atmospheric visibility 大气能见度
atmospheric volume extinction coefficient 大气体〔积〕消光系数
atmospheric vortex 大气涡旋
atmospheric water 大气水
atmospheric water budget 大气水分收支
atmospheric wave 大气波动【大气】
atmospheric window 大气窗【大气】
atmospheric window region 大气窗区
atmospherium ①大气层次②大气馆
ATN (Advanced TIROS-N) 先进泰罗斯-N卫星【大气】
ATOLL (analysis of tropical oceanic low-levels) 热带海洋地区低层分析(缩写词原义"环礁")
ATOLL Chart 热带海洋地区低层分析图
atom 原子
atomic absorption spectrometry 原子吸收谱测量
atomic bomb cloud 原子弹云
atomic clock 原子钟【天】
atomic cloud 原子〔爆炸〕烟云
atomic energy 原子能【物】

Atomic Energy Commission (AEC) 美国原子能委员会
atomic fluorescence spectrometry 原子荧光光谱法【化】
atomic mass unit 原子质量单位【物】
atomic model 原子模型【化】
atomic nuclear physics 原子核物理〔学〕【物】
atomic nucleus 原子核【物】
atomic number 原子序数【化】,原子序
atomic orbital 原子轨道【化】
atomic physics 原子物理〔学〕【物】
atomic spectrum 原子光谱【物】
atomic structure 原子结构【物】
atomic time(AT) 原子时【航海】
atomizer 雾化器【航海】
ATOVS (Advanced TIROS Operational Vertical Sounder) 先进泰罗斯业务垂直探测器
ATS(Applications Technology Satellite) 应用技术卫星【大气】
attached thermometer 附属温度表
attachment 附件
attachment coefficient 附着系数
attainable degree of supercooling 极限过冷度
attended station 有人测站
attenuation 衰减
attenuation coefficient 衰减系数【大气】
attenuation constant 衰减常数
attenuation cross-section 衰减截面【大气】
attenuation factor 衰减因子
attenuation length 衰减长度
attenuation of solar radiation 太阳辐射衰减
attitude angle 姿态角【航海】
attitude control 姿态控制
attitude determination 姿态确定
attitude error 姿态误差
attitude horizon sensor 姿态水平传感器
attitude measurement sensor 姿态测量传感器
attitude of satellite 卫星姿态
attitude precision 姿态精度
atto- 阿〔托〕(词头,10^{-18})
attraction ①吸引②引力
attractor 吸引子【数】
attribution 归因〔研究〕
audibility 能听度
audibility meter 听度表
audibility zone 可闻区
audible sound 可听声
audio frequency 声频
audio-modulated radiosonde 声频调制探空仪
auditory threshold 听阈,闻阈
Auger shower 俄歇簇射【地】
augmented matrix 增广矩阵【数】
Aura (=ora) 奥拉风(意大利加尔达湖的谷风)
aurassos 奥拉素风(法国罗讷河的西北强风)
aureole 华盖
auro 奥洛风(法国阿尔卑斯山的干热西南风)
aurora 极光【地】
aurora australis 南极光【地】
aurora borealis 北极光【地】
aurora ionization 极光电离
auroral arc 极光弧
auroral band 极光带【大气】
auroral belt 极光带【地】
auroral break-up 极光崩离
auroral bulge 极光隆起
auroral cap 极光盖
auroral cloud 极光云
auroral corona 极光冕【大气】
auroral curtains 极光幔
auroral drapery 极光幔
auroral electrojet 极光带电集流【地】
auroral electrojet (AE) index 极光带电集流指数(即高纬亚暴指数)

auroral excitation mechanisms 极光激发机理
auroral green line 极光绿谱线
auroral hiss 极光嘶声(0.5～500 kHz 天然电磁噪声)
auroral infrasonic wave 极光次声波
auroral ionosphere 极光带电离层
auroral(或 aurora) magnetic disturbance 极光磁扰
auroral oval 极光卵【大气】,极光卵形环【地】
auroral particle 极光粒子
auroral particle precipitation 极光粒子沉降
auroral rays 极光射线
auroral spectrum 极光光谱
auroral storm 极光暴
auroral substorm 极光亚暴
auroral zone 极光地带
aurora polaris 北极光
Austausch 交换(德语名)
austausch coefficient 交换系数
auster(=ostria) 奥斯特风(保加利亚沿海的干南风)
Australasian 澳洲界【地理】,大洋洲的
Australasian Mediterranean Water(AAMW) 大洋洲的地中海海水
Australia Current 澳大利亚海流
Australian Bureau of Meteorology(BOM) 澳大利亚气象局
Australian National Committee for Antarctic Research(ANCAR) 澳大利亚南极研究国家委员会
austral pole 南极
austru 焚风(罗马尼亚语名)
autan(=altanus) 奥唐风(法国南部的东南大风)
authorized station 正规测站
auto-alarm 自动报警装置
auto-analyzer 自动分析仪
autobarotropic 自〔动〕正压的
autobarotropic atmosphere 自〔动〕正压大气
autobarotropy 自〔动〕正压状态
autocode 自动编码
auto-collimator 准直望远镜,光学测角仪
auto command scanner 自动控制扫描器
autoconvection 自动对流
autoconvection gradient 自动对流梯度
autoconvective instability 自动对流不稳定度
autoconvective lapse rate 自动对流直减率
auto-conversion 自动转化(小水滴成对并合)
autocorrelation 自相关【大气】
autocorrelation coefficient 自相关系数
autocorrelation function (ACF)自相关函数【数】
autocorrelation spectrum 自相关谱
autocorrection 自动校正
autocorrelogram 自相关图
auto-covariance 自协方差
autocovariance spectrum 自协方差谱
auto-exhaust 汽车废气
auto-exhaust catalyst 汽车尾气催化剂【化】
autogenous electrification 自〔生〕起电
autographic instrument 自记仪器,自动图示仪器
autographic records 自记记录【大气】
autolevelling assembly 自动校平装置【航海】
automat(on) 自动装置
automated analysis 自动分析
automated logic inference 自动逻辑推理【计】
automated search 自动搜索,自动寻优
automated surface observing system(ASOS) 地面自动观测系统
automated verification system 自动验证系统【计】
automatically control 自动控制
automatic celestial navigation 自动天文导航
automatic check 自动检验【计】

automatic chloride analyzer　氯化物自动分析仪
automatic climatological recording equipment（ACRE）　自动气候测录装置
automatic cloud amount recorder　自动云量记录器
automatic conductivity analyzer　自动电导率分析仪
automatic data editing and switching system（ADESS）　资料自动编辑转接系统
automatic data exchange system　自动数据交换系统
automatic data processing　自动资料处理【大气】
automatic direction-finder　自动测向器（测天电方向用）
automatic error request equipment　自动误差校正装置
automatic evaporation pan（AUTOVAP）　自动蒸发皿
automatic evaporation station　自动蒸发站
automatic fluorescent particle counter　荧光粒子自动计数器
automatic following control　自动跟踪控制
automatic frequency control　自动频率控制【电子】
automatic gain control　自动增益控制【电子】
automatic meteorological observing station　自动气象观测站
automatic meteorological oceanographic buoy　〔自动〕海洋气象浮标站
automatic meteorological station　自动气象站【大气】
automatic multifrequency ionospheric recorder　自动多频电离层记录器
automatic paging　自动分页【计】
automatic picture transmission（APT）　自动图像传输【大气】
automatic precipitation（rain，snow）collector　自动降水（雨、雪）收集器
automatic precipitation sampler　自动降水取样器
automatic program control　自动程序控制
automatic programming　自动程序设计【计】
automatic radiometeorograph　自记无线电气象仪
automatic radio rain-gauge　自动无线电雨量计
automatic range tracking（A.R.T.）　自动远距离跟踪
automatic recorder　自记器
Automatic Relay Global Observation System（ARGOS）　全球观测自动中继系统（通过卫星收集环境资料并进行观测平台定位）
automatic sensibility control　自动灵敏度控制
automatic sequential precipitation sampler　顺序式自动降水取样器
automatic sprinkler system　自动洒水灭火系统
automatic standard magnetic observatory-remote　自动标准地磁遥测台
automatic standard magnetic observatory　自动标准地磁观测台
automatic station　自动测站
automatic station keeping　自动台站管理
automatic sunshine recorder　自动日照计
automatic suspended solids analyzer　自动浮粒分析仪
automatic test equipment　自动测试设备【电子】
automatic tide gauge　验潮器
automatic tracking　自动跟踪,自动追踪
automatic transmission　自动透射,自动传递
automatic turbidity analyzer　浑浊度自动分析仪
automatic volume control　自动音量控制

【电子】
automatic weather-data processing and communication control system (APCS) 自动天气资料加工和通信控制系统
automatic weather observing system (AWOS) 自动天气观测系统
automatic weather station (AWS) 自动气象站
automation gain control 自动增益控制
automation of field operation and services (AFOS) 〔美国〕业务和服务自动化系统
automobile emission 汽车废气排放
automobile exhaust 汽车废气
automonitor 自动监测仪
auto monitoring 自动检测
automotive emission 自动发射
autonomous system 自治系统【计】,自主系统
auto pollution 汽车污染
autoregression method 自回归法
autoregression model 自回归模型【数】
auto-regressive integrated moving average (ARIMA) 自回归积分滑动平均
autoregressive model 自回归模型【大气】
auto-regressive moving average (ARMA) 自回归滑动平均
autoregressive moving-average model 自回归滑动平均模型【数】
autoregressive process 自回归过程
autoregressive series 自回归级数
autospectrum 自〔乘〕谱
autotrack 自动跟踪
autumn 秋〔季〕
autumnal 秋天的
Autumnal Equinox 秋分【大气】【农】(节气)
autumn equinoctial period 秋分期
autumn equinox tide 秋分潮
autumn ice 秋冰

autumn rain 秋雨
auvergnasse (=auvergnac) 奥韦尔纳斯风(法国中央高原的西北冷风)
auxiliary agricultural meteorological station 辅助农业气象站
auxiliary chart 辅助图
auxiliary ship observation 辅助船舶观测【大气】
auxiliary ship station 辅助船舶站
auxiliary station 辅助站
auxiliary thermometer 辅助温度表
auxiliary variable 辅助变量【数】
availability 可用度【数】,可用性,有效性
available buoyant energy 可用浮〔力〕能,有效浮〔力〕能
available energy 有效能量
available head 可用水头,有效水头
available moisture of the soil 土壤中可用水分
available potential energy (APE) 有效位能【大气】
available precipitation amount 可用降水量,有效降水量
available soil moisture 土壤有效水分
available solar radiation 有效太阳辐射【大气】
available storage capacity 有效容量,可用库容
available water 有效水分【大气】
available wind energy 有效风能【大气】,可用风能
available wind velocity 有效风速【大气】
avalaison (=avalasse) 阿瓦莱松风(法国西部的持续西风)
avalanche 雪崩【大气】
avalanche wind 雪崩气浪
average 平均
average atmospheric atom number 平均大气原子序数
average departure 平均距平,平均偏差
average drag coefficient 平均曳引系数,

平均拖曳系数
average error 平均误差
average field 平均场
average geomagnetic level 平均〔地〕磁扰〔动〕程度
average integrator 平均积分器
average intensity 平均强度
average interstitial velocity 平均空隙流速(水文)
average lag 平均滞后
average life 平均寿命
average limit of ice 冰的平均边缘线
average net balance (冰体的)平均净收支,冰体单位面积净收支
average over the spectrum 谱平均
average ozone value 臭氧均值
average power 平均功率
average sidereal day 平恒星日
average solar day 平太阳日
average spectrum 平均谱
average value 平均值
average variability 平均变率
average velocity 平均速度
average wind velocity 平均风速【大气】
average year 常年,一般年份
averaging 求平均
averaging kernel 平均核
averaging operator 平均算符
avgongadaur 阿夫冈加达尔天气(英国法罗群岛无风晴朗的天气)
AVHRR(Advanced Very High Resolution Radiometer) 先进甚高分辨率辐射仪【大气】
AVHRR Polar Pathfinder(APP) 先进甚高分辨率辐射仪极地探索者〔计划〕
aviation 航空,飞行
aviation accident 空难,航空事件
aviation aera〔weather〕forecast 航空区域〔天气〕预报【大气】
aviation automated weather observation system(AV-AWOS) 航空自动气象观测系统
aviation climatology 航空气候学【大气】
aviation forecast 航空预报
aviation forecast zone 航空预报区
aviation medicine 航空医学
aviation meteorological code 航空气象电码【大气】
aviation meteorological element 航空气象要素【大气】
aviation meteorological information 航空气象信息【大气】
aviation meteorological observation 航空气象观测【大气】
aviation meteorological service 航空气象服务,航空气象勤务
aviation meteorological support 航空气象保障【大气】
aviation meteorology 航空气象〔学〕
aviation observation 航空观测
aviation〔weather〕forecast 航空〔天气〕预报【大气】
aviation weather hazard 航空天气灾害
aviation weather observation 航空天气观测
aviation weather report 航空气象报告
aviation weather service 航空气象服务
Avogadro constant 阿伏伽德罗常量【物】
Avogadro number 阿伏伽德罗常量【物】
Avogadro's hypothesis 阿伏伽德罗假设
Avogadro's law 阿伏伽德罗定律
avre 阿弗尔风(出现在法国德龙省吕克恩第乌地方暖冬及冷夏的风)
avulsion ①冲裂〔作用〕②破岸分流③改道(河流)④浪裂
Awakening from Hibernation 惊蛰【大气】【农】(节气)
aweather 迎风
AWIPS (Advanced Weather Interactive Processing System) 先进天气交互处理系统
AWS(automatic weather station) 自动

气象站
AXBT(airborne expendable bathythermograph) 机载投弃式温深仪
axial-flow anemometer 轴流风速表
axial force 轴向力
axialite 轴晶【化】
axially scattering spectrometer 轴向散射粒谱仪
axial root 主根【植】
axial symmetry 轴对称
axiom 公理【数】
axis(axes) ①轴径②轴
axis of anticyclone 反气旋轴
axis of contraction 收缩轴
axis of coordinates 坐标轴【数】
axis of depression 低压轴
axis of dilatation 展开轴
axis of inflow 内流轴
axis of jet stream 急流轴【大气】
axis of low 低压轴
axis of outflow 外流轴
axis of ridge 脊线
axis of rotation 旋转轴【数】
axis of symmetry 对称轴【物】
axis of trough 槽线
axisymmetric flow 轴对称流
axisymmetric trubulence 轴对称湍流
axisymmetric vortex 轴对称涡旋

ayalas 埃约拉斯风(法国中央高原热的东南大阵风, marin 海风的一种)
aygalas 阿加拉风(marin 海风的一种)
azel-scope 方位-高度指示器
aziab 阿齐阿勃风(红海上空闷热天气时的风)
azimuth 方位,方位角
azimuthal perturbation 方位小扰动
azimuthal quantum number 角量子数【物】
azimuthal wavenumber (= angular wavenumber) 角波数
azimuth angle 方位角
azimuth averaging 方位平均【大气】
azimuth distortion 方位畸变
azimuth-elevation 方位-仰角
azimuth error 方位误差
azimuth gauge 方位器
azimuth indicating goniometer 方位角指示仪
azimuth marker 方位标识器,方向标记
azimuth resolution 方位角分辨率【大气】
azimuth scale 方位标度,方位标尺,方位刻度〔盘〕
azo compound 偶氮化物【化】
Azores anticyclone 亚速尔反气旋
Azores Current 亚速尔海流
Azores high 亚速尔高压【大气】
azran 方位距离

B

Babinet point 巴比涅〔中性〕点
Babinet principle 巴比涅原理【物】
back-bent occlusion 后曲锢囚【大气】
back-door cold front 后门冷锋(美国大西洋沿岸)
back echo reflection 障碍物回波反射
back end (= late autumn) 晚秋,秋末
backflow 回流,反流

background 本底【物】
background air pollution 本底空气污染
Background Air Pollution Monitoring Network (BAPMoN) 大气本底污染监测网(1970 年代初由 WMO, WHO, UNEP 等发起组织)
background atmosplere 本底大气
background concentration 本底浓度【大气】

background errors 背景误差
background error covariance 背景误差协方差
background field 本底场,背景场
background field prediction 本底场预报
background level 本底水平,背景水准
background light sensor 背景光传感器
background luminance meter 背景亮度仪
background luminescence sensor 背景发光传感器
background monitor 本底监测器
background monitoring programme 本底监测计划
background noise 背景噪声
background pollution 本底污染,背景污染
background pollution observation 本底污染观测【大气】
background radiation 本底辐射,背景辐射
background station 本底〔观测〕站
background visibility 本底能见度
backing 逆转
backing wind 逆转风【大气】
backlash ①后退②齿隙
back-off system 补偿系统
backlobe 后〔波〕瓣
backpropagation 后向传播
back propagation (BP) network 反传网络【计】,前馈网络,BP 网络
back radiation 后向辐射
backscatter 后向散射
backscattered intensity 后向散射强度
backscattering 后向散射【大气】,背散射【物】
backscattering coefficient 后向散射系数
backscattering cross section 后向散射截面【大气】
backscattering differential cross section 微分后向散射截面
backscattering efficiency 后向散射效率
backscattering lidar 后向散射〔激〕光〔雷〕达

backscatter meter 后向散射表
backscatter measurement 后向散射测量
backscatter-to-extinction ratio 后向散射-消光比
backscatter ultraviolet spectrometer (BUV) 后向散射紫外光谱仪【大气】
backscatter ultraviolet technique 紫外辐射后向散射法【大气】
backscatter visibility sensor 后向散射能见度探测器
back-sheared anvil 后切云砧
backtracking ①返程,回程②反向跟踪,回溯〔法〕③系统搜索自动解题法
backup 备用元件,备用器件,备份【计】
backward difference 向后差分【数】,后向差分
backward flux density 后向流密度
backward scatter 后向散射,背散射【物】
backward scattering coefficient 后向散射系数
backward specific intensity 后向散射比强度
backward-tilting trough 后倾槽【大气】
backwash 回流,反溅,离岸流
backwater 回水,壅水
backwater curve 回水曲线,壅水曲线
bad-i-sad-o-bistroz(=seistan) 塞斯坦风(阿富汗与伊朗间的强季风)
bad visibility 恶劣能见度
bad weather 恶劣天气
bad weather approach 恶劣天气进场着陆
baffle 障板,隔板,导流板
baffling wind 无定向风
Bagrov's criterion 巴格罗夫判据
bag-type collector 袋式除尘器
baguio (= bagio, bagyo, vaguio, vario) 碧瑶风(菲律宾的强烈热带气旋)
Bahama Current 巴哈马海流
bahorok 巴霍洛风(苏门答腊 5~9 月间的一种焚风)
bai 沙霾,黄雾

bai-u 梅雨(日本语名)
baiu front 梅雨锋(日本用法)
balance ①平衡②差额③天平,秤
balanced baroclinic model 平衡斜压模式
balanced flow 平衡流
balanced gyroscope 平衡陀螺仪【航海】
balanced wind 平衡风
balance equation 平衡方程【大气】
balance height 平衡高度(高空气球的)
balance level 平衡面
balance meter 平衡表
balance of solar radiation 太阳辐射平衡,太阳辐射差额
balancer 平衡器
balance year 平衡年
balansometer 辐射平衡表,净辐射表
bali 鲍利风(意大利加尔达湖的风)
Bali Roadmap 巴厘岛路线图(联合国气候变化会议的一个部长级协议,2007年)
Bali wind 巴厘风(爪哇东部的一种东风)
ball ice 浮冰球
balling 雪聚成球
ballistic air density 弹道空气密度【大气】
ballistic density 弹道密度
ballistic meteorology 弹道气象学
ballistic particle 冲击粒子
ballistics ①弹道学②弹道特性
ballistic temperature 弹道温度【大气】
ballistic wind 弹道风【大气】
ball lightning(＝globe lightning) 球状闪电【大气】,球闪
ballonet 气室,副气囊
ballonet ceiling 气球升限
balloon 气球
balloon astronomy 气球天文学
balloon basket 气球吊篮
balloon bed 气球系留台
balloon-borne detector 球载探测器
balloon-borne laser radar 球载光〔雷〕达
balloon-borne reflector 球载反射器

balloon-borne sensor 球载传感器
balloon-borne turbulence probe 球载湍流探头
balloon ceiling 球测云底〔高度〕,气球云幂
balloon cover 气球罩
balloon drag 拖曳气球
balloon measurement 气球测量
balloon observation 气球观测
balloon satellite 气球卫星
balloon shroud 气球〔护〕罩
balloon-sonde 探测气球
balloon sounding 气球探测
balloon theodolite 测风经纬仪
balloon transit 测风经纬仪
ball pyranometer 球状全天空辐射计
ballute (ball与parachute两词合成的缩略词)球伞(供火箭探测用)
band 带,谱带
band absorption 带吸收
Banda Sea Water 班达海水
band cirrus 带状卷云
banded cloud system 带状云系【大气】
banded echo 带状回波【大气】
banded model 带模式【大气】
banded structure 带状结构
band gap 〔谱〕带〔间〕隙
band lightning 带状闪电【大气】
band model 带模式【大气】
band-pass 带通
bandpass-filtered data 带通滤波资料
band-pass filter 带通滤波〔器〕【大气】
band-pass fluctuation 带通振荡
band spectrum 带〔状〕光谱
bandspread 频带展宽
band stop filter 带阻滤波器【电子】
band width 〔频〕带宽〔度〕
bandwidth compression modem 带宽压缩调制解调器,带宽压缩调制解调技术
bandwidth limited white noise 有限带宽白噪声

ban-gull 班恩古尔风(苏格兰的夏季海风)
bank 〔存贮〕库
bankfull stage 齐岸水位,满槽水位
banking of the current 气流堆积〔现象〕
banking process (=pile-up process) 堆积过程
bank of clouds 云堤
bank storage 河岸调蓄,库岸调蓄
banner cloud 旗云【大气】
BAPMoN (Background Air Pollution Monitoring Network) 大气本底污染监测网
Baquios 南半球热带飓风
bar 巴(旧的气压单位)
barat 巴拉特飑(苏拉威西岛的一种烈飑)
barb 风速羽【大气】(风速填图符号,每根代表 4 m/s 或 10 海里①/时)
Barbados Oceanographic and Meteorological Experiment (BOMEX) 巴巴多斯海洋和气象试验
barbed arrow 风矢(风向风速填图符号)
barber 大风雪,冷风暴(美国和加拿大地区)
barchan (=barchane, barkhan) 新月形沙丘,新月形雪堆
bar code reader 条码阅读器【计】
bar code scanner 条码扫描器【计】
bare ice 裸冰【大气】
bare ice field 冰原
bare soil 裸地【大气】
baric analysis 气压分析
baric area 气压区
baric flow 压流
baric lapse rate 气压直减率
baric topography 气压形势
baric wind law 风压定律
barih (=shamal) 夏马风
barines 巴林风(委内瑞拉东部的一种西风)
barium cloud 钡云
barium fluoride film hygrometer 氟化钡膜湿度表
Barnes weighting function 巴恩斯〔加〕权函数
baro 气压表,气压〔计〕的
Barocap 压敏电容(芬兰维赛拉公司气压传感器产品名称)
barocell 气压膜盒,压敏元件
baroceptor 气压传感器,〔气〕压敏〔感〕元件
barocline 斜压
barocline state 斜压状态
baroclinic 斜压的
baroclinic adjustment 斜压适应
baroclinic annulus 斜压环带
baroclinic atmosphere 斜压大气【大气】
baroclinic boundary layer 斜压边界层
baroclinic condition 斜压条件,斜压状态(复数)
baroclinic disturbance 斜压扰动【大气】
baroclinic eddy 斜压涡
baroclinic flow 斜压气流
baroclinic fluid 斜压流体
baroclinic forecast 斜压预报
baroclinic instability 斜压不稳定【大气】
baroclinicity (=baroclinity, barocliny) 斜压性【大气】
baroclinic leaf 斜压叶
baroclinic mode 斜压模〔态〕【大气】,斜压波
baroclinic model 斜压模式【大气】
baroclinic motion 斜压运动
baroclinic numerical model 斜压数值模式
baroclinic process 斜压过程【大气】
baroclinic quasi-geostrophic flow 斜压准地转流
baroclinic torque vector 斜压转矩矢量
baroclinic wave 斜压波【大气】
baroclinic wave activity 斜压波活动
baroclinic zone 斜压带

---

① 1 海里=1.853 km

baroclinity　斜压性
barocyclometer　气压风暴表
barocyclonometer　气压风暴计,风暴位置测定仪
bar of the storm　风暴云堤
barogram　气压自记曲线
barograph　气压计【大气】
barograph trace　气压自记曲线
barometer　气压表【大气】
barometer box(或 case)　气压表匣
barometer cistern　气压表水银槽,测压腔
barometer column　水银柱
barometer constant　气压测高常数
barometer correction　气压表订正
barometer elevation　气压计海拔〔高度〕
barometer formula　气压测高公式
barometer level　气压表高度【大气】
barometer reading　气压表读数
barometer reduction　气压表订正
barometer tube　测压(气压)管
barometric　①气压〔的〕②气压计〔的〕
barometric altimeter　气压高度表
barometric altimetry (= barometric hypsometry)　气压测高法
barometric change　气压变化
barometric characteristic　气压倾向
barometric column　〔水银〕气压柱
barometric constant　气压测高常数
barometric correction　气压订正【大气】
barometric correction table　气压表器差订正表
barometric depression　低压
barometric disturbance　气压扰动
barometric effect　气压效应
barometric equation　气压测高公式
barometric errors　气压计误差
barometric fluctuation　气压起伏
barometric formula　压高公式
barometric gradient　气压梯度
barometric height　气压高度

barometric height formula　压高公式【大气】
barometric high　高压
barometric hypsometry (= barometric altimetry)　气压测高法
barometric leveling　气压高度测量
barometric low　低压
barometric maximum　气压最高值
barometric mean temperature　测高平均气温
barometric minimum　气压最低值
barometric pressure　气压〔表值〕
barometric rate　气压升降率
barometric reduction table　气压订正表
barometric ripple　气压微扰
barometric switch　气压开关
barometric tendency　气压倾向
barometric tube　水银气压表内管
barometric wave　气压波
barometrograph　气压自动记录仪,气压计
barometry　气压测定法
baromil　〔气〕压毫巴(旧的测气压单位)
baroreceptor　气压传感器
baroresistor　气压电阻
baroscope　验压器,气压测验器,气压计,大气浮力计
barosensor　气压传感器
barosphere　气压层(指 8 km 以上的大气层的旧称)
barostat　恒压器,气压调节器
baroswitch　气压开关【大气】(无线电探空仪上用)
barothermograph　气压温度计【大气】
barothermohygroanemograph　气压温度湿度风速计
barothermohygrograph　〔气〕压温〔度〕湿〔度〕计【大气】
barothermohygrometer　气压温度湿度表
barothermometer　压温表
barotron　气压传感器
barotropic　正压〔的〕

barotropic atmosphere 正压大气【大气】
barotropic condition 正压条件,正压状态(复数)
barotropic disturbance 正压扰动
barotropic eddy 正压涡流
barotropic equation 正压方程
barotropic equatorial wave 赤道正压波
barotropic fluid 正压流体
barotropic forecast 正压预报
barotropic instability 正压不稳定【大气】
barotropic mode 正压模〔态〕【大气】,正压波
barotropic model 正压模式【大气】
barotropic numerical model 正压数值模式
barotropic pressure function 正压压力函数
barotropic primitive equation 正压原始方程
barotropic process 正压过程
barotropic Rossby wave 正压罗斯贝波
barotropic state 正压状态
barotropic torque vector 正压转矩矢量
barotropic vorticity equation 正压涡度方程【大气】
barotropic wave 正压波【大气】
barotropic wave activity 正压波活动
barotropic zone 正压区
barotropy 正压性【大气】
barrens 瘠地,荒地,荒漠
barrier 壁垒,障壁
barrier berg 平板状冰山
barrier height 势垒高度【电子】
barrier iceberg 平顶冰山(海洋)
barrier jet 地形〔障碍〕急流【大气】
barrier layer 障碍层,阻挡层(海面混合层和温跃层之间,约 $30 \sim 80$ m 深处)
barrier theory 阻碍学说
barrier theory of cyclones 气旋的阻碍学说
barrier wind 屏障风
Barringer correlation spectrometer 巴林杰相关光谱仪(测 NO_2, SO_2 的)

barycenter 引力中心,质〔量中〕心
barye 微巴(旧的气压单位,dyn/cm^2)
basal fertilizer 基肥【土壤】
basal sliding 底冰滑动
basalt 玄武岩【地质】
base 碱【化】
base flow 基〔本径〕流
baseflow recession 基流消退,基流衰退,基流退水
baseflow recession curve 基流退水曲线
baseflow storage 基流储量(水文)
base level of erosion 侵蚀基准面(水文)
baseline ①原始的,基本的②原始资料③基线
baseline check ①基线校正②基值检定
baseline check box 基值检定箱,基线检定箱
baseline monitoring 基线监测
base map 底图
base plate 底板
base-pressure coefficient ①基线气压系数②基准压强系数
base spectrum 基本谱
base station 基本站,基准站
base surge 底散云(水下核爆炸所形成的雾、水和碎片混合云)
base temperature 基准温度
base width 底宽(水文)
BASIC(Beginner's All-purpose Symbolic Instruction Code) BASIC 语言【计】
basic data set (BDS) 基本资料组,基本资料库
basic equations 基本方程组
basic flow 基本气流【大气】
basic magnetospheric process 基本磁层过程
basic meridian 本初子午线
basic state 基本状态
basic system 基本系统
basic thermal radiation 宁静太阳热辐射
basin ①流域②盆地,海盆③水域

basin accounting 流域水量平衡,流域水量收支
basin lag 流域滞时,流域汇流时间
basin outlet 流域出口
basin recharge 流域〔再〕补给,流域回灌
basin response 流域响应
basis function 基函数【物】
batch ①批量(资料处理用语)②程序组
batch processing 批处理【计】
bathyal environment ①半深海环境②深海环境
bathymetric chart 水深图,等深线图
bathymetry 测〔水〕深法
bathythermograph (BT) 温深仪【大气】,深水温度仪
bathythermograph grid 温深仪〔读数〕网格
bathythermograph print 温深仪复制记录
bathythermograph slide 温深仪玻〔璃〕片
batticaloa kachchan 拜蒂克洛喀昌风(斯里兰卡的焚风)
battle ship 战列舰【航海】
baud 波特【计】(发报速率单位)
baud rate 波特率【地信】
Baur's solar index 鲍尔太阳指数
bayamo 巴亚莫飑(出现在古巴南海岸的一种雷飑)
Bayes analysis 贝叶斯分析【计】
Bayes decision function 贝叶斯决策函数【数】
Bayes estimate 贝叶斯估计【数】
Bayes formula 贝叶斯公式【数】
Bayesian decision rule 贝叶斯决策规则【计】
Bayesian inference 贝叶斯推理【计】
Bayesian probabilistic forecast 贝叶斯概率预报
Bayesian Processor of Output (BPO) 贝叶斯输出处理器
Bayesian theorem 贝叶斯定理【计】
bay ice 海湾冰

Bay of Bengal (BOB) 孟加拉湾
Bay of Bengal Water 孟加拉湾水
BCN (British Commonwealth of Nations) 英联邦
B-display B显示器【大气】
BDRF (bidirectional reflection function) 双向反射函数
beach ice 海滩冰
beacon 航标,信标
beaded lightning 串珠状闪电【大气】,珠状闪电【大气】
bead thermistor 珠状热敏电阻
beam 波束
beam angle 波束角,波束宽度
beam axis 束轴
beam broadening 波束加宽
beamcast 定向无线电传真
beam filling 光束充填〔量〕
beam filling coefficient 波束充塞系数【大气】
beam-forming 波束形成
beam illuminance 光束照度
beam irradiance 光束辐照度
beam-limited radar altimeter 限束雷达测高表
beam pattern 指向性图案(波束方向图)
beam size 波束大小
beam splitter 分束器【物】,分束镜,分光镜
beam spot wander 光斑漂移
beam spread 波束加宽
beam swinging 波束转动法(测风)
beam wander 波束漂移
beam wave 束状波
beam width 射束宽度,波束宽度
beam wind 航行侧风
bearing ①方位②支承
beat frequency 拍频
beat frequency oscillator 拍频振荡器【电子】
beating 拍频〔振荡〕

beat mode 拍频模
Beaufort force 蒲福风力
Beaufort weather notation 蒲福天气符号
Beaufort number 蒲福风力,蒲福数
Beaufort notation 蒲福天气符号
Beaufort〔wind〕scale 蒲福风级【大气】
Beaumont period 博蒙特时段(连续48小时内至少有46个小时气温≥10℃,相对湿度≥75%)
beavertail antenna 獭尾型天线
bed-e-simur 巴季锡穆尔风期(盛夏伊朗沙漠的热风季节)
bedrock 基岩【地质】
Beekley gauge 贝克来雨量计
Beer-Lambert law 比尔-朗伯〔吸收〕定律【物】
Beer's law 比尔定律
before Christ (B. C.) 公元前
before present (B. P. 或 BP,如 250 kaBP 表示距今 25 万年前) 距今
Beginner's All-purpose Symbolic Instruction Code BASIC 语言【计】
beginning height 出现点高度
Beginning of Autumn 立秋(节气)【大气】【农】
Beginning of Spring 立春(节气)【大气】【农】
Beginning of Summer 立夏(节气)【大气】【农】
beginning of the Meiyu period 入梅
Beginning of Winter 立冬(节气)【大气】【农】
beginning point 出现点(流星)
bel 贝〔尔〕(音强单位)
belat 皮拉脱风(阿拉伯南海岸的一种北风或西北风)
Bellamy method 贝拉米法(估测水平散度)
Bellani atmometer 贝拉尼蒸发表
Bellman's quasi-linearization 贝尔曼准线性化
bellows ①空盒,空盒组②波纹管
bell soliton 钟形孤波

bell taper 钟形圆锥〔法〕
below-cloud scavenging 云下清除
below minimums ①机场关闭天气②超极限天气条件
below normal 逊常,正常以下
beltane-ree 贝尔坦里天气(英国设得兰和奥克尼群岛上圣灵降临节前后的多风暴天气)
belt of calms 无风带
belt of convergency 辐合带
belt of fluctuation 脉动带
belt of maximum precipitation 最大降水带
Beltrami equation 贝尔特拉米方程【数】
Beltrami flow 贝尔特拉米流
Bemporad's formula 本波拉公式
Benard cell 贝纳胞【大气】,贝纳单体
Benard convection 贝纳对流【大气】
Benard heating approximation 贝纳加热近似
benchmark ①基准②基准点,基准站
benchmark station 基准〔气候〕站【大气】
bending of ray 射线弯曲
Benguela Current 本格拉海流
bent-back occlusion 后曲锢囚【大气】
benthic boundary layer 洋底边界层
benthos ①海底,洋底②底栖生物
bent-over plume 卷曲烟流
bent stem thermometer 曲管温度表
bentu de soli 邦蒂德索利风(意大利撒丁岛沿岸的东风)
benzene 苯(C_6H_6)
benzo-a-pyrene (=3,4-benzopyrene) 苯并〔a〕芘,3,4-苯并芘
benzo-pyrene emission 苯并芘排放
berber 巴巴风(美国圣劳伦斯湾的 boorga 风)
berg 冰山
Bergen School 卑尔根学派(挪威)
Bergeron classification 贝吉龙分类
Bergeron effect 贝吉龙效应

Bergeron-Findeisen process 贝吉龙-芬德森过程
Bergeron-Findeisen theory 贝吉龙-芬德森[冰晶]理论
Bergeron forcing 贝吉龙强迫[作用]
Bergeron mechanism 贝吉龙机制
Bergeron process 贝吉龙过程
berg wind 山风(南非南海岸的一种热风)
bergy bit 小冰山
Bering Slope Current 白令海斜坡流
Bermuda high 百慕大高压
Bernal-Fowler rules 伯纳耳-否勒定则(物理)
Bernard Price Institute of Geophysical Research (BPIGR) 伯纳德·普赖斯地球物理研究所
Bernoulli effect 伯努利效应
Bernoulli equation 伯努利方程【物】
Bernoulli number 伯努利数【数】
Bernoulli's law 伯努利定律
Bernoulli's theorem 伯努利定理【大气】
Bernoulli stream function 伯努利流函数
Berson winds (或 westerlies) 伯森西风带
beryllium copper 铍铜
Bessel function 贝塞尔函数【数】
Besson comb nephoscope 贝森梳状测云器
Besson nephoscope 贝森测云器
best approximation 最佳逼近【数】
best fit 最佳拟合【数】
best linear invariant estimate 最佳线性不变估计【数】
best linear unbiased estimate 最佳线性无偏估计【数】
Best number 最佳数
best track 最佳路径
beta drift β漂移,贝塔漂移
Beta effect β效应【大气】,贝塔效应
beta gyre β环流,贝塔环流,贝塔涡旋
beta particle β粒子,贝塔粒子
beta-plane β平面【大气】,贝塔平面
beta-ray (=β-ray) β射线,贝塔射线
Beta scale β尺度,贝塔尺度
beta spiral β螺线,贝塔螺线
beta(或 β)-plane approximation β平面近似,贝塔平面近似
Bethe formula 贝蒂公式
between layers 云层间[飞行]
bhut (=bhoot) 布特尘卷风(印度的一种尘卷风)
biannual 一年两次的
bias 偏倚【数】
biased estimate 有偏估计【数】
bias score 倾向性评分,系统性误差评分
Bickley jet 比克莱急流
bicoherence 双阶相干法
bicorrelation 双阶相关
bicubic spline functions 双三次样条函数
bicyclic elimination 双循环消去法
bidirectional diode 双向二极管【电子】
bi-directional reflectance 双向反射比
bidirectional reflectance factor 双向反射[比]因子【大气】
bidirectional reflection function 双向反射函数
bi-directional slip 双向滑动
bidirectional wind vane 双向风向标
biennial 两年[的]
biennial ice 二年冰
biennial oscillation 二年振荡
biennial plant 二年生植物【植】
biennial wind oscillation 二年风振荡
bifilar electrometer 双线静电表
bifurcation 分岔【物】
bifurcation point 歧点【数】,分岔点
bifurcation theory 分歧理论【数】,分岔理论
big bang cosmology 大爆炸宇宙论【天】,大爆炸宇宙【物】
Bigelow's evaporation formula 毕葛洛蒸发公式
Bigg chamber 毕格云室
bihemispherical reflectance 双半球反射比

bilinear interpolation 双线性内插【计】
billow cirrus clouds 波状卷云
billow cloud 浪云
billow wave （造成浪云的大气）波浪运动
bimetal element 双金属元件
bimetallic actinograph 双金属直接辐射计
bimetallic actinometer 双金属直接辐射表
bimetallic strip 双金属片
bimetallic thermograph 双金属片温度计【大气】
bimetallic thermometer 双金属温度表
bimetal thermometer 双金属温度表
bimodal distribution 双峰分布
bimodal spectrum 双峰谱【大气】
bimolecular reduction 双分子还原【化】
binary 二进制的
binary code 二进制代码
binary coded character 二进制编码字符
binary-coded decimal(BCD) 二-十进制码（二进制编码的十进制）
binary-coded octal 二-八进制码（二进制编码的八进制）
binary collision 二体碰撞【天】
binary cyclones 双气旋【大气】
binary digit 二进制数字【计】
binary image 二元图像
binary number 二进制数
binary prediction 两分类预报
binary system 二进制【计】
binary-to-decimal converter 二-十进制转换器
binary tree 二叉树
binary typhoons 双台风【大气】
Binary Universal Form for Representation of meteorological data (BUFR) 气象数据的二进制通用表示格式
bin card 卡片箱
binding energy 结合能【物】
binoculars 双筒望远镜
binomial 二项式【数】
binomial distribution 二项分布【数】
binomial smoothing 二项式平滑
binomial theorem 二项式定理【数】
bioacoustics 生物声学
bioactive peptide 活性肽【生化】
bioassay 生物测定
bioastronomy 生物天文学【天】
bio-battery 生物电池【电子】
biocenosis 生物群落
biochemical action 生化作用
biochemical oxygen demand(BOD) 生化需氧量【大气】
biochemistry 生物化学【生化】
biochore 〔副极地〕植物区气候（柯本气候分类之一）
bioclimate 生物气候
bioclimate zonation 生物气候分区
bioclimatic 生物气候的
bioclimatic law 生物气候律
bioclimatics 生物气候学
bioclimatograph 生物气候图
bioclimatology(=bioclimatics) 生物气候学【大气】
bioclock 生物钟【生】
biocoenosis 生物群落【生】
biocommunity 生物群落【生】
biocompatibility 生物适应性
biocycle 生物循环,生物带
biodegradable 可生物降解的
biodegradation 生物降解
biodiversity 生物多样性【植】
bioecology 生物生态学
bioelectricity 生物电【电子】
bioelectromagnetics 生物电磁学
bioelectronics 生物电子学【电子】
bioengineering 生物工程〔学〕
bioerosion 生物侵蚀【生】
biofeedback 生物反馈【电子】,生物回授
biofog 生物雾
biogenic greenhouse gas emission 生物温室气体排放
biogenic ice nucleus 生物冰核

biogenic trace gases 生物痕量气体
biogeochemical cycle 生物地球化学循环【大气】,生物地化循环【生】
biogeochemistry 生物地球化学
biogeography 生物地理学【植】
bioinorganic chemistry 生物无机化学【生化】
bioinstrumentation 生物测试仪器
biokinetic temperature limit 生物活力温度界限
biological clock 生物钟【植】
biological dating method 生物学定年法
biological decomposition 生物分解
biological degradation 生物降解
biological depollution 生物〔学〕去污〔染〕
biological diversity 生物多样性【生】
biological feedback 生物反馈
biological medium 生物介质
biological minimum temperature 生物学最低温度
biological nitrogen fixation (BNF) 生物固氮
biological productivity 生物生产力
biological rhythm 生物节律【植】
biological treatment 生物处理【土壤】
biological weathering 生物风化【土壤】
biological zero point 生物学零度【大气】
bioluminescence 生物发光【物】
biomacromolecule 生物大分子【生化】
biomass 生物量【植】
biomass burning 生物质燃烧【大气】
biome 生物群落,生物群系【植】
biometeorological index 生物气象指数【大气】
biometeorological time scale model 生物气象时间尺度模式
biometeorology 生物气象学【大气】
biometry 生物统计学
bionics 仿生学
biooptics 生物光学
bioorganic chemistry 生物有机化学【生化】
biophenology 生物物候学
biophysical chemistry 生物物理化学【生化】
biophysical model 生物物理模式
biophysics 生物物理〔学〕【物】
biopolymer 生物聚合体【生化】
bi-orthogonal wavelet transformation 双正交小波变换【计】
biosphere 生物圈【植】
biosphere-albedo feedback 生物圈-反照率反馈
biosphere-atmosphere interaction 生物圈-大气圈交互作用
biosphere reserves 生物圈保护区
biota 生物群
biotemperature 生物温度
biotope 群落生境【地理】
biotron 气候控制室,生物气候室
biphase 双阶位相
bipolar circulation pattern 偶极环流型
bipolar group 双极〔黑子〕群
bipolarized measurement 双极化测量
bipolar magnetic region 双极磁区
bipolar pattern 双极型(闪电)
bipolar transistor 双极晶体管【电子】
birainy 双雨季〔的〕
birainy climate 双雨季气候
birazon(=virazon) 比拉海风(西班牙)
bird burst 鸟群飞散〔回波〕
bird of migration(或 passage) 候鸟
bird's eye view map 鸟瞰图【测】
birefringence 双折射【物】
Birkeland currents 伯克兰电流【地】
bise(=bize) 比士风(法国南部山区的一种干冷风)
bise brume 比士雾风(瑞士和法国山区潮湿或有雾的西北风)
Bishop's corona (或 ring) 毕晓普光环【大气】
Bishop wave 毕晓普波(一种大气背风波)

bispectrum 双阶谱
bispectrum analysis 双阶谱分析
bissextile ①(年、月、日)闰的②闰年
bistability 双稳性
bistatic 双基地〔的〕,双站〔的〕,双点〔的〕
bistatic acoustic radar 双基地声〔雷〕达
bistatic acoustic sounder 双基地声学探测器,收发分置声学探测器
bistatic acoustic wind monitor system 双基地声学测风系统,收发分置声学测风系统
bistatic arrangement 双基地设置
bistatic cross section ①双基地截面②双向截面
bistatic Doppler acoustic sounder 双基地多普勒声学探测仪,收发分置多普勒声学探测仪
bistatic lidar 双基地激光雷达【大气】,双基地光〔雷〕达
bistatic radar 双基地雷达【电子】
bistatic radar cross section 双基地雷达截面
bistatic radar equation 双基地雷达方程
bistatic scattering of pulse 脉冲双基地散射
bit ①〔二进制〕位【数】②比特
bit error rate 比特误码率
biting wind 刺骨寒风
bit rate 位速率
bit string 位串(一串二进位信息)
bitter cold 严寒
bittern 盐卤,卤水
bivane 双向风标(能指风的水平及垂直方向的风向标)
bivariate distribution 二维分布,二元分布
bivariate time series 二元时间序列
Bjerknes circulation theorem (＝Bjerknes theorem of circulation) 皮叶克尼斯环流定理【大气】
Bjerknes feedback mechanism 皮叶克尼斯反馈机制
blaast 爆发风(苏格兰和瑞典的飑)
black and white bulb thermometer 黑白球温度表【大气】
black blizzard 黑尘暴(美国中南部的一种尘暴)
blackbody 黑体【大气】
blackbody emission 黑体发射
blackbody radiation 黑体辐射【大气】
blackbody spectrum 黑体光谱
black box 黑箱【计】
black box method 黑箱法,盲区法(降水预报用)
black box testing 黑箱测试【计】
black-bulb thermometer 黑球温度表【大气】
black buran 黑风暴
black carbon 黑碳
black carbon aerosol 黑碳气溶胶
black carbon concentration 黑碳浓度
Black current 黑潮【航海】
black fog 黑雾
black frost 黑霜,黑冻
black hole 黑洞【天】
black hole radiation 黑洞辐射【物】
black ice 黑冰【大气】,雨凇(俗名)
black lightning (＝dark lightning) 暗闪电
black northeaster 东北黑风(澳大利亚东南部的东北大风)
blackout (无线电)衰落〔(通讯)中断,停电
black rain 黑雨
black roller 黑滚轴风(美国西部的黑风暴)
black southeaster 东南黑风(非洲南部及澳大利亚的黑色东南风暴)
black squall 乌云飑
black storm 黑风暴(中亚细亚的一种沙尘暴)
black stratus 黑层云
Black stream 黑潮【航海】

blackthorn winds 刺李风(英国泰晤士河谷3、4月的干冷风)
Blackthorn wind 布拉桑冷风(英国的冷干风)
black wind(＝reshabar,rushabar) 黑风(伊朗南克尔迪斯坦的东北旱风的俗称)
blad 布莱德飑(苏格兰的雨飑)
blanced valve 平衡阀【航海】
blank 布蓝克飑(英国风暴天气时的突发性飑)
blank chart 空白图
blash 骤风
blast ①阵风②爆炸气浪
blast gas cloud 爆炸气体云
blast wave 爆炸波,冲击波
Blaton formula 白拉通公式
blaze 布莱士风(苏格兰的突发性干燥风)
Bleeker humidity diagram 白立克湿度图
blending height 混合高度
blending process 混合过程
blending region 混合区
bliffart(＝bluffart,bliffert) 布列法特飑(苏格兰的阵风或雪飑)
blight 枯萎病(植物所患)
blight weather 枯萎病天气
blind area 盲区
blind (flying) condition 盲目飞行条件,能见度极差的条件
blind drainage ①内流水系,闭流水系②暗沟排水
blind prognose 盲目预报
blind rollers 长涌,暗浪
blind speed 盲速
blink 反光,闪烁
blinter 布林特风(阵风的苏格兰方言名)
blip 回波,尖头信号
blirty 阵风(多变天气时的)
blizzard 雪暴
blizzard fatality 雪暴灾害
blizzard warning 雪暴预警

blizzard wind 暴风雪,凛风
blob (雷达屏上的)大气湍流斑,非均匀区
block 阻塞〔高压〕
block code 字组,分组码
block diagram 框图
blocking 封闭【化】,闭塞【通信】
blocking action 阻塞作用
blocking high(或 anticyclone) 阻塞高压【大气】,阻塞反气旋
blocking oscillator 间歇振荡器
blocking pattern 阻塞形势
blocking ridge 阻塞脊
blocking situation 阻塞形势【大气】
block iterative method 块迭代法,成组迭代法
block number 台站区号
blood-rain 血雨【大气】
blood-snow 血雪【大气】
blop 小尺度扰动(雷达用语)
blossom shower(＝mango shower) 芒果阵雨
blout (＝blouter,blowther,blowthir) 布劳特风暴(苏格兰的风暴,风雨或雹的突然来临)
blowby ①窜漏②气体喷出
blowing dust 高吹尘,扬尘
blowing sand 高吹沙【大气】,扬沙【大气】
blowing snow 高吹雪【大气】
blowing spray 高吹沫
blowland 风蚀地
Blown-ups theory 溃变理论
blowout 吹蚀【土壤】
bloxam 布洛克萨姆平均法(10～11 d的平均)(罕用词)
BLP(boundary layer profiler) 边界层廓线仪【大气】
blue band ①蓝带②蓝冰带
blue flash 蓝闪〔光〕
blue-green flame 蓝绿闪(天文)

blue haze 青霾
blue ice 蓝冰【大气】,纯净冰（冰愈纯洁色愈蓝）
blue-ice area 蓝冰区
blue jets 蓝〔放电〕急流
blue moon 蓝月亮
blue noise 蓝噪声
blue of the sky 天空蓝度【大气】
blue sky 碧空
blue sky scale 蓝天标〔度〕
blue sun 蓝太阳
blue thermal 晴空热泡
blunk 勃龙飑（英国）
bochorno 布秋诺热风（西班牙）
BNF(biological nitrogen fixation) 生物固氮
BOB(Bay of Bengal) 孟加拉湾
BOD(biochemical oxygen demand) 生化需氧量【大气】
bodily tides 固体潮
body force 〔彻〕体力【物】
body waves 体波
bog ①沼泽②泥炭地
bogus data 人造〔站〕资料【大气】,假想资料
bogus observation 虚拟观测
bogus vortex 虚拟涡旋
bohorok 巴霍洛风（苏门答腊5~9月间的一种焚风）
Bohr atom model 玻尔原子模型【物】
boil ①沸腾②泉涌,管涌,水涌
boiling point 沸点
boiling point thermometer 沸点温度表
bolide 火流星,火球
Bolling-Allerod 玻林-艾勒拉德暖期, BA暖期（14700—12700 BP）
bologram 热辐射仪自记曲线【大气】
bolograph 热辐射计
bolometer 热辐射仪【大气】
bolon 博朗风（北苏门答腊托巴湖的风）
bolt ①闪电②雷击

Boltzmann constant(＝Boltzmann's constant) 玻尔兹曼常量【物】,玻尔兹曼常数
Boltzmann equation 玻耳兹曼方程【数】
BOMEX（Barbados Oceanographic and Meteorological Experiment） 巴巴多斯海洋和气象试验
bond energy 键能【物】
bond structure 键结构【物】
bond valence 键价【化】
bonsai 盆景【林】
book-end vortices 书夹涡旋,两端涡旋
Boolean algebra 布尔代数【计】
Boolean operation 布尔运算【计】
boorga 布加风（阿拉斯加的冷风）
bora 布拉风（亚得里亚海东岸的一种干冷东北风）
boraccia 强布拉风
bora climatology 布拉风气候学
bora fog 布拉雾（布拉风引起的浓雾）
borasco（＝borasca, bourrasque） 布拉斯科雷暴（地中海上的一种雷暴或强飑）
bora scura 气旋式布拉风
bordelais 博德莱风（法国西南部的西风）
borderline science 边缘科学
border spring 边缘泉（冲积锥的）
border zone 边界区
bore ①涌潮②钻孔
Boreal 北方〔气候〕期
boreal climate 北部〔森林〕气候
boreal climatic phase 北方期（气候）
boreal forest 北方林区
boreal pole 北极
boreal region 北方区
boreal woodland 北方林地
boreal zone 北方区
boreas（＝borras） 北风（希腊语名）
borino 弱布拉风
bornan 布南风（日内瓦湖中央的一种风名）
Born approximation 玻恩近似【物】

boron trifluoride detector 三氟化硼检测器(测土壤含水量)
Bose-Einsten distribution 玻色-爱因斯坦分布
botanical garden 植物园【植】
botanical zone 植物带
botany 植物学【植】
both sideband 双边带
both-way communication 双向通信【电子】
bottleneck effect 瓶颈效应【化】
bottle post 漂流瓶
bottle thermometer 瓶式温度计
Bottlinger's rings 玻氏晕环,玻特林格晕环
bottom boundary condition 底边界条件
bottom current 底〔层〕流
bottom friction 底摩擦
bottom ice 底冰
bottom layer 底层
bottom temperature 底层水温,〔海〕底温〔度〕
bottom topography 水底地形
bottom-trapped wave 底拦截波
bottom water 底〔层〕水
Bouguer anomaly 布格异常【测】
Bouguer-Lambert law 布格-朗伯定律
Bouguer's halo 布格晕
Bouguer's law 布格定律
boulbie 布尔比埃风(法国阿里埃日省的北风)
Boulder Atmospheric Observatory(BAO) 博尔德大气观象台(美国)
boundary 边界,界限
boundary condition 边界条件
boundary currents 边界流
boundary-generated internal wave 边界内波
boundary layer 边界层
boundary layer climate 边界层气候【大气】
boundary layer dynamics 边界层动力学

boundary layer equations 边界层方程组
boundary layer jet stream 边界层急流【大气】
boundary layer meteorology 边界层气象学【大气】
boundary layer model(BLM) 边界层模式
boundary layer morphology 边界层形态学
boundary layer profiler(BLP) 边界层廓线仪【大气】
boundary layer pumping 边界层抽吸作用
boundary layer radar 边界层雷达【大气】
boundary layer radiosonde 边界层探空仪
boundary layer rolls 边界层滚〔动〕涡〔旋〕
boundary layer separation 边界层分离
boundary mixing 边界混合
boundary of saturation 饱和边界
boundary science 边缘科学
boundary surface 界面
boundary value condition 边值条件【数】
boundary value problem 边值问题【数】
boundary wave 边界波,界波
bound charge 束缚电荷【物】
bounded-derivative method 有界导数法
bounded weak echo region(BWER) 有界弱回波区
bound electron 束缚电子【电子】
Bourdon thermometer 巴塘温度表【大气】
Bourdon tube 巴塘管,布尔东管【大气】
Boussinesq approximation 布西内斯克近似【大气】
Boussinesq equation 布西内斯克方程【数】
Boussinesq fluid 布西内斯克流体
Boussinesq number 布西内斯克数
bow 虹
bow echo 弓形回波
Bowen ratio 鲍恩比【大气】
bow wave 头波
boxcar function 矩形波函数

boxcar integration method 矩形波串积分法
boxcar integrator 矩形波串积分器
boxcar window 矩形波串窗
box diagram 方块图
box kite 箱形风筝
box method 箱室法
box models 箱模式【大气】
box-whisker plot 盒须图
Boyden index 博伊登〔不稳定〕指数
Boyle law(＝Boyle's law) 玻意耳定律【物】
Boyle-Mariotte law 玻意耳-马略特定律
Boys camera 波伊思相机
BP (before present) 距今
BPI pan （美）作物局蒸发皿,BPI 蒸发皿
BPO (Bayesian Processor of Output) 贝叶斯输出处理器
brackish water 半咸水,苦咸水,微咸水
Brady array humidity sensor 布雷迪阵列湿敏元件
Bragg diffraction 布拉格衍射【物】
Bragg scattering 布拉格散射
branched function 分支函数
branching of lightning 闪电分支,闪电分岔
branching point 〔分〕支点【数】
brash ①碎浮冰〔堆〕②骤风,聚风暴（罕用）
brash ice 碎冰【航海】
brave west wind 咆哮西风带【大气】(见于 40°～65°N 和 36°～65°S 之间)
Brazil Current 巴西〔暖〕海流
BRDF (bidirectional reflectance distribution function) 双向反射比分布函数
breadth-first search 广度优先搜索
break 中断
breakaway depression 析离低压（脱离大低压而东移的副低压）
breakdown ①击穿,崩溃②疲竭
breakdown field 击穿电场

breakdown potential 击穿电位〔电压〕
breakdown strength 击穿强度【电子】
breakdown voltage 击穿电压【电子】
breaker 碎浪
breaker depth 碎浪深度,碎波深度
breaking-drop theory 水滴破碎理论【大气】
breaking-off process 切断过程
breaking wave （近堤）破波,碎波
breaking wind speed 破坏风速【大气】
break line 断裂线
break monsoon 季风中断【大气】
break of the Meiyu period 休梅期
break-period 中断期
break point 折点,断点,转效点
breaks in overcast 密云隙,准阴天（云量大于9,小于10）
break the cloud 穿云
breakthrough curve 穿透曲线
breakup ①解冻,冰破碎②(水滴)破碎③崩溃
breakup season 解冻季节,冰碎裂期
breakwater 防波堤
breather 热带飑
bred mode 繁育模
bred vector 繁育矢量
breeding method 繁育法
breeze 微风
Bremsstrahlung 轫致辐射【物】
Bremsstrahlung effect 轫致辐射效应
brenner 轻机枪风(见于英国海上的一种短时急阵风)
breva 白里伐风(意大利科摩湖的一种日风)
Brewer bubbler 布鲁尔起泡器（库伦法测臭氧用）
Brewster angle 布儒斯特角【物】
Brewster point 布儒斯特〔中性〕点
Brewster window 布儒斯特窗【物】
brickfielder 烧砖风(澳大利亚南海岸的干热风)
bridled anemometer 制动风速表

bridled-cup anemometer 制动风杯风速表
briefing 天气讲解(起飞前)
Brier score(BSCR) 布赖尔评分法
bright band 亮带
bright band echo 亮带回波
bright eruption 喷焰
brightness 亮度
brightness component 亮度分量
brightness contrast 亮度对比
brightness distribution 亮度分布
brightness level(=adaptation level) 适应亮度,场亮度
brightness temperature 亮度温度【大气】,亮温
bright network 亮网络区(色球层)
bright segment 曙暮光弧
bright sunshine 亮日照,环影日照
bright sunshine duration 亮日照时数
brine 卤水,盐水,浮雪
brine slush 松冰团
brisa(=briza) (西班牙语)布里沙海风
brise carabinee(=brise carabinera, carabine) 布里沙卡拉比内风(法国和西班牙的一种突发性暴风)
brisote 布里沙脱风(古巴的一种强东北海风)
British Antarctic Survey(BAS) 英国南极测量局
British Commonwealth of Nations(BCN) 英联邦
British National Committee on Antarctic Research(BNCAR) 英国南极研究委员会
British thermal unit 英制热〔量〕单位
broadband access 宽带接入【计】
broadband albedo 宽带反照率
broadband emissivity 宽带发射率
broadband flux emissivity 宽带通量发射率
broadband flux transmissivity 宽带通量透射率

broadband LAN 宽带局〔域〕网【计】
broadband pulse propagation 宽带脉冲传播
broadband radiation 宽带辐射
broadband ultraviolet radiation 宽带紫外辐射
broad beam transmitter 宽束发射机
broadcast center 广播中心
broad-crested weir 宽顶堰
broad leaved forest 阔叶林【林】
Brocken bow 峨眉宝光,布罗肯宝光
Brocken specter 峨眉宝光,布罗肯宝光,布罗肯幽灵
broeboe 布鲁贝风(印尼望加锡海峡南部的东风)
broiling 大寒天
broken 裂云〔天空〕(云量6～9)
brokenness 多云状况
broken rainbow 断虹
broken sky 多云(天空状况),裂云
bromine compounds 溴化合物
bromomethane 溴甲烷
brontides 轻微地震声(轻微雷声似的隆隆声)
brontograph 雷雨计
brontometer 雷雨表
brown cloud 棕云
brown fume 棕色尘雾
brown haze 棕色轻雾,棕霾
Brownian diffusion 布朗扩散
Brownian motion 布朗运动【物】
Brownian noise 布朗噪声
Brownian rotation 布朗旋转
Browning's severe right moving storm model 布朗宁右转强风暴模型
brown snow 棕雪
browser 浏览器【计】
Brückner cycle 布吕克纳周期,布氏周期
brüscha 白罗夏风(瑞士的一种东北风)
brubu 布鲁伯风(印度东部的一种飑)
brughierous 布律吉鲁斯风(法国努瓦尔

山区的南风）
bruma　布鲁马霾（见于智利沿海）
brume　雾（西班牙语名）
Brunt-Douglas isallobaric wind　布伦特-道格拉斯等变压风，B-D 等变压风
Brunt-Vaisala frequency　布伦特-维赛拉频率，B-V 频率，布伦特-韦伊塞莱频率【大气】
Brunt-Vaisala period　布伦特-维赛拉周期，B-V 周期
brush discharge　刷形放电
bryochore　①苔原区，冻原区②有生物区
B scope　B 显示器【大气】，B 示波器
bubble　气泡
bubble bursting　气泡破裂
bubble convection　气泡对流
bubble gauge　气泡水位计
bubble high　气泡高压，小高压，雷暴高压（直径约 80～480 km）
bubble nucleus　气泡核
bubble theory　气泡理论
bubbly ice　气泡冰
Buchan spells　巴肯冷暖期（苏格兰东南部九个不合时令的冷暖期）
bucket temperature　表面水温
bucket thermometer　表面水温表
Buckingham Pi theory　白金汉 π 理论
buckling　屈曲
buddy check　邻近点检验
budgets of atmospheric species　大气成分〔源汇〕收支
budget year　冰积年
Budyko number　布德科数
buffer　缓冲器【计】
buffer factor　缓冲因子
buffering　缓冲作用
buffer memory　缓冲存储器【计】，缓存
buffer storage　缓冲寄存器【计】
buffer zone　缓冲带
buffeting Mach number　（扰流）抖振马赫数

BUFR（Binary Universal Form for Representation of meteorological data）　气象数据的二进制通用表示格式
building climate　建筑气候【大气】
building climate demarcation　建筑气候区划【大气】
building climatology　建筑气候学
building sunshine　建筑日照
buildup index　火险等级指数
bulb　球部
bulbous cloud　球根状云
bulk　①容积，体积　②整体，总体
bulk aerodynamic drag formulation　整体气体动力学曳力公式
bulk aerodynamic method　整体空气动力学方法
bulk average　整体平均【大气】，体积平均
bulk boundary layer　整体边界层【大气】，粗边界层
bulk density of soil　土壤单位体积干重，土壤容量，土壤体积密度
bulk-drag scheme　整体曳力方式
bulk effect　体效应
bulk flow　整体流
bulk formula　整体公式
bulk heat flux　整体热通量
bulk memory　大容量存储器【计】
bulk method　整体法
bulk mixed layer model　整体混合层模式
bulk modulus　整体模数
bulk motion　整体运动
bulk parameterization　整体参数化
bulk phase　整体相，凝集相（固、液态同时存在）
bulk planetary boundary layer　整体行星边界层
bulk property　整体性质
bulk Richardson number　整体里查森数【大气】，粗里查森数
bulk stable boundary layer growth　整体稳定边界层增长

bulk transfer method 整体传输法
bulk transfer coefficient 整体输送系数
bulk transfer law 整体输送定律
bulk transport 整体输送
bulk turbulence scale 整体湍流尺度
bulk velocity 整体速度
bulk water 整体水
bulkwater (cloud) model 整体水〔云〕模式
bulkwater parameterization 整体水参数化
bulletin 公报,公告,会报
bullet rosettes 子弹花瓣状〔冰晶〕
bull's eye squall 牛眼云飑(晴天一小片云下的飑,见于南非沿海)
bummock 水下冰丘
bump 颠簸
bumpiness 颠簸性
bumpy air 颠簸空气
bumpy flight 颠簸飞行
buoy 浮标站
buoyancy 浮力
buoyancy effect 浮力效应
buoyancy factor 浮力因子
buoyancy fluctuation 浮力起伏
buoyancy flux 浮力通量
buoyancy force 浮力【大气】
buoyancy frequency 浮力频率
buoyancy length scale 浮力长度标尺
buoyancy lift 浮力,浮升
buoyancy oscillation 浮力振荡
buoyancy parameter 浮力参数
buoyancy production rate 浮力产生率
buoyancy subrange 浮力次区,浮力亚区
buoyancy velocity 浮力速度【大气】
buoyancy wave 浮力波
buoyancy wavenumber 浮力波数
buoyant convection 浮升对流
buoyant energy 浮力能
buoyant instability 浮升不稳定〔性〕
buoyant plume 浮升烟羽【大气】
buoyant subrange 浮升亚区
buoyant vortex 浮力涡旋

buoy observation 浮标观测,海上浮标气象观测
buoy weather station 浮标天气站
buran 布朗风(俄罗斯及中亚的一种寒冷东北风)
burga (见 boorga)布加风
Burger number 伯格数【大气】
Burgers' equation 伯格斯方程
Burgers vector 伯格斯矢〔量〕【物】
Burger's vortex 伯格涡旋
buria 布拉风(bora 的保加利亚称号)
burning index 火险指数
burn-off ①雾消(太阳加热消雾)②云消(太阳加热消云)③耗散热
burraxka silch 布拉卡雹暴(地中海马耳他附近的一种雹暴)
burrow microclimate (= hole microclimate)洞穴小气候
burst ①猝发②阵(降水间歇次数)③定向信号,相位信号④暴流
burst cloud band 爆发性云带
bursting height 爆裂高度(气球的)
bursting point 爆裂点,爆破点
bursting thickness 爆裂厚度
burst of monsoon 季风爆发
burst swath 下击暴流带
bus 汇流线,总线【计】
Busch lemniscate 中性线
bush 海龙卷尾,龙喷水(龙卷接触海面时向四周喷水花现象)
business as usual 照常排放情景【大气】,无控排放
Businger-Dyer relationship 布辛格-戴尔关系式,B-D 关系式(一种相似关系式)
bus network 总线网【计】
buster (=southerly buster, burster, southerly burster) 伯斯特风(澳洲猛烈的西北冷风)
busy tone 忙音
butadiene 丁二烯($CH_2CHCHCH_2$)
butterfly diagram 蝶形图(黑子在日面的

分布图)
butterfly effect 蝴蝶效应【物】
button cell 扁形电池【电子】,扣式电池
BUV（backscatter ultraviolet spectrometer） 后向散射紫外光谱仪【大气】
Buys Ballot's law 白贝罗定律【大气】,风压定律
BWER（bounded weak echo region） 有界弱回波区
by-pass 旁路【物】
Byram anemometer 拜拉姆风速表
byte 字节【计】

C

cabbeling(=cabbaling) 混合增密
cabin atmosphere 舱内大气
cable ①线缆【计】②电缆③有线电视的④现定义为0.1海里(185.2 m)历史上曾定义为百英寻(=600英尺,1829 m)
Cable News Network（CNN） 有线新闻网(美国)
cacaerometer 空气污染检查器
CACGP（Commission of Atmospheric Chemistry and Global Pollution, IAMAS） 大气化学和全球污染委员会(IAMAS)
cacimbo 加辛坡雾(非洲西南沿岸的一种浓雾)
CAD（computer-aided design） 计算机辅助设计【计】
cadastral attribute 地籍属地【地信】
cadastral information system 地籍信息系统【地信】
cadastral survey 地籍测量【测】
cadastre 地籍【测】
caesium clock 铯钟【天】
caju rains 贾如雨(巴西10月间的一种小阵雨)
cake ice 饼〔状〕冰
calcareous soil 石灰性土【土壤】
cal 卡〔路里〕的简写
Cal 千卡〔路里〕的简写,亦称大卡
calcite 方解石【化】
calculation 计算
calculus ①微积分〔学〕【数】②演算【数】
calefaction 热污染
calendar ①历②历法
calendar line 日界线【航海】
calendar year 历年【天】
calf 小块冰
calibrater (或 calibrator) 校准器
calibrating function 量规函数
calibration 校准【大气】,定标,检定
calibration curve 校准曲线【大气】
calibration of instrument 仪器标定,仪器校准
calibration tank 检定池,校准池
California Current 加利福尼亚〔冷〕海流
California method 加利福尼亚法(水文)
California norther 加利福尼亚北风(春夏季的强干而多尘的北风)
California plotting position 加利福尼亚标绘位置
calina 卡里纳霾(西班牙)
Callao painter (见 painter)卡劳雾
call lamp 信号灯
calm 零级风【大气】,静风
calm belt 无风带【大气】
calm central eye 无风眼
calm inversion pollution 静稳逆温污染,无风逆温污染
calm layer 无风层

calm night 静夜
calm sea 无浪(风浪零级)
calms of Cancer 北半球副热带无风带
calms of Capricorn 南半球副热带无风带
calm zone 无风带
caloradiance 热辐射强度
calorie(cal) 卡〔路里〕(热量的旧单位,=4.1868 J)
calorific value 热值
calorimeter ①量热器【物】②量能器③卡计
calorimetry 量热学【物】,量热法
calved ice 仔冰(崩离母冰后漂浮在水面的),崩裂冰
Calvin cycle 卡尔文循环【生化】
calving 裂冰〔作用〕
calvus (cal) 秃状(云)
camanchaca (=garua) 浓湿雾(见于南美洲西海岸)
cambial initial 形成层活动
Cambrian Period 寒武纪【地质】
Cambrian System 寒武系【地质】
camera 照相机,摄像机
Campbell-Stokes sunshine recorder 坎贝尔-斯托克斯日照计【大气】,玻璃球日照计,聚焦式日照计
campos 坎普草原(南美),热带草原(南美),坎普群落(生态)
camsine (见 khamsin) 喀新风
Canadian Advisory Committee on Remote Sensing(CACRS) 加拿大遥感咨询委员会
Canadian hardness-gauge 加拿大硬度器
Canadian Meteorological and Oceanographic Society(CMOS) 加拿大气象与海洋学会
Canadian Meteorological Centre(CMC) 加拿大气象中心
Canadian Meteorological Service (Can Met Ser) 加拿大气象局
canalization 狭管效应
canal theory 水槽理论
Canary Current 加那利海流(北大西洋流的南支)
cancel 取消,撤消
cancellation 消除
cancellation ratio 对消比【大气】
candela(cd) 坎〔德拉〕(发光强度单位)
candle 烛光
candle ice 烛〔状〕冰
Candlemas Eve winds (= Candlemas-crack) 圣烛节强风(英国 2、3 月的强风,圣烛节在 2 月 2 日)
canigonenc 坎尼沃南克风(法国比利牛斯省的干冷风)
canonical coordinates 正则坐标【物】
canonical coordinates of the first kind 第一类典范坐标【数】
canonical correlation 典型相关
canonical correlation analysis(CCA) 典型相关分析【数】
canonical equation 正则方程【物】,典型方程
canonical matrix 典范矩阵【数】
canonical regression 典型回归
canonical variable 典型变量
canopy ①林冠层【大气】,冠层② 森林覆盖③(天、顶、座舱)盖
canopy temperature 冠层温度【大气】
Canterbury northwester 坎特伯里西北大风(新西兰从南阿尔卑斯山吹来的强西北焚风)
canting angle 倾斜角
canyon 峡谷【地理】
canyon wind 峡谷风,下吹风,下降风
cap 云帽
capacitance rain gauge 电容雨量计
capacitive aneroid 电容空盒
capacitive thin-film sensor 薄膜电容〔测湿〕感应元件
capacitor (或 condenser) microphone 电容传声器
capacity correction 容量订正
capacity of the wind 风的挟带能力,风沙含量
cap cloud 山帽云【大气】

CAPE (convective available potential energy)　对流有效位能,对流可用位能
Cape Doctor　道克特角风(见于南非沿海的强东南风)
Cape Verde type　佛得角型(汉恩气候分类类型之一)
capillarity　毛细现象【物】,毛〔细〕管作用【大气】
capillarity correction　毛〔细〕管订正
capillary　毛〔细〕管
capillary action　毛细作用
capillary collector　毛管收集器
capillary conductivity (= unsaturated hydraulic conductivity)　毛〔细〕管传导性,毛管传导率
capillary depression　毛细下降
capillary diffusion (或 capillary movement)　毛〔细〕管水扩散〔作用〕
capillary electrometer　毛〔细〕管静电计
capillary forces　毛〔细〕管力
capillary gas chromatograph　毛细管气相色谱仪【化】
capillary head　毛〔细〕管水头
capillary hysteresis　毛〔细〕管滞后(作用,现象)
capillary interstice　毛〔细〕管孔隙
capillary moisture capacity　毛管持水量
capillary phenomenon　毛〔细〕管现象
capillary potential　毛〔细〕管位势
capillary pressure　毛〔细〕管压力
capillary ripple (= capillary wave)　毛〔细〕管涟波,界面波,表面张力波
capillary rise　毛〔细〕管上升〔高度〕
capillary rise of soil moisture　土壤毛〔细〕管上升水
capillary suction　毛〔细〕管吸力
capillary water　毛〔细〕管水〔分〕
capillary wave　表面张力波【物】,张力波,界面波
capillatus (cap)　鬃状(云)
capped column　冠柱〔冰〕晶(带帽的柱状冰晶)
CAPPI(constant-altitude plan position indicator)　等高平面位置显示器【大气】
capping inversion　覆盖逆温【大气】,冠盖逆温
capping inversion layer　覆盖逆温层
capping layer　冠盖〔逆温〕层
captain　船长【航海】
captive balloon　系留气球
captive balloon sounding　系留气球探测【大气】
capture　捕获
capture cross-section　捕获截面
capture rate　捕获率
carabine (或 brisa carabinera, brise carabinee)　卡拉宾风(法国和西班牙的强阵风)
carbacidometer　大气碳酸计
carbohydrate　碳水化合物【化】,糖类
carbon　碳【大气】
carbonaceous particle　含碳粒子
carbon assimilation　碳同化【大气】,碳〔素〕同化〔作用〕
carbonate　碳酸盐
carbonate analysis　碳酸盐分析
carbon-black seeding　碳黑催化
carbon bond mechanism　碳键机制
carbon budget　碳收支
carbon credit　碳信用,碳权
carbon cycle　碳循环【大气】
carbon-dating　碳定年法【大气】,碳-14定年法
carbon dioxide　二氧化碳【大气】
carbon dioxide atmospheric concentrations　二氧化碳大气浓度【大气】
carbon dioxide band　二氧化碳带【大气】
carbon dioxide equivalence　二氧化碳当量【大气】
carbon dioxide fertilization　二氧化碳施肥【大气】
carbon dioxide fertilizing effect　二氧化碳施肥效应
carbon dioxide laser　二氧化碳激光器【电子】,CO_2 激光器
carbon disulfide　二硫化碳(CS_2)

carbon emission 碳排放〔量〕
carbon-film hygrometer element 碳膜湿度〔表〕元件
carbon footprint 碳足迹
carbon hygristor 碳湿敏电阻
Carboniferous Period 石炭纪【地质】
carbon isotope 碳同位素
carbon monoxide 一氧化碳【大气】
carbon nutrition 碳营养【土壤】
carbon pool 碳池【大气】，碳库
carbon sequestration 碳封存
carbon sink 碳汇【大气】
carbon source 碳源【大气】
carbon tetrachloride 四氯化碳(CCl_4)
carbon trading 碳交易
carbonylation 羰基化【化】
carbonyl compounds 羰基化合物
carbonyl sulfide 硫化羰，氧硫化碳(COS)
carboxylation 羧基化【化】
carburetor icing 汽化器积冰
carcenet 卡塞内特风（东比利牛斯的寒冷谷风）
cardinal temperatures 基点温度
cardinal wind 主向风（即东、南、西、北风），正向风
careme 卡列姆风（见于小安的列斯群岛2—4月干季）
Caribbean Current 加勒比海流
Caribbean Meteorological Council (CMC) 加勒比气象理事会
Caribbean Meteorological Organization (CMO) 加勒比气象组织
Carnot cycle 卡诺循环【物】
Carnot efficiency 卡诺效率
Carnot engine 卡诺发动机
Carnot's cycle process 卡诺循环过程
Carnot's theorem 卡诺定理【物】
carriage return 回车【计】
carrier 载体【化】，载波，载流子
carrier balloon 运载气球
carrier balloon system 运载气球系统
carrier frequency 载频
carrier telephone 载波电话【电子】

carrier wave 载波
carry-over ①剩余物②转移，承前③滞后
carry-over effect 滞后效应（云催化的）
carry-over storage ①多年调节库容②多年调节水库
Cartesian coordinates 笛卡儿坐标【数】
Cartesian tensor 笛卡儿张量
cartogram 统计图
cartographical climatology 图示气候学
cartographic base 地理底图【地理】
cartographic compilation 地图编绘【地理】
cartographic hierarchy 制图分级【测】
cartographic information 地图信息【地理】
cartographic interpretation 地图判读【地理】
cartography 地图学【地理】
cartwheel satellite 滚轮式卫星
cascade ①喷流，小瀑布，急滩②级、级联，串级③梯阶
cascade decay 级联衰变【物】
cascade filtering 级联滤波
cascade hydroelectric station (= step hydroelectric station) 梯级水电站
cascade impactor 多级采样器【大气】
cascade of energy 能量串级
cascade process 级联过程【物】，串级过程
cascade shower 级联簇射【地】
cascade theory 串级理论
cascading glacier 悬冰川，湍降冰川，冰瀑
case-based reasoning 〔历史天气〕个例推理
Casella's siphon rainfall recorder 卡萨拉虹吸雨量计
case study 个例分析
Casimir effect 卡西米尔效应【物】
Casimir polynomial 开西米尔多项式【数】
Cassegrainian mirror 卡塞格伦反射镜
Cassegrainian telescope 卡塞格林望远镜
Cassegrain reflector antenna 卡塞格伦反射面天线【电子】

Cassegrain telescope 卡塞格林望远镜【天】
cassette (tape) 盒式磁带
castellanus (cas) 堡状(云)
castellatus 堡状(云)(旧名)
CAT (clear air turbulence) 晴空湍流【大气】
catadioptric telescope 折反射望远镜【天】
catafront (=katafront)下滑锋
Catalina eddy 卡塔利娜涡旋(美南加州海域上形成的中尺度气旋环流)
catalysis 催化【化】
catalyst 催化剂【化】
catalytic converter 催化转换器,催化式排气净化器(减少汽车排放的装置)
catalytic cycle 催化循环
cataphalanx 冷锋面
catastrophe 突变【物】,灾变
catastrophe geometry 突变几何形
catastrophe theory 灾变论【地理】,突变理论
catastrophic climatic change 灾变性气候变化
catastrophic event 灾变性事件
catastrophism 灾变论【地质】
catathermometer 卡他温度表,冷却率温度表
catchment area 汇水面积,集水面积
catchment glacier(= snowdrift glacier) 吹雪冰川,洼地冰川
catch 集雨量
categorical forecast 分类预报,分级预报
CATEX (China Abnormal Typhoon Scientific Experiment) 中国异常台风科学试验(1993—1994)
Cathaysia 华夏古大陆【地质】
cathetometer 高差表
cathode 阴极
cathode ray 阴极射线
cathode-ray direction-finder 阴极射线测向器
cathode-ray direction finding 阴极射线测向
cathode-ray oscillograph 阴极射线示波器(旧称,现不常用)

cathode-ray oscilloscope 阴极射线示波器
cathode-ray radiogoniometer 阴极射线定向器
cathode-ray tube (CRT) 阴极射线管
cathodic current 阴极电流【化】
cat ice 壳冰(水位下降后存留的)
cation 正离子,阳离子
cation exchange capacity 阳离子交换容量
cat's nose 猫鼻风(英国的一种西北冷风)
cat's paw 猫掌黑斑(阵风在水面产生的表面张力波引起的湖面或海面现象)
Cauchy mean value theorem 柯西中值定理【数】
Cauchy number (或 Hooke number) 柯西数(或胡克数)
Cauchy-Riemann equations 柯西-黎曼方程【数】
cauliflower cloud 花椰菜云
causality 因果性【物】,因果联系,因果规律
cavaburd (见 kavaburd)卡伐布大雪
cavaliers 骑士风期(法国 3—4 月盛吹 mistral 风的日子)
caver (见 kaver)卡佛风
cavitation 空腔作用
cavity 空腔,腔体
cavity pyrheliometer 腔体绝对日射表
cavity radiaton 〔空〕腔辐射
cavity radiometer 空腔辐射计【大气】
C band C 波段(雷达用, $4 \sim 8$ GHz 频率, $3.75 \sim 7.5$ cm 波长)
CBL (convective boundary layer) 对流边界层
CCL (convective condensation level) 对流凝结高度【大气】
CCN (cloud condensation nuclei) 云凝结核【大气】
cd (candela) 烛光,坎〔德拉〕(发光强度的单位)
CD (compact disc) 光碟【计】,光盘
CDF (cumulative distribution function) 累计分布函数
CDM(clean development mechanism) 清洁发展机制

CDMA(code division multiple access) 码分多址(一种数字通信技术)
CD-R(**CD-recordable**) 可录光碟【计】
CDR(circular depolarization ratio) 圆退偏〔振〕比
CD-ROM(**CD-read only memory**) 只读碟【计】
CD-RW(**CD-rewritable**) 可重写光碟【计】
ceiling ①云幂,云幕 ②云幂高度(美国用语) ③升限(飞机、上升气流等的)
ceiling alarm 云幂警告器
ceiling and visibility unlimited 晴朗,能见度极好
ceiling balloon 云幂气球【大气】,测云气球
ceiling classification 云幂分类,云幕分类
ceiling height 云幂高度
ceiling height indicator 云幂高度指示器
ceiling light 云幂灯【大气】
ceiling of convection 对流高度
ceiling projector 云幂灯【大气】,云幕灯
ceiling zero 零级云幂(云幂高度在15 m以下)
ceilograph 云幂计
ceilometer 云幂仪(测低云高)【大气】
ceilometry 云幂法
celerity 波速
celestial axis 天轴【航海】
celestial body 天体
celestial chart 天体图
celestial coordinate 天体坐标
celestial coordinate system 天球坐标系【天】
celestial equator 天赤道【天】,天〔球〕赤道
celestial fix 天文船位【航海】
celestial fixing 天文定位【航海】
celestial horizon 真地平圈【航海】
celestial horizon (=astronomical horizon) 天球地平圈,天文地平
celestial latitude 黄纬【天】
celestial longitude 黄经【天】
celestial meridian 天体子午圈

celestial navigation 天文导航【天】,天文航海【航海】
celestial pole 天极【天】,天〔球〕极
celestial sphere 天球【天】
celestial stem 天干【天】
cell ①环型,单体【大气】,单元 ②电池 ③细胞 ④元件 ⑤像元【地信】
cell cloud 细胞状云
cell echo 单体回波【大气】
Cellini's halo (见 Heiligenschein)露面宝光
cellophane hygrometer 胶膜湿度表,赛珞芬湿度表
cell outline 云单体轮廓线
cell resolution 像元分辨率【地信】
cell size 像元尺寸【地信】
cell theory 环型〔学〕说
cellular circulation 单体环流,细胞环流【大气】
cellular cloud pattern 〔细〕胞状云〔型〕
cellular convection 单体对流,细胞对流【大气】
cellular hypothesis 单体假说
cellular movement 环型运动
cellular pattern 细胞状云【大气】
cellular physiology 细胞生理学
cellular structure 环型构造
cellular vortex 环型涡旋
cellular wave 环型波
cellulosed paper 纤维素纸(测湿用)
Celsius temperature scale 摄氏温标【大气】
Celsius' thermometer 摄氏温度表
Celsius' thermometric scale 摄氏温标
Cenozoic Era 新生代【地质】
Cenozoic Erathem 新生界【地质】
centennial time-scale 百年时间尺度
center 中心点
centered difference 中心差分【数】
centered finite difference 中心有限差分
centered time difference 时间中央差

Center for Atmospheric Sciences (CAS, USA) 〔美国〕大气科学中心
Center for International Environment Information (CIEI) 国际环境新闻中心
center jump 〔低压〕中心跳跃
center of action 大气活动中心【大气】,活动中心
center of attraction 引力中心
center of buoyancy 浮心【航海】
center of disturbance 扰动中心
center of falls 降压中心
center of gravity 重心【物】
center of mass 质心
center-of-mass coordinate 质心坐标
center of pressure 压力中心
center of rises (= pressure-rise center) 升压中心
center of symmetry 对称中心
center of variation 变化中心
centi- 厘(词头, 10^{-2})
centibar 厘巴(=10 mb=10 hPa)
centigrade(℃) 百分度,摄氏度(温度单位)
centigrade temperature scale 摄氏温标
centigrade thermometer 摄氏温度表
centimeter-gram-second system (cgs system) 厘米·克·秒制
centimeter(cm) 厘米
centipoise 厘泊
central calm 中心无风区
central core 中心核
central data distribution facility (CDDF) 中央数据分发中心
central depression 中心气旋,中心低压
central difference 中心差分【数】
central eye 中心眼
central force 有心力,中心力,辏力
central forecasting office (国家或地区)预报中心
Centralized Storm Information System (CSIS) 风暴信息集中系统

central limit 中心极限
central limit theorem 中心极限定理【数】
central meridian passage 中心子午线中天
central moment 中心矩
Central Pacific Hurricane Center (CPHC) 中太平洋飓风中心(美国)
central plasma sheet 中心等离子体片
central pressure 中心气压
central processing unit (CPU) 中央处理器【计】,中央处理机
Central Standard Time 中部标准时间(美国)
central tendency 集中趋势(统计)
central water 中央水(深度150～800 m之间的海水)
centrifugal acceleration 离心加速度
centrifugal effect 离心效应
centrifugal force 离心力【物】
centrifugal instability 离心不稳定度
centrifugal potential 离心势
centrifugal stretching 离心伸长
centrifuge 离心机
centrifuge moisture equivalent 离心水分当量
centripetal acceleration 向心加速度【物】
centripetal force 向心力【物】
centroid 矩心,质心,形心【数】
CEO (Chief Executive Officer) 执行主管【计】,首席执行官
CEOF (complex empirical orthogonal function) 复经验正交函数
CEOS (Committee on Earth Observation Satellites) 卫星对地观测委员会
cepstrum 倒谱,对数倒频谱
cepstrum analysis 倒谱分析
ceramic relative humidity sensor 陶瓷相对湿度传感器
ceramic semiconductor 陶瓷半导体(元件)
ceramic tube psychrometer 瓷管干湿表
ceraunograph 雷电仪【大气】
ceraunometer 雷电仪【大气】

cerium titanate humidity transducer 钛酸铈湿度传感器
cers 西尔斯风(法国南部和西班牙东北部的强风)
certainty factor 确定性因子
cetane 十六烷($C_{16}H_{34}$),鲸蜡烷
CFCs (chlorofluorocarbons) 氯氟碳化物【大气】
C figure C 指数(地磁活动,旧称)
CFL (Courant-Friederichs-Lewy) condition 柯朗-弗里德里希斯-列维条件【数】,CFL 条件【大气】
CFLOS (cloud-free line of sight) 无云视线
CFR (VFR 的旧称)目视飞行规则
cfs (cubic feet per second) 立方英尺每秒(缩写)
cgs system 厘米·克·秒制
CGT (climatological global teleconnection) 全球气候遥相关
chaff 金属丝,金属箔(干扰雷达,或作反射体用)
chaff seeding 箔丝播撒【大气】
chaff wind technique 箔条测风法
chain length 链长【化】
chain lightning 链〔状〕闪〔电〕
chain reaction 链式反应【物】,连锁反应
chain rule 链式法则【数】,链规则
chalk 白垩【地质】
challiho 恰里荷风(印度 4 月前后的一种猛烈南来风)
chamsin (见 khamsin)喀新风
chandui (= chanduy) 昌杜伊凉风(厄瓜多尔 7—11 月干季期间)
change chart (或 tendency chart) 变化图(或趋势图)
change of phase 相变
change of state 状态变化
change of type 天气型更换
Changjiang-Huaihe cyclone 江淮气旋【大气】
Changjiang-Huaihe shear line 江淮切变线【大气】
Changjiang River 长江
channel ①通道【大气】,频道②信道【计】③闪道
channel control 河槽控制,测流控制断面
channeling 峡谷风
channel jet 通道急流
channel storage 河槽蓄水〔量〕,槽蓄〔量〕
chaos 混沌【数】
chaotic attractor 混沌吸引子【物】
chaotic dynamical system 混沌动力系统
chaotic motion 混沌运动【物】
chaotic sky 混乱天空【大气】
Chapman cycle 查普曼循环
Chapman layer 查普曼层【地】,D 层
Chapman mechanism 查普曼机制【大气】
Chapman region 查普曼区
Chapman theory 查普曼理论
Chappuis band 查普斯带
character ①特性 ②字符【计】
character figure 〔逐日〕磁变示数,磁情指数
characteristic coordinate 特征坐标
characteristic curve 特征曲线【数】
characteristic equation 特征方程
characteristic failure behavior 故障〔率〕特征曲线
characteristic frequency 特征频率
characteristic function 特征函数【数】,本征函数
characteristic humidity 特征湿度
characteristic length 特征长度
characteristic line 特征线
characteristic matrix 特征矩阵【数】
characteristic of the pressure tendency 气压倾向特征
characteristic passage time 特征通过时间
characteristic point 特性点

characteristic root　特征根
characteristics　①特性 ②特性曲线
characteristics of grain sizes　粒度特征
characteristic temperature　特征温度
characteristic time scale　特征时间尺度
characteristic value　特征值,本征值
characteristic-value problem　特征值问题,本征值问题
characteristic vector　特征向量【数】,特征矢量,本征矢量
characteristic velocity　特征速度
character representation form for data exchange（CREX）　用于数据交换的字符表示格式
characters per second　字符每秒【计】
character string　字符串【计】
character style　字体【计】
charge　①电荷 ②带电
charge balance　电荷平衡
charge conservation　电荷守恒【物】
charged cloud　荷电云
charged particle　带电粒子【物】
charged plate　荷电板
charge-mass ratio　荷质比【物】
charge separation　电荷分离
charging time constant　荷电时间常量【物】
Charles-Gay-Lussac law　查理-盖吕萨克定律
Charles law　查理定律【物】
Charner-Drazin theorem　查尼-德拉津定理,C-D定理
Charner-Stern theorem　查尼-斯特恩定理,C-S定理
Charney mode　查尼波,查尼模
Charnock's relation　查诺克关系式（流体力学中）
chart　图
chart datum　海图基准面【航海】
chart plotter　填图员
chart plotting　填图
chasing tornado　尾追龙卷

Chebyshev polynomial　切比雪夫多项式【数】
check　检验
checkerboard diagram　棋盘式图
check observation　校验观测
cheimophobia　寒冷恐怖症,惧寒症
chemical actinometer　化学日射表
chemical activation　化学活化【化】
chemical activity　化学活性【化】
chemical analysis　化学分析【化】
chemical bond　化学键【化】
chemical combination　化合【化】
chemical composition of precipitation　降水的化学成分
chemical energy　化学能【化】
chemical equilibrium　化学平衡
chemical fertilizer　化学肥料【土壤】
chemical geography　化学地理学【地理】
chemical hygrometer　化学湿度表【大气】
chemical laser　化学激光器【化】
chemically oxidizing atmosphere　化学氧化大气
chemical meteorological modeling　化学-气象模拟
chemical oxygen demand（COD）　化学需氧量【大气】
chemical physics　化学物理〔学〕【物】
chemical potential　化学势【化】
chemical reaction　化学反应【化】
chemical smoke　化学烟气
chemical tracer　化学示踪物
chemical weathering　化学风化【土壤】
chemiluminescence　化学发光【化】,化学荧光
chemiluminescent O_3 analyzer　化学荧光臭氧分析仪
chemi-luminescent sonde　化学荧光探空仪（测臭氧用）
chemisorption　化学吸附【物】
chemopause　光化层顶【大气】

chemosphere 光化层【大气】
chergui 秋尔古风(摩洛哥的一种东或东南风)
Chezy equation 谢齐方程(水文)
chibli(见 kibli) 基布利风
Chicago school 芝加哥学派
chichili 奇奇利风(阿尔及利亚南部的干热风)
chief executive officer(CEO) 执行主管【计】,首席执行官
chief mate 大副【航海】
chief officer 大副【航海】
Chile Current 智利海流
chili 奇利风(突尼斯的干热风)
chill 寒冷
chilled-mirror hygrometer 冷凝镜湿度表,露点湿度表
chill factor 风寒指数,寒冷因子
chill hour (蛰伏)寒冷时数
chilling 冷温[量],冷温需求
chilling damage 低温冷害
chilling hour (蛰伏)寒冷时数
chilling injury 寒害【大气】
chill unit (蛰伏)寒冷单位
chimney cloud 烟(囱)云
chimney current 烟囱式气流
chimney plume (烟囱排放)烟羽
chimonophilous plant 喜冬植物
China Abnormal Typhoon Scientific Experiment(CATEX) 中国异常台风科学试验(1993—1994)
China Coastal Current 中国海岸流
China Global Atmosphere Watch Baseline Observatory(CGAWBO) 中国全球大气监测网基准观象台(简称瓦里关本底站)
China Meteorological Administration(CMA) 中国气象局
China new generation weather radar (CINRAD) 中国新一代天气雷达
Chinese character card 汉卡【计】
Chinese Meteorological Society(CMS) 中国气象学会【大气】
Chinese National Antarctic Research Expedition(CHINARE) 中国南极考察队
chinook 钦诺克风(落基山东坡的一种干暖西南风)
chinook arch 钦诺克拱状云
chionophile 喜雪的
chionophilous plant 喜雪植物
chionophobous plant 嫌雪植物
chip 芯片【计】
chip log 拖板计程器,测程板(海洋)
chipware 芯件【计】
Chi-square distribution χ^2 分布【数】
Chi-square test χ^2 检验【数】
chlorine compounds 氯化物
chlorine dioxide 二氧化氯(OClO)
chlorine monoxide dimmer 一氧化氯二聚物(Cl_2O_2)
chlorine monoxide radical 一氧化氯基(ClO)
chlorine nitrate 硝酸氯($ClNO_3$,常写为$ClONO_2$)
chlorine oxides 氯[的]氧化物(ClO_x)
chlorinity 氯含量【大气】,氯度
chlorofluorocarbons(CFCs) 氯氟碳化物【大气】
chlorofluoromethane(CFM) 氟氯烷,氟利昂,氟冷剂
chloroform 氯仿,三氯甲烷($CHCl_3$)
chlorophyll 叶绿素【植】
chloroplast DNA 叶绿体 DNA【生化】
chloroplast RNA 叶绿体 RNA【生化】
chlorosity 氯度【大气】,体积氯度,氯容(等于样品氯度 chlorinity 乘以样品 20℃时的密度)
chocolate gale(=chocolatero) 巧克力北风(墨西哥湾北风不很强盛时的一种叫法)
chocollatta north 丘可拉塔北风(西印

度群岛的西北大风)
chom 科姆风(Sirocco 的北非方言)
chop ①乱短波,三角浪 ②有碎浪的水(海)面 ③断续,断层
chopper ①斩波器,限制器②断路器
choppy ①风向屡变的 ②波浪滔滔的
choroisotherm 〔地区〕等温线
chota barsat 小雨期(印度季风雨来前的一二天雨日,印度斯坦语名)
chou lao hu 秋老虎(中文)
Christoffel number 克里斯托弗尔数
chromatic scintillation 色闪烁
chromatogram 色〔层〕谱图【物】
chromatograph 色〔层〕谱仪【物】,色谱仪【化】
chromatography 色〔层〕谱法【物】
chromosphere 色球【天】
chronic pollution 慢性污染
chronoanemoisothermal diagram 时间风向温度图
chronograph ①时间记录器 ②〔精密〕记时计
chronoisotherm 时间等温线
chronology 年代学,年表
chronometer 天文钟【航海】,计时表
chronometer rate 日差【航海】
chronometric radiosonde 〔要素调制〕时距探空仪
chronothermometer 平均温度表,温度时间自记表
chubasco 丘巴斯科雷暴(见于中美西海岸雨季)
chun fung 春风(英拼中文)
chunyuh 春雨(英拼中文)
churada 楚那打飑(西太平洋上马里亚纳群岛上 1、2、3 月间的一种飑名)
churn 翻腾
cierzo 西尔左风(mistra 风的西班牙语名,常见于秋冬之际)
CIGRE (International Council on Large Electric Systems) 国际大电网会议
CIN (convective inhibition) 对流抑制能,对流约束能
C index C 指数(地磁活动)
cinetheodolite 电影经纬仪,高精度光学跟踪仪
CINRAD (China new generation weather radar) 中国新一代天气雷达
CIPAS (Climate Information Processing and Analysis System) 气候信息处理与分析系统
cipher 密码【电子】
cipher code 密码【电子】
cipher key 密钥【电子】
circle 圆
circle function 圆函数
circle of inertia 惯性圆
circuit 回路【数】,路径,电路
circular convolution 循环卷积
circular correlation 循环相关
circular cylindrical coordinates (= cylindrical coordinates) 〔圆〕柱坐标
circular depolarization ratio 圆退偏〔振〕比
circular frequency 圆频率【物】
circular hole diffraction 圆孔衍射【物】
circular motion 圆周运动
circular oscillation 圆振荡
circular polarization 圆偏振【物】
circular symmetry 圆对称
circular thermograph 圆形温度计
circular variable 圆变量
circular vortex 圆涡旋
circulation 环流
circulation cell 环流圈【大气】,环流单体
circulation in cyclone 气旋环流
circulation index 环流指数【大气】
circulation integral 环积分
circulation in wake 尾流环流
circulation model 环流模式
circulation pattern 环流型【大气】

circulation system 环流系统
circulation theorem 环流定理【大气】
circulation type 环流型
circumference ①圆周【数】②周长【数】
circumferential velocity 圆周速度
circumhorizontal arc 环地平弧【大气】, 日承
circumpolar circulation 绕极环流【大气】
circumpolar cyclone 绕极气旋
circumpolar map 环极地图
circumpolar vortex 绕极涡旋【大气】
circumpolar westerlies 绕极西风带【大气】
circumpolar whirl 绕极旋风
circumscribed halo 外切晕
circumsolar radiation 太阳周边辐射
circumsolar sky radiation 环日天空辐射
circumzenithal arc 环天顶弧【大气】,日载
cirque glacier 冰斗冰川
cirriform 卷云状〔的〕
cirrocumulus（Cc） 卷积云【大气】
cirrocumulus castellanus 堡状卷积云
cirrocumulus floccus 絮状卷积云
cirrocumulus lacunosus 网状卷积云
cirrocumulus lenticularies 荚状卷积云
cirrocumulus "mackerel" 鱼鳞状卷积云
cirrocumulus nebulosus 薄幕卷积云
cirrocumulus stratiformis 成层卷积云
cirrocumulus undulatus 波状卷积云
cirro-filum（或 thread） 纤缕卷云
cirro-macula 斑状卷云
cirro-nebula 薄幕卷云
cirro-ripples 绉纹卷云
cirrostratu cumulosus 积云状卷层云
cirrostratus（Cs） 卷层云【大气】
cirrostratus communis 普通卷层云
cirrostratus duplicatus 复卷积云
cirrostratus fibratus（Cs fib） 毛卷层云
cirrostratus filosus（Cs fil） 毛卷层云【大气】
cirrostratus nebulosus（Cs nebu） 薄幕卷层云【大气】

cirrostratus undulatus 波状卷层云
cirrostratus vittatus 带状卷层云
cirro-velum 缟状卷云
cirrus（Ci） 卷云【大气】
cirrus canopy 卷云幔
cirrus castellanus 堡状卷云
cirrus caudatus 尾状卷云
cirrus densus（Ci dens） 密卷云【大气】
cirrus duplicatus 复卷云
cirrus excelsus 高卷云
cirrus fibratus（Ci fib） 毛卷云
cirrus filosus（Ci fil） 毛卷云【大气】
cirrus floccus 絮状卷云
cirrus intortus 乱卷云
cirrus nothus（Ci not） 伪卷云【大气】
cirrus plume 卷云砧
cirrus radiatus 辐辏状卷云
cirrus sheet 卷云片
cirrus shield 盾状卷云层
cirrus spissatus（Ci spi） 密卷云
cirrus uncinus（Ci nuc） 钩卷云【大气】
cirrus ventosus（=windy cirrus） 风卷云
cirrus vertebratus 脊状卷云
CISK（conditional instability of the second kind） 第二类条件〔性〕不稳定【大气】
cissero 锡西罗风(法国罗讷河三角洲的热而有雨的东南 Sirocco 风)
cistern 〔水银〕槽
cistern barometer 水银槽气压表
city climate 城市气候
city ecological system 城市生态系统【地理】
city environment 城市环境
city fog 城市雾
city pollution 城市污染
civic landscape 城市景观
civil day 民用日【天】(从子夜开始计量的平太阳日)
civil time 民用时
civil twilight 民用晨昏朦影【航海】,民用曙暮光

CL(condensation level)　凝结高度【大气】
Clapeyron-Clausius equation(＝Clausius-Clapeyron equation)　克拉珀龙-克劳修斯方程,C-C 方程
Clapeyron's diagram　克拉珀龙图
clapotis　驻波(法文术语)
clarity〔of gas〕　纯洁度(气体的)
class A evaporation pan　A 级蒸发皿
class-A pan　A 级蒸发皿
classical coherence theory　经典相干理论
classical physics　经典物理〔学〕【物】
classical solution　经典解【数】
classification　分类〔法〕
classification analysis　分类分析【大气】
classification statistics　分类统计
class interval　(频数或次数分布的)组,组距
clathrate hydrate　水包合物(高压下存在的水合物)
Clausius-Clapeyron equation　克劳修斯-克拉珀龙方程【物】,C-C 方程
clausius(Cl)　克劳(熵的旧单位,＝1 卡/K)
Clausius equation of state　克劳修斯状态方程
clay atmometer　陶瓷蒸发计
Clayden effect　克雷登效应(闪电摄影)
clay soil　黏土〔土壤〕
clean air(＝pure air)　纯净空气,纯洁空气
clean development mechanism(CDM)　清洁发展机制
cleaning effect by rain　雨水清洁作用
clear　碧〔空〕
clear air　晴空
clear air echo　晴空回波【大气】
clear-air turbulence(CAT)　晴空湍流(颠簸)【大气】
clearance(＝clearing)　①云消 ②云消时间 ③云缝
clear-column radiance　晴空(气柱)辐射
clear day　晴天
clear ice　纯净冰

clear icing(＝clear ice)　纯净冰
clearing　①晴空化〔过程〕②清除
clearinghouse　变换站,交换中心
clear layer　透明层
clear line of sight(CLOS)　晴空视线
clear night　晴夜
clear-of clouds condition　云外飞行条件
clear radiance　晴空辐射
clear sky　晴天【大气】(总云量小于 $\frac{1}{8}$),碧空
clear sky precipitation　晴空降水
clear-weather approach　良好天气进场着陆
cleft　①磁隙区 ②极尖区
CliC(Climate and the Cryosphere)　气候与冰雪圈计划
cliff-eddy　悬壁涡流
climagram　气候图
climagraph　气候图表
CLIMAP(Climate: Long-range Investigation Mapping and Prediction)　气候长期调查测绘和预测计划(重建距今 18000 年的海面温度规划)
CLIMAT(report of monthly means and tatals from a land station)　陆地测站月平均与总量报告
CLIMAT broadcast　气候月报(每月 5 日前报告上月的)
climate　气候【大气】
climate abnormality　气候反常
climate accident　气候突变
climate analogs　气候相似,气候类比
Climate Analysis Centre(CAC,NOAA)　气候分析中心(NOAA)
Climate and the Cryosphere(CliC)　气候与冰雪圈计划
climate and human health(CHH)　气候与人类健康
climate anomaly　①气候距平 ②气候异常
Climate Application Programme(CAP)

气候应用计划
climate change(=climatic change) 气候变化
climate change detection (=climatic change detection, CCD) 气候变化探测,气候变化发现
climate change projection 气候变化展望
climate comfort 气候舒适〔度〕
climate control 气候控制〔方案〕
climate damage 气候灾害【大气】
Climate Data Programme (CDP) 气候资料计划
climate divide 气候分界【大气】
climate drift 气候漂移
climate fluctuations (=climatic fluctuations) 气候振动
climate forecast 气候预报
climate-friendly technique 气候友好技术,气候友善技术
Climate Impact Assessment Programme (CIAP) 气候影响评价计划
Climate Impact-Study Programme (CIP) 气候影响研究计划
Climate Information Processing and Analysis System (CIPAS) 气候信息处理与分析系统
climate melioration 气候改良
climate model 气候模式
climate modification 人工影响气候【大气】
climate monitoring 气候监测
climate noise 气候噪声【大气】
climate of eternal frost 永冻气候
climate periodicity 气候周期性【大气】
climate prediction 气候预测
climate projection 气候展望,气候预估
climate reconstruction 气候重建
climate regionalization 气候区划【大气】
climate resources 气候资源【大气】
climate sensitivity 气候敏感性
climate sensitivity experiment 气候敏感性实验【大气】
climate signal 气候信号
climate simulation 气候模拟
climate snow line (=climatic snow line) 气候雪线
climate state 气候状态
climate state vector (CSV) 气候状态矢量
climate stress load 气候应力荷载【大气】
climate system 气候系统【大气】
climate variability (=climatic variability) 气候变率
Climate Variability and Predictability (CLIVAR) 气候变率与可预测性〔研究计划〕
climatic adaptation 气候适应【大气】
climatic amelioration 气候改良
climatic amplitude 气候振幅
climatic analysis 气候分析【大气】
climatic anomaly 气候异常【大气】,气候距平
climatic assessment 气候评价【大气】
climatic atlas 气候图集【大气】
climatic barrier 气候障壁
climatic belt 气候带【大气】
climatic box 人工气候箱
climatic chamber 人工气候室
climatic change 气候变化【大气】
climatic characteristics 气候特征
climatic chart 气候图
climatic classification 气候分类【大气】
climatic climax 气候演替顶极
climatic coexistance 气候共存态
climatic comfort 气候舒适〔度〕
climatic condition 气候条件
climatic constraint 气候约束〔因子〕
climatic contrast 气候对比
climatic control 气候控制〔因子〕
climatic cultivation limit 气候栽培界限
climatic cycle 气候旋回(非严格的周期性)
climatic data 气候资料【大气】
climatic degeneration 气候恶化【大气】
climatic demand 气候要求

climatic deterioration 气候恶化【大气】
climatic determinism 气候决定论
climatic diagnosis 气候诊断【大气】
climatic diagram 气候图
climatic disaster 气候灾害
climatic discontinuity 气候不连续
climatic divide 气候分界
climatic domestication 气候驯化【大气】
climatic effect 气候效应
climatic element 气候要素【大气】
climatic ensemble 气候总体
climatic environment 气候环境
climatic evaluation of fire-danger 火险气候评价
climatic extreme 气候极值
climatic factor 气候因子
climatic feedback interaction 气候反馈作用
climatic feedback mechanism 气候反馈机制【大气】
climatic fluctuation 气候振动【大气】
climatic forecast 气候展望,气候预报
climatic impact 气候影响【大气】
climatic index 气候指数
climatic indicator 气候指示物,气候标志
climatic instability 气候不稳定性
climatic intransitivity 气候非传递性
climatic landscape 气候景观
climatic limit of crops 作物气候界限
climatic map 气候图【大气】
climatic modification 人工影响气候
climatic monitoring 气候监测【大气】
climatic noise 气候噪声
climatic non-periodic variation 气候非周期变化【大气】
climatic numerical modelling 气候数值模拟
climatic optimum 气候适宜〔期〕,气候最适期
climatic oscillation (= climatic cycle) 气候振荡【大气】

climatic pathology 气候病理学
climatic periodicity 气候周期性
climatic periodic variation 气候周期性变化【大气】
climatic phenomenon 气候现象
climatic persistence 气候持续性【大气】
climatic physiology 气候生理学
climatic plant formation 气候植物区系
climatic potential 气候潜力,气候潜势
climatic prediction 气候预测【大气】
climatic prediction of the first kind 第一类气候预测
climatic prediction of the second kind 第二类气候预测
climatic probability 气候概率【大气】
climatic problem 气候问题
climatic productivity index 气候生产力指数
climatic province 气候区
climatic psychology 气候心理学
climatic reconstruction 气候重建【大气】
climatic record 气候记录
climatic region 气候区【地理】
climatic regionalization 气候区划【地理】
climatic resources 气候资源
climatic revolution 气候演变【大气】
climatic rhythm 气候韵律
climatic risk analysis 气候风险分析【大气】
climatic scenario 气候背景
climatic sensitivity 气候敏感性【大气】
climatic simulation 气候模拟【大气】
climatic soil formation 气候性土壤形成
climatic soil type 气候土壤型
climatic state vector 气候状态矢量
climatic statistics 气候统计【大气】
climatic subdivision 气候副区
climatic system 气候系统
climatic teleconnection 气候遥相关
climatic time series 气候时间序列
climatic transition 气候转换
climatic transitivity 气候可传递性

climatic treatment 气候疗法
climatic trend 气候趋势【大气】
climatic turbulence 气候湍流
climatic type 气候型【大气】
climatic vacillation 气候波动
climatic value 气候值
climatic variability 气候变率【大气】,气候可变性
climatic variation 气候变迁【大气】
climatic year 气候年
climatic zonation 气候地带性
climatic zone 气候带【大气】
climatization 气候适应过程
climatizer 气候实验室
climatogenesis 气候生成
climatography 气候志【大气】
climatological atlas 气候图集
climatological background field 气候背景场
climatological chart 气候图
climatological condition 气候条件
climatological data 气候资料
climatological data bank 气候资料库
climatological diagram 气候图表
climatological division 气候分区
climatological forecast 气候〔学方法〕预报
climatological front 气候锋【大气】
climatological global teleconnection (CGT) 全球气候遥相关
climatological limit check 气候极值检验
climatological network 气候站网
climatological normals 气候平均值
climatological observation 气候观测【大气】
climatological pattern 气候型
climatological standard normals 气候标准平均值【大气】(1901—1930年,1931—1960年等30年年平均值)
climatological station 气候站
climatological station elevation 气候站海拔〔高度〕
climatological station network (= climatological network) 气候站网
climatological station pressure 气候站气压
climatological statistics 气候统计学【大气】
climatological substation 气候辅助站
climatological summary 气候概述
climatological transparent coefficient 气候透明系数
climatological year 气候年(通常与公历年相同,偶尔有从12月1日开始计算的12个月)
climatologist 气候工作者,气候学家
climatology 气候学【大气】
climatomesochore 中小气候情况,平均小气候情况
climatonomy 理论气候学
climatopathology 气候病理学【大气】
climatotherapy 气候治疗学,转地疗法
climax community 顶极群落【生】
climax soil 顶极土壤【土壤】
climogram (= climatic diagram) 气候图
climograph (= climatic diagram) 气候图
clinometer ①倾斜计,仰角器②磁倾计,倾角仪③象限仪
cliper 气候持续性
CLIVAR (Climate Variability and Predictability) 气候变率与可预测性〔研究计划〕
clo 克罗(衣着指数单位)
clock-star 校钟星
clock-wise 顺时针
clockwork anemometer 钟制风速表
clone 克隆【植】
CLOS (clear line of sight) 晴空视线
closed basin 闭合流域,内流流域,闭合盆地【地理】
closed〔cloud〕cells 封闭型细胞状云【大气】
closed cell 闭合环流,闭合胞
closed-cell stratocumulus 闭合胞层积云

closed circuit television 闭路电视
closed circulation 闭合环流
closed curve 闭合〔曲〕线
closed drainage 闭合流域,内流流域
closed equations 闭合方程组
closed form 闭形【物】
closed(geomagnetic)field line 闭〔磁〕力线
closed high 闭合高压
closed isobar 闭合等压线
closed lake 内陆湖,无出流湖
closed loop 闭〔合〕回路【物】
closed low 闭合低压
closed model 闭合模式
closed system 闭合系统【大气】,闭合方程组
closed weather （旧称,现常用 below minimums)机场关闭天气
closure assumption 闭合假设
closure problem 闭合问题
closure scheme 闭合方案
cloud 云【大气】
cloud absorption 云吸收
cloudage 云量
cloud albedo 云反照率【大气】
cloud amount 云量【大气】
cloud atlas 云图
cloud attenuation 云衰减【大气】
cloud band 云带【大气】
cloud bank 云堤【大气】
cloud banner(＝banner cloud) 旗〔状〕云
cloud bar 云带,云条
cloud base 云底【大气】
cloud base height 云底高度
cloud-base height balloon 云底高度探测气球
cloud-base recorder 云底高度测录器
cloudbow (亦称 fogbow, mistbow, white rainbow) 云虹
cloudbuster 破云器(一种播撒干冰用的机载设备)
cloud camera 摄云机

cloud canopy 云幔
cloud cap (＝cap cloud) 帽〔状〕云,云帽
cloud ceiling 云幂【大气】,云幕
cloud ceilometer 云幂仪
cloud chamber 云室【大气】
cloud chart 云图
cloud chemistry 云化学
cloud classification 云分类【大气】
cloud cluster 云团【大气】
cloud column 烟云柱
cloud computing 云计算
cloud condensation nuclei(CCN) 云凝结核【大气】
cloud condensation nucleus counter 云凝结核计数器
cloud contamination 云干扰
cloud cover 云量
cloud coverage 云覆盖区【大气】
cloud covered area 云覆盖区
cloud crest 云冠
cloud current 云〔中气〕流
cloud deck 云海
cloud decoupling 云际退耦〔作用〕
cloud density 云密度
cloud-detection radar 测云雷达【大气】
cloud diary 云〔的〕日志
cloud direction 云向
cloud discharge 云放电【大气】
cloud dispersal 消云
cloud dissipation 消云【大气】
cloud drop 云滴
cloud droplet 云滴【大气】
cloud droplet collector 云滴谱仪【大气】
cloud droplet-size distribution 云滴谱【大气】
cloud-drop sampler 云滴取样器
cloud due to volcanic eruption 火山云
cloud dynamics 云动力学【大气】
cloud echo 云回波【大气】
cloud electrification 云起电
cloud element 云素

cloud ensemble 云总体
cloud-environment system 云周围环境体系
cloud étage 云族【大气】
cloud feedback 云反馈【大气】
cloud field 云区
cloud flash 云闪
cloud flying 云中飞行
cloud forest 云林(云层上面多塔状突起的情形)
cloud form 云状【大气】
cloud formation 云的形成
cloud fraction 云量
cloud-free area 无云区
cloud-free line-of-sight 无云视线
cloud frequency 云〔的〕频率
cloud from fire 火灾云
cloud from waterfall 瀑布云
cloud genera (genus 的复数)云属【大气】
cloud groups 云簇
cloud gun 采云枪,云〔取样〕枪
cloud height 云高【大气】
cloud height indicator 云高指示器
cloud-height measurement method 云高测量法
cloud height meter 云幂器
cloud identification 云的识别
cloudier 人造云
cloud image animation 云图动画【大气】
cloudiness 云量
cloud layer 云层
cloudless 无云,全晴
cloudlet 小云块
cloud level 云高度,云层
cloud lightning discharge 云中闪电
cloud line 云线【大气】
cloud luminance 云辉光
cloud map 云图
cloud mass 云块
cloud microphysics 云微物理学【大气】
cloud microstructure 云的微〔观〕结构

cloud mirror 测云镜
cloud model 云模式【大气】
cloud modification 云的人工影响【大气】
cloud motion vector 云运动矢量【大气】
cloud-motion wind vector 云风矢
cloud nephoscope 测云器
cloud nucleus(cloud nuclei) (旧称)云〔粒子〕核
cloud observation 云观测
cloud optical depth 云光学路径(云顶和云底间的垂直光学厚度)
cloud parameterization 云参数化
cloud particle 云粒,云滴
cloud particle imager 云粒子成像器
cloud particle replicator 云粒子印模仪
cloud-particle sampler 云滴采样器【大气】
cloud pattern 云型
cloud-phase chart 云相图
cloud photography 摄云术
cloud physics 云物理学【大气】
cloud picture 云图
cloud population 云总数目,云密度
cloud radar 〔测〕云雷达
cloud-radiation interaction 云-辐射交互作用
cloud radiative forcing 云辐射强迫
cloud rake 梳状测云器
cloud reflector 测云镜
cloud resulting from explosion 爆炸云
cloud resulting from industry 工业污染云
cloud sea 云海
cloud searchlight 云幂灯
cloud sector 云区
cloud seeding 播云【大气】
cloud-seeding agent 播云剂【大气】,人工催化剂
cloud-setting chamber 沉降云室
cloud shadow 云影
cloud-sheet 云片

cloud shield 盾状云,云幛
cloud simulator 云室
cloud species 云种【大气】(以云形和云结构而分)
cloud stage 云的阶段
cloud street 云街【大气】
cloud structure 云结构【大气】
cloud symbol 云符号
cloud system 云系【大气】
cloud system of a depression 低压云系
cloud telemeter 云高遥测仪
cloud-to-cloud discharge 云际放电【大气】
cloud-to-ground discharge 云地〔间〕放电【大气】,霹雳
cloud-to-ground flash 云地闪电
cloud top 云顶【大气】
cloud-top entrainment instability 云顶夹卷不稳定〔性〕
cloud top height(CTH) 云顶高度【大气】,回波高度
cloud top height estimation system(CTHES) 云顶高度估计系统
cloud top temperature 云顶温度【大气】
cloud tracer 示踪云
cloud type 云型
cloud variety 云类【大气】(以云的排列和透明度而分)
cloud veil 云层,云盖,云纱
cloud water droplet sampler 液态云滴取样器
cloud wind estimation system 云风估计系统
cloud winds 云导风,云〔迹〕风
cloud with vertical development 直展云【大气】
cloudy 多云【大气】(在美国,指云量在 0.7 以上的天空)
cloudy day 多云日(该日大多数时间云量在 75% 以上)
cloudy sky 多云天空(总云量为 $\frac{3}{8} \sim \frac{5}{8}$)

cloudy weather 多云天气
cloud zone 云带
clutter 地物杂波
cluster 簇,束,群,团
cluster analysis 聚类分析【数】
cluster compound 簇合物【化】
cluster in squall line 飑线群
cluster ion 离子团
clutter rejection 地物杂波消除
CMAQ (Community Multi-scale Air Quality) (城市)多尺度空气质量模式
CME (coronal mass ejection) 日冕物质抛射【天】,日冕物质喷射【大气】
CMF(coherent memory filter) 相干存储滤波器【大气】
CNN(Cable News Network) 有线新闻网(美国)
cnoidal wave 椭圆余弦波【物】
COADS (Comprehensive Ocean-Atmosphere Data Set) 海洋-大气综合数据集
co-adsorption 共吸附【物】
coagulating sedimentation (= coagulating precipitation) 聚并沉降〔作用〕,混凝沉淀〔作用〕
coagulation 碰并【大气】,凝聚
coagulation coefficient 碰并系数
coalescence 并合(云滴的)【大气】
coalescence coefficient 并合系数
coalescence effect 并合作用
coalescence efficiency 并合系数【大气】
coalescence process 并合过程,碰并过程
coal gas 煤成气【地质】
coal geochemistry 煤地球化学【地质】
coalification 煤化作用【地质】
coamplitude line (= corange line) 等潮差线,同潮幅线
COARE (Coupled Ocean Atmosphere Response Experiment) 海气耦合响应试验
coarse-grained density 粗粒密度【物】

coarse-grained snow 粗粒雪
coarse-mesh 粗网格
coarse-mesh grid 粗网格
coarse particle mode 〔气溶胶〕粗粒
coarse particles 粗粒子(直径大于 2 μm 悬浮在空中的粒子)
coarse sand soil 粗砂土【土壤】
coastal boundary layer 海岸边界层
coastal climate 滨海气候【大气】,海岸带气候
coastal current 沿岸流【航海】
coastal effect 海岸效应
coastal fog 海岸雾
coastal front 海岸锋
coastal geomorphology 海岸地貌学【地理】
coastal jet 沿岸急流
coastal low 海岸低压
coastally trapped waves 海岸陷波,海岸俘获波,海岸拦截波
coastal marine environment 沿海海洋环境
coastal marsh 海滨沼泽【地理】
coastal meteorology 海岸气象学
coastal night fog 海岸夜雾
coastal terrace 海岸阶地【地理】
coastal upwelling 海岸涌升流,近岸上升流
coastal zone 海岸带【航海】
Coastal Zone Color Scanner (CZCS) 海岸带水色扫描仪【大气】,海色扫描仪
coasting lead 〔浅海〕测深锤
coastline 岸线【航海】
Coast Pilot ①沿海航行指南②沿海引水员
coaxial antenna 同轴天线【电子】
coaxial cable 同轴电缆【电子】,同轴线【物】
coaxiality 共轴性【物】
cobblestone turbulence 弱颠簸
COBOL (common business-oriented language) 面向商业的通用语言,COBOL 语言【计】

cockeyed bob 斜眼鲍勃飑(澳洲的一种飑)
cod 端点(轨迹弯曲时中心达到最西的一点)
COD (chemical oxygen demand) 化学需氧量
coda wave 尾波
CODAR (COrrelation Detection And Ranging) 相关探测和测距
codeclination ①余赤纬②极距
code division multiple access (CDMA) 码分多址(一种数字通信技术)
code figure 数码
code form 电码格式【大气】,电码型式
code format 代码格式
code generation 代码生成【计】
code generator 代码生成器【计】,代码生成程序
code group 电码组【大气】
code kind 电码种类【大气】
code letter 电码,字码
code name 电码名称
code section 电码段
code-sending radiosonde (= code-type radiosonde) 电码式探空仪
code specification 电码说明,编码规定
code symbol 电码符号
code table 电码表
code word 字码
codimension 余维〔数〕【物】
coding 编码
coefficient 系数
coefficient data base (CDB) 系数数据库
coefficient of barotropy 正压系数
coefficient of coherency 相干系数
coefficient of compressibility 可压缩〔性〕系数
coefficient of consolidation 固结系数,压实率
coefficient of continentality 大陆度系数
coefficient of correlation 相关系数
coefficient of diffusion (= diffusivity)

扩散系数【物】
coefficient of dynamic viscosity
　（＝dynamic viscosity）　动〔力〕黏〔滞〕系数
coefficient of eddy conduction（＝eddy conductivity）　涡旋传导系数
coefficient of eddy diffusion（＝eddy diffusivity）　涡旋扩散系数,涡动扩散系数
coefficient of eddy viscosity　涡旋黏滞系数,涡动黏滞系数
coefficient of excess　峰度系数,峭度系数
coefficient of exchange　交换系数
coefficient of expansion（＝coefficient of thermal expansion）　膨胀系数
coefficient of extinction　消光系数
coefficient of heat conduction（＝thermal conductivity）　热〔传〕导系数,热导率
coefficient of kinematic viscosity　运动黏滞系数
coefficient of molecular viscosity
　（＝dynamic viscosity）　分子黏滞系数【大气】
coefficient of multiple correlation　复相关系数
coefficient of mutual diffusion　互扩散系数
coefficient of partial correlation　偏相关系数
coefficient of piezotropy　压性系数
coefficient of polytropy　多元〔性〕系数
coefficient of regression　回归系数
coefficient of restitution　恢复系数【物】
coefficient of skewness　偏度系数
coefficient of skin friction　表面摩擦系数
coefficient of sliding friction　滑动摩擦系数【物】
coefficient of soil thermal conductivity　土壤导热系数
coefficient of soil thermometric conductivity　土壤导温系数
coefficient of static friction　静摩擦系数【物】
coefficient of tension　张力系数
coefficient of thermal conduction（＝thermal conductivity）　热〔传〕导系数,热导率
coefficient of thermal expansion　热胀系数
coefficient of transmission　透射系数
coefficient of transparency　透明度系数
coefficient of viscosity　黏滞系数,黏性系数【物】【大气】
coefficient of wind pressure　风压系数【大气】
coelostat　定天镜【天】
coercive force　顽〔磁〕力
coffin corner　危角
cognitive science　认知科学【计】
cognitive task analysis（CTA）　认知任务分析
coherence　①同调②相干性【大气】,相关性③黏附,附着④凝聚⑤谱相关
coherence bandwidth　相干带宽
coherence element　相干元素
coherence length　相干长度
coherence spectrum　相干谱
coherence time　相干时间
coherency　相干
coherent acoustic Doppler radar　相干多普勒声〔雷〕达
coherent detection　相干检测【数】,相干探测
coherent echo　相干回波【大气】
coherent field　相干场
coherent integration　相干积分
coherent intensity　相干强度
coherent light　相干光【物】
coherent light detection and ranging　相干光〔雷〕达
coherent memory filter（CMF）　相干存储滤波器【大气】
coherent optical adaptive technique（COAT）

相干光自适应技术
coherent optical radar 相干光〔雷〕达【物】
coherent oscillator 相干振荡器
coherent phase 相干位相
coherent radar 相干雷达【大气】
coherent radiation 相干辐射【物】
coherent scattering 相干散射【物】
coherent source 相干光源【物】
coherent state 相干态【物】
coherent structure 相干结构
coherent synthetic aperture imaging radar (CSAIR) 相干合成孔径成像雷达
coherent target 相干靶〔标〕
coherent-video signal 相干视频信号【大气】
coherent wave 相干波【物】
cohesion 内聚性,内聚力
COHMAP (Cooperative Holocene Mapping Project) 全新世测绘合作计划(用多途径的资料重建全球气候的计划)
coho (coherent oscillator 的缩写)相干振荡器
coincidence counting 符合计数【物】
coincidence error 重合误差
coincidence rate 符合率【物】
coincidental correlation 叠合相关
coincident spectral density 共谱密度
col 鞍型
colatitude 余纬
cold acclimatization 寒冷气候适应
cold advection 冷平流【大气】
cold air 冷空气
cold-air drop (= cold pool) 冷〔空气〕池,冷空气堆
cold-air injection 冷空气侵入
cold air mass 冷气团【大气】
cold-air outbreak 冷空气爆发,寒潮爆发
cold air pool ①闭合冷区(厚度场上的闭合低值区)②低空气堆
cold air sink 冷气汇
cold anticyclone 冷性反气旋【大气】
cold belt 冷带

cold box 冷箱(供估算空气中冻结核数目用)
cold cap 寒带
cold climate with dry winter 冬干寒冷气候
cold climate with moist winter 冬湿寒冷气候
cold cloud 冷云【大气】
cold conveyor belt 冷空气输送带
cold-core anticyclone 冷心反气旋
cold-core cyclone 冷心气旋
cold-core high 冷心高压
cold-core low 冷心低压
cold-core rings 冷心〔海洋〕涡流
cold core synoptic system 冷心天气系统
cold current 寒流【航海科技】(水温低于沿途海温的海流)
cold cyclone (= cold low) 冷性气旋【大气】
cold damage 寒害
cold desert (= arctic desert) 冷沙漠,寒漠
Cold Dew 寒露(节气)【大气】【农】
cold dome 冷堆
cold downdraft 冷下沉气流
cold drop (= cold pool) 冷池
cold element 寒冷要素
cold event 冷事件
cold fog 冷雾
cold front 冷锋【大气】
cold front cloud system 冷锋云系【大气】
cold front rain 冷锋雨
cold front thunderstorm 冷锋雷暴
cold-front type of occlusion 冷锋型锢囚
cold-front wave 冷锋波动
cold high 冷高压【大气】
cold island 冷岛
cold lake 冷湖
cold low 冷低压【大气】
cold ocean current 冷洋流
cold occluded front 冷性锢囚锋【大气】
cold occlusion 冷锢囚

cold-occlusion depression 冷性锢囚低压
cold of the late spring 倒春寒
cold-outburst 寒潮爆发【大气】
cold period 冷期
cold plasma flow 冷等离子体流
cold plasma 冷等离子体
cold pole 寒极【大气】
cold pool 冷池【大气】
cold resistance 抗寒性
cold season 冷季
cold sector 冷区
cold soak 冷浸
cold source 冷源
cold spell in spring 春寒,春寒期
cold start 冷启动【计】
cold tongue 冷舌【大气】
cold top 冷云顶
cold trough 冷槽【大气】
cold type occlusion 冷〔性〕锢囚
cold vortex 冷涡【大气】
cold wall 冷水壁
cold wave 寒潮【大气】
cold wedge 冷楔
cold wind 冷风
colla 可拉风(菲律宾有大雨时的南或西南风)
collada 可拉达风(加利福尼亚湾上的强风,可达 $16\sim22$ m/s)
collapse 崩塌【土壤】,崩溃,衰弱
collapsed cloud 崩溃云
collapsing 崩溃
collar cloud 项圈云
collection ①冲并,碰并②收集
collection efficiency 捕获系数【大气】,冲并效率
collective effect 集约效应
collector ①捕获器,收集器②集电器③集电极【电子】
colligative property 溶液特性
collimated beam 准直波束,平行波束
collimated intensity 准直强度

collimator 准直仪,平行光管
collision 碰撞【大气】
collisional quenching 碰撞猝灭【化】
collision broadening 碰撞〔谱线〕增宽【物】,碰撞加宽
collision-coalescence process 碰〔撞〕并〔合〕过程
collision cross-section 碰撞截面【物】
collision dynamics 碰撞动力学
collision efficiency 碰撞系数【大气】,碰撞率
collision frequency 碰撞频率【物】
collision heating 磁撞加热【物】
collision process 碰撞过程【物】
collision relaxation 碰撞弛豫【物】
collision relaxation time 碰撞张弛时间
collision theory 碰撞理论
collision time 碰撞时间【物】
collissionless plasma shock 无碰撞等离子体激波
colloid 胶体【化】,胶体(的)
colloidal dispersion 胶体分散【大气】,胶态弥散
colloidal instability 胶体不稳定性【大气】
colloidally stable 胶体稳定的
colloidal metastable 胶体亚平衡
colloidal solution 胶体溶液
colloidal stability 胶体稳定性
colloidal suspension 胶体悬浮
colloidal system 胶体系统【大气】
colloid chemistry 胶体化学【化】
Colorado low 科罗拉多低压
Colorado sunken pan 科罗拉多蒸发皿(约 1 m 见方 0.5 m 深)
color composite 彩色合成
colored precipitation 带色降水
colored rain 有色雨
colorimeter 比色计【化】
colorimetric method 比色法
colorimetric oxidant analyzer 氧化物比

色分析仪
colorimetry 比色法【化】,色度学【物】
color information 彩色信息
color look-up table（CLUT） 色标查算表
color parameter 彩色参数
color reaction 颜色反应
color space 彩色空间
color spin scan cloud camera 彩色自旋扫描摄云机
color temperature 色温【物】
col pressure field 鞍形气压场【大气】
column （气）柱
column abundance 〔气〕柱丰度【大气】
columnar crystal 柱状晶体
columnar resistance 柱电阻
columnar vortex 柱状涡旋
column model 柱模式【大气】
column number density 柱数密度
colure 二分圈,二至圈
comanchaca 科芒恰卡雾（南美由低的层云变成的浓雾）
COMBAR code（或 combat aircraft code）战斗机航空电码
combat aircraft meteorological report（COMBAR） 战斗机气象报告
comber ①深水白头浪②长周期碎浪③长浪（从大洋到达海岸）
combination band 组合带
combination coefficient 复合系数
combination line 组合〔谱〕线【物】
combination of lenses 透镜组【物】
combination principle 组合原理
comb nephoscope 梳状测云器
comb-type radiosonde 梳齿式探空仪
combustion dust 燃烧尘
combustion nucleus（combustion nuclei）燃烧核【大气】
comet 彗星
comfort chart 舒适度图
comfort current 舒适气流【大气】
comfort curve 舒适度曲线

comfort index 舒适指数【大气】
comfort standard（＝comfort zone）舒适标准
comfort temperature 舒适温度【大气】
comfort zone 舒适区
comma cloud 逗点〔状〕云
comma cloud system 逗点云系【大气】
comma head 逗点云头
command 命令【计】,指令
command and data acquisition station（CDA） 指令和资料接收站
command clock subsystem 指令钟子系统
command language 公共语言【计】
comma tail 逗点云尾
comminution 粉碎作用
Commission for Aeronautical Meteorology（CAeM,WMO） 航空气象学委员会（WMO）
Commission for Agricultural Meteorology（CAgM,WMO） 农业气象学委员会（WMO）
Commission for Atmospheric Sciences（CAS,WMO） 大气科学委员会（WMO）
Commission for Basic System（CBS,WMO） 基本系统委员会（WMO）
Commission for Bibliography and Publications（CBP,WMO） 目录和出版委员会（WMO）
Commission for Climatology（CCl,WMO）气候学委员会（WMO）
Commission for Climatology and Applications of Meteorology（CCAM,WMO）气候学和气象学应用委员会（WMO）
Commission for Hydrological Meteorology（CHM,WMO） 水文气象学委员会（WMO）
Commission for Hydrology（CHY,WMO）水文学委员会（WMO）
Commission for Instruments and Methods of Observation（CIMO,WMO） 仪器

和观测方法委员会(WMO)
Commission for Marine Meteorology (CMM, WMO)　海洋气象学委员会(WMO)
Commission for Polar Meteorology (CPM, WMO)　极地气象学委员会(WMO)
Commission for Special Applications of Meteorology and Climatology (CoSAMC, WMO)　气象学和气候学特殊应用委员会(WMO)
Commission for Synoptic Meteorology (CSM, WMO)　天气学委员会(WMO)
Commission on Atmospheric Chemistry and Global Pollution (CACGP, IAMAS)　大气化学和全球污染委员会(IAMAS)
Commission on Sustainable Development (CSD)　可持续发展委员会
Committee for Environmental Conservation(CoEnCo)　环境保护委员会
Committee for International Program in Atmospheric Sciences and Hydrology (CIPASH, UN)　国际大气科学及水文学计划委员会(UN)
Committee on Atmosphere and Oceans (CAO)　海洋与大气委员会(美国)
Committee on Atmospheric Measurements (CAM)　大气测量委员会(美国)
Committee on Atmospheric Sciences (CAS, NAS)　大气科学委员会(NAS)
Committee on Data for Science and Technology(CODATA, ICSU)　科学技术资料委员会(ICSU)
Committee on Earth Observation Satellites (CEOS)　卫星对地观测委员会
Committee on Fisheries(COFI, FAO)　渔业委员会(FAO)
Committee on Laser Atmospheric Studies (CLAS)　激光大气研究委员会(美国)
Committee on Scholarly Communication with the People's Republic of China(CSCPRC)　美中学术联络委员会(美国)
Committee on Solar-Terrestrial Research (CSTR)　日地研究委员会(美国)
Committee on Space Research (COSPAR, ICSU)　空间研究委员会(ICSU)
Common Business-Oriented Language COBOL　语言【计】
common earthing system　共用接地系统
common logarithm　常用对数【数】
common-path method　共路径法
common stratus　普通层云
Commonwealth Meteorology Research Centre(CMRC)　联邦气象研究中心(澳大利亚)
Commonwealth of Australia Bureau of Meteorology(CABM)　澳大利亚联邦气象局
Commonwealth Scientific and Industrial Research Organization (CSIRO)　澳大利亚联邦科学和工业研究组织
communication　通信【计】
communication centre　通信中心
communication channel　通信通道
communication interface　通信接口【计】
communication satellite(ComSat)　通信卫星
communication network　通信网络
communication software　通信软件【计】
communication station　通信站
community airborne waste　社区空气传播的废物
community atmosphere　城市大气
Community Multi-scale Air Quality (CMAQ)　(城市)多尺度空气质量模式
Community Radiation Transfer Model (CRTM)　共有辐射传输模式
commutator radiosonde　转换器式探空仪
compact difference scheme　紧差分格式【数】
compact differencing　紧差分
compact disc(CD)　光碟【计】,光盘

compact disc digital audio（CD-DA） 唱碟【计】
compact disc-read only memory（CD-ROM） 只读碟【计】
compact disc-recordable（CD-R） 可录光碟【计】
compaction 致密,压缩,压紧
compactness 密实度,致密度
compact radio source 致密射电源
comparative meteorology 比较气象学
comparative rabal 探空气球比较观测（目的是检测电子跟踪系统）
comparative simulation 比较模拟
comparison spectrum 比较光谱【物】
compartment 间隔,区划
compass 罗盘,指南针
compass theodolite 罗盘经纬仪【测】
compatibility 相容性【物】,兼容性
compatible computer 兼容计算机【计】
compensated air thermometer 补偿气温表
compensated pyrradiometer 补偿式全辐射表
compensated scale barometer 补偿式定标气压表
compensating pyrheliometer 补偿式绝对日射表
compensation 补偿作用
compensation current ①补偿气流 ②补偿〔海〕流
compensation of instrument 仪器补偿
compensation pressure 补偿气压
competence of the wind 风的挟带力
compete symmetry 全对称【物】
compiler 编译程序【计】,编译器
complacent series 满足序列
complement 补码【计】
complementary function 余函数
Complete Atmospheric Energetics Experiment（CAENEX, USSR） 综合大气能量学试验（前苏联）
complete conservation scheme 完全守恒格式【数】
complete convergence 全收敛【数】
complete elliptic integral of the first kind 第一类完全椭圆积分【数】
complete freeze-up 完全冻结（河流或湖泊中的水从上至下全部冻结）
complex 络合物【化】
complex climatology 综合气候学
complex conjugate 复共轭的【数】
complex demodulation 复解调
complex dielectric constant 复介电常数
complex empirical orthogonal function（CEOF） 复经验正交函数
complex envelope 复包络
complex group 复杂〔黑子〕群
complex hydrograph 复合〔水文〕过程线
complex index of refraction 复折射率【物】
complex ions 复离子
complex low 多中心低压
complex molecules 复分子
complex multiplication 复数乘法
complex number 复数【数】
complex quality control（CQC） 复质量控制
complex refractive index 复折射率
complex signal 复信号
complex terrain 复杂地形
complex variable 复变数【数】
component 部分,分量
component of force 分力
component velocity 分速度
composite flash 多次闪击
composite forecast chart 综合〔天气〕预报图
composite hydrograph 复合〔水文〕过程线
composite image 合成图像
composite map 综合图
composite medium 复合介质
composite observational system 综合观测系统

composite prognostic chart 综合〔形势〕预报图
composite reflectivity 综合反射率〔因子〕
composite rough surface 复合粗糙面
composite structure 合成结构
composite vertical cross-section 综合垂直剖面图
composite water sample 综合水样〔品〕
composite wave 合成波
composition 组成,成分
composition analysis of cosmic dust 宇宙尘成分分析
composition of atmosphere 大气组成
composition of forces 力的合成【物】
compound 化合物
compound centrifugal force (= Coriolis force) 复合离心力(即科里奥利力)
compound fertilizer 复合肥料【土壤】
compound function 复合函数【数】
compound hydrograph (= composite hydrograph, complex hydrograph) 复合〔水文〕过程线
compound lens 复合透镜【物】
compound matrix 复合矩阵【数】
compound pendulum 复摆【物】
compound probability 复〔合〕概率
Comprehensive Ocean-Atmosphere Data Set (COADS) 海洋-大气综合数据集
comprehensive treatment 综合治理
compressed digital transmission 压缩数字传输
compressed magnetosphere 受压磁层
compressed pulse radar altimeter 压缩脉冲雷达高度计
compressibility ①可压缩性【物】②压缩率【物】
compressibility wave 压缩波
compressible 可压缩的
compressible fluid 可压缩性流体
compressible model equation 可压缩模式方程

compression ①压缩②压力
compressional wave (= longitudinal wave) 压缩波
compression evaluation 压缩评价
compression heating 压缩增温
compression ratio 压缩比【电子】
compression wave 压缩波
compressive 可压缩的
Compton effect 康普顿效应【物】【大气】
Compton scattering 康普顿散射【物】【大气】
computabillity 可计算性【数】
computational dispersion 计算频散
computational domain 计算区域
computational instability 计算不稳定【大气】
computational mode 计算模〔态〕【大气】
computational stability 计算稳定性
computational stability criterion 计算稳定性判据
computer 计算机【计】
computer-aided design (CAD) 计算机辅助设计【计】
computer application 计算机应用【计】
computer control 计算机控制【计】
computer draft 计算机制图【计】
computer engineering 计算机工程【计】
computer-formatted terminal forecast (CFFT) 计算机制作的终点预报
computer hardware 计算机硬件【计】
computer instruction 计算机指令
computer language 计算机语言【计】
computer-output microfilm (COM) 计算机输出缩微胶片
computer-output microfilmer 计算机输出缩微摄影机
computer product 计算机产品
computer program 计算机程序【计】
computer program validation 计算机程序确认【计】

computer program verification　计算机程序验证【计】
computer reliability　计算机可靠性【计】
computer science　计算机科学【计】
computer security　计算机安全【计】
computer software　计算机软件【计】
computer storage　计算机存储器
computer system　计算机系统【计】
computer technology　计算机技术【计】
computer virus　计算机病毒【计】
computer virus counter-measure　计算机病毒对抗【计】
computer worded forecast(CWF)　计算机文字预报
computing method　计算方法【数】
computing technology　计算技术【计】
concave　①天空,苍穹②凹的
concave lens　凹透镜【物】
concave mirror　凹面镜【物】
concentration　①含量②浓度③富集
concentration basin　浓缩盆地,浓缩流域(蒸发大,淡水减少)
concentration gradient　浓度梯度
concentration variance　浓度方差
concentrator solar cell　聚光太阳电池【电子】
concentric beam　同心光束【物】
concentric circles　同心圆【数】
concentric eyewall cycle　同心眼墙循环,同心眼壁变化周期(大约一天)
concentric eyewalls　同心眼壁【大气】,同心眼墙
concentric lens　同心透镜【物】
conceptual model　概念模型【计】
conceptual picture　物理图像,概念图像
concrete minimum temperature　混凝土板最低温度
condensability　凝结性
condensate　冷凝物
condensation　凝结【大气】
condensation adiabat　湿绝热

condensation efficiency　凝结效率【大气】
condensation function　凝结函数
condensation level　凝结高度【大气】
condensation limit　凝结限度
condensation nuclei counter　凝结核计数器
condensation nucleus　凝结核【大气】
condensation parameterization　凝结参数化
condensation pressure　凝结压
condensation process　凝结过程【大气】
condensation stage　凝结阶段
condensation temperature　凝结温度
condensation trail(＝contrail)　凝结尾迹
condenser discharge anemometer　电容放电式风速表
conditional climatology　条件气候学,室内人造气候学
conditional conservatism　条件保守性
conditional distribution　条件分布【数】
conditional equilibrium　条件〔性〕平衡
conditional instability　条件〔性〕不稳定【大气】
conditional instability of the second kind (CISK)　第二类条件〔性〕不稳定【大气】
conditional linear regression　条件线性回归
conditionally unstable atmosphere　条件〔性〕不稳定大气
conditional mean　条件均值【数】
conditional probability　条件概率【数】
conditional sampling　条件取样
conditional stability　条件〔性〕稳定〔度〕
conditional statement　条件语句
conditional symmetric instability　条件对称不稳定〔性〕
conditional unstable　条件〔性〕不稳定的
conditions of readiness　(危险天气)警戒状态
conductance　电导【物】,传导

conduction 传导【大气】
conduction current 传导电流
conductive equilibrium 传导平衡
conductivity 电导率【物】,传导性,传导率
conductivity current 传导电流
conductivity-temperature-depth profiler (CTD) 电导率-温度-水深廓线仪
conductor 导体
cone ①圆锥体②风袋③风暴信号
cone angle 圆锥角
cone of depression (＝cone of influence) ①沉降锥②下降漏斗,浸润漏斗
cone of escape 逃逸圆锥(外대层中)
cone of visibility 能见圆锥
cone of vision 视界圆锥,视场圆锥
cone radiometer 圆锥辐射表
Conference of the Parties (COP)(UN Framework Convention on Climate Change) 缔约方大会(联合国气候变化框架公约)
confidence band (＝confidence interval) 置信区间
confidence coefficient 置信系数【数】
confidence degree 置信度【大气】(如95％)
confidence interval 置信区间【数】【大气】
confidence level 置信水平【数】【大气】(置信度的互补概率,95％置信度的置信水平为0.05)
confidence limit 置信限【数】
confidence region 置信区域【数】
confidence upper limit 置信上限【数】
configuration ①位形【物】②组态【物】③构形,图形,格局,布局
confined aquifer(＝artesian aquifer) 承压含水层,有压含水层
confined groundwater 承压地下水,承压潜水
confining bed (layer, stratum) 不透水层,隔水层,阻水层
confluence 合流,汇合,汇流【大气】(difluence 分流的反义)
confluent hypergeometric equation 汇合型超几何方程【数】
confluent hypergeometric function 合流超几何函数,合流超比函数
confluent thermal ridge 汇流型温度脊
confluent thermal trough 汇流型温度槽
conformal conic projection 正形圆锥投影
conformal map (＝isogonal map, orthomorphic map) 等角地图,正形地图
conformal projection 等角投影【测】,正形投影
confused sea 暴涛(道氏风浪九级)
congelation 冻结
congelifraction 冰冻风化
congeliturbation 融冻泥流作用,翻浆作用
congelont 等冰冻线
congestus (con) 浓(云)
conical beam 锥形射束
conical hail 锥形冰雹【大气】
conical-head thermometer 锥头温度表
conical marine raingauge 锥表船用雨量器
conical point 锥型针,象牙针
conical projection 圆锥投影
conical scanning 圆锥〔形〕扫描
conic section 圆锥曲线,圆锥截面
conifer 针叶树
coniferous forest 针叶林【林】
conimeter (＝konimeter) 测尘器
coning 圆锥形扩散
coniology 微尘学
coniscope 计尘仪
conjugate complex number 共轭复数【数】
conjugate gradient 共轭梯度
conjugate image 共轭图像
conjugate matrix 共轭矩阵【数】
conjugate point 共轭点
conjugate-power law 共轭指数律
conjugate region heating 共轭区域加热

conjunction 合【天】
conjunctive use （地表水与地下水）联合利用
connate water 原生水
connected field line 连通〔磁〕力线
connecting band 连接带
connection in parallel 并联【物】
connection in series 串联【物】
consecutive mean 动态平均【大气】
consensus average 集成平均【数】
consensus averaging 集成平均〔法〕
consensus forecast 集成预报
consequent 后项【数】,结果,后果
conservation law 守恒律【物】
conservation equation 守恒方程【数】
conservation of absolute angular momentum 绝对角动量守恒
conservation of absolute momentum 绝对动量守恒
conservation of absolute vorticity 绝对涡度守恒【大气】
conservation of angular momentum 角动量守恒【大气】
conservation of area 面积守恒
conservation of energy 能量守恒【大气】
conservation of mass 质量守恒【大气】
conservation of momentum 动量守恒
conservation of natural resources 自然资源保护【地理】
conservation of potential vorticity 位涡守恒【大气】
conservation of vorticity 涡度守恒
conservation scheme 守恒格式【大气】
conservatism 保守性
conservative difference scheme 守恒差分格式
conservative element 保守元素
conservative operator 保守算子【数】
conservative pollutants 持恒污染物
conservative property 保守性【大气】
conservative property of air mass 气团保守性【大气】
conservative scattering 守恒散射
conservatory 展览温室【林】
conserved parameter diagram (= conserved variable diagram) 守恒参量图
conserved variable diagram 守恒变量图
consistency 相容性【数】,协调性,和谐性,一致性
consistency check 一致性检验【计】,协调性检查
consistency constraint 一致性约束【计】
consistent estimate 相合估计【数】
consistent numerical scheme 相容数值格式,相容数值方案
console 操作台,控制台
console terminal 控制台终端
consolidated ice 密集厚冰,冻连冰群
consolidation ①固结〔作用〕②渗压
constancy 持续性,稳定性
constancy of angular momentum 角动量常定
constancy of climate 气候稳定性
constancy of winds 风的稳定性
constant ①常数②稳定〔的〕,常定〔的〕
constant absolute vorticity trajectory 等绝对涡度轨迹
constant altitude plan position indicator (CAPPI) 等高平面位置显示器【大气】
constant electron density surface 等电子密度面
constant f approximation f 常数近似
constant flux layer 等通量层,常值通量层
constant f model f 常数模式
constant-height balloon 定高气球
constant-height chart 等高面图
constant-height surface (= constant-level surface, isohypsic surface) 等高面
constant humidity 等湿
constant humidity line 等湿线
constant level balloon 定高气球【大气】

constant level chart 等高面图
constant-level surface 等高面
constant parameter linear system 常参数线性系统
constant pressure 定压
constant-pressure balloon 定压气球
constant pressure chart 等压面图
constant-pressure-pattern flight 等压面飞行
constant pressure surface 等压面
constant-rate dilution gauging 恒速溶液法测流〔量〕
constant temperature line 等温线
constant temperature surface 等温面
constant thickness line 等厚度线
constant voltage power supply 稳压电源【计】
constant volume balloon 等容气球
constant volume gas thermometer 等容气温表
constant volume sampling(CVS) 等容积取样
constant wind 稳定风
Constellation Observing System for Meteorology, Ionosphere and Climate (COSMIC) 气象、电离层和气候卫星探测系统
constituent 组元,组分
constituent of tides 分潮
constraint condition 约束条件【数】
constriction of the horizon 地形遮蔽图
constructive interference 相长干涉【物】
consultation ①咨询②会商,会诊
consumption ①消耗②(湍流)耗散
consumptive use 耗水量,需水量
contact 接触
contact anemometer 电接〔式〕风速表【大气】
contact angle 接触角,啮合角
contact cooling 接触冷却
contact fight 目视飞行

contact nucleus (contact nuclei) 接触核
contact potential difference 接触电势差【物】
contact weather 目视飞行天气
contaminant 污染物
contaminate 污染(动词)
contamination 污染【农】
contemporary correlation 同时相关
content 含量
contessa del vento 康德萨特凡托云(西西里岛依特拉山上的荚状云)
context analysis 语境分析【计】,上下文分析
continental aerosol 大陆性气溶胶
continental air mass 大陆气团【大气】
continental anticyclone 大陆性反气旋
continental borderland 大陆边缘地
continental climate 大陆性气候【大气】
continental cloud 大陆云
continental drift 大陆漂移【地质】
continental drift theory 大陆漂移说【地质】
continental environment 大陆环境
continental facies 陆相【地质】
continental glacier 大陆性冰川【地理】,大陆〔性〕冰川【地质】
continental hemisphere 陆半球
continental high 大陆高压
continental ice (= continental glacier) 大陆冰,陆冰
continentality 大陆度【大气】
continentality index (= coefficient of continentality, index of continentality) 大陆度指数【大气】
continental platform 大陆台地,大陆地台
continental polar air (mass) 极地大陆气团,极地大陆空气
continental shelf 大陆架【地理】,陆棚
continental shelf wave 大陆架波
continental slope 大陆坡【地理】
continental sphere 陆圈,陆界

continental tropical air（mass） 热带大陆气团,热带大陆空气
Continental Water Boundary 大陆水界
continental wind 大陆风
contingency table 列联表【数】【大气】,相关概率表
continuation 连续,延拓
continuing current 连续电流(回击之后)
continuity 连续性
continuity chart 连续性图
continuity equation 连续〔性〕方程【物】,连续方程【大气】
continuous absorption 连续吸收
continuous aerosol detector 气溶胶连续检测仪
continuous air monitoring program station 空气连续监测规划站
continuous curve 连续曲线
continuous distribution 连续分布【数】
continuous function 连续函数【数】
continuous heavy rain 连续性大雨
continuous leader 连续先导(闪电)
continuous medium 连续介质【物】
continuous moderate rain 连续性中雨
continuous monitoring 连续监测
continuous precipitation 连续性降水【大气】
continuous rain 绵雨,连续雨
continuous record 连续记录
continuous spectrum 连续〔频〕谱
continuous thunder and lightning 连续雷电
continuous wave 连续波【物】
continuous wave ionosonde 〔电离层〕连续波测高仪
continuous-wave radar(CW radar) 连续波雷达【大气】
continuous wave transmitter 连续波发射器【电子】
continuum 连续区【物】,连续谱【化】
continuum absorption 连续区吸收

contour chart 等压面图【大气】,等高线图
contour height 等高线高度,等压面高度
contour interval 等高线间隔,等值线间隔
contour line 等高线【大气】
contour microclimate 地形小气候【大气】
contraction 缩并【物】,缩并〔运算〕【数】,收缩
contraction axis 压缩轴
contraction coefficient 收缩系数
contraction of operators 算符缩并【物】
contrail（＝condensation trail）凝结尾迹
contrail-formation graph 〔凝结〕尾迹生成图
contra solem 反钟向(即气旋方向)
contrast ①差异②对比
contrast area 对比区
contraste 互逆风(航海用语,指直布罗陀海峡内相距不远的地方吹相反的风)
contrast enhancement 对比度增强
contrast of luminance 亮度对比【大气】
contrast point 对照点
contrast stretching 对比扩展(图像处理)
contrast telephotometer 对比遥测光度表
contrast threshold 视觉感阈【大气】,对比阈值
contributing region (回波)贡献区
contribution 贡献
control 控制
control analysis 对照分析
control area 对照区,对比区
control center 控制中心
control cloud 对照云
control day（＝key day）征兆日,特征日
control experiment 对照试验
control forecast 对照预报
controlled access system 受控访问系统【计】
controlled area 控制区,对照区

controlled system 受控系统【计】
control program(CP) 控制程序
control run 对照算程
control section 控制断面,控制河段,控制渠段
control system 控制系统
control-tower visibility 塔台能见度,控制塔能见度
control unit module 控制组件
control variable 控制变量
control well 控制井
convection 对流【大气】
convection adjustment 对流调整
convectional circulation 对流〔性〕环流
convectional rain 对流〔性〕雨
convectional theory 对流学说
convection cell 对流单体【大气】,对流胞
convection current ①对流气流②对流电流
convection depth 对流厚度
convection line 对流线
convection model 对流模式【大气】
convection over slope 过坡对流
convection process 对流过程
convection scheme 对流方案
convection theory of cyclones 气旋对流理论
convective acceleration 对流加速度
convective activity 对流活动
convective adjustment 对流调整【大气】
convective available potential energy(CAPE) 对流有效位能,对流可用位能
convective boundary layer(CBL) 对流边界层【大气】
convective cell 对流胞,对流单体
convective circulation 对流环流
convective cloud 对流云【大气】
convective cloud street 对流云街
convective cluster 对流云团
convective component 对流部分
convective condensation level 对流凝结高度【大气】
convective cooling 对流冷却
convective echo 对流回波【大气】
convective equilibrium 对流平衡
convective impulse 对流冲量
convective index 对流指数
convective inhibition(CIN) 对流抑制能,对流约束能
convective instability 对流不稳定【大气】
convective kinetic energy 对流动能
convectively unstable line 对流不稳定线
convective mass flux 对流质量通量
convective mixed layer 对流混合层
convective overturn 对流混合,对流翻腾(湖沼)
convective parameterization 对流参数化【大气】
convective plume 对流凝结羽
convective potential energy 对流势能
convective precipitation 对流性降水【大气】
convective rain 对流雨
convective region 对流区
convective Richardson number 对流里查森数
convective scale 对流尺度
convective shower 对流〔性〕阵雨
convective stability 对流〔性〕稳定度【大气】
convective storm 对流风暴
convective storm initiation mechanism 对流风暴发生机制
convective temperature 对流温度
convective theory 对流学说
convective theory of cyclogenesis 气旋生成的对流〔学〕说
convective thunderstorm 对流性雷暴【大气】
convective transport 对流输送
convective transport theory 对流输送〔学〕说

convective turbulence 对流湍流
convective velocity scale 对流速度尺度
conventional observation 常规观测【大气】
conventional radar 常规雷达
convergence 收敛〔性〕【数】,辐合【大气】
convergence band 辐合带
convergence field 辐合场
convergence line 辐合线【大气】
convergence trough 辐合槽
convergence zone 辐合带
convergent lens 会聚透镜【物】
convergent numerical scheme 收敛数值方案,收敛数值格式
convergent series 收敛级数【数】
conversational monitor system (CMS) 会话监控系统
converse proposition 逆命题【数】
conversion 转换
conversion factor 换算因子
converter 转换器
convex lens 凸透镜【物】
conveyance ①输送②水道输送能力③输水因子
conveyer(或 conveyor)belt 输送带
convolution ①卷积【数】②卷曲,回旋③脑回(解剖)
convoy 护航【航海】
cooking snow 湿雪
Cook reverse-flow thermometer 库克返流温度表
coolant 冷却剂
cool damage 冷害【大气】
cooling 冷却
cooling degree-day 冷却度日【大气】
cooling power 冷却率
cooling-power anemometer 冷却率风速表
cooling process 冷却过程
cooling rod 冷却杆(露点仪用)
cooling temperature 冷却温度
cool injury 冷害
coolometer 冷却率测定表

cool pool 冷池,冷空气潭
cool season 凉季【大气】
cool wave 凉波
Cooperative Institute for Mesoscale Meteorological Studies (CIMMS) 中尺度气象研究合作研究所(美国)
Cooperative Institute for Meteorological Satellite Studies (CIMSS) 气象卫星研究合作研究所(美国)
Cooperative Institute for Research of the Atmosphere (CIRA) 大气研究合作研究所(美国)
cooperative observer 合作观测者(美国的义务观测者称呼)
cooperative phenomenon 合作现象【物】
cooperative reflector 合作反射器
coordinate 坐标
σ-coordinate σ坐标【大气】
θ-coordinate 位温坐标【大气】
coordinate axis 坐标轴【数】
coordinate curves 坐标曲线【数】
Coordinated Observation and Prediction of the Earth System (COPES) 地球系统的协调观测和预报(WCRP 的 2005—2015 年战略框架)
coordinated universal time (UTC) 协调世界时【天】
coordinate line 坐标线
coordinate plane 坐标平面
coordinate surface 坐标曲面
coordinate system 坐标系【数】
coordinate transformation 坐标变换【数】
coordination bond 配位键【化】
coordination chemistry 配位化学【化】
COP 〔Conference of the Parties (UN Framework Convention on Climate Change)〕 缔约方大会(联合国气候变化框架公约)
Copenhagen Water 国际(哥本哈根)标准海水
COPES (Coordinated Observation and

Prediction of the Earth System) 地球系统的协调观测和预报
coplane scanning 共面扫描
copolarized signal 等偏振信号,等极化信号
coppice forest 矮林【林】
copy ①复制【计】,拷贝②副本【计】
corange line 等潮差线【航海】,同潮幅线
cordonazo 科多纳索风暴(墨西哥)
cordonazo de San Francisco 科多纳索秋分(西班牙水手对秋分的称法,因为他们认为那时风暴多)
core analysis 岩芯分析,岩样分析
coreless winter 平缓冬季(指南极冬季日平均气温变化不大)
core memory 磁芯存储
core metadata 核心元数据
core sample (岩、土、雪、冰)芯样品
Coriolis acceleration 科里奥利加速度【大气】,科氏加速度
Coriolis effect 科里奥利效应
Coriolis force 科里奥利力【大气】(又称地转偏向力),科氏力
Coriolis parameter 科里奥利参数【大气】,科氏参数
Coriolis term 科里奥利项
corner effect 隅角效应
corner-frequency method 隅角频率法
corner reflector 角反射器(测风雷达用)【大气】
corn heat unit 玉米热单位(生长度日的修改型)
cornice 雪檐(由于风的作用在山顶或绝壁下风方凝结的雪块)
corn snow 春天粒雪
coromell 康乐梅尔风(加利福尼亚湾拉巴士地方11—5月夜间盛行的陆风)
corona ①华【大气】(日华,月华) ②日冕
corona current 电晕电流
corona discharge 电晕放电【物】

coronagraph 日冕仪【天】
coronal hole 冕洞
coronal mass ejection (CME) 日冕物质抛射【天】,日冕物质喷射【大气】
coronal transient 日冕瞬变【天】
corotating magnetic field line 共旋磁力线
corotation field strength 共旋电场强度
corposant 电晕放电〔现象〕
corpuscle 微粒
corpuscular cosmic ray 微粒宇宙线
corpuscular heat source 微粒热源
corpuscular radiation 微粒辐射
corpuscular stream 微粒流
corpuscular theory 微粒说【物】
corpuscular theory of light 光的微粒说
corrasion 风蚀【大气】
corrected airspeed 订正空速
corrected altitude 订正高度
correcting lens 校正透镜【物】
correction 订正〔值〕,修正【物】,校正【物】
correction factor 订正系数
correlated-k 相关 k
correlation 相关【大气】,关联【物】
correlation analysis 相关分析【数】
correlation coefficient 相关系数【数】,关联系数【物】
COrrelation Detection And Ranging (CODAR) 相关探测和测距
correlation diagram 相关图
correlation distance 相关距离
correlation factor 相关因子,相关因素
correlation forecasting 相关预报【大气】
correlation function 相关函数【数】,关联函数【物】
correlation matrix 相关〔矩〕阵
correlation method 相关法
correlation peak 相关峰
correlation processor 相关处理机
correlation ratio 相关比
correlation slope method 相关斜率法

correlation spectrum 相关谱,关联谱【物】
correlation spectroscopy 相关光谱学
correlation synoptics 相关天气学
correlation table 相关表
correlation-transform method 相关变换法
correlation triangle 相关三角形
correlation variable 相关变数
correlative meteorology 相关气象学(罕用词,将统计相关用于多种气象观测)
correlogram 相关图
correspondence analysis 对应分析【数】
corresponding angles 同位角【数】
corresponding appraisal 函审鉴定
corresponding point 对应点
corrosion 溶蚀【土壤】,腐蚀,侵蚀
cosine law of illumination 照度余弦定律
cosine-tapered rectangular window 余弦矩形窗
cosine window 余弦窗
COSMIC (Constellation Observing System for Meteorology, Ionosphere and Climate) 气象、电离层和气候卫星探测系统
cosmic abundance 宇宙丰度【物】
cosmical constant 宇宙常数
cosmical meteorology 宇宙气象学
cosmic background radiation 宇宙背景辐射【地】
cosmic dust 宇宙尘【大气】
cosmic noise 宇宙噪声【物】
cosmic radiation 宇宙辐射
cosmic radio noise 宇宙射电噪声【地】
cosmic ray 宇宙线【大气】
cosmic-ray ionization 宇宙线电离
cosmic-ray shower 宇宙线簇射
cosmic ray storm 宇宙线暴【地】
cosmic velocity 宇宙速度
cosmism 宇宙〔演化〕论【物】
cosmochronology 宇宙年代学【地质】
cosmogenic radioisotopes 宇宙线产生的放射性同位素
cosmogenic radionuclides (CRNs) 宇宙成因的放射性核素
cosmogony 天体演化学【天】
cosmology 宇宙学【天】
cosmos 宇宙
cospectrum 协谱【大气】,共谱
cost-benefit analysis 成本效益分析【计】
cost-benefit approach 费用-收益法
cost-benefit ratio 费用-收益比
cost-effectiveness 效益
cost function 费用函数【数】,代价函数【计】
cost-loss ratio 费用-损失比
cotidal hour 等潮时,同潮时
cotidal line 等潮〔时〕线,同潮线
cotton ball cloud 棉球云
cotton-belt climate 棉〔花〕带气候(冬干夏雨的温暖气候)
cotton-region shelter 棉区百叶箱(广泛用于二级气象站)
Couette flow 库埃特流
coulomb(C) 库〔仑〕(电荷量单位)
Coulomb force 库仑力【物】
Coulomb law 库仑定律【物】
coulometric oxidant analyzer 电量法氧化剂分析仪
coulometry 电量测量法,库仑法
count area 计数区
counter ①计数器②反,逆
counter balance 抵偿,抗衡
counter-circulation 反环流
counter clock-wise 反时针
countercurrent ①逆流,回流②反向电流
counterglow 对日照【天】(夜空中与太阳相反方向处的一个弥漫椭圆形暗弱的亮区)
counter-gradient 反梯度
counter-gradient flux 反梯度通量

counter-gradient heat flux 反梯度热通量
countergradient wind 反梯度风【大气】
counter parhelia 反幻日,反假日
counter radiation 逆辐射
counter sun 反日
counter trade 反信风
countertwilight 反曙暮光
counting anemometer 计数风速表
counting efficiency 计数效率
counts 计数
couped atmospheric-hydrological system 耦合大气-水文系统
coupled difference equation 耦合差分方程
coupled magnetosphere-ionosphere system 磁层-电离层耦合体系
coupled model 耦合模式
coupled ocean-atmosphere general circulation model 海洋-大气环流耦合模式
Coupled Ocean Atmosphere Response Experiment(COARE) 海气耦合响应试验
couped ordinary differential equation 耦合常微分方程
coupled system 耦合系统
coupling 耦合,配合
coupling coefficient 耦合系数
coupling equation 耦合方程
coupling model 耦合模式
Courant condition 柯朗条件
Courant-Friederichs-Lewy（CFL） condition 柯朗-弗里德里希斯-列维条件【数】,CFL 条件
course 航向,航线,行程
covariability 协变性
covariable（＝covariate） 余变量
covariance 协方差【数】【大气】
covariance function 协方差函数
covariance matrix 协方差〔矩〕阵
covariant 协变量,协变
covariant tensor 协变张量【物】

covariate 余变量
covariation 协变
cover 覆盖
coverage ①范围 ②台站分布范围③覆盖区
coverage diagram 覆盖度图,有效区域图,作用范围图
covering fraction 覆盖系数(云滴取样)
cover of cloud 云量
cow quaker 英国 5 月风暴
cowshee(见 kaus) 考斯风
cow storm 牛暴风(加拿大埃尔兹米尔岛海上的大风)
cps 字符每秒【计】
CPU（central processing unit） 中央处理器【计】,中央处理机【计】
CQC（complex quality control） 复质量控制
crab angle 偏航角,偏流订正角
crabbing 偏航
Crachin 濛雨天气,克拉香天气【大气】(中国南部沿海和越南东部沿海地区毛毛雨和小雨天气)
crack ①裂隙,裂缝②冰间水路
crankcase ventilation 曲轴箱换气
creep ①蠕变【地】②蠕动,滑动
creeping flow 蠕动流,蠕流
creeping of aneroid barometer 空盒气压表的蠕动〔滞后〕
creepmeter 蠕变仪【地】
crepuscular arch 曙暮光弧
crepuscular ray 曙暮辉
crest 顶峰
crest cloud 盔云
crest gauge 最高水位水尺,峰顶水尺
crest of wave 波峰
crest profile 洪峰纵剖面
crest stage 洪峰阶段
Cretaceous Period 白垩纪【地质】
Cretaceous System 白垩系【地质】
crevasse（法） ①裂隙②冰隙③决口

crevasse hoar 冰隙白霜
CREX (character representation form for data exchange) 用于数据交换的字符表示格式
criador 克利亚德风(西班牙北部的一种西风)
CRISTA (Cryogenic Infrared Spectrometers and Telescopes for the Atmosphere) 大气低温红外光谱仪和望远镜
criteria pollutants 临界污染物(应加以控制的),关键污染物
criterion ①判据【物】②标准
critical angle 临界角【物】
critical area 警戒区
critical bursting point 临界破碎点
critical damping 临界阻尼【物】
critical day-length 临界光长【大气】
critical depth 临界水深,临界深度
critical depth control 临界水深控制点
critical discharge 临界流量
critical drop radius 临界水滴半径
critical flow 临界流〔动〕
critical frequency 临界频率
critical gradient 临界梯度
critical height 临界高度
critical lapse-rate 临界直减率
critical latitude 临界纬度
critical layer 临界层
critical level 〔逸的〕临界高度
critical level interaction 临界高度交互作用
critical level of escape 临界〔预报〕逃逸高度
critical liquid water content 临界液水含量
critical mountain height 临界山脉高度
critical period of growth 临界生长期
critical period of 〔crop〕 water requirement 〔作物〕需水临界期【大气】
critical point 临界点【物】
critical pressure 临界压力,临界压强【物】
critical receptor 临界感受器
critical region 临界区
critical Reynolds number 临界雷诺数
critical Richardson number 临界里查森数【大气】
critical saturation ratio 临界饱和比
critical speed 临界速率
critical state 临界状态,临界态【物】
critical success index (CSI) 临界成功指数
critical temperature 临界温度【物】
critical temperature of cold damage 寒害临界温度
critical tidal level 临界潮〔水〕位
critical value 临界值
critical velocity 临界速度【物】
critical wavelength 临界波长
critical wind speed 临界风速
crivetz(=crivat, krivu) 克里维茨风(罗马尼亚的一种寒冷东北风)
CRNs (cosmogenic radionuclides) 宇宙成因的放射性核素
crochet d'orage(=crochet de grain) 飑鼻,雷雨鼻
Cromwell Current 克伦威尔潜流(即太平洋赤道潜流)
crop calendar ①作物历②作物日程表
crop climate 作物气候
crop climatic adaptation 作物气候适应性
crop climatic ecotype 作物气候生态型
crop coefficient 作物系数
crop disease environment monitor (CDEM) 作物病害环境监测器
crop forecast 作物预测
crop microclimate 农田小气候
crop moisture index 作物水分指数
crop water requirement 作物需水量【大气】
crop-weather 作物天气
crop-weather model 作物-天气模式
cross ①十字晕 ②日柱

cross check 交叉检验
cross contamination 交叉污染
cross-correlation 互相关,交叉相关
【大气】
cross correlation function 互相关函数
【数】
cross-correlation spectrum 互相关谱
cross-correlogram 互相关图
cross-covariance 互协方差
cross dating 交叉定年法
cross-differentiation 交叉微分法
crossed beam 交叉波束
crossed vortex 交涡旋
cross-equatorial current 越赤道海流
cross-equatorial flow 越赤道气流【大气】
cross-front flow 穿锋流
crossing theory 交叉理论
cross-isobar angle 风矢等压线交角,越等压线角
cross-over 交叉
crossover experiment 交叉〔型〕实验
cross-over storm comma 交叉风暴逗点云系
"cross"parachute "十字形"降落伞
cross polarization 正交偏振,正交极化
cross-polarized signal 正交偏振信号
cross-power spectrum 互功率谱
cross product 向量积【数】,叉积
cross sea 横浪
cross-section 剖面【大气】,截面
cross-sectional analysis 剖面分析
cross-section diagram 剖面图【大气】
cross-section test 剖面检验
cross spectrum 互谱【数】,交叉谱【大气】
cross spectrum density 互谱密度
cross totals index 交叉总指数
cross-track scanner 跨轨扫描仪
cross validation 交叉核实【数】
cross-valley wind 横〔穿〕谷风
cross-wind 侧风【大气】

cross-wind diffusion 横风扩散,侧风扩散
crosswind gustiness 侧风阵性
crosswind sensor 侧风传感器
crosswise vorticity 正交涡度
crown flash 冕〔状〕闪〔电〕
CRT (cathode-ray tube) 阴极射线管
crude data 原始资料
cruiser 巡洋舰【航海】
cryobiochemistry 低温生物化学【生化】
cryochore 冰雪区
cryoconite hole 冰穴
cryoelectronics 低温电子学
cryogenic hygrometer 低温湿度计
Cryogenic Infrared Spectrometers and Telescopes for the Atmosphere(CRISTA) 大气低温红外光谱仪和望远镜
cryogenic period 冰雪时期,冰期
cryology 冷圈学【地理】,冰冻学【大气】
cryopedology 冻土学【地理】
cryopedometer 冻结仪
cryophilic 嗜寒性的
cryophyte 冰雪植物
cryoplanation 强霜冻侵蚀,冻蚀
cryoplankton 冰雪浮游生物
cryopump 低温抽气泵
cryosphere 冷圈【地理】,冰雪圈,冰冻圈【大气】,冰冻层,低温层
cryoturbation ①冻裂搅动作用,冻搅②冷冻风化作用③冷冻翻浆作用
cryptanalysis 密码分析【电子】
cryptoclimate 室内小气候【大气】
cryptoclimatology 室内气候学【大气】
cryptography 密码学【电子】
Cryptozoic Eon 隐生宙【地质】
crystal 晶体
crystal edge 晶棱【物】
crystal face 晶面【物】
crystal form 晶形【物】,晶状
crystal habit 晶体习性
crystal lattice 晶格【物】,晶体点阵
crystalline ①晶状〔的〕②结晶〔的〕

crystalline frost 晶体霜
crystalline lens 晶状体【物】
crystallinity 结晶度【化】,晶性【物】
crystallization 结晶
crystallization nucleus 结晶核
crystallographicc plane 晶面【物】
crystal structure 晶体结构【物】
CSAIR(coherent synthetic aperture imaging radar) 相干合成孔径成像雷达
CSD(Commission on Sustainable Development) 可持续发展委员会
CSI (conditional symmetric instability) 条件对称不稳定〔性〕
CSI (critical success index) 临界成功指数
CSTD monitoring system 电导率-盐度-温度-深度监测系统
CTA (cognitive task analysis) 认知任务分析
CTD(conductivity-temperature-depth profiler) 电导率-温度-水深廓线仪
ct DNA 叶绿体 DNA【生化】
CTH (cloud top height) 云顶高度【大气】
CTHES (cloud top height estimation system) 云顶高度估计系统
ct RNA 叶绿体 RNA【生化】
cube ①立方②立方体【数】
cubic equation 三次方程【数】
cubic system 立方〔晶〕系
cum sole 顺钟向(反气旋方向)
comulative concentration 累积浓度
cumulative distribution function (CDF) 累计分布函数
cumulative error 累积误差【数】
cumulative frequency 累积频率
cumulative histogram 累积直方图
cumulative spectrum 累积谱
cumulative temperature 积温
cumuliform (cuf) 积云状
cumuliform cloud 积状云【大气】
cumulogenitus (cug) 积云性
cumulonimbus (Cb) 积雨云【大气】
cumulonimbus arcus 弧状积雨云
cumulonimbus calvus(Cb calv) 秃积雨云【大气】
cumulonimbus capillatus(Cb cap) 鬃积雨云【大气】
cumulonimbus dynamics 积雨云动力学
cumulonimbus incus 砧状积雨云
cumulonimbus mammatus 悬球状积雨云(曾用乳房状积雨云)
cumulonimbus model 积雨云模式【大气】
cumulostratus 层积云
cumulus(Cu) 积云【大气】
cumulus base 积云底
cumulus congestus(Cu cong) 浓积云【大气】
cumulus convection 积云对流【大气】
cumulus fractus(Cu fra) 碎积云
cumulus heating 积云〔对流〕加热
cumulus humilis(Cu hum) 淡积云【大气】
cumulus-lenticularis 荚状积云
cumulus mediocris(Cu med) 中展积云【大气】
cumulus parameterization model 积云参数化模式
cumulus pileus 幞状积云
cumulus stage 积云阶段
cumulus undulatus 波状积云
Cunnane plotting position 库纳标绘位置
Cunningham slip correction 坎宁安滑移订正(斯托克斯定律的偏差)
cup anemometer 转杯风速表【大气】
cup-contact anemometer 电接转杯风速表
cup-counter anemometer 计数转杯风速表
cup crystal 杯状冰晶
cup current meter 转杯式海流计
cup-generator anemometer 磁感转杯风速表
curie (Ci) 居里(放射性的旧单位, $=3.7\times10^{10}$ Bq)

Curie point 居里点【物】
Curie temperature 居里温度(即居里点)
curl 旋度【数】
curl of vector 矢量旋度
current ①流②电流
current chart 海流图
current cross section 海流剖面,流速断面图
current curve 流速流向曲线
current density 流密度
current ellipse 潮流椭圆
current float 漂流杆,测流浮子
current function 流函数
current-limiting device 限流元件
current meter 测流表,流速表,流速仪
current pole 测流杆
current profiler 流速廓线仪
current rose 流玫瑰图,海流图
current tables 潮流〔预报〕表
current vortex 海流涡旋
current weather 现在天气
cursor ①光标【计】②游标【计】
curtain 幡
curtain aurora 帘状极光,极光帘
Curtis-Godson approximation 柯蒂斯-戈德森近似,C-G 近似
curvature 曲率【数】
curvature effect 曲率效应
curvature vorticity 曲率涡度【大气】
curve 曲线
curved isobar 弯曲等压线
curved-path error 大气折射误差
curve fitting 曲线拟合【数】【大气】
curve of growth 生长曲线
curvilinear coordinates 曲线坐标【数】
curvilinear motion 曲线运动【物】
curvilinear regression 曲线回归
cusp ①尖点②拐点
cusp catastrophe ①月尖型突变②尖拐型突变
custard winds 奶油蛋糕风(英国北部沿海的寒冷东风)
customer 客户【计】
cut-off angular frequency 截止角频率
cut-off frequency 截止频率
cut-off high 切断高压【大气】
cut-off low 切断低压【大气】
cut-off time 截止时间
cut-off voltage 截止电压【电子】
cut-off wavelength 截止波长【电子】
cutoff wavelength 截止波长【物】
cutting device 解脱器,切断装置
cutting-off process 切断过程
C weather 目视飞行天气(contact weather 的缩写)
CW radar (continuous-wave radar) 连续波雷达【大气】
cyanometer 天空蓝度测定仪【大气】
cyanometry 天空蓝度测定法
cybernation 〔计算机〕控制
cybernetics 控制论【数】
cycle 循环,变程
cycle per second (cps) 周每秒,周/秒(相当于赫兹)
cycle process 循环过程
cyclic boundary condition 循环边界条件
cyclic check 循环校验
cyclic code 循环码
cyclic peptide 环肽【生化】
cyclogenesis 气旋生成【大气】
cyclo-geostrophic current 气旋地转流
cyclohexane 环己烷(C_6H_{12})
cycloid 摆线【数】
cyclolysis 气旋消散【大气】
cyclone 气旋【大气】
cyclone cellar 防风窖,防龙卷掩体
cyclone collector 旋风集尘器(除去直径大于 3 μm 的粒子,或取样分析)
cyclone family 气旋族【大气】
cyclone model 气旋模式
cyclone path 气旋路径
cyclone precipitation 气旋降水

cyclone-prone area 多气旋区
cyclone track 气旋路径
cyclonette 小旋风
cyclone warning 气旋预警
cyclone wave 气旋波
cyclonic 气旋式的、气旋性的
cyclonic bora 气旋式布拉风
cyclonic centre 气旋中心
cyclonic circulation 气旋性环流【大气】
cyclonic curvature 气旋性曲率【大气】
cyclonic disturbance 气旋扰动
cyclonic extratropical storm（CYCLES）气旋性温带风暴
cyclonic phase 气旋期
cyclonic precipitation 气旋降水
cyclonic rain 气旋〔性〕雨
cyclonic rotation 气旋式旋转
cyclonic scale 气旋尺度
cyclonic shear 气旋性切变【大气】
cyclonic storm 气旋〔性〕风暴
cyclonic thunderstorm 气旋〔性〕雷暴
cyclonic vorticity 气旋性涡度【大气】
cyclonic wave 气旋波【大气】

cyclonic wind 气旋风
cyclostrophic 旋衡的
cyclostrophic balance 旋衡性平衡
cyclostrophic convergence 旋衡辐合
cyclostrophic divergence 旋衡辐散
cyclostrophic flow 旋衡流
cyclostrophic function 旋衡函数
cyclostrophic transport 旋衡输送
cyclostrophic wind 旋衡风【大气】
cyclotron 回旋加速器【物】
cyclotron frequency 回旋频率【物】
cyclotron radius 回旋半径【物】
cylindrical coordinates 柱面坐标【数】
cylindrical function 圆柱函数【数】
cylindrical net radiometer 圆柱形净辐射表
cylindrical polar coordinates 柱面极坐标
cylindrical projection 圆柱投影
cylindricity 圆柱度【航海】
cypher（＝cipher）①密码②数码
CZCS（Coastal Zone Color Scanner）海岸带水色扫描仪【大气】,海色扫描仪

D

DAB（digital audio broadcast）数字音频广播
dadur 达德风（印度恒河流域的顺水风）
dagpm（deka geopotential meters）十位势米（符号表示法）
dahatoe 达哈托风（北苏门答腊托巴湖的风）
daily amount 日量
daily amplitude 日振幅
daily average wind velocity 日平均风速
daily character figure 日磁情指数
daily course 日变程
daily extreme 日极值

daily forecast 每日预报
daily maximum temperature 日最高温度【大气】
daily maximum wind force 日最大风力
daily maximum wind velocity 日最大风速
daily mean 日平均【大气】
daily mean temperature 日平均温度【大气】
daily minimum temperature 日最低温度【大气】
daily motion 日运动【天】
daily precipitation 日降雨量【铁道】,日降水量
daily rainfall 日降雨量

daily rate 日差【航海】
daily range 日较差【大气】
daily range of temperature 温度日较差【大气】
daily regulation 日调节【水利】
daily storage 日储水量,日调节库容
daily temperature range 气温日较差
daily variation 日变化【大气】
daily weather chart (=map) 每日天气图
daisy world 雏菊世界(计算机模拟的假设世界)
daisyworld model 雏菊世界模式
d'Alembert formula 达朗贝尔公式【数】
d'Alembert's paradox 达朗贝尔佯谬【力】
Dalton's law 道尔顿定律【大气】
damage 灾害,损伤【力】
damage area 受灾面积
damage factor 污染系数【石油】
damage forecasting 灾害预报【医】
damage function 损耗函数
damming ①(冷空气的)堆积阻塞②筑坝,拦水,壅水
damp air 潮湿空气
damped collision 阻尼碰撞【物】
damped〔natural〕frequency 阻尼〔固有〕频率
damped oscillation 阻尼振荡
damped oscillator 阻尼振子【物】
damped vibration 阻尼振动【物】
damped wave 阻尼波
damp haze 湿霾
damping 阻尼【物】,挫抑
damping action 阻尼作用【物】
damping coefficient 阻尼系数【电工】
damping factor 阻尼因子
damping procedure 阻尼程序
damping radiation 阻尼辐射【天】
damping ratio 阻尼比(测风仪器响应特征)
damping scheme 阻尼格式
damping term 阻尼项

damp proofing course 防潮层【建筑】
damp tolerant plant 耐湿植物
D-analysis D-分析
dancing dervish 尘旋
dancing devil 尘旋
danger area 危险区【航空】
danger line 危险线(指洪水水位)
dangerous half 危险半圆
dangerous quadrant 危险象限【航海】
dangerous semicircle 危险半圆【航海】(在北半球热带气旋路径的右边,在南半球热带气旋路径的左边)
danger signal 危险信号
Dansgaard-Oeschger events 丹斯加德-厄施格事件(末次冰期中的冷暖振荡现象),D-O 事件
Darcian velocity 达西速度
Darcy 达西(多孔介质渗透率单位)
Darcy's law 达西定律
dark adaptation 暗适应【物】
dark band 暗带
darkening towards the limb 临边昏暗〔现象〕
dark field 暗视野
dark frost 黑霜【大气】
dark lane 暗带【天】
dark light 暗〔光〕线,暗光
dark lightning 暗闪电
dark limb 暗临边
dark line 暗线
dark matter 暗物质【天】
dark noise 暗噪声
dark radiation 暗辐射
dark ray 暗射线
dark segment 暗弧
Darling shower 达令尘暴(澳大利亚达令河附近)
dart leader 直窜先导(闪电)
dartsonde 标枪探空仪(火箭探空仪的一种)
dart-stepped leader 直串梯级先导

Darwin Station 达尔文站(澳大利亚)
dashed line 虚线
data(datum 的复数形式) ①数据②资料
data acquisition 资料收集,数据采集
data acquisition facility 资料接收中心
data acquisition system 数据采集系统【计】
data analysis center 资料分析中心
data analysis system 数据分析系统【计】
data assimilation 资料同化【大气】
data assimilation cycle 资料同化周期
data assimilation frequency 资料同化频率
data assimilation model 资料同化模式
data assimilation system 资料同化系统
data attribute 数据属性【计】
data bank 资料库,数据库
data base 数据库【计】
Database Control System(DBCS) 数据库控制系统
Database Management System(DBMS) 数据库管理系统
data capacity 信息容量
data capture 数据采集【地信】
data checking 资料检验
data classification system 数据分类系统
data collection and location system (DCLS) 资料收集和定位系统
data collection and platform location system(DCPLS) 资料收集和观测平台定位系统
data collection platform(DCP) 资料收集平台【大气】
data collection system(DCS) 资料收集系统【大气】
data communication system 数据通信系统
data compatibility 数据兼容性【地信】
data completeness 数据完整性【地信】
data compression 数据压缩【地信】,资料压缩
data conversion 数据转换【地信】
data coverage 资料覆盖范围,数据层【地信】
data directory 数据目录【计】
data display 数据显示【地信】
data dissemination 数据发布【地信】
data distribution 数据分发【地信】
data editing 数据编辑【地信】
data encoding 数据编码【地信】
data exchange system 数据交换系统
data extraction 数据提取,资料提取
data field 数据域【地信】,数据字段【地信】
data file 数据文件【地信】
data flow 数据流【计】
data format 资料格式,数据格式【计】
data frame 数据帧【计】
data fusion 数据融合【地信】,资料融合
data gathering network 资料收集站网
data handling 资料处理
data impact 资料冲击
data input 数据输入【地信】
data integration 数据集成【地信】
data interpretation 资料判读
data layout 数据〔打印〕格式
datalogger 资料记录〔输出〕器
data management 数据管理【地信】
Data Management Center(DMC) 数据管理中心
data management system(DMS) 数据管理系统【地信】
data modem 数据调制解调器,数据调制解调技术
data monitoring 资料监测
data monitoring system 资料监测系统
data network 数据网
data output 数据输出
data presentation 数据表达【地信】
data processing 数据处理【计】,信息加工
Data Processing Analysis Facility 资料分析加工中心

data quality 数据质量【地信】
data reality 数据真实性【地信】
data reconstitution 数据重组【计】
data reconstruction 数据重建【计】
data rejection 资料剔除
data representation 数据表示【地信】
data requirement 资料要求
data restoration 数据恢复【计】
data sample 样本资料
data selection 资料选择
data set 数据集
dataset catalog 数据集目录【地信】
dataset directory 数据集目录【地信】
data set documentation 数据集文档【地信】
data sharing 数据共享【计】
datasonde 数据探空仪(火箭探空仪的一种)
data source 数据源
data structure 数据结构【计】
data system test 资料系统试验
data tape 数据〔磁〕带
data transmission 数据传输【计】,资料传送
data type 数据类型【计】
data updating 数据更新【地信】
data utilization station 资料利用站
data validation 数据确认【计】
data validity 数据有效性【计】
data-void area 资料稀少地区
data volume 资料量
data window 数据窗
dateability 可定年性
date line 日界线【航海】,日期变更线
dating 定年
dating method 定年法
datum 基准【地信】,基〔准〕面
datum of chart 海图基准面【海洋】
datum level 基准面
datum line 基准线
datum static correction 基准面静校正【地】
daughter cell 子〔环流〕胞
daughter cloud 子体云
Davidson Current (= Davidson inshore current) 戴维森海流
Davies number 戴维斯数(计算雨滴末速度用的无因次量)
Davis weighing lysimeter 戴维斯称重式蒸散渗漏计
dawn 黎明
dawn chorus 晨噪(黎明时的无线电干扰)
dawn-dusk electric field 晨昏电场【地】
dawnside auroral oval 晨侧极光卵
dawnside magnetosphere 晨侧磁层
dawnside tail 晨侧磁尾
day 日
daybreak 黎明,破晓
day breeze 昼风,日风
day degree 日·度
day (= light) flight 昼间飞行【航空】
dayglow 白天气辉
daylength 昼长【大气】
daylight 白天光照
daylight-saving time (= summer time) 夏令时【天】(夏季把时钟拨快1小时)
daylight standard 日照标准【建筑】
daylight visibility 白天能见度
day mark 日标【航海】
day-night observation 连续观测
day of dry-hot wind 干热风日
day of rain 雨日
day of snow lying 积雪日(台站周围地面至少一半有雪的日子)
dayside aurora 昼侧极光
dayside auroral oval 昼侧极光卵
dayside magnetosphere 昼侧磁层
days with snow cover 积雪日数【大气】
day temperature 白天温度,日温
daytime 白天,白昼
daytime observations 白天观测
daytime visual range (= visual range)

白天能见距离
day-to-day change 日际变化,逐日变化
day without frost 无霜日
dBz(也用 dBZ)(decibel reflectivity factor) 分贝反射率因子【大气】,〔雷达〕反射强度(雷达反射率因子 Z 的对数尺度)
DCAPE (downdraft convective available potential energy) 下沉对流可用位能
DCP (data collection platform) 资料收集平台【大气】
DCS (data collection system) 资料收集系统【大气】
DCVZ (Denver convergence-vorticity zone) 丹佛辐合涡度带
DDA (depth-duration-area value) 〔降水〕时-深-面值
DDA value 时-深-面值
Deacon wind profile parameter 迪肯风速廓线参数
deactivation 减活作用
dead air 闭塞空气,不流通空气
dead end 端点【地信】
dead glacier 死冰川,停滞冰川
dead heart 枯心【农】
dead-line 截止期
deadly temperature 致死温度
dead reckoning 推算航行法
dead reservoir capacity 死库容【水利】
dead water 死水
dead water level 死水位【水利】
dead wind 逆风
deaister(＝doister, dyster) 戴斯特风暴(苏格兰的强烈风暴)
dealiasing 去混淆
Deardorff's model 迪尔多夫模式
Deardorff velocity 迪尔多夫速度(对流混合层中)
deasil 顺日向
debacle 解冻
debriefing 天气汇报(飞行员返航后的口头气象汇报)
debris 碎片
debris flow 泥石流【地质】
debris flow erosion 泥石流侵蚀【地质】
debugging failure 调试故障,初期故障
Debye shielding distance 德拜屏蔽距离
DEC (Digital Equipment Corporation) 数字设备公司, DEC 公司(美国)
deca- 十(词头), 10^1
decadal-to-centennial variability 十年至百年变化,十年至百年变率
decade ①十年②十进,十个构成的一组
decay 衰变,衰减
decay area 衰减区
decay cloud 伪装目标云
decay constant 衰变常数
decay distance 衰减距离
decaying stage 消散期
decaying wave 消散波
decaying mode 衰减模式
decay of waves 波衰减
decay rate 衰变率【物】
decelerate 减速
deci- 分(词头), 10^{-1}
decibar 分巴
decibel (dB) 分贝
decibel reflectivity factor (dBz) 分贝反射率因子【大气】
deciduous broadleaved forest 落叶阔叶林【林】
deciduous forest 落叶林【地理】
deciduous snow forest climate 落叶雪林气候
decile 十分位值,十分位数
decimal-binary conversion 十进二进制转换【数】
decimal coefficient of absorption 常用吸收系数
decimal coefficient of extinction 常用消光系数
decimal digit 十进制数字【计】
decimal system 十进制
decimeter (dm) 分米

decision analysis 决策分析【数】【大气】,判定分析
decision function 决策函数
decision making algorithm 决策算法
decision matrix 决策矩阵
decision tree 决策树【计】,决策树〔形图〕【大气】
decision under risk 风险决策【数】
decision under uncertainty 不确定决策【数】
deck(=lid) 〔云〕盖
declimatization 气候不适应〔症〕
declination ①偏角②磁偏角③赤纬【天】【大气】
decoder 译码器,解码器【计】
decoding 译码,解码【计】
deconvolution 反褶积,消卷积
decorrelation time (=coherence time) 抗相关时间
decoupling 解耦【数】,拆离【地质】,退耦合(的)
decrease 减少
decrescent 下弦月
deduction 推论,演绎
deep cloud band 深厚云带
deep convection 深对流【大气】
deep convection model 深对流模式
deep current 深海流【海洋】
deep easterlies 深厚东风带
deepening 加深
deepening cyclone 加深气旋
deepening of a depression 低压加深【大气】
deepening stage 加深阶段
deep freeze 深厚冰封
deep-moored instrument station 深海系泊仪器站
deep ocean 深海环流
deep ocean moored buoy 深海系泊浮标
deep ocean pressure gage 深海压力计
deep ocean sediment probe (DOSP) 深海沉积物探针
deep ocean survey vehicle (DOSV) 深海考察器
deep percolation 深层渗漏【水利】
deep research vehicle (DRV) 深海考察器
deep-sea 深海
deep-sea core 深海岩心
deep-sea lead 深海测深锤
deep sea sediment 深海沉积物
deep-seated soil temperature 深层地温
deep seepage 深层渗漏,深层渗透
deep sound channel 深水声道
deep space 外层空间,深空
deep space instrumentation facility (DSIF) 深空探测设备,太空〔飞行器〕控制与测量设备(设在地面上)
deep space probe 深空探测器
deep spring 晚春
deep trades 深厚信风
deep water 深层水【海洋】
deep-water ocean wave 深水海洋波
deep-water wave 深水波【大气】
deep well 深井【石油】
Defense Meteorological Satellite Program (DMSP) 国防气象卫星计划(美国)
deficit 亏值,不足量
definite integral 定积分【数】
definition ①定义 ②清晰度
Definitive Geomagnetic Reference Field (DGRF) 最终地磁参考场,确定的地磁参考场(过去地磁场的数学模式)
deflation 吹蚀【地理】
deflation hollow 风蚀洼地【地理】
deflecting force 偏向力,偏转力
deflecting force of earth rotation 地转偏向力
deflecting magnetic field 致偏磁场
deflection 偏向,偏转
deflection anemometer 偏转风速表
deflection force of earth rotation 地球自转偏向力
deflection front 偏转锋
deflection-modulated indicator (=amplitude-

modulated indicator） 调偏指示器
deflection of air flow 气流偏角
deflector 导风板,导流片
defocusing 散焦【物】
deforestation 毁林
deformation 形变【数】,变形【地质】
deformation axis 变形轴
deformation field 变形场【大气】
deformation flow 变形流
deformation gradient 变形梯度
deformation plane 变形面
deformation radius 变形半径
deformation thermograph 变形温度计
deformation thermometer 变形〔类〕温度表【大气】
deformation zone 变形带
deformation zone cloud system 变形场云系
deformed ice 变形冰
degeneracy 退化,简并【化】,简并度【物】
degenerate amphidrome 退化旋转潮波
degenerate elliptic equation 退化椭圆〔型〕方程【数】
degenerate hyperbolic equation 退化双曲〔型〕方程【数】
degenerate state 退化状态
degeneration 退化
deglaciation 冰川退缩【地理】,冰川减退【大气】,冰消〔作用〕,冰消期
degradation 降解【化】,普遍冲刷【水利】,退化
degradation of energy 能量递降
degradation of environment 环境退化
degradation of pollutant 污染物质降解【水利】
degree 度
degree-day 度日【大气】
degree-day correlation 度日相关
degree-hour 度时
degree of acidity 酸度【化学】
degree of coherence 相干度【物】

degree of confidence 〔置〕信度
degree of drought 干旱程度【林】
degree of freedom 自由度【物】
degree of humification 腐殖化程度【土壤】
degree of ionization 电离度【物】
degree of parallelism 并行度【计】
degree of polarization 偏振度【物】【大气】
degree of polymerization 聚合度【化】
degree of saturation 饱和度【水利】
degree of stability 稳定度
degree of subnet 子网度【计】
degree of superheat 过热程度
degree of vacuum 真空度【电子】
degree of water erosion 水蚀程度【林】
degree of wind erosion 风蚀程度【林】
degrees of frost 霜度
dehumidification 减湿〔作用〕
dehydration 脱水〔作用〕
dehydrogenation 脱氢【化】
deice 防冰,除冰
deicer 防冰器,除冰器
deicing 除冰
deionization 消离子作用【化】,消电离【电子】
dekad 十天,旬
delay 延迟,滞后
delayed automatic gain control 迟延自动增益控制
delayed（=stratospheric） fallout 延缓〔放射性〕沉降,平流层沉降
delayed report 迟到的报告,过时报告
delay line 延迟线【计】
delay picture transmission（DPT） 延时图像传输【大气】
delay time 滞后时间,延迟时间【计】
delinescope 幻灯
deliquescence 潮解【化】
Dellinger effect（=fadeout） 德林杰效应（短波通讯中断现象）

del-operator 向量微分算子
delta Eddington δ-E 技术,德耳塔-埃丁顿〔法〕(处理短波辐射传输的近似方法)
Delta function δ 函数
delta region ①三角洲 ②出口区
delta rule δ 规则,德耳塔规则(训练神经网络的方法)
deluge 大洪水,泛滥
DEM (digital elevation matrix) 数字高程矩阵【地信】
DEM (digital elevation model) 数字高程模型【地信】
demarcation 划界
demarcation line 界线
demodulation 解调
demultiplexer 分用器【电子】
dendrite 枝状
dendritic snow crystal 枝状雪晶
dendric snow crystals 枝状雪花晶体
dendritic crystal 枝状冰晶【大气】,载状晶体(雪花的一种)
dendrochore 树木景带,林木区
dendrochronology 年轮学
dendroclimatography 树木年轮气候志【大气】
dendroclimatology 树木年轮气候学【大气】,年轮气候学【地理】
dendroecology 年轮生态学
dendrograph 树木年轮测定器
denitrification 反硝化作用【土壤】,脱氮作用
dense distribution 稠密分布
dense fog 浓雾
dense upper cloud 浓密高云
densification 致密化【力】
densitometer ①比重计 ②显像密度计
density 密度【物】,浓度
density altitude 密度高度
density (=thickness) advection 厚度平流

density correction 密度订正
density channel 密度通道
density current 密度流
density fluctuation 密度涨落【物】
density function 密度函数
density distribution 密度分布
density evolution 密度演化【天】
density gradient 密度梯度
density inversion 密度逆增
density of dry air 干空气密度
density of moist air 湿空气密度
density of snow 积雪密度
density-of-snow gauge 〔量〕雪秤
density potential temperature 密度位温
density temperature 密度温度
densometer 密度计,〔纸张〕透气度测定仪
densus 浓密(云)
denterium content 重氢(氘)含量
denudation 剥蚀〔作用〕【地理】
denudation surface 剥蚀面【地理】
Denver convergence-vorticity zone (DCVZ) 丹佛辐合涡度带
Denver cyclone 丹佛气旋
deoxygenation 脱氧【化】
deoxyribonucleic acid (DNA) 脱氧核糖核酸【生化】
departure 偏差,距平【大气】
depegram 露点图
dependability 可靠性
dependence 相关,依赖
dependent meteorological office (DMO) 航空气象台(室)
dependent variable 应变数
depeq 德佩克风(西南季风时苏门答腊、东印度的强风)
depergelation 解冻
depletion curve 退水曲线
depolarization 退极化【物】,退偏振【物】
depolarization effect 退极化效应
depolarization ratio 退偏振比【大气】

deposit gauge 沉淀器(测量大气污染的沉淀作用)
deposition ①淀积〔作用〕②凝华【大气】③沉降
deposition nucleus 凝华核【大气】
deposition velocity 沉降速度
depression 低〔气〕压【大气】
depression angle 俯角
depression belt 低压带
depression of the dew point 〔温度〕露点差【大气】
depression of the wet bulb 干湿球温度差
depression storage 填洼【水利】,洼地蓄水〔量〕
depth ①深度,厚度 ②层次
depth-area-curve 〔降水〕深度-面积曲线
depth-area-duration analysis 雨量-面积-持续时间分析
depth-area formula 雨量-面积公式
depth contour 等深线【测】
depth-duration curve 雨量-持续时间曲线
depth-duration-frequency curve 雨量-持续时间-频数曲线
depth-first search 深度优先搜索【自动】
depth hoar 浓〔白〕霜
depth-integration sediment sampling 积深泥沙采样,全深集总泥沙采样
depth line 深度线
depth marker 深度标志器,深度标尺(测量冰或雪面被新降雪埋藏深度)
depth of compensation 补偿深度【地】(水中透光带下界)
depth of freezing 冻结深度【农】
depth of frictional influence 摩擦〔影响〕深度
depth of planetary bouandary layer 行星边界层厚度
depth of runoff 径流深【水利】
depth of snow 积雪深度
depth of spring (summer, autumn, winter) 仲春(盛夏,仲秋,隆冬)

depth-velocity integration method 水位-流速积分法
derecho 线状风暴,下击暴流族
DERF (dynamical extended range forecasting) 动力延伸预报
derivative 导数,微商【数】
derivative disaster 衍生灾害
derivative monitoring 导数法监测
derivative of convolution 卷积微分
derivative of high order 高次导数
derivative operator 微分算符
derived data 导出资料,统计性资料
derived gust velocity 导出阵风速度
derived unit 导出单位
desalination 脱盐〔作用〕,淡化〔作用〕
descaling 降尺度
Descartes ray 笛卡儿光线,笛卡儿射线
descendent 下降度(反梯度)
descending air 下沉空气
descending area 下沉区
descending cloudless region 下沉无云区
descending current 下降气流
descending node 降交点【天】
descent 下降
descent method 下降法
describing function 描述函数【电子】
descriptive climatology 描述性气候学
descriptive data 描述数据
descriptive meteorology 描述性气象学
descriptive model 〔描述性〕概念模式
descriptive text 文字说明
deseasonalizing 去季节化(消去季节性变化)
desert 沙漠【土壤】,荒漠【地理】
desert belt(=zone) 沙漠带
desert climate 沙漠气候亚类【大气】,荒漠气候【地理】
desert devil 沙卷风
desert historical geography 沙漠历史地理【地理】
desertification 荒漠化【大气】

desertization 荒漠化【大气】,沙漠化（自然因素）
desert meteorology 沙漠气象学
desert pavement 漠盖层【地理】
desert savanna 沙漠草原
desert soil 荒漠土壤【地理】
desert steppe 沙漠草原
desert wind 沙漠风(美国加利福尼亚州南部的干燥风)
desert wind squall 沙漠风飑
desert zone 沙漠带
desiccation 变旱,脱水【农】
design annual runoff 设计年径流量【水利】
design criteria for surface drainage 排涝设计标准
design factor 设计因子
design flood 设计洪水【水利】
design flood frequency 设计洪水频率【水利】
design flood level 设计洪水位【水利】
designing wind speed 设计风速
design low flow year 设计枯水年【水利】
design storm 设计暴雨【水利】
design storm pattern 设计暴雨雨型【水利】
design torrential rain 设计暴雨【大气】
design water level 设计水位
design wave 设计波〔浪〕
designated pollutant 指定污染物,标志污染物
desirable comfort zone 理想舒适区
despin control electronics (DSE) 消旋控制电子设备
destablization 去稳作用
destination 目的地,目标
destructive interference 相消干涉
destructiveness 破坏性,毁坏性
desulfurization 脱硫【化】
desuperheater 减温器【航海】
detached shock wave 离体激波
detached stratus 分裂层云
detail 细节,详图

detain 阻拦,扣留
detaining layer 阻挡层
detection ①探测②检测③检波
detection and attribution 探测与归因（气候变化的）
detectivity 探测率【电子】
detector 探测器【电子】
detemperature rate 降温率
detention period 滞留期,停留时间
detention storage 拦洪蓄水,拦洪库容
deterioration report 天气转坏报〔告〕
determinacy 确定性【数】
determinant 行列式【数】
determinate forecast 确定预报
determination ①测定②确定
determinism 决定论【物】
deterministic 确定性〔的〕,必然性〔的〕,定数论〔的〕
deterministic forecast 确定性预报【大气】
deterministic hydrology 确定性水文学
deterministic prediction 确定性预报
deterministic system 确定性系统【数】
detrainment 卷出,夹出
detrend 去倾
detrending 去倾
deuterium 氘【化】,重氢
deuteron 氘核【化】,重氢核
developed turbulence 发展的湍流
development ①发展,演变②拟定,研制③显影④发育【植】
development index 发展指数
deviating force 偏向力,偏转力
deviation ①偏差②偏角
devil（见 dust devil） 尘卷风
devil water 废水
Devonian period 泥盆纪【地质】
Devonian System 泥盆系【地质】
dew 露【大气】
Dewar flask 杜瓦〔真空保温〕瓶
dewbow 露虹【大气】
dew-cap 露冠

Dew cell 道氏元件(湿敏元件)
dew-drop 露滴
dew duration recorder 露持续时间记录器
dewfall 降露,结露
dewgauge 测露表【大气】,露量器
dewing 结露
dew plate 露水板
dew-point 露点【物】
dew-point〔temperature〕 露点〔温度〕
dew-point apparatus 露点测定器
dew-point deficit(=depression) 气温露点差
dew-point front 露点锋【大气】,干线
dewpoint formula 露点公式
dew-point hygrometer 露点湿度表【大气】
dew-point line 露点线
dew point meter 露点表
dew-point radiosonde 露点探空仪
dew-point recorder 露点记录仪
dew-point spread 气温露点差
dew point temperature 露点〔温度〕【大气】
DGRF (Definitive Geomagnetic Reference Field) 最终地磁参考场,确定的地磁参考场
diabatic change 非绝热变化
diabatic effect 非绝热效应
diabatic flow 非绝热流【力】
diabatic gradient 非绝热梯度
diabatic heating 非绝热加热
diabatic initialization 非绝热初始化
diabatic mesoscale circulation 非绝热中尺度环流
diabatic process 非绝热过程【大气】
diabatic term 非绝热项
diagnosis(=diagnose) ①诊断②分析
diagnostic analysis 诊断分析【大气】
diagnostic equation 诊断方程【大气】
diagnostic model 诊断模式【大气】
diagnostic study 诊断研究
diagnostic weather analysis 诊断天气分析
diagonal 对角线【数】

diagram 图〔解〕
diagrammatic sketch 示意图
diagram method 图解法,图示法
dial ①日晷仪②刻度盘
DIAL (differential absorption lidar) 微分吸收激光雷达【大气】
dial hygrometer 度盘湿度表
dial thermometer 度盘温度表
diameter 直径
diameter of icing 积冰直径
diamond dust 钻石尘(冰晶形的降水,小到悬浮在空气中)
diapause 冬眠状态,滞育【农】
diaphragm 膜片
diaphragm manometer 膜片压力表
diapycnal 法向(垂直于等密度面),垂直等密度面方向
diapycnal mixing 密度差异混合(穿过密度跃层的混合)
diatomic molecule 双原子分子【化】
diastrophism 地壳运动
diathermance(=diathermancy) 透辐射热性
diathermy 电热〔疗〕法,透热〔疗〕法
DIC (dissolved inorganic carbon) 溶解无机碳
dichotomy 两分〔法〕,二等分
dichroism 二向色性【物】
Dicke radiometer 狄克辐射计
diel- 昼夜(大气化学和土壤微气象学中常用此词代替 diurnal)
dielectric 〔电〕介质【电子】,电介质【物】
dielectric absorption 介电质吸收
dielectric breakdown 介质击穿【电子】
dielectric constant 介电常数
dielectric factor 介电因子(雷达方程中)
dielectric function(=relative permittivity) 介电函数(相对电容率)
dielectric gradient 介电梯度
dielectric isolation 介质隔离【电子】

dielectric loss 介电损耗
dielectric strength 介电强度【物】
dielectronic recombination 双电子复合
diffeomorphism 微分同胚
difference analogue 差分格式
difference coordinate 坐标差
difference equation 差分方程【数】
difference frequency 差频【物】
difference of higher order 高阶差分【数】
difference operator 差分算子【数】
difference scheme 差分格式【大气】
differential 微分【数】
differential absorption 微分吸收
differential absorption and scattering (DAS) 微分吸收和散射
differential absorption application 微分吸收应用
differential absorption cross section 微分吸收截面
differential absorption hygrometer 微分吸收湿度计【大气】
differential absorption lidar（DIAL) 微分吸收激光雷达【大气】
differential absorption lidar thermometer 微分吸收光〔雷〕达温度表
differential absorption technique 微分吸收法【大气】,微分吸收技术
differential actinometer 示差直接辐射表
differential advection 差动平流
differential air thermometer 示差气温表
differential analyser 微分分析仪
differential analysis 微分分析【大气】,差分分析【大气】,差值分析
differential attenuation 微分衰减
differential backscattering cross section 微分后向散射截面
differential ballistic wind 差动弹道风【大气】,分层弹道风
differential bathygraph 差动自记测深仪
differential coefficient 微分系数

differential chart 微分图
differential cross section 微分截面
differential equation 微分方程【数】
differential GPS 差分全球定位系统【航海】
differential heating 不均匀加热
differential kinematics 微分运动学
differential mobility analyzer 微分迁移率分析仪
Differential Omega 微分奥米加
differential operator 微分算子
differential optical absorption 微分光学吸收
differential phase function 微分相函数
differential phase shift 微分相〔位漂〕移
differential quotient 微商【数】
differential reflectivity 微分反射率
differential scattering cross section 微分散射截面
differential thermal advection 差动热平流(上下层不同的热平流)
differential thermometer 示差温度表
differential water capacity 微分持水量(持水量随土壤、水压的变化率)
differential wind 差动风【大气】,差异风
diffluence 分流【大气】
diffluent thermal ridge 分流型温度脊
diffluent thermal trough 分流型温度槽
diffracted ray 衍射线
diffracted wave 衍射波
diffraction 衍射【大气】
diffraction angle 衍射角【物】
diffraction coupled resonator（DCR) 衍射耦合谐振腔
diffraction fringe 衍射条纹
diffraction halo 衍射晕
diffraction pattern 衍射图样【物】
diffraction peak 衍射峰值
diffraction phenomenon 衍射现象
diffraction region 衍射区

diffraction spectrum 衍射光谱
diffraction zone 衍射区
diffuse 漫射,扩散
diffuse auroral patch 弥散极光亮斑
diffuse boundary 扩散界面
diffused light 漫射光
diffuse field 漫射场
diffuse front 扩散锋
diffuse illumination 漫射光照
diffuse incident intensity 漫射入射强度
diffuse intensity 漫射强度
diffuse length 弥散长度
deffuse light 漫射光
diffuser ①漫射体【物】②漫射器,汽化器雾化装置③扬声器纸盆④扩压管,扩散段
diffuse radiation 漫射辐射【大气】
diffuse reflection 漫反射【大气】
deffuse reflector 漫〔射〕反射体
diffuse scattering 漫散射
diffuse skylight 漫射天光
diffuse sky radiation 天空漫射辐射【大气】
diffuse solar radiation 漫射太阳辐射【大气】
diffusion ①扩散【大气】②漫射
diffusion approximation 漫射近似
diffusion chamber 扩散云室
diffusion coefficient ①扩散系数②漫射系数
diffusion denude technique 扩散解吸技术,扩散沉积技术
diffusion diagram 扩散图
diffusion equation 扩散方程【数】,漫射方程
diffusion hygrometer 扩散湿度表
diffusion layer 扩散层
diffusion level 扩散高度
diffusion model 扩散模式【大气】
diffusion pump 扩散泵【电子】
diffusion time 扩散时间

diffusion velocity 扩散速度
diffusiophoresis 扩散迁移
diffusiophoretic force 扩散迁移力
diffusiophoretic velocity 扩散迁移速度
diffusisphere 扩散层
diffusive convection 扩散对流
diffusive equilibrium 扩散平衡
diffusive force 扩散力
diffusive relaxation 扩散张弛法
diffusive separation 扩散分离
diffusivity 扩散率【大气】,扩散性
diffusivity equation 扩散方程
diffusivity factor 漫射率因子
diffusometer 扩散表
digifax 数模
digit 数〔字〕,〔十进数的〕位,号
digital 数字的,数字化的
digital audio broadcast（DAB） 数字音频广播
digital camera 数字照相机【计】
digital cartography 数字地图学【测】,数字地图制图【地信】
digital circuit 数字电路【计】
digital communication 数字通信【电子】
digital compression modem 数字压缩调制解调器
digital computer 数字计算机【计】
digital correlation technology 数字相关技术
digital data 数据【地信】,数字化资料
digital data acquisition system 数字资料获取系统
digital data processor 数字资料处理机
digital depth indicator 数字深度指示器
digital display 数字显示
digital Earth 数字地球【计】
digital elevation matrix（DEM） 数字高程矩阵【地信】
digital elevation model（DEM） 数字高程模型【地信】
digital facsimile recorder 数字传真记录

器,数字传真收片机
digital fax interface (DFI)　数字传真接口
digital filter　数字滤波器【农】
digital filtering　数字滤波
digital forecast　数字预报
digital image　数字图像【计】
digital image mosaic　数字图像镶嵌【地理】
digital image-processing　数字图像处理【测】
digitalization　数字化【计】
digital line system　数字有线系统【航海】
digital map　数字地图【地理】
digital mapping　数字制图【地信】
digital mercury barometer　数字水银气压表
digital orthoimage　数字正摄影像【测】,数字正射影像【地信】
digital orthophoto map　数字正摄影像图【测】,数字正射影像图【地信】
digital parallel input　数字并行输入
digital processing　数字处理
digital radar　数字〔化〕雷达
digital radar echo signal processor (DIREP)　数字〔化〕雷达回波〔信号〕处理器
digital radio system　数字无线系统【航海】
digital sampling　数字化取样
digital scan converter　数字扫描转换器
digital scanner　数字扫描器
digital signal　数字信号【电子】
digital signal processing　数字信号处理
digital signal processor　数字信号处理器【计】
digital simulation　数字模拟【地理】
digital smoke monitor　数字式烟雾监测仪
digital storage　数字贮存
digital sunshine duration recorder　数字式日照〔时间〕记录器
digital temperature sensor　数字式温度探测器
digital terrain model (DTM)　数字地形模型【地理】
digital theodolite　数字经纬仪
digital thermometer　数字式温度表
digital-to-analog conversion　数-模转换
digital-to-analog converter　数模转换器【计】
digital-to-voice computer　数字-话音计算机
digital-to-voice converter　数字-话音转换器
digital transmission　数字传输
digital video broadcast-satellite (DVB-S)　卫星数字电视广播
digital video disc　数字影碟【计】
digital video integrator and processor (DVIP)　数字视频积分处理器
digital voltmeter　数字电压表【电子】
digital weather chart recorder　数字式天气图记录器,数字式天气图传真收片机
digitization　数字化【地信】
digitized cloud map　数字化云图【大气】
digitized map　数字化地图【地信】
digitized radar experiment　数字雷达试验
digitizer　数字读出机,数字化装置,数字化仪【计】
digitizing　数字化
digitizing error　数字化误差
dihedral reflector　二面角反射器
dilatation　膨胀,伸展
dilatation axis　膨胀轴,伸展轴
dilatation field　膨胀场,伸展场
dilatometer　膨胀表
dilution　稀释,冲淡
Diluvial Epoch　洪积世
dimension　①大小②因次③度,维④量纲
dimensional analysis　量纲分析【物】,因次分析
dimensional equation　量纲方程
dimensionality reduction method　降维方法
dimensional parameter　量纲参数

dimensionless 无量纲
dimensionless equation 无量纲方程
dimensionless group 无量纲数群【化工】
dimensionless number 无量纲数
dimensionless parameter 无量纲参数
dimensions 量纲【物】,维〔数〕【数】
dimer 二聚体【化】
dimerization 二聚【化】
dimethyl disulfide 二甲基二硫(CH_3SSCH_3)
dimethyl sulfide (DMS) 二甲〔基〕硫【大气】(CH_3SCH_3)
dimethyl sulfoxide 二甲亚砜(($CH_3)_2SO$)
dimmerfoehn 迪默焚风(阿尔卑斯山南北气压差大于 12 hPa 时罕见的焚风)
Dines anemometer 丹斯测风表【大气】,达因测风表
Dines compensation 达因补偿〔作用〕
Dines compensation theorem 达因补偿定理
Dines float barograph 达因浮标气压计
Dines float manometer 达因流压表
Dines pressure anemograph 达因风压计
Dines radiometer 达因辐射计
Dines tilting-siphon rain-gauge 达因翻斗式虹吸雨量计
dinitrogen pentoxide 五氧化二氮(N_2O_5)
dinosaur 恐龙
diode 二极管
diode laser 二极管激光器
diopter 屈光度
dioxane 二氧杂环己烷($C_4H_8O_2$),二噁烷
dioxin 二噁英(一种致癌化合物)
dip ①倾角②倾斜
dip angle 倾角
dip circle 磁倾仪
dip equator 磁倾赤道【大气】
dip of the horizon 地平倾角
dipole 偶极子【力】
dipole antenna 偶极子天线
dipole anticyclone 偶极反气旋【大气】
dipole emission 偶极子发射
dipole flow pattern 偶极流型
dipole layer 偶极层【物】
dipole moment 偶极矩【物】
dip pole 磁极
Dirac delta function 〔狄拉克〕δ 函数
Dirac equation 狄拉克方程【数】
direct absorption 直接吸收【物】
direct access 直接存取【计】
direct aerosol effect 直接气溶胶效应
direct broadcast 直〔接广〕播
direct cell 直接〔环流〕胞
direct circulation 直接环流〔圈〕【大气】,正环流
direct crosswind 正侧风
direct current 直流电
direct detection 直接探测
direct flow 直接径流
direction ①方向 ②方位
directional absorptivity 定向吸收率
directional derivative 方向导数【数】
directional distribution 方向分布
directional emissivity 定向发射率
directional gain 指向性增益
directional-hemispherical reflectance 定向球面反射率
directional hydraulic conductivity 定向导水性,定向导水系数
directional shear 定向切变
directional spectrum 方向谱
direction cosine 方向余弦
direction finder 定向仪
direction finding 定向
direction-finding system 测向系统【电子】
direction of movement 移〔动方〕向
directivity 指向性【电子】
directivity factor 指向性因数
directivity function 指向性函数
directivity index 指向性指数
directivity pattern 指向性图案
direct lighting 直照

direct lightning flash 直击雷【电力】
direct lightning strike 直接雷击【电工】
direct method 直接法【物】
direct mode 直接〔观测〕方式
direct numerical simulation 直接数值模拟
directory 目录【地信】
direct product 直积【数】,内积
direct radiation 直接辐射【大气】
direct radiative forcing 直接辐射强迫
direct-reading instrument 直读式仪器
direct-reading thermometer 直读式温度表【大气】
direct read-out 直接读出
direct read-out ground station 直读式地面站【大气】
direct read-out image dissector (DRID) 直读式图像分析仪
direct read-out infrared radiometer (DRIR) 直读式红外辐射计
direct-recording bottom current meter 自记底层海流计
directrix 准线
direct route 最短航线
direct runoff 直接径流
direct segment method 直接分段法
direct solar radiation 直接太阳辐射
direct sound〔wave〕 直达声波
direct tide 顺潮
direct transmission system 直接传送系统
direct vision nephoscope 直视测云器
direct vision prism 直视棱镜
Dirichlet problem 狄利克雷问题【数】
dirigible balloon ascent 可控气球探测
discharge ①流量【地理】②放电【物】
discharge area 泄水区
discharge coefficient 流量系数【水利】,排放系数
discharge diagram 放电图
discharge-flow system 放电流动系统
discharge section line 测流断面
discipline 学科

discomfort 不舒适感
discomfort index 不适指数【大气】
discomfort zone 不舒适区
disconnection device 解脱器
discontinuity 不连续〔性〕【大气】
discontinuity of first order 一阶不连续性
discontinuity of zero order 零阶不连续性
discontinuity point 不连续点【数】
discontinuous turbulence 不连续湍流
discrepance 残留项,残余项
discrete data 离散数据【地信】
discrete distribution 离散分布【数】
discrete fluid 离散流体〔模型〕【力】
discrete Fourier transform 离散傅里叶变换【数】
discrete mesh 离散网格
discrete model 分离模式
discrete ordinates method 离散纵标法
discrete polar cap aurora 离散极盖区极光
discrete propagation 非连续传播
discrete set 离散集【数】
discrete space theory 分立(离散)空间理论
discrete spectrum 离散谱【大气】,断续光谱
discrete stochastic process 离散随机过程
discrete transform 离散变换
discretization 离散化【数】
discriminant 判别式【数】
discriminant analysis 判别分析【大气】
discriminate analysis 判别分析【数】
discriminatory analysis 判别分析
disdrometer 雨滴谱仪【大气】
dish 碟形天线(偶尔不正确地用于指所有雷达天线)
dish experiment 转盘实验
dishpan experiment 转盘实验
disintegration 分裂
disk hardness-gauge 盘式〔测雪〕硬度器
disk meter 盘式流量计
disk operating system (DOS) 磁盘操作系统

dislocation 错位
dispersed phase 弥散相态
dispersing medium 弥散剂,弥散介质
dispersion ①弥散【力】②色散【物】③频散【石油】④离散【数】
dispersion coefficient 弥散系数【土壤】,扩散系数
dispersion curve 色散曲线
dispersion diagram 频散图
dispersion equation 色散方程
dispersion medium 离散介质
dispersion of light 光的色散
dispersion power 色散本领【物】
dispersion relation 色散关系
dispersion relationship 频散关系【大气】
dispersive coefficient 分散系数【土壤】
dispersive flux 扩散通量
dispersive medium 色散介质【物】,色散媒质【电工】
dispersive wave 频散波
dispersivity 色散性,扩散性,扩散率(扩散系数除以速度)
displacement 位移
displacement distance 位移距离
displacement height 位移高度
displacement thickness 位移厚度
displacement current 位移电流
display 显示〔器〕
display format 显示格式【计】
display mode 显示方式【计】
display panel 显示板【电子】
display screen 显示屏【电子】
display terminal 显示终端【计】
dissemination ①分发 ②广播
dissipation 耗散【物】【大气】,消散
dissipation coefficient 耗散系数
dissipation constant 消散常数
dissipation factor 耗散因数【电子】
dissipation function 消散函数
dissipation integral 耗散积分
dissipation length scale 耗散长度尺度

dissipation of energy 能量消散
dissipation of waves 波耗散
dissipation power 耗散功率【电子】
dissipation range 耗散区
dissipation rate 耗散率【大气】
dissipation scale 耗散尺度
dissipative state 耗散态【数】
dissipation trail (= distrail) 消散尾迹(凝结尾迹的反意)
dissipative stage 消散阶段
dissipative structure 耗散结构【物】
dissipative system 耗散系统
dissociation 离散,分离,离解【物】
dissociative photoionization 离解光致电离【物】,离解光化电离作用
dissociative recombination 离解复合〔过程〕【物】
dissolution 溶解〔作用〕
dissolved inorganic carbon(DIC) 溶解无机碳
dissolved organic carbon(DOC) 溶解有机碳
dissolved organic compound(DOC) 溶解有机物
dissolved organic matter(DOM) 溶解有机物
distance 距离
distance constant 距离常数(风传感器的响应特征)
distance control 遥控
distance education 远程教育【计】
distance rainfall recorder 遥测雨量计
distance theodolite 测距经纬仪【测】
distance thermometer 遥测温度计
distant fishery 远洋渔业【航海】
distant flash 远闪,远电
distant-reading thermometer 遥测温度计
distillation 蒸馏
distillation method 蒸馏法【航海】
distortion 畸变,失真【电子】
distortional wave 畸变波

distortion correction 畸变校正【大气】
distortionless filter 无畸变滤波〔器〕
distrail 消散尾迹,耗散尾迹
distress 遇险【航海】
distress alerting 遇险报警【航海】
distributed capacitance 分布电容【电子】
distributed database 分布式数据库【计】
distributed target 分布目标【电子】【大气】,散布靶
distribution ①分布【物】②分配
distribution coefficient 分配系数【地质】
distribution function 分布函数【数】
distribution graph 分布图
distribution law 〔概率〕分布律
distributive law 分配律【数】
district forecast 区域预报
disturbance 扰动,摄动【天】
disturbance daily variation 〔地磁〕扰日变化
disturbance line 扰动线
disturbance polar current 扰动极区电流
disturbance ring 扰动环
disturbance selection of perturbation 扰动选择
disturbance variation 磁扰变化
disturbed body 受摄体【天】
disturbing body 摄动体【天】
disturbed boundary layer 扰动边界层
disturbed day 〔地磁〕扰日
disturbed quantity 扰动量
disturbed soil sample 扰动土样【土木工程】
disturbed sun 活动太阳
disturbed upper atmosphere 扰动高层大气
disturbed weather electricity 扰〔动〕天〔气〕电学,扰天电
diurnal 日(指一天内的),每日,周日
diurnal amplitude 日振幅
diurnal circulation 日环流
diurnal cooling 周日冷却(海洋中)
diurnal cycle 日循环
diurnal heating 周日加热(海洋中)
diurnal inequality 日差
diurnal pressure wave 日气压波
diurnal range 日较差【大气】
diurnal solar tide 周日太阳潮
diurnal tide 全日潮【航海】
diurnal variation 日变化【大气】
diurnal vertical migration 昼夜垂直移动【海洋】
diurnal wave 周日波
diurnal wind 周日变风
divective 散流〔的〕
divergence 辐散【大气】,散度【数】【大气】
divergence damping 散度阻尼
divergence equation 散度方程【大气】
divergence field 辐散场,散度场
divergence line 辐散线
divergence measurement 散度测定
divergence signature 辐散特征〔图型〕
divergence theorem 散度定理【大气】
divergence theory of cyclogenesis 气旋生成的辐散理论
divergence theory of cyclones 气旋的辐散理论【大气】
divergence-vorticity ratio 散度涡度比
divergent flow 辐散流
divergent wind 散度风
divergent wind component 辐散风分量
diverging beam 发散波束
diverging trough 疏散槽
diversification 多样化
diversion of water 水改道,分水,导水
DIVERSITAS（Biological Diversity Program） 生物多样性计划
diversity reception 多项接收
divers storm 多变风暴(埃及亚历山大港冬末雨季出现的北风风暴)
divide 分水岭
divide line 分水线
dividend 被除数【数】

dividing crest 分水岭
dividing ridge 分水岭
dividing streamline 分水流线,分隔流线
division ①除法【数】②分度
divisor 除数【数】
D-layer D层
DMO (dependent meteorological office) 航空气象台(室)
DMS (dimethyl sulfide) 二甲〔基〕硫【大气】(CH_3SCH_3)
DMSP (Defense Meteorological Satellite Program) 国防气象卫星计划(美国)
Dobson ozone spectrophotometer 陶普生臭氧分光光度计
Dobson spectrophotometer 陶普生分光光度计【大气】,多布森分光光度计【大气】
Dobson unit (DU) 陶普生单位【大气】(等于标准状态下 0.01 mm 臭氧层的厚度),多布森单位
DOC (1) dissolved organic carbon 溶解有机碳 (2) dissolved organic compound 溶解有机物
Doctor 保健风,郎中风①一种热带和副热带的海风,据说有益于健康②同 harmattan 风)
Documental period 文献记录时期
dog days 伏天
doister (见 deaister) 戴斯特风暴
doldrum equatorial trough 无风带赤道槽
doldrums 赤道无风带
DOM (dissolved organic matter) 溶解有机物
domain 域
domain average 区域平均
domain expert 领域专家【计】
domain knowledge 领域知识
domain of integration 积分域【数】
dome 堆,丘,圆顶,穹面〔式〕
domestic climatology 生活气候学

dominant scale ①主尺度(指空间域) ②主周期(指时间域)
dominant tree 优势木【林】
dominant wind 盛行风
Donghai Coastal Current 东海沿岸流【海洋】
Donghai Sea 东海【海洋】
Donora smog 多诺拉烟雾(美国)
donor 供体【生化】
donor impurity 施主杂质
D-operator D-算子
Doppler acoustic wind sensor 多普勒声学测风器
Doppler broadening 多普勒〔谱线〕增宽【物】【大气】,多普勒加宽
Doppler cloud sounding 多普勒测云法
Doppler effect 多普勒效应【物】
Doppler equation 多普勒方程
Doppler error 多普勒误差
Doppler frequency 多普勒频率
Doppler frequency shift 多普勒频移
Doppler laser radar 多普勒激光雷达
Doppler lidar 多普勒激光雷达【大气】
Doppler-limited spectroscopy 多普勒限光谱学
Doppler profile 多普勒线形【物】,多普勒廓线
Doppler radar 多普勒天气雷达【大气】
Doppler radar reflectivity 多普勒雷达反射率
Doppler radar spectrum 多普勒雷达波谱
Doppler radiosonde system 多普勒无线电探空仪系统
Doppler shift 多普勒频移【物】
Doppler-shifted frequency 多普勒频移频率
Doppler sodar 多普勒声〔雷〕达【大气】
Doppler spectral broadening 多普勒谱展宽
Doppler spectral moments 多普勒谱矩
Doppler spectral width 多普勒频谱宽度
Doppler spectrum 多普勒谱
Doppler spread 多普勒谱展宽

Doppler technique 多普勒技术
Doppler transceiver 多普勒收发机
Doppler velocity 多普勒速度【大气】
Doppler velocity spectra resolution 多普勒速度谱分辨率
Doppler weather radar 多普勒天气雷达
Doppler width 多普勒宽度
dormancy 休眠【植】,冬眠
dormant volcano 休眠火山【地质】,静火山
dose 剂量
dosimeter 紫外线表
dot dash line 点划线
dot line 点线
dot matrix printer 点阵打印机【计】
dot matrix size 点阵精度【计】
dot product 内积,点积
dotted curve 点线
dotted-dashed line 点划线
dots per inch 点每英寸【计】
dots per second 点每秒【计】
double angle formula 倍角公式【数】
double-beam spectrophotometer 双波束分光计
double-cell magnetometer 双元〔铷蒸气〕磁强计
double correlation 双相关,二重相关
double cusp 双尖点
double diffusive convection 双扩散对流
double ebb 双低潮【海洋】
double-ended system 双端系统
double eye structure 双眼墙结构
double front 双锋
double integral 二重积分【数】
Double-Kelvin wave 双开尔文波
double layer 双层【物】
double mass analysis 流量对照分析
double-mass curve 双累积曲线
double precision 双精度
double refraction 双折射
double register 双寄存器
double scattering 双散射

double-slidewire bridge 双滑线电桥
double stranded DNA 双链 DNA【生化】
double stranded RNA 双链 RNA【生化】
double sunspot cycle 黑子〔活动〕双周期
double-theodolite technique 双经纬仪法
double-theodolite observation 双经纬仪观测【大气】
double-theodolite technique 双经纬仪技术
double tide 双潮
double tropopause 双对流层顶
double tube thermometer 双管温度表
double vortex thunderstorm 双涡〔旋〕雷暴
double wave 双波
doubling method 倍加法
doubly stochastic matrix 重随机矩阵,双随机矩阵
Dove's law 杜维定律
downburst 下击暴流【大气】
downburst cluster ①下击暴流团②下击暴流云团
downburst swath 下击暴流带
downdraft 下曳气流【大气】
downdraft velocity 下曳气流速度
downdraught (= downdraft) 下曳气流【大气】,下沉气流
downglide motion 下滑运动
downgradient 顺梯度
down-gradient diffusion 顺梯度扩散
downgradient flux 顺梯度通量
downgradient transfer 顺梯度输送
down-gradient transport 顺梯度输送
download 下载【计】
downpour 倾盆大雨
downrush 强泻气流
downscale direction 降尺度方向（尺度减小）
downscaling 降尺度
downshear 顺切变
downslide 下滑
downslide surface 下滑面

downslope 下坡运动
downslope wind 下坡风【大气】
downslop windstorm 下坡风暴
downstream 下游,顺水【航海】
downstream development 上游〔对下游发展的〕效应
downstream effect 上游效应(指上游长波系统的变化逐渐影响到下游环流形势)
downstream shear lobe 下游切变瓣
down valley wind 山风
downward atmospheric radiation 向下大气辐射【大气】
downward continuation 向下延拓
downward flow 下沉气流【大气】
downward motion 向下运动
downward〔total〕radiation 向下〔全〕辐射【大气】
downward terrestrial radiation 大气向下辐射
downward viewing spectral method 俯视光谱法
downwash ①向下输送②污染物下沉
downwelling 下降流【航海】
downwind ①下风,顺风 ②下风方
dpi 点每英寸【计】
DPP（disaster prevention and preparedness） 防灾抗灾
dps 点每秒【计】
DPT（delay picture transmission） 延时图像传输【大气】
draft 小股铅直气流
drag 抗力,曳力,阻力
drag acceleration 阻力加速〔度〕
drag anemometer 阻尼风速表
drag coefficient 拖曳系数【大气】,滞凝系数
drag law 曳力定律
dragon 水龙卷
dragon's tail 测温链
drag plate 阻力板(地面边界层测量用)

drag sphere anemometer 阻尼球风速表
drag theory 阻力理论
drain 疏水【电力】,排水沟
drainage 排水
drainage area 流域,流泄区
drainage basin 流域
drainage density 河网密度【水利】
drainage divide 分水岭【水利】,流域分界线
drainage evapotranspirometer 渗漏式蒸散计
drainage gauge 渗漏计
drainage flow 下泻气流
drainage network 河网,排水网
drainage well 排水井【电力】
drainage wind 流泄风【大气】
dramundan 帷幕〔状〕极光,北极光幕
draught 过堂风(见于印度南部库马尔角)
drawdown 水位下降,地下水位降落,抽水降深
drawdown curve 下降曲线,地下水位下降曲线,抽水降深曲线
drawing velocity 引曳速度
drawoff 〔水库〕放水,泄水
dredger 挖泥船【航海】
D-region D区
dreikanter 三棱石
dribble 微雨
dried ice 无〔覆〕水冰,干白冰
drift 漂移【物】,漂流
drift bottle 漂浮瓶
drift card 漂流卡(密封在漂流瓶内)
drift-correction angle 偏流订正角
drift current 漂流,偏流
drift epoch 漂浮期
drift fishing boat 漂流鱼船【航海】
drifter 漂流物
drift ice 流冰【海洋】,浮冰
drifting 漂航【航海】
drifting buoy 漂流浮标
drifting dust 低吹尘

drifting sand 低吹沙【大气】
drifting snow 低吹雪【大气】
drift meter 偏流计
drift sight 漂流测示器
drift station 漂浮观测站
driller's log 钻井记录〔曲线〕
drilling platform 钻井平台【航海】,海上平台
drimeter 湿度计,含水量测定仪
drip 点,滴
drisk 毛毛雨雾
driven snow 吹雪【大气】
driver ①驱动器,②驱动程序【计】
driving force 驱动力【物】
driving frequency 激励频率
driving function 驱动函数
driving mechanism 驱动机制
driving rain 大风雨
driving rain index 风吹雨指数
driving rain on buildings 建筑物风吹雨
driving snow 大风雪
drizzle 毛毛雨【大气】,细雨
drizzle drop 毛毛雨滴
drizzle type 毛毛雨型(指极光粒子沉降)
drizzling fog 毛雨雾
drogue 浮标,海锚【航海】
dromi 得罗密风(叙利亚沿海的风暴)
drop 滴,水滴
drop breakup 雨滴破碎
drop collector 水滴集电器(测空中电位梯度用)
droplet 微滴
droplet collector 滴谱仪
droplet spectrum 滴谱
droppable pyrotechnic flare system 投掷焰弹系统
dropping flare 投掷焰弹
drop settling chamber 沉降云室
drop-size distribution 滴谱
drop-size distribution parameter 滴谱参数【大气】
drop-size meter 滴谱仪
drop-size spectrometer 〔自动〕滴谱仪
dropsonde (D/P) 下投式探空仪【大气】
dropsonde observation 探空仪观测
drop spectrum 滴谱【大气】
drop theory 微滴学说
dropwindsonde 下投式测风探空仪
drosograph 露量计
drosometer 测露表【大气】,露量表【大气】
drought 干旱【大气】
drought aggravation 干旱发展
drought alleviation 干旱缓和
drought area 干旱面积
drought damage 旱灾【大气】
drought catastrophe 大旱
drought frequency 干旱频数【大气】
drought index 干旱指数【大气】
drought resistance 抗旱性,耐旱性
drought-resistant variety 抗旱品种
drought severity index 干旱强度指数
drought-striken region 受旱地区
drouth (=drought) 干旱
drown 淹没
droxtal 过冷水冰滴(直径约 10～20 μm)
drum (of self-recording instrument) 鼓,钟,筒(自记仪器的)
drum recording 滚动记录
dry adiabat 干绝热线
dry adiabatic 干绝热的
dry-adiabatic atmosphere 干绝热大气
dry adiabatic change 干绝热变化
dry adiabatic cooling 干绝热冷却
dry adiabatic dynamics 干绝热动力学
dry adiabatic lapse rate 干绝热直减率【大气】
dry adiabatic process 干绝热过程【大气】
dry adiabatic rate 干绝热率
dry adiabatic temperature change 干绝热温度变化
dry air 干〔燥〕空气

dry and warm lid 干暖盖
dry adiabatic warming 干绝热增温
dry-and-wet-bulb hygrometer 干湿球湿度表
dry bulb 干球
dry-bulb temperature 干球温度【大气】
dry-bulb thermometer 干球温度表【大气】
dry climate 干燥气候
dry cold front 干冷锋【大气】
dry convection 干对流【大气】
dry convection adjustment 干对流调整
dry damage 旱灾
dry deposition 干沉降【大气】
dry enthalpy 干焓
dry farming 旱农【地理】
dry fog 干雾【大气】
dry freeze 干冻【大气】
dry growth 干增长【大气】
dry haze 干霾【大气】
dry hot wind(=hot-arid wind) 干热风【大气】
dry ice 干冰【大气】
dry-ice seeding 干冰催化
drying power 干燥率
dry instability 干不稳定性
dry inversion 干逆温
dry line 干线【大气】
dry model 干模式
dryness 干燥
dryness and wetness grades series 旱涝等级序列
dry period 干期
dry permafrost 干永冻区
dry season 干季【大气】
dry slot 干楔
dry snow 干雪【大气】
dry spell 干期【大气】
dry stage 干燥阶段
dry static energy 干静力能量
dry static total energy 干静力总能量
dry subhumid climate 干次湿气候

dryth 干风(英国干燥的北风或东风)
dry tongue 干舌【大气】
dry wind 干风
dry year 缺水年,旱年
dsDNA 双链DNA【生化】
dsRNA 双链RNA【生化】
DSS (Decision Support System) 决策支持系统
DTM (digital terrain model) 数字地形模型【地理】
DU (Dobson unit) 陶普生单位【大气】,多布森单位
dual catastrophe 双重突变
dual-cell magnetometer 双元〔铷蒸气〕磁强计
dual-channel radar 双通道雷达【大气】
dual cusp 双尖点
dual-Doppler analysis 双多普勒分析【大气】
dual frequency phase locked receiver 双频锁相接收机
dual-frequency radar 双频雷达
duality 对偶〔性〕【电子】
duality theorem 对偶定理
dual polarization radar 双极化雷达,双偏振雷达【大气】
dual solution 对偶解【数】
dual sounding 双探空仪探空
dual space 对偶空间【物】
dual traverse barograph 双杆气压计
dual-wavelength 双波长
dual wavelength Doppler radar 双波长多普勒雷达
dual wavelength radar 双波长雷达【大气】
duct ①波导【大气】②导管,管道
ducting 波导
duff ①落叶层,半腐层②煤粉〔屑〕
dull weather 阴沉〔的〕天气
dumbbell depression 孪生低气压
dumbbell distribution 哑铃形分布

dummy argument 虚变元,哑元
dummy statement 空语句
dummy variable 虚〔假〕变量
Dunmore-type humidity sensor 邓莫尔型湿度传感器
duplex 双工【电子】
duplexer 双工器,天线共用器,天线〔收发〕转换开关
duplicatus (du) 复(云)
Dupuit-Forchheimer assumptions 杜普特-福希海默尔假定,D-F 假定
durability 持久性,持续性
duration 持续时间
duration curve 历时曲线【水利】
duration of frost-free period 无霜期【大气】
duration of growing period 生长期
duration of possible sunshine 可照时数【大气】
duration of sunshine 日照时间
duration statistics 历时统计
dusenwind 杜森风(达达尼尔海峡的山口风,由达达尼尔海峡吹到爱琴海的东东北强风)
dusk 晨昏
duskside magnetosphere 昏侧磁层
duskside tail 昏侧磁尾
dust 浮尘
dust avalanche 干雪崩【大气】
Dust Bowl 沙尘暴碗形区(1930 年代对美国中南部五州:科罗拉多、新墨西哥、堪萨斯、得克萨斯、俄克拉何马,因受干旱和沙尘暴影响而得的名称)
dust cloud 尘云
dust collector 集(吸)尘器
dust content 含尘量
dust counter 计尘器
dust devil 尘卷风【大气】
dust-devil effect 尘卷风效应
dust electrification 粉尘起电
duster (= duststorm) 尘暴,喷粉机【农】

dust extinction 尘消光
dustfall 降尘【大气】
dust fall jars 尘降器,沉降器(用重力沉降法清除空气中较大质粒的器件)
dust (= sand) fog 尘雾,沙尘降雾
dust-free atmosphere 无尘大气
dust-haze 尘霾
dust horizon 尘埃层顶【大气】
dust-ladden air 含尘空气
dust loading 尘埃浓度,尘埃含量
dust particle 尘粒
dust rain 尘雨,雨土
dust sampling 尘埃取样
dust shower 尘阵雨
dustsonde 尘埃探空仪
duststorm 沙尘暴【大气】,尘暴
dust turbidity 尘埃浑浊度
dust veil 尘幔
dust veil index 尘幔指数
dust wall 尘壁
dust well 尘坑
dust whirl 尘旋【大气】
dust wind 尘风
duty cycle 工作比,占空因数【电子】
duty factor 工作比,占空因数【电子】
duty of water 灌溉额
duty ratio 占空比,负载比【电子】
Duvdevani dew gauge 杜德瓦尼露量器,杜氏露量器
D-value D 值(即 $Z-Z_P$,某点等压面高度与 ICAO 标准大气的等压面高度之差)
DVB-S (digital video broadcast-satellite) 卫星数字电视广播
Dvorak technique 德沃夏克技术
dwigh (= dwey, dwoy) 德怀风暴(加拿大纽芬兰的一种阵雨暴或雪暴)
dye laser 染料激光器【物】
dynamical climatology 动力气候学
dynamical convection 动力对流
dynamical extended range forecasting

(DERF) 动力延伸预报
dynamical feedback 动力反馈
dynamical forecast 动力预报
dynamical forecasting 动力预报
dynamical meteorology 动力气象学
dynamical model 动力学模式
dynamical predictability 动力可预报性
dynamical structure 动力学结构
dynamical sublayer 动力底层,动力副层
dynamical system 动力系统【数】,动态系统【力】
dynamical tide 动力潮汐
dynamic analysis 动态分析【计】
dynamic anticyclone 动力性高压
dynamic balance 动力平衡
dynamic boundary 动力边界
dynamic boundary condition 动力边界条件
dynamic bifurcation 动态分岔
dynamic climatology 动力气候学【大气】
dynamic cloud seeding 动力播云【大气】
dynamic coefficient of viscosity 动〔力〕黏〔滞〕系数
dynamic component 动力分量
dynamic convection 动力对流
dynamic cooling 动力冷却
dynamic cumulus 动力积云
dynamic depth 动力深度
dynamic display 动态显示【计】
dynamic equilibrium 动态平衡,动力平衡
dynamic flux 动力通量
dynamic forecasting 动力预报
dynamic geomorphology 动力地貌学
dynamic head 动压头【航海】
dynamic heating 动力加热
dynamic height 动力高度【大气】
dynamic-height anomaly 动力高度距平
dynamic initialization 动力初值化【大气】,动力初值处理
dynamic instability 动力不稳定〔性〕

【大气】
dynamic kilometer 动力千米
dynamic lift 动力举力
dynamic low 动力〔性〕低压【大气】
dynamic map 动态地图【测】
dynamic meter 动力米【大气】
dynamic microphone 电动传声器
dynamic meteorology 动力气象学【大气】
dynamic mode 动态模【物】
dynamic model 动力模式
dynamic oceanography 动力海洋学
dynamic parameter 动力参数
dynamic pressure 动〔力〕压强,动压力
dynamic pressure perturbation 动压力扰动
dynamic property 动力特性
dynamic range 动态范围
dynamics 动力学【物】
dynamic seeding 动力播撒
dynamic similarity 动力相似【大气】
dynamics of the atmosphere 大气动力学
dynamic stability 动力稳定〔性〕【大气】
dynamic state 动态【航海】
dynamic temperature change 温度动力变化
dynamic trough 动力槽
dynamic unit 动力单位
dynamic updraft 动力上曳气流
dynamic viscosity 动力黏性【大气】
dynamic viscosity coefficient 动力黏性系数,动力黏滞系数
dynamic warming 动力增暖
dynamo action 发电机作用
dynamo layer 发电机层
dynamo theory 发电机理论(地磁的)
dyne 达因
dyris 狄里斯风(小亚细亚南岸的强陆风)
dyster(见 deaister) 戴斯特风暴
dystrophication 河湖污染

E

Eady wave 伊迪波
eager(=ragre) 涌潮
EAPE(evaporative available potential energy) 有效蒸发势〔能〕
early fallout 早期沉降
early frost 早霜
early frost hidden 黑霜
Early Medieval Cool Period 中世纪初寒冷期
early snow 初雪,早雪
early spring 早春
early storm 早期风暴
early subboreal climatic phase 亚北方早期〔气候〕
early summer 初夏
early warning 预警
early warning radar 预警雷达【电子】
early warning system 预警系统
earlywood 早材
earth ①地球【天】②地【电工】
earth-air current 地-空电流
earth-atmosphere radiation budget 地气辐射收支
earth-atmosphere system 地气系统
earth axis 地轴【地质】
earth bow shock 地球弓形激波
earth coordinate 地球坐标【船舶】
earth crust 地壳【地质】
earth-current 大地电流,泄地电流【电工】
earth-current storm 大地电流暴,地电暴
earth curvature correction 地球曲率订正
earth ellipsoid 地球椭球【测】,地球椭圆体【航海】
earth environmental sciences 地球环境科学
earth equatorial plane 地球赤道平面
earth figure 地球形状【地信】
earth-fixed coordinate system 地固坐标系【地信】

earth flow 泥石流
earth gravity 地心引力
earth hummock 土坝,土堤,土丘
earth inductor 地磁感应仪
earthing conduct 接地导体
earthing electrode 接地体
earthing reference point 接地基准点
earthing resistance 接地电阻
earth-ionosphere waveguide 地球电离层波导
earth light(=earth shine) 地光,地球反照
Earthly Branches 地支
earth-moon system 地-月系统
earth mound 土坝,土堤,土丘
earth observation satellite 对地观测卫星【地信】
Earth Observing System(EOS) 地球观测系统【电子】【大气】
earth pole 地极
earth physics 地球物理学
earthquake 地震
earthquake flood 地震洪水,海啸
earthquake intensity 地震烈度【地】
earthquake magnitude 震级【地】
earthquake wave 地震波
earth radiation 地球辐射
earth radiation belt 地球辐射带【大气】
earth radiation budget(ERB) 地球辐射收支,地球辐射平衡
earth radiation budget experiment(ERBE) 地球辐射收支试验【大气】
Earth Radiation Budget Satellite(ERBS) 地球辐射收支卫星【大气】
earth radiation budget scanning radiometer 地球辐射收支扫描辐射仪
earth remote sensing satellite 地球遥感卫星
earth remote sensing system 地球遥感系统

Earth Resources Observation System (EROS)　地球资源观测系统【地信】
Earth Resources Satellite(ERS)　地球资源卫星
Earth Resources Technology Satellite (ERTS)　地球资源技术卫星【大气】
earth revolution　地球公转
earth rotation　地球自转
earth rotation parameter　地球自转参数【天】
earth's axis　地轴
earth science　地球科学【地质】,地学
earth's climate system　地球气候系统
earth's environment　地球环境
earth shadow　地〔球〕影
earth shine（见 earth light）　地光,地球反照
earth's magnetic field　地磁场
earth's magnetic field line　地磁场〔磁力〕线
earth's magnetism　地磁
earth's surface　地面
earth stabilization　地球稳定
earth station　卫星地面接收站
earth stripe　接地带【电工】
earth structure　地球结构
earth surface　地球表面【地理】
earth's vorticity　地球涡度
earth-synchronous orbit　地球同步轨道
earth system science　地球系统科学
Earth System Science Partnership(ESSP)　地球系统科学联盟
earth temperature　地温
earth thermometer　地温表
earth tide　固体潮【地质】,地潮
earthwatch　地球监视
east (E)　东
East African Coast Current (= Somali Current)　东非沿岸流（即索马里流）
East African jet　东非急流
East African Meteorological Department (EAMD)　东非气象局
East African time　东非时间
East Antarctic Ice Sheet　东〔半球〕南极冰板块
East Asia major trough　东亚大槽【大气】
East Asia monsoon　东亚季风【大气】
East Atlantic pattern (EA)　东大西洋型
East Auckland Current　东奥克兰海流（新西兰）
East Australia Current　东澳大利亚海流
east by north　东偏北
east by south　东偏南
East Cape Current　东好望角海流
East China Sea　东海【海洋】
easterlies　东风带【大气】
easterly belt　东风带【大气】
easterly disturbance　东风带扰动
easterly jet　东风急流
easterly trough　东风槽
easterly wave　东风波【大气】
Eastern Standard Time　东部标准时间（美国）
East Greenland Current　东格陵兰海流
eastern hemisphere　东半球
East Iceland Arctic Current　东冰岛北极流
East Indian Current　东印度洋海流
East Indian winter jet　东印度洋冬季涌流
East Kamchatka Current　东勘察加海流
East Korea Warm Current　朝鲜东部暖海流
east longitude　东经
East Madagascar Current　东马达加斯加海流
east northeast (ENE)　东东北
east southeast (ESE)　东东南
eastward　向东
ebb　落潮
ebb and flow　涨落潮
ebb current　落潮流【航海】
ebb interval　落潮流落潮间隙

ebb stream 落潮流【航海】
ebb strength 落潮强度
ebb tide 落潮【航海】
Ebert ion-counter 厄柏离子计数仪
EBM（energy balance model） 能量平衡模式【大气】
ebullition 沸腾
eccentric angle 偏正角
eccentric anomaly 偏近点角
eccentricity 偏心率,离心率【数】
Eccentric-orbiting Geophysical Observatory（EGO） 偏心轨道地球物理观测卫星
ECD（electron capture detection） 电子俘获检测,电子俘获探测
echelon cloud 梯状云
echo 回波
echo amplitude 回波振幅
echo analysis 回波分析【大气】
echo box 回波箱【电子】
echo character 回波特征【大气】
echo complex 回波复合体【大气】
echo contour 回波等值线,回波廓线
echo depth 回波厚度【大气】
echo distortion 回波畸变【大气】
echo-free vault 无回波穹
echo frequency 回波频率（可指回波幅度的变动频率或回波信号的多普勒频率）
echo identification 回波识别
echo intensity 回波强度
echo overhang 悬垂回波
echo power 回波功率
echo pulse 回波脉冲
echo radiosonde 回答式探空仪
echo signal 回波信号
echo signal simulator 回波信号模拟器
Echosonde 回声探测仪（美国雷廷公司双基地声雷达产品名称）
echo sounder 回声探测器
echo sounding apparatus 回声探测器
echo synthetic chart 回波综合图【大气】
echo wall 回波墙【大气】

ecidioclimate 微生态气候
eclipse 食（日,月）
eclipse weather 日食天气
eclipse wind 食风
eclipse year 食年
ecliptic 黄道【大气】
ecliptic coordinate system 黄道坐标系【天】
ecliptic latitude 黄纬【天】
ecliptic longitude 黄经【天】
ecliptic pole 黄极【航海】
ECMWF（European Center for Medium-range Weather Forecasts） 欧洲中期天气预报中心
e-ε closure e-ε 阶闭合,方程-耗散率闭合
ecnephias 埃克内菲斯飑（地中海的飑或雷暴）
ecoactivity 生态活动【自然】
eco-agriculture 生态农业
ecocatastrophe 生态灾难【自然】
ecoclimate 生态气候【动物】
ecoclimate forecasting 生态气候预测
ecoclimatic adaptation 生态气候适应
ecoclimatic forecast 生态气候预报
ecoclimatology 生态气候学【大气】
ecocline 生态梯度【植物】
ecodistrict 生态小区【地理】
ecogenetics 生态遗传学【生】
ecogeography 生态地理学【地理】
ecological agriculture 生态农业【农】
ecological balance 生态平衡【地理】
ecological benefit 生态效益【地理】
ecological biochemistry 生态生物化学【生】
ecological capacity 生态容量【水产】
ecological climatology 生态气候学
ecological complex 生态综合体【生】
ecological condition 生态条件
ecological crisis 生态危机【农】
ecological design 生态设计【农】
ecological engineering 生态工程【生】
ecological equilibrium 生态平衡

ecological environment 生态环境【大气】
ecological environment quality 生态环境质量
ecological factor 生态因子【植物】
ecological genetics 生态遗传学【生】
ecological monitoring 生态监测
ecological optimum 生态最适度【生】
ecological process 生态过程
ecological protection 生态保护
ecological stability 生态稳定性【生】
ecological strategy 生态对策【生】
ecological succession 生态演替【生】
ecological system 生态系统【植物】
ecological threshold 生态阈值【生】
ecological water requirement 生态需水
ecology 生态学【大气】
ecology environment 生态环境
ecolyzer 电化学-氧化碳分析仪
Economic and Social Council（ECOSOC, UN） 经济社会理事会(联合国)
Economic and Social Council of Asia and Pacific（ESCAP, UN） 亚洲和太平洋经济社会理事会(联合国)
Economic Commission for Africa（ECA, UN） 非洲经济委员会(联合国)
Economic Commission for Asia and the Far East（ECAFE, UN） 亚洲和远东经济委员会(联合国)
Economic Commission for Europe（ECE, UN） 欧洲经济委员会(联合国)
Economic Commission for Latin America（ECLA, UN） 拉丁美洲经济委员会(联合国)
economic conditions 经济状况
economic cost 经济费用
economic decision 经济决策
economic effect 经济效益
economic geography 经济地理学【地理】
economic impact 经济影响
economic principle 经济原则
economics of stormwater 暴雨水经济学
economic utility 经济效益
economic valuation 经济评价
economic yield of aquifer 〔含水层的〕经济抽水率,〔含水层的〕安全抽水率
economy of radiation 辐射收支
ecopedology 生态土壤学
ecoregion 生态区域【地理】
ecosphere 生态圈【农】,生态界
ecosystem 生态系〔统〕【生】,生态系统【大气】
ecosystem balance 生态系统平衡
ecosystem diversity 生态系统多样性【生】
ecotone 生态过渡带,群落交错区【动物】
ecotope 生态区【地理】
ecotype 生态型
ectotoxin（exotoxin） 外毒素【生化】
edapho-climate condition 风土条件
edapho-climatic condition 土壤气候条件,风土条件
edaphology 耕作土壤学【土壤】
Eddington approximation 埃丁顿近似法
eddy ①涡旋【大气】②涡动【力】
eddy accumulation 涡流累积〔法〕
eddy advection 涡动平流【大气】
eddy available potential energy 涡动有效位能
eddy coefficient 涡动系数
eddy conduction 涡动传导
eddy conduction coefficient 涡动传导系数
eddy continuum 涡动连续
eddy conductivity 涡动传导率【大气】
eddy-correlation evaporation sensor 涡动相关蒸发传感器
eddy correlation 涡动相关【大气】
eddy correlation method 涡动相关法
eddy covariance 涡动协方差
eddy current 涡流
eddy current coefficient 涡流系数
eddy current effect 涡流效应
eddy diffusion 涡动扩散【大气】

eddy diffusion coefficient　涡动扩散系数
eddy diffusivity　涡动扩散率
eddy energy　涡动能量
eddy equation　涡动方程
eddy exchange coefficient　涡动交换系数
eddy field　涡动场
eddy flow　涡流,紊流
eddy flux　涡动通量【大气】
eddy flux parameterization　涡动通量参数化
eddy friction　涡动摩擦
eddy heat flux　涡动热通量
eddy kinetic energy　涡动动能【大气】
eddy moisture flux　涡动水汽通量
eddy momentum flux　涡动动量通量
eddy momentum transfer　涡动动量输送
eddy motion　涡旋运动,涡动
eddy resistance　涡动阻力
eddy shearing stress　涡动切应力【大气】
eddy sink　涡汇
eddy size　涡旋尺度
eddy source　涡源
eddy spectrum　涡谱
eddy stress　涡动应力
eddy-stress tensor　涡动应力张量
eddy structure　涡旋结构
eddy tornado　涡动龙卷
eddy transfer coefficients　涡动传输系数
eddy transport　涡动输送
eddy velocity　涡动速度
eddy viscosity　涡动黏滞率【大气】,涡动黏滞性系数
edge wave　边缘波【大气】
EEOF (extended EOF)　扩展的经验正交函数
effect　①作用②效应【数】
β-effect　β效应【大气】
effective accumulated temperature　有效积温【大气】
effective abstractions　有效吸水量,有效降雨损失量

effective acoustic center　有效声中心
effective aperture　有效孔径
effective area　有效面积
effective atmosphere　〔光学〕有效大气〔路径〕
effective atmospheric transmission　有效大气透射【地】
effective attenuation factor　有效衰减系数
effective band width　有效带宽
effective boundary layer　有效边界层
effective brightness temperature　有效亮度温度
effective detection radius　有效探测半径
effective detection range　有效探测范围
effective diffusion coefficient　有效扩散系数
effective discharges　有效流量
effective earth radius　有效地球半径【大气】
effective evapotranspiration　有效蒸散【大气】
effective field　有效场
effective flood warning　有效洪水预警
effective flux　有效通量
effective gust velocity　有效阵风速度
effective focal length (EFL)　有效焦距
effective growth energy　有效生长能量
effective head　净压头,有效压头【航海】
effective height of anemometer　风速表有效高度
effective isotropic radiated power(EIRP)　有效各向同性辐射功率
effective length　有效长度
effective liquid water content　有效液水含量
effective load　有效荷载
effective moisture　有效水分
effectiveness of precipitation　降水效率
effective nocturnal radiation　有效夜间辐射
effective outgoing radiation　有效向外辐射
effective permeability　有效渗透率,有效

渗透性
effective porosity 有效孔隙度
effective potential temperature 有效位温
effective power 有效功率
effective precipitable water 有效可降水分
effective precipitation 有效降水【大气】
effective pulse length 有效脉冲长度
effective radiation 有效辐射【大气】
effective radiation layer 有效辐射层
effective radius 有效半径
effective radius of the earth 地球有效半径
effective rainfall 有效雨量【农】
effective rainfall hyetograph 有效雨量过程线
effective rainfall intensity 有效降雨强度,有效雨强
effective ray 有效射线
effective receiver aperture 有效接收孔径
effective receiver area 有效接收面积
effective roughness length 有效粗糙〔度〕长度
effective sample size 有效样本数
effective section 有效断面,有效截面
effective size parameter 有效尺度参数
effective snowmelt 有效融雪量
effective span 有效量程
effective stack height 有效烟囱高度【大气】
effective stress 有效应力【力】
effective temperature 有效温度【大气】
effective temperature index 有效温度指数
effective tensor 有效张量
effective terrestrial radiation 有效地球辐射,有效辐射
effective turbulent flux 有效湍流通量
effective use of water resources 有效水资源利用
effective value 有效值
effective velocity 〔出流〕有效速度(地下水的)
effective visibility 有效能见度【大气】

effective wave 有效波【地】
effective wavelength 有效波长
effective wind direction 有效风向
effective wind speed 有效风速【大气】
efficiency 效率
efficient statistic 有效统计量
effluent 流出物,排放物,喷射物
effluent channel 排污河槽
effluent discharge limits 排污极限
effluent gas 废气
effluent limitations 〔出流〕排放限制,污水排放限制
effluent plume 排放烟羽
effluent seepage 坡面渗流,渗出流(地下水)
effluent standards 排污标准
effluent stream 潜水补给河,地下水补给河,出渗河
efflux velocity 射流速度
effusion velocity 泻流速度
e-folding time e折减时间【大气】(常指变量衰减到e分之一所需的时间)
Egnell's law 伊格尼尔定律
Egyptian wind 埃及风(埃及冬季常伴有雾和尘的西风)
EHF (extremely high frequency) 极高频(30 G~300 GHz)
EHF communication 极高频通信【电子】,毫米波通信
eigenfunction 本征函数,特征函数
eigenperiod 固有周期
eigenvalue 本征值【物】
eigenvalue problem 本征值问题,特征值问题
eigenvector 本征向量【数】
eight-inch rain gauge 八英寸雨量计(集雨口径 20.3 cm)
eighteen-degree water 18℃水(在大西洋马尾藻海形成的18℃水就是副热带模式水的实例)
Einstein coefficient 爱因斯坦系数

Einstein equation 爱因斯坦方程【数】
Einstein's summation notation 爱因斯坦求和符号（常用于大气湍流的研究）
eissero 艾西罗风（尼罗河三角洲多暖雨的东风，sirocco 风）
ejectable radiosonde 弹射探空仪
ejection chamber 弹射器,弹射仓（施放投掷式探空仪用）
Ekman boundary conditions 埃克曼边〔界〕条件
Ekman convergence 埃克曼辐合
Ekman depth 埃克曼厚度
Ekman displacement 埃克曼位移
Ekman flow 埃克曼流【大气】
Ekman layer 埃克曼层【大气】
Ekman number 埃克曼数
Ekman profile 埃克曼廓线
Ekman pumping 埃克曼抽吸【大气】
Ekman reversing water bottle 埃克曼颠倒采水器
Ekman spiral 埃克曼螺线【大气】
Ekman suction 埃克曼〔层〕抽吸作用
Ekman transport 埃克曼输送
Ekman turning 埃克曼旋转
elaidic acid 反油酸【生化】
elastic backscattering 弹性后向散射
elastic barometer 变形气压表,弹性气压表
elastic body 弹性体【物】
elastic collision 弹性碰撞【物】
El Chichon 厄尔奇琼（1982 年喷发的一座墨西哥火山）
elasticity 弹性【物】
elastic modulus 弹性模量【物】
elastic wave 弹性波【力】
elastic scattering 弹性散射【物】
E-layer E 层
electrical aerosol size analyzer 电学〔法〕气溶胶尺度分析仪
electrical anemometer 电传风速表
electrical breakdown 电击穿
electrical calibration 电定标
electrical charge 电荷
electrical conductivity 电导率
electrical despun antenna（EDA） 电子消旋天线
electrical discharge 放电
electrical field effect 电场效应
electrical hygrometer 电测湿度计
electrically-heated thermometer 电热式温度表（风速测量仪器）
electrically scanning microwave radiometer（ESMR） 电子扫描微波辐射仪
electrical log 电测〔钻井〕记录曲线,电测井图
electrical mobility analyzer 电子迁移率分析仪
electrical phenomenon 电学现象
electrical storm 雷暴（thunderstorm 的英文俗称）
electrical substitution radiometer 电测辐射表
electrical switch 电子开关
electrical system 电气系统
electrical thermometer 电测温度表
electric breakdown 电击穿
electric-capacity moisture meter 电容式含水量测定仪
electric charge 电荷
electric conductivity ①电导率②导电性
electric conductivity raingauge 水导〔式〕雨量计【大气】
electric conductor 导〔电〕体
electric cup anemometer 电传风杯风速表
electric current 电流
electric currents in the atmosphere 大气电流
electric dipole transition 电偶极跃迁【物】
electric dipole 电偶极子【物】
electric discharge 放电
electric displacement 电位移【物】
electric disturbance 电扰

electric double layer 电偶层【物】
electric field 电场
electric field intensity 电场强度
electric field strength 电场强度
electric flux 电通量【物】
electric force 电力
electric hot bed 电热温床
electricity 电
electricity of precipitation 降水电学
electric lines of force(＝electric field line) 电力线
electric polarization 电极化
electric potential 电位【电子】,电势【物】
electric potential gradient 电位梯度
electric resistance thermometer 电阻温度计
electric thermometer 电测温度表
electrification 起电〔作用〕
electrification of ice nucleus 冰核带电
electrization 起电,带电
electroacoustics 电声学
electro-acoustic wave 电声波
electrochemical detector(ECD) 电化学检测器
electrochemical sonde 电化学探空仪(测臭氧用)【大气】
electrochemiluminescence 电化学发光【物】
electrochemistry 电化学【化】
electrode effect 电极效应
electrogram 大气电场自记曲线
electrohydrodynamics 电流体力学【物】
electrojet 电集流【地】【大气】,电急流(电离层中)
electroluminescence 电致发光【物】
electrolytic capacitor 电解电容器
electrolytic hydrogen generator 电解水制氢设备
electrolytic hydrogen plant 电解水制氢装置

electrolytic hygrometer 电解式湿度表
electrolytic thermometer 电解式温度表
electrolytic strip 电解片(测湿用)
electromagnetic acoustic probe 电磁-声探测器
electromagnetic compatibility(EMC) 电磁兼容【电子】,电磁兼容性
electromagnetic coupling 电磁耦合
electromagnetic echo sounding 电磁回波探测
electromagnetic energy 电磁能
electromagnetic equation 电磁方程
electromagnetic field 电磁场
electromagnetic induction 电磁感应【物】
electromagnetic interference 电磁干扰
electromagnetic radiation 电磁辐射
electromagnetics 电磁学【物】
electromagnetic screen 电磁屏蔽
electromagnetic shielding 电磁屏蔽
electromagnetic spectrum 电磁〔波〕谱
electromagnetic theory 电磁理论
electromagnetic transducer 电磁传感器
electromagnetic wave 电磁波
electromagnetism 电磁学【物】
electrometeor 大气电学现象
electrometer 静电计【物】
electron 电子【电子】
electron affinity 电子亲合势【电子】
electron atmosphere 电子云
electron attachment 电子附着
electron aurora 电子极光
electron avalanche 电子雪崩
electron beam 电子束【电子】,电子注
electron capture detection 电子俘获检测,电子俘获探测
electron collision frequency 电子碰撞频率
electron concentration 电子浓度
electron cooling rate 电子冷却率
electron density 电子密度

electron density profile 电子密度廓线
electron detector(MEPED) 中能质子和电子检测器
electron-electron energy transfer 电子间能量传输
electron energy balance 电子能量平衡
electron event 电子事件
electron heating rate 电子加热率
electronic charge 电子电荷
electronic commerce 电子商务【计】
electronic computer 电子计算机【计】,电脑
electronic current 电子流
electronic excitation 电子激发
electronic library system 电子资料库系统
electronic mail 电子信函【电子】,电子邮件
electronic psychrometer 电子干湿表
electronic publishing system 电子出版系统【计】
electronic readout equipment 电子读出装置
electronic signature 电子签名【计】
electronic temperature recorder 电子温度自记仪
electronic theodolite 电子经纬仪
electronic thermometer 电子干湿表
electronic transitions 电子跃迁
electronic weather station 电子气象站
electronic wind direction indicator 电子风向显示器
electron impact cross section 电子碰撞截面
electron-ion recombination 电子-离子复合
electron microscope 电子显微镜【物】
electron number density 电子数密度
electron optics 电子光学【物】
electron plasma frequency 电子等离子体频率
electron plasma oscillation 电子等离子体振荡

electron-positron collider 正负电子对撞机【物】
electron precipitation 电子沉降
electron production rate 电子产生速率
electron spectroscopy 电子能谱学
electron spin resonance 电子自旋共振【物】
electron synchrotron 电子同步加速器【物】
electron temperature profile 电子温度廓线
electron thermal conductivity 电子导热性
electron-volt(eV) 电子伏特
electro-optical meteorology 光电气象学
electro-osmosis 电渗〔透〕〔作用〕,电析
electrophoresis 电泳【化】
electroplating 电镀【化】
electropsychrometer 电测干湿表
electroscope 验电器【物】
electrosensitive〔recording〕paper 电敏〔记录〕纸
electrosonde 电要素探空仪
electrosphere 导电层(由几十千米高度延伸至电离层)
electrostatic aerosol sampler 静电式气溶胶取样器
electrostatic capacity 静电容
electrostatic charge 静电荷
electrostatic coalescence 静电并合
electrostatic earthing system 静电接地系统
electrostatic electrometer 静电计
electrostatic field 静电场【物】
electrostatic induction 静电感应【物】
electrostatic precipitator 静电沉降器
electrothermostat 电恒温器
element 要素,元素,元量,元件
elemental vortex 元量涡旋
elementary area 单位面积
elementary fluid mechanics 普通流体力学
elementary reaction 元反应【化】

elementary storm 元磁暴
elementary substance 单质【化】
elementary synoptic process 基本天气过程
elementary synoptic situation 基本天气形势
elementary volume 单元体积
elements of surface hydrology 地表水文学要素
elephanta 爱烈芬塔暴风雨(在印度马拉白沿海 9～10 月间西南季风终了时常发生伴有猛烈南风或东南风的暴雨)
elerwind 埃勒风(奥地利提罗耳的森谷中的风)
elevated convection 抬升对流
elevated echo 雨幡回波【大气】
elevated point source 高架源
elevated rain gauge 高架雨量计
elevated temperature psychrometer 增温式干湿表
elevation 高度,高程,仰角
elevation angle 仰角
elevation capacity curve 高程-库容关系曲线,库容曲线
elevation finder 测高仪
elevation of ivory point 象牙针高程(气压表的)
elevation of the zero point of barometer 气压表零点高度
elevation position indicator (EPI) 仰角位置指示器
elevation profile 垂直廓线
ELF (extremely low frequency) 极低频 (频率范围 3～30 Hz 之间)
ELF emission 极低频发射
Eliassen-Palm flux (E-P flux) E-P 通量【大气】,伊莱亚森-帕尔姆通量
elimination by substitution 代入消元法【数】
elimination method 消去法

ellipse 椭圆
ellipsoid of gyration 回转椭球
elliptical coordinates 椭圆坐标
elliptical depolarization ratio 椭圆退偏振比,椭圆退极化比
elliptical orbit 椭圆轨道
elliptic function 椭圆函数【数】
elliptic integral 椭圆积分【数】
elliptic polarization 椭圆偏振
elmo's fire (见 corposant) 爱尔摩火花
El Nino 厄尔尼诺【大气】
El Nino effect 厄尔尼诺效应
El Nino year 厄尔尼诺年
elongation 拉长,伸长
elongation stage 拔节期,伸长期
Elsasser absorption band 埃尔萨瑟吸收带
Elsasser's radiation chart 埃尔萨瑟辐射图
elve 气辉闪〔光〕
elvegust (见 sno) 斯诺飑
EMAC probe 电磁-声探测器
emagram 埃玛图【大气】,温度对数压力图
emanation 发射【计】,射气【化】
emanometer 测氡仪,射气仪
embata 恩巴塔风(大西洋加那利群岛背风坡的局地西南风)
embed cumulus 隐嵌积云
embed thunderstorm 隐藏性雷暴
embryo 胚,胚胎,胚粒
embryo curtain 胚胎帷幕
embryo formation region 雹胚生成区
EMC (electromagnetic compatibility) 电磁兼容【电子】,电磁兼容性
emergence angle 出射角【物】
emergency alerting network 紧急警戒站网
emergency drought impact areas 临时干旱影响区
emergency flood mitigation 应急减洪
emissary sky 预兆〔性〕天空

emission ①发射,放射②喷射,排放
emission factor 排放因子,排放系数
emission inventory 排放清单
emission layer 发射层
emission limit 排放限度
emission line 发射谱线
emission per capita 人均排放〔量〕
emission rate 排放率【大气】
emission spectroscopy 发射光谱学
emission spectrum 发射谱
emission standard 排放标准
emission trading (ET) （碳）排放交易
Emissions Trading Scheme (ETS) 排放贸易方案,排放贸易体制
emission theory 微粒说
emissive power 发射强度
emissivity 发射率【物】,比辐射率
emittance (M_e) 辐〔射〕出〔射〕率（W/m^2）
emittance of the earth's surface 地表辐射出射率
empirical coefficient 经验系数
empirical constant 经验常数
empirical curve 经验曲线【数】
empirical data 经验数据
empirical discriminant function 经验判别式函数
empirical eigenvector 经验特征向量
empirical expression 经验表达式
empirical flood formula 经验洪水公式
empirical formula 经验公式【数】
empirical mode decomposition 经验模态分解
empirical orthogonal function (EOF) 经验正交函数【大气】,自然正交函数
empirical regional relation 经验区域相关
empirical relationship 经验关系
empirical return periods 经验重现期
empirical risk minimization (ERM) 经验风险最小化
Empty Quarter 无人区
encipherning 加密【电子】

encode 编码
encoding method 编码方法【计】
encroachment method 浸和方法
end gas 尾气
endergonic reaction 吸能反应【生化】
ending of flood 洪水终止
ending of ice sheet 冰层消失
ending of Meiyu 出梅【大气】
ending of precipitation event 降水事件结束
ending of snow cover 积雪消失
ending of snowmelt 融雪终止
End of Heat 处暑【大气】
end of storm oscillation (EOSO) 雷暴电振荡终结
endorheic lake 内陆湖,内流湖
endothermic reaction 吸热反应
end-use efficiency 使用效率
energetics 力能学【大气】,能量学
energetics method 能量学方法
energetics of stratosphere 平流层能量学
energetics of the atmosphere 大气能量学
energy 能【化】,能量【物】
energy absorption 能量吸收
energy and enstrophy conserving model 能量和拟能守恒模式
energy balance 能量平衡
energy balance climate models 能量平衡气候模式
energy balance climatology 能量平衡气候学
energy balance equation 能量平衡方程
energy balance model (EBM) 能量平衡模式【大气】
energy band 能带【物】
energy budget 能量收支,能量平衡
energy cascade 能量串级【大气】
energy conservation 能量守恒【物】
energy-conserving model 能量守恒模式
energy conserving scheme 能量守恒格式
energy-containing eddies 含能涡旋
energy conversion 能量转换

energy conversion equation 能量转换方程
energy crisis 能源危机
energy cycle 能量循环
energy density 能量密度
energy density spectrum 能量密度谱【大气】
energy diagram 能量图
energy dispersion 能量频散,能量色散
energy dispersive spectrometer 散能分光计
energy dissipating rate 能量耗散率
energy dissipation 能量耗散
energy equation 能量方程
energy exchange 能量交换
energy flow 能流
energy flux 能流【物】,能量通量
energy flux density 能流密度【物】
energy front 能量锋
energy frontal zone 能量锋区
energy grade line 能坡线
energy identity 能量恒等式【数】
energy inequality 能量不等式【数】
energy level 能级【物】
energy level diagram 能级图【物】
energy loss 能量损失
energy method 能量法
energy norm 能量模
energy resolution 能量分辨率【物】
energy sources meteorology 能源气象学【大气】
energy spectral density 能[量]谱密度
energy spectrum 能谱【物】【大气】
energy sphere 能量球面
energy storage 能量积聚
energy transfer 能量传递
energy transformation 能量转换
energy transport 能[量]输送
engineering hydrology 工程水文学
engineering seismology 工程地震【地质】
engine-exhaust trail(=exhaust trail) 排气尾迹
enhanced cloud picture 增强云图【大气】
enhanced greenhouse effect 增强温室效应

enhanced image 增强云图
enhanced imagery 增强图像
enhanced infrared satellite cloud picture 增强显示红外卫星云图
enhanced mode 增强模式【地信】
enhanced network 增强网络【天】
enhanced picture 增强云图【大气】
enhanced satellite cloud picture 增强显示卫星云图
enhanced "V" 增强 V 型（卫星红外图像）
enhancement curve 增强曲线（处理数字图像用）
EnKF(ensemble Kalman filter) 集合卡尔曼滤波
enlarged partial-disc picture 部分放大[云]图
enormous flood 特大洪水
enrichment 富集【土壤】,浓缩
enrichment factor 富集因子
en route report 航线天气报告
en route weather 航线天气
ensemble 系综【物】,总体,群
ensemble average 集合平均【大气】
ensemble forecast 集合预报【大气】
ensemble Kalman filter (EnKF) 集合卡尔曼滤波
ensemble prediction system (EPS) 集合预报系统
ensemble spread ①系集离散②集合预报平均差值（集合离差）
ensemble theory 系综理论
ENSO(El Nino 与 Southern Oscillation 合成的缩略词) 恩索【大气】
ENSO event 恩索[天气]事件
ENSO indicator 恩索指示器
ENSO indices 恩索指数
ENSO phenomenon 恩索[天气]现象（由厄尔尼诺、拉尼娜和相关天气现象组成）
enstrophy 涡度拟能【大气】

enstrophy cascade 拟能串级
enstrophy conserving scheme 涡度拟能守恒方案
enstrophy norm 涡度拟能模
enthalpy 焓【物】
entity-type entrainment mixing 独立实体夹卷混合
entrainment ①夹卷【大气】②卷吸【海洋】
entrainment coefficient 夹卷系数
entrainment rate 夹卷率【大气】
entrainment velocity 夹卷速度
entrainment zone 夹卷区〔域〕
entrance region 入口区
entropy 熵【物】
entropy balance equation 熵平衡方程
entropy cascade 熵串级
entropy criterion 熵判据
entropy function 熵函数
entropy of information 信息熵【数】
entropy production 熵增量
envelope ①球皮②包络③波包
envelope function 波包函数
envelope hole 波包穴
envelope method 包络法
envelope shock 波包激波
envelope soliton 包络孤立子
envelope topography 包络地形
envelope velocity 包络速度,群速度
environics 环境学
environment 环境
environmental abnormality 环境异常【地理】
environmental acoustics 环境声学
environmental analysis 环境分析
environmental anomaly 环境异常
environmental appraisal 环境评价【农】,环境鉴定
environmental assimilating capacity 环境同化能力
environmental atmospheric quality monitoring 〔环境〕大气质量监测【大气】
environmental background 环境本底
environmental capacity 环境容量
environmental chamber 环境模拟室
environmental chemistry 环境化学
environmental climate 环境气候
environmental climatology 环境气候学【大气】
environmental condition 环境条件
environmental control 环境防治
environmental data 环境资料
Environmental Data and Information Service (EDIS) 环境资料信息部(NOAA)
environmental data buoy 环境资料浮标
Environmental Data Service (EDS) 环境资料局(美国)
environmental degradation 环境退化【地理】
environmental design 环境设计【地理】
environmental deterioration 环境恶化【农】
environmental disaster 环境灾难
environmental ecology 环境生态学
environmental effect 环境效应【地理】
environmental element 环境要素【地理】
environmental engineering 环境工程
environmental evaluation 环境评价【地质】
environmental event 环境事件
environmental evolution 环境演化【地理】
environmental factor 环境因子
environmental forecasting 环境预报【农】
environmental geochemistry 环境地球化学【地理】
environmental geography 环境地理学
environmental gradient 外界梯度
environmental greening 环境绿化【林】
environmental impact assessment 环境影响评价【地理】
environmental lapse rate 环境直减率

【大气】
environmental load 环境荷载
environmentally friendly 环境友好的
environmental mdicine 环境医学
environmental meteorology 环境气象学
environmental monitor 环境监测
environmental pollutant 环境污染物
environmental pollution 环境污染
Environmental Prediction Research Facility (EPRF) 环境预报研究中心(美国海军)
environmental protection 环境保护
Environmental Protection Agency (EPA) 美国环境保护局
environmental quality 环境质量【地理】
environmental quality comprehensive evaluation 环境质量综合评价【地理】
environmental quality evaluation 环境质量评价
environmental quality index 环境质量指数
environmental quality parameter 环境质量参数
environmental remote sensing 环境遥感【地信】
Environmental Research Laboratories (ERL, NOAA) 环境研究院(NOAA)
environmental resources 环境资源【农】
environmental risk 环境风险
environmental risk assessment 环境风险评价
environmental satellite 环境卫星
environmental sciences 环境科学
Environmental Science Services Administration (ESSA) 环境科学服务局
environmental self-purification 环境自净
environmental sensitivity 环境敏感性
environmental standard 环境标准
environmental stress 环境胁迫
environmental suitability 环境适应性

Environmental Survey Satellite (ESSA) 环境勘测卫星,艾萨卫星【大气】
environmental system 环境系统
environmental temperature 环境温度
environmental tracer 环境示踪剂,环境示踪物
environmental wind 环境风
environment flow 环境气流
enzyme 酶
Eocene Epoch 始新世【地质】
Eocene Series 始新统【地质】
EOF (empirical orthogonal function) 经验正交函数
eolation 风蚀〔作用〕
EOLE 风神(法国气象卫星名)
eolian action 风成作用
eon 宙【地质】
eonothem 宇【地质】
EOS (Earth Observing System) 地球观测系统【电子】【大气】,对地观测系统
Eozoic Era 始生代
epeirogenesis 造陆作用【地质】
epeirogeny 造陆运动【地质】
E-P flux (Eliassen-Palm flux) E-P通量,伊莱亚森-帕尔姆通量
ephemeral data 探测情景资料,补充资料
ephemeral lake 季节性湖泊
ephemeral rain 短暂降雨
ephemeral stream 季节性河流
ephemeris 历表【天】,星历【航空】
epicenter 震中
epicenter distribution 震中分布【地】
epicenter intensity 震中烈度【地】
epicycle ①本轮【天】②间升期
epifocus 震中【地】
epigeosphere 地球表层【地理】
epilimnion 温度跃层,变温水层
episode 节,段
epipycnal 等密度方向
epipycnal mixing 等密度面混合
episode criteria 〔污染〕事件标准

episodes of extreme rainfall 极端降水事件
epitaxy(＝epitaxis) 外延附生【物】
epoch 世【地质】,历元【天】
epoxide 环氧化合物【化】
Eppley normal incidence pyrheliometer 埃普利垂直入射绝对日射表
Eppley pyranometer 埃普利天空辐射表
Eppley pyrheliometer 埃普利绝对日射表
EPS (ensemble prediction system) 集合预报系统
equal altitude circle 等高度圈
equal area chart 等面积图
equal-area map 等面积地图
equal area transformation 等面积变换
equality 相等,等同,等式【数】
equalization 均衡
equal precipitation 等雨量
equal rainfall depth 等雨深,等雨量
equal surface projection 等面积投影
equal thickness interference 等厚干涉【物】
equation 方程【数】
ω-equation ω方程【大气】
equation for average field 平均场方程
equation of barotropy 正压方程
equation of constrain 约束方程
equation of continuity 连续方程
equation of dynamics 动力〔学〕方程【数】
equation of motion 运动方程【大气】
equation of piezotropy 压性方程
equation of radiation transfer 辐射传输方程
equation of state 状态方程【大气】
equation of static equilibrium 静力平衡方程
equation of thermal wind 热成风方程
equation of time 时差【天】
equation of transfer 传输方程
equation of vorticity 涡度方程
equator 赤道
equatorial acceleration 赤道加速度

equatorial air mass 赤道气团【大气】
equatorial anticyclone 赤道反气旋
equatorial beta-plane 赤道β平面
equatorial buffer zone 赤道缓冲带【大气】
equatorial bulge 赤道隆起【天】
equatorial calm belt 赤道无风带
equatorial calms 赤道无风带【大气】
equatorial cell 赤道〔铅直〕环流
equatorial climate 赤道气候【大气】
equatorial climate zone 赤道气候带
equatorial climatic region 赤道气候区
equatorial cold tongue 赤道冷舌
equatorial continental air mass 赤道大陆气团,赤道大陆空气
equatorial convergence belt 赤道辐合带【大气】
equatorial convergence zone 赤道辐合带
equatorial counter current 赤道逆〔海〕流
equatorial current 赤道流【海洋】
equatorial current system 赤道洋流系统
equatorial day 赤道日
equatorial deep jets 赤道深〔层急〕流
equatorial dry zone 赤道干旱带
equatorial easterlies 赤道东风带【大气】
equatorial electrojet 赤道电集流【地】
equatorial front 赤道锋
equatorial frontal zone 赤道锋区
Equatorial Intermediate Current 赤道中层洋流
equatorial jet 赤道急流
equatorial Kelvin wave 赤道开尔文波
equatorial low 赤道低压【大气】
equatorially-trapped wave 赤道陷波
equatorial maritime air mass 赤道海洋气团,赤道海洋空气
equatorial meteorology 热带气象学,赤道气象学
equatorial orbiting satellite 赤〔道〕轨〔道〕卫星
equatorial plane 赤道平面

equatorial Poincare wave 赤道庞加莱波【海洋】
equatorial precipitation belt 赤道多雨带
equatorial radius of deformation 赤道变形半径
equatorial rain forest 赤道雨林【大气】
equatorial rain forest climate 赤道雨林气候
equatorial rain forest ecosystem 赤道雨林生态系统
equatorial ring current 赤道环电流
equatorial stratospheric wave 赤道平流层波
equatorial tidal current 赤道潮流
equatorial tide 赤道潮汐
equatorial trough 赤道槽【大气】
Equatorial Under Current(EUC) 赤道潜流
equatorial upwelling 赤道涌升流
equatorial vortex 赤道涡旋
equatorial water 赤道水
equatorial wave 赤道波
equatorial waveguide 赤道波导
equatorial westerlies 赤道西风带【大气】
equatorial wind 赤道风
equatorial zone 赤道带【地理】
equiangular spiral 等角螺线
equidensen 等密度面
equideparture 等距平
equidiurnal effect 等日效应
equigeopotential surface 等重力位势面
equigravisphere 等引力带
equilateral triangle 等边三角形【数】
equilibrant 平衡力
equilibrant convection 平衡对流
equilibrant moment 平衡力矩
equilibration 平衡过程
equilibration time 平衡时间
equilibrium 平衡
equilibrium altitude 平衡高度(气球)
equilibrium climate 平衡态气候

equilibrium climate response 平衡态气候响应
equilibrium condition 平衡条件【力】
equilibrium condition of phase transition 相变平衡条件【物】
equilibrium drawdown 平衡降深(井抽地下水位)
equilibrium line 平衡线(冰川的)
equilibrium-line altitude 平衡线高度
equilibrium line of glacier 冰川〔积融〕平衡线
equilibrium paraboloid 平衡抛物面
equilibrium range 平衡区
equilibrium solar tide 平衡太阳潮
equilibrium spheroid 平衡球体
equilibrium state 平衡态
equilibrium system 平衡系统
equilibrium temperature 平衡温度
equilibrium theory 平衡理论
equilibrium tide 平衡潮
equilibrium vapour pressure 饱和水汽压
equilibrium well discharge 均衡井出水量
equinoctial colure 二分圈
equinoctial gale(=storm) 二分点风暴(春、秋分日附近常见的大风)
equinoctial rain 二分〔点〕雨【大气】
equinoctial storm 二分点风暴(英国和北美)
equinoctial storm effect 二分点风暴效应
equinoctial tide 分点潮【海洋】
equinoxes 二分点【大气】(春分点和秋分点)
equiparte(=equipatos) 伊奎帕特雨(墨西哥10—1月期间连续多日的大雨)
equipartition 均分
equipartition law 均分定律,匀布定律
equipluves 等雨量线【农】
equipment 设备,装置
Equipment Development Laboratory (EDL, NWS) 气象设备发展实验室

（美国）
equipotential bonding 等电位连接
equipotential bonding bar 等电位连接带
equipotential bonding network 等电位连接网
equipotential line 等势线【物】，等位线【电子】
equipotential surface 等势面【物】，等位面【电子】
equipotential surface of gravity 重力等位面【地】
equipotential temperature 等位温
equipressure 等压线
equipressure surface 等压面
equiscalar line 等值线
equiscalar surface 等值面
equisubstantial surface 等质面
equivalence principle 等效原理【物】
equivalence statement 等价语句
equivalence theorem 等价定理
equivalent ①等值，等价，等效 ②当量
equivalent altitude of aerodrome 等效机场高度【大气】
equivalent band width 等效带宽
equivalent barotropic atmosphere 相当正压大气【大气】
equivalent barotropic level 相当正压高度
equivalent barotropic model 相当正压模式【大气】
equivalent blackbody temperature (TBB) 等效黑体温度，辐射亮温
equivalent clear column radiance 等效晴空辐射率【大气】
equivalent constant wind 弹道风
equivalent cross section 等效截面【电子】
equivalent depth 相当深度
equivalent diameter 等效直径
equivalent duration 等效风时
equivalent fetch 等效风区

equivalent foot-candle 等效英尺-烛光
equivalent gradient wind 等效梯度风
equivalent gravity waves 等效重力波
equivalent head-wind 等效逆风
equivalent height 等值高度
equivalent isotropically radiated power (EIRP) 各向同性等效辐射功率
equivalent longitudinal wind 等效纵向风
equivalent normal incidence frequency 等值正射频率
equivalent of heat 热当量
equivalent path 等效路径
equivalent potential temperature 相当位温
equivalent precipitation 当量雨量，等效雨量
equivalent radiant temperature 等效辐射温度【电子】
equivalent radiative atmosphere 相当辐射大气
equivalent rain gauge density 等效雨量计密度
equivalent reflectivity factor 等效反射率因子
equivalent stationary rainstorm (= equivalent stationary storm) 等效静止暴雨
equivalent storms 等效暴雨
equivalent tail wind 等效顺风
equivalent temperature 相当温度【大气】
equivalent topography 等效地形学
equivalent width 等价宽度
ERBE (Earth Radiation Budget Experiment) 地球辐射收支试验【大气】
ERBS (Earth Radiation Budget Satellite) 地球辐射收支卫星【大气】
E-region E区
eremium 荒漠群落
eremophyte 荒原植物
erf(= error function) 误差函数
erg 尔格(功的旧单位)
ergodic condition 遍历条件

ergodicity 遍历性【数】,各态历经性
ergodic process 遍历过程【数】
ergodic system 遍历系统
ergonomics 工效学
ERM(empirical risk minimization) 经验风险最小化
eroded field 侵蚀区
EROS(Earth Resources Observation System) 地球资源观测系统【地信】
erosion 侵蚀〔作用〕【大气】,冲刷
erosion ridge 风蚀雪波
erratic 不规则性
erratic path 不规则路径
error 误差【数】
errror anahysis 误差分析【数】
error check 差错校验【数】
error contamination 误差污染,误差混淆
error correction 误差校正【数】,误差订正
error covariance 误差协方差
error covariance matrix 误差协方差矩阵
error-detecting code 误差检测电码
error detection 误差检验
error diagnosis 误差诊断
error distribution 误差分布【数】
error equation 差错方程【数】,误差方程
error estimate 误差估计【数】
error function 误差函数
error of digitization 数字化误差
error of interpolation 内插误差
error of mean square 均方误差
error of observation 观测误差
error of reading 读数误差
error propagation 误差传播【数】,差错传播【数】
error rate 差错率【数】
error variance 误差方差
ERSS(European Remote Sensing Satellites) 欧洲遥感卫星【大气】
Ertel potential vorticity E位势涡度,埃尔特尔位势涡度

ertor 欧托(臭氧层有效〔辐射〕温度)
ERTS(Earth Resources Technology Satellite) 地球资源卫星
erythemal dosimeter 红斑辐射剂量仪
erythemal spectrum 红斑谱(2900～3250 埃)
erythemal spectrum dosimeter 红斑谱辐射剂量仪
escape depth 逃逸深度【物】
escape radiation 外逸辐射
escape rate 外逸速率
escape speed 逃逸速度
escape velocity 外逸速度
Espy-Köppen theory 埃斯皮-柯本学说
ESSA(Environmental Survey Satellite) ①艾萨卫星【大气】②(Environmental Science Services Administration)环境科学服务局
essential amino acid 必需氨基酸【生化】
ESSP(Earth System Science Partnership) 地球系统科学联盟
estegram 埃斯特图
esterly wave 东风波
estimate 估计
estimated ceiling 估测云幂,估计云幂
estimate value 估计值
estimator 估计量,估计算子
estival(=aestival) 夏季的,夏令
estuary 河口【航海】,河口湾【地理】
ET(emission trading) （碳）排放交易
ETA(estimated time of arrival) 预计到达时间
eternal frost climate 永冻气候
eternal snow 粒雪
eternal water 永存水,原生水
Etesian climate 地中海气候
Etesians 地中海季风(地中海东部夏天的盛行偏北风)
e-text 电子文本【计】
ethane 乙烷(C_2H_6)

ethene 乙烯(C_2H_4)
ether ①以太【物】②醚【化】,乙醚
Ethernet 以太网【计】
ether wave 以太波
ether wind 以太风【物】
ethylene 乙烯,乙烯基
ethyne 乙炔(C_2H_2)
ETS(Emissions Trading Scheme) 排放贸易方案
e-type absorption e型吸收(双水分子吸收)
EU(European Union) 欧盟
EUC(Equatorial Undercurrent) 赤道潜流
Euclidean distance 欧几里得距离
Euclidean geometry 欧几里得几何〔学〕【数】,欧氏几何
Euclidean space 欧几里得空间【数】
EUCOS(European Union Composite Observing System) 欧洲综合观测系统
eugenics 优生学【生】
Euler backward scheme 欧拉后差格式
Eulerian change 欧拉变换
Eulerian coordinates 欧拉坐标【大气】
Eulerian correlation 欧拉相关
Eulerian current measurement 欧拉测流法
Eulerian equations 欧拉方程
Eulerian mean formulation 欧拉平均公式
Eulerian motion 欧拉运动
Eulerian time derivative 欧拉时间导数
Eulerian wind 欧拉风
Euler number 欧拉数【数】
Euler-Lagrange differential equation 欧拉-拉格朗日微分方程【数】
eupatheoscope 热损失估测仪
euphotic zone 光亮带,透光带
euraquilo(= euroaquilo,euroclydon) 尤拉奎洛风(阿拉伯和近东的暴风)
Eurasian continent 欧亚大陆
Eurasian pattern 欧亚型

eurithermic 适广温的(生物)
European Center for Medium-range Weather Forecasts (ECMWF) 欧洲中期天气预报中心
European Geophysical Society (EGS) 欧洲地球物理学会
European mean time 欧洲平时
European Meteorological Satellite (EuMetSat) ①欧洲气象卫星②欧洲气象卫星组织
European monsoon 欧洲季风
European Remote Sensing Satellites (ERSS) 欧洲遥感卫星【大气】
European Space Agency (ESA) 欧洲空间管理局
European Space Operations Centre (ESOC) 欧洲空间控制中心
European Space Research Organization (ESRO) 欧洲空间研究组织
European Telecommunication Satellites Organization (EUTELSAT) 欧洲通信卫星组织
European time 欧洲时间
European Union (EU) 欧盟
European Union Composite Observing System (EUCOS) 欧洲综合观测系统
Euros 带雨东南暴风(希腊语名)
eustasy 全球性海面升降【海洋】
eustatic movement 全球性海面升降【地理】
eustatic change 海面升降变化
eustatic sea level changes 海水升降的海平面变化
EUTELSAT (European Telecommunication Satellites Organization) 欧洲通信卫星组织
eutrophication 富营养化【土壤】
eutrophic mire 富营养沼泽【地理】
eutrophic system 富营养系统【土壤】
eutrophic water 富营养水【地理】
EUV (extreme ultraviolet radiation) 极紫外辐射

evanescent level 损耗层
evanescent wave (EW) 损耗波
evaporating dish 蒸发皿
evaporation 蒸发【大气】
evaporation capacity 蒸发量
evaporation coefficient 蒸发系数
evaporation curve 蒸发曲线
evaporation excess 过量蒸发
evaporation flux 蒸发通量
evaporation fog 蒸发雾【大气】
evaporation formula 蒸发公式
evaporation from ice 冰面蒸发
evaporation from land 陆面蒸发
evaporation from land surface 地面蒸发【农】
evaporation from phreatic water 潜水蒸发【农】
evaporation from soil 土壤蒸发【农】
evaporation from water surface 水面蒸发【农】
evaporation frost 蒸发霜
evaporation gauge 蒸发器
evaporation hook gauge 蒸发器
evaporation index 蒸发指数【农】
evaporation observation 蒸发观测
evaporation of water 水的蒸发
evaporation opportunity 蒸发可能率
evaporation pan 小型蒸发器【大气】,蒸发皿
evaporation power 蒸发能力,蒸发率
evaporation process 蒸发过程
evaporation rate 蒸发率
evaporation suppressor 蒸发抑制剂
evaporation tank 大型蒸发器【大气】
evaporation trail 蒸发尾迹
evaporative available potential energy 可用蒸发位能
evaporative capacity 蒸发能力,蒸发率
evaporative power 蒸发能力
evaporativity 蒸发率
evaporimeter 蒸发仪

evaporimetry 蒸发侧定法
evaporogram 蒸发量曲线
evaporograph 蒸发计【大气】
evaporometer 蒸发表
evapotranspiration 蒸散【大气】,总蒸发【地球】
evapotranspirometer 蒸散表
evapotron 涡动通量仪
even function 偶函数【数】
evening calm 夕静(黄昏海陆风交替时的静风现象)
evening tide 汐
even-odd check 奇偶校验【数】
event recorder 事件记录器
ever frozen soil 永冻土
everglade 湿地,轻度沼泽化低地
evergreen broad-leaved forest 常绿阔叶林【林】
Evershed effect 埃弗谢德效应
evidence of glaciation 冰川形迹,冰川痕迹
evogram 埃伏图
evolution 演化【物】,进化【遗传】
evolutional ecology 进化生态学【动物】
evolutionary botany 演化植物学【植】
evolutionary method 调优法
evolutionary power spectral density 调优功率谱密度
evolutionary power spectrum 调优功率谱
evolutionary spectrum 调优谱
evolutionary theory 进化论【植】
evolution of atmosphere 大气演化【大气】
evolution period 发展期
Ewing-Donn theory 尤因-唐﹝全球气候变化﹞理论
exact differential equation 恰当微分方程【数】
exact solution 精确解【数】
exceedance interval 超常间隔期
exceedance probability 超大概率,超过数概率

excellent visibility 最佳能见度
exceptional visibility 特佳能见度
excess attenuation 逾量衰减
excessive precipitation 非常降水量
excessive rain 久雨,霪雨
excessive rainfall 特大雨量,霪雨
excessive storm 特大暴雨
excess moisture injury 湿害
excess noise 过量噪声
excess pressure 超压,声压
excess snowfall 过量降雪
exchange 交换
exchange absorption 交换吸收
exchange capacity 交换量
exchange center 交换中心
exchange coefficient 交换系数
exchange of momentum 动量交换
excimer laser 准分子激光器【电子】
excitation 激发,激励
excitation mechanism 激发机理
excited atom 受激原子
excited ion 受激离子
excited molecule 受激分子
excited state 激发态,受激态
exciting force 扰动力
exclusive economic zone 专属经济区【航海】
executive committee（EC） 执行委员会
exergonic reaction 放能反应【生化】
exhalation 发散〔作用〕
exhaust air plume 排气烟羽【电力】
exhaust contrail 〔废气〕凝结尾迹【大气】
exhaust emission standard 废气排放标准
exhaust evaporation trail 〔废气〕蒸发尾迹【大气】
exhaust pollution 排放污染【航空】
exhaust port 排气口【航海】
exhaust temperature 排气温度【航海】
exhaust trail 排气尾迹
exhaust turbine generating set 废气涡轮发电机组【航海】
exhaust valve 排气阀【航海】
existing snowline 现存雪线,现在雪线
exit jet 流出急流,出口急流
exit region 出口区
ExKF（extended Kalman filter） 扩展的卡尔曼滤波
Exner function 埃克斯纳函数
exoatmosphere 外大气圈,外逸层
exobase 逸散层底【地】
exobiology 地外生物学【天】
exogenic influences 外源影响
exogenous electrification 外源起电
exopause 外逸层顶
exorheic〔basin,＝region〕 外流流域,外流区
exosphere 外〔逸〕层【大气】
exothermic reaction 放热反应
exotoxin 外毒素【生化】
Expanded Programme of Technical Assistance（EPTA,UN） 技术援助发展计划署（联合国）
expansibility 膨胀率
expansion ①展开式②膨胀
expansional cooling 膨胀冷却
expansion wave 膨胀波
expectance 期望
expectation 期望,期望值
expected loss 期望损失
expected utility 期望效益
expected value 期望值
expedition 探险,考察
Expendable Bathythermograph（XBT） 投弃式温深仪【大气】,消耗性温深仪
experiment 实验
experimental basin 实验流域,试验流域
experimental forecast 试验预报
experimental forecast center 试验预报中心
experimental meteorology 实验气象学
Experimental Meteorology Laboratory

（EML，NOAA） 实验气象学研究所（NOAA）
experiment design 试验设计
expertise 专门技术，专长
expert system 专家系统【计】【大气】
explained variance 解释方差，方差缩减
explicit time difference 显式时间差分
explicit difference scheme 显式差分格式
explicit integration method 显示积分法
explicit difference scheme 显式差分格式【数】
explicit parallelism 显式并行性【计】
explicit smoothing 显式平滑
explicit splitting technique 显式分解方法
Explorer 探险者（常用于空间探测卫星）
explosion cloud 爆炸云
explosion wave 爆炸波
explosive warming 爆发性增温
explosive wave 爆炸波
exponent 指数
exponential approximation 指数逼近【数】
exponential atmosphere 指数大气
exponential distribution 指数分布
exponential function 指数函数【数】
exponential integral 指数积分
exponential kernel approximation 指数核近似
exponential law 指数律【数】
exponential profile of turbulence 湍流指数廓线
exponential smoothing 指数平滑
exponential spectrum 指数谱
exponential sum fitting of transmission function（ESFT） 透过率函数指数和拟合
exponential window 指数窗
export 输出
exposure ①暴露②曝光〔量〕③开敞程度
exposure meter 曝光表

exposure of instruments 仪器露置
exsiccation 人为变旱
extended EOF（EEOF） 扩展的经验正交函数
extended forecast 延伸预报【大气】
extended Huygens-Fresnel principle 广义惠更斯-菲涅耳原理，广义 H-F 原理
extended Kalman filter（ExKF） 扩展的卡尔曼滤波
extended medium 广延介质
extended-range forecast（＝extended-term forecast） 延伸预报
extended source 扩展〔光〕源【物】
extensible balloon 弹性气球，膨胀气球
extensive air shower 广延大气簇射
extensive property 广度性质【化】
extensive quantity 广延量【物】
exterior angle 外角【数】
exterior ballistics 外弹道学
external force 外力
external forcing 外强迫【大气】，外界强迫作用
external galaxy 河外星系【物】
external gravity wave 重力外波【大气】
external lightning protection system 外部防雷装置
external mode 外模态
external parameter 外参数
external pressure 外压力
external reflection 外反射【物】
external Rossby radius of deformation 外罗斯贝变形半径
external water circulation 外水分循环（海洋大陆间的水分循环）
external wave 表面波，外波
external work 外功
extinction ①消光②〔放电的〕猝灭，猝熄③消亡
extinction coefficient（σ） 消光系数【大气】

extinction cross section 消光截面
extinction efficiency 消光效率
Extinction paradox 消光佯谬
extinct lake 干涸湖
extinct volcano 死火山【地质】
extraction method 抽样法
extraction of square root 开平方【数】
extra long-range〔weather〕forecast 超长期〔天气〕预报【大气】,短期气候预报
extra-long wave 超长波
extranet 外联网,外连网【计】
extraordinary refractive index 非〔寻〕常折射率【物】
extraordinary storm 特大暴雨
extraordinary wave 非常波【物】
extraordinary weather 异常天气
extrapolation 外推法,外推【数】
extrapolation method 外推法【大气】
extraterrestrial civilization 地外文明【天】
extra-terrestrial radiation 大气顶太阳辐射
extratropical belt (=zone) ①温带②热带外地区
extra-tropical circulation 温带环流
extra-tropical circulation numerical model 温带环流数值模式
extratropical cyclone 温带气旋【大气】
extra-tropical low 温带低压
extratropical planetary wave 温带行星波
extra-tropical severe local storm 温带局地强风暴
extra-tropical storm 温带风暴
extratropical tropopause 温带对流层顶
extreme arid environments 极端干燥环境
extreme climate 极端气候
extreme climatic event 极端气候事件
extreme climatic stability 极端气候稳定性
extreme crosswind 强侧风
extreme drought 极端干旱

extremely high frequency (EHF) 极高频
extremely hot and dry climate 极端干热气候
extremely low frequency (ELF) 极低频
extremely simple 简易的
extremely simple meteorological observing station (MOSES) 简易气象观测站
extremely wet climate 极端湿润气候
extreme precipitation 极端降水量
extreme precipitation process 极端降水过程
extreme precipitation quantities 极端降水量
extremes(＝extreme value) 极值
extreme severe sand and dust storm 特强沙尘暴【大气】
extremes of rainy years 极端雨年
extreme temperature 极端温度
extreme ultraviolet radiation (EUV) 极〔端〕紫外辐射
extreme value 极值
extreme value analysis 极值分析【大气】
extreme value distribution 极值分布
extreme wave height 极波高〔度〕
extreme weather events 极端天气事件 (洪水和干旱)
extreme wind speed 极大风速【大气】
extremum 极值
extremum check 极值检查
eye irritation 眼涩度(表示烟雾严重程度常用的定性指标)
eye observation 目测
eye of the storm 风暴眼
eye of the tropical cyclone 热带气旋眼
eye of typhoon 台风眼
eye of wind 风眼,风穴
eyepoint 视点【计】
eye wall 眼壁
eye-wall chimney 眼壁柱
e-zine 电子杂志【计】

F

Fabry-Perot etalon 法布里-珀罗标准具【物】
Fabry-Perot interferometer 法布里-珀罗干涉仪【物】
FACE (Free-Air CO_2 Enrichment) 自由大气中 CO_2 浓度富集
facet model 小平面模式
facility 设备
facsimile 传真
facsimile chart 传真图
facsimile copier equipment 传真图像设备
facsimile equipment 传真设备
facsimile transmission 传真发送
facsimile weather chart 传真天气图【大气】
facsimile weather chart recorder 天气图传真收片机
factor 因素,因子,因数
factor analysis 因子分析【大气】
factorial 阶乘【数】
factorial effect 因素效应【化】
factorial experiment 析因实验【化】
factoring 因式分解
factorization 因数分解【数】,因子分解
factors for climatic formation 气候形成因子【大气】
faculae 光斑
fade humidity 凋萎湿度
fadeout ①衰落,衰退②淡出【计】,中断
fading 衰落,凋萎
Fahrenheit temperature scale 华氏温标
Fahrenheit thermometer 华氏温度表
Fahrenheit thermometric scale 华氏温标【农】
failure ①失效【计】②故障【计】,失败【计】
failure data 失效数据【计】

failure rate ①故障率 ②失效率【电子】
fair 晴天
fairness 晴,好
fairway 航道【航海】
fair weather 晴天
fair-weather cirrus 晴天卷云
fair-weather cumulus 晴天积云
fair-weather current 晴天电流【大气】
fair-weather electric field 晴天电场【大气】
fair-weather electricity 晴天电学
fair-weather potential gradient 晴天电位梯度
fair wind 顺风
fait-spectrum source 平谱源【天】
Falkland Current 福克兰海流
fall ①降落②秋(美国)
fall equinox (见 Autumnal Equinox) 秋分
fall flood 秋汛洪水
fall frost 秋霜冻
fall growth period 秋季生长期
falling limb 退水段,落洪段
falling sphere 落球〔法〕
falling-sphere method 落球法
falling tide 落潮,退潮
fallout ①沉降物【大气】,沉降②放射性坠尘③放射性散落物
fallout front 沉降锋(带有放射性物质)
fallout plot 落尘图
fallout wind 沉降风【大气】(带有放射性散落物者)
fall seeding time 秋播期
fall speed 下落速度
fallstreak 雨(雪)幡
fallstreifen 雨(雪)幡(德语名)
fall velocity 下落速度

fall wind 下吹风
false alarm probability 虚警概率【电子】
false alarm rate 空报率
false alarm ratio(FAR) 虚警〔报〕率,空报率
false cirrus 伪卷云
false color 假彩色
false-color cloud picture 假彩色云图【大气】
false color image 假彩色图像
false echo 假回波【航海】
false front(=fictitious front) 假锋,伪锋
false reflection 假反射
false sun 幻日,假日
false warm sector 假暖区
false white rainbow 假白虹
family of chemical species 化学成分族
family of clouds 云族
family of cyclones 气旋族
family of depressions 低压族
family of downburst cluster 下击暴流族
family of tornadoes 龙卷族
fan filtering 扇形滤波
fanning 扇形扩散
fan scanning 扇形扫描
FAR(false alarm ratio) 空报率
farad(F) 法〔拉〕(电容单位)
faradaic current 法拉第电流【化】
Faraday effect 法拉第效应【物】
Faraday rotation 法拉第旋转【物】
Far East time 远东时间
far-end infrared frequency 超极高频
far field 远场
far infrared 远红外【化】
far infrared radiation 远红外辐射
far-infrared region 远红外区
farmer's proverb 农谚【大气】
farmer's year 农民年(英国,从每年最接近3月1日的星期日开始,共12个月,类似于我国农历)
farming season 农事季节

farm water requirement 田间需水量
farmyard manure 农家肥【土壤】
far ultraviolet radiation 远紫外辐射
fastest mile 最快每小时英里速
fastest mile wind 最快每小时英里风速
fast Fourier transform(FFT) 快速傅里叶变换【大气】
fast Fourier transform inverse(FFTI) 快速傅里叶逆变换
fast ice 固定冰
fast ion 快离子
fast mode equation 快模方程
fast response cup anemometer 快响应转杯风速表
fast-response O_3 detector 快〔速〕响应臭氧检测器
fast-response sensors 快响应传感器
fast scanning meteorological radar 快〔速〕扫描气象雷达
fata bromosa 幻雾蜃景
fatal humidity 致死湿度
fatal temperature 致死温度
fata morgana 复杂蜃景(意大利语名)
fathom 英寻(等于6英尺,约1.83 m)
fathom curve 等深线
fatigue break 疲劳断裂【海洋】
fault 故障【计】
fault diagnosis 故障诊断【航海】
fault earthquake 断层地震【地】
fault-tolerant computer 容错计算机【计】
favogn 焚风(瑞士语名)
favourable temperature 适宜温度
favourable weather 有利天气
favourable wind 顺风【航海】
fax(=facsimile) 传真
faxcasting 电视〔传真〕广播
fax chart(=fax map) 传真图
fax transmitter 传真发送机,传真发片机
FCL(free convection level) 自由对流高度【大气】

F distribution　F 分布【化】
FDMA (frequency division multiple access)　频分多址【电子】
FDP (Forecast Demonstration Project)　预报示范项目
feasibility　可行性
feasibility study　可行性研究
feather　风羽（风速填图符号，每根 4 m/s或10海里/时）
feather angle　羽角【海洋】
feathering　羽状移动【海洋】
feathery crystal　羽状晶体
feature　特征
Federal Aviation Administration (FAA)　〔美国〕联邦航空局
Federal Committee for Meteorological Services (FCMS)　联邦气象服务委员会（美国）
Federal Committee of Science and Technology (FCST)　联邦科学技术委员会（美国）
Federation of Astronomical and Geophysical Services (FAGS, ICSU)　天文与地球物理学联合会
feebly arid　轻度干旱
feed　馈〔给〕，供给
feedback　反馈
feedback system　反馈系统
feeder cloud　馈云，供水云
feedforward control　前馈控制【计】
feeding　①输进 ②〔风暴〕增强
feeling bottom　触底
Feigenbaum number　费根鲍姆数【物】
felled tree　伐倒木【林】
feller　伐木机【林】
felling　伐木【林】
felling machine　伐木机【林】
felt area　〔地震〕波及区
femto-　飞〔母托〕（词头，10^{-15}）
Fengyun 2　风云二号
Fermat principle　费马原理【物】

fermion　费米子【物】
Ferrel cell　费雷尔环流【大气】
Ferrel's law　费雷尔定律
Ferrel vortex　费雷尔涡旋
ferromagnetism　铁磁性【地质】
fertile soil　肥土【土壤】
festoon-cloud　花彩云
fetch　①风浪区【大气】②风浪区长度 ③行程
Feynmann diagram　F 图
FFT (fast Fourier transform)　快速傅里叶变换【大气】
fiard　低浅峡湾
fiber laser　光纤激光器【物】
fiber optics　纤维光学【物】
fiberoptic catheter　光导纤维管
fibratus (fib)　毛（云状）
fibre-glass unit　玻璃纤维测湿元件（测土壤含水量）
fibrous　纤维状〔的〕
fibrous ice　纤维状冰
Fickian diffusion　菲克扩散
Fickian diffusion equation　菲克扩散方程
Fickian equation　菲克方程
Fick law　菲克定律【物】
Fick's equation　菲克方程
Fick's law of diffusion　菲克扩散定律
fictitious front　假锋，伪锋
fictitious sun　虚太阳
fictitious viscosity　虚黏滞率
fidelity　保真性【物】
Fido (= fog intensive dispersal of)　消雾（用热驱散机场跑道上之雾）
fiducial interval　置信区间
fiducial limits　置信限
fiducial mark　基准标记
fiducial point　①参考点，基准点 ②置信点
fiducial temperature　基准温度
field　①场 ②现场 ③外场
field-aligned current　场向电流

field brightness 场亮度
field capacity 田间持水量【大气】
field changes 场变化
field ecosystem 农田生态系统【农】
field effect 场效应【化】
field elevation 场站海拔高度
field emission 场致发射【物】
field excursion associated with precipitation (FEAWP) 降水关联的电场极性反转
field experiment 外场试验,野外试验,现场试验
field ice 冰原冰
field intensity 场强
field ionization 场电离【化】
field jump 电场跃变【化】
field luminance 场亮度
field management 田间管理【农】
field maximum moisture capacity 田间最大持水量
field mill 电场强度计,转盘电场仪
field [moisture] capacity 田间持水量【大气】
field observation 外场观测【大气】,野外观测,现场观测
field of deformation 变形场
field of divergence 辐散场,散度场
field of pressure 气压场
field of solenoid 力管场,网格场
field of view 视场,视界,视野
field of vorticity 涡度场
field permeability 场致磁导率
field safeguarding forest 农田防护林【农】
field stop 视场光阑【物】,场阑
field strength 〔电〕场强〔度〕
field technique 田间技术【农】
field variables 场变数,场变量
field water-holding capacity 田间持水量
field water requirement 田间需水量【农】

FIFOR 航线预报(表示 flight forecast 的国际电码字)
figure ①数字②图
filament 暗条【天】
filamental flow 线流
filament current 灯丝电流
file ①文件【计】,档案②(外)存储器,(存储)资料
file access 文件存取【计】
file allocation 文件分配【计】
file directory 文件目录【计】
file memory 文件存储器【计】
file name 文件名【计】
fillet lightning 带状闪电
filling 填塞
filling balance 灌球平衡秤
filling barometer 充液气压计
filling cyclone 填塞气旋
filling of a depression 低压填塞【大气】
filling up ①填塞②充气
film ①薄层(冷空气)②薄膜③软片(摄影)
film crust 冰膜
film loop 环型胶卷
film renewal model 地表修复格式
film water 薄膜水
filosus 毛〔丝〕状〔云〕
filter ①滤波器②过滤器③滤光片
filter analysis 滤波分析
filter-captrue 过滤-捕获
filtered equations 滤波方程
filtered model 过滤模式【大气】,滤波模式
filter equation 滤波方程
filter function 滤波函数
filtering 滤波【数】
filtering approximation 滤波近似法
filtering function 滤波函数
filtering meteorological noise 过滤气象杂波,气象噪声滤波
filter photometer 滤光光度表

filter sampler 过滤取样器
filter wedge spectrometer（FWS） 滤波楔形光谱仪
filtration 过滤〔作用〕,渗透
final prediction error（FPE） 最终预报误差
final reading 终读数
final warming 最终增温
Findeisen-Bergeron nucleation process 芬德森-贝吉龙成核过程,F-B成核过程
fine 晴天
fine adjustment 细调【物】
fine day 晴日,晴天
fine-grained density 细粒密度【物】
fine line 细线
Fineman's nephoscope 法因曼测云器
fine mesh 细网格
fine-mesh grid 细网格【大气】
fine mesh model 细网格模式
fine particles 细粒子（直径小于$2\ \mu m$的粒子）
fine sand soil 细砂土【土壤】
finesse 细度【物】,细致因子
fine structure 精细结构【物】
fine weather 好天气
fine weather effect 晴天效应
fingering 指印现象
fingerprint method 指纹法
finger rafting 指状筏冰
finite amplitude 有限振幅【大气】
finite amplitude convection 有限振幅对流
finite amplitude disturbance 有限振幅扰动
finite amplitude instability 有限振幅不稳定度
finite bandwidth power loss 有限频宽功率损失
finite depth 有限厚度,有限深度
finite difference 有限差分
finite difference approximation 有限差分近似
finite difference equation 有限差分方程
finite difference method 有限差分法
finite difference model 差分模式【大气】
finite difference quotient 有限差〔分〕商
finite difference ratio 有限差比
finite differencing 有限差分
finite dimensional vector space 有限维向量空间【数】
finite element 有限元
finite element analysis 有限元分析【计】【数】
finite element method 有限元法【数】
finite element model 有限元模式
finite Fourier series 有限傅里叶级数
finite source method 有限源法
finite volume equation 有限体积方程
Finsterniswind 昏暗风（日全食时的风,德语名）
fiord 峡湾
fireball 火流星,火球
fire behavior 燃烧习性
fire belt 防火带【林】
fire-danger meter 火险标尺
fire hazard 〔自然〕火险（因天气而发生的火灾危险）
fireline intensity 隔火带强度【力学】
firestorm 火灾风暴（由森林火灾引起的风暴）
fire trench 防火沟【林】
fire triangle 禁火三角形【石油】
fire weather 火险天气
fire wind 火灾风（火灾风暴引起的风）
firing 点火,发射
firm yield ①稳定产量 ②可靠产水量
firn 永久积雪【大气】,永久冰雪（德语名）
firn field 陈年雪场
firn ice 陈年冰
firnification 永久积雪作用
firn limit 永久雪限

firn line 永久雪线【大气】
firn snow 永久积雪,冰川雪
firn spiegel 永久镜冰
firn wind(＝glacier wind) 冰川风
first 弗利斯特雨(苏格兰的突发性阵雨)
first aid at sea 海上急救【航海】
first come-first-serve(FCFS) 简单排队法(先来先服务)
first contact 初亏
first cosmic velocity 第一宇宙速度【物】
first frost 初霜【大气】
First Frost 霜降【大气】(节气)
First GARP Global Experiment(FGGE) 全球大气研究计划第一期全球试验
first guess 初估值【大气】,首次推定值
first-guess field 初估场
first gust 首次阵风
first-hop wave 第一反射波
first law of thermodynamics 热力学第一定律【物】
first moment 一阶矩
first order autocorrelation 一阶自相关
first order climatological station 一级气候站
first order closure 一阶闭合
first order closure model 一阶闭合模式
first order ecosystem 一级生态系统
first order multiple scattering approximation 一阶多次散射近似
first order perturbation solution 一阶扰动解
first order phase transition 一级相变【化】
first order reaction 一级反应【化】
first order smoothing 一阶平滑
first order smoothing approximation 一阶平滑近似
first order spectrum 一级图谱【化】
first order station 一级气象站
First Point of Aries 春分点
First Point of Libra 秋分点
first purple light 第一紫霞,第一紫光
first quarter 上弦【天】
first quarter (of the moon) 上弦(月的)
first radiation constant 第一辐射常数
first snowfall 初雪
first time step 起步
first year ice 一年冰【大气】,一冬冰
Fisher and Porter raingauge 费希尔-波特雨量计,F-P雨量计
Fisher discriminant function 费希尔判别函数【数】
Fisher (F) distribution F分布,费希尔分布
fisheries meteorology 渔业气象学
fisheries oceanography 渔业海洋学【海洋】
Fisher information function 费希尔信息函数【数】
fishery resources 水产资源【航海】,渔业资源【海洋】
fish reef 渔礁【航海】
fish resources 鱼类资源【海洋】
fission 裂变
fission chamber 裂变室
fission chemistry 裂变化学【化】
fission gas 裂变气体〔产物〕【化】
fission product 裂变物
fission product chemistry 裂变产物化学【化】
fission threshold 裂变阈【物】
fission track dating 裂变径迹定年法
fissure 裂缝,裂隙
fitness figure ①舒适度图②适合度因数
fitting 拟合,配曲线,调整
fitting method 拟合法
five and ten system 五进和十进风制
five channel scanning radiometer 五通道扫描辐射计
five day forecast 五日预报
five lens aerial camera 五镜头空中照相机

five-point method 五点法
fix 定位
fixed beam ceilometer (FBC) 固定光束〔式〕云幂仪
fixed boundary condition 固定边界条件
fixed cistern barometer 定槽式气压表
fixed cloud-top altitude (FCA) 固定云顶高度
fixed cloud-top temperature (FCT) 固定云顶温度
fixed dune 固定沙丘【地理】
fixed frequency sounder 固频探空仪
fixed level chart 等高面图
fixed platform 固定〔观测〕平台
fixed point 不动点,定点【数】
fixed point gauge 定点水位计
fixed point iteration 定点迭代
fixed point observation 定点观测【大气】
fixed point operation 定点运算
fixed sea platform 固定海上平台
fixed ship station 固定船舶站【大气】
fixed star 恒星
fixed time broadcast 定时广播
fixed time observation 定时观测
fixing 定影【物】
fixing of atmospheric gases 大气气体固定
fjord 峡湾
flag ①标记【地信】②标记位③旗帜
flake 雪片
flame collector 火焰集电器
flame ionization detection 火焰电离检测
flaming aurora 焰状极光
flan 弗兰风(苏格兰的突发性干阵风)
Flanders storm 佛兰德风暴(英国伴有南风的大降雪)
flanking line 侧翼阶梯云
flare 喷焰,辉斑
flare echo 耀斑回波
flare-radiation ionospheric effect 耀斑辐射电离层效应
flash 闪光

flash event 闪电事件
flash flood 暴洪(山洪暴发),突发性洪水
flash flood alarm system (FFAS) 暴洪预警系统
flash flood warning 暴洪预警
flash flood watch 暴洪监视
flashing 闪烁
flashlamp-pumped laser (FPL) 闪光灯抽运激光器
flash memory 闪速存储器【计】
flash photolysis systems 闪光光解系统【化】
flash rate 闪光率
flash spectrum 闪光谱【天】
flashy stream 暴洪河流
flask sampling 瓶采样
flat-bed scanner 平板扫描仪【计】
flat Bellani cup radiometer 贝拉尼板罩式辐射仪
flat copy facsimile scanner 平面拷贝传真扫描器
flat low 浅低压
flat maximum 平坦型最高(气温)
flat plate (cone) radiometer 平板(圆锥)辐射仪
flat spectrum 平谱【天】
flattened sun 扁日
flaw 缝口风(英国突发性阵风)
F layer F层
F_1-layer F_1层
F_2-layer F_2层
fleecy cloud 卷毛云
fleecy sky 卷毛云天
Fleet Broadcast 船队预警广播
Fleet Numerical Oceanography Center (FNOC) 舰队数值海洋学中心(美国海军)
Fleet Numerical Weather Central (FNWC) 舰队数值天气中心(美国海军)

Fleet Weather Central (FWC)　舰队天气中心(美国海军)
flexible disk　软磁盘【计】,软盘
flex point　拐点
flexural wave　挠曲波
flight briefing　航空简报
flight cross section　航空剖面〔图〕
flight documentation　飞行〔气象〕文件
flight dossier (= flight forecast-folder)　航空天气预报表格
flight forecast　航空天气预报
flight forecast-folder　航空气象要览
flight information centre　航空气象情报中心
flight information document　飞行文件
flight information region　气象情报区,飞行情报区
flight information service station　飞行情报服务台
flight level　飞行高度
flight plan　航空计划,飞行计划
flight service station　飞行服务站
flight train　探空设备系列
flight visibility　空中能见度【大气】,飞行能见度
flight weather briefing　航空天气简报
flight weather forecast bulletin　飞行天气预报表
flip-flop　触发器【计】
float barograph　浮子气压计
floating average　移动平均,滑动平均
floating balloon　漂移气尘
floating dust　飘尘【大气】,浮尘
floating ice　浮冰
floating lysimeter　浮动式渗漏计
floating pan　漂浮式蒸发皿【大气】
floating point　浮点
floating point operation　浮点运算
floating point processor　浮点处理机
floating population　流动人口【地理】
floating radiosonde　漂浮探空仪

float type raingauge　浮筒〔式〕雨量器
flocculus　谱斑【天】
floccus (flo)　絮状(云)
floe　浮冰块
floeberg　小冰山,浮冰丘
floe ice　浮冰
flood　①洪水②涨潮
flood catastrophe　水灾
flood channel　行洪河道
flood control　防洪,洪水调节
flood control reservoir　防洪水库
flood crest　洪〔水〕峰〔顶〕,最高洪水位
flood current　涨潮流【航海】
flood damage　水灾【地理】
flood discharge level　洪水位
flood erosion　洪水侵蚀【土壤】
flood fatality　洪灾
flood forecast　洪水预报
flood forecasting　洪水预报
flood freezing injury　冻涝害【农】
flood frequency　洪水频率
flood frequency distribution　洪水频率分布
flood gate　〔洪〕水闸
flood hydrograph　洪水涨落图
flood icing　冰泉,喷水冰层
flooding　洪涝
flooding ice　喷水冰层,冰泉
flooding irrigation　漫灌【农】
flooding routing　洪水演算,调洪演算
flood interval　洪水间隙
flood irrigation　漫灌,淹灌
flood-leading rainfall　致洪降水
flood marks　洪水标记,高潮标记
flood mitigation　减洪
flood peak　洪峰【航海】
flood period　汛期【大气】,洪水期
flood plain　①泛滥平原,洪泛平原②沃野
flood probability　洪水概率
flood proofing　防汛抢险

flood region 涝区,洪泛区
flood routing 洪水演算,调洪演算
floods-above-base series 超基准洪水系列
flood season 汛期
flood stage 洪水位
flood stream 涨潮流【航海】
flood strength 最大涨潮流速,洪水强度
flood tide 涨潮【大气】
flood warning 洪水警报
flood watch 洪水监测〔报告〕
flood-waterlogging damage 洪涝灾害
flood wave 洪水波
floppy disk 软磁盘,软盘【计】
floppy disk drive 软盘驱动器【计】,软驱
flora ①植物志②植物③植物群
Florida Current 佛罗里达海流
Florida Solar Energy Center（FSEC） 佛罗里达太阳能中心（美国）
flow 流〔动〕
flow chart 流程图,程序图,框图
flowchart 流程图【计】
flow concentration 汇流【地理】
flow diagram 流程图【计】,程序框图【化】
flow equilibrium 流动平衡
flower-bud appearing stage 现蕾期
flowering 开花【植】
flowering period 开花期
flowering stage 开花期【农】
flow law 流变定律
flow line 流线
flowmeter 流量表
flow net 流网
flow-off 径流
flow pattern 流型
flow separation ①气流分离 ②水流分离
flow splitting ①气流分支 ②水流分支
flow velocity 流速
fluctuating field 起伏场
fluctuating humidity 脉动湿度
fluctuating pressure 脉动气压
fluctuating temperature 脉动温度
fluctuating velocity 脉动速度
fluctuation ①变动,起伏②扰动,振荡 脉动
fluctuation velocity 脉动速度
flue gas 烟道气〔体〕
flue gas purifier 烟道气净化器
flue gas velocity 排烟速度
fluid 流体
fluid dynamics 流体动力学
fluid dynamics equation 流体动力学方程
fluidic wind sensor 射流测风仪,射流风传感器
fluid mechanics 流体力学
fluid parcel 流体块
flume 渡槽
fluorescence 荧光
fluorescence analysis 荧光分析
fluorescence chromatogram 荧光色谱
fluorescence quenching method 荧光猝灭法【化】
fluorescence spectrophotometer 荧光分光光度计
fluorescence spectrum 荧光谱,荧光光谱【化】
fluorescent light 荧光
fluorescent scattering 荧光散射
fluorescent tracer 荧光示踪物
flurry 阵雪
fluvial dynamics 河流动力学
fluvial morphology 河流形态学,河流地貌
flux 通量【大气】
flux adjustment 通量调整
flux aggregation 通量聚集
fluxatron 通量仪
flux-corrected transport 通量订正输送
flux correction 通量订正
flux density 通量密度
flux density threshold 通量密度低限
flux divergency 通量散度

flux emissivity   通量发射率
flux form   通量形式
fluxgate magnetometer   饱和式磁强计
fluxmeter   磁通计
flux of radiation   辐射通量
fluxplate   热流板,热通量板
flux-profile relationships   通量廓线关系
flux-ratio method   通量比方法
flux Richardson number   通量里查森数【大气】
flux transmissivity   通量透射率
flux transmittance   通量透射比
fly ash   飞灰【大气】,飞尘,浮尘
flying laboratory   飞行实验室
flying route   飞行路线,航线,航路
flying-spot   扫描点,飞点,浮动光点
flying spot recording technique   飞点记录法,浮动光点记录法
flyoff   蒸散(陈旧词)
FM-CW radar   调频连续波雷达
Fn (fracto-nimbus)   碎雨云【大气】
foam crust   泡沫雪
focal length   焦距
focal power   〔光〕焦度【物】
focus   焦点
focused beam   汇焦光束
focusing effect   聚焦效应
focus projection and scanning vidicon   聚焦投影与扫描视像管
foehn   焚风【大气】
foehn air   焚风气流
foehn bank   焚风〔云〕堤
foehn break   焚风〔云〕隙
foehn climatology   焚风气候学
foehn cloud   焚风云
foehn cyclone   焚风气旋
foehn effect   焚风效应【大气】
foehn gap   焚风〔云〕隙
foehn island   焚风岛
foehn nose   焚风鼻
foehn pause   ①焚风停顿 ②焚风界

foehn period   焚风期
foehn phase   焚风阶段
foehn storm   焚风风暴(德国巴伐利亚山10月份伴有强焚风的灾害性风暴)
foehn trough   焚风槽
foehn wall   焚风墙【大气】,焚风〔云〕壁
foehn wave   焚风波【大气】
foehn wind   焚风
fog   雾【大气】
fog bank   雾堤【大气】
fog bow   雾虹
fogbroom   散雾器
fog chamber   雾室
fog clearing   消雾
fog collector   雾收集器
fog crystal   雾晶
fog damage   雾害
fog day   雾日
fog deposit   冻雾覆盖层
fog detector   测雾仪(测雾天能见度)
fog dispersal   消雾
fog dispersal operation   消雾作业
fog dissipation   消雾【大气】
fog drip   (树上的)雾水滴落
fog-drop   雾滴【大气】
fog droplet   小雾滴
fog forest   〔高山〕雾林
fog-gauge   雾量计
foggy   有雾〔的〕,多雾〔的〕
fog horizon   雾层顶
fog investigation dispersal operation (FIDO)   研究性〔的〕消雾计划
fogmeter   雾量表(计)
fog patch   碎雾
fog point   露点
fog precipitation   雾降水
fog rain   雾雨
fog-region   雾区
fog scale   雾级标度
fog signal   雾号【航海】

fog visiometer 雾天能见度仪
fog warning 雾预警【航海】
fog water collector 雾滴收集器
fog wind 雾风
föhn 焚风
föhn air 焚风气流
föhn break（或 gap） 焚风云隙
föhn cloud 焚风云
föhn effect 焚风效应
föhn wall （或 bank） 焚风云壁（向风坡降水云的边缘）
föhn wave 焚风波
foil 反射薄片（高空测风用）
Fokker-Planck equation 福克尔-普朗克方程【物】
fold 褶皱【地质】
fold belt 褶皱带【地质】
fold catastrophe 褶皱型突变
folding 折叠（误差）
folding frequency 折叠频率
folding velocity 折叠速度
fold point 皱点
following wind 顺风
Food and Agricultural Organization of the United Nations(FAO) 联合国粮〔食及〕农〔业〕组织
foot-candle 英尺烛光
foot-lambert (=footlambert) 英尺朗伯
foot pound 英尺磅(=1.356 J)
foot-pound-second system 英尺-磅-秒制
footprint ①足迹,痕迹②星载探测仪器的瞬时视场
footprint modeling 足迹模拟
forano 弗拉诺风（意大利那不勒斯的海风）
forbidden band 禁带【物】
forbidden line 禁线
forbidden region 禁区（带电粒子运动的）
forbidden transition 禁戒跃迁【物】
forbidden zone of cloud formation 成云禁区（约高空 30～75 km）

force 力
forced cloud 强迫云
forced convection 强迫对流【大气】
forced oscillation 强迫振荡【大气】
forced vibration 强迫振动
forced wave 强迫波
forced wind circulation 强迫风环流
force-one wind 1 级风
force-restore method 强迫重建法
force three wind 3 级风
force two wind 2 级风
forcing 强迫,作用力,作用项
forcing function 强迫函数
forcing term 强迫项,外力项
forecast 〔天气〕预报
forecast accuracy 预报准确率【大气】
forecast amendment 订正预报【大气】
forecast-analysis cycle 预报分析循环
forecast anomaly 预报距平
forecast area 预报区【大气】
forecast bulletin 天气公报
forecast chart 预报图【大气】
Forecast Demonstration Project (FDP) 预报示范项目
forecast difference 预报差异
forecast district 预报区
forecast ensemble 预报集成
forecaster 预报员
forecast error 预报误差（预报值与验证值之差）
forecast evaluation 预报评估
forecast identification 预报鉴定
forecasting 预报〔方法〕
forecasting by mathematical statistics 数理统计预报方法
forecasting center 预报中心
forecasting criterion 预报判据
forecasting error 预报误差（预报值与观测值之差）
forecasting period 预报期
forecasting technique 预报技术

forecast lead time　预报提前时间
forecast period　预报时段
forecast-reversal test　反验证法,正反预报检验
forecast score　预报评分【大气】
forecast sensitivity　预报敏感性
forecast skill　预报技巧
forecast terminology　预报术语
forecast updating　更新预报
forecast verification　预报检验【大气】
forecast verification score　预报检验评分
forecast zone　预报区
foreign broadening　外加宽
foreign matter of the air　大气杂质
Forel scale　福来耳海色标度
forensic meteorology　法律气象学
forerunner　①征兆,预兆②先头波
foreshadow　预测
foreshortening　用透视法缩小绘制〔图〕
fore sky　锋区天空,前兆天空
forestation　造林,植林
forest biology　森林生物学【林】
forest border　林缘
forest canopy　林冠层【林】
forest climate　森林气候【大气】
forest conservation　森林保护【农】
forest ecology　森林生态学【林】
forest economics　林业经济学【林】
forest engineering　林业工程学【林】
forest-fire cloud　森林火灾云
forest fire-danger weather index　森林火险气象指数
forest fire-danger weather rating　森林火险天气等级
forest-fire meteorology　森林火灾气象学
forest-fire smoke　〔森〕林火〔灾〕烟云
forest-fire〔weather〕forecast　林火〔天气〕预报【大气】
forest hydrology　森林水文学【地理】
forest interior　森林内部
forest limit temperature　森林界限温度【大气】
forest line　森林线
forest meteorology　森林气象学【大气】
forest microclimate　森林小气候【大气】
forest mire　林地沼泽【地理】
forest plantation　人工林【林】
forest region　林区【地理】
forest resources　森林资源【林】
forestry　林学【林】
forest science　林学【林】
forest smoke　森林烟
forest soil　森林土壤【地理】
forest steppe　森林草原【地理】
forest-steppe climate　森林草原气候
forest wind　森林风
forked lightning　叉状闪电【大气】
formaldehyde　甲醛
formal solution　形式解【数】
format　①格式②信息编排
formation fluid　地层流体
formation function　形成函数【化】
form drag　形式曳力,形式应力
formic acid　蚁酸,甲酸
form of environmental substance　环境物质形态【地质】
formula　公式
formula for gain　增益公式
formulation　①立式,公式化②系统阐述③构想④配制
FORmula TRANslator　FORTRAN 语言【计】
Fortin barometer　福丁气压表【大气】,动槽式气压表
FORTRAN（formula translator 的缩写）FORTRAN 语言【计】
Forty Saint's storm　四十圣风暴(希腊强南风)
forward-backward scattering　向前-向后散射【化】,前向-后向散射
forward-backward scheme　向前向后格式
forward chaining　正向链接

forward difference 向前差分【数】
forward flux density 前向通量密度
forward integration 前向积分
forward integration scheme 前向积分格式
forward interpolation 前向内插
forward-marching techniques 时间前移法
forward overhang 前向悬垂回波
forward reaction 正向反应【化】
forward scatter(ing) 前向散射【大气】
forward scattering spectrometer 前向散射粒谱仪
forward scattering spectrometer probe (FSSP) 前向散射粒谱仪探头
forward scattering theorem 前向散射原理
forward scatter meter (FSM) 前向散射计
forward scatter visibility meter 前向散射能见度仪
forward specific intensity 前向比辐射强度
forward tilting 向前倾斜
forward-tilting trough 前倾槽【大气】
forward time difference 时间向前差
forward-upstream differencing 上游前差
forward visibility 前向能见度
fossil 化石
fossil fuels 矿物燃料,化石燃料
fossil ice 化石冰,埋藏冰
fossil permafrost 化石永久冻土
fossil water 化石水
Foucault pendulum 傅科摆【物】
foul weather 坏天气
fount 喷泉【地质】
fountain 喷泉【地质】
four-dimensional data assimilation 四维资料同化【大气】,四维资料分析
four dimensional spacetime 四维时空【物】
four dimensional spectral density 四维谱密度
four-dimensional variational assimilation 四维变分同化
four-dimensional variational data assimilation (4D-Var) 四维变分资料同化
four-dimensional optimal interpolation (4D-OI) 四维最优插值
four flux theory 四流理论
Fourier analysis 傅里叶分析【数】
Fourier coefficient 傅里叶系数【数】
Fourier integral 傅里叶积分【数】
Fourier integral operator 傅里叶积分算子【数】
Fourier kernel 傅里叶核
Fourier kernel window 傅里叶核窗
Fourier pair 傅里叶对
Fourier series 傅里叶级数【数】
Fourier space 傅里叶空间【化】
Fourier spectrum 傅里叶谱
Fourier synthesis 傅里叶合成
Fourier transform 傅里叶变换【数】
Fourier transform infrared spectrometer 傅里叶变换红外光谱计【化】
Fourier transform infrared spectroscopy 傅里叶变换红外光谱〔学〕【化】
Fourier transform mass spectrometer 傅里叶变换质谱计【化】
Fourier transform nuclear magnetic resonance 傅里叶变换核磁共振【化】
four-level system 四能级系统【物】
fourth contact 复圆(日食)
fourth order moment 四阶矩
foveal vision 黄斑中心视像,明视觉
FOV (field of view) 视场
fowan 福万风(大不列颠和马恩岛的干热风)
foyer 天电源地
f-plane approximation f平面近似
FPP scale (Fujita-Pearson scale) 藤田-皮尔逊尺度
fps 帧每秒【计】
fractal ①分形【计】②分形体【物】③非整维子
fractal analysis 分形分析【数】

fractal dimension 分形维数【物】,分维
fractal dimension theory 分〔数〕维〔数〕理论
fractals 分形分析【数】
fraction 分数【数】
fractional cloud amount 碎云量
fractional power law 分数幂定律
fractional step 小段步长
fractional step method 分步法【数】,分数步长法
fractional volume abundance 气体分容量
fractionation 分级【化】
fraction of saturation 饱和分数
fractocumulus(Fc) 碎积云【大气】
fractonimbus(Fn) 碎雨云【大气】
fractostratus(Fs) 碎层云【大气】
fractus(fra) 碎云
fragmentation 冰〔晶破〕碎
fragmentation nuclei 碎核
frame 幅,帧(图片等的)【计】
frame per second 帧每秒【计】
frame synchronizer 帧同步器
frank 平坦风(稳定风的航海术语)
fraternal twin integrations 不同模式双子积分〔实验〕
F ratio F 比〔方差比〕
Fraunhofer diffraction 夫琅禾费衍射【物】
Fraunhofer diffraction theory 夫琅禾费衍射理论
Fraunhofer line discriminator (FLD) 夫琅禾费线鉴别器
Fraunhofer's band 夫琅禾费谱带
Fraunhofer's line 夫琅禾费谱线
frazil 屑冰,潜冰,水内冰
frazil ice 片冰
freak wave 变异波,反常波
Fredholm integral equation 弗雷德霍姆积分方程【数】
free acidity 游离酸性
free aerostat 自由高空气球

free air 自由空气
free air anomaly 自由空气异常【地】
Free-Air CO_2 Enrichment Experiment (FACE) 自由大气中 CO_2 浓度富集试验
free air foehn 高空焚风(下沉形成的高空干热空气)
free atmosphere 自由大气【大气】
free balloon 自由气球
free board ①干舷〔高〕②出水高〔度〕,超高
free convection 自由对流【大气】
free convection level (FCL) 自由对流高度【大气】
free convection scaling 自由对流尺度
free convection scaling velocity 自由对流标定速度
free electron 自由电子【物】
free expansion 自由膨胀
free fall acceleration 自由落体加速度
free floating moored buoy 自由浮动式系留浮标
free foehn 高空焚风
free hydromagnetic oscillation 自由磁流振荡
free lift 净举力
free lifting force 净举力,自由上升力
free lift of balloon 气球净举力,气球自由上升力
free low frequency variability 自由低频变异度
free mode 自由〔波〕模
free nappe 自由水舌,通气水舌
free oscillation 自由振荡
free path 自由程【物】
free period 自由周期
free radical 自由基【化】,游离基
free Rossby wave 自由罗斯贝波
free-slip boundary condition 自由滑动边界条件
free space 自由空间【物】

free spectral range 自由〔光〕谱区
free streamline 自由流线
free-stream Mach number 自由气流马赫数
free surface 自由面
free surface approximation 自由表面近似【物】
free-surface condition 自由面条件
free turbulence 自由湍流
free water 自由水,自由波
free-water content 自由水含量
free wave 自由波【大气】
free wind 自由风
freeze (=freezing) 冻结
freeze drying 冷冻干燥【农】
freeze free period 无冻期,非冻结期
freeze-level chart 冻结高度图
freeze spray (=freezing spray) 冻沫
freeze-thaw cycle of water 水的冻融循环
freeze thaw cycles 冻融循环【地理】
freeze-thaw pattern 冻融〔土壤〕型
freeze up 封冻
freezing ①冻结【大气】②凝固
freezing damage 冻害【地理】
freezing day 冻日(最高温度0℃以下)
freezing degree-day 冻〔结〕度日
freezing drizzle 冻毛毛雨
freezing fog 冻雾【大气】
freezing hardiness 抗冻性
freezing index 冻结指数
freezing injury 冻害【大气】
freezing interface 冻结界面
freezing level 冻结高度,凝固高度
freezing-level chart 结冰高度图
freezing nucleus 冻结核【大气】
freezing nuclei spectrum 冻结核谱
freezing point 凝固点,冰点
freezing point depression 冰点温差
freezing point line 冰点线,冻结点线
freezing precipitation 冻〔结〕降水
freezing rain 冻雨【大气】

freezing season 冻季
freezing spray 冻沫
freezing temperature 冻结温度,凝固温度
F region F区,F层
F_1-region F_1层
F_2-region F_2层
freon 氟利昂
frequency 频率,频数
frequency agility 频率捷变
frequency analysis 频率分析
frequency band 频带
frequency conversion 变频【电子】
frequency converter 变频器【电子】
frequency-curve 频率曲线
frequency-direction spectrum 频率-方向谱
frequency discrimination 鉴频【电子】
frequency discriminator 鉴频器【电子】,甄频器
frequency distogram 频数直方图【电子】
frequency-distribution 频率分布
frequency diversity 频率分集
frequency division multiple access (FDMA) 频分多址【电子】
frequency domain 频率域,频域【计】
frequency-domain averaging 频域平均
frequency-doubled 倍频
frequency equation 频率方程
frequency filtering 频率滤波
frequency function 频率函数
frequency meter 频率计
frequency-meter anemometer 频率表式风速表
frequency-modulated continuous-wave radar (=FM-CW radar) 调频连续波雷达
frequency modulation (FM) 调频
frequency multiplication 倍频【电子】
frequency multiplier 倍频器【电子】
frequency polygon 频数多边形
frequency range 频段,波段,频率范围
frequency-ratio method 频率比法
frequency response 频率响应【大气】

frequency response function 频率响应函数
frequency shift 频移【物】
frequency shift keying 移频键控
frequency sounding method 频率测探法【地】
frequency spectrum 频谱【大气】
frequency splitting 频率分裂
frequency spread 频率展宽
frequency synthesizer 频率合成器
frequency-time analysis 频率-时间分析
frequency-time spectral density function 频率-时间谱密度函数
frequency-wavenumber filtering 频率-波数滤波
frequency-wavenumber migration 频率-波数偏移【地】
fresh air 新鲜空气
fresh air mass 新鲜气团
fresh breeze 5级风【大气】,清劲风
freshet ①洪水泛滥 ②春汛 ③(入海的)淡水河流
fresh gale 强风
Fresh Green 清明【大气】(节气)
fresh snow cover 新积雪
fresh water 淡水
fresh water lake 淡水湖【地理】
fresh weight 〔新〕鲜重〔量〕,初次称重
fresh wind 清劲风(5级风)
Fresnel diffraction 菲涅耳衍射【物】
Fresnel length 菲涅耳长度
Fresnel reflection 菲涅耳反射
Fresnel shrinkage 菲涅耳收缩
Fresnel size 菲涅耳尺度
Fresnel zone 菲涅耳带
friagem (= vriajem) 弗里阿格姆凉期(南美洲亚马孙河谷中上游和玻利维亚东部5、6月间5～6天的凉期)
friction 摩擦
frictional coefficient 摩擦系数【化】
frictional convergence 摩擦辐合【大气】
frictional dissipation 摩擦耗散
frictional divergence 摩擦辐散【大气】
frictional drag 摩擦曳力【大气】
frictional effect 摩擦效应
frictional force 摩擦力
frictional layer 摩擦层【大气】
frictional resistance 摩擦阻抗
frictional secondary flow 摩擦二次流,摩擦副流
frictional skin drag 表面摩擦曳力
frictional stress 摩擦应力
frictional sublayer 摩擦附属层,摩擦副层
frictional torque 摩擦力矩,摩擦转矩
friction head 摩擦水头
friction height 摩擦高度
friction layer 摩擦层
friction loss 摩擦损失
friction velocity 摩擦速度【大气】
friendly ice 亲和冰
frigid blasts 凛冽寒风
frigid weather 严寒天气,寒冷天气
frigid zone 严寒地区,寒带
frigofuge 畏寒植物,避寒植物
frigorigraph 冷却计
frigorimeter 冷却表
fringe region 边缘区,边缘层
fringe region of the atmosphere 大气边缘层
frog leap scheme 蛙跳格式
frog storm 蛙风暴,春风暴(春季的第一个风暴)
front 锋〔线〕【大气】
frontal action 锋面作用,锋面活动
frontal analysis 锋面分析【大气】
frontal band 锋带
frontal circulation 锋面环流
frontal cloud 锋面云
frontal collapse 锋面崩溃
frontal contour 锋面等高线
frontal contour chart 锋面等高线图
frontal convection 锋面对流

frontal cyclone	锋面气旋
frontal cyclone model	锋面气旋模式
frontal fog	锋面雾【大气】
frontal funnel effect	锋区漏斗效应
frontal inversion	锋面逆温【大气】
frontal lifting	锋面抬升
frontal line	锋线
frontal low	锋面低压
frontal mass	锋区气团
frontal model	锋面模式
frontal occlusion	锋面锢囚
frontal passage	锋面过境【大气】
frontal passage fog	锋面过境雾
frontal precipitation	锋面降水【大气】
frontal profile	锋区剖面
frontal rainfall	锋面雨
frontal shear	锋面切变
frontal slope	锋面坡度【大气】
frontal strip	锋区界条
frontal structure	锋面结构
frontal surface	锋面【大气】
frontal system	锋系
frontal theory	锋面理论
frontal thunder shower	锋面雷阵雨
frontal thunderstorm	锋面雷暴
frontal topography	锋面形势
frontal uplift	锋面抬升
frontal wave	锋面波动【大气】
frontal weather	锋面天气【大气】
frontal zone	锋区【大气】,锋带
front characteristics	锋的特征
frontogenesis	锋生【大气】
frontogenetical area	锋生区
frontogenetical front	锋生锋
frontogenetical function	锋生函数
frontogenetical sector	锋生区
frontology	锋面学,锋面学说
frontolysis	锋消【大气】
frontolytical area	锋消区
frontolytical front	消散中的锋
frontolytical sector	锋消区
front passage fog	锋际雾
front shear line	锋面切变线
front squall	锋面飑,线飑
frost	霜【大气】
frost action	冻裂作用,冰冻作用
frost belt	〔冰〕冻带
frostbite	霜害,冻伤
frost blister	冻土丘,冰丘
frost boil	①冻腾②冻凸地
frost churning	冻搅
frost climate	寒冷气候,冰原气候
frost damage	霜害
frost dam	冻堤
frost day	霜日【大气】
frost depth	霜〔透〕深〔度〕
frost detector	霜检测器
frost fan	防霜鼓风机,防霜风扇
frost feathers	霜羽
frost flakes	霜片
frost flower	霜花
frost fog	霜雾
frost free day	无霜日【农】
frost-free season	无霜期
frost hazard	霜害,冻害
frost haze	冻霾
frost heaving	冻胀丘,冻拔
frost heaving force	冻胀力【地理】
frost hole	霜眼,霜穴
frost hollow	霜坑,霜洼
frosting	结霜
frost injury	霜冻【大气】,霜害【农】
frost in the air	气霜,冰晶
frostless season	无霜期
frostless zone	无霜带,无霜区
frost-lifting	冻拔
frost line	霜线
frost mist	霜霭,冰晶霾
frost mound	冻丘
frost period	霜期【大气】

frost plants　冰雪浮游生物
frost pocket(或 hollow)　霜注
frost point　霜点【大气】
frost-point hygrometer　霜点湿度表
frost-point technique　霜点法
frost-point thermometer　霜点温度表
frost prevention　防霜【大气】
frost protection　防霜
frost ring　霜轮
frost riving　冻裂
froströk　冻烟
frost season　霜期
frost smoke　冻烟
frost snow　冻雪
frost splitting　冻裂
frost table　冻土面
frost weather　霜冻天气
frost weathering　寒冻风化【地理】
frost work　冻裂
frost zone　霜区
Froude number　弗劳德数【大气】
frozen dew　冻露
frozen fog　冻雾,冰雾
frozen ground　地冻,冻土【地理】
frozen-in-field　冻结场
frozen-in hypothesis　冻结假设
frozen-in magnetic field　冻结磁场
frozen-in random function　冻结随机函数
frozen precipitation　冻结降水
frozen rain　冻雨
frozen snow crust　冻结雪壳
frozen soil　冻土【大气】
frozen soil apparatus　冻土器【大气】
frozen turbulence　冻结湍流
frozen-turbulence approximation　冻结湍流近似
frozen-turbulence hypothesis　冻结湍流假定

fruit-frost meteorology　果树防冻气象学
frutex　(拉)灌木【植】
F-scale　F 尺度
F test　F 检验【农】
fuel moisture　可燃物含水量
Fuhjin　尘暴(日语名)
Fujita-Pearson scale（FPP）　藤田-皮尔逊等级(表征龙卷风强度的特征数)
Fujita scale　藤田等级
Fujiwara effect　藤原效应
fuketsu(＝wind cave)　风穴(日语名)
fulchronograph　闪电电流计,闪电(电流特性)记录仪
fulgurite　闪电熔岩(因闪电打击大地,使大地中的矿物熔成许多玻璃状管子)
full-disc cloud picture　全景圆盘云图【大气】
full-disc picture　全景圆盘〔云〕图
full-disk display　全景圆盘〔云〕图
full-disk image　全景〔圆盘〕图像
full moon　望【天】,望月
full resolution　全分辨率【大气】,原分辨率,最高分辨率
full-wave theory　全波理论【地】
fully-arisen sea（＝fully-developed sea）充分成长的风浪
fully compressible　〔完〕全压缩〔的〕
fully developed ocean wave　充分发展的海浪
fully developed turbulence　充分发展的湍流
fully ionized plasma　完全电离等离子体
fully penetrating well　全透水井
fully-rough flow　全粗糙流
fully turbulent　完全湍流的
fume　微粒污染物
fumigation　①熏烟法②烟下沉现象③稳定烟云

fumulus 缟状(云)(旧名)
function 函数
functional 泛函【数】
functional analysis 泛函分析【数】
functional derivative 函数导数,函数微商
function block 功能块【计】
function call 函数调用【计】
function fitting 函数拟合
function of frontogenesis 锋生函数【大气】
fundamental dimensions 基本维数
fundamental equations of hydrodynamics 流体动力学基本方程
fundamental frequency 基频
fundamental point 基本点
fundamental unit 基本单位
fundamental vibration 基本振动
fundamental wave 基波【物】
funnel(或 funnel column) 龙卷漏斗柱
funnel aloft 空中漏斗
funnel cloud 漏斗云【大气】
funnel column 漏斗云〔柱〕
funnelling 狭管效应

furacana(=furacan(e),furicane,furicano) 飓风
furiani 富里阿尼风(意大利波河附近的西南风)
Furious fifty 狂暴西风(指南半球 50°S 附近强西风)
further outlook 短期天气展望
fusion ①消融,融解②聚变
fusion chemistry 聚变化学【化】
fuzziness 模糊〔性〕
fuzzy 模糊〔性〕【数】
fuzzy classification 模糊分类
fuzzy differential equation 模糊微分方程
fuzzy information 模糊信息
fuzzy knowledge 模糊认识,弗晰认识
fuzzy logic 模糊逻辑
fuzzy mathematics 模糊数学【计】
fuzzy measure 模糊测度
fuzzy neartude 模糊贴近度
fuzzy set 模糊集【计】
fuzzy theory 模糊理论
fuzzy variable 模糊变量
F value F 值【农】

G

GA(genetic aglorithm) 遗传算法
gage(=gauge) 量器
Gaia hypothesis 盖娅假说
gaign 爬山风(意大利语名)
gain 增益,放大
gain factor 增益因子
gain function 增益函数
gaining stream 盈水河,地下水补给河
galactic cosmic ray 银河宇宙射线
galactic light 银河光

galactic radiation 银河系辐射,星系辐射
Galactic System 银河系【天】
galaxy 星系【天】
Galaxy 银河系【天】
gale 8 级风【大气】,大风
GALE(Genesis of Atlantic Lows Experiment) 大西洋低压起源试验
gale cone 大风信号
gale pollution 大风污染
Galerkin approximation 伽辽金近似

Galerkin finite element method 伽辽金有限元法
Galerkin method 伽辽金法
Galerkin-Petrov method 伽辽金-彼得罗夫法【数】
Galerkin spectral method 伽辽金波谱法
galerne(＝galerna, galerno) 加勒内风（英吉利海峡附近的西北狂风）
gale signal 大风信号
gale warning 大风警报【大气】
Galilean invariant 伽利略不变量
Galilei 伽(加速度单位,＝1 cm/s^2)
gallego 加耶果风（西班牙和葡萄牙的寒冷北风）
gallium 镓
gallium arsenide（GaAs） 砷化镓【物】
galvanic cell 原电池
Galvanic ozone monitor 伽伐尼臭氧监测仪
galvanometer 电流表
game theory 对策论【数】,博奕论
gamma distribution (＝γ distribution) 伽玛分布,γ分布
gamma function (＝γ function) 伽玛函数,γ函数
gamma radiation (＝γ radiation) 伽玛辐射,γ辐射
gamma-ray(＝γ-ray) γ射线,伽玛射线
gamma ray snow gauge (＝γ ray) γ射线雪量器
gamma scale (＝γ scale) 伽玛尺度,γ尺度
Ganges type 恒河型(温度)
Gans theory 甘斯理论
gap in spectrum 谱间隙
gap wind 狭道风,山口风
garbage 空间碎片（火箭或卫星的碎片）
garbin(＝garbi) 加宾风（法国西南部的一种海风）
garden city 田园城市【地理】
Gardner counter 加德纳计数器（测量空气中细微粒用）
gargal（见 grega） 格里加尔风
GARP（Global Atmospheric Research Program） 全球大气研究计划【大气】
GARP Atlantic Tropical Experiment (GATE) 全球大气研究计划大西洋热带试验
Garrett-Munk spectrum 加勒特-芒克谱,G-M谱
garua(＝camanchaca) 浓湿雾
gas 气体
gas cap 〔流星〕气冠,气顶
gas chromatograph 气相色谱仪
gas chromatography 气相色谱法【化】
gas chromatography-mass spectrometer system 气相色谱仪-质谱仪系统
gas coal 气煤【地质】
gas constant 气体常数【大气】
gas constant per molecule 分子气体常数【大气】
gas density balance 气体密度天平【化】
gas electrode 气体电极【化】
gaseous absorption loss 气体吸收损耗
gaseous diffusional separation 气体扩散分离〔法〕【化】
gaseous（electrical）**discharge** 气体放电
gaseous nebula 气体星云
gaseous phase 气相
gaseous pollutant 气体污染物
gaseous pollution 气体污染
gaseous polymerization 气相聚合【化】
gaseous state 气态
gaseous tracer technique 气态示踪物法（扩散测量用）
gases dissolved under pressure 加压溶解气体【航海】
gas field 气田【地质】
gas-filled detector 充气式探测器（中子法土壤含水量测量）
gas filter correlation spectroscopy 气体

过滤相关波谱学
gasification 气化
gas laser 气体激光器【物】
gas law 气体定律
gas liquid chromatography 气液色谱法【化】
gasol 液化气体
gasometer 气量表
gasometry 气体分析【化】
gas-particle distribution factor 气体粒子分布因子
gas phase 气态,气相
gas-phase kinetics 气相动力学
gas-phase polymerization 气相聚合【化】
gas phase reaction 气相反应
gas solid chromatography 气固色谱法【化】
gas-solid reactions 气固反应
gas-sphere(=air-sphere) 气界,气圈
gas theory 气体论
gas thermometer 气体温度表【大气】
gas-to-particle conversion (GTPC) 气体-粒子转化,气粒转化
gas transfer 气体输送
gate-to-gate azimuthal shear 距离选通方位切变
gateway 网关【计】
gateway city 要冲城市【地理】
gathering ground 聚水区
gating ①闸②选通
gauge(=gage) 量器
gauge height 规范高度,标准高度
gauge pressure 器示压力
gauge relation 水位关系,流量关系
gauge zero 规范零〔高度〕
gauging section 〔水文〕测量截面
gauging site 〔水文〕测量点
gauss(G) 高斯(磁通量密度单位)
Gauss divergence theorem 高斯散度定理
Gauss elimination 高斯消元法【数】
Gaussian correlation function 高斯相关函数
Gaussian curve 高斯曲线
Gaussian curvature 高斯曲率【数】
Gaussian dispersion model 高斯扩散模式
Gaussian distribution 高斯分布
Gaussian elimination method 高斯消元法
Gaussian grid 高斯网格
Gaussian latitude 高斯纬度
Gaussian lineshape 高斯线形
Gaussian model 高斯模式
Gaussian plume models 高斯烟流模式
Gaussian point 高斯点
Gaussian process 高斯过程【数】
Gaussian quadrature 高斯求积法
Gaussian random field 高斯随机场
Gaussian velocity fluctuation 高斯速度起伏
Gaussian wave packet 高斯波包
Gaussian weight 高斯权重
Gaussian white noise 高斯白噪声
Gauss precision weight 高斯精密度权重
Gauss quadrature formula 高斯求积公式
Gauss-Seidel iterative method 高斯-塞德尔迭代〔法〕【数】,G-S 迭代〔法〕
Gauss's theorem 高斯定理
Gauss window 高斯窗
GAW(Global Atmosphere Watch) 全球大气监测网
Gay-Lussac law 盖吕萨克定律【物】
Gazetteer period 方志时期(气候)
GCA① (ground-controlled approach) 引导着陆〔系统〕②(great circle arc) 大圆弧
GCA minimums GCA 低限
GCM (general circulation model) 大气环流模式【大气】
GCOS (Global Climate Observing System) 全球气候观则系统
GCP (Global Carbon Project) 全球碳计划
GCRP (Global Change Research Program) 全球变化研究计划(IGBP)

GCTE (Global Change and Terrestrial Ecosystem) 全球变化和地球生态系统(IGBP)
GDH (growing degree-hour) 生长度-时
GDP (gross domestic product) 国内生产总值
GDPS (Global Data Processing System) 全球资料处理系统
gebli (见 ghibli) 基布利风
GECAFS (Global Environmental Change and Food Systems) 全球环境变化与食物系统
GECHH (Global Environmental Change and Human Health) 全球环境变化与人类健康
GECHS (Global Environmental Change and Human Security) 全球环境变化与人类安全
GEF (Global Environment Facility) 全球环境基金(世界银行等1991年提议成立)
geg 杰克旋风(西藏地区的沙漠尘旋风)
gegenschein 反黄道光,对日照(德语名)
Geiger-Muller counter 盖革-米勒计数器【物】
Geiger-Muller tube 盖革-米勒计数管
gel 凝胶【物】
Geminids meteor shower 格米尼德期流星雨
gem mineralogy 宝石矿物学【地质】
gemology 宝石学【地质】
gemstone 宝石【地质】
gending 全丁风(爪哇的一种焚风)
gene 基因【生化】
general astronomy 普通天文学【天】
General Broadcast 通用〔气象〕广播
general cartography 普通地图学【地理】
general circulation 大气环流
general circulation model (GCM) 大气环流模式【大气】
general circulation of atmosphere 大气环流
general forecast 一般天气预报
general geographical name 地理通名【地理】
general geographic map 普通地理图【地理】
general geography 普通地理学【地理】
generalized absorption coefficient 广义吸收系数
generalized Bayes decision function 广义贝叶斯决策函数【数】
generalized Boolean algebra 广义布尔代数【数】
generalized coordinate 广义坐标
generalized directivity function 广义方向性函数
generalized displacement 广义位移
generalized hydrostatic equation 广义〔流体〕静力方程
generalized magnetic latitude 广义〔地〕磁纬〔度〕
generalized solution 广义解【数】
generalized transmission function 广义透射函数
generalized vertical coordinate 广义垂直坐标
general lighting 一般照明【航海】
general map 普通地图【地理】
general meteorology 普通气象学
general notice 一般飞行报告
general packet radio service (GPRS) 通用分组无线电业务系统
general physical geography 普通自然地理学【地理】
general radio communication 常规无线电通信【航海】
general solution 通解【数】
genera of cloud 云属
generate and test 生成检验〔法〕
generating area 生成区
generating cell 生成胞,生成单体

generating function　母函数【数】,生成函数
generation of landforms　地貌世代【地理】
generation rate　产生率【物】
generic rule　通用法则
genesis　发生,起源
Genesis of Atlantic Lows Experiment (GALE)　大西洋低压起源试验
genetic aglorithm(GA)　遗传算法
genetic engineering　基因工程【生化】
genetic classifcation of climate　气候形成分类法【大气】
genitus　衍生云
Genoa(Genova) cyclone　热那亚气旋
genome　基因组【生化】
gentle breeze　3级风【大气】,微风
GEO (Group on Earth Observations)　〔国际〕地球观测组织(2005年成立,已有60个国家和43个国际组织参加)
geoanalysis　地理分析【地理】
geobiochemistry　地球生物化学【生化】
geocentric coordinate　地心坐标【天】
geocentric gravitational constant　地心引力常数【天】
geocentric horizon　地心地平
geocentric latitude　地心纬度
geocentric longitude　地心经度
geocentric reference system　地心坐标系
geocentric system　地心体系【天】
geocentric zenith　地心天顶
geochemical environment　地球化学环境
geochemical gas survey　气体地球化学测量【地质】
geochemical landscape　地球化学景观【地理】
geochemical method　地球化学方法
geochemical model　地球化学模型【地质】
Geochemical Ocean Section Study (GEOSECS)　海洋断面地球化学研究计划
geochemistry　地球化学【地质】

geochemistry of element　元素地球化学【地】
geoclimatic drying power　地面干燥率
geocorona　地冕【地】
geocryology　冻土学【地理】
geodesic grid　测地网络
geodesic line　①测地线②〔最〕短程线
geodesy(=geodetics)　大地测量学
geodetic coordinate　地球坐标
geodetic curve　大地测量曲线
Geodetic Earth Orbiting Satellite(GEOS)　大地测量地球轨道卫星
Geodetic Satellite (GEOSAT)　大地测量卫星
geodetic survey　大地测量【航海】
geodynamic height　动力高度
geodynamic meter　动力米
geodynamics　地球动力学【地】
geoecology　地生态学【地理】
geographic accuracy　地理精度【地理】
geographical boundary　地理界线【地理】
geographical coordinate　地理坐标
geographical data bank　地理数据库【地理】
geographical data handling　地理数据采编【地理】
geographical distribution　地理分布【地理】
geographical environment　地理环境【地理】
geographical factors　地理因子【地理】
geographical information system (GIS)　地理信息系统【地理】
geographical latitude　地理纬度
geographical longitude　地理经度
geographically possible sunshine duration　地理可能日照时数
geographical name　地名【地理】
geographical name data bank　地名数据库【地理】
geographical process　地理过程【地理】
geographical remote sensing　地理遥感

【地理】
geographical sphere 地理圈【地理】
geographical survey 地理考察【地理】
geographical synthesis 地理综合【地理】
geographical system 地理学体系【地理】
geographical unit 地理单元【地理】
geographic base map 地理底图【地理】
geographic coordinate 地理坐标【地理】
geographic correlation 地理相关法【地理】
geographic effect on winds 地形对风的影响
geographic graticule 地理坐标网【地理】
geographic grid 地理网格
geographic horizon 地理地平
geographic identifier 地理标识符【地信】
geographic information system（GIS）地理信息系统【农】
geographic location 地理位置,地理定位
geographic mapping 地理制图【地理】
geographic meridian 地理子午线【地信】
geographic parallel 地理纬圈【地信】
geographic profile 地理剖面【地理】
geographic reference system 地理参照系【地信】
geographimetrics 地理计量学【地理】
geography 地理学
geography of communication 交通地理学【地理】
geoheat 地热【地】
geoid 大地水准面,地球体【地理】
geoidal height map 大地水准面高度图【航海】
geoidal surface 大地水准面
geoinformatics 地学信息技术
geoisotherm 等地温线
geological age 地质时代
geological chemistry 地质化学
geological climate 地质气候
geological era 地质代
geological oceanography 地质海洋学

【海洋】
geological period 地质时期(气候),地质纪
geologic environment 地质环境【地质】
geologic epoch 地质世
geologic era 地质代
geologic hazard 地质灾害【地质】
geologic thermometer 地质温度表
geologic time 地质时期【地质】
geologic time scale 地质年表【地质】
geology 地质〔学〕【地】
geomagnetic activity index 地磁活动指数
geomagnetically quiet time 〔地〕磁宁静期
geomagnetic coordinate 地磁坐标【地】
geomagnetic density response 大气密度对地磁的响应
geomagnetic dipole 地磁偶极子
geomagnetic disturbance 〔地〕磁扰
geomagnetic element 地磁要素
geomagnetic equator 地磁赤道
geomagnetic excursion 地磁漂移【地质】
geomagnetic field 地磁场
geomagnetic index 磁情指数
geomagnetic latitude 磁纬
geomagnetic meridian 〔地〕磁子午圈
geomagnetic pole 地磁极【航海】
geomagnetics 地磁学
geomagnetic storm 地磁暴
geomagnetic survey 地磁测量【地】,磁测
geomagnetic tide 地磁潮
geomagnetism 地磁〔学〕【地】
geomatics 测绘学【测】
geomedicine 地理〔环境〕医学
geometrical acoustics 几何声学
geometrical attenuation 几何衰减【农】
geometrical effect 几何效应【化】
geometrical horizon 几何地平
geometrical optics 几何光学
geometrical spreading 几何扩散
geometric correction 几何校正【农】,几

何订正
geometric cross section　几何截面
geometric height　几何高度
geometric mean　几何平均
geometric meter　几何米
geometric optical region　几何光学区
geometric series　等比级数【数】
geometry　几何学
geomorphochronology　地貌年代学【地理】
geomorphogenesis　地貌成因【地理】
geomorphological process　地貌过程【地理】
geomorphology　地貌学【大气】
geo-navigation　地文航海【航海】
geonomy　地学
geophysical anomaly　地球物理异常【地】
geophysical data　地球物理资料
geophysical day　地球物理日(类似地球物理年)
geophysical exploration　地球物理勘探【地】,物探
geophysical fluid dynamics　地球物理流体动力学
Geophysical Fluid Dynamics Laboratory (GFDL)　地球物理流体动力学实验室(美国)
geophysical modelling　地球物理学模拟
geophysical sciences　地球物理科学
Geophysical Year　地球物理年
geophysics　地球物理学【大气】
geopotential　大地位【地】,〔重力〕位势
geopotential constrained initialization　〔重力〕位势限制初始化
geopotential enstrophy　位势拟能
geopotential field　位势场
geopotential foot　位势英尺
geopotential height　位势高度【大气】
geopotential meter (gpm)　位势米【大气】
geopotential surface　等位势面
geopotential tendency equation　位势趋势方程
geopotential thickness　位势厚度
geopotential topography　位势〔等高〕图
geoprobe　地球物理探测火箭(探测约4000英里高空)
georeference system　地理〔坐标〕参照系【地信】
George's index　乔治指数(亦称 K 指数,度量雷暴潜势的稳定度指数)
GEOS (Geodetic Earth Orbiting Satellite)　大地测量地球轨道卫星
GEOSAT (Geodetic Satellite)　大地测量卫星
geoscience　地学【地质】
Geoscience Information Society (GIS)　地球科学信息学会(美国)
geoscientist　地学家
geosphere　地圈【地理】,陆界,陆圈
GEOSS (Global Earth Observing System of Systems)　全球综合地球观测系统
geostationary　对地静止
geostationary meteorological satellite (GMS)　地球静止气象卫星【大气】
geostationary meteorological satellite system (GMSS)　地球静止气象卫星系统(日本)
Geostationary Operational Environmental Satellite (GOES)　地球静止业务环境卫星
Geostationary Operational Meteorological Satellite (GOMS)　地球静止业务气象卫星(俄罗斯发射的卫星)
geostationary satellite　地球静止卫星
geostrophic　地转〔的〕
geostrophic adjustment　地转适应【大气】
geostrophic advection　地转平流【大气】
geostrophic advection scale　地转平流尺度
geostrophic approximation　地转近似
geostrophic assumption　地转假定
geostrophic balance　地转平衡
geostrophic coordinates　地转坐标

geostrophic current 地转流【大气】
geostrophic degeneracy 地转退化
geostrophic departure 地转偏差
geostrophic deviation 地转偏差【大气】
geostrophic distance 地转距离
geostrophic divider 地转风风速分析器
geostrophic drag 地转曳力
geostrophic drag coefficient 地转曳力系数
geostrophic equilibrium 地转平衡
geostrophic flow 地转流
geostrophic flux 地转通量
geostrophic front 地转锋
geostrophic increment 地转增量
geostrophic inertial instability 地转惯性不稳定
geostrophic momentum 地转动量
geostrophic momentum approximation 地转动量近似
geostrophic motion 地转运动【大气】
geostrophic normal mode 地转正交模,地转正交波
geostrophic potential vorticity 地转位势涡度
geostrophic Richardson number 地转里查森数
geostrophic shearing deformation 地转切变变形
geostrophic streamfunction 地转流函数
geostrophic stretching deformation 地转伸展变形
geostrophic transport 地转风输送
geostrophic turbulence 地转湍流
geostrophic value 地转风值
geostrophic vorticity 地转涡度【大气】
geostrophic wind 地转风【大气】
geostrophic wind flow 地转风流
geostrophic wind front 地转风锋
geostrophic wind level 地转风高度
geostrophic wind method 地转风方法
geostrophic wind scale 地转〔风〕风速标尺
geostrophy 地转状态
geosynchronous orbit 地球同步轨道
geosynchronous satellite 地球同步卫星【大气】
geosynchronous very high resolution radiometer(GVHRR) 地球同步甚高分辨率辐射仪
geosystem 地理系统【地理】
geotechnologie 地球工程学
geotherm 地热等温线
geothermal activity 地热活动【地质】
geothermal anomaly 地热异常【地质】
geothermal energy 地热能【地质】
geothermal field 地热田【地质】
geothermal gradient 地下增温率
geothermal heat 地热
geothermal phenomenon 地热现象【地质】
geothermal prospecting 地热勘探【地质】
geothermal reservoir 地热水库【地质】
geothermal resources 地热资源【地质】
geothermal survey 地热调查【地质】
geothermal system 地热系统【地质】
geothermic depth 地下增温深度
geothermics 地热学【地质】
geothermic step 单位深度地温差
geothermoenergy 地热能【农】
geothermometer 地温表【大气】,地球温度计【地】
geothermy 地热学
Gerdien aspirator 盖尔丁通风器(测离子用)
Gerdien conductivity apparatus 盖尔丁电导率测量仪
germination 萌发【植】
Gerstner wave 格斯特纳波(有限振幅的旋转重力波)
gestalt 完形,格式量
GEWEX(Global Energy and Water Cycle Experiment) 全球能量与水循环试验

geyser 间歇〔喷〕泉,间歇温泉
gharbi 加比风(摩洛哥西风)
gharra 格拉风(利比亚和非洲的东北风)
ghaziyah (＝ragut) 拉格特风(地中海伊斯肯德湾的飑性下沉风)
GHG (greenhouse gases) 温室气体【大气】
ghibli (＝chibli,gebli,gibleh,gibli,kibli) 基布利风(利比亚西北的干热风,类似焚风)
GHIS (GOES High-resolution Interferometer Sounder) 戈斯卫星高分辨率干涉仪探测器
ghost 反常回波,鬼怪波
ghost of Gouda 哥达鬼怪风(南非的局地飑)
giant elliptical galaxy 巨椭圆星系【天】
giant galaxy 巨星系【天】
giant molecule 巨分子【化】
giant nuclei 巨核
giant-pluse laser 巨脉冲激光器【物】
giant pulse technique 巨脉冲技术
giant star 巨星【天】
Gibbs free energy 吉布斯自由能【物】
Gibbs function 吉布斯函数【物】
Gibbs phenomenon 吉布斯现象
Gibbs stability concept 吉布斯稳定观念
giboule 吉布莱风(法国大西洋沿岸的西北风飑)
Gibraltar outflow water 直布罗陀外流水
GIFS (Global Interactive Forecast System) 全球交互式预报系统
giga- 吉〔咖〕(词头,10^9)
gigabit 十亿位,吉位【计】
gigabits per second 十亿位每秒【计】
gigabyte (GB) 十亿字节,吉字节【计】
gigabytes per second 十亿字节每秒【计】
gigahertz (GHz) 吉赫
Gill anemometer 吉尔风速表
gilles wind 吉尔斯风(非洲留尼汪岛的西风)

gimbal 平衡环,常平架(使气压计、罗盘等保持水平)
gimbal mounted raingauge 常平架雨量计器
GIS (geographical information system) 地理信息系统【地理】
GISC (Global Informction System Center) 全球信息系统中心
glacial ①冰川的②冰成的③冰的
glacial anticyclone 冰川反气旋
glacial anticyclone theory 冰川反气旋学说
glacial basin 粒雪盆,冰川盆地,冰川流域
glacial breeze 冰川风
glacial cycle 冰川循环
glacial drift 冰碛【物】
glacial epoch 冰川世,冰河期
glacial erosion 冰蚀作用【地质】
glacial fluctuation 冰川波动
glacial geology 冰川地质学【地理】
glacial high 冰原高压
glacial-interglacial cycle 冰期-间冰期旋回
glacial-interglacial variation 冰期-间冰期变化
glacial interstade 冰川间冰阶
glacial maximum 最大冰川期
glacial period (phase) 冰期
glacial rebound 冰消地面回升
glacial recession 冰川退缩,冰〔川消〕退
glacial record 冰川记录
glacial replenishment 冰川补给
glacial stage 冰阶,冰期【地质】
glacial stage snowline 冰期雪线
glacial stream 冰川河流【地理】
glacial termination 冰期结束期
glacial trough ①冰蚀槽②冰川槽
glacial varves 冰川纹泥,冰川融积层
glaciation 冰川作用【地理】
glaciatioin level 冰川作用程度
glaciation limit ①冰川极限②冰川最大进展(时间或位置)③冰川下限

glacier 冰川【地质】
glacier berg 〔冰川〕冰山
glacier breeze 冰川风【大气】
glacier discharge 冰川放水
glacier fall 冰川瀑布
glacier flood 冰川洪水
glacier front 冰川锋
glacier ice 冰川冰
glacierization 冰川化
Glacier Lake Outbreak Flood (GLOF) 冰川湖突发性洪水
glacier mass balance 冰川物质平衡
glacier plain 冰川平原
glacier tongue 冰川舌
glacier variation 冰川变迁
glacier wind 冰川风
glacigenic agent 成冰剂
glacioclimate 冰川气候【大气】
glacioclimatology 冰川气候学
glaciology 冰川学【地理】
glacon 小浮冰
glare 闪光
glare asthenopia 闪光视〔力减〕弱
glare ice 光滑冰
glass-house (＝greenhouse) climate 温室气候
glass ice 冰壳【航海】
glassy 静海,镜海
glaves(＝glave,glavis) 焚风(见于法罗群岛)
G-layer G 层
glaze 雨凇【大气】,冻雨
glazed frost 雨凇,冻雨
glaze ice 雨凇
glaze storm 雨凇暴
Gleissberg cycle 格莱斯堡黑子(90 年)周期
glide path 下滑航线,滑翔道
glime 半透明冰(介于 glaze 和 rime 之间),毛冰,瓷冰
glitter 闪〔光〕斑

global air pollution 全球性大气污染
global analysis ①全球分析②整体分析
global and diffuse solar radiation integrator 总辐射和太阳漫射辐射积分器
global area coverage(GAC) 全球覆盖
global association 全球性联系
Global Atmosphere Watch (GAW) 全球大气监测网
Global Atmospheric Research Program (GARP) 全球大气研究计划【大气】
Global baseline datasets 全球基本气候资料集
Global Carbon Project (GCP) 全球碳计划
global change 全球变化
Global Change and Terrestrial Ecosystem (GCTE,IGBP) 全球变化和地球生态系统(IGBP)
Global Change Research Program(GCRP,IGBP) 全球变化研究计划(IGBP)
Global Change System for Analysis, Research and Training (START) 全球变化的分析、研究和培训系统(IGBP 计划之一)
global circuit 全球电路
global circulation 全球环流
global circulation model 全球环流模式
global climate 全球气候【大气】
Global Climate Observing System(GCOS) 全球气候观测系统
global climate system 全球气候系统【大气】
global climatologic change 全球性气候变化
Global Data Assimilation System (GDAS) 全球资料同化系统
global data processing system (GDPS,WWW) 全球资料处理系统(WWW)
global distribution 全球分布
Global Earth Observing System of Systems

（GEOSS） 全球综合地球观测系统
global energy 全球能量
Global Energy and Water Cycle Experiment（GEWEX） 全球能量与水循环试验
global environment 全球环境
Global Environmental Change and Food Systems（GECAFS） 全球环境变化与食物系统
Global Environmental Change and Human Health（GECHH） 全球环境变化与人类健康
Global Environmental Change and Human Security（GECHS） 全球环境变化与人类安全
global environmental monitoring system（GEMS） 全球环境监测系统
Global Environment Facility（GEF） 全球环境基金
global fitting 整体拟合
global horizontal sounding technique（GHOST） 全球水平探测技术
global ice volume 全球总冰量
Global Information System Center（GISC） 全球信息系统中心
Global Interactive Forecast System（GIFS） 全球交互式预报系统
global internal energy 全球内能
globalization 全球化
global kinetic energy 全球动能
Global Land Project（GLP） 全球陆地计划（IGBP 内）
global mean energy balance 全球平均能量平衡
global mean temperature 全球平均温度
global meteorological data 全球气象资料
global-mode coverage 全球覆盖【航海】
global model 全球模式
Global Navigation Satellite System（GNSS） 全球导航卫星系统【地信】
Global Observing System（GOS，WWW） 全球观测系统（WWW）
Global Ocean Observing System（GOOS） 全球海洋观测系统【地信】
Global Positioning System（GPS） 全球定位系统【大气】
global potential energy 全球位能
global prediction 全球预报
global radiation 总辐射【大气】
Global/Regional Assimilation and Prediction Enhanced System（GRAPES） 全球/区域同化和预报增强系统（模式）（是中国自主研发的第一代数值预报模式）
global scale 全球尺度
Global Sea Level Observing System（GLOSS） 全球海平面观测系统
global spectral model 全球谱模式
global telecommunication system（GTS，WWW） 全球电信系统【大气】（WWW）
Global Terrestrial Observation System（GTOS） 全球陆地观测系统
global transport 全球运送，全球输送
global warming 全球增温
global warming potential（GWP） 全球增温潜势，全球变暖潜势
global warming rate 全球增温率
global water cycle 全球水循环
Global Water System Project（GWSP） 全球水系统计划（IGBP 内）
global weather experiment（GWE） 全球天气试验
global weather reconnaissance 全球天气侦察
global wind system 全球风系【地】
globe lightning（= ball lightning） 球〔状〕闪〔电〕
globe thermometer 〔黑〕球温〔度〕表（测环境辐射热用）
globular projection 球状投影
GLOF（Glacier Lake Ooutbreak Flood）

冰川湖突发性洪水
glomeratus　簇状(云)
glory　宝光[环]【大气】,峨眉宝光
glow discharge　辉光放电【大气】
glow-emitting layer　气辉层
GLP(Global Land Project)　全球陆地计划
glucose　葡萄糖【生化】
GMS(geostationary meteorological satellite)　地球静止气象卫星【大气】
GMST(Greenwich mean sidereal time)　格林尼治平恒星时【天】
gnomon　表【天】,圭表【天】
gnomonic projection　心射切面投影,球心投影【测】,日晷投影
GNP(gross national product)　国民生产总值
GNSS(Global Navigation Satellite System)　全球导航卫星系统【地信】
goal-directed reasoning　目标推理[法]
gobi　戈壁【地理】
Goddard Laboratory of Atmospheric Sciences(GLAS)　戈达德大气科学实验室(美国)
Goddard Space Flight Center(GSFC)　戈达德空间飞行中心(美国)
GOES(Geostationary Operational Environmental Satellite)　地球静止业务环境卫星,戈斯卫星
GOES High-resolution Interferometer Sounder(GHIS)　戈斯卫星高分辨率干涉仪探测器
Golay cell　高莱探测器
Goldbach problem　哥德巴赫问题【数】
goldbeater's skin　肠膜,肠衣
goldbeater's skin hygrometer　肠膜湿度表
golden section search　黄金分割搜索【数】
Gold slide　戈尔德滑尺(供船用水银气压表读数订正用)
golfada　戈尔法达风(地中海的烈风)

gondola　圆球宝,吊篮(气球的)
Gondwanaland　冈瓦纳古大陆【地质】
goniometer　①测向器②测角计
goniometer system　测向装置,测角系统
goniometry direction finding　测角术定向
goodness of fit　吻合度
good visibility　良好能见度
Goody absorption band　古迪吸收带
Goody band model　古迪带模式
gorge　峡谷【地理】
gorge wind　峡谷风【大气】
gorich　戈里奇风(伊朗的西南风)
gosling blast　小鹅风(英国突发性的阵雨或湿雪)
gosling storm　小鹅风暴(英国突发性的雨飑)
governing equation　控制方程【大气】,支配方程
gowk storm　布谷鸟风暴(英国4月底或5月初的烈风或风暴)
GOOS(Global Ocean Observing System)　全球海洋观测系统
GO_3OS(Global Ozone Observing System)　全球臭氧观测系统(WMO)
gpm(geopotential meter)　位势米
GPS(global positioning system)　全球定位系统【大气】
grab sampling　手选(简单)取样
gradex method　雨代法
gradient　梯度【数】
gradient current　梯度流
gradient flow　梯度流
gradient level　梯度风高度
gradient observation　梯度观测
gradient operator　梯度算子
gradient Richardson number　梯度里查森数
gradient transport theory　梯度输送理论
gradient velocity　梯度风速
gradient wind　梯度风【大气】
gradient wind equation　梯度风公式

gradient wind front 梯度风锋
gradient wind level 梯度风高度(约900 hPa)
grading 分级,粒度
gradually commencing type 缓始型(磁暴)
graduated glass cylinder 量杯(量雨用)
graduation 刻度,分度
graesy weather 雾天
Grain in Ear 芒种【大气】(节气)
Grain Rain 谷雨【大气】(节气)
grains of ice 冰粒,冰丸
gram 克
gram calorie 克·卡
gram-mole(= gram-molecule) 克分子〔量〕
Grand Maximum 中世纪〔太阳黑子〕极大期
Grand Vent(= plouazaou) 格朗旺风(引起法国奥佛尼持续阵雨的西风,法语名)
granite 花岗岩【地质】
grant 授权【计】
granular snow 春天粒雪,米雪
granulation 米粒组织【天】
granules 米粒组织
GRAPES (Global/Regional Assimilation and Prediction Enhanced System) 全球/区域同化和预报增强系统(模式)(中国自主研发的第一代数值预报模式)
graph 图,图表
graphical solution 图解【物】
graphical statics 图解静力学
graphic method 图解法【物】
graphic plotter 绘图机
graphic representation 图示
graphing board 绘图板
graph theory of molecular orbitals 分子轨道图形理论【化】
Grashof number 格拉斯霍夫数
grass 草波(雷达 A 型显示屏上的背景噪声,陈旧词)
grass-growing days 长草日数
grassland 草原
grassland climate 草原气候【农】
grassland farming 草地农业【农】
grass minimum〔temperature〕 最低草温
grass minimum thermometer 最低草温表
grass surface temperature 草面温度
grass temperature 草温
grass thermometer 草温表【大气】
grassy soil 草地
grating 光栅,栅格
graupel 霰【大气】,软雹,雪丸
gravel bar 砾石洲,砾石滩
gravest Rossby wave mode 最平缓罗斯贝波模
gravimeter 重力仪【地】
gravimetric analysis 重量〔分析〕法【化】
gravimetric factor 重量因子【化】
gravimetric hygrometry 重量测湿法,称重测湿法
gravimetric method 重量法【化】
gravimetric raingauge 重力雨量器
gravimetry 重力测量学【地】,重量〔分析〕法【化】
gravitation 引力【物】,重力
gravitational acceleration 重力加速度
gravitational attraction 地球引力
gravitational constant 引力常量【物】
gravitational convection 重力对流
gravitational equilibrium 重力平衡
gravitational field 引力场【物】,重力场
gravitational head 重力落差
gravitational instability 引力不稳定性【天】,重力不稳定
gravitational mode 重力模
gravitational moist-convective instability of the second kind 第二类重力湿对流不稳定度
gravitational potential 引力势【物】,重力位势

gravitational potential energy 引力势能，重力势能
gravitational pressure 重力压〔力〕
gravitational separation 重力分离
gravitational stability 重力稳定度
gravitational tide 引力潮【地】
gravitational water 重力水,自流水
gravitational wave 重力波
gravitational wind 重力风
gravity 重力【物】
gravity acceleration 重力加速度【地】
gravity correction 重力订正
gravity current 重力流【大气】,密度流
gravity field 重力场【地】
gravity flow 重力流,无压流,自流
gravity force 重力
gravity gradient stabilization 重力梯度稳定
gravity gradient survey 重力梯度测量【地】
gravity gradient zone 重力梯度带【地】
gravity gradiometer 重力梯度仪【地】
gravity measurement 重力测量【地】
gravity measurement at sea 海洋重力测量【地】
gravity model 引力模式【地理】
gravity potential 重力位【地】
gravity water 重力水,自流水
gravity wave 重力波【大气】
gravity wave breaking 重力波破碎
gravity wave drag 重力波拖曳【大气】
gravity wave drag parameterization 重力波拖曳参数化
gravity wind 重力风
gray absorber 灰色吸收体,灰体
gray atmosphere 灰体大气
gray body 灰体
Gray code 格雷码,二进制码
gray ice 灰冰
gray level 灰阶【物】
gray scale 灰〔色标〕度【大气】

grease ice 油脂状冰,冰脂
greasy weather 雾天
Great Basin high 大盆地高压（美国西部）
great circle 大圆
great circle arc (GCA) 大圆弧
great circle arc length 大圆弧长
great circle bearing 大圆方位【航海】
great circle chart 大圆海图【航海】
great circle course 大圆航向【航海】
great circle distance 大圆距离【航海】
great circle route 大圆路径
great circle sailing 大圆航线算法【航海】
great circle theory 大圆理论
great drought 大旱
Greater Cold 大寒【大气】(节气)
greater ebb tidal current 大潮落潮流
greater flood tidal current 大潮涨潮流
Greater Heat 大暑【大气】(节气)
greatest common divisor 最大公因数【数】
great pluvial 大雨期
great radiation belt 大辐射带（地球）
Great Red Spot 大红斑【天】
Great Whirl 〔索马里〕大涡流
grecale 格雷卡尔风（法国科西嘉的干冷风）
greco 格雷科风（地中海西部的东北风）
green belt 绿带（无霜带）
green belt city 绿带城市【地理】
green flash 绿闪〔光〕
Green function 格林函数【数】
greenhouse carbon dioxide enrichment 温室二氧化碳加浓【农】
greenhouse climate 温室气候【大气】
greenhouse effect 温室效应【大气】
greenhouse forcing 温室强迫〔作用〕
greenhouse gas emission 温室气体排放
greenhouse gases (GHG) 温室气体【大气】
greenhouse gas stabilization 温室气体稳定【化】
greenhouse heating 温室加热【农】
greenhouse management 温室管理【农】

greenhouse state 温室状态
greenhouse ventilation 温室通风【农】
greenhouse warming 温室〔效应〕增温
green iceberg 绿冰山
greening 绿化【林】
Greenland anticyclone 格陵兰反气旋
Greenland Current 格陵兰海流
Greenland high 格陵兰高压
Greenland Ice Sheet 格陵兰大冰原
Greenland Sea Deep Water 格陵兰海深水
green moon 绿月
green oxygen line 氧气绿线
green ray 绿射线
green revolution 绿色革命【地理】
green rim 绿边
green segment 绿弓形
Green's function 格林函数
green sky 绿天
green snow 绿雪
Green's theorem 格林定理
green sun 绿日
green thunderstorm 绿雷暴
Greenwich apparent civil time (GACT) 格林尼治视民用时
Greenwich apparent time (GAT) 格林尼治视〔太阳〕时
Greenwich civil time (GCT) 格林尼治民用时
Greenwich hour angle 格林〔尼治〕时角【航海】
Greenwich mean sidereal time (GMST) 格林尼治平恒星时【天】
Greenwich mean time (GMT) 格林尼治平时,世界时
Greenwich meridian 格林尼治子午线【天】,本初子午线
Greenwich sidereal time (GST) 格林尼治恒星时
Greenwich time (GT) 格林尼治时
grega (=gregale, gregau, gargal, guergal) 格里加尔风(地中海中西部和邻近欧洲地区的强东北风)
G region G 层
Gregorian calendar 格里历【天】
gregori wind 格里哥里风(奥地利提罗耳 3、4 月的东风)
grenade 榴弹(火箭测风用)
grenade method 榴弹测风法
grenade sounding 榴弹探测〔法〕
grey absorber 灰吸收体【大气】
grey atmosphere 灰大气【天】
grey body 灰体
grey body radiation 灰体辐射【大气】
grey level display 分层显示
grey scale 灰度
grid 网格【数】
grid data bank 格网数据库【地理】
gridding 画网格
gridding algorithm 方格坐标算法(加网格算法)
grid heading 网络航向
grid interval 格距
grid length 格距
grid map 网格地图【地理】
grid meridians 网格子午线
grid-mesh 格网
grid navigation 网格航行
grid nephoscope 栅状测云器
grid north 网格北
grid point 网格点【数】,格点
grid point method 〔网〕格点法
grid point model 格点模式
grid resolution 网格分辨率
grid spacing 格距
grid system 网格系
grid telescoping 网格嵌套
grid turbulence 栅屏湍流
grid variation 网络偏差,磁偏角
Gringorten plotting position 格兰戈腾图标位置
Gross-austausch 大型交换(德语名)

gross domestic product(GDP) 国内生产总值
gross error check 重大误差检验,极值检验
gross fog 浓雾
gross interception loss 总截留损失〔量〕
gross national product(GNP) 国民生产总值
gross photosynthetic intensity 总光合强度
gross primary production 总初级生产〔能〕力
gross saturator 预饱和器(双压法湿度发生器的)
gross weight 毛重,总重
Grosswetterlage 大型天气形势(德语名)
Grosswetter 大型天气
ground acquisition and command station (GA&CS) 地面接收和指令站
ground avalanche 地面雪崩
ground-based ionosonde 地基电离层探测仪
ground-based lidar 地基光〔雷〕达
ground-based measurement 地基测量
ground-based monitoring 地对空监测
ground-based observation 地基观测【大气】
ground check 基值(控空仪施放前的温、湿、压基值)
ground-check chamber 地面校正箱,地面校正室
ground clutter 地物杂波
ground-controlled approach (GCA) 引导着陆〔系统〕
ground control point 陆标
ground data 地面〔实况测量〕资料
ground discharge 云地放电,霹雳
ground echo 地物回波【大气】,地面回波
ground fallout plot 地面〔放射性〕落尘图
ground flash 地闪
ground flux 地〔面〕通量

ground fog 地面雾【大气】
ground frost 地面霜
ground generator 地面发生器(人工增雨防雹用)
Groundhog Day 美国土拨鼠日(2月2日圣烛节,民俗中的征兆日)
ground-hugging fog 贴地雾
ground ice ①底冰②地下冰,土内冰
ground ice mound 冰丘,冻丘
ground inversion 地面逆温
ground layer 近地层
ground level 地平〔面〕
ground level concentration 地面浓度
ground-level inversion 地面逆温
ground level source 地面源
ground line 搁浅线
ground magnetogram 地面磁照图
ground phenomenon 地面现象
ground receiving station 地面接收站【地理】
ground resolution 地面分辨率【地理】
ground speed 地速
ground state 基态【化】
ground station 地面站
ground streamer 地面流光
ground surface temperature 地表温度
ground swell ①长浪,长涌②地隆
ground target 地靶,地面目标
ground temperature 地温【大气】
ground test equipment 地面测试设备
ground-thermometer 地温表
ground-to-cloud discharge 地云闪电【大气】
ground track 地面轨迹
ground truth 地面真值,地面实况【地理】
ground truth data 〔认定〕实况资料
ground truth site 〔认定〕实况站
ground visibility 地面能见度
ground water 地下水
groundwater basin 地下水流域,地下水盆地

groundwater dam　地下水坝
groundwater dating　地下水定年(确定地下水存在的年龄)
groundwater depletion　地下水耗竭,地下水消落
ground-water depletion curve　地下水耗减曲线
groundwater divide　地下〔水〕分水界,潜水分水界
groundwater flow　地下水〔径流〕,基流
ground water hydrology(＝groundwater hydrology)　地下水文学【地理】,地下水水文学
ground water level(＝underground water level)　地下水面,潜水面
groundwater mining(＝groundwater overexploitation)　地下水超采,过度抽取地下水
groundwater mound　地下水隆起
groundwater recession　地下水消落,地下水亏耗
groundwater reserves　地下水储量【地质】
ground water resources　地面水资源
groundwater runoff　基流,地下水径流【地质】
groundwater station　地下水测站
groundwater storage　地下水储量,潜水储量
groundwater system　地下水系统【地质】
groundwater table　地下水面,地下水位,潜水位
ground wave　地表波【地质】,地波【航海】
ground-wave propagation　地波传播
group frequency　基团频率【化】
grouping　分类,组合
grouping of frequency distributions　频数分布组合
group of sunspots　〔太阳〕黑子群
group of waves　群波【航海】
Group on Earth Observations (GEO)　〔国际〕地球观测组织(2006年成立,已有60个国家和43个国际组织参加)
group spectrum　群谱
group velocity　群速度【物】,群速
grower's year　栽培年,作物年
growing degree-day　生长期有效积温(植物生长期内逐日温度与生物学下界温度的差之和)
growing degree-hour　生长度-时
growing mode　增长模
growing period　生长期【农】
growing season　生长期,生长季
growing wave　增长波
growler　碎浮冰(在海上,长度不到1 m)
growth　生长【植】
growth factor (GF)　生长因子【生化】
growth habit　生长习性【农】
growth habit of ice crystal　冰晶增长习性
growth index　生长指数
growth rate　生长率【农】
growth ring　年轮,生长轮
growth unit　生长单位
GRPS (general packet radio service)　通用分组无线电业务系统
G system　G系
GTOS (Global Terrestrial Observation System)　全球陆地观测系统
GTPC (gas-to-particle conversion)　气粒转化
GTS (global telecommunication system)　全球电信系统【大气】
guba　居巴雨飑(新几内亚海上的雨飑)
guess field　推测场
guidance terminal forecast (GTF)　指导〔性〕终点预报
guided wave　导波
guiding centre plasma　引导中心等离子体
guiding laser beam　制导激光束【物】
Guinea Current　几内亚〔暖〕海流
Gulf Stream　墨西哥湾流【大气】
Gulf Stream Extension　湾流延伸

Gulf Stream Rings 湾流涡旋
Gulf Stream System 湾流系统
gully-squally 峡谷飑(中美安第斯山西坡峡谷的强阵风)
Gumbel distribution 耿贝尔分布
Gunn-Bellani radiometer 冈恩-贝拉尼辐射计
Gunn diode 耿〔式〕二极管【电子】
Gunn effect 耿〔式〕效应【电子】
gunprobe 炮射探测系统
Gurley theodolite 格利经纬仪
gush 〔阵性〕暴雨
gust 阵风【大气】
gust amplitude 阵风振幅【大气】
gust and lulls 阵风阵息
gust component(=gustiness component) 阵风分量
gust decay time 阵风衰减时间(最大风速至阵风结束的时间)
gust duration 阵风持续时间【大气】
gust factor 阵风系数
gust formation time 降风形成时间(阵风开始至最大阵风的时间)
gust frequency 阵风频数
gust frequency interval 阵风频数时段
gust front 阵风锋【大气】,飑锋
gust-gradient distance 阵风梯度距离
gustiness 阵风性,阵风度
gustiness component 阵风分量
gustiness factor 阵风系数
gust load 阵风负荷
gust peak speed 阵风最大风速
gust probe 阵风探测器
gustsonde 阵风探空仪
gust vector 阵风矢量

gusty wind 阵风(风向屡变的风)
Guti weather 古蒂天气(南非 12 月至次年 3 月的飑和阵风)
guttation 吐水【植】,叶尖吐水(易误认为露)
guttra 伊朗 5 月飑
guxen 冷飑(见于阿尔卑斯山)
Guyana Current 圭亚那海流
guzzle 阵飑(见于设得兰群岛)
GWP(global warming potential) 全球增温潜势,全球变暖潜势
GWSP (Global Water System Project) 全球水系统计划
gypsum 石膏【地质】
gypsum unit 石膏测湿元件(测土壤含水量)
gyration center 回转中心
gyration frequency 回转频率
gyre(=gyral) ①环流②涡旋
gyres 环流,涡旋
gyro 陀螺仪【航海】
gyrocompass 陀螺罗经【航海】,电罗经
gyrocompass bearing 陀罗方位【航海】
gyrocompass course 陀罗航向【航海】
gyrocompass error 陀罗差【航海】
gyro erected optical navigation(GEON) 陀螺光学导航
gyro frequency 回旋频率
gyromagnetic frequency 回转磁频率
gyromotion 回转运动
gyro reference frequency 罗经参考频率
gyro reference system 陀螺参考系
gyroscope 陀螺仪【物】

H

haar 哈雾(苏格兰东部的一种海雾)
hababai 哈巴巴艾风(红海西岸 10—11 月干热的东风或东北风)
habitat 生境【地理】,栖息地【海洋】
haboob(＝habbub,haboub,hubbob, bhbbub) 哈布尘(沙)暴(见于埃及、阿拉伯沙漠地区)
Hadean Era 冥古代(距今 4600～3800 百万年)
Hadley cell 哈得来环流〔圈〕【大气】
Hadley energy cycle 哈得来能量循环
Hadley regime 哈得来域【大气】,哈得来对称流型(转盘实验的术语)
Hadley's principle 哈得来原理
Hagen-Poiseuille flow 哈根-泊肃叶流,H-P 流
Haidinger's brush 黑丁格刷子(一种极光形式)
hail 〔冰〕雹【大气】
hail cloud 〔冰〕雹云【大气】
hail core 雹心,雹核
hail damage 雹灾【大气】
hail embryo 雹核【大气】,雹胚,冰雹胚胎
hailfall 降雹
hail fallout zone 冰雹沉降区,降雹区
hail generation zone 冰雹生成区【大气】
hail growth 冰雹增长
hail growth zone 冰雹增长区
hail impactor 碰撞式测雹器
hail impression 雹痕
hail lobe 雹瓣【大气】
hail mitigation 消雹
hail-observing network 冰雹测站网
hailpad 测雹板【大气】,测雹盒
hail path 冰雹路径

hail pellet 雹粒【大气】
hail prevention 防雹
hail-rain separator 雹雨分离器【大气】
hail research experiment 冰雹研究实验
hail shooting 降雹
hail-shower 雹阵
hail squall 雹飑
hail stage 成雹阶段
hailstone 雹块【大气】
hail storm 雹暴【大气】
hailstorm recorder 雹暴记录器
hailstreak 雹击线
hail suppression 防雹【大气】
hail suppression program 防雹计划
hailswath 雹击带
hair hygrograph 毛发湿度计【大气】
hair hygrometer 毛发湿度表【大气】
halcyon days(＝alcyone days) 平安时期(冬至前后 7 天的风平浪静时期)
Hale cycle 黑尔周期
hale de mars 3 月干风(法国莫尔旺 3 月份干燥的 bise 风)
Hale telescope 海尔望远镜【天】
half-arc angle 半弧角
half band width 半峰宽【化】
half-life 半衰期【物】
half life period 半衰期【物】
half moon 上下弦,半规月
half-order closure 半阶闭合
half-period zone 半周期带【物】
half power beam width 半功率束宽
half-power point 半功率点【电子】
half-tide-level 半潮位
half-time 半衰期
half-value period 半值期
half value thickness 半值厚度(指吸收体

half width 半宽度【天】
half-yearly oscillation 半年振荡
haline water 高盐水
Hall current(of electrojet) 霍尔流(电急流的)
Hall effect 霍尔效应【物】
Hall-effect device 霍耳效应器件【电子】
Hallett-Mossop process 哈利特-莫索普过程,H-M过程(冰晶淞附繁生过程)
Halmahera Eddy 哈尔马赫拉涡流
halny wiatr 焚风(波兰山区)
halo 晕【大气】
halocarbons 卤化碳,卤烃
halocline 盐跃层
halogen 卤素【化】
halons 哈龙(用作阻燃剂的一组有机溴化物),氟氯碳化物的溴化物
halo of 22° 22度晕【大气】
halo of 46° 46度晕【大气】
halo of Hevelius 海韦留晕
Hamilton canonical equations 哈密顿典范方程【数】
Hamilton equation 哈密顿方程【数】
Hamilton-Jocobi equation 哈密顿-雅可比方程【数】
Hamming code 汉明码【数】
Hamming window 汉明窗【电子】
hamun 雨季湖
hand anemometer 手持风速表【大气】
hanging dam 浮冰坝
hanging glacier 悬冰川【地理】
hanging water 悬着水
Hankel function 汉克尔函数【数】
Hankel transform 汉克尔变换【数】
Hanning window 汉宁窗
hanning 修平〔处理〕
Hanzi 汉字【计】
Hanzi card 汉卡【计】
haracana 飓风
Harany discontinuity 哈朗不连续性(大气电学中的)

harbor oscillations 海港〔水位〕起伏
hard disk 硬磁盘【计】,硬盘
hard freeze 坚冻,酷寒
hard frost 黑霜,黑冻
hardiness 耐寒性,抗冻
hardness 硬度,硬性
hardpan ①硬质地层,硬盘 ②硬地,坚固的基础
hard radiation 硬辐射,贯穿辐射
hard rain 暴雨
hard rime 霜淞
hard science 硬科学(泛指自然科学)
hardware 硬件【计】
hardware model 硬件模型
harmattan(= harmatan, harmetan, hermitan) 哈麦丹风(同赫米丹风,西非的干燥多尘东北风)
harmonic 调和
harmonic analyser 调和分析器
harmonic analysis 调和分析【数】,谐波分析
harmonic dial 谐波标度盘
harmonic function 调和函数【大气】【数】
harmonic mean 调和平均【数】
harmonic oscillation 谐振荡
harmonic oscillator 谐振子【物】,正弦波振荡器
harmonic prediction 调和预测
harmonic product spectrum 调和乘积谱
harmonic radiation 谐振辐射
harmonics ①谐波 ②谐量
harmonic series 谐波级数,调和级数【数】
harmonic solution 调和解【数】
haroucana 飓风
harrua 浓湿度(见于厄瓜多尔、秘鲁、智利沿岸)
Hartley band 哈特莱带(O_3强吸收带)
Hartley region 哈特莱区
Hartmann number 哈特曼数

harvest moon 〔收〕获月(近秋分的满月)
hashing 散列
haster 英格兰猛烈雨暴
haud 豪德飑(英国苏格兰地区)
haugull(= havgul, havgula) 豪古尔风(苏格兰和挪威夏季的湿冷海风)
hauling capacity 牵引容量
Haurwitz-Helmholtz wave 豪威兹-亥姆霍兹波,H-H波
Haurwitz wave 豪威兹波【大气】
hay fever 干草热,花粉热
hazard analysis 危害分析【农】
hazard geography 灾害地理学【地理】
hazard index 危险指数
hazardous weather message 危险天气通报【大气】
hazardous weather warning 危险天气预警
haze 霾【大气】
haze aerosol 霾气溶胶
haze aloft 高空霾【大气】
haze droplet 霾滴
haze horizon 霾层顶
haze layer 霾层
haze level 霾面(即霾层顶)
haze line 霾线
hazemeter 测霾表,霾表(透明度表的别称)
Hazen method 哈森法
Hazen plotting position 哈森图标位置
HCMM (Heat Capacity Mapping Mission) Satellite 热容〔量〕测绘卫星
HDGEC (Human Dimensions of Global Environmental Change) 全球环境变化的人文因素
head ①蓄水高度,落差②水头,扬程
heading 航向,方向
heading line 航线
head-water point 水源地
headwaters ①上游〔水面〕②河源,水源③上游河道(常用复数)

head wind 逆风【大气】,顶风
healthy regionalism 健康区域制
heap cloud 直展云
heat 热
heat balance 热量平衡【大气】
heat budget 热量收支【大气】
heat budget instrument(HBI) 热量收支测量仪器
heat burst 热爆发
heat capacity 热容〔量〕【物】
heat capacity at constant pressure 等压热容【化】
heat capacity at constant volume 等容热容【化】
Heat Capacity Mapping Mission (HCMM) 热容〔量〕测绘卫星
heat capacity mapping radiometer (HCMR) 热容〔量〕测绘辐射仪
heat capacity method 热容〔量〕方法
heat-compensated net radiometer 热补偿净辐射表
heat condition forecasting 热量条件预报
heat conduction 热传导
heat conduction calorimeter 热导式热量计【化】
heat conductivity 导热性,热导率
heat content 焓
heat convection 热对流
heat cumulus 热成积云
heat cycle 热循环
heat damage 热害【地理】
heat dormancy 热眠,热休眠
heat drying 加温干燥【农】
heat(ed) air psychrometer 加热式空气干湿表
heat effect 热效应【化】
heat energy 热能
heat engine 热机
heat equator 热赤道【大气】
heat equivalent 热当量
heat exhaustion 热衰竭,中暑(医)

heat flow equation 热流量方程【大气】
heat flow measuring plate 热流测量板
heat flow meter 热流仪(测热通量)
heat-flow transducer 热流换能器
heat flux 热通量【大气】
heat flux budget 热通量收支
heat flux vector 热通量矢量
heat function 热函数
Heat Health Warning System(HHWS) 热浪健康预警系统
heat index 热指数
heat-induced 加热引起的
heating degree-day 采暖度日【大气】
heating function 加热函数
heating(process) 加热〔过程〕
heating rate 加热率,增温率
heating steam 加热蒸汽【航海】
heating water ratio 加热水倍率【航海】
heat input 热量输入
heat island 热岛【大气】
heat island effect 热岛效应【大气】
heat lightning 热闪【大气】
heat low 热低压
heat of adsorption 吸附热【物】
heat of combination 化合热
heat of combustion 燃烧热【化】
heat of condensation 凝结热
heat of crystallization 结晶热
heat of dilution 稀释热【化】
heat of dissociation 离解热【物】
heat of dissolution 溶解热
heat of evaporation 蒸发热【物】
heat of formation 生成热【化】
heat of freezing 凝固热
heat of fusion 熔化热【物】,融化热
heat of liquefaction 液化热【化】
heat of melting 融解热
heat of mixing 混合热【化】
heat of reaction 反应热【化】
heat of solidification 凝固热
heat of solution 溶解热【物】
heat of sublimation 升华热【物】
heat of vaporization 汽化热【物】
heat pollution 热污染
heat radiation 热辐射【物】
heat rate 耗热率【航海】
heat ray 热〔射〕线
heat reservoir 热库【物】
heat-resisting plant 耐热植物
heat resources 热量资源【大气】
heat sink 热汇【大气】
heat source 热源【大气】
heat storage 热储量
heat storm 热〔雷〕暴
heat stress 热应力
heat stress index 热应力指数
heat-stroke 中暑
heat thunderstorm 热雷暴【大气】
heat tolerance 耐热性【农】
heat transfer 传热【物】,热量输送【大气】
heat transfer coefficient 传热系数
heat unit 热〔量〕单位
heat-unit theory 热单位理论
heat-water balance 热水平衡【地理】
heat wave 热浪【大气】
heave ①平错,隆起②冻胀③波浪翻腾
heaven 天,天空
heavenly body 天体
Heavenly Stems 天干
heaving(＝frost heaving) 冻胀丘,冻拔
Heaviside function 赫维赛德函数【数】
Heaviside layer 赫维赛德层,E层
heavy clay soil 重黏土【土壤】
heavy fog 重雾
heavy hail 大雹
heavy ice 重冰
heavy ion 重离子【物】
heavy ion content 重离子含量
heavy line 粗线
heavy overcast 阴沉
heavy passing shower 大阵雨

heavy rain　大雨【大气】
heavy seeding　强撒播,强催化
Heavy Snow　大雪【大气】(节气)
heavy shower　大阵雨
heavy snow warning　大雪预警
heavy squall　烈飑
heavy water　重水【化】
heavy weather　阴天
heavy weather navigation operating mode management　大风浪航行工况管理【航海】
hecto-　百(词头,10^2)
hectopascal(hPa)　百帕(气压单位,1 hPa=1 毫巴)
height　高度
height above sea level　海拔〔高度〕【地理】
height analysis method　高度分析法
height-azimuth-range position indicator (HARPI)　高度方位距离位置显示器
height-change chart　变高图
height-change line　变高线
height difference　潮高差【航海】
height-finding radar　测高雷达
height gain　高度增益
height of centre of gravity　重心高度【航海】
height of cloud　云高
height of cloud base　云底高度
height of tide　潮高【航海】
height pattern　高度〔分布〕型〔式〕
height vertical coordinate　高度垂直坐标〔系〕
heigyne meteorology(=hygieno-meteorology)　卫生气象学
Heiligenschein　露面宝光,草地宝光(德语名)
Heinrich event　海因里希事件,H 事件
Heisenberg model　海森伯模型【物】
Heisenberg uncertainty principle　海森伯测不准原则
heliacal rising　晨出【天】

heliacal setting　夕没【天】
helical antenna　螺旋形天线
helical scanning　螺旋形扫瞄
helicity　螺旋度【物】,螺旋性【物】
helicoidal anemometer　螺旋桨式风速表
helicopter　旋翼式风速感应器
helicopter wake　直升机尾流
heliocentric coordinate　日心坐标【天】
heliocentric distance　日心距离【天】
heliocentric gravitational constant　日心引力常数【天】
heliocentric system　日心体系【天】
heliogeophysics　日地物理〔学〕
heliogramma　日照纸
heliograph　太阳光度计【大气】,日光仪,〔感光〕日照计
heliographic chart　日面图【天】
heliographic coordinate　日面坐标【天】
heliogreenhouse　日光温室
heliometric index　总辐射平均强度表指数
heliopause　太阳风顶【天】
heliophil(e)　喜阳植物
heliophobe　厌阳植物
helioplant　太阳能利用装置
helioscope　太阳〔望远〕镜,回照器,量日镜
helioseismology　日震学【天】
heliosis(=sunstroke)　日射病,中暑,灼伤
heliosphere　①日光层,日球层【地】②太阳风圈【天】,太阳域(太空中受太阳气体及磁场影响的领域)
heliostat　定日镜【天】
heliotechnics　太阳能技术
heliotrope　回光仪,日光反射信号器
heliotropic wind　日成风,日转风
helium(He)　氦〔气〕
helium neon laser　氦氖激光器【电子】
helix　螺〔旋〕线【数】
hellinis halo　露面宝光,草地宝光
Hellmann recording snow-gauge　赫尔曼

自记雪量器
helm 山头风(英国北部强阵性东北风)
helm bar 山头云堤
helm cloud 山头云
Helmholtz equation 亥姆霍兹方程【数】
Helmholtz free energy 亥姆霍兹自由能【化】
Helmholtz function 亥姆霍兹函数
Helmholtz gravitational wave 亥姆霍兹重力波
Helmholtz instability 亥姆霍兹不稳定【大气】
Helmholtz's theorem 亥姆霍兹定理
Helmholtz wave 亥姆霍兹波【大气】
helm wind 头盔风(北英格兰伊登河谷的寒冷东北强风)
hemisphere 半球
hemisphere broadcast 半球广播
hemisphere exchange center 半球交换中心
hemispherical and net radiometer 半球〔式〕净辐射计
hemispheric fine mesh 半球细网格
hemispheric flux 半球通量
hemispheric model 半球模式【大气】
hemispheric prediction 半球预报
hemispheric wave-number 半球波数
hemoglobin 血红蛋白【化】
hemoprotein 血红素蛋白【化】
He-Ne laser 氦氖激光器【物】
henry(H) 亨〔利〕(电感单位)
Henry's law 亨利定律【化】
Henyey-Greenstein phase function H-G 相函数,亨尼-格林斯坦相函数
herb 草本【植】
hericane 飓风
Herlofson diagram 赫洛弗森图
hermitan 赫米丹风(西非的干燥多尘东北风)
Hermite polynomial 埃尔米特多项式【数】

Hermitian matrix 埃尔米特矩阵【数】
Herschel's actinometer 赫舍尔日射计【大气】
hertz(Hz) 赫〔兹〕(频率单位)
Hertz's diagram 赫兹图
Hertz vector 赫兹矢量
Herzberg band 赫茨堡带
Herzberg continuum 赫茨堡连续吸收带
hesper 海斯潘风(地中海的西风)
heterodyne 外差
heterodyne detection 外差探测
heterogeneity ①不齐性②不均匀性,异质
heterogeneous 非均匀的,多相的,异质的
heterogeneous atmospheric chemical reaction 非均相大气化学反应
heterogeneous chemistry 多相化学,非均相化学
heterogeneous equilibrium 多相平衡【化】
heterogeneous fluid 非均匀流体
heterogeneous nucleation 异质成核,非均质成核〔作用〕
heterogeneous reaction 多相反应【化】
heterogeneous substance 多相〔物〕质
heterogeneous system 非均相系统【化】,多相系,非均匀系
heterojunction 异质结【电子】
heterojunction solar cell 异质结太阳电池【电子】
heterosphere 非均质层【大气】(90 km 以上)
heterostructure 异质结构【电子】
heterotropic 斜交〔的〕
hevelian halo 淡晕
Hevelius's parhelia 海韦留幻日
Hevel's halo 海韦尔晕
hexagonal column 六角柱体(冰晶)
hexagonal platelet 六角板状(冰晶)
hexagonal system 六方晶系
HFCs (hydrofluorocarbons) 氢氟碳化物
HF (high frequency) 高频
HHWS (Heat Health Warning System)

热浪健康预警系统
hibernal 冬天〔的〕,冬令
hibernation 冬眠,越冬
hidden layer 隐藏层
hidden periodicity 隐藏周期性
hierarchical model 层次模型【地信】
hierarchical sequence 层次序列【地信】
hierarchy 分层【数】,级别,谱系
hierarchy of closures 闭合体系
HIFOR 高空预报(high-level forecast)的国际电码缩写
hig(=**ig**) 伊格风暴(英格兰持续时间短的一种风暴)
high ①高②高压
high aloft 高空高压
high altitude 高空(按美国标准指1500~6000 m 的高空)
high altitude observatory 高山观象台
high-altitude station 高山〔测〕站,高原〔测〕站
high-analysis fertilizer 高浓度肥料【土壤】
high arctic climate 北极气候
high-based thunderstorm 高空雷暴(雷暴底部约在2400 m 以上)
high cloud 高云【大气】
high crosswind 大侧风
high data rate storage subsystem(**HDRSS**) 高速资料存储子系统
high definition TV (**HDTV**) 高清晰度电视【电子】
high density lipoprotein 高密度脂蛋白【生化】
high dimensional system 高维系统
high-energy astrophysics 高能天体物理【物】
high-energy physics 高能物理〔学〕【物】
high energy proton alpha telescope (**HEPAT**) 高能质子α粒子计数仪
high energy proton and alpha particle detector(**HEPAD**) 高能质子和α粒子检测器

higher high water 高高潮【航海】
higher low water 高低潮【航海】
higher moment 高阶矩
higher-order closure model 高阶闭合模式
higher-order finite difference 高阶有限差分
higher-order perturbation solution 高阶扰动解
higher order terms 高阶项
highest astronomical tide 最高天文潮
highest temperature 最高温度
high flow year 丰水年【大气】,多水年,湿润年
high foehn 高空焚风
high fog 高雾
high forest 乔木【林】
high frequency (**HF**) 高频
high-frequency amplifier (**HFA**) 高频放大器
high frequency discharge 高频放电【化】
high-frequency drying 高频干燥【农】
high-frequency fluctuation 高频振荡
high frequency oscillator (**HFO**) 高频振荡器
high-frequency radar 高频雷达
high frequency radio acoustic sounder 高频无线电声学探测器
high index 高指数【大气】
high impact weather 高影响天气
highland climate 高地气候
highland glacier 高地冰川,高原冰川
highland ice 高地冰盖,高原冰,山地冰
highland uplift effect 山体效应【地理】
high latitude 高纬度
high latitude substorm index 高纬亚暴指数(即极光电急流指数,简称 AE 指数)
high level anticyclone 高空反气旋
high-level cloud 高云
high-level cyclone 高空气旋
high level language 高级语言【计】

high-level ridge 高空脊
high-level thunderstorm 高空雷暴
high-level trough 高空槽
high mountain soil 高山土壤【地理】
high-order closure 高阶闭合
high-order language 高阶语言【计】
high overcast 高密云〔天空〕
high pass filter 高通滤波〔器〕
highpass filtering 高通滤波【计】
high-performance liquid chromatography (HPLC) 高效液相色谱法【化】
high polar glacier 高〔厚〕极地冰川
high-power ultrasensitive radar 大功率超灵敏雷达
high precision positioning system 高精度定位系统【航海】
high〔pressure〕 高〔气〕压【大气】
high-pressure barrier 高压坝【大气】
high pressure center 高压中心
high-pressure system 高压系统,高值系统
high priority 高优先级【计】
high-reference signal 声频参考信号
high-repetition-rate 高重复率
high resolution 高分辨率
High-Resolution Dynamics Limb Sounder (HIRDLS) 高分辨〔率〕动力学临边探测器(EOS)
HIgh-Resolution Imaging Spectrometer (HIRIS) 高分辨〔率〕成像光谱仪(EOS)
High Resolution Facsimile(HR-FAX) 高分辨率〔云图〕传真【大气】
High-Resolution Infrared Radiation Sounder (HIRS) 高分辨〔率〕红外辐射探测器【大气】
High Resolution Infrared Radiometer (HRIR) 高分辨〔率〕红外辐射计
High Resolution Infrared Sounder(HRIS) 高分辨〔率〕红外探测器
High Resolution Interferometric Sounder (HIS) 高分辨〔率〕干涉探测器【大气】
High-Resolution Microwave Spectrometer Sounder(HIMSS) 高分辨〔率〕微波波谱探测器
high resolution PBL 精细边界层
High Resolution Picture Transmission (HRPT) 高分辨〔率〕图像传输【大气】
high ridge 高空脊
high sea ①狂浪（风浪 6 级）②公海,远海,远洋
high-speed camera 高速照相机【物】,高速摄影机
high speed facsimile 高速传真
high speed plasma flow 高速等离子体流
high temperature 高温
high temperature corrosion 高温腐蚀【航海】
high-temperature plasma 高温等离〔子〕体【物】
high-temperature short-time processing 短时高温加工【农】
high tide 高潮【大气】,满潮
hight water time 高潮时【航海】
high vacuum 高度真空
high velocity induction air conditioning system 高速诱导空气调节系统【航海】
high volume air sampler 大容量空气取样器
high-volume filter 高容量过滤器
high-volume particulate sampler 大容量微粒取样器
high water 高水,高潮
high-water interval 高潮间隙
high water level 高水位,高潮位
high water mark 高水位,高潮线
high water time 高潮时【航海】
high wind 大风
high zonal circulation 强西风环流,强纬向环流
Hilbert space 希尔伯特空间【数】

hileia(＝hylea) 热带雨林
hill 丘陵【地理】
hill fog 山雾
hill wave cloud 山波云
HIMAWARI 《葵花》卫星（日本静止气象卫星名）
HIMSS(HIgh-resolution Microwave Spectrometer Sounder) 高分辨〔率〕微波波谱探测器
hindcast 历史回报
hindcasting technique 追算技术
hinterland 腹地【地理】
hippy "嘻皮士"传感器（测海洋表面重力波）
hiracano 飓风
HIRDLS (HIgh-Resolution Dynamics Limb Sounder) 高分辨〔率〕动力学临边探测器(EOS)
hi-reference signal 声频参考信号
HIRIS (HIgh-Resolution Imaging Spectrometer) 高分辨〔率〕成像光谱仪
HIRS (High-resolution Infrared Radiation Sounder) 高分辨率红外辐射探测器
HIS(High-resolution Interferometric Sounder) 高分辨〔率〕干涉探测器
histogram 直方图【大气】
historical atlas 历史地图集【地理】
historical climate 历史气候【大气】
historical climate series 历史气候序列
historical climatic data 历史气候资料
historical climatic record 历史气候记录
historical climatology 历史气候学
historical cumulative emission per capita 人均历史累计排放
historical geography 历史地理学【地理】
historical geology 地史学【地质】
historical sequence 历史序列
historical time 历史时期
hoar 白霜
hoar crystal 白霜〔冰〕晶
hoar-frost 白霜【大气】

Hobbs's theory 霍布斯理论
hodogram ①高空风分析图【大气】②速矢端迹图
hodograph ①高空风分析图【大气】②速矢端迹
hodograph analysis 高空风分析
hohlraum ①辐射腔,辐射空穴②黑体
Hoiland's circulation theorem 赫伊兰环流定理
holard 土壤总含水量
hole microclimate 洞穴小气候
hollerith card 霍尔瑞斯卡,洞卡（一种做数据处理用的穿孔卡）
hollerith system 洞卡装置
Holmboe instability 霍尔姆博不稳定度
Holocene climate 全新世气候
Holocene Epoch 全新世【地质】
Holocene Series 全新统【地质】
hologram 全息图【物】
holographic method 全息方法
holographic technique 全息术
holograph 全息照相【物】
holography 全息术【物】
holonomic constraint 完整约束
holophone 声音全息记录器
holophote 全光反射装置,全射镜
holoscope 全息照相机
holospheric temperature 全球温度,纬圈平均温度
holosteric barometer 固体气压表（即空盒气压表）
Holter system 动态心电图监护系统【电子】,霍尔特系统
homeotherm 恒温动物【动】
homobront 雷暴等时线,初雷等时线
homoclime 相同气候
homodyne 零差,零拍
homogeneity ①齐性②均匀性,同质
homogeneous 均一的,齐次的
homogeneous and isotropic random function 均匀各向同性随机函数

homogeneous (and) isotropic turbulence 均匀各向同性湍流
homogeneous atmosphere 均质大气【大气】
homogeneous broadening 均匀加宽
homogeneous complex random function 均匀复随机函数
homogeneous condensation 同质凝结,均质凝结
homogeneous differential equation 齐次微分方程【数】
homogeneous equation 齐次方程
homogeneous fluid 均质流体
homogeneous medium 均匀介质
homogeneous mixing 均匀混合
homogeneous nucleation 同质成核,均质成核
homogeneous ocean model 均质海洋模式
homogeneous sublimation 同质升华,均质升华
homogeneous system 均相系统【化】,单相系,均匀系
homogeneous terrain 均匀地形
homogeneous wave 均匀波【物】,均质波
homogenization 等质化,均〔一〕化〔作用〕
homojunction laser 同质结激光器【电子】
homojunction solar cell 同质结太阳电池【电子】
homological algebra 同调代数【数】
homologous ray 同系光线【物】
homologous turbulence 同调湍流
homology 同调【数】,同源性【生化】
homomorphic filter 同形滤波〔器〕
homopause 均质层顶【大气】
homoscedasticity 同方差性
homosphere 均质层【大气】
homosphere-heterosphere transition 均质层-非均质层过渡带(区)
homothermy 等温层,同温层,恒温层
hook echo 钩状回波【大气】
hook echo tornado 钩状回波龙卷
hook gauge 钩形水位计,钩规

Hooke law 胡克定律【物】
Hooke number 胡克数
Hopf bifurcation 霍普夫分岔【物】
Hopfield bands 霍普菲带
Hopkin's bioclimatic law 霍普金斯生物气候律
horizon ①地平圈②地平③水平④层顶
horizon brightening 地平亮光
horizon-crossing indicator 地平交互指示器
horizon distance 水平距离,地平距离
horizon sensor 地平传感器
horizontal advection 水平平流
horizontal angle 方位角
horizontal branch 水平支【天】
horizontal circle 地平圈
horizontal convective rolls 水平对流卷涡
horizontal convergence 水平辐合
horizontal coordinate system 地平坐标系【航海】
horizontal coupling 水平耦合
horizontal decoupling 水平退耦
horizontal diffusion 水平扩散
horizontal diffusion coefficient 水平扩散系数
horizontal divergence 水平散度【大气】,水平辐散
horizontal extent 水平范围
horizontal homogeneity 水平均一
horizontally floating balloon 水平漂浮气球
horizontal mixing 水平混合
horizontal momentum equation 水平动量方程
horizontal orientation particle 水平取向粒子
horizontal parallax 地平视差【航海】
horizontal plane 水平面,地平面
horizontal polarization 水平偏振,水平极化
horizontal precipitation 水平降水【地理】

horizontal pressure force 水平气压力
（单位质量的水平气压梯度）
horizontal pressure gradient force 水平气压梯度力
horizontal rainbow 水平虹
horizontal resolution 水平分辨率
horizontal roll convection 水平卷涡对流
horizontal roll vortices 水平卷涡涡列
horizontal scale 水平尺度
horizontal shear 水平切变
horizontal simplification 水平简化
horizontal sounding balloon 水平探测气球
horizontal sounding technique 水平探测技术
horizontal structure equation 水平结构方程
horizontal tangent arc 平珥
horizontal time section 水平时间剖面，水平时间截面
horizontal transfer 水平输送
horizontal transverse wave 水平横向波
horizontal visibility 水平能见度【大气】
horizontal wavelength 水平波长
horizontal wavenumber 水平波数
horizontal wind shear 水平风切变【大气】
horizontal wind vector 水平风矢量【大气】
horizontal zone 水平地带【地理】
horizon wave 水平波
hormone 激素【生化】
horn antenna 喇叭天线【电子】
horn card 角质风向盘
horn radiator 喇叭天线，喇叭辐射器
horse latitude high 马纬度高压
horse latitudes 马纬度【大气】（约南北纬30°～35°一带干热无风地区）
host machine 〔宿〕主机【计】
host model 主模式
host spot 热斑【天】
host system 宿主系统【计】

hot 热〔的〕
hot-air balloon 热〔空气〕气球
hot arid zone 炎热干旱区
hot belt 热〔地〕带（早期用作年平均温度超过20℃的地带）
hot cell 热室【化】
hot cloud 热云
hot damage 热害【大气】
hot day 热日（最高温度≥25℃）
hot-dry wind 干热风
hot-film anemometer 热膜风速表
hot-film probe 热膜探头（测风用）
hothouse effect 温室效应
hoting damage 热害
hot laboratory 热实验室【化】
hot magnetospheric plasma 磁层热等离子体
hot plasma 热等离子体
hot plume 热焰【地】
hot run 热试验【化】
hot season 热季【大气】
hot spot 热点【地】
hot spot effect 热点效应
hot spring 温泉
hot summer 酷暑
hot test 热试验【化】
hot tower 热塔
hot water pollution 热水污染
hot wave 热浪
hot wind 热风
hot-wire anemometer 热线风速表【大气】
hot-wire liquid water content meter 热线含水量仪
Hough function 霍夫函数
Hough transformation 霍夫变换【计】
hour 时
hour angle 时角
hourly distance scale 小时-距离尺
hourly observation 逐时观测
hourly precipitation 1小时降水量

hour of observation 观测时
hour-out line 等时线(航空)
house microclimate 室内小气候
Hovmoller diagram 槽脊图
Howling Fifties 50°风暴地带(纬度)
hPa 百帕
HPLC (High-Performance Liquid Chromatography) 高效液相色谱法
HR-FAX (high resolution facsimile) 高分辨率[云图]传真【大气】
HRIRS (high resolution infrared radiation sounder) 高分辨[率]红外辐射探测器【大气】
HRPT (High Resolution Picture Transmission) 高分辨[率]图像传输【大气】
Huanghai Coastal Current 黄海沿岸流【海洋】
Huanghai Cold Water Mass 黄海冷水团【海洋】
Huanghai Sea 黄海【海洋】
Huanghai Warm Current 黄海暖流【海洋】
Hubble law 哈勃定律【物】
Hubble space telescope (HST) 哈勃空间望远镜【天】
hubbob (见 haboob) 哈布风
hue 色相,色调
Huggins band(O_3) 哈金斯带(O_3)
hull method 船体[测水温]法
human bioclimatology 人类生物气候学【大气】
human biometeorology 人类生物气象学【大气】
human climate 人类气候
human climatology 人类[生物]气候学
human comfort 人体舒适
human-computer dialogue 人机对话【计】
Human Dimensions of Global Environmental Change (HDGEC) 全球环境变化的人文因素
human hearing range 人类听觉范围,人耳听阈
human-machine interface 人机界面
human ecology 人类生态学【地理】
human geography 人文地理学【地理】
human influence on climate 人类对气候影响
human-machine mix 人机结合
human map 人文地图【测】
Humboldt current 洪堡海流(Peru current 的别称)
humic acids 腐殖酸[类]【土壤】
Humicap 湿敏电容(芬兰瓦依萨拉公司湿度传感器产品名称)
humid 湿润
humid cliamte 湿润气候【大气】
humidification 增湿[作用],增湿作用【农】
humidifier 增(加)湿器,湿润器
humidiometer 湿度表
humidistat 恒湿器(箱),保湿箱,湿度调节器
humidity 湿度【大气】
humidity atmosphere producer 湿空气发生器
humidity coefficient 湿度系数
humidity effect 湿度效应
humidity element 湿度元件,湿度要素
humidity field 湿度场【大气】
humidity fluctuation 湿度起伏
humidity index 湿度指数
humidity lapse rate 湿度直减率
humidity mixing ratio 湿度混合比
humidity province 湿度区
humidity retrieval 湿度反演【大气】
humidity slide-rule 湿度计算尺
humidity strip 湿度片
humidity variance 湿度方差
humidizer 增湿剂
humid mesothermal climate 湿温气候
humid microthermal climate 湿润低温气候(即雪林气候)

humidometer 湿度表
humidor 保润盒,蒸汽饱和室
humidostat 湿度调节仪
humid region 湿润区
humid temperate climate 湿润温和气候
humid zone 湿润带
humification 腐殖化作用【土壤】
humilis(hum) 淡(云)
humistor 湿敏电阻
humiture 温湿度(华氏度数与相对湿度的和的一半)
hummock ①〔浮〕冰排②波状地
hummocked ice 冰丘冰,堆积冰
humus 腐殖质【土壤】
hupe 夜风(见于大洋洲塔希提岛)
huracano 飓风
hurleblast 飓风
hurlecano 飓风
hurly-burly 英格兰雷暴
hurricane 12级风【大气】,飓风【大气】
hurricane analog technique(HURRAN) 飓风路径相似方法
hurricane balloon 飓风探测气球
hurricane band 飓风〔回波〕带
hurricane bar 飓风云堤
hurricane beacon 飓风探测气球
hurricane cloud 飓风云
hurricane core 飓风核
hurricane current intensity number 飓风强度指数
hurricane dissipation 飓风消散
hurricane eye 飓风眼
hurricane fatality 飓风灾害
hurricane-force wind 飓风级风〔速〕
hurricane microseism 飓风微地震
Hurricane Modification Program (STORMFURY) 削台计划
hurricane modification 飓风人工影响
hurricane name 飓风名称
hurricane origin 飓风源地
hurricane path prediction 飓风路径预报

hurricane radar band 飓风雷达〔回波〕带
hurricane rain band 飓风雨带
hurricane seeding 飓风撒播
hurricane surge 飓风〔暴〕潮
hurricane surveillance radar 飓风监视雷达
hurricane tide 飓风潮
hurricane T number 飓风T指数
hurricane tracking 飓风跟踪
hurricane warning 飓风预警
hurricane warning system 飓风预警系统
hurricane watch 飓风监视
hurricane wave 飓风浪
hurricane wind 飓风
Huygens construction 惠更斯作图法【物】
Huygens-Fresnel method 惠更斯-菲涅耳方法,H-F方法
Huygens-Fresnel principle 惠更斯-菲涅耳原理【物】,H-F原理
Huygens principle 惠更斯原理【物】
Huygens's wavelets 惠更斯小波
hybrid ①混合波导联结②合成物③混合的
hybrid coordinate 混合坐标【大气】
hybrid model 混杂型模式
hybrid storm 混杂型风暴
hybrid vertical coordinate 混合垂直坐标
hydration 水合【化】
hydraulic analog 水力模拟
hydraulic conductivity 导水性,导水率
hydraulic flow 水力流(研究重力场影响下的密度分层流体的动力学)
hydraulic grade line 水力坡线,水力比降线
hydraulic gradient 液压梯度
hydraulic head 〔压力〕水头
hydraulic jump 水跃
hydraulic mean depth 水力平均深度
hydraulic radius 水力半径
hydraulic resistivity 水力阻抗性,水力阻抗率(导水率的倒数)
hydraulic routing 水力学法流量演算

hydraulics 水力学【地理】
hydraulic similarity 水力相似〔性〕
hydraulic soil evaporimeter 水力土壤蒸发器
hydraulic structures 水工建筑物,水工结构
hydraulic system 液压系统【航海】
hydroacoustics 水声学
hydrobiology 水生生物学【海洋】
hydrocarbon 碳氢化合物【化】
hydrochemical facies 水化学相
hydrochemistry 水化学【地理】
hydrochloric acid 盐酸,氢氯酸
hydrochloric acid gas 盐酸(氢氯酸)气体
hydrochlorofluorocarbons (HCFCs) 氢氯氟碳化物
hydroclimatology 水文气候学
hydrodynamic(al) dispersion 流体动力学扩散
hydrodynamic(al) equation 流体动力学方程
hydrodynamically rough surface 流体动力粗糙面
hydrodynamic instability 流体动力不稳定〔性〕
hydrodynamic noise 流体动力噪声【海洋】
hydrodynamic pressure 流体动压力
hydrodynamics 流体动力学
hydrodynamic stability 流体动力稳定〔性〕
hydrodynamic volume 流体力学体积【化】
hydroeconomics 水利经济学【地理】
hydroelectricity 水电
hydrogel 水凝胶【化】
hydrogen(H) 氢〔气〕
hydrogen atom 氢原子【物】
hydrogen chloride 氯化氢
hydrogen detection system 氢检测装置(气球充气安全装置)
hydrogen(-filled) balloon 氢气球
hydrogen generator 氢气发生器

hydrogen ion 氢离子
hydrogen-like atom 类氢原子【物】
hydrogen line 氢〔气谱〕线
hydrogen peroxide 过氧化氢,双氧水(H_2O_2)
hydrogen pot 氢气罐
hydrogen scale of temperature 氢温标
hydrogen sulfide 硫化氢
hydrogen thermometer 氢温〔度〕表
hydrogeochemistry 水文地球化学
hydrogeography 水文地理学【地理】
hydrogeological boundary 水文地质边界
hydrogeological condition 水文地质条件【地质】
hydrogeological division 水文地质分区【地质】
hydrogeology 水文地质学【地质】
hydrographical network 水系,河网
hydrographic 水文〔学〕的
hydrographic station 海洋观测站,海洋水文站
hydrograph ①水文图,水位曲线②自计水位仪
hydrography 水文地理学【大气】
hydrokinetics 流体动能学
hydrolapse 水分直减率
hydrologic accounting 水分收支,水量收支
hydrological basin 水域
hydrological cycle 水文循环【大气】,水分循环
hydrological drought 水文干旱
hydrological element 水文要素
hydrological forecasting 水文预报(制作水文预报的过程)
hydrological forecast 水文预报(陈述)
hydrological network 水文站网
hydrological routing 流量演算
hydrological warning 水文预警
hydrologic balance (或 budget) 水分平衡,水分收支,水量收支

hydrologic cycle 水〔文〕循环,水分循环
hydrologic equation 水文〔学〕方程
hydrologic index 水文指数
hydrologic properties 水文〔学〕性质,水文属性
hydrologic rainfall analysis project 水文雨量分析计划
hydrologic regime 水〔文〕情〔况〕
hydrologic year 水文年
hydrology 水文学【大气】
hydrolysable nitrogen 水解氮【土壤】
hydrolysis 水解【化】
hydromagnetic emissions 磁流体发射
hydromagnetics 磁流体力学
hydromagnetic wave 磁流波
hydromechanics 流体力学
hydrometeor 水凝物,水凝现象
hydrometeor charge 水凝物负荷
hydrometeor drag 水凝物曳力
hydrometeorological forecast 水文气象预报【大气】
Hydrometeorological Publishing House (HPH) 水文气象出版社(前苏联)
Hydrometeorological Research Center (HRC) 水文气象研究中心(前苏联)
Hydrometeorological Service(HMS) 水文气象局(前苏联)
hydrometeorologist 水文气象学家
hydrometeorology 水文气象学【大气】
hydrometer 比重表
hydrometric method 水文测定法
hydrometric station 测流站,水文站
hydrometry ①水文测量学②液体比重测定法
hydromorphic 水生形态〔的〕,水成〔的〕
hydrooptics 水光学
hydroperoxide 氢过氧化物【化】
hydrophilic 亲水〔的〕,喜水〔的〕,适水〔的〕
hydrophilic colloid 亲水胶体【化】
hydrophilic material 亲水性材料

hydrophobic 疏水〔的〕,拒水〔的〕
hydrophobic colloid 疏水胶体【化】
hydrophobic nuclei 疏水核
hydrophone calibrator 水听器校准器
hydrophotometer 水下光度表
hydrophyte 水生植物【植】
hydrophytic plant 水生植物
hydropower 水能【地理】
hydropsis 海洋预报〔学〕
hydroscopic nuclei 吸湿性核
hydrosol 水溶胶【化】
hydrosphere 水圈【大气】,水界
hydrostatic adjustment 流体静力调整
hydrostatic adjustment process 静力适应过程【大气】
hydrostatic approximation 流体静力近似【大气】
hydrostatic assumption 流体静力假设
hydrostatic balance 流体静力平衡
hydrostatic check 静力检查【大气】
hydrostatic defect 流体静力亏损
hydrostatic delay 流体静力延迟
hydrostatic equation 流体静力方程【大气】
hydrostatic equilibrium 流体静力平衡
hydrostatic instability 流体静力不稳定度【大气】
hydrostatic layer 流体静力层
hydrostatic model 流体静力模式
hydrostatic pressure 流体静压〔强〕【物】
hydrostatic stability 流体静力稳定〔性〕
hydrostatics 流体静力学【物】
hydrostatic wave 流体静力波
hydrothermal 热液的
hydrothermal activity 水热活动【地】
hydrothermal alteration 水热蚀变【地】
hydrothermal area 水热区【地质】
hydrothermal circulation 过液循环【海洋】,水热循环【地】
hydrothermal convection system 水热对流系统【地】

hydrothermal eruption 水热喷发【地】	hygrophyte 湿生植物【地理】
hydrothermal explosion 水热爆炸【地】	hygroscope 湿度器
hydrothermal field 水热田【地】	hygroscopic(al) ①吸湿〔的〕②收湿〔的〕
hydrothermal fluid 水热流体【地质】	hygroscopic coefficient 吸湿系数【农】
hydrothermal process 热液过程【地质】	hygroscopicity 吸湿率,吸湿性
hydrothermal reaction 水热反应【地质】	hygroscopic moisture 土壤水分,湿存水
hydrothermal resources 水热资源【地质】	hygroscopic nuclei 吸湿性核【大气】
hydrothermal system 水热系统【地质】	hygroscopic particle 吸湿性粒子
hydroxyl 羟基	hygroscopic water 吸湿水,吸着水
hydroxyl emission 羟基发射(光谱)	hygroscopic wettable nuclei 亲湿性核
hydroxyl ion 羟基离子	hygrostat 恒湿器,湿度检定箱
hydroxy radical 羟自由基【化】	hygrothermogram 温湿自记曲线
hyetal 降雨的,雨量的	hygrothermograph 温湿计
hyetal coefficient 雨量系数	hygrothermometer 温湿表
hyetal equator 雨量赤道	hygrothermoscope 温湿仪【大气】
hyetal region 雨区	hygrothermostat 恒温湿器(箱),温湿自动调节器(箱)
hyetograph 雨量计	
hyetographic curve 雨量计曲线	hygrotransducer 湿度传感器
hyetography 雨量分布学	hylea(=hylaea,hileia) 热带雨林
hyetology 降水量学	hyperbaroclinic zone 超斜压区
hyetometer 雨量表	hyperbar 高压点,高气压
hygieno-meteorology 卫生气象学	hyperbola 双曲线【数】
hygristor 湿敏电阻	hyperbolic attractor 双曲线吸引子
hygroautometer 湿度自记计	hyperbolic navigation system 双曲线导航系统【航海】
hygroclimate 降水气候	
hygrodeik 图示湿度表	hyperbolic point 双曲点【数】
hygrogram 湿度自记曲线	hyperbolic system 双曲系统
hygrograph 湿度计【大气】	hyperbolic type 双曲型
hygrokinematics 水物质运动学	hyperbolic umbilic 双曲脐〔型突变〕【物】
hygrology 湿度学	hyperbolic umbilic catastrophe 双曲脐型突变
hygrometer 湿度表【大气】	
hygrometric continentality 降水大陆度	hyperelliptic curve 超椭圆曲线【数】
hygrometric equation 湿度公式	hyperfunction 超函数【数】
hygrometric formula 湿度公式,湿度计算式	hypergeometric function 超几何函数【数】,超比函数
hygrometric state 湿态	hypergeometric function of confluent type 汇合型超几何函数【数】
hygrometric tables 湿度查算表	
hygrometry 测湿法	hypermedia 超媒体【计】
hygronics 电子湿度计	hyperon 超子【物】
hygronom 湿度仪	hypersonic 超声,高超音速的(运动速度大于5倍声速)
hygrophilous plant 喜湿植物	

hypertext 超文本【计】
hyperthermia 体温过高
hypervelocity 超高速〔度〕
hypocenter 震源【地】
hypochlorous acid 次氯酸
hypolimnion (=hypolymnion) 均温层，下层滞水带(湖、海的)
hypothermia 体温过低，低温
hypothesis 假说，假设
hypothesis test 假设检验【化】
hypothetical global climate 设想的全球气候
hypoxic layer 缺氧层
hypsithermal 高温期(气候)
hypsithermal interval 高温时段(冰期后)
hypsographic curve 陆高海深曲线
hypsography ①地势图 ②地形测绘图 ③等高线〔分布〕型〔式〕
hypsometer 沸点测高表
hypsometric equation 测高方程
hypsometric formula 测高公式
hypsometry 测高法，测高学
hyrracano 飓风
hysteresis ①滞后现象②磁滞现象
hysteresis of adsorption 吸附滞后【化】
hyther 温湿作用
hythergraph 温湿图【大气】

I

IACS (International Associaton of Cryospheric Sciences) 国际冰冻圈科学协会(IUGG 的第 8 个协会成员)
IAEA (International Atomic Energy Agency) 国际原子能机构
IAG(International Association of Geodesy) 国际大地测量学协会
IAGA (International Association of Geomagnetism and Aeronomy) 国际地磁和高层大气物理学协会
IAHS (International Association of Hydrological Sciences) 国际水文科学协会
IAMAP(International Association of Meteorology and Atmospheric Physics) 国际气象学和大气物理学协会(IAMAS前身)
IAMAS(International Association of Meteorology and Atmospheric Sciences) 国际气象学和大气科学协会【大气】(1993 年起用此名)
IAMAS Commission on Atmospheric Electricity 国际气象学和大气科学协会大气电学委员会
IAMAS Commission on Dynamic Meteorology (ICDM) 国际气象学和大气科学协会动力气象学委员会
IAMAS Commission on Meteorology of Upper Atmosphere 国际气象学和大气科学协会高层大气气象学委员会
IAMAS Commission on Polar Meteorology (ICPM) 国际气象学和大气科学协会极地气象学委员会
IAMAS Radiation Commission (IRC) 国际气象学和大气科学协会辐射委员会
IAMAS Ozone Commission (IOC) 国际气象学和大气科学协会大气臭氧委员会
I and Q Channels(in-phase and quadrature channels) 同相和 90°相差的通道
IAPSO (International Association of Physical Science of the Ocean) 国际海洋物理科学协会
IAS (indicated air speed) 表速【大气】，指示航速

IAU (International Astronomical Union) 国际天文学联合会【天】
Ibe wind 艾贝风(类似焚风的局地强风)
IBL (internal boundary layer) 内边界层
IBM(International Business Machines Corporation) 国际商业机器公司,IBM 公司(美国)
IBP (International Biological Programme) 国际生物学计划
ICAO (International Civil Aviation Organization) 国际民[用]航[空]组织
ICAO standard atmosphere 国际民航组织标准大气【大气】
IC card 智能卡【计】
ICDM (IAMAS Commission on Dynamic Meteorology) 国际气象学和大气科学协会动力气象学委员会
ice 冰
ice accretion 积冰
ice-accretion indicator 积冰指示器
ice age 冰期【大气】
ice age aridity 冰期干燥性
ICE Age Epoch 冰河时代
ice albedo 冰反照率
ice-albedo feedback 冰-反照率反馈
ice and snow albedo 冰雪反照率
ice and snow albedo-temperature feedback 冰雪反照率-温度反馈
ice apparatus 测冰器
ice atlas 冰图
ice avalanche 冰崩【航海】
ice band(=ice belt) 浮冰带
ice bay 冰凹湾(大块海冰边缘凹进部分)
ice-bearing current 冰凌
ice belt 浮冰带【大气】(宽度几千米至 100 km),流冰带
iceberg 冰山【地理】
iceberg calving 冰山崩解
ice bight(=ice bay) 冰凹湾
iceblink 冰映光
ice blister 喷水冰层,冰泉
ice boundary 冰区界限线【航海】
icebreaker 破冰船【航海】

ice breakup 解冻,开冻
ice-bulb temperature 冰球温度
ice cake 饼[状]冰,板冰,冰块
ice calorimeter 冰卡表
ice cap 冰盖,冰冠
ice-cap climate 冰盖气候,冰冠气候
ice cascade 冰瀑
ice cave 冰穴
ice clearing 冰[间]湖,冰[间]穴,冰裂
ice climate 冰雪气候
ice cloud 冰云【大气】
ice coating 冰膜,冰衣
ice concentration 海冰密集度
ice condition summary 冰情概况
ice content 含冰量【地理】
ice core 冰芯
ice cover 覆冰量
ice crust 冰壳,冰皮,薄层冰
ice crystal 冰晶【大气】
ice crystal cloud 冰晶云
ice crystal concentration 冰晶浓度
ice crystal effect 冰晶效应
ice crystal fog 冰晶雾
ice crystal haze 冰晶霾
ice crystal nucleus 冰晶核
ice crystal theory 冰晶学说
ice day 冰日
ice deposit 积冰
ice desert 冰漠
iced firn(=ice firn, firn ice) 冻陈雪,冻粒雪,粒雪冰
ice drift 流冰,漂冰,淌凌
ice drill 冰钻,钻冰机
ice edge 冰缘线【海洋】,冰缘,冰块间隙
ice effect 冰效应
ice fall 冰瀑
ice fat 油脂状冰,冰脂
ice feathers 冰羽,霜羽,冰凇
ice field 冰原【地理】(漂浮于海面的)
ice floe 浮冰
ice flow 冰流,冰水径流
ice flower 冰花
ice fog 冰雾【大气】

ice foot 冰脚,冰壁
ice forecast 冰情预报
ice-formation condition 结冰条件
ice-forming nuclei 成冰核
ice free 无冰区【航海】
ice friction theory 冰摩擦〔学〕说
ice fringe 冰条纹
ice formation 成冰作用【地理】
ice formation on aircraft 飞机结冰
ice front 冰川前沿,冰壁
icehouse climate 冰室气候,冰窖气候
icehouse state 冰室状态
ice hydrometeor 冰结物,冰冻物
ice island 冰岛,浮冰岛
ice jam 冰塞,冰坝
ice keel 冰底,冰龙骨
Icelandic low 冰岛低压【大气】
ice layer 冰层
ice lens 冰透镜体【地理】
ice limit 冰限(冰的边缘线)
ice line 冰线
ice-mantling injury 冰壳害
ice mixing ratio 冰〔水〕混合比
ice mound 冰丘
ice multiplication 冰晶繁生,冰晶增殖
ice needle 冰针【大气】
ice nuclei 冰核
ice nuclei counter 冰核计数器
ice nucleus 冰核【大气】
ice nucleus theory 冰核理论
ice-ocean feedback 冰海反馈
ice on runway 跑道结冰
ice-out 融冰(水面冰块)
ice pack 积冰,浮冰
ice particle 冰粒
ice pellet 冰丸【大气】(包括冰粒和小雹)
ice period 冰冻期【航海】,冰期【海洋】
ice phase 冰相
ice phase parameterization 冰相参数化
ice pillar 霜柱
ice point 冰点【大气】
ice pole 冰极
ice prism 冰晶柱,冰针

ice pyramid 冰锥
ice raft 浮冰排,冰筏
ice rain 冻雨
ice reconnaissance 冰情侦察
ice regime phase 冰情
ice ribbon 冰带,冰条纹
ice-rich permafrost 富冰冻土【地理】
ice rind 冰壳【海洋】,脆冰壳
ice rise 冰丘
ice run 淌凌,融冰流
Ice Saints 冰圣降临日(据说法国 5 月 11—13 日多霜)
ice-saturation 冰面饱和
icescape 冰景(尤指极地带风光)
ice-scope 电线积冰自记仪器(示冰器)
ice sheet 冰盖【地理】,大冰原(覆盖于陆地)
ice shelf 冰架
ice sky 冰照云光(有冰面反射的有云天空)
ice splinter 碎冰片
ice spicule 冰针
ice stalagmite 冰笋
ice storm 冰暴【大气】
ice stream ①冰川②冰河③浮冰流
ice strip 窄浮冰带
ice structure 冰结构
ice surge 冰涌
ice target 冰靶
ice thickness 冰厚【海洋】
ice tongue 冰舌
ice-up 结冰(雪或水)
ice warning 冰情警报【航海】
ice-water mixed cloud 冰水混合云【大气】
ice-water mixing ratio 冰水混合比
ice wedge 冰楔【地理】
ice wind 冰冷风
icicle 冰柱
icing 积冰【大气】
icing index 积冰指数

icing intensity 积冰强度
icing level 积冰高度
icing meter 积冰表
icing mound 冰丘
icing on runway 跑道积冰【大气】
icing-rate meter 积冰率表
iconoscope 光电摄像管
icosahedron 二十面体【数】
IC temperature transducer 集成电路温度传感器
ICPM (IAMAS Commission on Polar Meteorology) 国际气象学和大气科学协会极地气象学委员会
ICSU (International Council of Scientific Unions) 国际科学联盟理事会，国科联
ictus of sun 日射病
ICU (intensive care unit) 监护病室【电子】
icy wind 冰冷风
ideal climate 理想气候【大气】
ideal fluid 理想流体【大气】
ideal gas (= perfect gas) 理想气体【大气】
ideal-gas laws 理想气体定律
ideal horizon 理想地平
ideal liquid 理想液体
identical particles 全同粒子【物】
identical twin integrations 全同模式孪生子积分实验
identification (ID) 认证【物】,识别,判读,标ించ
identification of air mass 气团识别
identifier 标识符【计】
identity 恒等〔式〕【数】
identity matrix 单位矩阵【数】,幺矩阵
identity signal 识别信号【航海】
IDNDR (International Decade for Natural Disaster Reduction) 国际减灾十年计划
IDV (integrated data viewer) 整合资料阅读机
IEC (International Electrotechnical Commission) 国际电工(技术)委员会
IEEE (Institute of Electrical and Electronic Engineers) 美国电气和电子工程师学会
IF signal 中频信号
IF-THEN rule 如果-则定则
IFF (identification: friend or foe) 敌友识别装置
IFOV (instantaneous field of view) 瞬时视场【电子】
IFR (instrument flight rules) 仪表飞行规则
IFR flight 仪表飞行
IFR terminal minimums 仪表飞行最低着陆气象〔条件〕
IFR weather 仪表飞行天气
ig (= hig) 伊格风暴(英国的一种强短时风暴)
IGAC (International Global Atmospheric Chemistry Project) 国际全球大气化学研究计划(IGBP 内)
IGBP (International Geosphere Biosphere Programme) 国际地圈生物圈计划
IGFOV (instantaneous geometric field of view) 瞬时几何视场
igloo (= iglu) (爱斯基摩人的)圆顶冰屋
igneous meteor 大气放电现象
igneous rock 火成岩【地质】
ignitron 点火器
ignorosphere 未知层,忽略层
IGOS (Integrated Global Observing System) 全球综合观测系统【地信】
IGU (International Geographical Union) 国际地理联合会(ICSU 内)
IGY (International Geophysical Year) 国际地球物理年
IHDP (International Human Dimensions Program) 国际人文因素计划
ill-conditioned equation 病态方程【数】

ill-posedness 不适定性
ill-posed problem 不适定问题【数】
illuminance(*E*) 〔光〕照度【物】,照明【物】
illumination 〔光〕照度【物】
illumination climate(=light climate) 光照气候
illumination intensity 光照强度
illumination length 光照长度【大气】
illuminometer 照度计【物】
image 像【物】,〔映〕像【物】
image alignment 图像对准
image complex 图像复合【地理】
image composition 图像合成【农】
image compression 图像压缩【计】
image data processing system(IDAPS) 图像资料处理系统
image definition 图像清晰度【物】
image degradation 图像劣化【物】,图像退化
image digitization 图像数字化【物】
image digitizing 图像数字化【地理】
image dissector camera system(IDCS) 图像分析照相机系统
image element 像元【物】
image enhancement 图像增强【物】,影像强化
image-forming receiver 成像接收机
image geometry 图像几何测定【地理】
image gridding 图像网格定位
image intensifier 图像增强器
image interpretation 图像判读【农】,图像解译
image level 像〔强度〕级
image matching 图像匹配【地信】
image mutual coherence function 图像互相干函数
image navigation 图像导航
image plane 像平面【物】,图像平面【计】
image plane scanner 图像平面扫描器

image processing 图像处理【大气】
image processing system 图像处理系统【地信】
image projection transformation 图像投影变换【地理】
image quality 影像质量【地理】,图像质量【计】
image reconstruction 图像重建【物】
image registration 图像配准【地信】
image resolution(=imagery resolution) 图像分辨率【大气】,影像分辨率【地理】
image restoration 图像复原【计】
imager 成像器
imagery 图像,成像【物】
imagery recognition 图像识别【地理】
image scale factor 图像比例尺
image segmentation 图像分割【计】
image sharpening 图像增强,图像锐化【地信】
image simulation 图像模拟【地理】
image smoothing 图像平滑【地信】
image storage system 图像存储系统【地信】
image transform 图像变换【物】
imaginary number 虚数【数】
imaginary part 虚部【数】
imaging 成像【物】
imaging radar 成像雷达【地信】,测绘雷达
imaging spectrometer 成像光谱仪【地信】
imaging system 成像系统【地信】
imbibition 吸涨〔作用〕【植】
IMF① (interplanetary magnetic field) 行星际磁场【大气】②(intrinsic mode function) 本征模态函数
immiscible displacement 不〔可〕溶混位移
immobile wave 驻波
IMO(International Maritime Organization) 国际海事组织(在英国伦敦)
IMO (International Meteorological Organization) 国际气象组织(世界气象

组织前身)
impact ①碰撞,冲击②影响,效果
impact ice 碰撞冰
impactometer 撞击采样仪
impactor(—impactometer) 撞击采样器,碰撞器,云粒子取样器
impedance matching 阻抗匹配【物】
imperfect elasticity 非完全弹性学
imperfect forecast 不完全预报
impetus 冲击
impinger 撞击采样器(以液体代替固体表面收集样品)
impingement 冲击
implicit difference equation 隐式差分方程【数】
implicit difference scheme 隐式差分格式
implicit function 隐函数【数】
implicit integration scheme 隐式积分方案
implicit method 隐式法【化】,蕴含法【计】
implicit model 隐模式
implicit parallelism 隐式并行性【计】
implicit smoothing 隐式平滑
implicit time difference 隐式时间差分
implicit time scheme 隐式时间格式
imprecision 不精确性(状态),精密度不够
impressed trough 诱生槽
impression 印痕(小滴取样)
improved TIROS operational satellite (ITOS) 艾托斯卫星,改进型泰罗斯业务卫星
improvement report 天气好转报〔告〕
impulse ①冲量【物】②脉冲
impulse current 冲击电流
impulse response 脉冲响应
impulse response function 脉冲响应函数
impulsive force 冲力
impulsive grounding resistance 冲击接地电阻

impulsive warming 爆发性增暖
impurity 不纯性,不纯度
inaccuracy 不准确度
inactive front 不活跃锋
inadvertent climate modification 无意识气候影响
inadvertent cloud modification 无意识云影响
inadvertent weather modification 无意识天气影响
incendium 火焰风
inch(in) 英寸(≈ 2.54 cm)
inches of mercury 英寸汞柱高($=33.864$ hPa)
incidence 入射
incident angle 入射角【物】
incident radiation 入射辐射
incident ray 入射线【物】
incident Rossby wave 入射罗斯贝波
incident solar flux 入射太阳通量
incipient low 初生低压
inclination 倾角
inclination of the axis of a cyclone 气旋轴倾角
inclination of the axis of an anticyclone 反气旋轴倾角
inclination of the wind 风的偏差角(实测风与梯度风间的夹角)
inclinometer 磁倾计
in-cloud scavenging 云中清除
incoherence 非相干性【物】
incoherent echo 非相干回波【大气】
incoherent field 非相干场
incoherent intensity 非相干强度
incoherent radar 非相干雷达【大气】,常规雷达
incoherent scattering 非相干散射
incoherent scatter radar 非相干散射雷达
incoherent source 非相干源
incoming radiation 入射辐射【大气】
incoming solar radiation 入射太阳辐射

incompressibility 不可压缩性【物】
incompressibility condition 不可压缩〔性〕条件
incompressible atmosphere 不可压缩大气
incompressible flow 不可压缩流
incompressible fluid 不可压缩流体【大气】
incompressible motion 不可压缩运动
inconsistency 不一致性,不相容性
inconstant cirrus 易变卷云
incorporation with extra soil 客土〔土壤〕
IN counter 冰核计数器
increment 增量【数】
increment cone 生长锥
incursion 侵入
incus(inc) 砧状〔云〕
indefinite ceiling 不定云幕
indefinite integral 不定积分【数】
independence test 独立性检验
independently distributed 独立分布的
independent pixel approximation 独立像元近似
independent random variable 独立随机变量
independent sample 独立样本
independent sampling 独立取样
independent sideband 独立边带
independent trials 独立试验【数】
independent variable 自变量【数】
indestructibility 不灭性
indeterminacy 不确定性,不确定度
index ①指数,指标②索引
index arm(of sextant) 指标(六分仪的)
index correction 指标订正
index cycle 指数循环【大气】
indexed non-sequential file 倒排索引文件【地信】
index error ①指标差(水银气压表读数与标准仪表之差)②检索误差
index number 索引号码
index number of station 区站号

index of absorption 吸收指数
index of aircraft icing 飞机积冰指数
index of aridity 干燥指数
index of continentality 大陆度指数
index of cooperation(IOC) 合作系数
index of evaporating surface of horizontal area 水平区内蒸发面指数
index of refraction 折射率
index of stability 稳定度指数【大气】
index of thermal stress 热应力指数
index of warmth 温暖指数
index of wetness 湿润指数,年雨量指数,年水量指数
index trend 指数趋势
Indian and Burma trough 印缅低槽
Indian Deep Water 印度洋深水
Indian equatorial jet 印度洋赤道急流〔水〕
Indian low 印度低压【大气】
Indian Meteorological Department(IMD) 印度气象局
Indian monsoon 印度季风【大气】
Indian National Satellite System(INSAT) 印度国家卫星系统
Indian Ocean 印度洋【海洋】
Indian Ocean dipole(IOD) 印度洋偶极子
Indian Ocean Experiment(INOEX) 印度洋试验
Indian Ocean monsoon 印度洋季风
Indian Plate 印度洋板块【海洋】
Indian Space Research Organization(ISRO) 印度空间研究组织
Indian spring low water 印度洋平均大潮低潮面(印度洋基准面)
Indian summer 印第安夏【大气】(美国10、11月间的热期)
indicated air speed(IAS) 表速【大气】,指示气流速度,指示航速
indicated airspeed 表速,指示航速,指示气流速度,

indicated altitude (IA)　指示高度
indicating hygrometer　指示式湿度表
indication of a request　查询标志
indicator　①显示器,指示器②指标
indicator diagram　指示线图
indicator figure　指示码
indicator function　指示函数【数】
indicator lamp　指示灯
indifferent air mass　中性气团,变性气团(失掉发源地性质的气团)
indifferent equilibrium　随遇平衡,中性平衡
indifferent stability　中性稳定度
indirect aerological analysis　间接高空分析
indirect aerology　间接高空学
indirect aerosol effect　间接气溶胶效应
indirect cell　间接环流
indirect circulation　间接环流〔圈〕【大气】,逆环流,反环流
indirect insertion　间接插入
indirect lightning strike　感应雷击
indirect method　间接法
indirect static earthing　间接静电接地
individual derivative　个别微商
individual droplet　单滴
individual microburst　单个微下击暴流
individual nuclei　单核
individual variable　个体变元【数】
Indonesian Throughflow (ITF)　印度尼西亚〔上层海水〕输送流,印度尼西亚穿越流
indoor air velocity　室内气流速度
indoor climate　室内气候【大气】
indoor temperature　室内温度
induced current　感应电流
induced fission　诱发裂变【化】
induced reaction　诱导反应【化】
induced recharge　诱导回灌,人工回灌
inductance　电感【物】
induction　①感应②诱导③归纳(法)④进入⑤吸气

induction charging mechanism　感应起电机制
induction experiment　诱导试验
induction icing　进气口积冰
induction method　感应法
induction period　诱导期【化】
inductive effect　诱导效应【化】
inductive statistics　归纳统计学
industrial aerosol　工业气溶胶
industrial climate　工业气候【大气】
industrial climatology　工业气候学
industrial emission　工业排放
industrialization　工业化
industrial meteorology　工业气象学
industrial revolution　工业革命(始于18世纪后半叶)
ineffective temperature　无效温度
inelastic collision　非弹性碰撞
inelasticity　非弹性【物】
inelastic process　非弹性过程
inequality　不等式【数】
inert gases　惰性气体
inertia　惯性【物】
inertia fluctuation　惯性振荡
inertia frequency　惯性频率
inertia-gravitational wave　惯性重力波
inertia gravity wave　惯性重力波【大气】
inertial boundary current　惯性边界流
inertial circle　惯性圆【大气】
inertial-convective subrange　惯性-对流副区
inertial coordinate system　惯性坐标系
inertial current　惯性流
inertial-diffusive subrange　惯性-扩散副区
inertial flow　惯性流
inertial force　惯性力【大气】
inertial forecast　惯性预报【大气】
inertial frame　惯性框架
inertial frequency　惯性频率
inertial instability　惯性不稳定【大气】
inertial-gravity wave　惯性重力波

inertial lag 惯性滞后
inertial mode 惯性模
inertial motion 惯性运动
inertial navigation system 惯性导航系统
inertial oscillation 惯性振荡【大气】
inertial period 惯性周期
inertial range 惯性区
inertial reference frame 惯性参考系【物】,惯性参照系
inertial stability 惯性稳定度【大气】
inertial sublayer 惯性次层
inertial subrange 惯性副区
inertial system 惯性系
inertial torque 惯性〔力〕矩
inertia wave 惯性波【大气】
inertio-gravity oscillation 惯性重力振荡
inertio-gravity wave(IGW) 惯性重力波【大气】
inexactness 不确切性
inference 推理【计】,推论
inference engine 推理机
inference machine 推理机【计】
inferior atmospheric layer 低层大气
inferior mirage 下现蜃景
inferior planet 内行星【天】(水星和金星),地轨内行星
infertile soil 瘠土【土壤】
infiltration 入渗【大气】,渗润,渗透
infiltration capacity 入渗量【大气】,渗润量
infiltration gallery 集水廊道(地下水)
infiltration index 渗润指数,入渗指数
infiltration rate 入渗率【农】
infiltration routing 下渗演算,入渗演算
infiltration well 下渗井,入渗井
infiltrometer 渗透仪,下渗仪
infinite 无穷〔的〕【数】,无限〔的〕
infinite continued fraction 无限连分数
infinitesimal 无穷小【数】
inflection point 拐点【数】,转折点
inflection point instability 转折点不稳

定度
inflexion point 反曲点,转折点
in-flight calibration 飞行中校准
in-flight service 飞行中服务
inflight visibility 飞行能见度
inflow 内流,入流,进水
inflow angle 入流角
inflow boundary 流入边界
inflow-storage-discharge curve 入流-蓄水-出流曲线(进蓄出曲线,洪水演算用)
influence field 影响场
influence function 影响函数【数】
influence theory 感应学说,诱生学说
influent seepage 渗漏,渗透
influx 入流,流入
informatics 信息学【数】
information 信息,情报,资料,数据
information collection 信息采集【地信】
information compression 信息压缩【地理】
information content 信息量
information contents 信息内容【地信】
information efficiency 信息效率【化】
information extraction 信息提取【地理】
information format 信息格式【地信】
information fusion 信息融合【地信】
information industry 信息产业【计】
information profitability 信息效益【化】
information retrieval system 信息检索系统【地信】
information safety 信息安全【地信】
information science 信息科学【数】
information security 信息安全【地信】
information system 信息系统【地信】
information technology 信息技术【地信】
information technology equipment 信息技术设备
information theory 信息论【数】
informis 无定状(云)

infrabar 低气压
infralateral tangent arcs 下切晕弧
infrared(IR) 红外线
infrared absorption hygrometer 红外吸收湿度表
infrared absorption spectroscopy 红外吸收光谱学
infrared atmospheric band 红外大气光谱带
infrared cloud imagery 红外云图
infrared cloud picture 红外云图【大气】
infrared cooling rate 红外冷却率
infrared detector 红外检测器【化】
infrared digital data 红外数字资料
infrared Doppler lidar 红外多普勒光〔雷〕达
infrared Doppler system 红外多普勒系统
infrared drying 红外线干燥【农】
infrared heating 红外线加热【农】
infrared humidometer 红外湿度表
infrared hygrometer 红外湿度表
infrared image 红外图像
infrared imagery 红外图像
infrared interferometer spectrometer(IRIS) 红外干涉光谱仪
infrared passive(IRP) 红外被动遥感
infrared photochemistry 红外光化学【化】
infrared picture 红外图像
infrared polarizer 红外偏振器【化】
infrared radiation 红外辐射【大气】
infrared radiator 红外辐射器
infrared radiometry 红外辐射测量术
infrared ray 红外线【物】
infrared refractometer 红外折射仪
infrared remote sensing 红外遥感
infrared scanning 红外扫描【农】
infrared source 红外光源【化】
infrared spectral region 红外光谱区
infrared spectrometer 红外分光计
infrared spectrophotometer 红外分光光度计【物】
infrared spectroscopy 红外光谱学【物】,红外光谱法【化】
infrared telescope 红外望远镜
infrared temperature profile radiometer (ITPR) 红外温度廓线辐射仪
infrared thermometer 红外温度表
infrasonic 次声〔的〕
infrasonic observatory 次声观测台
infrasonic wave 次声波【物】
infrasound 次声
ingredients based forecasting methodology 配料法(基于构成要素的预报方法)
inhalational particulate matter 可吸入颗粒物
inhibition 抑制〔作用〕,阻碍〔作用〕
inhomogeneity 非均匀性,非齐次性
inhomogeneous medium 不均匀介质
inhomogeneous mixing 不均匀混合
inhomogeneous ocean water 非均匀海水
inhomogeneous path 非均匀路程
inhomogeneous polynomial 非齐次多项式【数】
inhomogeneous random medium 不均匀随机介质
inhomogeneous reaction 非均相反应【化】
inhomogeneous scattering medium 不均匀散射介质
inhomogeneous turbulence 非均匀湍流
initial 起始,初始
initial-boundary value problem 初〔值〕-边值问题【数】
initial condensation phase 凝结初相
initial condition 初〔始〕条件【物】,初始条件【大气】
initial data 原始数据
initial detention(=surface storage) 初始滞留,地面滞留
initial ensemble perturbation 初始集合扰动
initial field 初始场

initial geomagnetic storm phase 地磁暴初相
initialization 初始化【计】,初值化【大气】
initial perturbation 初始扰动
initial phase 初始位相
initial rainfall 初期降水
initial round-off error 初始舍入误差
initial reading 起始读数
initial state 初始状态【计】
initial step(见 first time step) 起步
initial survey 初次检验【航海】
initial temperature 初始温度
initial value problem 初值问题【数】
initial velocity 初速〔度〕【物】,起始速度
injection temperature 〔海水〕进入口水温
injury 伤害【农】
injury by hail 雹害
injury by warm winter 暖冬害【农】
inland 内陆
inland climate 内陆气候
inland desert 内陆沙漠
inland fog 内陆雾
inland ice 内陆冰
inland inundation 内涝
inlandity 内陆率
inland sea breeze 内陆海风
inner boundary 内边界
inner eyewall 内〔层〕眼壁
inner friction 内摩擦
inner iteration 内迭代【数】
inner planet 带内行星【天】
inner product 内积【大气】
inner radiation belt 内辐射带
inner radiation shield 防辐射内罩
inner region 内区
inner scale(microscale) 内尺度(微尺度)
inner scale of turbulence 湍流内尺度
inorganic chemistry 无机化学【化】
inotope hydrology 同位素水文学【化】
in phase 同相
in pool 渠化(河流上逐节筑坝拦水,形成航行渠化梯级)

input 输入
input-output control system (IOCS) 输入-输出控制系统
INQUA (International Union for Quaternary Research) 国际第四纪研究联合会
INSAT (Indian National Satellite System) 印度国家卫星系统
insert for snow 量雪尺
inshore current 向岸流,近海流,近滨流
inshore wind 向岸风
in situ 原地,实地,现场
in situ localized monitoring 现场定域监测
in situ measurement 原地测量【地】,实地测量
in-situ measurement techniques 实地测量方法
in situ monitoring 实地监测
in situ observation 实地观测,现场观测
in situ source monitoring 污染源现场监测
insolameter 日射表
insolation 日射
insolation duration 日照时间【大气】
insolation weathering 热力风化
insoluble wettable nuclei 憎湿性核
inspection 检查,目测
inspectional analysis 检验分析,检查分析
instability 不稳定性【物】,不稳定〔度〕
instability chart 不稳定〔度〕图
instability constant 不稳定常数【化】
instability energy 不稳定能量
instability index 不稳定指数
instability line 不稳定线【大气】
instability shower 不稳定阵雨
instabilizing factor 不稳定因子
installation 装置
instantaneous 瞬间的
instantaneous field of view (IFOV) 瞬时视场【电子】
instantaneous geometric field of view (IGFOV) 瞬时几何视场
instantaneous power 瞬时功率

instantaneous power spectral density
瞬时功率谱密度
instantaneous sampling 瞬时采样
instantaneous sensation 瞬时感〔觉〕
instantaneous spectrum 瞬时谱
instantaneous unit hydrograph 瞬时单位过程图
instantaneous velocity 瞬时速度【物】
instantaneous wind speed 瞬时风速
instant occlusion 瞬时锢囚
in-step condition 同步条件,同步状态
Institute for Atmospheric Optics and Remote Sensing(IFAORS) 大气光学和遥感研究所(美国)
Institute for Atmospheric Sciences(IAS) 美国大气科学研究所
Institute for Theoretical Meteorology (ITM) 理论气象研究所(丹麦)
Institute of Electrical and Electronic Engineers(IEEE) 美国电气和电子工程师学会
Institute of Geophysics and Planetary Physics(IGPP) 地球物理与行星物理研究所(美国)
Institute of Polar Studies(IPS) 极地研究所(美国)
Institutes for Environmental Research (IER) 环境研究协会(美国)
instruction 指令【计】
instruction address register 指令地址寄存器【计】
instrumental analysis 仪器分析【化】
instrumental drift 仪器漂移
instrumental error 仪器误差
instrumental neutron activation analysis 仪器中子活化分析【化】
instrumental period 仪器记录时期(气候)
instrumental response 仪器响应
instrumentation 仪器测定
instrumentation ship 仪表船
instrument correction 仪器订正

instrument exposure 仪器暴露
instrument flight 仪表飞行
instrument flight rules (IFR) 仪表飞行规则
instrument landing system 仪表着陆系统
instrument meteorological condition (IMC) 仪表飞行气象条件
instrument payload 仪器载荷
instrument response function 仪器响应函数
instrument screen 百叶箱
instrument shelter 百叶箱【大气】,仪器〔百叶〕箱
instrument weather 仪表飞行天气
insulated stream 隔离河,绝缘河(与饱和层隔绝的)
insulation 绝缘
in-tandem model 串联模式
integer 整数【数】
integral 整的【数】,积分【数】
integral accuracy 积分精度【化】
integral calculus 积分学【数】
integral constraint 积分约束
integral curve 积分曲线【数】
integral depth scale 积分厚度尺度
integral emissivity 积分发射率
integral equation 积分方程【数】
integral heat of adsorption 积分吸附热【化】
integral heat of solution 积分溶解热【化】
integral length scales 积分长度尺度
integral transform 积分变换【数】
integrand 被积函数【数】
integrated absorption 总吸收
integrated absorption method 积分吸收法【化】
integrated analysis 综合分析
integrated assessment 综合评估
integrated circuit(IC) 集成电路
integrated data viewer(IDV) 整合资料阅读机

Integrated Flood Observing and Warning System(IFLOWS) 洪水综合观测和警报系统
Integrated Global Obsering System (IGOS) 全球综合观测系统【地信】
Integrated Global Ocean Station System (IGOSS) 全球联合海洋台站网
integrated information system 集成信息系统【地信】
integrated monitoring network 综合监测网
integrated physical geography 综合自然地理学【地理】
integrated physicogeographical regionalization 综合自然区划【地理】
integrated service digital network (ISDN) 综合业务数字网【计】,ISDN 网
integrated sounding system(ISS) 综合探测系统
integrated spectrum 积分谱
integrated survey 综合考察【地理】
integration 积分
integration anemograph 积分风速计
integration by parts 分部积分法【数】
integration kinetic evaporation meter 积分动力蒸发仪
integration method 积分法,集成法
integration nephelometer 积分浑浊度表
integration water sampler 水平集总采水器
integrity 完整性
INTELSAT (International Telecommunications Satellite Consortium) 国际通信卫星组织
intensification 加深,加强
intensity 强度
intensity-duration formula 强度-时间公式
intensity fluctuation 强度起伏
intensity level 强度〔等〕级
intensity-modulated indicator 强度调制显示器(雷达),调强指示器

intensity modulation 强度调制
intensity of illumination 光照强度【农】
intensity of turbulence 湍流强度
intensive care unit (ICU) 监护病室【电子】
intensive observation 加密观测
intensive observing period (IOP) 加密观测期
intensive property 强度性质
intensive quantity 强度量
interaction 相互作用,交互作用
interactive color radar display(ICRAD) 交互式彩色雷达显示
interactive digital image manipulation system(IDIMS) 交互式数字图像处理系统
Interactive Flash Flood Analyzer(IFFA) 暴洪交互分析器
interactive forecast system 交互式预报系统
interactive image processing 交互式图像处理
interactive processing 交互式处理【计】
interactive research imaging system(IRIS) 交互式研究成像系统
interactive terminal interface 交互式终端接口
Interagency Committee for International Meteorological Programs (ICIMP) 国际气象计划机构间联合委员会
interannual change 年际变化
interannual climate variability 年际气候变率
interannual predictability 年际可预报性
interannual time scale 年际变化尺度
interannual variability 年际变率【大气】
intercalary day 闰日
intercardinal wind 间向风(即东北、东南、西南、西北风)
interception 阻止作用,截留

interception of precipitation 降水截留
interceptometer 受雨器
interchangeability 互换性
interchange coefficient 交换系数
intercirrus region 卷云间区
interclass variance 组间方差
intercloud discharge 云际放电【大气】
intercloud lightning 云际闪电
interconnection network 互联网〔络〕【计】,互连网〔络〕
intercumulus region 积云间区
interdecadal variability 年代际变率
interdisciplinary study 多学科研究
interdiscipline 边缘科学
interdiurnal 日际
interdiurnal pressure variation 气压日际变化
interdiurnal temperature variation 温度日际变化
interdiurnal variability 日际变率
interdiurnal variation 日际变化
interface ①界面【计】②接口【计】
interface card 接口电路板
interface exchange process 界面交换过程【海洋】
interface module 接口模板,接口组件
interface standard 接口标准
interfacial layer 内界面层
interfacial tension 界面张力
interfacial wave 界面波
interference 干涉【物】,干扰
interference color 干涉色【物】
interference-filter photometer 干涉滤光光度计
interference fringe 干涉条纹
interference lobe 干扰波瓣
interference region 干涉区
interferogram 干涉图
interferometer 干涉仪【物】
interferometry 干涉测量术【物】
interferon 干扰素【生化】

interflow ①壤中流,表层流,土内水流 ②间层流 ③混流
intergelisol(=pereletok) 残冻层,陈年冻层
interglacial 间冰期〔的〕
interglacial and glacial cycle 冰期间冰期循环
interglacial condition 间冰期状况
interglacial period(或 phase) 间冰期【大气】
interglacial stage 间冰期【地质】
Intergovernmental Maritime Consultative Organization(IMCO) 政府间海运咨询组织(现为 IMO)
Intergovernmental Oceanographic Commission(IOC) 政府间海洋委员会
Intergovernmental Panel on Climate Change(IPCC) 政府间气候变化专门委员会【大气】
interhemispheric mass exchange 南北半球质量交换
interhourly variability 时际变率
Interim Operational Test Facility(IOTF) 临时业务试验机构
interim reference psychrimeter(IRP) 暂用标准干湿表
interim reference sunshine recorder 暂用标准日照计
inter-ion momentum transfer 离子内动量输送
interior angle 内角【数】
interior ballistics 内弹道学
interior basin 内陆盆地【地】
interior chemistry of earth 地球内部化学【地】
interleaving 交错〔混合〕
interlocking phenomenon 连锁现象
intermediate anticyclone 中间尺度反气旋
intermediate cyclone 中间尺度气旋
intermediate field 中间场
intermediate form 过渡形式

intermediate frequency(IF)　中频
intermediate frequency amplifier　中频放大器【电子】
intermediate ion　中离子
intermediate layer　中间层【电子】,过渡层
intermediate scale　中间尺度
intermediate standard time　辅助天气时间(03,09,15 和 21 时世界时)
intermediate synoptic observation　辅助天气观测【大气】
intermediate water　①中层水②中间水域(用复数)
intermittency　间歇现象
intermittent chaos　间歇混沌
intermittent(data) assimilation　间歇〔资料〕同化
intermittent heavy rain　间歇性大雨
intermittent light rain　间歇性小雨
intermittent moderate rain　间歇性中雨
intermittent precipitation　间歇性降水【大气】
intermittent rain　间歇〔性〕雨
intermittent stream　间歇〔性〕河〔流〕,季节河〔流〕
intermittent turbulence　间歇湍流
inter-monthly pressure variation　气压月际变化
inter-monthly temperature variation　温度月际变化
inter-monthly variability　月际变率【大气】
inter-monthly variation　月际变化
intermountain basin　山间盆地【地理】
intermountain region　山间地区
internal boundary　内界面
internal boundary layer　内边界层
internal cold(calibration) target(ICT)　机内定标用冷靶
internal consistency　内在一致性
internal consistency check　〔要素间〕一致性检验,内部一致性检验
internal energy　内能
internal friction〔force〕　内摩擦〔力〕【大气】
internal gravity wave　重力内波【大气】
internal inertial gravity wave　惯性重力内波
internal latent heat　内潜热
internal lightning protection system　内部防雷系统,内部防雷装置
internal memory　内存储器
internal mode　内模
internal parameter　内参数
internal Rossby radius of deformation　罗斯贝变形内半径
internal sea　内海【航海】
internal stability parameter　内稳定性参数
internal tide　内潮(海洋)
internal variable　内变量
internal warm target(IWT)　机内热靶
internal water circulation　〔局地〕内水分循环
internal waters　内水【航海】
internal wave　内波【航海】,【大气】
Internal Wave Experiment(IWEX)　内波实验
International Aerological Commission(IAC)　国际高空气象学委员会
International Air Transport Association(IATA)　国际航空运输协会
international analysis code(IAC)　国际分析电码
International Antarctic Analysis Center(IAAC)　国际南极研究分析中心
International Antarctic Meteorological Research Committee(IAMRC)　国际南极气象研究委员会
International Association for Atmospheric Sciences(IAAS)　国际大气科学协会
International Association for Radiation Research(IARR)　国际辐射研究协

International Association of Cryospheric Sciences(IACS) 国际冰冻圈科学协会(IUGG 的第 8 个协会成员)
International Association of Geodesy(IAG) 国际大地测量学协会
International Association of Geomagnetism and Aeronomy(IAGA) 国际地磁和高层大气物理学协会
International Association of Hydrogeologists(IAH) 国际水文地质工作者协会
International Association of Hydrological Sciences(IAHS) 国际水文科学协会
International Association of Meteorology and Atmospheric Physics(IAMAP,IUGG) 国际气象学与大气物理学协会(IAMAS 的前身)
International Association of Meteorology and Atmospheric Sciences(IAMAS) 国际气象学和大气科学协会【大气】
International Association of Physical Science of the Ocean(IAPSO) 国际海洋物理科学协会
International Astronautical Federation(IAF) 国际宇宙航行联盟
International Astronomical Union(IAU) 国际天文学联合会【天】
International Atomic Energy Agency(IAEA) 国际原子能机构
International Atomic Time 国际原子时【天】
International Biological Programme(IBP) 国际生物学计划
International Business Machine(IBM) 国际商用机器公司(及其生产的计算机型号)
international candle 国际烛光
International Civil Aviation Organization(ICAO) 国际民[用]航[空]组织
International Climatological Commission(ICC) 国际气候学委员会

international cloud atlas 国际云图【大气】
international cloud year 国际云年
international code 国际电码
International Commission for Air Navigation(ICAN) 国际航空委员会
International Commission on Dynamic Meteorology(ICDM) 国际动力气象学委员会
International Commission on Irrigation and Drainage(ICID) 国际排灌委员会
International Commission on Polar Meteorology(ICPM,IUGG) 国际极地气象学委员会
International Commission on Snow and Ice(ICSI) 国际冰雪委员会
International Commission on the Atmospheric Chemistry and Global Pollution(ICACGP) 国际大气化学和全球污染委员会
International Commission on the Meteorology of the Upper Atmosphere(ICMUA) 国际高层大气气象学委员会
International Committee for Earth Sciences(ICES) 国际地球科学委员会
International Committee of Atmospheric Science(ICAS) 大气科学国际委员会
International Council for the Exploration of the Sea(ICES) 国际海洋考察委员会
International Council of Scientific Unions(ICSU) 国际科学联盟理事会,国科联
International Council on Large Electric Systems(CIGRE) 国际大电网会议
international custom and usage 国际惯例【航海】
International Date Line 国际日期变更线
International Decade for Natural Disaster Reduction(IDNDR) 国际减灾十年计划

International Decade of Ocean Exploration(IDOE) 国际海洋调查十年规划
International Development Association(IDA) 国际开发协会
International Electrotechnical Commission(IEC) 国际电工(技术)委员会
International Federation of Air Line Pilots Associations(IFALPA) 国际航空驾驶员协会
international flight operations weather service 国际飞行气象服务
International Geographical Union(IGU) 国际地理联合会
International Geomagnetic Reference Field(IGRF) 国际地磁参考场
International Geophysical Committee(IGC) 国际地球物理委员会
International Geophysical Co-operation(IGC) 国际地球物理协作年
International Geophysical Year(IGY) 国际地球物理年
International Geosphere-Biosphere Programme(IGBP) 国际地圈-生物圈计划
International Global Atmospheric Chemistry Project(IGAC) 国际全球大气化学研究计划(IGBP 内)
International Human Dimensions Program(IHDP) 国际人文因素计划
International Hydrological Decade(IHD) 国际水文十年
international index number 国际区站号
International Indian Ocean Expedition(IIOE) 国际印度洋考察
International Maritime Organization(IMO) 国际海事组织(设在英国伦敦)
International Meteorological Committee(IMC) 国际气象学委员会
International Meteorological Organization(IMO) 国际气象组织(WMO 的前身)
International Meteorological Telecommunication Network 国际气象电〔传通〕信网
International Meteorological Teleprinter Network 国际气象电传打字电报网
international mile(或 nautical mile) 海里(=1852 m)
international network evaporation pan(INEP) 国际〔气象观测〕站网用蒸发皿
international period 国际观测期
International Polar Year(IPY) 国际极地年
International Project in Dendroclimatology(IPID) 年轮气候学国际研究计划
international pyrheliometric scale(IPS) 绝对日射表国际标尺
international quiet day 国际宁静日
International Quiet Sun Year(IQSY) 国际宁静太阳年
International Radio Consultative Committee(CCIR,ITU) 国际无线电咨询委员会(ITU)
international reference atmosphere 国际参照大气
international reference ionosphere 国际参考电离层【地】
International Rice Research Institute(IRRI) 国际水稻研究所
International Satellite Cloud Climatology Project(ISCCP) 国际卫星云气候学计划
International Satellite Land Surface Climatology Project(ISLSCP) 国际卫星陆面气候学计划
International Scientific Vocabulary(ISV) 国际通用科技词汇
international sign 国际符号
International Snow Classification 国际雪〔花〕形〔状〕分类

International Society for Photogrammetry and Remote Sensing(ISPRS) 国际摄影测量和遥感学会(成立于1910年,初名ISP,1980年起改用此名)
International Society of Bioclimatology and Biometeorology(ISBB) 国际生物气候学及生物气象学会
International Society of Biometeorology(ISB) 国际生物气象学会
International Southern Ocean Studies(ISOS) 国际南大洋研究计划
international standard atmosphere(ISA) 国际标准大气
International Standard Book Number(ISBN) 国际标准书号
International Standard Serial Number(ISSN) 国际标准刊号
International Standardization Organization(ISO) 国际标准化组织
International Steam Table Calorie(IT cal.) 国际蒸汽表卡
international symbol 国际符号
international synoptic code 国际〔天气〕电码【大气】
international synoptic surface observation code 国际地面天气观测电码
International System of Units(SI) 国际单位制
International Table calorie(IT cal.) 国际卡(=4.1868J)
International Telecommunications Satellite Consortium(INTELSAT) 国际通信卫星组织
International Telecommunication Union(ITU) 国际电信联盟
International Telegraph and Telephone Consultative Committee(CCITT,ITU) 国际电报电话咨询委员会(ITU)
international temperature scale(ITS) 国际温标
International Time Bureau 国际时间局
International Tree Ring Data Bank(ITRDB) 国际年轮资料库
International Union for Conservation of Nature and Natural Resources(IUCN) 国际自然与自然资源保护联盟
International Union for Quaternary Research(INQUA) 国际第四纪研究联合会
International Union of Geodesy and Geophysics(IUGG) 国际大地测量与地球物理联合会【测】
International Union of Geological Sciences(IUGS) 国际地质科学联合会
International Union of Radio Science(URSI) 国际无线电科学联合会
International Union of Surveying and Mapping 国际测绘联合会【测】
International Units 国际单位
international visual storm warning signal 国际风暴警报〔目视〕信号
international weather code 国际〔天气〕电码
internet 互联网〔络〕【计】,互连网〔络〕
Internet 因特网【计】
Internet phone 因特网电话【计】,IP电话
internet protocol 网际协议【计】
interpermafrost water 冻结层间水【地质】,冻结层中水【地理】
interplanetary dust 行星际(宇宙)尘埃
interplanetary gas 行星际气体
interplanetary magnetic field(IMF) 行星际磁场【大气】
interplanetary space 行星际空间【天】
interplay 相互作用
interpluvial 间雨期
interpolation 插值【数】,插值法【大气】,内插
interpolation formula 插值公式【数】
interpolation polynomial 内插多项式
interpolation weight 内插权重

interpretation 判读,解释,释用
interpulse period(IPP) 脉冲间歇周期,脉冲间歇时间(等于脉冲重复频率的倒数)
inter-quartile range 四分位数间距
interrogation recording and location system(IRLS) 询问、记录和定位系统
interrogation sign 疑问号
interrupted stream 断续河,中断河,断续水道
interrupt vector 中断矢量
interseasonal variability 季〔节〕际变率
intersect 横切,相交
intersequential variability 序列变率
interstade 间冰阶,间冰段
interstadial 间冰阶〔的〕
interstadial period 间冰阶
interstellar dust 星际尘埃【天】,星际尘【大气】
interstellar gas 星际气体【天】
interstellar space 〔恒〕星际空间【地】
interstellar turbulence 星际湍流
interstitial CCN 填隙的云凝结核
interstitial water ①间隙水②底质
interstroke change 闪击间歇变化
interstroke channel 闪击间通道
intertidal zone 潮间带(海洋)
intertropical cloud zone 热带云区
intertropical confluence zone(或 band) 热带汇流区(带)
intertropical convergence zone(ITCZ) 热带辐合带【大气】
intertropical front 热带锋
intertropical trough 热带槽
Inter-Union Commission on Radio-Meteorology(IUCRM) 无线电气象学协会间委员会
Inter-Union Commission on Solar-Terrestrial Physics(IUCSTP) 日地物理学协会间委员会
interval ①间隔,区间②音程,频程③阵风最大递减时段

interval estimation 区间估计【化】
interval zone 间隔带【地质】
intervening atmosphere 居间大气层
intervening medium 居间媒质
intolerant tree species 喜光树种【林】,阳性树种
intortus(in) 乱(云)
intra-annual growth band 年内生长纹印
intracavity gas laser 内腔式气体激光器
intracloud discharge 云内放电【大气】
intracloud flash 云内闪电
intracloud lightning 云内闪电
intranet 内联网【计】
intranet security 内联网安全【计】
intransitive 非传递性〔的〕
intransitive system 不定转移系统,非可递系统
intransitivity 非可递性
intraseasonal 季〔节〕内〔的〕
intraseasonal low-frequency variability 季〔节〕内低频变率
intraseasonal oscillation(ISO) 季节内振荡
intraseasonal time scale 季节内时间尺度
intrinsic carrier 本征载流子【电子】
intrinsic equation 内禀方程【物】
intrinsic error 固有误差【电子】
intrinsic frequency 固有频率,本征频率
intrinsic mode function(IMF) 本征模态函数
intrinsic permeability ①内禀渗透率②内禀磁导率
intrinsic quality factor 固有品质因数【电子】
intrinsic resistivity 内禀电阻率【物】
intrinsic wave frequency 固有波频,内禀波频
intromission 插入〔流〕,输入〔流〕
intromission zone 输入区

intrusions 侵入〔流〕,侵袭
inundation 洪水,大水
invading air 入侵空气
invain 无效
invariable 不变
invariant 不变式【数】
invariant imbedding method 不变量嵌入法
invariant latitude 不变纬度
invasion (=intrusion) 侵入
invasion of air 气团侵入
inventories of anthropogenic pollutants 人为污染物清单,人为污染物编目
inventory 目录,清单
inverna 因韦尔纳风(意大利马焦雷湖上的东南风)
inverse algorithm 反演算法
inverse discrete Fourier transform(IDFT) 反演离散傅里叶变换
inverse Fourier transform 傅里叶逆变换【数】
inverse method 逆方法,反演方法
inverse problem 逆问题,反演问题
inverse-square law 平方反比定律【农】
inverse technique 反演法【大气】,反演技术
inverse transform 逆变换
inverse trigonometric function 反三角函数【数】
inversion ①逆转②逆增(指温度),逆温③逆减(指降水量)
inversion base 逆温层底
inversion break-up 逆温破坏
inversion cloud 逆温云
inversion fog 逆温雾
inversion freezing 逆温冻结
inversion haze 逆温霾
inversion height 逆温层高度
inversion layer 逆温层【大气】
inversion lid 逆温层顶,逆温盖
inversion method 反演法

inversion of landform 地貌倒置【地理】
inversion of precipitation 雨量逆减(沿山坡,降水随高度增大而减少)
inversion of rainfall 雨量逆减
inversion of satellite sounding 卫星探测反演【大气】
inversion wind 逆温风
inverted barometer effect 压高转换效应
inverted image 倒像【物】
inverted landform 倒置地貌【地理】
inverted trough 倒槽【大气】
inverted V-pattern 倒 V 型
invertibility 可逆性
invertibility principle 可逆性原理
invierno 热带美洲雨季
inviscid fluid 非黏性流体【大气】
invisible cirrus 不可见卷云
invisible radiation 不可见〔的〕辐射
inward flux 内向通量
IOC (Intergovernmental Oceanographic Commission) 政府间海洋学委员会
IOD (Indion Ocean Dipole) 印度洋偶极子
iodide 碘化物
iodine 碘
ion 离子
ion atmosphere 离子雾
ion-atom interchange 离子-原子互换
ion attachment 离子附着
ion beam 离子束【物】
ion-capture theory 离子捕获〔学〕说
ion chromatography 离子色谱法【化】
ion cloud 离子云
ion column 离子柱
ion concentration 离子浓度
ion counter 离子计数器【大气】
ion current 离子电流
ion density 离子密度
ion detachment 离子分离
ion drag 离子拖曳
ion energy balance 离子能量平衡

ion energy loss rate　离子能量损失速率
ion exchange　离子交换【地质】
ionflo　离子流风速表
ionic acidity　离子酸度
ionic activity　离子活度
ionic conduction　离子传导
ionic mobility　离子迁移率【化】
ionic reaction　离子反应【化】
ion-ion recombination　离子-离子复合
ionium dating　镍定年法
ionization　电离【物】,电离作用,离子化
ionization by collision　碰撞电离【物】
ionization chamber　电离室【物】
ionization chamber detector　电离室检测仪
ionization energy　电离能【化】
ionization equilibrium　电离平衡【化】
ionization layer　电离层
ionization potential　电离电势
ionization temperature　电离温度【天】
ionization threshold　电离阈值
ionized atmosphere　电离大气
ionized particle　电离粒子
ionized trail　〔流星〕电离余迹
ionizing efficiency　电离效率
ionizing radiation　致电离辐射
ion life　离子寿命
ion mean life　离子平均寿命
ion migration　离子迁移
ion mobility　离子迁移率
ion-molecule reaction　离子-分子反应【化】
ionogram　电离图【大气】
ionopause　电离层顶【地】
ionosonde　电离层测高仪【地】,电离层探测器(利用电离层反射的短波以确定和记录电离层高度的装置)
ionosphere　电离层【大气】
Ionosphere Sounding Satellite(ISS)　电离层探测卫星
ionospheric disturbance　电离层扰动
ionospheric layer　电离层
ionospheric level　电离层高度
ionospheric modification　人工影响电离层
ionospheric observatory　电离层观象台
ionospheric parameter　电离层参数
ionospheric plasma　电离层等离子体
ionospheric potential　电离层电位,电离层电势
ionospheric probing　电离层探测
ionospheric profile　电离层剖面
ionospheric propagation　电离层传播【电子】
ionospheric property　电离层特性
ionospheric radar　电离层雷达
ionospheric recorder　电离层记录器
ionospheric response　电离层响应
ionospherics　电离层物理学
ionospheric scatter communication　电离层散射通信【电子】
ionospheric scintillation　电离层闪烁
ionospheric sounding　电离层探测
ionospheric storm　电离层暴【大气】
ionospheric substorm　电离层亚暴
ionospheric sudden disturbance　电磁层突扰【大气】
ionospheric tide　电离层潮
ionospheric tilt　电离层倾斜
ionospheric trough　电离层槽
ionospheric wave　电离层波
ionospheric wind　电离层风
ion pair　离子对
ion trap　离子捕集器
IPCC(Intergovernmental Panel on Climate Change)　政府间气候变化专门委员会【大气】
IPCC Fourth Assessment Report(AR4)　IPCC第四次评估报告
IPS(International Pyrheliometric Scale)　绝对日射表国际标尺
IPY(International Polar Year)　国际极地年

IQSY (International Quiet Solar Year)
国际宁静太阳年
IR (infrared 或 infrared radiation)
①红外②红外辐射
iridescence ①虹彩②晕色
iridescent altocumulus 虹彩高积云
iridescent cloud 虹彩云
Iriminger Current 伊尔明格海流
iris 虹彩
irisation 虹彩
Irish Meteorological Service(IMS) 爱尔兰气象局
IRIS(Infrared Interferometer Spectrometer) 红外干涉光谱仪
iron wind 铁风(中美的东北风)
irradiance 辐照度【大气】(W/m²)
irradiation 辐照【大气】(J/m²)
irradiation chamber 辐照室【农】
irradiation processing 辐照加工【农】
irreducibility 不可约性
irreducible fraction 不可约分数【数】,简分数
irregular crystal 不规则冰晶,不规则雪晶
irregular high-frequency type atmospherics 无规高频型天电
irregular reflection 乱反射
irreversible adiabatic process 不可逆绝热过程【大气】
irreversible greenhouse effect 不可逆温室效应
irreversible process 不可逆过程【物】
irreversible reaction 不可逆反应【化】
irreversible wave 不可逆波【化】
irrigation 灌溉
irrigation erosion 灌溉侵蚀【土壤】
irrigation farming 灌溉农业【农】
irrigation requirement 灌溉需水量,灌水定额
irrotational 无旋〔的〕
irrotational flow 无旋流
irrotational motion 无旋运动【大气】
irrotational vector 无旋向量
irrotational vortex 势涡
irrotational wind(component) 无旋〔转〕风(分量)
irtron 红外光电管
isabnormal 等异常线
isabnormal line 等异常线,等偏差线
isactine 日射化学强度等值线
isalea 等日射线,等日射量线
isallo- 等变(词头)
isallobar 等变压线【大气】
isallobaric 等变压的
isallobaric analysis 等变压分析
isallobaric chart 等变压图
isallobaric effect 等变压效应
isallobaric gradient 等变压梯度
isallobaric high 正变压中心
isallobaric low 负变压中心
isallobaric maximum 最大正变压
isallobaric minimum 最小负变压
isallobaric wind 变压风【大气】
isallohypse 等变高线【大气】
isallohypsic wind 等变高风
isallotherm 等变温线【大气】
isametral 等偏差线
isametric 等偏差,等距平
isanabat 等上升速度线
isanabation 等〔上〕升速〔度〕
isanakatabar 等气压较差线(在高低气压系统经过时的)
isanakatabaric chart 等气压较差线图
isanamal 等偏差,等距平线
isanemone 等风速线
isanomal 等距平线
isanomalous 等距平线〔的〕
isanomalous line 等偏差线,等距平线
isanomaly 等距平【大气】
isanthesic line 等始花线(物候学)
isarithm 等值线
isasteric surface 等容面
isaurore 极光等频线

ISCCP(International Satellite Climatology Project) 国际卫星云气候学计划
ISDN(integrated service digital network) 综合业务数字网【计】,ISDN 网
isenthalpic process 等焓过程【化】
isentrope 等熵线
isentropic 等熵的
isentropic analysis 等熵分析【大气】
isentropic atmosphere 等熵大气
isentropic cap 等熵冠
isentropic change 等熵变化
isentropic chart 等熵面图【大气】
isentropic condensation level 等熵凝结高度【大气】
isentropic coordinates 等熵坐标
isentropic Hadley circulation 等熵哈得来环流
isentropic mass circulation 等熵质量环流
isentropic mass transport 等熵质量传送
isentropic mean meridional circulation 等熵平均经向环流
isentropic mixing 等熵混合
isentropic motion 等熵运动
isentropic potential vorticity(IPV) 等熵位涡
isentropic process 等熵过程
isentropics 等熵线
isentropic sheet 等熵薄层
isentropic surface 等熵面
isentropic thickness-chart 等熵厚度图
isentropic trajectory 等熵轨迹
isentropic vertical coordinate 等熵垂直坐标
isentropic vorticity 等熵涡度
isentropic Walker-type circulation 等熵沃克环流
isentropic weight-chart 等熵重量图
iseoric line 等年温较差线
isepire 等降水大陆度
island effect 岛屿效应

ISLSCP(International Satellite Land Surface Climatology Project) 国际卫星陆面气候学计划
ISO ①(International Standardization Organization) 国际标准化组织 ②(intraseasonal oscillation) 季节内振荡
iso- ①等②同(词头)
iso-abnormal 等异常线
iso-abnormal line 等距平线
iso-amplitude line 等振幅线
isoanabaric center 等升压中心
isoanth 等花期线
isoatmic 等蒸发线
isoaurore 极光等频〔率〕线
isobar 等压线【大气】
isobaric 等压的
isobaric analysis 等压分析
isobaric channel 等压线间
isobaric chart 等压面图
isobaric condition 等压条件
isobaric contour chart 等压高度图
isobaric cooling 等压冷却
isobaric coordinates 等压坐标
isobaric divergence 等压辐散
isobaric energy transport 等压面能量传送
isobaric equivalent temperature 等压相当温度【大气】
isobaric isosteric solenoid 压容管
isobaric isosteric tube 压容管
isobaric line 等压线
isobaric mixing 等压混合
isobaric motion 等压运动
isobaric process 等压过程【化】
isobaric surface 等压面【大气】
isobaric temperature change 等压温度变化
isobaric thickness chart 等压厚度图
isobaric topography(=height pattern) 等压高度型〔式〕
isobaric tube 等压管
isobaric vorticity 等压涡度

isobaric wet-bulb temperature　等压湿球温度
isobarometric line　等压线
isobar type　等压线型式
isobase　等基线
isobath　等〔水〕深线
isobathytherm　等温深度线,等温深度面
isobestic point　等消光点,等吸收点
isobront　等雷暴日数线,初雷等时线
isoceraunic(= isokeraunic) line　①年雷暴日百分频率等值线②等雷雨(次数和强度)线
isochasm　极光等频〔率〕线
isocheim(= isochimene)　等冬温线
isochion　等雪量线【大气】
isochoric　等容〔的〕
isochoric change　等容变化,恒容变化
isochoric process　等容过程【化】
isochrone　同时线,等时线
isochronism　同时性
isoclimatic line　等气候线
isoclinal(= isoclinic) line　等倾线
isoclinal fold　等斜褶皱【地质】
isoclines　①等斜线,等倾线【数】②等斜褶皱(地质)
iso-coefficient　等系数线
isocorrelation　等相关线
isocryme　最冷期等水温线
iso-D　等 D 值线(或面)
isodef　等亏率线
isodense　等密线
isodiaphore　等〔气候因素〕月变线
isodrosotherm　等露点线【大气】
isodynam　等〔磁〕力线
isodynamic　等〔磁〕力的
iso-echo　等回波强度线
iso-echo contour　等回波线【大气】
isoenergy chart analysis　等能面分析
isoenergy pipe　等能管
isoentrope　等熵线
isoentropic analysis　等熵分析

isoeral　等春温线
isofronts-preiso code　等锋等压电码
isogeotherm　等地温线
isogon　①等风向线【大气】②等〔磁〕偏线③等角多边形
isogonal map(= conformal map)　等角地图,正形地图
isogonic line　同向线
isogradient　等梯度线
isogram　等值线
isohaline　等盐度线
isoheight　等高线
isohel　等日照线【大气】
isohion　①等雪深线②等雪日线③等雪线高度线
isohume　等湿度线【大气】
isohyet　等雨量线【大气】
isohyetal map　等雨量线图
isohygrometric line　等湿度线
isohyle　森林线
isohypse(= contour)　等高线
isohypsic　等高〔度〕的
isohypsic chart　等高面图
isohypsic surface　等高面
isokatabaric center　等降压中心
isokatanabar　等气压较差线
isokeraunic　①等雷频〔的〕②等雷雨强度〔的〕
isokeraunic line　(见 isoceraunic line) 雷暴频数等值线,等雷雨(强度或频次)线
isokinetic　等动能线
isokinetic sampling　等动能取样法
isokinetic temperature　等动力学温度【化】
isokurtic curve　等峰态曲线
isolated cell　孤立单体【大气】
isolated conductor　孤立导体【物】
isolated echo　孤立回波体
isolated line　孤立线,单一线
isolated system　隔离系统【化】,孤立系统
isolated wave　孤立波

isolation	绝缘
isoline	等值线【大气】
isomagnetic line	等磁线
isomenal	月平均等值线(主要指气温而言)
isomentabole	气压日际等变线
isomer	等比值线(主要指月降水量对年降水量的百分率而言),等雨率线
isomeric line	等雨率线
isomeric value	降水百分率
isometabole	等逐日变差线
isometeoric line	气象要素等值线
isometeorograde	等级值线(此级值系由 C. Ritter 所提出),等气象度线
isometric	①等轴的②等体积的③等距离的,等径的
isometropal	等秋温线
isoneph	等云量线【大气】
isonif	等雪量线
isonival	等雪量的,等雪量线
isoombre	等蒸发线
iso-orthotherm	等正温线
isopach	等厚度线
isopag(ue)	等冻期线
isoparallage	等年温较差线
isopectrics	冰冻等时线
isoperibolic calorimeter	等环境热量计【化】
isophasm(of pressure)	等变压值线
isophene(=isophane)	等物候线【大气】
isophenological line	等物候线【农】
isophote	等发光率线,等照度线
isopiestic	等压线
isopipteses	同时出现线
isoplanatic condition	等晕条件
isoplanatic region	等晕区
isopleth	等值线
isopleth of brightness	等亮度线
isopleth of thickness	等厚度线【大气】
isopluvial	等雨量线
isopore	地磁等年变线
isoporic	等磁变线
isopotential	等位势线
isoprene	异戊二烯(分子式 C_5H_8)
isopycnal	等密度
isopycnal mixing	等密度混合
isopycnal surface(= isopycnic surface)	等密度面
isopycnic	等密度的
isopycnic line	等密度线
isoryme	最冷月份平均温度等值线
isosceles trapezoid	等腰梯形【数】
isoshear	等切变线
isosphere	等球体
isostasy	地壳均衡
isostath	等密度线
isostatic adjustment	均衡调整,均衡补偿(地质)
isostere	等比容线
isosteric	等比容的
isosteric surface	等比容面
isotac	同时解冻线,等解冻线
isotach	等风速线【大气】
isotach analysis	等风速线分析
isotach chart	等风速线图
isotalant	等年温较差线
isothene	等气压平衡线
isothere(=isotheral)	等夏温线
isotherm	等温线【大气】,恒温线
isothermal	等温〔线〕的
isothermal atmosphere	等温大气【大气】
isothermal change	等温变化
isothermal (cloud) chamber	等温云室
isothermal compression	等温压缩
isothermal equilibrium	等温平衡
isothermal expansion	等温膨胀
isothermal layer	等温层【大气】
isothermal line	等温线
isothermal process	等温过程【大气】
isothermal reactor	等温反应器【化】
isothermal surface	等温面
isothermobath	等水温线

isothermobrose 平均夏雨等值线
isothermohyps 等温线
isotherm ribbon 等温线密集区
isothyme 等蒸发量线
isotimic 等值线
isotimic line 等值线
isotimic surface 等值面
isotonicty 等渗压线
isotope 同位素【物】
isotope abundance 同位素丰度
isotope dating 同位素年代测定【化】
isotope geochemistry 同位素地球化学【化】
isotope geochronology 同位素地质年代学【化】
isotope geology 同位素地质学【化】
isotope ratio 同位素比率
isotopic abundance 同位素丰度【物】
isotopic analysis 同位素分析
isotopic equilibrium 同位素平衡【地质】
isotopic exchange equilibrium 同位素交换平衡【地质】
isotopic exchange reaction 同位素交换反应【地质】
isotopic geothermometer 同位素地热标【地质】,同位素地球温度计【地】
isotopic paleoclimatology 同位素古气候学
isotopic thermometry 同位素测温法【地质】
isotopic tracer 同位素示踪物
isotropic 各向同性〔的〕,迷向【数】
isotropic intensity 各向同性强度
isotropic line 迷向直线【数】
isotropic medium 各向同性介质
isotropic radiation 各向同性辐射
isotropic radiator 各向同性辐射器
isotropic random function 迷向随机函数
isotropic scattering 各向同性散射
isotropic target 各向同性靶〔标〕

isotropic temperature factor 各向同性温度因子【化】
isotropic turbulence 各向同性湍流【大气】
isotropy 各向同性【物】
isovel（＝isotach） 等风速线
ISPRS（International Society for Photogrammetry and Remote Sensing） 国际摄影测量和遥感学会
Isreal Meteorological Society（IMS） 以色列气象学会
ITCZ（intertropical convergence zone） 热带辐合带【大气】
iterated interpolation method 迭代插值法【数】
iteration 迭代〔法〕
iteration method 迭代法【数】
iterative analysis 迭代分析
iterative inversion technique 迭代反演法
iterative process 迭代过程
iterative scheme 迭代法
iterative solution 迭代解
iterative variational method 迭代变分法
ITF（Indonesian Throughflow） 印度尼西亚穿越流
Ito equation 伊藤方程
Ito process 伊藤过程【数】
ITOS（Improved TIROS Operational System） 艾托斯卫星【大气】,改进型泰罗斯业务系统
ITU（International Telecommunication Union） 国际电信联盟
IUCN（Internatonal Union for Conservation of Nature and Natural Resources） 国际自然与自然资源保护联盟
IUGG（International Union of Geodesy and Geophysics） 国际大地测量和地球物理联合会【测】
ivory point 象牙针尖

J

Jacobian 雅可比行列式【数】
Jacobian energy transformation 雅可比能量转换
Jacobian operator 雅可比算子
Jacobi iteration 雅可比迭代〔法〕【数】
Jacobi iterative method 雅可比迭代法
Jacobi's integral 雅可比积分【天】
Jacob's ladder 雅各布光象(日落辉在水中的倒影)
jade 玉石【地质】
jadeite 硬玉【地质】
jaloque(＝xaloque) 西罗科风
jansky (简写为 Jy)央(天体射电流量密度单位，＝ 10^{-26} W·m^{-2}·Hz^{-1})【天】
jansky noise 宇宙〔射电〕噪声
January thaw 1月暖期,1月解冻
Japan Current 日本海流【大气】,日本〔暖〕海流(黑潮)
Japanese Antarctic Research Expedition(**JARE**) 日本南极研究探险
Japanese Earth Resources Satellite(**JERS-1**) 日本地球资源卫星
Japanese Meteorological Society(**JMS**) 日本气象学会
Japanese Society of Limnology(**JSL**) 日本湖沼学会
Japan Meteorological Agency(**JMA**) 日本气象厅
Japan Sea Deep Water 日本海深层水(水深大于 200 m)
Japan Sea Middle Water 日本海中层水(水深 25～200 m)
Japan Sea Proper Water 日本海本征水(同深层水)
jauk(＝**jauch**) 焦克风(奥地利克拉根富特盆地的焚风)
Java Java 语言【计】
jet ①急流②喷射③喷注【物】
jet axis 急流轴
jet current 急流
jet-effect wind 急流效应风
jet flow 急流
jetlet 小急流
Jet Propulsion Laboratory(**JPL**) 喷气推进实验室(美国)
jet streak 急流
jet stream 急流【大气】
jet stream axis 急流轴
jet stream cirrus 急流卷云
jet stream cloud 急流云
jet stream cloud system 急流云系【大气】
jet stream core 急流核【大气】,急流心
jet stream front 急流锋
jetstream location 急流定位
jetstream pattern 急流型式
jetstream ridge 急流脊
jetstream trough 急流槽
Jevons effect 杰文斯效应(雨量器对测雨的影响)
Jimsphere balloon 棘面气球
job 作业
jochwinde 约克风(阿尔卑斯山陶埃恩山口的风)
jog-log 电磁海流计
Johnson-Williams liquid water meter(＝**J-W meter**) 约翰逊-威廉姆斯液态含水量仪,J-W 液态含水量仪
Johnson-Williams liquid water probe 约翰逊-威廉姆斯液态水探测器,J-W 液态水探测器
joint 节理【地质】

Joint Air-Sea Interation Experiment（JASIE） 海气交互作用联合试验
Joint Arctic Weather Stations（JAWS） 联合北极天气站
joint characteristic function 联合特征函数
Joint Climate Research Fund（JCRF） 联合气候研究基金
Joint Global Ocean Flux Study（with IGBP） 全球海洋通量联合研究（国际地圈-生物圈计划内）
Joint GOOS Scientific and Technical Committee（J-GOOS） 全球海洋观测系统联合科技委员会
jointing 节理状况
Joint Institute for Marine and Atmospheric Research（JIMAR） 海洋与大气联合研究所（美国NOAA环境研究院与夏威夷大学）
Joint Institute for the Study of the Atmosphere and the Ocean（JISAO） 大气与海洋联合研究所（美国NOAA环境研究院与华盛顿大学）
Joint Meteorological Radio Propagation Committee（JMRPC） 气象无线电传播联合委员会（英国）
Joint Meteorological Satellite Advisory Committee（JMASC） 联合气象卫星咨询委员会（美国）
Joint North Sea Wave Project（JONSWAP） 北海海浪联合研究计划
Joint Numerical Weather Prediction Unit（JNWPU） 联合数值天气预报中心
Joint Organizing Committee（JOC, WMO/ICSU） 联合组织委员会（WMO/ICSU）
Joint Planning Office（GCOS） 联合计划办公室（全球气候观测系统）
Joint Planning Staff（JPS） 联合计划专家
Joint Research Center（CEC） 联合研究中心（欧洲共同体委员会）
Joint Scientific Committee（JSC, WMO/ICSU） 联合科学委员会（WMO/ICSU）
Joint Typhoon Warning Center（JTWC, USAF and US Navy） 美国联合台风预警中心（美国空海军）
JONSWAP（Joint North Sea Wave Project） 北海海浪联合研究计划，琼斯韦普计划
JONSWAP spectrum 琼斯韦普海浪谱，受风区限制的海浪谱
jonvek 琼维克风（捷克的冷北风）
joran（=juran） 乔兰风（从瑞士汝拉山吹向日内瓦湖的冷风）
Jordan algebra 若尔当代数【数】
Jordan sunshine recorder 乔唐日照计【大气】，暗筒式日照计
joule（J） 焦[耳]（能量,热量和功的单位）
Joule heat 焦耳热【物】
Joule law 焦耳定律【物】
Joule's constant 焦耳常数
Jovian planet 类木行星【天】
J process J过程
judder 位移,不稳定
Julian calendar 儒略历【天】（历年平均长度为365.25日）
Julian day 儒略日
jumbo 大型喷气客机
jump ①突跳【物】,跳跃②跃迁③骤变
jump model 跳跃模式
junction condition 衔接条件
junction streamer 衔接闪流,连接流光
Junge aerosol size distribution 荣格气溶胶〔尺度〕谱
Junge law 荣格律
Junge（aerosol）**layer** 荣格〔气溶胶〕层（平流层中大粒子的高值层）
Junge model 荣格模式
Junge size distribution 荣格〔尺度〕谱
junk wind 行船风（亚洲的东南季风）
junta 琼泰风（南美洲安第斯山隘口的大风）
Jupiter 木星【天】

juran(＝joran)　乔兰风
Jurassic Period　侏罗纪【地质】
Jurassic System　侏罗系【地质】
Jurines phenomenon　侏里纳现象(北非沙漠中的一种光学现象)
jury problem　应急解题〔法〕,临时解题

justification　论证【数】
Juvenile water　(＝magmatic water)　岩浆水,初生水
J-W meter　J-W 液态含水量仪

K

kaavie　卡维雪(苏格兰的大雪)
ka BP　距今…千年
kachchan　卡契昌风(斯里兰卡山脉背风坡的焚风)
kahamsin　卡哈姆辛风(埃及的一种和 Sirocco 风同类的干热风)
Kaikias　东北风,北东北风(希腊语名,在冬季带来雷暴和冰雹的风)
Kal Baisakhi　卡贝萨基飑(孟加拉西南季风来临时的尘飑)
kalema　激浪(见于几内亚海岸)
Kalman-Bucy filter　(或简称 Kalman filter)　卡尔曼-布西滤波器【数】
Kalman filter　卡尔曼滤波器【数】
Kalman filtering　卡尔曼滤波【大气】
Kalman gain matrix　卡尔曼增益矩阵
Kalman weight matrix　卡尔曼权重矩阵
Kamchatka Current　堪察加海流
kamsin(＝khamsin)　喀新风
kanat(＝ghanat,ganat)　坎儿井,暗渠
kane-pu-ahio-hio　卡涅普阿希奥希奥风(夏威夷群岛的龙卷风)
kaolinite　高岭石【土壤】
kapalilua　卡帕里洛阿风(夏威夷群岛的海风)
karaburan　黑风暴(中亚东北风)
karajol(＝qarajel,quara)　卡拉乔尔风(保加利亚沿海的西风)

karbas　卡尔巴风(古腓尼基的东北风)
karema wind　卡雷马风(非洲坦噶尼喀湖的强东风)
karif(＝kharif)　卡里夫风(索马里亚丁湾的西南风)
Karman constant　卡门常数【大气】,卡曼常数
Karman spectrum　卡曼谱
Karman vortex　卡曼涡
Karman vortex street　卡门涡街【大气】,卡曼涡街
Karman vortex train　卡曼涡列
karst　喀斯特【地质】,岩溶
karstbora　喀斯特布拉风(南斯拉夫沿海的 bora 风)
karst facies　喀斯特相【地质】,岩溶相
karst geomorphology　喀斯特地貌学【地理】
karst hydrology　喀斯特水文学,岩溶水文学
karst landform　喀斯特地貌【地理】
katabaric(＝katallobaric)　降压〔的〕,负变压〔的〕
katabatic cold front　下滑锋【大气】
katabatic front　下滑锋
katabatic wind　下吹风
katafront(＝catafront)　下滑锋
kata-isallobar　等负变压线

katallobar 负变压线【大气】
katallobaric 负变压〔的〕
katallobaric center 负变压中心【大气】
kataphalanx(＝cataphalanx) 冷锋面
kata thermometer 卡他温度表,冷却温度表
kaus(亦写为 quas,cowshee,sharki) 考斯风(波斯湾冬半年带雨飑的东南风)
kavaburd(＝cavaburd) 卡伐布大雪(英国设得兰群岛的大雪)
kavaihae 卡瓦衣赫风(夏威夷群岛的阵风飑)
kaver(＝caver) 卡佛风(英国外赫布里底群岛的一种和风)
K band K 波段(雷达用,频率为 8～12 GHz 波长为 2.5～3.75cm)
K changes K 变化
K-ε closure K-ε 闭合,一阶半〔湍流〕闭合
KdV equation(＝Korteweg-de Vries equation) KdV 方程【数】
keel ①龙骨②脊棱
keen 凛寒
keily probe 凯里探头(测云滴谱用)
kelsher 英国大阵雨
Kelvin 开〔尔文〕(国际单位制的基本单位,记为 K)
Kelvin equation 开尔文公式【化】
Kelvin-Helmholtz billows 开尔文-亥姆霍兹大波,K-H 波状云
Kelvin-Helmholtz instability 开尔文-亥姆霍兹不稳定性【物】
Kelvin-Helmholtz wave 开尔文-亥姆霍兹波【大气】
Kelvin model 开尔文模型【化】
Kelvin's circulation 开尔文环流
Kelvin's circulation theorem 开尔文环流定理【大气】
Kelvin temperature 绝对温度
Kelvin temperature scale 绝对温标,开〔尔文〕温标
Kelvin wave 开尔文波【大气】
Kennedy Space Center(KSC) 〔美国〕肯尼迪空间中心
Kennelly-Heaviside layer 肯内利-赫维赛层,E 层
Kepler's equation 开普勒方程【天】
Kepler's law 开普勒定律【天】
keraunograph(＝ceraunograph) 天电仪,雷电计
kern counter 计核器
kernel function 核函数【数】
kernel ice 雾凇冰(一种飞机积冰)
Kerner's oceanicity index 柯尔纳海洋度指数
kerns arc 日承,对弧
Kessler type parameterization 凯斯勒型参数化
ketone ①酮②含酮基的
ketone body 酮体【生化】
Kew barometer 寇乌气压表
Kew pattern barometer 寇乌气压表【大气】,定槽式气压表
Kew photographic thermograph 寇乌照相式温度计
keyboard 键盘【计】
key day(＝control day) 征兆日,特征日
key group 指示组
keying 键入,打孔
key word 关键字〔词〕,电码字
keyword in context 上下文内关键字【计】
khamaseen(＝khamsin) 喀新风
khamsin 喀新风(埃及和红海上空的一种干热南或东南风)
kharif(＝karif) 卡里夫风
kharmatan 卡马塔(西非沙漠区冬季的多尘冷天气)
kibli 基布利风(利比亚地中海沿海岸的干热风)
killing freeze 严冬(影响生长季节的)
killing frost 严霜

killing temperature 致死温度
kilo- 千(词头,10^3)
kilobit 千比特
kilocalorie(＝kilogram calorie) 千卡,大卡
kilocycle 千周
kilogram(kg) 千克,公斤
kilogram calorie 千卡,大卡
kilohertz(kHz) 千赫
kilojoule 千焦耳
kilomegacycle 千兆周(旧称,现称吉周)
kilometer(km) 千米,公里
kilomole 千摩尔,千克分子(已陈旧,不用)
kilowatt-hour 千瓦时
kinase 激酶【生化】
K index K 指数【地】(雷暴稳定度)
kinematical analysis(＝kinematic analysis) 运动学分析
kinematical equation 运动学方程【物】
kinematical reference system 运动学参考系【天】
kinematic boundary condition 运动学边界条件【大气】
kinematic equator 运动学赤道
kinematic extrapolation 运动外延,运动学外推〔法〕
kinematic flux 运动学通量
kinematic front 运动锋
kinematic similarity 运动〔学〕相似性
kinematics 运动学【物】
kinematic viscosity 运动黏度【物】,运动黏滞性【大气】
kinematic viscosity coefficient 动黏滞系数
kinetheodolite 摄影经纬仪
kinetic cooling 动力致冷
kinetic energy 动能【物】
kinetic energy equation 动能方程
kinetic energy spectrum 动能谱
kinetic equation 动力〔学〕方程【物】,动力学方程
kinetic equilibrium 动力平衡

kinetic friction 动摩擦
kinetic parameter 动力学参数【化】
kinetic reaction rate 动力学反应速率
kinetics 动理学【物】,动力学(曾用名)
kinetic temperature 动力学温度,动理温度【物】
kinetic theory 〔分子〕动理学理论,〔分子〕运动论
kinetic theory of gases 气体动理〔学理〕论【物】,气体分子运动论
Kirchhoff approximation 基尔霍夫近似
Kirchhoff integral theorem 基尔霍夫积分定理【物】
Kirchhoff's equation 基尔霍夫方程
Kirchhoff's law 基尔霍夫定律【大气】
Kirchhoff vortex 基尔霍夫涡旋
kite 风筝
kite-ascent 风筝探测
kite balloon 系留气球
kite observation 风筝观测
kite sounding 风筝探测
kitoon(＝kytoon,kite balloon) 系留气球
kloof wind 克洛夫风(南非西蒙斯湾的西南冷风)
klydonograph 闪电特性仪
klystron 速调管【电子】
knik wind 尼克风(阿拉斯加的强东南风)
Knollenberg optical array probe 诺伦贝格光阵式探测器
Knollenberg probe 诺伦贝格探测器
knot 海里/小时,节
knowledge acquisition 知识获取【计】
knowledge base 知识库【计】
knowledge-based consultation system 基于知识的咨询系统【计】
knowledge base system 知识库系统【计】
knowledge engineering 知识工程【计】
knowledge extraction 知识提取【计】
knowledge industry 知识产业【计】
Knudsen number 克努森数【物】
Knudsen's tables 克努森表

koembang 科厄姆班风(爪哇的焚风)
kogarashi 秋风(日语名)
kohala 科哈拉风(几内亚群岛的海陆风)
kohilo 科希洛风(几内亚群岛的飑)
Köhler equation 科勒方程(关于云滴增长的)
Kol Baisakgi 科尔贝萨吉风(西南季风开始时,在孟加拉湾北部沿岸发生的西北阵风)
Kollsman stagnation thermometer 科斯曼驻点温度表
Kolmogorov cascade 科尔莫戈罗夫串级
Kolmogorov-Chapman equation 科尔莫戈罗夫-查普曼方程【物】
Kolmogorov constant 科尔莫戈罗夫常数,科尔氏常数
Kolmogorov dissipation scale 科尔莫哥罗夫耗散尺度
Kolmogorov microscale 科尔莫哥罗夫微尺度
Kolmogorov-Obukhov inertial range 科尔莫哥罗夫-奥布霍夫惯性区,K-O惯性区
Kolmogorov-Obukhov inertial subrange 科尔莫哥罗夫-奥布霍夫惯性副区,K-O惯性副区
Kolmogorov-Obukhov length 科尔莫哥罗夫-奥布霍夫长度,K-O长度
Kolmogorov scale 科尔莫哥罗夫尺度,科尔氏尺度
Kolmogorov similarity 科尔莫哥罗夫相似性
Kolmogorov spectrum 科尔莫哥罗夫谱
Kolmogorov theory 科尔莫哥罗夫理论
Kolmogorov time scale 科尔莫哥罗夫时间尺度
Kolmogorov turbulence 科尔莫哥罗夫湍流
Kolmogorov velocity scale 科尔莫哥罗夫速度尺度
kona 科纳风(夏威夷西南方来的暴风雨)
kona cyclone (也称 kona storm)科纳气旋【大气】(冬季大型少移动的副热带气旋)
kona storm 科纳风暴(出现于夏威夷群岛)
konimeter(=conimeter) 计尘器
koniology(=coniology) 微尘学
koniscope(=coniscope) 计尘仪
konisphere(=staubosphere) 尘圈,尘层
Köppen classification 柯本分类
Köppen-Geiger climate 柯本-盖格气候
Köppen's climatic classification 柯本气候分类
Köppen-Supan line 柯本-苏潘等温线【大气】
Korteweg-de Vries equation (或 KdV equation) 考特维-德伏里斯方程,KdV方程【物】
Korteweg-de Vries wave (=KdV wave) 考特维-德伏里斯波,KdV波
Koschmieder's law 柯什密得定律
Koschmieder's theory 柯什密得理论(能见度计算)
Kosmos(=Cosmos) 宇宙卫星
kossava(=kosava,koschawa) 科萨瓦风(贝尔格莱德东南方多瑙河上的谷风)
K_p K_p 指数(地磁活动)
K process K过程(闪电)
Krakatao volcano (亦拼 Krakatau volcano)喀拉喀托火山(印度尼西亚境内,105°24′E,6°07′S,1883年发生震惊世界的火山大爆发事件)
Krakatao winds 喀拉喀托东风带(取名源自喀拉喀托火山爆发后发现的现象),高空信风(热带 18~24 km 高空的东风层)
Kriging 克里金插值〔法〕(邻近点加权内插);克里金法【地信】(一种求最优线性无偏内插估计量的方法)
krivu(=crivetz) 克里维茨风

Kronecker delta 克罗内克 δ,克氏符号
Kronecker symbol 克罗内克符号【数】
Kronecker tensor 克罗内克张量
kryptoclimate 室内小气候【大气】
kryptoclimatology 室内小气候学
krypton 氪〔气〕
krypton hygrometer 氪〔管〕湿度计
K theory K 理论
K theory of turbulence 湍流 K 理论【大气】
Kubelka-Munk two flux theory 库比尔卡-芒克二流理论
Kunming quasi-stationary front 昆明准静止锋【大气】
Kuo convection scheme 郭晓岚对流方案,郭氏对流方案
Kuo-Eliassen equatoin 郭-伊莱亚森方程,

K-E 方程
Kuo scheme 郭氏法,郭氏方案
Kurihara grid 栗原网格
Kurile Current 千岛海流,亲潮
Kuroshio 黑潮【大气】,台湾暖流
Kuroshio Countercurrent 黑潮逆流
Kuroshio Current 黑潮(日本暖流,台湾暖流)
Kuroshio Extension 黑潮延续体,黑潮续流
Kuroshio system 黑潮〔流〕系
kurtosis 峰态,峭度
kybernetics(=cybernetics) 控制论
kymatology 波浪学
Kyoto Protocol 京都议定书(1997 年通过)
kytoon(=kite balloon,kitoon) 系留气球

L

lab(=laboratory) ①实验室②研究所
labbe(=labe) 拉贝风(法国普罗旺斯的湿润西南风)
labech 拉贝契风(阿尔卑斯山的 labbe 风)
labile 不稳定〔的〕
labile phosphorus 活性磷【土壤】
lability 不稳定度,易变性
laboratory(=lab) ①实验室②研究所
Laboratory for Applications of Remote Sensing(LARS) 遥感应用实验室(美国珀杜大学)
Laboratory for Plannetary Atmosphere (LPA) 行星大气实验研究所(英国伦敦大学)
laboratory lightning 人造闪电
Laboratory of Atmospheric Sciences(LAS) 大气科学实验室(美国)
laboratory tank 实验池

laboratory tornado 模拟龙卷
Labrador current 拉布拉多〔冷〕海流
Labrador Sea Water 拉布拉多海水
lacunaris 网状(云)
lacunosus(la) 网状(云)
Lafond's Tables 赖芳德表
lag 落后,滞后
lag coefficient 滞后系数【大气】,后延系数
lag correlation 后延相关,自相关
lag correlation method 后延相关法
lag cross correlation 滞后交叉相关
lag effect 滞后效应【海洋】
lag error 滞差
lag of aneroid barometer 空盒气压表的滞后现象
lagoon 潟湖【地理】,氧化塘【农】
Lagrange expansion formula 拉格朗日展开公式【数】

Lagrange interpolation 拉格朗日插值
Lagrange interpolation formula 拉格朗日插值公式【数】
Lagrange mean value theorem 拉格朗日中值定理【数】
Lagrange motion 拉格朗日运动
Lagrange multiplier 拉格朗日乘子【物】
Lagrange's planetary equation 拉格朗日行星运动方程【天】
Lagrange turbulence 拉格朗日湍流【物】
Lagrangian advective scheme 拉格朗日平流格式
Lagrangian airshed models 拉格朗日空气(污染区)模式
Lagrangian change 拉格朗日变化
Lagrangian coordinates 拉格朗日坐标
Lagrangian correlation 拉格朗日相关
Lagrangian current measurement 拉格朗日测流〔法〕
Lagrangian description 拉格朗日描述
Lagrangian equations 拉格朗日方程
Lagrangian equations of motion 拉格朗日运动方程
Lagrangian float 拉格朗日浮筒,拉格朗日漂流标
Lagrangian function 拉格朗日函数【物】
Lagrangian hydrodynamic equations 拉格朗日流体动力学方程
Lagrangian mean current 拉格朗日平均流
Lagrangian mean formulation 拉格朗日平均公式
Lagrangian mean meridional circulation 拉格朗日平均经向环流
Lagrangian method 拉格朗日方法
Lagrangian similarity 拉格朗日相似
Lagrangian timescale 拉格朗日时间尺度
Lagrangian wave 拉格朗日波
lag time 滞后时间
Laguerre-Gauss quadrature 拉盖尔-高斯求积【数】
lag window 后延窗

laheimar 拉海玛飑(阿拉伯 10、11 月的飑)
lake 湖
lake basin 湖盆【地理】
lake breeze 湖风【大气】
lake breeze front 湖风锋
lake circulation 湖水环流【地理】
lake core 湖芯
lake current 湖流【地理】
lake effect 湖泊效应
lake-effect snow 湖泊效应降雪
lake effect snowstorm 湖泊效应雪暴
lake evaporation 湖泊蒸发
lake front 湖锋【大气】
lake hydrology 湖泊水文学【地理】
lake ice 湖〔泊〕冰
lake level 湖平面
lake sediment 湖相沉积
lake storage 湖泊蓄水量【地理】
lake (surface) temperature 湖面温度
LAM(limited area model) 有限区模式【大气】
lamb-blasts 英国小春雪
Lamber-Bouguer law 朗伯-布格定律【物】
lambert 朗伯(亮度单位)
Lambert-Beer law 朗伯-比尔定律【化】
Lambert conformal conic projection 兰勃特正形圆锥投影
Lambert conic projection 兰勃特圆锥投影
Lambert cosine law 朗伯余弦定律【物】
Lambertian scattering 朗伯散射
Lambertian surface 朗伯面
Lambert projection 兰勃特投影【测】
Lambert's formula 朗伯公式
Lambert's law 朗伯定律
Lambert's law of absorption 朗伯吸收定律
lambing storm (=lamb-storm) 小春雪
Lambrecht's polymeter 兰布雷奇特多能湿度表

Lamb shift 兰姆移位【物】
lamb-shower 小春雪
Lamb wave 兰姆波【大气】
lamellar flow 层流【物】,片流
lamellar vector 片式矢〔量〕,片向量
laminar boundary layer 层流边界层【大气】
laminar flow 片流
laminar plume 片状烟流
laminar sub-layer 片流下层
laminar vector 片式矢〔量〕
laminar vortex 片式涡流
laminated current 片流
Lamont-Doherty Earth Observatory of Columbia University 哥伦比亚大学拉蒙特-多尔蒂地球观象台
LAN (local area network) 局域网【计】
land and sea breezes 海陆风
land-atmosphere interaction 陆气交互作用
land-barometer 陆用气压表
land breeze 陆风【大气】
land breeze climatology 陆风气候学
land breeze depth 陆风厚度
land breeze front 陆风锋
land breeze speed 陆风风速
land breeze temperature 陆风温度
land capability 土地生产能力【地理】
land characteristics 土地特性【地理】
land classification 土地分类【地理】
land cover classification 土地覆盖分类【农】
landcreep 山崩
land degradation 土地退化
land effect 海岸效应,陆地效应【航海】
land element 土地要素【地理】
land environment 陆地环境【地质】
land evaluation 土地评价【地理】
land evaporation 陆面蒸发【地理】
landfall 着陆,登陆
landfast ice 岸冰

land fog 陆雾
landform 地貌【地理】
landform forming process 地貌形成作用【地理】
landform genesis 地貌成因【地理】
landforms series 地貌序列【地理】
land function 土地功能【地理】
land ice 陆冰【地理】
land improvement 土地改良【地理】
land information system 土地信息系统【地理】
landing〔weather〕forecast 着陆〔天气〕预报【大气】
land-lash 冲击大陆风暴(英格兰的狂风暴雨)
land management 土地管理【地理】
landmark 陆标,地标
landmark navigation 地标导航
land-origin ice 陆源冰【海洋】
land productivity 土地生产率【地理】
land quality 土地质量【地理】
land resources 土地资源【农】
land resources satellite 国土资源卫星【地理】
LANDSAT 陆地卫星【大气】
Land Satellite (＝Landsat) 陆地卫星【大气】
landscape 景观,自然景色
landscape architecture 园林学【林】
landscape climatology 景观气候学
landscape design 风景设计【林】
landscape ecology 景观生态学【地理】
landscape engineering 园林工程【林】
landscape geochemistry 景观地球化学【地理】
landscape planning 风景规划【林】
landscape science 景观学【地理】
land-sea contrast 海陆对比
land sky 陆照云光
landslide 滑坡【地理】,山崩
landslip 滑坡【地理】,山崩

landslip wind　山崩风
landspout　陆龙卷
land subsidence　地面沉降【地质】
land surface parameterization　陆面参数化
land surface process　陆面过程
land treatment system　土地处理系统【农】
landuse　土地利用【地信】
landuse change　土地利用变化
landuse classification　土地利用分类【地理】
landuse map　土地利用图【地信】
landuse system　土地利用制度【地理】
landuse type　土地利用类型【地理】
landutilization rate　土地利用率【农】
landward wind　海风
lane　冰巷
Langevin equation　朗之万方程
Langevin ion　朗之万离子(即重离子)
langkisau　兰基苏风(苏门答腊和东印度群岛白天强烈的焚风)
langley(ly)　兰(能通量旧单位)
Langley Research Center (LRC)　美国兰利研究中心
Langmuir cells　朗缪尔对流胞
Langmuir chain reaction　朗缪尔连锁反应
Langmuir circulation　朗缪尔环流
Langmuir layer　朗缪尔层
Langmuir number　朗缪尔数
Langmuir probe　朗缪尔探针【物】(量度等离子体密度用)
Langmuir probe measurement　朗缪尔测定
Langmuir streak　朗缪尔条纹
Langmuir trough　朗缪尔水槽(实验仪器)
Langmuir wave　朗缪尔波【物】
La Nina　拉尼娜【大气】,反厄尔尼诺
Laplace distribution　拉普拉斯分布【数】
Laplace equation　拉普拉斯方程
Laplace operator　拉普拉斯算子
Laplace tidal equation　拉普拉斯潮汐方程
Laplace transform　拉普拉斯变换
Laplacian operator　拉普拉斯算子
LAPS (local analysis and prediction system)　局地分析和预报系统(NOAA开发的)
lapse limit(=tropopause)　对流层顶
lapse line　直减率线
lapse rate　直减率,递减率
lapse rate of temperature　温度直减率
lard ice　脂状冰
large amplitude wave theory　大振幅波理论
large calorie　千卡,大卡
large eddy model　大涡旋模式
large eddy simulation　大涡流模拟
large eddy simulation model　大涡旋模拟模式
large halo　大晕(46°的晕)
large ion　大离子,重离子
large nuclei　大核
large particle　大粒子
large Reynolds number flow　大雷诺数流〔动〕
large rotation fluid annulus　大型流体转盘
large scale　大尺度
large-scale atmospheric motion　大尺度大气运动
large-scale circulation　大尺度环流【大气】
large-scale computer　大型计算机【计】
large-scale condensation　大尺度凝结
large-scale convection　大尺度对流
large-scale loop　大尺度〔磁力线〕环
large-scale motion　大尺度运动
large-scale planetary wave　大尺度行星波
large-scale turbulence　大尺度湍流
large-scale wave property　大尺度波特性
large-scale weather〔process〕　大尺度天气〔过程〕【大气】
large-scale weather situation　大尺度〔天气〕形势

Larmor frequency 拉莫尔频率【物】
larry 薄浆雾(英国特恩河口的浓雾)
laser(light amplification by simulated emission of radiation) 激光【物】,激光器【物】
laser absorption spectrometer(LAS) 激光吸收分光计
laser aligner 激光准直仪【测】
laser altimeter 激光测高仪【测】
laser anemometer 激光测风仪
laser beam 激光束【化】
laser beam recorder(LBR) 激光束记录仪
laser bonding 激光焊接【物】
laser boring 激光打孔【物】
laser cavity 激光〔共振〕腔【物】
laser ceilograph 激光云幂计
laser ceilometer 激光云幂仪【大气】
laser cloud indicator 激光云高指示器
laser crosswind system 激光侧风测量系统
laser diode 激光二极管【物】
laser display 激光显示【物】
laser Doppler anemometer 激光多普勒风速计
laser-excited radar 激光受激雷达
laser facsimile image recorder 激光传真图像记录器,激光传真云图收片机
laser fog nephelometer 激光雾能见度仪
laser guidance 激光制导【物】
laser gyro 激光陀螺【电子】
laser heterodyne radiometer 激光外差式辐射计
laser heterodyne spectrometer(LHS) 激光外差分光计
laser-induced fluorescence(LIF) 激光感生荧光【物】
laser interferometer 激光干涉仪【电子】
laser isotope separation 激光分离同位素【电子】
laser level 激光水准仪【测】
la serpe 拉塞尔佩云(长的云堤,见于意大利埃特纳火山南端)

laser present weather identifier 激光天气实况识别仪
laser printer 激光印刷机【计】
laser pumping 激光抽运【物】
laser radar 光〔雷〕达,激光雷达
laser radar equation 光〔雷〕达方程
laser radiation technology(LARAT) 激光辐射技术
laser remote sensing 激光遥测
laser resonator 激光〔共振〕腔【物】
laser spectrometer 激光分光计
laser spectroscopy 激光光谱〔学〕【电子】
laser typesetter 激光照排机【计】
laser velocimeter 激光速度仪
laser weather identifier 激光天气识别仪
LASG(State Key Laboratory of Numerical Modeling for Atmospheric Sciences and Geophysical Fluid Dynamics) 大气科学和地球流体力学数值模拟国家重点实验室(中科院大气物理所)
last contact 复圆(日食)【天】
last contact of umbra 复圆(月食)【天】
last date 终日
last deglaciation 末次冰消期
last frost 终霜【农】
last glacial 末次冰期(距今 15000~80000 年)
Last Glacial Maximum(LGM) 末次冰期冰盛期(距今约 2 万年),末次冰盛期
Last Glacial Termination(LGT) 末次冰期结束期
last hoarfrost 终白霜
last interglacial 末次间冰期(距今 115000~125000 年)
last killing frost 终杀霜
last quarter 下弦【天】
last snowfall 终降雪
last snow 终雪
late frost 晚霜
late glacial 晚冰期(在末次冰期之后,10—13 ka BP)

late glacial stage climate 晚冰期气候
latency 等待时间,取数时间
latent energy 潜能
latent heat 潜热【大气】
latent heat flux 潜热通量
latent heat of fusion 熔化潜热
latent heat of vaporization 汽化潜热
latent instability 潜在不稳定【大气】
latent instability index 潜在不稳定指数
latent value (= characteristic value) 特征值,本征值
latent vector (= characteristic vector) 特征矢量,本征矢量
lateral boundary 侧边界
lateral boundary condition 侧边界条件
lateral coupling 侧向耦合
lateral diffusion 侧向扩散
lateral entrainment 侧向卷入
lateral friction 侧向摩擦
lateral gustiness 侧阵风
lateral heterogeneity 侧向不均匀性
lateral inflow 侧向来水,旁侧入流
lateral mirage 侧现蜃景
lateral mixing 侧向混合【大气】
lateral refraction 侧向折射
lateral tangent arc 侧切晕弧
lateral wind 侧风
late spring 暮春
late spring coldness 倒春寒【农】
late spring cold 倒春寒【大气】
latest frost 终霜【大气】
late subboreal climatic phase 亚北方晚期〔气候〕
latewood 晚材
latitude 纬度
latitude barometer correction 气压表纬度订正〔值〕
latitude effect 纬度效应
latitude-longitude grid 经纬度网格
latitude-time distribution 纬度-时间分布
latitude-time section 纬度-时间剖面

latitudinal 经向〔的〕,纬度〔的〕
latitudinal fluctuation 南北摆动
latitudinal variation 经向变化
lattice 晶格,点阵
launching ①发射(火箭或卫星)②施放(气球、探空仪)
launch vehicle 运载火箭
laurence 闪烁景
Laurentide ice sheet (LIS) 劳仑泰冰盖
Lausan 劳桑风(大洋洲新赫布里底群岛附近的强东南信风)
lava flow 熔岩流【地质】
lava lake 熔岩湖【地质】
laveche (= leveche) 累韦切风
law 定律,法则
Law and Parson distribution 劳-帕森〔雨滴〕谱
law of conservation of energy 能量守恒定律【物】
law of darkening 暗度定律
law of dynamical similarity 动力相似性定律
law of error 误差定律
law of excluded middle 排中律【数】
law of mechanical equivalent of heat 热功当量定律【化】
law of rational indices 有理指数定律【化】
law of storms 风暴定律
law of thermodynamics 热力学定律
law of the sea 海洋法【海洋】
Lax equivalence theorem 拉克斯等价定律
Lax-Wendroff differencing scheme 拉克斯-温德罗夫差分格式,L-W 差分格式
Lax-Wendroff scheme 拉克斯-温德罗夫格式【数】,L-W 格式
layer 层
layer cloud 层云
layer depth 层厚
layered echo 层状回波【大气】
layered parallel plane medium 层状平行

平面介质
layer model 多层模式
layer of compensation 补偿层
layer of discontinuity 不连续层
layer of frictional influence 摩擦层
layer of no motion 无运动层
L band L波段(频率1~2 GHz,波长15~30 cm)
LCD 液晶显示【计】
LCL(lifting condensation level) 抬升凝结高度【大地】
leaching ①沥滤②淋洗作用【土壤】
leaching deposit 淋积矿床【地质】
lead 冰间水道,人工水道
leader 先导〔作用〕
leader〔streamer〕 先导〔流光〕【大气】
leader stroke 先导闪击【大气】
leading eigenvector 主特征向量
leading singular vector 主奇异向量
leading sunspot 前导黑子
leading wave 导式波
lead land 冰间水道【海洋】
leaf area index (LAI) 叶面积指数【农】
leaf temperature 叶温【大气】
leaf wetness 叶面湿润
leaf wetness duration 叶面湿润持续时间
league 里格(旧时长度单位,4828 m)
leakage 渗漏,泄漏
leakage coefficient ①渗漏系数,漏水系数②漏磁系数
leakage current 泄漏电流
leakance 电漏
leaky aquifer 漏水含水层,渗漏蓄水层
lean year 歉收年
leap-day 闰日(2月29日)
leapfrog differencing 蛙跃差分
leapfrog integration scheme 蛙跃式积分格式
leapfrog method 蛙跃法,时间中央差法
leap-frog scheme 蛙跳格式【数】,时间中央差格式

leap second 闰秒【天】
leap year 闰年【天】
learning machine 学习机
least significant difference 最小显著差数【农】
least square 最小二乘方
least square fitting 最小二乘法拟合【化】
least square method 最小二乘法
least square minimization 最小二乘方极小化
least-square procedure 最小二乘方过程
least squares approximation 最小二乘逼近【数】
least squares estimate 最小二乘估计【数】
least squares solution 最小二乘解【数】
least-time track 节时航线
LED display 发光二极管显示【计】
ledge 暗礁
lee cyclogenesis 背风旋生,背风坡气旋生成
lee depression 背风坡低压【大气】
lee eddy 背风坡涡旋
lee low 背风坡低压
lee side(=leeward) 背风面
leeside convergence 背风辐合
leeside trough 背风槽
lee tide 背风潮
lee trough 背风槽【大气】
Leeuwin Current 莱文海流
lee vortex 背风坡涡旋
leeward 背风面
leeward islands 背风列岛
leeward side 背风面
lee wave 背风波【大气】
lee wave cloud 背风波云
lee wave hydraulic jump 背风波水跃
lee-wave rotor 背风波转子
lee-wave separation 背风波分离
leeway 风压差
leeway angle 风压差【航海】
leeway coefficient 风压差系数【航海】

left-handed coordinates 左手[旋]坐标
left-handed coordinate system 左手坐标系
left-handed rotation 左转
legal hour 法定时
legal time 法定时间
legal year 法定年
legend 图例【测】
Legendre associated differential equation 勒让德连带微分方程【数】
Legendre equation 勒让德方程
Legendre function 勒让德函数
Legendre polynomial 勒让德多项式【数】
legislation 立法,法制,法规
Leibniz formula 莱布尼茨公式【数】
Leibniz's theorem of calculus 莱布尼茨微积分定理
lemma 引理
Lenard effect 列纳效应
length of record 记录长度
lenkonotos 白南风(见于古希腊,带来很热的天气)
lens 透镜
lenticular cloud 荚状云
lenticular cloud band 荚状云带
lenticularis(len) 荚状(云)
lenzbote 春汛风(阿尔卑斯山冬季焚风)
leptokurtic 高峰态
leptokurtosis 高狭峰,尖峰态
lepton 轻子
Lesser Cold 小寒【大气】(节气)
Lesser Fullness 小满【大气】(节气)
Lesser Heat 小暑【大气】(节气)
leste 累斯太风(北非及马特拉的一种干热东或东南风,西班牙语名)
LET(linear energy transfer) 传能线密度【农】,线性能量转换
lethal concentration 致死浓度
lethal dose 致死剂量
lethal ultraviolet radiation 致死紫外辐射
levant 累范特风(levante 风的法语名)
levante 累范特风(从法国南部沿海和岛屿吹向直布罗陀海峡的一种东或东南风,西班牙语名)

levanter 累范特风(levante 风的英语名)
levantera 累范太腊风(直布罗陀海峡中的一种持久东风)
levanto 累范托东南风(一种加那利群岛的热东南风)
leveche 累韦切风(sirocco 风的西班牙语名)
level ①水平面②水准③水准仪④层⑤高度
level ice 〔层状〕平滑冰
level model 多层模式
level of concern (污染物浓度)关注水平
level of escape 逃逸高度
level of free convection(LFC) 自由对流高度
level of free sink(LFS) 自由下沉高度
level of neutral buoyancy 中性浮力高度,平衡高度
level of no motion 无运动层
level of non-divergence 无辐散层
level of significance 显著性水平
level of significance of difference 差异显著平准【农】
level surface 重力位面
levent canards 鸭子风(法国加尔省的比士风,影响鸭子的移栖,法语名)
Lewis number 刘易斯数
Lewis Research Center(LeRC) 美国刘易斯研究中心
LFC(level of free convection) 自由对流高度
LF(low frequency) 低频(30 k～300 kHz)
LFM(limited area fine mesh model) 有限区细网格模式【大气】
LFO(low frequency oscillation) 低频振荡
LGM(Last Glacial Maximum) 末次[冰期]冰盛期
LGT(Last Glacial Termination) 末次冰期结束期
LI(lifting index) 抬升指数【大气】
Liapunov exponent 李雅普诺夫指数【数】

Liapunov exponent spectrum 李雅普诺夫指数谱
Liapunov function 李雅普诺夫函数
Liapunov index 李雅普诺夫指数
Liapunov stability 李雅普诺夫稳定性【数】
libeccio 西南风（意大利语）
liberator 利伯拉托风（直布罗陀海峡的西风）
library 〔程序〕库
lichen 地衣
lid ①〔云〕盖②逆温层顶
lidar (light detection and ranging) 激光雷达【大气】,光〔雷〕达
lidar ceilometer 光〔雷〕达云幂仪
lidar constant 光雷达常数
lidar-determined height 光〔雷〕达测高
lidar equation 光雷达方程
lidar meteorology 光〔雷〕达气象学
lidar ratio 光雷达比(体后向散射系数与体消光系数之比)
lidar reflectance 光〔雷〕达反射率
lifeboat 救生艇【航海】
lifebuoy 救生圈【航海】
life cycle 生命史
lifetime 寿命【化工】(化合物浓度降至原始值$1/e$所需的时间)
lifetime of satellite 卫星寿命
life zone 生物带
lift 升力
lifted index 抬升指数
lifting 抬升
lifting condensation level (LCL) 抬升凝结高度【大气】
lifting condensation level zone 抬升凝结高度区
lifting fog 抬升雾
lifting force 举力,上升力
lifting index (LI) 抬升指数【大气】
ligand 配体【化】
light 光

light air 1级风【大气】,软风
light amplification by stimulated emission of radiation (laser) 激光
light and temperature potential productivity 光温生产潜力【农】
light attenuation 光衰减
light barrier transducer 截光换能器
light boat 灯船【海洋】
light breeze 2级风【大气】,轻风
light clay soil 轻黏土【土壤】
light climate 光照气候
light compensation point 光补偿点【农】
light crosswind 小侧风
light detection and ranging (LIDAR) 光〔雷〕达
light dispersion 光弥散
light duration 光照延续时间
light emitting diode (LED) 发光二极管【物】
light emitting diode display (LED display) 发光二极管显示【计】
light energy 光能
light exposure 光照量
light filter 滤光片
light flow 光通量
light fog 轻雾
light freeze 轻冻
light frost 轻霜
light gate 光闸【物】
light gating 光选通【物】
light house 灯塔【海洋】
light intensity 光强度
light ion 轻离子
light-loving plant 喜光植物
light meter 光度计
light minimum 最低光量
lightning 闪电【大气】
lightning arrester 避雷器
lightning channel 闪电通道【大气】
lightning conductor 避雷针【大气】
lightning counter 闪电计数器

lightning current 闪电电流
lightning damage 闪电危害
lightning death 闪电猝灭
lightning detection and positioning system 闪电探测和定位系统
lightning detection and ranging(LDAR) 闪电探测和测距
lightning detection network 闪电探测网
lightning detection system 闪电探测系统
lightning diffusion 闪电扩散
lightning direction finder 闪电方向探测器
lightning disaster loss 雷电灾害损失
lightning discharge 闪电放电
lightning echo 闪电回波【大气】
lightning electromagnetic impulse 雷击电磁脉冲
lightning equipotential bonding 雷电等电位连接
lightning impulse 雷电冲击
lightning induction 雷电感应
lightning interference 雷电干扰
lightning fire 雷击火【林】
lightning flash 电闪
lightning-flash counter 闪电计数器
Lightning Imaging Sensor (LIS) 闪电成像传感器,闪电成像仪
lightning location data 闪电定位数据
lightning location method 闪电定位方法
lightning location system 闪电定位系统
Lightning Mapper Sensor（LMS） 闪电测绘仪,闪电测绘传感器
lightning mapping system 闪电探测系统
lightning outage 雷击跳闸
lightning overvoltage 闪电过电压
lightning protection system (LPS) 雷电保护系统,防雷装置
lightning protection zone (LPZ) 雷电保护区,防雷区
lightning protector plate 避雷板
lightning recorder 闪电记录器

lightning rod 避雷针【大气】
lightning spectrum 闪电光谱
lightning storm 雷暴
lightning striking distance 雷击距
lightning striking point 雷击点
lightning stroke 雷击
lightning suppression 人工雷电抑制【大气】
lightning surge 雷电浪涌
lightning warning set (LWS) 闪电警报仪
light of night sky 夜天光
light pen 光笔（一种笔型的小光电元件,用以把新资料储入电脑中）
light pillar 光柱（日柱）
light polarizing 光极化
light pressure 光压
light-pulse projector 〔云幂仪〕光脉冲发射器
light-pulse receiver 〔云幂仪〕光脉冲接收器
light pump 光泵【物】
light rain 小雨【大气】
light ray 光线
light requirement 需光量
light resources 光资源【大气】
light saturation 光饱和【海洋】
light saturation point 光饱和点【农】
light scattering 光散射
light scattering diagram 光散射图
light scattering table 光散射表
light sensitive index 感光指数
lightship code 灯船电码
lightship station 灯船测站
Light Snow 小雪【大气】(节气)
light source 光源
light squall 轻飑
light transmissivity 透光率
light treatment 光照处理【农】
light wave 光波【物】
light year 光年【物】
likelihood 似然【数】

likelihood function 似然函数【数】
LIM(linear inverse model) 线性转置模式
Liman Current 利曼海流(由鄂霍次克海流向日本海)
limb brightening 临边增亮【大气】
limb darkening 临边变暗【大气】
limb emission pressure modulated radiometer 临边发射压力调制辐射仪
limb emission radiometry 临边发射辐射测量法
limb infrared monitoring of the stratosphere(LIMS) 平流层临边红外监测
limb occultation interferometer 边蚀干涉仪
limb occultation laser heterodyne radiometry 边蚀激光外差辐射测量法
limb occultation photometer 边蚀光度计
limb radiance 临边辐射
limb radiance infrared radiometer(LRIR) 临边辐射红外辐射仪
limb radiance inversion radiometer (LRIR) 临边辐射反演辐射仪
limb radiance stratospheric sounder (LRSS) 临边辐射平流层探测器
limb retrieval 临边反演【大气】
limb scanning 临边扫描
limb scanning method 临边扫描法【大气】
limb-scanning pressure modulated radiometer 临边扫描压力调制辐射仪
limen 最低光限,(生理)阈限
limestone scrubbing 石灰石涤气,石灰面清洗(一种清除 SO_2 的工艺)
liminal contrast 对比阈
limit 极限,限度
limit circle 极限圆
limit cycle 极限环
limited area fine-mesh model(LFM) 有限区细网格模式【大气】
limited-area forecast model 有限区〔域〕预报模式

limited-area model 有限区模式【大气】
limited-area nested grid model(LNGM) 有限区〔域〕嵌套网格模式
limited-area prediction 有限区〔域〕预报
limited fine-mesh model 有限〔区域〕细网格模式
limit-gauge 限量雨量器
limiting angle 限角
limiting meteorological condition 气象限制条件
limiting velocity 极限速度
limiting wave 极限波
limit of audibility 能听极限
limit of convection(LOC) 对流限〔高〕
limit of the atmosphere 大气极限
limit of visibility 能见极限
limnological meteorology 湖沼生物气象学
limnology 湖沼学
line 线
line absorption 〔谱〕线吸收
linear 线性
linear absorption coefficient 线性吸收系数
linear action hygrometer 线性湿度表
linear advection equation 线性平流方程
linear algebra 线性代数【数】
linear computational instability 线性计算不稳定
linear constrained method 线性限制法
linear correlation 线性相关
linear depolarization ratio 线退偏比
linear differential equation 线性微分方程
linear differential equation of first order 一阶线性微分方程【数】
linear elliptic equation of higher order 高阶线性椭圆〔型〕方程【数】
linear energy transfer 线性能量转换【农】,传能线密度
linear entropy scale 线性熵标
linear equation 线性方程,一次方程
linear field of motion 线性运动场

linear field of temperature 线性温度场	linear wave 线性波
linear filtering 线性滤波【数】	line average 线平均
linear function 线性函数	line blow 强风
linear geostrophic balance 线性地转平衡	line breadth 线宽【物】,谱线宽度
linear global balance 线性全球平衡	line broadening 〔谱〕线加宽
linear homogeneous equation 线性齐次方程	line-by-line integration 逐线积分〔法〕
linear instability 线性不稳定(度)	line-by-line method 逐线法（一种大气透过率计算法）
linear interpolation 线性内插	line cloud 线状云
linear inverse model（LIM） 线性转置模式	line convection 线对流【大气】
linear inversion 线性反演【大气】	line echo 线状回波
linearity 线性度	line echo wave pattern（LEWP） 波型线状回波
linearization 线性化	line-end vortices 线端涡旋
linearized (differential) equation 线性化〔微分〕方程	line factor 谱线因子【化】
linear molecule 线形分子	line gale（= equinoctial gale） 二分点风暴（春、秋分日附近常见的大风）
linear momentum 线动量	line integral 线积分
linear mountain wave theory 线性山岳波理论	line intensity 线强
linear operator 线性算子	Line Island Experiment(LIE) 莱恩岛试验
linear partial differential equation 线性偏微分方程【数】	line of apsides 远近线
linear partial differential equation with variable coefficients 变系数线性偏微分方程【数】	line of constant wetness 等湿线
	line of discontinuity 不连续线
	line of echo 回波线
linear partial differential with constant coefficients 常系数线性偏微分方程【数】	line of force 力线
	line of frontogenesis 锋生线
linear programming 线性规划	line of position(L.O.P.) 位置线
linear regression 线性回归	line of sight 视线
linear scale 线性标度	line-of-sight propagation 视线传播
linear stability analysis 线性稳定度分析	line-of-sight range 视线范围
linear synthesis 线性合成【化】	line of storm 风暴线
linear system 线性系统	line parameter 〔谱〕线参数
linear tangent equations 线性正切方程	lineprinter 行式打印机
linear transformation 线性变换	line profile 谱线轮廓【化】
linear transition 线性推移	line scanning 行扫描【电子】
linear-trend removal 线性倾向去除	line shape function 线形函数
linear velocity 线速度	line source 线源
linear viscoelasticity 线性黏弹性【化】	line spacing 线间隔
linear water wave theory 线性水波理论	line spectrum 线状谱
	line-squall 线飑
	line-squall cloud 线飑云

line storm 线性风暴(二分点风暴〔equinoctial storm〕的别称)
line strength 线强
line thunderstorm 线雷暴
line vortex 线涡
line width 谱线宽度【化】,线宽【物】
Lineykin layer 林尼金层
Linke-Feussner actinometer 林克-福斯纳日射表
linker 连接程序
Linke-scale(＝Linke blue sky scale) 林克〔天空蓝度〕标
Linke turbidity factor 林克浑浊因子
lipid 脂质【生化】
lipoprotein 脂蛋白【生化】
Lips 李普斯风
liquefaction 液化〔作用〕
liquid 液体
liquid air 液态空气
liquid-bubble tracer 液泡追踪器
liquid chromatograph 液体色谱仪
liquid chromatography 液相色谱法
liquid-column barometer 液柱气压表
liquid crystal 液晶
liquid crystal display(LCD) 液晶显示【计】,液晶显示器
liquid-glass thermograph 玻管液体温度计
liquid-in-glass thermometer 玻管液体温度表
liquid(in-metal)**thermograph** 钢管水银土壤温度计
liquid limit 液〔体〕限〔度〕
liquid phase 液相
liquid phase reaction 液相反应【化】
liquid state 液态
liquid thermometer 液体温度表
liquid water 液态水
liquid-water content(LWC) 液态水含量
liquid water-holding capacity 液态水含量
liquid water loading 液态水载荷
liquid water mixing ratio 液态水混合比

liquid water path(LWP) 液态水光程
liquid water potential temperature 液态水位温
liquid water static energy 液态水静力能
liquified natural gas 液化天然气【地质】
LIS ①(Lightning Imaging Sensor) 闪电成像仪 ②(Laurntide ice sheet)劳仑泰冰盖
liter 公升,升
lithification 石化〔作用〕【地质】
lithium cell 锂电池【化】
lithium chloride 氯化锂
lithology 岩石学
lithometeor 大气尘粒【大气】
lithosphere 岩石圈【大气】
lithostratigraphy 岩石地层学【地质】
little brother 小弟气旋(南半球深厚扰动后面的稳定热带气旋)
Little Climatic Optimum 小气候适宜期
little ice age 小冰期【大气】
little sister 小妹气旋(北半球深厚扰动后面的稳定热带气旋)
littoral climate 海滨气候
littoral current 沿岸海流
littoral drift 沿岸泥沙流【海洋】
livestock meteorology 家畜气象学
livestock safety index(LSI) 家畜安全指数
living glacier 现存冰川
living resources conservation 生物资源保护【地理】
living space 生存空间【地理】
Livingstone atmometer 利文斯通蒸发表
Livingstone sphere 利文斯通球
lizard-balloon 拖尾气球
ljuka 焚风(见于奥地利卡林西亚)
llanos 蓝诺草原(即南美大草原,西班牙语名)
llebetjabo 卢贝特杰多风(西班牙卡塔卢尼亚的暴风)
llevant 利凡风
LLJ(low-level jet) 低空急流

llorano 冬雾（美国加利福尼亚湾沿岸因冷空气入侵所形成）
LLWS (low level wind shear) 低空风切变
LMS (Lightning Mapper Sensor) 闪电测绘仪，闪电测绘传感器
load ①负载②装入【计】
loading ①负荷②装载量③浓度，含量④〔程序〕装入
loading of air pollutant 空气污染物含量【大气】
load module 装配模块
loam 沃土，壤土（黏土、淤泥和沙的混和物）
loam soil 壤土【农】
loamy sand 壤砂土【土壤】
lobe 瓣，波瓣
lobe pattern 瓣型
lobe structure 裂片结构，瓣状结构
local 局部，局地，地方性
local acceleration 局部加速度
local action 局地作用
local AFOS MOS program (LAMP) 地方自动化模式输出统计预报计划
local analysis 局部分析
local analysis and prediction system (LAPS) 局地分析和预报系统（NOAA开发的）
local angular 局部角动量
local angular momentum 局地角动量
local apparent time 地方视时
local area coverage 局部区域范围，区域覆盖
local area network (LAN) 局域网【计】
local axis 局地轴【大气】
local change 局地变化
local circulation 局地环流【大气】，地方性环流
local civil time 地方民用时
local climate 局地气候【大气】
local climatology 局地气候学
local closure 局部闭合
local convergence 局部收敛【数】
local coordinate system 地方坐标系【地信】，独立坐标系
local derivative 局部导数【大气】
local effect 局部效应
local equilibrium 局域平衡【物】
local establishment (=lunitidal interval) 月潮间隙
local extra-observation 加点观测
local fallout 局部落尘
local field 局部域【数】
local fitting 局部拟合
local forecast 局地预报【大气】，当地预报，地方预报
local free convection 局地自由对流
local free-convection similarity 局地自由对流相似性
local heating 局部加热
local history 地方志
local horizon 局部地平
local hour angle 地方时角
local inflow 局部入流
local instability 局域不稳定性【物】
local isotropic (turbulence) 局地各向同性〔湍流〕
local isotropy 局部各向同性
locality 定域性【物】
localization 定域化【物】
localization principle 定域原理
local lifetime 局地存活期
local lightning-flash counter 本地闪电计数器
locally absent ring 部分欠缺年轮
locally frozen condition 局地冻结条件
locally generated sea 局地产生的风浪
locally homogeneous medium 局部均匀介质
locally homogeneous random function 局部均匀随机函数
locally homogeneous space 局部齐性空间
locally homogeneous turbulence 局部均匀湍流
local Mach number 局部马赫数
local mean time 地方平〔均〕时
local precipitation 地方性降水【大气】

local scale 局部比例尺【航海】
local-scale weather 局地天气
local (severe) storm 局部〔强〕风暴
local (severe) thunderstorm 局部〔强〕雷暴
local sideteal time 地方恒星时
local similarity 局域相似,局地相似性
local similarity hypothesis 局部相似假说
local solar time 地方太阳时
local spectrum 局地谱
local stability 局地稳定性
local standard time 地方标准时
local thermodynamic equilibrium (LTE) 局部热动平衡【天】,局地热力平衡
local thunderstorm 地方性雷暴
local time 地方时
local topography 方志【地理】
local true time 地方真时
local variation 局部变化,局地变化
local vorticity 局部涡度
local warning radar 局地警戒〔天气〕雷达
local water level 当地水位【航海】
local weather 地方性天气【大气】
local weather report 地方天气报告
local wind 地方性风【大气】
location 定位,位置,单元
location of lightning 闪电定位
Lockheed Missile and Space Company (LMSC) 美国洛克希德导弹和空间公司
lock-in 锁定
lock-in amplifier 锁相放大器
locus 轨迹
lodos 洛多斯风(保加利亚黑海沿岸的南风)
loess 黄土
loess landform 黄土地貌【地理】
lofting 垂直伸展
log 原木【林】
log amplitude fluctuation 对数振幅起伏
logarithm (log) 〔常用〕对数(以10为底)

logarithm table 对数表
logarithmic amplifier 对数放大器【电子】
logarithmic differentiation 对数微分
logarithmic normal distribution 对数正态分布【数】
logarithmic scale 对数标尺
logarithmic spectrum 对数谱
logarithmic velocity profile 风速对数廓线【大气】
log book 航海日志【航海】
logging 测井【地】,森林采运【林】
logical value 逻辑值
logic circuit 逻辑电路
logistical regression 逻辑斯谛回归【数】
logistic distribution 逻辑斯谛分布【数】
logn heavy swell 长狂涌【海洋】
lognormal cloud-size distribution 对数正态云尺度分布
log-normal distribution 对数正态分布
log-pressure coordinate 对数压力坐标
log transformation 对数变换【化】
lolly ice 片冰,海上浮冰
lombarde 伦巴德风(主要是沿法、意边境从伦巴迪亚吹来的东风)
Lomonosov Current 大西洋赤道潜流,罗蒙诺索夫海流【海洋】
London fog 伦敦雾
London (sulfurous) smog 〔含硫〕伦敦烟雾〔事件〕(1952年)
London Smog Incidents 伦敦烟雾事件
long-crested wave 长峰波(海面波)
long-day plant 长日照植物
long discharge 长放电
long duration precipitation recorder 长期降水记录器
long-exposure modulation transfer function 长曝光调制传递函数
longitude 经度
longitude effect 经度效应
longitude of ascending node 升交经度

longitudinal ①纬向〔的〕,经度〔线的〕 ②纵〔的〕,轴向〔的〕
longitudinally asymmetric heating 纬向非对称加热
longitudinal rolls 纵〔向〕滚涡
longitudinal roll vortices 纵〔向〕滚涡
longitudinal section 纵断面,纵切面
longitudinal variation 纬向变化
longitudinal vibration 纵振动
longitudinal wave 纵波【大气】,可压缩波
longitudinal wind 纵向风(顺风或逆风)
long-lasting precipitation 绵雨
long low swell 长轻涌【海洋】
long method 长法
long moderate swell 长中涌【海洋】
Longmont anticyclone 朗蒙特反气旋（美国科罗拉多州）
long optical path 长光程
long-path absorption 长光程吸收
long-path ambient air monitoring 长距离环境空气监测
long range and tactical navigation system (LORTAN) 罗坦系统,远程战术导航系统【电子】
long-range force 长程力【物】【化】
long range navigation (LORAN) 罗兰〔导航仪〕,远程〔无线电〕导航【电子】
long-range transport 长距离输送
long-range〔weather〕forecast 长期〔天气〕预报【大气】
longshore current (= littoral current) 沿岸海流
longshore drift 沿岸泥沙流【海洋】
longshore wind 沿岸风(印度马德拉斯的湿南风,斯里兰卡的东北夜风)
long spark 长火花
long-term climatic change 长期气候变化
long-term forecast ①长期预报②长时预报(航空上指2 h)
long-term hydrological forecast 长期水文预报

long-term model 长期模式
long-train atmospherics 长列〔型〕天电
long wave 长波
longwave approximation 长波近似
long-wave cloud radiative forcing (LWCF) 长波云辐射强迫〔作用〕
long-wave cutoff 长波截止
longwave cyclone 长波气旋
longwave feedback 长波反馈
long-wave formula 长波公式
long-wave radiation 长波辐射
long-wave ridge 长波脊
long-wave trough 长波槽【大气】
loo（或 lu, loo marna） 洛风(印度热西风)
Loofah 录发(第二次世界大战中对英国研发的极性浊度计的简称)
lookup table 查算表
looming 上现蜃景
loop ①回线②循环 ③〔波〕腹④打转路径
loop antenna 环形天线
Loop Current 环形海流(墨西哥湾)
loop hurricane 打转飓风
looping 波浪形扩散,环状扩展
loop prominence 圈状日珥
loop rating 水位流量关系绳套曲线
loop track 打转路径
loop typhoon 打转台风
loose avalanche 尘状雪崩
loose snow 疏松雪,尘状雪
LORAN (long range navigation) 罗兰〔导航仪〕,远程〔无线电〕导航【电子】
Loran-C sonde 罗兰 C 探空仪
Lorentz broadening 洛伦兹加宽
Lorentz energy cycle 洛伦兹能量循环
Lorentz force 洛伦兹力
Lorentzian lineshape 洛伦兹线型
Lorentz line 洛伦兹线
Lorentz Lorenz formule 洛伦兹-洛伦茨

公式【物】，L-L 公式
Lorentz profile 洛伦兹廓线
Lorentz transformation 洛伦兹变换
Lorenz attractor 洛伦兹吸引子【物】
Lorenz conversion 洛伦兹转换
Lorenz-Mie theory 洛伦兹-米理论，L-M理论
LORTAN (long range and tactical navigation system) 罗坦系统，远程战术导航系统【电子】
Los Angeles (photochemical) smog 洛杉矶光化学烟雾〔事件〕
Loschmidt number 洛施密特常量【物】，洛施密特数
loss function 损失函数
lost balloon 测风气球
loudness 响度
louvred screen 百叶箱
Love wave 勒夫波【地】，Q 波(简称)
low ①低②低压
low aloft 高空低压
low altitude radiosonde 低空〔无线电〕探空仪
low altitude surveillance radar 低空搜索雷达【电子】
low-analysis fertilizer 低浓度肥料【土壤】
low carbon economy 低碳经济
low cloud 低云【大气】
low coast 低平海岸【海洋】
low density lipoprotein 低密度脂蛋白【医学】
low dimensionality 低维数
low earth orbiting 低地球轨道
low-energy particle 低能粒子
low-energy plasma 低能等离子体
low energy proton alpha telescope (LEPAT) 低能质子α粒子计数仪
lower antitwilight 低反曙暮光，低反辉
lower arc 下珥
lower atmosphere 低层大气【大气】
lower atmosphere layer 低层大气

lower circumzenithal arc 下环天顶弧(圈)
lower cloud 低云
lower envelope 下包络(磁扰)
lower high water 低高潮【航海】
lower ionosphere 电离层下部
lower low water 低低潮【航海】
lower mirage 下现蜃景
lower reach 下游【航海】
lower stratosphere 平流层下部
lower tangential arc 下切弧(晕的)
lower thermosphere 热层下部
lower troposphere 对流层下部
lower tropospheric jet 低空急流
lowest astronomical tide 最低天文潮位
lowest temperature 最低温度
low-flow channel 枯水河槽，枯季行水河槽
low flow year 枯水年【大气】，少水年，干旱年
low frequency (LF) 低频
low-frequency fluctuation 低频振动
low frequency oscillation (LFO) 低频振荡
low-frequency regime 低频体系
low-frequency variability 低频变化
low index 低指数【大气】
Lowitz arc 日珥，罗维兹弧
lowland 低地【地理】
low latitude 低纬〔度〕
low-level air temperature 低层气温
low-level cloud 低云
low-level inversion 低层逆温
low-level jet 低空急流
low-level jet stream 低空急流【大气】
low-level nocturnal jet 夜间低空急流
low-level radiosonde 低空〔无线电〕探空仪
low-level shear 低空切变
low-level wind shear 低空风切变【大气】
low order model (LOM) 低阶模式
low parry arc 下偏珥

low pass filter 低通滤波〔器〕
low pass fluctuation 低通振荡
low〔pressure〕 低〔气〕压【大气】
low pressure area 低压区
low pressure center 低压中心
low-pressure system 低压系统,低值系统
low-reference signal 低参考信号
low resolution facsimile(LR-FAX) 低分辨率〔云图〕传真【大气】
low resolution infrared radiometer (LRIR) 低分辨率红外辐射仪
low resolution omnidirectional radiometer 低分辨率全向辐射仪
low salinity layer(LSL) 低盐度层
low speed plasma flow 低速等离子体流
low temperature damage in autumn 寒露风【大气】
low temperature drying 低温干燥【农】
low-temperature hygrometry 低温测湿术
low-temperature plasma 低温等离〔子〕体【物】
low tide 低潮
low topped cloud 矮云
low vacuum 低度真空
low-velocity layer 低速层
low water ①低水〔位〕②低潮
low water level 低水位
low-water line 低潮线
low-water mark ①低水〔位〕标志②低潮线
low-water plane(LWP) 低潮面
low water time 低潮时【航海】
low zonal circulation 弱西风环流,弱纬向环流
loxodrome 恒向线,方位线
LPS (lightning protection system) 雷电保护系统
LPZ (lightning protection zone) 雷电保护区,防雷区
LR-FAX(low resolution facsimile) 低分辨率〔云图〕传真

lu 洛风(见 loo)
lucimeter 总辐射平均强度表
luganot 卢加诺特风(意大利加尔达湖的强南风或南东南风)
lull 息静,暂静
lumber 成材【林】
lumen(lm) 流明(光通量单位)
luminance 亮度【物】,发光率,光亮度【大气】(单位:cd/m^2)
luminance contrast 光亮度对比
luminance temperature 亮〔度〕温〔度〕
luminescence 发光【物】
luminescence efficiency 发光效率【物】
luminosity 亮度【物】,发光度
luminous 发光
luminous cloud 发光云
luminous contrast 光亮度对比
luminous density 光密度
luminous efficiency 发光效率,光效能
luminous emittance 面发光率(单位:lm/m^2)
luminous emittance 光发射度【物】
luminous energy 光能
luminous exitance 发光率(单位:lm/m^2)
luminous exposure 光照量
luminous flux 光通量【物】(单位:lm)
luminous flux density 光通量密度
luminous intensity 发光强度【物】
luminous meteor 光象
luminous night cloud 夜光云
luminous power 光功率(单位:lm)
luminous vapour-trail 发光水汽尾迹
lumped chemical mechanism 集总化学机制
lunar atmospheric tide 大气太阴潮
lunar aureole 月华【大气】
lunar barometric variation 月致气压变化
lunar calendar 阴历
lunar corona 月华【大气】
lunar day 太阴日
lunar distance 太阴距离
lunar eclipse 月食【天】

lunar halo 月晕【大气】
lunar phase 月相【天】
lunar probe 月球探测器
lunar rainbow 月虹
lunar tide 太阴潮【大气】,月亮潮
lunar year 太阴年
lunation(=lunar month) 太阴月
lunitidal interval 满潮时距
lustrum 五年
luvside 向风面
lux (lx) 勒〔克斯〕(光照度单位, =1 lm/m²)
LWC(liquid water content) 液态含水量
Ly(langley) 兰〔利〕(1 g·cal/cm²)

Lyapunov dimension 李雅普诺夫维数
Lyapunov function 李雅普诺夫函数
Lyapunov stability 李雅普诺夫稳定性【物】
Lyman-α 莱曼α发射线(1215.67 Å)
Lyman α emission line 莱曼-α发射线
Lyman-α hygrometer 莱曼-α湿度表【大气】
Lyman series 莱曼系【物】
Lyman-α total water content meter 莱曼-α总含水量仪
lysimeter 蒸散量测定装置,蒸渗仪
lysocline 溶〔解〕跃面(大海中的一水层,某些化学物质在其中被溶解)

M

MAB (Man and Biosphere Program) 人与生物圈计划
Ma BP 距今…百万年
machine instruction 机器指令【计】
machine learning 机器学习【计】
machine translation 机器翻译【计】
Mach number 马赫数【大气】(速度与音速的比值)
Mach principle 马赫原理【天】
mackerel breeze(或 gale) 鲭风(适于捕鲭的强风)
mackerel sky 鱼鳞天【大气】
Macky effect 麦基效应
Maclaurin formula 麦克劳林公式【数】
Maclaurin series 麦克劳林级数
macroburst 大〔型〕下击暴流
macroclimate 大气候【大气】
macroclimatology 大气候学【大气】
macrofluidics 宏流控技术
macro instruction 宏指令
macrometeorology 大尺度气象学,大象学

macromolecule 高分子【化】,大分子
macro physics 宏观物理学
macro scale 大尺度,宏观尺度
macroscale circulation 大尺度环流
macroscopic ①大范围〔的〕②宏观〔的〕
macrosynoptic meteorology 大尺度天气学
macrosystem 宏观系统【物】
macroturbulence 宏观湍流
macroviscosity 宏观黏滞度【大气】,宏观黏性
maculosus 斑状(云)
Madden-Julian oscillation (MJO) 马登-朱利安振荡,M-J振荡
maelstorm 大涡旋
maestro 美斯屈罗风(亚得里亚海夏天的一种西北风)
MAG-CON thermo-regulator 磁接温度调节器
magma ①岩浆②悬浮体
magmatic water 岩浆水
magnesia method 氧化镁法
magnesium oxide method 氧化镁法

magnetically disturbed day 磁扰日
magnetically quiet day 磁静日
magnetic analyzer 磁分析器【化】
magnetic annual change 年差【航海】
magnetic bearing(M. B.) 磁方位〔角〕
magnetic bubble memory(MBM) 磁泡存贮器
magnetic card 磁性卡片(文献检索用)
magnetic character figure 磁性数
magnetic cooling 磁致冷【电子】
magnetic coordinate 磁坐标【地】
magnetic core 磁芯
magnetic coupling 磁耦合
magnetic crochet 磁鼻(地磁场的短时扰动)
magnetic crotchet 地磁钩扰
magnetic declination 磁偏角
magnetic deflection 磁偏转
magnetic dip 磁倾角【大气】
magnetic dipole 磁偶极子【物】
magnetic dipole field 磁偶极场
magnetic dipole moment 磁偶极矩
magnetic direction finding (MDF) 磁定向法
magnetic disk 磁盘
magnetic disk drive 〔磁〕盘驱动器【计】,盘驱
magnetic disturbance 磁扰
magnetic domain 磁畴【电子】
magnetic double refraction 磁性双折射
magnetic drum 磁鼓【电子】
magnetic element 地磁要素
magnetic equator 磁倾赤道【大气】,地磁赤道
magnetic field 磁场
magnetic field intensity 磁场强度
magnetic field line 磁力线
magnetic field strength 磁场强度
magnetic figure 磁力线图
magnetic flux 磁通量
magnetic gradlent 磁场梯度

magnetic gradiometer 磁力梯度仪【地】
magnetic head 磁头【电子】
magnetic hole 磁洞
magnetic hysteresis 磁滞现象【电子】
magnetic inclination 磁倾角【大气】
magnetic induction 磁感〔应〕强度【电子】
magnetic intensity 磁强度
magnetic isoanomalous line 等磁异常线【地】
magnetic latitude 磁纬
magnetic lines of force 磁力线
magnetic map 地磁图
magnetic meridian 磁子午线,磁子午圈
magnetic micropulsation 地磁微脉动
magnetic needle 磁针【物】
magnetic north 磁北
magnetic orientation control coil 磁性定向控制线圈
magnetic pole 磁极【物】
magnetic pole migration 磁极迁移
magnetic quantum number 磁量子数【化】
magnetic quiet-day solar daily variation 磁静日〔太阳日〕变化
magnetic sector boundary 磁扇形边界
magnetic south 磁南
magnetic spectrograph 磁谱仪
magnetic storm(=magstorm) 磁暴【大气】
magnetic storm time variation 磁暴时间变化
magnetic substorm 磁亚暴【地】
magnetic survey 地磁测量
magnetic susceptibility 磁化率【电子】
magnetic suspension 磁悬浮【电子】
magnetic tape 磁带
magnetic tape drive 磁带机
magnetic tape event recorder(MATER) 磁带式事件记录器
magnetic tape recorder 磁带记录器
magnetic variation 〔地〕磁变,磁倾角

magnetic wind direction 磁极风向
magnetism ①磁[性]②磁学
magnetization 磁化【物】
magnetized plasma 磁化等离[子]体【物】
magneto-anemometer 磁电风速表
magnetogram 磁照图,地磁记录图
magnetograph 磁强计,地磁记录仪
magnetohydrodynamics 磁流[体动]力学【物】
magnetohydrodynamic wave 磁流[体]动力[学]波
magneto-ionic theory 磁离子理论【地】
magnetometer 磁强计,磁力仪
magnetopause 磁层顶【大气】
magnetosheath 磁鞘【地】
magnetosphere 磁层【大气】(外逸层和大部分电离层的总称)
magnetospheric cleft 磁隙区(即极尖区)
magnetospheric convection 磁层对流
magnetospherics 磁层物理学
magnetospheric storm 磁层暴【地】
magnetospheric substorm 磁层亚暴
magnetospheric tail 磁[层]尾
magnetostratigraphy 磁性地层学【地】
magnetostriction 磁致伸缩【物】
magnetotail 磁尾【地】
magnetowind 磁风
magnetron 磁控管【电子】
magnification 放大率【物】
magnification factor 放大因子
magnification function 增益函数
magnifier 放大镜【物】
magnitude ①量,数量②星等
magnofango 马格诺凡果风(法国普罗旺斯的 mistral 风)
Magnus effect 马格努斯效应,马氏效应
Magnus force 马格努斯力,马氏力
Magnus formulas 马格努斯公式
magstorm(= magnetic storm) 磁暴
Mahalanobis distance 马哈拉诺比斯距离【数】

mail exploder 邮件分发器【计】
mailing list 邮件发送清单【计】
maillist 邮件发送清单【计】
Main Atlantic Climate Phase 主大西洋气候期
main current 主流,干流
main effect 主效应【化】
mainframe computer 主计算机
mainframe pulse 主机脉冲
Main Geophysical Observatory(MGO) 地球物理观象总台(苏联)
mainland climate 大陆气候
main lobe 主[波]瓣
main memory 主存储器
main meteorological office 气象总台
main precipitation core 主要降水中心
main pulse 主脉冲
main standard time 基本天气观测时间【大气】(格林尼治时间 00、06、12 和 18 时)
main stroke 主闪击
main telecommunications network(MTN) 主干电信网
maintenance 维持,维修
main thermocline 主[温]跃层【海洋】
main trunk circuit 主干线
maistrau(或 maistre) 密史脱拉风
maize rains 玉米雨季(东非 2—5 月)
major circulation 主[要]环流
major constituent 主成分【化】
major constituents of sea water 海水主要成分【海洋】
major cycle 黑子大周(指前导黑子极性向北时的 11 年周期)
major frame 主帧
major lobe 主[波]瓣(射束的)
major planet 大行星【天】
major ridge 主脊
major trough 主槽
major wave 主波
Maloja wind 马洛杰风(瑞士恩加迪的

顺水风)
Malvinas Current 马尔维纳斯海流(亦称福克兰海流)
mamatele(= mamaliti, mamatili) 马马特勒风(意大利西西里的西北风)
mamma(mam) 悬球状(云),乳房状(云)
mammato cloud 悬球状云,乳房状云
mammato-cumulus 悬球状积云,乳房状积云
mammatus 悬球状〔云〕
Mammoth 猛犸象
MAN(meteopolitan area network) 城域网【计】
management 管理
Man and Biosphere Program(MAB) 人与生物圈计划
Man-computer Interactive Data Access System(McIDAS) 人机对话数据存取系统
mandatory layer 标准等压层
mandatory level(或 surface) 标准等压面【大气】
mandatory surface 标准等压面
mangofango 曼果凡各风
mango shower 芒果阵雨(见于泰国中部沿海地区 2 月到 3 月)
manifold 流形,簇
man induced climatic change 人为气候变化
manipulated information rate processor(MIRP) 键控信息速率处理器
man-machine dialogue 人机对话【计】
man-machine interaction 人机交互【计】
man-machine mix 人机结合【大气】
man-machine weather forecast 人机结合天气预报【大气】
man-made dust 人工尘埃
man-made lake 人工湖【地理】
man-made source 人工源,人为源
manned balloon 载人气球
manned orbital space station 载人轨道空间站

manned space flight 载人空间飞行
Manning equation 满宁方程
manometer 测压器,压力表
manostat 恒压器,压力稳定器
man overboard 人员落水【航海】
manrope knot 握索结【航海】
man's impact on climate 人类影响气候
mantle ①罩,冲积层,浮土②地幔③等离子体幔
mantle convection 地幔对流【地】
manual control(M/C) 手控,手动,〔人工〕操纵
manual earthing electrode 人工接地体
manual gain control 人工增益调整
manual intervention 人工干预
manually digitized radar(MDR) 人工数字化雷达
manual observation 人工观测
many body problem 多体问题【天】
mao mao yuh 毛毛雨(英拼中文)
map ①图,地图【测绘】②映射
map correlation 天气图相关
map factor 地图〔放大〕因子【大气】
map interpretation statistics(MIS) 图像释用统计
map legend 图例【地信】
map name 图名【地信】
mapped data 图标资料,填图资料
mapping ①映射【数】②套图③绘图
mapping function 映射函数【数】
map plotting 填图
map projection 地图投影【大气】
map scale 地图比例尺【地信】
map specification 地图规范【地信】
map spotting 填图
map title 图名【地信】
marajos 马拉若斯风(巴西亚马孙河流域东北强飑)
marble crust 硬雪壳
marching problem 前进式问题
marching wave 行进波

march of temperature 温度变程
marenco 马朗科风(意大利马乔列湖的东东南风)
mares tail 马尾云(钩卷云的英国俗名)
Marfa front 马尔法锋(美国靠墨西哥边界的一种露点锋)
marginal effect 边际效应【大气】
marginal instability 边际不稳定〔性〕
marginally stable 边界稳定
marginal ray 边缘射线
marginal sea 边缘海【海洋】
marginal stability 边际稳定〔性〕
marginal visibility condition 最低能见度条件
marginal weather approach 极限气象条件进场着陆
Margules's equation 马古列斯方程
Margules' formula 马古列斯公式【大气】
Margules's model 马古列斯模式
marigram ①验潮图②验潮纪录
marigraph 验潮计
marin 海风(法国地中海沿岸和滨海阿尔卑斯山的东南风)
marinada 马林纳达风(西班牙卡塔卢尼亚和法国鲁西荣沿岸的海风)
marin blanc 白海风(无锋面的海风,法语名)
marine accident 海损事故【海洋】
marine acoustics 海洋声学
marine aerosol 海洋气溶胶
marine air fog 海洋气团雾
marine air (mass) 海洋气团,海洋空气
marine atlas 海〔洋〕图集【测】
marine atmosphere 海洋大气
Marine Atmospheric Boundary Layer Experiments (MABLES) 海洋大气边层试验
marine automatic meteorological observing station (MAMOS) 海洋自动气象观测站

marine barometer 船舶用气压表
marine bio-acoustics 海洋生物声学【海洋】
marine biochemical resource 海洋生化资源【海洋】
marine biochemistry 海洋生物化学【海洋】
marine biogeochemistry 海洋生物地球化学【海洋】
marine biological activity 海洋生物活动
marine biological noise 海洋生物噪声【海洋】
marine biology 海洋生物学
marine boundary layer 海洋边界层
marine bucket 测海水温吊桶
marine charting 海洋测绘【测绘】
marine chemical resource 海洋化学资源【海洋】
marine chemistry 海洋化学【海洋】
marine climate 海洋性气候【大气】
marine climatology 海洋气候学【大气】
marine current 海流
marine ecological investigation 海洋生态调查【航海】
marine ecology 海洋生态学
marine ecosystem 海洋生态系统
marine electric installation 船舶电气设备【航海】
marine engineering 海洋工程【航海】
marine environment 海洋环境【地理】
marine environmental assessment 海洋环境评价【海洋】
marine environmental capacity 海洋环境容量【海洋】
marine environmental chemistry 海洋环境化学【海洋】
marine enviornmental data information 海洋环境数据资料
marine environmental forecasting 海洋环境预报【海洋】
marine environmental monitoring 海洋环境监测【海洋】

marine environmental prediction　海洋环境预报【海洋】
marine environmental protection　海洋环境保护【海洋】
marine environmental quality　海洋环境质量【海洋】
marine environmental science　海洋环境科学【海洋】
marine erosion　海蚀作用【地理】
marine fog　海洋雾
marine forecast　海洋〔天气〕预报
marine geochemical exploration　海洋化探【地质】
marine geochemistry　海洋地球化学【地质】
marine geodesy　海洋大地测量学【测绘】
marine geology　海洋地质学【海洋】
marine geomorphology　海洋地貌学【海洋】
marine geophysics　海洋地球物理学【海洋】
marine gravity anomaly　海洋重力异常【海洋】
marine gravity survey　海洋重力调查【海洋】
marine hydrography　海洋水文学【海洋】
marine hydrology　海洋水文学【海洋】
marine inversion　海上逆温
marine layer　海水层
marine meteorological chart　海洋气象图【测】
marine meteorological code　海洋气象电码【大气】
marine meteorological service system (MMSS)　海洋气象服务系统
marine meteorology　海洋气象学【大气】
marine microorganism　海洋微生物【海洋】
marine navigation　航海学【航海】
marine navigation expert system　航海专家系统【航海】
marine observation　海洋〔天气〕观测
marine physics　海洋物理学【海洋】
marine pollutant　海洋污染物【海洋】
marine pollution　海洋污染【海洋】
marine psychology　航海心理学【航海】
marine radar　海洋雷达
marine radiofacsimile recorder　船用无线电传真记录器,船用无线电传真收片机
marine rainbow　海上虹
marine raingauge　船舶雨量计
Marine Reporting Station (MARS)　海洋数据报告站
marine resources　海洋资源【海洋】
marine science　海洋科学
marine science and technology　海洋科学技术【航海】
Marine Sciences Commission (MSC)　海洋科学委员会
marine search and rescue　海上搜救【航海】
marine sediment　海相沉积
marine stratigraphy　海洋地层学【海洋】
marine surveillance　海洋监视【航海】
marine thermodynamics　海洋热力学
marine thermometer　海水温度表
marine varve　海洋季候泥
marine weather data　海上气象数据【航海】
marine weather forecast　海洋天气预报【大气】
marine weather observation　海洋天气观测
Mariotte's law　马略特定律
maritime aerosol　海洋性气溶胶【大气】
maritime air　海洋空气
maritime air fog　海洋气团雾
maritime air mass　海洋气团【大气】
maritime climate　海洋气候
maritime cloud　海洋云
maritime coefficient　海洋系数
maritime continent　海洋陆地(指印度尼西亚等岛屿)
maritime continental contrast　海洋-大陆对比
maritime glacier　海洋性冰川【地理】
maritime law　海洋法
maritime meteorological observation

海洋气象观测
maritime meteorology 海洋气象学
maritime polar air（mass） 极地海洋气团,极地海洋空气
Maritime Province Cold Current 沿海地区冷海流
maritime satellite(MARISAT) 海事卫星
maritime satellite system(MARISAT) 海事卫星系统
maritime tropical air（mass） 热带海洋气团,热带海洋空气
maritime tundra 海洋性冻原
maritimity 海洋度,海洋对气候的影响程度
Markov chain 马尔可夫链
Markov deviation 马尔可夫偏差
Markovian 马尔可夫过程
Markov process 马尔可夫过程【数】
Mars 火星【天】
MARS（microwave atmospheric remote sensor） 微波大气遥感探测器
Marsden chart 马斯登图
Marsden square 马斯登方块（测绘）
marsh 沼泽
Marshall and Palmer distribution 马歇尔-帕尔默〔雨滴〕谱
Marshall-Palmer radar rainfall function 马歇尔-帕尔默雷达雨量函数
Marshall-Palmer relation 马歇尔-帕尔默关系,M-P关系
marsh gas 沼气
Martian atmosphere 火星大气
Martian year 火星年
Marvin sunshine recorder 马尔文日照计
Mascarene high 马斯克林高压
maser（microwave amplification by stimulated emission of radiation） ①微波激射器【物】②微波激射器【物】
masked front 隐锋
masked trough 隐槽
MASR（microwave atmospheric sounding radiometer） 微波大气探测辐射仪
mass absorption coefficient 质量吸收系数
mass accommodation coefficient 质量调节系数
Massachusetts Institute of Technology（MIT） 马萨诸塞州理工学院（美国）,麻省理工学院
mass balance 质量平衡【海洋】,质量〔收支〕平衡
mass budget 质量收支【海洋】
mass concentration 质量浓度(质量与体积之比,同密度)
mass conservation 质量守恒
mass convergence 质量辐合
mass curve 累计值曲线
mass curve of rainfall 降雨累计曲线
mass defect 质量亏损【物】
mass divergence 质量辐散
mass/energy constraint 质能约束
mass excess 质量过剩【物】
mass extinction 大灭绝,大量消亡
mass extinction coefficient 质量消光系数
mass extinction cross section 质量消光截面
mass flow adjustment 质量流调整
mass flux 质量通量
mass of atmosphere 大气质量
mass proportion 质量比例
mass rate of air flow 通风量
mass scattering coefficient 质量散射系数
mass spectrometer 质谱仪
mass storage 海量存储器【计】
mass storage device 海量存储设备（磁带机、磁盘机等）
mass-transfer method 质量迁移法
mass transport 质量输送,质量输运
mast （观测仪器）杆,桅杆
masterboard 母板【计】,主板
Master International Frequency List（MIFL） 国际频率总表
master program 主程序

master recession curve 主退水曲线
master routine 主程序
master seasonal trend 季节主趋势
master/slave computer 主从计算机【计】
master station 主站【计】
Matanuska wind 马塔牛斯加风（见于阿拉斯加的巴姆地方冬季）
match 匹配
matched filter 匹配滤波器
material boundary 物质边界
material coordinates 实质坐标
material curve 实质曲线
material derivative 实质导数
material point(=particle) 质点
material science 材料科学【物】
material surface 实质面
material volume 实质体积
maternal vortex 母涡旋
mathematical climate 数理气候【大气】
mathematical constraint 数学限制条件
mathematical expectation 数学期望【数】
mathematical filter 数值滤波
mathematical forecast 数值预报
mathematical forecasting (=numerical forecasting) 数值预报
mathematical meteorology 数理气象学
mathematical model 数学模型【数】
mathematical physics 数学物理学
mathematical simulation 数学模拟
mathematical statistics 数理统计〔学〕【数】
Matheson diagram 迈则松图
matinal 晨风（法国莫尔旺山地区夏天的一种东风）
matiniere 清晨风（阿尔卑斯山西麓的冬季下沉风）
matrix 矩阵【数】
matrix inversion 矩阵求逆【数】
matrix method 矩阵法
matrix method of stability analysis 矩阵法稳定度分析

Matsuno scheme 松野格式
matto grosso 热带草原（巴西）
mature phase(或 stage) 成熟阶段
mature soil 成熟土壤【土壤】
mauka breeze 莫卡风（夏威夷的清凉夜风）
Mauna Loa Observatory 冒纳罗亚观象台（夏威夷）
Maunder minimum 蒙德极小期【天】（英天文学家蒙德发现的1645—1715年间太阳活动持续处于低潮的时期）
Maurer sunshine chronograph 莫勒日照计
maxi-min criterion 极大极小判据
maxi-min strategy 极大极小策略
maximum 最大,最高,极大
maximum acceptable concentration 最大允许浓度
maximum and minimum thermometer 最高最低温度表
maximum contaminant level (MCL) 最大污染水平（饮水标准）
maximum depth-area-duration data 最大深-面-时资料
maximum depth of frozen ground 最大冻土深度【大气】
maximum depth of snow cover 最大雪深
maximum design wind speed 最大设计平均风速【大气】
maximum diffusion flow 最大扩散流〔动〕
maximum echo height 最大回波高度
maximum entropy method (MEM) 最大熵方法
maximum entropy spectrum 最大熵谱
maximum entropy spectrum analysis 最大熵谱分析
maximum growth temperature 最高生长温度【农】
maximum gust lapse 阵风最大递减
maximum gust lapse interval 阵风最大递减时段

maximum gust lapse time 阵风最大递减时间
maximum gust speed 最大阵风速
maximum height 最大高度【航海】
maximum height of lift 最大起升高度【航海】
maximum hygroscopicity 最大吸湿水【农】,最大吸湿量【土壤】
maximum instantaneous wind speed 最大瞬时风速【大气】
maximum likelihood criterion 最大似然判据
maximum likelihood estimate 最大似然估计【数】
maximum likelihood filter 最大似然滤波
maximum likelihood method 最大似然法
maximum measuring depth 最大测量深度【航海】
maximum possible precipitation 最大可能降水
maximum possible sunshine duration 最大可能日照时数
maximum precipitation 最大降水量【大气】
maximum pressure 最高气压
maximum principle 最大值原理【数】
maximum probable flood (= probable maximum flood) 最大可能洪水
maximum radiation thermometer 最高辐射温度表
maximum shelter distance 最大防护距离【大气】
maximum station pressure 最高本站气压
maximum temperature 最高温度【大气】
maximum thermometer 最高温度表【大气】
maximum total precipitation content 最大总降水量
maximum transport velocity 最大输送速度
maximum unambiguous range 最大不模糊距离【大气】
maximum unambiguous velocity 最大不模糊速度
maximum value 极大值,最大值
maximum vapour pressure 最大水汽压〔力〕
maximum visibility 最大能见度
maximum water-holding capacity 最大持水量
maximum water vapour tension 最大水汽张力
maximum wind 极值风,最大风力
maximum-wind and shear chart 最大风速和风切变图
maximum wind level 最大风速层【大气】
maximum wind pressure 最大风压【大气】
maximum wind speed 最大风速【大气】
maximum-wind topography 最大风高度型
maximum zonal westerlies 最大西风〔带〕
Maxwell-Boltzmann collision equation 麦克斯韦-玻耳兹曼碰撞方程
Maxwell-Boltzmann distribution 麦克斯韦-玻耳兹曼分布
Maxwell equations 麦克斯韦方程组【物】
Maxwell speed distribution 麦克斯韦速率分布【物】
Maxwell velocity distribution 麦克斯韦速度分布【物】
Maxwell's equation 麦克斯韦方程
Maxwell's law 麦克斯韦定律
maycnm 季风(阿拉伯语名)
Mayson's hygrometer 梅森湿度表
MCC (mesoscale convective complex) 中尺度对流复合体【大气】
McKenzie high 麦肯西高压
MCL (mixing condensation level) 混合凝结高度
M component M 分量(闪电)

MCS (mesoscale convective system) 中尺度对流系统【大气】
MDF (magnetic direction finding) 磁定向法
meadow 草甸【地理】
mean 均值【数】,平均
mean annual precipitation 年平均降水量
mean annual range of temperature 平均温度年较差【大气】
mean annual sea level (MASL) 年平均海平面
mean annual temperature 年平均温度
mean anomaly 平均距平
mean ascending current 平均上升气流
mean basin precipitation 流域平均降水量
mean chart 平均图
mean circulation 平均环流
mean climate characteristics 平均气候特征
mean cloud height 平均云高
mean cloudiness 平均云量
mean cosine of scattering angle 散射角平均余弦
mean daily maximum temperature 日最高温度平均值
mean daily maximum temperature for a month 月平均日最高温度
mean daily minimum temperature 日最低温度平均值
mean daily minimum temperature for a month 月平均日最低温度
mean daily temperature 日平均温度
mean day-to-day variation 平均逐日变化,平均日际变化
meander ①曲流②蜿蜒扩散(烟羽)
meandering course 蜿蜒路径
mean descending current 平均下沉气流
mean deviation 平均偏差【物】
mean difference 平均差
mean distance between earth and sun 平均日地距离

mean Doppler velocity 平均多普勒速度
mean effective diameter 平均有效直径
mean environmental wind 平均环境风
mean equator 平赤道【天】
mean equatorial day 平〔均〕赤道日
mean equinox 平春分点【天】
mean error 平均误差【数】
mean estimate histogram 平均估计值直方图
mean flow 平均〔气〕流
mean flux 平均通量
mean forecast error 平均预报误差
mean free path 平均自由程【物】
mean gust speed 平均阵风速
mean higher high water ①平均最高高水位②平均高高潮〔面〕
mean high water 平均高潮
mean high water interval 平均高潮间隙【航海】
mean high water springs 平均大潮高潮〔面〕
mean interdiurnal variability 日际变化平均值
mean isotherm 平均等温线
mean kinetic energy 平均动能
mean latitude 平均纬度【航海】
mean life 平均寿命
mean lifetime 平均寿命【物】
mean loss 平均损失
mean lower low water ①平均低低潮〔面〕②平均最低低水位
mean low water 平均低潮
mean low water interval 平均低潮间隙【航海】
mean low water springs 平均大潮低潮〔面〕
mean lunar day 平〔均〕太阴日
mean map 平均图
mean meridional circulation 平均经向环流
mean meridional stream function 平均

经向流函数
mean molecular weight　平均分子量
mean monthly maximum temperature
　月平均最高温度
mean monthly minimum temperature
　月平均最低温度
mean motion　平均运动【天】
mean noon　平正午【天】
mean parallax　平均视差【天】
mean pressure meter　平均压力计【航海】
mean radiant temperature　平均辐射温度
mean radius　平均半径
mean sea level（MSL）　平均海平面
mean sea level trends　平均海平面趋势
mean sensitivity　平均敏感度
mean sidereal day　平恒星日
mean sidereal time　平恒星时【天】
mean skin temperature　平均表层温度
mean solar day　平太阳日【天】
mean solar hour　平太阳时
mean solar time　平太阳时【大气】
mean solar year　平太阳年
mean square beam cross section　均方束截面
mean square deviation　均方差
mean square error　均方误差【数】
mean summer time（M. S. T）　平均夏令时
mean sun　平太阳【天】
mean synodic lunar month　平朔望月
mean temperature　平均温度
mean temperature of air column　气柱平均温度
mean tide level　平均潮位
mean time　平时【天】
mean time before failure（MTBF）　平均故障间隔
mean time to failure　平均故障时间
mean utility　平均效益
mean value　平均值
mean value theorem　中值定理【数】
mean vector　均值向量【数】

mean velocity　平均速度【物】
mean virtual temperature　平均虚温
mean-volume diameter　平均体积直径
mean-volume radius　平均体积半径
mean water level　平均水位,平均水平面
mean wind vector　平均风向量
mean wind velocity　平均风速
measure　度量,测量
measured ceiling　测定云高
measurement　①测量【物】②测量〔结果〕
measurement cell　测量单体,测量单元（雷达）
measurement errors　测量误差
measurement point　测点
measuring element　测量元件
mechanical despin antenna（MDA）　机械消旋天线
mechanical dispersion（亦称 hydraulic dispersion）　机械分散,水力分散
mechanical draft　机械通风
mechanical energy　机械能【物】
mechanical energy equation　机械能方程
mechanical equivalent of heat　热功当量【大气】
mechanical forcing　机械强迫作用
mechanical instability　机械性不稳定度
mechanical internal boundary layer（MIBL）　机械内边界层
mechanical mixing　机械混合
mechanical motion　机械运动【物】,力学运动
mechanical pressure　机械性压力
mechanical production rate　机械产生率
mechanical stability　机械稳定〔度〕
mechanical switch　机械开关
mechanical turbulence　机械湍流【大气】
mechanical unstable　机械性不稳定
mechanical weathering　机械风化〔作用〕【地理】
mechanics　力学【物】,机械学
mechanism　机理【物】,机制

mechanistic type crop growth simulator
机制型作物生长模拟装置
Meddy(Mediterranean Eddy) 地中海涡流
median 中位数【数】,中线【数】
median diameter 中位直径
median radius 中位半径
median volume diameter 中位体积直径
median volume radius 中位体积半径
medical climatology 医疗气候学【大气】
medical geography 医学地理学【地理】
medical meteorology 医疗气象学【大气】
Medieval Climate Optimum 中世纪气候适宜期
Medieval Maximum 中世纪极大值
Medieval Mild Phase 中世纪温暖期（气候）
Medieval Warm Epoch(MWE) 中世纪暖期（气候）
Medieval Warm Period 中世纪温暖期（约公元10—13世纪）
medina 梅迪纳风（西班牙卡迪斯冬季的风）
mediocris(med) 中（云）
Mediterranean climate 地中海气候【大气】
Mediterranean Deep Water 地中海深层水
Mediterranean front 地中海锋
Mediterranean lenses （亦称 Meddy）地中海透镜（亦称地中海涡流）
Mediterranean outflow 地中海出流水
Mediterranean regime 地中海型
Mediterranean Sea 地中海
Mediterranean type climate 地中海型气候【大气】
Mediterranean Water 地中海水
medium ①介质,载体②平均值③中等
medium angle camera 中角(78°)照相机
medium clay soil 中黏土【土壤】
medium cloud 中云
medium energy proton and electron detector (MEPED) 中能质子和电子检测器
medium field-of-view radiometer 中等视场辐射仪
medium frequency(MF) 中频
medium-level cloud 中云
medium-range oscillation 中期振荡
medium-range〔weather〕forecast 中期〔天气〕预报【大气】
medium resolution infrared radiometer (MRIR) 中分辨红外辐射仪
medium-scale 中间尺度
medium-scale data utilization station 中型〔卫星〕资料利用站
medium-term hydrological forecast 中期水文预报
meeting appraisal 会议鉴定
mega- 兆（词头,10^6）
megabarye 千毫巴,兆微巴,巴（旧的气压单位,$=10^5$ Pa）
megabyte(MB) 兆字节
megacycle 兆周
megadyne 兆达因
megafog 雾信号器
megahertz 兆赫（百万周/秒）
Megainterstadial 大间冰阶
Megamonsoon 超级季风
megatemperature 高温
megatherm 高温
megathermal 高温〔的〕
megathermal climate 高温气候
megathermal period 高温期,大暖期
megathermal type 高温植物型
megathermic 高温植物〔的〕
Meinel bands 迈内尔带（羟基发射带）
meiobar ①低压区②低压等值线（小于1000 hPa 的等压线）
meions 负距平中心
Meiyu(= plum rain) 梅雨【大气】
Meiyu front 梅雨锋【大气】
Meiyu period 梅雨期【大气】
Mellor-Yamada parameterization 梅勒-

山田参数化,M-Y 参数化
mellow soil 熟土【土壤】
melted soil 融土【地理】
meltem(或 meltemi) 梅尔特风(1.保加利亚沿海和博斯普鲁斯地区夏天的突发东风或东北风;2.季风同义语)
melting 融化
melting band 融化带
melting layer 融化层
melting level 融化层,融化高度
melting point 融〔化〕点【大气】
melt pond （海冰面上的)融水池
membership 隶属关系【数】
membership function 隶属函数
membership grade 隶属度
membrane filter 薄膜滤纸,滤膜
Memery period 梅默里周期
memory ①存贮②存贮器
memory protection 内存保护【计】
meniscus 弯液面,半月形
menu 选单【计】,菜单
MERCAST 商业无线电广播
Mercator mosaic 墨卡托拼图
Mercator projection 墨卡托投影【测】
mercurial barometer 水银气压表
mercurial thermometer 水银温度表
Mercury 水星【天】
mercury(Hg) 水银,汞
mercury barometer 水银气压表【大气】
mercury column 水银柱
mercury-in-glass psychrometer 玻璃水银干湿表
mercury-in-glass thermometer 玻璃水银温度表
mercury-in-steel thermograph 钢管水银温度计
mercury thermometer 水银温度表【大气】
merging cloud 合并云,混合云
mergozzo 梅尔果佐风(意大利马乔列湖的西北风)

meridian ①子午线【天】,经线②子午圈【天】
meridian circle ①子午圈②子午环【天】,子午仪
meridian plane 子午面
meridional 经向〔的〕
meridional cell 经向环流
meridional circulation 经向环流【大气】
meridional cross section 经向剖面
meridional exchange 经向交换
meridional flow 经向气流
meridional front 经圈锋
meridional gradient 经向梯度
meridional index 经向指数
meridionality 经向度,经向分布〔性〕
meridional overturn circulation (MOC) 经向翻转环流
meridional profile 经圈廓线
meridional wind 经向风
meromictic lake 半混合湖〔泊〕,局部混合湖
mesh scale 网格尺度,网格距
MESIS (Meteorological Services Information System) 决策气象服务〔信息〕系统
mesoanalysis 中〔尺度〕分析
mesoanticyclone 中〔尺度〕反气旋
mesochart 中尺度图
mesoclimate 中气候【大气】
mesoclimatology 中尺度气候学
mesocyclone 中气旋(与对流风暴有关的涡旋)
mesocyclone signature 中气旋回波特征
mesocyclone tornado 中气旋龙卷
mesohigh 中〔尺度〕高压
mesojet 中尺度急流
mesokurtic 常态峰的
mesokurtosis 常态峰
mesolateral arc 中侧弧
Mesolithic age 中石器时代
mesolow 中〔尺度〕低压

mesometeorology 中尺度气象学【大气】
meson 介子【物】
mesonet 中尺度网
mesonet station 中尺度观测网站
mesopause 中间层顶【大气】
mesopeak 中间层最高温度点
mesophyte 中生植物【植】
mesopic vision 黄昏黎明视觉
Mesoproterozoic Era 中元古代【地质】
Mesoproterozoic Erathem 中元古界【地质】
mesoscale 中尺度【大气】
meso-α scale α中尺度【大气】
meso-β scale β中尺度【大气】
meso-γ scale γ中尺度【大气】
mesoscale atmospheric motion 中尺度大气运动
mesoscale cellular convection (MCC) 中尺度单体对流
mesoscale convective [cloud] band 中尺度对流云带
mesoscale convective [cloud] cluster 中尺度对流云团
mesoscale convective complex (MCC) 中尺度对流复合体【大气】
mesoscale convective storm complex 中尺度对流风暴复合体
mesoscale convective system (MCS) 中尺度对流系统【大气】
mesoscale disturbance 中尺度扰动
mesoscale eddy 中尺度涡流
mesoscale elevated inversion 中尺度空中逆温
mesoscale entrainment instability 中尺度夹卷不稳定度
mesoscale high pressure area 中尺度高压区
mesoscale lee low pressure area 中尺度背风坡低压区
mesoscale lee vortex 中尺度背风坡涡旋
mesoscale low 中尺度低压【大气】

mesoscale low pressure area 中尺度低压区
mesoscale meteorological modeling 中尺度气象模拟
mesoscale meteorology 中尺度气象学
mesoscale model 中尺度模式【大气】
mesoscale motion 中尺度运动
mesoscale organization 中尺度组织
mesoscale precipitation area 中尺度降水区
mesoscale precipitation core 中尺度降水中心
mesoscale pressure system 中尺度气压系统
mesoscale rainband 中尺度雨带
mesoscale structure 中尺度结构
mesoscale synoptics 中尺度天气学
mesoscale system 中尺度系统【大气】
mesoscale unorganized convection(MUC) 中尺度无组织对流
mesoscale vortex street 中尺度涡街
mesoscale weather system 中尺度天气系统
mesoscopic structure 介观结构【物】
mesosphere 中间层【大气】(大气按热力性质分五层之第三层,约在50~85 km高度)
mesosphere-stratosphere-troposphere(MST) radar 中间层-平流层-对流层雷达,MST雷达
mesospheric circulation 中间层环流【大气】
mesospheric cloud 中间层云
mesospheric jet 中间层急流
mesospheric ozone 中间层臭氧
mesostructure 介观结构【物】
mesosynoptic meteorology 中尺度气象学
mesotherm 中温
mesothermal 中温[的]
mesothermal climate 中温气候【大气】
mesothermal type 中温植物型

mesothermic 中温植物〔的〕
mesovortex 中涡旋
Mesozoic Era 中生代【地质】
Mesozoic Erathem 中生界【地质】
message ①电报②信息
message addressing 报头,电报地址
message switching facility 转报设备
messenger ①信锤②先驱,信使
met ①气象学(meteorology)的口语简称②也可指气象工作者(meteorologist)
metabolic rate 新陈代谢率
metabolism 新陈代谢【生化】
metadata 元数据【计】,元资料
metadata element 元数据元素
metadata entity 元数据实体
metallic barometer 金属气压表(空盒气压表的同义语)
metallic thermometer 金属温度表
metallogenesis 成矿作用【地质】
metal-oxide-semiconductor structure 金属-氧化物-半导体结构【物】,MOS结构【物】
metal oxide varistor 金属氧化物压敏电阻
metamorphosis 变性,变质,变形
metamorphosis of snow 雪的形态变化
metapause 上大气层顶
METAR (Meteorological Terminal Air Report) 机场定时地面天气报告
metasphere 上大气层,氢原子优势圈
metastability 亚稳度
metastable 亚稳
metastable state 亚稳态【物】
meteor ①大气现象【大气】(云除外)②流星【天】,陨星
meteoric dust 流星尘埃
meteoric shower 流星雨
meteoric water 天落水,大气水
meteorite 陨星【天】,陨石
meteorite shower 陨石雨【地质】
meteoritic ablation 流星烧蚀
meteoritic dust 陨石尘埃

meteoritic material 流星体
meteoritics 陨石学【地质】,陨星学
meteorogram 天气实况演变图【大气】
meteorograph 气象计
meteoroid 流星体【天】
meteorological acoustics 气象声学
meteorological advisory 气象咨询报
Meteorological Agency 气象厅(日本)
meteorological aid 气象辅助设备
meteorological airborne data system (MADS) 航空气象资料系统
meteorological aircraft 气象飞机【大气】
meteorological airplane 气象飞机
meteorological analysis 气象分析
meteorological archive 气象档案
meteorological archive and retrieval system(MARS) 气象〔资料〕档案存取系统
Meteorological Authority 气象局(埃及)
meteorological balloon 气象〔探测〕气球
meteorological bomb 气象炸弹
meteorological broadcast 气象广播
Meteorological Bulletin 气象通报
meteorological car 气象车
meteorological center 气象中心
meteorological cloud chamber 气象云室
meteorological code 气象电码【大气】
meteorological communication 气象通信
Meteorological Communications Center (M.C.C.) 气象通信中心
meteorological constituent 气象潮分
meteorological data center 气象资料中心
meteorological data collecting and processing unit 气象数据收集与处理装置
meteorological data handling system (MDHS) 气象资料处理系统
meteorological datum plane 气象基准面
meteorological disaster 气象灾害
meteorological disease 气象官能症,天气疾病
meteorological district 气象区〔域〕

meteorological drought 气象干旱
meteorological drought index 气象干旱指数
meteorological dynamics 气象动力学
meteorological economic decision 气象经济决策
meteorological element 气象要素【大气】
meteorological element series 气象要素系列,气象要素序列
meteorological equator 气象赤道【大气】
meteorological equipment 气象设备
meteorological factor 气象因素,气象因子
meteorological forecast 气象预报
meteorological hazard mitigation 气象减灾
meteorological information 气象信息
meteorological information and dose acquisition system (MIDAS) 气象信息和辐射剂量获取系统
Meteorological Information Comprehensive Analysis and Processing System (MICAPS) 气象信息综合分析处理系统
Meteorological Institute ①气象研究所 ②气象学院
meteorological instrument 气象仪器【大气】
meteorological kinematics 气象运动学
meteorological log 气象日志,气象记录簿
meteorological loss 气象损失
meteorological magnetic event recorder 磁带式气象事件记录器
meteorological map 气象图
meteorological message 气象电报
meteorological minimum 最低气象条件【大气】
meteorological model 气象模式
meteorological modeling 气象模拟
meteorological navigation 气象导航【大气】
meteorological network 气象台站网
meteorological noise 气象噪声【大气】

meteorological observation 气象观测【大气】
meteorological observatory 气象台【大气】
meteorological observer 气象观测员
meteorological observing station 气象观测站
Meteorological Office 气象局(英国)
meteorological optical range 气象光学视距,气象光学视程
meteorological optics 气象光学
meteorological organization 气象组织
meteorological parameter 气象参数
meteorological phenomenon 气象现象
meteorological platform 气象观测平台【大气】,观测场
meteorological post 气象哨
meteorological proverb 气象谚语【大气】
meteorological radar 气象雷达【大气】
meteorological radar equation 气象雷达方程【大气】
meteorological radar station 气象雷达站
meteorological range 气象〔标准〕视距
meteorological realm 气象领域
meteorological reconnaissance flight 气象侦察飞行
meteorological record 气象记录
meteorological region 气象区
meteorological report 气象报告【大气】
meteorological reporting area 气象报告范围
meteorological research flight 气象研究飞行
meteorological rocket 气象火箭【大气】
meteorological rocket facility 气象火箭站
meteorological rocket network (MRN) 气象火箭网
meteorological rocket sonde 气象火箭探空仪
meteorological satellite 气象卫星【大气】
meteorological satellite ground station 气象卫星地面站【大气】
meteorological satellite laboratory (MSL)

气象卫星实验室
meteorological service 气象服务,气象业务
Meteorological Service 气象局(一般通用)
Meteorological Services Information System(MESIS) 决策气象服务〔信息〕系统
meteorological ship 气象观测船
meteorological shipping route 气象航线【大气】
meteorological society 气象学会
meteorological sounding rocket 气象探测火箭
meteorological stability 气象稳定性
meteorological statics 气象静力学
meteorological station 气象站
meteorological statistics 气象统计学
meteorological survey 气象测量
meteorological symbol 气象符号
meteorological table 气象表
meteorological telecommunication 气象通信
meteorological telecommunication hub 气象通信枢纽
meteorological telecommunication network 气象电〔传通〕信网
meteorological telegraph 气象〔有线〕电报
meteorological teleprinter network 气象电传通信网
meteorological thermodynamics 气象热力学
meteorological tide 气象潮【大气】(由于盛行风向的更替和水温的季节变化,引起海面的年或半年变化)
meteorological tower 气象塔
meteorological transmission 气象传送,气象〔情报〕传输
meteorological tropics 气象热带
meteorological visibility 气象能见度
meteorological visibility at night 夜间气象能见度
meteorological warfare 气象战
meteorological warning 气象预警
meteorological warning message 气象预警电报
meteorological watch office 气象监视台
meteorological weapon 气象武器
meteorological wind tunnel 气象风洞【大气】
meteorological yearbook 气象年鉴
meteorologist 气象工作者,气象学家
meteorology 气象学【大气】
meteorology broadcast 气象广播
meteorology echo 气象回波【航海】
meteorology history 气象史
meteorology of crops 作物气象【大气】
Meteorology Research, Inc.(MRI) 气象研究公司(美国)
meteoropathic reaction 气象生理反应
meteoropathology 气象病理学
meteoropathy 气象病【大气】
meteorotropic disease 气象病【大气】,气象官能症,天气疾病
meteorotropic effect 气候〔影响性〕反应
meteorotropism 气象官能症
meteor shower 流星雨
meteor stream(或 swarm) 流星群
meteor trail 流星尾迹
meteor trains 流星列
meteor trajectory 流星轨迹
meteor wake 流星瞬现尾迹,流星尾迹
Meteosat 欧洲同步气象卫星(1997年11月发射第一颗,定位于本初子午线上空即零度经度上空)
Meteosat Data Management Department(MDMD) 欧洲气象卫星资料管理部
meteotron 人工热流器,造云器
meter 米
meter-kilogram-second system(MKS) m-kg-s 单位制

meter-ton-second system of units m-t-s 单位制
methacrolein 异丁烯醛
methane 甲烷,沼气
methane bacteria 甲烷细菌【土壤】
methanesulfonic acid 甲基磺酸 (CH_3SO_3H)
methanol 甲醇 (CH_3OH)
methanol-water hydrogen generator 甲醇-水制氢设备
method of characteristics 特性法
method of finite difference 有限差分法
method of images 镜像法【物】,电像法
method of least squares 最小二乘法【数】
method of moments 矩量法
method of perturbations 扰动法,微扰法
method of reduction of order 降阶法【数】
method of separation of variables 分离变量法【数】
method of (small) perturbation 微扰法
method of successive 渐近法
method of undetermined coefficient 待定系数法【数】
method of variation of constant 常数变易法【数】
method of weighted least squares 加权最小二乘法【数】
methodology 方法学
methyl 甲基
methyl alcohol 甲醇
methyl bromide 甲基溴 (CH_3Br)
methyl chloride 甲基氯 (CH_3Cl)
methyl chloroform 甲基氯仿,三氯乙烷 (CH_3CCl_3)
methyl iodide 甲基碘
methylvinyl ketone 甲基·乙烯基(甲)酮,丁烯酮 ($CH_2:CH·CO·CH_3$)
metre(=meter) 米
metric coefficient 度量系数
metric force 度量力
metric system 米制
metric tensor 度量张量
metro 气象学的简称(军事气象服务中使用)
metropolitan area network (MAN) 城域网【计】
Metropolitan Meteorological Experiment (METROMEX) 都市气象试验
metrosonde 〔边界层〕气象探空仪
metsat 气象卫星(meteorological satellite)的缩写
Mexican hat wavelet 墨西哥帽小波
Mexican high 墨西哥高压
MF (medium frequency) 中频(300~3000 kHz)
MICAPS (Meteorological Information Comprehensive Analysis and Processing System) 气象信息综合分析处理系统
Michael-riggs (=rig) 米迦勒节捣乱风(英国9月底的大风)
Michelson actinograph 迈克耳孙直接辐射计
Michelson interferometer 迈克耳孙干涉仪【物】
micro- 微(词头,10^{-6})
micro analysis 微量分析【化】
microanalysis 小尺度分析
microbar 微巴(达因/厘米^2)
microbarm 微气压图
microbarogram 微〔气〕压记录图
microbarograph 微压计【大气】
microbarovariograph 微〔气〕压计,微变压计
microbiology 微生物学
microburst 微下击暴流【大气】
microburst line 微下击暴流线
microcircuit 微型电流
microclimate 小气候【大气】
microclimate in the fields 农田小气候【大气】
microclimatic factor 小气候因素
microclimatic heat island 小气候热岛
microclimatic measurement 小气候测量

microclimatic observation 小气候观测
microclimatology 小气候学【大气】
microcomputer 微〔型计算〕机
microcomputer data acquisition and processing system 微机数据采集加工系统【农】
microcomputer information system 微机信息系统【农】
microcoulomb ozone meter 臭氧微库仑分析仪
microcyclone 小气旋
microearthquake 微震【地】
microelectronics 微电子学【物】
microelement 微量元素【农】
microencapsulation technique 微封闭技术
microenvironment 人造环境,微环境
microfiche 缩微胶片
microfilm 缩微胶卷
microfilmer 缩微摄影机
microfilming 缩微摄影
microfront 微锋(数十厘米至数米宽度内达几摄氏度的温差)
microgeography 微观地理学,小地理学
microhabitat 小生境【地理】
microlayer ①微表层②微界面
micromanometer 微压表,微压计
micrometeorograph 微气象计
micrometeorology 微气象学【大气】
micrometer 测微表,微米
micron 微米(长度单位,10^{-6} m)
microphotography 缩微摄影【测】,显微摄影术
microphysical model 微物理模式
microphysical process 微观物理过程
microphysical property 微观物理特性
microphysics 微观物理学
microphysics parameterization 微观物理学参数化
micropluviometer 微雨量器
micro-power sunshine detector 微功率日照检测仪

micro-pressure gauge 微气压计
microprocessor 微〔信息〕处理机
micropulsation 微脉动【地】(地磁场 0.1 s 到 10 min 的周期脉动)
micro-Ringelmann chart 林吉曼小图
microscale 小尺度,微尺度
microscale effect 微尺度效应
microscale system 小尺度系统【大气】
microscale turbulence 微尺度湍流,小尺度湍流
microscale weather system 微尺度〔天气〕系统,小尺度天气系统
microscopic 微观〔的〕
microseismograph 微震仪
microseisms 微震
microstress 微应力
microstrip antenna 微带天线【电子】
microstructure 微结构【物】
microsynoptic map 小尺度天气图
microsynoptic meteorology 小天气学
microtherm 低温
microthermal 低温〔的〕
microthermal climate 低温气候【大气】
microthermal type 低温植物型
microthermic 低温植物〔的〕
microtremor 微振动
microturbulence 小尺度湍流
microvariation 微变化
microvariation of pressure 气压微变化
microviscosity 微黏度
microvortex 微涡,小尺度涡
microwave 微波【物】
microwave altimeter 微波高度计
microwave amplification by stimulated emission of radiation(MASER) 微波激射器,微波激射
microwave atmospheric remote sensor (MARS) 微波大气遥感探测器
microwave atmospheric sounding radiometer(MASR) 微波大气探测辐射仪
microwave backscattering 微波后向散射

microwave drying 微波干燥【农】
microwave image 微波图像【大气】
microwave-infrared sounding 微波-红外探测
microwave meteorology 微波气象学
microwave oven 微波炉
microwave probing 微波探测
microwave propagation 微波传播
microwave radar 微波雷达
microwave radiation 微波辐射
microwave radiometer 微波辐射仪【大气】
microwave radiometric measurement 微波辐射测量
microwave refractometer 微波折射计
microwave remote sensing 微波遥感【农】
microwave scatterometer 微波散射计
microwave sounding unit (MSU) 微波探测装置
microwave temperature-humidity profiler 微波温湿廓线〔探测〕仪
microwave ultrasonics 微波超声学
microwave wind scatterometer 测风微波散射计
microweather 小尺度天气
Mid-Atlantic ridge 大西洋洋中脊【海洋】
Mid-Atlantic trough (MAT) ①大西洋洋中槽②中大西洋海槽
midday 正午,中午
middle atmosphere 中层大气【大气】（对流层顶至 100 km 之间的大气）
Middle Atmosphere Programme Steering Committee (MAPSC, WMO/ICSU) 中层大气研究计划指导委员会 (WMO/ICSU)
Middle Atmospheric Programme (MAP) 中层大气研究计划
middle atmospheric physics 中层大气物理学【大气】
middle cloud 中云【大气】
middle Cretaceous 中白垩世

middle Devonian 中泥盆世
middle infrared 中红外（通常约为 4～15 μm）
middle Jurassic 中侏罗世
middle latitude 中纬度
middle-latitude air mass 中纬〔度〕气团
middle-latitude equable regime 中纬〔度〕平静型
middle-latitude high-pressure belt 中纬〔度〕高压带
middle-latitude system 中纬〔度〕系统
middle-latitude westerlies 中纬〔度〕西风〔带〕
middle-level cloud 中层云
middle Permian 中二叠世
middle Silurian 中志留世
middle stratosphere 平流层中部
Middle Subboreal climatic phase 亚北方中期〔气候〕
middle troposphere 中对流层
mideddy 中涡旋
midget tropical cyclone 小型热带气旋（半径约为 100～200 km）
midget tropical storm (= midget typhoon) 小型台风（直径 100 km 以内的台风）
Mid-Japan Sea Cold Current 中日本海冷海流
mid-latitude anticyclone 中纬度反气旋
mid-latitude cyclone 中纬度气旋
mid-latitude depression 中纬〔度〕低压
mid-latitude forcing 中纬强迫
mid-level cyclone 中层气旋
midnight 子夜,半夜
midnight sun 半夜太阳
midnight wind 子夜风（德国巴伐利亚高地维尔姆和阿梅尔河的偏南山风）
mid-ocean ridge 洋中脊【海洋】,中央海岭
mid-ocean rise 洋中隆【地质】
mid-ocean trough 大洋中部槽（简称洋中槽）
Mid-Pacific trough 太平洋洋中槽

mid-Proterozoic ice age　中元古代冰期
mid-range forecast　中期预报
mid-season month　季中月(1、4、7和10月)
midsection method　中断面法(推求断面流量)
midsummer　盛夏
mid-tropospheric cyclone　对流层中层气旋【大气】
midtropospheric front　中对流层锋
midtropospheric inversion　中对流层逆温
mid-value　中值
midwinter　隆冬
Mie backscattering　米后向散射
Mie backscattering efficiency　米后向散射系数
Mie extinction　米消光
miejour　米儒尔风(法国普罗旺斯南面吹来的湿热海风)
Mie scattering　米散射【大气】,粗粒散射
Mie scattering theory　米散射理论
Mie theory　米理论
migrating cyclone　迁移气旋
migration　①移动②位移
migration velocity　偏移速度【地】
migration velocity analysis　偏移速度分析【地】
migratory　移动[的],迁移[的]
migratory anticyclone　移动性反气旋
migratory bird　候鸟
migratory high　移动性高压
mil　①密耳(千分之一英寸)②密位(圆周的6400分之一)
Milankovitch cycle　米兰科维奇循环
Milankovitch hypothesis　米兰科维奇假说
Milankovitch oscillation　米兰科维奇振荡
Milankovitch Pleistocene climate variation　米兰科维奇更新世气候变化
Milankovitch solar radiation curve　米兰科维奇太阳辐射曲线,米兰氏太阳辐射曲线

Milankovitch theory　米兰科维奇理论,米兰氏理论
mild climate　温和气候
mild weather　温和天气
mile　英里(=1.609 km)
military climatography　军事气候志【大气】
military meteorological information　军事气象信息【大气】
military meteorological support　军事气象保障【大气】
military meteorology　军事气象学【大气】
milky ice　乳色冰
Milky Way　银河【天】
milky weather　乳白天空
Mill Creek culture　米尔河文化
Millennium Development Goals　千年发展目标
Miller's climatic classification　米勒气候分类法
millet rains　谷子雨(10—12月东非的一种大雨)
milli-　毫(词头,10^{-3})
millibar(mb)　毫巴(气压单位,1 mb=1 hPa)
millibar-barometer　毫巴[标尺]气压表
millibar scale　毫巴标尺
millimeter(mm)　毫米
millimeter wave　毫米波
millimetre of mercury　毫米汞柱(=1.333 hPa)
millipore filter　毫孔过滤器,毫孔滤膜
Mills period　(亦称 Mills infection period)米尔斯[传染]期
Milne problem　米尔恩问题
Minamata disease　水俣病【地理】
Mindanao Current　棉兰老海流
Mindanao Eddy　棉兰老涡流
mine[dangerous] area　水雷[危险]区【测】

mineral aerosol 矿物气溶胶
mineral deposit model 矿床模式【地质】
mineral economics 矿产经济学【地质】
mineral exploitation 矿产开发【地质】
mineral exploration 矿产勘察【地质】
mineralization 矿化作用【地质】
mineralization rate 矿化速率【土壤】
mineralized water 矿化水【地质】
mineralogy 矿物学【地质】
mineral spring 矿泉【地质】
mine-sweeping area 扫雷区【测绘】
miniature climatological station（MINICLIM） 小型〔自动〕气候站
miniature net radiometer 小型净辐射表
minicomputer 小型计算机
Minifax 小型气象传真机（美国奥尔登公司一种传真机产品名）
minimal flight 节时飞行
minimal flight path 节时航线
minimal headings 节时航向
minimax 极小化极大【数】
minimax criterion 极小化极大判据
minimax principle 极小化极大原理【数】
minimax strategy 极小极大策略【数】
minimum 最小,最低,极小
minimum altitude 低空（按美国标准指650 m以下）
minimum annual flow 最小年流量,年最小流量（无严格意义）
minimum detectable signal（MDS） 最小可探测信号
minimum deviation 最小偏向,最小偏距
minimum duration 最小风时【海洋】
minimum energy state 最低能级
minimum entropy exchange principle 极小熵交换原理
minimum flight altitude 最低飞行高度
minimum information method 最小信息法（卫星温度反演的一种方法）
minimum ionizing speed 电离最低速

〔率〕
minimum relative humidity 最小相对湿度
minimum station pressure 最低本站气压
minimums（＝operational weather limits） 飞行天气低限
minimum temperature 最低温度
minimum temperature prediction 最低温〔度〕预报
minimum thermometer 最低温度表【大气】
minimum-time track 节时航线
minimum value 最小值,极小值
minimum variance estimation 极小方差估计
minimum variance method 极小方差法
minisonde 小型探空仪
mini-supercell 小型超级单体
mini-tornado cyclone 小〔型〕龙卷气旋
mini-T-sonde 低空温度探空仪
minorant 弱函数
minor circulation 小〔型〕环流
minor cycle 黑子小周（指前导黑子极性向南时的11年周期,一般说来正极大值比负极大值的黑子数多）
minor frame 副帧
minor lobe 副瓣
minor ridge 次脊
minor sea breeze 小海风
minor trough 次槽
minor wave 次波
mintra 最低尾迹条件
minuano 米牛阿诺风（巴西南部6—9月间之一种寒冷西南风）
minute ①分②微小的
Miocene Epoch 中新世【地质】
Miocene Series 中新统【地质】
miothermic period 温和期（指间冰期）
mirage 蜃景【大气】,海市蜃楼,幻景
mirror nephoscope 测云镜
mirror point 镜点
mirror reflection 镜〔面〕反射【物】,镜

像反射【物】
miscellaneous data 杂项数据(资料)
miscible displacement 可溶混移动,可溶混驱替
misocyclone 小气旋(水平尺度小于4 km)
misoscale 小尺度(40 m～4 km)
missile 发射体(火箭)
missing ring 欠缺年轮
mist 轻雾【大气】,霭
mistbow 雾虹
mist droplet 霭滴
mistral(=mystral) 密史脱拉风(地中海北岸的一种干冷西北或北风)
mitgjorn 米若尔恩风(法国鲁西荣从比利牛斯山吹来的干燥微南风)
mitigation 减轻,减缓
mitochondrial DNA 线粒体 DNA【生化】
mitochondrial RNA 线粒体 RNA【生化】
mixed cloud 冰水混合云
mixed distribution 混合[频数]分布
mixed forest 混交林【林】
mixed icing condition 混合积冰条件
mixed layer 混合层
mixed-layer capping inversion 混合层冠盖逆温
mixed-layer depth 混合层厚度
mixed-layer evolution 混合层演变
Mixed Layer Experiment(MILE) 混合层实验
mixed-layer height 混合层高度
mixed layer homogeneous 混合层均一性
mixed-layer model 混合层模式
mixed-layer similarity 混合层相似性
mixed-layer spectra 混合层谱
mixed-layer top 混合层顶
mixed-layer venting 混合层排放,混合层通气
mixed model 混合模型【农】
mixed nucleus 混合核
mixed planetary-gravity wave 混合行星-重力波
mixed rain and snow 雨夹雪
mixed Rossby-gravity wave 混合罗斯贝重力波【大气】
mixed tide 混合潮【海洋】
mixing 混合
mixing chamber 混合云室
mixing cloud 混合云
mixing cloud top 混合云顶
mixing condensation level(MCL) 混合凝结高度【大气】
mixing cooling 混合冷却
mixing depth(或 height) 混合层深(高)度
mixing distribution 混合分布
mixing efficiency 混合效率
mixing fog 混合雾【大气】
mixing height 混合[层]高度
mixing layer 混合层【大气】
mixing length 混合长【大气】
mixing-length theory 混合长理论
mixing line 混合线
mixing-line structure 混合线结构
mixing path 混合路径
mixing potential 混合势
mixing ratio 混合比【大气】
mixing ratio indicator 混合比显示器
mixing turbulence 混合湍流
mixing zone 混合带
mixture 混合物
mizzle 混合轻雾(毛毛雨和雾同时出现)
mks system 米·千克·秒制
Mløler chart 密勒辐射图
ML(mixed layer) 混合层的缩写
M meter M 仪器(大气中液态水含量测定仪表的总称)
MMO (main meteorological office) 气象总台的缩写
moat 气旋壕(气旋眼壁外围的降水减少环带)
Moazagotl 焚风云(见于欧洲苏台德山

地区)
Moazagotl wind 过山强风(产生焚风云)
mobile ship 移动船舶
mobile ship station 移动船舶站【大气】
mobile source 移动[污染]源
mobile telemetering station 移动式遥测站
mobile weather station 流动气象站
mobility 流动性,迁移率
Moby Dick balloon(=skyhook balloon) 高层等高探测气球
MOC(meridional overturn circulation) 经向翻转环流
mock fog 假雾
mock moon(=paraselene) 幻月,假月
mock sun(=parhelion) 幻日,假日
mock sun ring (=parhelic circle) 幻日环
mode ①众数【数】②模态【大气】【物】③波型
mode eddies (亦称 mesoscale eddies) 中尺度涡流,中尺度涡旋
model 型式,模型
model atmosphere 标准大气,模式大气
model biases 模式偏差
model calibration 模式校正
model deficiency 模式缺陷
model experiment 模拟实验
model hierarchy 模式体系
modeling ①模拟②建模
modeling cirteria 模拟判据
mode-locking 锁模【物】
model output statistic prediction(MOS prediction) 模式输出统计预报【大气】
model output statistics(MOS) 模式输出统计
model resolution 模式分辨率【大气】
model stratosphere 模式平流层
model troposphere 模式对流层
modem 调制解调器【计】,调制解调装置
mode radius 众数半径
moderate 中度[的]

moderate baroclinity 中度斜压性
moderate breeze 4级风【大气】,和风
moderate crosswind 中[度]侧风
moderate fog 中雾
moderate gale 疾风,7级风
moderately nonlinear 中非线性
moderate rain 中雨【大气】
MODerate-resolution Imaging Spectroradiometer(MODIS) 中分辨[率]成像光谱辐射仪(EOS)
moderate sea 中浪(风浪3级)
moderate tropical storm 中等热带风暴
moderate visibility 中常能见度
mode water 模式水
MODIS (MODerate-resolution Imaging Spectroradiometer) 中分辨[率]成像光谱辐射仪(EOS)
modification ①控制,影响②变更,修正
modification of air mass 气团变性
modified Bessel function 修正贝塞耳函数
modified Gamma distribution 修正γ分布
modified refractive index 修正折射率
modified refractivity 修正折射率差
modified von Karman spectrum 修正冯·卡曼谱
modon 涡偶,偶极子,双涡
modon flow pattern 偶极子流型,偶极流型
modular automated weather system (MAWS) 模块式自动气象观测系统
modular flow 模式流
modulation 调制
modulational instability 调制不稳定[性]
modulation theorem 调制定理
modulation transfer function(MTF) 调制传递函数
modulator 调制器
module ①模【数】,模块【计】②模数,系数③微型组件
modulus ①模数【数】②模量【物】
modus ponens (affirmative mode) 肯定

法（拉丁语）
modus tollens（denial mode） 否定法（拉丁语）
Moeller chart 摩勒图
moist adiabat 湿绝热线【大气】
moist adiabatic lapse rate 湿绝热直减率【大气】
moist adiabatic process 湿绝热过程【大气】
moist air 湿空气【大气】
moist available energy(MAE) 湿有效能量
moist available potential energy 湿有效位能
moist climate 湿润气候
moist〔conservation〕equation 水汽〔守恒〕方程【大气】
moist convection 湿对流【大气】
moist convective adjustment 湿对流调整
moist enthalpy 湿焓
moistering 湿化
moist index 湿润指数【农】
moist instability 湿不稳定度
moist-labile energy 湿不稳定能量
moist-lability 〔潮〕湿不稳定性
moist model 湿模式
moist potential vorticity 湿位涡
moist process 湿过程
moist slantwise convection 湿倾斜对流
moist snow-flake 湿雪花
moist static energy 湿静力能
moist subhumid climate 潮亚湿气候
moist tongue 湿舌
moisture ①水分②水汽③湿气
moisture adjustment 水分调整
moisture anomaly index 水分距平指数
moisture apparatus 测湿器
moisture availability 水汽有效率(饱和率)
moisture available index 水汽可用指数
moisture budget 水汽收支
moisture burst 水汽猝发
moisture channel data 水汽通道资料
moisture coefficient 湿润系数

moisture content 水汽含量【大气】
moisture-continuity equation 水分连续方程
moisture convergence 水汽辐合
moisture deficit 水汽不足〔量〕
moisture equation 水汽方程
moisture equivalent 湿度当量(土壤的)
moisture factor 湿润因子
moisture flux 水汽通量
moisture index 湿润度【大气】
moisture indicator 湿度指示器
moisture inversion 逆湿【大气】
moisture pooling 水汽集中池
moisture profile 湿度廓线【大气】（土壤的）
moisture redistribution 水分再分布
moisture stress 水汽应力
moisture teller 水分〔快速〕测定仪
moisture-temperature index 温湿指数
moisture tension 水汽张力，水汽压
moisture tongue 湿〔度〕舌
moisture volume percentage 水分体积百分率
moisture weight percentage 水分重量百分率
molality(M/W) 〔重量〕摩尔浓度（每1000 g 溶液中溶质的摩尔数）
molan 莫兰风（由阿尔夫河吹向日内瓦湖的微风）
molar gas constant 摩尔气体常数【化】
molar heat capacity 摩尔热容【物】
molar internal energy 摩尔内能【化】
molarity(M/V) 〔体积〕摩尔浓度（每1升溶液中溶质的摩尔数）
molar mass average 摩尔质量平均【化】，分子量平均
molar specific heat capacity 摩尔比热容
molar volume 摩尔体积【物】
mold rain 梅雨，霉雨
mole 摩尔【化】
molecular absorption 分子吸收【化】

molecular absorption cross section
  分子吸收截面
molecular biology 分子生物学【生化】
molecular clock 分子钟【天】
molecular complex 分子络合物【化】
molecular conduction 分子传导
molecular diffusion 分子扩散【化】
molecular diffusivity 分子扩散率
molecular disease 分子病【生化】
molecular dissipation 分子耗散
molecular dynamics 分子力学【化】
molecular genetics 分子遗传学【生化】
molecular geometry 分子几何〔结构〕【化】
molecular heat 分子热
molecular heat conductivity 分子热传导率
molecular integral 分子积分【化】
molecular motion 分子运动
molecular optics 分子光学【物】
molecular pollutant 分子污染物
molecular reaction 分子反应【化】
molecular-scale temperature 分子〔尺度〕温度
molecular scattering 分子散射【化】
molecular sieve 分子筛【化】
molecular spectroscopy 分子光谱学【物】
molecular spectrum 分子光谱【物】
molecular structure 分子结构【物】
molecular theory 分子理论
molecular thermodynamics 分子热力学【化】
molecular viscosity 分子黏性【大气】
molecular viscosity coefficient 分子黏滞系数【大气】
molecular weight 分子量
molecule 分子
mole fraction 摩尔分数
mole fraction of water vapor 水汽摩尔分数
mole number 摩尔数

mole ratio method 摩尔比法【化】
möller chart 密勒辐射图
Moll-Gorczynski solarimeter 莫尔-戈辛斯基日射总量表
mollisol 活跃冻层
mollition 解冻
Moll thermopile 莫尔热电堆
Moltchanov board 莫尔恰诺夫盘
moment ①瞬间②矩③力矩
moment equation 矩方程
moment method 矩量法
moment of inertia 惯性矩
moment of momentum 角动量,动量矩
moment of rotation 转矩
momentum 动量
momentum anemometer 动量风速表
momentum conservation 动量守恒
momentum exchange 动量交换
momentum flux 动量通量
momentum transfer 动量传输
momentum-transport 动量传输
MONEX（Monsoon Experiment） 季风试验
Monge's phenomenon 蒙日现象（北非沙漠中的一种光学现象）
Mongolian anticyclone 蒙古反气旋
Mongolian cyclone 蒙古气旋【大气】
Mongolian low 蒙古低压【大气】
Monin-Obukhov equation 莫宁-奥布霍夫方程,M-O方程
Monin-Obukhov length 莫宁-奥布霍夫长度
Monin-Obukhov scaling length 莫宁-奥布霍夫尺度长度
Monin-Obukhov similarity theory 莫宁-奥布霍夫相似理论
monitor ①监视,监听②监视器,监听器,记录器③监控程序
monitoring 监测
monitoring car 监测车
monitoring net 监测网

monitoring network 监测网
monitoring ship 监测船
monitoring system 监测系统
monitor of ultraviolet solar energy（MUSE） 太阳紫外能监测仪
monitor of ultraviolet solar radiation（MUSR） 太阳紫外辐射监测仪
monitor well 监测井
monochromatic 单色的
monochromatic brightness 单色亮度
monochromatic equilibrium 单色平衡
monochromaticity 单相性【化】
monochromatic light 单色光【物】
monochromatic radiation 单色辐射【大气】
monochromatic transmittance 单色透射比
monochromator 单色仪【物】
monodisperse 单谱，单分散
monodispersed size distribution 单分散谱，均匀谱
monodispersion 单分散性【化】
monograph series 专题丛书，论丛
mono-lobe scanner 单瓣扫描器
monomer 单体【生化】
monomolecular film 单分子膜
monosaccharide 单糖【生化】
monostatic acoustic radar 单基地声〔雷〕达
monostatic acoustic sounder 单基地声学探测器，收发合置声学探测器
monostatic incocherent scatter radar 单基地非相干散射雷达
monostatic lidar 单基地光〔雷〕达
monostatic radar 单基地雷达【电子】
monotone difference scheme 单调差分格式【数】
monotonic 单调〔的〕
monotonicity 单调性
monsoon 季风【大气】
monsoon active period 季风活跃期
monsoon advance 季风前移，季风前进

monsoon air 季风气流
monsoon break 季风中断
monsoon burst 季风爆发【大气】
monsoon circulation 季风环流【大气】
monsoon circulation tube 季风流管
monsoon climate 季风气候【大气】
monsoon cloud cluster 季风云团【大气】
monsoon cluster 季风云团
monsoon convergence line 季风汇合线
monsoon current 季风洋流
monsoon cyclone 季风气旋
monsoon depression 季风低压【大气】
monsoon disturbance 季风扰动
monsoon dynamics 季风动力学
Monsoon Experiment（India）（MONEX） 季风试验（印度）
monsoon fog 季风雾
monsoon forest 季〔风〕雨林【地理】
monsoon gyre 季风涡旋
monsoon index 季风指数【大气】
monsoon low 季风弱低压
monsoon lull 季风间歇
monsoon meteorology 季风气象学
monsoon onset 季风建立【大气】
monsoon pulse 季风脉冲
monsoon rain 季风雨【大气】
monsoon rainfall 季风雨量
monsoon rainforest climate 季风雨林气候
monsoon regime 季风型
monsoon region 季风区【大气】,季风气候区
monsoon retreat 季风后退,季风撤退
monsoon season 季风季节
monsoon surge ①季风潮【大气】②季风爆发
monsoon trough 季风槽【大气】
monsoon weather 季风天气
monsoon wind 季风
monsoon zone 季风带
montagnere（＝montagneuse） 干冷风（见于法国南部）

Montana-monsoon 蒙大拿季风（美国蒙大拿州大草原的 chinook 风）
MONT code 山地电码（mountain code 的缩写）
Monte Carlo method 蒙特卡罗方法【大气】
Monte Carlo model 蒙特卡罗模式
monterpenes 单萜烯
Montgomery function 蒙哥马利函数
Montgomery potential 蒙哥马利势
Montgomery stream function 蒙哥马利流函数
month 月
monthly amount 月总量
monthly bulletin 月报,月刊
monthly climatological summary 月气候概述
monthly maximum temperature 月最高温度
monthly mean 月平均【大气】
monthly mean temperature 月平均温度【农】
monthly minimum temperature 月最低温度
monthly predictability 月可预报性
monthly record 月报
monthly sum 月总和
Montreal Protocol 蒙特利尔议定书（1987年,关于减少使用臭氧层耗减物质）
moon 月球【天】,月亮,太阴
moonbow 月虹
moon dog 幻月,假月
moon-earth relation 月-地关系
moon illusion 月径幻觉,月幻视
moonlet 小月
moonmist 月雾色
moon pillar 月柱
moonquake 月震【地质】
moonshine 月光
moonshot 月球探测器

moon's path 白道
moored acoustic buoy system（MABS） 系泊声浮标系统
moor-gallop 荒野飑（经过英国荒野的飑）
moraine 冰渍,冰川堆石
moraine terrace 冰渍阶地【地质】
morass 沼泽
morget（=morgeasson） 摩格特风（瑞士日内瓦湖的陆风）
Morlet wavelet 莫莱特小波
morning calm 晨静
morning glory 阵晨风（澳大利亚卡奔塔利亚湾冲破早晨寂静突然出现的风）
morning glow 晨辉
morning tide 早潮
morphology 形态学
morphology of polymer 聚合物形态学【化】
Morse code 莫尔斯电码
mosaic 镶嵌图,拼图
mososcale synoptics 中尺度天气学
MOS prediction（model output statistic prediction） 模式输出统计预报【大气】
Mossbauer effect 穆斯堡尔效应【物】
most frequent size 最可几大小
mother-cloud 母云
mother current 主流,母流,本流
mother of pearl cloud 珠母云
motion 运动
motion field 运动场
motivation 促动力,激发
motor aspirated temperature shield 马达通风测温防辐射罩
motorboating （低频寄生振荡的）汽船声,乘汽船
mountainados 山风暴（见于美国科罗拉多州）
mountain air 山地空气
mountain and valley breeze 山谷风

mountain and valley winds　山谷风
mountain anemograph　高山风速计
mountain barograph　高山气压计【大气】
mountain barometer　高山气压表【大气】
mountain barrier　山岳屏障,山地障碍
mountain breeze　山风【大气】
mountain cap cloud　山顶云
mountain chain　山链【地质】
mountain climate（亦称 highland climate）　山地气候【大气】
mountain climatology　山地气候学【大气】
mountain fog　山雾【大气】
mountain forcing　地形强迫作用,山脉强迫作用
mountain forms　山形【地理】
mountain-gap wind　山口风
mountain glacier　高山冰川
mountain lee wave　山岳背风波
mountain meteorology　山地气象学【大气】
mountain mire　山地沼泽【地理】
mountain observation　山地观测【大气】
mountain〔observation〕station　高山〔观测〕站【大气】
mountainous sea　怒涛（美制风浪8级）
mountain peak　山峰【地理】
mountain physiologic effect　高山生理反应
mountain-plain circulation　高山平原间环流
mountain-plains wind systems　山地平原风系
mountain range　山脉【地质】
mountain region farming　山区农业【农】
mountain sickness　高山病
mountain slope　山脉坡度
Mountain Standard Time　山地标准时间（美国）
mountain station　高山站,山地气象站
mountain system　山系【地质】
mountain torque　山脉力矩
mountain torrent　山洪
mountain tundra　高山冻原

mountain-valley breeze　山谷风【大气】
mountain-valley wind　山谷风
mountain-valley wind systems　山谷风系
mountain wave　地形波
mountain-wave cloud　山地波状云
mountain weather　山地天气
mountain wind　山风
mouse　鼠标〔器〕【计】
moutonnee　①冰川擦痕②羊背石
movable-area fine mesh　活动有限域细网格
movable fine-mesh model　可移动细网格模式【大气】
movable nested grid model　活动嵌套网格模式
movable platform　移动〔观测〕平台
movable-scale barometer　动标〔尺〕气压计
movie loop　环型胶卷
moving average　滑动平均
moving average（MA）model　移动平均模型
moving dune　流动沙丘
moving exploration through line　路线考察
moving fetch　移动风浪区
moving gravity wave　移动重力波
moving observation　流动观测
moving particle　运动粒子
moving random medium　运动随机介质
moving receiver　移动接收机
moving spectrum　滑动谱
moving-target indication（MTI）　移动目标指示（雷达显示）
moving target indicator　活动目标显示器
moving time window　滑动时间窗
moving-window method　流动窗方法
Mozambique Current　莫桑比克〔暖〕海流
MPSDM（multiple point source dispersion model）　多点源扩散模式
MPT（multiplier phototube）　光电倍增管
M-region　M区（太阳活动的某一未

知区）
MSI（multi-spectral image） 多波段图像【大气】,多光谱云图
MSIRR（multispectral infrared radiometer） 多波段红外辐射仪
MSSA（multichannel singular spectrum analysis） 多通道奇异谱分析
MST radar ①MST雷达【大气】(mesosphere-stratosphere-troposphere radar) ②中间层、平流层和对流层雷达
MSU（Microwave Sounding Unit） 微波探测装置
mt DNA 线粒体DNA【生化】
mt RNA 线粒体RNA【生化】
mts system 米·吨·秒单位制
MUC（mesoscale unorganized convection） 中尺度无组织对流
mud ball 泥球
mud crack 龟裂
mud flow 泥流【土壤】
mud rain 泥雨
mud rime 泥凇
mud-rock flow 泥石流
muerto 米尔托风（夏季墨西哥的强北风）
muggy 湿热的,闷热的（口语）
muggy weather 闷热天气
Muirhead Facsimile Copier Equipment（MUFAX） 米尔黑德传真图像复制设备
multiannual storage 多年库容,多年蓄水量
multi-band remote sensing 多波段遥感
multiband system 多频带系统（遥感）
multi-cell 多环流,多单体,复合单体
multicell convection 多单体对流
multicell convective storm 多单体对流风暴
multicell severe local storm 局地强风暴复合体,多单体局地强风暴
multi-cell storm 复合体风暴,多单体风暴

multicell thunderstorm 多胞雷暴,多单体雷暴
multichannel communication system 多路通信系统
multichannel infrared radiometer 多通道红外辐射仪
multichannel photometer 多通道光度计
multichannel pyrheliometer 多通道绝对日射表
multichannel sea surface temperature 多通道海面温度（遥感算法）
multichannel singular spectrum analysis（MSSA） 多通道奇异谱分析
multichannel system 多通道系统
multichannel window method 多通道窗区法
multicolor detection 多色探测
multi-color spin scan cloud camera（MSSCC） 彩色自旋扫描摄云机
multidimension 多维,高维
multi-dimensional Fourier series 多维傅里叶级数
multidisciplinary 多学科的,多学科性
multi-equilibrium states 多平衡态
multi-frequency microwave radiometer 多频微波辐射计
multigrid method 多重网格法
multilevel model 多层模式【大气】
multimedia 多媒体【计】
multimedia computer 多媒体计算机【计】
multimedia PC（MPC） 多媒体个人计算机【计】
multimodal spectrum 多峰谱
multi-nested mesh（或grid）model 多重嵌套网格模式
multiparameter radar 多参数雷达
multipath 多路径传播,多路径无线电传播（multiple-path radio propagation的创新字）
multipath transmission 多路传输
multi-phase atmospheric system 多相大

气系统
multi-phase method 多相方法
multi-phase model 多相模式
multi-photoelectron event 多光电子事件
multiple atmospheric parameter 多种大气参数
multiple-beam klystron(MBK) 多束速调管
multiple-cell echo 多单体回波【大气】
multiple corona 多重光晕
multiple correlation 复相关【大气】
multiple correlation coefficient 复相关系数【大气】
multiple curvilinear regression analysis 多重曲线回归分析
multiple discharge 多次放电
multiple discriminant analysis 多元判别分析
multiple drift corrections 多重偏流修正
multiple echo 多重回波
multiple effect evaporator 多效蒸发器
multiple equilibria 多平衡态
multiple equilibria system 多平衡态系统
multiple factorial experiment 多因子试验
multiple-filter method 多重滤波法
multiple flow equilibria 多流动平衡
multiple Fourier series 多重傅里叶级数【数】
multiple incursion theory 多次升降说
multiple-index 多重指标
multiple-layer inversion 多层逆温
multiple linear regression 多重线性回归
multiple linear regression analysis 多重线性回归分析
multiple nutrients compound fertilizer 多元〔复合〕肥料【土壤】
multiple-phase flow 多相流
multiple point 多重点
multiple point source dispersion model (MPSDM) 多点源扩散模式

multiple-purpose reservoir 多目标水库
multiple reflection 多次反射
multiple register 复合记录器
multiple regression 多元回归
multiple regression analysis 多元回归分析
multiple regression yield model 复回归产量模式
multiple scattering 多次散射【大气】
multiple scattering effect 多次散射效应
multiple scattering theory 多次散射理论
multiple stable layer 多重稳定层
multiple-stage evaporator 多级蒸发器
multiple stroke 多次闪击
multiple-time-scale 多时间尺度
multiple trip echo 多程回波
multiple-tropopause 复对流层顶
multiple vortex configuration 多重涡旋配置
multiple wave 多重波
multiple window channel method 多窗区频道法
multiplexer 复用器【电子】,多路调制器
multiplexing ①多路驱动【电子】②〔多路〕复用【电子】,复接
multiplication 乘法【数】
multiplier phototube(MPT) 光电倍增管
multi-point calibration 多点校准
multipole vibration 多极振动
multiprocessing 多重处理【计】
multiresolution 多分辨率【计】
multiscale analysis 多尺度分析【计】
multi-scale method 多尺度方法
multispectral imagery(MSI) 多波段云图
multi-spectral image(MSI) 多波段图像【大气】,多光谱云图
multispectral imaging 多波段成像
multispectral infrared radiometer(MSIRR) 多波段红外辐射仪
multispectral photography 多波段摄影
multispectral remote sensing 多波段遥感【地理】,多光谱遥感
multispectral scanner 多波段扫描器
multispectral sensor 多波段传感器

multi-spectrum 多维谱
multistatic radar 多基地雷达【电子】
multi-station Doppler radar 多站多普勒雷达
multi-station incoherent scatter radar 多站非相干散射雷达
multi-stylus recording system 多笔记录装置
multisystem ensemble 多系统集合
multiuser system 多用户系统【计】
multivariate analysis 多元分析
multivariate analysis of variance 多元方差分析【数】
multivariate analysis scheme 多元分析方案
multivariate hypergeometric distribution 多元超几何分布【数】
multivariate linear regression 多元线性回归
multivariate objective analysis 多元客观分析
multivariate optimum interpolation 多元最优插值
multivariate signal 多元信号
multivariate〔statistical〕analysis 多元〔统计〕分析【数】
multivariate time series 多元时间序列
multivariate transfer function 多元传递函数,多元转换函数
multiwavelength nephelometer 多波长浊度仪
multiwavelength sunphotometer 多波长太阳光度计
multiway-tree 多叉树
multiyear ice 多年冰
M-unit M 单位(折射率)
Munk boundary layer 芒克边界层(海洋)
Munsell color chart 芒塞尔色图
Munsell color system 芒塞尔色系【测】
Munsell soil charts 芒塞尔土色卡【土壤】
mushroom cloud 蘑菇云
muskeg ①泥炭沼泽②湿地植被
Muskingum method 马斯京根〔方〕法(洪水演算)
mutatus 转化云
mutual coherence 互相干〔性〕【物】
mutual coherence function 互相干函数
mutual correlation 互关联【物】,互相关
mutually exclusive events 不相容事件【数】
mutual spectral density 互谱密度【物】
muzzler 强逆风
Myers rating 迈尔斯检定法(比较不同流域最大流量的指标)
mylar balloon 聚酯纤维气球
mylar corner reflector 聚酯薄膜角形反射器
mystral(=mistral) 密史脱拉风

N

nabivnoy ice 多层冰,筏状冰
NACA(National Advisory Committee for Aeronautics, USA) 美国国家航空咨询委员会
NACA Standard Atmosphere NACA(美国国家航空咨询委员会)标准大气
nacreous cloud 珠母云【大气】
nadir 天底〔点〕
nadir angle 天底角
nadir solar backscatter meter 天底太阳后向散射计
naf hat 诺夫哈特风(阿拉伯的飑)
NAM(Nothern Annular Mode) 北半球环状模

NAND gate 与非门【计】
Nanhai Coastal Current 南海沿岸流【海洋】
Nanhai Sea 南海【海洋】
Nanhai Warm Current 南海暖流【海洋】
nano- 纳〔诺〕(词头,10^{-9})
nano-crystal 纳米晶体【物】
Nansen bottle 南森瓶(海水取样用)
Nansen cast (南森瓶)采水测温
NAO (North Atlantic Oscillation) 北大西洋涛动
Napierian logarithm(= natural logarithm) 自然对数(以 e 为底)
nappe ①水舌,水片(溢流)②叶(圆锥曲面的)③推覆体【地质】
narbonnais(= narbones) 纳榜内风(法语名)
narrow angle camera 窄角(12°)照相机
narrowband radiation 窄带辐射
narrow beam equation 窄束方程
narrow beam pulse 窄束脉冲
narrow beam radiogoniometer 无线电窄束定向器
narrow beam transmitter 窄束发射机
narrow channel approximation 狭管近似
narrow-field scanning system 窄视场扫描系统
narrow-sector recorders (NSR) 窄带记录器
NAS (National Academy of Sciences, USA) 美国国家科学院
NASA (National Aeronautics and Space Administration, USA) 美国国家航空与航天局【地信】,美国宇航局
NASA Scatterometer NASA 散射仪(星载微波雷达)
NASA standard atmosphere 美国国家航空与航天局标准大气
nascent cyclone 初生气旋
n'aschi (= nashi) 纳什风(波斯湾冬季东北风)
NAS Committee on Atmospheric Sciences (NASCAS) 美国国家科学院大气科学委员会
NAS Committee on Oceanography(NASCO) 美国国家科学院海洋委员会
nashi(=n'aschi) 东北风(阿拉伯语名)
nasslood 冰上冰(在旧冰面上形成的冰)
NAT (nitric acid trihydrate) 硝酸三水合物
natural background 自然本底,自然基准
National Academy of Sciences (NAS) 〔美国〕国家科学院
National Advisory Committee for Aeronautics (NACA) 〔美国〕国家航空咨询委员会
National Aeronautics and Space Administration (NASA) 美国国家航空与航天局【地信】
National Aeronautics and Space Council (NASC) 美国国家航空与航天理事会
National Ambient Air Quality Standards (NAAQS) (美国)国家环境空气质量标准
national and aviation meteorological facsimile network(NAMFAX) 国家与航空气象传真网
national atlas 国家地图集【地理】
National Aviation Weather Advisory (Unit)(NAWAU) 〔美国〕国家航空天气咨询处
National Bureau of Standards (UBS) 〔美国〕国家标准局
National Center for Air Pollution Control (NCAPC) 国家大气污染控制中心(美国)
National Center for Atmospheric Research (NCAR) 国家大气研究中心(美国)
National Centers for Environmental Prediction, USA (NCEP) 美国国家环境预报中心

National Climate Data Center（NCDC）
国家气候资料中心（美国）
National Climatic Center（NCC） 〔美国〕国家气候中心
National Committee for Antarctic Research（NCAR） 国家南极考察委员会（澳大利亚,新西兰）
National Committee for Clear Air Turbulence（NCCAT） 〔美国〕国家晴空湍流研究委员会
National Data Buoy Center（NDBC） 国家资料浮标中心（美国）
National Earth Satellite Service（NESS） 美国国家地球卫星局
National Environmental Satellite Center（NESC） 〔美国〕国家环境卫星中心
National Environmental Satellite, Data and Information Service（NESDIS） 〔美国〕国家环境卫星、资料和信息局
National Environmental Satellite Service（NESS,NOAA） 美国国家环境卫星局（NOAA）
National Geophysical Research Institute（NGRI）国家地球物理研究所（印度）
National Hurricane and Experimental Meteorology Laboratory（NHEML） 〔美国〕国家飓风和实验气象学研究所
National Hurricane Center（NHC,NOAA） 〔美国〕国家飓风中心（NOAA）
National Hurricane Research Laboratory（NHRL） 〔美国〕国家飓风实验室
National Institute of Polar Research（NIPR） 国立极地研究所（日本）
National Marine Data Inventory（NAMDI） 国家海洋资料目录（美国）
National Meteorological Center（NMC） 国家气象中心（WMO 的定名）
National Oceanic and Atmospheric Administration（NOAA） 〔美国〕国家海洋大气局
National Ocean Satellite System（NOSS） 〔美国〕国家海洋卫星系统
National Operational Meteorological Satellite System（NOMSS） 〔美国〕国家业务气象卫星系统
national park 国家公园【地理】,国家天然公园【林】
National Polar-orbiting Operational Environmental Satellite System（NPOESS）国家极轨业务环境卫星系统（美国）
National Research Center for Disaster Prevention（NRCDP） 国立防灾研究中心（日本）
National Research Council（NRC）〔美国〕国家研究理事会
National Science Foundation（NSF）〔美国〕国家科学基金会
National Scientific Balloon Facility（NSBF） 国家科学探测气球中心（美国）
National Severe Storm Forecast Center（NSSFC） 国家强风暴预报中心（美国）
National Severe Storms Laboratory（NSSL, NOAA） 国家强风暴实验室（NOAA）
national standard 国家标准
national standard barometer 国家标准气压表【大气】
National Tidal Datum Epoch 〔美国〕国家潮汐基准面时期
National Weather Record Center（NWRC）〔美国〕国家天气记录中心
National Weather Satellite Center（NWSC） 国家气象卫星中心（美国）
National Weather Service（NWS, USA）〔美国〕国家气象局
National Weather Service Forecast Office（NWSFO） 国家气象局预报台（美国）
National Weather Service Headquarter（NWSH） 国家气象局总部（美国）
National Weather Service Training Center（NWSTC） 国家气象局培训中心（美国）

nationwide natural disaster warning system（NADWARN） 全国自然灾害预警系统
native soil organic matter 土壤原有机质【土壤】
native vegetation 本地植物
natural boundary condition 自然边界条件
natural broadening 自然加宽
natural calamity 自然灾害
natural calendar 自然历【地理】
natural complex 自然综合体【地理】
natural control 自然控制,天然控制
natural convection 自然对流
natural coordinates 自然坐标〔系〕【大气】
natural disaster 自然灾害
natural-draftdrying 自然通风干燥【农】
natural environment 自然环境【地理】
natural feature 自然地貌【航海】
natural flow 天然径流,天然水流,未调节水流
natural frequency 自然频率
natural hail embryo 自然雹胚
natural hazard 自然灾害【地理】
natural illumination 自然光照
natural landscape 自然景观【地理】
natural law 自然法则
natural light 自然光【物】
natural linewidth 自然线宽【电子】
natural logarithm（ln） 自然对数（以 e 为底）
naturally ventilated psychrometer 自然通风干湿表
natural number 自然数【数】
natural orthogonal function 自然正交函数
natural oscillation 固有振动,固有振荡
natural period 自然周期,固有周期
natural polymer 自然高分子【化】
natural radioactivity 天然放射性【物】
natural radioelement 天然放射性元素【化】
natural release 自然释放【地质】
natural remanent magnetization（NRM） 天然剩磁【地】
natural reserve 自然保护区
natural resources 自然资源【地理】
natural resources of groundwater 地下水天然资源【地质】
natural science 自然科学
natural scientist 自然科学家
natural season 自然季节【农】
natural seasonal phenomenon 自然季节现象
natural selection 自然选择【农】
natural siphon rain-gauge 自然虹吸式雨量计
natural source 自然源
natural sulfur cycle 自然硫循环
natural synoptic period 自然天气周期【大气】
natural synoptic region 自然天气区【大气】
natural synoptic season 自然天气季节【大气】
natural variability of climate 气候的自然变率
natural vegetation 自然植被
natural wavelength 固有波长
naulu 瑙卢阵雨(夏威夷)
nautical almanac 航海天文历【航海】
nautical charts and publications 航海图书资料【航海】
nautical history 航海史【航海】
nautical instrument 航海仪器【航海】
nautical meteorology 航海气象【航海】
nautical mile 海里(＝1852 m)
nautical science 航海科学【航海】
nautical service 航海服务【航海】,航海保证【航海】
nautical system 航海制
nautical twilight 航海曙暮光

navaid 导(助)航设备
navaid dropwindsonde 下投式导航测风探空仪
navaid system 导航系统
navaid wind-finding 导航测风【大气】
Naval Environmental Prediction Research Facility(NEPRF) 海军环境预报研究所(美国)
Naval Weather Research Facility(NWRF, US Navy) 〔美国〕海军天气研究中心
Navier-Stokes equation 纳维-斯托克斯方程【数】
navigable semicircle 可航半圆(在北半球热带气旋路径的左边,在南半球热带气旋路径的右边)
navigable waters 通航水域【航海】
navigating in fog 雾中航行【航海】
navigating in narrow channel 狭水道航行【航海】
navigating in rocky water 岛礁区航行【航海】
navigation 航行
navigation aid in wind finding 导航测风(包括 Loran-C Omega 系统)
navigational aid 导航设备
navigational wind 航行风【大气】
navigation channel 航道【海洋】
navigation equipment 导航设备【海洋】
navigation facility 导航设备
navigation radar 导航雷达【电子】
navigation receiver 导航〔信号〕接收机
navigation satellite 导航卫星【电子】
navigation system of synchronous satellite 同步卫星导航系统【电子】
Navigation System Timing And Ranging (NAVSTAR) 时距导航系统【地信】
navigation table 航海表【航海】
NAVSTAR (Navigation System Timing And Ranging) 时距导航系统【地信】
navy oceanographic weather buoy(NOWEB) 海军海洋天气浮标
NCAR(National Center for Atmospheric Research) 国家大气研究中心(美国)
NCEP(National Centers for Environmental Prediction, USA) 美国国家环境预报中心
NCEP/NCAR reanalysis 美国国家环境预报中心/美国国家大气研究中心再分析资料
N-curve N 曲线
NDVI(Normalized Difference Vegetation Index) 归一化差分植被指数,植被指数
Nd:YAG laser Nd:YAG 激光器【物】,掺钕的钇铝石榴石激光器【物】
neap range 小潮差
neap rise 小潮升【航海】
neap tide 小潮【航海】
near-critical period 近关键期
near-equatorial jet 近赤道急流
near-equatorial ridge 近赤道脊
near-equatorial trough 近赤道槽
near field phenomenon 近场现象
near-field region 近场区【电子】
near gale 7级风【大气】,疾风
near-infrared 近红外【物】
near-infrared radiation 近红外辐射
near-polar-orbiting satellite 近极〔地〕轨道卫星
near-real-time 近实时
nearshore zone 近滨带【海洋】,近海,近岸带
near side of the moon 月球正面【天】
near-stationary magnetopause 近稳磁层顶(指接近稳定的磁层顶)
near surface reference temperature 近海面参考温度
near(或 quasi)polar sun-synchronous orbit 近极地(准极地)太阳同步轨道
nebelwind(= fog wind) 雾风(安底斯山)

nebula 星云
nebular hypothesis 星云假说
nebular theory 星云〔学〕说
NEBUL code NEBUL 电码
nebule 纳布尔(大气不透光度单位)
nebulosus(neb) 薄幕云
necessary and sufficient condition 充要条件【数】
necessary condition 必要条件【数】
NEC(North Equatorial Current) 北赤道海流
needle 冰针
needle ice 针冰
Neel temperature(=Curie temperature) 居里温度
negative 负,阴
negative accumulated temperature 负积温【农】
negative acknowledgement 否认【电子】
negative adsorption 负吸附【化】
negative anomaly 负异常【地】
negative area 负区
negative axis 负轴
negative charge 负电荷
negative cloud-to-ground lightning 负云地闪电(向地输送的负电荷减少)
negative correlation 负相关【化】
negative dissipation 负耗散
negative electricity 负电
negative feedback 负反馈
negative ground flash(=negative cloud-to-ground lightning) 负地闪
negative ion(=anion) 负离子,阴离子
negative ion-electron ratio 负离子-电子比率
negative ion mass spectrum 负离子质谱【化】
negative ion reaction 负离子反应
negative isothermal vorticity advection 负等温涡度平流
negative pole 负极,阴极

negative rain 带负电雨
negative refraction 负折射【大气】
negative surface tension 负表面张力【化】
negative temperature 负温度,零下温度
negative temperature coefficient 负温度系数【化】
negative viscosity 负黏性
negative vorticity advection (NVA) 负涡度平流
negentropy ①负熵②负平均信息量
nemere 尼默风(匈牙利的寒潮风)
Neogene Period 新近纪【地质】,新第三纪
Neogene System 新近系【地质】,新第三系
Neoglacial 新冰川期,新冰河期(距今3000—2000年期间)
neoglaciation 新冰川作用
Neolithic Age 新石器时代
neon 氖
neoprene balloon 氯丁橡胶气球
Neoproterozoic Era 新元古代【地质】
Neoproterozoic Erathem 新元古界【地质】
neper 奈培(衰减单位)
nephanalysis 云〔层〕分析【大气】
neph chart 云层分析图
nephcurve 云系分界线
nepheloid layer 雾状层(美国东岸数千米外发现的层状混浊海水)
nephelometer ①能见度测定表【大气】,浑浊度表②云量计
nephelometry 浊度测定法,测晕度法,测云速和方向法
nephelophyte (吸叶面)露水植物
nepheloscope ①云滴凝结器【大气】②测云器③气温改变指示器(由空气压缩或膨胀造成)
nephograph 摄云仪
nephohypsometer 云高度表
nephology 云学
nephometer 量云器,云量计

nephometry 测云法
nephoscope 测云仪【大气】,反射式测云器,云速计
nephsystem 云系
Neptune 海王星【天】
neritic zone 近海区【海洋】,浅海带【海洋】
Nernst equation 能斯特方程【化】
NESDIS (National Environmental Satellite, Data and Information Service)〔美国〕国家环境卫星、资料和信息局
nested grid 套网格【大气】
nested grid model (NGM) 〔嵌〕套网格模式
nested regional model 嵌套区域模式
nested tropical cyclone model 〔嵌〕套网格热带气旋模式
nesting problem 〔嵌〕套网格问题
net address 网络地址【计】
net balance 净平衡
net basin supply 净流域供水
net citizen 网民【计】
net exchange radiometer 净辐射表
net flow 净流〔量〕
netizen 网民【计】
netnews 网络新闻【计】
net outgoing IR (= net terrestrial radiation) 净射出红外辐射
net primary production 净初级生产量
net pyranometer 净辐射表【大气】,辐射平衡表
net pyrgeometer 大气净辐射表
net pyrradiometer 辐射平衡表
net radiation 净辐射【大气】
net radiation balance 净辐射平衡
net radiometer 净辐射表
net solar radiation 净太阳辐射
net source 净源
net storm rain 净暴雨量
net terrestrial radiation 净地面辐射
network analysis 网络分析

network density 站网密度
networking 连网【计】,联网
network management 网络管理【计】
network news 网络新闻【计】
network of meteorological station 气象台站网【大气】
network (of stations) 气象网,测站网,台站网
network radar 组网〔天气〕雷达
network software 网络软件
network worm 网络蠕虫【计】
Neuhoff diagram 诺霍夫图
Neumann boundary condition 诺伊曼边界条件
Neumann function 诺伊曼函数【数】
Neumayer screen 诺埃麦尔百叶箱
neural expert system 神经专家系统【计】
neural net 神经网络【电子】
neural network 神经网络【电子】
neural network model 神经网络模型【计】
neuron 神经元【计】
neuron network 神经元网络【计】
neutercane 变性热带气旋,中性热带气旋
neutral atmosphere 中性大气
neutral boundary layer 中性边界层
neutral cyclone 中性气旋,变性气旋
neutral diffusion coefficient 中性扩散系数
neutral drag instability 中性拖曳不稳定性
neutral energy balance 中性能量平衡
neutral equilibrium 中性平衡
neutral gas equilibrium 中性气体平衡
neutral ionization drift 中性电离漂浮
neutrality condition 中性条件,中性状态(复数)
neutralization 中和【化】
neutral line 中〔性〕线【物】
neutral mode 中性模
neutral occluded front 中性锢囚锋【大气】
neutral occlusion 中性锢囚

neutral oscillation 中性振荡
neutral particle 中性粒子,中性质点
neutral plane 中性平面
neutral point 中性点
neutral sheet 中性片
neutral soil 中性土【土壤】
neutral stability 中性稳定【大气】
neutral surfaces 中性面
neutral temperature 中和温度
neutral wave 中性波
neutral wind 中性风
neutrino 中微子【物】
neutron 中子【物】
neutron activation 中子激活
neutron activation analysis 中子活化分析【电子】
neutron logging 中子测井【地】
neutron moisture gauge 中子含水量测定器
neutron moisture meter 中子〔土壤〕含水量仪
neutron-scattering method 中子散射法
neutron soil moisture meter 中子土壤水分测定仪
neutron thermalization soil probe 中子〔热能〕慢化土壤湿度探头
neutron transport theory 中子输送理论
neutropause 中性层顶
neutrosphere 中性层(电离层底,约75 km以下的大气层)
nevada 涅瓦达风(一种冷风,西班牙语名)
neve(= firn) 永久冰雪,冰川雪(法语名)
neve frozen soil 永冻土
neve line 永久雪线
neve penitent 永久积雪
new candle 新烛光(旧的光照强度单位)
New Guinea Coastal Current 新几内亚海岸流
New Guinea Coastal Undercurrent 新几内亚海岸潜流

new ice 新〔成〕冰
Newman barometer 纽曼气压表
New Mixico Solar Energy Institute (NMSEI) 新墨西哥州太阳能研究所(美国)
new moon 朔【天】,新月
new snow 新雪(雪晶形状可辨认或24 h内新降的雪)
New Stone Age 新石器时代
newton (N) 牛〔顿〕(力的单位)
Newtonian cooling 牛顿冷却
Newtonian coordinates 牛顿坐标
Newtonian flow 牛顿流动【化】
Newtonian fluid 牛顿流体
Newtonian friction law 牛顿摩擦定律
Newtonian mechanics 牛顿力学
Newtonian reference frame 牛顿参考系
Newtonian shear viscosity 牛顿剪切黏度【化】
Newtonian speed of sound 牛顿声速
Newtonian viscosity 牛顿黏度【化】
Newton interpolation formula 牛顿插值公式【数】
Newton's formula for the stress 牛顿应力公式
Newton's law of cooling 牛顿冷却定律
Newton's laws of motion 牛顿运动定律
New Zealand Geophysical Society (NZGS) 新西兰地球物理学会
New Zealand Meteorological Service (NZMS) 新西兰气象局
NEXRAD(Next Generation Weather Radar) ①新一代〔多普勒〕气象雷达②下一代天气雷达计划的缩写词③WSR-88D天气雷达(波长10.5 cm,S频段)的俗称(美国国家天气局的业务运行雷达)
next generation upper air system 下一代高空探测系统
NGO(Non-Governmental Organization) 非政府组织
nieve penitente 融凝冰柱

night air-glow spectrum 夜辉光谱
night/day optical survey of lightning 日夜闪电光学测量
night-dew 夜露
night flight 夜间飞行
nightglow 夜间气辉,夜气辉
night jet 极夜急流
night-sky light 夜天光【大气】
night-sky luminescence 夜天发光
night sky radiation 夜天辐射
nighttime 夜间
nighttime ionosphere 夜间电离层
nighttime observations 夜间观测
night twilight 暮光
night visibility 夜间能见度
night visual range 夜间能见距离
night wind 夜风(即山风或岸风)
Nilsson rain-gauge 尼尔森雨量器
nimbostratus(Ns) 雨层云【大气】
NIMBUS 雨云卫星【大气】(美国实验气象卫星名)
nimbus(Nb) 雨云
nimbus-cumuliformis 积云状雨云
Nimbus Data Handling Facility(NDHF) 雨云卫星资料处理中心
NIMBUS E microwave spectrometer (NEMS) 《雨云-E》卫星微波波谱仪
Nimbus meteorological satellite 雨云气象卫星
nimbus radar 测雨雷达
nine-light indicator 九灯风向风速仪
Nipher shield 奈弗防护罩(雨量器用)
Nippon Electric Company(NEC) 日本电气公司
nirta 尼尔塔风(北苏门答腊托巴湖的风)
nitrate 硝酸盐
nitrate ion 硝酸根离子
nitrate radical 硝酸根
nitration 硝化【化】
nitric acid 硝酸

nitric acid trihydrate(NAT) 硝酸三水合物(极地平流层云内可能有此成分)
nitric oxide(NO) 一氧化氮【大气】
nitrification 硝化作用
nitrile 腈【化】
nitrile oxide 氧化腈【化】
nitrite-nitrogen 硝态氮【土壤】
nitrogen(N) 氮(气)
nitrogen cycle 氮循环
nitrogen dioxide(NO_2) 二氧化氮【大气】
nitrogen fixation 固氮作用【土壤】,固氮(作用)【化】
nitrogen-fixing plants 固氮植物
nitrogen oxides(NO_x) 氮氧化物(N_2O,NO,NO_2等)
nitrogen pentoxide 五氧化二氮(N_2O_5),过氧化氮
nitrophile 喜氮植物,适氮植物
nitrous acid 亚硝酸
nitrous oxide(N_2O) 氧化亚氮,一氧化二氮
nival 多雪的,积雪地区的
nival belt (=snow belt) 雪带
nival climate 冰雪气候【大气】
nivation (=snow erosion) ①雪蚀②霜蚀
nivometer 雪量器
NMHCs(nonmethane hydrocarbons) 非甲烷烃,非甲烷碳氢化物
NMI (normal mode initialization) 正规模(态)初值化【大气】
NOAA ①诺阿卫星【大气】(美国极轨气象业务卫星名)②美国国家海洋大气局
NOAA Data Buoy Center(NDBC) 〔美国〕国家海洋大气局数据浮标中心
NOAA Data Buoy Office(NDBO) 〔美国〕国家海洋大气局数据浮标处
NOAA satellite 诺阿卫星【大气】
NOAA weather wire system (NWWS) 〔美国〕国家海洋大气局有线气象通信系统

Noah's Ark 诺亚方舟卷云(辐辏状卷云条)
Noah's Deluge 诺亚洪水
noble gas 惰性气体
noctilucent cloud 夜光云【大气】
nocturnal boundary layer 夜间边界层
nocturnal convection 夜间对流
nocturnal cooling 夜间冷却
nocturnal drainage wind 夜间下泄风
nocturnal inversion 夜间逆温
nocturnal jet 夜间急流【大气】
nocturnal minimum temperature 夜间最低气温
nocturnal radiation 夜间辐射【大气】
nocturnal thunderstorm 夜间雷暴
nodal factors 节点因子
nodal increment ①截距②交点经度差
nodal line 交点线【天】,波节线
nodal longitudinal distance 交点经距
nodal period 交点周期
nodal regression rate 交点西退率
node ①节点【物】②交点【天】,结点
nodical month 交点月
no-frost zone 无霜区
noise 噪声【物】
noise background 噪声本底
noise control smoother 噪声控制平滑器,噪声控制平滑算子
noise equivalent radiance (NER) 噪声等效辐射率
noise equivalent radiance difference 噪声等效辐射差
noise equivalent temperature difference 噪声等效温度差
noise factor 噪声系数【电子】
noise figure 噪声指数
noise filtering 噪声过滤
noise index 噪声指数
noise level 噪声级
noise level of atmospherics 天电干扰高度
noise pollution 噪声污染
noise power 噪声功率
noise ratio 噪声比
noise reduction 减噪,噪声降低
noise spectrum 噪声谱
noise suppression 噪声抑制
noise temperature 噪声温度
noise threshold 噪声阈值
nomenclature 术语
nominal discharge current 标称放电电流
nominal voltage 标称电压【电子】
nomogram 列线图【大气】
nomograph 列线图
nomographic chart 列线图
nomography 列线图解法
nonadditivity 非〔叠〕加性
nonadditivity correction 非加性修正【物】
non-adiabatic cooling 非绝热冷却
non-adiabatic irreversible process 非绝热不可逆过程
non-adiabatic process 非绝热过程
non-adjustable siphon mercury barometer 不可调虹吸式水银气压表
non-coherence 非相干
non-coherent echo 非相干回波
noncoherent radar 非相干雷达
noncoherent target 非相干目标
non-conductor 非导体
nonconservation of momentum 动量非守恒性
nonconservative property 非保守性
nonconservative scattering 非守恒散射
nonconstant 非常数
non-contaminated atmosphere 未污染的大气
non-contributing area 不产流区,不产流面积
non-convective precipitation 非对流性降水【大气】
non-conventional observation 非常规观

测【大气】,非常规观测资料(复数)
non-deterministic 非确定性的,不确定〔的〕
non-deviative absorption 非偏移吸收【地】
non-dimensional coordinate 无量纲坐标
non-dimensional equation 无量纲方程【大气】,无因次方程
non-dimenstional number 无量纲数
nondimensionalization 无量纲化【物】
non-dimensional parameter 无量纲参数【大气】,无因次参数
non-dimensional quantity 无量纲量
non-dimensional spectrum 无量纲谱
nondimensional variable 无因次变量
nondispersed infrared analyzer 非色散红外分析仪
nondispersive analysis 非色散分析【化】
nondispersive infrared absorption 非色散红外吸收
nondispersive infrared specrtometry 非色散红外光谱法
non-divergence level 无辐散层【大气】,无辐散高度
non-divergent flow 无辐散流
nondivergent model 无辐散模式
nondivergent motion 无辐散运动【大气】
non-divergent wind(component) 无辐散风〔分量〕
nonelectrolyte 非电介质
non-electrostatic field 非静电场
nonequilibrium state 非平衡态【物】
nonequilibrium system 非平衡系统【化】
non-equilibrium thermodynamics 非平衡热力学
non-essential amino acid 非必需氨基酸【生化】
non-frontal squall line 无锋飑线
nonfrost period 无霜期

non-grey atmosphere 非灰大气【天】
Non-Governmental Organization(NGO) 非政府组织
nonharmonic tidal analysis 非调和潮汐分析
nonhomogeneous atmosphere 非均质大气
nonhomogeneous path 非均匀路径
nonhomogeneous terrain 非均匀地形
nonhydrostatic model 非流体静力模式
nonhydrostatic primitive equations 非流体静力原始方程组
nonhydrostatic wave regime 非流体静力波体系
non-hygroscopic nuclei 非吸湿性核
noninductive charging mechanism 非感应起电机制
non-irrigation period 非灌溉期间
nonisotropic turbulence 非各向同性湍流
non-lift balloon 无升力气球
nonlinear 非线性
nonlinear advection equation 非线性平流方程
nonlinear balance equation 非线性平衡方程
nonlinear computational instability 非线性计算不稳定
nonlinear equation 非线性方程
nonlinear instabililty 非线性不稳定〔性〕【大气】
nonlinear interaction 非线性交互作用
nonlinear iterative method 非线性迭代法
nonlinearity of the equation of state 状态方程的非线性关系
nonlinearity 非线性化
nonlinear mapping 非线性映射
nonlinear normal mode 非线性正交模
nonlinear normal-mode initializaton 非线性正交模初值化
nonlincar quasi-geostrophic flow 非线性

准地转流
nonlinear regression 非线性回归【数】
nonlinear saturation 非线性饱和
nonlinear Schrodinger equation 非线性薛定谔方程【物】
nonlinear trajectory 非线性轨迹
nonlinear viscoelasticity 非线性黏弹性【化】
nonlinear wave 非线性波
nonlinear wave-wave interactions 非线性波-波交互作用
nonlocal closure 非局地闭合
nonlocal flux 非局地通量
nonlocal mixing 非局地混合
nonlocal response 非局域响应【物】
nonlocal static stability 非局域静力稳定度
non-local thermodynamic equilibrium 非局部热动平衡【天】
nonmethane hydrocarbons（NMHCs） 非甲烷烃,非甲烷碳氢化物
non-Newtonian flow 非牛顿流动【化】
non-Newtonian reference frame 非牛顿参考系
nonnormal distribution 非正态分布【化】
non-parametric statistics 非参数统计
nonperiodic flow 非周期流
non-periodic variation 非周期变化
nonpolarized light 非偏振光【物】
non-real time 非实时
non-real time data 非实时资料【大气】
non-real-time data processing 非实时资料处理
non-recording rain gauge 雨量筒,非自记雨量器
non-representative 无代表性的
non-resettable current limiting 非恢复限流
nonreversibility 不可逆性
nonrotational field 无旋场

non-saturated air 未饱和空气
non-scattering atmosphere 无散射大气【大气】
nonseparable problem 不可分问题
non-squall cluster 非飑线云团
non-stable star 不稳定星【天】
non-stationary evolution 非稳定演变
nonsupercell tornado 非超级单体龙卷
non-synoptic data 非天气资料【大气】
non-synoptic observation 非定时天气观测
nonthermal radiation 非热辐射【天】
nontornadic storm 非龙卷风暴
nontranspiring canopy 无蒸腾森林覆盖
nonuniformity 不均匀性
non-viscous fluid 非黏性流体
nonvolcanic geothermal region 非火山地热区【地】
nonwetting liquid 不浸润流体
noon 中午,正午
Nordenskjold line 诺登舍尔德线,诺氏线
nor'easter 东北大风
NOR gate 或非门【计】
norm ①标准值②范数【数】③范【数】
normal ①正常〔的〕,标准〔的〕,正态〔的〕②垂直,正交,法线【数】
normal acceleration 法向加速度【物】
normal aeration 常态通气(土壤空气)
normal-air cabin 增压舱
normal atmosphere 正常大气
normal barometer 标准气压表【大气】
normal chart 标准图
normal circular distribution 标准圆分布
normal component 法向分量
normal curve 正态曲线【农】
normal curve of error 误差正态曲线
normal dispersion ①正常色散【物】②正态扩散,正频散【地】
normal distribution 正态分布【数】
normal equation 正规方程【数】,法方程

normal factor　正常因子
normal flow year　平水年【大气】,中水年,一般年
normal frequency　简正频率【物】,正态频率
normal function　正规函数【数】
normal galaxy　正常星系【天】
normal gradient　正常梯度,法〔线方〕向梯度
normal gravity　正常重力
normal gravity potential　正常重力位【地】
normal incidence　正入射【物】,垂直入射
normal incidence pyrheliometer(NIP)　垂直入射绝对日射表
normal isopleth　平均等值线
normalization　归一化【物】,正态化【化】,归一法【化】
normalization law of errors　误差正态〔分布〕律
normalization map　标准图
normalized autocorrelation　归一化自相关
normalized cross-correlation　归一化互相关
Normalized Difference Vegetation Index(NDVI)　归一化差分植被指数(常简称"植被指数")
normalized echo intensity　归一化回波强度【大气】
normalized height　归一化高度
normalized logarithmic spectrum　归一化对数谱
normalized power spectrum　归一化功率谱
normalized quadrature spectrum　归一化求积谱
normal law of errors　误差正态定律
normal map(= normal chart)　标准图
normal mode　正规模〔态〕【大气】,正交波

normal mode equation　正交模方程
normal mode initialization(NMI)　正规模〔态〕初值化【大气】
normal mode solution　正交模解
normal-plate amenometer　垂板风速表
normal population　正态总体
normal projection　正轴投影
normals　标准平均值,多年平均值
normal spiral galaxy　正常漩涡星系【天】
normal state　正常态【物】
normal stress　法向应力【物】
normal temperature and pressure(NTP)　标准温压($T=273.15$ K, $P=1013.25$ hPa)
normal value　正常值
normal water　标准海水
normal wind　正常风
Normand's theorem　诺曼德定理
noroet　诺罗埃风(英吉利海峡及比斯开湾的强北风或西北飑)
Norta　强北风(夏季智利沿岸)
nortada　北风(菲律宾)
norte　北风(①西班牙冬天的北风②墨西哥湾及中美洲的强冷北风)
north(N)　北
North American anticyclone　北美反气旋
North American high　北美高压
North American Plate　北美洲板块【海洋】
North Atlantic Current　北大西洋海流
North Atlantic Deep Water　北大西洋深层水【海洋】(1000~4000 m 深度)
North Atlantic drift　北大西洋漂流
North Atlantic high　北大西洋高压
North Atlantic Oscillation(NAO)　北大西洋涛动
North Atlantic Stream　北大西洋海流
North Atlantic Treaty Organization(NATO)　北大西洋公约组织
northbound　上轨道(对北半球而言)
North Brazil Current　北巴西海流

North China occluded front 华北锢囚锋【大气】
northeast (NE) 东北
Northeast China low 东北低压【大气】
northeaster 东北大风
northeastern storm 东北风暴
northeast monsoon 东北季风
northeast storm 东北风暴
northeast trade 东北信风
North Equatorial Countercurrent (NECC) 北赤道逆流
North Equatorial Current (NEC) 北赤道海流
norther 强北风(特别指美国南部的强冷风)
Northern Annular Mode (NAM) 北半球环状模
northern branch jet stream 北支急流【大气】
Northern Hemisphere 北半球
Northern Hemisphere data tabulations (NHDT) 北半球资料表
Northern Hemisphere Exchange Centre (NHEC) 北半球交换中心
northern latitude 北纬
northern lights 北极光
northern nanny 北方乳母狂风(英国一种带飑和雹的寒冷北风)
northern summer 北半球夏季
northern vernal equinox 春分,春分点
northern winter 北半球冬季
north foehn 北焚风
north frigid zone 北寒带,北极带
North Korean Current 北朝鲜海流
north magnetic pole 磁北极【地】
north northeast (NNE) 北东北(东北偏北)
north northwest (NNW) 北西北(西北偏北)
North Pacific Current 北太平洋海流
North Pacific Drift 北太平洋漂流

North Pacific Oscillation (NPO) 北太平洋涛动【大气】
north point 北方,北点
north polar region 北极区
North Pole 北极【地理】
North Polar Water 北极水【海洋】
north star 北极星
north temperate zone 北温带
north tropic 北热带
north tropic zone 北热带
northwest (NW) 西北
northwester(=nor'wester) 西北大风
Northwest Pacific anticyclone 西北太平洋反气旋
Northwest Pacific high 西北太平洋高压
Norwegian Current 挪威〔暖〕海流
Norwegian cyclone model 挪威气旋模式
Norwegian Polar Institute (NPI) 挪威极地研究所
Norwegian school 挪威学派
Norwegian Sea Deep Water 挪威海深层水
nor'wester 西北大风
no-slip boundary condition 无滑动边界条件
no-slip condition 无滑动条件
no swell 无涌【海洋】
NOT gate 非门【计】
notos 南风(希腊语名)
nova 新星【天】
nowaki 台风(日本古雅称)
nowcast 临近预报【大气】(常指未来数小时内的天气预报,一般是0~3 h,但0~6 h的天气预报也用此词)
nowcasting 临近预报,现时预报
NO_x (nitrogen oxides) 氮氧化物
Noxon cliff 诺克松峭壁(高层大气中发现冬季和早春时活性氮总气柱量锐减的现象)
NO_y 总活性氮(total reactive nitrogen)的符号标记

NPO (North Pacific Oscillation) 北太平洋涛动【大气】
NPOESS (National Polar-orbiting Operational Environmental Satellite System) 国家极轨业务环境卫星系统(美国)
NRM (natural remanent magnetization) 天然剩磁【地】
NRM wind scale NRM(北落基山脉)风级(风力和风速的对应关系同蒲福风级)
Ns(nimbostratus) 雨层云【大气】
NSCAT NASA 散射仪(NASA scatterometer 的缩写)
N.T.P (normal temperature and pressure) 标准温度和气压的缩写
nucleant(=nucleating agent) 成核剂
nuclear dust 核子尘
nuclear fuel 核燃料【化】
nuclear geochemistry 核地球化学【化】
nuclear magnetic resonance 核磁共振【化】
nuclear winter 核冬天【大气】
nucleation 核化【大气】,成核作用
nucleation threshold 成核温阈
nuclei counter 计核器
nucleon 核子【物】
nuclepore filter 核孔滤膜
nucleus 核,核心
nucleus counter 计核器,核子计数器
nudging 张弛逼近,纳近〔法〕,拉近〔法〕(一种四维资料同化方法)
nuée ardente 炽热火山云(法语名)
null hypothesis 解消假设
null layer (=zero layer) 零层
null voltage 零位电压【电子】
number concentration 数浓度
number density 数密度
number distribution function 数量分布函数【化】
number of range samples 距离采样数【大气】

number of vibration 振动数
numerical analysis 数值分析【数】
numerical calculation 数值计算【数】
numerical climatic classification 数值气候分类【大气】
numerical computation 数值计算【数】
numerical dispersion 数值分散
numerical experiment 数值实验【大气】
numerical forecast 数值预报
numerical forecasting 数值预报
numerical instability 数值不稳定
numerical integration 数值积分【大气】
numerical method 数值方法【数】
numerical model 数值模式
numerical modeling 数值模拟〔方法〕
numerical quadrature 数值求积
numerical scheme 数值方案
numerical simulation 数值模拟【大气】
numerical simulation experiment 数值模拟实验
numerical solution 数值解【数】
numerical stability 数值稳定性【数】
numreical step by step method 数值的逐步逼近法
Numerical Weather Center 数值天气中心
numerical weather forecast 数值天气预报
numerical weather prediction (NWP) 数值天气预报【大气】
N-unit N 单位
Nusselt number 努塞特数【航空】
nutation 章动【天】
nutation sensor 章动传感器
nutrient preserving capability 保肥性【土壤】
nutrient supplying capability 供肥性【土壤】
NVA (negative vorticity advection) 负涡度平流
N weather 仪表飞行天气(instrument weather)的缩写(罕用)
NWP (numerical weather prediction)

数值天气预报【大气】
n-year event   n年一遇事件
nylon unit   尼龙测湿元件（测土壤含水量）
Nyquist constraint   奈奎斯特约束
Nyquist criterion   奈奎斯特判据【电子】
Nyquist frequency   奈奎斯特频率
Nyquist interval   奈奎斯特间距
Nyquist sampling frequency   奈奎斯特采样频率【计】
Nyquist velocity   奈奎斯特速度
Nyquist wave number   奈奎斯特波数

O

oasis   绿洲（沙漠中的）
oasis cultivation   绿洲耕作【地理】
oasis effect   绿洲效应【大气】
ob   气象学中对天气观测的常用缩写
oberwind   奥伯风（奥地利萨尔茨卡默古特的风）
objective analysis   客观分析【大气】
objective forecast   客观预报【大气】
objective function   目标函数【数】
object space scanner   目标空间扫描器
oblateness   扁率
oblateness of the earth   地球扁率【天】
oblate spheroid   扁球
oblate symmetric top   扁对称陀螺
oblique Cartesian coordinates   斜角笛卡儿坐标
oblique coordinates   斜坐标【数】
oblique equidistant cylindrical projection   斜角等距圆柱投影
oblique incidence   斜射
oblique Mercator projection   斜角墨卡托投影
oblique projection   斜轴投影
oblique visibility   斜能见度
oblique visual range   斜视程
obliquity of the ecliptic   黄赤交角【天】
O'Brien cubic polynomial   奥布赖恩三次多项式
obscuration   ①朦胧〔天空〕②掩星
obscured sky cover   天空不明，朦胧天空
obscuring phenomenon   朦胧现象
observation   ①观测②〔观测〕记录③观察【物】
observational astrophysics   实测天体物理学【天】
observational data   观测记录，观测数据
observational day   观测日
observational error   观测误差【大气】，测量误差
observational evidence   观测证据
observational fact   观测事实
observational frequency   观测次数【大气】，观测频率
observational network   测站网
observational plot   观测小区
observational program   观测项目
observational result   观测结果
observational section   观测地段
observation equation   观测方程
observation landplane   陆地观测飞机
observation seaplane（OSP）   海上观测飞机
observation site   观测场【大气】
observation study   观测研究
observation systems simulation experiment   观测系统模拟实验
observation well   ①观测井②验潮井
observatory   气象台，观象台，观测台，天文台
observatory-remote   遥测台

observed altitude 观测高度【航海】
observed latitude 观测纬度【航海】
observed longitude 观测经度【航海】
observed value 观测值【化】,测定值
observing systems simulation experiment (OSSE) 观测系统模拟实验
observer 气象观测员【大气】
obstacle flow signature 障碍气流征兆（多普勒雷达观测中）
obstruction 障碍物
obstruction to vision 能见度变坏（大气现象所造成）,视程障碍
obtuse triangle 钝角三角形【数】
Obukhov length 奥布霍夫长度
Obukhov's criterion 奥布霍夫判据
occluded cyclone 锢囚气旋【大气】
occluded depression 锢囚低压
occluded front 锢囚锋【大气】
occlusion 锢囚
occlusion front 锢囚锋
occultation 掩【天】（一个天体被另一个角直径较大的天体（如月球或行星）所掩蔽的现象）
occultation method 掩星法【大气】
ocean 洋,海洋
ocean air(mass) 海洋气团,海洋空气
ocean-atmosphere coupled model 海洋大气耦合模式
ocean-atmosphere coupling 海洋大气耦合
ocean-atmosphere heat exchange 海洋大气热交换
ocean-atmosphere interaction 海洋大气交互作用
ocean-atmosphere model 海洋大气模式
ocean basin 海盆,洋盆
ocean bottom 海底,洋底
ocean circulation 海洋环流
ocean climate 海洋性气候
ocean color 海色
ocean color imager(OCI) 海色成像器

ocean conveyor belt 海洋输送带
ocean current 海流【航海】,洋流【大气】
ocean current chart 洋流图【航海】
ocean data buoy(ODB) 海洋资料浮标
ocean data station(ODS) 海洋资料站
ocean data transmitter(ODT) 海洋资料发送器
ocean deep 海渊
ocean drift 海洋漂流
ocean energy conversion 海洋能转换【海洋】
ocean energy resources 海洋能源【海洋】
ocean floor 海底
ocean floor drilling(OFD) 海底钻井
ocean front 海洋锋
ocean gyre 海洋环流单体
ocean heat transport 海洋热量输送
ocean surveillance system(OSS) 海洋监视系统
oceanic anticyclone 海洋反气旋
oceanic basin 洋盆【地理】
oceanic climate 海洋性气候
oceanic crust 洋壳【海洋】
oceanic front 海洋锋【海洋】
oceanic general circulation 海洋环流
oceanic general circulation model (OGCM) 海洋环流模式
oceanic heat transport 海洋热量输送
oceanic hemisphere 水半球
oceanic high 海洋高压
oceanicity (见 oceanity) 海洋性,海洋度
oceanic low 海洋低压
oceanic meteorology 海洋气象〔学〕
oceanic meteorological research 海洋气象研究
oceanic mixed layer 海洋混合层
oceanic moderate 海洋性温和〔气候〕
oceanic noise 海洋噪声,海鸣
oceanic optical remote sensing 海洋光学遥感【海洋】
oceanic plate 大洋板块【海洋】

oceanic ridge 海岭【地理】
oceanic surface mixed layer 海面混合层
oceanic tide 海洋潮
oceanic troposphere 大洋对流层【海洋】
oceanic turbulence 海洋湍流
oceanity(=oceanicity) 海洋度【大气】,海洋性
ocean mixed layer 海洋混合层
ocean mixing 海洋混合
ocean monitoring ship 海洋监测船【航海】
ocean navigation 大洋航行【航海】
ocean observation technology 海洋观测技术【海洋】
oceanographic equator 海洋学赤道(表面最高水温带)
oceanographic facility(OF) 海洋调查设备
oceanographic investigation 海洋调查【海洋】
oceanographic research buoy(ORB) 海洋研究浮标
oceanographic research equipment(ORE) 海洋研究设备
oceanographic research ship(ORS) 海洋调查船
oceanographic (research) vessel 海洋调查船
oceanographic survey 海洋调查【海洋】
oceanographic tower 海洋观测塔
oceanographic tracer 海洋示踪物
oceanographic winch 海洋观测绞车
oceanography 海洋地理学
oceanology 海洋学【海洋】
oceanophysics 海洋物理学
ocean optics 海洋光学【海洋】
Oceanospace Explorer(OSPER) 海洋空间探索计划
ocean sciences 海洋科学【海洋】
ocean space robot(OSR) 海上空间探测装置
ocean station 海洋观测站
ocean station vessel(OSV) 海洋天气船

ocean surveillance system(OSS) 海洋监测系统
ocean swell 涌浪
ocean temperature 海〔洋〕温〔度〕
ocean thermal energy 海洋温差能【海洋】
Ocean Topography Experiment (TOPEX) 海面形状试验计划(美法合作,1992年发射的海神卫星探测)
ocean turbulence 海洋湍流
ocean wave 海浪
ocean wave spectrum 海浪谱
ocean weather report 海洋气象报告【航海】
ocean weather ship 海洋天气船
ocean weather station 海洋气象站【大气】
ocean weather vessel 海洋天气船【航海】
ocean wind 海洋风
octa 八分之一(测云量用)
octagonal grid 八角形网格
octal system 八进制【计】
octant ①八分仪②卦限,八分之一(地面,天空等)
odd chlorine 奇氯〔族〕(包括 Cl, ClO, HCl 等)
odd function 奇函数【数】
odd hydrogen 奇氢〔族〕(包括 H, OH, HO_2 等)
odd nitrogen 奇氮〔族〕(包括 N, NO, NO_2 等)
odd oxygen 奇氧(O 和 O_3 浓度之和)
odd-oxygen system 奇氧系(主要包含 O 和 O_3)
ODF (ordinary dynamical forecast) 常规动力预报
ODS(ozone depleth substance) 臭氧耗减物质
oe 屋伊旋风(丹麦法罗群岛上)
OECD(Organization for Economic Cooperation and Development) 经济合作与发展组织

oersted 奥斯特(磁场强度单位)
off-airways 偏离航路
off-axis angle 偏轴角【物】
off-center PPI scope 偏心平面位置显示器(一种陈旧的雷达显示)
Office of Coastal Environment (OCE) 海岸环境局
off-ice wind 下冰风(从冰上向下吹的风)
official forecast 官方预报
off-level 非规定层上的
off-line 脱机【计】,离线
off-line analysis 离线分析【物】
off-line processing 脱机处理【计】
off-sea wind 向岸风
off-shore buoy observing equipment(OBOE) 近海浮标观测系统
off-shore current 远岸流
offshore navigation 近海航行【航海】
off-shore platform 海上平台
offshore survey 近海测量【航海】
off-shore wind 离岸风【大气】
offtime 非〔规〕定时
offtime report 非定时报〔告〕
off-year 小年
Ogasawara high 小笠原〔副热带〕高压
OGCM(oceanic general circulation model) 海洋环流模式
ogive 累积频率图
ogive curve 累积频率曲线,肩形曲线
ohmic current 欧姆电流
ohm(Ω) 欧〔姆〕(电阻单位)
oileic acid 油酸【生化】
oil field 油田【地质】
oilgotrophication 贫营养化【土壤】
oil pool 油藏【地质】
oil shale 油页岩【地质】
oil slick 浮油〔膜〕,〔海上〕油斑,油膜
Okhotsk blocking anticyclone 鄂霍次克阻塞反气旋
Okhotsk blocking high 鄂霍次克阻塞高压
Okhotsk high 鄂霍次克海高压【大气】
okta 八分之一(云量单位)
Older Dryas 较老仙女木事件
Oldest Dryas 老仙女木事件
old ice 陈冰,多年冰,多年雪
old snow 陈雪
old wives' summer 秋热,秋老虎(欧洲9月末、10月初的晴暖期)
olefine 烯烃
Oligocene Epoch 渐新世【地质】
Oligocene Series 渐新统【地质】
oligomer 寡聚体【生化】
oligopeptide 寡肽【生化】
Olland cycle 奥兰转换器
Olland principle 奥兰原则
OLR (outgoing long-wave radiation) 向外长波辐射【大气】,射出长波辐射,外逸长波辐射
ombrograph 〔微〕雨量计
ombrology 测雨学
ombrometer 微雨量器
ombrophile 喜雨植物
ombrophobe 厌雨植物
ombrophyte 喜阴植物
ombroscope 测雨器,报雨器
omega block Ω型阻塞
Omega equation (或 ω equation) 奥米伽方程
omega momentum flux 奥米伽动量通量
omega navigation system 奥米伽导航系统
Omega navigator 奥米伽导航仪【航海】
Omega Position Location Experiment 奥米伽定位试验
Omegasonde 奥米伽〔导航〕探测仪
Omega windfinding 奥米伽导航测风
omnidirectional 全向的,不定向的
omnidirectional air meter 全向气流表
omnidirectional antenna 全向天线【电子】

omnidirectional radiometer　全向辐射仪
on and off instruments　目视间或仪表飞行
on-board atmosphere　机内大气，舱内大气
on-board infrared Fourier spectrometer　舱内红外傅里叶光谱仪
on-board processing　舱上处理
ondometer　测波仪
one-and-a-half order closure　一阶半闭合
one-cell magnetometer　单元〔铷蒸气〕磁强计
one-dimensional cloud probe　一维云探测器
one-dimensional model　一维模式
one-parameter family　单参数族
one-sided difference　单侧差分
one-sided smoothing　单边光滑
one-tier model　一阶段模式
one-way approach　单向方法
one-way atmospheric transmittance　单程大气透射比
one-way attenuation　单程衰减【大气】
one-way entrainment process　单向夹卷过程
one-way grid nesting　单向〔嵌〕套网格
one-way nested model　单向嵌套模式
one-way interaction　单向影响
on-ice wind　上冰风（向冰上吹的风）
on instruments　仪表飞行
onion peeling method　逐层剥皮法
onlevel　规定层上的
on-line　联机【计】，在线
on-line analytical processing　联机分析处理【计】
on/off rain sensor　有无降雨传感器
onset　爆发，开始
onset of Meiyu　入梅【大气】
onset of monsoon　季风开始
onset vortex　〔季风〕爆发涡旋
on-shore wind　向岸风【大气】，离海风

ontime　〔规〕定时
on top　云上〔飞行〕
on-year　大年
opacity　不透明性，不透明度
opacus（op）　蔽光云
opalescence　乳白光
opalescence turbidity　乳白浑浊度
opalescent turbidity　乳白〔的〕浑浊度
opaque　不透明〔的〕
opaque approximation　不透明近似
opaque atmosphere　不透明大气
opaque ice　不透明冰
opaque layer　浑浊层，不透明层
opaque reflector　不透明反射体
opaque sky cover　蔽光云量
OPEC(Organization of the Petrolem Exporting Countries, Vienna, Austria)　石油输出国组织（奥地利维也纳）
open boundary　开边界
open boundary condition　开边界条件
open-cell cumulus　开口细胞状积云
open cellular convection　开口胞状对流
open channel flow　明渠流
open (cloud) cells　开口型细胞状云【大气】，开口云胞
open curve　开展曲线
opening angle　张角
open jet wind tunnel　开口式风洞
open ocean mining (OOM)　大洋采矿
open scale barograph　微〔气〕压计
open sea　①不冻海②公海
open system　开放系统【计】
open water　开阔水面，无冰水面
open wave　开放波
open-wire resistance thermometer　明线电阻温度表
operating sequence diagram　操作顺序图
operating system　操作系统【计】
operation　操作，运算，控制，运转，作业
operational forecast　业务预报【大气】
Operational Linescan System(OLS)

业务线扫描系统(美国国防气象卫星上装载的可见和红外成像仪)
operational meteorological condition 作战气象条件
operational model timing 业务模式(资料同化)时限
operational numerical model 业务数值模式
operational prediction 业务预报
operational research 运筹学【数】
operational risk 作业风险
Operational Software for Surface Meteorological Observation (OSSMO) 地面气象测报业务软件
operational use 业务使用
operational weather limits 飞行天气低限
operation center 控制中心,操作中心,作业中心
operation code 操作码
operation control key (OCK) 运算控制键
operation directive 操作指令
operation engaged in hydrogen 涉氢作业
operation order 操作,顺序,运算指令
operations research 运筹学【数】
operative temperature 操作温度
operator ①操作员②算子
opposing wind (= head wind) 逆风,顶头风
opposition 冲【天】
oppressive weather 闷热天气
optical active substance 旋光物质【物】
optical activity 光学活性【化】,旋光性【物】
optical air mass 大气光学质量
optical alignment 光学校直【物】
optical anemometer 光学风速仪
optical array precipitation spectrometer probe 光阵滴谱仪探头
optical astronomy 光学天文学【天】
optical axis 光轴【物】

optical black lacquer 光学黑色涂料
optical cavity 光学谐振腔【电子】
optical communication 光〔学〕通信【物】
optical component 光学元件
optical constant 光学常数【物】
optical correlation instrument 光学相关仪器
optical counter 光学计数器
optical crosswind profiler 光学横向风速廓线仪,光学侧风廓线仪
optical density 光〔学〕密度
optical density of a cloud 云中光学密度
optical depth 光学厚度【大气】
optical depth factor 光学厚度因子
optical disc 光碟【计】,光盘【物】
optical disc drive 光碟驱动器【计】
optical distance 光程
optical fiber 光纤【电子】
optical fiber cable 光缆【电子】
optical fiber communication 光纤通信【电子】
optical flowmeter 光学流量表(测云中含水量)
optical gate 光闸【物】
optical haze 光学霾,光霾
optical heterodyne 光〔学〕外差【物】
optical heterodyne radar 光外差雷达
optical horizon 光学地平
optical hygrometer 光学湿度表
optical image processing 光学图像处理【地理】
optical imaging probe 光学成像探测器
optically active gas 光学活性气体
optically denser medium 光密介质【物】
optically effective atmosphere 光学有效大气
optically homogeneous 光学均质的
optically-pumped laser 光泵激光器,光抽运激光器
optically smooth 光学平滑的

optically thick medium　光厚介质【天】
optically thin medium　光薄介质【天】
optically thinner medium　光疏介质【物】
optical mass　光学质量
optical meteor　大气光象
optical mixer　光学混频器
optical mode　光学模【物】
optical model　光学模型【物】
optical modulation　光调制【物】
optical observation　光学观测
optical ozonesonde system　光学臭氧探空仪系统
optical parametric oscillator（OPO）光学参量振荡器
optical particle probe　光学粒子探测器
optical path　光学路径,光程【物】
optical pathlength　光程长〔度〕
optical phenomenon　光学现象
optical principle　光学原理
optical probing of the atmosphere　大气光学探测
optical propagation　光传播
optical pulse propagation　光脉冲传播
optical pumping　光抽运【物】
optical pumping magne tometer　光泵磁强计
optical pyrometer　光测高温表
optical quenching　光猝灭【物】
optical rain gauge　光学雨量计
optical refractive index　光学折射率
optical remote sensing　光学遥感
optical rotation　旋光性【物】
optical scattering　光散射
optical scattering probe　光散射探测器（测量粒子或粒子群的前向散射）
optical shutter　光闸【物】
optical sizing device　光学粒子尺度测定装置（测气溶胶用）
optical slant range　光学斜距
optical snow depthmeter　雪深光学测量仪

optical-theodolite　光学经纬仪
optical thickness　光学厚度【物】
optical transfer function　光学传递函数
Optical Transient Detector（OTD）　光学瞬变探测器（早期装载于卫星上测闪电的仪器）
optical visibility　光学能见度
optical window　光学窗
optics　光学【物】
optimal climate normals　最优气候值,最优气候均态
optimal control　最优控制【航海】
optimal estimate　最优估计【化】
optimal network　最佳观测网
optimal perturbation　最优扰动
optimal temperature　最适温度【农】
optimal value　最优值【化】
optimal yield　最优出水量,最优产水量
optimization　最优化【数】
optimum climate　最佳气候
optimum control　最优控制【航海】
optimum decision　最优决策
optimum filter　最佳滤波
optimum flight　最佳飞行
optimum humidity　最佳湿度
optimum initialization　最佳初值化
optimum interpolation　最优内插〔法〕
optimum interpolation method　最优插值法【大气】
optimum interpolation quality control　最优内插质量控制
optimum observing system　最佳观测系统
optimum overrelaxation parameter　最佳超松弛参数
optimum resolution（OR）　最佳分辨率
optimum route　最佳航线【大气】
optimum solution　最优解【物】
optimum speed　最佳航速【航海】
optimum strategy　最优策略
optimum temperature　最适温度【大气】

optimum track line 最佳航线
optimum track route 最佳航线
optimum water need 最适需水
opto-acoustic detection 光声探测
ora (见 Aura)奥拉风(意大利加尔达湖的谷风)
orbit 轨道
orbital characteristics 地球轨道特征
orbital eccentricity 轨道偏心率
orbital eccentricity variation 轨道偏心率变化
orbital electron 轨道电子【物】
orbital element 轨道要素
orbital frequency 轨〔道〕频〔数〕
orbital inclination 轨道倾角
orbital latitude 轨道纬度
orbital longitude 轨道经度
orbital motion 轨道运动
orbital parameter 轨道参数
orbital plane 轨道面
orbital velocity 轨道速度
orbiting astronomical observatory(OAO) 轨道天文台【天】(1966—1972 年美国发射的 2 颗卫星)
orbiting geophysical observatory(OGO) 轨道地球物理观测站
orbiting solar observatory(OSO) 轨道太阳观测站【天】(1962—1975 年间美国发射的一组 8 枚研究太阳的卫星)
orbit number 轨道编号数
orbit plane 轨道平面
orchard heater 果园防霜炉
ordering relationship 量级关系
orderly human activities 有序人类活动
order of accuracy 准确度级
order of discontinuity 不连续阶
order of magnitude 数量级【数】,大小级次
order parameter 序参数
order statistics 顺序统计学
ordinal 序数【数】

ordinary agricultural meteorological station 一般农业气象站
ordinary average 普通平均
ordinary cell 寻常胞
ordinary climatological station 一般气候站【大气】
ordinary differential equation 常微分方程
ordinary dynamical forecast(ODF) 常规动力预报
ordinary light 寻常光【物】
ordinary radiation station 普通辐射站
ordinary ray 寻常光线,寻常射线
ordinary wave 寻常波【物】,常波
ordinary year 常年
ordinate 纵坐标
Ordovician Period 奥陶纪【地质】
organic acids 有机酸
organic aerosol 有机气溶胶
organic carbon 有机碳
organic chemistry 有机化学【化】
organic fertilizer 有机肥料【土壤】
organic geochemistry 有机地球化学【地】
organic nitrates 有机硝酸盐(通式 $RONO_2$)
organic peroxides 有机过氧化物(通式 ROOR 或 ROOH)
organic pollutant 有机污染物
Organization for Economic Cooperation and Development(OECD) 经济合作与发展组织
organized convection 有组织对流
organized large eddies 有组织大涡旋
organized magnetic field 规则磁场
OR gate 或门【计】
orientation ①取向②方位
orientational disorder 取向无序【化】
orientation fall 定向下落
orientation fluctuation 取向涨落【物】
orientation of ice crystal 冰晶取向
oriented overgrowth (=epitaxis)

晶体定向生长,外延生长
origin ①原点(坐标的)②来源,源地
original data 原始资料
original error 固有误差,原始误差
origin of atmosphere 大气起源
origin of coordinates 坐标原点
origin of element 元素起源【地质】
origin of species 物种起源【植】
origin time 起始时间
orogenesis 造山作用【地质】
orogenic cycle 造山循环
orogeny 造山运动【地质】
orographic 地形的,山地的
orographic anticyclone 地形反气旋
orographic cloud 地形云
orographic condition 地形条件
orographic depression 地形低压【大气】
orographic downward wind 地形下坡风
orographic drag 地形拖曳【大气】
orographic effect 地形效应
orographic fog 地形雾
orographic forcing 地形作用力,地形强迫〔项〕
orographic frontogenesis 地形锋生【大气】
orographic influence 地形影响
orographic leeside trough 地形背风槽
orographic lifting 地形抬升
orographic low 地形低压
orographic occluded front 地形锢囚锋【大气】
orographic occlusion 地形锢囚
orographic precipitation 地形降水【大气】
orographic rain 地形雨【大气】
orographic rainfall 地形雨量【大气】
orographic snowline 地形雪线【大气】
orographic stationary front 地形静止锋【大气】
orographic storm 地形风暴
orographic thunderstorm 地形雷暴【大气】
orographic uplift 地形抬升

orographic upward wind 地形〔性〕上坡风
orographic vortex 地形涡旋
orographic wave 地形波【大气】
orographic wave drag 地形波曳力
orographic wind flow 地形风气流
orography 山志学,山地形态学
oroshi 下滑风(日语名)
orsure 奥休尔风(法国利翁湾的烈风)
orthogonal 正交
orthogonal antennas 正交天线
orthogonal coefficient 正交系数【农】
orthogonal coordinates 正交坐标【数】
orthogonal curvilinear coordinates 正交曲线坐标【数】
orthogonal experiment 正交试验【农】
orthogonal function 正交函数【大气】
orthogonality 正交性
orthogonalization 正交化【化】
orthogonal lines 正交线
orthogonal polynomials 正交多项式【数】
orthogonal projection 正〔交〕投影【数】
orthogonal transformation 正交变换【数】
orthographic projection 正射投影【测】
orthomorphic map 正形图
orthonormality 标准正交性
orthonormal system 规范正交系【数】
orthophoto 正射像片【测】
orthophotomap 正射影像地图【测】
orthorhombic system 斜方晶体【地质】,正交晶系
oscillating body 振动体
oscillation 振荡【大气】【物】,振动【物】,涛动
oscillation equation 振荡方程
oscillation period 振荡周期
oscillation series 振荡级数
oscillator 振〔动〕子
oscillatory integral 摆动积分

oscillatory wave 振荡波
oscillograph 示波器【物】
oscilloscope 示波器【物】
osmometer 渗压表
osmosis 渗透〔作用〕【化】,渗透【物】
osmotic pressure 渗透压【化】
osmotic suction 渗透吸力【农】
osmotic water 渗透水
oso 熊风(美国加利福尼亚州的强西北风)
OSO (Orbiting Solar Observatory) 轨道太阳观测站【天】
Osos wind 奥索斯风(加利福尼亚的强西北风)
OSSMO(Operational Software for Surface Meteorological Observation) 地面气象测报业务软件
ostria(=auster) 奥斯特风
Ostrovski-Gauss formula 奥-高公式【数】
OTD (optical transient detector) 光学瞬变探测器
ouakhge 乌阿赫吉风(撒哈拉中部的尘暴)
ouari 瓦利风(索马里南风)
oued 枯水河
ounce (OZ) 盎司(常衡1盎司=28.35 g,金衡、药衡1盎司=31.103 g)
outbreak 爆发
outburst 外暴流
outdiffusion coefficient 外扩散系数
outer atmosphere 外大气层(外逸层的同义语)
outer boundary 外边界
outer eyewall 外眼墙
outer iteration 外迭代【数】
outer layer 外层(常指边界层顶层)
outer planet 外行星
outer product 外积〔向量积〕
outer radiation shield 防辐射外罩
outer region 外区

outer scale 外尺度
outer space 外层空间【地】
outer sphere 外层【化】,外界
outer vortex 外〔层〕涡旋
outflow 外流
outflow boundary 外流边界
outflow jet 外〔流〕急流
outflow spiral 外流螺旋(云或气流)
outflux ①出流②流出通量
outgassing 出气【电子】,释气〔作用〕,脱气
outgoing angle 出射角【物】
outgoing long-wave radiation (OLR) 向外长波辐射【大气】,射出长波辐射
outgoing radiation 向外辐射
out-growth 产生,生成
outlet glacier 出流冰川,注出冰川
outlier 离群值【数】
outline map 草图
outlook 展望,前景
outo 奥托风(西班牙类似于法国的autan风)
out-of-phase 异相〔位〕【物】,反相
output 输出【物】
output signal 输出信号
output spooling 脱机输出
oval-shaped belt 〔极光〕卵形带,极光椭圆区
overall dimension 总尺寸,外形(轮廓)尺寸
overall reaction 总反应【化】
over-all rate 优势反应率(大气电离平衡的)
overboat effect 过冲作用
overcast 阴天【大气】,阴
overcast circle 密云圈
overcast day 阴日
overcast sky 阴天
overcompensate 过分补偿
overcooling 过冷
overdamped system 过阻尼系统

overdamping 过阻尼【物】
overdetermined case 超定个例
overdetermined problem 超定问题
overdetermined system 超定组【数】
overdimension 超维
overestimate 估计过高
overfitting 过度拟合
overflow ①溢出②溢出数③上溢
overforecast 过度预报
overhang echo 悬垂回波(常指上升气流区上面的雷达回波)
overheat 过热
overheating layer 过热层
overland flow 地面径流,地表漫流
overlap 重叠,交叠,搭接部分
overlap integral 重叠积分【化】
overlapping average 滑动平均
overlapping band 重叠带
overlapping mean 滑动平均
overlayer 覆盖层【物】
overnormal wind 超常风
overpredicted 过度预报〔的〕
overreflection 超反射
overrelax 超松弛
overrelaxation 超松弛【数】
overrelaxation coefficient 超松弛系数
overrelaxation parameter 超松弛参数
overrunning 上滑
overrunning cold front 凌驾冷锋,上滑冷锋
over-seeding 过量播撒,过量催化
overshoot ①过辐射②逸出③超出规定④过调节⑤过渡特性上的上冲(峰突)⑥超值指示
overshooting cloud top 上冲云顶
overshooting top 上冲云顶
overshooting top temperature 上冲顶温度
overshooting updraft 冲顶上升气流
oversize 超差,加大(尺码)
overspecification 过定义,超定义
overstability 超稳定度(性)

over-the-nose visibility 飞机前方能见度
overtone 泛音【物】
overtone band 泛频带
overtrades (= Krakatao winds) 高空信风
overtrade wind 高空信风
overturn 翻转
overturning ①翻转,颠倒(常用于空气的迅速对流交换)②高低空风向相反
overweather flight 复杂气象飞行,恶劣天气飞行
overwinter 越冬
overwintering control 越冬防治【农】
overwintering crop 越冬作物【农】
overwrite 冲掉
overyearing 越冬,越年
Owen's dust recorder 欧文计尘器
owl-light 黄昏蒙影,曙暮光
oxidant 氧化剂【大气】
oxidation 氧化〔作用〕
oxidation-reduction indicator 氧化还原指示剂【化】
oxide 氧化物【化】
oxide catalyst 氧化物催化剂【化】
oxides of nitrogen 氮氧化物
oxides of sulfur 硫氧化物
oxidizing agent 氧化剂【化】
oxidizing capacity of atmosphere 大气的氧化能力
oxidizing environment 氧化环境【地】
oxidizing reaction 氧化反应
ox's eye 牛眼飓风(几内亚对飓风的称呼)
oxygen(O) 氧〔气〕
oxygen absorption band 氧吸收带
oxygen absorption spectrum 氧吸收谱
oxygen band 氧分子带($0.13\sim17\mu m$波长范围内对紫外辐射有强吸收)
oxygen coefficient 含氧系数【地】
oxygen content 氧气含量
oxygen cycle 氧循环
oxygen deficiency 氧亏欠
oxygen effect 氧效应【农】
oxygen-18 ice core record 氧18冰芯纪录

oxygen isotope ratio 氧同位素比
oxygen isotope record 氧同位素纪录
oxygen partial pressure 氧气分压
oxygen saturation 氧饱和
Oyashio 亲潮（千岛冷海流，日语名）
Ozmidov scale 奥斯米多大尺度〔湍流〕
ozone 臭氧【大气】
ozone budget 臭氧收支
ozone cloud 臭氧云
ozone concentration 臭氧浓度
ozone deficiency 臭氧亏欠
ozone degradation 臭氧渐衰〔作用〕
ozone-depleting potential（ODP） 臭氧耗减势
ozone depletion 臭氧耗竭【地】,臭氧耗减
ozone distribution 臭氧分布
ozone-ethylene chemiluminescent reaction 臭氧乙烯化学发光反应
ozone flux 臭氧通量
ozone formation 臭氧形成
ozone generator 臭氧发生器【航海】
ozone heating 臭氧加热
ozone heat source 臭氧热源

ozone hole 臭氧洞
ozone isopleth plot 臭氧等值线图
ozone layer 臭氧层
ozone photochemistry 臭氧光化学
ozone production 臭氧生成
ozone reaction 臭氧反应
ozone recovery 臭氧恢复
ozone shield 臭氧防护层
ozonesonde 臭氧探空仪
ozone spectrophotometer 臭氧分光光度计
ozone-temperature sensor 臭氧-温度传感器
ozonide 臭氧化物【化】
ozonization 臭氧化【化】
ozonogram 臭氧图（电化学探空仪得出的臭氧分布）
ozonograph 臭氧测定仪
ozonolysis 臭氧解【化】
ozonometer 臭氧计【大气】
ozonometry 臭氧测定术
ozonopause 臭氧层顶
ozonoscope 臭氧测量仪
ozonosphere 臭氧层【大气】

P

Pacific and North American（PNA）pattern 太平洋-北美型
Pacific Decadal Oscillation（PDO） 太平洋十年振荡
Pacific Deep Water 太平洋深层水（1000～3000 m 深度）
Pacific Equatorial Undercurrent 太平洋赤道潜流【海洋】,克伦威尔海流
Pacific high 太平洋高压【大气】
Pacific Marine Environmental Laboratory（PMEL） 太平洋海洋环境实验室
Pacific-North American Pattern 太平洋-北美型,PNA 型

Pacific Ocean 太平洋【海洋】
Pacific Plate 太平洋板块【海洋】
Pacific region 太平洋区域
Pacific Standard Time 太平洋标准时间（美国）
Pacific Subarctic Current 太平洋副北极海流
Pacific Time 太平洋时间（美国）
Pacific-type coastline 太平洋型岸线【海洋】
Pacific-type continental margin 太平洋型大陆边缘【海洋】
pack（＝pack ice） 块冰

package ①组装,封装 ②程序包
packed data format 压缩码资料格式
packed tower 填充塔(减少 SO_2 排放)
packet 包络,包裹
packet of electrons 电子包【物】
packet switching technique 包交换技术
pack ice 块冰,大块浮冰
packing 存储,合并,压缩
paesa 佩萨风(意大利加尔达湖的北东北大风)
paesano(见 aura) 奥拉风
page access time 页面存取时间【计】
PAGES(Past Global Changes) 历史全球变化计划(属国际地圈-生物圈研究计划 IGBP)
paging 分页【计】
pagoscope 测霜仪
PAH(polycyclic aromatic hydrocarbon) 多环芳烃
painter 不洁雾(秘鲁沿海的一种雾)
pair coupling 对耦合【物】
pair-distribution function 对分布函数【数】
pair production 对产生【物】
Palaeoclimate Modeling Intercomparison Project(PMIP) 古气候模式比较计划
palaeogeomagnetic equator 古地磁赤道【地】
palaeogeothermics 古地热学【地】
palaeolatitude 古纬度【地】
palaeolongitude 古经度【地】
palaeomagnetic field 古地磁场【地】
palaeomagnetic pole 古地磁极【地】
palaeomagnetism 古地磁〔学〕【地】
pal(a)eontology 古生物学,化石学
pal(a)eotemperature variation 古温〔度〕变化
pal(a)eotemperature analysis 古温〔度〕分析
pal(a)eotemperature curve 古温曲线
pal(a)eotemperature measure 古温测定,古温测量
pal(a)eotemperature record 古温记录
paleoanthropology 古人类学
paleobiochemistry 古生物化学【生化】
paleobiogeography 古生物地理学【地质】,生物古地理学【地质】
paleobiology 古生物学
paleobiometeorology 古生物气象学
paleobotany 古植物学【植】
paleoceanography 古海洋学【海洋】
Paleocene Epoch 古新世【地质】
Paleocene Series 古新统【地质】
paleoclimate 古气候【大气】
paleoclimatic evidence 古气候证据
paleoclimatic indicator 古气候指标
paleoclimatic reconstruction 古气候重建
paleoclimatic record 古气候记录
paleoclimatic sequence 古气候序列
paleoclimatology 古气候学【大气】
paleocrystic 长期冻结
paleocrystic ice 古结晶冰(至少10年以上),陈年海冰
paleocurrent 古海流【海洋】
paleoecology 古生态学【地质】
Paleogene Period 古近纪【地质】
Paleogene System 古近系【地质】
paleogeographic map 古地理图【地质】
paleogeography 古地理学【地理】
paleogeomorphology 古地貌学
paleogeophysics 古地球物理学
paleogeotemperature 古地温【地】
paleoglaciation 古冰川
paleoglaciation record 古冰川记录
paleohydrogeology 古水文地质学【地质】
paleohydrology 古水文学【地理】
paleolatitude 古纬度【地】
paleolith 旧石器时代
paleomagnetic epoch 古地磁时期
paleomagnetics 古地磁学
paleomagnetic time scale 古地磁时间尺度

paleomonsoon 古季风
paleontological record 化石记录
paleopalynology 古孢粉学
paleopedology 古土壤学【土壤】
Paleoproterozoic Era 古元古代【地质】
Paleoproterozoic Erathem 古元古界【地质】
paleosnowline 古雪线
paleosol 古土壤
paleotemperature 古温〔度〕
pal(a)eotemperature analysis 古温〔度〕分析
pal(a)eotemperature curve 古温曲线
pal(a)eotemperature measure 古温测定,古温测量
pal(a)eotemperature record 古温记录
pal(a)eotemperature stratification 古温层次,古温层位
paleotemperature variation 古温变化
Paleozoic Era 古生代【地质】
Paleozoic Erathem 古生界【地质】
paleozoology 古动物学
pallio-nimbus 层状雨云
pallium 层状雨云
Palmer drought severity 帕尔默干旱强度
Palmer Drought Severity Index (PDSI) 帕尔默干旱强度指数
palouser 尘暴(见于加拿大拉布拉多半岛)
Paluch diagram 帕卢奇图(用于研究云过程的大气热力学图)
palynological data 孢粉资料
palynology 孢粉学
pampas 南美大草原
pampero seco 无降水型帕姆佩罗风
pampero sucio 尘沙型帕姆佩罗风
pampero 帕姆佩罗风(南美拉普拉塔河口的寒冷南或西南风飑,冬季7—10月,多伴有大风雷暴)
PAN(peroxy acetyl nitrate) 过氧乙酰硝酸酯

Pan-American monsoon 泛美季风
panas oetara 帕纳索塔拉风(印度尼西亚2月份的强干北风)
pancake ice 饼状冰
panchromatic film 全色片【测】
panchromatic photography 全色摄影
pan coefficient 蒸发皿系数
pan evaporation 蒸发皿蒸发(由A类蒸发皿自由水面测得)
pan-ice(=floe-ice) 浮冰
pannus(pan) 碎片(云)
panorama camera 全景摄影机【测】
panoramic barograph 全量程气压计
panoramic camera 全景摄影机【测】
panoramic hygrograph 全量程湿度计
panoramic photography 全景摄影【测】
panoramic thermograph 全量程温度计
panoramic view 全景【物】
panzeractinometer 铠装日射表(林克-福斯纳温差电偶日射表)
papagaio 帕帕盖欧风(中美西海岸的强冷北风)
papagallos(=hurricane) 飓风(东北太平洋上用)
papagayo(=popogaio) 帕帕加约风(中美尼加拉瓜和危地马拉的太平洋沿岸的强东北风)
paper hygrometer 纸示湿度表
PAR(photosynthetically active radiation) 光合有效辐射【大气】
para-boen 帕拉-布恩风(巴西亚马孙河流域7—10月午后的飑)
parabola 抛物线【数】
parabolic antenna 抛物面天线【电子】
parabolic equation 抛物线方程
parabolic equation method 抛物线方程法
parabolic mirror 抛物柱面镜【物】
parabolic torus antenna 抛物环面天线【电子】
parabolic type 抛物型

parabolic window 抛物线窗
paraboloid 抛物面
paraboloidal mirror 抛物面镜【物】
paraboloidal reflector 抛物面反射器【电子】
parachute radiosonde 下投式探空仪,降落伞探空仪
paradox 佯谬【物】,悖论【数】
paraffins ①链烷烃,石蜡烃②石蜡
parafoveal vision 视外区视像,曙暮视觉
paraglider 滑翔降落伞,滑翔伞(宇宙飞船或火箭重返大气层时可展开以引导飞船降落,或助收回发射中的火箭)
parallactic angle 星位角【天】
parallax 视差【天】
parallax second(=parsec) 秒差距
parallel 平行线,纬圈
parallel band 平行〔谱〕带
parallel-beam approximation 平行光束近似
parallel-beam radiation 平行光束辐射
parallel bedding 平行层理【地质】
parallel computer 并行计算机【计】
parallel computing 并行计算【计】
parallel database 并行数据库【计】
parallelepiped 平行六面体
parallel flux 平行通量
parallel fold 平行褶皱【地质】
parallelogram 平行四边形【数】
parallel processing 并行处理
parallel processor operating system 并行处理机操作系统【计】
parallel shift 平行位移
parallel translation 平行移动【数】
paramagnetism 顺磁性【物】
parameter ①参数②参量
parameter estimation 参数估计【化】
parameterization 参数化【大气】,参量化【物】
parameterized cyclone wave 参数化气旋波

parameterized microphysics 参数化微物理学
parameterized model 参数化模式
parameterized turbulence 参数化湍流
parameterized vorticity flux 参数化涡度通量
parameterized vorticity transfer 参数化涡度输送
parameter test 参数检验【化】
parametric amplifier 参数放大器【电子】
parametric equation 参数方程
parametric model 参数模式
parametric oscillator 参量振荡器
paramito 帕拉密托风(哥伦比亚波哥大冬季7—9月冷东风)
paramp 参量放大器
paranthelion (复数为 paranthelia)远幻日【大气】,远假日
parantiselena 远幻月【大气】,远假月
paraselene 假月【大气】,幻月
paraselenic circle 幻月环【大气】
parasitic echo 寄生回波【电子】
parasitic fold 寄生褶皱【地质】
parasitic frequency 寄生频率
parasitic wave 寄生波
parcel 气块
parcel method 气块法
parcel theory 气块理论
parenchyma 薄壁组织
parent cloud 母云
parent material 母体物质
Pareto optimality 帕雷托最优性【数】
parhelic circle 幻日环【大气】(一种晕象)
parhelic ring 幻日环
parhelion(复数为 parhelia) 假日【大气】,幻日
parity 奇偶性【数】
parity check 奇偶校验【数】
parity conservation 宇称守恒【物】

parrey (=perry) 佩里飑
parry arc 内晕珥
parsec(=parallax second, pc) 秒差距【天】(天文距离单位,相当 3.258 光年)
Parscval cquality 帕塞瓦尔等式【数】
Parseval's theorem 帕塞瓦尔定理
part cloudy 少云
partial coherence 部分相干
partial correlation 偏相关
partial correlation coefficient 偏相关系数【化】
partial derviative 偏导数【数】,偏微商
partial differential coefficient 偏微分系数
partial differential equation 偏微分方程【数】
partial differential equation of higher order 高阶偏微分方程【数】
partial drought 轻旱
partial-duration series 部分历时序列
partial eclipse 偏食【天】
partial frequency 部分频率
partially penetrating well 不完全贯穿井,不完全渗水井,半透水井
partially polarized wave 部分偏振波
partial obscuration 部分天空不明
partial potential temperature 分位温
partial pressure 分压〖力〗
partial regression 偏回归【农】
partial ring 〔部分〕欠缺年轮
partial sensitivity 偏敏感性
partial tide 分潮
partial vacuum 未尽真空,部分真空
particle 质点【物】,粒子【物】
particle absorption 粒子吸收
particle camera 粒子摄影机
particle charge 粒子电荷
particle derivative 个别导数
particle detector 粒子探测器
particle flux 粒子通量

particle monitor 粒子监测仪
particle radiation 粒子辐射
particle size distribution 粒径分布,粒子谱
particle velocity 粒子速度
particular solution 特解【数】
particulate 微粒,悬浮微粒,粒子
particulate loading ①悬浮微粒②悬浮微粒的浓度(或含量)
particulate matter 颗粒物质
particulate pollutant 微粒污染物
partition function 分拆函数【数】,配分函数【物】
partition of energy 能量分配
partly cloudy 少云【大气】;部分有云(在美国常指云量在 0.3~0.6,也有时指 0.4~0.7,不严格)
partly occluded 部分锢囚〔的〕
parts per billion (ppb) 十亿分率(10^{-9})
parts per million (ppm) 百万分率(10^{-6})
parts per million by volume (ppmv) 体积百万分率(10^{-6}体积分数)
parts per trillion (ppt) 万亿分率(10^{-12})
Pascal (Pa) 帕〔斯卡〕(压力单位,$1\ Pa = 1\ N/m^2$)
Pascal law 帕斯卡定律【物】
Pasquill-Gifford stability category 帕斯奎尔-吉福德稳定度分类
Pasquill stability classes 帕斯奎尔稳定度分类
passage ①过境②星下点轨迹
pass-band 通带
passing shower 过境阵雨
passive anafront 被动上滑锋
passive cavity aerosol spectrometer probe 被动腔气溶胶分光计探测
passive cloud 被动〔隔断〕云
passive cloud collector 被动式云〔水〕收集器

passive front(= inactive front) 不活跃锋,被动锋
passive heterodyne radiometer 被动外差辐射计
passive katafront 被动下滑锋
passive microwave remote sensing 被动微波遥感
passive microwave sensor 被动微波传感器
passive permafrost 原永冻层
passive radiometer 无源辐射计
passive (remote sensing) system 被动〔遥感〕系统
passive remote sensing technique 被动遥感技术【大气】
passive satellite 无源卫星
passive steering 被动引导
passive technique 被动技术
pastagram 温高图
Past Global Changes(PAGES, IGBP) 历史全球变化计划(IGBP)
past hour 过去1小时
pastoral region 牧业区【地理】
past six hours 过去6小时
past three hours 过去3小时
past weather 过去天气【大气】
patch 碎片,斑点,斑纹
patchy turbulence 断续湍流,修补湍流,补丁湍流
patent ①专利,专利权②专利品
path ①路径【物】②路程【物】
path length 程长【物】,路径长度
path line 迹线【物】,路径〔曲〕线
path method 路径法
path of mixing 混合路径
pathological biometeorology 病理生物气象学
path radiance 路径辐射亮度
pathway 迹线
patrol landplane 陆地侦察机
patrol seaplane 海上侦察机
pattern 型式,分布,形势

pattern correlation 图型相关
patterned ground 图式地面,图形土
pattern recognition 图形识别,流型辨识法
pattern recognition technique 图形识别技术【大气】
Patterson barometer 帕特森〔水银〕气压表
Paulin aneroid 波林式空盒气压表
Pavlovsky's approximation 巴浦洛夫斯基近似(流体)
payload 净荷【计】,有效载荷,负载
PBL (planetary boundary layer) 行星边界层【大气】
PCA(polar cap absorption) 极盖吸收【航海】
PCA events 极盖吸收事件
PCBs(polychlorinated biphenyls) 多氯联苯
PCM(pulse code modulation) 脉码调制【电子】
P-coordinate P坐标【大气】
P-σ coordinate system P-σ坐标系
PDO (Pacific Decadal Oscillation) 太平洋十年振荡
PDSI (Palmer Drought Severity Index) 帕尔默干旱强度指数
peak ①峰【地理】②极大值,峰值
peak acceleration 峰值加速度【地】
peak current 峰值电流(闪电)
peak discharge 洪峰流量,最大流量
peak displacement 峰值位移【地】
peak flow method 洪峰水流法
peak frequency 峰值〔所在的〕频率
peak gust 最大阵风
peak power 峰值功率
peak value 峰值
peak velocity 峰值速度【地】
pearl lightning(= beaded lightning) 串珠状闪电【大气】,珠状闪电
pearl-necklace lightning 串珠状闪电【大气】,珠状闪电
Pearson distribution 皮尔逊分布

Pearson type distribution 皮尔逊型分布【数】
pea soup fog 浓雾(俗名)
Peclet number 佩克莱特数(雷诺数与普朗特数之积)
peculiar velocity 本动速度
pedestal (天线)基座
pedestal cloud 基柱〔状〕云
pedogeography 土壤地理学【地理】
pedohydrology 土壤水文学【地理】
pedology 发生土壤学【土壤】
pedosphere 土壤圈【地理】
peesash(= peshash, pisachee, pisachi) 皮萨希风(印度挟带尘埃的风)
peesweep storm (亦称 peasweep, peesweip, peewit, teuchit, swallow storm) 皮斯威普风暴(英国的早春风暴)
Peffer osmometer 佩弗渗压计
P-E index (precipitation-effectiveness index) 降水-有效性指数
Pelean cloud (见 nuée ardente) 炽热火山云
pellet 冰丸
pellicular water 薄膜水,附着水
Peltier effect 佩尔捷效应【物】,珀尔帖效应(旧译)
Peltier element 佩尔捷效应元件
penalty function 补偿函数
pendant cloud 漏斗云
pendule 摆涡(尾迹消失后形成的环状涡旋)
pendulum anemometer 摆式风速表
pendulum day 摆日(富科摆旋转一周的时间)
peneplain 准平原【地理】
penetation depth 渗透深度【农】,穿透深度【物】
penetrating radiation 穿透辐射
penetrating top 穿透云顶
penetration range 夜间视距
penetrative convection 穿透对流【大气】,贯透对流
penetrative downdraft 贯透下沉气流
penetrometer 硬度测量计,钻进深度自记仪
peninsula 半岛【地理】
penitent ice 融凝冰柱
penitent snow 融凝雪
penknife ice 刀〔状〕冰
Penman-Monteith method 彭曼-蒙蒂思法,P-M 法(确定水汽输送率或潜热通量的方法)
pennant 风三角【大气】(风速填图符号,代表 20 m/s 风速)
Penn State/NCAR Mesoscale Model (或 PSU/NCAR Mesoscale Model) 宾州〔大学〕/美国国家大气研究中心中尺度模式
pentad 候【大气】,五天
pentagon 五边形【数】
Pentagon 五角大楼(美国国防部)
penthemeron 候,五天
penumbra 〔黑子〕半影
PE (permanent echo) 固定回波(雷达中)
peptide 肽【生化】
per annum 每年(拉丁语)
P-E ratio(precipitation-evaporation ratio) 降水-蒸发比
per capita GDP 人均 GDP(国内生产总值)
percent 百分数,百分比(%)
percentage 百分法【数】,百分率
percentage error 百分误差【数】
percentage of possible sunshine 可照百分率
percentage of precipitation anomalies 降水〔量〕距平百分率
percentage of sunshine 日照百分率【大气】
percentile 百分位点
percent reduction 百分递减,还原率,衰减百分率

perched groundwater 栖留水,静滞地下水
perched stream 栖留河,滞水河
percolation 渗透作用【大气】
percolation rate 渗漏率【农】
percolation water 渗漏水【农】
per diem 每日(拉丁语名)
pereletok 多年冻层
perennial drought 多年性干旱
perennially frozen ground(＝permofrost) 永冻地
perennial plant 多年生植物【植】
perennial stream 常流河
perfect black body 理想黑体
perfect fluid 理想流体
perfect forecast 理想预报
perfect gas(＝ideal gas) 完全气体【物】,理想气体
perfect-gas law 完全气体定律,理想气体定律
perfectly diffuse radiator 理想漫射辐射体
perfect observation 理想观测
perfect prediction(PP) 完全预报【大气】,理想预报
perfectprog method(PPM) 完全预报法
perfect prognosis(PP) 完全预报法
perfect prognosis method 完全预报法,理想预报法
perfect prognostic 完全预报
perfect radiation 完全辐射,理想辐射
perfect radiator 完全辐射体,理想辐射体
perfect reflection wall 理想反射壁
perfluorocarbons(PFCs) 全氟碳化物(如 CF_4、C_2F_6 等)
performance index 性能指数
pergelation 永冻作用
pergelisol 永冻土【大气】,多年冻土
pergelisol table 永冻面
perhumid climate 过湿气候

pericenter 近心点
pericyclonic ring 气旋周环
perigean range 近地点潮幅,近地点潮差
pcrigean tide 近月潮
perigee ①近地点【大气】②〔弹道〕最低点
periglacial 冰川边缘
periglacial climate 冰缘气候
periglacial stage 冰缘期
perihelion ①近日点【大气】②〔弹道〕最高点
period 周期【大气】,时期
period average(或 mean) 长期平均【大气】(10 年以上)
period doubling 周期加倍(倍周期)
period doubling bifurcation 倍周期分岔【物】
periodic attractor 周期吸引子
periodic boundary condition 周期边界条件
periodic current 周期〔性海〕流
periodic function 周期函数【数】
periodicity 周期性【物】
periodic law of elements 元素周期律【化】
periodic motion 周期运动
periodic orbit 周期轨道【天】
periodic oscillation 周期振动,周期振荡
periodic signal 周期信号
periodic table of elements 元素周期表【化】
periodic variation 周期变化
periodic wind 周期风
periodization 周期化【数】
period mean 长期平均
period of drought 干旱周期
period of growth 生长〔时〕期
period of light variation 光变周期【天】
period of oscillation 振动周期
period of record 记录时期,记录年限

period of validity 预报时效
periodogram 周期图【大气】
periodogram analysis 周期图分析【数】
periodogram method 周期图法
periodogram vector 周期图矢量
period spectrum 周期谱
peripheral circulation 外周环流
peripheral device 外围设备【计】,外部设备,外设
peripheral equipment 外围设备【计】,外部设备,外设
peripherals 外围设备【计】,外部设备,外设
peripherization 边缘化【地理】
periphery 边缘地【地理】
perlucidus(pe) 漏隙(云)
permafrost 永冻土【大气】,永冻,常冻
permafrost aggradation 冻土进化【地理】
permafrost degradation 冻土退化【地理】
permafrost dynamics 冻土动力学【地理】
permafrost island 孤立永冻区
permafrost table 永冻面
permanent anticyclone 永久性反气旋【大气】
permanent aurora 夜气辉
permanent circulation 常定环流
permanent control 稳定控制(河道测量)
permanent current 永久〔性海〕流
permanent depression 永久性低压【大气】,常定低压
permanent echo 恒定回波
permanent gas 恒定气体
permanent high 永久性高压【大气】,常定高压
permanent low 常定低压
permanently frozen ground 永冻地
permanent magnet 永磁体【物】
permanent magnetization 永久磁化
permanent register 永久记录簿
Permanent Service for Mean Sea Level (PSMSL) 平均海平面资料常设局(英国)
permanent thermocline 永久性温跃层【海洋】
permanent wave 恒定波
permanent wilting point 永枯点
permanent wind 恒定风
permeability ①渗透性②磁导率
permeability coefficient 渗透〔性〕系数
per mensem 每月(拉丁语名)
Permian Period 二叠纪【地质】
Permian System 二叠系【地质】
per mille 千分率,千分比(‰)
permittivity 介电常数,电容率
Permo-Carboniferous era 石炭二叠纪
permutation 排列【数】,置换【数】
peroxide 过氧化物【化】
peroxy acetyl nitrate(PAN) 过氧乙酰硝酸酯
peroxy radical 过氧自由基
perpendicular ①垂直的【数】②垂线【数】
perpendicular band 垂直〔谱〕带
perpendicular incidence 直射
perpetual frost climate 永冻气候亚类【大气】,永冻气候
perpetual snow 积雪
perry(=parrey, parry, pirrie, pirry) 佩里飑(英国,常伴有大雨)
Persian Gulf Water 波斯湾海水
persistence 持续性【大气】
persistence forecast 持续性预报【大气】
persistence length 相关长度【化】
persistence organic pollutants (POPs) 持久性有机污染物,难降解有机污染物(如滴滴涕、六氯苯、多氯联苯、二噁英等)
persistence tendency 持续性趋势【大气】
persistency climatology 持续性气候学
persistent anomaly 持久性距平
persistent oscillation 持续振荡

personal and instrument equation 人仪差【测】
personal computer（PC） 个人计算机【计】
personal equation 人〔为误〕差（由个人造成的系统观测误差）
personal error 人为误差
personnel engaged in hydrogen 涉氢人员
perspective grid 透视网格
perspective projection 透视投影
Pers sunshine recorder 佩斯日照计,佩斯日照记录器
perturbation ①摄动【物】【大气】②微扰【物】
perturbation equation 扰动方程【大气】
perturbation method 小扰动法【大气】
perturbation motion 扰动运动,微扰动
perturbation pressure gradient force （PPGF） 扰动气压梯度力
perturbation quantity 扰动量
perturbation solution 扰动解
perturbation technique 微扰技术,微扰法
perturbation theory 微扰论【物】,微扰理论【化】
perturbed dimension 扰动尺寸【化】
perturbed ensemble forecast 扰动集合预报
perturbed model 扰动模式
Peru/Chile Current 秘鲁/智利海流
Peru Current 秘鲁〔冷〕海流
Peruvian paint 秘鲁色雾
peshash(＝peesash) 皮萨希风
peta- 拍〔它〕(词头,10^{15})
petit St. Bernard 小圣伯纳风(法国的一种山风)
petrochemical plant 石化工厂
Petterssen development equation 彼得森发展方程
Petterssen development theory 彼得森发展理论
PFCs（perfluorocarbons） 全氟碳化物
pH 酸碱度,pH值(表示氢离子浓度倒数的对数,$-\log[H^+]$)
Phanerozoic Eon 显生宙【地质】
phase ①相〔位〕【物】,位相②相【大气】③时段
phase angle 相角【物】
phase array 相控阵【电子】
phase averaging 相位平均
phase change 相变【大气】
phase coding 相位编码【航海】
phase coherence 相位相干
phase comparison 比相【航海】
phase constant 相位常量【物】,相常数
phase correlation function 相位相关函数
phased-array antenna 相控阵天线
phase delay 相位延迟【物】
phase diagram 相图
phase difference 相〔位〕差【物】
phase distortion 相畸变
phase equilibrium 相平衡【化】
phase fluctuation 相位涨落【物】,位相起伏
phase frequency spectrum 相位频率谱
phase front 波阵面,相前
phase function 相函数【大气】
phase instability 相态不稳定
phase lag 相位滞后
phase-lag spectrum 相位滞后谱
phase-locking 锁相【物】
phase matching 相位匹配【物】
phase matrix 相矩阵
phase modulation 调相【物】
phase of moon 月象,盈亏
phase parameter 相参数
phase path 相程
phase plane 相平面
phase portrait 相图
phase response 相位响应

phase rule　相律
phase screen theory　相位屏理论
phase sequence　相序【航海】
phase shift　相移【电子】
phase-shift keying　相移键控【电子】
phase space　相空间【物】
phase spectrum　位相谱【大气】
phase speed　相速
phase structure function　相位结构函数
phase time modulation　调相
phase transformation　态变,相变
phase transition　相变【物】
phase velocity　相速【物】
phase voltage　相电压【物】
phenodate　物候日【大气】
phenogram　物候图【大气】
phenological calendar　物候历
phenological chart　物候学图
phenological division　物候分区【大气】
phenological event　物候事件
phenological observation　物候观测【大气】
phenological phase　物候期
phenological relation　物候关系
phenological season　物候季节
phenological simulation　物候模拟
phenological simulation model　物候模拟模式
phenological spectrum　物候谱
phenology　物候学【大气】
phenology law　物候学定律【农】
phenomenal sea　骇浪,汹涛
phenomenological law　现象律
phenomenon　现象
phenometry　物候测定学
phenophase　物候期【大气】
phenoseason　物候季
phenospectrum　物候谱【大气】
Philippine Atmospheric, Geophysical and Astronomical Services Administration (PAGASA)　菲律宾大气、地球物理和天文服务管理局

Philippines Current　菲律宾海流
phloroglucinol　间苯三酚
pH meter　pH 计【化】,酸度计
phosphorescence　磷光
phot　辐透,厘米烛光(旧的照度单位)
photic zone　透光层
photoabsorption　光吸收
photoabsorption cross section　光吸收截面
photocathode　光阴极
photo-cell　①光电池【物】②光电管【物】
photochemical air pollution　光化学空气污染
photochemical dissociation　光化离解
photochemical equilibrium　光化平衡
photochemical model　光化模式
photochemical fog　光化学雾【化】
photochemical oxidant　光化〔学〕氧化剂
photochemical pollutant　光化〔学〕污染物
photochemical pollution　光化学污染【大气】
photochemical process　光化学过程【化】
photochemical reaction　光化学反应【大气】
photochemical smog　光化学烟雾【大气】
photochemistry　光化学【大气】
photochrome　彩色照片
photochromic glass　光致变色玻璃
photocoagulator　光致凝结器,光焊接器
photocolorimetry　光比色法
photoconductive cell　光敏电阻,光电导管,光电导电池
photoconductive detector　光〔电〕导探测器
photoconductive soild-state detector　光导型固态探测器
photoconductor　光电导体【物】,光敏电阻
photocounting　光子计数【物】

photodecomposition 光解【化】
photodegradation 光降解【化】
photodetachment 光致分离【物】
photodetection 光〔电〕探测【物】
photodetector 光探测器
photo-digitizing system 图像数字仪【天】
photodiode 光电二极管【物】
photodissociation 光解离【化】,光解作用
photoelectric astrolabe 光电等高仪【天】
photoelectric cell 光电池【物】
photoelectric effect 光电效应【物】
photoelectric hygrometer 光电湿度表
photoelectric photometer 光电光度计
photoelectric photometry 光电测光术
photoelectric transmittancemeter 光电透射表
photoelectrochemistry 光电化学【化】
photoelectron 光电子【物】
photoelectron collision 光电子碰撞
photoelectron energy spectrum 光电子能谱
photoelectron escape altitude 光电子外逸高度
photoelectron escape flux 光电子外逸通量
photoelectronic imaging 光电成像【天】
photofax 图片传真〔机〕
photogrammeter 摄影经纬仪
photogrammetry 摄影测量学【测】
photographic barograph 摄影气压计
photographic meteor 照相流星
photographic photometry 照相测光【天】
photographic projection 照相投影
photographic sunshine recorder 摄影日照计
photoionization 光致电离【物】
photoionization cross section 光致电离截面
photolabile 对光不稳定〔的〕
photology 光学
photolysis 光解作用【物】
photometeor 大气光学现象
photometer 光度计【物】
photometric observation 光度观测
photometric system 测光系统【天】
photometry 光度学【物】
photomicroscope 摄影显微镜
photomixing 光混频
photomultiplier 光电倍增管【物】
photon 光子【物】
photon activation analysis 光子活化分析【化】
photon counter 光子计数器
photon-counting 光子计数【天】
photon distribution function 光子分布函数
photon flow rate 光子流速率
photon flux 光子通量【物】
photon integrating nephelometer 光子积分浑浊度表
photon rocket 光子火箭
photooxidation 光氧化【化】,感光氧化作用
photoperiod 光周期【农】
photoperiodism 光周期〔性〕【大气】
photophase 光照阶段【大气】
photopic vision 眼窝正视
photopolarimeter 光偏振表
photo-recombination 光致复合
photoresistor 光敏电阻
photo-respiration 光呼吸【植】
photosensitive 光敏的
photosensitivity 光敏性【物】
photosensitization 光敏化【化】,光敏作用
photosphere 光球【天】
photostage 光照阶段【农】
photostationary state relation 光稳态关

系(决定对流层中 NO 和 NO_2 浓度比的关系式)
photosynthesis 光合作用【植】
photosynthetically active radiation(PAR) 光合有效辐射【大气】
photosynthetic efficiency 光合效率【农】
photosynthetic intensity 光合强度【农】
phototaxis 趋光性【农】
photovoltaic detector 光电探测器
photovoltaic device 光伏器件【电子】
photovoltaic effect 光生伏打效应【物】,光伏效应【电子】
photoxide 光氧化物
phreatic cycle 地下水涨落循环
phreatic line 地下水位线,渗流线
phreatic surface 地下水面
phreatic water 无压地下水,潜水
phreatic zone 地下水层,饱和层
phreatophyte 深根吸水植物
phugoid frequency 长周期振动频率,摆动频率
pH value pH 值【大气】,酸碱值
physical and chemical properties 物理化学特性
physical chemistry 物理化学【化】
physical climate 物理气候【大气】
physical climatology 物理气候学【大气】
physical constraint 物理限制条件
physical environment 自然环境
physical equation 物理方程
physical forecasting 物理预报
physical geochemistry 物理地球化学【地】
physical geodesy 物理大地测量学【测】,大地重力学
physical geography 自然地理学【地理】
physical geology 普通地质学【地质】
physical hydrodynamics 物理流体力学
physical initialization 物理初始化

physical limit check 物理合理性检验
physical mechanism 物理机理
physical memory 物理存储器
physical meteorology 物理气象学【大气】
physical mode 物理模〔态〕【人气】
Physical Oceanographic Real-Time System (PORTS) 物理海洋实时系统
physical oceanography 物理海洋学【海洋】
physical optics 物理光学【物】
physical weathering 物理风化【土壤】,物理风化〔作用〕【地理】
physico-statistic prediction 物理统计预报
physics 物理〔学〕【物】
physics of the atmosphere 大气物理学
physiography 地文学【地理】
physiological climatology 生理气候学
physiological disease 生理病害【农】
physiological drought 生理干旱【土壤】
physiological maturity 生理成熟【农】
phytochemistry 植物化学【植】
phytoclimate 植物〔小〕气候【大气】
phytoclimatology 植物气候学【大气】
phytoecology 植物生态学【植】
phytogeography 植物地理学【植】
phytological biometeorology 植物生物气象学
phytometer 蒸腾表
phytoplankton 〔海洋〕浮游植物
phytotoxic 毒害植物的,植物毒性的
phytotron 人工气候室【大气】,育苗室
pibal ①气球测风【大气】②测风气球
Piche evaporimeter 毕歇蒸发表
pickup 拾波器
pick-up coil 拾波线图【物】
pico- 皮〔可〕(词头,10^{-12})
picture 图像,照片
picture element 像元【物】,像素,像点
picture in picture (PIP) 画中画【计】

picture mosaic 拼图【大气】
piecewise approximation method 逐段逼近法
piecewise-uniform medium 分层均匀介质
pie chart 饼形图【计】
piedmont glacier 山麓冰川
piedmont plain 山麓平原【地理】
Pierson-Moskowitz spectrum 皮尔逊-莫斯科维茨海浪谱,P-M 海浪谱
pieze 厘巴,毕西(压强单位,＝1000 Pa)
piezocoefficient 压性系数
piezocrystal 压电晶体【物】
piezoelectric ceramic 压电陶瓷【电子】
piezoelectric crystal 压电晶体【物】
piezoelectric effect 压电效应【物】
piezoelectric sorption hygrometer 压电吸附式湿度表
piezometer 压强计【物】,测压计
piezometric head 测压水头,测压管水头
piezotropic 压性〔的〕
piezometric surface 测压管液面
piezotropic equation 压性方程
piezotropy 压性
pile-up process(＝banking process) 堆积过程,增水过程
pileus(pil) 幞状云
pilmer 皮尔默大阵雨(英国)
pilot balloon 测风气球
pilot balloon ascent 气球施放
pilot balloon observation 测风气球观测【大气】
pilot-balloon plotting board 测风绘图板【大气】
pilot balloon self-recording theodolite 测风气球自记经纬仪
pilot balloon slide-rule 气球高空风计算尺
pilot balloon station 气球测风站
pilot balloon theodolite 测风经纬仪
pilot briefing 飞行员天气简报
pilot chart 领航图
pilot meteorological report 飞行员气象报告【大气】
pilot report 飞行员天气报告
pilot streamer 先导流光(闪电)
pilot study 先期研究,预研究
pilot tube 测风管
Pinatubo, Mt. 皮纳图博火山(1991年曾喷发)
pinched lightning 钳〔状〕闪〔电〕
pinching frost 严寒,严霜
pinene 蒎烯($C_{10}H_{16}$)
piner 苦干风(英国北或东北强风)
pingo 大冻丘,冰核丘
pink snow 粉红雪
pin-point convection 单点对流
pin point typhoon 豆台风(小台风)
Pioneer 先驱者(行星和行星际探测器)
pip 点回波
PIP(picture in picture) 画中画【计】
piphigram 压熵图
piping ①管涌【地质】②管道系统③管道输水
pirani 皮拉尼(热线测压)
piston effect 活塞效应
pitch ①间距,节距②俯仰角③音调
pitch-angle 镜角
pitch axis 俯仰轴
pit gauge 地坑雨量器
pi-theorem π定理
Pitot probe 皮托管,总压测量管
Pitot-static tube 皮托管
Pitot tube 皮托管
Pitot tube anemometer 皮托管风速表
pivot arm anemometer 转臂风速计
pivoting high 枢轴高压
pixel 像元【物】,像素【大气】
pixel value 像素值
plage 谱斑
plain 平原

plain-language report 明语气象报告
【大气】
Planck constant 普朗克常量【物】
Planck function 普朗克函数
Planckian radiator 普朗克辐射体
Planck's law 普朗克定律
Planck's radiation law 普朗克辐射定律
Planck time 普朗克时间【天】
plant genetics 作物遗传〔学〕【农】
β-plane β平面【大气】
plane albedo 平面反照率
plane atmospheric wave 平面大气波
plane chart 平面图【航海】
plane-dendritic crystal 平面枝状晶体
plane flow 平面流
plane geometry 平面几何〔学〕【数】
plane of the ecliptic 黄道面
plane-parallel approximation 平面平行近似
plane-parallel assumption 平面平行假定
plane-parallel atmosphere 平面平行大气
plane parallel medium 平面平行介质
plane-parallel motion 平面平行运动【物】
plane polarization 平面偏振【物】
plane source 面源
planet 行星【天】
planetarium 天象仪【天】
planetary aberration 行星光行差【天】
planetary albedo 行星反照率【大气】,地球反射率
planetary atmosphere 行星大气【大气】
planetary boundary layer(PBL) 行星边界层【大气】
planetary boundary layer jet 行星边界层急流
planetary circulation 行星环流
planetary equilibrium temperature 行星平衡温度
planetary geodesy 行星大地测量学【测】
planetary-geostrophic 行星地转(近似)
planetary-gravity wave 行星-重力波
planetary motion 行星运动
planetary radio occultation experiment 行星无线电掩星试验
planetary scale 行星尺度【大气】
planetary-scale system 行星尺度系统【大气】
planetary temperature 行星温度【大气】
planetary vorticity 行星涡度【大气】
planetary vorticity effect 行星涡度效应【大气】
planetary vorticity gradient 行星涡度梯度
planetary wave 行星波【大气】
planetary-wave formula 行星波公式
planetary wind 行星风
planetary wind belt 行星风带
planetary wind system 行星风系【大气】,行星风带
planetesimal hypothesis 微星假说
planetoid 小行星
plane wave 平面波【大气】
plane wave pulse 平面波脉冲
planimeter 求积仪【测】
plankton 浮游生物
plan of land utilization 土地规划【农】
plan position indicator(PPI) 平面位置显示器【大气】
plan shear indicator(PSI) 平面切变显示器【大气】
plant biology 植物生物学【植】
plant breeding 作物育种〔学〕【农】
plant canopy 植〔物〕冠〔盖〕
plant climate 植物气候
plant climate zone 植物气候区
plant climatology 植物气候学【植】
plant community 植物群落【地理】
plant cover 植被

plant disease	植物病害【农】
plant ecology	植物生态学【植】
plant embryology	植物胚胎学【植】
plant geography	植物地理学【植】
planting density	种植密度【农】
planting season	种植期
plant morphology	作物形态〔学〕【农】
plant organography	植物器官学【植】
plant pathology	植物病理学【农】
plant phenology	植物物候学
plant physiology	植物生理学【植】
plant protection	植物保护【农】
plant science	植物学【植】
plant systematics	系统植物学【植】
plant temperature	植物温度
plasma	等离〔子〕体【物】
plasma cloud	等离子体云
plasma confinement	等离子体约束
plasma diffusive equilibrium	等离子体扩散平衡
plasma dispersion function	等离子体频散函数
plasma display	等离子〔体〕显示【计】
plasma drift velocity	等离子体漂移速度
plasma echo	等离子体回波
plasma flow	等离子体流
plasma frequency	等离子体频率
plasma mantle	等离子体幔
plasmapause	等离子体层顶【大气】
plasma physics	等离〔子〕体物理学【物】
plasma process	等离子体过程
plasmasphere	等离子体层【大气】
plasma temperature	等离子体温度
plasticity	〔可〕塑性
plastic package	塑封【电子】
plate anemometer	平板风速表
plateau	高原【地理】
plateau climate	高原气候【大气】
plateau glacier	高原冰川
plateau meteorology	高原气象学【大气】
plateau mire	高原沼泽【地理】
plateau monsoon	高原季风【大气】
plateau station	高原测站
plate crystal	板状晶体
platelet	小片冰晶
plate motion	板块运动【地质】
plate scale	底片比例尺【天】
plate tectonics	板块构造学【地质】
platform	①观测平台②台地【地理】
platform location and data relay	平台定位和资料中继
platinum resistance thermometer（PRT）	铂丝电阻温度表
platinum wire resistance total temperature probe	铂丝电阻全温探头
platinum wire thermometer	铂丝温度表
platometer	求积仪【测】
platykurtic	扁平峰〔的〕,低峰〔的〕
platykurtosis	低峰态
playa	干盐湖【地质】,干荒盆地
playback	重放,再现
pleiobar	①高值等压区②高值等压线（高于 1000 hPa 的等压线）
pleion	正偏平中心
Pleistocene Climate Optimum	更新世气候最适期
Pleistocene climate	更新世气候
Pleistocene Epoch	更新世【地质】(距今 250 万至 1 万年前)
Pleistocene glacial epoch	更新世冰期
Pleistocene glaciation	更新世冰川作用
Pleistocene ice age	更新世冰期
Pleistocene Series	更新统【地质】
pleochroic halo	五色晕
plidar(= polychromatic lidar)	多色光雷达
pliobar	高压等值线
Pliocene Epoch	上新世【地质】
Pliocene Series	上新统【地质】
plotter	绘图机【计】
plotting	填图

plotting board 填图板
plotting model 填图格式
plotting position 标绘位置
plotting symbol 填图符号【大气】
plouazaou(=Grand Vent) 格朗旺风
ploughing season 耕种期
plowshares ①犁铧②积雪表面消融时的犁铧形特征
plow wind(或 plough wind) 犁风(带有雷阵雨的直线移动的风)
plug and play 即插即用【计】
plug and play operating system 即插即用操作系统【计】
plume ①卷流,股流②烟云,烟羽,云羽③云砧
plume height 烟羽高度【大气】
plume rise 烟羽抬升【大气】
plume(rise)model 烟流〔上升〕模式
plume type 烟羽类型【大气】
plum rains(=Meiyu) 梅雨【大气】
plunging breaker 卷浪
Pluto 冥王星【天】
pluvail index 雨量指数
pluvail lake 雨成湖
pluvail period 多雨期【大气】,雨季
pluvail region 多雨地区
pluvial 雨期
pluviograph 雨量计【大气】
pluviometer 雨量器
pluviometer-association 累计雨量器
pluviometric coefficient 雨量系数
pluviometric quotient 雨量比
pluviometry 雨量测定学【大气】,雨量测定法
pluvioscope 雨量器
pluviothermic ratio 雨温比
PMC(polar mesospheric clouds) 极地中间层云
PMIP (Palaeoclimate Modeling Intercomparison Project) 古气候模式比较计划
PMP(probable maximum precipitation) 可能最大降水【大气】,可能最大暴雨
PM-2.5 细颗粒物(空气中悬浮的粒径等于小于2.5 μm的颗粒物)
PM-10 可吸入颗粒物(空气中悬浮的粒径等于小于10 μm的颗粒物)
PMSE (polar mesosphere summer echoes) 极地中间层夏季回波
PNA(Pacific-North American pattern) 太平洋-北美型
pneumatic bridge hygrometer 桥臂压降式湿度表
pneumatology 气体力学
pocket anemometer 轻便风速表
pocket aspiration psychrometer 轻便通风干湿表
pocky cloud 悬球状云,乳房状云
pocky hygrometer 轻便湿度表
POD (probability of detection) 识别率,报准率,探测概率
podsol 灰壤
poechore 草生区
POES(Polar-Orbiting Operational Environmental Satellite) 极轨业务环境卫星
POGO (Polar-Orbiting Geophysical Observatory) 极轨地球物理观测卫星
pogonip 冻雾
Poincare formula 庞加莱公式【数】
Poincare section 庞加莱截面【物】
Poincare wave 庞加莱波
point ①点(澳大利亚雨量单位,1/100英寸)②方位点(罗盘上)
point attractor 点吸引子
point data 点〔测〕资料
point discharge 尖端放电【大气】
point discharge current 尖端放电电流
pointed lightning protector 避雷针
pointed summer 短夏(指极地夏季甚短)
pointer ①指极星②记录笔尖
pointer barometer 指针式气压表
point gauge 测针,针形水尺
point of neutral buoyancy 中性浮力点

point of occlusion　锢囚点【大气】
point of saturation　饱和点
point of symmetry　对称点
point precipitation　〔站〕点降水
point rainfall　单点雨量
point receiver　点接收机
point source　点源,点光源【物】
point spread function　点扩展函数
point target　点目标(如飞机之类的雷达目标)
point-to-multipoint communication　点-多点通信【电子】
point-to-point communication　点-点通信【电子】,点对点通信
point-to-point transmission　点对点传送
point vortex　点涡【大气】
poise　泊(旧的黏度单位)
Poiseuille flow　泊肃叶流
Poiseuille-Hagen law　泊肃叶-哈根定律
Poiseuille law　泊肃叶定律【物】
Poisson constant　泊松常数
Poisson distribution　泊松分布
Poisson equation　泊松方程【数】
Poisson random variable　泊松随机变量
polacke(＝polake)　波拉克风(阿尔巴尼亚波希米亚的下沉冷风)
polar air depression　极地气团低压
polar air haze　极地〔空气〕霾
polar air mass　极地气团【大气】
polar amplification　极地放大现象(气候模拟中)
polar angle　极角
polar anticyclone　极地反气旋【大气】
polar Atlantic Water　极地大西洋海水
polar aurora　极光
polar axis　极轴
polar band　极带,极光谱带
polar belt　极带
polar blackout　极地〔无线电〕衰落
polar cap　极冠,冠盖
polar cap absorption　极冠吸收

polar cap absorption events　极盖吸收事件
polar-cap ice　极冠冰
polar cell　极地胞,极地单体
polar circle　极圈
polar climate　极地气候【大气】,极地候类【大气】,寒带气候
polar continental air　极地大陆空气【大气】
polar continental high　极地大陆高压
polar coordinate method　极坐标法【航海】
polar coordinates　极坐标【数】
polar corona　极光冕
polar current　极地气流
polar cyclone　极地气旋【大气】
polar day　极昼【地理】
polar distance　极距
polar easterlies　极地东风〔带〕【大气】
polar-easterlies index　极地东风指数
polar elementary storm　极区元磁暴
Polar Experiment(POLEX, USSR)　极地试验(前苏联)
polar front　极锋【大气】
polar frontal zone　极锋区
polar-front jet(stream)　极锋急流
polar front theory　极锋理论【大气】,极锋学说
polar glacier　极地冰川【地理】
polar glow　极〔地气〕辉
polar high　极地高压【大气】
polar ice　极地冰
polar ice cap　极地冰冠,极地冰盖
polar ice sheet　极地冰原
polarimeter　偏振计【物】
polarimetric radar　偏振雷达
polar invasion(＝outbreak)　寒潮
polaris(＝polar star)　北极星
polariscope　偏〔振〕光镜【物】;偏振仪
polarity　①极性②配极〔变换〕③截然相反

polarity transition 极性过渡【地】
polarizability 极化率【物】
polarizability tensor 偏振张量
polarization ①极化【物】②极化度【物】③偏振(多用于光学)
polarization curve 极化曲线
polarization diversity 极化分集,散极化
polarization-diversity radar 多偏振雷达
polarization electric field 偏振电场
polarization filtering 偏振滤波
polarization isocline 偏振等倾线
polarization lidar 偏振光雷达
polarization matrix 极化矩阵
polarization of light 光的偏振
polarization spectroscory 偏振光谱学【物】
polarized light 偏振光
polarized plane 偏振面
polarizer 起偏器【物】
polarizing radar 极化雷达
polarizing radiometer 偏振辐射仪
polar jet 极地急流
polar light 极光
polar low 极地低压【大气】
polar low pressure 极地低压
polar magnetic disturbance 极区磁扰
polar magnetic storm 极区磁暴
polar magnetic substorm 极区磁亚暴
polar magnetogram 极区磁照图
polar marine air (mass) 极地海洋气团,极地海洋空气
polar maritime air 极地海洋气团,极地海洋空气
polar mesosphere summer echoes (PMSE) 极地中间层夏季回波(高纬夏季收到中间层顶来的异常强雷达回波)
polar mesospheric clouds (PMC) 极地中间层云
polar meteorology 极地气象学【大气】
polar monsoon zone 极地季风带
polar motion 极移【天】

polar navigation 极区航行【航海】
polar night 极夜【地理】
polar night jet (stream) 极夜急流【大气】
polar night vortex 极夜涡旋
polarography 极谱〔分析〕法
Polar Orbiting Geophysical Observatory (POGO) 极地轨道地球物理观测卫星
polar-orbiting meteorological satellite 极轨气象卫星【大气】
Polar-Orbiting Operational Environmental Satellite (POES) 极轨业务环境卫星
polar orbiting satellite (POS) 极轨卫星【大气】
polar outbreak 寒潮爆发
polar phase shift 极相漂移【地】
polar plume 极羽【天】(太阳两极区域的日冕羽毛状明亮结构)
polar projection 极投影
polar rain 极雨(指小通量的极光粒子沉降)
polar region 极地
polar shower 极地簇射
polar squall 极飑(指大通量的极光粒子沉降)
polar star 北极星
polar stereographic map 极射赤面地图
polar stereographic mosaic 极射赤面拼图
polar stratospheric cloud (PSC) 极地平流层云
polar-tropical coupling 极地-热带耦合
polar tropopause 极地对流层顶
polar trough 极地槽
polar type 极地型
polar vortex 极涡【大气】,极地涡旋
polar vortex index 极涡指数
polar wandering 极〔地迁〕移
polar wave 极波
polar westerlies 极地西风带

polar wind 极风
polar wind belt 极风带
polar wind divide 极地风界
polar year 极年
polar zone 极区
polder 围圩,围垦,围海造田
pole 极
pole of inaccessibility 难近冰极
pole-star recorder 北极星记录器
pole tide 极点潮
poleward 向极的
pollen 花粉
pollen analysis 孢粉分析
pollen density 花粉浓度
pollen diagram 孢粉图式
pollen indicator 花粉指标
pollen rain 花粉雨
pollen record 花粉记录
pollen zone 花粉带
pollutant 污染物
pollutant emission 污染物排放
pollutant index 污染物指数
pollutant monitoring 污染监测
pollutants standard index(PSI) 空气污染标准指数
pollutant transport 污染物迁移
polluted air 污染空气
pollution 污染【农】,浑浊
pollutional equivalent 污染当量
pollution control 污染控制
pollution disease 污染病
pollution dose 污染剂量【地理】
pollution flux 污染〔物〕通量
pollution index 污染指数【地质】
pollution level 污染水平【农】
pollution meteorology 污染气象学【大气】
pollution prediction 污染预报
pollution source 污染源【地质】
pollution warning system 污染预警系统
Polycene 多新世
Polychlorinated biphenyls(PCBs) 多氯联苯
polychloroprene balloon 氯丁橡胶气球
polychromatic lidar 多色光〔雷〕达
polychromatic transmittance 多色透射比
polyconic projection 多圆锥投影
polycyclic aromatic hydrocarbons (PAH) 多环芳烃
polycylidrical projection 多圆柱投影
polydisperse 多分散,多谱
polyethylene balloon 聚乙烯气球
polygon 多边形【数】
polyhedric projection 多面投影
polyhedron 多面体
polymer 多聚体【生化】,聚合物【化】
polymer chemistry 高分子化学【化】
polymer crystal 高分子晶体【化】
polymer electret 聚合物驻极体【化】
polymerization 聚合〔反应〕【化】
polymeter 多能湿度表
polymological data 孢粉资料
polynomial 多项式【数】
polynomial fitting 多项式拟合
polynomial regression 多项式回归【农】
polynya 冰间湖,冰隙,冰穴(水面未冰处,俄语名)
polypeptide 多肽【生化】
polyspectrum 多阶谱
polytetrafluoroethylene (PTFE) 聚四氟乙烯【化】
polytropic atmosphere 多元大气【大气】
polytropic process 多元过程【大气】
polyvinyl chloride (PVC) 聚氯乙烯【化】
pondage ①蓄水②蓄水量③调节容量
ponente 波南兑风(地中海的一种西风)
Ponentino 波能蒂诺风(意大利西海岸上的西海风)
poniente 波尼恩特风(直布罗陀海峡的西风)
pontias 蓬提阿风(法国尼昂的山风)
pooling 集中,汇集

pool of cold air 冷空气堆
poorga(=purga) 普加风(buran 风的别名)
poor visibility 不良能见度
POP(principal oscillation pattern) 主振荡型
popcorn cluster 爆米花状云团
popcorn cumulonimbi 爆米花状积雨云群
popogaio(见 papagayo) 帕帕加约风(中美尼加拉瓜和危地马拉的太平洋沿岸的强东北风)
POPs(persistence organic pollutants) 持久性有机污染物,难降解有机污染物(如滴滴涕、六氯苯、多氯联苯、二噁英等)
PoPT(probability of precipitation type) 降水[类]型[出现]概率
population ①种群【地理】②总体【数】③全域④群体
population density ①群体密度【农】②人口密度【地理】
population dynamics 种群动态【海洋】
population ecology 种群生态【海洋】
population geography 人口地理学【地理】
population improvement 群体改良【农】
population of droplet 滴群
population of nuclei 核群
pop-up thunderstorm 突发雷暴
poriaz 波里阿兹风(黑海的猛烈东北风)
porlezzina 波勒齐纳风(意大利罗加诺湖的东风)
porosity 孔隙度【地质】,孔积率,多孔性
porous ice 多孔冰晶
porous medium 多孔介质
porous porcelain atmometer 多孔陶瓷蒸发仪
porous sponge condition 泛海绵[侧]边界条件

portable automated mesonet (PAM) 可移动的自动中尺度观测网
portable automatic meteorological (PAM) network 可移动的自动气象站网
portable cup anemometer 轻便风杯风速表
portable life-support system (PLSS) 〔太空人〕轻便维生系统
portable mesonet stations 可移动的中尺度观测网站
portable wind profile recording system 可移动的风廓线记录系统
port meteorological liaison officer 港口气象联络官
port of departure 出发港【航海】
port of destination 目的港【航海】
POS(polar-orbiting satellite) 极轨卫星【大气】
position indicator 位置显示器
position vector 位矢
positive 正,阳
positive anomaly 正异常【地】
positive area 正区
positive axis 正轴
positive charge 正电荷
positive cloud-to-ground lightning 正云地闪电(由云向地输送正电荷)
positive correlation 正相关
positive definite 正定[的]
positive definite advection scheme 正定平流方案
positive difinite matrix 正定矩阵【数】
positive discharge 正放电
positive feedback 正反馈
positive ground flash 正地闪
positive ion (或 cation) 正离子,阳离子
positive isothermal vorticity advection 正等温涡度平流
positively charged ionosphere 〔带〕正〔电的〕电离层
positive particle 正粒子

positive pole 正极,阳极
positive rain 带正电雨
positive semi-definite 半正定〔的〕
positive semi-definite matrix 正半定矩阵【数】
positive static stability 正静力稳定
positive tilt 正斜槽
positive vorticity advection（PVA） 正涡度平流【大气】
positron 正电子【物】
possible error 可能误差
possible global radiation 可能总辐射
post analysis 事后分析
post-colouring 后色化
posterior distribution 后验分布【数】
posterior probability 后验概率【数】
post-frontal fog 锋后雾
post-frontal precipitation band 锋后降水带
postglacial age 冰后期
post-glacial climate 冰后期气候【大气】
post-glacial period 冰后期【地质】
post-processing 后处理
potamology 河流学,河流水文学
potassium 钾
potassium-argon dating 钾-氩定年法
potassium dihydrogen phosphate（KDP） 磷酸二氢钾【电子】
potential ①势【物】,位②潜在的,可能的③电势,电位
potential box 潜势区
potential buoyancy 潜在浮力
potential centre of crystallization 潜在结晶中心
potential density 位密度
potential dew-point temperature 露点位温
potential drop 电势降〔落〕【物】,位势降〔落〕
potential due to hydrostatic pressure 流体静压位势
potential economic effectiveness 潜在经济效益
potential energy 势能【物】,位能【大气】
potential energy diagram 势能图
potential enstrophy 位涡度拟能
potential enstrophy conserving scheme 位涡度拟能守恒格式
potential equivalent temperature 相当位温
potential evaporation 潜在蒸发【大气】,蒸发力,可能蒸发率
potential evaporation rate 潜在蒸发率
potential evapotranspiration 潜在蒸散【大气】,可能蒸散,蒸散势
potential evapotranspiration ratio 潜在蒸散比
potential flow 位〔势〕流
potential flow theory 势流理论
potential function 势函数【物】
potential gradient 位势梯度
potential index of refraction 位折射率,位折射指数
potential instability 位势不稳定【大气】
potential instability index 位势不稳定指数
potential loss 潜在失水
potential method 潜势法
potential predictability 潜在可预报性
potential pseudo-equivalent temperature 假相当位温
potential pseudo-wet-bulb temperature 假湿球位温
potential rate 潜在率
potential recharge 潜在再充水,持水能力
potential refractive index 位折射率
potential refractivity 潜折射本领
potential runoff 潜在径流
potential synergism 最大协同作用
potential temperature 位温【大气】
potential temperature coordinate 位温坐标,等熵坐标
potential temperature gradient 位温梯度

potential transpiration 可能蒸腾
potential visibility 有效能见度
potential vloume 位容
potential vorticity 位势涡度【大气】,位涡
potential vorticity equation 位势涡度方程
potential wet-bulb temperature 湿球位温
potentiometer 电位器【电子】,电位表
potentiometric surface(=piezometric surface) 静水压面,测压管液面
potometer 透明度表,蒸腾表,散发仪
poudrin 漂落冰针
Pouillet pyrheliometer 普约绝对日射表
poultry stress index (PSI) 家禽受热指数(指数大,家禽死亡率上升)
pound 磅(常衡1磅=453.592 g,金衡、药衡1磅=373.242 g)
powder snow 粉雪
powdery snow (=powder snow) 粉末〔状〕雪
power ①功率②率③幂【数】,乘方
power density 功率密度
power 〔density〕 spectrum 功率〔密度〕谱【物】
power fluctuation 功率起伏
power flux 功率流
power frequancy grounding resistance 工频接地电阻
power function 幂函数【数】
power law 〔乘〕幂〔定〕律
power-law profile 幂律廓线
power reflection coefficient 功率反射系数
power series 幂级数【数】
power source 电源【物】
power spectral method 功率谱方法
power spectrum ①功率谱【数】②能谱
power spectrum density 功率谱密度【数】
power subsystem 能源〔子〕系统,电能〔子〕系统,能源部分
power supply 电源【物】
power system 电力系统
power transfer function 功率传递函数
power window 幂窗
Poynting vector 玻印亭矢量【物】,能流密度矢〔量〕
ppb (parts per billion) 十亿分率 (10^{-9})
PPI (plan position indicator) 平面位置显示器【大气】
PPI scope 平面位置显示器
ppm (parts per million) 百万分率 (10^{-6})
ppmv (parts per million by volume) 体积百万分率
ppt (parts per trillion) 万亿分率 (10^{-12})
practical analysis 实用分析
practical salinity units (PSU) 实用盐度单位
praecipitatio(pra) 降水性(云)
prairie 草原(加拿大南部和美国中、北部)
prairie climate 草原气候【大气】
Prandtl mixing length 普朗特混合长
Prandtl mixing-length theory 普朗特混合长理论【大气】
Prandtl number 普朗特数【大气】
preactivation 预活化,预激活
preamplifier 前置放大器【电子】
preanalysis 事前分析,预分析
prebaratic chart 地面天气形势预报图
prebiotic chemistry 前生命化学【生化】
Precambrian 前寒武纪【地质】
precast 〔天气〕预报
precaution 预防
preceding hour 观测前一小时
preceding sunspot 前导黑子
precession 岁差【天】,进动,旋进
precession parameter 进动参数

precession rate　进动率
precipitable water　可降水量【大气】
precipitable water index　可降水量指数
precipitable water vapor（＝ precipitable water）　可降水量
precipitation　①降水【大气】,降水量 ②沉降
precipitation acidity　降水酸度【大气】
precipitation aloft　高空降水
precipitation anomalies　降水〔量〕距平
precipitation area　降水区
precipitation attenuation　降水衰减【大气】
precipitation band　降水带
precipitation budget　降水收支
precipitation cell　降水中心
precipitation ceiling　降水云幕
precipitation chart　降水量图【大气】
precipitation chemistry　降水化学【大气】
precipitation collector　降水收集器
precipitation convection　降水性对流
precipitation cumulus model　降水积云模式
precipitation current　降水电流
precipitation day　降水日【大气】
precipitation detector　降水检测器
precipitation duration　降水持续时间【大气】
precipitation echo　降水回波【大气】
precipitation effectiveness　降水有效度,降水有效性
precipitation effectiveness index　降水有效指数
precipitation effectiveness ratio　降水有效度比
precipitation efficiency　降水效率
precipitation electricity　降水电学
precipitation element　降水元
precipitation enhancement program（PEP）　增水计划
precipitation evaporation index　降水蒸发指数

precipitation evaporation（PE）ratio　降水-蒸发比
precipitation flux　降水通量
precipitation gauge　雨量筒
precipitation-generating element　降水生成胞,降水生成单体
precipitation index　降水指数
precipitation intensity　降水强度【大气】
precipitation inversion　降水逆减【大气】
precipitation-measuring radar　测雨雷达
precipitation mechanism　降水机制
precipitation particle spectrometer（PPS）　降水粒子谱仪
precipitation physics　降水物理学【大气】
precipitation process　降水过程
precipitation rate　降水率
precipitation recorder　降水记录器
precipitation regime　降水季节特征【大气】
precipitation rose　降水频率图
precipitation scavenging　降水清除
precipitation shadow　雨影
precipitation shield　盾形降水区
precipitation spectrometer probe（PSP）　降水粒子谱仪探头
precipitation "staircase"　降水"阶梯"（云及其降水形成的步骤）
precipitation statics　降水静电
precipitation station　雨量站
precipitation stimulation　人工影响降水,人工催化降水
precipitation surge　降水骤增区
precipitation trails　雨幡
precipitation trajectory　降水轨迹
precipitation variation　降水变率,雨量变率
precipitation within sight　视区内有降水
precipitator　除尘器
precipitous sea　怒涛【海洋】（道氏风浪8级）
precipitus　降水性云

precision 精密度【物】,精〔密〕度【大气】
precision altimeter 精密高度表
precision aneroid barometer 精密空盒气压表
precision optics 精密光学〔系统〕【物】
precision theodolite 精密经纬仪
pre-cold frontal squall line 冷锋前飑线
pre-cooling water reservoir 预冷水槽
precursor ①前体【生化】,前体物②前兆【地】
precursor gas 原〔起〕始气体
precursor signal 前兆信号
predecessor 前身,前兆物
predictability 可预报性【大气】
predictability limit 可预报性极限
predictand 预报量【大气】(要预报的变量或概率)
prediction 预报,预测
predictive equation 预报方程
predictor ①预估〔式〕【数】②预报因子【大气】
predissociation 预离解
predominant radius 优势半径
predominant wind 主导风,盛行风
predominant wind direction 主导风向【大气】
preexponential factor 指数前的系数
preferred cloud coverage 优势云覆盖区
preferred period 优势周期
prefigurance 前估
prefix ①电报报头②词头
pre-flight calibration 施放前检定
prefrontal fog 锋前雾
prefrontal instability line 锋前不稳定度线
prefrontal squall line 锋前飑线
prefrontal thunderstorm 锋前雷暴
prehistoric period 史前期
prehistory 史前考古学
prehurricane squall line 飓风前飑线

preirradiation 前辐照【农】
preliminary phase 初相
prelims 预设置值(质量检查用)
preprocessing 预处理
pre-selected tolerance 预选容许误差
present climate 现代气候
present monitoring 现时监测
present position 现在位置
present weather 现在天气【大气】
pressure 压力,压强
pressrue drag 气压曳力
pressure altimeter 气压测高表【大气】
pressure altitude 气压高度
pressure anemometer 压力风速表
pressure anomaly 气压距平
pressure averaging method 气压平均法
pressure broadening 压致〔谱线〕增宽【物】,压力〔谱线〕增宽【大气】
pressure capsule 气压表空盒
pressure center 气压中心
pressure change 气压变化
pressure-change chart 气压变化图
pressure chart 气压图
pressure coefficient 压力系数
pressure coordinates 气压坐标
pressure coordinate system 气压坐标系【大气】
pressure correlation 气压相关
pressure deepening 气压变低,气压加深
pressure distribution 气压分布
pressure dome 气压丘
pressure drag 气压曳力
pressure drop 气压降落,气压跌落
pressure equation 气压方程
pressure-fall centre 降压中心
pressure field 气压场【大气】
pressure fluctuation spectrum 气压起伏谱
pressure gauge 压力器,压强器
pressure gradient 气压梯度【大气】
pressure-gradient force 气压梯度力【大气】

pressure head 压力[水]头
pressure height 气压高度,压力高度
pressure ice 压力冰
pressure jump 气压涌升,压力突跃
pressure jump detector 气压涌升探测器
pressure jump sensor 气压涌升传感器
pressure melting 〔加〕压融〔化〕
pressure modulated cell(PMC) 压力调制池(压力调制辐射仪的)
pressure modulated infrared radiometer (PMIR) 压力调制红外辐射仪
pressure modulated radiometer(PMR) 压力调制辐射仪
pressure nose 气压鼻
pressure of radiation 辐射压
pressure of wind 风压
pressure pattern 气压[分布]型[式]
pressure pattern flight 气压型飞行
pressure-plate anemometer 压板风速表
pressure profile 气压廓线
pressure recorder 气压记录器
pressure reduction 〔海平面〕气压换算【大气】,气压订正
pressure ridge 高压脊
pressure-rise centre 升压中心
pressure scaling 压力换算
pressure-sensor array 气压传感器阵列
pressure spectrum 气压谱
pressure-sphere anemometer 压力球风速计
pressure stress 压应力
pressure surge 气压跃升,气压涌升
pressure surge line 气压涌升线【大气】
pressure system ①气压系统【大气】②气压坐标系
pressure-temperature correlation 压温相关
pressure-temperature humidity sounding 压温湿探测
pressure tendency 气压倾向【大气】
pressure-tendency chart 气压倾向〔分布〕图

pressure-tendency equation 气压倾向方程
pressure tide gauges 压力验潮仪
pressure topography 等压面形势
pressure torque 气压力矩
pressure transducer 气压传感器,压强换能器
pressure trough 气压槽
pressure tube anemometer 压管风速表
pressure variation 气压变量【大气】
pressure variograph 气压变量计
pressure variometer 气压变量表
pressure velocity 气压(铅直)速度
pressure vertical coordinates 气压垂直坐标
pressure wave 气压波【大气】
pressure work 压力功
prester 普雷斯特龙卷(地中海和希腊出现的带闪电的旋风或水龙卷)
prestress 预应力【地质】
pretreatment 预处理【电子】
pretty breeze 惠风
prevailing visibility 盛行能见度,主导能见度
prevailing westerlies 盛行西风带
prevailing westerly winds 盛行西风
prevailing wind 盛行风【大气】
prevailing wind direction 盛行风向
prewhitening 预白化
PRF(pulse repetition frequency) 脉冲重复频率【电子】
Price meter 普莱斯流速计
primary absolute cavity radiometer (PACRAD) 一级绝对腔体辐射仪
primary air 原生空气,一次空气
primary bow(=primary rainbow) 主虹
primary circulation 主级环流【大气】,一级环流
primary color 基色
primary cosmic ray 初级宇宙线
primary cyclone 主气旋

primary data 原始资料
primary data user station (PDUS) 一级资料使用站
primary data utilization station(PDUS) 〔卫星〕一次资料利用站,〔卫星〕原始资料利用站
primary depression 主低压
primary disaster 原生灾害
primary front 主锋
primary low 主低压
primary maximum 主极大
primary minimun 主极小
primary pollutant 原生污染物【大气】,原发性污染物,一次污染物
primary process 初级过程【化】
primary radar 一次雷达【电子】
primary radar technique 一次雷达技术
primary rainbow 虹【物】,主虹
primary scattering 一次散射【大气】
primary source of infection 初侵染源【农】
primary standard 一级标准【化】,基准物
primary standard barometer 一级标准气压表
primary standard pyrheliometer 一级标准绝对日射表
primary wind-finding radar 一次测风雷达
prime meridian 本初子午线【天】
primitive earth 原始地球
primitive equation 原始方程【大气】
primitive equation model 原始方程模式【大气】
primitive farming 原始农业【农】
primitive soil 原始土壤【土壤】
primordial atmosphere 初生大气
principal agricultural meteorological station 基本农业气象站
principal axis 主轴【数】
principal band 主〔螺旋〕带(气旋内)

principal climatological station 基本气候站
principal component analysis (PCA) 主成分分析【数】
principal front 主锋【大气】
principal land station 基本陆地站
principal oscillation pattern (POP) 主振荡型
principal synoptic observation 基本天气观测【大气】
principal weight 主要权重
principle 原理,原则
principle of agroclimatic analogy 农业气候相似原理【农】
principle of circulation 环流原理
principle of conservation of energy 能量守恒原理
principle of geometric association 几何组合原理
principle of historical sequence 历史顺序原则
principle of indeterminism 测不准原理
principle of invariance 不变性原理
principle of superposition 重叠原理
principle of virtual displacement 虚位移原理
principle of virtual work 虚功原理
principle plane 主面
principle point 主点
principle section 主截面
printed (circuit) board 印刷电路板
printer ①打印机 ②印刷机【计】
prior distribution 先验分布【数】
priori 先验〔的〕
prior information 先验信息
priority 优先权【数】,优先级【计】
prior probability 先验概率【数】
prism 棱镜【物】
privacy 保密性
probabilistic process 概率过程
probability 概率【大气】,几率

probability density 概率密度【化】
probability density function 概率密度函数
probability distribution 概率分布【大气】
probability distribution function 概率分布函数
probability ellipse 概率椭圆
probability forecast 概率预报【大气】
probability function 概率函数【数】
probability generation function 概率母函数【数】
probability integral 概率积分
probability levels 概率水准
probability model 概率模式
probability of detection 识别率,报准率,探测概率
probability of lightning strike damage 雷击损害概率
probability of precipitation type(PoPT) 降水〔类〕型〔出现〕概率
probability of reality 实在概率
probability paper 概率〔坐标〕纸【数】
probability score(PS) 概率评分
probability theory 概率论【数】
probable error 概率误差
probable maximum flood (或 maximum possible flood) 可能最大洪水【大气】
probable maximum precipitation(PMP) 可能最大降水【大气】,可能最大暴雨
probe 探测,探针
procedural error 非技术错误(质量检验用)
procedure body 过程体
procedure signal 程序信号【航海】
procedure statement 过程语句
process 过程
process lapse rate 过程直减率
processor 处理器【计】,处理机,处理单元
process of large scale atmosphere 大尺度大气过程

product-moment 积矩
production rule 产生式规则【计】
production system 产生式系统【计】
productivity of land 土地生产力【农】
product rule 乘积法则
profile 纵断面图【测】,廓线【大气】,剖面
profile matching 廓线匹配
profiler 断面仪【测】,廓线仪【大气】
profile similarity 廓线相似〔性〕
profiles 剖面图【测】
prog 预报图(prognostic chart)的常简称
prog chart 〔形势〕预报图
prognosis 形势预报【大气】
prognostic chart 预报图
prognostic clouds 预报云〔方程组〕
prognostic contour chart 等压面预报图
prognostic equation 预报方程
prograde orbit 前进轨道
program(或 programme) ①程序【计】②项目③计划
program block 程序块【计】
program communication block 程序通信块
program design 程序设计【计】
program development 程序开发
program library 程序库【计】
programmed control 程序控制【航海】
Programme on Long-range Forecasting Research(PLFR) 长期预报研究计划
Programme on Short and Mid-range Prediction(PSMP) 短、中期预报研究计划
programmer ①程序员②程序写入器
programming 编程【计】,程序设计
programming language 程序设计语言【计】
Program of Regional Observation and Forecast Service(PROFS) 区域观测和预报业务计划

program specification bolck 程序说明块
progression of the monsoon 季风推进
progressive motion 前进运动
progressive wave 前进波【大气】
projcetion 投影【物】,投射【数】
Project on Research in Tropical Meteorology(PRTM) 热带气象研究计划
projector 云幕灯,探照灯
prolate cycloidal track 扁长旋转路径
prolate spheroid 长球〔体〕
prolate symmetric top 扁长对称陀螺
prolonged frontal rain 持续锋面雨
prominence(=solar prominence) 日珥
promojna 冰面开裂(冰面上因水流而成的开裂,俄语名)
prontour chart 高空预报图
propagating (Rossby) wave 移行〔罗斯贝比〕波
propagation 传播
propagation constant 传播常数【电子】
propagation effect 传播效应
propane 丙烷(C_3H_8)
propeller 螺旋桨,推进器
propeller anemometer 螺旋桨〔式〕风速表【大气】
propeller-type current meter 旋桨式流速仪
propeller-vane anemonmeter 旋桨式风标风速计
propeller vane 〔螺〕旋桨〔式〕风标
propene 丙烯(C_3H_6)
proper time 固有时【物】,原时
proper vibration 固有振动
proportion 比例【数】
protectable loss 可预防损失
protected reversing thermometer 闭端颠倒海水温度表
protected thermometer 闭端保护温度表(一种颠倒温度表)
protection forest 防护林【林】
protection of water resources 水资源保护【地理】

protein engineering 蛋白质工程【生化】
Proterozoic Eon 元古宙【地质】
Proterozoic Eonothem 元古宇【地质】
Proterozoic Era 元古代
protium 氕【化】
protobiochemistry 原始生物化学【生化】
protocol 协议【计】,规约
protomer 原聚体【生化】
proton 质子【物】
protonosphere 质子层【地】
proton precipitation 质子雨(来自外空间的高能质子通量进入高层大气)
proton-proton chain 质子-质子链式反应
proton-proton reaction 质子-质子反应【天】(4个氢核聚变为1个氦核的一系列热核反应过程)
prototype 原型【计】,样机
prototype regional observing and forecasting service(PROFS) 典型区域观测和预报服务
proxy climate indicator 代用气候指标
proxy climatic record 代用气候记录
proxy data 代用资料
PSC (polar stratospheric cloud) 极地平流层云
pseudo-adiabat 假绝热〔线〕
pseudo-adiabatic ①假绝热〔的〕②假绝热〔线〕
pseudo-adiabatic chart (或 diagram) 假绝热图
pseudo-adiabatic convection 假绝热对流
pseudo-adiabatic diagram 假绝热图【大气】
pseudo-adiabatic expansion 假绝热膨胀
pseudo-adiabatic lapse rate 假绝热直减率【大气】
pseudo-adiabatic process 假绝热过程【大气】
pseudo-cirrus 伪卷云
pseudo cold-front 假冷锋

pseudo color 伪彩色【电子】
pseudo-color cloud picture 伪彩色云图【大气】
pseudocolor enhancement 伪彩色增强
pseudocolor image 伪彩色图像
pseudocoloring 伪彩色(不同色彩代表不同的灰度)
pseudo-colour enhancement 伪彩色处理显示
pseudo diffusion 假扩散
pseudo-equivalent potential temperature 假相当位温【大气】
pseudo-equivalent temperature 假相当温度
pseudo front 假锋
pseudo-geostrophic approximation 准地转近似
pseudo-latent instability 假潜不稳定
pseudomomentum 假动量
pseudo noise 伪噪声
pseudo noise code 伪噪声码【电子】
pseudo-potential temperature 假位温
pseudo potential vorticity equation 假位涡度方程
pseudo-random number 伪随机数【数】
pseudospectral approximation 伪谱近似
pseudospectral method 假谱方法【大气】,伪谱法
pseudo-spectral model 假谱模式
pseudo-spectrum 伪谱
pseudo-wet-bulb potential temperature 假湿球位温
pseudo-wet-bulb temperature 假湿球温度
PSI(plan shear indicator) 平面切变显示器【大气】
PSMSL(Permanent Service for Mean Sea Level) 平均海平面资料常设局(英国)
psophometer 噪声计
psychrograph 干湿计

psychrometer 干湿表【大气】,玻璃水银干湿表
psychrometer difference 干湿球温度差值
psychrometric calculator 湿度计算器
psychrometric chart 湿度计算图
psychrometric coefficient 湿度系数,测湿系数
psychrometric constant 干湿表常数
psychrometric formula 测湿公式【大气】,湿度计算式
psychrometric table 湿度查算表
psychrometry 测湿学
psychrosphere 湿气圈,水汽圈
p system(=pressure system) 气压坐标系
PTFE(polytetrafluoroethylene) 聚四氟乙烯【化】
public forecast 公众预报
public nuisance 公害
pucker point 皱点
puddle 冰上融水坑
puelche 帕尔希风(吹过南美安第斯山的东风)
puff ①阵风②烟团
pulg-in 插入【计】
pulsar 脉冲星【物】
pulsating aurora 脉动极光
pulsating auroral glow 脉动状极光辉
pulsation 脉动
pulse 脉冲【电子】
pulse actinometer 脉冲直接辐射表
pulse-amplitude-code modulation 脉冲幅度编码调制
pulse-amplitude modulation 脉幅调制【电子】
pulse broadening 脉冲加宽
pulse ceilometer 脉冲云幂仪
pulse code modulation (PCM)(=pulse coded modulation) 脉码调制【电子】

pulse coding 脉冲编码
pulse compression 脉积压缩
pulse-counting sonde 脉冲计数式探空仪
pulsed laser 脉冲激光器【物】
pulsed-light cloud-height indicator 脉冲光云高指示器
pulsed light 脉冲光
pulse Doppler radar 脉冲多普勒雷达【电子】
pulsed radar 脉冲雷达【电子】
pulse duration 脉冲持续时间, 脉冲宽度
pulse duration modulation 脉宽调制【电子】
pulse frequency 脉冲频率
pulse frequency modulation 脉冲调频
pulse integration 脉冲积分
pulse integrator 脉冲积分器
pulse ionosonde 脉冲电离层探测装置
pulse ionosound 脉冲电离层探测器
pulse length 脉冲长度(线距)
pulse length modulation 脉冲长度调制
pulse-limited radar altimeter 限幅雷达测高计
pulse modulation 脉冲调制【电子】
pulse number modulation 脉冲频度调制, 脉冲数调制
pulse-pair processing 脉冲对处理
pulse-pair processor (PPP) 脉冲对处理器
pulse period 脉冲周期
pulse phase modulation 脉冲调相
pulse-position modulation 脉位调制
pulse propagation 脉冲传播
pulse radar 脉冲雷达【大气】
pulse recurrence (或 repetition) frequency 脉冲重复频率
pulse repetition time (PRT) 脉冲重复时间
pulse sounder 脉冲探测仪
pulse-time-modulated radiosonde 脉冲时间调制探空仪
pulse time modulation 脉时调制【电子】
pulse volume 脉冲体积
pulse width 脉冲宽度【电子】, 脉冲持续时间
pulse width modulation 脉宽调制【电子】
pulse xenon lamp 脉冲氙灯【物】
pumping 抽运【物】, 抽吸作用, 泵作用
pumping head 抽运水头
pumping (of barometer) 振荡(气压表的)
pumping water level 抽水水位
pumpkin-shaped balloon 南瓜型气球
punched card 穿孔卡
punched tape 穿孔纸带
punching machine 打孔机, 穿孔机
punos 普诺斯风(南美普纳地区干冷的南风或西南风)
pure air 纯洁空气
pure deformation 纯粹变形
purga (=poorga) 普加风
purification 净化〔作用〕
purl 盘旋飞行路径
purple glow 紫辉
purple light ①紫霞 ②紫光
purple spot 紫〔辉光〕斑(清晨或黄昏时)
purple twilight 曙暮紫光
PVA (positive vorticity advection) 正涡度平流【大气】
PVC (polyvinyl chloride) 聚氯乙烯【化】
pycnocline 密度跃层【大气】
pyranogram 天空辐射自记曲线
pyranograph 天空辐射计

pyranometer 总辐射表【大气】,净辐射表
pyranometry 天空辐射测量〔法〕
pyrgeometer 大气辐射表,地球辐射表,红外辐射表
pyrheliogram 绝对日射自记曲线
pyrheliograph 绝对日射计
pyrheliometer 直接辐射表【大气】
pyrheliometric scale 绝对日射表标尺
pyrheliometry 直接日射测量学【大气】,太阳直接辐射测量学
pyrhenerwind 佩韩纳风(奥地利阿尔卑斯山焚风)
pyrocumulus 热〔成〕积云
pyroelectric effect 热释电效应【物】
pyroelectric detector 热电探测器
pyroelectric turbidity meter 热电式浑浊度表
pyrolysis 热解〔作用〕
pyrometer 高温表
pyrometry 测高温法
pyrotechnic flare 焰弹
pyrotechnic generator 焰弹发生器
pyrradiometer 全辐射表

Q

qarajel (= karajol, qarajel, quara) 卡拉乔风(保加利亚沿海的西风)
qaus (= kaus) 考斯风
Q band Q 带(即 Ku 雷达频带,频率为 12~18 GHz,波长为 1.67~2.5 cm)
QBO (quasi-biennial oscillation) 准两年振荡【大气】
Q burst Q 爆发(地磁瞬变)
QBWO (quasi-biweekly oscillation) 准双周振荡
Q channel (quadrature channel) Q 通道,正交通道
Q code Q 电码(航空)
Qiantang River tidal bore 钱塘〔江〕涌潮【海洋】
qibla (= ghibli) 朝南风(摩洛哥热风)
Qinghai-Xizang high 青藏高压【大气】
Qinghai-Xizang Plateau Meteorological Science Experiment (QXPMEX) 青藏高原气象科学试验(1979 年 5—8 月)
Qinghai-Xizang Plateau monsoon 青藏高原季风
Qinghai-Xizang trough 青藏低槽
Qinghai-Xizang Plateau meteorology 青藏高原气象学
Q noise Q 噪声(闪电发出的甚高频辐射中)
QPF (quantitative precipitation forecast) 定量降水预报【大气】
Q-switch Q 开关,光量开关
Q-switched ruby laser Q 开关红宝石激光器
Q-switching Q 开关【电子】,光量开关
quadrant 象限〔数〕,象限仪
quadrant analysis 象限分析
quadrant angle 象限角,射角
quadrant electrometer 象限静电计【物】
quadrant visibility 象限能见度
quadratic conserving scheme 二次守恒格式
quadratic equation with one unknown 一元二次方程【数】
quadratic form 二次形
quadratic transformation 二次变换【数】
quadrature ①正交②上(下)弦③求

积分
quadrature formula 求积公式
quadrature spectral density 求积谱密度
quadrature spectrum 求积谱【大气】
quadrilateral element 四边形元
quadruple recorder 四用自记计
quad-spectrum 求积谱
qualification 鉴定【计】
qualification requirements 鉴定需求
【计】
qualified product 合格品【电子】
qualitative analysis 定性分析
qualitative observation 定性观测
qualitative spectrometric analysis 光谱定性分析【化】
quality 质,质量
quality control 质量控制【电子】
quality evaluation 质量评价【农】
quality management 质量管理【电子】
quality of snow 雪质（冰晶所占重量百分比）
quality of water 水质
quality standard 质量标准【农】
quantification theory 数量化理论
quantile 分位数【数】
quantitative analysis 定量分析
quantitative forecasting 定量预报方法
quantitative precipitation estimation 定量降水估计
quantitative precipitation forecast(QPF) 定量降水预报【大气】
quantitative spectrometric analysis 光谱定量分析【化】
quantity 量【数】,数量
quantity of illumination 光照量(lx·s)
quantity of radiant energy 辐射能量
quantization 量子化【物】,量化【数】
quantized signal 量化信号【大气】
quantometer 光量计
quantum 量子【物】
quantum computer 量了计算机【计】

quantum efficiency 量子效率
quantum electronics 量子电子学【电子】
quantum mechanics 量子力学【物】
quantum number 量子数【物】
quantum predictability analysis 量子可预测性分析
quantum theory 量子理论【物】
quara（见 karajol） 卡拉乔风
quartile 四分位数
quartile deviation 四分差
quartz 石英【物】
quartz-glass pyrheliometer 石英玻璃绝对日射表
quartz-glass thermometer 石英玻璃温度表
quartz thermometer 石英温度计
quasi- 准（词头）
quasi-barotropic fluid 准正压流体
quasi-biennial oscillation（QBO） 准两年振荡【大气】
quasi-biennial period 准两年周期
quasi-biweekly oscillation（QBWO） 准双周振荡
quasi-center of action 准活动中心
quasi-conservative 准保守性的
quasi-discontinuity 准不连续性
quasi-equilibrium 准平衡【物】
quasi-geostrophic（＝quasigeostrophic） 准地转
quasi-geostrophic approximation 准地转近似
quasi-geostrophic current 准地转流
quasi-geostrophic equations 准地转方程
quasi-geostrophic equilibrium 准地转平衡
quasigeostrophic filtering 准地转滤波
quasi-geostrophic flow 准地转流
quasi-geostrophic model 准地转模式【大气】
quasi-geostrophic motion 准地转运动【大气】
quasi-geostrophic omega equation 准地

转 Ω 方程
quasi-geostrophic potential vorticity 准地转位涡度
quasi-geostrophic potential vorticity equation 准地转位涡度方程
quasigeostrophic scaling 准地转定标
quasi-geostrophic system 准地转系统
quasi-geostrophic theory 准地转理论
quasi-geostrophic turbulence 准地转湍流
quasi-gray atmosphere 准灰体大气
quasi-homogeneous 准均匀的
quasi-horizontal motion 准水平运动
quasi-horizontal transport 准水平输送
quasi-hydrostatic approximation 准静力近似
quasi-hydrostatic system 准静力系统
quasi-isobaric process 准等压过程
quasi-Lagrangian coordinates 准拉格朗日坐标
quasi-Lagrangian nested grid model(QNGM) 准拉格朗日[嵌]套网格模式
quasi-linear convective system 准线状对流系统
quasi-linear partial differential equation 拟线性偏微分方程【数】
quasi-linear wave train 准线性波列
quasi-molecule 准分子【物】
quasi-monotonic scheme 准单调格式
quasi-Newton method 拟牛顿法【数】
quasi-nondivergence 准无辐散【大气】
quasi-nondivergent 准无辐散[的]
quasi-nondivergent flow 准无辐散气流
quasi-periodic 准周期性【大气】
quasi-permanent low 准常定低压,准永久性低压
quasi-polar sun-synchronous orbit 准极地太阳同步轨道
quasi-reflection 准反射
quasi-static process 准静力过程

quasi-stationary 准静止
quasi-stationary front 准静止锋【大气】
quasi-stationary perturbation 准静止扰动
quasi-stationary time series 准稳定时间序列
quasi-steady 准稳定
quasi steady state 准稳定态
quasi-triennial oscillation 准三年振荡
quasi-two weeks oscillation 准两周振荡
Quaternary 第四纪
Quaternary climate 第四纪气候【大气】
Quaternary climatic record 第四纪气候记录
Quaternary Ice Age 第四纪冰期【地质】
Quaternary low sea level 第四纪低海平面
quaternary low sea-level period 第四纪低海面期
Quaternary Period 第四纪【地质】
Quaternary System 第四系【地质】
quench frequency 淬频
quenching 淬火【物】,猝灭
quenching effect 淬灭效应【电子】,猝灭效应
queue 队列
queueing theory 排队论【数】
queueing time 排队时间【数】
"Quick Look" data 速阅资料
quicksand 流沙
quicksilver ①水银,汞锡合金 ②水银似的
quiescent prominence 宁静日珥【天】
quiet level 宁静水平
quiet solar radio 宁静太阳射电
quiet sun 宁静太阳【天】
QuikSCAT 快速散射计卫星(1999年发射,测量海上风速风向的极轨卫星)
quintics 五次曲线【数】
quintile 五分位数
quotient ①商【数】②份额 ③降水蒸发商
quotiety 率,系数
Q vector Q向量(准地转和半地转运动中出现)

R

rabal 无线电测风,无线电探空气球
RACOON （RAdiation-Controlled balloON）（缩写词字面义"浣熊"）〔辐射控制〕零压气球,辐射平衡气球
radar（radio detection and ranging） 雷达【电子】
radar algorithm 雷达算法【大气】
radar altimeter 雷达高度表,雷达测高仪【测】
radar altitude 雷达高度
radar antenna 雷达天线
radar band 雷达波带
radar beacon 雷达信标【航海】,雷康
radar beam 雷达波束
radar calibration 雷达标定【大气】,雷达校准
radar chaff 雷达示踪物
radar climatology 雷达气候学【大气】
radar cloud-base and cloud-top indicator 云底云顶雷达指示器
radar coded message 雷达编码电报
radar colored display 雷达彩色显示器
radar constant 雷达常数
radar cross section 雷达截面积【电子】,雷达截面
radar cross section per unit area 单位面积雷达截面积
radar data 雷达资料
radar data assimilation 雷达资料同化
radar database 雷达数据库【电子】
radar data processor 雷达数据处理器
radar detection gap 雷达探测空隙
radar detection of storm 雷达风暴探测
radar diagnosis of severe-storm 雷达强风暴识别,雷达强风暴诊断
radar digital display 雷达数字显示〔器〕
radar digitizer 雷达数字转换器

radar displays 雷达显示〔图〕
radar dome 雷达天线罩
radar duct 雷达波导
radar duplexer 雷达转换开关
radar echo 雷达回波【大气】
radar echo composite map 雷达回波综合图
radar echo fluctuation 雷达回波起伏
radar echo linear extrapolation forecasts 雷达回波线性外推预报
radar echo pillar profile 雷达回波直柱廓线
radar echo signal simulator 雷达回波信号模拟器
radar equation 雷达方程【电子】
radar equivalent reflectivity factor 雷达等效反射率因子【大气】
radar facsimile transmission（RAFAX） 雷达图像传真发送
radar foreshortening 雷达透射收缩
radar frequency band 雷达〔频率〕波段
radar hail detector 雷达冰雹探测器
radar hole 雷达〔回波〕洞
radar horizon 雷达视线水平,雷达地平线【电子】
radar hydrology 雷达水文学
radar hygrometer 雷达湿度计
radar image animation 雷达动态图像
radar image geometric distortion 雷达图像几何失真
radar imagery 雷达图像
radar indicating moving target 移动目标显示雷达
radar indicator 雷达显示器【电子】
radar interferometry 雷达相干测量法
radarman 雷达操作员
radar measurement of precipitation

雷达测量降水
radar meteorological observation 雷达气象观测【大气】
radar meteorology 雷达气象学【大气】
radar microwave link 雷达微波中继装置
radar mile 雷达英里(陈旧词,雷达脉冲至1英里目标往返一次所需要的时间)
radar navigation 雷达导航【航海】
radar navigation chart 雷达引航图【航海】
radar network composite 雷达组网【大气】
radar noise 雷达噪声
radar obscuring 雷达遮挡
radar [meteorological] observation 雷达[气象]观测
radar observation (ROB) 雷达观测[报告]
rader overlay 雷达覆盖区【测】
radar polarimetry 雷达偏振测量[法]
radar precipitation echo 雷达降水回波
radar rainfall integrator 雷达雨量积分器【大气】
radar raingauge 雷达雨量计
radar-raingauge comparison 雷达雨量计比对
radar range 雷达有效距离,雷达探测距离【电子】
radar range equation 雷达测距方程
radar receiver 雷达接收机
radar reflectivity 雷达反射率【大气】
radar reflectivity factor 雷达反射率因子【大气】
radar reflector 雷达反射器【航海】
radar remote sensing 雷达遥感【地理】
radar report (RAREP) 雷达报告
Radar Research and Development Establishment (RRDE) 雷达研究发展中心(英国)
radar resolution 雷达分辨率,雷达分辨力【电子】
radar resolution volume 雷达分辨体积【大气】
radar return 雷达回波
rader room 雷达室【船舶】
radar scatterometer 雷达散射计
radar scope 雷达显示器【电子】
radar screen 雷达荧光屏
radar sensitivity 雷达灵敏度
radar shadow 雷达盲区
radar signal spectrograph (RASAPH) 雷达信号摄谱仪
radar site (或 station) 雷达站
radarsonde 雷达测风仪【大气】
radar sounding 雷达探空【大气】
radar station 雷达站【电子】
radar storm detection 雷达风暴探测【大气】
radar theodolite 雷达经纬仪
radar tracking 雷达跟踪【电子】
radar transponder 雷达应答器【航海】
radar transmitter 雷达发射机
radar triangulation 雷达三角法
radar volume 雷达[照射]体积
radar wave band 雷达波段
radar weather observation 雷达天气观测
radar weather station 雷达气象站
radar wind finding 雷达测风
radar wind observation (或 sounding) 雷达(无线电)测风
radar wind system 雷达测风系统
radial acceleration 径向加速度【物】
radial inflow 径向流入【大气】
radial Reynolds number 径向雷诺数
radial spectrum 径向谱
radial symmetry 径向对称,辐射对称【建筑】
radial velocity 径向速度【物】,视向速度【天】
radial wavenumber 径向波数
radial wind 径向风【大气】
radian 弧度【数】
radiance 辐射率【大气】,辐射亮度

【物】
radiance temperature 辐射亮温,辐射温度
radiance unit 辐射率单位
radiant 辐射的,辐射点【天】
radiant density 辐射密度
radiant emittance 辐射度,辐射发射度【物】
radiant energy 辐射能
radiant energy density 辐射能密度【物】
radiant-energy thermometer 辐射能温度表
radiant exitance 辐射度,辐射出射度【物】
radiant exposure 辐照量【大气】,辐射曝光量【电工】
radiant flux 辐射通量(W)【大气】,辐射功率
radiant flux density 辐射通量密度
radiant heat 辐射热
radiant intensity 辐射温度(W/sr)【大气】
radiant point 辐射点
radiant power 辐射功率
radiation ①辐射【物】②放射
radiational index of dryness 干燥辐射指数
radiation balance 辐射平衡【大气】,净辐射,辐射差额
radiation balance meter 辐射平衡表
radiation balance of earth's surface 地表辐射差额
radiation balance of the earth-atmosphere system 地气系统辐射差额,地球辐射平衡
radiation balance of underlying surface 下垫面辐射平衡
radiation belt 辐射带【地】
radiation belts of the earth 地球辐射带
radiation biology 辐射生物学【农】
radiation boundary condition 辐射边界条件
radiation budget 辐射收支【地】
radiation center 辐射中心

radiation chart 辐射图【大气】
radiation climate 辐射气候【大气】
radiation-cloud-aerosol interaction 辐射-云-气溶胶交互作用
radiation coefficient 辐射系数【电力】
radiation-controlled balloon (RACOON) 辐〔射〕控〔制〕气球,零压气球(飞行在对流层顶之上)，
radiation cooling 辐射冷却【大气】
radiation correction 辐射订正
radiation damage 辐射损伤【物】
radiation damping 辐射阻尼【物】
radiation efficiency 辐射效率【电子】
radiation dry index 辐射干燥度
radiation equilibrium 辐射平衡
radiation error 辐射误差
radiation error of radiosonde 探空仪辐射误差
radiation feedback 辐射反馈
radiation filter 辐射滤波器
radiation flux 辐射通量【物】
radiation flux divergence meter 辐射通量散度计
radiation fog 辐射雾【大气】
radiation frost 辐射霜【大气】
radiation frost injury 辐射霜冻害
radiation heating 辐射加热
radiation imbalance 辐射不平衡
radiation intensity 辐射强度【物】
radiation instruments 辐射仪器
radiation inversion 辐射逆温【大气】
radiation laws 辐射〔定〕律
radiation length 辐射长度【物】
radiation loss 辐射亏损
radiation model 辐射模式【大气】
radiation night 强辐射夜(指无云无风的夜间)
radiation of heat 热辐射
radiation of light 光辐射
radiation of visual ray 可见光辐射
radiation pattern 辐射型,辐射〔方向〕

图【物】
radiation pattern avalanche 辐射型雪崩
radiation point 辐射点
radiation pressure 辐射压〔强〕【物】
radiation protection 辐射防护【物】
radiation recorder 辐射记录器
radiation reference scale 辐射基准标尺
radiation ridge 辐射脊
radiation sensor 辐射传感器【测】
radiation shield 防辐射罩,辐射屏蔽【物】
radiation shield for thermometer 温度表防辐射罩
radiation shield for radiosonde 探空仪防辐射罩
radiation sonde 辐射探空仪
radiation source 辐射源【物】
radiation station 辐射站
radiation temperature 辐射温度
radiation thermometer 辐射温度表【大气】
radiation torque 辐射转矩
radiation transfer 辐射传输,辐射传递
radiation transfer equation 辐射传输方程【大气】
radiation transfer model 辐射传输模式
radiation transition 辐射跃迁【机械】
radiative absorption 辐射吸收【物】
radiative-convective equilibrium 辐射-对流平衡
radiative-convective model 辐射-对流模式
radiative cooling 辐射冷却
radiative damping 辐射阻尼
radiative diffusivity 辐射漫射率
radiative equilibrium 辐射平衡
radiative equilibrium temperature 辐射平衡温度
radiative feedback process 辐射反馈过程
radiative flux 辐射通量
radiative flux density 辐射通量密度
radiative flux divergence 辐射通量散度

radiative forcing 辐射强迫〔作用〕
radiative half-life 辐射半衰期
radiative heat exchange 辐射热交换
radiative heat flux 辐射热通量
radiative heating 辐射加热
radiatively active gas 辐射活跃气体
radiative recombination 辐射复合
radiative sky temperature 辐射天空温度
radiative transfer equation 辐射传输方程
radiative transfer model 辐射传输模式
radiative transfer theory 辐射传输理论
radiative-turbulent equilibrium 辐射湍流平衡
radiator ①辐射体,辐射器【电子】②散热体【电工】
radiatus(ra) 辐辏状〔云〕
radio 无线电,射电【天】
radioactive aerosol 放射性气溶胶【化】
radioactive air monitoring system 放射性空气监测系统
radioactive carbon 放射性碳
radioactive collector 放射性集电器(测空中电位梯度用)
radioactive dating 放射性鉴年法【物】
radioactive decay 放射性衰变【化】
radioactive element 放射性元素【化】
radioactive fallout 放射性尘降物【化】,放射性散落物
radioactive fallout forecast 放射性尘降预报
radioactive fallout plot 放射性尘降区
radioactive gas 放射性气体【冶金】
radioactive isotope 放射性同位素【物】
radioactive nuclide 放射性核素【化】
radioactive particle ionization 放射性粒子电离
radioactive pollutant 放射性尘降物
radioactive pollution 放射性污染【地质】
radioactive precipitation 放射性降水
radioactive rain 放射性雨

radioactive snow gauge 放射性雪量计
radioactive tracer 放射性示踪剂
radioactive waste 放射性废物【化】
radioactivity 放射性【化】
radio acoustic sounding system (RASS) 无线电-声学探测系统,RASS 系统
radio altimeter 无线电高度表【航空】
radioassay 放射性测量,放射性检测【化】
radio astrometry 射电天体测量学【天】
radio astronomy 射电天文学【天】
radio atmometer 辐射蒸发表
radio aurora 射电极光
radio beacon 无线电信标【航海】
radio-beacon buoy 无线电浮标【电子】
radio beam 无线电射束
radiobearing 无线电定向
radio blackout 无线电通信中断
radio burst 射电暴
radiocarbon 放射性碳
radiocarbon age 放射性[碳]年代
radiocarbon concentration 放射性碳含量
radiocarbon dating 放射性碳定年【地质】,碳定年法【大气】,碳-14 定年法
radio ceiling 无线电云幂器
radiochemistry 放射化学【化】
radiocinemograph 射电摄像仪(记录天电次数)
radio climatology 无线电气候学
radio compass 无线电罗盘【电子】
radio direction finder 无线电定向仪
radio direction finding 无线电测向
radio direction finding radar 无线电测向雷达
radio distance finder 无线电测距器
radio duct 无线电波导
radioelement 放射性元素【化】
radio energy 无线电能
radioenvironmental chemistry 放射环境化学【化】
radiofacsimile converter 无线电传真变换器

radio fadeout 无线电衰减
radio frequency (RF) 无线电频率,射频
radio frequency amplifier 射频放大器【电子】
radio frequency band 射频波段
radio frequency range 无线电[航向]信标
radio frequency spectrum 射频[波]谱【物】
radio goniograph 无线电测向仪
radio goniometer 无线电测向仪
radio goniometry 无线电定向法
radio hole 无线电洞
radio horizon 无线电地平
radioisotope 放射性同位素
radioisotope snow-gauge 射线雪量计
radio link 无线电线路【航海】
radiolocator 无线电定位器
radiological air pollution 放射性空气污染
radiological chemistry 放射化学
Radiological Society of North America (RSNA) 北美辐射学会
radiomaximograph 天电强度仪,射电极值计(记录天电电场极大值)
radio meteor 无线电流星
radiometer 辐射计【电子】
radio meteorology 无线电气象学【大气】
radiometer sonde 无线电探空仪
radiometeorograph 无线电气象计
radiometer amplifier 辐射仪放大器
radiometersonde 辐射探空仪
radiometric correction 辐射订正
radiometric quantity 辐射量
radiometric thermosonde 测温辐射仪
radiometry 辐射测量法,辐射度量学【物】
radio mirage 无线电蜃景
radio nebula 射电星云【物】
radio navigation 无线电导航【船舶】
radio navigation aid 无线电导航设备

radio noise　无线电噪声【铁道】
radio nova　射电新星【物】
radionuclide　放射性核素【化】
radio occultation technique　无线电掩星法
radio oceanography　无线电海洋学
radiophotograph　无线电传真〔照片〕
radio paging　无线电寻呼【电子】
radio pilot〔balloon〕　无线电测风气球
radio-range finding　无线电测距
radio range goniometer　无线电导航测向器
radio-range receiver　无线电测距接收机
radio refraction correction　大气折射订正
radio refractive index　大气折射指数
radio-refractometer　无线电折射仪【大气】
radio-sondage　无线电探空
radiosonde　无线电探空仪【大气】
radiosonde and radiowind station（RWS）　无线电探空测风站
radiosonde balloon　探空气球【大气】
radiosonde calibration　探空仪校准
radiosonde commutator　探空仪转换器
radiosonde data digitizer　探空〔仪〕资料数字转换器
radiosonde encipher　探空仪编码器
radiosonde ground equipment　探空仪地面设备
radiosonde modulator　探空仪调制器
radiosonde network　探空站网
radiosonde observation（RAOB）　〔无线电〕探空〔仪〕观测〔记录〕
radiosonde-radiowind system　无线电探空测风系统
radiosonde receiver-recorder　无线电探空〔仪〕接收记录器
radiosonde recorder　探空仪记录器
radiosonde reference signal　探空仪参考信号
radiosonde report　探空报告
radiosonde station　探空站,无线电探空站
radiosonde station recorder　探空〔仪〕台站记录器

radiosonde system　〔无线电〕探空仪系统
radiosonde temperature　探空温度
radiosonde transmitter　探空仪发报机
radiosonde unwinder　探空仪退绕器
radio sounding　无线电探空【大气】
radiosounding operation　探空业务
radio source　射电源【天】,辐射源
radioteletype receiver　无线电传打字电报接收机
radioteletype transmitter（RTT）　无线电传打字电报发报机
radioteletypewriter　无线电传打字机
radio telescope　射电望远镜【天】
radio theodolite　无线电经纬仪【大气】
radio theodolite recording receiver　无线电经纬仪记录式接收机
radiotherapy　放射治疗【物】
radiotracer　放射性示踪物
radio time signal　无线电时号【航海】
radio-tracked pilot balloon　无线电跟踪测风气球
radio tracking　无线电追踪,无线电跟踪法【动】
radio transmission　无线电传送
radio wave　无线电波
radio wave absorption by clouds　无线电波的云层吸收
radio wave propagation　无线电波传播
radio wave ray　无线电波射线
radio wave ray curvature　无线电波射线曲率
radio-wave sounding　无线电波探空
radio weather service　无线电气象业务【航海】
radio weather warning system　无线电天气预警系统
radio whistler　无线电哨声
radio wind finding　无线电测风
radiowind observation　无线电测风观测【大气】

radio window 射电窗
radiowind station 无线电测风站
radium 镭
radius 半径【数】
radius of convergence 收敛半径【数】
radius of curvature 曲率半径【数】
radius of deformation 变形半径
radius of influence 影响半径
radius of maximum wind 最大风速半径
radius of protection 避雷半径
radius vector 径矢【数】,径向量【数】
radix ①基,基数②根【动】
radix layer 〔边界层〕底层
RADOB(report of ground radar weather observation) 地面雷达天气观测报告
radome 天线罩【电子】,〔雷达〕天线罩【大气】
radon(Rn) 氡【大气】
radtrack system 雷达跟踪系统
RAFAX(radar facsimile transmission) 雷达图像传真发送
raffiche(=refoli) 拉菲希阵风(地中海地区从山又吹来的阵风,阵性的 bora 风)
RAFOS technology RAFOS 技术〔是 Sofar(声发)的倒拼,也是 Sofar 的逆运用,是一种声学观测海洋内层洋流的拉格朗日方法〕
rafraichometer 体温表
rafted ice 多层冰
rageas 布拉风
ragged ceiling 碎云幂
raggiatura 强陆风(意大利语名)
ragut(=rageas,ghaziyah) 拉格特风(地中海伊斯肯德湾的飑线下沉风)
rain 雨【大气】
rain and snow 雨夹雪
rain and snow mixed 雨夹雪
rain and snow shower 阵性雨夹雪
rain area 雨区
rain attenuation 雨衰减

rainband 雨带【大气】
rain belt 雨带
rainbow 虹【大气】
rain cell 雨胞
rain cloud 雨云
rain cloud polarization 雨云极化
rain cover 防雨罩【机械】
rain crust 雨斑壳,雨水冻结壳
rain day 雨日【大气】
rain detector 感雨器
rain drop 雨滴【大气】
raindrop disdrometer 雨滴谱仪【大气】
raindrop erosion 雨滴侵蚀【土壤】
raindrop freezing 雨滴冻结
rain droplet collector 雨滴谱仪
raindrop kinetic energy 雨滴动能【土壤】
raindrop recorder 雨滴记录器
raindrop size 雨滴大小【土壤】
raindrop size distribution 雨滴谱【大气】
raindrop size distribution of Marshall and Palmer 马歇尔-帕尔默雨滴谱
raindrop sorting collector 雨滴选集器
raindrop spectrograph 雨滴谱仪
raindrop spectrum 雨滴谱
rainer 降雨装置【机械】
rain erosion 雨蚀【大气】
rain factor 降水因子
rainfall ①雨量【航海】②降雨
rainfall〔amount〕 雨量【大气】
rainfall amount during Meiyu period 梅雨量
rainfall area 降雨区
rainfall chart 雨量图【大气】
rainfall density 雨量强度
rainfall distribution 雨强分布,雨量分布【农】
rainfall duration 降雨持续时间【农】
rainfall effectiveness 降雨效率
rainfall erosion 雨蚀〔作用〕【农】,降雨

侵蚀【土壤】
rainfall excess 超渗雨量
rainfall factor 降水因子
rainfall frequency 降雨频率
rainfall hour 降雨时数
rainfall information 降雨信息
rainfall intensity 降雨强度【水利】
rainfall intensity pattern 雨强分布型
rainfall intensity recorder 雨强计【大气】
rainfall intensity return period 雨强重现周期
rainfall inversion 雨量逆减
rainfall loss 雨量损失
rainfall message 雨量报
rainfall pattern 雨型
rainfall phone-line network 电话应答式雨量站
rainfall potential 降雨潜势
rainfall rate 降雨率
rainfall recorder 雨量计,自计雨量计【水利】
rainfall regime 雨量型
rainfall reliability 降雨量可靠性
rainfall simulation 人工降雨【土壤】
rainfall runoff relation 降雨径流关系【水利】
rainfall station 雨量站
rainfall variability 雨量变率
rainfed cropland 雨养农田
rainfed farming 雨养农业【农】
rain field 雨区,雨量场
rain forest 雨林【植物】
rainforest climate 雨林气候
rain frequency 降雨频率
rain gauge 雨量器【大气】
rain gauge for radar calibration 雷达标定雨量计
rain gauge shield 雨量器防护罩
rain gauge wind shield 雨量器防风罩,雨量器风档

rain gush(或 gust) 暴雨
rain gust 暴雨
rain hour 降雨时数
raininess 雨〔量〕强〔度〕
rain intensity 〔降〕雨强〔度〕
rain intensity gauge 降水强度器
rainless region 无雨区
rain-making 人造雨,人工降雨【农】
rain-measuring radar 测雨雷达
rain-out 雨洗【大气】(指雨从大气中清除尘埃等),雨冲刷
rain penetration 雨水渗透【建筑】
rain-producing system 致雨系统
rain-rate gauge 雨强计
rain salinity 雨水含盐度
rain shadow 雨影【大气】
rainshadow effect 雨影效应
rain shower 阵雨
rain simulator 降雨模拟机,模拟降雨装置【土壤】
rain spell 雨期
rain squall 雨飑
rain stage 成雨阶段
rain stimulation by artificial means 人工影响降水,人工催化降水
rainstorm 暴雨【水利】,雨暴
rainstorm combination 暴雨组合
rainstorm erosion 暴雨侵蚀【土壤】
rainstorm flood 暴雨洪水
rainstorm intensity 暴雨强度【公路】
rainstorm investigation 暴雨调查
rainstorm runoff 暴雨径流【公路】
rainstorm transposition 暴雨移置
rainstorm warning 暴雨预警
rainstorm with strong wind 暴风雨
raintrees 雨林
rain virga 雨幡【大气】
rainwash 雨蚀
rain washout 降雨冲刷,雨洗
rainwater 雨水,软水
Rain Water 雨水【大气】(节气)

rainy climate 多雨气候
rainy day 雨日【农】
rainy season 雨季【大气】
rainy tropics 多雨热带
raising erosion 扬蚀【土壤】
RAM (random access memory) 随机存储器【计】
Raman effect 拉曼效应【电子】
Raman frequency shift 拉曼频移
Raman lidar 拉曼光达
Raman lidar detection 拉曼光雷达探测
Raman scattering 拉曼散射
Raman spectral analysis 拉曼光谱分析
RAMOS (remote automatic meteorological observing system) 自动遥感气象观测系统
ramp function 斜坡函数
ramp structure 斜坡结构
rams 冰角
RAMS (random access measurement system) 随机通道测量系统
ramsonde 冰雪硬度器
random 随机〔的〕【数】
random access 随机存取【计】,随机接入【数】
random access measurement system (RAMS) 随机通道测量系统
random access memory (RAM) 随机存储器【计】
random amplitude 随机振幅
random decision 随机决策
random Elsasser band model 随机爱尔沙色带模式【大气】
random error 随机误差【数】
random fluctuation 随机起伏
random forcing 随机强迫
random forecast 随机预报【大气】
random function 随机函数【数】
randomization 随机化
randomized cross experiment 随机交叉试验
randomized experiment 随机化试验
randomized experiment at single domain 单区随机试验
randomized seeding trial 随机播云试验
randomized test 随机化检验【数】
random medium 随机介质
random mixing 随机混合
random model 随机模式
randomness 随机性
random noise 随机噪声【电工】
random number 随机数【数】
random observing error 随机观测误差
random phase approximation 随机相位近似,无规相〔位〕近似【物】
random process 随机过程【自动】
random sample 随机样本【数】
random sampling 随机取样【地质】
random sea 随机海〔况〕
random signal 随机信号
random variable 随机变量【数】
random variate 随机变量
random walk 随机游动【数】,无规行走【物】
range ①极差【数】②距离,范围③量程【电子】,射程【物】④波段⑤发射场
range aliasing 距离混淆〔现象〕
range ambiguity 距离模糊【大气】
range analysis 极差分析
range and range rate 距离及距离变化率
range attenuation 距离衰减【大气】
range averaging 距离平均【大气】
range bin 距离库【大气】
range delay 射程延迟
range distortion 射程畸变
range elevation indicator 距离仰角显示器
range error 射程误差
range finder 测距仪【航海】
range folding 距离折错,距离混淆〔现象〕

range gate 距离门(雷达)
range gating 距离选通
range-height indicator (RHI) 距离高度显示器【大气】
range marker 距离标志【电子】,固定距标【航海】
range normalization 距离归一化,距离订正
range of tide 潮差
range resolution 距离分辨率【大气】
range strobe 距离标
range-velocity display (RVD) 距离速度显示【大气】
range wind 射程风,纵风
rank correlation 秩相关,等级相关
Rankine temperature scale 兰氏温标(冰点为 491.69 °R,沸点为 671.69 °R,中间分为 180°,同华氏分度)
Rankine vortex 兰金涡旋
rank of matrix 矩阵的秩【数】
RAOB (radiosonde observation) 〔无线电〕探空〔仪〕观测〔记录〕
Raoult effect 拉乌尔效应(亲水沾污物使露点仪镜面视在露点下降)
Raoult's law 拉乌尔定律
rapid city flood 城市湍急洪水
rapid distortion theory 快速畸变理论,陡变理论
rapid interval imaging 快速间隔取像,短间隔取像
rapid interval scan 快速〔间隔〕扫描,短间隔扫描
rapidly moving aurora 速移极光
rapidly moving cold front 快行冷锋【大气】,第二型冷锋
rapid scan mode 快速扫描方式
rare earth element 稀土元素【化】
rare earth metal 稀土金属【化】
rare element 稀有元素【化】
rarefaction wave 稀疏波【物】
rarefied gas dynamics 稀薄气体动力学【力】
rare gase 稀有气体【化】
rare metal 稀有金属【化】
rare optical phenomenon 奇异光学现象,罕见光象
RAREP (radar report) 〔航空天气代码〕雷达报告
RASAPH (radar signal spectrograph) 雷达信号摄谱仪
RASS (radio acoustic sounding system) 无线电-声学探测系统
raster line 光栅线
rate coefficient 速率系数
rate constant 速率常数【化】,速率系数
rated output 标定功率【航海】
rated power 标定功率【航海】,额定功率【电力】
rate equation 速率方程【化】
rate of accretion 〔碰冻〕增长率
rate of change 变率
rate-of-climb indicator 爬升指示器
rate of convergence 收敛速率【数】,辐合率
rate of ice accretion 积冰率
rate of ion production 离子产生率
rate of Lagrangian to Eulerian time scale 拉格朗日-欧拉时间尺度比
rate of precipitation 降水率
rate-of-rainfall gauge 雨强计,降雨率记录器
rate-of-rainfall recorder 雨强计,降雨率记录器
rate of response 响应速率
rate of speed 额定风速
rating curve 率定曲线,检定曲线
rating flume 标定槽,检定槽
ratio 比【数】,比率
rational distribution of factories 厂房合理布局
rational function 有理函数【数】
rational horizon 合理地平
rational method 推理〔方〕法,利用率法

【林】
rational number 有理数【数】
ratio of evaporation to transpiration 蒸发蒸腾比
rattler 猛烈雷暴,倾盆大雨
ravine wind 峡谷风,沟谷风
raw data 原始资料,原始数据【计】
raw data of wind-finding record 测风原始记录
rawin(=radio wind finding) 无线电测风仪
rawin target 无线电测风靶
rawinsonde 无线电探空测风仪【大气】
rawinsonde observation 无线电探空测风观测
rawinsonde station 无线电探空测风站
raw moment 原力矩
raw radio sounding record 探空原始记录
raw spectral density estimate 原始谱密度估计量
raw soil 生土【土壤】
raw water 原水【土木】
ray 射线【数】,放射
ray bending 射线弯曲
ray equation 射束方程
Rayleigh atmosphere 瑞利大气
Rayleigh-Benard convection 瑞利-贝纳对流
Rayleigh-Benard convection rotation 瑞利-贝纳对流旋转
Rayleigh-Benard instability 瑞利-贝纳不稳定性【物】
Rayleigh cell 瑞利单体
Rayleigh-Debye scattering 瑞利-德拜散射
Rayleigh distribution 瑞利分布
Rayleigh friction 瑞利摩擦
Rayleigh hypothesis 瑞利假设
Rayleigh instability 瑞利不稳定
Rayleigh-Jeans formula 瑞利-金斯公式【物】,R-J 公式
Rayleigh-Jeans law 瑞利-金斯定律,R-J 定律
Rayleigh number 瑞利数【大气】
Rayleigh optical depth 瑞利光学厚度
Rayleigh parameter 瑞利参数
Rayleigh phase function 瑞利相函数
Rayleigh scattering 瑞利散射【大气】,分子散射
Rayleigh's formula 瑞利公式
Rayleigh's scattering law 瑞利散射定律
Rayleigh smoothness criterion 瑞利平滑判据
Rayleigh wave 瑞利波
rayon cloud 人造云
ray optics 射线光学
ray tracing 光线追迹【物】
RDF (radio direction finding) 无线电测向
reach ①河段②到达距离③区域
reaction kinetics 反应动力学【化工】
reaction order 反应级数【化】
reaction rate 反应速率【化】,反应率【物】
reactivation 再活化,复活【地质】
reactive power 无功功率【物】
read-in 读入
reading 读数
reading error 读数误差
read microscope 读数显微镜
read only memory (ROM) 只读存储器【计】
read out 读出
read-out station 指令和资料接收站
read-out time 读出时间【计】
read-write cycle 读写周期【计】
real air temperature 实际气温
real atmosphere 实际大气
real horizon 真地平
realizability 可实现性【数】
realized energy 实在能量
real latent instability 真潜〔在〕不稳定〔度〕
real number 实数【数】

real part 实部【数】
real root 实根【数】
real time 实时【计】
real time analysis 实时分析
real time analyzer 实时分析仪
Real Time Computing Center (RTCC) 实时计算中心
real time data 实时资料【大气】
real time display 实时显示【大气】
real-time ice particle size analysis system 冰粒尺度实时分析系统
real-time measurement 实时测量,实时测定
real-time meteorological data 实时气象资料
real-time prediction 实时预报
real-time process 实时过程
real-time processing 实时处理【计】
real time signal processing 实时信号处理【电子】
real-time system 实时系统【计】
real-time transmission 实时传送
real-time transmission system (RTTS) 实时传送系统
real-time video 实时视频【计】
real video on demand 真点播电视【计】
reanalysis 再分析〔资料〕
rear 后部
rear-inflow jet 后向流入急流,尾部流入急流
rear of depression 低压后部,低压尾
Reaumur temperature scale 列氏温标
Reaumur thermometer 列氏温度表
rebat 里巴风(日内瓦湖的微风)
reboyo 里包约风(巴西沿海的风暴)
recco "飞机气象侦查(reconnaissance)"的常用缩写
RECCO code 飞机气象侦查编码
received power 接收功率
receiver ①接收机,接收器【计】②输入元件

receiver gain 接收机增益
receiver〔noise〕temperature 接收机〔噪声〕温度【大气】
receiving aerial 接收天线
receiving antenna 接收天线
receiving antenna of meteorological satellite 气象卫星接收天线
receiving bucket 受雨桶
receiving cross section 接收截面
recent drizzle 观测时有毛毛雨
recent freezing rain 观测时有冻雨
recent hail 观测时有雹
recent rain 观测时有雨
recent rain and snow 观测时有雨夹雪
recent snow 观测时有雪
recent snow shower 观测时有阵雪
recent thunderstorm 观测时有雷暴
receptor 受体【生化】
recession 消退
recession curve 消退曲线,退水曲线【地理】
recharge ①回灌,补给②再装③再充电
recharge area 补给区(地下水),回灌区,供水区【石油】
recharge capacity 回灌容量,补水容量
reciprocal spreading 互反展开
reciprocity 互反性
reciprocity law 互反律【数】
reciprocity principle 互反原理
reciprocity theorem 倒易定理【物】,互易定理【电子】
recirculation 再循环【医】,闭合环流
recognition mode 识别模式
recombination 复合【物】
recombination coefficient 复合系数
recombination energy 复合能
recombination rate 复合率【物】
recommended route 推荐航线
reconnaissance 飞行侦察(常简写为 recco 或 recon)
reconnaissance summary 飞行侦察简报

（台风的）
reconnection 重联
reconstruction 重建,复原
record data line 记录日界
recorder 记录器
recording albedometer 自记反照率表
recording anemometer 〔自记〕风速计
recording arm 记录杆
recording barometer 〔自记〕气压计
recording ceilometer 自记云幂仪
recording clock 自记钟
recording cylinder 记录筒,钟筒
recording drum 记录筒,钟筒
recording equipment 自记设备
recording frigorimeter 〔自记〕冷却计
recording instrument 自记仪器
recording lever 记录杆
recording observation 记录观测
recording ombrometer 自记微雨量计
recording pan 自记蒸发皿
recording pen 自记笔
recording pluviometer 自记雨量计
recording pointer 自记笔
recording potentiometer 自记电位计
recording psychrometer 自记干湿计
recording raingauge 雨量计【大气】
recording theodolite 自记经纬仪
recording visibility meter 自记能见度仪
record observation 时次观测
records and documents of meteorological science and technology 气象科技文献资料
record with data-missing 记录缺测
recovery 复苏现象
recovery factor 订正因子
recovery test 复原检验
recovery time 恢复时间【电子】,恢复周期【电工】
rectangular Cartesian coordinate 直角笛卡儿式坐标
rectangular coordinate 直角坐标【数】,正交坐标
rectangular curvilinear coordinate 正交曲线坐标
rectangular function 矩形函数
rectangular grid 长方形网格
rectangular waveguide 矩形波导
rectangle 矩形,长方形【数】
rectification 整流【物】
rectifier 整流器【物】
rectilinear current 往复〔海〕流
rectilinear wind shear 线性垂直风切变
recurrence 回复,重现
recurrence formula 递推公式,循环公式
recurrence frequency 重现频率
recurrence interval 重现期
recurrence relation 递推关系
recurrence surface 重复面
recurrent flow 回流
recurrent magnetic storm 重现磁暴
recurrent period 重现周期
recurring decimal 循环小数【数】
recurring weather 重现天气
recursion 循环
recursiveness 递归性【数】
recurvature 转向
recurvature latitude 转向纬度
recurvature of storm 风暴转向
recurvature of typhoon 台风转向
recurved occlusion 后曲锢囚
recycled water 再循环水
red flash 红闪
red noise 红噪声
red noise phenomenon 红噪声现象
red-noise type 红噪声型
redox 氧化还原反应,氧化还原〔作用〕【化】
redox cell 氧化还原电池【电子】
red rain 红雨
Red Sea Water 红海水
red-shift 红移【物】
red snow 红雪

Red Spot Hollow 红斑穴
red tide 赤潮【航海】
reduced 订正过的
reduced equation 简化方程【数】
reduced frequency 约化频率,简约频率【力】
reduced form 约化公式
reduced gravity 约化重力
reduced grid 约化网格
reduced incident intensity 约化入射强度
reduced kernel 约化核【数】
reduced mass 约化质量
reduced pressure 订正气压
reduced temperature 订正温度
reduced wave number 约化波数
reducing agent 还原剂
reducing atmosphere 还原性大气
reduction 订正,整理,换算,简化,归约【数】
reduction factor 订正因子
reduction of a fraction 约分【数】
reduction of pressure to a standard level 标准等压面气压订正
reduction of temperature to mean sea-level 平均海平面温度订正
reduction of variance 方差缩减
reduction table 订正表
reduction to sea level 海平面订正
reductive perturbation 约化摄动
redundancy(＝redundance) 冗余度
redundant 多余观测
red water 赤潮【航海】
Reech number 里奇数(弗罗德数的倒数)
re-entry 再入,重返(大气层)
reference 参考【计】,参考文献
reference atmosphere 参考大气【大气】
reference circuit 参考电路【电子】
reference climatological station 基准气候站
reference condition 标准条件【电子】
reference crop evaporation 参考作物蒸发〔量〕
reference data 参考资料
reference ellipsoid 参考椭球体
reference frame 参考系【物】
reference information 参照信息(质量检验程序中用)
reference junction 参比接点【电工】
reference level 参考层,参考水平【化】
reference material 标准物质【化】,参比物质【海洋】
reference measurement 测量准值,基准测量
reference pressure 参考气压,参考压力【煤炭】
reference radiosonde 基准无线电探空仪,标准无线电探空仪
reference sonde 基准探空仪,标准探空仪
reference sound power 参考声功率【机械】,基准声功率
reference sound pressure 参考声压【机械】,基准声压
reference standard 参照标准
reference standard barometer 基准气压表
reference temperature 参考温度
reference thermometer 基准温度表
referral 检索
reflectance 反射比【物】
reflectance measurement 反射比测量
reflected fohn 反射焚风
reflected global radiation 地球反射辐射
reflected radiation 反射辐射【大气】
reflected ray 反射线
reflected solar radiation 反射太阳辐射【大气】
reflected sunlight 反射太阳光
reflected sunlight correction 反射太阳光订正
reflected terrestrial radiation 向上反射地球辐射
reflecting nephoscope 反射测云器
reflecting power 反射能力,反射率

reflecting telescope 反射望远镜【天】
reflection 反射【物】
reflection coefficient 反射系数
reflection function 反射函数
reflection nephoscope 反射测云器
reflection of light 光的反射
reflection rainbow 反射虹
reflective index 反射率
reflective power 反射能力,反射率
reflective wave 反射波
reflectivity 反射率【大气】
reflectivity coating 反射膜层
reflectivity factor 反射率因子
reflectometer 反射表,反射计【电子】,反射辐射表
reflector 反射器,反射镜,反射望远镜【天】
reflex klystron 反射速调管【电子】
reflexion(=reflection) 反射
refoli(=raffiche) 拉菲希阵风
reforecast 再预报
refracted ray 折射线
refracting telescope 折射望远镜【天】
refraction 折射【物】,蒙气差
refraction coefficient 折射系数
refraction correction 折射修正
refraction diagram 折射图
refraction error 折射误差
refraction haze 折射霾
refraction index 折射率【物】
refraction index of radio wave 电波折射率
refraction law 折射定律【物】
refraction modulus 折射模数
refraction of water waves 水波折射
refraction prediction 折射预测
refractive index 折射率【物】
refractive index factor 折射率因子
refractive index fluctuation 折射率起伏
refractive index structure constant 折射率结构常数

refractive modulus 折射模数
refractivity ①折射模数(=refractive modulus)②折射本领
refractometer 折射计【物】,折光仪
refractor 折射望远镜【天】
refrigerant 制冷剂【航海】
refrigerator 制冷机,冷气机,电冰箱
reforestation 再造林
Refsdal diagram 列夫斯达图
reg 砾〔质沙〕漠
regelation 再冻〔作用〕,复冰现象
regeneration 再生
regeneration of a depression 低压再生【大气】
regeneration of typhoon 台风再生
regime ①体系②季节变化特征③状况
regime channel 稳定河槽,不冲不淤渠道
regional agroclimatic delimitation 地方农业气候区划,区域农业气候区划
regional air pollution 区域性空气污染
Regional Air Pollution Study(RAPS) 区域性空气污染研究
regional analysis 区域分析
regional analysis and forecasting system (RAFS) 区域分析和预报系统
Regional Association (RA) 区域协会(世界气象组织的)
Regional Association for Africa (RA Ⅰ) 世界气象组织第一(非洲)区域协会
Regional Association for Asia (RA Ⅱ) 世界气象组织第二(亚洲)区域协会
Regional Association for Europe (RA Ⅵ) 世界气象组织第六(欧洲)区域协会
Regional Association for North and Central America (RA Ⅳ) 世界气象组织第四(北、中美)区域协会
Regional Association for South America (RA Ⅲ) 世界气象组织第三(南美)区域协会
Regional Association for the South-West

Pacific(RA Ⅴ) 世界气象组织第五(西南太平洋)区域协会
regional auxiliary meteorological telecommunication network 地区气象辅助电信网
regional basic synoptic network 区域基本天气网
regional broadcast 区域广播
Regional Center for Tropical Meteorology(RCTM) 热带气象区域中心(美国)
regional classification of aviation climate 航空气候区划
regional climate 区域气候【大气】
regional climate modeling 区域气候模拟
regional climatology 区域气候学
regional ensemble forecast 区域集合预报
regional forecast 区域预报【大气】
regional geology 区域地质学
regionalization map 区划图
Regional Meteorological Centre(RMC) 区域气象中心(世界气象组织)
regional meteorological office 区域气象中心
regional meteorological subcentre 区域气象副中心
Regional Meteorological Telecommunication Network(RMTN) 区域气象电信网
regional model 区域模式【大气】
regional modeling 区域模拟
regional precipitation 区域性降水
regional pollution 区域性污染
regional radiation center(RRC) 区域辐射中心
regional regression randomized experiment 区域回归随机试验
regional scale model 区域尺度模式
regional spectral model 区域谱模式
regional standard 区域标准器

regional standard barometer 区域标准气压表【大气】
Regional Telecommunication Hub(RTH,WWW) 区域电传通信枢纽(WWW)
region of escape 外气层
region of influence 影响区
region of thunderstorm activity 雷暴活动区
register 自计器,寄存器【计】
registering balloon 记录气球
registering rain gauge 自记雨量计
registering weather vane 风向计
registering wind 自计风标
register pen 记录笔
registrar 记录员
regolith 风化层【地理】,土被
regreening 返青
regreening stage 返青期
regressand 回归值
regression 回归
regression analysis 回归分析【数】
regression coefficient 回归系数【数】
regression diagnostics 回归诊断【数】
regression equation 回归方程
regression estimate 回归估计【数】
regression function 回归函数【数】
regression line 回归线
regression matrix 回归矩阵
regression method 回归法
regression of the node 交点退行【天】(天体轨道升交点经度不断减少的现象)
regression prediction equation 回归预报方程
regressor 回归量
regular advection 正则平流
regular band model 规则带模式【大气】
regular broadcast 定时广播
regular boundary 正则边界【数】
regular forecast 定时预报
regular observation 常规观测

regular peaked and long train atmospherics 规律尖峰长列型天电
regular perturbation 正则摄动
regular reflection 单向反射
regular reflector 单向反射体
regular smooth atmospherics 规律平稳型天电
regular synoptic report 常规天气〔图〕报告
Regular World Days（RWD） 预定世界日（国际地球物理年期间事先选定的某些工作日，每月约 3~4 天，以便同时观测）
regulating device 调准装置
regulation of verification 检定规程
regulator 调节器
reinforcement 增强
rejection region 拒绝区域【数】,否定区域
relative age 相对时代
relative angular momentum 相对角动量
relative aperture 相对孔径【物】
relative error 相对误差【数】
relative concentration 相对浓度
relative contour 相对等高线
relative coordinate system 相对坐标系
relative current 相对流
relative density 相对密度
relative divergence 相对辐散,相对散度
relative evaporation 相对蒸发
relative evaporation gauge 相对蒸发器
relative frequency 相对频率
relative gradient 相对梯度
relative gradient current 相对梯度流
relative height 相对高度
relative homogeneity 相对均匀性
relative humidity 相对湿度【大气】
relative humidity of moist air with respect to ice 湿空气冰面相对湿度
relative humidity of moist air with respect to water 湿空气水面相对湿度
relative humidity probe 相对湿度探头
relative humidity with respect to ice 冰面相对湿度
relative hypsography 相对高度型
relative ionospheric opacity-meter（riometer） 电离层电磁吸收测定仪
relative isohypse 等厚度线,等高度线
relative luminous efficiency 相对发光效率
relative mean variability 平均相对变率
relative moisture of the soil 土壤相对湿度
relative momentum 相对动量
relative motion 相对运动【物】
relative number of sunspot 黑子相对数【大气】
relative ozone concentration 相对臭氧浓度
relative permeability 相对磁导率【物】
relative permittivity 相对电容率【物】
relative phase change 相对相位变化
relative pluviometric coefficient 相对雨量系数
relative reduction 相对递减率
relative response 相对响应
relative sea level 相对海平面
relative scattering function 相对散射函数
relative scattering intensity 相对散射强度
relative soil moisture 土壤相对湿度【大气】
relative spectrum 相对谱
relative stability 相对稳定〔度〕
relative sunshine 相对日照
relative sunspot number 相对太阳黑子数
relative topography 相对形势
relative velocity 相对速度
relative visibility 相对能见度
relative vorticity 相对涡度【大气】
relative water content of soil 土壤相对含水量
relative wind 相对风
relativity 相对论【物】
relaxation 张弛,弛豫【物】
relaxation method 松弛法【数】,张弛法

【大气】
relaxation of a trough 低槽张弛
relaxation oscillator 张弛振荡器
relaxation time 弛豫时间【物】,松弛时间
relaxed algorithm 松弛算法【计】
relaxed eddy accumulation 张弛涡动累积
relay-interplanting 套种
relay station 中继站【计】,接力站【电子】
releasable energy 可释放能量
reliability 可靠性,可靠率
reliability test 可靠性检验
reliable weight 可靠权重
relief ①地形,地势②地形模型
relocation 再定位,浮动
relocation diffusion 迁移扩散
remainder 余数【数】
remaining green 贪青
remanent magnetism 剩磁
remote access 远程访问【计】
remote access server 远程访问服务器【计】
remote automatic meteorological observing system（RAMOS） 自动遥测气象观测系统
remote control equipment 遥控装置
remote controlled air sampler 遥控空气取样器
remote database access 远程数据库访问【计】
remote forcing 远距离作用
remote heterodyne detection 远距离外差探测
remote infrared atmospheric profiling system 大气廓线红外遥测系统
remote instruction 远程教学【计】
remotely sensed data 遥感资料
remote measurement 遥测
remote meteorological station 遥测气象站
remote mode 遥测方式
remote picture 遥测图像

remote range-resolved hygrometry 远距离分辨测湿法
remote reading psychrometer 遥测干湿表
remote response 遥响应
remote sensing（RS） 遥感【地信】
remote sensing data 遥感数据
remote sensing information 遥感信息
remote sensing instrument 遥感仪器
remote sensing of the atmosphere 大气遥感
remote sensing platform 遥感平台
remote sensing technique 遥感技术
remote sensor 远距离传感仪
remote sounding 遥〔测〕探〔空〕
remote temperature diagnostic 远距离温度诊断
remote terminal 远程终端【计】
remote thermometer 遥测温度计
removal correction 移动订正
renewable energy source 可再生能源
renewable resources 可再生资源
renormalization group 重正规化群
reoccurrence of Meiyu 倒黄梅
REOF（rotated EOF analysis） 旋转EOF分析
repeatability 重复性
repetition frequency 重复频率
repeller 排斥子,反射极
report-form examination 报表审核
reporting point 报告点
report of ground radar weather observation（RADOB） 地面雷达天气观测报告
representability 可表示性【数】
representative basin 代表流域【水文】
representative meteorological observation 代表性气象观测
representativeness 代表性
representative period 再生期
reproducibility 复现性【电子】,可重复性【物】

reproduction 复制
reproductive growth 生殖增长
reradiation 再辐射
re-radiator 再辐射器,反射靶
Research and Development Board of America (RDB) 美国研究发展局
Research Flight Facility (RFF) 〔大气〕研究飞行中心
research paper 研究论文
research report 研究报告
research vessel 调查船,考察船
reseau 测候网(法语名)
reseau mondial 世界〔台站〕网(法语名)
resemblance 相似性
reservoir ①水库②蓄水池③水箱④储油层
reservoir compounds 气藏化合物(如 $CINO_3$, HO_2 NO_2 等)
reservoir operation 水库控制运行,水库业务运行
reset 调整,复位【计】,清除
reshabar(=rushabar) 黑风(土耳其拜尔德斯坦的 bora 风)
residence half-time 半滞留期
residence time 滞留期,停留时间【化】
residual 残差【数】
residual ageostrophic motion 剩余非地转运动
residual atmosphere 剩余大气
residual circulation 剩余环流
residual depression storage 剩余洼地蓄水量
residual layer 残留层【大气】
residual loss 剩余损失,剩余损耗【物】
residual mass curve 残留〔质〕量曲线
residual mean circulation 剩余平均环流
residual mean meridional circulation 余差平均经向环流,剩余平均经向环流
residual meridional circulation 余差经向环流,剩余经向环流
residual stream function 余差流函数,剩余流函数
residual sum of squares 残差平方和【数】
residual variance 剩余方差
residue 剩余【数】,残数,留数
resilience 复原【动】,回弹力【物】
resin 树脂【化】
resinous electricity 阴电
resistance ①阻力②电阻【物】
resistance psychrometer 电阻式干湿表
resistance thermometer 电阻温度表
resistance-type mercury barograph 电阻式水银气压计
resistivity 电阻率【物】
resolution ①分解,分辨,判定②预解【数】③分辨率【物】
resolution cell 分辨单元
resolution of facsimile equipment 传真机分辨率,传真设备分辨率
resolution of imagery 图像分辨率
resolution of power of optical instrument 光学仪器的分辨率
resolution of velocity 速度分解
resolution volume 分辨体积
resolving power 分辨能力
resonance 共振【物】
resonance absorption 共振吸收【物】
resonance broadening 共振加宽
resonance cascading 共振级联
resonance fluorescence 共振荧光【物】
resonance fluorescence lidar 共振荧光光〔雷〕达
resonance frequency 共振频率
resonance interaction 共振交互作用
resonance interaction of Rossby waves 罗斯贝波共振交互作用
resonance oscillation 共振振动
resonance radiation 共振辐射
resonance Raman scattering 共振拉曼散射
resonance scattering 共振散射【物】
resonance theory 共振理论【大气】,共振

学说
resonance trough 共振槽
resonance waves over hills 过山共振波
resonance waves over thermals 热泡共振波
resonant cavity 共振腔【物】
resonant frequency 共振频率
resonant spectrophone 共振光谱测声器
resonant triad 三波共振子
resonator 共振器【物】
resource information system 资源信息系统
resource-sharing 资源共享
respiration 呼吸〔作用〕
respiratory chain 呼吸链【生化】
responder 回答器
response ①特性曲线②灵敏度③反响,响应
response analysis 特性曲线分析
response distance 响应距离,距离常数
response envelope spectrum (RES) 反应包络谱
response function 响应函数【物】
response length 响应距离
response spectrum 响应谱
response time 响应时间
responsible forecasting area 预报责任区【大气】
responsive parameterization 响应参数化
responsivity 响应率
rest frame 静止参考系
resting stage in autumn 秋季停止生长期
rest mass 静质量【物】
restoring 回复,重建
restoring force 回复力
resultant 合量,合成量,结式【数】
resultant error 总误差,合成误差
resultant force 合力【物】
resultant of forces 力的合成
resultant temperature 合成温度
resultant temperature index 合成温度指数
resultant wind 合成风【大气】
resultant wind direction 合成风向
resultant wind velocity 合成风速
result of every other year 隔年结果
result of measurement 测量结果
resumed operation 恢复作业
retardation 减速
retarded flow 减速流,阻滞流
retarding basin 滞洪区
retarding effect 减速效应
retention curve 持水曲线
retention factor 保留因子
retention storage 持水量,拦蓄量
retraction 收缩【数】
retransmitter 中继〔转发〕发射机
retreater 退回
retrieval ①检索②恢复(信息的)③反演,还原
retro-correlation 反相关
retroflection(=retroflexion) 反曲率
retrograde depression 后退低压
retrograde orbit 后退轨道
retrograde system 后退系统
retrograde wave 后退波【大气】
retrogression 退行【大气】,逆行,退化【动】,后退
retrogression-reflection 强回声〔区〕
retrogressive wave 后退波
retroreflection 返射【物】
retroreflector 返射器
return convection 回返对流,冷对流
return current 回流
return echo 回波
returned flash 回闪
return-flow wind tunnel 回流式风洞,闭式回流风洞
returning air-mass 回流气团
returning flow 回流
returning flow weather 回流天气【大气】
returning polar air (mass) 极地回流气

团,极地回流空气
returning polar maritime air 极地回流海洋气团
return lightning 回〔返〕闪〔电〕
return period 重现期
return signal 回波信号
return streamer 回流,回返闪流
return stroke 回击【大气】
returnto zero 归零制,回零
reverberation 混响
reversal film 反转片【测】
reversal of the monsoon 季风转换
reversal temperature 转换温度
reverse cell 逆环流圈
reversed tide 逆潮
reverse flow thermometer (RFT) 回流式温度表
reverse flow thermometer housing 回流式温度表罩
reverse of phenological sequence 物候倒置
reverse-oriented monsoon trough 反向季风槽
reversibility 可逆性,可易性
reversible adiabatic process 可逆绝热过程【大气】
reversible change 可逆变化
reversible circulation 可逆环流
reversible cycle 可逆循环
reversible engine 可逆机
reversible machine 可逆机械
reversible Markov process 可逆马尔可夫过程【数】
reversible moist-adiabatic process 可逆湿绝热过程
reversible process 可逆过程【物】
reversible saturation adiabatic process 可逆饱和绝热过程
reversing current 往复〔潮〕流
reversing layer 反变层
reversing thermometer 颠倒温度表
reversing tidal current 往复潮流

reverted warm front 倒暖锋
review ①回顾②评述
revised local reference 本站订正基准面
revolution ①旋转②公转【天】
revolution of the earth 地球公转
revolving fluid 旋转流体
revolving storm 旋转风暴
Reynolds analogy 雷诺相似
Reynolds averaging 雷诺平均
Reynolds condition 雷诺条件
Reynolds effect 雷诺效应
Reynolds number (Re) 雷诺数【大气】
Reynolds number similarity 雷诺数相似性
Reynolds stress 雷诺应力【大气】
RF (radio frequency) 射频
RF signal 射频信号
rheology 生物流变学【力】,流变学
RHI (range-height indicator) 距离高度显示器【大气】
RHI scope 距离高度显示器
rhizobia 根瘤菌【土壤】
rhomboidal truncation 菱形截断
rhumb line 等角线(航行用)
rhumb line course 等角航线
rhythm 韵律
ribbon ice 带状冰
ribbon lightning 带〔状〕闪〔电〕
ribut 里巴特飑(马来西亚5—11月持续时间短的飑)
ribonucleic acid 核糖核酸【生化】
Riccati differential equation 里卡蒂微分方程【数】
rice blast induced by coldness 冷稻瘟病
Richardson extrapolation 理查森外推〔法〕【数】
Richardson grid 里查森网格
Richardson number 里查森数【大气】
Richard thermograph 理查德温度计
Richter scale 里氏〔地震等级〕表(分为1到10级)
Richter magnitude 里氏震级【地】

ridge 高压脊【大气】,脊
ridge aloft 高空脊
ridge cultivating 垄作
ridge line 脊线【大气】
ridge of high pressure 高压脊
ridge point 脊点
ridge regression 岭回归
ridging 脊生
riefne 马耳他强风暴
Riemannian geometry 黎曼几何[学]【数】
Riemann-Hilbert problem 黎曼-希尔伯特问题
rift in clouds 云隙,云中裂缝
rig（见 Michael-riggs） 米迦勒节捣乱风
right angle 直角【数】
right ascension 赤经【大气】
right-handed coordinates 右手〔旋转〕坐标
right-handed rotation 右手式旋转
right-handed rectangular coordinates 右手直角坐标系
right triangle 直角三角形【数】
rigid body 刚体【物】
rigid boundary 刚体边界
rigid boundary condition 刚体边界条件【大气】
rigid disk 硬磁盘,硬盘【计】
rigid-lid 刚盖
rigid rotation 刚体旋转
rigid system 刚体系
rime 雾凇【大气】,树挂,不透明冰
rime fog 冰雾
rime ice 雾凇冰,粗冰
rime rod 雾凇探棒
rime sensor 雾凇测量器,雾凇传感器
rime-splinter process 凇碎过程
riming 结凇,凇附
rind ice 壳冰
Rinehart projection 莱因哈特投影
ring densitometry 年轮密度测定

Ringelmann chart 林格曼图
Ringelmann number 林格曼数
Ringelmann shades 林格曼阴影
ring halo 环晕
ringing 振铃【电子】
ring vortex 环状涡旋
ring-width index 轮宽指数
riometer 电离层电磁吸收测定仪
rip 裂流,裂浪(两股水流相冲击形成的巨澜)
rip current 离岸流,裂流
ripe 软雪,粗粒状雪
ripening of snow 雪的软化(熟化)
ripe snow 软雪
ripple wave 涟〔波〕,皱波【物】
rip tide 离岸潮
rise 上升
rise time 上升时间
rising limb 涨水段,涨洪段
rising tide 涨潮
risk assessment 风险评估【计】
risk function 风险函数【数】
risk rate 风险率
river 河
river basin 流域
river basin planning 流域规划
river bed 河床
river fog 河雾
river forecast 流域预报
river forecast center（RFC） 河流〔水情〕预报中心
river gauge 水位尺,水位标
river ice 河冰
river network 河网
river stage 水位
river system 水系
river〔surface〕temperature 河面温度
R-meter R-计
R-mode factor analysis R-型因子分析
RMSE（root-mean-square error） 均方根误差【大气】,均方差

RMTN(Regional Meteorological Telecommunication Network) 区域气象电信网
Rn(radon) 氡【大气】
road damped by snow 雪阻
roaring forties 咆哮西风
ROB(radar observation) 雷达观测〔报告〕
Robin balloon 罗宾气球（气象火箭上携带的一种充气探测气球）
Robin constant 罗宾常数【数】
Robin Hood's wind 罗宾汉风（温度近零度的湿阵风）
robin snow 春天小雪
Robinson's anemometer 罗宾孙风速表，转杯风速表
Robin sphere 罗宾球
Robitzsch actinograph 罗别茨直接辐射计
Robitzsch bimetallic pyranograph 罗别茨双金属天空辐射计
robot 机器人【计】
robot device 自动装置
robot weather station 自动气象站
robust control 鲁棒控制【数】,稳健控制
robustness 稳健性【数】,鲁棒性
robustness regression 稳健回归，鲁棒回归【数】
rockair 机载探空火箭
rocket 火箭
rocket balloon instrument(RBI) 火箭气球装置（膨胀球）
rocket-borne detector 箭载探测器
rocket-borne laser radar system 箭载光达系统
rocket-borne photometric ozone and temperature measuring sonde 箭载光度法臭氧与温度探空仪
rocket exhaust cloud 火箭废气云
rocket exhaust trail 火箭排烟尾迹
rocket-grenade method 火箭榴弹法（测高空风用）
rocket-launching site 火箭发射点

rocket lightning 火箭状闪电【大气】
rocket meteorograph 火箭气象计
rocket ozonesonde 火箭臭氧探空仪
rocket projectile 火箭弹
rocketsonde 火箭探空仪
rocket sounding 火箭探测【大气】
rocket-triggered lightning 火箭激发闪电
rockoon 气球火箭（由气球上发射）
rockoon sounding 气球火箭探测
rodada 车辙风（24小时内从西北方向顺时针变动约360°的风,西班牙语名）
rod thermometer 棒状温度表
ROFOR(route forecast) 航线预报编码报头组（沿一条空中航线）
ROFOT(route forecast) 航线预报编码报头组（以英制单位）
rogue wave 奇异波,怪异波
roll ①（复数 rolls）滚轴涡旋,滚涡②滚动,横摇,摇摆,卷,绕
roll axis 滚轴
roll cloud 滚轴云
roll cumulus 滚轴积云
roller ①滚〔轴〕云,飑云②长涌
roll instability 滚流不稳定性【物】
roll vortex 滚轴涡旋【大气】
ROMET(route forecast) 航线预报编码报头组（以公制单位）
rondada 车辙风,郎达达风（西班牙航海术语,表示风向日变化按西北—北—东—南—西类似车轮行走方式的风）
Rontgen ray 伦琴射线,X射线
room climate 室内气候
rooted histogram 含根直方图
root-mean-square(RMS) 均方根
root-mean-square error 均方根误差【数】【大气】,均方差,标准差
root system 根系【植】
rope cloud 索状云
ropes of Maui 茂宜岛曙暮辉
rosau 罗索风（法国罗纳河谷的西或西

南昼风）
Rosemount temperature housing 罗式温〔度表〕罩
Rossby critical velocity 罗斯贝临界速度
Rossby diagram 罗斯贝图解【大气】
Rossby energy cycle 罗斯贝能量循环
Rossby formula 罗斯贝公式
Rossby gravity wave 罗斯贝重力波
Rossby-Haurwitz wave 罗斯贝-豪威茨波
Rossby manifold 罗斯贝流型
Rossby number 罗斯贝数【大气】
Rossby number similarity 罗斯贝数相似
Rossby parameter 罗斯贝参数【大气】
Rossby plane 罗斯贝平面
Rossby profile 罗斯贝廓线
Rossby radius 罗斯贝半径
Rossby radius of deformation 罗斯贝变形半径
Rossby regime 罗斯贝域【大气】
Rossby retardation 罗斯贝延迟〔量〕
Rossby term 罗斯贝项
Rossby wave 罗斯贝波【大气】
Rossby wave equation 罗斯贝波方程
Rossby-wave radiation 罗斯贝波辐射
rotary current 旋转流
rotary joint 旋转关节
rotary tidal current 旋转潮流
rotary vector series 旋转向量序列
rotated eigenvector 旋转特征向量
rotated EOF analysis （REOF） 旋转EOF分析
rotating adaptation process 旋转适应过程
rotating anemometer 旋转风速表
rotating annulus （或 dishpan） 转盘
rotating-beam ceilometer （BBC） 旋转光束〔式〕云幂仪
rotating coordinate frame 旋转坐标系
rotating dishpan experiment 转盘实验【大气】
rotating fluid 旋转流体
rotating fluid dynamics 旋转流体动力〔学〕
rotating joint 旋转关节
rotating mirror Q-switching 转镜式Q开关
rotating reflection sunshine recorder 回转反射式日照计
rotating Reynolds number 旋转雷诺数
rotating scanning differential polarimeter 旋转扫描示差偏振仪
rotation ①转动②自转③旋转,回④旋度【数】
rotational 转动,旋转
rotational axis of earth 地球自转轴
rotational energy 转动动能
rotational energy level 转动能级
rotational field 旋转场
rotational flow 旋转流
rotational instability 旋转不稳定〔度〕
rotational inertia 转动惯性
rotational Mach number 旋转马赫数
rotational mode 旋转模
rotational model 旋转模型
rotational motion 旋转运动
rotational similarity criterion 旋转相似性判断
rotational spectrum 旋转谱
rotational wave 旋转波【地】
rotational wind〔component〕 旋转风〔分量〕
rotational wind constrained initialization 旋转风约束初始化
rotation anemometer 旋转风速表
rotation band 转动谱带【大气】
rotation moment 转矩
rotation of crops 换茬
rotation-vibration spectrum 转振光谱
rotenturm wind 红塔风（南喀尔巴阡山罗田土尔姆山口的火热南风）
rotor 转子

rotor clouds　滚轴云【大气】
rotor effect　转子效应
rotor flow　转子气流
rotorod sampler　旋杆取样器
rotor streaming　转子气流
rotoscope　旋转镜
rotten ice　溶冰
rouergue　（见 arouergue)阿洛厄尔格风
rough air　颠簸空气
rough earth　粗糙地面
rough flow　粗糙流
roughness　粗糙度
roughness coefficient　粗糙度系数
roughness element　粗糙元
roughness layer　粗糙层
roughness length　粗糙度长度【大气】
roughness parameter　粗糙度参数【大气】
roughness sublayer　粗糙次层
rough sea　大浪(风浪4级)
rough surface　粗糙面
rough surface scattering　粗糙面散射
rough weather　狂风暴雨,恶劣天气
rounding　约数
rounding error　舍入误差【数】
round-off error　舍入误差【大气】
route　①航线,航路【航海】,②路由【计】
route component　航线分速
route cross section　航线剖面图
route forecast　航线预报
router　路由器【计】
routine　①程序②常规③日常业务
routine sonde　常规探空仪
row　行【计】
row address　行地址【计】
Rowe osmometer　罗伍渗压计
Royal Meteorological Society (RMS)　英国皇家气象学会
Royal Netherlands Meteorological Institute (RNMI)　荷兰皇家气象研究所
Royal Observatory (R.O.)　皇家观象台(英国)
Royal Society (RS)　〔英国〕皇家学会
Royco counter　罗伊库计数器(测量空气中粗微粒用)
RS (remote sensing)　遥感【地信】
R-scope　R型显示器(与A显类似,但带有扫描扩展和精密定时装置)
rubber balloon　橡胶气球
rubber ice　弹性冰
rubidium clock　铷钟【天】
ruby laser　红宝石激光器【物】
rule　①规则,定则②尺子
rule-based system　〔基于〕规则系统
ruled surface　直纹〔曲〕面【数】
rule of thumb　启发式规则,试探规则
rules of inference　推理规则
runaway process　失稳过程,失控过程
Runge-Kutta method　龙格-库塔法【数】
run mode　运行方式【计】
running average　滑动平均【大气】
running difference　滑动差分
running mean　滑动平均
running mean model　滑动平均模型
running spectrum　滑动谱
running time　运行时间【计】
running transform　滑动变换
run-off=(runoff)　径流
runoff coefficient　径流系数
runoff cycle　径流循环
runoff plot　径流图
runoff ratio　径流比
run-off-wind　风程
run-off-wind anemometer　风程风速表
runway　跑道
runway elevation　跑道〔海拔〕高度
runway observation　跑道观测
runway puddle　跑道积水
runway temperature　跑道温度
runway visibility　跑道能见度
runway visual range (RVR)　跑道能见度【大气】,跑道视程【航空】

rush 阵发(风的)
rushabar(＝reshabar) 黑风
rush of snowmelt flood 融雪洪水
Russell cycle 拉塞尔周期

RVD(range velocity display) 距离速度显示【大气】
RVR(runway visual range) 跑道能见度【大气】,跑道视程【航空】

S

S(siemens) 西〔门子〕(电导单位,1 S＝1 A/V)
sack-cloud 袋状云
saddle 鞍型低〔气〕压
saddle-back 鞍状无云区(浓积云、积雨云和平坦低云区之间的无云天空)
saddle point 鞍点
saddle-point equilibrium 鞍点平衡
saddle point integration method 鞍点积分法
safe net 安全网【计】
safesonde 安全探空仪(不影响航线安全的)
safe temperature 安全温度
safety 安全
safety communication 安全通信【航海】
safety fairway 安全航路【航海】
safety speed 安全航速
Saffir-Simpson hurricane scale 赛福尔-辛普森飓风等级,SS飓风等级
SAGE(Stratospheric Aerosol and Gas Experiment) 平流层气溶胶和气体试验
Saharan dust 撒哈拉尘
sahel ①萨赫勒风(摩洛哥沙漠尘风)②荒漠草源(指非洲撒哈拉沙漠以南的干旱少雨地带)
sahelian drought 萨赫勒干旱
sahil 岸风(北非的干热风)
sailing directions 航海指南
Saint Elmo's fire 〔圣〕爱尔摩火〔花〕,电晕放电
Saint Martin's summer 圣马丁热期

salgaso(＝agua enferma) 秘鲁赤潮
saline intrusion 盐碱侵入,盐碱化
salinity 盐度
salinity bridge 盐度电桥
salinity of sea ice 海冰盐度
salinity-temperature-depth(STD) 盐度、温度和深度
salinity tongue 盐舌
salinization 盐化,脱盐【农】,盐渍化
salinometer 盐度计
saloon(见 satellite launched from a balloon) 气球发射卫星
saltation ①跃移〔作用〕【地质】②流沙③沙暴
salt content 含盐量【物】,盐分
salt desert 盐漠【地理】
salt efflorescence 盐霜
salt fingering 盐指【海洋】,盐舌
salt haze 盐霾
salt intrusion 盐侵
salt lake 盐湖【地理】
salt line 盐线
salt nucleus 盐核【大气】
salt particle 盐粒
salt-seeding 盐粉播撒【大气】
salt tolerant crop 耐盐作物
saltwater 卤水
saltwater intrusion 盐侵【海洋】,海水入侵
salty wind damage 盐风灾害
SAM ①(stratospheric aerosol measurement) 平流层气溶胶测量 ②(Southern Annular Mode) 南半球环状模

samiyel 沙米耶利风(土耳其的 simoon 风)
samoon(＝samun) 沙蒙风(撒哈拉北部夏季的一种焚风)
sample 样本【数】
sample capacity 样本〔容〕量
sample deviation 样本偏差
sample function 样本函数【数】
sample likelihood function 样本似然函数
sample mean 样本均值【数】
sample median 样本中位数【数】
sample percentile 样本百分位数【数】
sample quantile 样本分位数【数】
sample quartile 样本四分位数【数】
sampler 取样器
sample size 样本量【数】,样本大小
sample space 样本空间【数】
sampling 抽样【数】,采样
sampling error 采样误差,抽样误差【自然】
sampling frequency 采样频率【计】
sampling interval 采样间隔【大气】
sampling method 采样方法
sampling station 采样站,观测站
SAMS (stratospheric and mesospheric sounder) 平流层和中间层探测器
samun(见 samoon) 沙蒙风
sand and dust storm 沙尘暴【大气】
sand and dust storm weather 沙尘暴天气【大气】
sand and dust weather 沙尘天气【大气】
sand auger 沙卷(见于加利福尼亚死谷)
sand-bearing wind 含沙风
sand blink 沙照云光
sand devil 沙卷
sand dune 沙丘【地理】
sand-dust storm 沙尘暴
sand fog 沙雾
sand haze 沙霾【大气】
sand mist 沙霭
sand pillar 沙柱,沙卷
sand snow 沙性雪
sand soil 砂土【土壤】

sandstorm 沙尘暴【大气】
sand tornado 沙龙卷
sand wall 沙壁【大气】,沙墙
sand wave 沙波
sand whirl 沙旋
sandwich 夹层结构
sandy desert 沙漠【地理】
sandy loam soil 砂壤土【土壤】
sansan 云冠
sansar(＝sarsar, shamshir) 桑萨风(伊朗的东北风)
Santa Ana 圣安娜风(焚风的地方名称,见于美国加利福尼亚)
Santa Maria 圣玛丽亚风(美国加利福尼亚州的焚风)
Santa Rosa 圣罗萨风(阿根廷风暴)
SAO (semiannual oscillation) 半年振荡【大气】
saoet 萨奥特风(北苏门答腊托巴湖的风)
SAR ①(Synthetic Aperture Radar) 合成孔径雷达【大气】②(IPCC Second Assessment Report) IPCC 第二次评估报告
sarande-aghii 萨兰达·艾伊风(希腊春分日前的偏南风暴)
sarca 萨卡风(意大利加尔达湖的强北风)
Sargasso Sea 马尾藻海〔域〕【海洋】
sarsar(见 sansar) 桑萨风
sastruga 单个雪面波纹脊
sastrugi ①雪面波纹【大气】②波状沙层
satellite 卫星【天】,伴随体【动物】
satellite altimetry 卫星测高
Satellite and Mesometeorological Research Project (SMRP) 卫星和中尺度气象研究计划
satellite based navigational system 星基导航系统
satellite-borne ionosonde 星载〔电离层〕测高仪
satellite-borne lightning mapper sensor 星载闪电摄像仪

satellite-borne weather radar　星载气象雷达
satellite climatology　卫星气候学【大气】
satellite cloud picture　卫星云图【大气】
satellite cloud photograph　卫星云图
satellite cloud picture analysis　卫星云图分析【大气】
satellite cloud-tracked wind　卫星云迹风
satellite communication　卫星通信【电子】
satellite coverage　卫星覆盖区【航海】
satellite data　卫星资料
satellite day (SD)　卫星白天
satellite density gauge　卫星〔大气〕密度计
satellite derived wind　卫星云迹风【大气】
satellite disturbed orbit　卫星摄动轨道【航海】
satellite Doppler positioning　卫星多普勒定位
satellite ephemeris　〔卫星〕星历
satellite field service station (SFSS)　卫星区域站
satellite geodesy　卫星大地测量学
satellite image　卫星影像
satellite image navigation　卫星影像导航
satellite infrared radiometer spectrometer (SIRS)　卫星红外辐射光谱仪
satellite infrared spectrometer (SIRS)　卫星红外光谱仪
satellite interpretation message (SIM)　卫星判释信息
satellite laser radar　卫星光〔雷〕达
satellite laser radar probing　卫星光〔雷〕达探测
satellite launched from a balloon (Saloon)　气球发射卫星
satellite launching vehicle (SLV)　卫星运载火箭
satellite lightning sensor　卫星闪电探测器
satellite meteorology　卫星气象学【大气】
satellite monitoring　卫星监测

satellite navigation system　卫星导航系统
satellite navigator　卫星导航仪
satellite night (SN)　卫星黑夜
satellite observation (SATOB)　卫星观测
satellite oceanography　卫星海洋学
satellite orbit　卫星轨道
satellite perspective　卫星资料剖析
satellite photograph　卫星照片
satellite photographic study　卫星照片判读
satellite photo receiver　卫星图片接收器
satellite picture　卫星云图
satellite picture navigation　卫星云图导航
satellite picture recorder　卫星云图传真收片机，卫星云图〔传真〕记录器
satellite radiance　卫星辐射率
satellite relay of data　卫星转播资料
satellite remote sensing observation　卫星遥感探测
satellite sounding　卫星探测【大气】
satellite subpoint　卫星星下点
satellite temperature (SATEM)　①卫星温度②温度遥感卫星
satellite temperature retrieval　卫星温度反演
satellite town　卫星城
satellite tracking antenna　卫星跟踪天线
satellite weather bulletin　卫星天气报告
satellite weather data collection platform　卫星中继的天气资料收集平台
satellite weather information system (SWIS)　卫星天气信息系统
satellite wind estimate　卫星风估计
satellite winds　卫星风
satellite zenith angle　卫星天顶角
SATEM　卫星探空（资料）
satin ice　丝状冰
SATOB (satellite observation)　卫星观测
saturated adiabatic　饱和绝热线
saturated adiabatic process　饱和绝热过程
saturated air　饱和空气【大气】
saturated humidity　饱和湿度【农】

saturated soil 饱和土【水利】
saturated vapour 饱和水汽
saturated virtual temperature 饱和虚温
saturated zone 饱和区
saturation 饱和【物】
saturation air temperature 饱和空气温度【建筑】
saturation adiabat 湿绝热线
saturation-adiabatic lapse rate 湿绝热直减率
saturation-adiabatic process 湿绝热过程
saturation-adiabatics 湿绝热线
saturation curve 饱和曲线
saturation deficit（或 deficiency） 饱和差【大气】
saturation equivalent potential temperature 饱和相当位温
saturation hygrometer 饱和湿度计
saturation level（SL） 饱和高度
saturation mixing ratio 饱和混和比
saturation mixing ratio with respect to ice 冰面饱和混合比【大气】
saturation mixing ratio with respect to water 水面饱和混合比【大气】
saturation moisture capacity 饱和持水量【大气】
saturation of scintillation 闪烁饱和
saturation point 饱和点
saturation pressure 饱和气压
saturation signal 饱和信号
saturation specific humidity 饱和比湿【大气】
saturation static energy 饱和静力能
saturation temperature 饱和温度
saturation total temperature 饱和总温度
saturation vapour pressure 饱和水汽压【大气】
saturation vapour pressure in the pure phase with respect to ice 纯冰面饱和水汽压【大气】
saturation vapour pressure in the pure phase with respect to water 纯水面饱和水汽压【大气】
saturation vapour pressure of moist air with respect to ice 湿空气冰面饱和水汽压【大气】
saturation vapour pressure of moist air with respect to water 湿空气水面饱和水汽压【大气】
saturation water vapour pressure 饱和水汽压
saturator 饱和器(湿度发生器用)
Saturn 土星【天】
Saussure's hygrometer 毛发湿度表
savanna ①萨瓦纳【地理】②热带稀树草原
savanna climate 萨瓦纳气候【地理】,热带稀树草原气候【农】
Savart polariscope 沙伐尔脱偏振仪
savet 萨维特风(北苏门答腊托巴湖的风)
saw-tooth function 锯齿形函数
saw-tooth waveform 锯齿波形【电子】
S-band S 波段(频率 2～4 吉赫,波长 7.5～15 cm)
SBL（stable boundary layer） 稳定边界层
SBLI（surface-based lifted index） 地面抬升指数
SBL strength 稳定边界层强度
SBUV（Solar Backscatter Ultraviolet Radiometer） 太阳后向散射紫外辐射仪
Sc（stratocumulus） 层积云【大气】
scalability 可扩充性
scalar 标量【物】,纯量
scalar computer 标量计算机【计】
scalar field 标量场【数】
scalar function 标量函数
scalarization 标量化【数】
scalar magnetometer 标量磁强计
scalar potential 标位
scalar product 标量积【数】,纯积
scalar quantity 标量,纯量
scale ①尺度②标尺③比例尺

scale analyses 尺度分析【大气】
scale collapse 尺度崩溃
scale dependence 尺度依存性
scale dependence instability 尺度依存不稳定度
scale factor 地图〔放大〕因子【大气】,缩尺比数,标度因子【化工】
scale height 〔大气〕标高,标高【天】
scale line 比例尺线,标度线
scale of thermometer 温标
scale of turbulence 湍流尺度
scale of wind-force 风〔力等〕级
scale parameter 尺度参数【数】
scale relation 尺度关系
scale selective 尺度选择
scale separation 尺度分离
scales interaction 尺度交互作用
scale value 标值
scaling ①尺度分析【大气】②定标〔标度〕③标定【物】④标度无关性【物】
scaling parameter 尺度参数
scaling up 升尺度
scaling variable 标度无关变量【物】
SCAMS（scanning microwave spectrometer） 微波扫描波谱仪
scan 扫描
scan circle 扫描圆
scan line 扫描线
scan mirror 扫描镜
scanner 扫描器【物】,扫描仪【计】
scanning 扫描
scanning detector 扫描探测器
scanning imaging spectrophotometer（SIS） 扫描成像分光光度计
scanning laser Doppler velocimeter system（SLDVS） 扫描激光多普勒速度计系统
scanning lidar 扫描光〔雷〕达
scanning microwave spectrometer（SCAMS） 微波扫描波谱仪
scanning multichannel microwave radiometer（SMMR） 多通道微波扫描辐射仪【大气】
scanning radiometer 扫描辐射仪【大气】
scanning radiometer data manipulator（SRDM） 扫描辐射仪资料处理器
scanning spectrometer 扫描波谱仪
scanning visible/infrared radiometer（SV/IRR） 可见光和红外扫描辐射仪
scan radius 扫描半径
SCAPE（slantwise convective available potential energy） 斜对流可用位能
scarf 云幞
scarf cloud 幞状云
scatter ①散射【电子】②散布
scatter angle 散射角
scatter communication 散射通信【电子】
scatter diagram 点聚图【大气】,散布图
scattered light 散射光【物】
scattered light photometer 散射光光度计
scattered power 散射功率
scattered radiation 散射辐射【大气】
scattered ray 散射线
scattered shower 零散阵雨
scattered sky 少云,疏云〔天空〕
scattergram 散布图
scattering 散射【物】
scattering amplitude 散射振幅【物】
scattering angle 散射角【物】
scattering area coefficient 散射面积系数
scattering area ratio 散射面积比
scattering coefficient 散射系数【大气】
scattering coefficient meter 散射系数仪
scattering cross-section 散射截面【大气】
scattering efficiency 散射效率
scattering efficiency factor 散射效率因子
scattering function 散射函数
scattering in the atmosphere 大气散射
scattering matrix 散射矩阵【物】
scattering phase function 散射相函数
scattering power 散射能力
scattering ratio 散射比

scatterometer 散射计【电子】
scatter propagation 散射传播
scatter radar 单站非相干散射雷达
scavenging 清除
scavenging air 清净空气
scavenging by precipitation 降水清除,雨洗
scavenging coefficient 清除系数
Sc cast 堡状层积云【大气】
Sc cug 积云性层积云【大气】
Sc du 复层积云
scenarios 情景,构想
scentometer 呼吸测污计
scharnitzer 寒北风(奥地利提罗耳的持续寒冰北风)
schedule 时间表,目录表
schematic diagram ①略图,示意图②原理图【机械】
scheme ①方案,格式②概形【数】
schema concept 模式概念【计】
schlieren 异离气块,异密层
Schmidt lines 施密特线【物】
Schmidt number (Sc) 施密特数
Schottky barrier diode 肖特基势垒二极管【电子】
Schottky barrier 肖特基势垒【电子】
Schrodinger equation 薛定谔方程【数】
Schwarzschild coordinates 施瓦氏坐标【物】
Schumann-Ludlam limit 舒曼-卢德兰极限
Schumann resonance 舒曼共振
Schumann-Runge band 舒曼-龙格吸收带(氧的)
Schwarzchild-Milne integral equation S-M 积分方程
Schwarzchild solution 施瓦氏解【物】
Scientific and Technological Activities Committee (STAC, AMS) 科学技术活动委员会(AMS)
Scientific Committee on Antarctic Research (SCAR, ICSU) 南极研究科学委员会(ICSU)
Scientific Committee on Oceanic Research (SCOR, ICSU) 海洋研究科学委员会(ICSU)
scientific paper 科学论文
scientific research 科学研究
scintillation 闪烁【天】
scintillation counter 闪烁计数器【物】
scintillation crystal 闪烁晶体【物】
scintillometer 闪烁计数器
scirocco 西罗科风
sciron 史凯隆(西北风的希腊名称)
scissors effect 剪刀效应
SCIT (Storm Cell Identification and Tracking) 风暴单体识别与追踪
Sc la 网状层积云
Sc lent 荚状层积云【大气】
SCMR (surface composition mapping radiometer) 表面成分图像辐射仪
Scofield-Oliver technique S-O 技术,S-O 法(由卫星红外云图估计降水)
Sc op 蔽光层积云【大气】
scorching 酷暑
Scorer parameter 斯科勒参数(大气重力波波动方程)
Scotch mist 苏格兰雾(有强毛毛雨的浓雾)
scotopic vision 适暗视觉【物】,微光视觉
Scottish Meteorological Society(SMS) 苏格兰气象学会(英国)
scourring 冲蚀【土壤】,冲刷【铁道】
Sc pl 漏隙层积云
SCR (selective chopper radiometer) 选择调制辐射仪
Sc ra 辐状层积云
screen ①百叶箱【大气】②屏幕③筛选
screened pan 网罩蒸发皿
screening 筛选
screening layer 屏蔽层
screening multiple linear regression 筛选多重线性回归
screening procedure 筛选步骤
screen psychrometer 百叶箱干湿表
Scripps Institution of Oceanography(SIO) 斯克里普斯海洋研究所

scrubber 洗涤器,涤气器,脱硫系统
SCSMEX (South China Sea Monsoon Experiment) 南海季风试验
Sc str 成层状层积云
Sc tra 透光层积云【大气】
scud 碎雨云,飞云
Sc un 波状层积云
S-curve 〔method〕 S 曲线〔法〕
SCV (submesoscale coherent vortices) 次中尺度相干涡旋
sea ①海②海况
sea air 海洋〔性〕空气
Sea-Air Interaction Laboratory (SAIL) 海-气交互作用实验室(美国)
sea and land breeze circulation 海陆风环流
sea-barometer 船用气压表
sea bank 海堤【航海】
sea-biosphere interface 海水-生物圈界面
sea breeze 海风【大气】
sea breeze convergence 海风辐合
sea breeze convergence zone 海风辐合区
sea breeze depth 海风厚度
sea breeze front 海风锋【大气】
sea breeze of the second kind 似冷锋海风,第二类海风
sea chart 海图
sea climate 海洋气候
sea clutter 海面回波
sea coast 海岸
sea condition 海况【航海】
sea disturbance 海面扰动
sea echo 海浪回波
sea fog 海雾【大气】
sea fret 海涌雾(英格兰东北沿海的一种海雾)
sea horizon 海地平
sea ice 海冰【航海】
sea-ice frequency 海冰频率
sea-land breeze 海陆风【大气】
sea-land interface 海陆界面
sealed mercury barometer 密封水银气压表
sea level 海平面
sea-level change 海平面变化
sea-level chart 海平面图
sea level correction 海平面订正
sea level elevation 海拔高度
sea-level horizon 海地平
sea level measurements 海平面测量(资料)
sea level pressure (SLP) 海平面气压【大气】
sea level pressure chart 海平面气压图
sea-level record 海平面记录
sea level rise 海平面上升
sea level synoptic chart 海平面天气图
sea map 海图
sea mist 蒸汽轻雾,海霭
seamless forecast 无缝〔隙〕预报
seamless steel pipe 无缝钢管
sea of cloud 云海
sea of fog 雾海
sea-particle interface 海水-颗粒物界面
sea rainbow 海虹
search ①搜索【计】②到处侵入(风、冷风)
search algorithm 搜索算法【计】
search and rescue satellite system 搜救卫星系统【航海】
search and rescue service 搜救业务【航海】
search light ①探照灯【航海】②云幕灯
searchlighting ①探照②水平扫描
search radar 搜索雷达
search space 查找空间,搜索空间
seaquake 海底地震
sea return 海面回波
sea-salt nucleus 海盐核【大气】
SEASAT 海洋卫星
sea satellite (SEASAT) 海洋卫星(美国)
Seasat-A scatterometer system (SASS) 海洋卫星-A 上的散射计系统

SEASAT surface wind 海洋卫星地面风
sea scale 风浪等级
sea smoke 海面蒸汽雾【大气】
season 季节【大气】
seasonal adjustment 季节调整
seasonal character 季节性
seasonal change 季节变化
seasonal current 季节性海流
seasonal disease 季节病
seasonal drought 季节性干旱
seasonal effect on appetite 食欲的季节反应
seasonal forecast 季节预报【大气】
seasonal genesis parameter 季节成因参数
seasonality 季节性【大气】
seasonal lag 季节性滞后
seasonal lake 季节性湖泊
seasonal predictability 季节可预报性
seasonal thermocline 季节性温跃层
seasonal time span 季节跨度
seasonal transition 季节过渡
seasonal variation 季节性变化
seasonal weather 季节性天气
seasonal weather forecasting 季节天气预报
seasonal wind 季风,季节性风
season of growth 生长季〔节〕
season of sowing 播种季〔节〕
sea spray 海沫
Sea Star 海〔事卫〕星
sea state 海况
sea-state scale 海况标尺
sea station 海洋站
sea surface albedo 海面反照率【大气】
sea surface radiation 海面辐射【大气】
sea surface roughness 海面粗糙度
sea-surface temperature (SST) 海面温度【大气】
sea surface temperature anomaly 海面温度异常,海温异常
sea surface wind speed 海面风速
sea temperature 海水温度【航海】

sea truth measurement 海洋真值观测
sea turn 海来风
Sea Use Program (SUP) 海洋利用规划
Sea-Viewing Wide Field-of-View Sensor (SeaWiFS) 海向广角遥感器
seawall 海堤【航海】
seawater 海水
seawater desalination 海水淡化
seawater intrusion 海水入侵
seawater pollution 海水污染
seawater thermometer 海水温度表
sea wave 海浪
SEC (South Equatorial Current) 南赤道海流
seca 塞卡干风(巴西)
secant projection 正割投影
SECC (South Equatorial Countercurrent) 南赤道逆流,南赤道反海流
Secchi disk(或 disc) 透明度板,赛克板(测水透明度)
sechard 塞查德风(瑞士日内瓦湖的焚风)
seclusion 闭塞〔过程〕
second ①秒②第二
secondary air 二次空气
secondary anticyclone 次生反气旋
secondary bands 次生〔飑〕带
secondary bow 霓,副虹
secondary circulation 二级环流
secondary cold front 副冷锋【大气】
secondary cyclone 次生气旋【大气】
secondary data utilization station (SDUS) 〔卫星〕二次资料利用站
secondary depression(或 low) 次生低压【大气】,副低压
secondary diffraction 二次衍射,二次绕射
secondary disaster 次生灾害
secondary electron emission 次级电子发射【电子】
secondary emission 次级发射
secondary flow 二次流,副流

secondary forest 次生林【林】
secondary front 副锋【大气】
secondary hydrometric station 二等水文站,辅助水文站
secondary ice crystal 次生冰晶
secondary instrument 补助仪器,二级仪器
secondary low 副低压,次生低压
secondary maximum 次极大【物】
secondary pollutant 次生污染物【大气】,继发性污染物
secondary pyrheliometer 二级绝对日射表
secondary radar 二次雷达【电子】
secondary rainbow 霓【大气】
secondary scattering 二次散射
secondary standard barometer 二级标准气压表
second contact 食既【天】
second cosmic velocity 第二宇宙速度【天】
second-foot 立方英尺/秒,每秒立方英尺(英制流量单位)
second-foot day 每秒立方英尺日(相当于86400立方英尺水量,英水量单位)
second generation computer 第二代计算机【计】
second law of thermodynamics 热力学第二定律
second moment 二阶矩,二次力矩
second moment equation 二阶矩方程
second-order climatological station 二级气候站
second-order closure 二阶闭合
second order finite difference 二阶有限差分
second order model 次级模式
second order perturbation solution 二阶扰动解
second order reaction 二级反应
second-order station 二级站
second purple light 第二紫光,第二紫霞
second radiation constant 第二辐射常数

Second Tibetan Plateau Meteorological Experiment (TIPEX) 第二次青藏高原大气科学试验(1998年)
second trip echo 二次回波【大气】
second-year ice 二年冰
sectional autocorrelogram 局部自相关图
sector ①扇形②区,段,部分
sectorized cloud picture 分区云图【大气】
sectorized picture 分区〔云〕图
sector 扇形【数】
sector scan 扇形扫描
sector scan indicator 扇形扫描显示器
sector wind 〔航路〕分段风
secular 世纪〔的〕,多年〔的〕
secular change 世纪变化,长期变化
secular trend 世纪趋势,长期趋势
secular trend in climate 长期气候趋势【大气】
secular variation 世纪变化,长期变化
sediment 沉积物【地质】,泥沙【土壤】
sedimentation ①沉积【力】②沉降【大气】③沉淀
sedimentation boundary 沉降界面
sedimentation coefficient 沉降系数
sedimentation diameter 〔等价〕沉积〔物〕直径
sedimentation equilibrium 沉降平衡
sedimentation potential 沉降势
sedimentation velocity 沉降速度
sediment-carrying capacity 挟沙能力,挟沙量
sediment concentration 沉积〔物〕浓度
sediment density meter 沉降物浓度测量仪
sediment discharge rating 输沙量-流量关系〔曲线〕
sediment grading 沉积物分类,泥沙颗粒分级
sedimentology 沉积学【石油】,泥沙学
sediment sampler 泥沙取样器
sediment transport monitor 沉降物输送监测器

sediment trap	沉降物捕集器
sediment yield	泥沙产量,泥沙清除量
seedability	可催化性,可播撒性
seeder cloud	播种云
seeder-feeder	播馈机制
seed fertilizer	种肥【土壤】
seeding	播撒,催化
seeding agent	云催化剂【大气】
seeding device	播撒装置
seeding location	播撒位置,播撒部位
seeding material	播撒物质
seeding rate	催化率
seeding subroutine	催化子程序
seed-time	播种时期,催化时间
seektime	寻找时间
seepage	渗漏(土壤中的),渗流【力】
seepage spring	渗水泉,渗流泉
seepage velocity	渗流速度
seesaw structure	跷跷板结构,涨落结构
segment	部分,段
seiche	假潮【海洋】,湖面波动,湖震
seism	地震
seismicity	地震流动性【地】,地震度
seismic sea wave	海啸【地】
seismic sea wave warning system (SSWWS) 地震海浪预警系统	
seismic source	震源【地】
seismically active belt	地震活动带【地】
seismically active zone	地震活动区【地】
seismic-tectonic zone	地震构造带【地质】
seismic wave	地震波【地】
seismogeology	地震地质学【地质】
seismogram	地震图【地】
seismograph	地震仪【地】
seismology	地震学【地】
seismometer	地震计【地】
seismometry	测震学【地】
seismotectonics	地震构造学【地质】
Seistan wind	十二旬风(伊朗赛斯登夏季的一种强北风,可持续120天之久,波斯语名)
selatan(＝slatan)	塞拉坦风(印度尼西亚的强干南风)
selected pressure surface	特选气压面,特选气压场
selected ship station	特选船舶站
selection hypothesis	选择假设
selection rule	选择定则【物】
selective absorption	选择吸收【物】
selective action	选择作用
selective adsorption	选择吸附
selective chopper radiometer (SCR) 选择调制辐射仪	
selective damping	选择性阻尼
selective filtering	选择滤波
selective instability	选择不稳定〔度〕
selective scattering	选择散射
selective stability	选择稳定〔度〕
self-adjoint operator	自伴算子【数】
self-broadening	自加宽
self-calibrating	自标定
self-consistency	自洽性【物】,自身一致性
self-cleaning	自净【化】
self-cleaning process	自净过程
self coherence	自相干【物】
self-collection	自碰并
self correlation	自相关
self correlation function	自相关函数
self-development	自〔身〕加强
self-exited oscillation	自发性振荡
self-generated chaos	自生混沌
self-generating anemometer	自发电风速表
self-heating	自热
self-heating dewpoint sensor	自热式露点传感器
self-healing of atmosphere	大气自复〔原〕〔效应〕
self-organization	自组织
self-organization system	自组织系统
self-oscillating Rb-vapor magnetometer 自振铷蒸气磁强计	

self-reciprocal function 自反函数【数】
self-recorder 自记器
self-recording barometer 自记气压计
self-recording hygrometer 自记湿度计
self-recording rain gauge 自记雨量计
self-registering anemometer 自记风速计
self-registering barometer 自记气压计
self-registering thermometer 自记温度计
self-scavenging ability 自清能力
selsyn ①自动同步〔系统〕②同步测风仪
selvas 南美热带雨林（葡萄牙语名）
SEM（space environmental monitor） 空间环境监测器
semantic network 语义网络【计】
semiannual oscillation(SAO) 半年振荡【大气】
semiannual temperature oscillation 半年温度振荡
semiannual wind oscillation 半年风振荡
semi-arid 半干旱
semi-arid climate 半干旱气候【大气】
semi-arid region 半干旱区
semi-arid zone 半干旱带
semi-automatic apparatus 半自动仪器
semi-automatic balloon launcher（SABL） 半自动气球施放器
semi-average method 半平均法
semi-circle theorem 半圆定理
semiconductor 半导体【物】
semiconductor laser 半导体激光器【物】
semiconductor detector 半导体探测器【电子】
semiconductor element 半导〔感应〕元件
semiconductor thermocouple 半导体温差电偶
semiconductor thermoelement 半导体热电元件
semiconductor thermometer 半导体温度表
semiconfined aquifer 半承压含水层,漏水含水层
semi-desert 半荒漠

semidiurnal solar tide 半日太阳潮
semidiurnal tide 半日潮
semidiurnal variation 半日变化
semidiurnal wave 半日波【大气】
semi-empirical theory of turbulence 湍流半经验理论
semi-explicit scheme 半显格式
semi-fixed dune 半固定沙丘
semi-fixed platform 半固定平台
semi-geostrophic equations 半地转方程【大气】
semi-geostrophic flow 半地转流
semi-geostrophic frontogenesis 半地转锋生
semi-geostrophic height-tendency equation 半地转高度倾向方程
semi-geostrophic motion 半地转运动【大气】
semi-geostrophic space 半地转空间
semi-geostrophic theory 半地转理论【大气】
semi-grey radiation 半灰辐射
semi-humid 半湿润
semi-humid region 半湿润区
semi-implicit method 半隐式法【数】
semi-implicit scheme 半隐式格式【大气】
semi-infinite atmosphere 半无限大气
semi-infinite medium 半无限介质
semi-interquartile range 四分位差
semi-Lagrangian method 半拉格朗日法
semi-Lagrangian model 半拉格朗日模式
semi-Lagrangian scheme 半拉格朗日方案,半拉格朗日格式
semi-logarithmic scale 半对数尺度
semiperiod 半周期
semi-permanent action center 半永久性活动中心
semi-permanent anticyclone 半永久性反气旋
semi-permanent depression 半永久性低

压【大气】
semi-permanent high 半永久性高压【大气】
semi-permanent low 半永久性低压
semiprognosis 不完全预报
semi-spectral method 半谱方法
semi-submersible barge 半潜〔式〕钻探站,半潜〔式〕平台
semi-tropical 副热带的
semi-tropical cyclone 副热带气旋
sensible heat 感热【大气】,显热【航空】
sensible heat flow 感热流〔量〕
sensible heat flux 感热通量【农】
sensible horizon 显著地平
sensible temperature 感觉温度【大气】
sensing element 感应元件
sensing system 测定系统
sensitive series 敏感序列
sensitive surveillance radar 灵敏监视雷达
sensitivity 灵敏度【物】,敏感性【计】
sensitivity analysis 灵敏度分析【数】
sensitivity experiment 敏感性实验
sensitivity ratio 灵敏度比
sensitivity study 敏感性研究
sensitivity test 敏感性检验
sensitivity time control 灵敏度时间控制
sensitometer 感光计【物】
sensitometry 感光测定法
sensor 传感器【物】,感应元件
separability 可分性
separating density effect 分离〔大气〕密度效应
separating of charge 电荷分离
separation 分离
separation constant 分离常数
separation line 分离线
separation of charge 电荷分离
separation of flow 气流分离
separation of variables 变数分离
separation point 分离点

separation vortex 分离涡旋
separation wake 分离尾流
separator ①分离器②除尘器
separatrix 分型线,分隔号
sequence 顺序,序列【数】
sequence estimation 序列估计
sequence of circulation 环流序列
sequence of weather 天气历史顺序
sequential access 顺序存取【计】
sequential algorithm 序列算法
sequential analysis 序贯分析【数】
sequential assimilation 序列同化
sequential computer 串行计算机【计】
sequential processing 顺序处理【计】
sequential relaxation 顺序张弛
sequestration ①隐蔽作用②汇集,隔绝③(多元)结合
serac 冰塔,冰雪柱
serein 晴空雨【大气】(法语名)
serene 晴空,静海
serial communications 串行通信
serial correlation 序列相关
serial correlation coefficient 序列相关系数
serial mouse 串行鼠标〔器〕【计】
serializability 可串行性【计】
serial number 流水号(电报),序号
serial processing 串行处理
serial station 定点观测站【海洋】,海洋水文站
series ①序列②级数【数】
series expansion method 级数展开法
server 伺服器,服务器【计】
service bit rate 服务比特率【计】
service ceiling 飞行高度
services system 服务系统
service telegram 公务电报
servo amplifier 伺服放大器【电子】
servo barometer 伺服气压表
servo loop 伺服环〔路〕
servo-motor 伺服电〔动〕机【电子】

servo system　伺服系统【电子】
SESAME（Severe Environmental Storm and Mesoscale Experiment）　强环境风暴和中尺度试验(1979年)
sesquiterpene　倍半萜【化】
seston　浮游物,悬浮物【海洋】
set　集〔合〕【数】
set-function　集函数【数】
setting　拨定装置
setting-in of bai-u　入梅(日本语名)
settled weather　晴天
settlement dust counter　沉淀尘埃计数器
settling cloud chamber（SCC）　沉降云室
set-top box　机顶盒【计】
setup　设置,建立【计】
setup time　建立时间【计】
severe cold　严寒
severe contamination　严重污染
severe convection　强对流
severe convective storm　强对流风暴
severe drought　大旱,严旱
severe drought year　大旱年,严旱年
Severe Environmental Storm and Mesoscale Experiment（SESAME）　强环境风暴和中尺度试验(1979年)
severe frost　严霜
severe gale　厉风
severe line squall　强线飑
severe local storm　局地强风暴
Severe Local Storm Forecast Center（SLSFC）　局地强风暴预报中心(美国)
severely shear　强切变
severe right moving storm　右移强风暴
severe sand and dust storm　强沙尘暴【大气】
severe storm　强风暴
severe storm observation　强风暴观测
Severe Storm Observation Satellite（SSOS）　强风暴观测卫星
severe thunderstorm　猛烈雷暴,强雷暴
severe thunderstorm watch　强雷暴监视

severe tropical storm　强热带风暴(风力10～11级)【大气】
severe typhoon（STY）　强台风【大气】
severe weather　灾害性天气【大气】,恶劣天气,灾害天气
severe weather automatic nowcast system（SWAN）　灾害天气短时临近预报系统
severe weather threat index　剧烈天气征兆指数,强天气征兆指数,灾害性天气征兆指数【大气】
severe weather warning　危险天气预警【大气】
severe winter　严冬
severity　严酷〔性〕
sewage irrigation　污水灌溉【土壤】
SFAZI（spherics azimuth）code　天电方位编码
SFAZU code　测站天电方位编码
sferics　①天电②远程雷电③天电学④天电测定法
sferics fix　天电定位
sferics network　天电观测网
sferics observation　天电观测
sferics receiver　天电接收机
sferics recorder　天电记录器
sferics source　天电源
sferix　低频天电
shaded area　阴影区
shade ratio　遮光率
shades of gray　层次,灰度
shade temperature　阴温
shading coefficient　遮阳系数
shading effect ratio　阴影有效比
shadow　盲区(雷达)
shadow band pyranometer　遮光罩天空辐射计
shadow effect　阴影效应【物】
shadow of the earth　地影
shadow radiometer　遮光罩辐射计
shadow zone　影区【地】
shaft　风矢杆

shahali 撒哈拉风（阿尔及利亚撒哈拉和 scirocco 同类的干热风）
shaitan 尘旋
shale ice 页状冰，页片冰
shallow boundary layer 浅边界层
shallow cellular circulation 浅层单体环流
shallow cloud 浅云
shallow cloud street 浅云街
shallow convection 浅对流【大气】
shallow convection parameterization 浅对流参数化【大气】
shallow convective system 浅对流系统
shallow fog 浅雾
shallow layer 薄层
shallow low 浅低压【大气】
shallow water approximation 浅水近似
shallow water effect 浅水作用
shallow water equation 浅水方程
shallow water model 浅水模式
shallow water theory 浅水理论
shallow water wave 浅水波【大气】
shaluk 沙卢克风（北非沙漠的热风）
shamal(＝shemaal, shimal, shumal) 夏马风（美索不达米亚的一种西北风）
shamshir（见 sansar） 桑萨风
Shannon entropy 香农熵【电子】
Shannon theory 香农理论【数】
shape function 形状函数
shape of the sky 天空形状
shape operator 形状算子【数】
shape parameter 形状参数【数】
sharav 大陆酷热（希伯来语名）
sharki（＝sherki, shuquee, shurgee, 见 kaus） 考斯风
sharp-edged gust 突发阵风
sharp front 陡锋
sharttering 破碎，吹散
shear 剪切【物】，切变
shear axis 切变轴
shear energy production 切变能量产生
shear flow 切变流

shear-gravity wave 切变重力波
shear hodograph 切变速矢端迹图
shearing deformation 切变
shearing instability 切变不稳定【大气】
shearing strength 抗切强度
shearing stress 切应力
shearing vorticity 切变涡度【大气】
shearing wave 切变波【大气】
shear instability 切变不稳定
shear layer 切变层【大气】
shear line 切变线【大气】
shear of wind 风切变
shear production 切变产生〔量〕
shear ridge 切变脊
shear strain 切应变，剪应变
shear stress 剪应力【物】
shear term 切变项
shear vector 切变矢量
shear wave 切变波
sheathed thermometer 双管温度表
shedding 泄离
sheet cloud 层状云
sheet erosion 片蚀【土壤】
sheet flow 片流
sheet frost 片霜
sheet ice 片冰
sheet lightning 片状闪电【大气】
sheet wash 片蚀，表面冲蚀
shelf cloud 板架云
shelf ice 陆架冰，陆缘冰
shelf waves 〔大〕陆架地形波
shell ice 壳冰（水位下降后遗留的）
shelter 百叶箱
shelter-belt 防风带
sheltering coefficient 遮蔽系数
shelter thermometer 百叶箱温度表
shemaal 夏马风
sherki 考斯风
Sherwood number 舍伍德数【化】
SHF（super high frequency） 超高频
SHF communication 超高频通信【电子】，

厘米波通信
shield 防风罩(雨量器上用),防辐射罩(测温用)
shield cloud 盾状云【大气】
shielding effect 屏蔽效应
shielding effectiveness 屏蔽效能
shielding layer 屏蔽层
shift 位移【医】,变移,漂移
shift-in 移入【计】
shifting level 变移高度
shifting sand 流沙
shifting wind 不定风,无定向风
shift-out 移出【计】
shimmer 闪晃
shine 露面宝光
ship barometer 船用气压表【大气】
shipboard expendable bathythermograph (SXBT) 船用投弃式深度温度计
shipboard raingauge 船用雨量器
shipboard theodolite 船用经纬仪
shipboard wind plotter 船上真风填算盘
shipborne radar 船载雷达【电子】
ship Doppler (SDOP) 船用多普勒系统
ship motion 船舶运动,船体运动
ship observation 船舶观测【大气】
ship report 船舶报〔告〕
ship routing 船舶航线〔安排〕
ship synoptic code 船舶天气电码
ship weather report 船舶气象报告
ship wind 航行风
shoal 浅滩【测】,险滩【水利】
shoaling 变浅,浅水作用
shock 激波
shock seeding 雷冲击波催化
shock solution 激波解
shock wave 激波,〔冲〕击波【物】
shooting star 流星
shore current 岸流
shore ice 岸冰
shore lead 冰岸水道
shore meteorology 海岸气象,滨海气象学

shore wind 〔海〕岸风
short circuit 短路【物】
short circuit current 短路电流
short-crested wave 短峰波
short-day plant 短日照植物
short-exposure modulation transfer function 短曝光调制传递函数
short method 短法
short rainbow 短虹
short-range〔weather〕forecast 短期〔天气〕预报【大气】
short-term beam spread 短周期波束加宽
short-term forecast 短期预报
short-term forest fire danger forecast 短期森林火险预报
short-term hydrologic forecast 短期水文预报
short wave 短波
short wave albedo 短波反照率
short wave cloud radiative forcing 短波云辐射强迫作用
short wave cutoff 短波截止
short wave radiation 短波辐射
short wave trough 短波槽
shot noise 散粒噪声【物】
Showalter index 肖沃特〔稳定度〕指数
Showalter stability index 肖沃特稳定〔度〕指数
shower 阵雨
shower cloud 阵雨云
shower formula 阵雨公式
shower of ash 阵尘
shower street 阵雨带
showery precipitation 阵性降水【大气】
showery rain 阵雨【大气】
showery snow 阵雪【大气】
shred cloud 碎片状云
shrieking sixties 咆哮西风(指南半球60°S附近的强西风)
shroud 〔气〕球罩
shrub 灌木【植】

shuga 白松冰团,冰花
shuga ice 雪泥冰
shuttle 航天飞机
shuttle imaging radar（SIR） 航天飞机成像雷达
Siberian anticyclone 西伯利亚反气旋
Siberian high 西伯利亚高压【大气】
SID（sudden ionospheric disturbance） 突发电离层骚扰【地】,电离层突扰【天】
sideband 边带【电子】,旁带
side-band instability 边带不稳定
side effect 边缘效应【电子】
side flash 侧闪电
side lobe 旁瓣【电子】
side lobe echo 旁瓣回波
side looking aircraft（或 airborne）radar（SLAR） 机载侧视雷达
side-looking radar 侧视雷达【电子】
sidereal day 恒星日【天】
sidereal month 恒星月【天】
sidereal time 恒星时
sidereal year 恒星年【天】
side-scanning sonar 旁扫描声呐
sidetone 侧音
sidetones ranging 侧音测距【电子】
side view 侧视图
side wind 侧风
siemens（S） 西〔门子〕（电导单位）
siffanto 西范托风（意大利亚德里亚海的西南风）
si giring giring 西吉林吉林风（北苏门答腊托巴湖的风）
sigma-coordinate system σ坐标系
sigma meter 累加装置
sigma vertical coordinate σ垂直坐标
signal 信号【电子】
signal generator 信号发生器【电子】
signal intensity 信号强度
signal needle code 莫尔斯电码
signal strength 〔无线电〕信号强度
signal to clutter ratio 信号杂波比,信杂比【电子】
signal to jamming ratio 信号干扰比,信干比【电子】
signal-to-noise problem 信噪问题
signal-to-noise ratio（SNR） 信噪比【物】
signal variability processor（SVP） 信号变率处理器
signal velocity 信号速度
signature ①特征,信号②符号差【物】
signature analysis techniques 特征分析技术,特征分析法
significance 显著性
significance level 显著性水平【大气】（如0.05,0.01等）
significance test 显著性检验【数】
significant cloud 特殊云,重要云层
significant difference 显著性差异
significant digit 有效数字【数】
significant figure 有效数字【数】
significant level 特性层【大气】
significant meteorological information 重要气象信息【大气】
significant point 特殊(性)点
significant wave 有效波
significant wave height 有效波高
significant weather 重要天气【大气】
significant weather chart 重要天气图【大气】
significant weather report 重要天气报告
sign test 符号检验【数】
sigua 西格风（菲律宾8级以上的季风）
silicon pressure sensor 硅压敏元件
silicon solar cell 硅太阳电池
silicosis 矽肺
Sil intensity gauge 西尔〔降雨〕强度计
sill ①海底山脊②岩床【地质】③基古
sill depth 海槛深度（海盆和大海之间可以水平通讯的最大深度）
silt 粉尘,粉砂（直径在 $3.9 \sim 62.5\ \mu m$ 的沉积颗粒）
silt-discharge rating 输沙量-流量关系

〔曲线〕
silty clay soil 粉〔砂〕黏土【土壤】
Silurian Period 志留纪【地质】
Silurian System 志留系【地质】
Siluro-Devonian ice age 志留-泥盆纪冰期
silver analysis 含银量分析
silver-disk pyrheliometer 银盘绝对日射表
silver frost 银光霜
silver iodide 碘化银
silver iodide seeding 碘化银〔云〕催化【大气】
silver storm 银光风暴
silver thaw 雨凇
similarity (＝equality, resemblance) 相似性【计】
similarity hypothesis 相似性假设
similarity law 相似〔定〕律
similarity relationship 相似〔理论〕关系〔式〕
similarity theory 相似理论
similarity theory of turbulence 湍流相似理论【大气】
similarity transformation 相似转换
simoom(＝simm) 西蒙风(非洲和阿拉伯沙漠区的干热风,阿拉伯语名)
simple average 简单平均
simple climate model 简单气候模式
simple correlation 单相关
simple harmonic motion type 简谐运动型
simple harmonic wave 简谐波【物】
simple linear correlation 简单线性相关
simple pendulum 单摆【物】
simple reflection 单反射
simple reflector 单反射体
simple vortex 简涡
simple wave 简波
Simpson dirt trap 辛普森挡污器(水银气压表用)
simulated climate 模拟气候
simulated rainfall 模拟降水
simulation 模拟【物】

simulation computer 仿真计算机【计】
simulation correlation 同时相关
simulation test 模拟试验【航海】
simultaneity 同时性【物】
simultaneous auroral multiballoons observation (SAMBO) 多〔气〕球同步极光观测
simultaneous equation 联立方程
simultaneous observation 同步观测,同时观测
simultaneous physical retrieval method 同时物理反演法
simultaneous pressure 同时气压
simultaneous relaxation 联合张弛
simultaneous temperature 同时温度
sine curve 正弦曲线【数】
sine function 正弦函数
sine-function expansion 正弦函数展开
sine galvanometer 正弦电流表
sine window 正弦窗
single-board computer 单板计算机【计】
single cell severe local storm 局地强风暴单体
single-cell storm 单体风暴
single-chip computer 单片计算机【计】,单片机
single-column manometer 单管压力表
single Doppler radar 单多普勒雷达
single drift correction 单偏航订正
single-grid-heading navigation 单网格航向导航
single-heading navigation 单航向导航
single-ended remote probing 单端遥测
single-ended system 单端系统
singleion 单离子
single observer forecast 单站预报,单点预报
single scattering 一次散射,单次散射
single scattering albedo 单次散射反照率
single scattering approximation 单次散射近似

single sheath 单鞘(等离子体的)
single sideband 单边带
single-sideband modulation 单边带调制【电子】
single space mean 一次空间平均
single station analysis 单站分析
single station〔weather〕forecast 单站〔天气〕预报【大气】
single-step iteration 单步迭代【数】
single-theodolite observation 单经纬仪观测
singlet oxygen 纯态氧
single-unit transceiver 收发〔合一〕两用机
single-wavelength lidar 单波长光〔雷〕达
singular advection 奇异平流
singular corresponding point 奇异对应点
singular integral equation 奇异积分方程【数】
singularity ①奇点【数】②奇异性【物】③独特性
singular perturbation 奇异扰动【物】,奇异摄动
singular point 奇点【数】
singular spectrum analysis(SSA) 奇异谱分析
singular value 奇异值【数】
singular value decomposition(SVD) 奇异值分解【数】
singular vectors 奇异矢量
Sinian Period 震旦纪【地质】
Sinian System 震旦系【地质】
sink 汇,下沉,汇点【数】
sinkhole ①灰岩坑,落水洞②污水井
sinking mirage 下现蜃景【大气】
sinking motion 下沉运动
sink processes for trace gases 痕量气体移除过程
sintering phenomenon 晶化现象
sinuosity 小曲折
sinusoid 正弦式
sinusoidal pattern 正弦型

sinusoidal wave 正弦波
siphon barograph 虹吸气压计
siphon barometer 虹吸气压表【大气】
siphon cistern barometer 虹吸槽式气压表
siphon rainfall recorder 虹吸〔式〕雨量计【大气】
siphon raingauge 虹吸〔管〕雨量器
siphon recording barometer 虹吸式〔自记〕气压计
sirocco 西罗科风(欧洲南部的焚风)
sirocco di levante 希腊西南焚风
SIRS(satellite infrared spectrometer) 卫星红外光谱仪
SIRS control unit module(SCUM) 卫星红外光谱仪控制组件
SIRS instrument power supply(SIPS) 卫星红外光谱仪电源
SIS(scanning imaging spectrophotometer) 扫描成像分光光度计
site 位置,场,站点【计】,网站
site characterization 测站表征
site of a station 测站位置
siting 选址
SI unit 国际〔制〕单位
Six's thermometer 悉克斯〔最高最低〕温度表
size 尺寸,大小
size distribution 尺度分布,谱分布
size parameter 尺度参数
size spectrum 尺度谱
skare 雪面壳,雪结皮(瑞典语名)
skavler 雪面波纹(sastrugi的挪威语名)
skew 偏斜
skew curve 偏斜曲线
skewness 偏度【数】
skew-symmetric determinant 反称行列式【数】
skew-symmetric tensor 斜称张量【数】
skew-symmetry 斜对称
skew T-log P diagram 温度对数气压斜交图表(即赫洛弗森热力图表)

skill score 技巧〔评〕分【大气】
skin current 表流
skin drag 表面拖曳
skin effect 趋肤效应【物】
skin-friction 表面摩擦
skin-friction coefficient 表面摩擦系数
skin-surface temperature 表面温度
skip distance 越程,跳跃距离
skip effect 超越效应
Skiron 西北风(冬天寒冷,夏天干热,希腊语名)
skirt region 磁〔尾〕裙
sky 天空,天空状况
sky background 天空背景
sky background noise 天空背景噪声
sky brightness 天空亮度【大气】
sky brightness temperature 天空亮度温度
sky clear 碧空,晴天
sky condition 天空状况【大气】
sky cover 云量
skyhook balloon 高层等高探测气球
skylab(＝sky laboratory) 空间实验室
skylight ①天光②天窗【建筑】
skyline 地平线
sky luminance 天空亮度【大气】
sky map 云底亮度图(地面反光引起的云底亮度分布)
sky mirror 天空反射镜
sky polarization 天空偏振
sky radiation 天空辐射【大气】
sky radiometer (＝diffusometer) 天空辐射表【大气】
sky slightly clouded 少云天空(总云量为八分之一、二)
sky sweeper 天扫风
sky wave 天波【电子】
sky wave correction 天波改正量【航海】
sky wave delay 天波延迟【航海】
sky wave radar 天波雷达
slab approximation 平板近似
slab layer 平板层

slab like motion 平板式运动
slab model 平板模式
slab of scatterer 散射元薄层
slab symmetry 平板对称
slab thickness 平板厚度(电离层的)
slack current channel 缓流航道【航海】
slack water 憩潮
slake 〔石灰〕熟化,减弱
slant path 倾斜路径
slant path molecular absorption 斜距分子吸收
slant range 斜距
slant visibility 倾斜能见度【航空】,斜能见度
slantwise convection 倾斜对流
slantwise convection available potential energy 斜对流可用位能
slantwise instability 倾斜不稳定性
slatan (见 selatan) 塞拉坦风
slave station 从〔属〕站
SLDVS (scanning laser Doppler velocimeter system) 扫描激光多普勒速度计系统
sleet 雨夹雪【大气】
sleet and ice 冰凌
SLF communication 超低频通信【电子】,超长波通信
slice method 薄片法【大气】
slideing scale 滑尺
slide rule method 计算尺法
slight air 软风
slight breeze 轻风
slight hail 轻雹
slight haze 轻霾
Slight Heat 小暑(节气)
slightly stable layer 弱稳定层
slight rain 微雨
slight sea 轻浪(风浪2级)
slight shower of rain 小阵雨
slight thunderstorm 小雷暴
sling psychrometer 手摇干湿表【大气】
sling thermometer 手摇温度表

slip 滑坡【土壤】
slip stream 滑流
slit 狭缝【物】
slit function 狭缝函数
slob ice 海面乱冰
slope 斜率、斜度,坡度
slope area method 比降-面积法,坡降-面积法
slope current 坡度流
slope flow 坡度流
slope function 斜率函数【数】
slope method 斜率法
slope of a front 锋面坡度
slope of an isobaric surface 等压面坡度【大气】
slope wind 坡风【大气】
slope wind circulation 坡风环流
slope windstorm 下坡风暴
sloping overhang echo 伸悬回波
slow adjustment 慢调整
slow climate system 缓慢气候系统
slow ion 慢离子
slowly moving cold front 慢行冷锋【大气】,第一型冷锋
slow neutron 慢中子【物】
slowly varying wave train 缓变波列
slow manifold 慢波簇,慢流形
slow mode 慢模【物】
slow mode constrained initialization 慢模约束初始化
slow mode equation 慢模方程
slow wave 慢波
slow tail 慢尾
SLP (sea level pressure) 海平面气压【大气】
sludge ①雪泥冰,软冰②泥浆③淤渣
sludging ①污泥【农】②融冻泥流③泥流〔作用〕
sluff 小滑雪
slug test 导水率检验,斯勒格检验
sluice (见 flood gate) 水闸,泄水道

slush ①雪浆、温雪②泥浆
slush icing 飞机沾雪,飞机湿雪,密集冰花
Smale horsehoe 斯梅尔马蹄【物】
Smale horseshoe transform 斯梅尔马蹄变换
small amplitude 小振幅
small amplitude theory 小振幅理论
small angle approximation 小角近似
small angle scattering 小角散射【物】
small circle 小圆〔航路〕(球面与不通过球心的平面的交线)
small-craft warning 小船预警
small eddy closure 小涡闭合
small eddy theory 小涡理论
small hail 小冰雹,小雹
small halo 〔小〕晕(22°的晕)
small ion 轻离子
small ion combination 轻离子结合
small orifice raingauge 小集水口雨量筒
small parameter method 小参数法
small perturbation 小扰动
small-scale atmospheric dynamics 小尺度大气动力学
small-scale convection 小尺度对流
small-scale data utilization station (SDUS) 小型〔卫星〕资料利用站
small-scale turbulence 小尺度湍流
smaze 烟霾(smoke 与 haze 的合成缩略词)
Smithsonian pyrheliometric scale 斯密森绝对日射表标尺
Smithsonian silver disk pyrheliometer 斯密森银盘绝对日射表
Smithsonian water-flow pyrheliometer 斯密森水柱式绝对日射表
SMMR (scanning multifrequency microwave radiometer) 多通道微波扫描辐射仪【大气】
SMO (supplementary meteorological office) 补充气象台
smog 烟雾【大气】(smoke 与 fog 的合

成缩略词)
smog aerosol 烟雾气溶胶【大气】
smog chamber 〔光化学〕烟雾〔实验〕室
smog front 烟雾锋
smog horizon 烟雾层顶,烟雾亮度
smog index 烟雾指数
smog layer 烟雾层
smoke 烟【大气】
SMOKE (Sparse Matrix Operator Kernel Emission) 稀疏矩阵算子核心排放(排放源处理程序)
smoke aloft 高空烟
smoke chart 烟尘图
smoke cloud 烟云【大气】
smoke density indicator 烟雾浓度显示仪,烟雾浓度指示仪
smoke fly ash 烟灰
smoke fog 烟雾
smoke haze 烟霾
smoke horizon 烟尘层顶
smoke injury 烟害
smokeless zone 无烟区
smokemeter 烟尘计
smoke pall 厚烟层【大气】
smoke plume 烟羽【大气】,烟流
smoke plume density 烟羽密度
Smokes 白浓雾(非洲几内亚海岸干季常见)
smoke screen 烟幕【大气】
smoke shade 烟雾色调
smoke trail 烟迹
smoke volatility index 烟雾挥发度指数
smoky fog 烟雾
smooth function 光滑函数【数】
smoothing 平滑【数】,修约
smoothing coefficient 平滑系数
smoothing operator 平滑算子
smoothing parameter 平滑参数
smoothing process 平滑过程
smooth perturbation method(SPM) 平缓扰动法

smooth sea 微浪(风浪 1 级)
smooth surface 平滑面
SMS (synchronous meteorological satellite) 同步气象卫星
smudging 熏烟(果园防霜的一种措施)
S/N (signal to noise ratio) 信噪比
snap 短冷期
snappy 寒冷
sneak out 渐隐,淡出(电视图像的)
Snell law 斯涅耳定律【物】,折射定率
snifter 暴风,暴风雨
sno 斯诺飑(挪威峡湾北部的寒冷下滑飑)
snorter 强风,暴风雨(雪)
snow 雪【大气】
snow accumulation 积雪
snow albedo-temperature feedback 雪反照率-温度反馈
snow and rain 雨夹雪
snowball 雪球
snowball Earth hypothesis 雪球假说【古生物】
snowbank 雪堤
snow banner 雪旗(常指由山脊吹起的雪)
snow barrier 雪障
snowberg 雪山,覆雪冰山
snow bin 量雪箱
snow blindness 雪盲【大气】
snow blink 雪照云光
snow board 测雪板
snowbreak 雪障
snow-broth 雪水,新融的雪
snow burn 雪灼
snow cap 雪冠
snow climate 雪原气候
snow cloud 雪云(下雪的云)
snow concrete 坚雪
snow course 测雪路线,雪径(采雪样路线)
snow cover 积雪【大气】,雪盖
snow cover chart 积雪图
snow covered area 积雪面积,雪被面积

snow covered ice 雪盖冰
snow cover imdex 积雪指数
snow cover line 积雪线
snow-cover meter 雪量计
snowcreep 雪移,雪蠕变
snowcrete 坚雪,凝固雪
snow crust 〔冰〕雪壳
snow crystal 雪晶【大气】
snow crystal concentration meter 雪晶浓度计
snow damage 雪灾【大气】
snow day 雪日【大气】
snow density 雪密度【大气】
snow depth 雪深【大气】
snow depthmeter 雪深测定仪
snow depth scale 量雪尺【大气】
snow-depth telemeter 遥测雪深计
snow drift 〔吹〕雪堆
snowdrift glacier 吹雪冰川
snowdrift ice 吹雪冰
snow-eater 融雪风
snow effect 雪效应
snow erosion 雪蚀【土壤】
snowfall 〔amount〕 雪量【大气】
snow fence 〔防〕雪栅〔栏〕
snowfield 雪原
snowflake 雪花【大气】
snow flurry 阵雪,雪阵
snow forest climate 雪林气候
snow garland 雪花环
snow gauge 量雪器
snow generating cell 雪生单体
snow generating level 生雪层,降雪层
snow geyser 雪喷
snow grains 米雪【大气】
snow hydrology 积雪水文学【地理】
snow ice 雪冰
Snow, Ice and Permafrost Research Establishment (SIPRE) 冰雪与永久冻土研究所(美国)
snow igloo 小雪丘

snow injury 雪害
snow level 量雪尺
snow line 雪线【大气】
snow load 雪荷载【大气】
snow lying 雪掩,积雪覆盖
snowmaker 人工造雪机
snow mat 雪席(用于标记新雪和陈雪界面的器具)
snow measuring plate 测雪板
snow melt 雪融,解冻〔水〕
snowmelt erosion 融雪侵蚀【土壤】
snowmelt flood 融雪洪水,春汛
snowmelt runoff 融雪径流
snow mist 雪霭(＝ice crystals)
snowout 雪散落物,雪沉降物
snowpack 积雪总量【大气】,〔积〕雪场,雪堆
snowpack yield 积雪〔水当〕量
snow patch 雪斑
snow pellets 雪丸(包括软雹和霰)
snow pillow 雪枕(测量雪压)
snow plume 雪羽
snow precipitation line 降雪线
snow pressure 雪压【大气】
snowquake (＝snow tremor) 雪震,雪崩
snow reflectivity 雪反射率
snow regime 积雪型式
snow ripple 雪纹
snow roller 雪卷
snow sampler 雪柱收集器,取雪管
snow-sampling equipment 取雪装置
snow scale 量雪尺【大气】
snow shed 避雪栅(山坡上遮护铁道的)
snow shield 防雪设备,雪障
snow shower 阵雪
snow sky (＝snow blink) 雪映天(地面有积雪时,地平线上的彩色天空)
snow slide 雪崩
snow sludge 雪泥,雪浆
snow slush 雪泥(指大量积雪为雨水粘结而成的雪块)

snow smoke 雪烟
snow squall 雪飑
snow stage 成雪阶段【大气】
snow stake 测雪桩
snow static 雪花静电
snow storm 雪暴【大气】
snow stratification 积雪成层〔作用〕,积雪分层
snow survey 测雪
snow thunderstorm 雪雷暴
snow totalizer 雪量累计器
snow tower 测雪架
snow trail 雪迹
snow tremor 〔积〕雪震〔落〕
snow tube 取雪管,雪柱收集器
snow virga 雪幡【大气】
snow water 雪水
snow-water equivalent 雪水当量
SNR（signal-to-noise ratio） 信噪比
SO（Southern Oscillation） 南方涛动【大气】
soaking downpour 滂沱大雨
soaking rain 透雨
social environment 社会环境
social geography 社会地理学
Society of Agricultural Meteorology of Japan（SAMJ） 日本农业气象学会
socio-economic factor 社会经济因素
socked in 机场关闭天气
sodalime glass dome 钠钙玻璃罩（辐射仪器用）
sodar（sound detection and ranging） 声〔雷〕达【大气】
sodium 钠（Na）
sodium cloud 钠云
sodium iodide detector 碘化钠探测器
sodium layer 钠层【地】
sodium spectrum 钠谱
sodium vapor tail wind technique 钠蒸气尾迹测风法（高层风测量用）
sofar （或 SOFAR, sound fixing and ranging） 声发（声定位测距仪）
SOFAR（sound fixing and ranging）channel 声发通道
SOFAR technology 声发技术（声定位测距仪技术）
soft air （见 moist air）湿空气
soft decision 软决策【数】
soft hail 软雹
soft-land 软着陆（宇宙飞船等）
soft particle 软粒子（低能粒子）
soft radiation 软辐射
soft rime 雾凇
soft science 软科学（泛指社会科学）
software 软件
software agent 软件主体【计】
software configuration 软件配置【计】
software development cycle 软件开发周期【计】
software engineering 软件工程【计】
software library 软件库【计】
software package 软件包【计】
soft water 软水
SOHO（Solar and Heliospheric Observatory） 索贺【天】(太阳和日球层观测站,1995年发射）
SOI（southern oscillation index） 南方涛动指数
soil 土壤【土壤】
soil acidity 土壤酸度
soil aeration 土壤通气〔性〕【土壤】
soil air 土壤空气
soil air capacity 土壤空气容量【土壤】
soil air composition 土壤空气组成【土壤】
soil air exchange 土壤空气交换【土壤】
soil air diffusion 土壤空气扩散【土壤】
soil air regime 土壤空气状况
soil acidification 土壤酸化作用【土壤】
soil alkalization 土壤碱化作用【土壤】
Soil and Water Assessment Tool（SWAT） 水土评价工具
soil and water conservation 水土保持

【土壤】
soil and water conservation 水土保持学【林】
soil areation 土壤通气〔性〕【土壤】
soil atmosphere 土壤空气
soil auger 〔取〕土钻
soil biochemistry 土壤生物化学【土壤】
soil bio-geochemistry 土壤生物地球化学【土壤】
soil chemistry 土壤化学【土壤】
soil climate 土壤气候【大气】
soil climatology 土壤气候学
soil colloid chemistry 土壤胶体化学【土壤】
soil conductivity 土壤传导率
soil creep 土壤蠕动【土壤】,土体蠕动【地理】
soil degradation 土壤退化
soil depth thermometer 土壤温度表
soil development 土壤发育【土壤】
soil drought 土壤干旱
soil dust 土壤尘埃,土壤尘
soil ecosystem 土壤生态系统【土壤】
soil environment 土壤环境【土壤】
soil environment capacity 土壤环境容量【土壤】
soil environment quality 土壤环境质量【土壤】
soil environment quality assessment 土壤环境质量评价【土壤】
soil emissivity 土壤发射率
soil environment quality assessment 土壤环境质量评价【土壤】
soil erosion 土壤侵蚀〔学〕【土壤】,土壤流失,土蚀
soil evaporation 土壤蒸发【大气】
soil evaporation pan 土壤蒸发皿
soil evaporimeter 土壤蒸发表
soil family 土族【土壤】
soil fertility 土壤肥力【土壤】
soil feasibility 土宜【土壤】

soil fertility monitoring 土壤肥力监测【土壤】
soil field capacity 土壤田间持水量
soil flow 土石流
soil gas 土壤气体
soil geochemistry 土壤地球化学【土壤】
soil geography 土壤地理〔学〕【土壤】
soil heat flux 土壤热通量
soil heat flux plate 土壤热通量〔测量〕板
soil heat regime 土壤热状况【土壤】
soil horizon 土〔壤化育〕层
soil humidity monitor 土壤湿度监测仪
soil humus chemistry 土壤腐殖质化学【土壤】
soil hygrograph 土壤湿度计
soil hygrometry 土壤测湿法
soil mechanics 土力学【地质】
soil mineral chemistry 土壤矿物化学【土壤】
soil model 土壤模式
soil moisture 土壤水〔分〕【土壤】,土壤湿度
soil moisture content 土壤水〔分〕含量
soil moisture deficit 土壤水分差值
soil moisture profile 土壤湿度廓线
soil moisture tension 土壤水分张力
soil organic matter 土壤有机质【土壤】
soil pan 土壤蒸发皿
soil physics 土壤物理〔学〕【土壤】
soil-plant-atmosphere continuum (SPAC) 土壤-植物-大气系统,土壤-植物-大气连续体
soil pollution 土壤污染【土壤】
soil pollution chemistry 土壤污染化学【土壤】
soil porosity 土壤孔〔隙〕度【土壤】
soil productivity 土壤生产力【土壤】
soil productivity grading 土壤生产力分级【土壤】
soil purification 土壤净化【土壤】
soil redox system 土壤氧化还原体系

【土壤】
soil regionalization 土壤区划【土壤】
soil resources 土壤资源
soil science 土壤学【土壤】
soil self-purification function 土壤自净化功能【土壤】
soil sodication 土壤钠质化作用【土壤】
soil structure 土壤结构【土壤】
soil stripe 冻埂
soil surface thermometer 地表温度表
soil texture 土壤质地【土壤】
soil temperature 土壤温度【大气】
soil thermal exchange 土壤热状况【土壤】
soil thermograph 土壤温度计
soil thermometer 土壤温度表【大气】
soil tillage 土壤耕作【土壤】
soil tilth 土壤耕性【土壤】
soil tube 取土管
soil utilization 土壤利用【土壤】
soil-vegetation-atmosphere transfer [SVAT] model 土壤-植被-大气传输模式
soil water 土壤水【分】
soil water balance 土壤水分平衡【大气】
soil water content 土壤含水量【大气】
soil water potential 土壤水势【大气】,土水势【土壤】
soil water pressure 土壤水压
soil zone 土壤带
solaire (= soulédras, soulédre) 日出风（法国中、南部的东风，来自日出方向，法语名）
sol-air temperature 等价日射气温
solano 沙拉拿风（西班牙东南海岸夏天的一种东风）
solar active region 太阳活动区
solar activity 太阳活动【大气】
solar air mass 日射大气光程,日射大气质量
solar altitude 太阳高度,太阳高度角

solar altitude angle 太阳高度角
solar altitude diagram 太阳高度查算图
Solar and Heliospheric Observatory (SOHO) 索贺【天】(1995年发射的太阳和日球层观测站卫星)
solar annual [apparent] motion 太阳周年视运动
solar atmosphere 太阳大气
solar azimuth (angle) 太阳方位角
solar atmospheric tide 〔太阳〕大气潮
solar aureole 日晕（轮）,环日散射辐射
solar-backscattered ultraviolet/total ozone mapper system (SBUV/TOMS) 太阳后向散射紫外线/臭氧总量测绘系统
solar backscatter ultraviolet radiometer (SBUV) 太阳后向散射紫外辐射仪
solar backscatter ultraviolet spectrometer (SBUVS) 太阳后向散射紫外分光计
solar barometric variation 日致气压变化
solar beam 太阳波束
solar blind region 太阳盲区
solar calendar 阳历【天】
solar cell 太阳〔能〕电池
solar chromosphere 太阳色球〔层〕
solar climate 太阳气候【大气】,天文气候
solar constant 太阳常数【大气】
solar corona ①日华【大气】②日冕【天】
solar corpuscle 太阳微粒
solar corpuscular emission 太阳微粒发射【地】
solar corpuscular radiation 太阳微粒辐射
solar corpuscular ray 太阳微粒射线
solar corpuscular theory 太阳微粒〔学〕说
solar cross 太阳十字〔晕〕
solar crown 日华
solar cycle 太阳活动周【天】,太阳活动周期【大气】
solar day 太阳日【天】
solar daily variation 太阳日变化
solar declination 太阳赤纬
solar disk 日面【天】

solar distance 日〔地〕距
solar disturbance 太阳扰动
solar ebb 太阳活动低潮
solar eclipse 日食【天】
solar electron event 太阳电子事件【地】
solar elevation 太阳仰角
solar elevation angle 太阳仰角(等于太阳高度角,与太阳天顶角互余)
solar energy 太阳能
solar energy demarcation 太阳能区划【大气】
solar energy resources 太阳能资源【大气】
Solar Energy Society of America (SESA) 美国太阳能学会
solar facula 太阳光斑
solar flare 太阳耀斑【大气】
solar flare activity 〔太阳〕耀斑活动
solar flare disturbance 太阳耀斑扰动
solar flood(见 solar tide) 太阳潮
solar flux density 太阳通量密度
solar granule 太阳米粒
solar halo 日晕【大气】
solar heat 太阳热
solar heating rate 太阳〔辐射〕加热率
solar heterodyne radiometer 太阳外差辐射仪
solar high energy particle 太阳高能粒子
solar house 温室
solarigram 总日射自记曲线
solarigraph 日射总量计
solarimeter 日射总量表
solar infrared 太阳红外线
solar irradiance 太阳辐照度【天】
solarization 日曝
solar limb 日面边缘
solar luminosity 太阳光度【天】(光度单位,等于 3.826×10^{26} J/s)
solar magnetic cycle 太阳磁周【天】
solar maximum 太阳〔黑子〕极大
solar maximum temperature 最高日射温度
solar minimum 太阳〔黑子〕极小

solar-meteorological relationship 太阳-气象关系
solar occultation 掩日法
solar optical telescope 太阳光学望远镜
solar paddle(或 panel) 太阳电池帆板,太阳翼
solar particle emission 太阳粒子发射
solar particle event 太阳粒子事件
solar particle stream 太阳微粒流
solar physics 太阳物理学【天】
solar pointer 太阳指向器(跟踪太阳的)
solar prominence 日珥【天】
solar proton event 太阳质子事件【天】
solar proton monitor (SPM) 太阳质子监测仪
solar radiation 太阳辐射【大气】
solar radiation correction 日射订正
solar radiation error 太阳辐射误差
solar radiation observation 日射观测,太阳辐射观测
solar radiation pressure 日射压
solar-radiation thermometer 日射温度表
solar radio emission 太阳电亡[发射]
solar radiometer 太阳辐射表
solar radio radiation 太阳射电
solar radio telescope 太阳射电望远镜
solar radius 太阳半经【天】(长度单位,$=6.9599 \times 10^5$ km)
solar short-wave radiation 太阳短波辐射
solar signal 太阳信号
solar spectrograph 太阳摄谱仪
solar spectrum 太阳光谱
solar spicule 日暑,日针
solar synchronous meteorological satellite 太阳同步气象卫星【大气】,极轨气象卫星
solar system 太阳系【天】
solar telescope 太阳望远镜【天】
solar temperature 太阳温度
solar term 节气【天】
solar-terrestrial disturbance 日地扰动

solar-terrestrial physics　日地物理学【地】
solar-terrestrial relationship　日地关系【天】
solar-terrestrial space　日地空间【地】
solar thermometer　日射温度表
solar tide　太阳潮【地】
solar time　太阳时
solar-topographic theory　太阳地文说
solar-type star　太阳型星【天】
solar ultraviolet radiation　太阳紫外辐射
solarure（＝solore）索洛尔风（吹向法国德龙河的夜间山风）
solar variability　太阳变异度
solar wind　太阳风【天】,日射微粒流
solar wind boundary　太阳风边界
solar year　太阳年
solar zenith angle　太阳天顶角
SOLAS（Surface Ocean-Lower Atmosphere Study）表层海洋-低层大气研究
solenoid　力管,螺线管【物】
solenoidal　力管〔的〕
solenoidal index　力管指数,网络指数
solenoidal vector　管状矢量
solenoid effect　力管效应
solenoid field　力管场,零散度场【电工】
solenoid term　力管项
sol-gel process　溶胶-凝胶过程【化】
solid angle　立体角【数】
solid boundary　刚体边界
solid carbon dioxide（＝dry ice）固体二氧化碳,干冰
solid geometry　立体几何〔学〕【数】
solidification　凝固【大气】
solidify　固化
solidifying point　凝固点【物】
solid line　实线
solid phase　固态,固相【石油】
solid rotation　刚体转动
solid state　固态,固体
solid-state electrometer　固体静电计

solid state electronics　固体电子学【物】
solid-state imaging sensor　固体成像传感器
solid state laser　固体激光器【物】
solid state plasma　固体等离〔子〕体【物】
solifluction　解冻泥流,泥流【土壤】
solitary wave　孤立波【大气】
soliton　孤〔立〕子【物】,孤〔立〕波,单极子,单波型
Solomon's paper　索洛蒙纸（测湿用的）
solore（见 solarure）索洛尔风
solstices　二至点（夏至点和冬至点）
solstitial colure　二至圈
solstitial tide　至点潮
solubility　溶〔解〕度
solubility parameter　溶度参数【化】
solum　风化层
solute　溶质
solute effect　溶质效应
solution　溶液【化】
solvate　溶剂合物【化】
solvent　溶剂【化】
solvent effect　溶剂效应【化】
Somali Current　索马里海流
Somali jet　索马里急流
somatic temperature　体温
sonar（sound navigation and ranging）声呐【航海】
sonar image　声呐图像【测】
sonar remote sensing　声呐遥感【地理】
sonar sweeping　声呐扫海【测】
sonde　①探空仪【大气】②探测气球③探头,探针
sonde error　探空仪误差
sondo（见 zonda）桑达风
sonic anemometer　声〔学〕风速表
sonic anemometer-thermometer（SAT）声〔学〕风速温度表
sonic boom　声爆
sonic energy　声能
sonic grenade　声榴弹（测风、温用）

sonics 声能学
sonic speed 声速
sonic thermometer 声学温度表
sonic wave 声波
Sonntag pyrheliometer 桑塔格绝对日射表
Sonora 索诺拉雷暴
Sonora storm 索诺拉风暴(美国加利福尼亚南部山地及沙漠中夏季的雷暴)
Sonora weather 索诺拉天气(阵雨条件下适于飞行的天气,美国俚语)
Sonora weather 索诺拉雷暴天气
soot 煤烟(燃料不完全燃烧而产生的以碳为主要成分的微粒),烟灰,烟炱
soot concentration 煤烟浓度
soot content 煤烟含量
soot deposit 煤烟沉积
sootfall 煤烟沉降
sopero(=Sover) 索佩罗风(意大利加尔达湖附近的夜间微风)
sophisticated 先进的,高级的,复杂的
soroche 山岳症(西班牙语名)
sorption 吸收,吸附
sorting machine 分类机
SOS ①(save our ship)国际通用的(船舶、飞机等)呼救信号,无线电求救信号②(synchronous orbit satellite)同步轨道卫星
sou'easter (southeaster 的缩写)东南大风
soulaire 苏莱拉雷风(法国莫尔旺山的南风,法语名)
sound 声
sound absorption 吸声,声吸收
sound absorption coefficient 吸声系数
sound beam 声束
sound burst 猝发声
sound channel 声通道
sound energy density 声能密度
sound energy flux 声能通量
sounder 探空仪,探测器
sound fixing and ranging (SOFAR) 声发(声定位测距仪)

sound frequency 声频
sounding 探空【大气】,探测
sounding balloon 探空气球【大气】
sounding meteorograph 探空气象仪
sounding pole 测深杆
sounding rocket 探空火箭【大气】
sounding weight 测深锤
sound intensity 声强
sound level 声级【物】
sound level meter 声级仪
sound lump echoes 声回波团块结构
sound navigation and ranging(sonar) 声呐【航海】
sound pollution 噪音污染,声污染
sound power level 声能级
sound pressure integrator 声压积分仪
sound pressure level 声压级
sound pressure level meter 声压等级仪
sound propagation 声音传播
sound radar 声〔雷〕达
sound ray 声线
sound shadow 声影
sound velocity 声速【物】
sound wave 声波【物】
source 源〔泉〕,源地,源点【数】
source emission 污染源排放物
source function 源函数
source function coefficient 源函数系数
source of an atmospheric 天电源
source of light 光源
source of thunderstorm activity 雷暴活动源地【大气】
source program 源程序【计】
source region 源地
source-sink flow 源汇流
source strength 源地强度
source terms 源项
south (S) 南
South Asia high 南亚高压【大气】
South Asian monsoon 南亚季风
South Atlantic Central Water 南大西洋

中央水
South Atlantic convergence zone　南大西洋辐合带
South Atlantic Current　南大西洋海流
South Atlantic high　南大西洋高压
southbound　下轨道（对北半球而言）
South China quasi-stationary front　华南准静止锋【大气】
South China Sea　南海
South China Sea anticyclone　南海反气旋
South China Sea depression　南海低压【大气】
South China Sea high　南海高压
South China Sea monsoon　南海季风
South China Sea Monsoon Experiment (SCSMEX)　南海季风试验
southeast (SE)　东南
Southeast Asia Weather Center (SEAWECEN)　东南亚天气中心
southeaster　东南大风
southeast monsoon　东南季风
southeast-storm warning　东南风暴预警
southeast trades　东南信风
South Equatorial Countercurrent (SECC)　南赤道逆流，南赤道反洋流
South Equatorial Current (SEC)　南赤道海流
South Equatorial Drift Current　南赤道漂流
souther　南大风
southerly burster　南寒风（澳大利亚）
Southern Annular Mode (SAM)　南半球环状模
southern branch jet stream　南支急流【大气】
Southern Hemisphere　南半球
Southern Hemisphere Exchange Centre (SHEC)　南半球交换中心
southern latitude　南纬
southern lights　南极光
Southern Ocean　南大洋，南极海

Southern Oscillation (SO)　南方涛动【大气】
Southern Oscillation Index (SOI)　南方涛动指数
Southern summer　南半球夏季
Southern winter　南半球冬季
south frigid zone　南寒带，南极带
south foehn　南焚风
South Indian Ocean Current　南印度洋海流
South Java Current　南爪哇海流
south magnetic pole　磁南极【地】
South Pacific convergence zone (SPCZ)　南太平洋辐合带
South Pacific Current　南太平洋海流
south point　南（英语系罗盘上用语）
South Pole　南极
south southeast (SSE)　南东南
south southwest (SSW)　南西南
south temperate zone　南温带
south tropic　南回归线
south tropical calm zone　南热带无风带
southward　向南
southwest (SW)　西南
Southwest China vortex　西南〔低〕涡【大气】
southwester (=sou'wester)　西南大风
southwesterlies　西南风带
southwesterly flow　西南气流
southwest monsoon　西南季风
Southwest Monsoon Current　西南季风海流
Sover (见 sopero)　索佩罗风
Soviet Antarctic Expedition (SAE)　（前）苏联南极考察队
Soya Warm Current　宗谷暖洋流
Soya Warm water　宗谷暖流
soyokaze　微风（日语名）
SPAC (soil-plant-atmosphere continuum)　土壤-植物-大气连续体
space　空间【物】，太空
Space and Upper Atmosphere Research Committee (SUPARCO)　空间与高层

大气研究委员会(巴基斯坦)
space averaging 空间平均
space-based observation 天基观测【大气】
space-based observation system 天基观测系统
space-based subsystem 天基子系统【大气】
spaceborne meteorological radar 星载气象雷达
space change 空间变化
space charge 空间电荷【物】
space consistency check 空间一致性检验
space correlation 空间关联【物】,空间相关
spacecraft 空间飞行器(如卫星、宇宙飞船等)
space difference scheme 空间差分格式
space domain 空[间]域【物】
space environment 空间环境
space environmental monitor (SEM) 空间环境监视器
space filtering 空间滤波
space flight 空间飞行
space function 空间函数
space harmonic representation 空间简谐表达式
space integration 空间积分
space mean chart 空间平均图
space medicine 宇宙医学
space meteorology 空间气象学【大气】
space platform 空间观测平台
space physics ①空间物理学【地】②空间物理【地】
space physics exploration 空间物理探测
space probe 宇宙探测器
space remote sensing 航天遥感
space research 空间研究
space resonance 空间共振
space scale 空间尺度
space shuttle 航天飞机【测】
space smoothing 空间平滑【大气】

spacetime 时空【物】
spacetime coordinates 时空坐标【物】
space-time correlation 时空相关
space-time sampling 时空取样
space tracking 空间跟踪
space tracking and data acquisition network (STADAN) 空间跟踪和资料接收站网
space truncation error 空间截断误差
space weather 空间天气【大气】
space weather disaster 空间天气灾害
space weather mechanism 空间天气机制
spacing ①植距②间隔,空隙
span 量距
SPARC (Stratospheric Processes and Their Role in Climate) 平流层过程及其在气候中的作用
spark ①火花②电花
spark discharge 火花放电【物】
sparse data area 资料稀少区域
Sparse Matrix Operator Kernel Emission (SMOKE) 稀疏矩阵算子核心排放(排放源处理程序)
spate 洪水,突然性的大暴雨,倾盆大雨(苏格兰语)
spatial and temporal distribution 时空分布
spatial and time resolution 时空分辨率
spatial average 空间平均
spatial coherence 空间相干
spatial correlation 空间相关
spatial crystal 立体冰晶
spatial dendrite 立体枝状冰晶
spatial derivative 空间导数,空间微商
spatial distribution 空间分布
spatial factor 空间因子
spatial filter function 空间滤波函数
spatial filtering 空间滤波
spatial representativeness 空间代表性
spatial resolution 空间分辨率【大气】
spatial scale 空间尺度

spatial-temporal variability 时空变率
spatial variation 空间变化
SPCZ (South Pacific convergence zone) 南太平洋辐合带
SPD (surge protection device) 电涌保护器
special aerodrome report 特殊机场报告
Special Experiment Concerning Typhoon Recurvature and Unusual Movement (SPECTRUM) 台风转向和异常运动特别试验
special function 特殊函数【数】
specialized meteorological service 专业气象服务
special observation 特殊观测
special sensor microwave imager 特种传感器微波成像仪
special sensor microwave temperature 特种传感器微波温度〔仪〕
special weather report 特殊天气报告
special weather statements 特殊天气综述
Special World Intervals (SWI) 特殊世界时段
species of clouds 云种
specific ①特定的,专门的②具体的,明确的③比的,单位的
specific absorption 吸收率(井水回灌)
specification ①规范,规格②说明书③规程,技术条件,技术要求
specific attenuation 比衰减,单位衰减
specific capacity ①比容量②单位(井)出水量
specific conductance 电导率,比电容
specific conductivity 比传导率
specific density 比密
specific discharge 比流量,单位〔面积〕流量
specific energy 比能
specific enthalpy 比焓
specific entropy 比熵
specific flux 比通量,单位〔面积〕通量

specific gas constant 比气体常数
specific gravity 比重【物】
specific heat〔capacity〕 比热〔容〕【物】
specific heat at constant pressure 定压比热【物】
specific heat at constant volume 定容比热,定体〔积〕比热【物】
specific heat capacity 比热容
specific humidity 比湿【大气】
specific humidity line 比湿线
specific intensity 辐射单位强度,比强度
specificity ①特异性,特征②专一性【海洋】
specific precipitation density 特定降水密度
specific retention 比持水量,单位持水量
specific storage 单位贮水量(地下水)
specific volume 比体积【物】,比容
specific-volume anomaly 比容异常
specific weight 比重
specific yield 比出水量
specified isobaric surface 规定等压面
specified level 规定层
speckle 斑点,污迹
spectral ①光谱的②分谱的③单色的
spectral absorption 谱吸收
spectral albedo 光谱反照率
spectral analysis 谱分析【数】
spectral blocking 〔波〕谱阻塞
spectral broadening 谱加宽
spectral channel 光谱通道,频谱通道
spectral concentration of a radiometric quantity 辐射量谱密度
spectral decomposition 谱分解【数】
spectral decomposition of stationary process 平稳过程谱分解【数】
spectral density 谱密度
spectral density function 谱密度函数【数】
spectral density tensor 谱密度张量
spectral detectivity 单色探测率

spectral diffusivity 谱扩散率
spectral distribution function 谱分布函数
【数】
spectral evolution 谱展开
spectral filter function 谱滤波函数
spectral frequency 谱频率
spectral function 谱函数
spectral gap 谱〔间〕隙
spectral hygrometer 光谱湿度表
spectral integral 谱积分【数】
spectral interference 光谱干扰
spectral interval 谱间隔
spectral irradiance 谱辐照度
spectral line 〔光〕谱线
spectral line intensity 谱线强度
spectral line interference 谱线干扰
spectral mapping 谱映射
spectral measure 谱测度【数】
spectral method 谱方法【数】
spectral model 谱模式【大气】
spectral numerical analysis 谱数值分析
spectral numerical prediction 谱数值预报
spectral pattern 谱型
spectral photoconductivity 谱光电导
spectral processing 谱处理
spectral pyranometer 分光天空辐射表
spectral radiance 谱辐射亮度
spectral representation 谱表达式
spectral resolution 谱分辨率【数】
spectral-response characteristics 谱响应特征〔性〕
spectral separation 谱分离
spectral set 谱集
spectral similarity 谱相似〔性〕
spectral space 谱空间
spectral transfer function 谱传递函数
spectral transform model 谱变换模式
spectral width 谱宽
spectral window 谱窗【数】
spectrobologram 分光变阻测热图
spectrobolometer 分光变阻测热计

spectrograph 光谱计,摄谱仪【化】
spectroheliograph 日射光谱计
spectrometer 〔光〕谱仪【物】,分光计,分光仪
spectrometry 分光测量法
spectrophone 光声计
spectrophone detection 光声探测
spectrophotometer 分光光度表【物】
spectrophotometric dosimeter 紫外分光光度表
spectrophotometry 分光光度学
spectropluviometer 分谱雨量计
spectropyrheliometer 绝对日射光谱仪
spectroscope 分光镜【物】
spectroscopic hygrometer 分光湿度表
spectroscopic laser radar technology 分光光〔雷〕达技术
spectroscopic method 光谱法
spectroscopic term 光谱项【化】
spectroscopy ①光谱学【物】②分光学
spectrum ①谱【物】②光谱③频谱④波谱⑤功率谱
SPECTRUM (Special Experiment Concerning Typhoon Recurvature and Unusual Movement) 台风转向和异常运动特别试验(1990年)
spectrum abscissa 谱的横坐标
spectrum amplification 谱放大率
spectrum analysis 谱分析
spectrum area 谱区
spectrum banding 谱带
spectrum broadening 谱加宽
spectrum calculation 谱计算
spectrum characteristic 谱特性,谱特征
spectrum coordinate 谱坐标
spectrum curvature 谱曲率
spectrum density 谱密度
spectrum distribution 谱分布
spectrum distribution function 谱分布函数
spectrum energy 谱能
spectrum-energy spread 波谱能量扩展

spectrum folding　波谱折叠
spectrum function　谱函数
spectrum gap　谱〔间〕隙
spectrum index　谱指数
spectrum inflection　谱的拐点
spectrum interval　谱间隔
spectrum level　谱级,谱水平
spectrum maximum　谱的极大值
spectrum minimum　谱的极小值
spectrum model　谱模式
spectrum moment　谱矩
spectrum of lateral velocity　横向速度谱
spectrum of longitudinal velocity　纵向速度谱
spectrum of temperature　温度谱
spectrum of turbulence　湍流谱
spectrum of velocity　速度谱
spectrum ordinate　谱的纵坐标
spectrum parameter　谱参数
spectrum pattern　谱图
spectrum peak frequency　谱峰频率
spectrum reliability　谱的可靠性
spectrum slope　谱斜率
spectrum tensor　谱张量
spectrum terminology　谱术语
spectrum theory　波谱理论
spectrum width　谱宽
spectrum window　谱窗
specular reflection　镜反射
specular reflector　镜反射体
speech analysis　语言分析
speech recognition　语音识别【计】
speech synthesis　语音合成【计】
speed　速率【物】,速度
speed of cloud movement　云速
speed of light　光速
speed of sound　声速
speed over ground　实际航速【航海】
speed regulation by frequency variation　变频调速【航海】
speed regulation by pole changing　变极调速【航海】
speed-up height　加速高度
speed-up ratio　加速比
speed-up wind　加速风
spelaeo-meteorology　洞穴气象学
speleothem　洞穴沉积层,洞穴堆积物
sphere　①球②圈
sphere calibration　球校准
spheric aerostat　球形气球
spherical albedo　球面反照率
spherical Bellani cup radiometer　贝拉尼球罩式辐射仪
spherical coordinates　球面坐标【数】
spherical coordinate system　球面坐标系
spherical curvature　球面曲率
spherical function　球函数
spherical grid　经纬度网格
spherical harmonic analysis　球〔面〕谐〔波〕分析
spherical harmonics　球面调和函数【数】,球谐函数
spherical harmonic wave　球〔面〕谐波
spherical incident wave　球面入射波
spherical lens　球面透镜【物】
spherical mirror　球面镜【物】
spherical net　球面网格
spherical projection　球面投影【物】
spherical pyrgeometer　球形地面辐射表
spherical pyrradiometer　球形全辐射表
spherical shell　球壳【力】,球皮
spherical shell apparatus　球壳装置
spherical symmetry　球对称
spherical top　球陀螺
spherical trigonometry　球面三角学【数】
spherical wave　球面波
spherics　远程雷电,天电
spherics azimuth (SFAZI) code　天电方位编码
spheric state　远程雷电状况,天空状况
spheroid　球状体,椭球体
spheroidal oscillation　球型振荡

spicule 冰针
spider lightning 蛛网状闪电
spike ①尖峰(仪器记录图中的)②穗状花序【植】③穗(指禾本科)
spikelet 小穗【生】
spiky phenomenon 尖劈现象
spillway 溢洪道
spillway capacity 溢洪道容量,溢洪道泄洪能力
spillway design flood 溢洪道设计洪水,校核洪水
spin 旋转,自旋【物】
spin axis 自旋轴
spin axis point 旋轴点
spin-down 消转
spin-down effect 消转效应
spin-down process 消转过程
spindown time 消转时间【大气】,旋转减弱时间
spin labeling 自旋标记【生化】
spinning bowl LWC meter 转杯液水含量仪
spinning coil magnetometer 自旋线圈磁强计
spin quantum number 自旋量子数【物】
spin-rotation band 自旋-转动带
spin scan cloud camera (SSCC) 自旋扫描摄云机
spin stabilization 自旋稳定
spin-up 起转【航空】
spinup time 起转时间【大气】
spin-up problem 起转问题
spin-up process 起转过程
spin wave 自旋体【物】
spiral 螺线【数】
spiral band 螺旋带
spiral cirrus 螺旋状卷云
spiral cloud band 螺旋云带【大气】
spiral equilibrium 螺线平衡
spiral layer 螺线层
spiral pattern 螺旋状云型

spiral rain-band 螺旋雨带
spiral rain-band echo 螺旋雨带回波【大气】
spiral scanning 螺旋扫描
spiral structure 螺旋[状]结构
spiral wave 螺旋波
spirit-in-glass thermometer 酒精[玻璃]温度表
spirit thermometer 酒精温度表
spissatus (spi) 密云
spit 微量降水
Spitsbergen Current 斯匹次卑尔根海流
splashdown 溅落
splash erosion 溅蚀[土壤]
spline 样条[函数]【数】
spline fitting 样条拟合【数】
spline function 样条[函数]
spline-function expansion 样条函数展开
spline interpolation 样条插值
splinter 冻离[冰]屑
splintering 碎splitting
split cold front 分离冷锋
split-explicit scheme 分离显式格式【大气】
split-explicit method 分离显式方法
splitting convective storm 分离对流风暴
splitting factor 分离因子
splitting frequency 分频
splitting method 分离法,分裂法
splitting of westerlies 西风带分支
split window 分离窗[区],分裂窗
split window technique 分窗法
sponge boundary condition 海绵边界条件【大气】
sponge growth 海绵生长
sponge layer 海绵层
sponge zone 海绵带
spongy hail 海绵状雹
spongy ice 海绵冰
spontaneous condensation 自发凝结
spontaneous convection 自发对流
spontaneous emission 自发发射【物】

spontaneous freezing 自发冻结
spontaneous nucleation 自发核化
spontaneous sublimation 自发凝华
sporadic E 散见E层【地】
sporadic E-region （或 Es-layer）散见E层
sporadic wind 时现风
spore-pollen analysis 孢粉分析
Sporrer minimum 斯波勒极小【天】（1400—1510年期间太阳活动持续处于低水平）
spot group 〔太阳〕黑子群
spotting ①填图②确定准确位置
spot wind 定点风
spout 龙卷
spray ①飞沫②日喷【天】
spray electrification 水沫起电
spray region 〔大气〕边缘区
spread 展形,散布
spread E layer 扩展E层
spread F layer 扩展F层
spread of the ensemble 集合扩展,集合扩散
spread skill correlation 扩展技巧相关
spread spectrum singal 扩频信号
spring 春〔季〕
Spring Beginning 立春（节气）【大气】
spring crust 春雪壳
Spring Equinox 春分（节气）【大气】
spring flood 春汛
spring frost 春霜,晚霜【农】
spring land 湿润土地
spring-neap cycle 春潮周期
springness 春性
spring rise 大潮升【航海】
spring sludge 融冰
spring snow 春雪
spring tapping 泉水引流,引泉
spring tide 大潮【海洋】,春潮
springwood 早材【植物】,春材
sprinkle 微阵雨
sprite 幽〔灵〕闪〔电〕,滚闪
Sprung barograph 斯普朗气压计
spurious correlation 伪相关

spurious gravity-wave 伪重力波
sputnik 人造卫星（俄语名）
squall 飑【大气】
squall cloud 飑〔线〕云【大气】
squall cluster 飑线云簇
squall front 飑锋
squall line 飑线【大气】
squall line echoes 飑线回波
squall line thunderstorm 飑线雷暴
squall storm 飑暴
squall surface 飑面
squall wind 飑性风
squally weather 飑性天气
square mesh 〔正〕方形网格【数】
square net 〔正〕方形网格【数】
square root absorption 平方根吸收
square wave 正方波
square-wave generator 方波发生器【电子】
sr(steradian) 立体弧度,球面〔角〕度
SR（scanning radiometer） 扫描辐射仪【大气】
SREH（storm-relative environmental helicity） 风暴对环境的螺旋度
SSA（singular spectrum analysis） 奇异谱分析
SSOA（Severe Storm Observation Satellite） 强风暴观测卫星
SSP（subsatellite point） 卫星星下点
SSM/I（special sensor microwave imager） 特种传感器微波成像仪
SSM/T（special sensor microwave temperature） 特种传感器微波温度〔仪〕
SSU（stratospheric sounding unit） 平流层探测装置
SST（sea surface temperature） 海面温度【大气】
St（stratus） 层云【大气】
stability 稳定〔度〕,稳定性【物】
stability analysis 稳定度分析,稳定性分

析【物】
stability categories 稳定度分类
stability chart 稳定度图
stability condition 稳定〔度〕条件
stability criterion 稳定〔度〕判据
stability index 稳定度指数
stability length 稳定度长度
stability parameter 稳定度参数【大气】
stability theory 稳定度理论
stabilization 稳定作用
stabilization system 稳定系统
stabilizing factor 稳定因子
stabilizing force 稳定力
stable air 稳定空气【大气】
stable air mass 稳定气团【大气】
stable antenna platform 稳定天线平台
stable atmosphere 稳定大气
stable boundary layer 稳定边界层
stable equilibrium 稳定平衡态
stable lamina 稳定片〔流〕层
stable local oscillator 稳定本机振荡器
stable manifold 稳定流形
stable motion 稳定运动
stable node 稳定节点
stable oscillation 稳定振荡
stable profile 稳定廓线
stable stratification 稳定层结
stable type 稳定型
stable wave 稳定波
staccato lightning flash 不连续闪电
stack ①堆栈【计】②烟囱【电力】
stack effect 烟囱效应
stack effluent 烟囱排放物
stack emission 烟囱排放
stack height 烟囱高度
stacking 空中待命〔降落〕
stadia technique 视距尺方法
staff gauge 水位标,水尺
stage ①阶段,期 ②水位【水力】
stage discharge relation 水位流量关系
stage relation 水位关系

stage wise regression procedure 逐段回归方法
staggered grid 交错网格【大气】,跳点网格
staggered mesh 交错网格【数】
staggering of horizontal grid 水平网格点交错
stagger scheme 交错格式【大气】,跳点格式
stagnant air 停滞空气
stagnant glacier 停滞冰川
stagnant ground water 静止地下水
stagnated air mass 停滞气团
stagnation area 滞留区
stagnation point 滞点
stagnation pressure 停滞压
stalling Mach number 失速马赫数【航空】
stalo（stable local oscillator） 稳定本〔机〕振〔荡〕器
stamukhi 搁浅冰山（俄语名）
stand 平潮
standard ①标准②标准的
standard artillery atmosphere 标准弹道大气
standard artillery zone 标准弹道层
standard aspirated psychrometer 标准通风干湿表【大气】
standard atmosphere 标准大气【大气】
standard atmosphere pressure 标准大气压【大气】
standard atmospheric condition 标准大气状况
standard ballistic atmosphere 标准弹道大气
standard barometer 标准气压表
standard coordinate 标准坐标
standard density 标准密度
standard density altitude 标准密度高度【航空】
standard depth 标准深度

standard deviation 标准偏差【物】,标准差【大气】
standard eight-inch rain gauge 标准八英寸雨量计
standard error 标准误差【数】
standard error of estimate 估计标准误差
standard geopotential meter 标准位势米
standard gravity 标准重力
standard instrument 标准仪器
standard isobaric surface 标准等压面【大气】
standardization 标准化,规格化
standardized variate 标准化变量
standard level 标准层
standard meridian 标准子午线
standard meteorological screen 标准气象百叶箱
standard observation time 标准观测时间
standard operative temperature 标准作业温度
standard pan 标准蒸发器【大气】
standard pressure 标准气压
standard pressure altitude 标准气压高度
standard〔pressure〕level 标准层【大气】,规定层
standard precipitation index 标准降水指数
standard project flood 标准预估洪水
standard propagation 标准传播
standard psychrometer 标准干湿表
standard pyrheliometer 标准日射强度计
standard raingauge 标准雨量计【大气】
standard refraction 标准折射
standard sea level 基本水准面
standard sea water 标准海水
standard snow tube 标准取雪筒
standard station screen 标准百叶箱
standard sunshine recorder 标准日照计
standard target 标准目标
standard temperature 标准温度
standard temperature and pressure (STP) 标准温压
standard test procedure 标准测试程序
standard thermometer 标准温度表
standard time 标准时
standard time of observation 标准观测时间【大气】
standard unit 标准单位,标准单元【计算机】
standard visibility 标准能见度
standard visual range 标准视程
standby 备用【计】
standing cloud 驻云【大气】
standing eddies 驻涡
standing wave 驻波【大气】
standpipe storage precipitation gauge 立柱式贮存雨量器
Stanford University Microwave Laboratory (SU-ML) 斯坦福大学微波实验室(美国)
Stanton number 斯坦通数(普朗特数的倒数)
star ①星②恒星【天】
Stark effect 斯塔克效应【物】
starshine recorder 星照计
START (Global Change System for Analysis, Research and Training) 全球变化的分析、研究和培训系统(IGBP 计划之一)
starting plume 初始烟羽
starute 缓降伞(供火箭探测用)
state 态【物】,状态【数】
state curve 状态曲线
State Key Laboratory of Numerical Modeling for Atmospheric Science and Geophysical Fluid Dynamics (LASG) 大气科学和地球流体力学数值模拟国家

重点实验室
state of ground 地面状态【大气】,地表状况
state-of-sea scale 海况等级
state of the sea 海﹝面状﹞况
state of the sky 天空状况
state parameter 状态参数
state space 态空间【物】
state variable 状态变量
static ①天电②静力的
statical meteorology 静力气象学
static analysis 静态分析【计】
static bifurcation 静态分岔
static ceiling 气球平衡高度
static electricity ①静电②静电学
static energy 静力能
static equilibrium 静力平衡
static head 静压头【航海】
static initialization 静力初值化【大气】
static instability 静力不稳定〔度〕【大气】
static instability parameter 静力不稳定参数
static level 静力水位
static method 静态方法
static pressure 静压【物】
static pressure port 静压孔
static process 静力过程
statics 静力学【物】
static seeding 静力播撒云,静力催化
static stability 静力稳定度【大气】
static vent 静压孔
station 台,站
stationary 静止,稳定,滞留,驻立,定常
stationary automated mesonet (SAM) 永久性自动中尺度观测网
stationary cyclone 静止气旋,滞留气旋
stationary eddy 常定涡旋,驻涡
stationary ensemble 定态系综【物】
stationary front 静止锋【大气】
stationary function 平稳函数【数】

stationary Gaussian process 平稳高斯过程
stationary Gaussian time series 平稳高斯时间序列
stationary medium 平衡介质
stationary motion 常定运动
stationary orbit 平稳轨道
stationary phase 平稳期
stationary planetary wave 行星驻波
stationary process 平稳过程【数】
stationary random function 平稳随机函数
stationary random process 平稳随机过程
stationary real random function 平稳实随机函数
stationary Rossby wave 罗斯贝驻波
stationary source 平稳﹝信﹞源
stationary satellite 静止卫星
stationary state 定态【物】,平稳态
stationary state hypothesis 定态假设
stationary stochastic process 平稳随机过程
stationary time series 平稳时间序列【数】
stationary wave 定常波【大气】,定态波【物】
stationary wavelength 定常波长
station barometer 台站气压表
station circle 站圈【大气】(天气图上用小圈表示的测站位置)
station continuity chart 单站气象要素变化图
station designator (= station identifier) 区站号(5位数字)
station elevation 台站海拔
station identifier 区站号
station index 〔台〕站号,台站索引
station index number 区站号【大气】
station location 站址【大气】
station model 填图格式
station pressure 本站气压【大气】
station rainfall 站点雨量

station-ratio method 台站比法
station weather forecast 测站天气预报
station-year method 台站年〔分析〕法
statistic 统计量【数】
statistical 统计的
statistical analysis 统计分析【数】
statistical assimilation method 统计同化法
statistical band model 统计带模式【大气】
statistical characteristics 统计特征
statistical climate 统计气候
statistical climatology 统计气候学【大气】
statistical data 统计数据,统计资料
statistical dependence 统计相关〔性〕
statistical description of turbulence 湍流统计描述
statistical-dynamic model (SDM) 统计动力模式【大气】
statistical-dynamic prediction 统计动力预报【大气】
statistical forecast 统计预报【大气】
statistical hypothesis 统计假设【数】
statistical independence 统计独立〔性〕
statistical information 统计信息
statistical interpolation 统计内插
statistical interpolation method 统计插值法【大气】
statistical interpretation 统计释用,统计解译,统计判读
statistical inversion method 统计反演法
statistical inversion technique 统计反演方法
statistically based crop-weather analysis model 作物-天气分析统计模式
statistically homogeneous rough surface 统计均匀的粗糙面
statistical method 统计方法
statistical model 统计模式【大气】,统计模型【水利】

statistical-numerical prediction 统计-数值预报
statistical parameter 统计参数
statistical physics 统计物理〔学〕【物】
statistical power 统计功率
statistical regulation 统计控制法
statistical significance test 统计显著性检验
statistical test 统计检验
statistical thermodynamics 统计热力学
statistical weather forecast 统计天气预报
statistical weight 统计权重
statistics 统计,统计〔学〕【数】
statoscope 微动气压器,灵敏气压表
staubosphere 尘圈,尘层
steadiness 定常度
steady climate 定常气候
steady flow 定常流【物】
steady motion 定常运动
steady rain 连续雨,绵雨
steady state 定常态【物】,稳态【电工】
steady-state hypothesis 定态假设
steady-state solution 稳定解,定态解
steady state storm 稳定风暴,稳态风暴
steady system 定常系统
steady velocity 常定速度
steady vortex 定常涡旋
steady wind pressure 稳定风压【大气】
steam 蒸汽
steam devil 水汽暴
steam fog 蒸汽雾
steam mist 蒸汽轻雾
steepness 斜度
steering 引导
steering concept 引导概念
steering current 引导气流
steering flow 引导气流【大气】
steering law 引导法则
steering level 引导高度【大气】

steering line 驶线
steering surface 引导层
steering wind 引导风
Stefan-Boltzmann constant 斯蒂芬-玻耳兹曼常数
Stefan-Boltzmann law 斯蒂芬-玻耳兹曼辐射律,斯特藩-玻尔兹曼定律【物】
stefan problem 施特藩问题【数】
stellar crystal 星状结晶
stellar lightning 星〔状〕闪〔电〕
stellar scintillation 星光闪烁
stellar wind 星风【天】(类似太阳风)
St. Elmo's fire 圣爱尔摩火
stem flow 树冠〔水〕流
stem thermometer 棒状温度表
step 跃阶,单位高度气压差,单位深度温度差
step-by-step (或 suceessive) approximation 逐步近似〔法〕
step curve 阶梯曲线
step function 阶梯函数【数】,分层〔作用〕
step function model 阶梯函数模式
step length 步长【数】
step-over method 跃步法
steppe 〔干〕草原
steppe climate 草原气候〔亚类〕【大气】
stepped function 阶梯函数
stepped leader 梯级先导(闪电)
stepped motor 步进电〔动〕机【电子】
steppe soil 草原土壤
step response function 阶梯反应函数
step size 步长【数】
step voltage 跨步电压
step width 步长【数】
stepwise 逐步
stepwise discriminatory analysis 逐步判别分析法
stepwise regression 逐步回归【数】
stepwise regression analysis 逐步回归分析

stepwise sequential test 逐步序列检验
steradian 球面度(立体角单位)
stereo image 立体图像
stereogram 立体图
stereographic image 立体图像
stereographic projection 极射赤面投影,赤平投影,球面投影【测】
stereopair 立体像对【测】
stereophotogrammetry 立体摄影测量【测】
stereoscopic pair of photographs 立体照片对(测云用)
steric anomaly 比容异常
Stern climate 严酷气候
Stevenson screen 〔斯蒂文森〕百叶箱
St fra 碎层云【大气】
sticking coefficient 黏性系数
sticky weather 湿热天气
stiff (或 strong) crosswind 大侧风
stiffness 刚度,劲度【地】
stiffness matrix 刚度矩阵【数】
stiff system 刚性系统,刚性组【数】
Stikine wind 斯提金风(加拿大斯提金河附近的阵性东东北风)
stilb 熙提(旧的亮度单位)
still pond 静水器(观测蒸发池内水位用)
still-water level 静水位
still well 静水井(观测蒸发池内水位用)
still-well gauges 稳水器
stimulated radiation 受激辐射【物】
stirring 扰动
stirring length scale 扰动长度尺度
St. Luke's summer 圣路克热期(10月18日前后的晴暖期)
St. Martin's summer 圣马丁热期(11月11日前后的晴暖期)
St nebu 薄幕层云
St op 蔽光层云
stochastic 随机的

stochastic analysis 随机分析【数】
stochastic amplitude 随机振幅
stochastic coalescence 随机并合
stochastic coalescence equation 随机并合方程
stochastic differential equation 随机微分方程【数】
stochastic dynamic prediction 随机动力预报
stochastic dynamic model 随机动力模式
stochastic equation 随机方程
stochastic events 随机事件
stochastic forcing 随机强迫
stochastic hydrology 随机水文学
stochastic initial perturbation 随机初始扰动
stochastic model 随机模型,随机模式
stochastic noise 随机噪声
stochastic particle modeling 随机粒子模拟
stochastic perturbation 随机扰动
stochastic process 随机过程【数】
stochastic response 随机响应
stochastic sampling 随机取样
stochastic variable 随机变量
stoichiometric concentration 化学计量浓度【化】
stoichiometric ratio 化学计量比
stoke (St) 斯(动力黏度单位, cm^2/s)
Stokes fluorescence 斯托克斯荧光
Stokes formula 斯托克斯公式【数】
Stokesian wave 斯托克斯波
Stokes law 斯托克斯定律
Stokes line 斯托克斯线【物】
Stokes matrix 斯托克斯矩阵
Stokes number 斯托克斯数
Stokes parameter 斯托克斯参数
Stokes's drift 斯托克斯漂移
Stokes's drift velocity 斯托克斯漂移速度
Stokes's stream function 斯托克斯流函数
Stokes's theorem 斯托克斯定理

Stokes wave 斯托克斯波
stomatal resistance 气孔阻力
stone ice 石冰
stooping 光折
storage 存储,存储器【计】,库容
storage curve 库容曲线,储量曲线
storage effect 蓄水效应
storage equation 蓄洪方程,储量方程,水量平衡方程
storage function 蓄水函数
storage gauge 蓄水式雨量器,累积雨量器
storage ratio 蓄水率,蓄水系数
storage routing 蓄水演算,流量演算
storage term ①蓄水期②存储项
storativity 蓄水率,储水系数【土木工程】,给水度(=含水层厚度×单位储量)
storm 10级风【大气】,狂风,风暴
STORM (stormscale operational and research meteorology) 风暴尺度业务和研究气象学〔计划〕
storm belt 风暴区
stormburst 风暴猝发
storm cell 风暴单体
storm cell identification and tracking (SCIT) 风暴单体识别与追踪
storm center 风暴中心【大气】
storm chaser 风暴追踪
storm cloud 风暴云
storm cone 风暴信号
storm core 风暴核心
storm cyclone 风暴气旋
storm detection equipment 风暴探测装置
storm detection radar (SDR) 风暴探测雷达
storm duration 风暴持续时间,暴雨历时
storm eddy 风暴轴涡动
storm-feeder band 风暴旋入带
storm flow 暴雨径流
storm gage 风暴强度计
storm gale 暴风
storminess 风暴度

storm initiator 风暴引发
storm lane 风暴路径
storm loss 暴雨损失〔量〕
storm model 风暴模式
storm observation satellite (SOS) 风暴观测卫星
storm of Bay of Bengal 孟加拉湾风暴【大气】
storm path 风暴路径
storm precipitation 风暴降水量
storm-proof 抗风暴的,防暴风雨的
storm radar data processor (STRADAP) 风暴雷达数据处理机
storm rainfall 暴雨
storm-relative environmental helicity 风暴相对〔环境〕螺旋度
stormscale operational and research meteorology (STORM) 风暴尺度业务和研究气象学〔计划〕
stormscope 风暴探测器
storm signal 风暴信号
storm smear 暴雨强度分配(按地区的)
storm splitting 风暴分裂
storm spotter 风暴雷达
storm sudden commencement 磁暴急始
storm surge 风暴潮【大气】
storm swell 风暴涌〔浪〕
stormter 暴风雨中心
storm tide 风暴潮(因风暴所引起的海洋波浪)
storm time 暴时(磁暴开始起算的时间)
storm track 风暴路径,风暴轴
storm transposition 雨区移置【水利】(洪水设计用)
storm warning 风暴警报【大气】,暴风警报【航海】
storm-warning signal 风暴预警信号
storm-warning tower 风暴预警塔
stormwater 暴雨水
storm-wave 风暴浪
stormy weather 风暴天气

storm wind 暴风
STP (standard temperature and pressure) 标准温压
ST radar 平流层对流层雷达,ST 雷达
straight blow 直击强风
straight fertilizer 单元肥料【土壤】
straight-line gale 〔长程〕直击大风
straight-line wind 直线风,直行风
straight wind 直线风
strain 应变【物】,胁应
strain-gage anemometer 应力型风速表
straining motion 形变运动
strain tensor 应变张量
strange attractor 奇怪吸引子【大气】,怪引子
stratification 层结
stratification curve 层结曲线【大气】
stratified atmosphere 层结大气
stratified echoes 层状回波
stratified fluid 分层流体,层状流体
stratified fluid dynamics 层状流体动力学
stratified shear flow 层状切变流
stratiform 层状〔的〕
stratiform cloud 层云【大气】
stratiformis (str) 层状〔云〕
stratiform recipitaton 层状云降水
stratiform precipitation area 层状云降水区
stratigraphy 地层学
stratocumulus (Sc) 层积云【大气】
stratocumulus castellanus (Sc cast) 堡状层积云【大气】
stratocumulus cumulogenitus (Sc cug) 积云性层积云【大气】
stratocumulus duplicatus (Sc du) 复层积云
stratocumulus floccus (Sc flo) 絮状层积云
stratocumulus lacunosus (Sc la) 网状层积云

stratocumulus lenticularis (Sc lent) 荚状层积云【大气】
stratocumulus mammatus 旋球状层积云
stratocumulus opacus (Sc op) 蔽光层积云【大气】
stratocumulus perlucidus 漏隙层积云
stratocumulus radiatus (Sc ra) 辐辏状层积云
stratocumulus stratiformis (Sc str) 层状层积云
stratocumulus translucidus (Sc tra) 透光层积云【大气】
stratocumulus undulatus (Sc un) 波状层积云
stratocumulus vesperalis (Se ves) 向晚层积云
stratonull layer 平流层零层
stratopause 平流层顶【大气】
stratosphere 平流层【大气】
stratosphere coupling 平流层耦合作用
stratosphere jet flow 平流层急流
stratosphere radiation 平流层辐射
stratosphere-troposphere exchange 平流层-对流层间交换
stratospheric aerosol 平流层气溶胶【大气】
stratospheric aerosol and gas experiment (SAGE) 平流层气溶胶和气体试验
stratospheric aerosol layer 平流层气溶胶层
stratospheric aerosol measurement (SAM) 平流层气溶胶测量
stratospheric and mesospheric sounder (SAMS) 平流层和中间层探测器
stratospheric chemistry 平流层化学
stratospheric circulation index 平流层环流指数
stratospheric cloud 平流层云
stratospheric compensation 平流层补偿
stratospheric coupling 平流层耦合作用
【大气】
stratospheric fallout 平流层〔放射性物〕沉降
stratospheric inversion 平流层逆温
stratospheric monsoon 平流层季风
stratospheric oscillation 平流层振荡
stratospheric ozone 平流层臭氧
stratospheric photochemistry 平流层光化学
stratospheric polar vortex 平流层极〔地〕涡〔旋〕
stratospheric pollution 平流层污染【大气】
Stratospheric Processes and Their Role in Climate (SPARC) 平流层过程及其在气候中的作用
stratospheric sounding unit (SSU) 平流层探测装置
stratospheric steering 平流层引导
stratospheric sudden warming 平流层爆发〔性〕增温【大气】
stratospheric sulfate layer 平流层硫酸盐层
stratospheric super-pressure balloon 平流层超压气球
stratospheric warming 平流层增温
stratostat 平流层气球
stratus (St) 层云【大气】
stratus communis (= common stratus) 普通层云
stratus fractus (Fs) 碎层云
stratus lenticularis 荚状层云
stratus maculosus 斑状层云
stratus nebulosus (St nebu) 薄幕层云
stratus opacus (St op) 蔽光层云
stratus translucidus (St tra) 透光层云
stratus undulatus (St un) 波状层云
strays 天电
streak 条纹,条痕

streak lightning 条状闪电【大气】
streakline 纹线
stream 河流【水利】,气流,流
stream devil 蒸汽暴
streamer ①云幡,云线②流光,光繇
streamer echoes 雨幡回波
streamers 帐帷状极光
stream flow 河川流量
stream flow routing 〔河道〕流量演算
stream frequency 河流频数(单位流域面积的河流分支数)
stream function 流函数
stream gauge 流量计
stream gauging 流量测定,测流
stream hydrograph 流量曲线
streamline 流线【大气】
streamline analysis 流线分析
streamline chart 流线图【大气】
streamline field 流线场
streamline flow 流线流
streamline pattern 流线型
streamline vane 流线型风标
stream network 河网,水系
stream order 河流等级,支流等级
stream piracy 河流袭夺,夺流
stream profile 河流纵断面,河流纵剖面
stream segment 河流分支,河道分支
stream tube 流管【物】
streamwise vorticity 流线式涡度,流线涡〔度〕
Streeter-Phelps equation 斯特里特-费尔普斯方程(生化需氧量),S-P 方程
streetlamp halo 路灯晕
street tree 行道树【林】
stress 应力【物】,胁迫【植物】
stress effect 应力效应
stress tensor 应力张量【物】
stretched data 时展资料【大气】
stretched grid 展宽网格
stretched gridded data 展宽网格资料

stretched horizontal coordinates 伸展水平坐标
stretched image 展宽〔云〕图
stretched-out field 伸展磁场
stretched variable 伸展变量
stretched VISSR data 展宽 VISSR(可见光和红外自旋扫描辐射仪)资料
stretching deformation 伸展变形
stretching term 伸展项
striation 纹线,纹理
strictly stationary process 强平稳过程【数】
string model 强模型【物】
strip chart 拼条图
strip chart recorder 条带记录器
stroke 闪击
stroke density 闪击密度
strong breeze 6级风【大气】,强风
strong constraint 强限制
strong downdraft 强下沉气流
strong evolution 强演变
strong fluctuation 强起伏
strong fluctuation region 强起伏区
strong gale 9级风【大气】,烈风
strong-line approximation 强线近似【大气】
strong-line limit 强线极限
strongly coupled system 强耦合系统【数】
strongly stationary process 强平稳过程【数】
strong wind anemograph 强风仪【大气】
strontium cloud 锶云
Strouhal number 斯特劳哈尔数
structural stability 结构稳定性
structural symmetry 结构〔性〕对称【物】
structure coefficient 结构系数
structure constant 结构常数
structure function 结构函数【大气】
structure sonde 〔低空〕大气结构探空仪

Struma fall wind 斯特鲁玛下沉风(见于保加利亚)
strut thermometer 翼架温度表
St. Swithin's Day 圣斯威辛日(英国民俗,7月15日如下雨,其后40天均雨)
student distribution t 分布【数】
student's t-test t 检验
sturmpause 焚风歇
Stuve diagram 施蒂威图
STY(severe typhoon) 强台风
Suahili 苏希利风(波斯湾的强西南风)
sub-adiabatic state 亚绝热状态
subantarctic front 副南极锋
subantarctic mode water 副南极模式水
subantarctic upper water 副南极高层水
subantarctic zone 副南极带
subarctic climate 副极地气候【大气】
Subarctic Current 副北冰洋流
Subarctic Intermediate Water 亚北极中层水
Subarctic Upper Water 副北极高层水
Subarctic Water 亚北极水
subarctic zone 副极带
subarid 半干〔的〕,半干燥
subatlantic 大西洋〔气候〕亚期
subatomic particle 亚原子粒子
subauroral zone 副极光带
subboreal climatic phase 亚北方〔气候〕期
sub-bottom depth recorder (SDR) 海底浅层记录仪
subcarrier 副载波
subcloud ①云下②云下层
subcloud layer 云下气层
subcooling 过冷
sub-critical 亚临界
subcritical bifurcation 亚临界分岔
subcritical flow 亚临界流
subcritical stability 亚临界稳定度
subduction 潜没,消减〔地〕,沉降【海洋】
subequatorial belt 副赤道带
subequatorial ridge 副赤道脊
subfrigid zone 副寒带
subfrontal cloud 锋下云【大气】
subgelisol 不冻下层
subgeostrophic wind 次地转风【大气】
subgradient wind 次梯度风【大气】
subgrid 次网格
subgrid condensation 次网格凝结
subgrid scale 次网格尺度
subgrid-scale eddy 次网格尺度涡旋
subgrid-scale motions 次网格尺度运动
subgrid-scale orography 次网格尺度地形
subgrid-scale parameterization 次网格〔尺度〕参数化【大气】
subgrid-scale process 次网格尺度过程【大气】
subgrid scale turbulence 次网格尺度湍流
subhumid 半湿的
subhumid climate 半湿气候
subimage 子图像
sub-inversion layer 逆温下气层
subjective analysis 主观分析
subjective assessment 主观估计,主观评价
subjective forecast 经验预报【大气】,主观预报
sublayer 云下气层
sublimation ①升华【大气】②凝华
sublimation curve 升华曲线
sublimation nucleus 凝华核
sublimation of ice 冰升华
sublinear 次线性的【数】,亚线性的
sub-Lorentzian spectral lines 亚洛伦兹谱线
submanifold 子流形【数】
submarine climate 副海洋性气候
submarine geomorphology 海底地貌【测】
submatrix 子矩阵【数】
submesoscale coherent vortices (SCV)

次中尺度相干涡旋
submillimerter 亚毫米波【电子】
subminimum 次极小【数】
submodel 子模型【数】,子模式
subnormal 次常,次正规
subnormal refraction 半正常折射
suborbital track 星下轨迹
subpoint 星下点
subpoint track 星下点轨迹
subpolar 副极地
subpolar airmass 副极地气团
subpolar anticyclone 副极地反气旋
subpolar cap 副极地冷气冠
subpolar glacier 副极地冰川
subpolar gyre 副极地转子
subpolar high 副极地高压
subpolar low 副极地低压
subpolar low-pressure belt 副极地低压带
subpolar westerlies 副极地西风带
subpolar zone 副极带
subpolar zone ionosphere 副极地带电离层(即极光带电离层)
subprogram 子程序【计】
subrange 子区域
subrefraction 次折射
subregional broadcast 次区域广播
subroutine 子程序
sub-satellite point (SSP) 〔卫星〕星下点【大气】
sub-satellite point track 星下点路径
subscript 下标
subseasonal oscillation 次季节振荡
subseasonal time scale 次季节尺度
subset 子集【数】
subshrub 半灌木【植】
subsidence 下沉
subsidence flow 下沉气流
subsidence inversion 下沉逆温【大气】
subsidiary vortex 副涡旋
subsistence emission 社会生存排放
subsoil 底土

subsoil ice 底冰
subsolar point 日下点
subsonic 亚声速〔的〕,次声速〔的〕
subsonic motion 亚声〔速〕运动
substandard propagation 次标准传播
substandard refraction 次标准折射
substantial derivative(= individual derivative) 实质导数
substantial sheet 物质薄层
substantial surface 物质面
substorm 亚暴
substrate 下垫面
substratosphere 次平流层
subsurface flow 伏流
subsun 日下晕
subsurface current 地下水流
subsurface float 水下浮标,深水浮标
subsurface flow 〔次〕表层流,壤中流【水利】
subsurface storm flow 〔次〕表层暴雨流
subsurface water 地下水,次表层水【海洋】
subsynoptic advection model (SAM) 次天气平流模式
subsynoptic scale 次天气尺度(中小尺度的通称)
subsynoptic scale system 次天气尺度系统【大气】,中间尺度天气系统
subsynoptic Ω system 次天气尺度 Ω 系统
subsynoptic update model (SUM) 次天气〔尺度〕更新模式
subsystem ①部件②子系统【数】
subterranean ice 地下冰
subterranean stream 伏流,地下河
subterranean water 地下水,暗河【水利】
subthermocline 温跃层以下海层
subtrack 星下点轨迹【电子】
subtraction 减法【数】
subtropical 副热带〔的〕
subtropical air ①副热带空气②副热带

气团
subtropical anticyclone 副热带反气旋【大气】
subtropical belt（或 zone） 副热带,亚热带
subtropical calms 副热带无风带【大气】
subtropical cell 副热带高压单体
subtropical climate 副热带气候【大气】
subtropical convergence 副热带辐合带
subtropical countercurrent 副热带逆流
subtropical cyclone 副热带气旋【大气】
subtropical easterlies 副热带东风带【大气】
subtropical easterlies index 副热带东风指数
subtropical front 副热带锋
subtropical gyre 副热带转子
subtropical high 副热带高压【大气】
subtropical high pressure belt 副热带高压带
subtropical high pressure zone 副热带高压区
subtropical jet 副热带急流
subtropical jet stream 副热带急流【大气】
subtropical monsoon zone 副热带季风区
subtropical mode water 副热带模式水
subtropical ridge 副热带高压脊
subtropical westerlies 副热带西风带【大气】
subtropical wind belt 副热带风带
subtropical zone 副热带,亚热带【地理】
subtropics 副热带,亚热带
subzero temperature 零下温度
sub-zero weather 近零天气,低云和低能见度天气
succession 顺序,序列
successive approximation 逐步近似〔法〕
successive correction 逐次订正
successive correction analysis 逐次订正〔分析〕法【大气】

successive elimination 逐次消元法
successive filtering 逐项滤波
successive iteration 逐次迭代
successive order of scattering 逐级散射
successive over-relaxation 逐次超松弛【数】
successive regression 逐步回归
suchovei（＝sukhovei） 苏霍威风（俄罗斯南部草原的多尘干热风）
suction anemometer 吸管式风速表
suction spot 抽吸点
suction tube 吸管
suction vortex 抽吸性涡旋【大气】
suction zone 吸气区
Sudan type 苏丹式（汉恩气候分类之一类型）
sudden change report 突变报〔告〕
sudden commencement〔geomagnetic〕 急始磁暴【地】
sudden-commencement magnetic storm 急始磁暴【地】
sudden enhancement of atmospherics 天电突增【天】
sudden ionospheric disturbance(SID) 电离层突扰【天】,突发电离层骚扰【地】
sudden onset 急始(空间物理现象的)
sudden stratospheric warming 平流层爆发性增温
sudden warming 突然增温
sudois 萨多伊斯风(瑞士日内瓦湖上的西南风)
suer 修尔风(意大利加尔达湖强的北北风)
suestado（＝sudestades） 苏埃斯塔德风暴(南美洲阿根廷及乌拉圭沿海的东南风大风暴)
sufficient condition 充分条件【数】
sugar berg 多孔冰山
sugar snow 雪中白霜
sukhovei（见 suchovei） 苏霍威风

sulfaric acid aerosol 硫酸气溶胶
sulfate 硫酸盐
sulfate aerosol 硫酸盐气溶胶
sulfur hexafluoride 六氟化硫(SF_6)
sulfur cycle 硫循环
sulfurization 硫化【化】
sulfurous smog 含硫烟雾
sulphur(=sulfur,下同)cycle 硫循环
sulphur dioxide 二氧化硫(SO_2)
sulphur dust 硫尘
sulphuric acid 硫酸(H_2SO_4)
sulphur monoxide 一氧化硫(SO)
sulphur oxides 硫氧化物(SO_x)
sulphur rain 黄雨【大气】,硫化雨
sulphur mist 硫酸雾
sultriness 闷热
sum 和【数】,总数
sumatra 苏门答腊风暴(马六甲海峡西南季风期中的一种强风暴)
sum check 和校验
summation notation 求和约定
summation principle 总和原则
summer 夏〔季〕
summer convective precipitation 夏季对流降水
summer day 夏日,热日
summer dry region 夏干区
summer fog 夏雾
summer half year 夏半年
summer hemisphere 夏半球
summer lightning 热闪
summer monsoon 夏季风【大气】
summer of St. Martin 圣马丁热期
summer solstice 夏至点
Summer Solstice 夏至(节气)【大气】
summer time(或 daylight saving time) 夏令时【天】
summerwood 晚材,夏材
sun 太阳【天】,日
sunbeam 日光
sunburning spectrum 日灼谱

sun cross(=cross) 十字晕
sun crust 再冻雪壳(地面雪晶被太阳加热融化后再次冻结形成)
sundial 日晷【天】
sun dog 近幻日,近假日
sun drawing water 云隙辉
sun-earth relationship 日-地关系
sun glint 日照亮斑【大气】(卫星云图上),太阳耀斑
sunglow 霞〔光〕
sunlight 日照
sunlit aurora 日耀极光
sunny slope 阳坡【地理】
sunpath diagram 日径图
sunphotometer 太阳光度计
sunpillar 日柱【大气】
sun pointing 太阳标定(雷达)
sun proton monitor 太阳质子监视仪
sunrise 日出
sunrise colours 曙光,早霞
sunrise glow 朝霞
sunrise time 日出时刻
sun's altitude 太阳高度
sunscald 日灼
sun's corona 日冕
sun sensor 太阳传感器,太阳感应器
sunset 日没,日落
sunset colours 夕阳
sunset effect 日没效应,日落效应
sunset glow 晚霞
sunset time 日落时刻
sunshine 日照【大气】
sunshine autograph 日照仪
sunshine card 日照记录纸
sunshine duration 日照时数【大气】
sunshine duration recorder 日照时间记录器
sunshine-hour 日照时数【农】
sunshine integrator 日照时数计算器
sunshine record 日照记录
sunshine recorder 日照计【大气】

sunshine switch 日照开关
sunshine time 日照时间
sunspot 太阳黑子【天】,日斑
sunspot cycle 太阳黑子周期【大气】
sunspot group 太阳黑子群【天】
sunspot maximum 太阳黑子极大
sunspot minimum 太阳黑子极小
sunspot number 太阳黑子数
sunspot period 太阳黑子周期
sunspot periodicity 太阳黑子周期,日斑周期
sunspot relative number 太阳黑子相对数
sunstroke 日射病,中暑,日灼伤
sun-synchronous orbit 太阳同步轨道
sun-synchronous satellite 太阳同步卫星
sunward side 向日面
sun-weather correlation 太阳-天气相关
sun-weather effect 太阳天气效应
superadiabatic lapse rate 超绝热直减率【大气】
superadiabatic state 超绝热状态
superbolt 超级闪电
supercell 超级单体【大气】
supercell severe local storm 局地强风暴超〔级〕单体
supercell storm 超单体风暴
supercell structure 超单体结构
supercell thunderstorm 超单体雷暴
supercell tornado 超单体龙卷
supercomputer 巨型计算机【计】,超级计算机
supercooled cloud 过冷云【大气】
supercooled cloud droplet 过冷云滴【大气】
supercooled fog 过冷却雾【大气】
supercooled rain 过冷却雨【大气】
supercooled water 过冷〔却〕水
supercooled water droplet 过冷〔却〕水滴
supercooling 过〔度〕冷〔却〕

supercooling phenomenon 过冷现象【物】
supercritical bifurcation 超临界分岔
supercritical convection 超临界对流
supercritical flow 超临界流
super-ensemble 超集合
superfluous water vapour 过剩水汽
super-fusion 过熔
supergeostrophic wind 超地转风【大气】
supergradient 超梯度
supergradient wind 超梯度风【大气】
super-growth of singular vector 奇异向量超增长
superheated vapor 过热水汽
superheating 过热〔的〕
superheating phenomenon 过热现象【物】
super high frequency (SHF) 超高频
superimposed field 叠加场
superimposed ice 层叠冰
superior air ①高层空气②高空〔下降〕空气
superior mirage 上现蜃景【大气】
superior planet 外行星【天文】
superlinear convergence 超线性收敛【数】
super-minicomputer 超级小型计算机【计】
supernormal refraction 超常折射
supernova 超新星【物】
supernumerary bow 附属虹(主虹之内和副虹之外,色彩不明之虹带)
supernumerary rainbow 附属虹【大气】
superposed-epoch method 时段叠加法
superposition 叠加【计】,叠合
superposition principle 叠加原理【数】
super-pressure balloon 超压气球
superrefraction 超折射【大气】
superrefraction echo 超折射回波【大气】

supersaturated air 过饱和空气【大气】
supersaturation 过饱和〔度〕
supersaturation with respect to ice 冰面过饱和
supersaturation with respect to water 水面过饱和
superscript 上标
supersonic 超声速〔的〕
supersonic motion 超声〔速〕运动
superstandard propagation 超标传播
superstandard refraction 超标折射
superstorm 超级风暴
superstratosphere 超平流层(平流层以上的气层)
superturbulent flow 超湍流
super TY 超强台风【大气】
super typhoon 超强台风【大气】
supervisor 管理程序【计】
supplemental observations 补充观测
supplementary feature〔of clouds〕〔云的〕附加特征
supplementary〔weather〕forecast 补充〔天气〕预报【大气】,订正预报
supplementary illumination 辅助光照
supplementary land station 补充陆地站
supplementary meteorological office 补充气象台
supplementary observation 补充观测【大气】
supplementary point 辅助点
supplementary ship station 补充船舶站
supplementary station 补充站
supply current 补充流,补充电流
support program 支持程序【计】
support effect 载体效应【化】
support vector machines (SVM) 支持向量机
suppress 抑制,控制
supragelisol 永冻上层
supralateral tangent arcs 上珥
suprapermafrost layer 永冻土活动层
sur 苏尔风(巴西冷风)

suracon 苏拉松风(玻利维亚寒风)
surazo 苏拉祖风(巴西强南风或西南风)
surf ①碎波,激浪②拍岸浪
surf beat 波击
surface ①表面【物】②地面
surface air observation 地面〔气象〕观测
surface air temperature 地面气温
surface albedo 地面反照率
surface analysis 地面分析
surface avalanche 表层雪崩
surface-based lifted index (SBLI) 地面抬升指数
surface-based observation system 地面观测系统
surface boundary condition 地面边界条件
surface boundary layer 地面边界层【大气】
surface chart 地面〔天气〕图【大气】
surface composition mapping radiometer (SCMR) 表面成分图像辐射仪
surface condensation 表面凝结
surface current 表层流
surface data 地面资料【大气】
surface detention 表层滞水
surface drag 表面拖曳
surface electrical conductivity recorder 地面大气电导率记录器
surface emissivity 表面发射率
surface energy 表面能【物】
surface energy balance 地表能量平衡
surface energy budget 地表能量收支
surface erosion 面蚀【土壤】
surface float 水面浮标
surface flow 地面流
surface flux 地面通量
surface force 表面力
surface forecast chart 地面预报图【大气】
surface free energy 表面自由能【物】
surface freezing-index 地面冻结指数
surface friction 地面摩擦

surface front 地面锋【大气】
surface geothermometer 地面温度表【大气】,零厘米温度表
surface gravity wave 表面重力波
surface hail sensor 地面测雹器
surface hoar ①雪面白霜②表面白霜
surface humidity 地面湿度
surface integral 〔曲〕面积分【数】
surface intensity 表面强度
surface inversion 地面逆温【大气】
surface layer 近地层【大气】
surface layer temperature scale 近地面层温度尺度
surface layer velocity scale 近地面层速度尺度
surface map 地面图
surface observation 地面观测【大气】
Surface Ocean-Lower Atmosphere Study (SOLAS) 表层海洋-低层大气研究（IGBP 计划）
surface of constant amplitude 恒振幅面
surface of constant phase 恒相位面
surface of discontinuity 不连续面
surface of no motion 无运动面
surface of separation 分隔面
surface of subsidence 下沉面
surface ozone 地表臭氧
surface ozone concentration recorder 地表臭氧浓度记录器
surface potential gradient recorder 地表电位梯度记录器
surface pressure 地面气压【大气】
surface radiation balance 地表辐射平衡
surface radiation budget 地表辐射收支【大气】
surface renewal model 地表修复模式,地表更新模式
surface resistance 地面阻力
surface retention 地面蓄水
surface roughness 地面粗糙度【大气】
surface roughness length 地面粗糙度

〔长度〕
surface runoff 地表径流
surface skin temperature 表层温度
surface stress 地面切应力,表面应力【物】
surface storage 地面滞留
surface synoptic chart 地面天气图
surface synoptic station 地面天气站
surface temperature 地面温度【大气】
surface temperature mapping 地表温度分布测绘
surface temperature of water 水面温度
surface tension 表面张力【物】
surface thawing-index 地面融化指数
surface thermometer 地面温度表
surface traction 表面曳引力
surface trough 地面槽【大气】
surface turbulence 地面湍流
surface velocity 表面速度
surface visibility 地面能见度【大气】
surface water 地表水
surface water hyrology 地表水文学
surface wave 地面波,表面波【物】
surface weather chart 地面天气图
surface weather observation 地面天气观测
surface wetness 叶面湿润度
surface wetness duration 叶面湿润期
surface wet static energy analysis 地面湿静能量分析
surface wind 地面风【大气】
surface wind stress 地面风应力
surfactant film 表面活性剂薄膜
surficial creep 表层蠕动
surf zone 碎波带【海洋】,碎浪带
surge ①气压波②涌浪,强风潮③湍振【力】
surge current 涌流
surge line ①风速突变线②气压波线③湍振边界【航空】
surge protection device（SPD）电涌保护器

surge-tide interaction 浪-潮相互作用
suroet 苏罗特风(法国西海岸上带雨的西南风)
surplus area 积雪区
survey datum 测量基准面
surveying 测量学【测】
surveying and mapping 测绘学【测】
survival emission 生理生存排放
susceptibility 敏感度【计】
suspended ash 悬浮灰分
suspended dust 浮尘【大气】
suspended load 悬浮载荷
suspended matter 悬浮物,悬浮体【海洋】
suspended particle 悬浮粒子
suspended particulate 悬浮颗粒物
suspended phase 悬浮相
suspended phase 悬浮态
suspended solids monitor 浮粒监测仪
suspended sonde 悬挂式探空仪
suspending phase 悬浮态
suspension 悬浮
suspensoid 悬浮体【化】,悬浮液【地质】
sustainable development 可持续发展
sustained convection 持续对流
sustained rainfall 持续降雨
sustained wind 持续风(指美国风速测值是用目测的一分钟平均值)
Sutton's diffusion formula 萨顿扩散公式
S-values S值
SVAT model (soil-vegetation-atmosphere transfer) model 土壤-植被-大气传输模式
SVD (singular value decomposition) 奇异值分解【数】
Sverdrup relation 斯维尔德鲁普关系
Swallow float 斯瓦罗浮子(海洋)
swallow storm (=peasweep storm) 燕暴(英国的早春雷暴)
swallow-tail catastrophe 燕尾突变
swamp 湿地,沼泽

SWAN (Severe Weather Automatic Nowcast System) 灾害天气短时临近预报系统
swash 流溅
SWAT (Soil and Water Assessment Tool) 水土评价工具
swath ①扫描带②行迹
swath width 行迹宽度
sway 倾侧,横荡【海洋】
sweat 凝结水
SWEAT index (severe weather threat index) 强天气T指标,强天气威胁指标
Swedish Meteorological and Hydrological Institute (SMHI) 瑞典气象与水文研究所
sweep 扫描【电子】,拂掠
sweep length 拂掠长度
swell 涌浪
swell scale 涌〔浪等〕级
swept frequency interferometer 扫频干涉仪【电子】
SWI (special world intervals) 特殊世界时段
SWIFT (Severe Weather Integrated Forecasting Tools) 强天气综合预报工具(用于临近预报系统)
swinging plate anemometer 压板风速仪
swirl ratio 弯曲比
Swiss Meteorological Institute (SMI) 瑞士气象研究所
switching anemometer 开关风速表,选通风速表
switching circuit 开关电路【电子】
switching diode 开关二极管【电子】
sylphon 空盒
sylvanshine 植物夜光
symbol ①符号②预兆
symbolic coding 符号编码【数】
symbolic figure 指示码
symbolic figurc group 指示组

symbolic language 符号语言【数】
symmetric filtering 对称滤波
symmetric matrix 对称矩阵【数】
symmetric top 对称陀螺【化】
symmetry 对称〔性〕【物】
symmetry instability 对称不稳定
symmetry point 对称点
symmetry principle 对称原理【数】
sympathetic discharge 引发闪电
sympiesometer 甘油气压表
synchronization 同步【电子】
synchronized parallel algorithm 同步并行算法【计】
synchronizer/data buffer (S/DB) 同步数据缓冲器
synchronous change 同步变化
synchronous communication 同步通信
synchronous detection 同步探测
synchronous meteorological satellite (SMS) 同步气象卫星
synchronous orbit satellite (SOS) 同步轨道卫星
synchronous pulse-gated photon counting 同步脉冲选通光子计数
synchronous single-photoelectron counting 同步单光子计数
synchronous teleconnection (或 teleconnexion) 同步遥相关
synchronous transmission 同步传输【计】
synchror 同步
synergetics 协同学【物】
synergism 协同〔作用〕
syngeothermal line 地面同时等温线
synodic month 朔望月【天】
synodic motion 会合运动
SYNOP code 天气电码（缩写）
synopsis ①天气表格②梗概③说明书
synoptic ①天气图的,天气尺度的②综观的
synoptic analysis 天气分析【大气】
synoptic background 天气背景

synoptic chart 天气图【大气】
synoptic climatology 天气气候学【大气】
synoptic code 天气电码【大气】
synoptic data 天气资料【大气】
synoptic forecast 天气预报
synoptic forecasting 天气预报
synoptic hour 天气观测时间【大气】
synoptic map 天气图
synoptic meteorology 天气学【大气】
synoptic model 天气学模式
synoptic observation 天气观测【大气】
synoptic process 天气过程【大气】
synoptic regime 天气〔学〕体系,天气实体
synoptic report 天气报告【大气】
synoptic scale 天气尺度【大气】
synoptic scale system 天气尺度系统【大气】
synoptic situation 天气形势【大气】
synoptic station 天气站
synoptic system 天气系统【大气】
synoptic type 天气型【大气】
synoptic update model 天气更新模式
synoptic wave-chart 〔海洋〕波浪图
synoptic weather code 天气电码
synoptic weather observation 综合天气观测
synthetic aperture imaging radar 合成孔径成像雷达
synthetic aperture radar (SAR) 合成孔径雷达【大气】
synthetic climatology 综合气候学
synthetic hydrograph 综合水文图
synthetic seamless rubber balloon 无缝合成橡胶气球
syphon 虹吸【物】
syphon barometer 见 siphon barometer
systematic analysis 系统分析【数】
systematic error 系统误差【数】
systematic observations 系统观测
systematic sampling 系统抽样【数】

system compatibility 系统兼容性【计】
systeme international (SI) 国际单位制（法语名）
system engineering 系统工程
system for nuclear auxiliary power 核子辅助电源系统
system function 系统函数
system of coordinates 坐标系
system of equations 方程组
system of homogeneous linear equations 齐次线性方程组【数】
system of winds 风系
system reset 系统复原
system of partial differential equations 偏微分方程组【数】
syzygical tide 朔望潮
systems science 系统科学【数】
systems theory 系统理论【数】
syzygy ①朔望②西齐基风（新几内亚与澳大利亚间海上的西风，在夏季风前盛行）

T

tabetisol 不冻地
table 表〔格〕
tablecloth 桌布云
table iceberg 桌状冰山
tableland 台地【地理】
Table Mountain southeaster 桌山东南大风（南非）
Table of Standard Atmosphere 标准大气表
tabular crystal 片状晶体
tabular iceberg 平顶冰山
tabulating machine 制表机
tabulation 列表【数】,制表
tachometer ①转速度②流速计
tachometry 转速测量〔法〕
Tahiti Island 塔希提岛
taiga 泰加林【地理】(西伯利亚针叶林区)
taiga climate 泰加林气候,副极地气候
tail cloud 尾云
tail Doppler radar 机尾多普勒雷达（装在机尾,常用作螺旋形扫描）
tailwater 尾水,下游
tail wind 顺风【大气】
taino 泰诺风（大安的列斯群岛的飓风）
Taiwan Area Mesoscale Experiment (TAMEX) 台湾地区中尺度试验
Taiwan low 台湾低压
Taiwan Warm Current 台湾暖流
takeoff forecast 起飞预报
Taku wind 塔谷风（阿拉斯加的强东北风）
talik 融区（在多年冻土上的,俄语名）
tall grass 茂草
tamboen 塔姆贝恩风（北苏门答腊托巴湖的风）
TAMEX (Taiwan Area Mesoscale Experiment) 台湾地区中尺度试验
tandem radiosonde 串列探空仪
tangent ①切线②正切【数】
tangent arc 正切弧
tangent arc to 22° halo 22°晕切弧,珥
tangent arc to 46° halo 46°晕切弧,珥
tangential acceleration 切向加速度
tangential stresses 切向应力【物】
tangential velocity 切向速度
tangential wind 切向风【大气】
tangent linear approximation 切线性近似
tangent linear equation 切线性方程
tangent linear model 切线性模式

tape ①纸带②磁带
tape mark 〔磁带〕带标
tape recorder ①磁带录音机②磁带记录器
tapioca snow 雪丸
taproot 直根【植】
TAR(IPCC Third Assessment Report) IPCC第三次评估报告
tarantata 塔兰塔塔风(地中海区的强西北风)
target 靶,目标
target area 目标区【大气】,作业区
target code 目标〔代〕码【计】
targeted observations (或 adaptive observations) 目标观测
target-control cross-over 目标区-对照区交叉
target function 目标函数【电子】
target position indication 目标位置显示
target position indicator 目标位置显示器
target signal 目标信号
target volume (= pulse volum) 照射体积
taryn 多季性陆地积冰(俄语名)
TAS (ture airspeed) 真空速〔度〕
task description 任务说明书
Taylor column 泰勒柱
Taylor diagram 泰勒图
Taylor effect 泰勒效应
Taylor〔series〕expansion 泰勒〔级数〕展开
Taylor formula 泰勒公式【数】
Taylor microscale 泰勒微尺度
Taylor number 泰勒数
Taylor polynomial 泰勒多项式
Taylor-Proudman theorem 泰勒-普罗德曼定理
Taylor series 泰勒级数
Taylor's diffusion theorem 泰勒扩散定理
Taylor's hypothesis 泰勒假设
Taylor's theorem 泰勒定理
Taylor vortex 泰勒涡
TBB (black body temperature equivalent) 辐射亮温,〔相当〕黑体温度
TBSS (three-body scatter spike) 三体散射长钉〔状回波〕(雷达显示屏上表示有大冰雹出现的虚假回波)
TCLV(tropical cyclone-like vortices) 热带气旋类低涡
TDMA(time division multiple access) 时分多址【电子】
TDR (time domain reflectometry) 时域反射法(测土壤湿度)
TDWR (terminal Doppler weather radar) 机场多普勒天气雷达
teardrop balloon 泪珠状气球
technical commission 技术委员会
technical data 技术数据
technical note (TN) 技术文稿
technical progress report 技术进展报告
technical regulation 技术规则,技术规范
technical report 技术报告
technique ①方法②技术③技术装备
technology assessment 技术评定
tecton 构造底
tectonics 构造学【地质】
teeth of the gale 大风口(旧航海术语,指风的来向)
tegenwind 捷振风(荷兰的东北寒风)
tehuano 特华诺风(西班牙语,表示强的近海南风)
tehuantepecer 特万特佩克风(墨西哥特万特佩克湾冬季的强北风)
T-E index 温效指数
telecommunication 电信【电子】,远程通信
telecommunication hub 电〔传通〕信枢纽
teleconnection(= teleconnexion) 遥相关【大气】
teleconnection pattern 遥相关型
telecontrol 遥控
telecopier 电传复印机
telegraph exchange 〔普通〕电报

telegraphic equation 电报方程
telemeteorograph 遥测气象计
telemeteorography 遥测气象仪器学
telemeteorometry 遥测气象仪器制造学
telemeter 遥测表
telemetering automatic weather station 遥测自动气象站
telemetering pluviograph 遥测雨量计【大气】
telemetering precipitation gauge 遥测雨量计
telemetering thermometer 遥测温度计
telemetric receiver 遥测接收机
telemetry 遥测【电子】
teleoperator 遥控机械装置
telephotometer 遥测光度表
telephotometry 光度遥测法
teleprinter 电传机【电子】
telepsychrometer 遥测干湿表
tele-recording 遥测记录
telesat 通信卫星
telescope 望远镜
teletext 图文电视【电子】,广播型图文
telethermometer 遥测温度表【大气】
telethermoscope 遥测温度器
teletype（见 teleprinter） 电传机
teletype printout 电传机打印输出
television (TV) 电视
Television and Infrared Observation Satellite (TIROS) 电视和红外辐射观测卫星,泰罗斯卫星
television picture 电视图像
television system 电视系统
telex 用户电报【电子】,电传
telluric line 大气吸收线,大气谱线
telophase 末期
TEMPAL code 等温线〔传输〕编码
temperate belt 温带【地理】
temperate climate 温带气候【大气】
temperate climate with summer rain 温带夏雨气候

temperate climate with winter rain 温带冬雨气候
temperate glacier 温带冰川
temperate latitude 中纬度
temperate rainforest 温带雨林
temperate rainy climate 温带多雨气候
temperate westerlies 温带西风带【大气】
temperate westerlies index 温带西风带指数
temperate zone 温带【地理】
temperature 温度
temperature advection 温度平流【大气】
temperature altitude chart 温度高度图
temperature anomaly 温度异常,温度距平
temperature average 平均温度
temperature belt 温度带
temperature coefficient 温度系数
temperature compensation 温度补偿
temperature contrast 温度对比
temperature control 温度控制
temperature control vessel 恒温器,恒温槽
temperature correction 温度订正【大气】
temperature departure 气温距平
temperature-dewpoint spread 温度露点差
temperature-difference quotient 温差商数
temperature effect 温度效应
temperature（或 thermal）efficiency 温度效率,热效率
temperature-efficiency index (T-E index) 温效指数
temperature-efficiency ratio (T-E ratio) 温效比
temperature-entropy chart 温熵图
temperature extremes 温度极值
temperature field 温度场【大气】
temperature fluctuation 温度脉动
temperature gradient 温度梯度【大气】
temperature history 温度随时间的变化
temperature humidity index 温湿指数
temperature humidity infrared radiometer (THIR) 温湿红外辐射仪
temperature inversion 逆温【大气】

temperature inversion layer　逆温层
temperature lapse rate　气温直减率【大气】
temperature microscale　温度微尺度
temperature-moisture diagram　温湿图
temperature-moisture index　温湿指数
temperature of the soil surface　土壤表面温度【大气】
temperature pressure curve　温压曲线
temperature profile　温度廓线【大气】
temperature profile recorder (TPR)　温度廓线记录器
temperature province　温度区域,温度分区
temperature range　温度较差【大气】
temperature recorder　温度记录仪
temperature reduction　温度订正
temperature reference sonde　温度基准探空仪
temperature retrieval　温度反演【大气】
temperature-salinity curve　温盐曲线
temperature-salinity diagram　温盐图
temperature scale　温标
temperature-sensing element　感温元件
temperature-specific mortality ratio　温度特定死亡率比
temperature structure coefficient　温度结构系数
temperature structure function　温度结构函数
temperature variance　温度方差
temperature variation　温度变化
temperature wave　温度波
temperature zone　温度带
tempest　恶劣天气,风暴天气
tempestada　可拉风(见 colla,菲律宾一种带暴雨的强烈西南风)
temporal coherence　时间相干性【物】
temporal correlation function　时间相关函数
temporal distribution　时间分布
Temporale　坦波拉尔风(中美太平洋沿岸的强西南风,西班牙语名)
temporal factor　时间因子
temporal fluctuation　时间起伏
temporal frequency spectrum　时间频谱
temporal resolution　时间分辨率【大气】
temporal scale　时间尺度
temporal standard deviation　时间标准差
temporal variation　时间变化
temporary current　短暂流
ten-days agrometeorological bulletin　农业气象旬报
ten-days average　旬平均
tendency　倾向,趋势
tendency chart　趋势图
tendency correlation　倾向相关
tendency equation　倾向方程【大气】
tendency interval　趋势时距
tendency method　倾向法,趋势法
tendency profile　倾向廓线,趋势廓线
tenebrescence　曙暮光,光吸收,变色荧光
tenggara　(见 tongara)〔印尼〕汤加拉风
tensiometer　土壤湿度计,液体表面张力计,伸长计
tension　张力【物】
tension saturated zone　张力饱和区
tensor　张量【物】
tensor algebra　张量代数【数】
tensor product　张量积【数】
tented ice　棚冰
tenuity factor　稀薄因子
tephigram　温熵图【大气】
tephrochronology　火山灰年代学
T-E ratio (temperature-efficiency ratio)　温效比
tercentesimal thermometric scale　近似绝对温标
tercile　百分位点
terdiurnal tide　三日潮汐
tereno　(见 terrenho)特伦诺风
term hour　定时观测时〔间〕
terminal　终端

terminal area 终止区域
terminal device 终端设备,终端【计】
terminal Doppler weather radar 机场多普勒天气雷达
terminal equipment 终端设备,终端【计】
terminal fall velocity 下落末速
terminal forecast 降落机场预报
terminal point 终点
terminal velocity 沉降末速【煤炭】,终极速度【力】
terminator 晨昏线
terpenes 萜【化】,萜烯【林】
terpenoids 萜类化合物【林】,类萜【生物】
terrain effect 地形效应
terrain factor 地形因子【地信】
terrain features 地形特性【地信】
terrain-following coordinate 〔随〕地形坐标
terrain-following radar 地形跟踪雷达【电子】
terrain-induced system 地致系统
terral 坦拉尔风(安底斯山脉沿岸强烈东南海风)
terral levante 偏东陆风(西班牙语名)
Terra Satellite 泰拉卫星(EOS 上午轨道星)
terre altos 高空陆风(里约热内卢的下沉阵风)
terrenho 特伦诺风(印度干冷陆风)
terrestrial branch 地支【天】
terrestrial coordinate system 地球坐标系
territorial development 国土开发【地理】
terrestrial ecosystem 陆地生态系统
terrestrial globe 地球
terrestrial heat 地热
terrestrial infrared radiation 地球红外辐射
terrestrial magnetism 地磁

territorial management 国土整治【地理】
terrestrial object 陆标【航海】
terrestrial planet 类地行星【天】
territorial planning 国土规划【地理】
terrestrial plant 陆生植物【植】
terrestrial pole 地极【航海】
terrestrial radiation 地球辐射(包括大气辐射)【大气】
terrestrial radiation balance 地球辐射平衡
terrestrial radiation thermometer 地面辐射温度表
terrestrial refraction 地球折射
terrestrial scintillation 地球闪烁
territorial sea 领海【航海】
terrestrial sphere 地球圆球体
territorial sky 领空【地理】
terrestrial surface radiation 地球表面辐射(不包括大气辐射)
territorial water 领海【航海】
terrestrial whirl wind 陆旋风
terrestrial wind 陆成风
territorial broadcast 地区广播
territorial transmission 地区传送
territory 领土【地理】
Tertiary 第三纪
tertiary circulation 三级环流【大气】,局部环流
Tertiary climate 第三纪气候
test ①测试②检验【数】
test and evaluation laboratory (T&EL) 测试鉴定实验室
test center 试验中心
test data 试验数据
test of significance (significance test) 显著性检验
test paper 试纸【化】
test range 试验场
test solution 试液【化】
Tetens's formula 特滕斯公式(饱和水

汽压)
tethered balloon 系留气球
tethered balloon profiler 系留气球大气廓线仪
tethersonde 系留气球探空仪
tethersonde profiler 系留气球大气廓线仪
tetroon 等容气球【大气】,定高气球
teuchi (见 peesweep storm)皮斯威普风暴(英国早春雷雨)
Teweles-Wobus index 图韦尔-汶布斯指数,T-W 指数(预报验证系统)
texture 纹理
Th 钍射气【大气】
thalwind 泰尔风(法国阿尔萨斯的冷风)
thaw ①融化②解冻【大气】
thawing index 融化指数【地理】
thawing season 解冻季节
thaw-water 解冻水
THC(thermohaline circulation) 温盐环流【海洋】,热盐环流
thematic 专题【地信】
thematic cartography 专题地图学【地理】
thematic image 专题影像【地信】
thematic mapper 专题制图仪【地理】
thematic map 专题地图【测】,专题图【地信】
The Observing System Research and Predictability Experiment (THORPEX) 全球观测系统研究与可预报性试验
theodolite 经纬仪【测】
theorem 定理【数】
theoretical meteorology 理论气象学【大气】
theory 理论【物】,学说
theory of cyclone 气旋学说
theory of origin species 物种起源说【植】
theory of similarity 相似理论
theory of stratospheric steering 平流层引导学说
thermal ①热泡【大气】②热的,热力的

③温度的
thermal advection 温度平流
thermal analysis 热分析
thermal balance 热量平衡
thermal band 热辐射带
thermal belt 高温带(山坡)
thermal blooming 热晕(热畸变)
thermal boundary layer 热力边界层
thermal calibration 热定标
thermal capacity 热容量
thermal circulation 热环流
thermal climate 热型气候
thermal comfort 热适应
thermal conduction 热传导
thermal conductivity ①导热性②导热系数,热导率【物】
thermal conductivity of soil 土壤热导率
thermal conductivity profile 热导率廓线
thermal constant 需热常数(植物生长的)
thermal continentality 温度大陆度
thermal contrast 热对比,温度对比
thermal convection 热对流【大气】
thermal core 热泡核
thermal current 热流
thermal day 热日
thermal death point 致死温度【农】
thermal decomposition 热分解
thermal degradation 热降解
thermal depression 热低压
thermal diffusion 热扩散【物】
thermal diffusion factor 热扩散因子
thermal diffuse scattering 热漫散射【物】
thermal diffusivity 热扩散率
thermal distortion 热成畸变
thermal downslope wind 热下坡风
thermal effect 热效应
thermal efficiency ①热效率【航空】②温〔度〕效率
thermal-efficiency index 温效指数

thermal-efficiency ratio　温效比
thermal emission　热发射
thermal energy　热能
thermal enthalpy　热焓
thermal equation of energy　热能方程
thermal equator　热赤道
thermal equilibrium　热平衡【物】
thermal excitation　热激发【物】
thermal expansion　热膨胀【物】
thermal flux　热通量
thermal fog dispersal system　消暖雾系统
thermal forcing　热力强迫作用,热成型【化】
thermal gradient　热梯度,温度梯度【公路】
thermal-gradient diffusion chamber　热梯度扩散云室
thermal high　热高压【大气】,热成高压
thermal hygrometer　量热湿度表
thermal hysteresis　温〔度〕滞〔后〕,热滞
thermal imager　热象仪【电子】
thermal-inertia　热惯性
thermal infrared　热红外
thermal instability　热力不稳定
thermal internal boundary layer　热力内边界层
thermal inversion　热逆温
thermal island〔effect〕　热岛〔效应〕
thermal jet　热成〔风〕急流
thermal load　热负荷,热荷载
thermal low　热低压【大气】,热成低压
thermally direct circulation　热力直接环流,热力正环流
thermally indirect circulation　热力间接环流,热力反环流
thermally stratified medium　热力分层介质
thermal motion　热运动【物】
thermal mountain effect　热丘效应
thermal neutrality　热力中性
thermal particle　热粒子

thermal plume　热〔力烟〕羽
thermal pollution　热污染【大气】
thermal potential　热势
thermal precipitator　热力沉淀仪
thermal production rate　热力产生率
thermal quality of snow　融雪量
thermal radiation　热辐射【物】
thermal regime　热状况
thermal relaxation　热力张弛
thermal resistance　热敏电阻
thermal resources　热量资源
thermal ridge　温度脊
thermal Rossby number　热力罗斯贝数
thermal roughness　热粗糙度【大气】
thermal scanning　热〔辐射〕扫描
thermal slope　热坡度
thermal slope wind　热坡度风
thermal steering　热引导【大气】
thermal stimulus　热刺激
thermal stratification　热力层结【大气】
thermal thunder shower　热雷阵雨
thermal tide　热力潮(由于大气温度日变化引起的大气压力变化)
thermal time constant　热力滞后系数,加热时间常数
thermal trough　温度槽,热力槽
thermal turbulence　热湍流
thermal unit　热单位
thermal vortice　热涡旋
thermal vorticity　热涡度【大气】
thermal vorticity advection　热涡度平流
thermal wave　热力波
thermal wind　热成风【大气】
thermal-wind equation　热成风方程【大气】
thermal wind steering　热成风引导
thermal zone　高温带【大气】
thermasonde　热感探空仪
thermautostat　自动恒温箱
thermic anomaly　气温异常
thermic cumulus　热积云

thermion 热离子
thermionics 热离子学
thermisopleth 等变温线
thermistor 热敏电阻【电子】
thermistor anemometer 热敏电阻风速表【大气】
thermistor bolometer 热敏电阻热辐射仪
thermistor thermometer 热敏电阻温度表
thermo-alcoholometer 酒精温度表
thermo-anemometer 温差电偶风速表,热线风速表
thermobarograph 温压计
thermobarometer 温压表(根据水的沸点测定高度)
Thermocap 温敏电容(芬兰维萨拉公司气温传感器产品名称)
thermochemical cycle 热化学循环【化】
thermochemistry 热化学【化】
thermocline 斜温层,温跃层
thermoclinic 斜温层的,温跃层的
thermoclinicity 斜温性
thermocouple 热电偶【电力】,温差电偶【物】
thermocouple anemometer 热电偶风速表
thermocouple dew point hygrometer 热电偶露点湿度表
thermocouple psychrometer 热电偶式干湿表
thermocouple thermometer 热电偶温度表
thermocyclogenesis 热气旋生成
thermodetector 测温计,温差探测器,热(温差电)检波器
thermoduct 热波导(逆温形成的大气层波导)
thermodynamical function 热力〔学〕函数
thermodynamical potential 热力学势【物】
thermodynamic chart 热力学图
thermodynamic circulation 热力环流
thermodynamic dew-point temperature 热力学露点温度

thermodynamic diagram 热力学图【大气】
thermoelectric effect 热电效应【电子】,温差电效应【电子】
thermodynamic efficiency 热力学效率
thermodynamic energy 热力学能
thermodynamic energy equation 热力学能量方程
thermodynamic equation 热力学方程【大气】
thermodynamic equilibrium 热力学平衡【物】
thermodynamic frost-point temperature 热力学霜点温度【大气】
thermodynamic function 热力学函数【物】
thermodynamic function of state 热力学状态函数
thermodynamic ice-bulb temperature 热力学冰球温度【大气】
thermodynamic method for mixed layer growth 热力学混合层增长方法
thermodynamic parameter 热力学参数
thermodynamic probability 热力学概率【物】
thermodynamic process 热力学过程【物】
thermodynamic property 热力学性质
thermodynamic potential 热力势
thermodynamics 热力学【物】
thermodynamic scale of temperature 热力学温标【物】
thermodynamic speed limit 热力学速度极限
thermodynamic stratification 热力层结
thermodynamic wet-bulb temperature 热力学湿球温度【大气】
thermoelectric actinograph 热电直接辐射计
thermoelectric actinometer 热电直接辐射表
thermoelectric effect 温差电效应【物】

thermoelectric hot-wire anemometer 热电式热线风速表
thermoelectric thermometer 热电偶温度表
thermo-element 热电偶,温差电偶
thermogram 温度自记曲线【大气】
thermograph 温度计【大气】
thermograph correction card 温度计订正表
thermography 测温法
thermohaline circulation (THC) 温盐环流【海洋】,热盐环流
thermohyet diagram 温湿综合图
thermohygrogram 温湿自记曲线【大气】
thermohygrograph 温湿计【大气】
thermohygrometer 温湿表【大气】
thermo-hygrostat 恒温恒湿箱
thermo-integrator 土壤积热仪
thermo-isodrome 等温差商数线
thermo-isodromic quotient 等温差商数
thermo-isohyp 实际温度等值线
thermoisopleth 等温线
thermo-junction thermometer 热偶温度计
thermokarst topography 〔热〕喀斯特地貌,热石灰岩地貌
thermoluminescence 热致发光【物】
thermoluminescence dating 热发光定年法
thermometer 温度表【大气】
thermometer bulb 温度表球部
thermometer column 温度表柱
thermometer correction 温度表订正〔值〕
thermometer float 测温浮标
thermometer-screen 百叶箱(温度表的)
thermometer shelter 百叶箱(温度表的)
thermometer support 温度表架
thermometric conductivity 热导率
thermometric constant 需热量(植物生长的)
thermometric element 感温元件
thermometric scale 温标
thermometrograph 温度记录器
thermometry 测温学
thermoneutrality 热力中性
thermoneutral zone 〔热力〕中性带
thermo-osmosis 热渗透
thermopause 热层顶【大气】
thermoperiod 温周期
thermoperiodicity 温周期性
thermoperiodism 温周期〔性〕【大气】
thermophoretic velocity 热迁移速度
thermopile 热电堆,温差电堆【物】
thermoprobe 测温探针
thermopsych 加热致冷器
thermosalinograph 温度盐度计
thermoscope 验温器
thermoscreen 百叶箱(温度表的)
thermosonde 热感探测仪
thermosphere 热层【大气】
thermospheric circulation 热层环流
thermostat 恒温器【物】,恒温箱
thermosteric anomaly 比容异常
thermotopographic effect 地形热效应
thermotropic 正温
thermotropic model 正温〔大气〕模式【大气】
thermowell 热电偶(温度计)套管
theta coordinate 位温坐标系,θ坐标系
theta (或 θ) coordinate system θ坐标系
thetagram 位温高度图,温压图
THI (temperature-humidity index) 温湿指数
THI (time-height indicator) 时高显示器
thick fog 浓雾
thickness 厚度
thickness advection 厚度平流
thickness chart 厚度图
thickness line 厚度线【大气】
thickness pattern 厚度场型
thick squall 强飑
thick-thin chart (= isentropic thickness chart) 等熵厚度图

thick weather 雾天,阴天
Thiessen polygon method 蒂森多边形法（降雨分析）
thin drizzle 细毛毛雨
thin film humidity sensor 薄膜湿度传感器
thin fog 轻雾
thin line echo 窄线回波
thin screen thoery 薄屏理论
thioformaldehyde 硫代甲醛
THIR (temperature-humidity infrared radiometer) 温湿红外辐射仪
third cosmic velocity 第三宇宙速度【天】
third harmonic 三次谐波【物】
third law of thermodynamics 热力学第三定律【物】
third-order climatological station 三级气候站
third-order closure 三阶闭合
thirty-day forecast 三十天预报
Thomson scattering 汤姆孙散射【物】
Thornthwaite evapotranspirometer 桑斯威特蒸散计
Thornthwaite moisture index 桑斯威特湿度指数
Thornthwaite's climatic classification 桑思韦特气候分类【大气】,桑斯威特气候分类
thoron 钍射气【大气】
THORPEX (THe Observing system Research and Predictability EXperiment) 全球观测系统研究与可预报性试验
THORPEX Interactive Grand Global Ensemble (TIGGE) 全球交互式大集合预报系统
thousand-grain weight 千粒重
threatening 险恶的(天气现象)
Threat score T评分
three-axis stabilized 三轴稳定
three-body interaction 三体相互作用
three-body reaction 三体反应

three-body scatter spike (TBSS) 三体散射长钉〔状回波〕
three-cell〔meridional〕circulation 三圈〔经向〕环流【大气】
three-component propeller anemometer 三分量螺旋桨式风速表
three-cup anemometer 三杯风速表
three-dimensional convection mode 三维对流模式
three-dimensional Fourier transform 三维傅里叶转换
three-dimensional model 三维模式
three-dimentional structure 三维结构
three-dimensional teleconnection (teleconnexion) 三维遥相关
three dimensional turbulence 三维湍流
three-dimensional variational analysis 三维变分分析
three-dimensional variational assimilation 三维变分同化
three dimensional wind observation 三维风观测
three primary colors 三原色【物】
three-time level scheme 三时次法
threshold 阈值,临界值
threshold autoregression 阈值自回归
threshold condition 阈值条件【物】
threshold contrast 视觉感阈【大气】,低限〔光〕对比,反衬阈值
threshold depth 低限深度
threshold illuminance 低限照度
threshold limit value 阈限度,最低限值
threshold of audibility 闻阈,听阈
threshold of discomfort 刺耳低限
threshold of hearing 闻阈,听阈
threshold of nucleation 成核临界温度,核化阈〔值〕
threshold of odor 气味阈值
threshold of pain 刺耳低限
threshold sensor 阈值感应元件
threshold signal 阈值信号

threshold temperature 阈温,临界温度
threshold temperature of ice nucleation 成冰阈温【大气】
threshold value 门限值,阈值【数】
threshold voltage 阈值电压【电子】
threshold wavelength 【电子】阈值波长
throughfall 贯穿降雨〔量〕(穿过植被而到达地面的降雨部分)
thrust-anemometer 推力风速表
Thumba Equatorial Rocket Launching Station (TERLS) 印度顿巴赤道火箭发射站
thunder 雷【大气】
thunderbolt 霹雳
thunder clap 霹雳
thundercloud 雷雨云
thunderhead 雷暴云砧,雷雨云砧
thunder shower 雷阵雨【大气】
thundersquall 雷飑【大气】
thunderstorm 雷暴【大气】
thunderstorm cell 雷暴单体【大气】
thunderstorm charge 雷暴电荷
thunderstorm charge separation 雷暴电荷分离
thunderstorm cirrus 雷暴卷云
thunderstorm complex 雷暴群
thunderstorm day 雷暴日【大气】
thunderstorm dipole 雷暴偶极子
thunderstorm downdraft 雷暴下击流
thunderstorm dynamics 雷暴动力学
thunderstorm electrification 雷暴起电
thunderstorm electricity 雷暴电学
thunderstorm high 雷暴高压【大气】
thunderstorm initiation mechanism 雷暴启动机制,雷暴触发机制
thunderstorm locating device 雷暴定位装置
thunderstorm low 雷暴低压
thunderstorm modification 雷暴改造
thunderstorm monitor 雷暴监测装置
thunderstorm monitoring 雷暴监测【大气】
thunderstorm observing station 雷暴观测站
thunderstorm outflow 雷暴泄流【大气】
thunderstorm rain 雷暴雨
thunderstorm recorder 雷暴记录器
thunderstorm sign 雷暴先兆
thunderstorm tripole 雷暴三极子
thunderstorm turbulence 雷暴湍流【大气】,雷暴〔飞机〕颠簸
thunderstorm wind 雷暴风
thunderstroke 雷击
thundery cloud system 雷雨云系【大气】
thundery front 雷雨锋
thundery precipitation 雷雨云降水【大气】
thundery sky 雷雨天空
Tianshan quasi-stationary front 天山准静止锋【大气】
Tibetan anticyclone 青藏反气旋
Tibetan high 青藏高压
Tibetan Plateau 青藏高原
TID (travelling ionospheric disturbance) 电离层行扰【地】【大气】
tidal age 潮龄
tidal bore 涌潮
tidal breeze 潮〔汐〕风
tidal component 分潮
tidal constituent 分潮
tidal crack 潮汐冰裂【海洋】(潮汐作用引起的冰裂缝)
tidal creek 潮沟【地理】
tidal current 潮流【航海】
tidal datum 潮高基准面【航海】
tidal day 潮日
tidal equation 潮〔汐〕方程
tidal excursion 潮程
tidal factor 潮汐因子
tidal force 潮汐力
tidal glacier 入海冰川,有潮冰川
tidal marsh 潮沼,盐泽
tidal meter 验潮仪,测潮表
tidal motion 潮汐运动

tidal oscillation 潮汐振荡
tidal period 潮汐周期,潮汐调和常数【航海】
tidal〔gravitational〕potential 引潮〔重力〕位势【海洋】
tidal prism 进潮量【海洋】
tidal range 潮差【航海】
tidal river 有潮河,潮汐河流【铁道】
tidal stream 潮流【航海】
tidal wave 潮〔汐〕波【海洋】,海啸【地】
tidal wind 潮〔汐〕风
tide 潮汐【航海】
tide amplitude 潮振幅
tide blow 激潮
tide gauge 测潮仪【海洋】
tide gauge benchmark 测潮器基准
tide-generating force 引潮力【地】
tide-generating potential 引潮位【地】
tide level 潮面【航海】
tide-predicting machine 潮汐计算机
tide-producing force 起潮力
tide-producing potential 起潮势
tide rip 潮急浪
tide table 潮汐表〔格〕
tierra caliente 高山暖温带(西班牙语名)
tierra fria 高山凉温带(西班牙语名)
tierra templada 高山中温带(西班牙语译名)
TIGGE (THORPEX Interactive Grand Global Ensemble) 全球交互式大集合预报系统(是全球各国和地区的业务数值预报中心的联合行动)
tillage 耕地;耕种;耕作物
tillage dynamics 耕作动力学【土壤】
tilted trough 斜槽
tilting bucket raingauge 翻斗〔式〕雨量计【大气】
tilting effect 倾斜效应
tilting term 倾斜项
tilting updraft 倾斜上击流
timber line 森林线,树木线
time and space scale 时空尺度

time-area-depth curve 时-面-深曲线
time average 时间平均
time-averaged measurement 时间平均观测〔值〕
time averaging 时间平均
time base 时基【电子】,时基线
time consistency check 时间一致性检验
time constant 时间常量【物】,时间常数【自动化】,滞后系数
time correlated scattering cross section 时间相关散射截面
time correlation 时间相关,时间关联【物】
time-correlation scattering amplitude 时间相关散射振幅
time cross-section 时间剖面图【大气】
time curve(=time front) 走时曲线(航空)
time delay 时延【天】,时滞
time dependent 不定常【数】,时间相关【数】
time-dependent boundary condition 时变边界条件
time-dependent flow 时变流
time-dependent perturbation 含时微扰【物】
time derivative 时间导数
time-distance graph 时程图
time division multiple access(TDMA) 时分多址【电子】
time domain 时域【数】
time-domain averaging 时域平均
time domain reflectometry(TDR) 时域反射法(测土壤湿度)
time en route 途中飞行时间
time filter 时间滤波器
time filtering 时间滤波
time front 走时曲线(航空)
time function 时间函数
time-height indicator 时间-高度显示器
time-height section 时间-高度剖面
time-independent boundary condition

固定边界条件
time integration 时间积分
time interval 时间间隔,时段
time-interval radiosonde 时距探空仪
time lag 时间滞差,时滞【数】
time lapse 时间推移
time-lapse film 连续拍摄的胶卷
time level 时次
time-marking device 时标装置
time mean chart 时间平均图
time-mean flow 时间平均〔气〕流
time normalization 时间归一化法
time-of-arrival location 时差定位【电子】
time-of-arrival technique 〔到达〕时差定位法,TOA法
time of concentration 集中时间,汇流时间【土木工程】
time of observation 观测时〔间〕
time parameter 时间参数
time representativeness 时间代表性
time-resolution 时间分辨率
time response 时间响应
time scale ①时标②时间尺度【数】
time scheme 时间格式
time section 时间剖面
time series 时间序列【数】
time series analysis 时间序列分析【数】
time share 分时
time sharing 分时【计】
time-sharing monitor system 分时监控系统【计】
time-sharing running mode 分时运行方式【计】
time smoothing 时间平滑【大气】
Time Standard 时间标准
time step 时〔间〕步〔长〕
time-temperature curve 温时曲线
time-to-space conversion 时空转换
time-variable filter 时变滤波〔器〕
time-variation 随时〔间〕的变化

time-varying power spectrum 时变功率谱
time-varying random medium 时变随机介质
time-varying rough surface 时变粗糙面
time-varying scattering cross section 时变散射截面
time-varying spatial spectral density 时变空间谱密度
time-varying specific intensity 时变辐射单位强度
time-varying spectral density 时变谱密度
time-varying spectrum 时变谱
time-varying system 时变系统【数】
time-varying transfer function 时变传递函数
time window 时间窗
time zone 时区
TIPEX（the second Tibetan Plateau Meteorological Experiment） 第二次青藏高原大气科学试验(1998年)
tipping-bucket rain gauge 翻斗式雨量计
tipping term 倾斜项
TIROS（television and infrared observation satellite） 泰罗斯卫星（电视及红外观测卫星）【大气】
TIROS Atmospheric Radiance Module（TARM） 泰罗斯大气（晴空）辐射模块(美国第三代极轨气象业务卫星垂直探测资料处理系统中的)
TIROS Information Processor（TIP） 泰罗斯信息处理机（美国第三代极轨气象业务卫星上的资料处理机）
TIROS-N 泰罗斯-N卫星【大气】
TIROS-N Operational Vertical Sounder（TOVS） 泰罗斯-N卫星业务垂直探测器
TIROS operational system 泰罗斯卫星业务系统
TIROS Operational Satellite（TOS） 泰罗斯业务卫星(美国第一代极轨气象业务卫星名,即艾萨卫星)

TIROS Operational Vertical Sounder
(TOVS) 泰罗斯业务垂直探测器(美国第三代极轨气象业务卫星上探测大气垂直结构的)
TIROS Stratospheric Mapper(TSM) 泰罗斯平流层探测资料映射模块(美国第三代极轨气象业务卫星垂直探测资料处理系统中的)
Titan ①泰坦(希腊神话中的巨人族之任何一员)②太阳神
titrand 被滴定物【化】
titrant 滴定剂【化】
titration 滴定【化】
titrimetric analysis 滴定〔分析〕法【化】
titrimetry 滴定〔分析〕法【化】
tivano 提旺诺风(意大利哥莫湖的夜风)
tjäle 地冻,冻土(瑞典语名)
TKE(turbulence kinetic energy) 湍流动能
T-ln$p$ diagram 温度-对数压力图【大气】
Toeplitz operator 特普利茨算子【数】
tofan(=tufon,tufan) 托范风暴(印尼山区的春天风暴)
TOGA(Tropical Ocean and Global Atmosphere) 热带海洋全球大气试验计划
token variable 标记变量【计】
tolerance 公差,容许误差,容限【物】
tolerance error 容许误差【数】
TOM(total organic matter) 总有机物
tomography 层析成像技术
TOMS(total ozone mapping spectrometer) 臭氧总量测绘分光计,臭氧总量测绘光谱仪
ton(=tonne) ①吨②美吨(=2000磅)③英吨(=2240磅)
tone bar 色调带
tone pattern 色调显示
tongara 汤加拉风(印尼望加锡海峡的东南风)
tonotron 雷达显示管
tons per day(TPD) 吨/日

tons per year(TPY) 吨/年
top ①顶部②陀螺【物】
top-down/bottom-up diffusion 边界层顶向下/地表向上漫射
top dressing 追肥【土壤】
topoclimate 地形气候【大气】
topoclimatology 地形气候学【大气】
TOPEX ①(Typhoon Operational Experiment)台风业务试验②(ocean topography experiment)海面形状试验计划(美法合作)
top of boundary layer 边界层顶
topographical amplification factor 地形放大因子
topographical effect 地形效应
topographically forced flow 地形强迫流
topographically generated wave 地形波动
topographic backscatter 地形后向散射体
topographic forcing 地形强迫,地形作用力
topographic frontogenesis 地形锋生〔作用〕
topographic isobar 地形等压线【大气】
topographic map 地形图【地理】
topographic Rossby wave 地形罗斯贝波
topographic target 地形目标
topographic trough 地形槽【大气】
topographic vortex 地形涡旋
topographic waves 地形波
topography 地形学
topological structure 拓扑结构
topology 拓扑〔学〕【数】
topomicroclimate 地形小气候
topothermogram 局部温度自记曲线
topside sounder 顶视探测仪【地】
tormento 托门图风(阿根廷的风,类似pampero风)
tornadic phase 多龙卷时段
tornadic storm 龙卷风暴
tornadic thunderstorm 龙卷雷暴

tornadic vortex signature（TVS） 龙卷涡旋信号
tornado 龙卷【大气】,陆龙卷
tornado alley 龙卷通道
tornado belt 龙卷区
tornado cave 龙卷躲避所
tornado cellar 防龙卷掩体,防风窖
tornado cyclone 龙卷气旋
tornado damage 龙卷灾害
tornado echo 龙卷回波【大气】
tornado forecast 龙卷预报
tornado hook 龙卷钩
tornado outbreak 龙卷爆发
tornado prominence 龙卷日珥
tornado swatch 龙卷行迹
tornado track 龙卷路径
tornado vortex 龙卷涡旋
tornado vortex signature（TVS） 龙卷涡旋特征
tornado warning 龙卷预警
tornado watch 龙卷监测
torque 转矩【物】
torr 托（1毫米汞柱高）
torrent 山洪【力】,激流
torrential rain 暴雨【大气】
Torricelli's tube 托里拆利管
Torricelli's vacuum 托里拆利真空
torrid zone 热带
torsion 扭转【物】
torsional moment 扭矩【物】
torsion hygrometer 扭转湿度表
TOS（TIROS Operational System） 泰罗斯卫星业务系统
tosca 托斯卡风（意大利加尔达湖的西南风）
total acidity 总酸度
total active temperature 活动积温
total attenuation coefficient 总衰减系数
total carbon 总碳〔量〕
total cloud cover 总云量【大气】
total column ozone measurements 臭氧柱总量测量
total conductivity 总电导率
total correlation factor 全相关因子
total cross section 总截面【物】
total derivative 全导数
total dew point 总露点
total differential 全微分【数】
total direct radiation 总直接辐射
total eclipse 全食【天】
total effective temperature 有效积温
total emissive power 总发射〔功〕率
total energy 总能量
total energy detector（TED） 总能检测器
total energy equation 总能量方程
total evaporation 总蒸发
total gradient current 总梯度流
total head line 总能头线
total head 总水头,总能头
total hemispherical radiometer 半球体总辐射计
total intensity 总强度
totality 全食【天】
totalizer（=totalizator） 累积计算器
totalizer for rain 累积雨量器
totalizing rain gauge 累计雨量器
total lift 全升力
total lift of a balloon 气球全升力
total organic matter（TOM） 总有机物
total ozone 臭氧总量【大气】
total ozone mapping spectrometer（TOMS） 臭氧总量测绘分光计
total ozone mapping system 臭氧总量测绘系统
total ozone monitoring satellite 臭氧总量监测卫星
total potential energy 总势能
total pressure 全压力
total probability 全概率
total radiation 全辐射【大气】(太阳辐射和地球辐射之和,所有波长辐射之和)
total reactive nitrogen 总活性氮

total reflection 全反射【物】
total residue 总残留量
total scattering coefficient 总散射系数
total scattering cross section 总散射截面
total sky cover 天空遮蔽总量
total solar eclipse 全日蚀
total solar energy monitor (TSEM) 总太阳能监测仪
total solar irradiance (TSI) 全日射（在地球大气外界,日地平均距离上与入射辐射垂直的单位表面上所接受的太阳辐射量,常取值为 1368 W/m^2）
total storm precipitation 天气过程总降水量
total suspended particulates (TSP) 总悬浮粒子〔数〕
total temperature 总温度
total-totals index TT 指数,全总量指数（垂直总量指数与交叉总量指数之和,一种稳定度指数）
total variance 总方差
total vorticity 全涡度
total water content 总含水量
total water content instrument (TWCI) 总含水量测量仪
total-water mixing ratio 总水分混合比
touriello 托利埃罗风（法国比利牛斯的焚风型南风）
tourism 旅游业【地理】
tourist point 旅游点【地理】
tourist resources 旅游资源【地理】
tourist site 游览地
Toussaint's formula 图森特公式
TOVS (TIROS Operational Vertical Sounder) 泰罗斯业务垂直探测器
towering 伸长蜃景
towering cumulus 塔状积云
tower layer 塔层
tower micrometeorology 塔层微气象学
tower of the winds 风塔（希腊雅典时代建立）
tower-produced fog 塔成雾
tower sensor 塔用传感器
tower visibility 塔台能见度
town fog 城市雾
Townsend discharge 汤森放电【电子】
Townsend support 汤森支架（最高温度表用）
toxicity 毒性
toxicology 毒理学
toxic pollutants 毒性污染物
toxin 毒素【生化】
trace ①痕量②微量③迹【物】
trace atmospheric constituents 大气痕量组分
torsional moment 扭矩【物】
trace chemistry 痕量化学【化】
trace constituent 痕量成分【化】
trace element 痕量元素【大气】
trace gas 痕量气体,示踪气体
trace of rainfall 雨迹
tracer 示踪物,示踪剂【化】
tracer analysis 示踪分析
tracer diffusion experiment 示踪扩散实验【大气】
trace recorder 微雨量计
tracer technique 示踪技术
trace substance 示踪物
track ①痕迹,路径②径迹【物】
track distance 轨距角
tracking 跟踪,追踪
tracking and data system (T&DS) 跟踪和资料系统
tracking radar 跟踪雷达
tracking radar echoes by correlation (TREC) 雷达回波相关跟踪法【大气】
tracking system 跟踪系统
track of a depression 低压路径【大气】
track radar 跟踪雷达
track wind 航线风

traction 曳引力
tractionability 曳引度
tractive force 曳引力,牵引力
trade 信风
trade air 信风空气
trade cumulus 信风积云
trade cumulus equilibrium 信风积云平衡
trade inversion 信风逆温
trade-off 协调,适应,一致,折中,比较评定
trade surge 信风潮
trade-wind belt 信风带【大气】
trade-wind cell 信风圈
trade-wind circulation 信风环流【大气】
trade-wind cumulus 信风积云
trade-wind desert 信风沙漠
trade-wind equatorial trough 信风赤道槽
trade-wind front 信风锋【大气】
trade-wind inversion 信风逆温【大气】
trade-winds 信风【大气】,贸易风
trade-wind zone 信风带
traditional approximation 惯用近似
traersu 特雷欧苏风(意大利加尔达湖的东风)
traffic ①报量②运输量
trail ①尾迹②[轨]迹【数】③退曳【航空】
trailing flare 曳光弹
trailing front 曳式锋【大气】
trailing wave 尾迹波【物】
trails of precipitation 雨幡,降水迹
trails of rain 雨幡
train effect 列车效应(飑线的移动速度矢量基本平行于其走向,使得飑线中的强降水单体依次经过同一地点,产生了最大的累积降水量)
trajectory ①轨迹②轨道【物】
trajectory technique 轨迹法
tramontana 脱拉蒙塔那风(地中海沿岸的一种干冷北风)
transceiver 收发器【计】
transcription ①转录②抄写③副本④改编
transcritical 跨临界
transducer ①换能器【船舶】②传感器【计】
transfer 输送,传递
transfer function 传递函数【数】,转换函数【机械】
transfer function model 传递函数模型【自动】
transfer of water vapor 水汽输送【大气】
transfer property 传递性质
transfer velocity 传输速度
transformation ①变换②变性③转化
transformation of air mass 气团变性
transformation point 相变点
transformation relation 变换关系
transformed air mass 变性气团【大气】
transformed Eulerian mean system 变换欧拉平均系统
trans-frontal 穿锋的
transient 瞬变〔的〕
transient analysis 瞬态分析【计】
transient climate response 瞬变气候响应
transient data 瞬变数据
transient eddy 瞬变涡动【大气】
transient fluctuation 瞬变振荡
transient motion ①暂态运动【物】,瞬态运动②过渡运动【数】
transient planetary wave 瞬变行星波
transient problem 瞬态问题
transient thermocline 瞬变温跃层
transient turbulence closure 过渡湍流闭合
transient variation 瞬变
transient wave 瞬变波【大气】
transilient matrix 跳跃矩阵
transilient turbulenle theory 湍流跳跃理论

transistor 晶体管【物】
transistor thermal resistance 晶体管热敏电阻
transit ①经纬仪【测】②中天【天】
transition ①转换,过渡②跃迁【物】
transitional flow 过渡气流【大气】
transition area 过渡带,过渡区
transition curve 过渡曲线,临界曲线
transition layer 过渡层【大气】
transition of circulation 环流型过渡
transition period 过渡时期
transition season 过渡季节【大气】
transition state theory 跃迁态理论
transition to turbulence 过渡湍流
transition zone 过渡带,过渡区
transitive system 确定转移系统,可递系统
transitivity 传递性【数】
transitory variation 暂时变化
translation 平移【物】,移动,位移
translational energy 平移能
translational motion 平移运动
translation flow 平移流
translatory field 平移场
translatory velocity 平移速度
translatory wave 平移波
translucency (=translucence) 半透明度
translucent 半透明〔的〕
translucidus (tr) 透光(云)
transmissibility 可转移性
transmissibility function 转移函数
transmission ①传输【电子】,发送②透射【物】
transmission coefficient 透射系数【物】
transmission curve 透射曲线
transmission density 透射密度
transmission function 透射函数
transmission matrix 透射矩阵
transmission line 传输线【计】
transmission loss 透射损耗
transmission range 透射距离

transmissivity (τ) 透射率【物】
transmissormeter 透射表
transmissormetry 透射测量技术,透射测定法
transmit-receive tube 收发转换开关
transmittance 透射比【物】
transmittance meter 透射表
transmittance monitor 透射比监测仪,透光度监测仪
transmitted power 发射功率
transmitter 发报机,发射机
transmitter-responder 应答器(见transponder)
trans-Neptunian planet 海外行星【天】
transonic 跨声速
transosonde 平移探空仪
transparency 透明度
transparency of atmosphere 大气透明度
transparent 透明〔的〕
transparent cloud 透明云
transparent coefficient 透明系数
transparent ice 透明冰
transparent sky cover 透光云量
transpiration 蒸腾【大气】
transpiration coefficient 蒸腾系数
transpiration efficiency 蒸腾效率
transpiration rate 蒸腾速率
transpiration ratio 蒸腾率
transpiring canopy 森林冠层蒸腾
trans-Plutonian planet 冥外行星【天】
transponder (=transmitter-responder) 应答器【电子】
transport 输送,输运
transport cross section 输运截面
transport mechanism 输送机制
transport processes 输送过程,输运过程
transport theory 输运理论
transpose of matrix 〔矩〕阵转置
transversal vibration 横振动
transverse circulation 横向环流
transverse cirrus banding 横向卷云带

transverse cloud band（或 line） 横云带
transverse velocity 横向速度【物】
transverse wave 横波【物】
trap 陷阱【物】
trapezoid 梯形【数】
trapezoidal scheme 梯形法
trapezoidal window 梯形窗
trapped wave 俘获波,陷波【海洋】
travelling anticyclone 移动性反气旋
travelling ionospheric disturbance（TID）
 电离层行扰【大气】
travelling standard barometer 巡检用标准气压表
travelling wave 行波【电子】
travelling-wave tube(TWT) 行波管【电子】
travel time 走时
traverse 横穿风（法国中部的西风,法语名）
traversia 海上西风（南美海员俚语）
traversier 凶险风（地中海的一种危险的入港风）
TREC（tracking radar echoes by correlation） 雷达回波相关跟踪法【大气】
tree algorithm 树算法【数】
tree climate 树木气候
tree graph 树图【数】
tree line 森林线,树木线
tree-line movement 树线移动
tree representation 树表示【数】
tree ring 年轮
tree-ring analysis 年轮分析
tree-ring climatology 年轮气候学
tree-ring record 年轮记录
tremor 微震
trend analysis 趋势分析
trend curve 趋势曲线
trend chart 趋势图
trend line 趋势线
trend method 趋势法
trial 试验,试探

trial-and-error method 尝试法【物】
trial solution 试探解【数】
triangular function 三角函数
triangular truncation 三角截断
triangular window 三角形窗
triangulation 三角法,三角剖分【数】
triangulation balloon 三角测量气球
Triassic Period 三叠纪【地质】
Triassic System 三叠系【地质】
triaxial accelerometer 三轴加速度仪
triboelectrification 摩擦起电
tributary 支流【地理】
tricellular model 三胞模式
tridiagonal matrix 三角对线矩阵
trihedral reflector 三面角反射器
trigger 触发器【物】
trigger action 激发作用,触发作用
triggered flash 触发闪电
trigger effect 触发效应
trigger mechanism 触发机制
trigonometric function 三角函数【数】
trigonometric interpolation 三角插值【数】
trigonometric series 三角级数【数】
trilateration 三边测量,长距三角测量
trilateration range and range rate（TRRR） 三点测距
trimer 三聚体【化】
trimodal acoustic radar receiver 三模态声〔雷〕达接收机
trimodal distribution 三峰分布
triode 三极管【物】
triple collision 三体碰撞【天】
triple correlation 三重相关
triple-Doppler analysis 三维多普勒分析
triple integral 三重积分【数】
triple point 三态点,三相点【物】
triple register 三相自记器
triple scalar product 三重数积
triple state 三相
tritiation 氚化【化】

tritide 氚化物【化】
tritium 氚【大气】
tritium ratio 氚比【化】
triton 氚核【物】
TRMM（tropical rainfall measuring mission） 热带降雨测量〔卫星〕,热带测雨任务〔卫星〕
trivane 三叶风向标
tromba 特隆巴风（马耳他旋风）
tropic ①回归线②热带
tropical air fog 热带气团雾【大气】
tropical air mass 热带气团【大气】
tropical anticyclone 热带反气旋,热带高压
tropical belt 热带
tropical calm zone 热带无风带
tropical cell 热带〔经向〕环流
tropical circulation 热带环流【大气】
tropical climate 热带气候【大气】
tropical climatology 热带气候学【大气】
tropical cloud cluster 热带云团【大气】
tropical constant-level balloon system （TCLBS） 热带定高气球系统
tropical continental air（mass） 热带大陆气团,热带大陆空气
tropical convergence 热带辐合带
tropical crop 热带作物
tropical cyclone 热带气旋【大气】
tropical cyclone clssification system from satellite imagery 卫星图像热带气旋分类
tropical cyclone-like vortices（TCLV） 热带气旋类涡旋
tropical cyclone program（TCP） 热带气旋计划
tropical cyclone twins 孪生热带气旋,双热带气旋
tropical day 酷热日
tropical depression 热带低压【大气】
tropical disturbance 热带扰动【大气】
tropical easterlies 热带东风带【大气】

tropical easterlies jet 热带东风急流【大气】
tropical environment 热带环境
tropical experiment（TROPEX） 热带试验
Tropical Experiment Board（TEB,WMO/ICSU） 热带试验局（WMO/ICSU）
Tropical Experiment Council（TEC,WMO/ICSU） 热带试验委员会（WMO/ICSU）
tropical front 热带锋
tropical high 热带高压【大气】
tropical marine air mass 热带海洋气团【大气】
tropical meteorology 热带气象学【大气】
tropical monsoon 热带季风【大气】
tropical monsoon climate 热带季风气候
tropical monsoon rain climate 热带季风雨气候亚类【大气】
tropical month 分至月
Tropical Ocean and Global Atmosphere （TOGA） 热带海洋与全球大气试验计划
tropical rain climate 湿热多雨气候
tropical rainfall measuring mission （TRMM） 热带降雨测量〔卫星〕,热带测量任务〔卫星〕
tropical rainforest 热带雨林【大气】
tropical rainforest climate 热带雨林气候【大气】
tropical rainy climate 热带常湿气候亚类【大气】
tropical revolving strom 热带风暴
tropical savanna climate 热带草原气候
tropical severe local storm 热带局地强风暴
tropical squall line 热带飑线
tropical steppe climate 热带〔稀树〕草原气候【大气】
tropical storm 热带风暴【大气】
tropical superior air mass 热带高空气团
tropical system 热带系统

tropical upper-tropospheric trough （TUTT） 热带对流层高空槽【大气】，洋中槽
tropical wave 热带波
tropical wave disturbance 热带波扰动
tropical wet and dry climate 热带干湿气候
tropical wet climate 热带潮湿气候
Tropical Wind Energy Conversion and Reference Level Experiment (TWERLE) 热带风能转换和参考层试验
tropical wind observation ship (TWOS) 热带风观测船
tropical winter dry climate 热带冬干气候亚类【大气】
tropical year 回归年【天】,分至年
tropical zone 热带【地理】
tropic higher high water (TcHHW) 回归高高潮
tropic lower low water (TcLLW) 回归低低潮
Tropic of Cancer 北回归线【地理】
Tropic of Capricorn 南回归线【地理】
tropics 热带
tropic tide 回归潮
tropogram 对流层高空气象图
tropopause 对流层顶【大气】
tropopause break 对流层顶断层
tropopause break-line 对流层顶断裂线
tropopause bump 对流层顶颠簸
tropopause chart 对流层顶图
tropopause discontinuity 对流层顶不连续
tropopause fold 对流层顶折叠,多对流层顶
tropopause funnel 对流层顶漏斗【大气】
tropopause inversion 对流层顶逆温
tropopause layer 对流层顶层
tropopause leaf 〔单叶〕对流层顶
tropopause wave 对流层顶波动【大气】
troposcatter 对流层散射

troposphere 对流层【大气】
troposphere-stratosphere exchange 对流层-平流层交换
tropospheric chemistry 对流层化学
tropospheric fallout 对流层沉降
tropospheric folding 对流层折叠
tropospheric ozone 对流层臭氧【大气】
tropospheric propagation 对流层传播
tropospheric reflection 对流层反射
tropospheric refraction 对流层折射【大气】
tropospheric scatter(=troposcatter) 对流层散射
tropospheric transport 对流层输送
trough 低压槽【大气】
trough aloft 高空槽
trough cyclone 槽〔式〕气旋
troughing 槽生
trough in westerlies 西风槽
trough line 槽线【大气】
trough line orientation 槽线走向
trough of low pressure 低压槽
trovadoes (=hurricane) 飓风（马达加斯加附近的用语）
trowal (=trowell) 高空暖舌(加拿大)
TR-tube (transmit-receive tube) 收发转换开关
true airspeed 真空速〔度〕
true altitude 真高度
true anomaly 真近点角【天】
true equator 真赤道【天】
true equinox 真春分点【天】
true error 真误差
true freezing point 真冻结温度
true force 真力
true gravity 真重力
true horizon 真地平【航海】
true latent heat 真潜热

true mean	真平均
true mean temperature	真平均温度
true noon	真午
true north	真北
true solar	真太阳
true solar day	真太阳日
true solar hour	真太阳时
true solar time	真太阳时
true value	真值【物】
true wind	真风【大气】
true wind direction	真风向
truncated distribution	截断分布
truncated normal distribution	截断常态分布
truncated spectral model	截谱模式
truncated spectrum	截谱
truncation error	截断误差【大气】,舍项误差
trunk	龙卷漏斗柱
trust territory	托管地区【地理】
trustworthiness	可靠性
truth-function	真值函数
T-S curve	温盐曲线
T-S diagram	温盐图
T-S relation	温盐关系
TSP(total suspended particulates)	总悬浮粒子〔数〕
Tsugaru Warm Current	津轻暖流
tsunami	海啸【海洋】,地震海浪
Tsushima Current	对马海流
tsuyu	(见 bai-u)梅雨(日语名)
TTE(typhoon tracking experiment)	台风跟踪试验
t-test	t 检验
tuba (tub)	管状(云)
tube-anemometer	管状风速表
tube photometer	管式光度计
tube-typed geothermometer	直管地温表【大气】
tube wind tunnel	管式风洞,简易风洞
tufan	(见 tofan)托范风暴
tufon	(见 tofan)托范风暴
tule fog	吐尔雾(美国加利福尼亚沼泽低地上的浓雾)
Tulipan radiometer	图里潘辐射计
tumble	翻滚
tunable dye laser	可调谐染料激光器
tunable gas laser	可调谐气体激光器
tunable laser	可调谐激光【化】
tunable laser diode spectrometer (TLDS)	可调谐激光二极管分光计
tunable laser spectroscopy	可调谐激光光谱学
tunable spin-flip Raman laser	可调谐自旋反转拉曼激光器
tundra	冻原【植物】,苔原【地理】
tundra climate	苔原气候亚类【大气】
tundra desert	冻原荒漠
tunnel	风洞
tunnelling effect	隧道效应【化】
turbid atmosphere	浑浊大气
turbidimeter	浑浊度表,浊度计【机械】
turbidity	浑浊度,浊度【物】
turbidity coefficient	浑浊系数,浊度系数
turbidity factor	浑浊因子【大气】
turbidity maximum	浑浊度极大〔区〕
turbid medium	浑浊媒质
turboclair	涡轮喷气消雾机(法国奥利和戴高乐机场用)
turbonada	大雷雨,狂风(西班牙语名)
turbopause	湍流层顶【地】
turbosphere	湍流层【地】
turbulence	湍流【物】,颠簸【航空】
turbulence body	湍流体
turbulence closure	湍流闭合
turbulence closure scheme	湍流闭合法
turbulence cloud	湍流云【大气】
turbulence component	湍流分量

turbulence condensation level 湍流凝结高度【大气】
turbulence effect 湍流效应
turbulence energy 湍流能量【大气】
turbulence factor 湍流因子
turbulence intensity 湍流强度
turbulence inversion 湍流逆温【大气】
turbulence kinetic energy 湍流动能
turbulence length scales 湍流长度尺度
turbulence Reynolds stress 湍流雷诺应力
turbulence severity 湍流度
turbulence shear stress 湍流切应力
turbulence spectrum 湍流谱【大气】
turbulence time scale 湍流时间尺度
turbulence velocity scale 湍流速度尺度
turbulency 湍流〔度〕
turbulent boundary layer 湍流边界层【大气】
turbulent convection 湍流对流
turbulent crosswind 阵侧风
turbulent diffusion 湍流扩散【大气】
turbulent diffusion coefficient 湍流扩散系数
turbulent diffusivity 湍流扩散率
turbulent dissipation 湍流消散
turbulent Ekman layer 湍流埃克曼层
turbulent energy 湍流能量
turbulent energy dissipation rate 湍流能量耗散率
turbulent energy flux 湍流能通量
turbulent entrainment 湍流夹卷
turbulent exchange 湍流交换【大气】
turbulent field 湍流场
turbulent flow 湍流【物】
turbulent fluctuation 湍流脉动
turbulent flux 湍流通量【大气】
turbulent friction velocity 湍流摩擦速度
turbulent gust 湍流阵风
turbulent heat transfer 湍流热输送
turbulent inertial sub-range 湍流惯性次区
turbulent intensity 湍流强度
turbulent intermittency 湍流间歇性
turbulent inversion 湍流逆温
turbulent kinetic energy 湍流动能
turbulent medium 湍流媒质
turbulent microscale 湍流微尺度
turbulent mixing 湍流混合
turbulent motion 湍流运动
turbulent plume 湍流烟流
turbulent process 湍流过程
turbulent Reynolds number 湍流雷诺数
turbulent scale 湍流尺度
turbulent scale stress 湍流尺度应力
turbulent shear stress 湍流切应力
turbulent similarity theory 湍流相似理论
turbulent statistical theory 湍流统计理论
turbulent structure 湍流结构
turbulent time scale 湍流时间尺度
turbulent transfer 湍流传输
turbulent transfer coefficients 湍流传输系数
turbulent transport 湍流输运
turbulent velocity vector 湍流速度矢量
turbulent vortex 湍流涡旋
turbulent wake 湍流尾流
turbulent zone 湍流区
turbulivity 湍流度,湍流系数
turf wall 草皮墙(雨量器防风用)
turn around ranging station (TARS) 往返测距站(测静止卫星精确位置)
Turner angle 特纳角
turning latitude 转向纬度
turning point 转向点
turnover frequency 反转频率
turnover rate 转换率
turnover time 转换时间
turreted cloud 塔状云
TUTT (tropical upper-tropospheric trough) 热带对流层高空槽【大气】
TVS (tornadic vortex signature) 龙卷涡

旋信号
TWCI(total water content instrument)
总含水量测量仪
TWECRLE(Tropical Wind Energy Conversion and Reference Level Experiment) 热带风能转换和参考层试验
tweeks 大气干扰
Twenty-four Solar Terms 二十四节气【大气】
twilight 曙暮光【大气】
twilight arch 曙暮光弧
twilight colours 霞【大气】
twilight correction 曙暮光时订正
twilight flash(或 glow) 曙暮辉
twilight glow 曙暮辉
twilight phenomena 曙暮光现象
twilight spectrum 曙暮光谱
twin channel radiometer amplifier 双通道辐射仪放大器
twin flights 双施放(探空仪)
twin-gauge station 双水位测站
twinkling 闪烁(星光)
twin radiosonde 双探空仪
twin soundings 双探空仪探空
twin water-flow pyrheliometer 双水流绝对日射表
twister(=tornado) 陆龙卷(美国口语)
twisting term 扭曲项
two-axis fluidic wind sensor 射流式双轴测风传感器
two concentric eyes 同心双眼(台风)
two-dimensional autocorrelation 二维自相关
two-dimensional cloud probe 二维云粒子探测器
two-dimensional crosscorrelation 二维互相关
two-dimensional detrending 二维去倾
two-dimensional eddies 二维涡旋
two-dimensional FFT 二维快速傅里叶变换

two-dimensional filtering 二维滤波
two-dimensional flow 二维流
two-dimensional Fourier series 二维傅里叶级数
two-dimensional model 二维模式
two-dimensional optical cloud probe (2D-C) 二维光学云探测器
two-dimensional power-wavenumber spectrum 二维功率-波数谱
two-dimensional precipitation probe 二维降水粒子探测器
two-dimensional spatial filtering 二维空间滤波
two-dimensional turbulence 二维湍流【大气】
two-flux theory 双通量理论
two-frequency correlation function 双频相关函数
two-frequency mutual coherence function 双频互相干函数
two-grid-interval noise 二倍格距噪声
two-gridlength wave 二倍格距波
two-photon absorption 双光子吸收【物】
two-photon process 双光子过程
two-pressure principle 双压法(湿度发生器的)
two-sided spectrum 双边谱
two satndard parallel conic projection 双标准纬线圆锥投影
two-step photon interaction 二阶光子交互作用
two-stream 双流〔法〕(处理大气中单色辐射传输复杂性的近似方法)
two-stream approximation 双流近似
two-stream instability 双流不稳定性【物】(等离子)
two-thirds law 三分之二律
two-time level scheme 二时次法
two-way approach 双向法
two-way attenuation 双程衰减【大气】
two-way interaction 双向影响,双向交互作用

two-way nesting 双向嵌套,双重嵌套
TWT（travelling wave tube） 行波管【电子】
T-year event 年温度跃变〔事件〕
Tyndall flower 丁铎尔花（出现在冰体内的六面形冰穴）
Tyndall phenomenon 丁铎尔现象【化】
Tyndallometer 丁铎尔测尘器（根据散射光强度测量大气中尘埃粒子数）
Tyndall scattering 丁铎尔散射
type 型,型式,类型
type-α leader α型前导
type-β leader β型前导
type map 类型图
typical year 标准年,代表年
typhoon 台风【大气】

typhoon bar 台风〔云〕墙
typhoon center 台风中心
Typhoon Committee（TC） 台风委员会
typhoon development 台风发展
typhoon eye 台风眼【大气】
typhoon genesis 台风生成
Typhoon Operational Experiment（TOPEX）台风业务试验
typhoon rain 台风雨
typhoon squall 台风飑
typhoon tide 台风潮
typhoon track 台风路径
typhoon tracking experiment（TTE）台风跟踪试验
typhoon warning 台风警报【大气】
typhoon watch 台风监测

U

uala-andhi 瓦拉安地尘暴（见于印度旁遮普）
UARS（Upper Atmosphere Research Satellite） 高层大气研究卫星
UAV（unmanned aerial vehicle） 无人航空飞行器,无人驾驶飞机
ubac 山阴,阴坡【地理】
UCAR（University Cooperation for Atmospheric Research） 〔美国〕大气研究大学联合会
udometer 雨量器
udomograph 自记雨量计
U figure（＝U index） U指数
UFO（unidentified flying object） 不明飞行物,幽浮
UGIS（urban geographic information system） 城市地理信息系统【地信】
ugly sky 阴沉天空
ugly weather 阴沉天空
UHF（ultra-high frequency） 特高频【电子】

UHF communication 特高频通信【电子】,分米波通信
UHF Doppler radar 超高频雷达【大气】
U index U指数（有关地磁场水平分量的指数）
ukiukiu 乌基乌基乌风（夏威夷群岛的东北信风）
UKMO（United Kingdom Meteorological Office） 英国气象局
ULF（ultra-low frequency） 特低频【电子】
Ulloa's circle（或 ring） 邬洛亚环（不常见景象）
Ulloa's ring 邬洛亚环,邬氏环
ultimate infiltration capacity 基本渗透量
ultra-fine mesh 超细网格
ultra-fine particle 超微颗粒【物】,微粒,超细粒子（直径≤10 nm）
ultra-high frequency（UHF） 特高频【电子】,超高频（300～3000 MHz）
ultra-long wave 超长波【大气】
ultra-low frequency（ULF） 特低频

ultrared ray 红外线
ultra-short wave 超短波
ultrasonic 超声〔的〕
ultrasonic anemometer 超声测风仪【大气】
ultra-sonic anemometer/thermometer 超声风速温度测量仪
ultrasonic holography 超声全息术【物】
ultrasonic property 超声特征
ultrasonics 超声学
ultrasonic snow data logger 超声波雪深测录器
ultraviolet (UV) 紫外线
ultraviolet absorbance monitor 紫外吸收监测仪
ultraviolet catastrophe 紫外灾难
ultraviolet differential absorption lidar (UV-DIAL) 紫外差分吸收光〔雷〕达
ultraviolet index (UVI) 紫外线指数
ultraviolet laser 紫外激光器
ultraviolet light 紫外光
ultraviolet photometry 紫外光度测定法
ultraviolet radiation (UV) 紫外辐射(按波长由短至长常分为三区,记为UV-A,UV-B,UV-C)
ultraviolet radiation meter 紫外辐射表
ultraviolet ray 紫外线【物】
ultraviolet remote sensing 紫外〔线〕遥感
ultraviolet spectrometer 紫外光谱仪
ultraviolet spectrum 紫外光谱
umbra ①本影【物】②本影【天】(太阳黑子的中央暗黑部)
umbral eclipse 本影食【天】
umbrella effect 阳伞效应
UMIS (urban management information system) 城市管理信息系统【地信】
Umkehr effect 回转效应【大气】
unambiguous range interval 不模糊距离〔时间〕区间
unambiguous velocity 不模糊速度【大气】
unambiguous velocity interval 不模糊速度〔时间〕间隔
unavailable potential energy 不可用位能
unbalance 不平衡【力】,紊乱
unbiased estimate 无偏估计【数】
unbiased test 无偏检验【数】
uncentred differencing 非中央差分
uncertainty 不确定度【物】,不定度【数】
uncertainty principle 不确定原理【物】,测不准原理
uncinus (unc) 钩状(云)
unconditional regression 无条件回归
unconfined aquifer 非承压含水层【地质】,自由含水层
uncorrelated samples 不相关样本【数】
uncorrelated scattering channel 非相关散射通道
UNCSD (United Nations Commission on Sustainable Development) 联合国可持续发展委员会
undamped oscillation 无阻尼振荡
undercast 机下云障
undercooling (见 supercooling)欠冷却,冷却不足
undercurrent 下层海流,潜流【海洋】
undercutting cold front 下切冷锋
underdamped system 欠阻尼系统
underdamping 欠阻尼【物】
underdetermined problem 欠定问题
underdetermined system 欠定组【数】,亚定组
underestimate 估计过低
underfitting 低度拟合
underflow 下溢,潜流
underforecast 不足预报
underground ice 地下冰
underground water 地下水
underground water level 地下水面,潜水面
underlying earth's surface 下垫面
underlying surface 下垫面
undermelting 冰融

under-moon 下幻月
under-parhelion 下幻日,下日
under predicated 预报偏低(少)的
underrelaxation 低松弛,亚松弛【数】
underrun 潜流
undershoot 低差
undersun 下日(一种晕的现象,太阳下面的假日)
undertow 退波流
underwater acoustic scattering 水下声散射
undetermined coefficient 待定系数
undistorted 未失真的
undisturbed motion 无扰运动
undisturbed solar condition 未扰动太阳状况
undisturbed sun 非扰动太阳
undograph 测波计
UNDP (United Nations Development Programme) 联合国开发计划署
undular bore 波状潮
undulation 波动
undulatus (un) 波状(云)
UNEP (United Nations Environment Programme) 联合国环境计划署
uneven grid 非均匀网格
unexplained variance 无释方差
UNESCO (United Nations Educational, Scientific and Cultural Organization) 联合国教科文组织
UNFCCC (United Nations Framework Convention on Climate Change) 联合国气候变化框架公约
unfiltered model 未滤波模式
unflyable weather 禁飞天气【大气】
unfree water 死水
unfreezing 不冻土,冻融作用
unidentified flying object (UFO) 不明飞行物,幽浮
unidirectional flux 单向通量
unidirectional vertical wind shear 单向垂直风切变

unifilar electrometer 单丝静电表
unified field theory 统一场论【物】
uniform convergence 一致收敛性【数】
uniform distribution 均匀分布【数】
uniform flow 均流
uniform illumination 均匀照明【物】
uniformitarianism 均变说
uniformity of spacetime 时空均匀性【物】
uniformization 单值化【数】,均匀化
uniform layer 均匀层
uniformly magnetized sphere 均匀磁化球
uniform magnetoionic medium 均匀磁离子介质
uniform midlatitude positive perturbation 均匀中纬正扰动
uniform motion 匀速运动【物】
uniform space 一致空间【数】
uniform velocity 等速度
unimodal distribution 单峰分布【数】
unimodal spectrum 单峰谱
unipolar electrical charge 单极电荷
unipolar group 单极黑子群
unipolar sunspot 单极黑子
uniformization 单值化【数】
unit 单位【物】
unitary filter 单通滤波〔器〕
unit distribution 单位常态分布
United Kingdom Meteorological Office (UKMO) 英国气象局
United Nations (UN) 联合国
United Nations Commission on Sustainable Development (UNCSD) 联合国可持续发展委员会
United Nations Conference on Desertification (UNCD) 联合国沙漠化问题会议
United Nations Development Programme (UNDP) 联合国开发计划署
United Nations Educational, Scientific and Cultural Organization (UNESCO)

联合国教育、科学和文化组织,联合国教科文组织
United Nations Environment Programme (UNEP) 联合国环境计划署
United Nations Framework Convention on Climate Change (UNFCCC) 联合国气候变化框架公约
United Nations Millennium Development Goals 联合国千年发展目标
United Nations Research Institute for Social Development (UNRISD) 联合国社会发展研究所
United Nations Scientific Committee on the Effects of Atomic Radiation (UNSCEAR) 联合国原子辐射效应科学委员会
United Nations World Food Conference (UNWFC) 联合国世界粮食会议
United States Air Force Environmental Technical Application Center (USAFETAC) 美国空军环境技术应用中心
United States Coast Pilot 美国海岸导航指南
United States Standard Atmosphere (USSA) 美国标准大气
United States Weather Bureau (USWB) 美国气象局
unit function 单位函数【数】
unit hydrograph 单位水位线
unit matrix 单位矩阵,幺矩阵【数】
unit normal distribution 单位正态分布
unit step function 单位阶梯函数
unit storm 单体雷暴
unit vector 单位矢量
univariate 单变数
univariate optimum interpolation 一元最优内插
univariate time series 一元时间序列
universal abundance 宇宙丰度
universal algebra 泛代数【数】
universal constant 普适常量【物】

universal day 国际日
universal decimal classification 国际十进位分类法
universal equilibrium hypothesis 普适平衡假说
universal filter 全通滤波〔器〕
universal function 普适函数,通用函数【数】
universal gas constant 普适气体常量【物】,普适气体常数【大气】
universal gravitational constant 万有引力常数
universal gravitational law 万有引力定律
universal rain gauge 通用雨量计
universal time (UT) 世界时【天】,格林尼治民用时
universal time coordinated (UTC) 协调世界时
universal transmission function 通用透射函数
universe ①宇宙【天文】②全域【数】③底集【数】
universe evolution 宇宙演化【物】
University Cooperation for Atmospheric Research (UCAR) 〔美国〕大气研究大学联合会
unlimited ceiling 无限云幂高
unlimited-unlimited 云幂和能见度无限,晴朗,能见度极好(系英语口语)
unmagnetized plasma 无磁等离子体
unmanned aerial vehicle (UAV) 无人航空飞行器,无人驾驶飞机
unmanned recovery vehicle 无人回收运载工具
unmanned scientific satellite 无人科学卫星
unmanned spacecraft 无人飞行器
unpacked data format 非压缩资料格式
unperturbed atmosphere 未扰〔动〕大气
unperturbed model 未扰〔动〕模式
unperturbed stratospheric aerosol

平流层静稳气溶胶,未扰平流层气溶胶
unpolarized light 非偏振光【物】
unpolarized wave 非极化波,非偏振波
unpredictability 不可预报性,不可预测性【物】
unpredictable loss 不可预防损失
unprotected reversing thermometer 开端颠倒温度表
unprotected thermometer 开管温度表,无保护颠倒温度表
unrestricted visibility 无限远能见度
unsaturated air 未饱和空气
unsaturated downdraft 未饱和下沉气流
unsaturated flow 未饱和气流
unsaturated hydraulic conductivity 未饱和导水性,未饱和导水率
unsaturated zone 未饱和区
unsaturation 未饱和
unsettled weather 多变天气,不稳定天气
unsharp masking 模糊遮光法
unstability 不稳定〔度〕
unstable air 不稳定空气【大气】
unstable air mass 不稳定气团【大气】
unstable atmosphere 不稳定大气
unstable channel 不稳定河床,不稳定河槽
unstable condition 不稳定条件,不稳定状态
unstable cyclone wave 不稳定气旋波
unstable equilibrium 不稳定平衡
unstable equilibrium point 不稳定平衡点
unstable manifold 不稳定流型
unstable motion 不稳定运动
unstable node 不稳定〔波〕节
unstable oscillation 不稳定振荡
unstable stratification 不稳定层结
unstable system 不稳定系统
unstable time series 不稳定时间序列
unstable wave 不稳定波【大气】
unsteady flow 不稳定水流

unterwind 昂特风(温地利萨尔茨卡默古特湖上的微风)
unusual solar activity 异常太阳活动〔性〕
unusual synoptic pattern 异常天气型
unusual typhoon movement 异常台风移动
unusual weather 异常天气
unutan 乌努坦风(夏威夷群岛的强旋风)
upbank thaw 上游解冻,上方增温
updating 更新,修改,校正
updraft(＝updraught) 上曳气流【大气】
updraft curtains 上升气流幕,上升气流薄层
updraft velocity 上曳气流速度
upglide cloud 上滑云【大气】
upglide motion 上滑运动
upglide surface 上滑面
upgoing electron 上升电子
upgoing wave 上传波
upgradient 逆梯度
upgradient flux 反梯度通量
upgradient transport 逆梯度传送
upland mire 山地沼泽
up-link radio experiment 对空无线电通信试验
upper air ①高空②高层空气
upper air analysis 高空分析【大气】
upper air anticyclone 高空反气旋
upper air chart 高空〔天气〕图【大气】
upper air circulation 高空环流
upper air climatology 高空气候学
upper air current 高空气流
upper-air cyclone 高空气旋
upper air data 高空资料【大气】
upper air disturbance 高空扰动
upper air divergence 高空辐散
upper air fluorescence 高层大气荧光
upper air front 高空锋
upper air inversion 高空逆温
upper air mass 高空气团
upper air meteorological research 高空

气象研究
upper-air observation 高空观测【大气】
upper-air ridge 高空脊
upper-air sounding 高空探测
upper-air station 高空站
upper-air structure 高层大气结构
upper-air synoptic station 高空气象站
upper-air trough 高空槽
upper anticyclone 高空反气旋
upper arcs 上珥
upper atmosphere 高层大气【大气】
upper atmosphere research 高层大气研究
Upper Atmosphere Research Satellite （UARS） 高层大气研究卫星
upper atmospheric phenomenon 高层大气现象
upper atmospheric physics 高层大气物理学
upper bright-band 上亮带
upper circumzenithal arc 上环天顶弧〔圈〕
upper cloud 高空云（对流层以上）
upper cold front 高空冷锋
upper cyclone 高空气旋
upper envelope 上包络（磁扰）
upper flight information region 高空飞行信息区
upper forest limit 森林上限
upper frictional region 高空摩擦区
upper front 高空锋【大气】
upper high 高空高压
upper inversion 高空逆温
upper ionosphere 电离层上部
upper layer station 高空站
upper level 高层,高能层
upper-level anticyclone 高空反气旋【大气】
upper-level chart 高空图
upper-level cyclone 高空气旋【大气】
upper-level disturbance 高空扰动
upper-level high 高空高压
upper-level inversion 高空逆温
upper-level jet stream 高空急流【大气】
upper-level low 高空低压

upper-level rain band 高空雨带
upper-level ridge 高空脊【大气】
upper-level trough 高空槽【大气】
upper-level wind 高空风【大气】
upper limit of the atmosphere 大气层上限
upper low 高空低压
upper ridge 高空脊
upper mirage 上现蜃景
upper mixing layer 高空混合层
upper perturbation 高空微扰
upper stratosphere 平流层上部
upper symmetric regime 高空对称型
upper tangential arc 上切弧（晕圈的）
upper troposphere 对流层上部
upper troposphere and lower stratosphere （UTLS） 上对流层—下平流层（200 hPa 至 30 hPa 区间）
upper tropospheric cold vortex 对流层高空冷涡
upper tropospheric humidity （UTH） 对流层上部湿度
upper trough 高空槽
upper warm front 高空暖锋
upper wind 高空风
upper-wind chart 高空风图
upper-wind observation 高空风观测
upper-wind report 测风报告
uprush 冲升气流,垂直急流
upscale direction 逆尺度（尺度增加）方向
upscale transfer 逆尺度（向较大尺度）转换
upshear 逆切变
upslide surface 上滑面
upslope fog 上坡雾【大气】
upslope wind 上坡风
upstream 迎风,上游
upstreaming ion 上流离子
upstream scheme 迎风差格式【大气】,迎风格式【数】

upstream shear lobe 上游切变瓣
upstream shock transition 上游激波跃迁（指太阳风遇地磁场时速度减低）
up-valley wind 谷风
upward atmospheric radiation 向上大气辐射
upward continuation 向上延拓
upward current 上升气流
upward flow 上升气流【大气】
upward heat transport 热量的向上输送
upward ion current 上流离子流
upward motion 向上运动
upward 〔total〕 radiation 向上〔全〕辐射【大气】
upward streaming ions 上流离子
upward terrestrial radiation 向上地球辐射
upwelling 涌升，上升流【航海】
upwelling current 涌升流
upwelling pocket 涌升水域
upwelling radiation 上行辐射
upwind 上风向
upwind difference 迎风差分【数】
upwind-downwind ratio 逆风顺风比
upwind effect 上风效应
upwind range 逆风航程
Uranus 天王星【天】
Ural blocking high 乌拉尔山阻塞高压【大气】
urban air 城市空气
urban air pollution 城市空气污染【大气】
urban atmosphere 城市大气
urban boundary layer 城市边界层
urban canopy 城市冠层
urban canopy layer 城市冠层
urban canyon 城市峡谷
urban circulation 都市环流
urban climate 城市气候【大气】
urban climatology 城市气候学【大气】
urban ecology 城市生态学

urban effects 城市效应
urban environmental pollution 城市环境污染
urban fringe 城市边缘【地理】
urban geographic information system （UGIS） 城市地理信息系统【地信】
urban geography 城市地理学【地理】
urban heat island 城市热岛【地理】
urban heat island effect 城市热岛效应
urban hydrology 城市水文学
urban implosion 城市内外（在既定的城区范围内废弃旧有功能，重新设计布局新功能的过程）
urbanism 城市化
urbanization 城市化【地理】
urbanization effect 城市化效应
urban management information system （UMIS） 城市管理信息系统【地信】
urban plume 城市烟羽
urban runoff 城市径流
urban-rural circulation 城乡环流【大气】
urban weather 城市天气【大气】
urban water logging 城市内涝
urea seeding 尿素撒播
U. S. airways code 美国航空〔天气〕代码，美国航线电码
U. S. Committee for the GARP (USC-GARP) 全球大气研究计划美国委员会
useful forecast 可用预报
user area 用户区【计】
user-end information 用户端信息
user key 用户密钥【计】
user log-in 用户注册【计】
user log-off 用户注销【计】
user log-on 用户注册【计】
user-oriented 用户导向的
user requirement 用户需求
U. S. Extension to the Standard ICAO Atmosphere 美国国际民航组织补充标

准大气
USGS(US Geological Survey) 美国地质调查局【地信】
U. S. Standard Atmosphere 美国标准大气
UT（universal time） 世界时【天】
UTC（ universal time coordinated） 协调世界时【天】
utility function 效益函数
UTLS（upper troposphere and lower stratosphere） 上对流层—下平流层（200 hPa 至 30 hPa 区间）
UV（ultraviolet radiation） 紫外辐射
UV-A A波段紫外辐射（最短的紫外波长）
UV and IR hygrometers 紫外和红外湿度计
UV-B B波段紫外辐射
UV detector 紫外检测仪
UV-C C波段紫外辐射
UV dosimeter 紫外线表
UV hygrometer 紫外湿度表
UV radiometer 紫外辐射仪
UV spectroirradiometer（UVS） 紫外分光辐照计
UVW anemometer 三分量风速仪

V

vacillation 振荡,摆动
vacillation cycle 振荡周期
vacillation phenomena 振荡现象
vacuum 真空【物】
vacuum correction 真空订正
vacuum electron device 真空电子器件【电子】
vacuum microeletronics 真空微电子学【电子】
vacuum-tube electrometer 真空管静电表
VAD（velocity azimuth display） 速度方位显示【大气】
vadose water 渗流水
vadose zone 渗水区
VAD wind profile 速度方位显示风廓线
vagueness 含糊性
vaguio（见 baguio, vario） 碧瑶风
valais wind 瓦莱风（由日内瓦湖上端沿罗讷河谷吹的谷风）
validation 确认【计】,验证,检验
valid character 有效字符
validity 有效性【数】
valid time 〔预报〕时效
valley breeze 谷风【大气】
valley exit jet 谷口急流
valley fog 谷雾
valley glacier 谷冰川
valley inversion 山谷逆温
valley outflow jet 山谷泄流
valley storage 河谷贮水〔量〕
valley wind 谷风
valley wind circulation 谷风环流
Vallot heliothermometer 瓦劳特日光温度表
valuation 赋值【数】
Van Allen radiation belt 范艾伦辐射带【大气】,地球内辐射带
Vandenberg Air Force Base, California（VAFB） 美国范登堡空军基地（加利福尼亚州）
van der Pol's equation 范德波尔方程【物】
van der Waals equation 范德瓦耳斯方程【物】
van der Waals radius 范德华半径【化】
vane 风向标,风信旗
vane anemometer 风杯风速计【大气】
vanishing point 合点【测】,灭点【计】

van't Hoff's law 范托夫定律【化】
VAP (velocity-azimuth processing) 速度方位处理
vaporization 汽化【物】
vapor scarf 凝结尾迹
vapour(=vapor) 水汽,汽,蒸气【物】
vapour concentration 水汽浓度
vapour density 水汽密度
vapour flux 水汽通量
vapour line 水汽线
vapour pressure ①水汽压②蒸气压【物】
vapour pressure curve 水汽压曲线
vapour pressure deficit 水汽压亏值
vapour pressure hygrograph 水汽压湿度计
vapour pressure thermometer ①水汽压温度表②蒸气压低温计【物】
vapour source 水汽源
vapour tension 水汽张力
vapour to liquid ratio 气态与液态之比
vapour trail 水汽尾迹
varatrazo 瓦拉特拉佐风(马达加斯加努西贝岛的陆风)
VARD (velocity azimuth range display) 速度方位距离显示【大气】
vardar(=vardarac) 瓦尔达尔风(由希腊瓦尔达尔河谷吹向萨洛尼卡海湾的下沉风)
vardarac 瓦尔达尔风(见 vardar)
variability 变率
variable ①变量【物】②变化〔的〕
variable area 可变区域
variable-audio-frequency radiosonde 变声频〔无线电〕探空仪
variable ceiling 多变云幕高
variable direction 风向多变
variable-radio-frequency radiosonde 变射频〔无线电〕探空仪
variable resolution model 可变分辨率模式

variable-slit impactor 可变缝隙取样器
variable visibility 多变能见度
variable wind 不定风【大气】
variance 方差【数】
variance analysis 方差分析【人气】
variance component 方差分量【数】
variance equation 变分方程【数】
variance ratio 方差比
variance reduction 方差缩减
variance spectrum 方差谱
variate 变量
variation ①变化②变分【数】
variational assimilation 变分同化
variational Doppler radar analysis system (VDRAS) 多普勒雷达变分分析系统
variational method 变分法
variational normal mode initialization 变分正模初始化
variational objective analysis 变分客观分析【大气】
variational optimization analysis 变分最佳分析
variational principle 变分原理【数】
variational quality control 变分质量控制
varied flow 非均匀流,变流
Vario (见 baguio)碧瑶风
variograph 变压计
variometer ①可变电感器【电子】,变感器②爬升速度仪
varsha 雨季(印度语名)
varves 纹泥【地理】,季候泥,冰湖季泥
VAS (VISSR atmospheric sounder) VISSR 大气探测器
vaudaire(=vauderon) 焚风(见于日内瓦湖)
vault 穹窿,拱顶(有界弱回波区)
V-band V 波段
V depression V 形低压
VDRAS (variational Doppler radar analysis system) 多普勒雷达变分分析系统
vectopluviometer 定向测雨器

vector 矢量,向量【数】
vector analysis 矢量分析
vector computer 向量计算机【计】
vector-diagram 矢量图
vector editing technology 矢量编辑技术
vector equation 矢量方程
vector field 矢量场
vector Fortran language 矢量 Fortran 语言
vector function 矢量函数
vector gauge 定向测雨器,向风雨量器
vectorial integration system 矢量积分系统
vectorial pluviometer 定向雨量器,矢量雨量器
vector potential 矢势【物】
vector product 矢积【物】
vector vane 矢量风标
veering 顺转
veering wind 顺转风【大气】
vegetation 植被
vegetation index 植被指数【大气】
vegetation season 植物生长季
vegetation type 植被型
vegetative period 植物生长期
vehicle 运载工具,车辆
vehicular emission 车辆排放
veil 幕,幔
velium 云幔
velocity 速度【物】
velocity aliasing 速度混淆
velocity ambiguity 速度模糊【大气】
velocity-area method 流速-面积法
velocity azimuth display (VAD) 速度方位显示【大气】
velocity-azimuth processing (VAP) 速度方位处理
velocity azimuth range display (VARD) 速度方位距离显示【大气】
velocity circulation 速度环流
velocity-contour method 等〔流〕速线方法
velocity curve 速度曲线
velocity defect 风速亏损
velocity-defect law 速度亏损定理
velocity distribution 速度分布
velocity divergence 速度辐散
velocity entrainment 速度夹卷
velocity field 速度场
velocity filtering 速度滤波
velocity fluctuation 速度起伏
velocity folding 速度折错
velocity-frequency filtering 速度-频率滤波
velocity gradient 速度梯度
velocity head 〔流〕速〔水〕头
velocity-height ratio 速度高度比
velocity indication coherent integrator (VICI) 速度指示相干积分器
velocity indicator 速度指示器
velocity of approach 接近流速,渐近流速
velocity of escape 逃逸速度【物】
velocity of recession 退行速度
velocity of sound 声速
velocity potential 速度势
velocity pressure 风压
velocity profile 速度廓线
velocity rod 测速杆,浮杆
velocity shear 速度切变
velocity spectrum 速度谱
velocity variance 速度方差
velocity-wavenumber filtering 速度-波数滤波
velopause 零风速层顶【大气】,速度顶
velum (vel) 缟状(云)
vena contracta 射流紧缩,收缩断面
vendaval(es) 文达瓦尔风(直布罗陀海峡的一种强西南风)
Venn diagram 文恩图
ventania 阿尼亚风(葡萄牙近海岸的飑)

vent da Mut 木特风(意大利加尔达湖的湿风)
vent des dames 达梅斯风(法国罗讷三角洲东部地中海沿岸的海风,法语名)
vent du midi 米迪风(法国中央高原中部和赛文山南部的南风,法语名)
vent hole 通气孔(水银气压表的)
ventifact 风棱石,风磨石
ventilated psychrometer 通风干湿表
ventilated thermometer 通风温度表
ventilation 通风
ventilation coefficient 通风系数
venting mixed layer air 通风混合层空气
vento di sotto 索托风(意大利加尔达湖的微风)
Venturi aircraft thermometer 文图里飞机温度表
Venturi effect 文图里效应
Venturi tube 流速管
Venus 金星【天】
Venus atmosphere 金星大气
veranillo 范拉尼罗旱期(中美洲六月底短期停雨的日子,西班牙语名)
verano 范拉诺旱期(美洲热带仲冬久旱期)
verdant zone 无霜带
vergence 聚散度
verglas 雨凇
verification 验证【物】
verification sample 验证样本
verification statistics 验证统计
vernal 春天的
vernal equinox 春分点【天】
Vernal Equinox 春分【大气】
vernalization 春化〔作用〕【农】,春化处理
vernier 游标【物】
vernier scale 游标尺

versatile information processor 通用信息处理器
vertebratus (ve) 脊状(云)
vertex angle 顶角【数】
vertical acceleration 垂直加速度
vertical advection 垂直平流
vertical anemometer 垂直风速表
vertical anemoscope 垂直风速仪
vertical angle 仰角,对顶角【数】
vertical arcs of the 22° parhelia 罗维兹弧
vertical-axis anemometer 垂直风速表
vertical-beam radar 垂直射束雷达【大气】
vertical beam ceilometer 垂直光束云幂仪
vertical climatic zone 垂直气候带【大气】
vertical circle 地平经圈【天】,垂直圈【航海】
vertical circulation 垂直环流
vertical continuation 垂直延伸
vertical convection 垂直对流
vertical convergence 垂直辐合
vertical coordinate system 垂直坐标系统
vertical correction factor 垂直订正因数
vertical coupling 垂直耦合
vertical cross section 垂直剖面图【大气】
vertical-current recorder 〔大气〕垂直电流记录仪
vertical decoupling 垂直退耦
vertical-development cloud 直展云
vertical differencing scheme 垂直差分法
vertical differential chart 垂直变差图
vertical diffusion 垂直扩散
vertical diffusion coefficient 垂直扩散系数
vertical displacement 垂直移动,垂直位移

vertical distribution 垂直分布
vertical divergence 垂直辐散
vertical division 垂直分层
vertical eddy temperature flux 垂直涡动温度通量
vertical extent of a cloud 云厚度【大气】
vertical flux 垂直通量
vertical gustiness 垂直阵风
vertical-gust recorder (v. g. recorder) 垂直阵风记录器
vertical heat transport 热量垂直输送
vertical incidence Doppler radar (VID) 垂直入射多普勒雷达
vertical instability 垂直不稳定度
verticality 铅垂度
vertical jet 垂直急流
vertical line 铅垂线
vertically integrated liquid (VIL) 累积液态含水量
vertically pointing Doppler radar 垂直指向多普勒雷达
vertically propagating wave 垂直传播波
vertical mass flux 垂直质量通量
vertical mixing 垂直混合
vertical modes 垂直模
vertical momentum flux 垂直动量通量
vertical motion 垂直运动
vertical polarization 垂直偏振,垂直极化
vertical pressure gradient force 垂直气压梯度力
vertical profile 垂直廓线【大气】
vertical ray 垂直射线
vertical resolution 垂直分辨率
vertical scale 垂直尺度
vertical section 垂直剖面
vertical shear 垂直切变
vertical shearing force 垂直切变力
vertical shrinking 垂直收缩
vertical simplification 垂直简化
vertical sounder 〔大气〕垂直〔结构〕探测器
vertical speed indicator 垂直速度显示器
vertical stability 垂直稳定度
vertical stretching 垂直伸展
vertical structure equation 垂直结构方程
vertical tangent arc 直切珥
vertical temperature gradient 温度垂直梯度
vertical temperature profile radiometer (VTPR) 温度垂直廓线辐射仪【大气】
vertical tilt 垂直倾斜
vertical time section 垂直时间剖面
vertical totals index 垂直总〔量〕指数
vertical transfer 垂直输送
vertical transform 垂直转换
vertical transverse wave 垂直横波
vertical velocity 垂直速度
vertical velocity variance 垂直速度方差
vertical visibility 垂直能见度【大气】
vertical vorticity 垂直涡度
vertical wave 垂直波
vertical wave length 垂直波长
vertical wave number 垂直波数
vertical wind error 垂直风误差
vertical wind shear 风的垂直切变【大气】
vertical wind velocity 垂直风速【大气】
vertical zonality 垂直地带性
vertical zone 垂直地带
very cloudy sky 多云天空(总云量为八分之六或七)
very high frequency (VHF) 甚高频 (30～300 MHz)
very high frequency direction finder (VHFDF) 甚高频测向器
very high frequency radar 甚高频雷达
very high resolution radiometer (VHRR) 甚高分辨率辐射仪【大气】

very high sea 狂涛(风浪7级)
very large eddy simulation(VLES) 甚大〔尺度〕涡旋模拟
very low frequency(VLF) 甚低频(<30 kHz)
very rough sea 巨浪(风浪5级)
very short-range〔weather〕forecast 甚短期〔天气〕预报【大气】(未来6 h内)
Very Small aperture terminal(VSAT) 甚小〔孔径〕地球站【电子】
vesine 韦西纳风(法国德龙省的谷风)
vessel of the rain gauge 雨量桶,雨量计容器
vestigial sideband(或 band) 残留边带
VFR(visual flight rules) 目视飞行规则【电子】
VFR between layers 云层间目视飞行规则
VFR flight 目视飞行
VFR on top 云顶目视飞行规则
VFR terminal minimums 目视飞行规则最低条件
VFR weather 目视飞行规则天气
VHF(very high frequency) 甚高频(30~300 MHz)
VHF radar 甚高频雷达【大气】
VHF source 甚高频源
VHRR(very high resolution radiometer) 甚高分辨率辐射仪【大气】
vibrating diaphragm transducer 振膜式传感器
vibration 振动【物】,颤动
vibration band 振动带
vibration energy 振动能
vibration-rotation band 振转带
vibration-rotation spectra 振转谱
vibration spectra 振动谱
vicissitude 变迁,变化,交替
video 视频
video frequency(=vision frequency) 视频

video gain 视频增益
videograph 视程计(一种后向散射能见度仪)
video integrator and processor(VIP) 视频积分处理器
video mapping 视像映射
videometer 视程表(测大气透光度用)
video scanner 视频扫描器
video signal 视频信号
video tape recorder 磁带录像机
video terminal 显示终端
videotex 视频数据系统
vidicon 视像管【电子】
vidicon camera 光导摄像机
vidicon camera system 光导相机系统
viento zonda(见 zonda) 松达风(阿根廷热风)
viewing angle correction 视角订正
vigil basins 警戒池,观察性流域
viking "海盗"火箭(美国高空研究火箭)
VIL(vertically integrated liquid) 垂直积分液水量
vinessa 文内萨风(意大利加尔达湖的风)
violence 狂暴
violent storm 11级风【大气】,暴风
VIP(video integrator and processor) 视频积分处理器
virazon(=birazon) 维拉海风(西班牙和葡萄牙语名)
virazones 维拉海陆风(西班牙和葡萄牙的语名)
virga(vir) ①幡状(云)②雨幡
virgin forest 原始林【林】
virtual access method 虚拟存取法
virtual computer 虚拟计算机【计】
virtual displacement 虚位移【物】
virtual gravity 虚重力
virtual height ①有效高度②虚高【地】
virtual image 虚像【物】

virtual impactor 虚拟冲击器
virtual machine 虚拟机【计】
virtual map 虚拟地图【地信】
virtual memory 虚拟存储〔器〕【计】
virtual potential temperature 虚位温
virtual pressure 虚压
virtual reality 虚拟现实【计】
virtual stress 虚应力
virtual sublayer 虚副层
virtual temperature 虚温【大气】
virtual work 虚功【物】
viscosity 黏滞性,黏度【物】,黏性【物】
viscosity coefficient 黏滞〔性〕系数
viscosity resistance 黏滞〔性〕阻力
viscous body 黏性体
viscous-convective subrange 黏性对流副区
viscous damping 黏性阻尼
viscous dissipation 黏滞耗散【大气】
viscous drag 黏滞曳力
viscous fluid 黏性流体【物】
viscous force 黏〔性〕力【物】
viscous stress 黏应力
viscous sublayer 黏滞副层
visentina 文散提纳风（意大利加尔达湖的强烈东东北到东风）
visibility 能见度【大气】,可见度【物】
visibility in cloud 云中能见度
visibility index 视程指数
visibility marker（或 object） 能见度目标物【大气】
visibility meter 能见度表【大气】
visibility object 能见度目标物
visibility ratio（＝luminosity） 发光率
visibility recorder 能见度表
visibility reduction 能见度降低
visible albedo 可见光反照率
visible and infrared spin scan radiometer（VISSR） 可见光和红外自旋扫描辐射仪
visible cloud picture 可见光云图【大气】
visible horizon 可见地平
visible image 可见光图像
visible-IR radiometer 可见光和红外辐射仪
visible light 可见光【大气】
visible plume tracer technique 可见烟羽示踪法（扩散测量）
visible 〔light〕 radiation 可见光辐射
visible spectrum 可见光谱
visiometer 能见度表【大气】
vision frequency 视频
VISSR（visible and infrared spin scan radiometer） 可见光和红外自旋扫描辐射仪
VISSR atmospheric sounder（VAS） VISSR 大气探测器
visual angle 视角
visual-aural range 视听距离
visual contact height 目视进场高度
visual Doppler indicator（VDI） 多普勒雷达目视显示器
visual extinction meter 可见光消光计（测能见度）
visual field 视场
visual flight 目视飞行
visual flight rules（VFR） 目视飞行规则【电子】
visualization 可视化【计】
visual meteorological condition 目视飞行气象条件
visual observation 目测【大气】
visual photometer 目测光度表
visual range 视程
visual range formula 视程公式
visual range of light 灯光能见距离【大气】
visual storm signal 风暴信号【大气】
visual storm warning 风暴预警标志
vital role 维生作用
vital temperature 生机温度
viuga 浮加风（俄罗斯草原带的一种偏北风风暴）

vivosphere 生物圈
VLES (very large eddy simulation) 甚大〔尺度〕涡旋模拟
VLF (very low frequency) 甚低频(小于30 kHz)
VOCs(volatile organic compounds) 挥发性有机化合物
voice-frequency 音频
voicespondence 录音通信
void ratio 空隙比
Voigt profile 沃伊特廓线,混合线型
volatile organic compounds (VOCs) 挥发性有机化合物
volatiles 挥发物
volatilization 挥发
volatilization method 挥发法【化】
volcanic activity 火山活动
volcanic aerosol 火山气溶胶
volcanic arc 火山弧
volcanic ash 火山灰【大气】
volcanic bomb 火山弹
volcanic cloud 火山云
volcanic dust 火山灰【大气】,火山尘
volcanic eruption 火山喷发
volcanic gases 火山气〔体〕
volcanic lake 火山湖【地理】
volcanic lightning 火山闪电
volcanic plume 火山烟流
volcanic sand 火山砂
volcanic sediment 火山沉积
volcanic storm 火山暴
volcanic thunder 火山雷鸣
volcanic wind 火山风
volcanic winter 火山冬天
volcanism 火山作用【地质】
volcano 火山【地理】
volcanology 火山学【地质】
VOLMET broadcast 对空气象广播【大气】
volt (V) 伏〔特〕(电位,电压和电动势单位)

voltage 电压【物】,伏特数
Volterra integral equation 沃尔泰拉积分方程【数】
volume 容积【物】,体积
volume absorption coefficient 体吸收系数
volume angular scattering coefficient 体角散射系数
volume average 体积平均【大气】
volume element 体积元【数】
volume expansivity 体膨胀率【物】
volume extinction coefficient 体消光系数
volume median diameter 体积中数直径
volume scan 体积扫描
volume scattering 体积散射
volume scattering coefficient 体散射系数
volume scattering function 体散射函数【大气】
volume target 体积目标【大气】
volume total scattering coefficient 体总散射系数
volume transport 体积输送
volumetric lysimeter 容积式蒸散计
volumetric snow sampler 容积式雪取样器
Voluntary Assistance Programme (VAP, WWW) 自愿援助计划(WWW)
Voluntary Co-operation Programme (VCP) 志愿协作计划(世界气象组织的)
voluntary observer 志愿观测者
Volz turbidity factor 沃斯混浊因子
von Karman constant 冯·卡曼常数
von Karman law 冯·卡曼定律
von Karman vortex 冯·卡曼涡旋
von Neumann algebra 冯·诺伊曼代数【数】
von Neumann method 冯·诺伊曼方法
vortex 低涡【大气】,涡旋
vortex breakdown 涡旋崩溃
vortex cloud street 涡旋云街
vortex cloud system 涡旋云系【大气】

vortex cooling tube 涡旋冷却管
vortex dew-point hygrometer 涡管露点湿度表
vortex filament 涡〔旋〕丝
vortex flow 涡流
vortex line 涡〔旋〕线
vortex merging 涡旋合并
vortex motion 涡旋运动
vortex pair 涡偶
vortex rain 涡旋雨
vortex-rings 涡〔旋〕环
vortex roll 转涡,涡卷
vortex Rossby waves 涡旋罗斯贝波
vortex scale 涡旋尺度
vortex shedding 涡旋曳离
vortex sheet 涡层
vortex shrinking 涡旋收缩
vortex signature 涡旋特征
vortex sink 涡汇
vortex source 涡源
vortex speed sensor 涡旋式风速传感器
vortex streets 涡〔旋〕列
vortex stretching 涡旋伸展
vortex thermometer 涡管温度表
vortex trail 涡旋轨迹
vortex train 涡列
vortex tube 涡管

vortex twinning 孪生涡旋
vorticity 涡度【大气】
vorticity advection 涡度平流【大气】
vorticity center 涡度中心
vorticity equation 涡度方程【大气】
vorticity forcing 涡度强迫
vorticity lobe 涡度瓣
vorticity path 涡度路径
vorticity transfer 涡度输送
vorticity transport hypothesis 涡度输送假设
vorticity transport theory 涡度输送理论
Voss polariscope 沃斯偏光器
V-point (见 radiant point) 辐射点
vriajem (见 friagem) 弗里阿吉姆干冷期
VSAT(very small aperture terminal) 甚小〔孔径〕地球站【电子】
V-shaped depression V形低压【大气】
V-shaped isobars V形等压线【大气】
VTPR (vertical temperature profile radiometer) 温度垂直廓线辐射仪【大气】
vulnerability 脆弱性
vuthan 乌丹暴(见于南美南部)
VWP (VAD wind profile) 速度方位显示风廓线

W

wadi 干枯河道,干河床,旱谷
wailer 怒号(风的)
wake 尾流【大气】
wake axis 尾流轴
wake capture 尾流捕捉
wake depression 尾流低压【大气】
wake effect 尾流效应
wake-following typhoon 尾随台风
wake low 尾低压

wake stream 尾流
wake turbulence 尾流湍流
wake vortex 尾涡
wake vorticity 尾涡度
waldsterben 森林衰亡,森林枯萎(德语,酸雨造成的)
Walker cell 沃克环流〔圈〕【大气】
Walker circulation 沃克环流
Walk-in environmental chamber 可进入

的大型环境模拟室
wall cloud 云墙
wall effect 壁效应(电子,离子)
wall of sand 沙壁
wall region 壁区
Walsh sequential spectrum 沃尔什序贯谱
Walsh transform 沃尔什变换【物】
wam-andai 瓦姆安代风(新几内亚岛西部冬季强西北阵风)
wam Braw 万布罗风(几内亚北部的干热风)
WAN (wide area network) 广域网【计】
wandering dune 流动沙丘【地理】
warm 暖,温暖
warm advection 暖平流【大气】
warm air 暖空气
warm air advection 暖平流
warm-air drop 暖池
warm air mass 暖气团【大气】
warm air wedge 暖空气楔
warm anticyclone 暖性反气旋【大气】,暖高压
warm braw 焚风,暖风
warm climate with dry summer 夏干温暖气候
warm climate with dry winter 冬干温暖气候
warm cloud 暖云【大气】
warm conveyor belt 暖输送带
warm-core anticyclone 暖心反气旋
warm-core cyclone 暖心气旋
warm-core disturbance 暖心扰动
warm-core high 暖心高压
warm-core low 暖心低压
warm-core rings 暖心涡旋
warm-core system 暖心系统
warm current 暖〔海〕流
warm cyclone 暖性气旋【大气】,暖低压
warm drop 暖池
warm eddy 暖涡
warm event 暖事件

warm fog 暖雾【大气】
warm front 暖锋【大气】
warm front cloud system 暖锋云系【大气】
warm front precipitation 暖锋降水
warm front rain 暖锋雨
warm front surface 暖锋面
warm front type 暖锋型
warm front type occlusion 暖锋型锢囚
warm-front wave 暖锋波动
warm half year 暖半年
warm high 暖高压【大气】,暖性反气旋
warm lake 热湖【地理】
warm low 暖低压【大气】,暖性气旋
warm night 暖夜(最低温度≥20℃)
warm occluded front 暖性锢囚锋【大气】
warm occlusion 暖性锢囚
warm pool 暖池
warm rain 暖雨【大气】
warm rain process 暖雨过程
warm ridge 暖脊【大气】
warm season 暖季
warm sector 暖区【大气】
warm start 热启动【计】
warmth index 温量指数
warm tongue 暖舌【大气】
warm-tongue steering 暖舌引导
warm-type occlusion 暖型锢囚
warm up time 预热时间
warm vortex 暖涡【大气】
warmward jet 暖侧急流
warm water fog 暖水雾
warm water mass 暖水团
warm water sphere 暖水层
warm wave 暖浪
warm-wet climate 湿热气候【大气】
warm wind 暖风
warning 预警
warning stage ①预警阶段②警戒水位【地理】
Wasatch winds 沃萨奇风(美国沃萨奇山区的强东风)
wash 冲刷

wash-leather bag 〔水银气压表〕皮囊
Washoe zephyr 瓦肖焚风（美国内华达一带的 chinook 风）
washout 冲洗,冲刷,清除
washout coefficient 冲洗系数,冲刷系数
washout efficient 冲洗效率
waste recycling 废物再循环
wastewater 废水
wastewater irrigation 污水灌溉
wastewater treatment 废水处理
water 水
water-activated battery 浸水活化电池
water and soil conservation 水土保持
water and soil erosion 水土流失
water atmosphere 水汽圈
water balance 水平衡
water bleeding 崩冰水,冰隙水
water body 水体【地理】
water budget 水平衡【地理】,水分收支【大气】,水分差额
water circulation coefficient 水循环系数【大气】
water cloud 水云【大气】
water compensation 水分补偿
water condensation index 水凝结指数
water conservation 水分保持
water consumption 耗水量
water content 含水量
water content of cloud 云含水量【大气】
water cycle 水循环【大气】
water deficit 水分亏缺【植物】
water drop 水滴
water droplet 微水滴,小水滴
water dropper (= water dropping collector) 水滴集电器（测空中电位梯度用）
water emissivity 水发射率
water equivalent 水当量
water equivalent of snow 雪水当量【大气】
water equivalent of snow cover 雪盖水当量
water erosion 水蚀【土壤】

waterfall 瀑布【地理】
waterfall effect 瀑布效应
water-flow pyrheliometer 水流式绝对日射表
water-flow recorder 水流记录器
water fog 水态雾,水雾
water hammer 水击,水锤
water head 水头,水位差
water-hot coefficient 水热系数
water inventory 总水量
water in soil 土壤水分
water level 水位,水准面
water-level recorder 水位计
waterlogged 涝的,多水的
waterlogging 涝【土壤】
waterlogging tolerance 耐渍性
waterlogging tolerant crop 耐涝作物
water loss 水耗
water management 水分管理
water mass 水团【地理】
water mass formation 水团构成
water mass transformation 水团变性
water meter 水表,水位表
water molecule 水分子
water need 需水量
water pollution 水污染【地理】
water pollution control 水污染控制
water pollution monitoring 水污染监测
water potential 水势【植物】
water preserving capability 保水性【土壤】
water quality 水质【地理】
water quality analysis 水质分析
water quality assessment 水质评价
water regime 水分状况
water requirement 需水量【植物】
water reservoir ①水槽（干湿表用）②水库
water resources 水资源【大气】
water resources assessment 水资源评估
water retention capacity 保水能力

watershed ①流域②集水区③分水岭
【地理】
water sky 水照云光
water smoke 蒸气雾
water snow 湿雪
watersonde 云滴谱探测仪
watershed morphology 流域形态
【地理】
waterspout 水龙卷【大气】,海龙卷
water spray seeding 喷水撒播
water spreading 水流扩展
water-stage recorder 水位计
water-stir pyrheliometer 搅水式绝对日射表
water stress effect 水应力作用
water structure 水结构
water substance 水质素
water supply 给水,供水【地理】
water-supply forecast 供水量预报,径流量预报
water supplying capability 供水性【土壤】
water supply sensitivity 供水敏感性
water surface evaporation 水面蒸发
water surplus 水盈值,过剩水量
water system 水系
water table 潜水面,地下水位【地理】
water table aquifer 非承压含水层,自由含水层
water temperature 水温
water-thermometer 水温表
watertight stratum 土壤不透水层
water type 水型
water-use coefficient 水分利用系数
water-use ratio (= transpiration ratio) 蒸腾率
water vapor 水蒸气【大气】,水汽
water-vapor absorption 水汽吸收
water-vapor bands 水汽带【大气】
water-vapor content 水汽含量
water vapor continuum 水汽连续带
water vapor convergence 水汽辐合

water-vapor correction 水汽订正
water-vapor density 水汽密度
water-vapor feedback 水汽反馈
water-vapor flux 水汽通量
water vapor-greenhouse effect 水汽-温室效应
water vapor imagery 水汽图像
water vapor meter 水汽测量仪
water vapor mixing ratio 水汽混合比
water vapor pressure 水汽压【大气】
water vapor profile 水汽廓线【大气】
water vapor retrieval 水汽反演【大气】
water-vapor spectroscope 水汽吸收分光仪
water vapor winds 水汽〔推导〕风
water wave 水波
water year 水文年度
watt (W) 瓦〔特〕
wave ①波【物】②波浪【航海】
wave action 波作用量
wave activity 波活动性
wave age ①波龄【海洋】②波速风速比
wave amplification 波幅增大
wave amplitude 波振幅
wave attenuation 波减弱
wave average 波平均
wave blocking 波阻塞
wave breaking 波〔浪破〕碎
wave celerity 波速(波脊或波谷的传播速度)
wave-CISK 波致第二类条件不稳定
wave climate 波〔浪气〕候(波浪的长期统计状态)
wave cloud 波状云【大气】
wave crest 波顶,波峰【海洋】
wave-current interaction 波流交互作用
wave cyclone 波动性气旋
wave depression 锋面低压,波状低压
wave diffraction 波衍射
wave dispersion 波扩散
wave dissipation 波消

wave disturbance 波状扰动
wave drag 波阻
wave dynamometer 波力计（波压观测装置）
wave energy density 波能密度
wave energy flux 波能通量
wave ensemble 波群、波包
wave equation 波动方程【物】
wave forcing 波动〔强迫〕作用
wave forecasting 波浪预报
waveform 波形
waveform analysis 波形分析
waveform recorder 波形记录仪（测天电用）
wave frequency 波动频率
wave front 波前【物】,波阵面【物】
wave-front method 波前法
wave generation 波生
wave group 波群【物】
waveguide 波导【物】
wave height 波高
wave interaction 波间交互作用
wave in the easterlies 东风波
wavelength 波长【物】
wavelet 小波,子波【物】
wavelet analysis 小波分析【数】
wavelet transform 小波变换【数】
wave-mean flow interaction 波和平均气流的交互作用
wave modeling 波动模拟
wave motion 波动
wave number 波数【物】
wave-number domain 波数域
wavenumber filtering 波数滤波
wavenumber-frequency filtering 波数-频率滤波
wave number-frequency spectrum 波数-频率谱
wave-number regime 波数域
wavenumber spectrum 波数谱
wave of condensation and rarefaction 疏密波
wave of oscillation 振荡波
wave of translation 平移波
wave packet 波包【物】
wave period 波动周期
wave pole 测波杆
wave profile 波廓线
wave propagation 波传播
Wave Propagation Laboratory（WPL, NOAA） 波传播实验〔研究〕室（NOAA）
wave range 波段
wave ray 波射线
wave recorder 波浪计
wave ridge 波脊,波峰【海洋】
wave set-up/set-down 波涨/落
wave source 波源
wave spectrum 波谱
wave speed 波速
wave staff 测波杆
wave steepness 波动坡度
wave system 波系
wave theory 波动理论,波动说【物】
wave theory of cyclogenesis 气旋生成的波动理论【大气】
wave theory of cyclones 气旋波动理论
wave theory of light 光波动理论
wave train 波列【物】
wave transience 波瞬态
wave trough 波槽,波谷【物】
wave turbulence 波状湍流
wave-type disturbance 波型扰动
wave velocity 波速
wave vector 波矢
wave-wave interaction 波间交互作用,波—波交互作用
wave-zonal flow interaction 波和纬向气流的交互作用
WCASP（World Climate Applications and Services Program） 世界气候应用和服务计划

WCED（World Commission on Environment and Development） 世界环境与发展委员会
WCIRP（World Climate Impact and Response Strategies Program） 世界气候影响和应对策战略计划
WCP（World Climate Program） 世界气候计划
WCRP（World Climate Research Program） 世界气候研究计划
weak absorption limit 弱吸收极限
weak cell 弱单体【大气】
weak constraint 弱限制
weak echo region（WER） 弱回波区
weak echo vault 弱回波穹窿【大气】
weakening 变弱
weak evolution 弱演变
weak-field dynamo mode 弱场动力模式
weak fluctuation 弱起伏
weak front 弱锋
weak-line approximation 弱线近似【大气】
weak-line limit 弱线极限
weakly nonlinear 弱非线性
weakly nonlinear theory 弱非线性理论
weak monsoon 弱季风
weak solution 弱解
weather 天气【大气】
weather above minimum 适航天气
weather advisory 天气警示报,天气提醒
weather analysis 天气分析
weather anomaly 天气异常
weather below minimum 禁航天气
weather-bound 禁航天气
weather breeder 风暴来临前晴朗天气
weather broadcast 天气广播
weather bureau 气象局
weather chart 天气图
weather chart analysis 天气图分析
weather chart drum scanner 滚筒式天气图传真扫描器
weather chart recorder 天气图传真收片机,天气图记录器
weather cock 风标
weather code 天气电码
weather condition 天气状况
weather control 天气控制
weather cycle 天气循环
weather data processing 天气资料处理
weather derivatives 天气合同
weather diary 天气日记
weather divide 天气区界线
weather echo 天气回波【大气】
weathered ice 风化冰
weather element 天气要素
weather erosion 天气侵蚀
weather event 天气事件
weather facsimile（WEFAX） 天气图传真【大气】
weather forecast 天气预报【大气】
weather forecaster 天气预报员
weather forecast model 天气预报模式
weather gauge （见 barometer）气压计
weather glass 晴雨计
weather grade of forest-fire 森林火险天气等级【大气】
weather hazard 天气灾害
weather information network and display system（WINDS） 气象信息网及显示系统
weathering 风化〔作用〕【大气】
weathering index 风化指数【土壤】
weathering intensity 风化强度【土壤】
weathering product 风化产物【土壤】
weathering residue 风化残余物【土壤】
weather instrumentation tower 气象观测塔
weather interpretation 天气释用,天气解译
weather lore 天气谚语
weatherman 气象员,天气预报员
weather map 天气图
weather-map type 天气图类型

weather message 气象电报
weather minimum 最低气象条件
weather modification 人工影响天气【大气】
weather monitoring 天气监测
weather moise 天气噪声
weather observation 天气观测
weather outlook 天气展望【大气】
weather parameter 天气参数
weather patrol ship 气象侦察艇
weather phenomenon 天气现象【大气】
weather plane 天气侦察飞机
weather prediction 天气预报
weather-prognostics 天气形势预报
weatherproof 全天候的
weather prospect 天气展望
weather proverb 天气谚语
weather radar 天气雷达【大气】
weather radar data processor and analyzer 天气雷达资料处理分析器
weather radar observation 天气雷达观测
weather radar signal integrator 天气雷达信号积分器
weather reconnaissance〔flight〕 天气侦察〔飞行〕
weather record center 天气记录中心
weather recurrence 天气重现
weather regime 天气实体
weather report 天气报告
Weather Research and Forecasting Model (WRF) 天气研究和预报模式
weather resistance 气候适应能力,耐风雨能力
weather routing 气象定线【航海】,气象导航【大气】
weather satellite 气象卫星
weather satellite facsimile receiver 气象卫星传真收片机
weather satellite recorder 气象卫星资料记录器

weather sequence 天气序列
weather service 天气服务【大气】
weather service bill 天气业务条例
Weather Service Forecast Office (WSFO) 〔美国〕气象局预报台
Weather Service Meteorological Observatory (WSMO) 〔美国〕气象局气象站
weather severity indicator 天气严重指数
weather shack 气象哨
weather ship 〔海洋〕天气观测船
weather-side 上风的,迎风的
weather sign 天气符号
weather simulation 天气模拟
weather situation 天气形势
weathersphere 天气层
Weather Squadron, U.S.A. 美国天气中队
weather station 天气站
weather surveillance radar (WSR) 天气监视雷达
weather symbol 天气符号【大气】
weather system 天气系统
weather tide 气象潮,迎风潮
weather tower 气象塔
weather trend 天气趋势
weather type 天气型
weather typing 天气分类
weather vane 风向标
weather warfare 气象战
weather warning 天气警报【大气】
weather warning bulletin 天气预警公报
weather warning system 天气预警系统
weather watch 天气监视
weather wave 天气波
weather window 天气〔安全期〕窗
weather wire service 天气有线广播
weber (Wb) 韦〔伯〕（磁通量单位）
Weber-Fechner law 韦伯-费希纳定律（心理物理学）
Weber number 韦伯数

Weddell Deep Water 威德尔深层水
Weddell Gyre Boundary 威德尔旋涡边界
wedge 〔高压〕楔
wedge isobar 楔形等压线
wedge line 楔线,脊线
WEFAX(weather facsimile) 天气图传真【大气】
Wegener-〔Bergeron〕-Findeisen theory 韦格纳-〔贝吉龙〕-芬德森理论
Weger aspirator 韦格通风离子测定仪
Weibull distribution 韦布尔分布【数】
Weibull plotting position 韦布尔测绘点
weighed average 加权平均
weighed mean 加权平均【物】
weighed residual method 加权剩余法
weighing 加权,权重
weighing barometer 称重式气压表【大气】
weighing factor 加权因子
weighing function 加权函数
weighing hygrometer 称重式湿度表
weighing lysimeter 称重式渗漏计
weighing mercury barometer 称重式水银气压表
weighing raingauge 称重式雨量器【大气】
weighing snow-gauge 称雪器【大气】
weight 重量【物】,权【测绘】
weight barograph 称重式气压计
weight barometer 称重式气压表
weight chart 气柱重量图
Weight distribution 权重分布【数】
weighted average 加权平均
weighted error 加权误差【数】
weighted mean(或 average) 加权平均
weighted spectrum 加权谱
weight factor 权重因子
weight function 权〔重〕函数【物】
weight snow-gauge 称雪器
weir ①堰②坝
well logging 测井【地】

well-mixed layer 完全混合层
well-posedness 适定性
well-posed problem 适定问题【数】
WER(weak echo region) 弱回波区
west(W) 西
WEST(wind energy simulation toolkit) 风能资源数值模拟软件
West African tornado 西非龙卷
West Antarctic Ice Sheet 南极西半球冰原
West Atlantic pattern(WA) 大西洋西部型
West Australia Current 西澳大利亚海流
west coast climate 西岸气候(大陆西岸受海洋影响比较温和的气候)
west-east transport 〔西〕东向输送(指由西向东方向的输送)
westerlies 西风带【大气】
westerlies rain belt 西风多雨带
westerly belt 西风带
westerly jet 西风急流【大气】
westerly trough 西风槽【大气】
westerly vortex 西风涡
westerly wave 西风波
westerly wind burst 西风爆发
westerly zone 西风带
western boundary current 西边界流
Western North Atlantic Central Water(WNACW) 西北大西洋中央水
Western North Pacific Central Water(WNPCW) 西北太平洋中央水
Western Test Range(WTR,VAFB) 〔美国〕西部试验场(范登堡空军基地)
West Greenland Current 西格陵兰海流
west northwest(WNW) 西西北
West Pacific Oscillation(WPO) 西太平洋涛动
West Pacific pattern(WP) 太平洋西部型

West-Siberian Regional Research Hydrometeorological Institute（WSRRHMI） 西伯利亚西部地区水文气象研究所（前苏联）
west southwest（WSW） 西西南
west wind drift 西风漂流【大气】
wet adiabat 湿绝热线
wet adiabatic 湿绝热〔的〕
wet adiabatic change 湿绝热变化
wet-adiabatic lapse rate 湿绝热直减率
wet adiabatic temperature difference 湿绝热温度差
wet air 湿空气
wet and dry bulb hygrometer 干湿球湿度表
wet bulb 湿球
wet-bulb depression 干湿球温差【大气】
wet-bulb potential temperature 湿球位温【大气】
wet-bulb pseudo-potential temperature 假湿球位温【大气】
wet-bulb pseudo-temperature 假湿球温度【大气】
wet-bulb temperature 湿球温度【大气】
wet-bulb temperature difference 湿球温度差值
wet-bulb thermometer 湿球温度表【大气】
wet climate 湿润气候
wet damage 湿害【大气】,渍害
wet day 雨日
wet delay 湿延迟
wet deposition 湿沉降【大气】
wet energy solenoid 湿能管
wet enthalpy 湿焓
wet-equivalent potential temperature 湿相当位温
wet fog 湿雾【大气】
wet growth 湿生长
wet index 湿指数
wet injury 湿害
wet instability 湿不稳定性

wet katathermometer 湿球冷却温度表
wetland 湿地【地理】
wet-line correction 湿线订正
wet microburst 湿微暴流
wet precipitation 湿沉降
wet removal process 除湿过程
wet season 湿季
wet snow 湿雪【大气】
wet solenoid 湿力管
wet spell 雨期
wet static energy 湿静力能
wet static total energy 湿静力总能量
wettability 可湿性【力】,可湿度
wetted area 过水断面面积
wetted perimeter 湿界,过水周界
wetterlage 天气型（德语名）
wetting front 湿锋,湿润界面
wet tongue 湿舌【大气】
wet unstable 湿不稳定的
wet year 湿年,多水年,丰水年
whaleback cloud 延展荚状云
wheel-barometer 轮形气压表
whip-poor-will storm 春暴（回春后第一次恶劣天气）
whirl ①旋风,涡旋,旋流②涡动【力】
whirlies 小风暴
whirling echo 涡旋状回波【大气】
whirling psychrometer 手摇干湿表【大气】
whirling thermometer 手摇温度表
whirlpool 旋涡,旋流
whirlwind（=whirl） 旋风,旋流
whistler 哨声【地】
whistling wind 呼啸风
white band 冰延层
white body 白体
white-bulb thermometer 白球温度表
white buran 白布冷风（寒冷吹雪的buran风）
whitecap 浪沫
White cell 怀特〔吸收〕池

white dew 冻露【大气】
White Dew 白露(节心)【大气】
white earth 白地,深厚冰封
white frost 白霜
white horizontal circle 假日环
white ice 白冰
white noise 白噪声【物】
whiteout 乳白天空,白朦天,白化〔天气〕
white rainbow 雾虹
white smog 白雾(指强烈阳光与空气中悬浮汽车废物作用而产生的光化学烟雾)
white spectrum 白谱【物】
white squall 无形飑(热带天空无云时的飑),白飑
WHO（World Health Organization） 世界卫生组织
whole gale 狂风(十级风的旧名)
whole-gale warning 狂风预警
wide angle camera 广角〔照〕相机
wide angle radiometer 广角辐射仪
wide area network（WAN） 广域网【计】
wideband radar 宽带雷达
wide-bodied jet 广体急流
wide field of view radiometer 宽视场辐射仪
wide field scanning system 宽视场扫描系统
wide sense stationary channel 广义平稳通道
wide sense stationary uncorrelated scattering channel 广义平稳非相关散射通道
widespread precipitation 大范围降水
width 宽度
Wien's displacement law 维恩位移律【物】
Wiener-Hopf integral equation 维纳-霍普夫积分方程【数】,W-H 积分方程
Wien's distribution law 维恩单波分配定律
Wien's law 维恩定律
Wien's law of radiation 维恩辐射定律
Wigand visibility meter 维甘德视程计
WIGOS（WMO Integrated Global Observing Systcm） 世界气象组织全球综合观测系统
Wild fence 栅栏(雨量器四周围用)
Wild's evaporimeter 维尔德蒸发器
wild snow 松雪
Wilks λ-statistic 威尔克斯 λ 统计量【数】
williwaw（＝willywaw, willwau willie-wa, willy-waa） 威利瓦飑(见于南美麦哲伦海峡)
willy-willy 畏来风(见于澳大利亚西部)
Wilson cloud chamber 威尔逊云室【物】
wilting 萎蔫【植物】
wilting coefficient 凋萎系数
wilting moisture 凋萎湿度【大气】
wilting point 萎蔫点【大气】,凋萎点
wind 风【大气】
wind action 风力作用
windage effect 风阻影响【大气】
wind aloft 高空风【大气】
wind arrow 风矢【大气】(风向风速填图符号)
wind avalanche 风雪崩
wind averaging system 风值平均系统
wind barb 风羽
wind-barometer table 风速气压换算表
wind belt 风带
windblast ①阵风②爆炸气浪
wind-break 风障【大气】,风灾【地理】
wind-break forest 防风林
windbreak net 防风网
windburn 〔植物〕风灼伤
wind-channel 风道
wind-chill 风寒
wind-chill factor 风寒因子
wind-chill index 风寒指数【大气】

wind circulation zone 风力环流带
wind-cock 风向标
wind component measuring system 风分量测量系统
wind cone 风向袋
wind corrosion 风蚀
wind crust 风成雪壳
wind current 风带
wind damage 风害【大气】,风灾【地理】
wind daily run 风日程,每日风行程
wind deflection 风向偏转
wind direction 风向【大气】
wind direction indicator 风向指示器
wind direction recorder 风向自记器
wind-dirction shaft 风矢杆(风向填图符号)
wind divide 风向界线
wind drag 风曳
wind drift 漂流
wind-driven barotropic model 风致正压模式
wind-driven current 漂流
wind driven ocean circulation 风〔驱〕动洋流
wind eddies 风涡
wind effect 风效应
wind energy 风能【大气】
wind energy content 风能资源储量【大气】
wind energy conversion 风能转换
wind energy demarcation 风能区划【大气】
wind energy density 风能密度【大气】
wind energy metering and recording system 风能测量和记录系统
wind energy potential 风能潜力【大气】
wind energy resources 风能资源【大气】
wind energy rose 风能玫瑰〔图〕【大气】
wind energy simulation toolkit (WEST) 风能资源数值模拟软件
wind engineering 风力工程
wind equation 风方程

wind erosion 风蚀【大气】
wind erosion hazard 风蚀灾害【土壤】
wind erosion prediction 风蚀预报【土壤】
wind factor 风因子
wind farm 风力田【大气】,风电场
wind field 风场【大气】
wind-finding 高空测风
wind-finding radar 测风雷达【大气】
wind flurry 风飑
wind force 风力【大气】
wind force scale 风级【大气】
wind form coast 离岸风
wind friction 风摩擦
wind gap 风口【地理】
wind-gauge 风速器
wind-generated current 风生海流
wind-generated inertial oscillation 风生惯性振荡
wind-generated wave 风成波
wind generator 风力发电机
windicator 指示式测风仪
wind induced oscillation 风振【大气】
wind-induced surface heat exchange 风生表面热交换
wind layer 风层
wind load 风荷载【大气】
wind lull 风速暂减
wind map 风图
wind measurement from satellite 卫星测风
wind mill 风力机
windmill anemometer 风车式风速表
windmill generator 风车发电机
windmill switch 风车式开关(探空仪用)
wind minder anemometer 指示式风速仪
wind minder vane 指示式风向仪
wind mixing 风混合
windness 多风
wind of 120 days (见 Seistan wind) 十二旬风

wind-over-deck　甲板风
window　①大气窗②视窗
window channel　窗区频道
window frost　窗霜
window function　窗函数【电子】
window hoar　窗霜
window ice　窗冰
window length　窗长度
window region　大气窗区
window-thermometer　窗外温度表
window type　窗型,窗类型
wind power plant　风力发电厂,风电厂
wind power potential　风能潜力
wind pressure　风压【大气】
wind profile　风速廓线
wind profiler　风廓线仪
wind-profiling radar(WPR)　风廓线〔探测〕雷达
wind radar　测风雷达
wind-recorder　自记测风器【大气】
wind regime　风况,风系
wind resistance　抗风性
wind reversal　风向逆转
wind ridge　风脊,风分界
wind ripple　风雪纹
wind rose　风玫瑰〔图〕【大气】
windrow cloud(=billow cloud)　波状云(英语俗称)
wind run　风程
wind run integrator　风程积分器
winds aloft　高空风
winds aloft observation　高空风观测
winds-aloft plotting board　高空风计算盘
WINDSAT　测风卫星(美国的一种气象卫星名)
wind scale　风级,风速标尺
wind scoop　风窝
wind sea　风浪
wind setup　风增水,气象潮

wind shadow　静风区
wind shaft　风矢杆【大气】(风向填图符号)
wind shear　风切变【大气】
wind shear and gust front warning system　风切变及阵风锋预警系统
wind shear detection system　风切变探测系统
wind shield　遮风板
wind shift　风向转变
wind-shift line　风向突变线【大气】
wind site assessment　风场评价【大气】
wind slad　风成雪壳
wind sleeve　风向袋【大气】
wind sock　风〔向〕袋
wind spectrum　风谱
wind speed　风速【大气】
wind speed counter　风速计数器
wind speed profile　风速廓线【大气】
wind speed threshold sensor　风速阈值传感器
wind-spun vortex　风动涡旋
wind squalls　风飑
windstorm　风暴
wind strata　风层
wind stress　风应力【大气】
wind stress curl　风应力旋度
wind structure　风的结构
wind system　风系
wind teeth　风向齿(航用齿形风向器)
wind throw　风倒(大风将树木连根拔起)
wind toward coast　向岸风
wind tunnel　风洞
wind turbine　风力涡轮
wind turbine generator　风力发电机
wind vane　风向标【大气】
wind vector　风矢量【大气】,风向量
wind velocity　风速【大气】
wind velocity counter　风速计数器
wind velocity equation　风速方程
wind velocity fluctuation　风速脉动

【大气】
wind veloclty indicator 风速指针
wind velocity profile 风速廓线
wind velocity recorder 风速计
wind vibration 风振
windward 迎风〔的〕,向风〔的〕,上风向
windward side 迎风面
wind wave 风浪
windy cirrus(＝cirrus ventosus) 风卷云
wind zone 风带
wing aerogenerator 翼式风力发电机
WINTEM (aviation forecast of winds and temperatures aloft at specific points) 定点风温航空预报
winter 冬〔季〕
winter cold and rainy climate 冬寒常湿气候亚类【大气】
winter cold and winter dry climate 冬寒冬干气候亚类【大气】
winter cold climate 冬寒气候类【大气】
winter day 冬日,寒日
winter dry moderate 冬干温和〔气候〕
winter fog 冬雾
winter half year 冬半年
winter hemisphere 冬半球
winter ice 严冰
winterization 过冬准备
winter moderate and rainy climate 冬温常湿气候亚类【大气】
winter moderate and summer dry climate 冬温夏干气候亚类【大气】
winter moderate and winter dry climate 冬温冬干气候亚类【大气】
winter moderate climate 冬温气候类【大气】
winter monsoon 冬季风【大气】
winterness 冬性【植物】
winterness index 冬性指数
winter season 冬季
winter severity index 冬性指数
Winter Solstice 冬至(节气)【大气】
winter time 冬令时
winter warm center 冬季暖核〔中心〕
wire icing 电线积冰【大气】
wiresonde 系留探空仪
wire-weight gauge 悬锤式水尺,悬链式水尺
WIS (WMO Information System) 世界气象组织信息系统
Wisconsin-Würm 威斯康星-玉木(冰期)
WISHE (wind-induced surface heat exchange) 风生表面热交换
wisperwind 威兹珀风(法国威兹珀河河谷的寒冷夜风)
withershins 对日〔照〕
Witterung 天候(一个短时间内的天气总合,德语名)
WKB(或 WKBJ) approximation WKB(或 WKBJ)近似
WKB method WKB方法
WMO (World Meteorological Organization) 世界气象组织
WMO Information System (WIS) 世界气象组织信息系统
WMO Integrated Global Obseving System (WIGOS) 世界气象组织全球综合观测系统
WMO Member 世界气象组织成员
WNACW (Western North Atlantic Central Water) 西北大西洋中央水
WNPCW (Western North Pacific Central Water) 西北太平洋中央水
WOCE(World Ocean Circulation Experiment) 世界海洋环流试验
Wolf number 沃尔夫数【天】,〔相对〕太阳黑子数
Wollaston prism 沃拉斯顿棱镜【物】
Woods Hole Oceanographic Institution 伍兹霍尔海洋研究所(美国马萨诸塞州伍兹霍尔)
wood technology ①木材学 ②木材工艺

学【林】
woolpack cloud 卷毛云
work 功【物】
workability of soil 适耕性【土壤】
work function 功函数,逸出功【物】
working memory 智能库,工作存储器
working standard barometer 工作〔级〕标准气压表
Workman-Reynolds effect 沃克曼-雷诺效应(一种雷暴起电机制),W-R效应
world climate 世界气候【大气】
World Climate Applications and Services Program(WCASP) 世界气候应用和服务计划
World Climate Conference(WCC) 世界气候会议
World Climate Impact and Response Strategies Program(WCIRP) 世界气候影响和应对策略计划
World Climate Programme (WCP) 世界气候计划
World Climate Research Program (WCRP) 世界气候研究计划
World Climatic Data Information Referral Service(INFOCLIMA) 世界气候资料信息检索服务
World Commission on Environment and Development(WCED) 世界环境与发展委员会
World Data Centre(WDC,ICSU) 世界资料中心(ICSU)
World Health Organization (WHO) 世界卫生组织
World Meteorological Centre(WMC,WMO) 世界气象中心(WMO)
World Meteorological Congress 世界气象组织代表大会
World Meteorological Day 世界气象日
World Meteorological Intervals(WMI) 世界气象时段
World Meteorological Organization (WMO) 世界气象组织
world meteorology 世界气象学
World Ocean Circulation Experiment (WOCE) 世界海洋环流试验
World Occanic Organization(WOO) 世界海洋组织
World Power Conference(WPC) 世界动力会议
World Radiation Center(WRC) 世界辐射中心
world radiometric reference(WRR) 世界辐射测量基准
World Trade Organization(WTO) 世界贸易组织
world weather 世界天气
World Weather Research Programme (WWRP) 世界天气研究计划
World Weather Watch(WWW) 世界天气监视网
world-wide fallout 全球放射性沉降
World Wide Fund for Nature(WWF) 世界自然基金会
world-wide natural disaster warning system 全球自然灾害预警系统
world wide web(WWW) 万维网【计】
WPO(West Pacific Oscillation 西太平洋涛动
WPR(wind profiling radar) 风廓线雷达
WRC(World Radiation Center) 世界辐射中心
WSR-88D (weather surveillance rader-1988 Doppler) 88型天气监测多普勒雷达
WTO(World Trade Organization) 世界贸易组织
WWF(World Wide Fund for Nature) 世界自然基金会
WWRP(World Weather Research Programme) 世界天气研究计划
WWW ①(world wide web)万维网

【计】②(World Weather Watch)世界天气监视网

X

xaloque(=xaloc,xaloch) 西罗科风（sirocco风的西班牙语名）
xaroco 西罗科风(sirocco风的葡萄牙语名)
X-band X带,X波段(8～12吉赫频率,2.5～3.75 cm波长)
XBT（expendable bathythermograph) 投弃式温深仪【大气】,消耗性温深仪
xenon 氙〔气〕(Xe)
xenon spark lamp 闪光氙灯
xerasium 旱涝演替
xerochore 沙漠区
xerophobous 嫌旱的,避旱的
xerophobous plant 适旱植物
xerophilous tree species 旱年树种【林】
xerophyte 旱生植物【地理】
xerothermal index 干热指标
xerothermal period 干热〔气候〕期
Xigaze Group 日喀则群【地质】
XML（extensible markup language） 可扩展置标语言【计】
Xondar 声〔雷〕达(美国声纳克斯公司双基地声雷达产品名称）
X-ray X射线
X-ray absorption spectrometry X射线吸收光谱法【化】
X-ray crystallography X射线晶体学【化】
X-ray fluorescence analysis X射线荧光分析【物】
X-ray spectrograph X射线摄谱仪【物】
X-weather 禁航天气
xylem 木质部【植物】

Y

Yagi antenna 八木天线【电子】
Yagi-Uda antenna 八木-有田天线
Yaglou thermometer 亚格鲁〔加热式〕温度表(测风用)
yalca 耶尔卡雪暴(见于秘鲁北部安第斯山）
Yamamoto radiation chart 山本辐射图
Yanai wave 柳井波,混合波
Yangtze River 长江
yard 码(=91.44 cm)
yaw 横摇
yaw angle 航摆角【航海】,偏航角
yaw axis 偏航轴
year 年
year-book 年鉴
year-climate 年气候
year temperature difference 年温度差
year-to-year pressure difference 年际气压差
yellow rain 黄雨【大气】
Yellow Sea 黄海
Yellow Sea Warm Current 黄海暖流
yellow snow 黄雪(伴有黄色花粉的雪)
yellow wind 黄〔土〕风
yield forecasting 产量预报(方法)
yield prediction 产量预测,收获预测【林】
Yin Dynasty ruin 殷墟

youg 尤格风（地中海地区夏天的热风）
Younger Dryas 新仙女木时期（公元前 10800—前 9600 年）
Younger Dryas event 新仙女木事件
young ice 新冰
Yttrium Aluminum Garnet (YAG) Laser 钇铝石榴石激光器【电子】

Z

Zanzibar Current 桑给巴尔海流
zastrugi 雪面波纹（俄语名）
z-coordinate z 坐标
Zebra time (Z) 世界时,零时区时间
Zeeman effect 塞曼效应【物】
Zeldovich mechanism 泽尔多维奇机制（由分子氮产生活性氮的化学机制）
Zener diode 齐纳二极管【电子】
zenith 天顶【天】
zenithal rains 天顶期雨（类似二分点雨）
zenith angle 天顶角
zenith distance 天顶距【天】
Zephyros 西风（古希腊代风名）
zero-activity atmosphere 宁静大气
zero curtain 土壤零温度层
zero-dimension 零维
zero-dimensional modeling 零维模似
zero-dimensional system 零维系统【物】
zero gravity 失重,零重力
zero isotherm 零度等温线
zero layer (=null layer) 零层
zero-level 零水平（指磁扰前宁静时水平）
zero line 零线【物】,基准线
zero-order closure 零阶闭合
zero-plane displacement 零平面位移
zero point ①零点②致死临界温度
zero-pressure balloon 零压气球
zero-temperature level 零温度层
zeroth law of thermodynamics 热力学第零定律【化】
zero-zero 零零条件（云层低且能见度极差）
zero-zero weather 零零天气
zero zone 零时区
zigzag lightning 锯齿〔状〕闪〔电〕
zigzag pattern 锯齿型,曲折型
zigzag track 蛇行路径
zobaa 佐巴尘旋风（埃及）
zodiac 黄道,黄道带【天】
zodiacal counterglow 黄道对日照
zodiacal light 黄道光【天】
zodiac band 黄道带
zodiac cone 黄道锥
zombie turbulence 残余湍流
zonal 纬向的
zonal available potential energy 纬圈平均有效位能,纬向可用位能
zonal circulation 纬向环流【大气】
zonal climatologic temperature 纬向气候温度
zonal distribution 纬向分布
zonal energy 纬向能量
zonal flow 纬向气流
zonal index 纬向指数
zonality 成带分布,地带性【地理】
zonality of hydrological phenomena 水文现象分带性,水文现象分区性
zonal kinetic energy 纬圈平均动能
zonally averaged models 纬向平均模式
zonally symmetric circulation 纬向对称环流
zonal mean 纬向平均
zonal mean characteristics 纬向平均特性
zonal mean mesospheric circulation 中间层纬向平均环流

zonal mean motion 纬向平均运动
zonal momentum equation 纬向动量方程
zonal pressure gradient 纬向气压梯度
zonal wavenumber 纬向波数
zonal westerlies 纬向西风带
zonal wind 纬向风【地】,带状气流
zonal wind profile 纬向风廓线
zonal wind-speed profile 纬向风速廓线
zonation 带状排列
zonda(=sondo) ①北风(阿根廷)②干热焚风(阿根廷安第斯山东坡的西风)
zone 区,带,地带
zone description ZD 时区号【航海】
zone of abnormal audibility 异常能听区
zone of aeration 通气层
zone of audibility 能听区,可闻区
zone of constant temperature 等温带
zone of discontinuity 不连续带
zone of evaporative regulation 蒸发调节范围

zone of maximum precipitation 最大降水带
zone of recurvature 转向区
zone of saturation 饱和区
zone of silence 静区,无声区
zone of thermal neutrality 热中性区
zone of trade wind 信风带
zone of transition 过渡地带
zone of wastage 消融区
zone of variable wind 不定风带
zone〔standard〕time 〔地〕区时【天】
zone wind 区域风
zoogeography 动物地理学【地理】
zoom 变焦镜头【物】
zooming 变焦【物】
Z-R relation Z-R 关系【大气】(雷达反射率因子和雨强关系)
Z time Z 时,世界时
Zurich number 苏黎世数【天】(太阳黑子)

汉英大气科学词汇

A

阿贝不变量【物】 Abbe invariant
阿贝数 Abbe number
阿贝折射计【物】 Abbe refractometer
阿波罗 Apollo
阿尔卑斯山试验 Alps Experiment (ALPEX)
阿尔及利亚海流 Algerian Current
阿尔梅里亚-奥兰锋(西地中海锋区) Almeria-Oran Front
阿尔文波(磁流体动力波)【力】 Alfven wave
阿尔文层【地】 Alfven layer
阿伏伽德罗常量【物】 Avogadro number, Avogadro constant
阿伏伽德罗定律 Avogadro's law
阿伏伽德罗假设 Avogadro's hypothesis
阿贡〔火山〕(印尼) Agung
阿基米德浮力 Archimedean buoyant force
阿基米德原理【物】 Archimedes principle
阿卡斯-洛宾火箭探空系统 ARCAS-ROBIN system
阿卡斯气象火箭(全能大气探测火箭) all-purpose rocket for collecting atmospheric soundings (ARCAS)
阿拉果点【大气】 Arago point
阿拉果距 Arago distance
阿拉果中性点 Arago's neutral point
阿拉斯加高压 Alaskan high
阿拉斯加海流 Alaska Current
阿拉斯加〔续〕海流 Alaskan Stream
阿利索夫气候分类〔法〕 Alisov's classification of climate
阿留申低压【大气】 Aleutian low
阿留申海流 Aleutian Current

阿伦尼乌斯表达式 Arrhenius expression
阿伦尼乌斯电离理论【化】 Arrhenius theory of electrolytic dissociation
阿伦尼乌斯方程【化】 Arrhenius equation
阿纳德尔海流 Anadyr Current
阿普尔顿层 Appleton layer
阿普尔顿异常【地】 Appleton anomaly
阿斯曼干湿表【大气】 Assmann psychrometer
阿斯曼通风干湿表 Assmann ventilated psychrometer
阿〔托〕(词头,10^{-18}) atto-
锕射气【大气】 actinon
埃(长度单位,10^{-8} cm) Angström
埃丁顿近似法 Eddington approximation
埃尔米特多项式 Hermite polynomial
埃尔萨瑟辐射图 Elsasser's radiation chart
埃尔萨瑟吸收带 Elsasser absorption band
埃弗谢德效应 Evershed effect
埃伏图 evogram
埃及风 Egyptian wind
埃克曼边〔界〕条件 Ekman boundary conditions
埃克曼抽吸【大气】 Ekman suction, Ekman pumping
埃克曼层【大气】 Ekman layer
埃克曼颠倒采水器 Ekman reversing water bottle
埃克曼辐合 Ekman convergence
埃克曼廓线 Ekman profile
埃克曼流【大气】 Ekman flow
埃克曼螺线【大气】 Ekman spiral

埃克曼输送　Ekman transport
埃克曼数　Ekman number
埃克曼位移　Ekman displacement
埃克曼旋转　Ekman turning
埃克斯纳函数　Exner function
埃玛图　emagram
埃普利垂直入射绝对日射表　Eppley normal incidence pyrheliometer
埃普利绝对日射表　Eppley pyrheliometer
埃普利天空辐射表　Eppley pyranometer
埃斯皮-柯本学说　Espy-Köppen theory
埃斯特朗补偿直接日射表　Ångström compensation pyrheliometer
埃斯特朗浑浊度系数　Ångström turbidity coefficient
埃斯特朗净大气辐射表　Ångström pyrgeometer
埃斯特图　estegram
矮林【林】　coppice forest
矮云　low topped cloud
霭滴　mist droplet
艾贝风　Ibe wind
艾里斑【物】　Airy disk
艾里波　Airy wave
艾里函数　Airy function
艾萨卫星【大气】　ESSA (Environmental Survey Satellite)
艾托斯卫星【大气】　ITOS (Improved TIROS Operational System)
爱尔兰气象局　Irish Meteorological Service (IMS)
爱根核　Aitken nuclei
爱根计尘器　Aitken nucleus counter, Aitken dust counter
爱根粒子　Aitken particle
爱因斯坦方程【数】　Einstein equation
爱因斯坦求和符号　Einstein's summation notation
爱因斯坦系数　Einstein coefficient
安布尔图　Amble diagram
安的列斯海流　Antilles Current

安哥拉-本格拉〔海流〕锋（海温、盐度）　Angola-Benguela Front
安哥拉海流　Angola Current
安全　safety
安全飞行最大高度　aircraft ceiling
安全航路【航海】　safety fairway
安全航速　safety speed
安全探空仪　safesonde
安全通信【航海】　safety communication
安全网【计】　safe net
安全温度　safe temperature
氨　ammonia
氨化〔作用〕　ammonification
氨基酸【化】　amino acid
氨基酸定年法　amino-acid dating
氨态氮【土壤】　ammonium-nitrogen
氨循环　ammonia cycle
鞍点　saddle point
鞍点积分法　saddle point integration method
鞍点平衡　saddle-point equilibrium
鞍形气压场【大气】　col pressure field
鞍型　col
鞍型低〔气〕压　saddle
鞍状无云区　saddle-back
铵离子　ammonium ion
岸冰　shore ice, landfast ice
岸流　shore current
岸线【航海】　coastline
暗带　dark band
暗带【天】　dark lane
暗度定律　law of darkening
暗辐射　dark radiation
暗〔光〕线　dark light
暗弧　dark segment
暗礁　ledge
暗临边　dark limb
暗闪电　dark lightning, black lightning
暗射线　dark ray
暗视野　dark field
暗适应【物】　dark adaptation

暗条【天】 filament
暗物质【天】 dark matter
暗线 dark line
暗噪声 dark noise
凹的 concave
凹面镜【物】 concave mirror
凹透镜【物】 concave lens
奥布霍布判据 Obukhov's criterion
奥布霍夫长度 Obukhov length
奥布赖恩三次多项式 O'Brien cubic polynomial
奥-高公式【数】 Ostrovski-Gauss formula
奥兰原则 Olland principle
奥兰转换器 Olland cycle
奥米伽导航测风 Omega windfinding
奥米伽〔导航〕探测仪 Omegasonde
奥米伽导航系统 omega navigation system
奥米伽导航仪【航海】 Omega navigator
奥米伽定位试验 Omega Position Location Experiment
奥米伽动量通量 omega momentum flux
奥米伽方程 Omega equation（或 ω equation）
奥斯米多夫尺度（湍流） Ozmidov scale
奥斯特 oersted
奥陶纪【地质】 Ordovician Period
澳大利亚海流 Australia Current
澳大利亚联邦科学和工业研究组织 Commonwealth Scientific and Industrial Research Organization（CSIRO）
澳大利亚联邦气象局 Commonwealth of Australia Bureau of Meteorology（CABM）
澳大利亚南极研究国家委员会 Australian National Committee for Antarctic Research（ANCAR）
澳大利亚气象局 Australian Bureau of Meteorology（BOM）

B

八分仪 octant
八分之一 okta, octa
八角形网格 octagonal grid
八进制【计】 octal system
八木天线【电子】 Yagi antenna
八木-有田天线 Yagi-Uda antenna
八英寸雨量计 eight-inch rain gauge
巴巴多斯海洋和气象试验 Barbados Oceanographic and Meteorological Experiment（BOMEX）
巴比涅原理【物】 Babinet principle
巴比涅〔中性〕点 Babinet point
巴恩斯〔加权〕函数 Barnes weighting function
巴格罗夫判据 Bagrov's criterion
巴哈马海流 Bahama Current
巴厘岛路线图（联合国气候变化会议的一个部长级协议,2007 年） Bali Roadmap
巴林杰相关光谱仪 Barringer correlation spectrometer
巴浦洛夫斯基近似（流体） Pavlovsky's approximation
巴塘管【大气】 Bourdon tube
巴塘温度表【大气】 Bourdon thermometer
巴西〔暖〕海流 Brazil Current
〔拔海〕高度 elevation
拔节期 elongation stage
靶 target
白贝罗定律【大气】 Buys Ballot's law
白冰 white ice
白道 moon's path
白地 white earth
白垩【地质】 chalk

白垩纪【地质】 Cretaceous Period
白垩系【地质】 Cretaceous System
白金汉π理论 Buckingham Pi theory
白拉通公式 Blaton formula
白立克湿度图 Bleeker humidity diagram
白令海斜坡流 Bering Slope Current
白露【大气】 White Dew
白谱【物】 white spectrum
白球温度表 white-bulb thermometer
白霜【大气】 hoar-frost, white frost, hoar
白霜〔冰〕晶 hoar crystal
白松冰团 shuga
白体 white body
白天光照 daylight
白天能见度 daylight visibility
白天能见距离 daytime visual range, visual range
白天气辉 dayglow
白天温度 day temperature
白雾 white smog
白噪声【物】 white noise
白昼 daytime
百分递减 percent reduction
百分法【数】 percentage
百分率 percent
百分位点 percentile, tercile
百分误差【数】 percentage error
百慕大高压 Bermuda high
百年时间尺度 centennial time-scale
百帕(气压单位,1 hPa＝1 毫巴) hectopascal(hPa)
百万分率(10^{-6}) ppm (parts per million)
百叶箱 thermoscreen, louvred screen, shelter, thermometer shelter, thermometer-screen, instrument screen
百叶箱【大气】 screen, instrument shelter
百叶箱干湿表 screen psychrometer
百叶箱温度表 shelter thermometer

摆动积分 oscillatory integral
摆日 pendulum day
摆式风速表 pendulum anemometer
摆脱天气影响 above the weather
摆涡 pendule
摆线【数】 cycloid
拜拉姆风速表 Byram anemometer
班达海水 Banda Sea Water
斑点 speckle
斑状(云) maculosus
斑状层云 stratus maculosus
斑状高层云 altostratus maculosus
斑状卷云 cirro-macula
板架云 shelf cloud
板块构造学【地质】 plate tectonics
板块运动【地质】 plate motion
板状晶体 plate crystal
半潮位 half-tide-level
半承压含水层 semiconfined aquifer
半导体【物】 semiconductor
半导体〔感应〕元件 semiconductor element
半导体激光器【物】 semiconductor laser
半导体热电元件 semiconductor thermoelement
半导体探测器【电子】 semiconductor detector
半导体温差电偶 semiconductor thermocouple
半导体温度表 semiconductor thermometer
半岛【地理】 peninsula
半地转方程【大气】 semi-geostrophic equations
半地转锋生 semi-geostrophic frontogenesis
半地转高度倾向方程 semi-geostrophic height-tendency equation
半地转空间 semi-geostrophic space
半地转理论 semi-geostrophic theory
半地转流 semi-geostrophic flow

半地转运动【大气】	semi-geostrophic motion
半对数尺度	semi-logarithmic scale
半峰宽【化】	half band width
半腐层	duff
半干旱	semi-arid
半干旱带	semi-arid zone
半干旱气候【大气】	semi-arid climate
半干旱区	semi-arid region
半干燥	subarid
半功率点【电子】	half-power point
半功率束宽	half power beam width
半固定平台	semi-fixed platform
半固定沙丘	semi-fixed dune
半灌木【植】	subshrub
半弧角	half-arc angle
半荒漠	semi-desert
半灰辐射	semi-grey radiation
半混合湖〔泊〕	meromictic lake
半阶闭合	half-order closure
半径【数】	radius
半宽度【天】	half width
半拉格朗日法	semi-Lagrangian method
半拉格朗日方案	semi-Lagrangian scheme
半拉格朗日模式	semi-Lagrangian model
半年风振荡	semiannual wind oscillation
半年温度振荡	semiannual temperature oscillation
半年振荡	semi-annual oscillation, half-yearly oscillation
半平均法	semi-average method
半谱方法	semi-spectral method
半潜〔式〕平台	semi-submersible barge
半球	hemisphere
半球波数	hemispheric wave-number
半球广播	hemisphere broadcast
半球交换中心	hemisphere exchange center
半球模式	hemispheric model
半球〔式〕净辐射计	hemispherical and net radiometer
半球体总辐射计	total hemispherical radiometer
半球通量	hemispheric flux
半球细网格	hemispheric fine mesh
半球预报	hemispheric prediction
半日变化	semidiurnal variation
半日波	semidiurnal wave
半日潮	semidiurnal tide
半日太阳潮	semidiurnal solar tide
半深海环境	bathyal environment
半湿的	subhumid
半湿气候	subhumid climate
半湿润	semi-humid
半湿润区	semi-humid region
半衰期	half-time
半衰期【物】	half life period, half-life
半透明〔的〕	translucent
半透明冰	glime
半透明度	translucency (=translucence)
半无限大气	semi-infinite atmosphere
半无限介质	semi-infinite medium
半咸水,苦咸水,微咸水	brackish water
半显格式	semi-explicit scheme
半夜太阳	midnight sun
半隐式法【数】	semi-implicit method
半隐式格式【大气】	semi-implicit scheme
半永久性低压【大气】	semi-permanent depression, semi-permanent low
半永久性反气旋	semi-permanent anticyclone
半永久性高压【大气】	semi-permanent high
半永久性活动中心	semi-permanent action center
半圆定理	semi-circle theorem
半正常折射	subnormal refraction
半正定〔的〕	positive semi-definite
半值厚度(指吸收体)	half value thickness
半值期	half-value period
半滞留期	residence half-time

半周期　semiperiod
半周期带【物】　half-period zone
半自动气球施放器　semi-automatic balloon launcher（SABL）
半自动仪器　semi-automatic apparatus
伴随　adjoint
伴随法　adjoint method
伴随方程　adjoint equation
伴随矩阵【数】　adjoint matrix
伴随灵敏度　adjoint sensitivity
伴随模式　adjoint model
伴随同化　adjoint assimilation
伴随微分方程【数】　adjoint differential equation
瓣　lobe
瓣型　lobe pattern
棒状温度表　stem thermometer, rod thermometer
磅　pound
包交换技术　packet switching technique
包络　packet, envelope
包络地形　envelope topography
包络法　envelope method
包络速度　envelope velocity
孢粉分析　spore-pollen analysis, pollen analysis
孢粉图式　pollen diagram
孢粉学　palynology
孢粉资料　polymological data
雹瓣【大气】　hail lobe
雹暴【大气】　hail storm
雹暴记录器　hailstorm recorder
雹飑　hail squall
雹害　injury by hail
雹痕　hail impression
雹击带　hailswath
雹击线　hailstreak
雹块【大气】　hailstone
雹粒【大气】　hail pellet
雹胚【大气】　hail embryo
雹胚生成区　embryo formation region

雹心　hail core
雹雨分离器【大气】　hail-rain separator
雹灾【大气】　hail damage
雹阵　hail-shower
饱和【物】　saturation
饱和比湿　saturation specific humidity
饱和边界　boundary of saturation
饱和差【大气】　saturation deficit（或 deficiency）
饱和持水量【大气】　saturation moisture capacity
饱和点　saturation point, point of saturation
饱和度【水利】　degree of saturation
饱和分数　fraction of saturation
饱和高度　saturation level（SL）
饱和混和比　saturation mixing ratio
饱和静力能　saturation static energy
饱和绝热线　saturated adiabatic
饱和空气【大气】　saturated air
饱和空气温度【建筑】　saturation air temperature
饱和气压　saturation pressure
饱和器（湿度发生器用）　saturator
饱和区　zone of saturation, saturated zone
饱和曲线　saturation curve
饱和湿度【农】　saturated humidity
饱和湿度计　saturation hygrometer
饱和式磁强计　fluxgate magnetometer
饱和水汽　saturated vapour
饱和水汽压【大气】　saturation water vapour pressure, saturation vapour pressure, equilibrium vapour pressure
饱和土【水利】　saturated soil
饱和温度　saturation temperature
饱和相当位温　saturation equivalent potential temperature
饱和信号　saturation signal
饱和虚温　saturated virtual temperature
饱和总温度　saturation total temperature

宝光〔环〕【大气】 glory
宝石【地质】 gemstone
宝石矿物学【地质】 gem mineralogy
宝石学【地质】 gemology
保肥性【土壤】 nutrient preserving capability
保留因子 retention factor
保密性 privacy
保润盒,蒸汽饱和室 humidor
保守算子【数】 conservative operator
保守性 conservative property, conservatism
保守元素 conservative element
保水能力 water retention capacity
保水性【土壤】 water preserving capability
保真性【物】 fidelity
堡状(云) castellanus (cas)
堡状层积云【大气】 stratocumulus castellanus (Sc cas)
堡状高积云【大气】 altocumulus castellanus(Ac cas)
堡状卷积云 cirrocumulus castellanus
堡状卷云 cirrus castellanus
报表审核 report-form examination
报告点 reporting point
报头 message addressing
鲍恩比【大气】 Bowen ratio
鲍尔太阳指数 Baur's solar index
暴风 storm wind, storm gale, snifter
暴风雪 blizzard wind
暴风雨 rainstorm with strong wind
暴风雨中心 stormter
暴洪 flash flood
暴洪河流 flashy stream
暴洪监视 flash flood watch
暴洪交互分析器 Interactive Flash Flood Analyzer(IFFA)
暴洪预警 flash flood warning
暴洪预警系统 flash flood alarm system (FFAS)

暴时 storm time
暴涛(道氏风浪九级) confused sea
暴雨 storm rainfall, rain gust, hard rain
暴雨【大气】 torrential rain
暴雨【水利】 rainstorm
暴雨调查 rainstorm investigation
暴雨洪水 rainstorm flood
暴雨径流 storm flow
暴雨径流【公路】 rainstorm run off
暴雨强度【公路】 rainstorm intensity
暴雨强度分配 storm smear
暴雨侵蚀【土壤】 rainstorm erosion
暴雨水 stormwater
暴雨水经济学 economics of stormwater
暴雨损失〔量〕 storm loss
暴雨移置 rainstorm transposition
暴雨预警 rainstorm warning
暴雨组合 rainstorm combination
曝光表 exposure meter
曝光〔量〕 exposure
爆发 outbreak
爆发,开始 onset
爆发性云带 burst cloud band
爆发性增暖 impulsive warming, explosive warming
爆裂点 bursting point
爆裂高度(气球的) bursting height
爆裂厚度 bursting thickness
爆米花状积雨云群 popcorn cumulonimbi
爆米花状云团 popcorn cluster
爆炸波 explosive wave, blast wave
爆炸气体云 blast gas cloud
爆炸云 explosion cloud, cloud resulting from explosion
杯状冰晶 cup crystal
北 north(N)
北巴西海流 North Brazil Current
北半球 Northern Hemisphere
北半球冬季 northern winter
北半球副热带无风带 calms of Cancer

北半球交换中心　Northern Hemisphere Exchange Centre（NHEC）
北半球夏季　northern summer
北半球资料表　Northern Hemisphere data tabulations（NHDT）
北冰洋〔烟〕雾【大气】　arctic（sea）smoke
北部〔森林〕气候　boreal climate
北朝鲜海流　North Korean Current
北赤道海流　North Equatorial Current（NEC）
北赤道逆流　North Equatorial Countercurrent（NECC）
北大西洋高压　North Atlantic high
北大西洋公约组织　North Atlantic Treaty Organization（NATO）
北大西洋海流　North Atlantic Stream, North Atlantic Current
北大西洋漂流　North Atlantic drift
北大西洋深层水【海洋】　North Atlantic Deep Water
北大西洋涛动　North Atlantic Oscillation（NAO）
北东北　north northeast（NNE）
北方，北点　north point
北方林地　boreal woodland
北方林区　boreal forest
北方期（气候）　boreal climatic phase
北方〔气候〕期　Boreal
北方区　boreal zone, boreal region
北焚风　north foehn
北海海浪联合研究计划　Joint North Sea Wave Project（JONSWAP）
北寒带　north frigid zone
北回归线【地理】　Tropic of Cancer
北极　boreal pole, Arctic Pole
北极【地理】　North Pole
北极霭　arctic mist
北极表层水　Arctic Surface Water
北极冰动力学联合试验　Arctic Ice Dynamics Joint Experiment（AIDJEX）

北极陈冰（两年以上析出盐分的冰，厚度在25 m以上）　arctic pack
北极大陆气团　arctic continental air（mass）
北极带　arctic zone, north frigid zone
北极底层水　Arctic Bottom Water
北极对流层顶　arctic tropopause
北极反气旋【大气】　arctic anticyclone
北极风　arctic wind
北极锋【大气】　arctic front, Arctic Polar Front
北极浮冰群【大气】　arctic pack
北极高压　arctic high
北极光　northern lights, aurora polaris
北极光【地】　aurora borealis
北极海流　Arctic Current
北极化　arcticization
北极荒漠　arctic desert
北极林木线　arctic tree line
北极霾【大气】　arctic haze
北极平流层涡旋　arctic stratospheric vortex
北极气候　high arctic climate
北极气候【大气】　arctic climate
北极气团【大气】　arctic air mass
北极区　north polar region
北极圈　Arctic Circle
北极乳白天空现象　arctic whiteout
北极水【海洋】　North Polar Water
北极涛动　AO（Arctic Oscillation）
北极天气站　arctic weather station
北极无线电衰失现象　arctic blackout
北极雾　arctic fog
北极锋【大气】　arctic front
北极星　polaris, polar star, north star
北极星记录器　pole-star recorder
北极增温　arctic warming
北极中层水　Arctic Intermediate Water
北美反气旋　North American anticyclone
北美辐射学会　Radiological Society of North America（RSNA）

北美高压　North American high
北美洲板块【海洋】　North American Plate
北热带　north tropic zone, north tropic
北太平洋海流　North Pacific Current
北太平洋漂流　North Pacific Drift
北太平洋涛动　North Pacific Oscillation (NPO)
北纬　northern latitude
北温带　north temperate zone
北西北　north northwest(NNW)
北支急流【大气】　northern branch jet stream
卑尔根学派　Bergen School
贝蒂公式　Bethe formula
贝尔曼准线性化　Bellman's quasi-linearization
贝尔特拉米方程【数】　Beltrami equation
贝尔特拉米流　Beltrami flow
贝〔尔〕（音强单位）　bel
贝吉龙分类　Bergeron classification
贝吉龙-芬德森〔冰晶〕理论　Bergeron-Findeisen theory
贝吉龙-芬德森过程　Bergeron-Findeisen process
贝吉龙过程　Bergeron process
贝吉龙机制　Bergeron mechanism
贝吉龙强迫〔作用〕　Bergeron forcing
贝吉龙效应　Bergeron effect
贝克来雨量计　Beekley gauge
贝拉米法（估测水平散度）　Bellamy method
贝拉尼板罩式辐射仪　flat Bellani cup radiometer
贝拉尼球罩式辐射仪　spherical Bellani cup radiometer
贝拉尼蒸发计　Bellani atmometer
贝纳胞【大气】　Benard cell
贝纳对流【大气】　Benard convection
贝纳加热近似　Benard heating approximation
贝塞尔函数【数】　Bessel function
贝森测云器　Besson nephoscope
贝森梳状测云器　Besson comb nephoscope
贝塔尺度　Beta scale
贝塔粒子　beta particle
贝塔螺线　beta spiral
贝塔漂移　beta drift
贝塔平面　beta-plane
贝塔平面近似　beta(或 β)-plane approximation
贝塔射线　beta-ray (=β-ray)
贝塔涡旋　beta gyre
贝塔效应　Beta effect
贝叶斯定理【计】　Bayesian theorem
贝叶斯分析【计】　Bayes analysis
贝叶斯概率预报　Bayes probabilistic forecast
贝叶斯公式【数】　Bayes formula
贝叶斯估计【数】　Bayes estimate
贝叶斯决策规则【计】　Bayesian decision rule
贝叶斯决策函数【数】　Bayes decision function
贝叶斯输出处理器　Bayes processor of output (BPO)
贝叶斯推理【计】　Bayesian inference
备降机场　alternate airport
备降〔机〕场天气预报　alternate forecast
备用【计】　standby
备用元件　backup
背风　alee
背风波【大气】　lee wave
背风波分离　lee-wave separation
背风波水跃　lee wave hydraulic jump
背风波云　lee wave cloud
背风波转子　lee-wave rotor
背风槽　leeside trough
背风槽【大气】　lee trough
背风潮　lee tide

背风辐合　leeside convergence
背风列岛　leeward islands
背风面　leeward side, leeward, lee side
背风坡低压　lee low
背风坡低压【大气】　lee depression
背风坡涡旋　lee vortex, lee eddy
背风坡气旋生成　lee cyclogenesis
背景发光传感器　background luminescence sensor
背景光传感器　background light sensor
背景亮度仪　background luminance meter
背景误差　background errors
背景误差协方差　background error covariance
背景噪声　background noise
钡云　barium cloud
倍半萜【化】　sesquiterpene
倍加法　doubling method
倍角公式【数】　double angle formula
倍频　frequency-doubled
倍频【电子】　frequency multiplication
倍频器【电子】　frequency multiplier
倍周期分岔【物】　period doubling bifurcation
悖论【数】　paradox
被除数【数】　dividend
被滴定物【化】　titrand
被动〔隔断〕云　passive cloud
被动技术　passive technique
被动腔气溶胶分光计探测　passive cavity aerosol spectrometer probe
被动上滑锋　passive anafront
被动式云〔水〕收集器　passive cloud collector
被动外差辐射计　passive heterodyne radiometer
被动微波传感器　passive microwave sensor
被动微波遥感　passive microwave remote sensing

被动下滑锋　passive katafront
被动遥感方法　passive remote sensing technique
被动〔遥感〕系统　passive (remote sensing) system
被动引导　passive steering
被积函数【数】　integrand
本波拉公式　Bemporad's formula
本初子午线　basic meridian
本初子午线【天】　prime meridian
本底【物】　background
本底场　background field
本底场预报　background field prediction
本底辐射, 背景辐射　background radiation
本底〔观测〕站　background station
本底监测计划　background monitoring programme
本底监测器　background monitor
本底空气污染　background air pollution
本底能见度　background visibility
本底浓度【大气】　background concentration
本底水平　background level
本底污染　background pollution
本底污染观测【大气】　background pollution observation
本地闪电计数器　local lightning-flash counter
本地植物　native vegetation
本动速度　peculiar velocity
本格拉海流　Benguela Current
本轮【天】　epicycle
本影【物】　umbra
本影食【天】　umbral eclipse
本站订正基准面　revised local reference
本站气压【大气】　station pressure
本征函数　eigenfunction
本征向量【数】　eigenvector
本征载流子【电子】　intrinsic carrier
本征值【物】　eigenvalue

本征值问题　eigenvalue problem
苯　benzene
苯并〔a〕芘,3,4-苯并芘　benzo-a-pyrene
　（＝3,4-benzopyrene）
苯并芘排放　benzo-pyrene emission
崩溃　collapsing
崩溃云　collapsed cloud
崩塌【土壤】　collapse
逼近【数】　approximation
逼近方程　approximation equation
逼近误差【数】　approximation error
比【数】　ratio
比持水量　specific retention
比出水量　specific yield
比传导率　specific conductivity
比尔定律　Beer's law
比尔-朗伯〔吸收〕定律【物】　Beer-Lambert law
比焓　specific enthalpy
比降-面积法　slope area method
比较光谱【物】　comparison spectrum
比较模拟　comparative simulation
比较气象学　comparative meteorology
比克莱急流　Bickley jet
比例【数】　proportion
比流量　specific discharge
比密　specific density
比能　specific energy
比气体常数　specific gas constant
比热　specific heat
比热容　specific heat capacity
比容距平　anomaly of specific volume
比容量　specific capacity
比容异常　steric anomaly, thermosteric anomaly, specific-volume anomaly
比色法　colorimetric method
比色法【化】　colorimetry
比色计【化】　colorimeter
比熵　specific entropy
比湿【大气】　specific humidity
比湿线　specific humidity line

比特误码率　bit error rate
比体积【物】　specific volume
比通量　specific flux
比相【航海】　phase comparison
比重　specific weight
比重【物】　specific gravity
比重表　hydrometer
比重计　densitometer
彼得森发展方程　Petterssen development equation
彼得森发展理论　Petterssen development theory
必需氨基酸【生化】　essential amino acid
必要条件【数】　necessary condition
毕格云室　Bigg chamber
毕葛洛蒸发公式　Bigelow's evaporation formula
毕歇蒸发表　Piche evaporimeter
毕晓普光环【大气】　Bishop's corona
　（或 ring）
闭〔磁〕力线　closed (geomagnetic) field line
闭端保护温度表　protected thermometer
闭端颠倒海水温度表　protected reversing thermometer
闭合胞层积云　closed-cell stratocumulus
闭合等压线　closed isobar
闭合低压　closed low
闭合方案　closure scheme
闭合方程组　closed equations
闭合高压　closed high
闭合环流　recirculation, closed circulation, closed cell
闭〔合〕回路【物】　closed loop
闭合假设　closure assumption
闭合冷区（厚度场上的闭合低值区）　cold air pool
闭合流域　closed drainage
闭合模式　closed model
闭合盆地【地理】　closed basin

闭合〔曲〕线　closed curve
闭合体系　hierarchy of closures
闭合问题　closure problem
闭合系统【大气】　closed system
闭路电视　closed circuit television
闭塞　blocking【通信】
闭塞〔过程〕　seclusion
闭塞空气,不流通空气　dead air
闭形【物】　closed form
碧〔空〕　clear
碧空　blue sky
碧空【大气】　clear sky
蔽光层积云【大气】　stratocumulus opacus（Sc op）
蔽光层云　stratus opacus（St op）
蔽光高层云【大气】　altostratus opacus（As op）
蔽光高积云【大气】　altocumulus opacus（Ac op）
蔽光云　opacus（op）
蔽光云量　opaque sky cover
壁垒　barrier
壁区　wall region
壁效应　wall effect
避旱的　xerophobous
避雷板　lightning protector plate
避雷半径　radius of protection
避雷器　lightning arrester
避雷针　pointed lightning protector
避雷针【物】　lightning rod, lightning conductor
避雪栅　snow shed
边带【电子】　sideband
边带不稳定　side-band instability
边际不稳定〔性〕　marginal instability
边际稳定〔性〕　marginal stability
边际效应【大气】　marginal effect
边界,界限　boundary
边界波,界波　boundary wave
边界层　boundary layer
边界层抽吸作用　boundary layer pumping
〔边界层〕底层　radix layer
边界层顶　top of boundary layer
边界层顶向下/地表向上漫射　top-down/bottom-up diffusion
边界层动力学　boundary layer dynamics
边界层方程组　boundary layer equations
边界层分离　boundary layer separation
边界层滚〔动〕涡〔旋〕　boundary layer rolls
边界层急流【大气】　boundary layer jet stream
边界层廓线仪【大气】　boundary layer profiler（BLP）
边界层雷达【大气】　boundary layer radar
边界层模式　boundary layer model（BLM）
边界层气候【大气】　boundary layer climate
〔边界层〕气象探空仪　metrosonde
边界层气象学【大气】　boundary layer meteorology
边界层探空仪　boundary layer radiosonde
边界层形态学　boundary layer morphology
边界混合　boundary mixing
边界流　boundary currents
边界内波　boundary-generated internal wave
边界区　border zone
边界条件　boundary condition
边界稳定　marginally stable
边蚀干涉仪　limb occultation interferometer
边蚀光度计　limb occultation photometer
边蚀激光外差辐射测量法　limb occultation laser heterodyne radiometry
边缘波【大气】　edge wave
边缘地【地理】　periphery
边缘海【海洋】　marginal sea
边缘化【地理】　peripherization
边缘科学　interdiscipline, boundary

science, borderline science
边缘区　fringe region
边缘泉(冲积锥的)　border spring
边缘射线　marginal ray
边缘效应【电子】　side effect
边值条件【数】　boundary value condition
边值问题【数】　boundary value problem
编程【计】　programming
编码　encode, coding
编码方法【计】　encoding method
编译程序【计】　compiler
扁长对称陀螺　prolate symmetric top
扁长旋转路径　prolate cycloidal track
扁对称陀螺　oblate symmetric top
扁率　oblateness
扁平峰〔的〕　platykurtic
扁球　oblate spheroid
扁日　flattened sun
扁形电池【电子】,扣式电池　button cell
变分【数】　variation
变分法　variational method
变分方程【数】　variance equation
变分客观分析　variational objective analysis
变分同化　variational assimilation
变分原理【数】　variational principle
变分正模初始化　variational normal mode initialization
变分质量控制　variational quality control
变分最佳分析　variational optimization analysis
变高风　allohypsic wind
变高图　height-change chart
变高线　height-change line
变高线型　allohypsography
变旱　desiccation
变化图(或趋势图)　change chart (或 tendency chart)
变化中心　center of variation
变换　transformation
变换关系　transformation relation
变换欧拉平均系统　transformed Eulerian mean system
变换站　clearinghouse
变极调速【航海】　speed regulation by pole changing
变焦【物】　zooming
变焦镜头【物】　zoom
变量　variate
变量【物】　variable
变率　variability, rate of change
变频【电子】　frequency conversion
变频调速【航海】　speed regulation by frequency variation
变频器【电子】　frequency converter
变迁　vicissitude
变浅,浅水作用　shoaling
变弱　weakening
变射频〔无线电〕探空仪　variable-radio-frequency radiosonde
变声频〔无线电〕探空仪　variable-audio-frequency radiosonde
变数分离　separation of variables
变系数线性偏微分方程【数】　linear partial differential equation with variable coefficients
变形半径　radius of deformation, deformation radius
变形冰　deformed ice
变形场　field of deformation
变形场【大气】　deformation field
变形场云系　deformation zone cloud system
变形带　deformation zone
变形流　deformation flow
变形面　deformation plane
变形气压表　elastic barometer
变形梯度　deformation gradient
变形温度表【大气】　deformation thermometer
变形温度计　deformation thermograph
变形轴　deformation axis
变性　metamorphosis

变性气团　indifferent air mass
变性气团【大气】　transformed air mass
变性热带气旋　neutercane
变压〔的〕　allobaric
变压场【大气】　allobaric field
变压风【大气】　allobaric wind
变压计　variograph
变压区　allobar
变移高度　shifting level
变异波　freak wave
遍历过程【数】　ergodic process
遍历条件　ergodic condition
遍历系统　ergodic system
遍历性【数】　ergodicity
标称电压【电子】　nominal voltage
标称放电电流　nominal discharge current
标定功率【航海】　rated power, rated output
标度线　scale line
标度因子【化工】　scale factor
标绘位置　plotting position
标记【地信】　flag
标记变量【计】　token variable
标量　scalar quantity
标量【物】　scalar
标量场【数】　scalar field
标量磁强计　scalar magnetometer
标量函数　scalar function
标量化【数】　scalarization
标量积【数】　scalar product
标量计算机【计】　scalar computer
标枪探空仪　dartsonde
标识符【计】　identifier
标位　scalar potential
标值　scale value
标志污染物　designated pollutant
标准　standard
标准八英寸雨量计　standard eight-inch rain gauge
标准百叶箱　standard station screen

标准测试程序　standard test procedure
标准层　standard level
标准层【大气】　standard pressure level
标准传播　standard propagation
标准大气【大气】　standard atmosphere
标准大气表　Table of Standard Atmosphere
标准大气压【大气】　standard atmosphere pressure
标准大气状况　standard atmospheric condition
标准单位　standard unit
标准弹道层　standard artillery zone
标准弹道大气　standard ballistic atmosphere, standard artillery atmosphere
标准等压层【大气】　mandatory layer
标准等压面　mandatory surface, mandatory level
标准等压面【大气】　standard isobaric surface
标准等压面气压订正　reduction of pressure to a standard level
标准干湿表　standard psychrometer
标准观测时间　standard observation time
标准观测时间【大气】　standard time of observation
标准海水　standard sea water, normal water
标准化　standardization
标准化变量　standardized variate
标准降水指数　standard precipitation index
标准密度　standard density
标准密度高度【航空】　standard density altitude
标准目标　standard target
标准能见度　standard visibility
标准年　typical year
标准偏差【物】　standard deviation
标准平均值,多年平均值　normals

标准气候平均值【大气】(1901—1930年,1931—1960年等30年年平均值) climatological standard normals
标准气象百叶箱　standard meteorological cal screen
标准气压　standard pressure
标准气压表　standard barometer
标准气压表【大气】　normal barometer
标准气压高度　standard pressure altitude
标准取雪管　standard snow tube
标准日射强度计　standard pyrheliometer
标准日照计　standard sunshine recorder
标准深度　standard depth
标准时　standard time
标准视程　standard visual range
标准条件【电子】　reference condition
标准通风干湿表【大气】　standard aspirated psychrometer
标准图　normalization map, normal map, normal chart
标准位势米　standard geopotential meter
标准温度　standard temperature
标准温度表　standard thermometer
标准温压　standard temperature and pressure (STP), normal temperature and pressure (NTP)
标准物质【化】　reference material
标准误差【数】　standard error
标准仪器　standard instrument
标准雨量计【大气】　standard raingauge
标准预估洪水　standard project flood
标准圆分布　normal circular distribution
标准折射　standard refraction
标准蒸发器【大气】　standard pan
标准正交性　orthonormality
标准重力　standard gravity
标准子午线　standard meridian
标准作业温度　standard operative temperature

标准坐标　standard coordinate
飑【大气】　squall
飑暴　squall storm
飑锋【大气】　squall front
飑面　squall surface
飑线【大气】　squall line
飑线回波　squall line echoes
飑线雷暴　squall line thunderstorm
飑线群　cluster in squall line
飑线云簇　squall cluster
飑性风　squall wind
飑性天气　squally weather
飑云【大气】　squall cloud
表【天】　gnomon
表层海洋-低层大气研究(IGBP 计划) Surface Ocean-Lower Atmosphere Study (SOLAS)
表层流　surface current
表层蠕动　surficial creep
表层温度　surface skin temperature
表层雪崩　surface avalanche
表层滞水　surface detention
表〔格〕　table
表流　skin current
表面白霜　surface hoar
表面波　external wave
表面成分图像辐射仪　surface composition mapping radiometer (SCMR)
表面发射率　surface emissivity
表面活性剂薄膜　surfactant film
表面力　surface force
表面摩擦　skin-friction
表面摩擦系数　skin-friction coefficient, coefficient of skin friction
表面摩擦曳力　frictional skin drag
表面能【物】　surface energy
表面凝结　surface condensation
表面强度　surface intensity
表面水温　bucket temperature
表面水温表　bucket thermometer
表面速度　surface velocity

表面拖曳　surface drag, skin drag
表面温度　skin-surface temperature
表面曳引力　surface traction
表面应力【物】　surface stress
表面张力【物】　surface tension
表面张力波【物】　capillary wave
表面重力波　surface gravity wave
表面自由能【物】　surface free energy
宾州〔大学〕/美国国家大气研究中心中尺度模式　Penn State/NCAR Mesoscale Model（或 PSU/NCAR Mesoscale Model）
滨海气候【大气】　coastal climate
滨海气象学　shore meteorology
冰　ice
冰岸水道　shore lead
冰凹湾　ice bight, ice bay
冰靶　ice target
冰坝　ice jam
冰雹测站网　hail-observing network
冰雹沉降区　hail fallout zone
〔冰〕雹【大气】　hail
冰雹路径　hail path
冰雹生成区【大气】　hail generation zone
冰雹研究实验　hail research experiment
〔冰〕雹云【大气】　hail cloud
冰雹增长　hail growth
冰雹增长区　hail growth zone
冰暴【大气】　ice storm
冰崩【航海】　ice avalanche
冰层　ice layer
冰层消失　ending of ice sheet
冰川　ice stream
冰川【地质】　glacier
冰川边缘　periglacial
冰川变迁　glacier variation
冰川冰　glacier ice
〔冰川〕冰山　glacier berg
冰川波动　glacial fluctuation
冰川补给　glacial replenishment
冰川擦痕　moutonnee

冰川槽　glacial trough
冰川地质学【地理】　glacial geology
冰川反气旋　glacial anticyclone
冰川反气旋学说　glacial anticyclone theory
冰川放水　glacier discharge
冰川风　glacier wind
冰川风　firn wind（＝glacier wind）
冰川风【大气】　glacier breeze
冰川锋　glacier front
冰川河流【地理】　glacial stream
冰川洪水　glacier flood
冰川湖突发性洪水　Glacier Lake Outbreak Flood（GLOF）
冰川化　glacierization
冰川〔积融〕平衡线　equilibrium line of glacier
冰川极限　glaciation limit
冰川记录　glacial record
冰川间冰期　glacial interstade
冰川盆地　glacial basin
冰川平原　glacier plain
冰川瀑布　glacier fall
冰川期雪线　glacial stage snowline
冰川气候【大气】　glacioclimate
冰川气候学　glacioclimatology
冰川前沿　ice front
冰川舌　glacier tongue
冰川世　glacial epoch
冰川退缩　glacial recession
冰川退缩【地理】　deglaciation
冰川纹泥　glacial varves
冰川物质平衡　glacier mass balance
冰川形迹　evidence of glaciation
冰川学【地理】　glaciology
冰川循环　glacial cycle
冰川作用【地理】　glaciation
冰川作用程度　glaciation level
冰带　ice ribbon
冰岛　ice island
冰岛低压【大气】　Icelandic low
冰的平均边缘线　average limit of ice

冰底　ice keel
冰点【大气】　ice point
冰点温差　freezing point depression
冰点线　freezing point line
〔冰〕冻带　frost belt
冰冻等时线　isopectics
冰冻风化　congelifraction
冰冻期【航海】　ice period
冰冻圈　cryosphere
冰冻学【大气】　cryology
冰冻作用　frost action
冰斗冰川　cirque glacier
冰反照率　ice albedo
冰-反照率反馈　ice-albedo feedback
冰盖　ice cap
冰盖【地理】　ice sheet
冰盖气候　ice-cap climate
冰海反馈　ice-ocean feedback
冰核　ice nuclei
冰核【大气】　ice nucleus
冰核带电　electrification of ice nucleus
冰核计数器　ice nuclei counter, IN counter
冰核理论　ice nucleus theory
冰后期　postglacial age
冰后期【地质】　post-glacial period
冰后期气候【大气】　post-glacial climate
冰厚【海洋】　ice thickness
冰花　ice flower
冰积年　budget year
冰积〔作用〕【大气】　accumulation
冰极　ice pole
冰架　ice shelf
冰〔间〕湖　ice clearing
冰间水道　lead
冰间水道【海洋】　lead land
冰角　rams
冰脚　ice foot
冰结构　ice structure
冰结物　ice hydrometeor
冰晶【大气】　ice crystal
冰晶繁生　ice multiplication

冰晶核　ice crystal nucleus
冰晶霾　ice-crystal haze
冰晶浓度　ice crystal concentration
冰〔晶破〕碎　fragmentation
冰晶取向　orientation of ice crystal
冰晶雾　ice crystal fog
冰晶效应　ice crystal effect
冰晶学说　ice crystal theory
冰晶云　ice crystal cloud
冰晶增长习性　growth habit of ice crystal
冰晶柱　ice prism
冰景　icescape
冰卡表　ice calorimeter
冰壳,冰皮,薄层冰　ice crust
冰壳【海洋】　ice rind, glass ice
冰壳害　ice-mantling injury
冰冷风　icy wind, ice wind
冰粒　ice particle, grains of ice
冰粒尺度实时分析系统　real-time ice particle size analysis system
冰凌　ice-bearing cuttent, sleet and ice
冰流　ice flow
冰面饱和　ice-saturation
冰面饱和混合比【大气】　saturation mixing ratio with respect to ice (γ_i)
冰面过饱和　supersaturation with respect to ice
冰面开裂　promojna
冰面相对湿度　relative humidity with respect to ice
冰面蒸发　evaporation from ice
冰漠　ice desert
冰膜　film crust
冰膜　ice coating
冰摩擦〔学〕说　ice friction theory
冰瀑　ice fall, ice cascade
冰期　glacial period (phase)
冰期【大气】　ice age
冰期【地质】　glacial stage
冰期干燥性　ice age aridity
冰期间冰期循环　interglacial and glacial

cycle, glacial-interglacial cycle
冰期结束期【大气】 glacial termination
冰碛【物】 glacial drift
冰情 ice regime phase
冰情概况 ice condition summary
冰情警报【航海】 ice warning
冰情预报 ice forecast
冰情侦察 ice reconnaissance
冰丘 ice rise, icing mound, ice mound
冰丘冰 hummocked ice
冰丘,冻丘 ground ice mound
冰球温度 ice-bulb temperature
冰区界限线【航海】 ice boundary
冰泉 flood icing
冰日 ice day
冰融 undermelting
冰山 berg
冰山【地理】 iceberg
冰山崩解 iceberg calving
冰上融水坑 puddle, pucker point
冰舌 ice tongue
冰升华 sublimation of ice
冰室状态 icehouse state
冰蚀作用【地质】 glacial erosion
冰室气候 icehouse climate
冰水混合比 ice water mixing ratio
冰〔水〕混合比 ice mixing ratio
冰水混合云 mixed cloud
冰水混合云【大气】 ice-water mixed cloud
冰笋 ice stalagmite
冰塔 serac
冰条纹 ice fringe
冰透镜体【地理】 ice lens
冰图 ice atlas
冰丸 pellet
冰丸【大气】 ice pellet
冰雾 rime fog
冰雾【大气】 ice fog
冰隙 polynya
冰隙白霜 crevasse hoar

冰隙水 water bleeding
冰限 ice limit
冰线 ice line
冰相 ice phase
冰相参数化 ice phase parameterization
冰消地面回升 glacial rebound
冰消〔作用〕,冰(雪)面消融,融蚀 ablation
冰效应 ice effect
冰楔【地理】 ice wedge
冰芯 ice core
冰穴 ice cave, cryoconite hole
冰雪反照率 ice and snow albedo
冰雪反照率-温度反馈 ice and snow albedo-temperature feedback
冰雪浮游生物 cryoplankton, frost plants
〔冰〕雪壳 snow crust
冰雪气候 ice climate
冰雪气候【大气】 nival climate
冰雪区【大气】 cryochore
冰雪圈【大气】 cryosphere
冰雪时期 cryogenic period
冰雪硬度器 ramsonde
冰雪与永久冻土研究所(美国) Snow, Ice and Permafrost Research Establishment (SIPRE)
冰雪植物 cryophyte
冰延层 white band
冰映光 iceblink
冰涌 ice surge
冰羽 ice feathers
冰原 bare ice field
冰原【地理】 ice field
冰原冰 field ice
冰原高压 glacial high
冰原气候 frost climate
冰缘期 periglacial stage
冰缘气候 periglacial climate
冰缘线【海洋】 ice edge
冰云【大气】 ice cloud

冰照云光　ice sky
冰针　ice spicule, spicule, ice needle
冰柱　icicle
冰锥　ice pyramid
冰渍　moraine
冰渍阶地【地质】　moraine terrace
冰钻　ice drill
丙酮　acetone
丙烷(C_3H_8)　propane
丙烯(C_3H_6)　propene
丙烯醛　acrolein
饼形图【计】　pie chart
饼状冰　pancake ice
饼〔状〕冰　ice cake, cake ice
并合（云滴的）【大气】　coalescence
并合过程　coalescence process
并合系数【大气】　coalescence efficiency
并合系数　coalescence coefficient
并合作用　coalescence effect
并联【物】　connection in parallel
并行处理　parallel processing
并行处理机操作系统【计】　parallel processor operating system
并行度【计】　degree of parallelism
并行计算【计】　parallel computing
并行计算机【计】　parallel computer
并行数据库【计】　parallel database
病理生物气象学　pathological biometeorology
病态方程【数】　ill-conditioned equation
拨定装置　setting
波包【物】　wave packet
波包孤立子　envelope soliton
波包函数　envelope function
波包激波　envelope shock
波包穴　envelope hole
波长【物】　wavelength
波传播　wave propagation
波传播实验〔研究〕室（NOAA）　Wave Propagation Laboratory（WPL，NOAA）
波导　ducting, duct

波导【物】　waveguide
波动　wave motion, undulation
波动方程【物】　wave equation
波动理论　wave theory
波动模拟　wave modeling
波动频率　wave frequency
波动坡度　wave steepness
波动〔强迫〕作用　wave forcing
波动性气旋　wave cyclone
波动周期　wave period
波段　wave range
波峰　crest of wave
波峰【海洋】　wave ridge, wave crest
波幅增大　wave amplification
波高　wave height
波谷【物】　wave trough
波耗散　dissipation of waves
波和平均气流的交互作用　wave-mean flow interaction
波和纬向气流的交互作用　wave-zonal flow interaction
波活动性　wave activity
波击　surf beat
波间交互作用　wave-wave interaction, wave interaction
波减弱　wave attenuation
波扩散　wave dispersion
波廓线　wave profile
波浪【航海】　wave
波浪计　wave recorder
波〔浪破〕碎　wave breaking
波〔浪气〕候　wave climate
波浪形扩散，环状扩展　looping
波浪学　kymatology
波浪预报　wave forecasting
波浪运动　billow wave
波列【物】　wave train
波林式空盒气压表　Paulin aneroid
波龄【海洋】　wave age
波流交互作用　wave-current interaction
波能密度　wave energy density

波力计　wave dynamometer
波能通量　wave energy flux
波平均　wave average
波谱　wave spectrum
波谱理论　spectrum theory
波谱能量扩展　spectrum-energy spread
波谱折叠　spectrum folding
〔波〕谱阻塞　spectral blocking
波前【物】　wave front
波前法　wave-front method
波群　wave ensemble
波群【物】　wave group
波射线　wave ray
波生　wave generation
波矢　wave vector
波束　beam
波束充塞系数【大气】　beam filling coefficient
波束大小　beam size
波束加宽　beam spread, beam broadening
波束角　beam angle
波束漂移　beam wander
波束形成　beam-forming
波束转动法(测风)　beam swinging
波数【物】　wave number
波数滤波　wavenumber filtering
波数-频率滤波　wavenumber-frequency filtering
波数-频率谱　wave number-frequency spectrum
波数谱　wavenumber spectrum
波数域　wave-number regime, wave-number domain
波衰减　decay of waves
波瞬态　wave transience
波斯湾海水　Persian Gulf Water
波速　wave celerity, wave speed, wave velocity, celerity
波特【计】(发报速率单位)　baud
波特〔率〕　baud rate

波纹管　bellows
波系　wave system
波消　wave dissipation
波形　waveform
波形分析　waveform analysis
波形记录仪　waveform recorder
波型扰动　wave-type disturbance
波型线状回波　line echo wave pattern (LEWP)
波衍射　wave diffraction
波伊思相机　Boys camera
波源　wave source
波涨/落　wave set-up/set-down
波阵面　phase front
波振幅　wave amplitude
波致第二类条件不稳定　wave-CISK
波状　undulatus (un)
波状层积云　stratocumulus undulatus (Sc un)
波状层云　stratus undulatus (St un)
波状潮　undular bore
波状高层云　altostratus undulatus
波状高积云　altocumulus undulatus
波状积云　cumulus undulatus
波状卷层云　cirrostratus undulatus
波状卷积云　cirrocumulus undulatus
波状卷云　billow cirrus clouds
波状扰动　wave disturbance
波状湍流　wave turbulence
波状云【大气】　wave cloud
波阻　wave drag
波阻塞　wave blocking
波作用量　wave action
玻恩近似【物】　Born approximation
玻尔原子模型【物】　Bohr atom model
玻尔兹曼常量【物】　Boltzmann constant (=Boltzmann's constant)
玻耳兹曼方程【数】　Boltzmann equation
玻管液体温度表　liquid-in-glass thermometer

玻管液体温度计　liquid-glass thermograph
玻璃水银干湿表　mercury-in-glass psychrometer
玻璃水银温度表　mercury-in-glass thermometer
玻璃纤维测湿元件(测土壤含水量)　fibreglass unit
玻林-艾勒拉德暖期,BA 暖期(14700—12700 BP)　Bolling-Allerod
玻色-爱因斯坦分布　Bose-Einsten distribution
玻氏晕环　Bottlinger's rings
玻意耳定律【物】　Boyle law(＝Boyle's law)
玻意耳-马略特定律　Boyle-Mariotte law
玻印亭矢量【物】　Poynting vector
剥蚀面【地理】　denudation surface
剥蚀〔作用〕【地理】　denudation
播馈机制　seeder-feeder
播撒　seeding
播撒位置　seeding location
播撒物质　seeding material
播撒装置　seeding device
播云【大气】　cloud seeding
播云剂【大气】　cloud-seeding agent
播种季〔节〕　season of sowing
播种时期　seed-time
播种云　seeder cloud
伯格数【大气】　Burger number
伯格斯方程　Burgers' equation
伯格斯矢〔量〕【物】　Burgers vector
伯格涡旋　Burger's vortex
伯克兰电流【地】　Birkeland currents
伯纳耳-否勒定则〔物理〕　Bernal-Fowler rules
伯努利定理【大气】　Bernoulli's theorem
伯努利定律　Bernoulli's law
伯努利方程【物】　Bernoulli equation
伯努利流函数　Bernoulli stream function
伯努利数【数】　Bernoulli number
伯努利效应　Bernoulli effect
伯森西风带　Berson winds（或 westerlies）
泊　poise
泊松常数　Poisson constant
泊松方程【数】　Poisson equation
泊松分布　Poisson distribution
泊松随机变量　Poisson random variable
泊肃叶定律【物】　Poiseuille law
泊肃叶-哈根定律　Poiseuille-Hagen law
泊肃叶流　Poiseuille flow
铂丝电阻全温探头　platinum wire resistance total temperature probe
铂丝电阻温度表　platinum resistance thermometer（PRT）
铂丝温度表　platinum wire thermometer
博尔德大气观象台(美国)　Boulder Atmospheric Observatory（BAO）
博伊登〔不稳定〕指数　Boyden index
箔条测风法　chaff wind technique
箔丝播撒【大气】　chaff seeding
薄壁组织　parenchyma
薄层　shallow layer，film
薄浆雾　larry
薄膜电容〔测湿〕感应元件　capacitive thin-film sensor
薄膜滤纸　membrane filter
薄膜湿度传感器　thin film humidity sensor
薄膜水　pellicular water，film water
〔薄膜〕吸附水　adhesive water
薄幕层云　stratus nebulosus (St neb)
薄幕卷层云【大气】　cirrostratus nebulosus(Cs neb)
薄幕卷积云　cirrocumulus nebulosus
薄幕卷云　cirro-nebula
薄幕云　nebulosus(neb)
薄幕状高积云　altocumulus nebulosus
薄片法【大气】　slice method

薄屏理论　thin screen thoery
补偿层　layer of compensation
补偿函数　penalty function
补偿气流　compensation current
补偿气温表　compensated air thermometer
补偿气压　compensation pressure
补偿深度【地】　depth of compensation
补偿式定标气压表　compensated scale barometer
补偿式绝对日射表　compensating pyrheliometer
补偿式全辐射表　compensated pyrradiometer
补偿系统　back-off system
补偿作用　compensation
补充船舶站　supplementary ship station
补充观测【大气】　supplementary observation
补充流　supply current
补充陆地站　supplementary land station
补充气象台　supplementary meteorological office (SMO)
补允〔天气〕预报【大气】　supplementary 〔weather〕forecast
补充站　supplementary station
补充资料　ephemeral data
补码【计】　complement
补助仪器　secondary instrument
捕获　capture
捕获截面　capture cross-section
捕获率　capture rate
捕获器　collector
捕获系数【大气】　collection efficiency
不变　invariable
不变量嵌入法　invariant imbedding method
不变式【数】　invariant
不变纬度　invariant latitude
不变性原理　principle of invariance
不产流区　non-contributing area
不纯性　impurity
不等动能取样法　anisokinetic sampling
不等式【数】　inequality
不等压的　anisobaric
不定常【数】　time dependent
不定风【大气】　variable wind, shifting wind
不定风带　zone of variable wind
不定积分【数】　indefinite integral
不定云幕　indefinite ceiling
不定转移系统　intransitive system
不动点,定点【数】　fixed point
不冻地　tabetisol
不冻海　open sea
不冻下层　subgelisol
不对称波结构　asymmetric wave structure
不对称参数　asymmetry parameter
不对称环流型　asymmetric circulation pattern
不对称陀螺【化】　asymmetrical top
不对称性　asymmetry
不对称因子　asymmetry factor
不规则冰晶　irregular crystal
不规则路径　erratic path
不规则性　erratic
不活跃锋　passive front, inactive front
不浸润流体　nonwetting liquid
不精确性　imprecision
不均匀混合　inhomogeneous mixing
不均匀加热　differential heating
不均匀介质　inhomogeneous medium
不均匀散射介质　inhomogeneous scattering medium
不均匀随机介质　inhomogeneous random medium
不均匀性　nonuniformity
不可调虹吸式水银气压表　non-adjustable siphon mercury barometer
不可分问题　nonseparable problem
不可见〔的〕辐射　invisible radiation
不可见卷云　invisible cirrus
不可逆波【化】　irreversible wave

不可逆反应【化】 irreversible reaction
不可逆过程【物】 irreversible process
不可逆绝热过程【大气】 irreversible adiabatic process
不可逆温室效应 irreversible greenhouse effect
不可逆性 nonreversibility
不〔可〕溶混位移 immiscible displacement
不可压缩大气 incompressible atmosphere
不可压缩流 incompressible flow
不可压缩流体【大气】 incompressible fluid
不可压缩〔性〕条件 incompressibility condition
不可压缩性【物】 incompressibility
不可压缩运动 incompressible motion
不可用位能 unavailable potential energy
不可预报性 unpredictability
不可预防损失 unpredictable loss
不可约分数【数】 irreducible fraction
不可约性 irreducibility
不利气象条件 adverse weather condition
不利天气 adverse weather
不连续层 layer of discontinuity
不连续带 zone of discontinuity
不连续点【数】 discontinuity point
不连续阶 order of discontinuity
不连续面 surface of discontinuity
不连续闪电 staccato lightning flash
不连续湍流 discontinuous turbulence
不连续线 line of discontinuity
不连续〔性〕【大气】 discontinuity
不良能见度 poor visibility
不灭性 indestructibility
不明飞行物 unidentified flying object (UFO)
不模糊距离〔时间〕区间 unambiguous range interval
不模糊速度【大气】 unambiguous velocity
不模糊速度〔时间〕间隔 unambiguous velocity interval
不平衡【力】 unbalance
不齐性,不均匀性,异质 heterogeneity
不确定度【物】 uncertainty
不确定决策【数】 decision under uncertainty
不确定性 indeterminacy
不确定原理【物】 uncertainty principle
不确切性 inexactness
不适定问题【数】 ill-posed problem
不适定性 ill-posedness
不适指数【大气】 discomfort index
不舒适感 discomfort
不舒适区 discomfort zone
不同模式双子积分〔实验〕 fraternal twin integrations
不透辐射热〔的〕 athermous
不透辐射热性 athermancy
不透明冰 opaque ice
不透明大气 opaque atmosphere
不透明〔的〕 opaque
不透明反射体 opaque reflector
不透明近似 opaque approximation
不透明性 opacity
不透水层 confining bed (layer, stratum)
不透水层【地质】 aquifuge
不完全贯穿井 partially penetrating well
不完全预报 semiprognosis, imperfect forecast
不稳定波【大气】 unstable wave
不稳定〔波〕节 unstable node
不稳定层结 unstable stratification
不稳定常数【化】 instability constant
不稳定大气 unstable atmosphere
不稳定〔的〕 labile
不稳定度 unstability, 2lability
不稳定〔度〕图 instability chart
不稳定河床 unstable channel

不稳定空气【大气】 unstable air
不稳定流型 unstable manifold
不稳定能量 instability energy
不稳定平衡 unstable equilibrium
不稳定平衡点 unstable equilibrium point
不稳定气团【大气】 unstable air mass
不稳定气旋波 unstable cyclone wave
不稳定时间序列 unstable time series
不稳定水流 unsteady flow
不稳定天气 unsettled weather
不稳定条件 unstable condition
不稳定系统 unstable system
不稳定线【大气】 instability line
不稳定星【天】 non-stable star
不稳定性【物】 instability
不稳定因子 instabilizing factor
不稳定运动 unstable motion
不稳定阵雨 instability shower
不稳定振荡 unstable oscillation
不稳定指数 instability index
不相关样本【数】 uncorrelated samples
不相容事件【数】 mutually exclusive events
不一致性 inconsistency
不准确度 inaccuracy
不足预报 underforecast
布德科数 Budyko number
布尔代数【计】 Boolean algebra
布尔运算【计】 Boolean operation
布格定律 Bouguer's law
布格-朗伯定律 Bouguer-Lambert law
布格异常【测】 Bouguer anomaly
布格晕 Bouguer's halo
布拉风气候学 bora climatology
布拉格散射 Bragg scattering
布拉格衍射【物】 Bragg diffraction
布拉雾(布拉风引起的浓雾) bora fog
布赖尔评分法 Brier score (BSCR)
布朗扩散 Brownian diffusion
布朗宁右转强风暴模型 Browning's severe right moving storm model

布朗旋转 Brownian rotation
布朗运动【物】 Brownian motion
布朗噪声 Brownian noise
布雷迪阵列湿敏元件 Brady array humidity sensor
布鲁尔起泡器(库伦法测臭氧用) Brewer bubbler
布吕克纳周期 Brückner cycle
布伦特-道格拉斯等变压风 Brunt-Douglas isallobaric wind
布伦特-维赛拉频率【大气】 Brunt-Vaisala frequency
布伦特-维赛拉周期 Brunt-Vaisala period
布罗肯宝光 Brocken bow
布罗肯幽灵 Brocken specter
布儒斯特窗【物】 Brewster window
布儒斯特角【物】 Brewster angle
布儒斯特〔中性〕点 Brewster point
布西内斯克方程【数】 Boussinesq equation
布西内斯克近似【大气】 Boussinesq approximation
布西内斯克流体 Boussinesq fluid
布西内斯克数 Boussinesq number
布辛格-戴尔关系式 Businger-Dyer relationship
步长【数】 step length, step size, step width
步进电〔动〕机【电子】 stepped motor
部分放大〔云〕图 enlarged partial-disc picture
部分 component
部分锢囚〔的〕 partly occluded
部分历时序列 partial-duration series
部分偏振波 partially polarized wave
部分频率 partial frequency
部分欠缺年轮 locally absent ring, partial ring
部分天空不明 partial obscuration
部分相干 partial coherence
部分真空 partial vacuum

C

材料科学【物】 material science
采暖度日【大气】 heating degree-day
采样方法 sampling method
采样间隔【大气】 sampling interval
采样频率【计】 sampling frequency
采样站 sampling station
采云枪 cloud gun
彩色参数 color parameter
彩色合成 color composite
彩色空间 color space
彩色信息 color information
彩色照片 photochrome
彩色自旋扫描摄云机 multi-color spin scan cloud camera（MSSCC），color spin scan cloud camera
参比接点【电工】 reference junction
参考层 reference level
参考大气【大气】 reference atmosphere
参考电路【电子】 reference circuit
参考气压 reference pressure
参考声功率【机械】 reference sound power
参考声压【机械】 reference sound pressure
参考椭球体 reference ellipsoid
参考温度 reference temperature
参考文献 reference
参考系【物】 reference frame
参考资料 reference data
参考作物蒸发〔量〕 reference crop evaporation
参量放大器 paramp
参量振荡器 parametric oscillator
参数 parameter
参数方程 parametric equation
参数放大器【电子】 parametric amplifier

参数估计【化】 parameter estimation
参数化【大气】 parameterization
参数化模式 parameterized model
参数化气旋波 parameterized cyclone wave
参数化湍流 parameterized turbulence
参数化微物理学 parameterized microphysics
参数化涡度输送 parameterized vorticity transfer
参数化涡度通量 parameterized vorticity flux
参数检验【化】 parameter test
参数模式 parametric model
参照标准 reference standard
参照信息 reference information
残差【数】 residual
残差平方和【数】 residual sum of squares
残冻层 intergelisol，pereletok
残留边带 vestigial sideband（或 band）
残留层【大气】 residual layer
残留项 discrepancy
残留影像 after-image
残留〔质〕量曲线 residual mass curve
残余湍流 zombie turbulence
舱内大气 cabin atmosphere
舱内红外傅里叶光谱仪 on-board infrared Fourier spectrometer
舱上处理 on-board processing
操作 operation
操作码 operation code
操作顺序图 operating sequence diagram
操作台，控制台 console
操作温度 operative temperature
操作系统【计】 operating system
操作指令 operation directive
槽脊图 Hovmoller diagram

槽生　troughing
槽〔式〕气旋　trough cyclone
槽线　axis of trough
槽线【大气】　trough line
槽线走向　trough line orientation
草本【植】　herb
草波　grass
草地　grassy soil
草地农业【农】　grassland farming
草甸【地理】　meadow
草面温度　grass surface temperature
草生区　poechore
草图　outline map
草温　grass temperature
草温表【大气】　grass thermometer
草原　grassland
草原气候【大气】　prairie climate
草原气候【大气】　steppe climate
草原气候亚类【大气】　grassland climate
草原土壤　steppe soil
侧边界　lateral boundary
侧边界条件　lateral boundary condition
侧风　side wind, lateral wind
侧风【大气】　cross-wind
侧风传感器　crosswind sensor
侧风阵性　crosswind gustiness
侧滚角　angle of roll
侧切晕弧　lateral tangent arc
侧闪电　side flash
侧视雷达【电子】　side-looking radar
侧视图　side view
侧现蜃景　lateral mirage
侧向不均匀性　lateral heterogeneity
侧向混合【大气】　lateral mixing
侧向卷入　lateral entrainment
侧向扩散　lateral diffusion
侧向来水　lateral inflow
侧向摩擦　lateral friction
侧向耦合　lateral coupling
侧向折射　lateral refraction

侧翼阶梯云　flanking line
侧音　sidetone
侧音测距【电子】　sidetones ranging
侧阵风　lateral gustiness
测雹板【大气】　hailpad
测冰器　ice apparatus
测波杆　wave staff, wave pole
测波计　undograph
测波仪　ondometer
测不准原理　principle of indeterminism
测潮表　tidalmeter
测潮器基准　tide gauge benchmark
测潮仪【海洋】　tide gauge
测尘器　conimeter, konimeter
测地网络　geodesic grid
测地线　geodesic line
测点　measurement point
测定　determination
测定系统　sensing system
测定云高　measured ceiling
测氡仪　emanometer
测风报告　upper-wind report
测风法　anemometry
测风杆　anemometer mast
测风管　pilot tube
测风绘图板【大气】　pilot-balloon plotting board
测风经纬仪　pilot balloon theodolite, balloon theodolite
测风经纬仪【大气】　aerological theodolite
测风雷达　wind radar
测风雷达【大气】　wind-finding radar
测风气球　pilot balloon
测风气球观测【大气】　pilot balloon observation
测风气球自记经纬仪　pilot balloon self-recording theodolite
测风器　anemoscope
测风塔　anemometer tower
测风微波散射计　microwave wind scat-

terometer
测风学　anemology
测风原始记录　raw data of wind-finding record
测高法　hypsometry
测高方程　hypsometric equation
测高公式　hypsometric formula
测高雷达　height-finding radar
测高平均气温　barometric mean temperature
测高术　altimetry
测高仪　elevation finder
测高温法　pyrometry
测光系统【天】　photometric system
测海水温吊桶　marine bucket
测绘学【测】　surveying and mapping, geomatics
测角术定向　goniometry direction finding
测井【地】　well logging
测距经纬仪【测】　distance theodolite
测距仪【航海】　range finder
测量　measure
测量【物】　measurement
测量单体　measurement cell
测量基准面　survey datum
测量结果　result of measurement
测量误差　measurement errors
测量学【测】　surveying
测量元件　measuring element
测流表　current meter
测流断面　discharge section line
测流杆　current pole
测露表【大气】　drosometer
测霾表　hazemeter
测深锤　sounding weight
测深杆　sounding pole
测湿法　hygrometry
测湿公式【大气】　psychrometric formula
测湿器　moisture apparatus
测湿学　psychrometry
测试　test

测试鉴定实验室　test and evaluation laboratory（T&EL）
测霜仪　pagoscope
测〔水〕深法　bathymetry
测微表　micrometer
测温法　thermography
测温浮标　thermometer float
测温辐射仪　radiometric thermosonde
测温计　thermodetector
测温链　dragon's tail
测温探针　thermoprobe
测温学　thermometry
测雾仪　fog detector
测向器　goniometer
测向系统【电子】　direction-finding system
测向装置,测角系统　goniometer system
测雪　snow survey
测雪板　snow measuring plate, snow board
测雪架　snow tower
测雪路线　snow course
测雪桩　snow stake
测压管液面　piezometric surface
测压（气压）管　barometer tube
测压器　manometer
测压水头　piezometric head
测雨雷达　precipitation-measuring radar, rain-measuring radar, nimbus radar
测雨器　ombroscope
测雨学　ombrology
测云法　nephometry
测云镜　mirror nephoscope, cloud reflector, cloud mirror
测云雷达【大气】　cloud-detection radar, cloud radar
测云仪【大气】　nephoscope, nepholoscope, cloud nephoscope
测站表征　site characterization
测站天气预报　station weather forecast
测站网　observational network
测站位置　site of a station

测针　point gauge
测震学【地】　seismometry
层　layer
层次模型【地信】　hierarchical model
层次序列【地信】　hierarchical sequence
层叠冰　superimposed ice
层厚　layer depth
层积云【大气】　stratocumulus（Sc）
层结　stratification
层结大气　stratified atmosphere
层结曲线　stratification curve
层流　lamellar flow
层流边界层【大气】　laminar boundary layer
层析成像技术　tomography
层云　layer cloud
层云【大气】　stratus（St）
层状　stratiform
层状（云）　stratiformis（str）
层状层积云　stratocumulus stratiformis（Sc str）
层状回波　stratified echoes
层状回波【大气】　layered echo
层状流体　stratified fluid
层状流体动力学　stratified fluid dynamics
〔层状〕平滑冰　level ice
层状平行平面介质　layered parallel plane medium
层状切变流　stratified shear flow
层状雨云　pallium, pallio-nimbus
层状云　sheet cloud
层状云【大气】　stratiform cloud
层状〔云〕降水区　stratiform precipitation area
叉状闪电【大气】　forked lightning
差错方程【数】　error equation
差错率【数】　error rate
差错校验【数】　error check
差动弹道风【大气】　differential ballistic wind
差动风【大气】　differential wind
差动平流　differential advection
差动热平流　differential thermal advection
差动自记测深仪　differential bathygraph
差分方程【数】　difference equation
差分格式　difference analogue
差分格式【大气】　difference scheme
差分模式【大气】　finite difference model
差分全球定位系统【航海】　differential GPS
差分算子【数】　difference operator
差频【物】　difference frequency
差异显著平准【农】　level of significance of difference
差值分析　differential analysis
插入【计】　pulg-in
插入〔流〕　intromission
插入管（测土壤含水量用）　access tube
插值【数】　interpolation
插值法【大气】　interpolation
插值公式【数】　interpolation formula
查理定律【物】　Charles law
查理-盖吕萨克定律　Charles-Gay-Lussac law
查尼-德拉津定理　Charner-Drazin theorem
查尼模　Charney mode
查尼-斯特恩定理　Charner-Stern theorem
查诺克关系式（流体力学中）　Charnock's relation
查普曼层【地】　Chapman layer
查普曼机制【大气】　Chapman mechanism
查普曼理论　Chapman theory
查普曼区　Chapman region
查普曼循环　Chapman cycle
查普斯带　Chappuis band
查算表　lookup table
查询标志　indication of a request
产量预报（方法）　yield forecasting

产量预测　yield prediction
产生　out-growth
产生率【物】　generation rate
产生式规则【计】　production rule
产生式系统【计】　production system
长波　long wave
长波槽【大气】　long-wave trough
长波调整【大气】　adjustment of long wave
长波反馈　longwave feedback
长波辐射【大气】　long-wave radiation
长波公式　long-wave formula
长波脊　long-wave ridge
长波截止　long-wave cutoff
长波近似　longwave approximation
长波气旋　longwave cyclone
长波云辐射强迫〔作用〕　long-wave cloud radiative forcing (LWCF)
长程力【化】　long-range force
〔长程〕直击大风　straight-line gale
长法　long method
长方形【数】　rectangle
长方形网格　rectangular grid
长放电　long discharge
长峰波(海面波)　long-crested wave
长光程　long optical path
长光程吸收　long-path absorption
长火花　long spark
长江　Yangtze River, Changjiang River
长距离环境空气监测　long-path ambient air monitoring
长距离输送　long-range transport
长狂涌【海洋】　logn heavy swell
长浪　ground swell
长列〔型〕天电　long-train atmospherics
长曝光调制传递函数　long-exposure modulation transfer function
长期变化　secular variation, secular change
长期冻结　paleocrystic
长期降水记录器　long duration precipitation recorder
长期模式　long-term model
长期平均【大气】　period average, period mean
长期气候变化　long-term climatic change
长期气候趋势【大气】　secular trend in climate
长期趋势　secular trend
长期水文预报　long-term hydrological forecast
长期〔天气〕预报【大气】　long-range 〔weather〕forecast
长期预报　long-term forecast
长期预报研究计划　Programme on Long-range Forecasting Research(PLFR)
长轻涌【海洋】　long low swell
长球〔体〕　prolate spheroid
长日照植物　long-day plant
长涌　blind rollers
长中涌【海洋】　long moderate swell
长周期振动频率　phugoid frequency
肠膜　goldbeater's skin
肠膜湿度表　goldbeater's skin hygrometer
尝试法【物】　trial-and-error method
常参数线性系统　constant parameter linear system
常定低压　permanent low
常定环流　permanent circulation
常定速度　steady velocity
常定运动　stationary motion
常规　routine
常规动力预报　ordinary dynamical forecast (ODF)
常规观测　regular observation
常规观测【大气】　conventional observation
常规雷达【大气】　incoherent radar, conventional radar
常规探空仪　routine sonde
常规天气〔图〕报告　regular synoptic report
常规无线电通信【航海】　general radio communication

常流河 perennial stream
常绿阔叶林【林】 evergreen broad-leaved forest
常年 ordinary year, average year
常平架雨量计器 gimbal mounted raingauge
常数 constant
常数变易法【数】 method of variation of constant
f 常数近似 constant f approximation
f 常数模式 constant f model
常态峰 mesokurtosis
常态峰的 mesokurtic
常态通气 normal aeration
常微分方程 ordinary differential equation
常系数线性偏微分方程【数】 linear partial differential with constant coefficients
常用对数【数】 common logarithm
常用吸收系数 decimal coefficient of absorption
常用消光系数 decimal coefficient of extinction
场 field
场变化 field changes
场变量 field variables
场电离【化】 field ionization
场亮度 field luminance, field brightness
场面气压【大气】 airdrome pressure
场强 field intensity
场向电流 field-aligned current
场效应【化】 field effect
场站海拔高度 field elevation
场致磁导率 field permeability
场致发射【物】 field emission
唱碟【计】 compact disc digital audio (CD-DA)
超标传播 superstandard propagation
超标折射 superstandard refraction
超差 oversize
超长波 extra-long wave

超长波【大气】 ultra-long wave
超长期〔天气〕预报【大气】 extra long-range〔weather〕forecast
超常风 overnormal wind
超常间隔degree exceedance interval
超常折射 supernormal refraction
超常 above normal
超大概率 exceedance probability
超单体风暴 supercell storm
超单体结构 supercell structure
超单体雷暴 supercell thunderstorm
超单体龙卷 supercell tornado
超地转风【大气】 supergeostrophic wind
超低频通信【电子】 SLF communication
超定个例 overdetermined case
超定问题 overdetermined problem
超定组【数】 overdetermined system
超短波 ultra-short wave
超反射 overreflection
超高频【电子】 super high frequency (SHF)
超高频雷达【大气】 UHF Doppler radar
超高速〔度〕 hypervelocity
超函数【数】 hyperfunction
超几何函数【数】 hypergeometric function
超基准洪水系列 floods-above-base series
超级单体【大气】 supercell
超级风暴 superstorm
超级闪电 superbolt
超级小型计算机【计】 super-minicomputer
超极高频 far-end infrared frequency
超集合 super-ensemble
超绝热直减【大气】 superadiabatic lapse rate
超绝热状态 superadiabatic state
超临界对流 supercritical convection
超临界分岔 supercritical bifurcation
超临界流 supercritical flow
超媒体【计】 hypermedia

超平流层　superstratosphere
超强台风【大气】　super typhoon, super TY
超渗雨量　rainfall excess
超声　hypersonic
超声波雪深测录器　ultrasonic snow data logger
超声测风仪【大气】　ultrasonic anemometer
超声〔的〕　ultrasonic
超声风速温度测量仪　ultra-sonic anemometer/thermometer
超声全息术【物】　ultrasonic holography
超声速〔的〕　supersonic
超声〔速〕运动　supersonic motion
超声特征　ultrasonic property
超声学　ultrasonics
超松弛　overrelax
超松弛【数】　overrelaxation
超松弛参数　overrelaxation parameter
超松弛系数　overrelaxation coefficient
超梯度　supergradient
超梯度风【大气】　supergradient wind
超湍流　superturbulent flow
超椭圆曲线【数】　hyperelliptic curve
超微颗粒【物】　ultra-fine particle
超维　overdimension
超文体【计】　hypertext
超稳定度(性)　overstability
超细网格　ultra-fine mesh
超线性收敛【数】　superlinear convergence
超斜压区　hyperbaroclinic zone
超新星【物】　supernova
超压　excess pressure
超压气球　super-pressure balloon
超越效应　skip effect
超折射【大气】　superrefraction
超折射回波【大气】　superrefraction echo
超子【物】　hyperon
朝鲜东部暖海流　East Korea Warm Current
潮差　range of tide
潮差【航海】　tidal range
潮程　tidal excursion
潮高【航海】　height of tide
潮高差【航海】　height difference
潮高基准面【航海】　tidal datum
潮沟【地理】　tidal creek
潮急浪　tide rip
潮间带　intertidal zone
潮解【化】　deliquescence
潮龄　tidal age
潮流【航海】　tidal stream, tidal current
潮流椭圆　current ellipse
潮流〔预报〕表　current tables
潮面【航海】　tide level
潮日　tidal day
〔潮〕湿不稳定性　moist-lability
潮湿空气　damp air
潮汐【航海】　tide
潮汐表〔格〕　tide table
潮汐冰裂【海洋】　tidal crack
潮〔汐〕波【海洋】　tidal wave
潮〔汐〕方程　tidal equation
潮〔汐〕风　tidal wind, tidal breeze
潮汐河流【铁道】　tidal river
潮汐计算机　tide-predicting machine
潮汐力　tidal force
潮汐因子　tidal factor
潮汐运动　tidal motion
潮汐振荡　tidal oscillation
潮汐周期　tidal period
潮亚湿气候　moist subhumid climate
潮振幅　tide amplitude
〔彻〕体力【物】　body force
尘埃层顶【大气】　dust horizon
尘埃含量　dust loading
尘埃浑浊度　dust turbidity
尘埃取样　dust sampling
尘埃探空仪　dustsonde

尘暴　duststorm
尘壁　dust wall
尘层　staubosphere
尘风　dust wind
尘降【大气】　dust fall
尘降器　dust fall jars
尘卷风【大气】　dust devil
尘卷风效应　dust-devil effect
尘坑　dust well
尘粒　dust particle
尘霾　dust-haze
尘幔　dust veil
尘幔指数　dust veil index
尘圈,尘层　konisphere(＝staubosphere)
尘雾　dust(＝sand)fog
尘消光　dust extinction
尘旋【大气】　dust whirl, dancing devil, dancing dervish, shaitan, ash devils
尘雨　dust rain
尘云　dust cloud
尘状雪崩　loose avalance
尘阵雨　dust shower
沉淀尘埃计数器　settlement dust counter
沉淀器　deposit gauge
沉积物【地质】　sediment
沉积物分类　sediment grading
沉积〔物〕浓度　sediment concentration
沉积学【石油】　sedimentology
沉降【化】　sedimentation
沉降风【大气】　fallout wind
沉降锋　fallout front
沉降界面　sedimentation boundary
沉降末速【煤炭】　terminal velocity
沉降平衡　sedimentation equilibrium
沉降势　sedimentation potential
沉降速度　sedimentation velocity, deposition velocity
沉降物捕集器　sediment trap
沉降物【大气】　fallout
沉降物浓度测量仪　sediment density meter

沉降物输送监测器　sediment transport monitor
沉降系数　sedimentation coefficient
沉降云室　settling cloud chamber（SCC）, drop settling chamber, cloud-setting chamber
沉降锥　cone of depression, cone of influence
陈冰　old ice
陈年冰　firn ice
陈年雪场　firn field
陈雪　old snow
晨侧磁层　dawnside magnetosphere
晨侧磁尾　dawnside tail
晨侧极光卵　dawnside auroral oval
晨出【天】　heliacal rising
晨辉　morning glow
晨昏　dusk
晨昏电场【地】　dawn-dusk electric field
晨昏线　terminator
晨静　morning calm
晨噪　dawn chorus
称雪器　weight snow-gauge
称雪器【大气】　weighing snow-gauge
称重式气压表【大气】　weighting barometer
称重式气压计　weight barograph
称重式渗漏计　weighing lysimeter
称重式湿度表　weighing hygrometer
称重式水银气压表　weighing mercury barometer
称重式雨量器【大气】　weighing rain-gauge
成雹阶段　hail stage
成本效益分析【计】　cost-benefit analysis
成冰核　ice-forming nuclei
成冰剂　glacigenic agent
成冰阈温【大气】　threshold temperature of ice nucleation
成冰作用【地理】　ice formation
成材【林】　lumber

成层高积云	altocumulus stratiformis

成层高积云　altocumulus stratiformis
成层卷积云　cirrocumulus stratiformis
成层状层积云　stratocumulus stratiformis(Sc str)
成核剂　nucleant, nucleating agent
成核临界温度　threshold of nucleation
成核温阈　nucleation threshold
成核作用　nucleation
成熟阶段　mature phase（或 stage）
成熟土壤【土壤】　mature soil
成像【物】　imaging
成像光谱仪【地信】　imaging spectrometer
成像接收机　image-forming receiver
成像雷达　imaging radar
成像器　imager
成像系统【地信】　imaging system
成雪阶段【大气】　snow stage
成雨阶段　rain stage
成云禁区　forbidden zone of cloud formation
承压地下水　confined groundwater
承压含水层　confined aquifer(＝artesian aquifer)
城市边界层　urban boundary layer
城市边缘【地理】　urban fringe
城市大气　urban atmosphere, community atmosphere
城市地理信息系统【地信】　urban geographic information system (UGIS)
城市地理学【地理】　urban geography
城市冠层　urban canopy layer, urban canopy
城市管理信息系统【地信】　urban management information system (UMIS)
城市化【地理】　urbanization
城市化效应　urbanization effect
城市环境　city environment
城市环境污染　urban environmental pollution
城市景观　civic landscape
城市径流　urban runoff
城市空气　urban air
城市空气污染【大气】　urban air pollution
城市内爆　urban implosion
城市内涝　urban waterlogging
城市气候　city climate
城市气候【大气】　urban climate
城市气候学【大气】　urban climatology
城市热岛【地理】　urban heat island
城市热岛效应　urban heat island effect
城市生态系统【地理】　city ecological system
城市生态学　urban ecology
城市水文学　urban hydrology
城市天气【大气】　urban weather
城市湍急洪水　rapid city flood
城市污染　city pollution
城市雾　town fog, city fog
城市峡谷　urban canyon
城市效应　urban effects
城市烟羽　urban plume
城乡环流【大气】　urban-rural circulation
城域网【计】　metropolitan area network (MAN)
乘法【数】　multiplication
乘积法则　product rule
〔乘〕幂〔定〕律　power law
程长【物】　path length
程序【计】　program(me)
程序开发　program development
程序控制【航海】　programmed control
程序库【计】　program library
程序块【计】　program block
程序设计【计】　program design
程序设计语言【计】　programming language
程序说明块　program specification bolck
程序通信块　program communication block
程序信号【航海】　procedure signal

程序员 programmer
弛豫时间【物】 relaxation time
弛豫【物】 relaxation
迟延自动增益控制 delayed automatic gain control
持恒污染物 conservative pollutants
持久性距平 persistent anomaly
持久性有机污染物 persistence organic pollutants（POPs）
持水量 retention storage
持水能力 potential recharge
持水曲线 retention curve
持续对流 sustained convection
持续风 sustained wind
持续锋面雨 prolonged frontal rain
持续降雨 sustained rainfall
持续时间 duration
持续性【大气】 persistence, durability, constancy
持续性气候学 persistency climatology
持续性趋势【大气】 persistence tendency
持续性预报【大气】 persistence forecast
持续振荡 persistent oscillation
尺寸 size
尺度 scale
尺度崩溃 scale collapse
尺度参数 size parameter, scaling parameter
尺度参数【数】 scale parameter
尺度分布 size distribution
尺度分离 scale separation
尺度分析【大气】 scale analyss, scaling
尺度关系 scale relation
尺度交互作用 scales interaction
尺度谱 size spectrum
尺度选择 scale selective
尺度依存不稳定度 scale dependence instability
尺度依存性 scale dependence
赤潮【航海】 red water, red tide
赤道 equator
赤道变形半径 equatorial radius of deformation
赤道波 equatorial wave
赤道波导 equatorial waveguide
赤道槽【大气】 equatorial trough
赤道潮流 equatorial tidal current
赤道潮汐 equatorial tide
赤道大陆气团 equatorial continental air mass
赤道带【地理】 equatorial zone
赤道低压【大气】 equatorial low
赤道电集流【地】 equatorial electrojet
赤道东风带【大气】 equatorial easterlies
赤道多雨带 equatorial precipitation belt
赤道反气旋 equatorial anticyclone
赤道风 equatorial wind
赤道锋 equatorial front
赤道锋区 equatorial frontal zone
赤道辐合带 equatorial convergence zone
赤道辐合带【大气】 equatorial convergence belt
赤道干旱带 equatorial dry zone
赤〔道〕轨〔道〕卫星 equatorial orbiting satellite
赤道海洋气团 equatorial maritime air mass
赤道环电流 equatorial ring current
赤道缓冲带【大气】 equatorial buffer zone
赤道急流 equatorial jet
赤道加速度 equatorial acceleration
赤道开尔文波 equatorial Kelvin wave
赤道冷舌 equatorial cold tongue
赤道流【海洋】 equatorial current
赤道隆起【天】 equatorial bulge
赤道逆〔海〕流 equatorial counter current
赤道庞加莱波【海洋】 equatorial Poincare wave
赤道平流层波 equatorial stratospheric wave

赤道平面　equatorial plane
赤道β平面　equatorial beta-plane
赤道气候【大气】　equatorial climate
赤道气候带　equatorial climate zone
赤道气候区　equatorial climatic region
赤道气团【大气】　equatorial air mass
赤道〔铅直〕环流　equatorial cell
赤道潜流　Equatorial Undercurrent (EUC)
赤道日　equatorial day
赤道深〔层急〕流　equatorial deep jets
赤道水　equatorial water
赤道涡旋　equatorial vortex
赤道无风带　equatorial calm belt, doldrums
赤道无风带【大气】　equatorial calms
赤道西风带【大气】　equatorial westerlies
赤道陷波　equatorially-trapped wave
赤道洋流系统　equatorial current system
赤道涌升流　equatorial upwelling
赤道雨林【大气】　equatorial rain forest
赤道雨林气候　equatorial rain forest climate
赤道雨林生态系统　equatorial rain forest ecosystem
赤道正压波　barotropic equatorial wave
赤道中层洋流　Equatorial Intermediate Current
赤经【天】　right ascension
赤纬【天】　declination
充分成长的风浪　fully-arisen sea(=fully-developed sea)
充分发展的海浪　fully developed ocean wave
充分发展的湍流　fully developed turbulence
充分条件【数】　sufficient condition
充沛降水　ample rainfall
充气式探测器　gas-filled detector
充要条件【数】　necessary and sufficient condition
充液气压计　filling barometer

重叠　overlap
重叠带　overlapping band
重叠积分【化】　overlap integral
重叠原理　principle of superposition
重放　playback
重复面　recurrence surface
重复频率　repetition frequency
重复性　repeatability
重合误差　coincidence error
重建　reconstruction, restoring
重现　recurrence
重现磁暴　recurrent magnetic storm
重现频率　recurrence frequency
重现期　recurrence interval, return period
重现天气　recurring weather
重现周期　recurrent period
重正规化群　renormalization group
冲【天】　opposition
冲并效率　collection efficiency
冲掉　overwrite
冲顶上升气流　overshooting updraft
冲击　impingement, impetus
〔冲〕击波【物】　shock wave
冲击电流　impulse current
冲击接地电阻　impulsive grounding resistance
冲击粒子　ballistic particle
冲积含水层　alluvial aquifer
冲积扇【地理】　alluvial fan
冲积土〔层〕　alluvial
冲积物【地理】　alluvium, alluvial deposit
冲力　impulsive force
冲裂〔作用〕　avulsion
冲升气流　uprush
冲蚀【土壤】　scourring
冲刷　washout, wash
冲刷系数　washout coefficient
冲洗效率　washout efficient
抽水降深　drawdown
抽水降深曲线　drawdown curve

抽水水位 pumping water level
抽吸点 suction spot
抽吸性涡旋【大气】 suction vortex
抽象 abstraction
抽象测试方法【地信】 abstract test method
抽样【数】 sampling
抽样法 extraction method
抽样误差【自然】 sampling error
抽运【物】 pumping
抽运水头 pumping head
稠密分布 dense distribution
臭氧【大气】 ozone
臭氧测定术 ozonometry
臭氧测定仪 ozonograph, ozonoscope
臭氧层 ozone layer
臭氧层【大气】 ozonosphere
臭氧层顶 ozonopause
臭氧等值线图 ozone isopleth plot
臭氧洞 ozone hole
臭氧发生器【航海】 ozone generator
臭氧反应 ozone reaction
臭氧防护层 ozone shield
臭氧分布 ozone distribution
臭氧分光光度计 ozone spectrophotometer
臭氧光化学 ozone photochemistry
臭氧耗减势 ozone-depleting potential (ODP)
臭氧耗减物质 ozone depleth substance (ODS)
臭氧耗竭【地】 ozone depletion
臭氧化【化】 ozonization
臭氧化物【化】 ozonide
臭氧计 ozonometer
臭氧加热 ozone heating
臭氧渐衰〔作用〕 ozone degradation
臭氧解【化】 ozonelysis
臭氧均值 average ozone value
臭氧亏欠 ozone deficiency
臭氧浓度 ozone concentration
臭氧热源 ozone heat source
臭氧生成 ozone production
臭氧收支 ozone budget
臭氧探空仪 ozonesonde
臭氧通量 ozone flux
臭氧图 ozonogram
臭氧微库仑分析仪 microcoulomb ozone meter
臭氧-温度传感器 ozone-temperature sensor
臭氧形成 ozone formation
臭氧乙烯化学发光反应 ozone-ethylene chemiluminescent reaction
臭氧云 ozone cloud
臭氧柱总量测量【大气】 total column ozone measurements
臭氧总量 total ozone
臭氧总量测绘分光计 total ozone mapping spectrometer (TOMS)
臭氧总量测绘系统 total ozone mapping system
臭氧总量监测卫星 total ozone monitoring satellite
出发港【航海】 port of departure
出口区 exit region
出流冰川 outlet glacier
出梅【大气】 ending of Meiyu
出气【电子】 outgassing
出射角【物】 outgoing angle, emergence angle
出现点(流星) beginning point
出现点高度 beginning height
初次检验【航海】 initial survey
初估场 first-guess field
初估值【大气】 first guess
初级过程【化】 primary process
初级宇宙线 primary cosmic ray
初亏 first contact
初期降水 initial rainfall
初侵染源【农】 primary source of infection
初生大气 primordial atmosphere

初生低压　incipient low
初生气旋　nascent cyclone
初始场　initial field
初始化【计】　initialization
初始集合扰动　initial ensemble perturbation
初始扰动　initial perturbation
初始舍入误差　inital round-offerror
初始条件【大气】　initial condition
初始位相　initial phase
初始温度　initial temperature
初始烟羽　starting plume
初始滞留　initial detention(＝surface storage)
初始状态【计】　initial state
初霜【大气】　first frost
初速〔度〕【物】　initial velocity
初夏　early summer
初相　preliminary phase
初雪　first snowfall, early snow
初〔值〕-边值问题【数】　initial-boundary value problem
初值化【大气】　initialization
初值问题【数】　initial value problem
雏菊世界（计算机模拟的假设世界）　daisy world
除冰　deicing, deice
除冰器　deicer
除尘器　precipitator
除法【数】　division
除湿过程　wet removal process
除数【数】　divisor
处理器【计】　processor
处暑【大气】　End of Heat
触底　feeling bottom
触发机制　trigger mechanism
触发器【计】　flip-flop
触发器【物】　trigger
触发闪电　triggered flash
触发效应　trigger effect
触发作用　trigger action

氚【大气】　tritium
氚比【化】　tritium ratio
氚核【物】　triton
氚化【化】　tritiation
氚化物【化】　tritide
穿锋的　trans-frontal
穿锋流　cross-front flow
穿孔卡　punched card
穿孔纸带　punched tape
穿透对流【大气】penetrative convection
穿透辐射　penetrating radiation
穿透曲线　breakthrough curve
穿透云顶　penetrating top
穿云　break the cloud
传播　propagation
传播常数【电子】　propagation constant
传播效应　propagation effect
传导【大气】　conduction
传导电流　conductivity current, conduction current
传导平衡　conductive equilibrium
传递函数【数】　transfer function
传递函数模型【自动】　transfer function model
传递性【数】　transitivity
传递性质　transfer property
传感器【物】　sensor
传能线密度【农】　linear energy transfer (LET)
传热【物】　heat transfer
传热系数　heat transfer coefficient
传输方程　equation of transfer
传输速度　transfer velocity
传输线【计】　transmission line
传真　fax, facsimile
传真发送　facsimile transmission
传真发送机　fax transmitter
传真机分辨率　resolution of facsimile equipment
传真设备　facsimile equipment
传真天气图【大气】　facsimile weather

chart
传真图　fax chart(=fax map), facsimile chart
传真图像设备　facsimile copier equipment
船舶报〔告〕　ship report
船舶电气设备【航海】　marine electric installation
船舶观测【大气】　ship observation
船舶航线〔安排〕　ship routing
船舶气象报告　ship weather report
船舶天气电码　ship synoptic code
船舶用气压表　marine barometer
船舶雨量计　marine raingauge
船舶运动　ship motion
船长【航海】　captain
船队预警广播　Fleet Broadcast
船上真风填算盘　shipboard wind plotter
船体〔测水温〕法　hull method
船用多普勒系统　ship Doppler (SDOP)
船用经纬仪　shipboard theodolite
船用投弃式深度温度计　shipboard expendable bathythermograph (SXBT)
船用气压表　sea-barometer
船用气压表【大气】　ship barometer
船用无线电传真记录器　marine radio-facsimile recorder
船用雨量器　shipboard raingauge
船载雷达【电子】　shipborne radar
串级　cascade
串级理论　cascade theory
串联模式　in-tandem model
串联【物】　connection in series
串列探空仪　tandem radiosonde
串行处理　serial processing
串行计算机【计】　sequential computer
串行鼠标〔器〕【计】　serial mouse
串行通信　serial communications
串珠状闪电【大气】　pearl-necklace lightning, beaded lightning
窗冰　window ice
窗长度　window length

窗函数【电子】　window function
窗区频道　window channel
窗霜　window hoar, window frost
窗外温度表　window-thermometer
窗型　window type
吹蚀【地理】　deflation
吹蚀【土壤】　blowout
吹雪【大气】　driven snow
吹雪冰　snowdrift ice
吹雪冰川　catchment glacier, snowdrift glacier
〔吹〕雪堆　snow drift
垂板风速表　normal-plate amenometer
垂线【数】　perpendicular
垂直变差图　vertical differential chart
垂直波　vertical wave
垂直波长　vertical wave length
垂直波数　vertical wave number
垂直不稳定度　vertical instability
垂直差分法　vertical differencing scheme
垂直尺度　vertical scale
垂直传播波　vertically propagating wave
垂直地带　vertical zone
垂直地带性　vertical zonality
垂直订正因数　vertical correction factor
垂直动量通量　vertical momentum flux
垂直对流　vertical convection
垂直分辨率　vertical resolution
垂直分布　vertical distribution
垂直分层　vertical division
垂直风速【大气】　vertical wind velocity
垂直风速表　vertical-axis anemometer, vertical anemometer
垂直风速仪　vertical anemoscope
垂直风误差　vertical wind error
垂直辐合　vertical convergence
垂直辐散　vertical divergence
垂直光束云幂仪　vertical beam ceilometer
垂直横波　vertical transverse wave
垂直环流　vertical circulation

垂直混合　vertical mixing
垂直积分液水量　vertically integrated liquid(VIL)
垂直急流　vertical jet
垂直加速度　vertical acceleration
垂直简化　vertical simplification
垂直结构方程　vertical structure equation
垂直扩散　vertical diffusion
垂直扩散系数　vertical diffusion coefficient
垂直廓线【大气】　vertical profile, elevation profile
垂直模　vertical modes
垂直能见度【大气】　vertical visibility
垂直耦合　vertical coupling
垂直偏振　vertical polarization
垂直平流　vertical advection
垂直剖面　vertical section
垂直剖面图【大气】　vertical cross section
垂直〔谱〕带　perpendicular band
垂直气候带【大气】　vertical climatic zone
垂直气压梯度力　vertical pressure gradient force
垂直切变　vertical shear
垂直切变力　vertical shearing force
垂直倾斜　vertical tilt
垂直入射多普勒雷达　vertical incidence Doppler radar(VID)
垂直入射绝对日射表　normal incidence pyrheliometer(NIP)
垂直射束雷达【大气】　vertical-beam radar
垂直射线　vertical ray
垂直伸展　vertical stretching
垂直时间剖面　vertical time section
垂直收缩　vertical shrinking
垂直输送　vertical transfer
垂直速度　vertical velocity
垂直速度方差　vertical velocity variance
垂直速度显示器　vertical speed indicator
垂直通量　vertical flux
垂直退耦　vertical decoupling
垂直稳定度　vertical stability
垂直涡动温度通量　vertical eddy temperature flux
垂直涡度　vertical vorticity
垂直延伸　vertical continuation
垂直运动　vertical motion
垂直阵风　vertical gustiness
垂直阵风记录器　vertical-gust recorder (v. g. recorder)
垂直指向多普勒雷达　vertically pointing Doppler radar
垂直质量通量　vertical mass flux
垂直转换　vertical transform
垂直总〔量〕指数　vertical totals index
σ垂直坐标　sigma vertical coordinate
垂直坐标系统　vertical coordinate system
春暴　whip-poor-will storm
春潮周期　spring-neap cycle
春分　Spring Equinox, northern vernal equinox
春分【天】　Vernal Equinox
春分点　First Point of Aries
春分点【天】　vernal equinox
春寒　cold spell in spring
春化〔作用〕【农】　vernalization
春〔季〕　spring
春天的　vernal
春天粒雪　corn snow
春天小雪　robin snow
春性　springness
春雪　spring snow
春雪壳　spring crust
春汛　spring flood, snowmelt flood
纯冰面饱和水汽压【大气】　saturation vapor pressure in the pure phase with respect to ice
纯粹变形　pure deformation

纯洁度(气体的)　clarity〔of gas〕
纯洁空气　pure air
纯净冰　clear icing, clear ice
纯净空气　clean air, pure air
纯水面饱和水汽压【大气】　saturation vapor pressure in the pure phase with respect to water
纯态氧　singlet oxygen
词头　prefix
瓷管干湿表　ceramic tube psychrometer
磁暴　magstorm
磁暴【大气】　magnetic storm(＝magstorm)
磁暴急始　storm sudden commencement
磁暴时间变化　magnetic storm time variation
磁北　magnetic north
磁北极【地】　north magnetic pole
磁鼻　magnetic crochet
磁层暴【地】　magnetospheric storm
磁层【大气】　magnetosphere
磁层-电离层耦合体系　coupled magnetosphere-ionosphere system
磁层顶【大气】　magnetopause
磁层对流　magnetospheric convection
磁层热等离子体　hot magnetospheric plasma
磁〔层〕尾　magnetospheric tail
磁层物理学　magnetospherics
磁层亚暴　magnetospheric substorm
磁场　magnetic field
磁场强度　magnetic field intensity, magnetic field strength
磁场梯度　magnetic gradient
磁畴【电子】　magnetic domain
磁带　magnetic tape, tape
〔磁带〕带标　tape mark
磁带机　magnetic tape drive
磁带记录器　tape recorder, magnetic tape recorder
磁带录像机　video tape recorder

磁带式气象事件记录器　meteorological magnetic event recorder
磁带式事件记录器　magnetic tape event recorder(MATER)
磁电风速表　magneto-anemometer
磁定向法　magnetic direction finding (MDF)
磁洞　magnetic hole
磁方位〔角〕　magnetic bearing(M. B.)
磁分析器【化】　magnetic analyzer
磁风　magnetowind
磁感风杯风速表　cup-generator anemometer
磁感〔应〕强度【电子】　magnetic induction
磁鼓【电子】　magnetic drum
磁化【物】　magnetization
磁化等离〔子〕体【物】　magnetized plasma
磁化率【电子】　magnetic susceptibility
磁极　dip pole
磁极风向　magnetic wind direction
磁极迁移　magnetic pole migration
磁极【物】　magnetic pole
磁接温度调节器　MAG-CON thermoregulator
磁静日　magnetically quiet day
磁静日〔太阳日〕变化　magnetic quiet-day solar daily variation
磁控管【电子】　magnetron
磁离子理论【地】　magneto-ionic theory
磁力梯度仪【地】　magnetic gradiometer
磁力线　magnetic lines of force, magnetic field line
磁力线图　magnetic figure
磁量子数【化】　magnetic quantum number
磁流波　hydromagnetic wave
磁流〔体〕动力〔学〕波　magnetohydrodynamic wave
磁流〔体动〕力学【物】　magnetohydro-

dynamics
磁流体发射　hydromagnetic emissions
磁流体力学　hydromagnetics
磁南　magnetic south
磁南极【地】　south magnetic pole
磁偶极场　magnetic dipole field
磁偶极矩　magnetic dipole moment
磁偶极子【物】　magnetic dipole
磁耦合　magnetic coupling
磁盘　magnetic disk
磁盘操作系统　disk operating system（DOS）
〔磁〕盘驱动器【计】　magnetic disk drive
磁泡存贮器　magnetic bubble memory（MBM）
磁偏角　magnetic declination
磁偏转　magnetic deflection
磁谱仪　magnetic spectrograph
磁强度　magnetic intensity
磁强计　magnetometer, magnetograph
磁鞘【地】　magnetosheath
磁倾赤道【大气】　dip equator, magnetic equator
磁倾计　inclinometer
磁倾角【大气】　magnetic inclination, magnetic dip
磁倾仪　dip circle
磁情指数　geomagnetic index
磁扰　magnetic disturbance
磁扰变化　disturbance variation
磁扰日　magnetically disturbed day
磁扇形边界　magnetic sector boundary
磁通计　fluxmeter
磁通量　magnetic flux
磁头【电子】　magnetic head
磁尾【地】　magnetotail
磁〔尾〕裙　skirt region
磁纬　magnetic latitude, geomagnetic latitude
磁隙区　magnetospheric cleft, cleft
磁芯　magnetic core
磁芯存储　core memory
磁性地层学【地】　magnetostratigraphy
磁性定向控制线圈　magnetic orientation control coil
磁性卡片　magnetic card
磁性数　magnetic character figure
磁性双折射　magnetic double refraction
磁悬浮【电子】　magnetic suspension
磁学　magnetism
磁亚暴【地】　magnetic substorm
磁照图　magnetogram
磁针【物】　magnetic needle
磁致冷【电子】　magnetic cooling
磁致伸缩【物】　magnetostriction
磁滞现象【电子】　magnetic hysteresis
磁撞加热【物】　collision heating
磁子午线　magnetic meridian
磁坐标【地】　magnetic coordinate
次标准传播　substandard propagation
次标准折射　substandard refraction
〔次〕表层暴雨流　subsurface storm flow
〔次〕表层流【水利】　subsurface flow
次波　minor wave
次槽　minor trough
次地转风【大气】　subgeostrophic wind
次级电子发射【电子】　secondary electron emission
次级发射　secondary emission
次级模式　second order model
次极大【物】　secondary maximum
次极小【数】　subminimum
次脊　minor ridge
次季节尺度　subseasonal time scale
次季节振荡　subseasonal oscillation
次氯酸　hypochlorous acid
次平流层　substratosphere
次区域广播　subregional broadcast
次生〔飑〕带　secondary bands
次生冰晶　secondary ice crystal
次生低压　secondary low
次生低压【大气】　secondary depression

（或 low）
次生反气旋　secondary anticyclone
次生林【林】　secondary forest
次生气旋【大气】　secondary cyclone
次生污染物【大气】　secondary pollutant
次生灾害　secondary disaster
次声　infrasound
次声波【物】　infrasonic wave
次声〔的〕　infrasonic
次声观测台　infrasonic observatory
次梯度风【大气】　subgradient wind
次天气尺度　subsynoptic scale
次天气〔尺度〕更新模式　subsynoptic update model（SUM）
次天气尺度 Ω 系统　subsynoptic Ω system
次天气尺度系统【大气】　subsynoptic scale system
次天气平流模式　subsynoptic advection model（SAM）
次网格　subgrid
次网格尺度　subgrid scale
次网格尺度参数化【大气】　subgrid-scale parameterization
次网格尺度地形　subgrid-scale orography
次网格尺度过程　subgrid-scale process
次网格尺度湍流　subgrid scale turbulence
次网格尺度涡旋　subgrid-scale eddy
次网格尺度运动　subgrid scale motions
次网格凝结　subgrid condensation
次线性的【数】　sublinear
次折射　subrefraction
次正规　subnormal
次中尺度相干涡旋　submesoscale coherent vortices（SCV）
刺耳低限　threshold of pain, threshold of discomfort
刺骨寒风　biting wind
从〔属〕站　slave station
粗糙层　roughness layer

粗糙次层　roughness sublayer
粗糙地面　rough earth
粗糙度　roughness
粗糙度参数【大气】　roughness parameter
粗糙度长度【大气】　roughness length
粗糙度系数　roughness coefficient
粗糙流　rough flow
粗糙面　rough surface
粗糙面散射　rough surface scattering
粗糙元　roughness element
粗粒密度【物】　coarse-grained density
粗粒散射【大气】　Mie scattering
粗粒雪　coarse-grained snow
粗粒子（直径大于 2 μm 悬浮在空中的粒子）　coarse particles
粗砂土【土壤】　coarse sand soil
粗网格　coarse-mesh grid, coarse-mesh
粗线　heavy line
猝发　burst
猝发声　sound burst
簇　cluster
簇合物【化】　cluster compound
簇状高积云　altocumulus glomeratus
簇状（云）　glomeratus
窜漏，气体喷出　blowby
催化【化】　catalysis
催化剂【化】　catalyst
催化率　seeding rate
催化循环　catalytic cycle
催化转换器，催化式排气净化器（减少汽车排放的装置）　catalytic converter
催化子程序　seeding subroutine
脆弱性　vulnerability
淬火【物】　quenching
淬灭效应【电子】　quenching effect
存储　packing
存储器【计】　storage
存取　access
存取时间【计】　access time
存贮　memory

〔存贮〕库　bank

错位　dislocation

D

达尔文站(澳大利亚)　Darwin Station
达朗贝尔公式【数】　d'Alembert formula
达朗贝尔佯谬【力】　d'Alembert's paradox
达西定律　Darcy's law
达西速度　Darcian velocity
达因　dyne
达因补偿定理　Dines compensation theorem
达因补偿〔作用〕　Dines compensation
达因翻斗式虹吸雨量计　Dines tilting-siphon rain-gauge
达因风速表　Dines anemometer
达因风压计　Dines pressure anemograph
达因浮标气压计　Dines float barograph
达因辐射计　Dines radiometer
达因流压表　Dines float manometer
打孔机,穿孔机　punching machine
打印机　printer
打转飓风　loop hurricane
打转路径　loop track
打转台风　loop typhoon
大雹　heavy hail
大爆炸宇宙论【天】　big bang cosmology
大侧风　high crosswind, stiff (或 strong) crosswind
大潮【海洋】　spring tide
大潮落潮流　greater ebb tidal current
大潮升【航海】　spring rise
大潮涨潮流　greater flood tidal current
大尺度　macro scale, large scale
大尺度波特性　large-scale wave property
大尺度〔磁力线〕环　large-scale loop
大尺度大气过程　process of large scale atmosphere
大尺度大气运动　large-scale atmospheric motion
大尺度对流　large-scale convection
大尺度环流　macroscale circulation
大尺度环流【大气】　large-scale circulation
大尺度凝结　large-scale condensation
大尺度气象学　macrometeorology
大尺度天气〔过程〕　large-scale weather 〔process〕
大尺度〔天气〕形势　large-scale weather situation
大尺度天气学　macrosynoptic meteorology
大尺度湍流　large-scale turbulence
大尺度行星波　large-scale planetary wave
大尺度运动　large-scale motion
大地测量【航海】　geodetic survey
大地测量地球轨道卫星　Geodetic Earth Orbiting Satellite(GEOS)
大地测量曲线　geodetic curve
大地测量卫星　Geodetic Satellite (GEOSAT)
大地测量学　geodesy(＝geodetics)
大地电流暴　earth-current storm
大地水准面　geoidal surface, geoid
大地水准面高度图【航海】　geoidal height map
大冻丘,冰核丘　pingo
大范围〔的〕　macroscopic
大范围降水　widespread precipitation
大风【大气】　gale high wind
大风警报【大气】　gale warning

大风浪航行工况管理【航海】 heavy weather navigation operating mode management
大风污染 gale pollution
大风信号 gale signal, gale cone
大风雪 driving snow
大风雨 driving rain
大辐射带 great radiation belt
大副【航海】 chief officer, chief mate
大功率超灵敏雷达 high-power ultra-sensitive radar
大寒【大气】 Greater Cold
大寒天 broiling
大旱 severe drought, great drought, drought catastrophe
大旱年 severe drought year
大核 large nuclei
大红斑【天】 Great Red Spot
大间冰阶 Megainterstadial
大浪 rough sea
大雷诺数流〔动〕 large Reynolds number flow
大离子 large ion
大粒子 large particle
大陆边缘地 continental borderland
大陆冰 continental ice, continental glacier
大陆度【大气】 continentality
大陆度系数 coefficient of continentality
大陆度指数【大气】 continentality index, coefficient of continentality, index of continentality
大陆风 continental wind
大陆高压 continental high
大陆环境 continental environment
大陆架波 continental shelf wave
大陆架【地理】 continental shelf
〔大〕陆架地形波 shelf waves
大陆漂移【地质】 continental drift
大陆漂移说【地质】 continental drift theory
大陆坡【地理】 continental slope
大陆气候 mainland climate
大陆气团【大气】 continental air mass
大陆水界 Continental Water Boundary
大陆台地 continental platform
大陆性冰川【地理】 continental glacier
大陆性反气旋 continental anticyclone
大陆性气候【大气】 continental climate
大陆性气溶胶 continental aerosol
大陆云 continental cloud
大灭绝 mass extinction
大年 on-year
大盆地高压 Great Basin high
大片目标 volume target
大气【大气】 atmosphere
大气本底污染监测网 Background Air Pollution Monitoring Network (BAPMoN)
大气本底〔值〕【大气】 atmospheric background
大气边界层【大气】 atmospheric boundary layer
大气边缘层 fringe region of the atmosphere
〔大气〕边缘区 spray region
大气变性 atmospheric metamorphism
〔大气〕标高【大气】 scale height
大气波 atmospheric billow
大气波导【大气】 atmospheric duct
大气波动【大气】 atmospheric wave
大气不透明度 atmospheric opacity
〔大气〕不稳定度【大气】 〔atmospheric〕 instability
大气测量委员会(美国) Committee on Atmospheric Measurements(CAM)
大气测湿法 atmospheric hygrometry
大气层【大气】 atmosphere, atmospheric layer
大气层(包围地球的) atmospheric envelope
大气层爆炸 atmospheric explosion

大气层结【大气】 atmospheric stratification
大气层上限 upper limit of the atmosphere
大气长波【大气】 atmospheric long wave
大气潮汐【大气】 atmospheric tide
大气潮汐振荡 atmospheric tidal oscillation
大气尘埃 atmospheric dust
大气尘粒【大气】 lithometeor
大气成分【大气】 atmospheric composition
大气成分〔源汇〕收支 budgets of atmospheric species
大气臭氧【大气】 atmospheric ozone
大气传输 atmospheric transport
大气传输模式【大气】 atmospheric transmission model
大气窗【大气】 atmospheric window
大气窗区 atmospheric window region, window region
〔大气〕垂直电流记录仪 vertical-current recorder
〔大气〕垂直〔结构〕探测器 vertical sounder
大气簇射【地】 air shower
大气单分散粒径分布 atmospheric monodispersion
大气的氧化能力 oxidizing capacity of atmosphere
大气地面层 atmospheric surface layer
大气电场【大气】 atmospheric electric field
大气电场自记曲线 electrogram
大气电导率 air conductivity
大气电导率【大气】 atmospheric electric conductivity
大气电离〔作用〕 atmospheric ionization
大气电流 electric currents in the atmosphere
大气电位图 atmospheric electric potential chart
大气电学【大气】 atmospheric electricity
大气电学现象 electrometeor
大气顶太阳辐射 extra-terrestrial radiation
大气订正 atmospheric correction
大气动力学 dynamics of the atmosphere
大气动力学【大气】 atmospheric dynamics
大气多分散粒径分布 atmospheric polydispersion
大气惰性气体 atmosphere noble gas
大气发电机 atmospheric dynamo
大气反馈机制 atmospheric feedback mechanism
大气反演问题 atmospheric retrieval problem
大气放电现象 igneous meteor
大气放射性【大气】 atmospheric radioactivity
大气分层【大气】 atmospheric subdivision
大气浮游生物 air plankton, aerial plankton
大气辐射【大气】 atmospheric radiation
大气辐射表,地球辐射表 pyrgeometer
大气辐射收支 atmospheric radiation budget
大气干旱 atmospheric drought
大气干扰 tweeks
大气更新剂 atmospheric regenerant
大气光化学【大气】 atmospheric photochemistry
大气光解〔作用〕【大气】 atmospheric photolysis
大气光谱 atmospheric spectrum
大气光谱【大气】 atmospheric optical spectrum
大气光象 optical meteor
大气光学【物】 atmospheric optics

大气光学和遥感研究所(美国) Institute for Atmospheric Optics and Remote Sensing(IFAORS)
大气光学厚度【大气】 atmospheric optical thickness, atmospheric optical depth
大气光学探测 optical probing of the atmosphere
大气光学现象 photometeor
大气光学现象【大气】 atmospheric optical phenomena
大气光学质量 optical air mass
大气光学质量【大气】 atmospheric optical mass
大气-海洋混合层模式 atmosphere-mixed layer ocean model
大气海洋交互作用 atmosphere-ocean interaction
大气-海洋-陆面耦合模式,气海陆耦合模式 atmosphere-ocean-land coupled model
大气和陆面过程计划 Atmospheric and Land Surface Processes Project (ALSPP)
大气痕量分子光谱仪 atmospheric trace molecular spectroscopy (ATMOS)
大气痕量气体【大气】 atmospheric trace gas
大气痕量组分 trace atmospheric constituents
大气红外探测器 Atmosphenic Infrared Sounder(AIRS)
大气后向散射 atmospheric backscattering
大气后向散射系数 atmospheric backscattering coefficient
大气候【大气】 macroclimate
大气候学【大气】 macroclimatology
大气化学【大气】 atmospheric chemistry
大气化学成分 atmospheric chemical composition
大气化学和全球污染委员会(IAMAS) Commission on Atmospheric Chemistry and Global Pollution (CACGP, IAMAS)
大气环境 atmospheric environment
大气环境局(加拿大) Atmospheric Environment Service (AES)
大气环境评价【大气】 assessment of atmospheric environment
大气环境容量【大气】 atmospheric environment capacity
大气环流 general circulation of atmosphere, general circulation
大气环流【大气】 atmospheric circulation
大气环流模式 atmospheric circulation model, atmospheric general circulation model (AGCM)
大气环流模式【大气】 general circulation model (GCM)
大气浑浊度【大气】 atmospheric turbidity
大气活动中心【大气】 atmospheric center of action
大气基本方程 atmospheric basic equation
大气及降水监测网 air and precipitation monitoring network
大气极限 limit of the atmosphere
大气技术部(NCAR) Atmospheric Technology Division (ATD, NCAR)
大气加热源 atmospheric heating source
大气监测 air monitoring
大气监测网 air monitoring network
大气监测系统 atmosphere monitoring system
大气监测站 air monitoring station
大气降水 atmospheric precipitation
大气结构 atmospheric structure
大气净辐射表 net pyrgeometer
大气净化 atmospheric scavenging

大气净化【大气】 atmospheric cleaning
大气科学【大气】 atmospheric science
大气科学国际委员会 International Committee of Atmospheric Science(ICAS)
大气科学和地球流体力学数值模拟国家重点实验室 State Key Laboratory of Numerical Modeling for Atmospheric Sciences and Geophysical Fluid Dynamics(LASG)(IAP, CAS)
大气科学实验室(美国) Laboratory of Atmospheric Sciences(LAS)
大气科学委员会(NAS) Committee on Atmospheric Sciences (CAS, NAS)
大气科学委员会(WMO) Commission for Atmospheric Sciences(CAS, WMO)
大气可预报性 atmospheric predictability
大气扩散【大气】 atmospheric diffusion
大气扩散方程【大气】 atmospheric diffusion equation
大气廓线红外遥测系统 remote infrared atmospheric profiling system
大气累积〔作用〕 atmospheric accumulation
大气离子【大气】 atmospheric ion
大气粒子 atmospheric particle
大气流体动力学 atmospheric hydrodynamics
大气霾 atmospheric haze
大气弥散胶体 atmospheric dispersoid
大气密度【大气】 atmospheric density
大气密度对地磁的响应 geomagnetic density response
大气模式 atmospheric model
大气模式比较计划 Atmospheric Models Intercomparison Project(AMIP)
大气钠 atmospheric sodium
大气能见度 atmospheric visibility
大气能量学 energetics of the atmosphere, atmospheric energetics
大气逆辐射【大气】 atmospheric counter radiation

大气逆温 atmospheric inversion
大气排除 atmospheric removal
大气偏振【大气】 atmospheric polarization
大气品味【大气】 air quality
大气屏蔽高度 atmospheric screening height
大气气溶胶 atmospheric aerosol
大气气体固定 fixing of atmospheric gases
大气起源 origin of atmosphere, atmospheric origin
大气强迫【大气】 atmospheric forcing
大气清洁机制 atmospheric cleaning mechanism
大气圈〔层〕 atmospheric shell, atmospheric region
大气扰动【大气】 atmospheric disturbance
大气热机 atmospheric engine
大气热力学【大气】 atmospheric thermodynamics
大气热流〔动〕 atmosphere heat flow
大气热平衡 atmospheric heat balance
大气瑞利散射截面 atmospheric Rayleigh scattering cross section
大气散射 scattering in the atmosphere
大气色散 atmospheric dispersion
大气闪晃 atmospheric boil
大气闪烁 atmospheric shimmer
大气上界 aeropause
大气声学【大气】 atmospheric acoustics
大气湿度 atmospheric moisture, atmospheric humidity
大气视宁度【天】 atmospheric seeing
大气衰减 air attenuation
大气衰减【大气】 atmospheric attenuation
大气衰减系数 atmospheric attenuation coefficient
大气水 atmospheric water

大气水分收支　atmospheric water budget
大气随机噪声　atmospheric stochastic noise
大气太阴潮　lunar atmospheric tide
大气探测【大气】　atmospheric sounding and observing, atmospheric probing
VISSR 大气探测器　VISSR atmospheric sounder(VAS)
大气碳酸计　carbacidometer
大气特性　atmospheric property
大气体〔积〕消光系数　atmospheric volume extinction coefficient
大气同化〔作用〕　atmospheric assimilation
大气同位素成分　atmosphere isotopic composition
大气透明窗　atmospheric transparency window
大气透明度　transparency of atmosphere
大气透明度【大气】　〔atmospheric〕 transparency
大气透射　atmospheric transmittance, atmospheric transmission
大气透射率【大气】　atmospheric transmissivity
大气湍流【大气】　air turbulence
大气湍流【大气】　atmospheric turbulence
大气湍流和扩散实验室（美国）　Atmospheric Turbulence and Diffusion Laboratory（ATDL）
大气微粒含量　atmospheric particulate content
大气温度　atmospheric temperature
大气温度测量　atmospheric temperature measurement
〔大气〕稳定度【大气】　〔atmospheric〕 stability
大气涡旋　atmospheric vortex
大气污染　atmospheric contamination
大气污染【大气】　atmospheric pollution

大气污染监测【大气】　atmospheric pollution monitoring
大气〔污染〕监测仪　air monitoring instrument
大气污染气体　atmospheric pollutant gas
大气污染物【大气】　atmospheric pollutant
大气污染源【大气】　atmospheric pollution sources
大气物理和大气化学实验室（NOAA）　Atmospheric Physics and Chemistry Laboratory（APCL, NOAA）
大气物理学　physics of the atmosphere
大气物理学【大气】　atmospheric physics
大气吸收【大气】　atmospheric absorption
〔大气〕吸收率【大气】　〔atmospheric〕 absorptivity
大气稀释　atmospheric dilution
大气现象　atmospheric phenomenon
大气现象【大气】　meteor
大气向下辐射　downward terrestrial radiation
大气消光【大气】　atmospheric extinction
大气消光系数　atmospheric extinction coefficient
大气效应　atmospheric effect
大气行为　atmospheric behaviour
大气需水量　atmospheric demand
大气悬浮物【大气】　atmospheric suspended matter
〔大气〕研究飞行中心　Research Flight Facility（RFF）
大气研究合作研究所（美国）　Cooperative Institute for Research of the Atmosphere（CIRA）
大气盐度　atmospheric salinity
大气掩星　atmospheric obscurant
大气演化　atmospheric evolution

大气演化【大气】 evolution of atmosphere
大气氧化剂 atmospheric oxidant
大气遥感 remote sensing of the atmosphere
大气遥感【大气】 atmospheric remote sensing
大气与海洋联合研究所(美国 NOAA 环境研究院与华盛顿大学) Joint Institute for the Study of the Atmosphere and the Ocean (JISAO)
大气云物理实验室(NASA) Atmospheric Cloud Physics Laboratory (ACPL,NASA)
大气杂质 foreign matter of the air
大气杂质【大气】 atmospheric impurity
大气噪声【大气】 atmospheric noise
大气折射〔差〕 astronomical refraction
大气折射【大气】 atmospheric refraction
大气折射订正 radio refraction correction
大气折射误差 curved-path error
大气折射指数 radio refractive index
大气振荡 atmospheric oscillation, atmospheric fluctuation
大气质量 mass of atmosphere
大气质量【大气】 atmospheric mass
大气质量标准【大气】 air quality standard
大气质量法 Air Quality Act
大气质量管理 air quality management
大气质量监测仪 air quality monitor
大气质量监视网 air quality surveillance network
大气质量控制区 air quality control region
大气质量判据(标准) air quality criteria
大气质量指数 air quality index
大气致癌物 atmospheric carcinogen
大气自复〔原〕〔效应〕 self-healing of atmosphere
大气棕色云 Atmospheric Brown Cloud (ABC)
大气总臭氧 atmospheric total ozone
大气阻塞 atmospheric blocking
大气组成 composition of atmosphere
大气组分 atmospheric constituents
大容量存储器【计】 bulk memory
大容量空气取样器 high volume air sampler
大容量微粒取样器 high-volume particulate sampler
大暑【大气】 Greater Heat
大涡流模拟 large eddy simulation
大涡旋 maelstorm
大涡旋模拟模式 large eddy simulation model
大涡旋模式 large eddy model
大西洋 Atlantic Ocean
大西洋低压起源试验 Genesis of Atlantic Lows Experiment(GALE)
大西洋副热带偶极子 Atlantic subtropical dipole
大西洋海洋气象实验室 Atlantic Oceanographic and Meteorological Laboratories (AOML)
大西洋〔气候〕亚期 subatlantic
大西洋时间 Atlantic time
大西洋水 Atlantic Water
大西洋西部型 West Atlantic pattern (WA)
大西洋信风试验 Atlantic Tradewind Experiment (ATEX)
大西洋洋中槽 Mid-Atlantic trough (MAT)
大西洋洋中脊〔海洋〕 Mid-Atlantic ridge
大行星【天】 major planet
大型计算机【计】 large-scale computer
大型交换 Gross-austausch
大型流体转盘 large rotation fluid annulus

大〔型〕下击暴流　macroburst
大型蒸发器【大气】　evaporation tank
大雪【大气】　Heavy Snow
大雪预警　heavy snow warning
大洋板块【海洋】　oceanic plate
大洋采矿　open ocean mining（OOM）
大洋对流层【海洋】　oceanic troposphere
大洋航行【航海】　ocean navigation
大洋中部槽　mid-ocean trough
大洋洲的地中海海水　Australasian Mediterranean Water（AAMW）
大雨【大气】　heavy rain
大雨期　great pluvial
大圆　great circle
大圆方位【航海】　great circle bearing
大圆海图【航海】　great circle chart
大圆航线算法【航海】　great circle sailing
大圆航向【航海】　great circle course
大圆弧　great circle arc（GCA）
大圆弧长　great circle arc length
大圆距离【航海】　great circle distance
大圆理论　great circle theory
大圆路径　great circle route
大晕(46°的晕)　large halo
大阵雨　heavy shower, heavy passing shower
大振幅波理论　large amplitude wave theory
代表流域【水文】　representative basin
代表性　representativeness
代表性气象观测　representative meteorological observation
代码格式　code format
代码生成【计】　code generation
代码生成器【计】　code generator
代入消元法【数】　elimination by substitution
代数方程【数】　algebraic equation
代数函数　algebraic function
代数学【数】　algebra
代数语言【计】　algebraic language
代数整数【数】　algebraic integer
代烷基　alkylperoxy radicals
代用气候记录　proxy climatic record
代用气候指标　proxy climate indicator
代用资料　proxy data
带　band
带电粒子【物】　charged particle
带负电雨　negative rain
带宽压缩调制解调器　bandwidth compression modem
带模式【大气】　band model, banded model
带内行星【天】　inner planet
带色降水　colored precipitation
带通　band-pass
带通滤波〔器〕【大气】　band-pass filter
带通滤波资料　bandpass-filtered data
带通振荡　band-pass fluctuation
带吸收　band absorption
〔带〕正〔电的〕电离层　positively charged ionosphere
带正电雨　positive rain
带状冰　ribbon ice
带〔状〕光谱　band spectrum
带状回波【大气】　banded echo
带状结构　banded structure
带状卷层云　cirrostratus vittatus
带状卷云　band cirrus
带状排列　zonation
带〔状〕闪〔电〕　ribbon lightning, fillet lightning, band lightning
带状闪电【大气】　band lightning
带状云系【大气】　banded cloud system
带阻滤波器【电子】　band stop filter
殆周期函数【数】　almost periodic function
殆周期解【数】　almost periodic solution
待定系数　undetermined coefficient
待定系数法【数】　method of undetermined coefficient

袋式除尘器　bag-type collector
袋状云　sack-cloud
戴维森海流　Davidson Current, Davidson inshore current
戴维斯称重式蒸散渗漏计　Davis weighing lysimeter
戴维斯数　Davies number
丹佛辐合涡度带　Denver convergence-vorticity zone(DCVZ)
丹佛气旋　Denver cyclone
丹斯加德-厄施格事件　Dansgaard-Oeschger events
单摆【物】　simple pendulum
单板计算机【计】　single-board computer
单瓣扫描器　mono-lobe scanner
单边带　single sideband
单边带调制【电子】　single-sideband modulation
单边光滑　one-sided smoothing
单变数　univariate
单波长光〔雷〕达　single-wavelength lidar
单步迭代【数】　single-step iteration
单参数族　one-parameter family
单侧差分　one-sided difference
单程大气透射比　one-way atmospheric transmittance
单程衰减【大气】　one-way attenuation
单次散射反照率　single scattering albedo
单次散射近似　single scattering approximation
单滴　individual droplet
单点对流　pin-point convection
单点雨量　point rainfall
单调差分格式【数】　monotone difference scheme
单调〔的〕　monotonic
单调性　monotonicity
单端系统　single-ended system
单端遥测　single-ended remote probing
单多普勒雷达　single Doppler radar
单反射　simple reflection
单反射体　simple reflector
单分散谱　monodispersed size distribution
单分散性【化】　monodispersion
单分子膜　monomolecular film
单峰分布【数】　unimodal distribution
单峰谱　unimodal spectrum
单个微下击暴流　individual microburst
单个雪面波纹脊　sastruga
单管压力表　single-column manometer
单航向导航　single-heading navigation
单核　individual nuclei
单基地非相干散射雷达　monostatic incoherent scatter radar
单基地光〔雷〕达　monostatic lidar
单基地雷达【电子】　monostatic radar
单基地声〔雷〕达　monostatic acoustic radar
单基地声学探测器　monostatic acoustic sounder
单极电荷　unipolar electrical charge
单极黑子　unipolar sunspot
单极黑子群　unipolar group
单经纬仪观测　single-theodolite observation
单离子　singleion
单片计算机【计】　single-chip computer
单偏航订正　single drift correction
单谱　monodisperse
单鞘　single sheath
单区随机试验　randomized experiment at single domain
单色的　monochromatic
单色辐射【大气】　monochromatic radiation
单色光【物】　monochromatic light
单色亮度　monochromatic brightness
单色平衡　monochromatic equilibrium
单色探测率　spectral detectivity
单色透射比　monochromatic transmit-

tance
单色仪【物】 monochromator
单丝静电表 unifilar electrometer
单糖 monosaccharide【生化】; cell【大气】
单体 monomer【生化】; cell【大气】
单体对流 cellular convection
单体风暴 single-cell storm
单体环流 cellular circulation
单体回波【大气】 cell echo
单体假说 cellular hypothesis
单体雷暴 unit storm
单通滤波〔器〕 unitary filter
单网格航向导航 single-grid-heading navigation
单位【物】 unit
单位常态分布 unit distribution
单位函数【数】 unit function
单位阶梯函数 unit step function
单位矩阵【数】 identity matrix
单位面积 elementary area
单位面积雷达截面积 radar cross section per unit area
单位深度地温差 geothermic step
单位矢量 unit vector
单位衰减 specific attenuation
单位水位线 unit hydrograph
单位正态分布 unit normal distribution
单位贮水量 specific storage
单相关 simple correlation
单相性【化】 monochromaticity
单向垂直风切变 unidirectional vertical wind shear
单向反射 regular reflection
单向反射体 regular reflector
单向方法 one-way approach
单向夹卷过程 one-way entrainment process
单向嵌套模式 one-way nested model
单向〔嵌〕套网格 one-way grid nesting
单向通量 unidirectional flux

单向影响 one-way interaction
〔单叶〕对流层顶 tropopause leaf
单元肥料【土壤】 straight fertilizer
单元〔铷蒸气〕磁强计 one-cell magnetometer
单元体积 elementary volume
单站分析 single station analysis
单站气象要素变化图 station continuity chart
单站〔天气〕预报【大气】 single station 〔weather〕forecast
单站预报 single observer forecast
单值化【数】 uniformization
单质【化】 elementary substance
淡积云【大气】 cumulus humilis (Cu hum)
淡水 fresh water
淡水湖【地理】 fresh water lake
淡(云) humilis(hum)
淡晕 hevelian halo
弹道风 equivalent constant wind
弹道风【大气】 ballistic wind
弹道空气密度【大气】 ballistic air density
弹道密度 ballistic density
弹道气象学 ballistic meteorology
弹道温度【大气】 ballistic temperature
弹道学 ballistics
弹射器 ejection chamber
弹射探空仪 ejectable radiosonde
弹性【物】 elasticity
弹性冰 rubber ice
弹性波【力】 elastic wave
弹性后向散射 elastic backscattering
弹性模量【物】 elastic modulus
弹性碰撞【物】 elastic collision
弹性散射【物】 elastic scattering
弹性体【物】 elastic body
蛋白质工程【生化】 protein engineering
氮(气) nitrogen (N)
氮循环 nitrogen cycle

氮氧化物　nitrogen oxides(NO_x)
当地水位【航海】　local water level
档案　archive
刀〔状〕冰　penknife ice
氘【化】　deuterium
氘核【化】　deuteron
导波　guided wave
导出单位　derived unit
导出阵风速度　derived gust velocity
导出资料　derived data
导电层　electrosphere
导〔电〕体　electric conductor
导电性　electric conductivity
导风板　deflector
导航测风　navigation aid in wind finding
导航测风【大气】　navaid wind-finding
导航雷达【电子】　navigation radar
导航设备　navigation facility, navigational aid
导航设备【海洋】　navigation equipment
导航卫星【电子】　navigation satellite
导航系统　navaid system
导航〔信号〕接收机　navigation receiver
导热率廓线　thermal conductivity profile
导热性　thermal conductivity
导式波　leading wave
导数法监测　derivative monitoring
导水率　hydraulic conductivity
导水率检验　slug test
导体　conductor
导(助)航设备　navaid
岛礁区航行【航海】　navigating in rocky water
岛屿效应　island effect
倒槽【大气】　inverted trough
倒春寒　cold of the late spring
倒春寒【大气】　late spring cold
倒春寒【农】　late spring coldness
倒黄梅　reoccurrence of Meiyu
倒暖锋　reverted warm front
倒排索引文件【地信】　indexed non-sequential file
倒谱　cepstrum
倒像【物】　inverted image
倒V型　inverted V-pattern
倒易定理【物】　reciprocity theorem
倒置地貌【地理】　inverted landform
到达角　angle of arrival
〔到达〕时差定位法　time-of-arrival technique
到〔达〕时差法　arrival-time difference technique
到时【地】　arrival time
到时差【地】　arrival time difference
道尔顿定律【大气】　Dalton's law
道氏元件　Dew cell
德拜屏蔽距离　Debye shielding distance
德林杰效应　Dellinger effect, fadeout
德沃夏克技术　Dvorak technique
灯船【海洋】　light boat
灯船测站　lightship station
灯船电码　lightship code
灯光能见距离【大气】　visual range of light
灯丝电流　filament current
灯塔【海洋】　light house
登陆　landfall
等比级数【数】　geometric series
等比容的　isosteric
等比容面　isosteric surface
等比容线　isostere
等比值线　isomer
等边三角形【数】　equilateral triangle
等变(词头)　isallo-
等变高风　isallohypsic wind
等变高线【大气】　isallohypse
等变温线【大气】　isallotherm
等变压的　isallobaric
等变压分析　isallobaric analysis
等变压风　isallobaric wind
等变压梯度　isallobaric gradient
等变压图　isallobaric chart

等变压线【大气】 isallobar
等变压效应 isallobaric effect
等变压值线 isophasm(of pressure)
等冰冻线 congelont
等差级数【数】 arithmetic series
等潮差线 coamplitude line, corange line
等潮差线【航海】 corange line
等潮时 cotidal hour
等潮〔时〕线 cotidal line
等春温线 isoeral
等磁变线 isoporic
等〔磁〕力的 isodynamic
等〔磁〕力线 isodynam
等磁线 isomagnetic line
等磁异常线【地】 magnetic isoanomalous line
等待时间 latency
等地温线 isogeotherm, geoisotherm
等电子密度面 constant electron density surface
等冬温线 isocheim, isochimene
等动力学温度【化】 isokinetic temperature
等动能取样法 isokinetic sampling
等动能线 isokinetic
等冻期线 isopag(ue)
等风速线 isovel, isotach, isanemone
等风速线【大气】 isotach
等风速线分析 isotach analysis
等风速线图 isotach chart
等风向线【大气】 isogon
等峰态曲线 isokurtic curve
等锋等压电码 isofronts-preiso code
等负变压线 kata-isallobar
等高〔度〕的 isohypsic
等高度圈 equal altitude circle
等高面 isohypsic surface, constant-level surface, constant-height surface
等高面图 fixed level chart, isohypsic chart, constant level chart, constant-height chart
等高平面位置显示器【大气】 constant altitude plan position indicator (CAPPI)
等高线 isohypse, contour, isoheight
等高线【大气】 contour line
等高线〔分布〕型〔式〕 hypsography
等高线高度 contour height
等高线间隔 contour interval
等高仪【测】 astrolabe
等焓过程【化】 isenthalpic process
等厚度线 isopach, constant thickness line
等厚度线【大气】 isopleth of thickness
等厚干涉【物】 equal thickness interference
等花期线 isoanth
等环境热量计【化】 isoperibolic calorimeter
等回波强度线 iso-echo
等回波线【大气】 iso-echo contour
等基线 isobase
等级值线 isometeorograde
等价 equivalent
〔等价〕沉积〔物〕直径 sedimentation diameter
等价定理 equivalence theorem
等价宽度 equivalent width
等价日射气温 sol-air temperature
等价语句 equivalence statement
等降水大陆度 isepire
等降压中心 isokatabaric center
等角地图 conformal map, isogonal map, orthomorphic map
等角航线 rhumb line course
等角螺线 equiangular spiral
等角投影【测】 conformal projection
等角线 rhumb line
等解冻线 isotac
等距平 equideparture
等距平线 iso-abnormal line, isanomal

等距平线【大气】 isanomaly
等距平线〔的〕 isanomalous
等绝对涡度轨迹 constant absolute vorticity trajectory
等亏率线 isodef
等雷暴日数线 isobront
等雷频〔的〕 isokeraunic
等雷雨线 isoceraunic, isokeraunic line
等离〔子〕体【物】【大气】 plasma
等离〔子〕体层【物】【大气】 plasmasphere
等离〔子〕体层顶【物】【大气】 plasmapause
等离子体过程 plasma process
等离子体回波 plasma echo
等离子体扩散平衡 plasma diffusive equilibrium
等离子体流 plasma flow
等离子体幔 plasma mantle
等离子体漂移速度 plasma drift velocity
等离子体频率 plasma frequency
等离子体频散函数 plasma dispersion function
等离子体温度 plasma temperature
等离〔子〕体物理学【物】 plasma physics
等离子〔体〕显示【计】 plasma display
等离子体约束 plasma confinement
等离子体云 plasma cloud
等亮度线 isopleth of brightness
等〔流〕速线方法 velocity-contour method
等露点线【大气】 isodrosotherm
等密度 isopycnal
等密度的 isopycnic
等密度方向 epipycnal
等密度混合 isopycnal mixing
等密度面 isopycnal surface, isopycnic surface, equidensen
等密度面混合 epipycnal mixing
等密度线 isostath, isopycnic line
等密线 isodense

等面积变换 equal area transformation
等面积地图 equal-area map
等面积投影 equal surface projection
等面积图 equal area chart
等能管 isoenergy pipe
等能面分析 isoenergy chart analysis
等年温较差线 isotalant, isoparallage, iseoric line
等偏差 isametric
等偏差线 isanomalous line, isanamal, isametral
等偏振信号 copolarized signal
等气候线 isoclimatic line
等〔气候因素〕月变线 isodiaphore
等气压较差线 isokatanabar, isanakatabar
等气压较差线图 isanakatabaric chart
等气压平衡线 isothene
等切变线 isoshear
等倾线 isoclinal(=isoclinic) line
等秋温线 isometropal
等球体 isosphere
等日射线 isalea
等日效应 equidiurnal effect
等日照线【大气】 isohel
等容变化 isochoric change
等容〔的〕 isochoric
等容过程【化】 isochoric process
等容积取样 constant volume sampling (CVS)
等容面 isasteric surface
等容气球 constant volume balloon, tetroon【大气】
等容气温表 constant volume gas thermometer
等容热容【化】 heat capacity at constant volume
等熵变化 isentropic change
等熵薄层 isentropic sheet
等熵垂直坐标 isentropic vertical coordinate

等熵大气　isentropic atmosphere
等熵的　isentropic
等熵分析【大气】　isoentropic analysis
等熵冠　isentropic cap
等熵轨迹　isentropic trajectory
等熵过程　isentropic process
等熵哈得来环流　isentropic Hadley circulation
等熵厚度图　thick-thin chart, isentropic thickness chart
等熵厚度图　isentropic thickness-chart
等熵混合　isentropic mixing
等熵面　isentropic surface
等熵面分析　isentropic analysis
等熵面图【大气】　isentropic chart
等熵凝结高度【大气】　isentropic condensation level
等熵平均经向环流　isentropic mean meridional circulation
等熵位涡　isentropic potential vorticity（IPV）
等熵涡度　isentropic vorticity
等熵沃克环流　isentropic Walker-type circulation
等熵线　isoentrope, isentropics, isentrope
等熵运动　isentropic motion
等熵质量传送　isentropic mass transport
等熵质量环流　isentropic mass circulation
等熵重量图　isentropic weight-chart
等熵坐标　isentropic coordinates
等〔上〕升速〔度〕　isanabation
等上升速度线　isanabat
等深线　fathom curve
等深线【测】　depth contour
等渗压线　isotonicty
等升压中心　isoanabaric center
等湿　constant humidity
等湿度线　isohygrometric line
等湿度线【大气】　isohume
等湿线　line of constant wetness, constant humidity line
等时线　isochrone
等时线（航空）　hour-out line
等始花线（物候学）　isanthesic line
等式【数】　equality
等势面【物】　equipotential surface
等势线【物】　equipotential line
等〔水〕深线　isobath
等水温线　isothermobath
等速度　uniform velocity
等梯度线　isogradient
等通量层　constant flux layer
等位势面　geopotential surface
等位势线　isopotential
等位温　equipotential temperature
等温变化　isothermal change
等温层　homothermy
等温层【大气】　isothermal layer
等温差商数　thermo-isodromic quotient
等温差商数线　thermo-isodrome
等温大气【大气】　isothermal atmosphere
等温带　zone of constant temperature
等温反应器【化】　isothermal reactor
等温过程【大气】　isothermal process
等温面　isothermal surface, constant temperature surface
等温膨胀　isothermal expansion
等温平衡　isothermal equilibrium
等温深度线　isobathytherm
等温线　isothermal line, isothermohyps, thermoisopleth, constant temperature line
等温线【大气】　isotherm
等温〔线〕的　isothermal
等温线密集区　isotherm ribbon
等温压缩　isothermal compression
等温云室　isothermal（cloud）chamber
等物候线【大气】　isophene(=isophane)
等物候线【农】　isophenological line
等系数线　iso-coefficient

等夏温线 isothere(=isotheral)	等压高度图 isobaric contour chart

等夏温线　isothere(=isotheral)
等相关线　isocorrelation
等消光点　isobestic point
等效暴雨　equivalent storms
等效带宽　equivalent band width
等效地形学　equivalent topography
等效反射〔率〕因子【大气】　equivalent reflectivity factor
等效风区　equivalent fetch
等效风时　equivalent duration
等效辐射温度【电子】　equivalent radiant temperature
等效黑体温度　equivalent blackbody temperature (TBB)
等效机场高度【大气】　equivalent altitude of aerodrome
等效截面【电子】　equivalent cross section
等效静止暴雨　equivalent stationary rainstorm (=equivalent stationary storm)
等效路径　equivalent path
等效逆风　equivalent head-wind
等效晴空辐射率【大气】　equivalent clear column radiance
等效顺风　equivalent tail wind
等效梯度风　equivalent gradient wind
等效英尺-烛光　equivalent foot-candle
等效雨量　equivalent precipitation
等效雨量计密度　equivalent rain gauge density
等效原理【物】　equivalence principle
等效重力波　equivalent gravity waves
等效纵向风　equivalent longitudinal wind
等斜线　isoclines
等斜褶皱【地质】　isoclinal fold
等雪量线　isonival, isonif
等雪量线【大气】　isochion
等雪深线　isohion
等压的　isobaric
等压分析　isobaric analysis
等压辐散　isobaric divergence

等压高度图　isobaric contour chart
等压高度型〔式〕　isobaric topography (=height pattern)
等压管　isobaric tube
等压过程【化】　isobaric process
等压厚度图　isobaric thickness chart
等压混合　isobaric mixing
等压冷却　isobaric cooling
等压面　equipressure surface, constant pressure surface
等压面【大气】　isobaric surface
等压面飞行　constant-pressure-pattern flight
等压面能量传送　isobaric energy transport
等压面坡度【大气】　slope of an isobaric surface
等压面图　isobaric chart, constant pressure chart
等压面图【大气】　contour chart
等压面形势　pressure topography
等压面预报图　prognostic contour chart
等压热容【化】　heat capacity at constant pressure
等压湿球温度　isobaric wet-bulb temperature
等压条件　isobaric condition
等压温度变化　isobaric temperature change
等压涡度　isobaric vorticity
等压线　isopiestic, isobaric line, isobarometric line, equipressure
等压线【大气】　isobar
等压线间　isobaric channel
等压线型式　isobar type
等压相当温度【大气】　isobaric equivalent temperature
等压运动　isobaric motion
等压坐标　isobaric coordinates
等盐度线　isohaline
等腰梯形【数】　isosceles trapezoid

等异常线　isabnormal line, iso-abnormal, isabnormal
等引力带　equigravisphere
等雨量　equal precipitation
等雨量线　isopluvial
等雨量线【大气】　isohyet
等雨量线【农】　equipluves
等雨量线图　isohyetal map
等雨率线　isomeric line
等雨深　equal rainfall depth
等云量线【大气】　isoneph
等晕区　isoplanatic region
等晕条件　isoplanatic condition
等照度线　isophote
等振幅线　iso-amplitude line
等蒸发量线　isothyme
等蒸发线　isoombre, isoatmic
等正变压线　anisallobar
等正温线　iso-orthotherm
等值高度　equivalent height
等值面　isotimic surface, equiscalar surface
等值线　isotimic, isotimic line, isogram, isopleth, isarithm, equiscalar line
等值线【大气】　isoline
等D值线（或面）　iso-D
等值正射频率　equivalent normal incidence frequency
等质化　homogenization
等质面　equisubstantial surface
等重力位势面　equigeopotential surface
等轴的　isometric
等逐日变差线　isometabole
邓莫尔型湿度传感器　Dunmore-type humidity sensor
低参考信号　low-reference signal
低槽张弛　relaxation of a trough
低层大气　lower atmosphere layer, inferior atmospheric layer
低层大气【大气】　lower atmosphere
低层逆温　low-level inversion

低层气温　low-level air temperature
低差　undershoot
低潮　low tide
低潮面　low-water plane(LWP)
低潮时【航海】　low water time
低潮线　low-water line
低吹尘　drifting dust
低吹沙【大气】　drifting sand
低吹雪【大气】　drifting snow
低低潮【航海】　lower low water
低地【地理】　lowland
低地球轨道　low erath orbiting
低度拟合　underfitting
低度真空　low vacuum
低反曙暮光　lower antitwilight
低分辨率红外辐射仪　low resolution infrared radiometer (LRIR)
低分辨率全向辐射仪　low resolution omnidirectional radiometer
低分辨率〔云图〕传真【大气】　low resolution facsimile(LR-FAX)
低峰态　platykurtosis
低高潮【航海】　lower high water
低阶模式　low order model(LOM)
低空　minimum altitude
〔低空〕大气结构探空仪　structure sonde
低空风切变【大气】　low-level wind shear(LLWS)
低空急流　lower tropospheric jet
低空急流【大气】　low-level jet stream, low-level jet(LLJ)
低空切变　low-level shear
低空搜索雷达【电子】　low altitude surveillance radar
低空温度探空仪　mini-T-sonde
低空〔无线电〕探空仪　low-level radiosonde, low altitude radiosonde
低密度脂蛋白【医学】　low density lipoprotein
低能等离子体　low-energy plasma
低能粒子　low-energy particle

低能质子α粒子计数仪　low energy proton alpha telescope（LEPAT）
低浓度肥料【土壤】　low-analysis fertilizer
低频　low frequency（LF）
低频变化　low-frequency variability
低频体系　low-frequency regime
低频天电　sferix
低频振荡　low frequency oscillation（LFO）
低频振动　low-frequency fluctuation
低平海岸【海洋】　low coast
低气压　infrabar
低〔气〕压【大气】　low〔pressure〕, depression
低浅峡湾　fiard
低水〔位〕　low water
低水位　low water level
低水〔位〕标志　low-water mark
低速层　low-velocity layer
低速等离子体流　low speed plasma flow
低碳经济　low carbon economy
低通滤波〔器〕　low pass filter
低通振荡　low pass fluctuation
低纬〔度〕　low latitude
低维数　low dimensionality
低温　microtherm
低温测湿术　low-temperature hygrometry
低温抽气泵　cryopump
低温〔的〕　microthermal
低温等离〔子〕体【物】　low-temperature plasma
低温电子学　cryoelectronics
低温干燥【农】　low temperature drying
低温冷害　chilling damage
低温气候【大气】　microthermal climate
低温生物化学【生化】　cryobiochemistry
低温湿度计　cryogenic hygrometer
低温植物〔的〕　microthermic
低温植物型　microthermal type
低涡【大气】　vortex

低限〔光〕对比　threshold contrast
低限深度　threshold depth
低限照度　threshold illuminance
低压　low, barometric low
低压槽　trough of low pressure
低压槽【大气】　trough
低压带　depression belt
低压后部　rear of depression
低压加深【大气】　deepening of a depression
低压路径【大气】　track of a depression
低压区　meiobar, low pressure area
低压填塞【大气】　filling of a depression
低压系统　low-pressure system
低压云系　cloud system of a depression
低压再生【大气】　regeneration of a depression
低压中心　low pressure center
〔低压〕中心跳跃　center jump
低压轴　axis of low, axis of depression
低压族　family of depressions
低盐度层　low salinity layer（LSL）
低云　low-level cloud, lower cloud
低云【大气】　low cloud
低指数【大气】　low index
滴　drop
滴定【化】　titration
滴定〔分析〕法【化】　titrimetry, titrimetric analysis
滴定剂【化】　titrant
滴谱　drop-size distribution, droplet spectrum
滴谱【大气】　drop spectrum
滴谱参数【大气】　drop-size distribution parameter
滴谱仪　drop-size meter, droplet collector
滴群　population of droplet
狄克辐射计　Dicke radiometer
狄拉克方程【数】　Dirac equation
〔狄拉克〕δ函数　Dirac delta function

狄利克雷问题【数】 Dirichlet problem
迪尔多夫模式 Deardorff's model
迪尔多夫速度 Deardorff velocity
迪肯风速廓线参数 Deacon wind profile parameter
敌友识别装置 identification：friend or foe（IFF）
笛卡儿射线 Descartes ray
笛卡儿张量 Cartesian tensor
笛卡儿坐标【数】 Cartesian coordinates
抵偿 counter balance
底板 base plate
底边界条件 bottom boundary condition
底冰 ground ice，subsoil ice，bottom ice
底冰滑动 basal sliding
底层 bottom layer
底〔层〕流 bottom current
底〔层〕水 bottom water
底层水温 bottom temperature
底宽（水文） base width
底拦截波 bottom-trapped wave
底摩擦 bottom friction
底片比例尺【天】 plate scale
底栖生物 benthos
底散云（水下核爆炸所形成的雾、水和碎片混合云） base surge
底图 base map
底土 subsoil
地标导航 landmark navigation
地表臭氧 surface ozone
地表臭氧浓度记录器 surface ozone concentration recorder
地表电位梯度记录器 surface potential gradient recorder
地表辐射差额 radiation balance of earth's surface
地表辐射出射率 emittance of the earth's surface
地表辐射平衡 surface radiation balance
地表辐射收支【大气】 surface radiation budget

地表径流 surface runoff
地表能量平衡 surface energy balance
地表能量收支 surface energy budget
地表水 surface water
地表水文学 surface water hyrology
地表水文学要素 elements of surface hydrology
（地表水与地下水）联合利用 conjunctive use
地表温度 ground surface temperature
地表温度表 soil surface thermometer
地表温度分布测绘 surface temperature mapping
地表修复格式 film renewal model, surface renewal model
地波【航海】 ground wave
地波传播 ground-wave propagation
地层流体 formation fluid
地层学 stratigraphy
地磁 terrestrial magnetism，earth's magnetism
地磁暴 geomagnetic storm
地磁暴初相 initial geomagnetic storm phase
〔地〕磁变 magnetic variation
地磁测量 magnetic survey
地磁测量【地】 geomagnetic survey
地磁场 geomagnetic field，earth's magnetic field
地磁场〔磁力〕线 earth's magnetic field line
地磁潮 geomagnetic tide
地磁赤道 magnetic equator，geomagnetic equator
地磁等年变线 isopore
地磁感应仪 earth inductor
地磁钩扰 magnetic crotchet
地磁活动指数 geomagnetic activity index
地磁极【航海】 geomagnetic pole
〔地〕磁宁静期 geomagnetically quiet time
地磁偶极子 geomagnetic dipole

地磁漂移【地质】 geomagnetic excursion
〔地〕磁扰 geomagnetic disturbance
〔地磁〕扰动日 active day, disturbed day
〔地磁〕扰日变化 disturbance daily variation
地磁图 magnetic map
地磁微脉动 magnetic micropulsation
地磁学 geomagnetics
地磁〔学〕【地】 geomagnetism
地磁要素 magnetic element, geomagnetic element
〔地〕磁子午圈 geomagnetic meridian
地磁坐标【地】 geomagnetic coordinate
地带性【地理】 zonality
地对空监测 ground-based monitoring
地方标准时 local standard time
地方恒星时 local sideteal time
地方民用时 local civil time
地方平〔均〕时 local mean time
地方时 local time
地方时角 local hour angle
地方视时 local apparent time
地方太阳时 local solar time
地方天气报告 local weather report
地方性风【大气】 local wind
地方性降水【大气】 local precipitation
地方性雷暴 local thunderstorm
地方性天气【大气】 local weather
地方真时 local true time
地方志 local history
地方自动化模式输出统计预报计划 local AFOS MOS program（LAMP）
地固坐标系【地信】 earth-fixed coordinate system
地基电离层探测仪 ground-based ionosonde
地基观测【大气】 ground-based observation
地基光〔雷〕达 ground-based lidar
地极 earth pole
地极【航海】 terrestrial pole

地籍【测】 cadastre
地籍测量【测】 cadastral survey
地籍属地【地信】cadastral attribute
地籍信息系统【地信】 cadastral information system
地壳【地质】 earth crust
地壳均衡 isostasy
地壳运动 diastrophism
地坑雨量器 pit gauge
地-空电流 earth-air current
地理标识符【地信】 geographic reference system
地理单元【地理】 geographical unit
地理底图【地理】 geographic base map, cartographic base
地理地平 geographic horizon
地理分布【地理】 geographical distribution
地理分析【地理】 geoanalysis
地理过程【地理】 geographical process
地理环境【地理】 geographical environment
地理〔环境〕医学 geomedicine
地理计量学【地理】 geographimetrics
地理界线【地理】 geographical boundary
地理经度 geographical longitude
地理精度【地理】 geographic accuracy
地理考察【地理】 geographical survey
地理可能日照时数 geographically possible sunshine duration
地理剖面【地理】 geographic profile
地理圈【地理】 geographical sphere
地理数据采编【地理】 geographical data handling
地理数据库【地理】 geographical data bank
地理通名【地理】 general geographical name
地理网格 geographic grid
地理位置 geographic location
地理纬度 geographical latitude

地理纬圈【地信】 geographic parallel
地理系统【地理】 geosystem
地理相关法【地理】 geographic correlation
地理信息系统【地理】 geographical information system（GIS）
地理学 geography
地理学体系【地理】 geographical system
地理遥感【地理】 geographical remote sensing
地理因子【地理】 geographical factors
地理制图【地理】 geographic mapping
地理子午线 geographic meridian
地理综合【地理】 geographical synthesis
地理坐标 geographical coordinate
地理坐标【地理】 geographic coordinate
地理坐标网【地理】 geographic graticule
地幔对流【地】 mantle convection
地貌【地理】 landform
地貌成因【地理】 landform genesis, geomorphogenesis
地貌倒置【地理】 inversion of landform
地貌过程【地理】 geomorphological process
地貌年代学【地理】 geomorphochronology
地貌世代【地理】 generation of landforms
地貌形成作用【地理】 landform forming process
地貌序列【地理】 landforms series
地貌学【地理】 geomorphology
地冕【地】 geocorona
地面 surface, earth's surface
地面边界层【大气】 surface boundary layer
地面边界条件 surface boundary condition
地面波 surface wave

地面槽【大气】 surface trough
地面测雹器 surface hail sensor
地面测试设备 ground test equipment
地面沉降【地质】 land subsidence
地面磁照图 ground magnetogram
地面粗糙度【大气】 surface roughness
地面粗糙度〔长度〕 surface roughness length
地面大气电导率记录器 surface electrical conductivity recorder
地面冻结指数 surface freezing-index
地面发生器 ground generator
地面反照率 surface albedo
地面〔放射性〕落尘图 ground fallout plot
地面分辨率【地理】 ground resolution
地面分析 surface analysis
地面风【大气】 surface wind
地面风应力 surface wind stress
地面锋【大气】 surface front
地面辐射温度表 terrestrial radiation thermometer
地面干燥率 geoclimatic drying power
地面观测【大气】 surface observation
地面观测系统 surface-based observation system
地面轨迹 ground track
地面接收和指令站 ground acquisition and command station（GA&CS）
地面接收站【地理】 ground receiving station
地面径流 overland flow
地面雷达天气观测报告 report of ground radar weather observation（RADOB）
地面流 surface flow
地面流光 ground streamer
地面摩擦 surface friction
地面目标 ground target
地面能见度 ground visibility
地面能见度【大气】 surface visibility
地面逆温 ground-level inversion, ground

inversion
地面逆温【大气】 surface〔temperature〕inversion
地面浓度 ground level concentration
地面气温 surface air temperature
地面气象测报业务软件 Operational Software for Surface Meteorological Observation（OSSMO）
地面〔气象〕观测 surface air observation
地面气压【大气】 surface pressure
地面融化指数 surface thawing-index
地面湿度 surface humidity
地面湿静力能量分析 surface wet static energy analysis
地面〔实况测量〕资料 ground data
地面霜 ground frost
地面水资源 ground water resources
地面抬升指数 surface-based lifted index（SBLI）
地面天气观测 surface weather observation
地面天气图 surface weather chart, surface synoptic chart
地面〔天气〕图【大气】 surface chart
地面天气形势预报图 prebaratic chart
地面天气站 surface synoptic station
地面通量 surface flux, ground flux
地面同时等温线 syngeothermal line
地面图 surface map
地面湍流 surface turbulence
地面温度【大气】 surface temperature
地面温度表 surface thermometer
地面温度表【大气】 surface geothermometer
地面雾【大气】 ground fog
地面现象 ground phenomenon
地面校正箱 ground-check chamber
地面蓄水 surface retention
地面雪崩 ground avalanche
地面预报图【大气】 surface forecast chart

地面源 ground level source
地面站 ground station
地面真值 ground truth
地面蒸发【农】 evaporation from land surface
地面滞留 surface storage
地面状态【大气】 state of ground
地面资料【大气】 surface data
地面自动观测系统 automated surface observing system（ASOS）
地面阻力 surface resistance
地名【地理】 geographical name
地名数据库【地理】 geographical name data bank
地平传感器 horizon sensor
地平交互指示器 horizon-crossing indicator
地平经圈【天】 vertical circle
地平经纬仪, 高度方位仪 altazimuth
地平亮光 horizon brightening
地平〔面〕 ground level
地平倾角 dip of the horizon
地平圈 horizontal circle, horizon
地平视差【航海】 horizontal parallax
地平纬圈 almucantar
地平线 skyline
地平坐标系【航海】 horizontal coordinate system
地气辐射收支 earth-atmosphere radiation budget
地气系统 earth-atmosphere system
地气系统反照率【大气】 albedo of the earth-atmosphere system
地气系统辐射差额 radiation balance of the earth-atmosphere system
地球 earth, terrestrial globe
地球扁率【天】 oblateness of the earth
地球表层【地理】 epigeosphere
地球表面【地理】 earth surface
地球表面辐射（不包括大气辐射） terrestrial surface radiation

地球赤道平面　earth equatorial plane
地球电离层波导　earth-ionosphere waveguide
地球动力学【地】　geodynamics
地球反射辐射　reflected global radiation
地球反照　earth light，earth shine
地球反照率　albedo of the earth
地球辐射【大气】　terrestrial radiation, earth radiation
地球辐射带　radiation belts of the earth
地球辐射带【大气】　earth radiation belt
地球辐射平衡　terrestrial radiation balance
地球辐射收支　earth radiation budget (ERB)
地球辐射收支扫描辐射仪　earth radiation budget scanning radiometer
地球辐射收支试验【大气】　earth radiation budget experiment (ERBE)
地球辐射收支卫星【大气】　Earth Radiation Budget Satellite (ERBS)
地球弓形激波　earth bow shock
地球工程学　geotechnologie
地球公转　revolution of the earth, earth revolution
地球观测系统【电子】【大气】　earth observing system (EOS)
地球轨道特征　orbital characteristics
地球红外辐射　terrestrial infrared radiation
地球化学【地质】　geochemistry
地球化学方法　geochemical method
地球化学环境　geochemical environment
地球化学景观【地理】　geochemical landscape
地球化学模型【地质】　geochemical model
地球环境　earth's environment
地球环境科学　earth environmental sciences
地球监视　earthwatch

地球结构　earth structure
地球静止气象卫星【大气】　geostationary meteorological satellite (GMS)
地球静止气象卫星系统（日本）　geostationary meteorological satellite system (GMSS)
地球静止卫星　geostationary satellite
地球静止业务环境卫星　Geostationary Operational Environmental Satellite (GOES)
地球静止业务气象卫星（俄罗斯发射的卫星）　Geostationary Operational Meteorological Satellite (GOMS)
地球科学【地质】　earth science
地球科学信息学会（美国）　Geoscience Information Society (GIS)
地球内部化学【地】　interior chemistry of earth
地球气候系统　earth's climate system
地球曲率订正　earth curvature correction
地球闪烁　terrestrial scintillation
地球生物化学【生化】　geobiochemistry
地球同步轨道　geosynchronous orbit, earth-synchronous orbit
地球同步甚高分辨率辐射仪　geosynchronous very high resolution radiometer (GVHRR)
地球同步卫星【大气】　geosynchronous satellite
地球椭球【测】　earth ellipsoid
地球稳定　earth stabilization
地球涡度　earth's vorticity
地球物理观象总台（前苏联）　Main Geophysical Observatory (MGO)
地球物理勘探【地】　geophysical exploration
地球物理科学　geophysical sciences
地球物理流体动力学　geophysical fluid dynamics
地球物理流体动力学实验室（美国）

Geophysical Fluid Dynamics Laboratory (GFDL)
地球物理年　Geophysical Year
地球物理日　geophysical day
地球物理探测火箭　geoprobe
地球物理学　earth physics
地球物理学【地】　geophysics
地球物理学模拟　geophysical modelling
地球物理异常【地】　geophysical anomaly
地球物理与行星物理研究所（美国）　Institute of Geophysics and Planetary Physics（IGPP）
地球物理资料　geophysical data
地球系统的协调观测和预报（WCRP 的 2005—2015 年战略框架）　Coordinated Observation and Prediction of the Earth System（COPES）
地球系统科学　earth system science
地球系统科学联盟　Earth System Science Partnership（ESSP）
地球〔行星〕反照率　albedo of the earth planetary
地球形状【地信】　earth figure
地球遥感卫星　earth remote sensing satellite
地球遥感系统　earth remote sensing system
地球引力　gravitational attraction
地〔球〕影　earth shadow
地球有效半径　effective radius of the earth
地球圆球体　terrestrial sphere
地球折射　terrestrial refraction
地球资源观测卫星【地信】　Earth Resources Observation Satellite（EROS）
地球资源观测系统【地信】　Earth Resource Observation System（EROS）
地球资源技术卫星【大气】　Earth Resource Technology Satellite（ERTS）
地球资源卫星　Earth Resources Satellite（ERS）

地球自转　earth rotation
地球自转参数【天】　earth rotation parameter
地球〔自转〕角速度　angular velocity of the earth
地球自转偏向力　deflection force of earth rotation
地球坐标　geodetic coordinate
地球坐标【船舶】　earth coordinate
地球坐标系　terrestrial coordinate system
地区传送　territorial transmission
〔地区〕等温线　choroisotherm
地区广播　territorial broadcast
地区气象辅助电信网　regional auxiliary meteorological telecommunication network
〔地〕区时【天】　zone〔standard〕time
地圈【地理】　geosphere
地热　terrestrial heat, geothermal heat
地热【地】　geoheat
地热等温线　geotherm
地热调查【地质】　geothermal survey
地热活动【地质】　geothermal activity
地热勘探【地质】　geothermal prospecting
地热能【地质】　geothermal energy
地热能【农】　geothermoenergy
地热水库【地质】　geothermal reservoir
地热田【地质】　geothermal field
地热系统【地质】　geothermal system
地热现象【地质】　geothermal phenomenon
地热学　geothermy
地热学【地质】　geothermics
地热异常【地质】　geothermal anomaly
地热资源【地质】　geothermal resources
地闪　ground flash
地生态学【地理】　geoecology
地史学【地质】　historical geology
地速　ground speed
地图比例尺　map scale

地图编绘【地理】 cartographic compilation
地图〔放大〕因子【大气】 map factor, scale factor
地图规范【地信】 map specification
地图判读【地理】 cartographic interpretation
地图投影【地理】 map projection
地图信息【地理】 cartographic information
地图学【地理】 cartography
地外生物学【天】 exobiology
地外文明【天】 extraterrestrial civilization
地温 earth temperature
地温【大气】 ground temperature
地温表 ground-thermometer, earth thermometer
地温表【大气】 geothermometer
地文航海【航海】 geo-navigation
地文学【地理】 physiography
地物回波【大气】 ground echo
地物杂波 ground clutter, clutter
地物杂波消除 clutter rejection
地下冰 underground ice, subterranean ice
地下河 subterranean stream
地下水 subsurface water, subterranean water, underground water, ground water
地下水坝 groundwater dam
地下水补给河 effluent stream
地下水测站 groundwater station
地下水层,饱和层 phreatic zone
地下水超采 groundwater mining (＝groundwater overexploitation)
地下水储量 groundwater storage
地下水储量【地质】 groundwater reserves
地下水定年 groundwater dating
地下〔水〕分水岭 groundwater divide
地下水耗减曲线 ground-water depletion curve
地下水耗竭 groundwater depletion

地下水〔径流〕 groundwater flow
地下水流 subsurface current
地下水流域 groundwater basin
地下水隆起 groundwater mound
地下水面 phreatic surface, underground water level, groundwater table, ground water level
地下水天然资源【地质】 natural resources of groundwater
地下水位【地理】 water table
地下水位线 phreatic line
地下水文学【地理】 ground water hydrology(＝groundwater hydrology)
地下水系统【地质】 groundwater system
地下水消落 groundwater recession
地下水涨落循环 phreatic cycle
地下增温率 geothermal gradient
地下增温深度 geothermic depth
地心地平 geocentric horizon
地心经度 geocentric longitude
地心体系【天】 geocentric system
地心天顶 geocentric zenith
地心纬度 geocentric latitude
地心引力 earth gravity
地心引力常数【天】 geocentric gravitational constant
地心坐标【天】 geocentric coordinate
地心坐标系 geocentric reference system
地形 relief
地形背风槽 orographic leeside trough
地形波 topographic waves, mountain wave
地形波【大气】 orographic wave
地形波动 topographically generated wave
地形波曳力 orographic wave drag
地形槽【大气】 topographic trough
地形的 orographic
地形等压线【大气】 topographic isobar
地形低压 orographic low
地形低压【大气】 orographic depression
地形对风的影响 geographic effect on

winds
地形反气旋　orographic anticyclone
地形放大因子　topographical amplification factor
地形风暴　orographic storm
地形风气流　orographic wind flow
地形锋生【大气】　orographic frontogenesis
地形锋生〔作用〕　topographic frontogenesis
地形跟踪雷达【电子】　terrain-following radar
地形锢囚　orographic occlusion
地形锢囚锋【大气】　orographic occluded front
地形后向散射体　topographic backscatter
地形降水【大气】　orographic precipitation
地形静止锋【大气】　orographic stationary front
地形雷暴【大气】　orographic thunderstorm
地形罗斯贝波　topographic Rossby wave
地形目标　topographic target
地形气候【大气】　topoclimate
地形气候学【大气】　topoclimatology
地形强迫流　topographically forced flow
地形强迫作用　mountain forcing
地形热效应　thermotopographic effect
地形抬升　orographic uplift
地形特性【地信】　terrain features
地形条件　orographic condition
地形图【地理】　topographic map
地形拖曳【大气】　orographic drag
地形涡旋　topographic vortex，orographic vortex
地形雾　orographic fog
地形下坡风　orographic downward wind
地形小气候【大气】　topomicroclimate，contour microclimate
地形效应　terrain effect，topographical effect，orographic effect

地形〔性〕上坡风　orographic upward wind
地形学　topography
地形雪线【大气】　orographic snowline
地形因子【地信】　terrain facter
地形影响　orographic influence
地形雨【大气】　orographic rain
地形雨量【大气】　orographic rainfall
地形云　orographic cloud
地形〔障碍〕急流【大气】　barrier jet
地形遮蔽图　constriction of the horizon
地形作用力　topographic forcing，orographic forcing
地学　geonomy
地学【地质】　geoscience
地学家　geoscientist
地学信息技术　geoinformatics
地衣　lichen
地影　shadow of the earth
地-月系统　earth-moon system
地云闪电【大气】　ground-to-cloud discharge
地震　seism，earthquake
地震波　earthquake wave
地震波【地】　seismic wave
〔地震〕波及区　felt area
地震地质学【地质】　seismogeology
地震构造带【地质】　seismic-tectonic zone
地震构造学【地质】　seismotectonics
地震海浪预警系统　seismic sea wave warning system（SSWWS）
地震活动带【地】　seismically active belt
地震活动区【地】　seismically active zone
地震计【地】　seismometer
地震烈度【地】　earthquake intensity
地震流动性【地】　seismicity
地震图【地】　seismogram
地震学【地】　seismology
地震仪【地】　seismograph
地支　Earthly Branches

地支【天】　terrestrial branch
地址标志　address marker
地址码　address code
地质代　geologic era, geological era
地质海洋学【海洋】　geological oceanography
地质化学　geological chemistry
地质环境【地质】　geologic environment
地质年表【地质】　geologic time scale
地质气候　geological climate
地质时代　geological age
地质时期【地质】　geologic time
地质时期（气候）　geological period
地质世　geologic epoch
地质温度表　geologic thermometer
地质〔学〕【地】　geology
地质灾害【地质】　geologic hazard
地致系统　terrain-induced system
地中海　Mediterranean Sea
地中海出流水　Mediterranean outflow
地中海锋　Mediterranean front
地中海季风　Etesians
地中海气候　Mediterranean climate, Etesian climate
地中海深层水　Mediterranean Deep Water
地中海水　Mediterranean Water
地中海透镜　Mediterranean lenses（亦称 Meddy）
地中海涡流　Meddy (Mediterranean Eddy)
地中海型　Mediterranean regime
地中海型气候【大气】　Mediterranean type climate
地轴　earth's axis
地轴【地质】　earth axis
地转〔的〕　geostrophic
地转动量　geostrophic momentum
地转动量近似　geostrophic momentum approximation
地转风【大气】　geostrophic wind
地转风方法　geostrophic wind method
地转〔风〕风速标尺　geostrophic wind scale
地转风风速分析器　geostrophic divider
地转风锋　geostrophic wind front
地转风高度　geostrophic wind level
地转风流　geostrophic wind flow
地转〔风气〕流　geostrophic current
地转风输送　geostrophic transport
地转风值　geostrophic value
地转锋　geostrophic front
地转惯性不稳定　geostrophic inertial instability
地转海洋学实时观测阵　Array for Real-time Geostrophic Oceanography（ARGO）
地转假定　geostrophic assumption
地转近似　geostrophic approximation
地转距离　geostrophic distance
地转里查森数　geostrophic Richardson number
地转流【大气】　geostrophic current
地转流函数　geostrophic streamfunction
地转偏差　geostrophic departure
地转偏差【大气】　geostrophic deviation
地转偏向力　deflecting force of earth rotation
地转平衡　geostrophic equilibrium, geostrophic balance
地转平流【大气】　geostrophic advection
地转平流尺度　geostrophic advection scale
地转切变变形　geostrophic shearing deformation
地转伸展变形　geostrophic stretching deformation
地转适应【大气】　geostrophic adjustment
地转通量　geostrophic flux
地转湍流　geostrophic turbulence
地转退化　geostrophic degeneracy
地转位势涡度　geostrophic potential vorticity
地转涡度【大气】　geostrophic vorticity
地转曳力　geostrophic drag

地转曳力系数　geostrophic drag coefficient
地转运动【大气】　geostrophic motion
地转增量　geostrophic increment
地转正交模　geostrophic normal mode
地转状态　geostrophy
地转坐标　geostrophic coordinates
递归性【数】　recursiveness
递推公式　recurrence formula
递推关系　recurrence relation
第二次青藏高原大气科学试验（1998）　Second Tibetan Plateau Meteorological Experiment（TIPEX）
第二代计算机【计】　second generation computer
第二辐射常数　second radiation constant
第二类海风　sea breeze of the second kind
第二类气候预测　climatic prediction of the second kind
第二类条件〔性〕不稳定【大气】　conditional instability of the second kind（CISK）
第二类重力湿对流不稳定度　gravitational moist-convective instability of the second kind
第二型冷锋【大气】　rapidly moving cold front（快行冷锋）
第二宇宙速度【天】　second cosmic velocity
第二紫霞　second purple light
第三纪　Tertiary
第三纪气候【大气】Tertiary climate
第三宇宙速度【天】　third cosmic velocity
第四纪　Quaternary
第四纪【地质】　Quaternary Period
第四纪冰期【地质】　Quaternary Ice Age
第四纪低海面期　quaternary low sea-level period
第四纪低海平面　Quaternary low sea level
第四纪气候【大气】　Quaternary climate
第四纪气候记录　Quaternary climatic record
第四系【地质】　Quaternary System
第一反射波　first-hop wave
第一辐射常数　first radiation constant
第一类典范坐标【数】　canonical coordinates of the first kind
第一类气候预测　climatic prediction of the first kind
第一类完全椭圆积分【数】　complete elliptic integral of the first kind
第一型冷锋【大气】　slowly moving cold front（慢行冷锋）
第一宇宙速度【物】　first cosmic velocity
第一紫霞　first purple light
蒂森多边形法　Thiessen polygon method
缔约方大会（联合国气候变化框架公约）　Conference of the Parties（COP）（UN Framework Convention on Climate Change）
颠簸　bump
颠簸飞行　bumpy flight
颠簸空气　rough air，bumpy air
颠簸性　bumpiness
颠倒温度表　reversing thermometer
典范矩阵【数】　canonical matrix
典型变量　canonical variable
典型回归　canonical regression
典型区域观测和预报服务　prototype regional observing and forecasting service（PROFS）
典型相关　canonical correlation
典型相关分析【数】　canonical correlation analysis（CCA）
点〔测〕资料　point data
点-点通信【电子】　point-to-point communication
点对点传送　point-to-point transmission
点-多点通信【电子】　point-to-multipoint communication

点划线　dotted-dashed line, dot dash line
点回波　pip
点火　firing
点火器　ignitron
点积　dot product
点接收机　point receiver
点聚图【大气】　scatter diagram
点扩展函数　point spread function
点每秒【计】　dots per second, dps
点每英寸【计】　dots per inch, dpi
点目标　point target
点涡【大气】　point vortex
点吸引子　point attractor
点线　dotted curve, dot line
点源　point source
点阵打印机【计】　dot matrix printer
点阵精度【计】　dot matrix size
碘　iodine
碘化钠探测器　sodium iodide detector
碘化物　iodide
碘化银　silver iodide
碘化银〔云〕催化【大气】　silver iodide seeding
电　electricity
电报　message
电报方程　telegraphic equation
电波折射率　refraction index of radio wave
电测辐射表　electrical substitution radiometer
电测干湿表　electropsychrometer
电测湿度仪　electrical hygrometer
电测温度表　electric thermometer, electrical thermometer
电测〔钻井〕记录曲线　electrical log
电场　electric field
电场强度　electric field strength, field strength, electric field intensity
电场强度计　field mill
电场效应　electrical field effect
电场跃变【化】　field jump

电传风杯风速表　electric cup anemometer
电传风速表　electrical anemometer
电传复印机　telecopier
电传机　teletype（见 teleprinter）
电传机【电子】　teleprinter
电传机打印输出　teletype printout
电〔传通〕信枢纽　telecommunication hub
电磁波　electromagnetic wave
电磁〔波〕谱　electromagnetic spectrum
电磁场　electromagnetic field
电磁方程　electromagnetic equation
电磁辐射　electromagnetic radiation
电磁干扰　electromagnetic interference
电磁感应【物】　electromagnetic induction
电磁海流计　jog-log
电磁回波探测　electromagnetic echo sounding
电磁兼容【电子】　electromagnetic compatibility (EMC)
电磁理论　electromagnetic theory
电磁能　electromagnetic energy
电磁耦合　electromagnetic coupling
电磁屏蔽　electromagnetic screen
电磁-声探测器　electromagnetic acoustic probe, EMAC probe
电磁学【物】　electromagnetism, electromagnetics
电导【物】　conductance
电导率　specific conductance, electrical conductivity
电导率【物】,传导性　conductivity
电导率-温度-水深廓线仪　conductivity-temperature-depth profiler (CTD)
电导率-盐度-温度-深度监测系统　CSTD monitoring system
电定标　electrical calibration
电动传声器　dynamic microphone
电动风速计　anemocinemograph
电镀【化】　electroplating

电感【物】 inductance
电荷 electrical charge, electric charge, charge
电荷分离 separation of charge, charge separation
电荷平衡 charge balance
电荷守恒【物】 charge conservation
电恒温器 electrothermostat
电化学【化】 electrochemistry
电化学发光【物】 electrochemiluminescence
电化学检测器 electrochemical detector (ECD)
电化学探空仪【大气】 electrochemical sonde
电化学-氧化碳分析仪 ecolyzer
电话应答式雨量站 rainfall phone-line network
电击穿 electric breakdown, electrical breakdown
电极化 electric polarization
电极效应 electrode effect
电集流【大气】 electrojet
电接风杯风速表 cup-contact anemometer
电接〔式〕风速表【大气】 contact anemometer
电解电容器 electrolytic capacitor
电解片 electrolytic strip
电解式湿度表 electrolytic hygrometer
电解式温度表 electrolytic thermometer
电解水制氢设备 electrolytic hydrogen generator
电解水制氢装置 electrolytic hydrogen plant
电介质【物】 dielectric
电离【物】 ionization
电离层 ionospheric layer, ionization layer
电离层【大气】 ionosphere
电离层暴【地】 ionospheric storm
电离层波 ionospheric wave

电离层参数 ionospheric parameter
电离层槽 ionospheric trough
电离层测高仪 ionosonde
电离层潮 ionospheric tide
电离层传播【电子】 ionospheric propagation
电离层等离子体 ionospheric plasma
电离层电磁吸收测定仪 relative ionospheric opacity-meter (riometer)
电离层电位 ionospheric potential
电离层顶【地】 ionopause
电离层风 ionospheric wind
电离层高度 ionospheric level
电离层观象台 ionospheric observatory
电离层记录器 ionospheric recorder
电离层雷达 ionospheric radar
〔电离层〕连续波测高仪 continuous wave ionosonde
电离层剖面 ionospheric profile
电离层倾斜 ionospheric tilt
电离层扰动 ionospheric disturbance
电离层散射通信【电子】 ionospheric scatter communication
电离层闪烁 ionospheric scintillation
电离层上部 upper ionosphere
电离层探测 ionospheric sounding, ionospheric probing
电离层探测卫星 Ionosphere Sounding Satellite (ISS)
电离层特性 ionospheric property
电离层突扰【大气】ionospheric sudden disturbance
电离层物理学 ionospherics
电离层下部 lower ionosphere
电离层响应 ionospheric response
电离层行扰【地】 travelling ionospheric disturbance (TID)
电离层亚暴 ionospheric substorm
电离大气 ionized atmosphere
电离电势 ionization potential
电离度【物】 degree of ionization

电离粒子　ionized particle
电离能【化】　ionization energy
电离平衡【化】　ionization equilibrium
电离室【物】　ionization chamber
电离室检测仪　ionization chamber detector
电离图【地】　ionogram
电离温度【天】　ionization temperature
电离效率　ionizing efficiency
电离阈值　ionization threshold
电离最低速〔率〕　minimum ionizing speed
电力　electric force
电力系统　power system
电力线　electric lines of force(＝electric field line)
电量测量法　coulometry
电量法氧化剂分析仪　coulometric oxidant analyzer
电流　electric current
电流表　galvanometer
电流体力学【物】　electrohydrodynamics
电漏　leakance
电码　code letter
电码表　code table
电码段　code section
电码符号　code symbol
电码格式【大气】　code form
电码名称　code name
电码式探空仪　code-sending radiosonde, code-type radiosonde
电码说明　code specification
电码型式【大气】　code form
电码种类【大气】　code kind
电码种类名称【大气】　code kind
电码组【大气】　code group
电敏〔记录〕纸　electrosensitive〔recording〕paper
电偶层【物】　electric double layer
电偶极跃进【物】　electric dipole transition

电偶极子【物】　electric dipole
电扰　electric disturbance
电热〔疗〕法　diathermy
电热式温度表　electrically-heated thermometer
电热温床　electric hot bed
电容传声器　capacitor（或 condenser）microphone
电容放电式风速表　condenser discharge anemometer
电容空盒　capacitive aneroid
电容式含水量测定仪　electric-capacity moisture meter
电容雨量器　capacitance rain gauge
电闪　lightning flash
电渗〔透〕〔作用〕　electro-osmosis
电声波　electro-acoustic wave
电声学　electroacoustics
电势降〔落〕【物】　potential drop
电视　television (TV)
电视〔传真〕广播　faxcasting
电视和红外辐射观测卫星　Television and Infrared Observation Satellite (TIROS)
电视图像　television picture
电视系统　television system
电通量【物】　electric flux
电位【电子】　electric potential
电位器【电子】　potentiometer
电位梯度　electric potential gradient
电位移【物】　electric displacement
电线积冰【大气】　wire icing
电线积冰自记仪器　ice-scope
电信【电子】　telecommunication
电学〔法〕气溶胶尺度分析仪　electrical aerosol size analyzer
电学现象　electrical phenomenon
电压【物】　voltage
电要素探空仪　electrosonde
电影经纬仪　cinetheodolite
电泳【化】　electrophoresis
电涌保护器　surge protection device

电源【物】 power supply, power source
电晕电流 corona current
电晕放电【物】 corona discharge
电晕放电〔现象〕 corposant
电致发光【物】 electroluminescence
电子【电子】 electron
电子包【物】 packet of electrons
电子产生速率 electron production rate
电子沉降 electron precipitation
电子出版系统【计】 electronic publishing system
电子导热性 electron thermal conductivity
电子等离子体频率 electron plasma frequency
电子等离子体振荡 electron plasma oscillation
电子电荷 electronic charge
电子读出装置 electronic readout equipment
电子风向显示器 electronic wind direction indicator
电子伏特 electron-volt (eV)
电子俘获检测 electron capture detection (ECD)
电子附着 electron attachment
电子干湿表 electronic psychrometer
电子光学【物】 electron optics
电子激发 electronic excitation
电子极光 electron aurora
电子计算机【计】 electronic computer
电子加热率 electron heating rate
电子间能量传输 electron-electron energy transfer
电子经纬仪 electronic theodolite
电子开关 electrical switch
电子冷却率 electron cooling rate
电子-离子复合 electron-ion recombination
电子流 electronic current
电子密度 electron density
电子密度廓线 electron density profile

电子能量平衡 electron energy balance
电子能谱学 electron spectroscopy
电子浓度 electron concentration
电子碰撞截面 electron impact cross section
电子碰撞频率 electron collision frequency
电子气象站 electronic weather station
电子迁移率分析仪 electrical mobility analyzer
电子签名【计】 electronic signature
电子亲合势【电子】 electron affinity
电子扫描微波辐射仪 electrically scanning microwave radiometer (ESMR)
电子商务【计】 electronic commerce
电子湿度计 hygronics
电子事件 electron event
电子束【电子】 electron beam
电子数密度 electron number density
电子同步加速器【物】 electron synchrotron
电子温度表 electronic thermometer
电子温度廓线 electron temperature profile
电子温度自记仪 electronic temperature recorder
电子文本【计】 e-text
电子显微镜【物】 electron microscope
电子消旋天线 electrical despun antenna (EDA)
电子雪崩 electron avalanche
电子邮件 electronic mail
电子跃迁 electronic transitions
电子云 electron atmosphere
电子杂志【计】 e-zine
电子资料库系统 electronic library system
电子自旋共振【物】 electron spin resonance
电阻率【物】 resistivity
电阻式干湿表 resistance psychrometer
电阻式水银气压计 resistance-type mercury barograph

电阻温度表　resistance thermometer
电阻温度计　electric resistance thermometer
凋萎　fading
凋萎湿度　fade humidity
凋萎湿度【大气】　wilting moisture
凋萎系数　wilting coefficient
吊篮　gondola
迭代变分法　iterative variational method
迭代插值法【数】　iterated interpolation method
迭代〔法〕　iteration
迭代法　iterative scheme
迭代法【数】　iteration method
迭代反演法　iterative inversion technique
迭代分析　iterative analysis
迭代过程　iterative process
迭代解　iterative solution
叠合相关　coincidental correlation
叠加【计】　superposition
叠加场　superimposed field
叠加原理【数】　superposition principle
碟形天线　dish
蝶形图（黑子在日面的分布图）　butterfly diagram
丁铎尔测尘器　Tyndallometer
丁铎尔花　Tyndall flower
丁铎尔散射　Tyndall scattering
丁铎尔现象【化】　Tyndall phenomenon
丁二烯（$CH_2CHCHCH_2$）　butadiene
顶部，陀螺【物】　top
顶峰　crest
顶级群落【地理】　climax
顶极土壤【土壤】　climax soil
顶角【数】　vertex angle
顶视探测仪【地】　topside sounder
订正　reduction
订正表　reduction table
订正高度　corrected altitude
订正过的　reduced

订正空速　corrected airspeed
订正气压　reduced pressure
订正温度　reduced temperature
订正系数　correction factor
订正因子　reduction factor, recovery factor
订正预报【大气】　forecast amendment
定槽式气压表　fixed cistern barometer
定槽式气压表【大气】　Kew pattern barometer
定常波【大气】　stationary wave
定常波长　stationary wavelength
定常度　steadiness
定常流【物】　steady flow
定常气候　steady climate
定常态【物】　steady state
定常涡旋　steady vortex
定常系统　steady system
定常运动　steady motion
定点迭代　fixed point iteration
定点风　spot wind
定点风温航空预报　WINTEM (aviation forecast of winds and tempemtures aloft at specific points)
定点观测【大气】　fixed point observation
定点观测站【海洋】　serial station
定点水位计　fixed point gauge
定点运算　fixed point operation
定高气球　constant-height balloon
定高气球【大气】　constant level balloon
定积分【数】　definite integral
定理【数】　theorem
π定理　pi-theorem
定量分析　quantitative analysis
定量降水估计　QPE (quantitative precipitation estimation)
定量降水预报【大气】　QPF (quantitative precipitation forecast)
定量预报方法　quantitative forecasting
定律　law

定年　dating
定年法　dating method
定日镜【天】　heliostat
定时观测　fixed time observation
定时观测时〔间〕　term hour
定时广播　regular broadcast, fixed time broadcast
定时预报　regular forecast
定态【物】　stationary state
定态假设　steady-state hypothesis, stationary state hypothesis
定态解　steady-state solution
定态系综【物】　stationary ensemble
定体〔积〕比热【物】　specific heat at constant volume
定天镜【天】　coelostat
定位　location, fix
定向　direction finding
定向测雨器　vector gauge, vectopluviometer
定向导水性　directional hydraulic conductivity
定向发射率　directional emissivity
定向切变　directional shear
定向球面反射率　directional-hemispherical reflectance
定向无线电传真　beamcast
定向吸收率　directional absorptivity
定向下落　orientation fall
定向仪　direction finder
定向雨量器　vectorial pluviometer
定性分析　qualitative analysis
定性观测　qualitative observation
定压　constant pressure
定压比热【物】　specific heat at constant pressure
定压气球　constant-pressure balloon
定义　definition
定影【物】　fixing
定域化【物】　localization
定域性【物】　locality
定域原理　localization principle
东　east（E）
东奥克兰海流（新西兰）　East Auckland Current
东澳大利亚海流　East Australia Current
东半球　eastern hemisphere
东〔半球〕南极冰板块　East Antarctic Ice Sheet
东北　northeast（NE）
东北大风　northeaster
东北低压【大气】　Northeast China low
东北风暴　northeast storm, northeastern storm
东北季风　northeast monsoon
东北信风　northeast trade
东冰岛北极流　East Iceland Arctic Current
东部标准时间（美国）　Eastern Standard Time
东大西洋型　East Atlantic pattern（EA）
东东北　east northeast（ENE）
东东南　east southeast（ESE）
东非急流　East African jet
东非气象局　East African Meteorological Department（EAMD）
东非时间　East African time
东非沿岸流　East African Coast Current（＝Somali Current）
东风波　esterly wave, wave in the easterlies
东风波【大气】　easterly wave
东风槽　easterly trough
东风带　easterly belt, easterlies
东风带扰动　easterly disturbance
东风急流　easterly jet
东格陵兰海流　East Greenland Current
东海【海洋】　East China Sea, Donghai Sea
东海沿岸流【海洋】　Donghai Coastal Current
东好望角海流　East Cape Current

东经　east longitude
东勘察加海流　East Kamchatka Current
东马达加斯加海流　East Madagascar Current
东南　southeast（SE）
东南大风　southeaster
东南风暴预警　southeast-storm warning
东南季风　southeast monsoon
东南信风　southeast trades
东南亚天气中心　Southeast Asia Weather Center（SEAWECEN）
东偏北　east by north
东偏南　east by south
东西效应（宇宙线）　asymmetric effect
东亚大槽【大气】　East Asia major trough
东亚季风【大气】　East Asia monsoon
东印度洋海流　East Indian Current
冬半年　winter half year
冬半球　winter hemisphere
冬干寒冷气候　cold climate with dry winter
冬干温和〔气候〕　winter dry moderate
冬干温暖气候　warm climate with dry winter
冬寒常湿气候亚类【大气】　winter cold and rainy climate
冬寒冬干气候亚类【大气】　winter cold and winter dry climate
冬寒气候类【大气】　winter cold climate
冬〔季〕　winter
冬季　winter season
冬季风【大气】　winter monsoon
冬季暖核〔中心〕　winter warm center
冬令时　winter time
冬眠　hibernation
冬日　winter day
冬湿寒冷气候　cold climate with moist winter
冬天〔的〕,冬令　hibernal
冬温常湿气候亚类【大气】　winter moderate and rainy climate
冬温冬干气候亚类【大气】　winter moderate and winter dry climate
冬温气候类【大气】　winter moderate climate
冬温夏干气候亚类【大气】　winter moderate and summer dry climate
冬雾　winter fog
冬性【植物】　winterness
冬性指数　winterness index, winter severity index
冬至【大气】　Winter Solstice
氡【大气】　radon（Rn）
动标〔尺〕气压计　movable-scale barometer
动槽式气压表　adjustable cistern barometer
动理学【物】　kinetics
动理〔学〕方程【物】　kinetic equation
动力边界　dynamic boundary
动力边界条件　dynamic boundary condition
动力播撒　dynamic seeding
动力播云【大气】　dynamic cloud seeding
动力不稳定〔度〕　dynamic instability
动力参数　dynamic parameter
动力槽　dynamic trough
动力潮汐　dynamical tide
动力初值化【大气】　dynamic initialization
动力单位　dynamic unit
动力底层　dynamical sublayer
动力地貌学　dynamic geomorphology
动力对流　dynamic convection, dynamical convection
动力反馈　dynamical feedback
动力分量　dynamic component
动力高度【大气】　dynamic height, geodynamic height
动力高度距平　dynamic-height anomaly
动力海洋学　dynamic oceanography
动力积云　dynamic cumulus

动力加热　dynamic heating
动力举力　dynamic lift
动力可预报性　dynamical predictability
动力冷却　dynamic cooling
动力米　dynamic meter, geodynamic meter
动力模式　dynamic model
动力黏性【大气】　dynamic viscosity
动力黏性系数　dynamic viscosity coefficient
动〔力〕黏〔滞〕系数　dynamic coefficient of viscosity, coefficient of dynamic viscosity (=dynamic viscosity)
动力平衡　kinetic equilibrium, dynamic balance
动力气候学　dynamical climatology
动力气候学【大气】　dynamic climatology
动力气象学　dynamical meteorology
动力气象学【大气】　dynamic meteorology
动力千米　dynamic kilometer
动力上曳气流　dynamic updraft
动力深度　dynamic depth
动力特性　dynamic property
动力通量　dynamic flux
动力稳定〔性〕【大气】　dynamic stability
动力系统【大气】【数】　dynamical system
动力相似　dynamic similarity
动力相似性定律　law of dynamical similarity
动力性高压　dynamic anticyclone
动力学【物】　dynamics
动力学参数【化】　kinetic parameter
动力学反应速率　kinetic reaction rate
动力学方程　kinetic equation
动力〔学〕方程【数】　equation of dynamics
动力学结构　dynamical structure
动力学模式　dynamical model
动力学温度　kinetic temperature
动〔力〕压强　dynamic pressure
动力延伸预报　dynamical extended range forecasting (DERF)
动力预报　dynamical forecasting, dynamic forecasting, dynamical forecast
动力增暖　dynamic warming
动力致冷　kinetic cooling
动量　momentum
动量传输　momentum-transport, momentum transfer
动量非守恒性　nonconservation of momentum
动量风速表　momentum anemometer
动量交换　momentum exchange, exchange of momentum
动量守恒　momentum conservation, conservation of momentum
动量通量　momentum flux
动摩擦　kinetic friction
动能【物】　kinetic energy
动能方程　kinetic energy equation
动能谱　kinetic energy spectrum
动黏滞系数　kinematic viscosity coefficient
动态【航海】　dynamic state
动态地图【测】　dynamic map
动态范围　dynamic range
动态分岔　dynamic bifurcation
动态分析【计】　dynamic analysis
动态模【物】　dynamic mode
动态平衡　dynamic equilibrium
动态平均【大气】　consecutive mean
动态显示【计】　dynamic display
动态心电图监护系统【电子】　Holter system
动物地理学【地理】　zoogeography
动物界【动】　animal kingdom
动物群落【地理】　animal community
动物生态学【动】　animal ecology
动物物候学　animal phenology
动物雾　animal fog
动压力扰动　dynamic pressure perturbation
动压头【航海】　dynamic head

冻拔　frost-lifting
冻疮　acrocyanosis
冻堤　frost dam
冻附过程　adfreezing
冻埂　soil stripe
冻害【大气】　freezing injury
冻害【地理】　freezing damage
冻季　freezing season
冻搅　frost churning
冻结　freeze, freezing, congelation
冻结【大气】　freezing
冻结层间水【地质】　interpermafrost water
冻结场　frozen-in-field
冻结磁场　frozen-in magnetic field
冻〔结〕度日　freezing degree-day
冻结高度　freezing level
冻结高度图　freeze-level chart
冻结核【大气】　freezing nucleus
冻结核谱　freezing nuclei spectrum
冻结假设　frozen-in hypothesis
冻结降水　frozen precipitation
冻结界面　freezing interface
冻结深度【农】　depth of freezing
冻结随机函数　frozen-in random function
冻结湍流　frozen turbulence
冻结湍流假定　frozen-turbulence hypothesis
冻结湍流近似　frozen-turbulence approximation
冻结温度,凝固温度　freezing temperature
冻结雪壳　frozen snow crust
冻结仪　cryopedometer
冻结指数　freezing index
冻涝害【农】　flood freezing injury
冻离〔冰〕屑　splinter
冻粒雪　iced firn(＝ice firn, firn ice)
冻裂　frost work, frost splitting, frost riving

冻裂搅动作用,冻搅　cryoturbation
冻露　white dew【大气】, frozen dew
冻霾　frost haze
冻毛毛雨　freezing drizzle
冻沫　freezing spray, freeze spray
冻丘　frost mound
冻日　freezing day
冻融〔土壤〕型　freeze-thaw pattern
冻融循环【地理】　freeze thaw cycles
冻融作用　unfreezing
冻腾　frost boil
冻土【大气】　frozen soil
冻土【地理】　frozen ground
冻土动力学【地理】　permafrost dynamics
冻土进化【地理】　permafrost aggradation
冻土面　frost table
冻土器【大气】　frozen soil apparatus
冻土丘,冰丘　frost blister
冻土退化【地理】　permafrost degradation
冻土学【地理】　geocryology, cryopedology
冻雾　frozen fog, pogonip, freezing fog【大气】
冻雾覆盖层　fog deposit
冻雪　frost snow
冻烟　frost smoke, froströk
冻雨　frozen rain, ice rain
冻雨【大气】　freezing rain
冻原荒漠　tundra desert
冻胀力【地理】　frost heaving force
冻胀丘　heaving, frost heaving
洞卡装置　hollerith system
洞穴沉积层,洞穴堆积物　speleothem
洞穴气象学　spelaeo-meteorology
洞穴小气候　burrow microclimate, hole microclimate
都市环流　urban circulation
都市气象试验　Metropolitan Meteorological Experiment (METROMEX)

陡度　abruptness
陡锋　sharp front
陡坡　abrupt slope
豆台风　pin point typhoon
逗点云头　comma head
逗点云尾　comma tail
逗点云系【大气】　comma cloud system
逗点〔状〕云　comma cloud
毒害植物的　phytotoxic
毒理学　toxicology
毒素【生化】　toxin
毒性　toxicity
毒性污染物　toxic pollutants
独立边带　independent sideband
独立分布的　independently distributed
独立取样　independent sampling
独立实体夹卷混合　entity-type entrainment mixing
独立试验【数】　independent trials
独立随机变量　independent random variable
独立像元近似　independent pixel approximation
独立性检验　independence test
独立样本　independent sample
读出　read out
读出时间【计】　read-out time
读入　read-in
读数　reading
读数误差　reading error, error of reading
读数显微镜　read microscope
读写周期【计】　read-write cycle
杜德瓦尼露量器　Duvdevani dew gauge
杜瓦〔真空保温〕瓶　Dewar flask
杜维定律　Dove's law
度　degree
度量力　metric force
度量系数　metric coefficient
度量张量　metric tensor
度盘湿度表　dial hygrometer
度盘温度表　dial thermometer

度日【大气】　degree-day
度日相关　degree-day correlation
度时　degree-hour
22度晕【大气】　22° halo
46度晕【大气】　46° halo
渡槽　flume
端点【地信】　dead end
短波　short wave
短波槽　short wave trough
短波反照率　short wave albedo
短波辐射　short wave radiation
短波截止　short wave cutoff
短波云辐射强迫作用　short wave cloud radiative forcing
短法　short method
短峰波　short-crested wave
短虹　short rainbow
短冷期　snap
短路【物】　short circuit
短路电流　short circuit current
短曝光调制传递函数　short-exposure modulation transfer function
短期森林火险预报　short-term forest fire danger forecast
短期水文预报　short-term hydrologic forecast
短期〔天气〕预报【大气】　short-range 〔weather〕 forecast
短期天气展望　further outlook
短期预报　short-term forecast
短日照植物　short-day plant
短时高温加工【农】　high-temperature short-time processing
短夏　pointed summer
短暂降雨　ephemeral rain
短暂流　temporary current
短、中期预报研究计划　Programme on Short and Mid-range Prediction (PSMP)
短周期波束加宽　short-term beam spread
断层地震【地】　fault earthquake
断虹　broken rainbow

断裂线　break line
断续河　interrupted stream
断续湍流　patchy turbulence
堆积过程　pile-up process, banking process
堆积阻塞　damming
队列　queue
对比　contrast
对比度增强　contrast enhancement
对比扩展（图像处理）　contrast stretching
对比区　contrast area
对比遥测光度表　contrast telephotometer
对比阈　liminal contrast
对比阈值　contrast threshold
对策论【数】　game theory
对产生【物】　pair production
对称不稳　symmetry instability
对称点　symmetry point, point of symmetry
对称矩阵【数】　symmetric matrix
对称滤波　symmetric filtering
对称陀螺【化】　symmetric top
对称〔性〕【物】　symmetry
对称原理【数】　symmetry principle
对称中心　center of symmetry
对称轴【物】　axis of symmetry
对地观测卫星【地信】　earth observation satellite
对地静止　geostationary
对分布函数【数】　pair-distribution function
对光不稳定〔的〕　photolabile
对华（天象）　anticorona
对角线【数】　diagonal
对空气象广播【大气】　VOLMET broadcast
对空无线电通信试验　up-link radio experiment
对流【大气】　convection
对流胞，对流单体　convective cell
对流边界层　convective boundary layer（CBL）
对流不稳定【大气】　convective instability
对流不稳定线　convectively unstable line
对流部分　convective component
对流参数化【大气】　convective parameterization
对流层【大气】　troposphere
对流层沉降　tropospheric fallout
对流层臭氧【大气】　tropospheric ozone
对流层传播　tropospheric propagation
对流层顶【大气】　tropopause
对流层顶波动【大气】　tropopause wave
对流层顶不连续　tropopause discontinuity
对流层顶层　tropopause layer
对流层顶颠簸　tropopause bump
对流层顶断层　tropopause break
对流层顶断裂线　tropopause break-line
对流层顶漏斗【大气】　tropopause funnel
对流层顶逆温　tropopause inversion
对流层顶图　tropopause chart
对流层顶折叠，多对流层顶　tropopause fold
对流层反射　tropospheric reflection
对流层高空冷涡　upper tropospheric cold vortex
对流层高空气象图　tropogram
对流层化学【大气】　tropospheric chemistry
对流层-平流层交换　troposphere-stratosphere exchange
对流层散射　tropospheric scatter, troposcatter
对流层上部　upper troposphere
对流层上部湿度　upper tropospheric humidity (UTH)
对流层输送　tropospheric transport
对流层下部　lower troposphere
对流层折叠　tropospheric folding
对流层折射【大气】　tropospheric refraction

对流层中层气旋【大气】 mid-tropospheric cyclone (MTC)
对流尺度 convective scale
对流冲量 convective impulse
对流单体【大气】 convection cell
对流动能 convective kinetic energy
对流翻腾(湖沼) convective overturn
对流方案 convection scheme
对流风暴 convective storm
对流风暴发生机制 convective storm initiation mechanism
对流高度 ceiling of convection
对流过程 convection process
对流厚度 convection depth
对流环流 convective circulation
对流回波【大气】 convective echo
对流混合层 convective mixed layer
对流活动 convective activity
对流加速度 convective acceleration
对流冷却 convective cooling
对流里查森数 convective Richardson number
对流模式【大气】 convection model
对流凝结高度【大气】 convective condensation level(CCL)
对流凝结羽 convective plume
对流平衡 convective equilibrium
对流气流 convection current
对流区 convective region
对流势能 convective potential energy
对流输送 convective transport
对流输送〔学〕说 convective transport theory
对流速度尺度 convective velocity scale
对流调整【大气】 convective adjustment
对流湍流 convective turbulence
对流温度 convective temperature
对流限〔高〕 limit of convection(LOC)
对流线 convection line
对流〔性〕环流 convectional circulation
对流性降水【大气】 convective precipitation
对流性雷暴【大气】 convective thunderstorm
对流〔性〕稳定度【大气】 convective stability
对流〔性〕雨 convectional rain
对流〔性〕阵雨 convective shower
对流学说 convective theory, convectional theory
对流抑制能 convective inhibition (CIN)
对流有效位能 convective available potential energy(CAPE)
对流雨 convective rain
对流云【大气】 convective cloud
对流云街 convective cloud street
对流云团 convective cluster
对流指数 convective index
对流质量通量 convective mass flux
对马海流 Tsushima Current
对偶定理 duality theorem
对偶解【数】 dual solution
对偶空间【物】 dual space
对偶〔性〕【电子】 duality
对耦合【物】 pair coupling
对日点 anti-solar point
对日〔照〕 withershins
对日照【天】 counterglow
对数变换【化】 log transformation
对数标尺 logarithmic scale
对数表 logarithm table
对数放大器【电子】 logarithmic amplifier
对数风速廓线 logarithmic velocity profile
对数谱 logarithmic spectrum
对数微分 logarithmic differentiation
对数压力坐标 log-pressure coordinate
对数振幅起伏 log amplitude fluctuation
对数正态分布 log-normal distribution
对数正态分布【数】 logarithmic normal

对数正态云尺度分布 lognormal cloud-size distribution
对消比【大气】 cancellation ratio
对应点 corresponding point
对应分析【数】 correspondence analysis
对照点 contrast point
对照分析 control analysis
对照区 controlled area, control area
对照试验 control experiment
对照算程 control run
对照预报 control forecast
对照云 control cloud
吨/年 tons per year (TPY)
吨/日 tons per day (TPD)
钝角三角形【数】 obtuse triangle
盾形降水区 precipitation shield
盾状卷云层 cirrus shield
盾状云 cloud shield
盾状云【大气】 shield cloud
多胞雷暴 multicell thunderstorm
多笔记录装置 multi-stylus recording system
多边形【数】 polygon
多变风暴 divers storm
多变能见度 variable visibility
多变云幂高 variable ceiling
多波长浑浊度仪 multiwavelength nephelometer
多波长太阳光度计 multiwavelength sunphotometer
多波段成像 multispectral imaging
多波段传感器 multispectral sensor
多波段红外辐射仪 multispectral infrared radiometer (MSIRR)
多波段扫描器 multispectral scanner
多波段摄影 multispectral photography
多波段图像【大气】 multi-spectral image (MSI)
多波段遥感 multi-band remote sensing
多波段遥感【地理】 multispectral remote sensing
多波段云图 multispectral imagery (MSI)
多参数雷达 multiparameter radar
多层冰 rafted ice, nabivnoy ice
多层模式 multilevel model, level model【大气】, layer model
多层逆温 multiple-layer inversion
多叉树 multiway-tree
多程回波 multiple trip echo
多尺度方法 multi-scale method
多尺度分析【计】 multiscale analysis
多尺度空气质量模式 Community Multi-scale Air Quality (CMAQ)
多重波 multiple wave
多重处理【计】 multiprocessing
多重点 multiple point
多重傅里叶级数【数】 multiple Fourier series
多重光晕 multiple corona
多重回波 multiple echo
多重滤波法 multiple-filter method
多重偏流修正 multiple drift corrections
多重嵌套网格模式 multi-nested mesh (或 grid) model
多重曲线回归分析 multiple curvilinear regression analysis
多重网格法 multigrid method
多重稳定层 multiple stable layer
多重涡旋配置 multiple vortex configuration
多重线性回归分析 multiple linear regression analysis
多重指数 multiple-index
多窗区频道法 multiple window channel method
多次反射 multiple reflection
多次放电 multiple discharge
多次散射【大气】 multiple scattering
多次散射理论 multiple scattering theory
多次散射效应 multiple scattering effect
多次闪击 multiple stroke, composite

flash
多次升降说　multiple incursion theory
多单体　multi-cell
多单体对流　multicell convection
多单体对流风暴　multicell convective storm
多单体回波【大气】　multiple-cell echo
多点校准　multi-point calibration
多点源扩散模式　multiple point source dispersion model(MPSDM)
多分辨率【计】　multiresolution
多风　windness
多峰谱　multimodal spectrum
多光电子事件　multi-photoelectron event
多光谱云图【大气】　multi-spectral image
多环芳烃　polycyclic aromatic hydrocarbons（PAH）
多基地雷达【电子】　multistatic radar
多级采样器【大气】　cascade impactor
多级蒸发器　multiple-stage evaporator
多极振　multipole vibration
多阶谱　polyspectrum
多孔冰晶　porous ice
多孔冰山　sugar berg
多孔介质　porous medium
多孔陶瓷蒸渗仪　porous porcelain atmometer
多流动平衡　multiple flow equilibria
多龙卷时段　tornadic phase
多路传输　multipath transmission
多路径传播　multipath
多路驱动【电子】　multiplexing
多路通信系统　multichannel communication system
多氯联苯　Polychlorinated biphenyls(PCBs)
多媒体【计】　multimedia
多媒体个人计算机【计】　multimedia PC（MPC）
多媒体计算机【计】　multimedia computer
多面体　polyhedron
多面投影　polyhedric projection
多目标水库　multiple-purpose reservoir
多能湿度表　polymeter
多年变化尺度　interannual time scale
多年冰　multiyear ice
多年〔的〕　secular
多年调节库容　carry-over storage
多年冻层　pereletok
多年冻土　pergelisol
多年库容　multiannual storage
多年生植物【植】　perennial plant
多年性干旱　perennial drought
多诺拉烟雾(美国)　Donora smog
多偏振雷达　polarization-diversity radar
多频带系统(遥感)　multiband system
多频微波辐射计　multi-frequency microwave radiometer
多平衡态　multiple equilibria, multi-equilibrium states
多平衡态系统　multiple equilibria system
多普勒测云法　Doppler cloud sounding
多普勒方程　Doppler equation
多普勒激光雷达　Doppler laser radar
多普勒激光雷达【大气】　Doppler lidar
多普勒技术　Doppler technique
多普勒宽度　Doppler width
多普勒雷达【大气】　Doppler radar
多普勒雷达变分分析系统　Variational Doppler Radar Analysis System（VDRAS）
多普勒雷达波谱　Doppler radar spectrum
多普勒雷达反射率　Doppler radar reflectivity
多普勒雷达目视显示器　visual Doppler indicator（VDI）
多普勒频率　Doppler frequency
多普勒频移　Doppler frequency shift
多普勒频移【物】　Doppler shift
多普勒频移频率　Doppler-shifted frequency
多普勒谱　Doppler spectrum

多普勒谱矩　Doppler spectral moments
多普勒〔谱线〕增宽【物】　Doppler broadening
多普勒谱展宽　Doppler spread, Doppler spectral broadening
多普勒声〔雷〕达【大气】　Doppler sodar
多普勒声学测风器　Doppler acoustic wind sensor
多普勒收发机　Doppler transceiver
多普勒速度【大气】　Doppler velocity
多普勒速度谱分辨率　Doppler velocity spectra resolution
多普勒天气雷达　Doppler weather radar, Doppler radar【大气】
多普勒无线电探空仪系统　Doppler radiosonde system
多普勒误差　Doppler error
多普勒限光谱学　Doppler-limited spectroscopy
多普勒线形【物】　Doppler profile
多普勒效应【物】　Doppler effect
多谱　polydisperse
多〔气〕球同步极光观测　simultaneous auroral multiballoons observation（SAMBO）
多气旋区　cyclone-prone area
多色光雷达　plidar(=polychromatic lidar)
多色探测　multicolor detection
多色透射比　polychromatic transmittance
多时间尺度　multiple-time-scale
多束速调管　multiple-beam klystron（MBK）
多水年【大气】　high flow year
多肽【生化】　polypeptide
多体问题【天】　many body problem
多通道窗区法　multichannel window method
多通道光度计　multichannel photometer
多通道海面温度（遥感算法）　multichannel sea surface temperature
多通道红外辐射仪　multichannel infrared radiometer

多通道绝对日射表　multichannel pyrheliometer
多通道奇异谱分析　multichannel singular spectrum analysis（MSSA）
多通道微波扫描辐射仪【大气】　scanning multichannel microwave radiometer（SMMR）
多通道系统　multichannel system
多维　multidimension
多维傅里叶级数　multi-dimensional Fourier series
多维谱　multi-spectrum
多系统集合　multisystem ensemble
多相反应【化】　heterogeneous reaction
多相方法　multi-phase method
多相流　multiple-phase flow
多相模式　multi-phase model
多相平衡【化】　heterogeneous equilibrium
多相〔物〕质　heterogeneous substance
多项接收　diversity reception
多项式【数】　polynomial
多项式回归【农】　polynomial regression
多项式拟合　polynomial fitting
多效蒸发器　multiple effect evaporator
多新世　Polycene
多学科的,多学科性　multidisciplinary
多学科研究　interdisciplinary study
多雪的　nival
多样化　diversification
多因子试验　multiple factorial experiment
多用户系统【计】　multiuser system
多余观测　redundant
多雨地区　pluvail region
多雨期,雨季　pluvail period【大气】
多雨气候　rainy climate
多雨热带　rainy tropics
多元超几何分布【数】　multivariate hypergeometric distribution
多元大气【大气】　polytropic atmosphere

多元方差分析【数】 multivariate analysis of variance
多元分析 multivariate analysis
多元分析方案 multivariate analysis scheme
多元〔复合〕肥料【土壤】 multiple nutrients compound fertilizer
多元过程【大气】 polytropic process
多元回归 multiple regression
多元回归分析 multiple regression analysis
多元客观分析 multivariate objective analysis
多元判别分析 multiple discriminant analysis
多元时间序列 multivariate time series
多元〔统计〕分析【数】 multivariate 〔statistical〕 analysis
多元线性回归 multivariate linear regression
多元信号 multivariate signal
多元〔性〕系数 coefficient of polytropy

多元转换函数 multivariate transfer function
多元最优插值 multivariate optimum interpolation
多圆柱投影 polycylidrical projection
多圆锥投影 polyconic projection
多云【大气】 cloudy
多云日(该日大多数时间云量在75%以上) cloudy day
多云天空(总云量为 $\frac{3}{8} \sim \frac{5}{8}$) cloudy sky, very cloudy sky
多云天气 cloudy weather
多云状况 brokenness
多站多普勒雷达 multi-station Doppler radar
多站非相干散射雷达 multi-station incoherent scatter radar
多中心低压 complex low
多种大气参数 multiple atmospheric parameter
惰性气体 noble gas, inert gas

E

峨眉宝光【大气】 glory
俄歇簇射【地】 Auger shower
额定风速 rate of speed
厄柏离子计数仪 Ebert ion-counter
厄尔尼诺【大气】 El Nino
厄尔尼诺年 El Nino year
厄尔尼诺效应 El Nino effect
厄尔奇琼(火山) El Chichon
厄加勒斯海流 Agulhas Current (＝Agulhas Stream)
恶劣能见度 bad visibility
恶劣天气 rough weather, tempest, bad weather
恶劣天气进场着陆 bad weather approach
恶劣天气预报图【大气】 significant weather chart
鄂霍次克海高压【大气】 Okhotsk high
鄂霍次克阻塞反气旋 Okhotsk blocking anticyclone
鄂霍次克阻塞高压 Okhotsk blocking high
恩索【大气】 ENSO (El Nino 与 Southern Oscillation 合成的缩略词)
恩索〔天气〕事件 ENSO (EI Nino-Southern Oscillation) event
恩索〔天气〕现象 ENSO (EI Nino-Southern Oscillation) phenomenon

恩索指示器　ENSO indicator
恩索指数　ENSO indices
珥　arc of contact of halo
二-八进制码（二进制编码的八进制）　binary-coded octal
二倍格距波　two-gridlength wave
二倍格距噪声　two-grid-interval noise
二叉树　binary tree
二重积分【数】　double integral
二次变换【数】　quadratic transformation
二次回波【大气】　second trip echo
二次空气　secondary air
二次雷达【电子】　secondary radar
二次散射　secondary scattering
二次守恒格式　quadratic conserving scheme
二次形　quadratic form
二次衍射　secondary diffraction
二等水文站,辅助水文站　secondary hydrometric station
二叠纪【地质】　Permian Period
二噁英　dioxin
二分点【大气】　Equinoxes
二分点风暴　line gale, equinoctial gale, equinoctial, storm
二分点风暴效应　equinoctial storm effect
二分点雨【大气】　equinoctial rain
二分圈　equinoctial colure, colure
二级标准气压表　secondary standard barometer
二级反应　second order reaction
二级环流　secondary circulation
二级绝对日射表　secondary pyrheliometer
二级气候站　second-order climatological station
二级站　second-order station
二极管　diode
二极管激光器　diode laser
二甲基二硫（CH_3SSCH_3）　dimethyl disulfide
二甲〔基〕硫（CH_3SCH_3）【大气】　dimethyl sulfide（DMS）
二甲亚砜（$(CH_3)_2SO$）　dimethyl sulfoxide
二阶闭合　second-order closure
二阶光子交互作用　two-step photon interaction
二阶矩　second moment
二阶矩方程　second moment equation
二阶扰动解　second order perturbation solution
二阶有限差分　second order finite difference
二进制【计】　binary system
二进制编码字符　binary coded character
二进制代码　binary code
二进制的　binary
二进制码　Gray code
二进制数　binary number
二进制数字【计】　binary digit
〔二进制〕位【数】　bit
二聚【化】　dimerization
二聚体【化】　dimer
二硫化碳（CS_2）　carbon disulfide
二面角反射器　dihedral reflector
二年冰　second-year ice, biennial ice
二年风振荡　biennial wind oscillation
二年生植物【植】　biennial plant
二年振荡　biennial oscillation
二-十进制码（二进制编码的十进制）　binary-coded decimal（BCD）
二-十进制转换器　binary-to-decimal converter
二十面体　icosahedron
二十四节气【大气】　Twenty-four Solar Terms
二时次法　two-time level scheme
二体碰撞【天】　binary collision
二维傅里叶级数　two-dimensional Fourier series
二维功率-波数谱　two-dimensional power-wavenumber spectrum

二维光学云探测器　two-dimensional optical cloud probe（2D-C）
二维互相关　two-dimensional crosscorrelation
二维降水粒子探测器　two-dimensional precipitation probe
二维空间滤波　two-dimensional spatial filtering
二维快速傅里叶变换　two-dimensional FFT
二维流　two-dimensional flow
二维滤波　two-dimensional filtering
二维模式　two-dimensional model
二维去倾　two-dimensional detrending
二维湍流【大气】　two-dimensional turbulence
二维涡旋　two-dimensional eddies
二维云粒子探测器　two-dimensional cloud probe
二维自相关　two-dimensional autocorrelation
二向色性【物】　dichroism
二项分布【数】【大气】　binomial distribution
二项式【数】　binomial
二项式定理【数】　binomial theorem
二项式平滑　binomial smoothing
二氧化氮【大气】　nitrogen dioxide（NO_2）
二氧化硫（SO_2）　sulphur dioxide
二氧化氯（$OClO$）　chlorine dioxide
二氧化碳【大气】　carbon dioxide
二氧化碳大气浓度【大气】　carbon dioxide atmospheric concentrations
二氧化碳当量【大气】　carbon dioxide equivalence
二氧化碳激光器【电子】　carbon dioxide laser
二氧化碳〔谱〕带【大气】　carbon dioxide band
二氧化碳施肥【大气】　carbon dioxide fertilization
二氧化碳施肥效应　carbon dioxide fertilizing effect
二氧杂环己烷，二噁烷　dioxane
二元分布　bivariate distribution
二元时间序列　bivariate time series
二元图像　binary image
二至点　solstices
二至圈　solstitial colure

F

发电机层　dynamo layer
发电机理论（地磁的）　dynamo theory
发电机作用　dynamo action
发光　luminous
发光【物】　luminescence
发光二极管【物】　light emitting diode（LED）
发光二极管显示【计】　light emitting diode display（LED display）
发光率　visibility ratio（＝luminosity），luminous exitance
发光率对比　contrast of luminance
发光强度【物】　luminous intensity
发光水汽尾迹　luminous vapour-trail
发光效率　luminous efficiency
发光效率【物】　luminescence efficiency
发光云　luminous cloud
发散束　diverging beam
发散〔作用〕　exhalation
发射　launching

发射【计】 emanation
发射层 emission layer
发射功率 transmitted power
发射光谱学 emission spectroscopy
发射机 transmitter
发射率【物】【大气】 emissivity（ε）
发射谱 emission spectrum
发射谱线 emission line
发射强度 emissive power
发射体（火箭） missile
发生 genesis
发声器 acoustical generator
发展的湍流 developed turbulence
发展期 evolution period
发展指数 development index
伐倒木【林】 felled tree
伐木【林】 felling
伐木机【林】 felling machine，feller
法布里-珀罗标准具【物】 Fabry-Perot etalon
法布里-珀罗干涉仪【物】 Fabry-Perot interferometer
法定年 legal year
法定时 hour
法定时间 legal time
法〔拉〕 farad（F）
法拉第电流【化】 faradaic current
法拉第效应【物】 Faraday effect
法拉第旋转【物】 Faraday rotation
法律气象学 forensic meteorology
法〔线方〕向梯度 normal gradient
法向分量 normal component
法向加速度【物】 normal acceleration
法向应力【物】 normal stress
法因曼测云器 Fineman's nephoscope
幡 curtain
幡状 virga（vir）
翻斗式雨量计 tipping-bucket rain gauge
翻斗〔式〕雨量计【大气】 tilting bucket raingauge
翻滚 tumble
翻腾 churn
翻转 overturning, overturn
繁育法 breeding method
繁育模 bred mode
繁育矢量 bred vector
反变层 reversing layer
反常传播 anaprop
反常低压 anomalous low
反常高压 anomalous high
反常回波 ghost
反常色散 anomalous dispersion
反常消光 anomalous extinction
反常衍射 anomalous diffraction
反称矩阵【数】 anti-symmetric matrix
反称行列式【数】 skew-symmetric determinant
反传网络〔计〕，前馈网络，BP 网络 back propagation（BP）network
反电子 antielectron
反对称〔性〕【物】 antisymmetry（=anti-symmetry）
反对数【数】 anti-logarithm
反厄尔尼诺【大气】 anti El Nino
反谷风 anti-valley wind
反海风 anti-sea breeze
反环流 counter-circulation, indirect circulation【大气】
反季风 antimonsoon
反假日 counter parhelia, antihelion
反假日【大气】 anthelion
反假月 antiselene
反馈 feedback
反馈系统 feedback system
反粒子【物】 antiparticle
反气旋【大气】 anticyclone
反气旋焚风 anticyclone foehn
反气旋环流【大气】 anticyclonic circulation
反气旋脊，高压脊 anticyclonic ridge
反气旋龙卷 anticyclonic tornado, anticyclone tornado

反气旋逆温　anticyclonic inversion
反气旋期　anticyclonic phase
反气旋气流　anticyclonic flow
反气旋生成【大气】　anticyclogenesis
反气旋式旋转　anticyclonic rotation
反气旋涡度【大气】　anticyclonic vorticity
反气旋涡度平流　anticyclonic vorticity advection（AVA）
反气旋下沉　anticyclone subsidence
反气旋消散【大气】　anticyclolysis
反气旋〔性〕辐散　anticyclonic divergence
反气旋〔性〕切变【大气】　anticyclonic shear
反气旋〔性〕曲率【大气】　anticyclonic curvature
反气旋〔性〕涡旋　anticyclonic vortex, anticyclonic eddies
反气旋移动　anticyclone movement
反气旋阴沉天气　anticyclonic gloom
反气旋轴　axis of anticyclone
反气旋轴倾角　inclination of the axis of an anticyclone
反曲点　inflexion point
反曲率　retroflection, retroflexion
反日　counter sun, anthelion【大气】
反日点　anthelic point
反日弧　anthelic arc
反日柱　anthelic pillar
反三角函数【数】　inverse trigonometric function
反山风　anti-mountain wind
反射【物】　reflection
反射靶　re-radiator
反射比【物】　reflectance
反射比测量　reflectance measurement
反射表　reflectometer
反射波　reflective wave
反射薄片　foil
反射测云器　reflection nephoscope
反射焚风　reflected fohn

反射辐射【大气】　reflected radiation
反射函数　reflection function
反射虹　reflection rainbow
反射极　repeller
反射角　angle of reflection
反射率　reflective index
反射率【物】【大气】　reflectivity
反射率因子　reflectivity factor
反射膜层　reflectivity coating
反射能力　reflective power, reflecting power
反射器　reflector
反射速调管【电子】　reflex klystron
反射太阳辐射【大气】　reflected solar radiation
反射太阳光　reflected sunlight
反射太阳光订正　reflected sunlight correction
反射望远镜【天】　reflecting telescope
反射系数　reflection coefficient
反射线　reflected ray
反时针　counter clock-wise, anticlockwise
反曙暮光　countertwilight, anti-twilight
反曙暮光弧　anti-twilight arch, anticrepuscular arch
反曙暮辉　anticrepuscular ray
反斯托克斯线　anti-Stokes line
反斯托克斯荧光　anti-Stokes fluorescence
反梯度　counter-gradient
反梯度风【大气】　counter-gradient wind
反梯度热通量　counter-gradient heat flux
反梯度通量　upgradient flux, counter-gradient flux
反相关，负相关　anticorrelation
反相〔位〕【物】　antiphase
反向跟踪　backtracking
反向季风槽　reverse-oriented monsoon trough
反硝化作用【土壤】,脱氮作用

denitrification
反信风　counter trade
反信风【大气】　anti-trade
反压流　antibaric flow
反演　retrieval
反演法　inversion method
反演法【大气】　inverse technique
反演方法　inverse method
反演离散傅里叶变换　inverse discrete Fourier transform(IDFT)
反演算法　inverse algorithm
反演问题　inverse problem
反验证法,正反预报检验　forecast-reversal test
反应包络谱　response envelope spectrum (RES)
反应动力学【化工】　reaction kinetics
反应级数【化】　reaction order
反应率【物】　reaction rate
反应热【化】　heat of reaction
反油酸【生化】　elaidic acid
反照率【大气】,反射率　albedo
反照率表【大气】　albedometer
反照率计　albedograph
反照中子　albedo neutron
反褶积　deconvolution
反质子　antiproton
反中微子　antineutrino
反中子　antineutron
反钟向(即气旋方向)　contra solem
反转片【测】　reversal film
反转频率　turnover frequency
返青　regreening
返青期　regreening stage
返射【物】　retroreflection
返射器　retroreflector
泛代数【数】　universal algebra
泛海绵〔侧〕边界条件　porous sponge condition
泛函【数】　functional
泛函分析【数】　functional analysis

泛滥　deluge
泛滥平原　flood plain
泛美季风　Pan-American monsoon
泛频带　overtone band
泛音【物】　overtone
范艾伦辐射带【大气】　Van Allen radiation belt, Allen radiation belt
范德波尔方程【物】　van der Pol's equation
范德华半径【化】　van der Waals radius
范德瓦耳斯方程【物】　van der Waals equation
范数【数】　norm
范托夫定律【化】　van't Hoff's law
范围　coverage
方案　scheme
方波发生器【电子】　square-wave generator
方差【数】　variance
方差比　variance ratio
方差分量【数】　variance component
方差分析【大气】　variance analysis
方差分析【数】　analysis of variance
方差谱　variance spectrum
方差缩减　variance reduction, reduction of variance
方程【数】　equation
KdV方程【数】　KdV equation (=Korteweg-de Vries equation)
方程组　system of equations
方法学　methodology
方格坐标算法　gridding algorithm
方解石【化】　calcite
方块图　box diagram
方位　bearing
方位标度　azimuth scale
方位标识器,方向标记　azimuth marker
方位-高度指示器　azel-scope
方位畸变　azimuth distortion
方位角　horizontal angle, azimuth
方位角分辨率【大气】　azimuth resolution

方位角指示仪　azimuth indicating goniometer
方位距离　azran
方位平均【大气】　azimuth averaging
方位器　azimuth gauge
方位误差　azimuth error
方位小扰动　azimuthal perturbation
方位-仰角　azimuth-elevation
方向　direction
方向导数【数】　directional derivative
方向分布　directional distribution
方向谱　directional spectrum
方向余弦　direction cosine
方志【地理】　local topography
芳炔【化】　arene
芳(香族)烃　aromatic hydrocarbon
防雹　hail prevention
防雹【大气】　hail suppression
防雹火箭　antihail rocket
防雹计划　hail suppression program
防雹炮　antihail gun
防波堤　breakwater
防潮层【建筑】　damp proofing course
防冻　anti-freezing
防风带　shelter-belt
防风窨　cyclone cellar
防风林　wind-break forest
防风网　windbreak net
防风罩　shield
防辐射　antiradiance
防辐射内罩　inner radiation shield
防辐射外罩　outer radiation shield
防洪　flood control
防洪水库　flood control reservoir
防护林【林】　protection forest
防火带【林】　fire belt
防火沟【林】　fire trench
防积冰　anti-icing
防龙卷掩体　tornado cellar
防霜　frost protection
防霜【大气】　frost prevention

防霜鼓风机　frost fan
防雾剂　antifoggant
防雪设备　snow shield
〔防〕雪栅〔栏〕　snow fence
防汛抢险　flood proofing
防雨罩【机械】　rain cover
防灾抗灾　disaster prevention and preparedness（DPP）
防止空气污染系统　anti-air-pollution system
仿射变换　affine transformation
仿生学　bionics
仿真计算机【计】　simulation computer
放大　amplification
放大镜【物】　magnifier
放大矩阵　amplification matrix
放大率【物】　magnification
放大器　amplifier
放大因子　magnification factor
放大因子【数】　amplification factor
放电　electric discharge, electrical discharge
放电【物】　discharge
放电流动系统　discharge-flow system
放电图　discharge diagram
放能反应【生化】　exergonic reaction
放热反应　exothermic reaction
放射化学　radiological chemistry
放射化学【化】　radiochemistry
放射环境化学【化】　radioenvironmental chemistry
放射性【化】　radioactivity
放射性尘降区　radioactive fallout plot
放射性尘降物　radioactive pollutant
放射性尘降物【化】　radioactive fallout
放射性尘降预报　radioactive fallout forecast
放射性废物【化】　radioactive waste
放射性核素【化】　radionuclide, radioactive nuclide

放射性集电器 radioactive collector
放射性检测【化】 radioassay
放射性鉴年法【物】 radioactive dating
放射性降水 radioactive precipitation
放射性空气监测系统 radioactive air monitoring system
放射性空气污染 radiological air pollution
放射性粒子电离 radioactive particle ionization
放射性气溶胶【化】 radioactive aerosol
放射性气体【冶金】 radioactive gas
放射性示踪剂 radioactive tracer
放射性示踪物 radiotracer
放射性衰变【化】 radioactive decay
放射性碳 radiocarbon, radioactive carbon
放射性碳定年【地质】 radiocarbon dating
放射性碳定年法【大气】 carbon dating, radio carbon dating
放射性碳含量 radiocarbon concentration
放射性〔碳〕年代 radiocarbon age
放射性同位素 radioisotope
放射性同位素【物】 radioactive isotope
放射性污染【大气】 active pollution
放射性污染【地质】 radioactive pollution
放射性雪量计 radioactive snow gauge
放射性雨 radioactive rain
放射性元素【化】 radioelement, radioactive element
放射治疗【物】 radiotherapy
飞尘 fly ash
飞点记录法 flying spot recording technique
飞灰【大气】 fly ash
飞机报告 air report
飞机测温法 aircraft thermometry
飞机颠簸【大气】 aircraft bumpiness
飞机观测 airplane observation, aircraft observation

飞机〔机体〕除冰 airframe deicing
飞机积冰【大气】 aircraft icing, aircraft ice accretion
飞机积冰指数 index of aircraft icing
飞机结冰 ice formation on aircraft
飞机累积资料系统 aircraft integration data system (AIDS)
飞机气象计 airplane meteorograph
飞机气象探测【大气】 airplane meteorological sounding
飞机气象站(设在飞机内的气象站) aircraft meteorological station
飞机气象侦查编码 RECCO code
飞机〔气象〕侦察飞行 aircraft reconnaissance flight
飞机起电 aircraft electrification
飞机前方能见度 over-the-nose visibility
飞机事故 aircraft hazard
飞机探测【大气】 aircraft sounding
飞机天气侦察【大气】 aircraft weather reconnaissance
飞机天线 aeroplane antenna
飞机通信寻址与报告系统 ACARS (Aircraft Communications Addressing and Reporting System)
飞机湍流 aircraft turbulence
飞机卫星资料中继 aircraft to satellite data relay (ASDAR)
飞机尾迹【大气】 aircraft trail
飞机尾流【大气】 aircraft wake
飞机仪器观测 apob
飞机迎面气流 air against airplane
飞机沾雪 slush icing
飞沫 spray
飞艇,飞船 airship
飞行报〔告〕 aircraft report
飞行服务站 flight service station
飞行高度 service ceiling, flight level
飞行路线 flying route
飞行能见度 flight visibility, inflight visibility

飞行〔气象〕文件 flight documentation
飞行情报服务台 flight information service station
飞行实验室 flying laboratory
飞行探空仪,无人机探空 aerosonde
飞行天气低限 minimums(=operational weather limits)
飞行天气预报表 flight weather forecast bulletin
飞行文件 flight information document
飞行员气象报告【大气】 pilot meteorological report
飞行员天气报告 pilot report
飞行员天气简报 pilot briefing
飞行侦察 reconnaissance
飞行侦察简报 reconnaissance summary
飞行中服务 in-flight service
飞行中校准 in-flight calibration
非保守性 nonconservative property
非必需氨基酸【生化】 non-essential amino acid
非飑线云团 non-squall cluster
非参数统计 non-parametric statistics
非常波【物】 extraordinary wave
非常规观测【大气】 non-conventional observation
非常降水量 excessive precipitation
非常数 nonconstant
非超级单体龙卷 nonsupercell tornado
非承压含水层【地质】 unconfined aquifer
非传递性〔的〕 intransitive
非弹性【物】 inelasticity
非弹性过程 inelastic process
非弹性碰撞 inelastic collision
非导体 non-conductor
非地转风【大气】 ageostrophic wind
非地转风分量 ageostrophic wind component
非地转环流 ageostrophic circulation
非地转加速度 ageostrophic acceleration
非地转流 ageostrophic flow

非地转平流 ageostrophic advection
非地转运动【大气】 ageostrophic motion
非电介质 nonelectrolyte
非调和潮汐分析 nonharmonic tidal analysis
非〔叠〕加性 nonadditivity
非定时报〔告〕 offtime report
非定时天气观测 non-synoptic observation
非对流性降水【大气】 non-convective precipitation
非感应起电机制 noninductive charging mechanism
非各向同性湍流 nonisotropic turbulence
非灌溉期间 non-irrigation period
非规定层上的 off-level
非〔规〕定时 offtime
非规定天气观测 asynoptic observation
非规定天气观测时间 asynoptic time
非灰大气【天】 non-grey atmosphere
非恢复限流 non-resettable current limiting
非火山地热区【地】 nonvolcanic geothermal region
非极化波 unpolarized wave
非技术错误(质量检验用) procedural error
非加性修正【物】 nonadditivity correction
非甲烷烃,非甲烷碳氢化物 nonmethane hydrocarbons(NMHCs)
非晶霜 amorphous frost
非晶〔形〕的 amorphous
非静电场 non-electrostatic field
非局部热动平衡【天】 non-local thermodynamic equilibrium
非局地闭合 nonlocal closure
非局地混合 nonlocal mixing
非局地通量 nonlocal flux

非局域静力稳定度　nonlocal static stability
非局域响应【物】　nonlocal response
非绝热变化　diabatic change
非绝热不可逆过程　non-adiabatic irreversible process
非绝热初始化　diabatic initialization
非绝热过程　non-adiabatic process
非绝热过程【大气】　diabatic process
非绝热加热　diabatic heating
非绝热冷却　non-adiabatic cooling
非绝热流【力】　diabatic flow
非绝热梯度　diabatic gradient
非绝热项　diabatic term
非绝热效应　diabatic effect
非绝热中尺度环流　diabatic mesoscale circulation
非均相大气化学反应　heterogeneous atmospheric chemical reaction
非均相反应【化】　inhomogeneous reaction
非均相化学　heterogeneous chemistry
非均相系统【化】　heterogeneous system
非均匀的　heterogeneous
非均匀地形　nonhomogeneous terrain
非均匀海水　inhomogeneous ocean water
非均匀流　varied flow
非均匀流体　heterogeneous fluid
非均匀路程　inhomogeneous path, nonhomogeneous path
非均匀湍流　inhomogeneous turbulence
非均匀网格　uneven grid
非均匀性　inhomogeneity
非均质层【大气】　heterosphere
非均质大气　nonhomogeneous atmosphere
非可递性　intransitivity
非连续传播　discrete propagation
非流体静力波体系　nonhydrostatic wave regime
非流体静力模式　nonhydrostatic model
非流体静力原始方程组　nonhydrostatic primitive equations
非龙卷风暴　nontornadic storm
非门【计】　NOT gate
非黏性流体　non-viscous fluid, inviscid fluid【大气】
非〔黏〕滞流体　inviscid fluid
非牛顿参考系　non-Newtonian reference frame
非牛顿流动【化】　non-Newtonian flow
非偏移吸收【地】　non-deviative absorption
非偏振光【物】　unpolarized light, nonpolarized light
非平衡热力学　non-equilibrium thermodynamics
非平衡态【物】　nonequilibrium state
非平衡系统【化】　nonequilibrium system
非齐次多项式【数】　inhomogeneous polynomial
非确定性的　non-deterministic
非扰动太阳　undisturbed sun
非热辐射【天】　nonthermal radiation
非色散分析【化】　nondispersive analysis
非色散红外分析仪　nondispersed infrared analyzer
非色散红外光谱法　nondispersive infrared specrtometry
非色散红外吸收　nondispersive infrared absorption
非生物的　abiotic
非生物悬浮物　abioseston
非实时　non-real time
非实时资料【大气】　non-real time data
非实时资料处理　non-real-time data processing
非守恒散射　nonconservative scattering
非天气图定时资料　asynoptic data
非天气资料【大气】　non-synoptic data
非同步遥相关　asynchronous teleconnection

非完全弹性学	imperfect elasticity
非稳定演变	non-stationary evolution
非吸湿性核	non-hygroscopic nuclei
非线性	nonlinear
非线性饱和	nonlinear saturation
非线性波	nonlinear wave
非线性波-波交互作用	nonlinear wave-wave interactions
非线性不稳定〔性〕【大气】	nonlinear instabililty
非线性迭代法	nonlinear iterative method
非线性方程	nonlinear equation
非线性轨迹	nonlinear trajectory
非线性化	nonlinearity
非线性回归【数】	nonlinear regression
非线性计算不稳定	nonlinear computational instability
非线性黏弹性【化】	non-linear viscoelasticity
非线性平衡方程	nonlinear balance equation
非线性平流方程	nonlinear advection equation
非线性交互作用	non-linear interaction
非线性薛定谔方程【物】	nonlinear Schrodinger equation
非线性映射	non-linear mapping
非线性正交模	non-linear normal mode
非线性正交模初值化	nonlinear normal-mode initializaton
非线性准地转流	nonlinear quasi-geostrophic flow
非相干	non-coherence
非相干场	incoherent field
非相干回波	non-coherent echo
非相干回波【大气】	incoherent echo
非相干雷达	noncoherent radar
非相干雷达【大气】	incoherent radar
非相干目标	noncoherent target
非相干强度	incoherent intensity
非相干散射	incoherent scattering
非相干散射雷达	incoherent scatter radar
非相干性【物】	incoherence
非相干源	incoherent source
非相关散射通道	uncorrelated scattering channel
非谐振子【物】	anharmonic oscillator
非〔寻〕常折射率【物】	extraordinary refractive index
非压缩码资料格式	unpacked data format
非正态分布【化】	nonnormal distribution
非中央差分	uncentred differencing
非周期变化	non-periodic variation
非周期解	aperiodic solution
非周期流	nonperiodic flow, aperiodic flow
非周期信号	aperiodic signal
非周期性【物】	aperiodicity
非周期振动	aperiodic oscillation
非洲东风急流	African easterly jet (AEJ)
非洲急流	African jet
非洲经济委员会(联合国)	Economic Commission for Africa (ECA, UN)
菲克方程	Fick's equation, Fickian equation
菲克扩散	Fickian diffusion
菲克扩散定律	Fick's law of diffusion
菲克扩散方程	Fickian diffusion equation
菲律宾大气、地球物理和天文服务管理局	Philippine Atmospheric, Geophysical and Astronomical Services Administration (PAGASA)
菲律宾海流	Philippines Current
菲涅耳长度	Fresnel length
菲涅耳尺度	Fresnel size
菲涅耳带	Fresnel zone
菲涅耳反射	Fresnel reflection
菲涅耳收缩	Fresnel shrinkage
菲涅耳衍射【物】	Fresnel diffraction
肥土【土壤】	fertile soil

废气　effluent gas
〔废气〕凝结尾迹【大气】　〔exhaust〕 contrail
废气排放标准　exhaust emission standard
废气涡轮发电机组【航海】　exhaust turbine generating set
〔废气〕蒸发尾迹【大气】　〔exhaust〕 evaporation trail
废水　wastewater
废水　devil water
废水处理　wastewater treatment
废物再循环　waste recycling
沸点　boiling point
沸点测高表　hypsometer
沸点温度表　boiling point thermometer
沸腾　ebullition, boil
费根鲍姆数【物】　Feigenbaum number
费雷尔定律　Ferrel's law
费雷尔环流【大气】　Ferrel cell
费雷尔涡旋　Ferrel vortex
费马原理【物】　Fermat principle, Fermat's principle【大气】
费米子【物】　fermion
费希尔-波特雨量计　Fisher and Porter raingauge
费希尔判别函数【数】　Fisher discriminant function
费希尔信息函数【数】　Fisher information function
费用函数【数】　cost function
费用-收益比　cost-benefit ratio
费用-收益法　cost-benefit approach
费用-损失比　cost-loss ratio
分　minute, deci-
分巴　decibar
分贝　decibel (dB)
分贝反射率因子【大气】　decibel reflectivity factor, dBz
分辨单元　resolution cell
分辨率【物】　resolution
分辨能力　resolving power

分辨体积　resolution volume
分布【物】　distribution
F分布　Fisher (F) distribution
F分布【化】　F distribution
分布电容【电子】　distributed capacitance
分布函数【数】　distribution function
分布目标【电子】　distributed target
χ^2分布【数】　chi-square distribution
t分布【数】　student distribution
分布式数据库【计】　distributed database
分布法【数】分数步长法　fractional step method
分布图　distribution graph
分部积分法【数】　integration by parts
分层【数】　hierarchy
分层弹道风　differential ballistic wind
分层均匀介质　piecewise-uniform medium
分层显示　grey level display
分岔【物】【大气】　bifurcation
分拆函数【数】　partition function
分潮　tidal component, tidal constituent, partial tide, constituent of tides
分窗法　split window technique
分点潮【海洋】　equinoctial tide
分隔面　surface of separation
分光变阻测热计　spectrobolometer
分光变阻测热图　spectrobologram
分光测量法　spectrometry
分光光度表【物】　spectrophotometer
分光光度学　spectrophotometry
分光光〔雷〕达技术　spectroscopic laser radar technology
分光计　spectrometer
分光镜【物】　spectroscope
分光湿度计　spectroscopic hygrometer
分光天空辐射表　spectral pyranometer
分级　grading
分级【化】　fractionation
分类　grouping

分类〔法〕 classification
分类分析【大气】 classification analysis
分类机 sorting machine
分类统计 classification statistics
分类预报 categorical forecast
分离 separation
分离变量法【数】 method of separation of variables
分离常数 separation constant
分离窗〔区〕 split window
分离〔大气〕密度效应 separating density effect
分离点 separation point
分离对流风暴 splitting convective storm
分离法 splitting method
分离冷锋 split cold front
分离模式 discrete model
分离器 separator
分离尾流 separation wake
分离涡旋 separation vortex
分离显式方法 split-explicit method
分离显式格式【大气】 split-explicit scheme
分离线 separation line
分离因子 splitting factor
分力 component of force
分立(离散)空间理论 discrete space theory
分裂 disintegration
分裂层云 detached stratus
分流【大气】 diffluence
分流型温度槽 diffluent thermal trough
分流型温度脊 diffluent thermal ridge
分米 decimeter(dm)
分配律【数】 distributive law
分配系数【地质】 distribution coefficient
分频 splitting frequency
分谱雨量计 spectropluviometer
分歧理论【数】 bifurcation theory
分区〔云〕图 sectorized picture

分区云图【大气】 sectorized cloud picture
分散系数【土壤】 dispersive coefficient
分时 time share
分时【计】 time sharing
分时监控系统【计】 time-sharing monitor system
分时运行方式【计】 time-sharing running mode
分束器【物】 beam splitter
分数【数】 fraction
分数幂定律 fractional power law
分〔数〕维〔数〕理论 fractal dimension theory
分水岭 dividing ridge, dividing crest, divide
分水岭【水利】 drainage divide
分水流线 dividing streamline
分水线 divide line
分速度 component velocity
分位数【数】 quantile
分位温 partial potential temperature
分析 analysis
D-分析 D-analysis
分析程序 analyzer
分析初始化循环 analysis initialization cycle
分析化学【化】 analytical chemistry
分析模型误差 analysis model error
分析图 analysed chart (或 map)
分析中心 analysis center
分形【计】 fractal
分形分析【数】 fractals, fractal analysis
分形维数【物】 fractal dimension
分型线, 分隔号 separatrix
分压〔力〕 partial pressure
分页【计】 paging
分用器【电子】 demultiplexer
〔分〕支点【数】 branching point
分支函数 branched function
分至月 tropical month

分子 molecule
分子病【生化】 molecular disease
分子〔尺度〕温度 molecular-scale temperature
分子传导 molecular conduction
分子反应【化】 molecular reaction
分子光谱【物】 molecular spectrum
分子光谱学【物】 molecular spectroscopy
分子光学【物】 molecular optics
分子轨道图形理论【化】 graph theory of molecular orbitals
分子耗散 molecular dissipation
分子积分【化】 molecular integral
分子几何〔结构〕【化】 molecular geometry
分子结构【物】 molecular structure
分子扩散【化】 molecular diffusion
分子扩散率 molecular diffusivity
分子理论 molecular theory
分子力学【化】 molecular dynamics
分子量 molecular weight
分子络合物【化】 molecular complex
分子黏性【大气】 molecular viscosity
分子黏滞系数【大气】 molecular viscosity coefficient
分子黏性系数【大气】 coefficient of molecular viscosity（＝dynamic viscosity）
分子气体常数【大气】 gas constant per molecule
分子热 molecular heat
分子热传导率 molecular heat conductivity
分子热力学【化】 molecular thermodynamics
分子散射【化】 molecular scattering
分子筛【化】 molecular sieve
分子生物学【生化】 molecular biology
分子污染物 molecular pollutant
分子吸收【化】 molecular absorption
分子吸收截面 molecular absorption cross section
分子遗传学【生化】 molecular genetics
分子运动 molecular motion
〔分子〕运动论 kinetic theory
分子钟【天】 molecular clock
分组码 block code
芬德森-贝吉龙成核过程 Findeisen-Bergeron nucleation process
焚风 foehn wind, warm braw, föhn
焚风【大气】 foehn
焚风鼻 foehn nose
焚风波 foehn wave【大气】, föhn wave
焚风槽 foehn trough
焚风岛 foehn island
焚风风暴 foehn storm
焚风阶段 foehn phase
焚风界 foehn pause
焚风期 foehn period
焚风气候学 foehn climatology
焚风气流 foehn air, föhn air
焚风气旋 foehn cyclone
焚风墙【大气】 foehn wall
焚风效应 föhn effect
焚风效应【大气】 foehn effect
焚风歇 sturmpause
焚风云 foehn cloud, föhn cloud
焚风云壁 föhn wall
焚风〔云〕壁 foehn wall
焚风〔云〕堤 foehn bank
焚风〔云〕隙 foehn gap, foehn break
焚风云隙 föhn break（或 gap）
粉尘 silt
粉尘起电 dust electrification
粉红雪 pink snow
粉末〔状〕雪 powdery snow（＝powder snow）
粉〔砂〕黏土【土壤】 silty clay soil
粉碎作用 comminution
粉雪 powder snow
丰产年 abundant year
丰度 abundance
丰度值 abundance value

丰水年　wet year
丰水年【大气】　high flow year
风【大气】　wind
风暴　windstorm, storm【大气】
风暴潮【大气】　storm surge, storm tide
风暴持续时间　storm duration
风暴尺度业务和研究气象学〔计划〕　stormscale operational and research meteorology（STORM）
风暴猝发　stormburst
风暴单体　storm cell
风暴单体识别与追踪【大气】　SCIT（Storm Cell Identification and Tracking）
风暴定律　law of storms
风暴度　storminess
风暴对环境的螺旋度　storm-relative environmental helicity（SREH）
风暴分裂　storm splitting
风暴观测卫星　storm observation satellite（SOS）
风暴核心　storm core
风暴降水量　storm precipitation
风暴警报【大气】　storm warning
风暴来临前晴朗天气　weather breeder
风暴浪　storm-wave
风暴雷达　storm spotter
风暴雷达数据处理机　storm radar data processor（STRADAP）
风暴路径　storm path, storm lane, storm track
风暴模式　storm model
风暴气旋　storm cyclone
风暴强度计　storm gage
风暴区　storm belt
风暴探测雷达　storm detection radar（SDR）
风暴探测器　stormscope
风暴探测装备　storm detection equipment
风暴天气　stormy weather
风暴线　line of storm
风暴信号　visual storm signal【大气】, storm signal, storm cone
风暴信息集中系统　Centralized Storm Information System（CSIS）
风暴旋入带　storm-feeder band
风暴眼　eye of the storm
风暴引发　storm initiator
风暴涌〔浪〕　storm swell
风暴预警　storm warning
风暴预警标志　visual storm warning
风暴预警塔　storm-warning tower
风暴预警信号　storm-warning signal
风暴云　storm cloud
风暴云堤　bar of the storm
风暴中心【大气】　storm center
风暴轴涡动　storm eddy
风暴转向　recurvature of storm
风暴追踪　storm chaser
风杯风速计　vane anemometer【大气】, cup anemometer
风标　weather cock
风飑　wind squalls, wind flurry
风布植物【植】　anemochore
风层　wind strata, wind layer, aeolosphere
风场【大气】　wind field
风场评价【大气】　wind site assessment
风车发电机　windmill generator
风车式风速表　windmill anemometer
风车式开关　windmill switch
风成波　wind-generated wave
风成的　aeolian
风成雪壳　wind slad, wind crust
风〔成〕音, 风吹声　aeolian tones
风成作用　eolian action
风程　wind run
风程　run-off-wind
风程风速表　run-off-wind anemometer
风程积分器　wind run integrator
风吹声　aeolian sounds
风吹雨指数　driving rain index

风带 wind zone, wind belt
风倒 wind throw
风道 wind-channel
风的垂直切变【大气】 vertical wind shear
风的结构 wind structure
风的偏差角 inclination of the wind
风的稳定性 constancy of winds
风的挟带力 competence of the wind
风的挟带能力 capacity of the wind
风电场【大气】 wind farm
风动涡旋 wind-spun vortex
风洞 wind tunnel, tunnel
风方程 wind equation
风分量测量系统 wind component measuring system
风干 air-dry
风害【大气】 wind damage
风寒 wind-chill
风寒因子 wind-chill factor
风寒指数【大气】 wind-chill index
风寒指数,寒冷因子 chill factor
风荷载【大气】 wind load
风化冰 weathered ice
风化残余物【土壤】 weathering residue
风化层 solum
风化层【地理】 regolith
风化产物【土壤】 weathering product
风化强度【土壤】 weathering intensity
风化指数【土壤】 weathering index
风化作用【土壤】 weathering
风混合 wind mixing
风级 wind scale
风级【大气】 wind force scale
风脊 wind ridge
风景规划【林】 landscape planning
风景设计【林】 landscape design
风卷云 cirrus ventosus, windy cirrus
风口【地理】 wind gap
风况 wind regime
风廓线〔探测〕雷达 wind-profiling radar(WPR)
风廓线仪 wind profiler
风浪 wind wave, wind sea
风浪等级 sea scale
风浪区【大气】 fetch
风棱石 ventifact
风力【大气】 wind force
风〔力等〕级 scale of wind-force
风力发电厂 wind power plant
风力发电机 wind turbine generator, wind generator
风力工程 wind engineering
风力环流带 wind circulation zone
风力机 wind mill
风力减弱 abatement of wind
风力田【大气】 wind farm
风力涡轮 wind turbine
风力自记曲线 anemogram
风力作用 wind action
风玫瑰〔图〕【大气】 wind rose
风媒【植】 anemophily
风媒植物 anemophilae
风媒植物【植】 anemophilous plant
风摩擦 wind friction
风能【大气】 wind energy
风能测量和记录系统 wind energy metering and recording system
风能玫瑰〔图〕【大气】 wind energy rose
风能密度【大气】 wind energy density
风能潜力 wind power potential
风能潜力【大气】 wind energy potential
风能区划【大气】 wind energy demarcation
风能转换 wind energy conversion
风能资源【大气】 wind energy resources
风能资源储量【大气】 wind energy content
风能资源数值模拟软件 wind energy simulation toolkit (WEST)
风谱 wind spectrum
风切变 shear of wind

风切变【大气】 wind shear
风切变及阵风锋预警系统 wind shear and gust front warning system
风切变探测系统 wind shear detection system
风〔驱〕动洋流 wind driven ocean circulation
风日程 wind daily run
风三角【大气】 pennant
风生表面热交换 wind-induced surface heat exchange(WISHE)
风生惯性振荡 wind-generated inertial oscillation
风生海流 wind-generated current
风生涡旋效应 anemogenic curl effect
风蚀 wind corrosion, corrasion
风蚀【地理】【大气】 wind erosion
风蚀程度【林】 degree of wind erosion
风蚀地 blowland
风蚀洼地【地理】 deflation hollow
风蚀雪波 erosion ridge
风蚀〔作用〕 eolation
风矢【大气】 wind arrow
风矢等压线交角 cross-isobar angle
风矢(风向风速填图符号) barbed arrow
风矢杆 wind-dircetion shaft, wind shaft 【大气】, shaft
风矢量【大气】 wind vector
风速【大气】 wind velocity, wind speed
风速表 anemotachometer
风速表【大气】 anemometer
风速表高度 anemometer level
风速表系数(风速与风杯切线速度之比) anemometer factor
风速表有效高度 effective height of anemometer
风速测定法【大气】 anemometry
风速顶 velopause
风速对数廓线【大气】 logarithmic velocity profile

风速方程 wind velocity equation
风速计 wind velocity recorder
风速计【大气】 anemograph
风速计数器 wind velocity counter, wind speed counter
风速亏损 velocity defect
风速廓线 wind velocity profile, wind profile
风速廓线【大气】 wind speed profile
风速脉动【大气】 wind velocity fluctuation
风速气压表 anemobarometer
风速气压换算表 wind-barometer table
风速器 wind-gauge
风速羽【大气】 barb
风速阈值传感器 wind speed threshold sensor
风速暂减 wind lull
风速指针 wind veloclty indicator
风图 wind map
风土条件 edapho-climate condition
风涡 wind eddies
风窝 wind scoop
风系 wind system, system of winds
风险函数【数】 risk function
风险决策【数】 decision under risk
风险率 risk rate
风险评估【计】 risk assessment
风向【大气】 wind direction
风向标 wind-cock, weather vane, vane
风向标【大气】 wind vane
风向齿 wind teeth
风向袋 wind sleeve【大气】, wind cone
风〔向〕袋 wind sock
风向多变 variable direction
风向风速表【大气】 anemorumbometer
风向风速风压计 anemometrograph
风向风速计【航海】 anemorumbograph
风向风速器 anemovane
风向风速仪 aerovane
风向界线 wind divide

风向量【大气】 wind vector
风向屡变的 choppy
风向逆转 wind reversal
风向偏转 wind deflection
风向突变线【大气】 wind-shift line
风向指示器 wind direction indicator
风向转变 wind shift
风向自记器 wind direction recorder
风效应 wind effect
风斜表 anemoclinometer
风斜计 anemoclinograph
风雪崩 wind avalanche
风雪纹 wind ripple
风压 velocity pressure, pressure of wind
风压【大气】 wind pressure
风压差 leeway
风压差【航海】 leeway angle
风压差系数【航海】 leeway coefficient
风压定律 baric wind law, Buys Ballots' low【大气】
风压系数【大气】 coefficient of wind pressure
风眼 eye of wind
风曳 wind drag
风因子 wind factor
风应力【大气】 wind stress
风应力旋度 wind stress curl
风羽 wind barb, feather, barb（风速填图符号，每根代表 4 m/s 或 10 海里①/时）
风云二号 Fengyun 2, FY-2
风障【大气】 wind-bread
风振【大气】 wind induced oscillation, wind vibration
风筝 kite
风筝观测 kite observation
风筝探测 kite sounding, kite-ascent
风值平均系统 wind averaging system
风致正压模式 wind-driven barotropic model
风阻影响【大气】 windage effect

封闭型细胞状云【大气】 closed 〔cloud〕 cells
封冻 freeze up
峰【地理】 peak
峰度系数 coefficient of excess
峰态 kurtosis
峰值 peak value
峰值电流 peak current
峰值功率 peak power
峰值加速度【地】 peak acceleration
峰值速度【地】 peak velocity
峰值〔所在的〕频率 peak frequency
峰值位移【地】 peak displacement
锋带 frontal band
锋的结构 frontal structure
锋的特征 front characteristics
锋的作用 frontal action
锋后降水带 post-frontal precipitation band
锋后雾 post-frontal fog
锋际雾 front passage fog
锋面【大气】 frontal surface
锋面崩溃 frontal collapse
锋面飑 front squall
锋面波动【大气】 frontal wave
锋面等高线 frontal contour
锋面等高线图 frontal contour chart
锋面低压 frontal low, wave depression
锋面对流 frontal convection
锋面分析【大气】 frontal analysis
锋面锢囚 frontal occlusion
锋面过境【大气】 frontal passage
锋面过境雾 frontal passage fog
锋面环流 frontal circulation
锋面降水【大气】 frontal precipitation
锋面雷暴 frontal thunderstorm
锋面雷阵雨 frontal thunder shower
锋面理论 frontal theory
锋面模式 frontal model

---

① 1 海里＝1.853 km

锋面逆温【大气】 frontal inversion
锋面坡度 slope of a front
锋面坡度【大气】 frontal slope
锋面气旋 frontal cyclone
锋面气旋模式 frontal cyclone model
锋面切变 frontal shear
锋面切变线 front shear line
锋面抬升 frontal uplift, frontal lifting
锋面天气【大气】 frontal weather
锋面雾【大气】 frontal fog
锋面形势 frontal topography
锋面学 frontology
锋面雨 frontal rainfall
锋面云 frontal cloud
锋前飑线 prefrontal squall line
锋前不稳定度线 prefrontal instability line
锋前雷暴 prefrontal thunderstorm
锋前雾 prefrontal fog
锋区【大气】 frontal zone
锋区界条 frontal strip
锋区漏斗效应 frontal funnel effect
锋区剖面 frontal profile
锋区气团 frontal mass
锋生【大气】 frontogenesis
锋生锋 frontogenetical front
锋生函数 frontogenetical function
锋生函数【大气】 function of frontogenesis
锋生区 frontogenetical sector, frontogenetical area
锋生线 line of frontogenesis
锋系 frontal system
锋下云【大气】 subfrontal cloud
锋线 frontal line
锋〔线〕【大气】 front
锋消【大气】 frontolysis
锋消区 frontolytical sector, frontolytical area
冯·卡曼常数 von Karman constant
冯·卡曼定律 von Karman law
冯·卡曼涡旋 von Karman vortex

冯·诺伊曼代数【数】 von Neumann algebra
冯·诺伊曼方法 von Neumann method
佛得角型（汉恩气候分类类型之一） Cape Verde type
佛罗里达海流 Florida Current
佛罗里达太阳能中心（美国） Florida Solar Energy Center（FSEC）
否定法 modus tollens（denial mode）
否认【电子】 negative acknowledgement
夫琅禾费谱带 Fraunhofer's band
夫琅禾费谱线 Fraunhofer's line
夫琅禾费线鉴别器 Fraunhofer line discriminator（FLD）
夫琅禾费衍射【物】 Fraunhofer diffraction
夫琅禾费衍射理论 Fraunhofer diffraction theory
弗劳德数【大气】 Froude number
弗雷德霍姆积分方程【数】 Fredholm integral equation
伏流 subsurface flow
伏〔特〕 volt（V）
伏天 dog days
拂掠长度 sweep length
服务比特率【计】 service bit rate
服务器【计】 server
服务系统 services system
氟化钡膜湿度表 barium fluoride film hygrometer
氟利昂 freon
氟氯烷 chlorofluoromethane（CFM）
浮标观测 buoy observation
浮标天气站 buoy weather station
浮标站 buoy
浮冰 floe ice, ice floe, floating ice
浮冰坝 hanging dam
浮冰带 ice belt【大气】, ice band
浮冰块 floe
浮冰排 ice raft, hummock
浮冰球 ball ice

浮尘【大气】 dust
浮点 floating point
浮点处理机 floating point processor
浮点运算 floating point operation
浮动 relocation
浮动式渗漏计 floating lysimeter
浮杆 velocity rod
浮力 buoyancy force【大气】,buoyancy
浮力波 buoyancy wave
浮力波数 buoyancy wavenumber
浮力参数 buoyancy parameter
浮力产生率 buoyancy production rate
浮力长度标尺 buoyancy length scale
浮力次区 buoyancy subrange
浮力能 buoyant energy
浮力频率 buoyancy frequency
浮力起伏 buoyancy fluctuation
浮力速度【大气】 buoyancy velocity
浮力通量 buoyancy flux
浮力涡旋 buoyant vortex
浮力效应 buoyancy effect
浮力因子 buoyancy factor
浮力振荡 buoyancy oscillation
浮粒【大气】 airborne particulate
浮粒监测仪 suspended solids monitor
浮升 buoyancy lift
浮升不稳定〔性〕 buoyant instability
浮升对流 buoyant convection
浮升亚区 buoyant subrange
浮升烟羽【大气】 buoyant plume
浮筒〔式〕雨量器 float type raingauge
浮心〔航海〕 center of buoyancy
浮游生物 plankton
浮子气压计 float barograph
符号编码【数】 symbolic coding
符号检验【数】 sign test
符号语言【数】 symbolic language
符合计数【物】 coincidence counting
符合率【物】 coincidence rate
幅 frame
辐辏状 radiatus（ra）

辐辏状层积云 stratocumulus radiatus（Sc ra）
辐辏状高层云 altostratus radiatus
辐辏状高积云 altocumulus radiatus
辐辏状卷云 cirrus radiatus,Abraham's tree
辐合【大气】 convergence
辐合槽 convergence trough
辐合场 convergence field
辐合带 convergence band,convergence zone,belt of convergency
辐合渐近线 asymptote of convergence
辐合率 rate of convergence
辐合线【大气】 convergence line
辐散【大气】 divergence
辐散场 field of divergence
辐散风分量 divergent wind component
辐散渐近线 asymptote of divergence
辐散理论 divergence theory
辐散流 divergent flow
辐散特征〔图型〕 divergence signature
辐散线 divergence line
辐射【物】 radiation
辐射半衰期 radiative half-life
辐射边界条件 radiation boundary condition
辐射不平衡 radiation imbalance
辐射差额【大气】 radiation balance
辐射长度【物】 radiation length
辐射出射度【物】 radiant exitance
辐〔射〕出〔射〕率(W/m^2) emittance
辐射传感器〔测〕 radiation sensor
辐射传输 radiation transfer
辐射传输方程 radiative transfer equation,equation of radiation transfer
辐射传输方程【大气】 radiation transfer equation
辐射传输理论 radiative transfer theory
辐射传输模式 radiative transfer model,radiation transfer model
辐射带【地】 radiation belt

辐射单位强度　specific intensity
辐射点　radiation point, radiant point
辐射点【天】　radiant
辐射订正　radiometric correction, radiation correction
辐射〔定〕律　radiation laws
辐射度量学【物】　radiometry
辐射-对流模式　radiative-convective model
辐射-对流平衡　radiative-convective equilibrium
辐射发射度【物】　radiant emittance
辐射反馈　radiation feedback
辐射反馈过程　radiative feedback process
辐射〔方向〕图【物】　radiation pattern
辐射防护【物】　radiation protection
辐射复合　radiative recombination
辐射干燥度　radiation dry index
辐射功率　radiant power
辐射活跃气体　radiatively active gas
辐射基准标尺　radiation reference scale
辐射脊　radiation ridge
辐射计【电子】　radiometer
辐射记录器　radiation recorder
辐射加热　radiative heating, radiation heating
辐〔射〕控〔制〕气球　radiation-controlled balloon (RACOON)
辐射亏损　radiation loss
辐射冷却　radiative cooling
辐射冷却【地】【大气】　radiation cooling
辐射亮温　radiance temperature
辐射量　radiometric quantity
辐射量谱密度　spectral concentration of a radiometric quantity
辐射滤波器　radiation filter
辐射率【大气】　radiance
辐射率单位　radiance unit
辐射漫射率　radiative diffusivity

辐射密度　radiant density
辐射模式【大气】　radiation model
辐射能　radiant energy
辐射能量　quantity of radiant energy
辐射能密度【物】　radiant energy density
辐射能温度表　radiant-energy thermometer
辐射逆温【大气】　radiation inversion
辐射平衡　radiative equilibrium, radiation equilibrium
辐射平衡【大气】　radiation balance
辐射平衡表　radiation balance meter, net pyrradiometer, balansometer
辐射平衡气球　racoon
辐射平衡温度　radiative equilibrium temperature
辐射屏蔽【物】　radiation shield
辐射气候【大气】　radiation climate
辐射器【电子】　radiator
辐射腔　hohlraum
辐射强度【大气】　radiant intensity
辐射强度【物】　radiation intensity
辐射强迫〔作用〕　radiative forcing
辐射热　radiant heat
辐射热交换　radiative heat exchange
辐射热通量　radiative heat flux
辐射生物学【农】　radiation biology
辐射收支　economy of radiation
辐射收支【地】　radiation budget
辐射霜【大气】　radiation frost
辐射霜冻害　radiation frost injury
辐射损伤【物】　radiation damage
辐射探空仪　radiometersonde, radiation sonde
辐射天空温度　radiative sky temperature
辐射通量　radiative flux【大气】, flux of radiation
辐射通量(W)【电工】　radiant flux
辐射通量【物】　radiation flux
辐射通量密度　radiative flux density,

radiant flux density
辐射通量散度　radiative flux divergence
辐射通量散度计　radiation flux divergence meter
辐射图【大气】　radiation chart
辐射湍流平衡　radiative-turbulent equilibrium
辐射温度　radiation temperature
辐射温度表　radiation thermometer
辐射误差　radiation error
辐射雾【大气】　radiation fog
辐射吸收【物】　radiative absorption
辐射系数【电力】　radiation coefficient
辐射效率【电子】　radiation efficiency
辐射型雪崩　radiation pattern avalanche
辐射压　pressure of radiation
辐射压〔强〕【物】　radiation pressure
辐射仪放大器　radiometer amplifier
辐射仪器　radiation instruments
辐射源【物】　radiation source
辐射跃迁【机械】　radiation transition
辐射-云-气溶胶交互作用　radiation-cloud-aerosol interaction
辐射站　radiation station
辐射蒸发表　radio atmometer
辐射中心　radiation center
辐射转矩　radiation torque
辐射阻尼　radiative damping
辐射阻尼【物】　radiation damping
辐照【大气】(J/m^2)　irradiation
辐照度【大气】(W/m^2)　irradiance
辐照加工【农】　irradiation processing
辐照量【电工】【大气】　radiant exposure
辐照室【农】　irradiation chamber
福丁气压表【大气】　Fortin barometer
福克尔-普朗克方程【物】　Fokker-Planck equation
福克兰海流　Falkland Current
福来耳海色标度　Forel scale
幞状积云　cumulus pileus
幞状云　scarf cloud, pileus (pil)
俯角　depression angle
俯角【数】　angle of depression
俯视光谱法　downward viewing spectral method
俯仰角　angle of pitch
俯仰轴　pitch axis
辅助变量【数】　auxiliary variable
辅助船舶观测【大气】　auxiliary ship observation (ASO)
辅助船舶站　auxiliary ship station
辅助点　supplementary point
辅助光照　supplementary illumination
辅助农业气象站　auxiliary agricultural meteorological station
辅助天气观测【大气】　intermediate synoptic observation
辅助天气时间　intermediate standard time
辅助图　auxiliary chart
辅助温度表　auxiliary thermometer
辅助站　auxiliary station
腐殖化程度【土壤】　degree of humification
腐殖化作用【土壤】　humification
腐殖酸〔类〕【土壤】　humic acids
腐殖质【土壤】　humus
负　negative
负变压〔的〕　katallobaric
负变压线【大气】　katallobar
负变压中心　isallobaric low
负变压中心【大气】　katallobaric center
负表面张力【化】　negative surface tension
负等温涡度平流　negative isothermal vorticity advection
负地闪　negative ground flash (= negative cloud-to-ground lightning)
负电　negative electricity
负电荷　negative charge
负反馈　negative feedback
负耗散　negative dissipation

负荷　loading
负积温【农】　negative accumulated temperature
负极　negative pole
负距平中心　meions
负离子　negative ion(＝anion)
负离子-电子比率　negative ion-electron ratio
负离子反应　negative ion reaction
负离子,阴离子　anion
负离子质谱【化】　negative ion mass spectrum
负黏性　negative viscosity
负偏差中心　antipleion
负区　negative area
负熵　negentropy
负温度　negative temperature
负温度系数【化】　negative temperature coefficient
负涡度平流　nagative vorticity advection (NVA)
负吸附【化】　negative adsorption
负相关【化】　negative correlation
负异常【地】　negative anomaly
负云地闪电　negative cloud-to-ground lightning
负载　load
负载比【电子】　duty ratio
负折射【大气】　negative refraction
负轴　negative axis
附件　attachment
附录　appendix
附属虹　supernumerary rainbow
【大气】, supernumerary bow
附属温度表　attached thermometer
附属云【大气】　accessory cloud
附着力　adhesion
附着系数　attachment coefficient
附着效率　adhesion efficiency
复摆【物】　compound pendulum
复包络　complex envelope
复变数【数】　complex variable
复层积云　stratocumulus duplicatus (Sc du)
复对流层顶　multiple-tropopause
复分子．complex molecules
复高层云　altostratus duplicatus
复高积云　altocumulus duplicatus
复共轭的【数】　complex conjugate
复合【物】　recombination
复合粗糙面　composite rough surface
复合肥料【土壤】　compound fertilizer
复〔合〕概率　compound probability
复合函数【数】　compound function
复合记录器　multiple register
复合介质　composite medium
复合矩阵【数】　compound matrix
复合离心力(即科里奥利力)　compound centrifugal force (＝Coriolis force)
复合率【物】　recombination rate
复合能　recombination energy
复合〔水文〕过程线　compound hydrograph (＝composite hydrograph, complex hydrograph)
复合体风暴　multi-cell storm
复合透镜【物】　compound lens
复合系数　recombination coefficient, combination coefficient
复回归产量模式　multiple regression yield model
复解调　complex demodulation
复介电常数　complex dielectric constant
复经验正交函数　complex empirical orthogonal function (CEOF)
复卷积云　cirrostratus duplicatus
复卷云　cirrus duplicatus
复离子　complex ions
复数【数】　complex number
复数乘法　complex multiplication
复苏现象　recovery
复位【计】　reset
复相关【数】【大气】　multiple correlation

复相关系数　coefficient of multiple correlation
复相关系数【数】【大气】　multiple correlation coefficient
复信号　complex signal
复用器【电子】　multiplexer
复原【动】　resilience
复原检验　recovery test
复圆　fourth contact(日食)
复圆【天】　last contact(日食), last contact of umbra(月食)
复(云)　duplicatus (du)
复杂地形　complex terrain
复杂〔黑子〕群　complex group
复杂气象飞行　overweather flight
复折射率　complex refractive index
复折射率【物】　complex index of refraction
复制　reproduction
复制【计】　copy
复质量控制　complex quality control (CQC)
副瓣　minor lobe
副北冰洋流　Subarctic Current
副北极高层水　Subarctic Upper Water
副赤道带　subequatorial belt
副赤道脊　subequatorial ridge
副锋【大气】　secondary front
副海洋性气候　submarine climate
副寒带　subfrigid zone
副极带　subpolar zone, subarctic zone
副极地　subpolar
副极地冰川　subpolar glacier
副极地带电离层　subpolar zone ionosphere
副极地低压　subpolar low
副极地低压带　subpolar low-pressure belt
副极地反气旋　subpolar anticyclone
副极地高压　subpolar high
副极地冷气冠　subpolar cap
副极地气候【大气】　subarctic climate
副极地气团　subpolar airmass

副极地西风带　subpolar westerlies
〔副极地〕植物区气候(柯本气候分类之一)　biochore
副极地转子　subpolar gyre
副极光带　subauroral zone
副冷锋【大气】　secondary cold front
副流　secondary flow
副南极带　subantarctic zone
副南极锋　subantarctic front
副南极高层水　subantarctic upper water
副南极模式水　subantarctic mode water
副气旋【大气】　secondary cyclone
副热带〔的〕　subtropical
副热带的　semi-tropical
副热带东风带【大气】　subtropical easterlies
副热带东风指数　subtropical easterlies index
副热带反气旋【大气】　subtropical anticyclone
副热带风带　subtropical wind belt
副热带锋　subtropical front
副热带辐合带　subtropical convergence
副热带高压【大气】　subtropical high
副热带高压带　subtropical high pressure belt
副热带高压单体　subtropical cell
副热带高压脊　subtropical ridge
副热带高压区　subtropical high pressure zone
副热带急流　subtropical jet
副热带急流【大气】　subtropical jet stream
副热带季风带　subtropical monsoon zone
副热带空气　subtropical air
副热带模式水　subtropical mode water
副热带逆流　subtropical countercurrent
副热带气候【大气】　subtropical climate
副热带气旋　semi-tropical cyclone
副热带气旋【大气】　subtropical cyclone

副热带无风带【大气】 subtropical calms
副热带西风带【大气】 subtropical westerlies
副热带,亚热带 subtropics, subtropical belt（或 zone）
副热带,亚热带【地理】 subtropical zone
副热带转子 subtropical gyre
副涡旋 subsidiary vortex
副帧 minor frame
赋值【数】 valuation
傅科摆【物】 Foucault pendulum
傅里叶变换【数】 Fourier transform
傅里叶变换核磁共振【化】 Fourier transform nuclear magnetic resonance
傅里叶变换红外光谱计【化】 Fourier transform infrared spectrometer
傅里叶变换红外光谱〔学〕【化】 Fourier transform infrared spectroscopy
傅里叶变换质谱计【化】 Fourier transform mass spectrometer
傅里叶对 Fourier pair
傅里叶分析【数】【大气】 Fourier analysis
傅里叶合成 Fourier synthesis
傅里叶核 Fourier kernel
傅里叶核窗 Fourier kernel window
傅里叶积分【数】 Fourier integral
傅里叶积分算子【数】 Fourier integral operator
傅里叶级数【数】 Fourier series
傅里叶空间【化】 Fourier space
傅里叶逆变换【数】 inverse Fourier transform
傅里叶谱 Fourier spectrum
傅里叶系数【数】 Fourier coefficient
富冰冻土【地理】 ice-rich permafrost
富集【土壤】 enrichment
富集因子 enrichment factor
富营养化【土壤】 eutrophication
富营养水【地理】 eutrophic water
富营养系统【土壤】 eutrophic system
富营养沼泽【地理】 eutrophic mire
腹地【地理】 hinterland
覆冰量 ice cover
覆盖 cover
覆盖层【物】 overlayer
覆盖度图 coverage diagram
覆盖逆温【大气】 capping inversion
覆盖区 area of coverage
覆盖系数（云滴取样） covering fraction

G

伽 Galilei（1 Gal＝1 cm/s^2）
伽伐尼臭氧监测仪 Galvanic ozone monitor
伽利略不变量 Galilean invariant
伽辽金-彼得罗夫法【数】 Galerkin-Petrov method
伽辽金波谱法 Galerkin spectral method
伽辽金法 Galerkin method
伽辽金近似 Galerkin approximation
伽辽金有限元法 Galerkin finite element method
伽玛尺度 gamma scale（＝γ scale）
伽玛分布 gamma distribution（＝γ distribution）
伽玛辐射 gamma radiation（＝γ radiation）
伽玛函数 gamma function（＝γ function）
盖尔丁通风器（测离子用） Gerdien aspirator

盖革-米勒计数管　Geiger-Muller tube
盖革-米勒计数器【物】　Geiger-Muller counter
盖吕萨克定律【物】　Gay-Lussac law
盖娅假说　Gaia hypothesis
概率【物】【大气】　probability
概率分布【数】【大气】　probability distribution
概率分布函数　probability distribution function
〔概率〕分布律　distribution law
概率过程　probabilistic process
概率函数【数】　probability function
概率积分　probability integral
概率论【数】　probability theory
概率密度【化】　probability density
概率密度函数　probability density function
概率模式　probability model
概率母函数【数】　probability generation function
概率评分　probability score(PS)
概率水准　probability levels
概率椭圆　probability ellipse
概率误差　probable error
概率预报【大气】　probability forecast
概率〔坐标〕纸【数】　probability paper
概念模型【计】　conceptual model
概念图像　conceptual picture
干冰　solid carbon dioxide, dry ice【大气】
干冰催化　dry-ice seeding
干不稳定性　dry instability
干草热,花粉热　hay fever
〔干〕草原　steppe
干沉降【大气】　dry deposition
干次湿气候　dry subhumid climate
干冻【大气】　dry freeze
干对流调整　dry convection adjustment
干对流【大气】　dry convection
干风　dry wind

干谷、旱谷,干涸河道　arroyo
干焓　dry enthalpy
干旱　drouth(＝drought)
干旱【大气】　drought
干旱程度【林】　degree of drought
干旱带　arid zone
干旱带水文学　arid-zone hydrology
干旱年【大气】　low flow year
干旱频数【大气】　drought frequency
干旱气候【大气】　arid climate
干旱强度指数　drought severity index
干旱指数【大气】　drought index
干旱周期　period of drought
干涸湖　extinct lake
干季【大气】　dry season
干静力能量　dry static energy
干静力总能量　dry static total energy
干绝热变化　dry adiabatic change
干绝热大气　dry-adiabatic atmosphere
干绝热的　dry adiabatic
干绝热动力学　dry adiabatic dynamics
干绝热过程【大气】　dry adiabatic process
干绝热冷却　dry adiabatic cooling
干绝热率　dry adiabatic rate
干绝热温度变化　dry adiabatic temperature change
干绝热线　dry adiabat
干绝热增温　dry adiabatic warming
干绝热直减率【大气】　dry adiabatic lapse rate
干空气密度　density of dry air
干枯河道　wadi
干冷锋【大气】　dry cold front
干霾【大气】　dry haze
干模式　dry model
干逆温　dry inversion
干暖盖　dry and warm lid
干期　dry period
干期【大气】　dry spell
干球　dry bulb

干球温度【大气】 dry-bulb temperature
干球温度表【大气】 dry-bulb thermometer
干扰波瓣 interference lobe
干扰素【生化】 interferon
干热风 hot-dry wind
干热风【大气】 dry hot wind(=hot-arid wind)
干热〔气候〕期 xerothermal period
干热指标 xerothermal index
干舌【大气】 dry tongue
干涉【物】 interference
干涉测量术【物】 interferometry
干涉滤光光度计 interference-filter photometer
干涉区 interference region
干涉色【物】 interference color
干涉条纹 interference fringe
干涉图 interferogram
干涉仪【物】 interferometer
干生长 dry growth
干湿表【大气】 psychrometer
干湿表常数 psychrometric constant
干湿计 psychrograph
干湿球湿度表 wet and dry bulb hygrometer, dry-and-wet-bulb hygrometer
干湿球温差 wet-bulb depression【大气】, depression of the wet bulb, psychrometer difference
干雾【大气】 dry fog
干舷〔高〕 free board
干线 dry line【大气】, dew-point front
干楔 dry slot
干雪【大气】 dry snow
干雪崩【大气】 dust avalanche
干盐湖【地质】 playa
干永冻区 dry permafrost
干燥 dryness
干燥〔的〕 arid
干燥度【大气】 aridity
干燥度指数【大气】 aridity index

干燥辐射指数 radiational index of dryness
干燥阶段 dry stage
干〔燥〕空气 dry air
干燥率 drying power
干燥气候类【大气】 dry climate
干〔燥〕区〔域〕 arid region
干燥系数 aridity coefficient
干燥因子 aridity factor
干燥指数 index of aridity, aridity index
干燥周期 arid cycle
干增长【大气】 dry growth
甘斯理论 Gans theory
甘油气压表 sympiesometer
感光测定法 sensitometry
感光计【物】 sensitometer
感光指数 light sensitive index
感觉温度【大气】 sensible temperature
感热【大气】 sensible heat
感热流〔量〕 sensible heat flow
感热通量【农】 sensible heat flux
感温元件 thermometric element, temperature-sensing element
感应电流 induced current
感应法 induction method
感应雷击 indirect lightning strike
感应起电机制 induction charging mechanism
感应学说 influence theory
感应元件 sensing element
感雨器 rain detector
冈恩-贝拉尼辐射计 Gunn-Bellani radiometer
刚度矩阵【数】 stiffness matrix
刚盖 rigid-lid
刚体【物】 rigid body
刚体边界 solid boundary, rigid boundary
刚体边界条件【大气】 rigid boundary condition
刚体系 rigid system
刚体旋转 rigid rotation, solid rotation

刚性系统　stiff system
钢管水银土壤温度计　liquid (in-metal) thermograph
钢管水银温度计　mercury-in-steel thermograph
港口气象联络官　port meteorological liaison officer
高层　upper level
高层大气　aeronomosphere
高层大气【大气】　upper atmosphere
高层大气结构　upper-air structure
〔高层〕大气探测火箭　atmospheric sounding projectile
高层大气物理学　upper atmospheric physics
高层大气现象　upper atmospheric phenomenon
高层大气研究　upper atmosphere research
高层大气研究卫星　Upper Atmosphere Research Satellite (UARS)
高层大气荧光　upper air fluorescence
高层等高探测气球　skyhook balloon, Moby Dick balloon
高层积云　altostratocumulus
高层空气　superior air
高层云【大气】　altostratus (As)
高差表　cathetometer
高潮【大气】　high tide
高潮间隙　high-water interval
高潮时【航海】　high water time
高吹尘　blowing dust
高吹沫　blowing spray
高吹沙【大气】　blowing sand
高吹雪【大气】　blowing snow
高次导数　derivative of high order
高低潮【航海】　higher low water
高地冰川　highland glacier
高地冰盖　highland ice
高地气候　highland climate
高度　height, altitude

高度表　altimeter
高度表拨定〔值〕【大气】　altimeter setting
高度表拨正指示器　altimeter setting indicator
高度表订正　altimeter correction
高度表方程　altimeter equation
高度垂直坐标〔系〕　height vertical coordinate
高度订正　altitude correction
高度方位距离位置显示器　height-azimuth-range position indicator(HARPI)
高度〔分布〕型〔式〕　height pattern
高度分析法　height analysis method
高度计　altigraph
高度廓线测量　altitude profile measurement
高度增益　height gain
高度真空　high vacuum
高分辨率　high resolution
高分辨〔率〕成像光谱仪(EOS)　High-Resolution Imaging Spectrometer(HIRIS)
高分辨〔率〕动力学临边探测器(EOS)　High-Resolution Dynamics Limb Sounder (HIRDLS)
高分辨〔率〕干涉探测器【大气】　High Resolution Interferometric Sounder (HIS)
高分辨率红外辐射计　High Resolution Infrared Radiometer(HRIR)
高分辨〔率〕红外辐射探测器【大气】　High-Resolution Infrared Radiation Sounder (HIRS)
高分辨〔率〕红外探测器　High Resolution Infrared Sounder(HRIS)
高分辨率图像传输【大气】　High Resolution Picture Transmission(HRPT)
高分辨〔率〕微波波谱探测器　High-Resolution Microwave Spectrometer Sounder(HIMSS)
高分辨率〔云图〕传真【大气】　High

Resolution Facsimile(HR-FAX)
高分子【化】 macromolecule
高分子化学【化】 polymer chemistry
高分子晶体【化】 polymer crystal
高峰态 leptokurtic
高高潮【航海】 higher high water
高〔厚〕极地冰川 high polar glacier
高积云【大气】 altocumulus (Ac)
高级研究计划局网络【计】 Advanced Research Project Agency Network (ARPANET)
高级语言【计】 high level language
高架雨量计 elevated rain gauge
高架源 elevated point source
高阶闭合 high-order closure
高阶闭合模式 higher-order closure model
高阶差分【数】 difference of higher order
高阶矩 higher moment
高阶偏微分方程【数】 partial differential equation of higher order
高阶扰动解 higher-order perturbation solution
高阶线性椭圆〔型〕方程【数】 linear elliptic equation of higher order
高阶项 higher order terms
高阶有限差分 higher-order finite difference
高阶语言【计】 high-order language
高精度定位系统【航海】 high precision positioning system
高卷云 cirrus excelsus
高空 upper air, high altitude
高空报表 aerological table
高空病 altitude disease, aeroembolism
高空槽 upper-level trough【大气】, upper trough, trough aloft, upper-air trough, high-level trough
高空测风 wind-finding
高空大气学【大气】 aeronomy
高空低压 upper-level low, upper low, low aloft
高空电位计 alti-electrograph
高空对称型 upper symmetric regime
高空反气旋 upper-level anticyclone【大气】, upper air anticyclone, upper anticyclone, high level anticyclone
高空飞行信息区 upper flight information region
高空分析 upper-air analysis【大气】, aerological analysis
高空焚风 free foehn, high foehn, free air foehn
高空风 winds aloft, upper-level wind【大气】, upper wind, aloft wind
高空风分析 hodograph analysis
高空风分析图 hodogram
高空风分析图【大气】 hodograph
高空风观测 winds aloft observation, upper-wind observation
高空风计算盘 winds-aloft plotting board
高空风图 upper-wind chart
高空锋 upper air front
高空锋【大气】 upper front
高空辐散 upper air divergence
高空高压 upper-level high, upper high, high aloft
高空观测【大气】 upper-air observation
高空观测日 aerological days
高空〔观测〕站 aerological station
高空观象台 aerological observatory
高空环流 upper air circulation
高空混合层 upper mixing layer
高空急流【大气】 upper-level jet stream
高空脊 upper-level ridge【大气】, upper ridge, ridge aloft, upper-air ridge, high ridge, high-level ridge
高空降水 precipitation aloft
高空雷暴 high-level thunderstorm, high-based thunderstorm
高空冷锋 upper cold front

高空霾【大气】 haze aloft
高空摩擦区 upper frictional region
高空逆温 upper-level inversion, upper inversion, upper air inversion
高空暖锋 upper warm front
高空气候学 upper air climatology
高空气候学【大气】 aeroclimatology
高空气流 upper air current
高空气体化学 aerochemistry
高空气团 upper air mass
高空气象计 aerometeorograph, aerograph
高空气象图 aerographical chart
高空气象学【大气】 aerology
高空气象研究 upper air meteorological research
高空气象站 upper-air synoptic station
高空气旋 upper cyclone, upper-level cyclone【大气】, upper-air cyclone, high-level cyclone
高空扰动 upper-level disturbance, upper air disturbance
高空生物学 aerobiology
高空探测 aerological sounding, upper-air sounding, aerological ascent, aerial exploration
高空〔天气〕图【大气】 upper air chart
高空图 upper-level chart, aerological diagram, aerogram
高空微扰 upper perturbation
高空信风 overtrade wind, overtrades (=Krakatao winds)
高空烟 smoke aloft
高空仪器 aerological instrument
高空雨带 upper-level rain band
高空预报(high-level forecast)的国际电码缩写 HIFOR
高空预报图 prontour chart
高空云 upper cloud
高空站 upper layer station, upper-air station

高空资料【大气】 upper air data
高莱探测器 Golay cell
高岭石【土壤】 kaolinite
高密度脂蛋白【生化】 high density lipoprotein
高密云〔天空〕 high overcast
高能天体物理【物】 high-energy astrophysics
高能物理〔学〕【物】 high-energy physics
高能质子和α粒子检测器 high energy proton and alpha particle detector (HEPAD)
高能质子α粒子计数仪 high energy proton alpha telescope (HEPAT)
高浓度肥料【土壤】 high-analysis fertilizer
高频 high frequency (HF)
高频放大器 high-frequency amplifier (HFA)
高频放电【化】 high frequency discharge
高频干燥【农】 high-frequency drying
高频雷达 high-frequency radar
高频无线电声学探测器 high frequency radio acoustic sounder
高频振荡 high-frequency fluctuation
高频振荡器 high frequency oscillator (HFO)
高〔气〕压【大气】 high 〔pressure〕
高清晰度电视【电子】 high definition TV
高容量过滤器 high-volume filter
高山冰川 mountain glacier, alpine glacier
高山病 mountain sickness
高山〔测〕站 high-altitude station
高山冻土【地理】 alpine permafrost
高山冻原 mountain tundra, alpine tundra
高山风速计 mountain anemograph
高山观象台 high altitude observatory

高山〔观测〕站【大气】 mountain 〔observation〕 station
高山辉,染山霞 alpine glow, alpenglow (＝alpengluhen)
高山平原间环流 mountain-plain circulation
高山气候 alpine climate
高山气压表【大气】 mountain barometer
高山气压计【大气】 mountain barograph
高山生理反应 mountain physiologic effect
高山土壤【地理】 high mountain soil
〔高山〕雾林 fog forest
高山站 mountain station
高山植物【林】 alpine plant
高水 high water
高水位 high water level
高斯 gauss (G)
高斯白噪声 Gaussian white noise
高斯波包 Gaussian wave packet
高斯窗 Gauss window
高斯点 Gaussian point
高斯定理 Gauss's theorem
高斯分布 Gaussian distribution
高斯过程【数】 Gaussian process
高斯精密度权重 Gauss precision weight
高斯扩散模式 Gaussian dispersion model
高斯模型 Gaussian model
高斯求积法 Gaussian quadrature
高斯求积公式 Gauss quadrature formula
高斯曲率【数】 Gaussian curvature
高斯曲线 Gaussian curve
高斯权重 Gaussian weight
高斯-塞德尔迭代〔法〕【数】 Gauss-Seidel iterative method
高斯散度定理 Gauss divergence theorem
高斯速度起伏 Gaussian velocity fluctuation

高斯随机场 Gaussian random field
高斯网格 Gaussian grid
高斯纬度 Gaussian latitude
高斯线形 Gaussian lineshape
高斯相关函数 Gaussian correlation function
高斯消元法 Gaussian elimination method
高斯消元法【数】 Gauss elimination
高斯烟流模式 Gaussian plume models
高速传真 high speed facsimile
高速等离子体流 high speed plasma flow
高速诱导空气调节系统【航海】 high velocity induction air conditioning system
高速照相机【物】,高速摄影机 high-speed camera
高速资料存储子系统 high data rate storage subsystem (HDRSS)
高通滤波【计】 highpass filtering
高通滤波〔器〕 high pass filter
高维系统 high dimensional system
高纬度 high latitude
高纬亚暴指数 high latitude substorm index
高温 megatemperature, megatherm, high temperature
高温表 pyrometer
高温带 thermal zone, thermal belt
高温〔的〕 megathermal
高温等离〔子〕体【物】 high-temperature plasma
高温腐蚀【航海】 high temperature corrosion
高温期 megathermal period, hypsithermal
高温气候 megathermal climate
高温时段 hypsithermal interval
高温植物〔的〕 megathermic
高温植物型 megathermal type
高雾 high fog
高效液相色谱法【化】 high-performance liquid chromatography (HPLC)

高〔气〕压【大气】 high〔pressure〕
高压 high, barometric high
高压坝【大气】 high-pressure barrier
高压等值线 pliobar
高压点 hyperbar
高压脊 ridge of high pressure, pressure ridge
高压脊【大气】 ridge
高压区 area of high pressure
高压系统 high-pressure system
〔高压〕楔 wedge
高压中心 high pressure center
高盐水 haline water
高影响天气 high impact weather
高优先级【计】 high priority
高原【地理】 plateau
高原冰川 plateau glacier
高原测站 plateau station
高原季风【大气】 plateau monsoon
高原气候【大气】 plateau climate
高原气象学【大气】 plateau meteorology
高原沼泽【地理】 plateau mire
高云 high-level cloud
高云【大气】 high cloud
高值等压区 pleiobar
高指数【大气】 high index
高重复率 high-repetition-rate
缟状 velum (vel)
缟状卷云 cirro-velum
戈壁【地理】 gobi
戈达德大气科学实验室(美国) Goddard Laboratory of Atmospheric Sciences (GLAS)
戈达德空间飞行中心(美国) Goddard Space Flight Center (GSFC)
戈尔德滑尺(供船用水银气压表读数订正用) Gold slide
戈斯卫星高分辨率干涉仪探测器 GOES High-resolution Interferometer Sounder(GHIS)
哥德巴赫问题【数】 Goldbach problem

搁浅线 ground line
格点模式 grid point model
格距 grid spacing, grid length, grid interval
格拉斯霍夫数 Grashof number
格莱斯堡黑子(90年)周期 Gleissberg cycle
格兰戈腾图标位置 Gringorten plotting position
格里历【天】 Gregorian calendar
格利经纬仪 Gurley theodolite
格林定理 Green's theorem
格林函数 Green's function
格林函数【数】 Green function
格林尼治恒星时 Greenwich sidereal time (GST)
格林尼治民用时 Greenwich civil time (GCT)
格林尼治平恒星时【天】 Greenwich mean sidereal time(GMST)
格林尼治平时, 世界时 Greenwich mean time(GMT)
格林尼治时 Greenwich time (GT)
格林〔尼治〕时角【航海】 Greenwich hour angle
格林尼治视民用时 Greenwich apparent civil time (GACT)
格林尼治视〔太阳〕时 Greenwich apparent time(GAT)
格林尼治子午线【天】, 本初子午线 Greenwich meridian
格陵兰大冰原 Greenland Ice Sheet
格陵兰反气旋 Greenland anticyclone
格陵兰高压 Greenland high
格陵兰海流 Greenland Current
格陵兰海深水 Greenland Sea Deep Water
格米尼德期流星雨 Geminids meteor shower
格式, 信息编排 format
格斯特纳波 Gerstner wave

格网　grid-mesh
格网数据库【地理】　grid data bank
隔火带强度【力学】　fireline intensity
隔离河　insulated stream
隔离系统【化】　isolated system
隔年结果　result of every other year
隔水层　aquiclude
个别导数　particle derivative, individual derivative
个例分析　case study
个人计算机【计】　personal computer (PC)
个体变元【数】　individual variable
各向同性【物】　isotropy
各向同性靶〔标〕　isotropic target
各向同性等效辐射功率　equivalent isotropically radiated power (EIRP)
各向同性辐射　isotropic radiation
各向同性辐射器　isotropic radiator
各向同性介质　isotropic medium
各向同性强度　isotropic intensity
各向同性散射　isotropic scattering
各向同性湍流【大气】　isotropic turbulence
各向同性温度因子【化】　isotropic temperature factor
各向异性　aeolotropism
各向异性【物】　anisotropy
各向异性尺度分析　anisotropic scaling
各向异性介质　anisotropic medium
各向异性散射　anisotropic scattering
各向异性随机介质　anisotropic random medium
各向异性湍流　anisotropic turbulence
各向异性因子　anisotropic factor
各种云天条件　all-sky condition
根瘤菌【土壤】　rhizobia
根系【植】　root system
跟踪　tracking
跟踪和资料系统　tracking and data system (T&DS)

跟踪雷达　track radar, tracking radar
跟踪系统　tracking system
更新世【地质】　Pleistocene Epoch
更新世冰川作用　Pleistocene glaciation
更新世冰期　Pleistocene ice age, Pleistocene glacial epoch
更新世气候　Pleistocene climate
更新世气候最适期　Pleistocene Climate Optimum
更新统【地质】　Pleistocene Series
更新预报　forecast updating
耕地;耕种;耕作物　tillage
耕种期　ploughing season
耕作动力学【土壤】　tillage dynamics
耕作土壤学【土壤】　edaphology
耿贝尔分布　Gumbel distribution
耿〔式〕二极管【电子】　Gunn diode
耿〔式〕效应【电子】　Gunn effect
工程地震【地质】　engineering seismology
工程水文学【地质】　engineering hydrology
工频接地电阻　power frequency grounding resistance
工效学　ergonomics
工业革命(始于18世纪后半叶)　industrial revolution
工业化　industrialization
工业排放　industrial emission
工业气候【大气】　industrial climate
工业气候学　industrial climatology
工业气溶胶　industrial aerosol
工业气象学　industrial meteorology
工业污染云　cloud resulting from industry
工作〔级〕标准气压表　working standard barometer
弓形回波　bow echo
公报　bulletin
公差,容限【物】　tolerance
公共语言【计】　command language
公海　open sea
公害　public nuisance
公理【数】　axiom

公升,升　liter
公式　formula
公务电报　service telegram
公元(拉丁语名)　Anno Domini (A. D.)
公元前　before Christ (B. C.)
公制单位区域预报国际电码　ARMET
公众预报　public forecast
公转【天】　revolution
功【物】　work
功函数　work function
功率　power
功率传递函数　power transfer function
功率反射系数　power reflection coefficient
功率流　power flux
功率密度　power density
功率〔密度〕谱【物】　power〔density〕spectrum
功率谱【数】　power spectrum
功率谱方法　power spectral method
功率谱密度【数】　power spectrum density
功率起伏　power fluctuation
功能块【计】　function block
供肥性【土壤】　nutrient supplying capability
供水【地理】　water supply
供水量预报　water-supply forecast
供水敏感性　water supply sensitivity
供水区【石油】　recharge area
供水性【土壤】　water supplying capability
供体【生化】　donor
拱线　apsidal line
拱状云飑　arched squall
共轭点　conjugate point
共轭复数【数】　conjugate complex number
共轭矩阵【数】　conjugate matrix
共轭区域加热　conjugate region heating
共轭梯度　conjugate gradient
共轭图像　conjugate image

共轭指数律　conjugate-power law
共路径法　common-path method
共面扫描　coplane scanning
共谱【大气】　cospectrum
共谱密度　coincident spectral density
共吸附【物】　co-adsorption
共旋磁力线　corotating magnetic field line
共旋电场强度　corotation field strength
共有辐射传输模式　Community Radiation Transfer Model (CRTM)
共振【物】　resonance
共振槽　resonance trough
共振辐射　resonance radiation
共振光谱测声器　resonant spectrophone
共振级联　resonance cascading
共振加宽　resonance broadening
共振交互作用　resonance interaction
共振拉曼散射　resonance Raman scattering
共振理论【大气】　resonance theory
共振频率　resonant frequency, resonance frequency
共振器【物】　resonator
共振腔【物】　resonant cavity
共振散射【物】　resonance scattering
共振吸收【物】　resonance absorption
共振学说　resonance theory
共振荧光【物】　resonance fluorescence
共振荧光光〔雷〕达　resonance fluorescence lidar
共振振动　resonance oscillation
共轴性【物】　coaxiality
贡献　contribution
钩卷云【大气】　cirrus uncinus(Ci nuc)
钩形水位计　hook gauge
钩状　uncinus (unc)
钩状回波【大气】　hook echo
钩状回波龙卷　hook echo tornado
构造地质　tecton
构造学【地质】　tectonics

估测云幂　estimated ceiling
估计　estimate
估计标准误差　standard error of estimate
估计过低　underestimate
估计过高　overestimate
估计算子　estimator
估计值　estimate value
孤立波　isolated wave
孤立波【大气】　solitary wave
孤立单体【大气】　isolated cell
孤立导体【物】　isolated conductor
孤立回波体　isolated echo
孤立线　isolated line
孤立永冻区　permafrost island
孤〔立〕子【物】　soliton
古孢粉学　paleopalynology
古冰川　paleoglaciation
古冰川记录　paleoglaciation record
古磁学　archaeomagnetism
古代的　archaic
古迪带模式　Goody band model
古迪吸收带　Goody absorption band
古地磁场【地】　palaeomagnetic field
古地磁赤道【地】　palaeogeomagnetic equator
古地磁极【地】　palaeomagnetic pole
古地磁时间尺度　paleomagnetic time scale
古地磁时期　paleomagnetic epoch
古地磁学　paleomagnetics
古地磁〔学〕【地】　palaeomagnetism
古地理图【地质】　paleogeographic map
古地理学【地理】　paleogeography
古地貌学　paleogeomorphology
古地球物理学　paleogeophysics
古地热学【地】　palaeogeothermics
古地温【地】　paleogeotemperature
古动物学　paleozoology
古海流【海洋】　paleocurrent
古海洋学【海洋】　paleoceanography
古季风　paleomonsoon

古结晶冰　paleocrystic ice
古近纪【地质】　Paleogene Period
古近系【地质】　Paleogene System
古经度【地】　palaeolongitude
古气候【大气】　paleoclimate
古气候记录　paleoclimatic record
古气候序列　paleoclimatic sequence
古气候学【大气】　paleoclimatology
古气候证据　paleoclimatic evidence
古气候指标　paleoclimatic indicator
古气候重建　paleoclimatic reconstruction
古人类学　paleoanthropology
古生代【地质】　Paleozoic Era
古生界【地质】　Paleozoic Erathem
古生态学【地质】　paleoecology
古生物地理学【地质】　paleobiogeography
古生物化学【生化】　paleobiochemistry
古生物气象学　paleobiometeorology
古生物学　paleobiology
古水文地质学【地质】　paleohydrogeology
古水文学【地理】　paleohydrology
古土壤　paleosol
古土壤学【土壤】　paleopedology
古纬度【地】　paleolatitude, palaeolatitude
古温变化　paleotemperature variation
古温测定　pal(a)eotemperature measure
古温层次　pal(a)eotemperature stratification
古温〔度〕　paleotemperature
古温〔度〕变化　pal(a)eotemperature variation
古温〔度〕分析　pal(a)eotemperature analysis
古温记录　pal(a)eotemperature record
古温曲线　pal(a)eotemperature curve
古新世【地质】　Paleocene Epoch
古新统【地质】　Paleocene Series
古雪线　paleosnowline

古元古代【地质】 Paleoproterozoic Era
古元古界【地质】 Paleoproterozoic Erathem
古植物学【植】 paleobotany
谷冰川 valley glacier
谷风 valley wind, up-valley wind
谷风【大气】 valley breeze
谷风环流 valley wind circulation
谷口急流 valley exit jet
谷雾 valley fog
谷雨【农】【大气】 Grain Rain
固氮植物 nitrogen-fixing plants
固氮作用【土壤】 nitrogen fixation
固定边界条件 time-independent boundary condition, fixed boundary condition
固定冰 fast ice
固定船舶站【大气】 fixed ship station
固定〔观测〕平台 fixed platform
固定光束〔式〕云幂仪 fixed beam ceilometer (FBC)
固定海上平台 fixed sea platform
固定回波(雷达中) PE (permanent echo)
固定沙丘【地理】 fixed dune
固定云顶高度 fixed cloud-top altitude (FCA)
固定云顶温度 fixed cloud-top temperature (FCT)
固化 solidify
固结系数 coefficient of consolidation
固结〔作用〕 consolidation
固频探空仪 fixed frequency sounder
固态 solid state
固体潮 bodily tides
固体潮【地质】 earth tide
固体成像传感器 solid-state imaging sensor
固体等离〔子〕体【物】 solid state plasma
固体电子学【物】 solid state electronics
固体激光器【物】 solid state laser
固体静电计 solid-state electrometer
固体气压表(即空盒气压表) holosteric barometer
固相【石油】 solid phase
固有波长 natural wavelength
固有波频,内禀波频 intrinsic wave frequency
固有频率 intrinsic frequency
固有品质因数【电子】 intrinsic quality factor
固有时【物】 proper time
固有误差 original error
固有误差【电子】 intrinsic error
固有振荡 natural oscillation
固有振动 proper vibration
固有周期 natural period, eigenperiod
故障【计】 fault
故障率 failure rate
故障〔率〕特征曲线 characteristic failure behavior
故障诊断【航海】 fault diagnosis
锢囚 occlusion
锢囚低压 occluded depression
锢囚点【大气】 point of occlusion
锢囚锋 occlusion front
锢囚锋【大气】 occluded front
锢囚气旋【大气】 occluded cyclone
寡聚体【生化】 oligomer
寡肽【生化】 oligopeptide
拐点 flex point
拐点【数】 inflection point
关键字〔词〕 key word
观测 observation
观测场【大气】 observation site
观测次数【大气】 observational frequency
观测地段 observational section
观测方程 observation equation
观测高度【航海】 observed altitude
观测记录 observational data
观测结果 observational result
观测经度【航海】 observed longitude
观测井 observation well
观测平台 platform

观测前一小时　preceding hour
观测日　observational day
观测时　hour of observation
观测时〔间〕　time of observation
观测时有雹　recent hail
观测时有冻雨　recent freezing rain
观测时有雷暴　recent thunderstorm
观测时有毛毛雨　recent drizzle
观测时有雪　recent snow
观测时有雨　recent rain
观测时有雨夹雪　recent rain and snow
观测时有阵雪　recent snow shower
观测事实　observational fact
观测纬度【航海】　observed latitude
观测误差　error of observation
观测误差【大气】　observational error
观测系统模拟实验　observing systems simulation experiment（OSSE）
观测项目　observational program
观测小区　observational plot
观测研究　observation study
（观测仪器）杆　mast
观测员　observer
观测证据　observational evidence
观测值【化】　observed value
观察性流域　vigil basins
官方预报　official forecast
冠〔层〕温〔度〕　canopy temperature
冠层【大气】　canopy
冠层温度【大气】　canopy temperature
冠盖〔逆温〕层　capping layer
冠盖逆温【大气】　capping inversion
冠柱〔冰〕晶（带帽的柱状冰晶）　capped column
管理　management
管理程序【计】　supervisor
管式风洞　tube wind tunnel
管式光度计　tube photometer
管涌【地质】　piping
管状风速表　tube-anemometer
管状矢量　solenoidal vector

管状（云）　tuba（tub）
贯穿降雨〔量〕　throughfall
贯透对流　penetrative convection
贯透下沉气流　penetrative downdraft
惯性【物】　inertia
惯性边界流　inertial boundary current
惯性模　inertial mode
惯性波【大气】　inertia wave
惯性不稳定【大气】　inertial instability
惯性参考系【物】　inertial reference frame
惯性次层　inertial sublayer
惯性导航系统　inertial navigation system
惯性-对流副区　inertial-convective subrange
惯性副区　inertial subrange
惯性矩　moment of inertia
惯性框架　inertial frame
惯性-扩散副区　inertial-diffusive subrange
惯性〔力〕矩　inertial torque
惯性力【物】【大气】　inertial force
惯性流　inertial flow, inertial current
惯性频率　inertial frequency, inertia frequency
惯性区　inertial range
惯性稳定度【大气】　inertial stability
惯性系　inertial system
惯性预报【大气】　inertial forecast
惯性圆　inertial circle【大气】, circle of inertia
惯性运动　inertial motion
惯性振荡　inertial oscillation【大气】, inertia fluctuation
惯性滞后　inertial lag
惯性重力波　inertio-gravity wave【大气】（IGW）, inertial-gravity wave, inertia-gravitational wave
惯性重力内波　internal inertial gravity wave

惯性重力振荡　inertio-gravity oscillation
惯性周期　inertial period
惯性坐标系　inertial coordinate system
惯用近似　traditional approximation
灌溉　irrigation
灌溉额　duty of water
灌溉农业【农】　irrigation farming
灌溉侵蚀【土壤】　irrigation erosion
灌溉需水量　irrigation requirement
灌木【植】　shrub, frutex（拉丁）
灌球平衡秤　filling balance
光　light
光斑　faculae
光斑漂移　beam spot wander
光饱和【海洋】　light saturation
光饱和点【农】　light saturation point
光泵【物】　light pump
光泵磁强计　optical pumping magnetometer
光泵激光器　optically-pumped laser
光比色法　photocolorimetry
光笔　light pen
光变周期【天】　period of light variation
光标　cursor
光波【物】　light wave
光波动理论　wave theory of light
光薄介质【天】　optically thin medium
光补偿点【农】　light compensation point
光测高温表　optical pyrometer, ardometer
光程　optical distance
光程长〔度〕　optical pathlength
光抽运【物】　optical pumping
光传播　optical propagation
光猝灭【物】　optical quenching
光导摄像机　vidicon camera
光导纤维管　fiberoptic catheter
光导型固态探测器　photoconductive soild-state detector
光导照相机系统　vidicon camera system
光的反射　reflection of light

光的偏振　polarization of light
光的色散　dispersion of light
光的微粒说　corpuscular theory of light
光电倍增管　multiplier phototube（MPT）
光电倍增管【物】　photomultiplier
光电测光术　photoelectric photometry
光电成像【天】　photoelectronic imaging
光电池【物】　photoelectric cell, photo-cell
光电〔]导探测器　photoconductive detector
光电导体【物】　photoconductor
光电等高仪【天】　photoelectric astrolabe
光电二极管【物】　photodiode
光电光度计　photoelectric photometer
光电化学【化】　photoelectrochemistry
光电气象学　electro-optical meteorology
光电摄像管　iconoscope
光电湿度表　photoelectric hygrometer
光〔]探测【物】　photodetection
光电探测器　photovoltaic detector
光电透射表　photoelectric transmittancemeter
光电效应【物】　photoelectric effect
光电子【物】　photoelectron
光电子能谱　photoelectron energy spectrum
光电子碰撞　photoelectron collision
光电子外逸高度　photoelectron escape altitude
光电子外逸通量　photoelectron escape flux
光调制【物】　optical modulation
光碟【计】　optical disc, compact disc(CD)
光碟驱动器【计】　optical disc drive
光度观测　photometric observation
光度计　light meter
光度计【物】　photometer
光度学【物】　photometry
光度遥测法　telephotometry
光发射度【物】　luminous emittance
光辐射　radiation of light
光合强度【农】　photosynthetic intensity

光合效率【农】 photosynthetic efficiency
光合有效辐射【大气】 photosynthetically active radiation(PAR)
光合作用【植】【大气】 photosynthesis
光厚介质【天】 optically thick medium
光呼吸【植】 photo-respiration
光滑冰 glare ice
光滑函数【数】 smooth function
光化 actinic
光化层【大气】 chemosphere
光化层顶【大气】 chemopause
光化反应【大气】 photochemical reaction
光化辐射【物】 actinic radiation
光化离解 photochemical dissociation
光化模式 photochemical model
光化平衡 photochemical equilibrium
光化射线 actinic ray
光化通量 actinic flux
光化吸收 actinic absorption
光化学【化】 photochemistry
光化学反应【大气】 photochemical reaction
光化学过程【化】 photochemical process
光化学空气污染 photochemical air pollution
光化学污染【大气】 photochemical pollution
光化〔学〕污染物 photochemical pollutant
光化学雾【化】 photochemical fog
〔光化学〕烟雾〔实验〕室 smog chamber
光化〔学〕氧化剂 photochemical oxidant
光化学烟雾【大气】 photochemical smog
光混频 photomixing
光极化 light polarizing
光降解【化】 photodegradation
〔光〕焦度【物】 focal power
光解【化】 photodecomposition
光解离【化】 photodissociation
光解作用【物】 photolysis

光缆【电子】 optical fiber cable
光〔雷〕达 light detection and ranging (LIDAR), laser radar
光雷达比 lidar ratio
光〔雷〕达测高 lidar-determined height
光雷达常数 lidar constant
光〔雷〕达反射率 lidar reflectance
光雷达方程 lidar equation, laser radar equation
光〔雷〕达气象学 lidar meteorology
光〔雷〕达云幂仪 lidar ceilometer
光亮度【大气】 luminance
光亮度对比 luminance contrast
光亮对比 luminous contrast
光量计 quantometer
光脉冲传播 optical pulse propagation
光弥散 light dispersion
光密度 luminous density
光密介质【物】 optically denser medium
光敏的 photosensitive
光敏电阻 photoresistor, photoconductive cell
光敏化【化】 photosensitization
光敏性【物】 photosensitivity
光能 luminous energy, light energy
光能测定仪 actinometer
光年【物】 light year
光偏振表 photopolarimeter
光谱的 spectral
光谱定量分析【化】 quantitative spectrometric analysis
光谱定性分析【化】 qualitative spectrometric analysis
光谱法 spectroscopic method
光谱反照率 spectral albedo
光谱干扰 spectral interference
光谱湿度表 spectral hygrometer
谱通道 spectral channel
〔光〕谱线 spectral line
光谱项【化】 spectroscopic term
光谱学【物】 spectroscopy

光强度　light intensity
光球【天】　photosphere
光散射　optical scattering, light scattering
光散射表　light scattering table
光散射探测器　optical scattering probe
光散射图　light scattering diagram
光生伏打效应【物】　photovoltaic effect
光声计　spectrophone
光声探测　spectrophone detection, optoacoustic detection
光疏介质【物】　optically thinner medium
光束充填〔量〕　beam filling
光束辐照度　beam irradiance
光束照度　beam illuminance
光衰减　light attenuation
光速　speed of light
光探测器　photodetector
光通量　luminous power, light flow
光通量【物】　luminous flux
光通量密度　luminous flux density
光外差雷达　optical heterodyne radar
光温生产潜力【农】　light and temperature potential productivity
光稳态关系　photostationary state relation
光吸收　photoabsorption
光吸收截面　photoabsorption cross section
光纤【电子】　optical fiber
光纤激光器【物】　fiber laser
光纤通信【电子】　optical fiber communication
光线　light ray
光线追迹【物】　ray tracing
光象　luminous meteor
光选通【物】　light gating
光学　photology
光学【物】　optics
光学参量振荡器　optical parametric oscillator (OPO)
光学常数【物】　optical constant
光学成像探测器　optical imaging probe
光学臭氧探空仪系统　optical ozonesonde system
光学传递函数　optical transfer function
光学窗　optical window
光学地平　optical horizon
光学风速仪　optical anemometer
光学观测　optical observation
光学黑色涂料　optical black lacquer
光学横向风速廓线仪　optical crosswind profiler
光学厚度【大气】　optical depth
光学厚度【物】　optical thickness
光学厚度因子　optical depth factor
光学混频器　optical mixer
光学活性【化】　optical activity
光学活性气体　optically active gas
光学计数器　optical counter
光学经纬仪　optical-theodolite
光学均质的　optically homogeneous
光学粒子尺度测定装置　optical sizing device
光学粒子探测器　optical particle probe
光学流量表　optical flowmeter
光学路径　optical path
光学霾　optical haze
光〔学〕密度　optical density
光学模【物】　optical mode
光学模型【物】　optical model
光学能见度　optical visibility
光学平滑的　optically smooth
光学湿度表　optical hygrometer
光学瞬变探测器　Optical Transient Detector (OTD)
光学天文学【天】　optical astronomy
光〔学〕通信【物】　optical communication
光学图像处理【地理】　optical image processing

光〔学〕外差【物】 optical heterodyne
光学现象 optical phenomenon
光学相关仪器 optical correlation instrument
光学校直【物】 optical alignment
光学斜距 optical slant range
光学谐振腔【电子】 optical cavity
光学遥感 optical remote sensing
光学仪器的分辨率 resolution of power of optical instrument
光学有效大气 optically effective atmosphere
〔光学〕有效大气〔路径〕 effective atmosphere
光学雨量计 optical rain gauge
光学元件 optical component
光学原理 optical principle
光学折射率 optical refractive index
光学质量 optical mass
光压 light pressure
光氧化【化】 photooxidation
光氧化物 photoxide
光阴极 photocathode
光源 source of light, light source
光闸【物】 optical shutter, optical gate, light gate
光栅 grating
光栅线 raster line
光照长度【大气】 illumination length
光照处理【农】 light treatment
〔光〕照度【物】 illumination, illuminance (E)
光照阶段【大气】 photophase
光照阶段【农】 photostage
光照量 quantity of illumination, luminous exposure, light exposure
光照气候 light climate, illumination climate
光照强度 illumination intensity
光照强度【农】 intensity of illumination
光照延续时间 light duration

光折 stooping
光阵滴谱仪探头 optical array precipitation spectrometer probe
光致变色玻璃 photochromic glass
光致电离【物】 photoionization
光致电离截面 photoionization cross section
光致分离【物】 photodetachment
光致复合 photo-recombination
光致凝结器 photocoagulator
光周期【农】 photoperiod
光周期〔性〕【大气】 photoperiodism
光轴【物】 optical axis
光柱(日柱) light pillar
光资源【大气】 light resources
光子【物】 photon
光子分布函数 photon distribution function
光子活化分析【化】 photon activation analysis
光子火箭 photon rocket
光子积分浑浊度表 photon integrating nephelometer
光子计数【天】 photon-counting
光子计数【物】 photocounting
光子计数器 photon counter
光子流速率 photon flow rate
光子通量【物】 photon flux
广播 dissemination
广播中心 broadcast center
广度性质【化】 extensive property
广度优先搜索 breadth-first search
广角辐射仪 wide angle radiometer
广角〔照〕相机 wide angle camera
广体急流 wide-bodied jet
广延大气簇射 extensive air shower
广延介质 extended medium
广延量【物】 extensive quantity
广义贝叶斯决策函数【数】 generalized Bayes decision function
广义布尔代数【数】 generalized Boolean

algebra
广义垂直坐标 generalized vertical coordinate
广义〔地〕磁纬〔度〕 generalized magnetic latitude
广义方向性函数 generalized directivity function
广义解【数】 generalized solution
广义〔流体〕静力方程 generalized hydrostatic equation
广义平稳非相关散射通道 wide sense stationary uncorrelated scattering channel
广义平稳通道 wide sense stationary channel
广义透射函数 generalized transmission function
广义位移 generalized displacement
广义吸收系数 generalized absorption coefficient
广义惠更斯-菲涅耳原理 extended Huygens-Fresnel principle
广义坐标 generalized coordinate
广域网【计】 wide area network (WAN)
归纳统计学 inductive statistics
归一化【物】 normalization
归一化差分植被指数 Normalized Difference Vegetation Index (NDVI)
归一化对数谱 normalized logarithmic spectrum
归一化高度 normalized height
归一化功率谱 normalized power spectrum
归一化互相关 normalized cross-correlation
归一化回波强度【大气】 normalized echo intensity
归一化求积谱 normalized quadrature spectrum
归一化自相关 normalized autocorrelation
归因〔研究〕 attribution

圭亚那海流 Guyana Current, Guiana Current
龟裂 mud crack
规定层 specified level
规定层上的 onlevel
规定等压面 specified isobaric surface
〔规〕定时 ontime
规范 specification
规范高度 gauge height
规范零〔高度〕 gauge zero
规范正交系【数】 orthonormal system
规律尖峰长列型天电 regular peaked and long train atmospherics
规律平稳型天电 regular smooth atmospherics
规则 rule
规则磁场 organized magnetic field
规则带模式【大气】 regular band model
硅太阳电池 silicon solar cell
硅压敏元件 silicon pressure sensor
轨道 orbit
轨道编号数 orbit number
轨道参数 orbital parameter
轨道地球物理观测站 orbiting geophysical observatory (OGO)
轨道电子【物】 orbital electron
轨道经度 orbital longitude
轨道面 orbital plane
轨道偏心率 orbital eccentricity
轨道偏心率变化 orbital eccentricity variation
轨〔道〕频〔数〕 orbital frequency
轨道平面 orbit plane
轨道倾角 orbital inclination
轨道速度 orbital velocity
轨道太阳观测站【天】 orbiting solar observatory (OSO)
轨道天文台【天】 orbiting astronomical observatory (OAO)
轨道纬度 orbital latitude
轨道要素 orbital element

轨道运动　orbital motion
轨迹　trajectory, locus
轨迹法　trajectory technique
轨距角　track distance
鬼波【大气】　angel echo
滚动记录　drum recording
滚流不稳定性【物】　roll instability
滚轮式卫星　cartwheel satellite
滚筒式天气图传真扫描器　weather chart drum scanner
滚涡　rolls
滚轴　roll axis
滚轴积云　roll cumulus
滚轴涡旋【大气】　roll vortex
滚轴云　rotor cloud【大气】, roll cloud
郭氏法,郭氏方案　Kuo scheme
郭晓岚对流方案,郭氏对流方案　Kuo convection scheme
国防气象卫星计划(美国)　Defense Meteorological Satellite Program (DMSP)
国际标准大气　international standard atmosphere (ISA)
国际标准化组织　International Standardization Organization (ISO)
国际标准刊号　International Standard Serial Number (ISSN)
国际标准书号　International Standard Book Number (ISBN)
国际冰冻圈科学协会(IUGG 的第 8 个协会成员)　International Association of Cryospheric Sciences (IACS)
国际冰雪委员会　International Commission on Snow and Ice (ICSI)
国际参考电离层【地】　international reference ionosphere
国际参照大气　international reference atmosphere
国际测绘联合会【测】　International Union of Surveying and Mapping
国际大地测量学协会　International Association of Geodesy (IAG)
国际大地测量与地球物理联合会【测】　International Union of Geodesy and Geophysics (IUGG)
国际大电网会议　International Council on Large Electric Systems (CIGRE)
国际大气化学和全球污染委员会　International Commission on the Atmospheric Chemistry and Global Pollution (ICACGP)
国际大气科学及水文学计划委员会 (UN)　Committee for International Program in Atmospheric Sciences and Hydrology (CIPASH, UN)
国际大气科学协会　International Association for Atmospheric Sciences (IAAS)
国际单位　International Units
国际单位制　International System of Units (SI)
国际地磁参考场　International Geomagnetic Reference Field (IGRF)
国际地磁和高层大气物理学协会　International Association of Geomagnetism and Aeronomy (IAGA)
国际地理联合会　International Geographical Union (IGU)
国际地面天气观测电码　international synoptic surface observation code
[国际]地球观测组织(2006年成立,已有60 个国家和 43 个国际组织参加)　Group on Earth Observations (GEO)
国际地球科学委员会　International Committee for Earth Sciences (ICES)
国际地球物理年　International Geophysical Year (IGY)
国际地球物理委员会　International Geophysical Committee (IGC)
国际地球物理协作年　International Geophysical Co-operation (IGC)
国际地圈-生物圈计划　International Geosphere-Biosphere Programme (IGBP)

国际地质科学联合会　International Union of Geological Sciences（IUGS）
国际第四纪研究联合会　International Union for Quaternary Research（INQUA）
国际电报电话咨询委员会（ITU）　International Telegraph and Telephone Consultative Committee（CCITT，ITU）
国际电工（技术）委员会　International Electrotechnical Commission（IEC）
国际电码　international code
国际电信联盟　International Telecommunication Union（ITU）
国际动力气象学委员会　International Commission on Dynamic Meteorology（ICDM）
国际飞行气象服务　international flight operations weather service
国际分析电码　international analysis code（IAC）
国际风暴预警〔目视〕信号　international visual storm warning signal
国际符号　international symbol, international sign
国际辐射研究协会　International Association for Radiation Research（IARR）
国际高层大气气象学委员会　International Commission on the Meteorology of the Upper Atmosphere（ICMUA）
国际高空气象学委员会　International Aerological Commission（IAC）
国际（哥本哈根）标准海水　Copenhagen Water
国际观测期　international period
国际惯例【航海】　international custom and usage
国际海洋调查十年规划　International Decade of Ocean Exploration（IDOE）
国际海洋考察委员会　International Council for the Exploration of the Sea（ICES）
国际海洋物理科学协会　International Association of Physical Science of the Ocean（IAPSO）
国际航空驾驶员协会　International Federation of Air Line Pilots Associations（IFALPA）
国际航空委员会　International Commission for Air Navigation（ICAN）
国际航空运输协会　International Air Transport Association（IATA）
国际环境新闻中心　Center for International Environment Information（CIEI）
国际极地气象学委员会　International Commission on Polar Meteorology（ICPM，IUGG）
国际极地年　International Polar Year（IPY）
国际减灾十年计划　International Decade for Natural Disaster Reduction（IDNDR）
国际卡　International Table calorie（IT cal.）
国际开发协会　International Development Association（IDA）
国际科学联盟理事会，国科联　International Council of Scientific Unions（ICSU）
国际民航组织标准大气【大气】　ICAO standard atmosphere
国际民〔用〕航〔空〕组织　International Civil Aviation Organization（ICAO）
国际南大洋研究计划　International Southern Ocean Studies（ISOS）
国际南极气象研究委员会　International Antarctic Meteorological Research Committee（IAMRC）
国际南极研究分析中心　International Antarctic Analysis Center（IAAC）
国际年轮资料库　International Tree Ring Data Bank（ITRDB）
国际宁静日　international quiet day
国际宁静太阳年　International Quiet Sun Year（IQSY）

国际排灌委员会 International Commission on Irrigation and Drainage(ICID)
国际频率总表 Master International Frequency List(MIFL)
国际气候学委员会 International Climatological Commission(ICC)
国际气象电传打字电报网 International Meteorological Teleprinter Network
国际气象电〔传通〕信网 International Meteorological Telecommunication Network
国际〔气象观测〕站网用蒸发皿 international network evaporation pan(INEP)
国际气象计划机构间联合委员会 Interagency Committee for International Meteorological Programs(ICIMP)
国际气象学和大气科学协会 IAMAS (International Association of Meteorology and Atmospheric Sciences)【大气】
国际气象学和大气科学协会大气臭氧委员会 IAMAS Ozone Commission(IOC)
国际气象学和大气科学协会大气电学委员会 IAMAS Commission on Atmospheric Electricity
国际气象学和大气科学协会动力气象委员会 IAMAS Commission on Dynamic Meteorology(ICDM)
国际气象学和大气科学协会辐射委员会 IAMAS Radiation Commission(IRC)
国际气象学和大气科学协会高层大气气象学委员会 IAMAS Commission on Meteorology of Upper Atmosphere
国际气象学和大气科学协会极地气象学委员会 IAMAS Commission on Polar Meteorology(ICPM)
国际气象学委员会 International Meteorological Committee(IMC)
国际气象学与大气物理学协会(IAMAS 的前身) International Association of Meteorology and Atmospheric Physics(IAMAP,IUGG)
国际气象组织(WMO 的前身) International Meteorological Organization(IMO)
国际区站号 international index number
国际全球大气化学研究计划(IGBP 内) International Global Atmospheric Chemistry Project(IGAC)
国际人文因素计划 International Human Dimensions Program(IHDP)
国际日 universal day
国际日期变更线 International Date Line
国际商用机器公司(及其生产的计算机型号) International Business Machine(IBM)
国际摄影测量和遥感学会 International Soiety for Photogrammetry and Remote Sensing(ISPRS)
国际生物气候学及生物气象学会 International Society of Bioclimatology and Biometeorology(ISBB)
国际生物气象学会 International Society of Biometeorology(ISB)
国际生物学计划 International Biological Programme(IBP)
国际十进位分类法 universal decimal classification
国际时间局 International Time Bureau
国际水稻研究所 International Rice Research Institute(IRRI)
国际水文地质工作者协会 International Association of Hydrogeologists(IAH)
国际水文科学协会 International Association of Hydrological Sciences(IAHS)
国际水文十年 International Hydrological Decade(IHD)
国际〔天气〕电码 international weather code
国际〔天气〕电码【大气】 international synoptic code
国际天文学联合会【天】 International

Astronomical Union(IAU)
国际通信卫星组织 International Telecommunications Satellite Consortium (INTELSAT)
国际通用的(船舶、飞机等)呼救信号,无线电求救信号 SOS
国际通用科技词汇 International Scientific Vocabulary(ISV)
国际卫星陆面气候学计划 International Satellite Land Surface Climatology Project(ISLSCP)
国际卫星云气候学计划 International Satellite Cloud Climatology Project (ISCCP)
国际温标 international temperature scale (ITS)
国际无线电科学联合会 International Union of Radio Science (URSI)
国际无线电咨询委员会(ITU) International Radio Consultative Committee (CCIR,ITU)
国际雪〔花〕形〔状〕分类 International Snow Classification
国际印度洋考察 International Indian Ocean Expedition (IIOE)
国际宇宙航行联盟 International Astronautical Federation(IAF)
国际原子能机构 International Atomic Energy Agency(IAEA)
国际原子时【天】 International Atomic Time
国际云年 international cloud year
国际云图【大气】 international cloud atlas
国际蒸汽表卡 International Steam Table Calorie(IT cal.)
国际〔制〕单位 SI unit
国际烛光 international candle
国际自然与自然资源保护联盟 International Union for Conservation of Nature and Natural Resources (IUCN)

国家标准 national standard
国家标准气压表【大气】 national standard barometer
国家大气污染控制中心(美国) National Center for Air Pollution Control (NCAPC)
国家大气研究中心(美国) National Center for Atmospheric Research(NCAR)
国家地球物理研究所(印度) National Geophysical Research Institute (NGRI)
国家地图集【地理】 national atlas
国家公园【地理】,国家天然公园【林】 national park
国家海洋资料目录(美国) National Marine Data Inventory(NAMDI)
(国家或地区)预报中心 central forecasting office
国家极轨业务环境卫星系统(美国) National Polar-orbiting Operational Environmental Satellite System (NPOESS)
国家科学探测气球中心(美国) National Scientific Balloon Facility(NSBF)
国家南极考察委员会(澳大利亚,新西兰) National Committee for Antarctic Research (NCAR)
国家气候资料中心(美国) National Climate Data Center (NCDC)
国家气象局培训中心(美国) National Weather Service Training Center (NWSTC)
国家气象局预报台(美国) National Weather Service Forecast Office(NWSFO)
国家气象局总部(美国) National Weather Service Headquarter(NWSH)
国家气象卫星中心(美国) National Weather Satellite Center (NWSC)
国家气象中心(WMO 的定名) National Meteorological Center(NMC)
国家强风暴实验室(NOAA) National Severe Storms Laboratory(NSSL,NOAA)

国家强风暴预报中心(美国)　National Severe Storm Forecast Center (NSSFC)
国家与航空气象传真网　national and aviation meteorological facsimile network(NAMFAX)
国家资料浮标中心(美国)　National Data Buoy Center (NDBC)
国立防灾研究中心(日本)　National Research Center for Disaster Prevention (NRCDP)
国立极地研究所(日本)　National Institute of Polar Research(NIPR)
国民生产总值　gross national product (GNP)
国内生产总值　gross domestic product (GDP)
国土规划【地理】　territorial planning
国土开发【地理】　territorial development
国土整治【地理】　territorial management
国土资源卫星【地理】　land resources satellite
果树防冻气象学　fruit-frost meteorology
果园防霜炉　orchard heater
过饱和〔度〕　supersaturation
过饱和空气【大气】　supersaturated air
过程　process
过程体　procedure body
过程语句　procedure statement
过程直减率　process lapse rate
过冲作用　overboat effect
过定义　overspecification
过冬准备　winterization
过〔度〕冷〔却〕　supercooling
过度拟合　overfitting
过度预报　overforecast
过度预报〔的〕　overpredicted
过渡层【大气】　transition layer
过渡带　transition zone
过渡地带　zone of transition

过渡季节【大气】　transition season
过渡气流【大气】　transitional flow
过渡区　transition area
过渡曲线　transition curve
过渡时期　transition period
过渡湍流　transition to turbulence
过渡湍流闭合　transient turbulence closure
过渡形式　intermediate form
过分补偿　overcompensate
过辐射　overshoot
过脊气流　airflow over ridge
过境　passage
过境阵雨　passing shower
过冷　subcooling, overcooling, supercooling
过冷〔却〕水　supercooled water
过冷〔却〕水滴　supercooled water droplet
过冷却雾【大气】　supercooled fog
过冷却雨【大气】　supercooled rain
过冷水冰滴　droxtal
过冷现象【物】　supercooling phenomenon
过冷云【大气】　supercooled cloud
过冷云滴【大气】　supercooled cloud droplet
过量播撒　over-seeding
过量降雪　excess snowfall
过量噪声　excess noise
过量蒸发　evaporation excess
过滤-捕获　filter-captrue
过滤模式【大气】　filtered model
过滤气象杂波　filtering meteorological noise
过滤取样器　filter sampler
过滤〔作用〕　filtration
过坡对流　convection over slope
过去天气【大气】　past weather
过去6小时　past six hours
过去3小时　past three hours
过去1小时　past hour
过热　overheat

过热层　overheating layer
过热程度　degree of superheat
过热〔的〕　superheating
过热水汽　superheated vapour
过热现象【物】　superheating phenomenon
过熔　super-fusion
过山共振波　resonance waves over hills
过剩水汽　superfluous water vapour

过湿气候　perhumid climate
过时报告　delayed report
过水断面面积　wetted area
过氧化氢　hydrogen peroxide
过氧化烷基原子团　peroxyalkyl radical
过氧化物【化】　peroxide
过阻尼【物】　overdamping
过阻尼系统　overdamped system

H

哈勃定律【物】　Hubble law
哈勃空间望远镜【天】　Hubble space telescope(HST)
哈得来环流〔圈〕【大气】　Hadley cell
哈得来能量循环　Hadley energy cycle
哈得来域【大气】　Hadley regime
哈得来原理　Hadley's principle
哈尔马赫拉涡流　Halmahera Eddy
哈根-泊肃叶流　Hagen-Poiseuille flow
哈金斯带(O_3)　Huggins band(O_3)
哈朗不连续性　Harany discontinuity
哈利特-莫索普过程　Hallett-Mossop process
哈密顿典范方程【数】　Hamilton canonical equations
哈密顿方程【数】　Hamilton equation
哈密顿矩阵　Hermitian matrix
哈密顿-雅可比方程【数】　Hamilton-Jacobi equation
哈森法　Hazen method
哈森图标位置　Hazen plotting position
哈特莱带　Hartley band
哈特莱区　Hartley region
哈特曼数　Hartmann number
海岸　sea coast
海岸边界层　coastal boundary layer
海岸带【航海】　coastal zone

海岸带气候【大气】　coastal climate
海岸带水色扫描仪【大气】　Coastal Zone Color Scanner (CZCS)
海岸低压　coastal low
海岸地貌学【地理】　coastal geomorphology
〔海〕岸风　shore wind
海岸锋　coastal front
海岸环境局　Office of Coastal Environment (OCE)
海岸阶地【地理】　coastal terrace
海岸气象学　coastal meteorology
海岸雾　coastal fog
海岸陷波　coastally trapped waves
海岸效应　coastal effect
海岸夜雾　coastal night fog
海岸涌升流　coastal upwelling
海拔高度　sea level elevation, above sea level(ASL)
海拔〔高度〕【地理】　altitude, height above sea level
海滨气候　littoral climate
海滨沼泽【地理】　coastal marsh
海冰【航海】　sea ice
海冰密集度　ice concentration
(海冰面上的)融水池　melt pond
海冰频率　sea-ice frequency

海冰盐度　salinity of sea ice
海堤【航海】　seawall, sea bank
海底　ocean floor, ocean bottom
海底地貌【测】　submarine geomorphology
海底地震　seaquake
海底浅层记录仪　sub-bottom depth recorder(SDR)
海底钻井　ocean floor drilling(OFD)
海地平　sea-level horizon, sea horizon
海尔望远镜【天】　Hale telescope
海风　landward wind
海风【大气】　sea breeze
海风锋【大气】　sea breeze front
海风辐合　sea breeze convergence
海风辐合区　sea breeze convergence zone
海风厚度　sea breeze depth
海港〔水位〕起伏　harbor oscillations
海虹　sea rainbow
海槛深度　sill depth
海军海洋天气浮标　navy oceanographic weather buoy(NOWEB)
海军环境预报研究所(美)　Naval Environmental Prediction Research Facility (NEPRF)
海况　sea state
海况【航海】　sea condition
海况标尺　sea-state scale
海况等级　state-of-sea scale
海来风　sea turn
海浪　sea wave, ocean wave
海浪回波　sea echo
海浪谱　ocean wave spectrum
海里　nautical mile, international mile
海里/小时　knot
海量存储器【计】　mass storage
海量存储设备　mass storage device
海岭【地理】　oceanic ridge
海流　marine current
海流【航海】　ocean current
海流剖面　current cross section
海流图　current chart

海流涡旋　current vortex
海陆对比　land-sea contrast
海陆风　land and sea breezes
海陆风【大气】　sea-land breeze
海陆风环流　sea and land breeze circulation
海陆界面　sea-land interface
海锚【航海】　drogue
海绵边界条件【大气】　sponge boundary condition
海绵冰　spongy ice
海绵层　sponge layer
海绵带　sponge zone
海绵生长　sponge growth
海绵状雹　spongy hail
海面粗糙度　sea surface roughness
海面反照率【大气】　sea surface albedo
海面风速　sea surface wind speed
海面辐射【大气】　sea surface radiation
海面广角遥感器　Sea-Viewing Wide Field-of-View Sensor(SeaWiFS)
海面回波　sea return, sea clutter
海面混合层　oceanic surface mixed layer
海面乱冰　slob ice
海面扰动　sea disturbance
海面升降变化　eustatic change
海面温度【大气】　sea-surface temperature (SST)
海面形状试验计划　Ocean Topography Experiment(TOPEX)
海〔面状〕况　state of the sea
海面蒸汽雾【大气】　sea smoke
海沫　sea spray
海盆　ocean basin
海平面　sea level
海平面变化　sea-level change
海平面测量　sea level measurements
海平面订正　sea level correction, reduction to sea level
海平面记录　sea-level record
海平面气压【大气】　sea level pressure

（SLP）
海平面气压订正【大气】 pressure reduction
〔海平面〕气压换算【大气】 pressure reduction
海平面气压图 sea level pressure chart
海平面上升 sea level rise
海平面天气图 sea level synoptic chart
海平面图 sea-level chart
海气边界过程 air-sea boundary process
海气交互作用联合试验 Joint Air-Sea Interation Experiment（JASIE）
海-气交互作用实验室（美国） Sea-Air Interaction Laboratory（SAIL）
海气交换【大气】 air-sea exchange
海气界面【大气】 air-sea interface
海气耦合模式 air-sea coupled model, air-ocean coupled model
海气温差 air-sea temperature difference
海气相互作用【大气】 air-sea interaction（ASI）
海色 ocean color
海色成像器 ocean color imager（OCI）
海色扫描仪【大气】 coastal zone color scanner（CZCS）
海森伯测不准原则 Heisenberg uncertainty principle
海森伯模型【物】 Heisenberg model
海上观测飞机 observation seaplane（OSP）
海上虹 marine rainbow
海上急救【航海】 first aid at sea
海上空间探测装置 ocean space robot（OSR）
海上逆温 marine inversion
海上平台 off-shore platform
海上气象数据【航海】 marine weather data
海上搜救【航海】 marine search and rescue
海上侦察机 patrol seaplane

海蚀作用【地理】 marine erosion
海市蜃楼【大气】 mirage
海〔事卫〕星 Sea Star
海事卫星 maritime satellite（MARISAT）
海事卫星 maritime satellite system（MARISAT）
海水 seawater
海水层 marine layer
海水淡化 seawater desalination
〔海水〕进入口水温 injection temperature
海水-颗粒物界面 sea-particle interface
海水入侵 seawater intrusion
海水升降的海平面变化 eustatic sea level changes
海水-生物圈界面 sea-biosphere interface
海水温度表 marine thermometer, seawater thermometer
海水温度【航海】 sea temperature
海水污染 seawater pollution
海水主要成分【海洋】 major constituents of sea water
海损事故【海洋】 marine accident
海滩冰 beach ice
海图 sea map, sea chart
海图基准面【航海】 datum of chart
海图基准面【海洋】 chart datum
海外行星【天】 trans-Neptunian planet
海湾冰 bay ice
海王星【天】 Neptune
海韦尔晕 Hevel's halo
海韦留幻日 Hevelius's parhelia
海韦留晕 halo of Hevelius
海雾【大气】 sea fog
海相沉积 marine sediment
海啸 earthquake flood
海啸【地】 seismic sea wave
海啸【海洋】 tsunami
海盐核【大气】 sea-salt nucleus
海洋边界层 marine boundary layer

〔海洋〕波浪图　synoptic wave-chart
海洋测绘【测】　marine charting
海洋潮　oceanic tide
海洋大地测量学【测】　marine geodesy
海洋-大陆对比　maritime continental contrast
海洋大气　marine atmosphere
海洋大气边界层试验　Marine Atmospheric Boundary Layer Experiments (MABLES)
海洋-大气环流耦合模式　coupled ocean-atmosphere general circulation model
海洋大气交互作用　ocean-atmosphere interaction
海洋大气模式　ocean-atmosphere model
海洋大气耦合　ocean-atmosphere coupling
海洋大气耦合模式　ocean-atmosphere coupled model
海洋-大气耦合研究试验　Coupled Ocean-Atmosphere Research Experiment (COARE)
海洋大气热交换　ocean-atmosphere heat exchange
海洋-大气综合数据集　Comprehensive Ocean-Atmosphere Data Set (COADS)
海洋低压　oceanic low
海洋地层学【海洋】　marine stratigraphy
海洋地理学　oceanography
海洋地貌学【海洋】　marine geomorphology
海洋地球化学【地质】　marine geochemistry
海洋地球物理学【海洋】　marine geophysics
海洋地质学【海洋】　marine geology
海洋调查【海洋】　oceanographic investigation, oceanographic survey
海洋调查船　oceanographic (research) vessel, oceanographic research ship (ORS)
海洋调查设备　oceanographic facility (OF)
海洋度　maritimity, oceanicity
海洋度【大气】　oceanity
海洋断面地球化学研究计划　Geochemical Ocean Section Study (GEOSECS)
海洋法　maritime law
海洋法【海洋】　law of the sea
海洋反气旋　oceanic anticyclone
海洋风　ocean wind
海洋锋　ocean front
海洋锋【海洋】　oceanic front
〔海洋〕浮游植物　phytoplankton
海洋高压　oceanic high
海洋工程【航海】　marine engineering
海洋观测技术【海洋】　ocean observation technology
海洋观测绞车　oceanographic winch
海洋观测塔　oceanographic tower
海洋观测站　ocean station, hydrographic station
海洋光学【海洋】　ocean optics
海洋光学遥感【海洋】　oceanic optical remote sensing
海洋化探【地质】　marine geochemical exploration
海洋化学【海洋】　marine chemistry
海洋化学资源【海洋】　marine chemical resource
海洋环境保护【海洋】　marine environmental protection
海洋环境【地理】　marine environment
海洋环境化学【海洋】　marine environmental chemistry
海洋环境监测【海洋】　marine environmental monitoring
海洋环境科学【海洋】　marine environmental science
海洋环境评价【海洋】　marine environmental assessment
海洋环境容量【海洋】　marine environmental capacity

海洋环境数据资料　marine enviornmental data information
海洋环境预报【海洋】　marine environmental prediction, marine environmental forecasting
海洋环境质量【海洋】　marine environmental quality
海洋环流　oceanic general circulation, ocean circulation
海洋环流单体　ocean gyre
海洋环流模式　oceanic general circulation model（OGCM）
海洋混合　ocean mixing
海洋混合层　oceanic mixed layer
海洋季候泥　marine varve
海洋监测船【航海】　ocean monitoring ship
海洋监测系统　ocean surveillance system（OSS）
海洋监视【航海】　marine surveillance
海洋科学　marine science
海洋科学【海洋】　ocean sciences
海洋科学技术【航海】　marine science and technology
海洋科学委员会　Marine Sciences Commission（MSC）
海洋空间探索计划　Oceanospace Explorer（OSPER）
海洋空气　maritime air
海洋雷达　marine radar
海洋利用规划　Sea Use Program（SUP）
海洋陆地　maritime continent
海洋能源【海洋】　ocean energy resources
海洋能转换【海洋】　ocean energy conversion
海洋漂流　ocean drift
海洋气候　sea climate, maritime climate
海洋气候学【大气】　marine climatology
海洋气溶胶【大气】　marine aerosol
海洋气团　marine air（mass）, ocean air（mass）
海洋气团【大气】　maritime air mass
海洋气团雾　maritime air fog, marine air fog
海洋气象报告【航海】　ocean weather report
海洋气象电码【大气】　marine meteorological code
海洋气象服务系统　marine meteorological service system（MMSS）
海洋气象观测　maritime meteorological observation
海洋气象图【测】　marine meteorological chart
海洋气象〔学〕　oceanic meteorology, maritime meteorology
海洋气象学【大气】　marine meteorology
海洋气象学委员会（WMO）　Commission for Marine Meteorology（CMM, WMO）
海洋气象研究　oceanic meteorological research
海洋气象研究咨询委员会（WMO）　Advisory Committee on Oceanic Meteorological Research（ACOMR, WMO）
海洋气象站【大气】　ocean weather station
海洋热力学　marine thermodynamics
海洋热量输送　oceanic heat transport, ocean heat transport
海洋生化资源【海洋】　marine biochemical resource
海洋生态调查【航海】　marine ecological investigation
海洋生态系统　marine ecosystem
海洋生态学　marine ecology
海洋生物地球化学【海洋】　marine biogeochemistry
海洋生物化学【海洋】　marine biochemistry
海洋生物活动　marine biological activity

海洋生物声学【海洋】 marine bio-acoustics
海洋生物学 marine biology
海洋生物噪声【海洋】 marine biological noise
海洋声学 marine acoustics
海洋示踪物 oceanographic tracer
海洋输送带 ocean conveyor belt
海洋数据报告站 Marine Reporting Station(MARS)
海洋水文学【海洋】 marine hydrology, marine hydrography
海洋天气船 ocean weather ship, ocean station vessel(OSV)
海洋天气船【航海】 ocean weather vessel
海洋天气观测 marine weather observation
海洋〔天气〕观测 marine observation
〔海洋〕天气观测船 weather ship
海洋〔天气〕预报 marine forecast
海洋天气预报【大气】 marine weather forecast
海〔洋〕图集【测】 marine atlas
海洋湍流 ocean turbulence, oceanic turbulence
海洋微生物【海洋】 marine microorganism
海洋卫星 Seasat
海洋卫星(美国) sea satellite (SEASAT)
海洋卫星地面风 SEASAT surface wind
海洋卫星-A 上的散射计系统 Seasat-A scatterometer system(SASS)
海洋温差能【海洋】 ocean thermal energy
海〔洋〕温〔度〕 ocean temperature
海洋污染【海洋】 marine pollution
海洋污染物【海洋】 marine pollutant
海洋物理学 oceanophysics
海洋物理学【海洋】 marine physics
海洋雾 marine fog

海洋系数 maritime coefficient
海洋性 oceanicity (见 oceanity)
海洋性冰川【地理】 maritime glacier
海洋性冻原 maritime tundra
海洋〔性〕空气 sea air
海洋性气候 oceanic climate, ocean climate
海洋性气候【大气】 marine climate
海洋性气溶胶【大气】 maritime aerosol
海洋性温和〔气候〕 oceanic moderate
海洋学【海洋】 oceanology
海洋学赤道(表面最高水温带) oceanographic equator
海洋研究浮标 oceanographic research buoy(ORB)
海洋研究科学委员会 Scientific Committee on Oceanic Research(SCOR, ICSU)
海洋研究设备 oceanographic research equipment(ORE)
海洋与大气联合研究所 Joint Institute for Marine and Atmospheric Research(JIMAR)
海洋预报〔学〕 hydropsis
海洋云 maritime cloud
海洋噪声 oceanic noise
海洋站 sea station
海洋真值观测 sea truth measurement
海洋重力测量【地】 gravity measurement at sea
海洋重力调查【海洋】 marine gravity survey
海洋重力异常【海洋】 marine gravity anomaly
海洋资料发送器 ocean data transmitter(ODT)
海洋资料浮标 ocean data buoy(ODB)
海洋资料站 ocean data station(ODS)
海洋资源【海洋】 marine resources
海洋资源研究咨询委员会(FAO) Advisory Committee on Marine Resources

Research (ACMRR, FAO)
海洋自动气象观测站　marine automatic meteorological observing station (MAMOS)
海渊　ocean deep
亥姆霍兹波【大气】　Helmholtz wave
亥姆霍兹不稳定【大气】　Helmholtz instability
亥姆霍兹定理　Helmholtz's theorem
亥姆霍兹方程【数】　Helmholtz equation
亥姆霍兹函数　Helmholtz function
亥姆霍兹重力波　Helmholtz gravitational wave
亥姆霍兹自由能【化】　Helmholtz free energy
骇浪　phenomenal sea
氦氖激光器【电子】　helium neon laser
氦氖激光器【物】　He-Ne laser
氦〔气〕　helium(He)
含冰量【地理】　ice content
含尘空气　dust-ladden air
含尘量　dust content
含根直方图　rooted histogram
含糊性　vagueness
含灰空气　ash air
含量　content
〔含硫〕伦敦烟雾〔事件〕(1952年)　London (sulfurous) smog
含硫烟雾　sulfurous smog
含能涡旋　energy-containing eddies
含沙风　sand-bearing wind
含时微扰【物】　time-dependent perturbation
〔含水层的〕经济抽水率　economic yield of aquifer
含水层【地质】　uifer
含水层检验　aquifer test
含水层系　aquifer system
含水量　water content
含水量探空仪　aquasonde
含酸煤烟　acid-containing soot

含盐量【物】　salt content
含氧系数【地】　oxygen coefficient
含银量分析　silver analysis
函审鉴定　corresponding appraisal
函数　function
δ函数　Delta function
函数导数　functional derivative
函数调用【计】　function call
函数拟合　function fitting
焓　heat content
焓【物】【大气】　enthalpy
寒潮　polar invasion(＝outbreak)
寒潮【大气】　cold wave
寒潮爆发　polar outbreak
寒潮爆发【大气】　cold-outburst
寒带　cold cap, frigid zone
寒带气候【大气】　polar climate
寒冻风化【地理】　frost weathering
寒〔海〕流　cold current
寒害　cold damage
寒害【大气】　chilling injury
寒极【大气】　cold pole
寒极高山区　arctic-alpine
寒冷　chill, snappy
寒冷恐怖症　cheimophobia
寒冷气候适应　cold acclimatization
寒冷天气　frigid weather
寒冷要素　cold element
寒露(节气)【农】【大气】　Cold Dew
寒露风【大气】　low temperature damage in autumn
寒武纪【地质】　Cambrian Period
寒武系【地质】　Cambrian System
汉卡【计】　Chinese character card, Hanzi card
汉克尔变换【数】　Hankel transform
汉克尔函数【数】　Hankel function
汉明窗【电子】　Hamming window
汉明码【数】　Hamming code
汉宁窗　Hanning window
汉字【计】　Hanzi

旱地　arid land
旱涝等级序列　dryness and wetness grades series
旱涝演替　xerasium
旱年树种【林】　xerophilous tree species
旱农【地理】　dry farming
旱生植物【地理】　xerophyte
旱灾　dry damage
旱灾【大气】　drought damage
航摆角【航海】　yaw angle
航测海洋学　airborne oceanography
航道【海洋】　navigation channel
航道【航海】　fairway
航海表【航海】　navigation table
航海服务【航海】　nautical service
航海科学【航海】　nautical science
航海气象【航海】　nautical meteorology
航海日志【航海】　log book
航海史【航海】　nautical history
航海曙暮光　nautical twilight
航海天文历【航海】　nautical almanac
航海图书资料【航海】　nautical charts and publications
航海心理学【航海】　marine psychology
航海学【航海】　marine navigation
航海仪器【航海】　nautical instrument
航海指南　sailing directions
航海制　nautical system
航海专家系统【航海】　marine navigation expert system
航空　aviation
航空安全　air traffic safety
航空病　air sickness
航空放射性测量【地】　airborne radioactivity survey
航空观测　aviation observation
航空-航天的　air-space
航空计划　flight plan
航空简报　flight briefing
航空农〔艺〕学　aeroagronomy
航空剖面〔图〕　flight cross section

航空气候区划　regional classification of aviation climate
航空气候区划【大气】　aeronautical climate regionalization
航空气候学　aeronautical climatology
航空气候学【大气】　aviation climatology
航空气候志【大气】　aeronautical climatography
航空气象保障【大气】　aviation meteorological support
航空气象报告　aviation weather report
航空气象电码【大气】　aviation meteorological code
航空气象服务　aeronautical meteorological service, aviation weather service, aviation meteorological service
航空气象观测【大气】　aviation meteorological observation
航空气象情报中心　flight information centre
航空气象台（室）　dependent meteorological office（DMO）
航空气象信息【大气】　aviation meteorological information
航空气象〔学〕　aviation meteorology
航空气象学【大气】　aeronautical meteorology
航空气象学委员会（WMO）　Commission for Aeronautical Meteorology（CAeM, WMO）
航空气象要览　flight forecast-folder
航空气象要素【大气】　aviation meteorological element
航空气象站　aeronautical meteorological station
航空气象资料系统　meteorological airborne data system（MADS）
航空区域〔天气〕预报【大气】　aviation area〔weather〕forecast
航空摄谱仪【测】　aerial spectrograph
航空摄影　aeroplane photography

航空摄影【测】 aerial photography
航空摄影测量【测】 aerophotogrammetry
航空摄影学 aerophotography
航空声学 aeroacoustics
航空天气电码 AERO code
航空天气订正预报【大气】 amendment of aviation weather forecast
航空天气观测 aviation weather observation
航空天气简报 flight weather briefing
航空天气预报 flight forecast, aeromancy
航空〔天气〕预报【大气】 aviation 〔weather〕forecast
航空天气预报表格 flight dossier, flight forecast-folder
航空物理学 aerophysics
航〔空〕线 airway
航空学 aeronautics
航空遥感【地理】 aerial remote sensing
航空医学 aviation medicine, aeromedicine
航空预报 aviation forecast
航空预报区 aviation forecast zone
航空侦察 aerial reconnaissance
航空重力测量【测】 airborne gravity measurement
航空专用电传通信网 Aeronautical Fixed Telecommunications Network(AFTN)
航空自动气象观测系统 aviation automated weather observation system (AVAWOS)
〔航路〕分段风 sector wind
航摄照片 aerial photograph
航天飞机 shuttle
航天飞机【测】 space shuttle
航天飞机成像雷达 shuttle imaging radar（SIR）
航天遥感 space remote sensing
航线 air route, route
航线分速 route component

航线风 track wind
航线观测 airways observation
航线〔观测〕百叶箱 airways shelter
航线监视雷达【电子】 air route surveillance radar
航线剖面图 route cross section
航线天气 airway weather, en route weather
航线天气报告 en route report
航线〔天气〕预报【大气】 air route 〔weather〕forecast
航线预报 route forecast, FIFOR, airways forecast
航线预报〔方法〕 airway forecasting
航线运输指挥中心 Air Route Traffic Control Center（ARTCC）
航向 heading line, heading, course
航行 navigation
航行侧风 beam wind
航行风 ship wind
航行风【大气】 navigational wind
毫巴(气压单位,1 mb=1 hPa) millibar（mb）
毫巴标尺 millibar scale
毫巴〔标尺〕气压表 millibar-barometer
毫孔过滤器 millipore filter
毫米 millimeter(mm)
毫米波 millimeter wave
毫米汞柱(=1.333 hPa) millimetre of mercury
豪威兹波【大气】 Haurwitz wave
豪威兹-亥姆霍兹波 Haurwitz-Helmholtz wave
好天气 fine weather
耗热率【航海】 heat rate
耗散【物】【大气】 dissipation
耗散长度尺度 dissipation length scale
耗散尺度 dissipation scale
耗散功率【电子】 dissipation power
耗散积分 dissipation integral
耗散结构【物】 dissipative structure

耗散率【大气】 dissipation rate
耗散区 dissipation range
耗散态【数】 dissipative state
耗散系数 dissipation coefficient, dissipative system
耗散因数【电子】 dissipation factor
耗水量 water consumption, consumptive use
合【天】 conjunction
合并云 merging cloud
合成波 composite wave
合成风【大气】 resultant wind
合成风速 resultant wind velocity
合成风向 resultant wind direction
合成结构 composite structure
合成孔径成像雷达 synthetic aperture imaging radar
合成孔径雷达【电子】【大气】 synthetic aperture radar (SAR)
合成图像 composite image
合成温度 resultant temperature
合成温度指数 resultant temperature index
合成物 hybrid
合成误差 resultant error
合点【测】 vanishing point
合格品【电子】 qualified product
合理地平 rational horizon
合力【物】 resultant force
合量 resultant
合流【大气】 confluence
合流超几何函数 confluent hypergeometric function
合流渐近线 asymptote of confluence
合作反射器 cooperative reflector
合作观测者(美国的义务观测者称呼) cooperative observer
合作系数 index of cooperation (IOC)
合作现象【物】 cooperative phenomenon
和【数】 sum
和风【大气】 moderate breeze
和校验 sum check
河 river
河岸调蓄 bank storage
河冰 river ice
河槽控制 channel control
河槽蓄水〔量〕 channel storage
河川流量 stream flow
河床 river bed
〔河道〕流量演算 stream flow routing
河谷贮水〔量〕 valley storage
河湖污染 dystrophication
河口【航海】 estuary
河流等级 stream order
河流动力学 fluvial dynamics
河流分支 stream segment
河流频数 stream frequency
河流袭夺 stream piracy
河流形态学 fluvial morphology
河流学 potamology
河流预报中心 river forecast center (RFC)
河流纵断面 stream profile
河面温度 river〔surface〕temperature
河外星系【物】 external galaxy
河网 river network, stream network, drainage network
河网密度【水利】 drainage density
河雾 river fog
荷电板 charged plate
荷电时间常量【物】 charging time constant
荷电云 charged cloud
荷兰皇家气象研究所 Royal Netherlands Meteorological Institute (RNMI)
荷质比【物】 charge-mass ratio
核 nucleus
核磁共振【化】 nuclear magnetic resonance
核地球化学【化】 nuclear geochemistry
核冬天【大气】 nuclear winter
核函数【数】 kernel function

核化【大气】 nucleation
核孔滤膜 nuclepore filter
核群 population of nuclei
核燃料【化】 nuclear fuel
核糖核酸【生化】 ribonucleic acid
核子【物】 nucleon
核子尘 nuclear dust
核子辅助电源系统 system for nuclear auxiliary power
盒式磁带 cassette (tape)
盒须图 box-whisher plot
赫茨堡带 Herzberg band
赫茨堡连续吸收带 Herzberg continuum
赫尔曼自记雪量器 Hellmann recording snow-gauge
赫洛弗森图 Herlofson diagram
赫舍尔日射计【大气】 Herschel's actinometer
赫维赛德层 Heaviside layer
赫维赛德函数【数】 Heaviside function
赫伊兰环流定理 Hoiland's circulation theorem
赫〔兹〕 hertz(Hz)
赫兹矢量 Hertz vector
赫兹图 Hertz's diagram
黑白球温度表【大气】 black and white bulb thermometer
黑冰【大气】 black ice
黑层云 black stratus
黑潮 Kuroshio【大气】,Kuroshio Current
黑潮【航海】 Black stream
黑潮〔流〕系 Kuroshio system
黑潮逆流 Kuroshio Countercurrent
黑潮延续体 Kuroshio Extension
黑丁格刷子 Haidinger's brush
黑洞【天】 black hole
黑洞辐射【物】 black hole radiation
黑尔周期 Hale cycle
黑盒(即组装式电子学单元) black box
黑盒法 black box method

黑球温度表【大气】 black-bulb thermometer
〔黑〕球温〔度〕表(测环境辐射热用) globe thermometer
黑霜 hard frost, early frost hidden, black frost
黑霜【大气】 dark frost
黑碳 black carbon
黑碳气溶胶 black carbon aerosol
黑体【大气】 blackbody
黑体发射 blackbody emission
黑体辐射【大气】 blackbody radiation
黑体光谱 blackbody spectrum
黑雾 black fog
黑箱测试【计】 black box testing
黑雨 black rain
黑子大周 major cycle
黑子〔活动〕双周期 double sunspot cycle
黑子相对数【大气】 relative number of sunspot
黑子小周 minor cycle
痕量 trace
痕量成分【化】 trace constituent
痕量化学【化】 trace chemistry
痕量气体 trace gas
痕量气体移除过程 sink processes for trace gases
痕量元素【化】【大气】 trace element
亨〔利〕(电感单位) henry(H)
亨利定律【化】 Henry's law
恒等〔式〕【数】 identity
恒定波 permanent wave
恒定风 permanent wind
恒定回波 permanent echo
恒定气体 permanent gas
恒河型(温度) Ganges type
恒湿器 hygrostat
恒湿器(箱) humidistat
恒速溶液法测流〔量〕 constant-rate dilution gauging
恒温槽 temperature control vessel

| 恒温动物【动】 homeotherm
| 恒温恒湿箱 thermo-hygrostat
| 恒温器【物】 thermostat
| 恒温湿器（箱） hygrothermostat
| 恒相位面 surface of constant phase
| 恒向线 loxodrome
| 恒星 fixed star
| 恒星【天】 star
| 〔恒〕星际空间【地】 interstellar space
| 恒星年【天】 sidereal year
| 恒星日【天】 sidereal day
| 恒星时 sidereal time
| 恒星月【天】 sidereal month
| 恒压器 barostat, manostat
| 恒振幅面 surface of constant amplitude
| 横波【物】 transverse wave
| 横〔穿〕谷风 cross-valley wind
| 横荡【海洋】 sway
| 横风扩散 cross-wind diffusion
| 横浪 cross sea
| 横切 intersect
| 横向环流 transverse circulation
| 横向卷云带 transverse cirrus banding
| 横向速度【物】 transverse velocity
| 横向速度谱 spectrum of lateral velocity
| 横摇 yaw
| 横云带 transverse cloud band（或 line）
| 横振动 transversal vibration
| 横坐标【数】 abscissa
| 红斑辐射剂量仪 erythemal dosimeter
| 红斑谱 erythemal spectrum
| 红斑谱辐射剂量仪 erythemal spectrum dosimeter
| 红斑穴 Red Spot Hollow
| 红宝石激光器【物】 ruby laser
| 红海水 Red Sea Water
| 红闪 red flash
| 红外 IR(infrared 或 infrared radiation)
| 红外被动遥感 infrared passive(IRP)
| 红外大气光谱带 infrared atmospheric band
| 红外多普勒光〔雷〕达 infrared Doppler lidar
| 红外多普勒系统 infrared Doppler system
| 红外分光光度计【物】 infrared spectrophotometer
| 红外分光计 infrared spectrometer
| 红外辐射【大气】 infrared radiation
| 红外辐射测量术 infrared radiometry
| 红外辐射器 infrared radiator
| 红外干涉光谱仪 infrared interferometer spectrometer(IRIS)
| 红外光电管 irtron
| 红外光化学【化】 infrared photochemistry
| 红外光谱区 infrared spectral region
| 红外光谱学【物】 infrared spectroscopy
| 红外光源【化】 infrared source
| 红外检测器【化】 infrared detector
| 红外冷却率 infrared cooling rate
| 红外偏振器【化】 infrared polarizer
| 红外扫描【农】 infrared scanning
| 红外湿度表 infrared hygrometer, infrared humidometer
| 红外数字资料 infrared digital data
| 红外图像 infrared picture, infrared imagery, infrared image
| 红外望远镜 infrared telescope
| 红外温度表 infrared thermometer
| 红外温度廓线辐射仪 infrared temperature profile radiometer(ITPR)
| 红外吸收光谱学 infrared absorption spectroscopy
| 红外吸收湿度表 infrared absorption hygrometer
| 红外线 ultrared ray, infrared(IR)
| 红外线【物】 infrared ray
| 红外线干燥【农】 infrared drying
| 红外线加热【农】 infrared heating
| 红外遥感 infrared remote sensing
| 红外云图 infrared cloud imagery

红外云图【大气】 infrared cloud picture
红外折射仪 infrared refractometer
红雪 red snow
红移【物】 red-shift
红雨 red rain
红噪声 red noise
红噪声现象 red noise phenomenon
红噪声型 red-noise type
宏观黏(滞)度【大气】 macroviscosity
宏观湍流 macroturbulence
宏观物理学 macro physics
宏观系统【物】 macrosystem
宏流控技术 macrofluidics
宏指令 macro instruction
虹 bow
虹【大气】 rainbow
虹【物】 primary rainbow
虹彩 irisation, iris, iridescence
虹彩高积云 iridescent altocumulus
虹彩云 iridescent cloud
虹的艾里理论 Airy theory of rainbow
虹吸【物】 syphon
虹吸槽式气压表【大气】 siphon cistern barometer
虹吸〔管〕雨量器 siphon raingauge
虹吸气压表【大气】 siphon barometer
虹吸气压计 siphon barograph
虹吸〔式〕雨量计【大气】 siphon rainfall recorder
虹吸式〔自记〕气压计 siphon recording barometer
洪堡海流 Humboldt current
洪峰【航海】 flood peak
洪峰阶段 crest stage
洪峰流量 peak discharge
洪峰水流法 peak flow method
洪峰纵剖面 crest profile
洪积世 Diluvial Epoch
洪涝 flooding
洪涝灾害 flood-waterlogging damage
洪水 inundation, flood
洪水标记 flood marks
洪水波 flood wave
洪水泛滥 freshet
洪〔水〕峰〔顶〕 flood crest
洪水概率 flood probability
洪水间隙 flood interval
洪水监测〔报告〕 flood watch
洪水警报 flood warning
洪水频率 flood frequency
洪水频率分布 flood frequency distribution
洪水侵蚀【土壤】 flood erosion
洪水位 flood stage, flood discharge level
洪水演算 flood routing, flooding routing
洪水预报 flood forecasting, flood forecast
〔洪〕水闸 flood gate
洪水涨落图 flood hydrograph
洪水终止 ending of flood
洪水综合观测和警报系统 Integrated Flood Observing and Warning System (IFLOWS)
洪灾 flood fatality
后〔波〕瓣 backlobe
后处理 post-processing
后切云砧 back-sheared anvil
后倾槽【大气】 backward-tilting trough
后曲锢囚【大气】 bent-back occlusion, recurved occlusion, back-bent occlusion
后色化 post-colouring
后退 backlash
后退波 retrogressive wave
后退波【大气】 retrograde wave
后退低压 retrograde depression
后退轨道 retrograde orbit
后退系统 retrograde system
后向传播 backpropagation
后向辐射 back radiation
后向流密度 backward flux density
后向流入急流 rear-inflow jet

后向散射　backscatter
后向散射【大气】　backscattering
后向散射【物】　backward scatter
后向散射比强度　backward specific intensity
后向散射表　backscatter meter
后向散射〔激〕光〔雷〕达　backscattering lidar
后向散射截面【大气】　backscattering cross section
后向散射能见度探测器　backscatter visibility sensor
后向散射强度　backscattered intensity
后向散射系数　backward scattering coefficient, backscattering coefficient
后向散射-消光比　backscatter-to-extinction ratio
后向散射效率　backscattering efficiency
后向散射紫外光谱仪【大气】　backscatter ultraviolet spectrometer (BUV)
后项【数】　consequent
后延窗　lag window
后延系数　lag coefficient
后延相关　lag correlation
后延相关法　lag correlation method
后验分布【数】　posterior distribution
后验概率【数】　posterior probability
厚度　thickness
厚度场型　thickness pattern
厚度平流　thickness advection, density (=thickness) advection
厚度图　thickness chart
厚度线【大气】　thickness line
厚烟层【大气】　smoke pall
候　penthemeron
候【大气】　pentad
候鸟　migratory bird, bird of migration (或 passage)
呼吸测污计　scentometer
呼吸链【生化】　respiratory chain
呼吸〔作用〕　respiration

呼啸风　whistling wind
弧　arc
弧度【数】　radian
弧光放电　arc discharge
弧状积雨云　cumulonimbus arcus
弧状（云）　arcus（arc）
弧状云【大气】　arc cloud
弧状云线【大气】　arc cloud line
胡克定律【物】　Hooke law
胡克数　Hooke number
湖　lake
湖〔泊〕冰　lake ice
湖泊水文学【地理】　lake hydrology
湖泊效应　lake effect
湖泊效应降雪　lake-effect snow
湖泊效应雪暴　lake effect snowstorm
湖泊蓄水量【地理】　lake storage
湖泊蒸发　lake evaporation
湖风【大气】　lake breeze
湖锋【大气】　lake front
湖流【地理】　lake clake coreurrent
湖面温度　lake (surface) temperature
湖盆【地理】　lake basin
湖平面　lake level
湖水环流【地理】　lake circulation
湖相沉积　lake sediment
湖芯　lake core
湖沼生物气象学　limnological meteorology
湖沼学　limnology
蝴蝶效应【物】　butterfly effect
互反律【数】　reciprocity law
互反性　reciprocity
互反原理　reciprocity principle
互反展开　reciprocal spreading
互功率谱　cross-power spectrum
互关联【物】　mutual correlation
互换性　interchangeability
互扩散系数　coefficient of mutual diffusion
互联网〔络〕【计】　internet, interconnec-

tion network
互谱【数】 cross spectrum
互谱密度 cross spectrum density
互谱密度【物】 mutual spectral density
互相干函数 mutual coherence function
互相干〔性〕【物】 mutual coherence
互相关【大气】 cross-correlation
互相关函数【数】 cross correlation function
互相关谱 cross-correlation spectrum
互相关图 cross-correlogram
互协方差 cross-covariance
护航【航海】 convoy
护面层 armouring
花彩云 festoon-cloud
花粉带 pollen zone
花粉记录 pollen record
花粉浓度 pollen density
花粉雨 pollen rain
花粉指标 pollen indicator
花岗岩【地质】 granite
花椰菜云 cauliflower cloud, causality
华【大气】(日华,月华) corona
华北锢囚锋【大气】 North China occluded front
华南准静止锋【大气】 South China quasi-stationary front
华氏温标 Fahrenheit temperature scale
华氏温标【农】 Fahrenheit thermometric scale
华氏温度表 Fahrenheit thermometer
滑尺 slideing scale
滑动变换 running transform
滑动差分 running difference
滑动摩擦系数【物】 coefficient of sliding friction
滑动平均 running average【大气】, moving average, overlapping average, overlapping mean, running mean, floating average
滑动平均模型 running mean model
滑动谱 running spectrum, moving spectrum
滑动时间窗 moving time window
滑流 slip stream
滑坡【地理】 landslip, landslide
滑坡【土壤】 slip
滑翔降落伞 paraglider
化合【化】 chemical combination
化合热 heat of combination
化合物 compound
化石 fossil
化石冰 fossil ice
化石记录 paleontological record
化石水 fossil water
化石永久冻土 fossil permafrost
化学成分族 family of chemical species
化学地理学【地理】 chemical geography
化学发光【化】 chemiluminescence
化学反应【化】 chemical reaction
化学肥料【土壤】 chemical fertilizer
化学分析【化】 chemical analysis
化学风化【土壤】 chemical weathering
化学活化【化】 chemical activation
化学活性【化】 chemical activity
化学激光器【化】 chemical laser
化学计量比 stoichiometric ratio
化学计量浓度【化】 stoichiometric concentration
化学键【化】 chemical bond
化学能【化】 chemical energy
化学平衡 chemical equilibrium
化学-气象模拟 chemical meteorological modeling
化学日射表 chemical actinometer
化学湿度表【大气】 chemical hygrometer
化学示踪物 chemical tracer
化学势【化】 chemical potential
化学物理〔学〕【物】 chemical physics
化学吸附【物】 chemisorption
化学需氧量【大气】 chemical oxygen

demand (COD)
化学烟气　chemical smoke
化学氧化大气　chemically oxidizing atmosphere
化学荧光臭氧分析仪　chemiluminescent O_3 analyzer
化学荧光探空仪　chemi-luminescent sonde
划界　demarcation
画网格　gridding
画中画【计】　picture in picture (PIP)
怀特〔吸收〕池　White cell
坏天气　foul weather
还原剂　reducing agent
还原性大气　reducing atmosphere
环　annulus
环地平弧【大气】　circumhorizontal arc
环积分　circulation integral
环极地图　circumpolar map
环己烷(C_6H_{12})　cyclohexane
环境　environment
环境保护　environmental protection
环境保护委员会　Committee for Environmental Conservation(CoEnCo)
环境本底　environmental background
环境标准　environmental standard
环境场　ambient field
环境大气　ambient atmosphere, ambient air
环境大气标准　ambient air standard
环境大气监测　ambient air monitoring
环境大气质量　ambient air quality
环境大气质量标准　ambient air quality standard
〔环境〕大气质量监测【大气】　〔environmental〕atmospheric quality monitoring
环境地理学　environmental geography
环境地球化学【地理】　environmental geochemistry
环境恶化【农】　environmental deterioration
环境防治　environmental control

环境分析　environmental analysis
环境风　environmental wind
环境风险　environmental risk
环境风险评价　environmental risk assessment
环境荷载　environmental load
环境化学　environmental chemistry
环境监测　environmental monitor
环境勘测卫星【大气】　Environmental Survey Satellite (ESSA)
环境科学　environmental sciences
环境科学服务局　Environmental Sciences Services Administration (ESSA)
环境颗粒浓度　ambient particulate concentration
环境空气取样　ambient air sampling
环境空气取样仪　ambient air sampling instrument
环境绿化【林】　environmental greening
环境敏感性　environmental sensitivity
环境模拟室　environmental chamber
环境评价【地质】　environmental evaluation
环境评价【农】　environmental appraisal
环境气候　environmental climate
环境气候学【大气】　environmental climatology
环境气流　environment flow
环境气象学　environmental meteorology
环境气压　ambient pressure
环境容量　environmental capacity
环境设计【地理】　environmental design
环境生态学　environmental ecology
环境声学　environmental acoustics
环境示踪剂　environmental tracer
环境适应性　environmental suitability
环境适应症　adaptive disease
环境条件　environmental condition
环境同化能力　environmental assimilating capacity
环境退化　degradation of environment

环境退化【地理】 environmental degradation
环境卫星 environmental satellite
环境温度 environmental temperature, ambient temperature
环境污染 environmental pollution
环境污染负荷 ambient pollution burden
环境污染物 environmental pollutant
环境物质形态【地质】 form of environmental substance
环境系统 environmental system
环境胁迫 environmental stress
环境效应【地理】 environmental effect
环境学 environics
环境研究协会(美国) Institutes for Environmental Research(IER)
环境研究院(NOAA) Environmental Research Laboratories (ERL, NOAA)
环境演化【地理】 environmental evolution
环境遥感【地信】 environmental remote sensing
环境要素【地理】 environmental element
环境液〔态〕水含量 ambient liquid water content
环境异常 environmental anomaly
环境异常【地理】 environmental abnormality
环境医学 environmental medicine
环境因子 environmental factor
环境影响评价【地理】 environmental impact assessment
环境预报【农】 environmental forecasting
环境预报研究中心(美国海军) Environmental Prediction Research Facility(EPRF)
环境灾难 environmental disaster
环境噪声 ambient noise
环境噪声场 ambient noise field

环境噪声级 ambient noise level
环境直减率【大气】 environmental lapse rate
环境质量【地理】 environmental quality
环境质量标准 ambient quality standard
环境质量参数 environmental quality parameter
环境质量评价 environmental quality evaluation
环境质量指数 environmental quality index
环境质量综合评价【地理】 environmental quality comprehensive evaluation
环境资料 environmental data
环境资料浮标 environmental data buoy
环境资料局(美国) Environmental Data Service(EDS)
环境资料信息部(NOAA) Environmental Data and Information Service(EDIS)
环境资源【农】 environmental resources
环境自净 environmental self-purification
环流 circulation, gyres, gyre(=gyral)
环流定理【大气】 circulation theorem
环流模式 circulation model
环流圈【大气】 circulation cell
环流调整【大气】 adjustment of circulation
环流系统 circulation system
环流型 circulation type
环流型【大气】 circulation pattern
环流型过渡 transition of circulation
环流序列 sequence of circulation
环流原理 principle of circulation
环流指数【大气】 circulation index
环日天空辐射 circumsolar sky radiation
环蚀 angular eclipse
环肽【生化】 cyclic peptide
环天顶弧【大气】 circumzenithal arc
环形海流 Loop Current

环形天线　loop antenna
环型　cell
环型波　cellular wave
环型构造　cellular structure
环型胶卷　movie loop, film loop
环型涡旋　cellular vortex
环型〔学〕说　cell theory
环型运动　cellular movement
环氧化合物【化】　epoxide
环宇能见度　all-round visibility
环晕　ring halo
环状涡旋　ring vortex
锾定年法　ionium dating
缓变波列　slowly varying wave train
缓冲存储器【计】　buffer memory
缓冲带　buffer zone
缓冲寄存器【计】　buffer storage
缓冲器【计】　buffer
缓冲因子　buffer factor
缓冲作用　buffering
缓降伞（供火箭探测用）　starute
缓流航道【航海】　slack current channel
缓慢气候系统　slow climate system
缓始型（磁暴）　gradually commencing type
幻灯　delinescope
幻日　mock sun, parhelion, false sun
幻日环　parhelic ring, parhelic circle【大气】, mock sun ring
幻雾蜃景　fata bromosa
幻月　moon dog, mocd moon, paraselene
幻月环【大气】　paraselenic circle
换茬　rotation of crops
换能器【船舶】　transducer
换气　aeration
换气率　air change rate
换算因子　conversion factor
荒川-雅可比近似　Arakawa jacobian
荒漠【地理】　desert
荒漠化【大气】　desertization, desertification
荒漠群落　eremium

荒漠土壤【地理】　desert soil
荒原植物　eremophyte
皇家观象台（英国）　Royal Observatory (R. O.)
黄斑中心视像　foveal vision
黄赤交角【天】　obliquity of the ecliptic
黄道【大气】　ecliptic
黄道带　zodiac band
黄道带【天】　zodiac
黄道对日照　zodiacal counterglow
黄道光【天】　zodiacal light
黄道面　plane of the ecliptic
黄道锥　zodiac cone
黄道坐标系【天】　ecliptic coordinate system
黄海　Yellow Sea
黄海【海洋】　Huanghai Sea
黄海冷水团【海洋】　Huanghai Cold Water Mass
黄海暖流　Yellow Sea Warm Current
黄海暖流【海洋】　Huanghai Warm Current
黄海沿岸流【海洋】　Huanghai Coastal Current
黄昏黎明视觉　mesopic vision
黄极【航海】　ecliptic pole
黄金分割搜索【数】　golden section search
黄经【天】　ecliptic longitude, celestial longitude
黄土　loess
黄土地貌【地理】　loess landform
黄〔土〕风　yellow wind
黄纬【天】　ecliptic latitude, celestial latitude
黄雪　yellow snow
黄雨【大气】　yellow rain, sulfur rain
灰冰　gray ice
灰尘沉降　ash fall
灰大气【天】　grey atmosphere
灰度　shades of gray, grey scale【大气】
灰分【航海】　ash content

灰阶【物】　gray level
灰壤　podsol
灰〔色标〕度【大气】　gray scale
灰色光　ash-grey light
灰色吸收体　gray absorber
灰体　grey body, gray body
灰体大气　gray atmosphere
灰体辐射【大气】　grey body radiation
灰吸收体【大气】　grey absorber
挥发　volatilization
挥发法【化】　volatilization method
挥发物　volatiles
挥发性有机化合物　volatile organic compounds (VOCs)
恢复时间【电子】　recovery time
恢复系数【物】　coefficient of restitution
恢复作业　resumed operation
辉光放电【大气】　glow discharge
回波　return echo, echo
回波等值线　echo contour
回波分析【大气】　echo analysis
回波复合体【大气】　echo complex
回波覆盖区域　area covered with echoes
回波功率　echo power
(回波)贡献区　contributing region
回波厚度【大气】　echo depth
回波畸变【大气】　echo distortion
回波脉冲　echo pulse
回波密实覆盖区域　area solidly covered with echoes
回波频率　echo frequency
回波强度　echo intensity
回波墙【大气】　echo wall
回波特征【大气】　echo character
回波线　line of echo
回波箱【电子】　echo box
回波信号　return signal, echo signal
回波信号模拟器　echo signal simulator
回波振幅　echo amplitude
回波综合图【大气】　echo synthetic chart
回车【计】　carriage return
回答器　responder
回答式探空仪　echo radiosonde
回返对流　return convection
回〔返〕闪〔电〕　return lightning
回复力　restoring force
回灌　recharge
回灌容量　recharge capacity
回光仪　heliotrope
回归　regression
回归潮　tropic tide
回归低低潮　tropic lower low water (TcLLW)
回归法　regression method
回归方程　regression equation
回归分析【数】　regression analysis
回归高高潮　tropic higher high water (TcHHW)
回归估计【数】　regression estimate
回归函数【数】　regression function
回归矩阵　regression matrix
回归量　regressor
回归年【天】　tropical year
回归系数　coefficient of regression
回归系数【数】　regression coefficient
回归线　tropic, regression line
回归预报方程　regression prediction equation
回归诊断【数】　regression diagnostics
回归值　regressand
回击【大气】　return stroke
回零　return-to zero
回流　return streamer, recurrent flow, return current, returning flow, backwash, backflow
回流气团　returning air-mass
回流式风洞　return-flow wind tunnel
回流式温度表　reverse flow thermometer (RFT)
回流式温度表罩　reverse flow thermometer housing
回流天气【大气】　returning flow weather

回路【数】 circuit
回闪 returned flash
回声探测 acoustic echo sounding
回声探测器 echo sounder, echo sounding apparatus, acoustic echo sounder
回声探测仪 Echosonde
回水 backwater
回水曲线 backwater curve
回线 loop
回旋半径【物】 cyclotron radius
回旋加速器【物】 cyclotron
回旋频率 gyro frequency
回旋频率【物】 cyclotron frequency
回转磁频率 gyromagnetic frequency
回转反射式日照计 rotating reflection sunshine recorder
回转频率 gyration frequency
回转椭球 ellipsoid of gyration
回转效应【大气】 Umkehr effect
回转运动 gyromotion
回转中心 gyration center
汇 sink
汇编程序 assembly program
汇编语言【计】 assembly language
汇合型超几何方程【数】 confluent hypergeometric equation
汇合型超几何函数【数】 hypergeometric function of confluent type
汇焦光束 focused beam
汇流【地理】 flow concentration
汇流【大气】 confluence
汇流型温度槽 confluent thermal trough
汇流型温度脊 confluent thermal ridge
汇水面积 catchment area
会合运动 synodic motion
会话监控系统 conversational monitor system（CMS）
会聚透镜【物】 convergent lens
会议鉴定 meeting appraisal
绘图板 graphing board
绘图机 graphic plotter

绘图机【计】 plotter
彗星 comet
惠风 pretty breeze
惠更斯-菲涅耳方法 Huygens-Fresnel method
惠更斯小波 Huygens's wavelets
惠更斯原理【物】 Huygens principle
惠更斯作图法【物】 Huygens construction
毁林 deforestation
昏侧磁层 duskside magnetosphere
昏侧磁尾 duskside tail
浑浊层 opaque layer
浑浊大气 turbid atmosphere
浑浊度极大〔区〕 turbidity maximum
浑浊度自动分析仪 automatic turbidity analyzer
浑浊媒质 turbid medium
浑浊因子【大气】 turbidity factor
混沌【数】【大气】 chaos
混沌动力系统 chaotic dynamical system
混沌吸引子【物】 chaotic attractor
混沌运动【物】 chaotic motion
混合 mixing
混合比【大气】 mixing ratio
混合比显示器 mixing ratio indicator
混合层 mixed layer(ML)
混合层【大气】 mixing layer
混合层顶 mixed-layer top
混合长【大气】 mixing length
混合〔层〕高度 mixing height
混合层高度 mixed-layer height
混合层冠盖逆温 mixed-layer capping inversion
混合层厚度 mixed-layer depth
混合层均一性 mixed layer homogeneous
混合层模式 mixed-layer model
混合层排放 mixed-layer venting
混合层谱 mixed-layer spectra
混合层深(高)度 mixing depth(或height)

混合层实验　Mixed Layer Experiment（MILE）
混合层相似性　mixed-layer similarity
混合层演变　mixed-layer evolution
混合长【大气】　mixing length
混合长理论　mixing-length theory
混合潮【海洋】　mixed tide
混合垂直坐标　hybrid vertical coordinate
混合带　mixing zone
混合分布　mixing distribution
混合高度　blending height
混合过程　blending process
混合核　mixed nucleus
混合积冰条件　mixed icing condition
混合冷却　mixing cooling
混合路径　path of mixing, mixing path
混合罗斯贝重力波【大气】　mixed Rossby-gravity wave
混合模型【农】　mixed model
混合凝结高度【大气】　mixing condensation level（MCL）
混合〔频数〕分布　mixed distribution
混合轻雾　mizzle
混合区　blending region
混合热【化】　heat of mixing
混合势　mixing potential
混合湍流　mixing turbulence
混合物　mixture
混合雾【大气】　mixing fog
混合线　mixing line
混合线结构　mixing-line structure
混合效率　mixing efficiency
混合行星-重力波　mixed planetary-gravity wave
混合云　mixing cloud
混合云顶　mixing cloud top
混合云室　mixing chamber
混合增密　cabbeling, cabbaling
混合坐标【大气】　hybrid coordinate
混交林【林】　mixed forest

混乱天空【大气】　chaotic sky
混凝土板最低温度　concrete minimum temperature
混响　reverberation
混淆空间频率　aliased spatial frequency
混淆误差　aliasing error
混淆〔现象〕　aliasing
混杂型风暴　hybrid storm
混杂型模式　hybrid model
活冰川　active glacier
活动　activity
活动层　active layer
活动光带（极光的）　activated band
活动积温　total active temperature
活动积温【大气】　active accumulated temperature
活动〔积状〕云　active cloud
活动极光　active aurora
活动面　active surface
活动目标显示器　moving target indicator
活动嵌套网格模式　movable nested grid model
活动日珥【天】　active prominence
活动太阳　disturbed sun, active sun
活动温度【大气】　active temperature
活动文件【计】　active file
活动永冻土　active permafrost
活动有限域细网格　movable-area fine mesh
活动中心　center of action【大气】, active center, action center
活度系数　activity coefficient
活化【化】　activation
活化分析　activation analysis
活化络合物理论　activated complex theory
活化能【化】【大气】　activation energy
活化自由能　activation free energy
活火山　active volcano
活塞效应　piston effect
活性部位　active site
活性氮　active nitrogen

活性反应　active reaction
活性磷【土壤】　labile phosphorus
活性肽【生化】　bioactive peptide
活性炭【化】　active carbon
活跃冻层　mollisol
活跃锋　active front
活跃季风【大气】　active monsoon
火成岩【地质】　igneous rock
火花放电【物】　spark discharge
火箭　rocket
火箭臭氧探空仪　rocket ozonesonde
火箭弹　rocket projectile
火箭发射点　rocket-launching site
火箭废气云　rocket exhaust cloud
火箭激发闪电　rocket-triggered lightning
火箭榴弹法　rocket-grenade method
火箭排烟尾迹　rocket exhaust trail
火箭气球装置　rocket balloon instrument（RBI）
火箭气象计　rocket meteorograph
火箭探测【大气】　rocket sounding
火箭探空仪　rocketsonde
火箭状闪电【大气】　rocket lightning
火流星　fireball, bolide
火山【地理】　volcano
火山暴　volcanic storm
火山尘　volcanic dust
火山沉积　volcanic sediment
火山弹　volcanic bomb
火山冬天　volcanic winter
火山风　volcanic wind
火山弧　volcanic arc
火山湖【地理】　volcanic lake
火山灰【大气】　volcanic ash
〔火山〕灰　ash
火山灰暴　ash shower
火山灰年代学　tephrochronology
火山活动　volcanic activity
火山雷鸣　volcanic thunder

火山喷发　volcanic eruption
火山气溶胶　volcanic aerosol
火山气〔体〕　volcanic gases
火山砂　volcanic sand
火山闪电　volcanic lightning
火山学【地质】　volcanology
火山烟流　volcanic plume
火山云　volcanic cloud, cloud due to volcanic eruption
火山作用【地质】　volcanism
火险标尺　fire-danger meter
火险等级指数　buildup index
火险天气　fire weather
火险指数　burning index
火星【天】　Mars
火星大气　Martian atmosphere
火星年　Martian year
火焰电离检测　flame ionization detection
火焰风　incendium
火焰集电器　flame collector
火灾风　fire wind
火灾风暴　firestorm
火灾云　cloud from fire
或非门【计】　NOR gate
或门【计】　OR gate
霍布斯理论　Hobbs's theory
霍尔流（电急流的）　Hall current (of electrojet)
霍尔姆博不稳定度　Holmboe instability
霍尔瑞斯卡　hollerith card
霍尔效应【物】　Hall effect
霍耳效应器件【电子】　Hall-effect device
霍夫变换【计】　Hough transformation
霍夫函数　Hough function
霍普菲带　Hopfield bands
霍普夫分岔【物】　Hopf bifurcation
霍普金斯生物气候律　Hopkin's bioclimatic law

J

击穿 breakdown
击穿电场 breakdown field
击穿电位〔电压〕 breakdown potential
击穿电压【电子】 breakdown voltage
击穿强度【电子】 breakdown strength
机场 airfield
机场标高 airport height
机场等效高度 equivalent altitude of aerodrome
机场标高〔高度〕 airport elevation
机场定时地面天气报告 Meteorological Terminal Air Report(METAR)
机场多普勒天气雷达 terminal Doppler weather radar(TDWR)
机场关闭天气 socked in, below minimums
机场海拔〔高度〕 aerodrome elevation
机场监视雷达【电子】 airport surveillance radar
机场色标 airfield color code
机场特别天气报告 airdrome special weather report
机场特殊天气报告【大气】 aerodrome special weather report
机场〔天气〕预报 airdrome forecast
机场天气预报 aerodrome forecast
机场危险天气警报【大气】 aerodrome hazardous weather warning
机场预报 airport forecast
机场预警 airdrome warning
机场预约天气报告【大气】 appointed aerodrome weather report
机场最低气象条件【大气】 aerodrome meteorological minima（或 minimums）
机顶盒【计】 set-top box
机理【物】 mechanism
机内大气 on-board atmosphere

机内定标用冷靶 internal cold (calibration) target(ICT)
机内热靶 internal warm target(IWT)
机器翻译【计】 machine translation
机器人【计】 robot
机器学习【计】 machine learning
机器指令【计】 machine instruction
机体积冰 airframe icing
机尾多普勒雷达 tail Doppler radar
机下云障 undercast
机械产生率 mechanical production rate
机械分散 mechanical dispersion（亦称 hydraulic dispersion）
机械风化〔作用〕【地理】 mechanical weathering
机械混合 mechanical mixing
机械开关 mechanical switch
机械内边界层 mechanical internal boundary layer（MIBL）
机械能【物】 mechanical energy
机械能方程 mechanical energy equation
机械强迫作用 mechanical forcing
机械通风 mechanical draft
机械湍流【大气】 mechanical turbulence
机械稳定〔度〕 mechanical stability
机械消旋天线 mechanical despin antenna(MDA)
机械性不稳定 mechanical unstable
机械性不稳定度 mechanical instability
机械性压力 mechanical pressure
机械运动【物】 mechanical motion
机载侧视雷达 side looking aircraft（或 airborne）radar（SLAR）
机载测量〔仪器〕 aircraft measurement
机载分光仪 airborne spectrometer
机载辐射温度仪 airborne radiation

thermometer(ART)
机载光〔雷〕达　airborne laser radar
机载雷达　airborne radar
机载染料光〔雷〕达　airborne dye lidar
机载搜索雷达　airborne search radar
机载探空火箭　rockair
机载天气雷达【大气】　airborne weather radar
机载投弃式温深仪　airborne expendable bathythermograph（AXBT）
〔机载〕投弃式温深仪　air-dropped（或airborne）expendable bathythermograph（AXBT）
机载下投式测风探空仪系统　aircraft dropwindsonde system（ACDWS）
机载云〔含水量〕收集器　airborne cloud collector
机载云凝结核粒谱仪　airborne CCN spectrometer
机载直接辐射表　aircraft actinometer
机载撞击〔取样〕器　aircraft impactor
机制型作物生长模拟装置　mechanistic type crop growth simulator
迹线　pathway
迹线【物】　path line
奇氮〔族〕　odd nitrogen
奇函数【数】　odd function
奇氯〔族〕　odd chlorine
奇偶校验【数】　parity check, even-odd check
奇偶性【数】　parity
奇氢〔族〕　odd hydrogen
奇氧　odd oxygen
奇氧系　odd-oxygen system
积冰　ice pack, ice deposit, ice accretion
积冰【大气】　icing
积冰表　icing meter
积冰高度　icing level
积冰率　rate of ice accretion
积冰率表　icing-rate meter
积冰强度　icing intensity

积冰指示器　ice-accretion indicator
积冰指数　icing index
积分　integration
积分【数】　integral
积分变换【数】　integral transform
积分长度尺度　integral length scales
积分动力蒸发仪　integration kinetic evaporation meter
积分发射率　integral emissivity
积分法　integration method
S-M 积分方程　Schwarzchild-Milne integral equation
积分方程【数】　integral equation
积分风速计　integration anemograph
积分厚度尺度　integral depth scale
积分浑浊度表　integration nephelometer
积分精度【化】　integral accuracy
积分谱　integrated spectrum
积分曲线【数】　integral curve
积分溶解热【化】　integral heat of solution
积分吸附热【化】　integral heat of adsorption
积分吸收法【化】　integrated absorption method
积分学【数】　integral calculus
积分域【数】　domain of integration
积分约束　integral constraint
积矩　product-moment
积深泥沙采样　depth-integration sediment sampling
积温　cumulative temperature
积温【大气】　accumulated temperature
积温曲线　accumulated temperature curve
积雪　snow accumulation, perpetual snow
积雪【大气】　snow cover
积雪成层〔作用〕　snow stratification
积雪覆盖　snow lying
积雪密度　density of snow
积雪面积　snow covered area
积雪区　surplus area
积雪日　day of snow lying

积雪日数【大气】 days with snow cover
积雪深度 depth of snow
积雪〔水当〕量 snowpack yield
积雪图 snow cover chart
积雪线 snow cover line
积雪消失 ending of snow cover
积雪型式 snow regime
〔积〕雪震〔落〕 snow tremor
积雪总量【大气】 snow pack
积雨云【大气】 cumulonimbus (Cb)
积雨云动力学 cumulonimbus dynamics
积雨云模式【大气】 cumulonimbus model
积云【大气】 cumulus(Cu)
积云参数化模式 cumulus parameterization model
积云底 cumulus base
积云对流【大气】 cumulus convection
积云〔对流〕加热 cumulus heating
积云间区 intercumulus region
积云阶段 cumulus stage
积云性 cumulogenitus (cug)
积云性层积云【大气】 stratocumulus cumulogenitus (Sc cug)
积云性高积云【大气】 altocumulus cumulogenitus(Ac cug)
积云状 cumuliform (cuf)
积云状卷层云 cirrostratu cumulosus
积云状雨云 nimbus-cumuliformis
积状云【大气】 cumuliform cloud
基本磁层过程 basic magnetospheric process
基本单位 fundamental unit
基本点 fundamental point
基本方程组 basic equations
基〔本径〕流 base flow
基本陆地站 principal land station
基本农业气象站 principal agricultural meteorological station
基本谱 base spectrum
基本气候站 principal climatological station
基本气流【大气】 basic flow
基本渗透量 ultimate infiltration capacity
基本水准面 standard sea level
基本天气观测【大气】 principal synoptic observation
基本天气观测时间【大气】 main standard time
基本天气过程 elementary synoptic process
基本天气形势 elementary synoptic situation
基本维数 fundamental dimensions
基本系统 basic system
基本系统委员会(WMO) Commission for Basic System(CBS,WMO)
基本站 base station
基本振动 fundamental vibration
基本状态 basic state
基本资料库 basic data set (BDS)
基波【物】 fundamental wave
基点温度 cardinal temperatures
基尔霍夫定律【大气】 Kirchhoff's law
基尔霍夫方程 Kirchhoff's equation
基尔霍夫积分定理【物】 Kirchhoff integral theorem
基尔霍夫近似 Kirchhoff approximation
基尔霍夫涡旋 Kirchhoff vortex
基肥【土壤】 basal fertilizer
基函数【物】 basis function
基流 groundwater runoff
基流储量(水文) baseflow storage
基流退水曲线 baseflow recession curve
基流消退 baseflow recession
基频 fundamental frequency
基色 primary color
基数 radix
基态【化】 ground state
基团频率【化】 group frequency
基线 baseline
基线监测 baseline monitoring
基线气压系数 base-pressure coefficient

基线校正　baseline check
基岩【地质】　bedrock
基因【生化】　gene
基因工程【生化】　genetic engineering
基因组【生化】　genome
〔基于〕规则系统　rule-based system
基于知识的咨询系统【计】　knowledge-based consultation system
基值　ground check
基值检定箱　baseline check box
基柱〔状〕云　pedestal cloud
基准　benchmark
基准标记　fiducial mark
基准测量　reference measurement
基准面　datum level
基准面静校正【地】　datum static correction
基准气候站　reference climatological station
基准〔气候〕站【大气】　benchmark station
基准气压表　reference standard barometer
基准探空仪　reference sonde
基准温度　fiducial temperature, base temperature
基准温度表　reference thermometer
基准无线电探空仪　reference radiosonde
基准线　datum line
畸变波　distortional wave
畸变校正【大气】　distortion correction
激波　shock
激波解　shock solution
激潮　tide blow
激发　motivation, excitation
激发机理　excitation mechanism
激发态　excited state
激光【物】　laser (light amplification by simulated emission of radiation)
激光侧风测量系统　laser crosswind system

激光测风仪　laser anemometer
激光测高仪【测】　laser altimeter
激光抽运【物】　laser pumping
激光传真图像记录器　laser facsimile image recorder
激光打孔【物】　laser boring
激光多普勒风速计　laser Doppler anemometer
激光二极管【物】　laser diode
激光分光计　laser spectrometer
激光分离同位素【电子】　laser isotope separation
激光辐射技术　laser radiation technology (LARAT)
激光干涉仪【电子】　laser interferometer
激光感生荧光【物】　laser-induced fluorescence (LIF)
激光〔共振〕腔【物】　laser resonator, laser cavity
激光光谱〔学〕【电子】　laser spectroscopy
激光焊接【物】　laser bonding
激光雷达【大气】　lidar
激光受激雷达　laser-excited radar
激光束【化】　laser beam
激光束记录仪　laser beam recorder (LBR)
激光水准仪【测】　laser level
激光速度仪　laser velocimeter
激光天气识别仪　laser weather identifier
激光天气实况识别仪　laser present weather identifier
激光陀螺【电子】　laser gyro
激光外差分光计　laser heterodyne spectrometer (LHS)
激光外差式辐射计　laser heterodyne radiometer
激光雾能见度仪　laser fog nephelometer
激光吸收分光计　laser absorption spectrometer (LAS)
激光显示【物】　laser display

激光遥测　laser remote sensing
激光印刷机【计】　laser printer
激光云高指示器　laser cloud indicator
激光云幂计　laser-ceilograph
激光云幂仪【大气】　laser ceilometer(LC)
激光照排机【计】　laser typesetter
激光制导【物】　laser guidance
激光准直仪【测】　laser aligner
激励频率　driving frequency
激酶【生化】　kinase
激素【生化】　hormone
吉布斯函数【物】　Gibbs function
吉布斯稳度观念　Gibbs stability concept
吉布斯现象　Gibbs phenomenon
吉布斯自由能【物】　Gibbs free energy
吉尔风速表　Gill anemometer
吉赫　gigahertz（GHz）
吉〔咖〕（词头，10^9）　giga-
0 级风【大气】　calm
1 级风　force-one wind
1 级风【大气】　light air
2 级风　force-two wind
2 级风【大气】　light breeze
3 级风　force-three wind
3 级风【大气】　gentle breeze
4 级风【大气】　moderate breeze
5 级风【大气】　fresh breeze
6 级风【大气】　strong breeze
7 级风【大气】　near gale
8 级风【大气】　gale
9 级风【大气】　strong gale
10 级风【大气】　storm
11 级风【大气】　violent storm
12 级风【大气】　hurricane
级联簇射【地】　cascade shower
级联过程【物】　cascade process
级联滤波　cascade filtering
级联衰变【物】　cascade decay
级数展开法　series expansion method
A 级蒸发皿　class-A pan, class A evaporation pan

极　pole
极飑　polar squall
极波　polar wave
极波高〔度〕　extreme wave height
极差【数】　range
极差分析　range analysis
极大风速【大气】　extreme wind speed
极大极小策略　maxi-min strategy
极大极小判据　maxi-min criterion
极大值　maximum value
极带　polar belt, polar band
极低频　extremely low frequency（ELF）
极低频发射　ELF emission
极地　polar region
极地胞　polar cell
极地冰　polar ice
极地冰川【地理】　polar glacier
极地冰冠　polar ice cap
极地冰原　polar ice sheet
极地槽　polar trough
极地簇射　polar shower
极地大陆高压　polar continental high
极地大陆空气【大气】　polar continental air, continental polar air（mass）
极地大西洋海水　polar Atlantic Water
极地低压　polar low pressure
极地低压【大气】　polar low
极地东风〔带〕　polar easterlies
极地东风指数　polar-easterlies index
极地对流层顶　polar tropopause
极地反气旋【大气】　polar anticyclone
极地放大现象　polar amplification
极地风界　polar wind divide
极地高压【大气】　polar high
极地轨道地球物理观测卫星　polar orbiting geophysical observatory（POGO）
极〔地〕轨〔道〕卫星　polar-orbiting satellite（POS）
极地海洋气团　polar marine air（mass），maritime polar air（mass）
极地回流海洋气团　returning polar

mari-time air
极地回流气团　returning polar air（mass）
极地急流　polar jet
极地季风带　polar monsoon zone
极地〔空气〕霾　polar air haze
极地平流层云　polar stratospheric cloud（PSC）
极地气候【大气】　polar climate
极〔地气〕辉　polar glow
极地气流　polar current
极地气团【大气】　polar air mass
极地气团低压　polar air depression
极地气象学【大气】　polar meteorology
极地气象学委员会（WMO）　Commission for Polar Meteorology（CPM，WMO）
极地气旋【大气】　polar cyclone
极〔地迁〕移　polar wandering
极地-热带耦合　polar-tropical coupling
极地〔无线电〕衰落　polar blackout
极地西风带　polar westerlies
极地型　polar type
极地研究所（美国）　Institute of Polar Studies（IPS）
极地中间层夏季回波　polar mesospheric summer echoes（PMSE）
极地中间层云　polar mesospheric clouds（PMC）
极点潮　pole tide
极端干旱　extreme drought
极端干热气候　extremely hot and dry climate
极端干燥环境　extreme arid environments
极端降水过程　extreme precipitation process
极端降水量　extreme precipitation quantities，extreme precipitation
极端降水事件　episodes of extreme rainfall
极端气候　extreme climate
极端气候事件　extreme climatic event

极端气候稳定性　extreme climatic stability
极端湿润气候　extremely wet climate
极端天气事件　extreme weather events
极端温度　extreme temperature
极端雨年　extremes of rainy years
极端紫外辐射　extreme ultraviolet radiation
极风　polar wind
极风带　polar wind belt
极锋【大气】　polar front
极锋急流　polar-front jet(stream)
极锋区　polar frontal zone
极锋理论【大气】，极锋学说【大气】　polar front theory
极盖吸收【航海】　polar cap absorption（PCA）
极盖吸收事件　polar cap absorption events，PCA events
极高频　extremely high frequency（EHF）
极冠　polar cap
极冠冰　polar-cap ice
极冠吸收　polar cap absorption
极光　polar light，polar aurora
极光【地】【大气】　aurora
极光暴　auroral storm
极光崩离　auroral break-up
极光磁扰　auroral（或 aurora）magnetic disturbance
极光次声波　auroral infrasonic wave
极光带【大气】　auroral band
极光带【地】　auroral belt
极光带电集流【地】　auroral electrojet
极光带电集流指数　auroral electrojet（AE）index
极光带电离层　auroral ionosphere
极光等频〔率〕线　isochasm，isoaurore
极光地带　auroral zone
极光电离　aurora ionization
极光盖　auroral cap
极光光谱　auroral spectrum

极光弧　auroral arc
极光激发机理　auroral excitation mechanisms
极光粒子　auroral particle
极光粒子沉降　auroral particle precipitation
极光隆起　auroral bulge
极光绿谱线　auroral green line
〔极光〕卵形带　oval-shaped belt
极光卵形环【地】,极光卵【大气】　auroral oval
极光幔　auroral drapery, auroral curtains
极光冕　polar corona, auroral corona
极光射线　auroral rays
极光嘶声　auroral hiss
极光椭圆区【大气】　auroral oval
极光亚暴　auroral substorm
极光云　auroral cloud
极轨地球物理观测卫星　Polar-Orbiting Geophysical Observatory（POGO）
极轨气象卫星　solar synchronous meteorological satellite
极轨气象卫星【大气】　polar-orbiting meteorological satellite
极轨卫星【大气】　polar orbiting satellite（POS）
极轨业务环境卫星　Polar-Orbiting Operational Environmental Satellite（POES）
极化【物】　polarization
极化率【物】　polarizability
极化分集　polarization diversity
极化矩阵　polarization matrix
极化雷达　polarizing radar
极化曲线　polarization curve
极角　polar angle
极距　polar distance
极年　polar year
极谱〔分析〕法　polarography
极区　polar zone
极区磁暴　polar magnetic storm
极区磁扰　polar magnetic disturbance

极区磁亚暴　polar magnetic substorm
极区磁照图　polar magnetogram
极区航行【航海】　polar navigation
极区元磁暴　polar elementary storm
极圈　polar circle
极射赤面地图　polar stereographic map
极射赤面拼图　polar stereographic mosaic
极投影　polar projection
极涡【大气】　polar vortex
极涡指数　polar vortex index
极限　limit
极限波　limiting wave
极限过冷度　attainable degree of supercooling
极限环　limit cycle
极限气象条件进场着陆　marginal weather approach
极限速度　limiting velocity
极限圆　limit circle
极相漂移【地】　polar phase shift
极小方差法　minimum variance method
极小方差估计　minimum variance estimation
极小化极大【数】　minimax
极小化极大判据　minimax criterion
极小化极大原理【数】　minimax principle
极小极大策略【数】　minimax strategy
极小熵交换原理　minimum entropy exchange principle
极性　polarity
极性过渡【地】　polarity transition
极夜【地理】　polar night
极夜急流【大气】　polar night jet （stream）, night jet
极夜涡旋　polar night vortex
极移【天】　polar motion
极羽【天】　polar plume
极雨　polar rain
极值　extremes, extreme value, extremum

极值分布　extreme value distribution
极值分析【大气】　extreme value analysis
极值风　maximum wind
极值检查　extremum check
极轴　polar axis
极昼【地理】　polar day
极坐标【数】　polar coordinates
极坐标法【航海】　polar coordinate method
即插即用操作系统【计】　plug and play operating system
即插即用【计】　plug and play
急流　jet, jet current, jet flow, jet streak
急流【大气】　jet stream
急流槽　jetstream trough
急流定位　jetstream location
急流锋　jet stream front
急流核【大气】　jet stream core
急流脊　jetstream ridge
急流卷云　jet stream cirrus
急流效应风　jet-effect wind
急流型式　jetstream pattern
急流云　jet stream cloud
急流云系【大气】　jet stream cloud system
急流轴　jet axis, jet stream axis
急流轴【大气】　axis of jet stream
急始　sudden onset
急始磁暴【地】　sudden-commencement magnetic storm, sudden commencement〔geomagnetic〕
急性污染　acute pollution
疾风【大气】　near gale
棘面气球　Jimsphere
集成电路　integrated circuit（IC）
集成电路温度传感器　IC temperature transducer
集成平均【数】　consensus average
集成平均〔法〕　consensus averaging
集成信息系统【地信】　integrated information system
集成预报　consensus forecast
集函数【数】　set-function

集〔合〕【数】　set
集合卡尔曼滤波　ensemble Kalman filter（EnKF）
集合扩展, 集合扩散　spread of the ensemble
集合平均【大气】　ensemble average
集合预报【大气】　ensemble forecast
集合预报系统　ensemble prediction system（EPS）
集水廊道　infiltration gallery
集（吸）尘器　dust collector
集雨量　catch
集约效应　collective effect
集中　pooling
集中趋势（统计）　central tendency
集中时间　time of concentration
集总化学机制　lumped chemical mechanism
瘠地　barrens
瘠土【土壤】　infertile soil
几何地平　geometrical horizon
几何高度　geometric height
几何光学　geometrical optics
几何光学区　geometric optical region
几何截面　geometric cross section
几何扩散　geometrical spreading
几何米　geometric meter
几何平均　geometric mean
几何声学　geometrical acoustics
几何衰减【农】　geometrical attenuation
几何校正【农】　geometric correction
几何效应【化】　geometric effect
几何学　geometry
几何组合原理　principle of geometric association
几内亚〔暖〕海流　Guinea Current
脊点　ridge point
脊生　ridging
脊线　axis of ridge
脊线【大气】　ridge line
脊状卷云　cirrus vertebratus

脊状(云) vertebratus (ve)
计尘器 dust counter, konimeter, conimeter
计尘仪 coniscope, koniscope
计核器 nuclei counter, kern counter, nucleus counter
pH 计【化】 pH meter
计时【地质】 age dating
计数 counts
计数风杯风速表 cup-counter anemometer
计数风速表 counting anemometer
计数器 counter
计数区 count area
计数效率 counting efficiency
计算 calculation
计算不稳定【大气】 computational instability
计算尺法 slide rule method
计算方法【数】 computing method
计算机【计】 computer
计算机安全【计】 computer security
计算机病毒【计】 computer virus
计算机病毒对抗【计】 computer virus counter-measure
计算机产品 computer product
计算机程序【计】 computer program
计算机程序确认【计】 computer program validation
计算机程序验证【计】 computer program verification
计算机存储器 computer storage
计算机辅助设计【计】 computer-aided design(CAD)
计算机工程【计】 computer engineering
计算机技术【计】 computer technology
计算机科学【计】 computer science
计算机可靠性【计】 computer reliability
〔计算机〕控制 cybernation
计算机控制【计】 computer control
计算机软件【计】 computer software
计算机输出缩微胶片 computer-output microfilm(COM)
计算机输出缩微摄影机 computer-output microfilmer
计算机文字预报 computer worded forecast(CWF)
计算机系统【计】 computer system
计算机应用【计】 computer application
计算机硬件【计】 computer hardware
计算机语言【计】 computer language
计算机指令 computer instruction
计算机制图【计】 computer draft
计算机制作的终点预报 computer-formatted terminal forecast (CFFT)
计算技术【计】 computing technology
计算模〔态〕【大气】 computational mode
计算频散 computational dispersion
计算区域 computational domain
计算稳定性 computational stability
计算稳定性判据 computational stability criterion
记录笔 register pen
记录长度 length of record
记录杆 recording lever, recording arm
记录观测 recording observation
记录气球 registering balloon
记录器 recorder
记录缺测 record with data-missing
记录日界 record data line
记录时期 period of record
记录筒 recording drum, recording cylinder
记录员 registrar
技巧〔评〕分【大气】 skill score
技术 technique
S-O 技术 Scofield-Oliver technique
δ-E 技术 delta Eddington
技术报告 technical report
技术规范 technical regulation
技术进展报告 technical progress report
技术评定 technology assessment

技术数据　technical data
技术委员会　technical commission
技术文稿　technical note（TN）
技术援助发展计划署（联合国）　Expanded Programme of Technical Assistance（EPTA,UN）
季风　monsoon wind
季风【大气】　monsoon
季风爆发　burst of monsoon
季风爆发【大气】　monsoon burst
〔季风〕爆发涡旋　onset vortex
季风槽【大气】　monsoon trough
季风潮【大气】　monsoon surge
季风带　monsoon zone
季风低压【大气】　monsoon depression
季风动力学　monsoon dynamics
季风后退　monsoon retreat
季风环流【大气】　monsoon circulation
季风汇合线　monsoon convergence line
季风活跃　active monsoon
季风活跃期　monsoon active period
季风季节　monsoon season
季风间歇　monsoon lull
季风建立【大气】　monsoon onset
季风开始　onset of monsoon
季风流管　monsoon circulation tube
季风脉冲　monsoon pulse
季风气候【大气】　monsoon climate
季风气流　monsoon air
季风气象学　monsoon meteorology
季风气旋　monsoon cyclone
季风前移　monsoon advance
季风区【大气】,季风气候区【大气】　monsoon region
季风扰动　monsoon disturbance
季风弱低压　monsoon low
季风试验　MONEX（Monsoon Experiment）
季风试验（印度）　Monsoon Experiment（India）（MONEX）
季风天气　monsoon weather
季风推进　progression of the monsoon
季风涡旋　monsoon gyre
季风雾　monsoon fog
季风型　monsoon regime
季风洋流　monsoon current
季风雨【大气】　monsoon rain
季风雨量　monsoon rainfall
季〔风〕雨林【地理】　monsoon forest
季风雨林气候　monsoon rainforest climate
季风云团　monsoon cluster
季风云团【大气】　monsoon cloud cluster
季风指数【大气】　monsoon index
季风中断　monsoon break
季风中断【大气】　break monsoon
季风转换　reversal of the monsoon
季候泥　varves
季节【大气】　season
季节变化　seasonal change
季节病　seasonal disease
季节成因参数　seasonal genesis parameter
季节风【大气】　anniversary wind
季节调整　seasonal adjustment
季节过渡　seasonal transition
季〔节〕际变率　interseasonal variability
季节降水特征　precipitation regime
季节可预报性　seasonal predictability
季节跨度　seasonal time span
季〔节〕内〔的〕　intraseasonal
季〔节〕内低频变率　intraseasonal low-frequency variability
季〔节〕内时间尺度　intraseasonal time scale
季〔节〕内振荡　intraseasonal oscillation
季节天气预报　seasonal weather forecasting
季节性　seasonal character
季节性【大气】　seasonality
季节性变化　seasonal variation
季节性风　seasonal wind
季节性干旱　seasonal drought

季节性海流　seasonal current
季节性河流　ephemeral stream
季节性湖泊　seasonal lake, ephemeral lake
季节性天气　seasonal weather
季节性温跃层　seasonal thermocline
季节性滞后　seasonal lag
季节预报【大气】　seasonal forecast
季节主趋势　master seasonal trend
季中月　mid-season month
剂量　dose
寄存器【计】　register
寄生波　parasitic wave
寄生回波【电子】　parasitic echo
寄生频率　parasitic frequency
寄生褶皱【地质】　parasitic fold
继发性污染物【大气】　secondary pollutant
加成反应【化】　addition reaction
加德纳计数器　Gardner counter
加点观测　local extra-observation
加法【数】　addition
加勒比海流　Caribbean Current
加勒比气象理事会　Caribbean Meteorological Council (CMC)
加勒比气象组织　Caribbean Meteorological Organization (CMO)
加勒特-芒克谱　Garrett-Munk spectrum
加密【电子】　encipherning
加密观测　intensive observation
加密观测期　intensive observing period (IOP)
加拿大气象局　Canadian Meteorological Service (Can Met Ser)
加拿大气象与海洋学会　Canadian Meteorological and Oceanographic Society (CMOS)
加拿大气象中心　Canadian Meteorological Centre (CMC)
加拿大遥感咨询委员会　Canadian Advisory Committee on Remote Sensing (CACRS)
加拿大硬度器　Canadian hardness-gauge
加那利海流　Canary Current
加强　intensification
加权　weighing
加权函数　weighing function
加权平均　weighted average, weighed average, weighted mean
加权平均【物】　weighed mean
加权谱　weighted spectrum
加权余量法【数】　weighted residual method
加权误差【数】　weighted error
加权因子　weighing factor
加权最小二乘法【数】　method of weighted least squares
加热〔过程〕　heating (process)
加热函数　heating function
加热率　heating rate
加热式空气干湿表　heat(ed) air psychrometer
加热水倍率【航海】　heating water ratio
加热引起的　heat-induced
加热蒸汽【航海】　heating steam
加热致冷器　thermopsych
加深　deepening
加深阶段　deepening stage
加深气旋　deepening cyclone
加速比　speed-up ratio
加速度　acceleration
加速度表　accelerometer
加速度谱　acceleration spectrum
加速度势　acceleration potential
加速风　speed-up wind
加速高度　speed-up height
加速器　accelerator
加速蚀损〔作用〕　accelerated erosion
加速运动　accelerated motion
加温干燥【农】　heat drying
加性噪声　additive noise
加压溶解气体【航海】　gases dissolved

under pressure
〔加〕压融〔化〕 pressure melting
夹层结构 sandwich
夹卷【大气】 entrainment
夹卷率【大气】 entrainment rate
夹卷区〔域〕 entrainment zone
夹卷速度 entrainment velocity
夹卷系数 entrainment coefficient
家禽受热指数 poultry stress index (PSI)
家畜安全指数 livestock safety index(LSI)
家畜气象学 livestock meteorology
镓 gallium
荚状层积云 Sc lent
荚状层积云【大气】 stratocumulus lenticularis (Sc lent)
荚状层云 stratus lenticularis
荚状高层云 altostratus lenticularis
荚状高积云【大气】 altocumulus lenticularis(Ac lent)
荚状积云 cumulus-lenticularis
荚状卷积云 cirrocumulus lenticularies
荚状（云） lenticularis(len)
荚状云 lenticular cloud
荚状云带 lenticular cloud band
甲板风 wind-over-deck
甲醇(CH_3OH) methyl alcohol, methanol
甲醇-水制氢设备 methanol-water hydrogen generator
甲基 methyl
甲基碘 methyl iodide
甲基磺酸(CH_3SO_3H) methanesulfonic acid
甲基氯(CH_3Cl) methyl chloride
甲基氯仿（CH_3CCl_3） methyl chloroform
甲基溴(CH_3Br) methyl bromide
甲基·乙烯基(甲)酮,丁烯酮($CH_2:CH·CO·CH_3$) methylvinyl ketone
甲醛 formaldehyde

甲烷 methane
甲烷细菌【土壤】 methane bacteria
钾 potassium
钾-氩定年法 potassium-argon dating
假白虹 false white rainbow
假边缘 artificial edge
假彩色 false color
假彩色图像 false color image
假彩色云图【大气】 false-color cloud picture
假潮【海洋】 seiche
假动量 pseudomomentum
假反射 false reflection
假锋 pseudo front
假回波【航海】 false echo
假绝热〔的〕 pseudo-adiabatic
假绝热对流 pseudo-adiabatic convection
假绝热过程【大气】 pseudo-adiabatic process
假绝热膨胀 pseudo-adiabatic expansion
假绝热图【大气】 pseudo-adiabatic diagram
假绝热图 pseudo-adiabatic chart
假绝热〔线〕 pseudo-adiabat
假绝热直减率【大气】 pseudo-adiabatic lapse rate
假扩散 pseudo diffusion
假冷锋 pseudo cold-front
假暖区 false warm sector
假谱方法【大气】 pseudospectral method
假谱模式 pseudo-spectral model
假潜不稳定 pseudo-latent instability
假日【大气】 parhelion（复数为 parhelia)
假日环 white horizontal circle
假设 assumption
假设检验【化】 hypothesis test
假湿球位温 pseudo-wet-bulb potential temperature, potential pseudo-wet-bulb temperature
假湿球位温【大气】 wet-bulb pseudo-

假湿球温度 wet-bulb pseudo-temperature【大气】, pseudo-wet-bulb temperature
假说 hypothesis
假位温 pseudo-potential temperature
假位涡度方程 pseudo potential vorticity equation
假雾 mock fog
假相当位温 potential pseudo-equivalent temperature
假相当位温【大气】 pseudo-equivalent potential temperature
假相当温度 pseudo-equivalent temperature
假月【大气】 paraselene
尖点 cusp
尖端放电【大气】 point discharge
尖端放电电流 point discharge current
尖峰 spike
尖峰态 leptokurtosis
尖角括号 angle bracket
尖劈现象 spiky phenomenon
尖头信号 blip
坚雪 snowcrete, snow concrete
间苯三酚 phloroglucinol
间冰阶 interstadial period, interstade
间冰期【大气】 interglacial period (或 phase)
间冰期【地质】 interglacial stage
间冰期〔的〕 interglacial
间冰期状况 interglacial condition
间隔 spacing, interval, compartment
间隔带【地质】 interval zone
间接插入 indirect insertion
间接法 indirect method
间接高空分析 indirect aerological analysis
间接高空学 indirect aerology
间接环流 indirect cell
间接环流〔圈〕【大气】 indirect circulation
间接静电接地 indirect static earthing
间接气溶胶效应 indirect aerosol effect
间距 pitch
间距式探空仪 chronometric radiosonde
间隙水 interstitial water
间向风 intercardinal wind
间歇混沌 intermittent chaos
间歇〔喷〕泉 geyser
间歇湍流 intermittent turbulence
间歇现象 intermittency
间歇性大雨 intermittent heavy rain
间歇〔性〕河〔流〕 intermittent stream
间歇性降水【大气】 intermittent precipitation
间歇性小雨 intermittent light rain
间歇〔性〕雨 intermittent rain
间歇性中雨 intermittent moderate rain
间歇振荡器 blocking oscillator
间歇〔资料〕同化 intermittent (data) assimilation
间雨期 interpluvial
监测 monitoring
监测车 monitoring car
监测船 monitoring ship
监测井 monitor well
监测网 monitoring network, monitoring net
监测系统 monitoring system
监护病室【电子】 intensive care unit (ICU)
监视 monitor
兼容计算机【计】 compatible computer
检查 inspection
检定【大气】 calibration
检定槽 rating flume
检定池 calibration tank
检定规程 regulation of verification
检索 referral
检验 check, validation
t 检验 t-test, Student's t-test

检验分析　inspectional analysis
F 检验【农】　F test
χ^2 检验【数】　chi-square test
剪刀效应　scissors effect
剪应力【物】　shear stress
减法【数】　subtraction
减洪　flood mitigation
减活作用　deactivation
减轻　mitigation
减少　decrease
减湿〔作用〕　dehumidification
减速　retardation, decelerate
减速流　retarded flow
减速效应　retarding effect
减退　abatement
减温器【航海】　desuperheater
减噪　noise reduction
简波　simple wave
简单排队法　first come-first-serve（FCFS）
简单平均　simple average
简单气候模式　simple climate model
简单线性相关　simple linear correlation
简化方程【数】　reduced equation
简略船舶电码　abbreviated ship code
简涡　simple vortex
简谐波【物】　simple harmonic wave
简谐运动型　simple harmonic motion type
简易的　extremely simple
简易气象观测站　extremely simple meteorological observing station（MOSES）
简约频率【力】　reduced frequency
碱【化】　base
碱度【化】　alkalinity
碱性尘雾　alkali fume
碱性土【土壤】　alkaline soil
建立【计】　setup
建立时间【计】　setup time
建筑气候【大气】　building climate
建筑气候区划【大气】　building climate demarcation
建筑气候学　building climatology

建筑气象学　architectural meteorology
建筑日照　building sunshine
建筑物风吹雨　driving rain on buildings
健康区域制　healthy regionalism
舰队数值海洋学中心（美国海军）　Fleet Numerical Oceanography Center（FNOC）
舰队数值天气中心（美国海军）　Fleet Numerical Weather Central（FNWC）
舰队天气中心（美国海军）　Fleet Weather Central（FWC）
渐近法　method of successive
渐近分析　asymptotic analysis
渐近解　asymptotic solution
渐近理论　asymptotic theory
渐近流速　velocity of approach
渐近线【数】　asymptote
渐近展开　asymptotic expansion
渐近值【数】　asymptotic value
渐新世【地质】　Oligocene Epoch
渐新统【地质】　Oligocene Series
溅落　splashdown
溅蚀【土壤】　splash erosion
鉴定【计】　qualification
鉴定需求【计】　qualification requirements
鉴频【电子】　frequency discrimination
鉴频器【电子】　frequency discriminator
键价【化】　bond valence
键结构【物】　bond structure
键控信息速率处理器　manipulated information rate processor（MIRP）
键能【物】　bond energy
键盘【计】　keyboard
键入　keying
箭载光达系统　rocket-borne laser radar system
箭载光度法臭氧与温度探空仪　rocket-borne photometric ozone and temperature measuring sonde
箭载探测器　rocket-borne detector
江淮气旋【大气】　Changjiang-Huaihe

cyclone
江淮切变线【大气】 Changjiang-Huaihe shear line
降雹 hail shooting, hailfall
降尘【大气】 dustfall
降尺度 descaling, downscaling
降尺度方向（尺度减小） downscale direction
降风形成时间 gust formation time
降交点【天】 descending node
降阶法【数】 method of reduction of order
降解【化】 degradation
降落机场预报 terminal forecast
降水【大气】 precipitation
降水百分率 isomeric value
降水变率 precipitation variation
降水持续时间【大气】 precipitation duration
降水大陆度 hygrometric continentality
降水带 precipitation band
降水的化学成分 chemical composition of precipitation
降水电流 precipitation current
降水电学 precipitation electricity, electricity of precipitation
降水定量预报 quantitative precipitation forecast(QPF)
降水关联的电场极性反转 field excursion associated with precipitation (FEAWP)
降水轨迹 precipitation trajectory
降水过程 precipitation process
降水化学【大气】 precipitation chemistry
降水回波【大气】 precipitation echo
降水机制 precipitation mechanism
降水积云模式 precipitation cumulus model
降水记录器 precipitation recorder
降水检测器 precipitation detector
降水"阶梯" precipitation "staircase"

降水截留 interception of precipitation
降水静电 precipitation statics
降水〔类〕型〔出现〕概率 probability of precipitation type(PoPT)
降水粒子谱仪 precipitation particle spectrometer(PPS)
降水粒子谱仪探头 precipitation spectrometer probe(PSP)
降水量【大气】 amount of precipitation
降水〔量〕距平 precipitation anomalies
降水〔量〕距平百分率 percentage of precipitation anomalies
降水量图【大气】 precipitation chart
降水量学 hyetology
降水率 rate of precipitation, precipitation rate
降水逆减【大气】 precipitation inversion
降水频率图 precipitation rose
降水气候 hygroclimate
降水强度【大气】 precipitation intensity
降水强度器 rain intensity gauge
降水清除 precipitation scavenging
降水区 precipitation area
降水日【大气】 precipitation day
〔降水〕深度-面积曲线 depth-area-curve
降水生成单体 precipitation-generating element
〔降水〕时-深-面值 depth-duration-area value(DDA)
降水事件结束 ending of precipitation event
降水收集器 precipitation collector
降水收支 precipitation budget
降水衰减【大气】 precipitation attenuation
降水酸度【大气】 precipitation acidity
降水通量 precipitation flux
降水物理学【大气】 precipitation physics
降水效率 precipitation efficiency, effectiveness of precipitation
降水性对流 precipitation convection

降水性高层云　altostratus precipitus
降水性云　precipitus
降水性(云)　praecipitatio(pra)
降水因子　rainfall factor, rain factor
降水有效度比　precipitation effectiveness ratio
降水有效性　precipitation effectiveness
降水-有效性指数　precipitation-effectiveness index (P-E index)
降水有效指数　precipitation effectiveness index
降水元　precipitation element
降水云幕　precipitation ceiling
降水-蒸发比　precipitation evaporation (PE) ratio
降水蒸发指数　precipitation evaporation index
降水指数　precipitation index
降水中心　precipitation cell
降水骤增区　precipitation surge
降维方法　dimensionality reduction method
降温率　detemperature rate
降雪层　snow generating level
降雪线　snow precipitation line
降压〔的〕　katabaric, katallobaric
降压中心　pressure-fall centre, center of falls
降雨持续时间【农】　rainfall duration
降雨的　hyetal
降雨径流关系【水利】　rainfall runoff relation
降雨累计曲线　mass curve of rainfall
降雨量可靠性　rainfall reliability
降雨率　rainfall rate
降雨频率　rain frequency, rainfall frequency
降雨潜势　rainfall potential
〔降〕雨强〔度〕　rain intensity
降雨强度【水利】　rainfall intensity
降雨区　rainfall area
降雨时数　rain hour, rainfall hour
降雨效率　rainfall effectiveness
降雨信息　rainfall information
降雨装置【机械】　rainer
较老仙女木事件　Older Dryas
交叉　cross-over
交叉波束　crossed beam
交叉定年法　cross dating
交叉风暴逗点云系　cross-over storm comma
交叉核实【数】　cross validation
交叉检验　cross check
交叉理论　crossing theory
交叉谱【大气】　cross spectrum
交叉微分法　cross-differentiation
交叉相关【大气】　cross-correlation
交叉污染　cross contamination
交叉〔型〕实验　crossover experiment
交叉总指数　cross totals index
交错单位张量　alternating unit tensor
交错格式【大气】　stagger scheme
交错〔混合〕　interleaving
交错网格【大气】　staggered grid, stagged mesh
交错张量　alternating tensor
交点经距　nodal longitudinal distance
交点西退率　nodal regression rate
交点线【天】　nodal line
交点月　nodical month
交点周期　nodal period
交互式彩色雷达显示　interactive color radar display(ICRAD)
交互式处理【计】　interactive processing
交互式数字图像处理系统　interactive digital image manipulation system (IDIMS)
交互式图像处理　interactive image processing
交互式研究成像系统　interactive research imaging system (IRIS)
交互式预报系统　interactive forecast system

交互式终端接口　interactive terminal interface
交换　exchange
交换量　exchange capacity
交换吸收　exchange absorption
交换系数　interchange coefficient, exchange coefficient, coefficient of exchange, austausch coefficient
交换中心　exchange center
交流电　alternating current
交替方向法　alternate direction method
交通地理学【地理】　geography of communication
交涡旋　crossed vortex
胶膜湿度表　cellophane hygrometer
胶态弥散　colloidal dispersion
胶体【化】　colloid
胶体分散【大气】　calloidal dispersion
胶体化学【化】　colloid chemistry
胶体不稳定性【大气】　colloidal instability
胶体溶液　colloidal solution
胶体稳定的　colloidally stable
胶体稳定〔度〕　colloidal stability
胶体系统【大气】　colloidal system
胶体悬浮　colloidal suspension
胶体亚平衡　colloidal metastable
焦点　focus
焦〔耳〕　joule(J)
焦耳常数　Joule's constant
焦耳定律【物】　Joule law
焦耳热【物】　Joule heat
焦距　focal length
角波数　azimuthal wavenumber, angular wavenumber, angular wave number
角程长度　angular path length
角动量　moment of momentum
角动量【物】【大气】　angular momentum
角动量常定　constancy of angular momentum
角动量平衡【大气】　angular momentum balance
角动量平面　angular momentum plane
角动量守恒【大气】　conservation of angular momentum
角度传感器【航海】　angular position sensor
角度谱　angular spectrum
角度校正　angularity correction
角反射器（测风雷达用）【大气】　corner reflector
角分辨率【物】【大气】　angular resolution
角分布【物】　angular distribution
角幅　angular width
角回旋频率　angular gyro-frequency
角加速度【物】　angular acceleration
角量子数【物】　azimuthal quantum number
角滤波函数　angular filter function
角偏振　angular polarization
角频率【物】　angular frequency
角散度　angular divergence
角速度【物】　angular velocity
角位移【物】　angular displacement
角系数　angular coefficient
角型　angular pattern
角展宽【大气】　angular spreading
角展〔宽〕因子　angular spreading factor
角质风向盘　horn card
搅水式绝对日射表　water-stir pyrheliometer
校验观测　check observation
校正　updating
校正透镜【物】　correcting lens
校钟星　clock-star
校准　calibration【大气】, adjustment
校准器　calibrater（或 calibrator）
校准曲线【大气】　calibration curve
阶乘【数】　factorial
阶梯反应函数　step response function

阶梯函数　stepped function
阶梯函数【数】　step function
阶梯函数模式　step function model
阶梯曲线　step curve
接触　contact
接触电势差【物】　contact potential difference
接触核　contact nucleus(contact nuclei)
接触角　contact angle
接触冷却　contact cooling
接地带【电工】　earth stripe
接地导体　earthing conduct
接口标准　interface standard
接口电路板　interface card
接口模板　interface module
接纳体【生化】　acceptor
接收　acquisition
接收范围　acquisition range
接收功率　received power
接收机　receiver
接收机〔噪声〕温度【大气】　receiver 〔noise〕temperature
接收机增益　receiver gain
接收截面　receiving cross section
接收天线　receiving antenna, receiving aerial
接受区〔域〕　acceptance region
节点【物】　node
节点因子　nodal factors
节理【地质】　joint
节理状况　jointing
节气【天】　solar term
节时飞行　minimal flight
节时航线　minimal flight path, minimum-time track, least-time track
节时航向　minimal headings
杰文斯效应　Jevons effect
结冰　ice-up
结冰高度图　freezing-level chart
结冰条件　ice-formation condition
结构常数　structure constant
结构函数【大气】　structure function
结构稳定性　structural stability
结构系数　structure coefficient
结构〔性〕对称【物】　structural symmetry
结合能【物】　binding energy
结晶　crystallization
结晶度【化】　crystallinity
结晶核　crystallization nucleus
结晶热　heat of crystallization
结露　dewing, dewfall
结霜　frosting
结凇　riming
截断常态分布　truncated normal distribution
截断分布　truncated distribution
截断误差【数】【大气】　truncation error
截光换能器　light barrier transducer
截距　nodal increment
截留地形波　arrested topographic wave
截谱　truncated spectrum
截谱模式　truncated spectral model
截止波长【电子】　cut-off wavelength
截止波长【物】　cutoff wavelength
截止电压【电子】　cut-off voltage
截止角频率　cut-off angular frequency
截止频率　cut-off frequency
截止期　dead-line
截止时间　cut-off time
解调　demodulation
解冻　mollition, depergelation, ice breakup, debacle, breakup
解冻【农】【大气】　thaw
解冻季节　thawing season, breakup season
解冻泥流　solifluction
解冻水　thaw-water
解码【计】　decoding
解码器【计】　decoder
解耦【数】　decoupling
解释方差　explained variance
解脱器　disconnection device
解析泛函【数】　analytic functional

解析函数【数】 analytic function
解析解 analytical solution
解析理论 analytical theory
解析误差 analytical error
解析延拓【数】 analytic continuation
解消假设 null hypothesis
介电常数 dielectric constant, permittivity
介电函数 dielectric function
介电强度【物】 dielectric strength
介电损耗 dielectric loss
介电梯度 dielectric gradient
介电因子 dielectric factor
介电质吸收 dielectric absorption
介观结构【物】 mesostructure, mesoscopic structure
介质 medium
介质隔离【电子】 dielectric isolation
介质击穿【电子】 dielectric breakdown
介子【物】 meson
界面 boundary surface
界面【计】 interface
界面波 interfacial wave
界面交换过程【海洋】 interface exchange process
界面张力 interfacial tension
界线 demarcation line
金星【天】 Venus
金星大气 Venus atmosphere
金属空盒 aneroid capsule
金属气压表 metallic barometer
金属丝 chaff
金属丝撒播 chaff seeding
金属温度表 metallic thermometer
金属-氧化物-半导体结构【物】 metal-oxide-semiconductor structure
金属氧化物压敏电阻 metal oxide varistor
津轻暖流 Tsugaru Warm Current
紧差分 compact differencing
紧差分格式【数】 compact difference scheme

紧急警戒站网 emergency alerting network
进潮量【海洋】 tidal prism
进动参数 precession parameter
进动角【物】 angle of precession
进动率 precession rate
进化论【植】 evolutionary theory
进化生态学【动物】 evolutionary ecology
进气口积冰 induction icing
〔进入稠密〕大气层飞行 atmospheric penetration
近滨带【海洋】 nearshore zone
近场区【电子】 near-field region
近场现象 near field phenomenon
近赤道槽 near-equatorial trough
近赤道急流 near-equatorial jet
近赤道脊 near-equatorial ridge
（近堤）破波 breaking wave
近地层 ground layer
近地层【大气】 surface layer
近地点【天】【大气】 perigee
近地点潮差 perigean range
近地面层速度尺度 surface layer velocity scale
近地面层温度尺度 surface layer temperature scale
近点周期 anomalistic period
近关键期 near-critical period
近海测量【航海】 offshore survey
近海浮标观测系统 off-shore buoy observing equipment（OBOE）
近海航行【航海】 offshore navigation
近海面参考温度 near surface reference temperature
近海区【海洋】 neritic zone
近红外【物】 near-infrared
近红外辐射 near-infrared radiation
近幻日 sun dog
近极〔地〕轨道卫星 near-polar-orbiting satellite
近极地(准极地)太阳同步轨道 near（或

quasi) polar sun-synchronous orbit
近零天气　sub-zero weather
近日点【天】【大气】　perihelion
近实时　near-real-time
近似表示【数】　approximate representation
近似解【数】　approximate solution
近似绝对温标　tercentesimal thermometric scale, approximate absolute temperature scale
近似值【数】　approximate value
近稳磁层顶　near-stationary magnetopause
近心点　pericenter
近月潮　perigean tide
劲度【地】　stiffness
浸和方法　encroachment method
浸水活化电池　water-activated battery
禁带【物】　forbidden band
禁飞天气【大气】　unflyable weather
禁航天气　weather-bound, X-weather, weather below minimum
禁火三角形【石油】　fire triangle
禁戒跃迁【物】　forbidden transition
禁区　forbidden region
禁线　forbidden line
京都议定书(1997年通过)　Kyoto Protocol
经典解【数】　classical solution
经典物理〔学〕【物】　classical physics
经典相干理论　classical coherence theory
经度　longitude
经度效应　longitude effect
经济地理学【地理】　economic geography
经济费用　economic cost
经济合作与发展组织　Organization for Economic Cooperation and Development (OECD)
经济决策　economic decision
经济评价　economic valuation
经济社会理事会(联合国)　Economic and Social Council (ECOSOC, UN)
经济效益　economic utility, economic effect
经济影响　economic impact
经济原则　economic principle
经济状况　economic conditions
经圈锋　meridional front
经圈廓线　meridional profile
经纬度网格　spherical grid, latitude-longitude grid
经纬仪【测】　transit, theodolite
经向变化　latitudinal variation
经向〔的〕　meridional
经向度　meridionality
经向翻转环流　meridional overturn circulation (MOC)
经向风　meridional wind
经向环流　meridional cell
经向环流【大气】　meridional circulation
经向交换　meridional exchange
经向剖面　meridional cross section
经向气流　meridional flow
经向梯度　meridional gradient
经向指数　meridional index
经验表达式　empirical expression
经验常数　empirical constant
经验重现期　empirical return periods
经验风险最小化　empirical risk minimization
经验公式【数】　empirical formula
经验关系　empirical relationship
经验洪水公式　empirical flood formula
经验判别式函数　empirical discriminant function
经验区域相关　empirical regional relation
经验曲线【数】　empirical curve
经验数据　empirical data
经验特征向量　empirical eigenvector
经验系数　empirical coefficient
经验预报【大气】　subjective forecast
经验正交函数【大气】　empirical

orthogonal function (EOF)
惊蛰【农】【大气】(节气) Awakening from Hibernation
晶格　lattice
晶格【物】　crystal lattice
晶化现象　sintering phenomenon
晶棱【物】　crystal edge
晶面【物】　crystal face, crystallographicc plane
晶体　crystal
晶体定向生长　oriented overgrowth (=epitaxis)
晶体管热敏电阻　transistor thermal resistance
晶体管【物】　transistor
晶体结构【物】　crystal structure
晶体霜　crystalline frost
晶体习性　crystal habit
晶形【物】　crystal form
晶状〔的〕　crystalline
晶状体【物】　crystalline lens
腈【化】　nitrile
精密度【物】【大气】　precision
精密高度表　precision altimeter
精密光学〔系统〕【物】　precision optics
精密经纬仪　precision theodolite
精密空盒气压表　precision aneroid barometer
精确解【数】　exact solution
精细边界层　high resolution PBL
精细结构【物】　fine structure
景观　landscape
景观地球化学【地理】　landscape geochemistry
景观气候学　landscape climatology
景观生态学【地理】　landscape ecologyg
景观学【地理】　landscape science
警戒区　critical area
警戒区域　alerting zone
径流　run-off, flow-off
径流比　runoff ratio
径流深【水利】　depth of runoff
径流图　runoff plot
径流系数　runoff coefficient
径流循环　runoff cycle
径向波数　radial wavenumber
径向对称　radial symmetry
径向风【大气】　radial wind
径向加速度【物】　radial acceleration
径向雷诺数　radial Reynolds number
径向量【数】　radius vector
径向流入【大气】　radial inflow
径向内流　radial inflow
径向谱　radial spectrum
径向速度【物】　radial velocity
净暴雨量　net storm rain
净初级生产量　net primary production
净地面辐射　net terrestrial radiation
净辐射【大气】　net radiation
净辐射表　net pyranometer【大气】, net radiometer, net exchange radiometer
净辐射平衡　net radiation balance
净荷【计】　payload
净化〔作用〕　purification
净举力　free lifting force, free lift
净流〔量〕　net flow
净流域供水　net basin supply
净平衡　net balance
净射出红外辐射　net outgoing IR, net terrestrial radiation
净太阳辐射　net solar radiation
净天空辐射表　net pyranometer
净源　net source
静电　static electricity
静电并合　electrostatic coalescence
静电场【物】　electrostatic field
静电沉降器　electrostatic precipitator
静电感应【物】　electrostatic induction
静电荷　electrostatic charge
静电计　electrostatic electrometer
静电计【物】　electrometer
静电容　electrostatic capacity

静电式气溶胶取样器　electrostatic aerosol sampler
静风【大气】　calm
静风区　wind shadow
静海　glassy
静力播撒云　static seeding
静力不稳定参数　static instability parameter
静力不稳定〔性〕【大气】　static instability
静力初值化【大气】　static initialization
静力过程　static process
静力检查【大气】　hydrostatic check
静力能　static energy
静力平衡　static equilibrium
静力平衡方程　equation of static equilibrium
静力气象学　statical meteorology
静力适应过程【大气】　hydrostatic adjustment process
静力水位　static level
静力稳定度【大气】　static stability
静力学【物】　statics
静摩擦系数【物】　coefficient of static friction
静区　zone of silence
静水井　still well
静水器　still pond
静水位　still-water level
静水压面　potentiometric surface, piezometric surface
静态方法　static method
静态分岔　static bifurcation
静态分析【计】　static analysis
静稳逆温污染　calm inversion pollution
静压孔　static vent, static pressure port
静压头【航海】　static head
静压【物】　static pressure
静夜　calm night
静止参系　rest frame
静止地下水　stagnant ground water
静止锋【大气】　stationary front
静止气旋　stationary cyclone
静止卫星　stationary satellite
静质量【物】　rest mass
镜点　mirror point
镜反射　specular reflection
镜反射体　specular reflector
镜角　pitch-angle
镜〔面〕反射【物】　mirror reflection
镜像法【物】　method of images
九灯风向风速仪　nine-light indicator
酒精〔玻璃〕温度表　spirit-in-glass thermometer
酒精温度表　alcohol thermometer, spirit thermometer, thermo-alcoholometer, alcohol-in-glass thermometer
旧石器时代　paleolith
救生圈【航海】　lifebuoy
救生艇【航海】　lifeboat
居间大气层　intervening atmosphere
居间媒质　intervening medium
居里（放射性的旧单位，$=3.7 \times 10^{10}$ Bq）　curie（Ci）
居里点【物】　Curie point
居里温度　Neel temperature, Curie temperature
局部　local
局部比例尺【航海】　local scale
局部闭合　local closure
局部变化　local variation
局部导数【大气】　local derivative
局部地平　local horizon
局部分析　local analysis
局部各向同性　local isotropy
局部加热　local heating
局部加速度　local acceleration
局部角动量　local angular
局部均匀介质　locally homogeneous medium
局部均匀随机函数　locally homogeneous random function
局部均匀湍流　locally homogeneous

turbulence
局部落尘　local fallout
局部马赫数　local Mach number
局部拟合　local fitting
局部齐性空间　locally homogeneous space
局部〔强〕风暴　local（severe）storm
局部〔强〕雷雨　local（severe）thunderstorm
局部区域范围　local area coverage（LAC）
局部入流　local inflow
局部收敛【数】　local convergence
局部温度自记曲线　topothermogram
局部涡度　local vorticity
局部相似假说　local similarity hypothesis
局部效应　local effect
局部域【数】　local field
局部自相关图　sectional autocorrelogram
局部坐标　local coordinates
局地变化　local change
局地产生的风浪　locally generated sea
局地存活期　local lifetime
局地冻结条件　locally frozen condition
局地分析和预报系统（NOAA 开发的）　local analysis and prediction system（LAPS）
局地各向同性〔湍流〕　local isotropic（turbulence）
局地环流【大气】　local circulation
局地角动量　local angular momentum
局地警戒〔天气〕雷达　local warning radar
〔局地〕内水分循环　internal water circulation
局地谱　local spectrum
局地气候【大气】　local climate
局地气候学　local climatology
局地强风暴　severe local storm
局地强风暴超〔级〕单体　supercell severe local storm

局地强风暴单体　single cell severe local storm
局地强风暴复合体　multicell severe local storm
局地强风暴预报中心（美国）　Severe Local Storm Forecast Center（SLSFC）
局地热力平衡　local thermodynamic equilibrium（LTE）
局地天气　local-scale weather
局地稳定性　local stability
局地预报【大气】　local forecast
局地轴【大气】　local axis
局地自由对流　local free convection
局地自由对流相似性　local free-convection similarity
局地作用　local action
局域不稳定性【物】　local instability
局域平衡【物】　local equilibrium
局域网【计】　local area network（LAN）
局域相似　local similarity
矩　moment
矩方程　moment equation
矩量法　moment method, method of moments
矩心　centroid
矩形波串窗　boxcar window
矩形波串积分法　boxcar integration method
矩形波串积分器　boxcar integrator
矩形波导　rectangular waveguide
矩形波函数　boxcar function
矩形函数　rectangular function
矩阵【数】　matrix
矩阵的秩【数】　rank of matrix
矩阵法　matrix method
矩阵法稳定度分析　matrix method of stability analysis
矩阵求逆【数】　matrix inversion
〔矩〕阵转置　transpose of matrix
巨分子【化】　giant molecule
巨核　giant nuclei

巨浪（风浪5级） very rough sea
巨脉冲激光器【物】 giant-pluse laser
巨脉冲技术 giant pulse technique
巨椭圆星系【天】 giant elliptical galaxy
巨星【天】 giant star
巨星系【天】 giant galaxy
巨型计算机【计】 supercomputer
拒绝区域【数】 rejection region
剧烈天气征兆指数 severe weather threat index
距今 before present（B. P. 或 BP，如 250 kaBP 表示距今 25 万年前）
距离 distance
距离标 range strobe
距离标志【电子】 range marker
距离采样数【大气】 number of range samples
距离常数 distance constant
距离分辨率【大气】 range resolution
距离高度显示器 range-height indicator（RHI）【大气】，RHI scope
距离归一化 range normalization
距离混淆[现象] range aliasing
距离及距离变化率 range and range rate
距离库【大气】 range bin
距离门 range gate
距离模糊【大气】 range ambiguity
距离平均【大气】 range averaging
距离衰减【大气】 range attenuation
距离速度显示【大气】 range-velocity display（RVD）
距离选通 range gating
距离选通方位切变 gate-to-gate azimuthal shear
距离仰角显示器 range elevation indicator
距离折错 range folding
距平【大气】 departure
距平相关 anomaly correlation
飓风 hurricane wind，hurricane【大气】，haracana
飓风[暴]潮 hurricane surge

飓风潮 hurricane tide
飓风等级 SS hurricane scale
飓风跟踪 hurricane tracking
飓风核 hurricane core
飓风[回波]带 hurricane band
飓风级风[速] hurricane-force wind
飓风监视 hurricane watch
飓风监视雷达 hurricane surveillance radar
飓风浪 hurricane wave
飓风雷达[回波]带 hurricane radar band
飓风路径相似方法 hurricane analog technique(HURRAN)
飓风路径预报 hurricane path prediction
飓风名称 hurricane name
飓风前飑线 prehurricane squall line
飓风强度指数 hurricane current intensity number
飓风人工影响 hurricane modification
飓风撒播 hurricane seeding
飓风探测气球 hurricane beacon，hurricane balloon
飓风微地震 hurricane microseism
飓风消散 hurricane dissipation
飓风眼 hurricane eye
飓风雨带 hurricane rain band
飓风预警 hurricane warning
飓风预警系统 hurricane warning system
飓风源地 hurricane origin
飓风云 hurricane cloud
飓风云堤 hurricane bar
飓风灾害 hurricane fatality
飓风 T 指数 hurricane T number
锯齿波形【电子】 saw-tooth waveform
锯齿形函数 saw-tooth function
锯齿型 zigzag pattern
锯齿[状]闪[电] zigzag lightning
聚变 fusion

聚变化学【化】 fusion chemistry
聚并沉降〔作用〕 coagulating sedimentation, coagulating precipitation
聚光太阳能电池【电子】 concentrator solar cell
聚合【大气】 aggregation
聚合度【化】 degree of polymerization
聚合〔反应〕【化】 polymerization
聚合物【化】 polymer
聚合物形态学【化】 morphology of polymer
聚合物驻极体【化】 polymer electret
聚积模【大气】 accumulation mode
聚焦式日照计【大气】 campbell-stokes sunshine recorder
聚焦投影与扫描视像管 focus projection and scanning vidicon
聚焦效应 focusing effect
聚类分析【数】 cluster analysis
聚氯乙烯 polyvinyl chloride（PVC）
聚散度 vergence
聚水区 gathering ground
聚四氟乙烯【化】 poly(tetrafluoroethylene)
聚乙烯气球 polyethylene balloon
聚酯薄膜角形反射器 mylar corner reflector
聚酯纤维气球 mylar balloon
卷层云【大气】 cirrostratus（Cs）
卷出 detrainment
卷积【数】 convolution
卷积微分 derivative of convolution
卷积云【大气】 cirrocumulus（Cc）
卷浪 plunging breaker
卷毛云 woolpack cloud, fleecy cloud
卷毛云天 fleecy sky
卷曲烟流 bent-over plume
卷云【大气】 cirrus（Ci）
卷云间区 intercirrus region
卷云幔 cirrus canopy
卷云片 cirrus sheet
卷云砧 cirrus plume

卷云状〔的〕 cirriform
决策分析【数】【大气】 decision analysis
决策矩阵 decision matrix
决策气象服务〔信息〕系统 Meteorological Services Information system（MESIS）
决策算法 decision making algorithm
决策树【计】【大气】 decision tree
决策支持系统 Decision Support System（DSS）
决定论【物】 determinism
绝对变率 absolute variability
绝对标尺 absolute scale
绝对标准气压表【大气】 absolute standard barometer
绝对不稳定【大气】 absolute instability
绝对参考系【物】 absolute reference frame
绝对大气质量 absolute air mass
绝对单位 absolute unit
绝对等高线 absolute isohypse
绝对定年法 absolute age determination
绝对动量 absolute momentum
绝对动量守恒 conservation of absolute momentum
绝对辐射标尺 absolute radiation scale
绝对干旱 absolute drought
绝对高度 absolute altitude
绝对高度表 absolute altimeter
绝对黑体【大气】 absolute black body
绝对极值 absolute extremum【数】, absolute extreme【大气】
绝对加速度 absolute acceleration
绝对角动量【大气】 absolute angular momentum
绝对角动量守恒 conservation of absolute angular momentum
绝对空腔辐射计 absolute cavity radiometer
绝对零点 absolute zero
绝对年较差 absolute annual range
绝对频率 absolute frequency

绝对气压　absolute pressure
绝对日射表　absolute pyrheliometer
绝对日射表标尺　pyrheliometric scale
绝对日射表国际标尺　international pyrheliometric scale(IPS)
绝对日射测量学　pyrheliometry
绝对日射光谱仪　spectropyrheilometer
绝对日射计　pyrheliograph
绝对日射自记曲线　pyrheliogram
绝对湿度【大气】　absolute humidity
绝对视差　absolute parallax
绝对速度　absolute velocity
绝对梯度流　absolute gradient current
绝对位涡　absolute potential vorticity
绝对温标　Kelvin temperature scale, absolute temperature scale【大气】
绝对温度　Kelvin temperature, absolute temperature
绝对温度极值　absolute temperature extremes
绝对稳定【大气】　absolute stability
绝对涡度【大气】　absolute vorticity
绝对涡度守恒【大气】　conservation of absolute vorticity
绝对误差【数】　absolute error
绝对线性动量　absolute linear momentum
绝对星等【天】　absolute magnitude
绝对形势　absolute topography
绝对月最低温度【大气】　absolute monthly minimum temperature
绝对月最高温度【大气】　absolute monthly maximum temperature
绝对云幂高　absolute ceiling
绝对运动【物】　absolute motion
绝对折射率　absolute refractive index, absolute index of refraction
绝对真空　absolute vacuum
绝对真空计【电子】　absolute vacuum gauge
绝对值【数】　absolute value
绝对坐标系　absolute coordinate system

绝热饱和点　adiabatic saturation point
绝热饱和气压　adiabatic saturation pressure
绝热饱和器(绝热干湿表用)　adiabatic saturator
绝热饱和温度　adiabatic saturation temperature
绝热闭合系统　adiabatic closed system, adiabatically enclosed system
绝热变化　adiabatic change
绝热不变量　adiabatic invariant
绝热大气【大气】　adiabatic atmosphere
绝热〔的〕　adiabat
绝热定律　adiabatic law
绝热方程　adiabatic equation
绝热干湿表　adiabatic psychrometer
绝热过程【大气】　adiabatic process
绝热检验【大气】　adiabatic trial
绝热近似　adiabatic approximation
绝热冷却【大气】　adiabatic cooling
绝热率　adiabatic rate
绝热模式【大气】　adiabatic model
绝热凝结　adiabatic condensation
绝热凝结点　adiabatic condensation point
绝热凝结气压【大气】　adiabatic condensation pressure
绝热凝结温度【大气】　adiabatic condensation temperature
绝热膨胀　adiabatic expansion
绝热平衡　adiabatic equilibrium
绝热平均　adiabatic meaning
绝热区　adiabatic region
绝热曲线　adiabatics, adiabatic curve
绝热上升【大气】　adiabatic ascending
绝热湿球温度　adiabatic wet-bulb temperature
绝热试验　adiabatic trial
绝热梯度　adiabatic gradient
绝热条件　adiabatic condition
绝热图　adiabatic chart
绝热图【大气】　adiabatic diagram

绝热温度变化　adiabatic temperature change
绝热温度梯度　adiabatic temperature gradient
绝热下沉【大气】　adiabatic sinking
绝热现象　adiabatic phenomenon
绝热相当温度【大气】　adiabatic equivalent temperature
绝热效应　adiabatic effect
绝热压缩　adiabatic compression
绝热液〔态〕水含量　adiabatic liquid water content
绝热运动　adiabatic motion
绝热增温　adiabatic warming
绝热增温【大气】　adiabatic heating
绝热直减率【大气】　adiabatic lapse rate
绝热状态　adiabatic state
绝缘　isolation, insulation
军事气候志【大气】　military climatography
军事气象保障【大气】　military meteorological support
军事气象信息【大气】　military meteorological information
军事气象学【大气】　military meteorology
均变说　uniformitarianism
均方差　mean square deviation
均方根　root-mean-square (RMS)
均方根误差【数】【大气】　root-mean-square error (RMSE)
均方束截面　mean square beam cross section
均方误差　error of mean square
均方误差【数】　mean square error
均分　equipartition
均分定律　equipartition law
均衡　equalization
均衡调整　isostatic adjustment
均衡井出水量　equilibrium well discharge
均流　uniform flow
均温层　hypolimnion (= hypolymnion)
均相系统【化】　homogeneous system
均一的　homogeneous
均一化　homogenization
均匀波【物】　homogeneous wave
均匀层　uniform layer
均匀磁化球　uniformly magnetized sphere
均匀磁离子介质　uniform magnetoionic medium
均匀地形　homogeneous terrain
均匀分布【数】　uniform distribution
均匀复随机函数　homogeneous complex random function
均匀各向同性随机函数　homogeneous and isotropic random function
均匀各向同性湍流　homogeneous (and) isotropic turbulence
均匀混合　homogeneous mixing
均匀加宽　homogeneous broadening
均匀介质　homogeneous medium
均匀性　homogeneity
均匀照明【物】　uniform illumination
均匀中纬正扰动　uniform midlatitude positive perturbation
均值【数】　mean
均值向量【数】　mean vector
均质层【大气】　homosphere
均质层顶【大气】　homopause
均质层-非均质层过渡带(区)　homosphere-heterosphere transition
均质成核　homogeneous nucleation
均质大气【大气】　homogeneous atmosphere
均质大气高度　atmospheric scale height
均质海洋模式　homogeneous ocean model
均质流体　homogeneous fluid
均质凝结　homogeneous condensation
均质升华　homogeneous sublimation

K

卡尔曼-布西滤波器【数】 Kalman-Bucy filter（或简称 Kalman filter）
卡尔曼滤波【数】【大气】 Kalman filtering
卡尔曼滤波器【数】 Kalman filter
卡尔曼权重矩阵 Kalman weight matrix
卡尔曼增益矩阵 Kalman gain matrix
卡尔文循环【生化】 Calvin cycle
卡〔路里〕（热量的旧单位，＝4.1868 J） calorie(cal)
卡门常数【大气】 Karman constant
卡曼谱 Karman spectrum
卡曼涡 Karman vortex
卡门涡街【大气】 Karman vortex street
卡曼涡列 Karman vortex train
卡诺定理【物】 Carnot's theorem
卡诺发动机 Carnot engine
卡诺效率 Carnot efficiency
卡诺循环【物】 Carnot cycle
卡诺循环过程 Carnot's cycle process
卡片箱 bin card
卡萨拉虹吸雨量计 Casella's siphon rainfall recorder
卡塞格林望远镜 Cassegrainian telescope
卡塞格林望远镜【天】 Cassegrain telescope
卡塞格伦反射镜 Cassegrainian mirror
卡塞格伦反射面天线【电子】 Cassegrain reflector antenna
卡他温度表 kata thermometer, catathermometer
喀斯特【地质】 karst
喀斯特地貌【地理】 karst landform
喀斯特地貌学【地理】 karst geomorphology
喀斯特水文学 karst hydrology
喀斯特相【地质】 karst facies
开边界 open boundary

开边界条件 open boundary condition
开端颠倒温度表 unprotected reversing thermometer
开尔文波【大气】 Kelvin wave
开尔文公式【化】 Kelvin equation
开〔尔文〕（国际单位制的基本单位，记为 K） Kelvin
开尔文-亥姆霍兹波【大气】 Kelvin-Helmholtz wave
开尔文-亥姆霍兹不稳定性【物】 Kelvin-Helmholtz instability
开尔文-亥姆霍兹大波 Kelvin-Helmholtz billows
开尔文环流 Kelvin's circulation
开尔文环流定理【大气】 Kelvin's circulation theorem
开尔文模型【化】 Kelvin model
开放波 open wave
开放系统【计】 open system
开关电路【电子】 switching circuit
Q 开关【电子】 Q-switching, Q-switch
开关二极管【电子】 switching diode
Q 开关红宝石激光器 Q-switched ruby laser
开管温度表 unprotected thermometer
开花【植】 flowering, anthesis
开花期 flowering period
开花期【农】 flowering stage
开口胞状对流 open cellular convection
开口式风洞 open jet wind tunnel
开口细胞状积云 open-cell cumulus
开口型细胞状云【大气】 open(cloud) cells
开阔水面 open water
开平方【数】 extraction of square root
开普勒定律【天】 Kepler's law
开普勒方程【天】 Kepler's equation
开展曲线 open curve

凯里探头(测云滴谱用) keily probe
凯斯勒型参数化 Kessler type parameterization
铠装日射表 panzeractinometer
铠装温度表(测海面温度用) armoured thermometer
堪察加海流 Kamchatka Current
坎〔德拉〕(发光强度单位) candela(cd)
坎儿井 kanat, ghanat, ganat
坎宁安滑移订正(斯托克斯定律的偏差) Cunningham slip correction
坎贝尔-斯托克斯日照计【大气】 Campbell-Stokes sunshine recorder
康普顿散射【物】【大气】 Compton scattering
康普顿效应【物】【大气】 Compton effect
抗冻性 freezing hardiness
抗风暴的 storm-proof
抗风性 wind resistance
抗寒性 cold resistance
抗旱品种 drought-resistant variety
抗切强度 shearing strength
抗生长素 antiauxin
抗相关时间 decorrelation time, coherence time
抗蒸腾的 anti-transpirant
抗重力 antigravity
考察 expedition
考察船 research vessel
考古时期 archaeological period
考古学 archaeology
考特维-德伏里斯波 Korteweg-de Vries wave (=KdV wave)
考特维-德伏里斯方程 Korteweg-de Vries equation (或 KdV equation)
柯本分类 Köppen classification
柯本-盖格气候 Köppen-Geiger climate
柯本气候分类【大气】 Köppen's climate classification
柯本-苏潘等温线【大气】 Köppen-Supan line

柯蒂斯-戈德森近似 Curtis-Godson approximation
柯尔纳海洋度指数 Kerner's oceanicity index
柯朗条件 Courant condition
柯朗-弗里德里希斯-列维条件【数】【大气】 Courant-Friederichs-Lewy (CFL) condition
柯什密得定律 Koschmieder's law
柯什密得理论(能见度计算) Koschmieder's theory
柯西-黎曼方程【数】 Cauchy-Riemann equations
柯西数(或胡克数) Cauchy number (或 Hooke number)
柯西中值定理【数】 Cauchy mean value theorem
科尔莫戈罗夫-查普曼方程【物】 Kolmogorov-Chapman equation
科尔莫戈罗夫常数 Kolmogorov constant
科尔莫戈罗夫串级 Kolmogorov cascade
科尔莫戈罗夫-奥布霍夫长度 Kolmogorov-Obukhov length
科尔莫戈罗夫-奥布霍夫惯性副区 Kolmogorov-Obukhov inertial subrange
科尔莫戈罗夫-奥布霍夫惯性区 Kolmogorov-Obukhov inertial range
科尔莫戈罗夫尺度 Kolmogorov scale
科尔莫戈罗夫耗散尺度 Kolmogorov dissipation scale
科尔莫哥罗夫理论 Kolmogorov theory
科尔莫哥罗夫谱 Kolmogorov spectrum
科尔莫哥罗夫时间尺度 Kolmogorov time scale
科尔莫哥罗夫速度尺度 Kolmogorov velocity scale
科尔莫哥罗夫湍流 Kolmogorov turbulence
科尔莫哥罗夫微尺度 Kolmogorov microscale

科尔莫哥罗夫相似性　Kolmogorov similarity
科勒方程　Köhler equation
科里奥利参数【大气】　Coriolis parameter
科里奥利加速度【大气】　Coriolis acceleration
科里奥利力【大气】　Coriolis force
科里奥利项　Coriolis term
科里奥利效应　Coriolis effect
科纳风暴【大气】　kona cyclone
科纳气旋【大气】　kona cyclone
科氏参数【大气】　Coriolis parameter
科氏加速度【大气】　Coriolis acceleration
科斯曼驻点温度表　Kollsman stagnation thermometer
科学技术活动委员会(AMS)　Scientific and Technological Activties Committee (STAC, AMS)
科学技术资料委员会(ICSU)　Committee on Data for Science and Technology (CODATA, ICSU)
科学论文　scientific paper
科学研究　scientific research
颗粒物质　particulate matter
壳冰　shell ice, rind ice, cat ice
可变电感器【电子】　variometer
可变分辨率模式　variable resolution model
可变缝隙取样器　variable-slit impactor
可变区域　variable area
可表示性【数】　representability
可播撒性　seedability
可持续发展　sustainable development
可持续发展委员会　Commission on Sustainable Development (CSD)
可串行性【计】　serializability
可递系统　transitive system
可调虹吸式水银气压表　adjustable siphon mercury barometer
可调谐激光二极管分光计　tunable laser diode spectrometer (TLDS)
可调谐激光光谱学　tunable laser spectroscopy
可调谐激光【化】　tunable laser
可调谐气体激光器　tunable gas laser
可调谐染料激光器　tunable dye laser
可调谐自旋反转拉曼激光器　tunable spin-flip Raman laser
可定年性　dateability
可分性　separability
可航半圆　navigable semicircle
可计算性【数】　computabillity
可见地平　visible horizon
可见光【大气】　visible light
可见光反照率　visible albedo
可见光辐射　visible〔light〕radiation, radiation of visual ray
可见光和红外辐射仪　visible-IR radiometer
可见光和红外扫描辐射仪　scanning visible/infrared radiometer (SV/IRR)
可见光和红外自旋扫描辐射仪　visible and infrared spin scan radiometer (VISSR)
可见光谱　visible spectrum
可见光图像　visible image
可见光消光计　visual extinction meter
可见光云图【大气】　visible cloud picture
可见烟羽示踪法　visible plume tracer technique
可降水量　precipitable water vapor
可降水量【大气】　precipitable water
可降水量指数　precipitable water index
可进入的大型环境模拟室　Walk-in environmental chamber
可靠权重　reliable weight
可靠性　reliability, dependability, trustworthiness
可靠性检验　reliability test
可控气球探测　dirigible balloon ascent

可扩充性　scalability
可扩展置标语言【计】　extensible markup language（XML）
可录光碟【计】　compact disc-recordable（CD-R）
可能误差　possible error
可能蒸散【大气】　potential evapotranspiration
可能蒸腾　potential transpiration
可能总辐射　possible global radiation
可能最大暴雨【大气】　probable maximum precipitation
可能最大洪水【大气】　probable maximum flood（或 maximum possible flood）
可能最大降水【大气】　probable maximum precipitation（PMP）
可逆饱和绝热过程　reversible saturation adiabatic process
可逆变化　reversible change
可逆过程【物】　reversible process
可逆环流　reversible circulation
可逆机　reversible engine
可逆机械　reversible machine
可逆绝热过程【大气】　reversible adiabatic process
可逆马尔可夫过程【数】　reversible Markov process
可逆湿绝热过程　reversible moist-adiabatic process
可逆性　invertibility，reversibility
可逆性原理　invertibility principle
可逆循环　reversible cycle
可燃物含水量　fuel moisture
可溶混移动　miscible displacement
可生物降解的　biodegradable
可湿性【力】　wettability
可实现性【数】　realizability
可视化【计】　visualization
可释放能量　releasable energy
〔可〕塑性　plasticity
可听声　audible sound

可闻区　audibility zone
可吸入颗粒物　inhalable particulate matter，PM-10
可行性　feasibility
可行性研究　feasibility study
可压缩的　compressive，compressible
可压缩模式方程　compressible model equation
可压缩性【物】　compressibility
可压缩性流体　compressible fluid
可压缩〔性〕系数　coefficient of compressibility
可移动的风廓线记录系统　portable wind profile recording system
可移动的中尺度观测网站　portable mesonet stations
可移动的自动气象站网　portable automatic meteorological（PAM）network
可移动的自动中尺度观测网　portable automated mesonet（PAM）
可移动细网格模式【大气】　movable fine-mesh model
可用度【数】　availability
可用浮〔力〕能　available buoyant energy
可用降水量　available precipitation amount
可用水头　available head
可用预报　useful forecast
可用蒸发位能　evaporative available potential energy
可预报性【大气】　predictability
可预报性极限　predictability limit
可预防损失　protectable loss
可再生能源　renewable energy source
可再生资源　renewable resources
可照百分率　percentage of possible sunshine
可照时数【大气】　duration of possible sunshine
可重复性【物】　reproducibility
可重写光碟【计】　CD-RW（CD-rewritable）

可转移性　transmissibility
克　gram
克分子〔量〕　gram-mole(＝gram-molecule)
克·卡　gram calorie
克拉珀龙-克劳修斯方程　Clapeyron-Clausius equation，Clausius-Clapeyron equation
克拉珀龙图　Clapeyron's diagram
克拉香天气【大气】　Crachin
克劳修斯-克拉珀龙方程【物】　Clausius-Clapeyron equation
克劳修斯状态方程　Clausius equation of state
克雷登效应(闪电摄影)　Clayden effect
克里金插值〔法〕　kriging；克里金法【地信】(一种求最优线性无偏内插估计量的方法)
克里斯托弗尔数　Christoffel number
克隆【植】　clone
克伦威尔潜流(即太平洋赤道潜流)　Cromwell Current
克罗内克 δ　Kronecker delta
克罗内克符号【数】　Kronecker symbol
克罗内克张量　Kronecker tensor
克罗(衣着指数单位)　clo
克努森表　Knudsen's tables
克努森数【物】　Knudsen number
刻度　graduation
客观分析【大气】　objective analysis
客观预报【大气】　objective forecast
客户【计】　customer
客土〔土壤〕　incorporation with extra soil
氪〔管〕湿度计　krypton hygrometer
氪〔气〕　krypton
肯定法　modus ponens (affirmative mode)
肯内利-赫维赛层　Kennelly-Heaviside layer
空白图　blank chart
空报率　false alarm rate

空-地传导电流【大气】　air-earth conduction current
空-地电流【大气】　air-earth current
空调　air conditioning
空管雷达【电子】　air traffic control radar
空盒　sylphon
空盒高度表　aneroid altimeter
空盒气压表　aneroid
空盒气压表【大气】　aneroid barometer
空盒气压表的蠕动〔滞后〕　creeping of aneroid barometer
空盒气压表的滞后现象　lag of aneroid barometer
空盒气压计　aneroidograph
空盒气压计【大气】　aneroid barograph
空盒气压曲线　aneroidogram
空基观测【大气】　space-based observation
空基观测系统　space-based observation system
空基子系统【大气】　space-based subsystem
空间【物】　space
空间变化　spatial variation, space change
空间差分格式　space difference scheme
空间尺度　spatial scale, space scale
空间代表性　spatial representativeness
空间电荷【物】　space charge
空间飞行　space flight
空间飞行器　spacecraft
空间分辨率【大气】　spatial resolution
空间分布　spatial distribution
空间跟踪　space tracking
空间跟踪和资料接收站网　space tracking and data acquisition network (STADAN)
空间共振　space resonance
空间观测平台　space platform
空间函数　space function
空间环境监视器　space environmental

monitor（SEM）
空间积分　space integration
空间简谐表达式　space harmonic representation
空间截断误差　space truncation error
空间滤波　spatial filtering，space filtering
空间滤波函数　spatial filter function
空间平滑【大气】　space smoothing
空间平均　spatial average，space averaging
空间平均图　space mean chart
空间气象学【大气】　space meteorology
空间实验室　skylab，sky laboratory
空间碎片（火箭或卫星的碎片）　garbage
空间天气【大气】　space weather
空间微商　spatial derivative
空间物理【地】　space physics
空间相干　spatial coherence
空间相关　spatial correlation，space correlation
空间研究　space research
空间研究委员会（ICSU）　Committee on Space Research（COSP AR，ICSU）
空间一致性检验　space consistency check
空间因子　spatial factor
空间与高层大气研究委员会（巴基斯坦）　Space and Upper Atmosphere Research Committee（SUPARCO）
空〔间〕域【物】　space domain
空军大气模式（美国）　Air Force atmospheric model
空军基地　air force base（A.F.B）
空军天气局（美国）　Air Weather Service（AWS）
空气　air
空气称重器　air poise
空气传染　air infection
〔空气〕颠簸　air bump
空气动力凝结尾迹　aerodynamic contrail
空气动力实验室　aerodynamic laboratory
空气动力尾迹　aerodynamic trail
空气动力学【大气】　aerodynamics
空气动力学方法　aerodynamic method
空气动力学平衡　aerodynamic balance
空气冻结指数　air freezing index
空气毒素　air toxins
空气管路腔　air-line well
空气管路倾斜订正　air line correction
空气管路探测　air-line sounding
空气光公式　airlight formula
〔空〕气光（悬浮物散射光）【大气】　airlight
空气轨迹　air travel，air trajectory
空气环流　air circulation
空气急流　air torrent
空气夹卷　air entrainment
空气检测　aerial detection
空气解冻指数　air thawing index
〔空〕气阱　air trap
空气净化　air purification
空气净化设备　air cleaning facility
空气静力平衡　aerostatic balance
空气静力学　aerostatics
空气连续监测规划站　continuous air monitoring program station
空气流泄　air drainage
空气密度　air density
空气取样器　air sampler
空气取样装置　air sampling rig
空气燃料比　air-fuel ratio
空气热力化学　aerothermochemistry
空气停滞模式　air stagnation model
空气污染　aerial contamination
空气污染【大气】　air pollution
空气污染标准　air pollution standard
空气污染标准指数　pollutants standard index（PSI）
空气污染法　air pollution law
空气污染法规　air pollution code
空气污染观测站　air pollution

observation station
空气污染化学　air pollution chemistry
空气污染监测　air pollution surveillance
空气污染检查器　cacaerometer
空气污染控制　air pollution control
空气污染立法　air pollution legislation
空气污染模拟【大气】　air pollution modeling
空气污染模型【大气】　air pollution model
空气污染气象学　air pollution meteorology
空气污染潜势　air pollution potential
空气〔污染区〕模式　airshed model
空气污染事故　air pollution disaster
空气污染事件　air pollution episode
空气污染诉讼　air pollution complaint
空气污染条例　air pollution regulation
空气污染物　air pollutant, aerial contaminant
空气污染物含量【大气】　loading of air pollutant
空气污染物排放【大气】　air pollutant emission
空气污染物排放标准【大气】　air pollutant emission standard
空气污染预报　air pollution forecasting
空气污染预警　air pollution alert
空气污染源　air pollution source
空气污染指数　air pollution index
空气雾化器　air atomizer
空气曳力　air drag
空气运动　air motion
空气质量【大气】　air quality
〔空气中〕二氧化碳测定法　anthracometry
〔空气中〕二氧化碳测定仪　anthracometer
空气中悬浮的粒径大于 2.5 μm 的微粒质量　PM-2.5
空气中悬浮的粒径大于 10 μm 的微粒质量　PM-10
空气轴承　air bearing
空气资源　air resource
空气阻力　air resistance
空腔　cavity
〔空〕腔辐射　cavity radiaton
空腔辐射计【大气】　cavity radiometer
空腔作用　cavitation
空速度　airspeed
空隙比　void ratio
空语句　dummy statement
空域　airspace
空中待命〔降落〕　stacking
空中放电【大气】　air discharge
空中浮游的　airborne
空中交通管制【电子】　air traffic control
空中漏斗　funnel aloft
空中能见度【大气】　flight visibility
空中透视　aerial perspective
空中位置显示器　air-position indicator (API)
空中悬浮微粒【大气】　airborne particulate
孔径　aperture
孔径角　angle of aperture
孔径平均　aperture averaging
孔径修匀效应　aperture smooth effect
孔隙度【地质】　porosity
恐雷〔电〕感　astraphobia
恐龙　dinosaur
控制　modification, control
控制变量　control variable
控制程序　control program(CP)
控制断面　control section
控制方程【大气】　governing equation
控制井　control well
控制论　kybernetics
控制论【数】　cybernetics
控制塔能见度　control-tower visibility
控制台终端　console terminal
控制系统　control system

控制中心 control center, operation center
控制组件 control unit module
寇乌气压表【大气】 Kew pattern barometer
寇乌照相式温度计 Kew photographic thermograph
枯水河 oued
枯水河槽 low-flow channel
枯水年【大气】 low flow year
枯萎病天气 blight weather
枯萎病(植物所患) blight
枯心【农】 dead heart
库埃特流 Couette flow
库比尔卡-芒克二流理论 Kubelka-Munk two flux theory
库克返流温度表 Cook reverse-flow thermometer
库〔仑〕(电荷量单位) coulomb(C)
库仑定律【物】 Coulomb law
库仑力【物】 Coulomb force
库纳标绘位置 Cunnane plotting position
库容曲线 storage curve, elevation capacity curve
酷寒 hard freeze
酷热日 tropical day
酷暑 scorching, hot summer
跨步电压 step voltage
跨临界 transcritical
跨声速 transonic
块冰 pack, pack ice
块迭代法 block iterative method
快离子 fast ion
快模方程 fast mode equation
快速傅里叶变换【大气】 fast Fourier transform (FFT)
快速傅里叶逆变换 fast Fourier transform inverse (FFTI)
快速畸变理论 rapid distortion theory
快速间隔取像 rapid interval imaging
快速间隔扫描 rapid interval scan
快速扫描方式 rapid scan mode
快速散射计卫星 Quik SCAT
快〔速〕扫描气象雷达 fast scanning meteorological radar
快〔速〕响应臭氧检测器 fast-response O_3 detector
快响应传感器 fast-response sensors
快响应转杯风速表 fast response cup anemometer
快行冷锋【大气】 rapidly moving cold front
宽带发射率 broadband emissivity
宽带反照率 broadband albedo
宽带辐射 broadband radiation
宽带接入【计】 broadband access
宽带局〔域〕网【计】 broadband LAN
宽带雷达 wideband radar
宽带脉冲传播 broadband pulse propagation
宽带通量发射率 broadband flux emissivity
宽带通量透射率 broadband flux transmissivity
宽顶堰 broad-crested weir
宽度 width
宽视场辐射仪 wide field of view radiometer
宽视场扫描系统 wide field scanning system
宽束发射机 broad beam transmitter
狂暴 violence
狂风 whole gale, storm【大气】
狂风预警 whole-gale warning
狂浪 high sea
狂涛(风浪7级) very high sea
矿产经济学【地质】 mineral economics
矿产开发【地质】 mineral exploitation
矿产勘察【地质】 mineral exploration
矿床模式【地质】 mineral deposit model
矿化水【地质】 mineralized water

矿化速率【土壤】 mineralization rate
矿化作用【地质】 mineralization
矿泉【地质】 mineral spring
矿物气溶胶 mineral aerosol
矿物燃料 fossil fuels
矿物学【地质】 mineralogy
框图 block diagram
亏年 annus deficiens
亏值 deficit
盔云 crest cloud
《葵花》卫星 HIMAWARI
溃变理论 Blown-ups theory
馈〔给〕 feed
馈云 feeder cloud
昆明准静止锋【大气】 Kunming quasi-stationary front
扩频信号 spread spectrum singal
扩散【大气】 diffusion
扩散泵【电子】 diffusion pump
扩散层 diffusisphere, diffusion layer
扩散对流 diffusive convection
扩散方程 diffusivity equation
扩散方程【数】 diffusion equation
扩散分离 diffusive separation
扩散锋 diffuse front
扩散高度 diffusion level
扩散解吸技术 diffusion denude technique
扩散界面 diffuse boundary
扩散力 diffusive force
扩散率【大气】 diffusivity

扩散模式【大气】 diffusion model
扩散平衡 diffusive equilibrium
扩散迁移 diffusiophoresis
扩散迁移力 diffusiophoretic force
扩散迁移速度 diffusiophoretic velocity
扩散湿度表 diffusion hygrometer
扩散时间 diffusion time
扩散速度 diffusion velocity
扩散通量 dispersive flux
扩散图 diffusion diagram
扩散系数 diffusion coefficient
扩散系数【物】【大气】 coefficient of diffusion (= diffusivity)
扩散性 diffusivity
扩散云室 diffusion chamber
扩散张弛法 diffusive relaxation
扩展 F 层 spread F layer
扩展 E 层 spread E layer
扩展的经验正交函数 EEOF (extended EOF)
扩展的卡尔曼滤波 extended Kalman filter (ExKF)
扩展〔光〕源【物】 extended source
扩展技巧相关 spread skill correlation
阔叶林【林】 broad leaved forest
廓线【大气】 profile
廓线匹配 profile matching
廓线仪【大气】 profiler
廓线相似〔性〕 profile similarity

L

拉布拉多海水 Labrador Sea water
拉布拉多〔冷〕海流 Labrador Current
拉丁美洲经济委员会(联合国) Economic Commission for Latin America (ECLA, UN)
拉盖尔-高斯求积【数】 Laguerre-Gauss quadrature
拉格朗日变化 Lagrangian change
拉格朗日波 Lagrangian wave
拉格朗日测流〔法〕 Lagrangian current measurement
拉格朗日插值 Lagrange interpolation

拉格朗日插值公式【数】 Lagrange interpolation formula
拉格朗日乘子【物】 Lagrange multiplier
拉格朗日方程 Lagrangian equations
拉格朗日方法 Lagrangian method
拉格朗日浮筒 Lagrangian float
拉格朗日空气(污染区)模式 Lagrangian airshed models
拉格朗日流体动力学方程 Lagrangian hydrodynamic equations
拉格朗日描述 Lagrangian description
拉格朗日-欧拉时间尺度比 rate of Lagrangian to Eulerian time-scale
拉格朗日平均公式 Lagrangian mean formulation
拉格朗日平均经向环流 Lagrangian mean meridional circulation
拉格朗日平均流 Lagrangian mean current
拉格朗日平流格式 Lagrangian advective scheme
拉格朗日时间尺度 Lagrangian time scale
拉格朗日湍流【物】 Lagrange turbulence
拉格朗日相关 Lagrangian correlation
拉格朗日相似 Lagrangian similarity
拉格朗日行星运动方程【天】 Lagrange's planetary equation
拉格朗日运动 Lagrange motion
拉格朗日运动方程 Lagrangian equations of motion
拉格朗日展开公式【数】 Lagrange expansion formula
拉格朗日中值定理【数】 Lagrange mean value theorem
拉格朗日坐标 Lagrangian coordinates
拉克斯等价定律 Lax equivalence theorem
拉克斯-温德罗夫差分格式 Lax-Wendroff differencing scheme
拉克斯-温德罗夫格式【数】 Lax-Wendroff scheme
拉曼光达 Raman lidar
拉曼光雷达探测 Raman lidar detection
拉曼光谱分析 Raman spectral analysis
拉曼频移 Raman frequency shift
拉曼散射 Raman scattering
拉曼效应【电子】 Raman effect
拉莫尔频率【物】 Larmor frequency
拉尼娜【大气】 La Nina
拉普拉斯变换 Laplace transform
拉普拉斯潮汐方程 Laplace tidal equation
拉普拉斯方程 Laplace equation
拉普拉斯分布【数】 Laplace distribution
拉普拉斯算子 Laplacian operator, Laplace operator
拉塞尔周期 Russell cycle
拉乌尔定律 Raoult's law
拉乌尔效应 Raoult effect
喇叭天线 horn radiator
喇叭天线【电子】 horn antenna
莱布尼茨公式【数】 Leibniz formula
莱布尼茨微积分定理 Leibniz's theorem of calculus
莱恩岛试验 Line Island Experiment (LIE)
莱曼-α发射线 Lyman-α, Lyman α emission line
莱曼-α湿度表【大气】 Lyman-α hygrometer
莱曼系【物】 Lyman series
莱曼-α总含水量仪 Lyman-α total water content meter
莱文海流 Leeuwin Current
莱因哈特投影 Rinehart projection
赖芳德表 Lafond's Tables
濑尾型天线 beavertail antenna
兰勃特投影【测】 Lambert projection

兰勃特圆锥投影　Lambert conic projection
兰勃特正形圆锥投影　Lambert conformal conic projection
兰布雷奇特多能湿度表　Lambrecht's polymeter
兰金涡旋　Rankine vortex
兰姆波【大气】　Lamb wave
兰姆移位【物】　Lamb shift
兰氏温标　Rankine temperature scale
拦洪蓄水　detention storage
蓝冰【大气】　blue ice
蓝冰区　blue-ice area
蓝带　blue band
蓝〔放电〕急流　blue jets
蓝绿闪〔天文〕　blue-green flame
蓝闪〔光〕　blue flash
蓝太阳　blue sun
蓝天标〔度〕　blue sky scale
蓝月亮　blue moon
蓝噪声　blue noise
朗伯(亮度单位)　lambert
朗伯-比尔定律【化】　Lambert-Beer law
朗伯-布格定律【物】　Lamber-Bouguer law
朗伯定律　Lambert's law
朗伯公式　Lambert's formula
朗伯面　Lambertian surface
朗伯散射　Lambertian scattering
朗伯吸收定律　Lambert's law of absorption
朗伯余弦定律【物】　Lambert cosine law
朗缪尔波【物】　Langmuir wave
朗缪尔测定　Langmuir probe measurement
朗缪尔层　Langmuir layer
朗缪尔对流胞　Langmuir cells
朗缪尔环流　Langmuir circulation
朗缪尔连锁反应　Langmuir chain reaction
朗缪尔数　Langmuir number
朗缪尔水槽　Langmuir trough
朗缪尔探针【物】　Langmuir probe
朗缪尔条纹　Langmuir streak
朗之万方程　Langevin equation
朗之万离子(即朗离子)　Langevin ion
浪-潮相互作用　surge-tide interaction
浪沫　whitecap
浪云　billow cloud
劳-帕森〔雨滴〕谱　Law and Parson distribution
老化　aging
老仙女木事件　Oldest Dryas
涝【土壤】　waterlogging
涝的　waterlogged
涝区　flood region
勒夫波【地】　Love wave
勒让德多项式【数】　Legendre polynomial
勒让德方程　Legendre equation
勒让德函数　Legendre function
勒让德连带微分方程【数】　Legendre associated differential equation
雷【大气】　thunder
雷暴　lightning storm
雷暴【大气】　thunderstorm
雷暴触发机制　thunderstorm initiation mechanism
雷暴单体【大气】　thunderstorm cell
雷暴等时线　homobront
雷暴低压　thunderstorm low
雷暴电荷　thunderstorm charge
雷暴电荷分离　thunderstorm charge separation
雷暴电学　thunderstorm electricity
雷暴电振荡终结　end of storm oscillation (EOSO)
雷暴定位装置　thunderstorm locating device
雷暴动力学　thunderstorm dynamics
雷暴风　thunderstorm wind
雷暴改造　thunderstorm modification
雷暴高压【大气】　thunderstorm high

雷暴观测站　thunderstorm observing station
雷暴活动区　region of thunderstorm activity
雷暴活动源地【大气】　source of thunderstorm activity
雷暴记录器　thunderstorm recorder
雷暴监测【大气】　thunderstorm monitoring
雷暴监测装置　thunderstorm monitor
雷暴卷云　thunderstorm cirrus
雷暴偶极子　thunderstorm dipole
雷暴频数等值线　isokeraunic line, isoceraunic line
雷暴起电　thunderstorm electrification
雷暴群　thunderstorm complex
雷暴日【大气】　thunderstorm day
雷暴三极子　thunderstorm tripole
雷暴湍流【大气】　thunderstorm turbulence
雷暴下击流　thunderstorm downdraft
雷暴先兆　thunderstorm sign
雷暴泄流【大气】　thunderstorm outflow
雷暴雨　thunderstorm rain
雷飑【大气】　thunder squall
雷冲击波催化　shock seeding
雷达【电子】　radar
雷达报告　radar report (RAREP)
雷达编码电报　radar coded message
雷达标定【大气】　radar calibration
雷达标定雨量计　rain gauge for radar calibration
雷达冰雹探测器　radar hail detector
雷达波带　radar band
雷达波导　radar duct
雷达波段　radar wave band
雷达波束　radar beam
雷达彩色显示器　radar colored display
雷达操作员　radarman
雷达测风　radar wind sounding【大气】, radar wind finding, radar wind observation (或 sounding)
雷达测风系统　radar wind system
雷达测风仪【大气】　radarsonde
雷达测距方程　radar range equation
雷达测量降水　radar measurement of precipitation
雷达常数　radar constant
MST雷达【大气】　MST radar
雷达导航【航海】　radar navigation
雷达等效反射率因子【大气】　radar equivalent reflectivity factor
雷达地面天气观测报告　RADOB (report of ground radar weather observation)
雷达地平线【电子】　radar horizon
雷达动态图像　radar image animation
雷达发射机　radar transmitter
雷达反射率【大气】　radar reflectivity
雷达反射率因子【大气】　radar reflectivity factor
雷达反射器【航海】　radar reflector
〔雷达〕反射强度　dBZ (常用 dBz)
雷达方程【电子】　radar equation
雷达分辨率　radar resolution
雷达分辨体积【大气】　radar resolution volume
雷达风暴探测　radar storm detection【大气】, radar detection of storm
雷达覆盖区【测】　rader overlay
雷达高度　radar altitude
雷达高度表　radar altimeter
雷达跟踪【电子】　radar tracking
雷达跟踪系统　radtrack system
雷达观测　ROB
雷达回波【电子】【大气】　radar echo
雷达〔回波〕洞　radar hole
雷达回波起伏　radar echo fluctuation
雷达回波线性外推预报　radar echo linear extrapolation forecasts
雷达回波相关跟踪法【大气】　tracking radar echoes by correlation (TREC)
雷达回波信号模拟器　radar echo signal

simulator
雷达回波直柱廓线　radar echo pillar profile
雷达回波综合图　radar echo composite map
雷达降水回波　radar precipitation echo
雷达接收机　radar receiver
雷达截面积【电子】　radar cross section
雷达经纬仪　radar theodolite
雷达灵敏度　radar sensitivity
雷达盲区　radar shadow
雷达偏振测量〔法〕　radar polarimetry
雷达〔频率〕波段　radar frequency band
(雷达屏上的)大气湍流斑　blob
雷达气候学【大气】　radar climatology
雷达〔气象〕观测　radar〔meteorological〕observation
雷达气象观测【大气】　radar meteorological observation
雷达气象学【大气】　radar meteorology
雷达气象站　radar weather station
雷达强风暴识别　radar diagnosis of severe-storm
雷达三角法　radar triangulation
雷达散射计　radar scatterometer
雷达湿度计　radar hygrometer
雷达示踪物　radar chaff
雷达室【船舶】　rader room
雷达数据处理器　radar data processor
雷达数据库【电子】　radar database
雷达数字显示〔器〕　radar digital display
雷达数字转换器　radar digitizer
雷达水文学　radar hydrology
雷达算法【大气】　radar algorithm
雷达探测距离【电子】　radar range
雷达探测空隙　radar detection gap
雷达探空【大气】　radar sounding
雷达天气观测　radar weather observation
雷达天线　radar antenna
雷达天线罩　radar dome, radome

【大气】
雷达透射收缩　radar foreshortening
雷达图像　radar imagery
雷达图像传真发送　radar facsimile transmission（RAFAX）
雷达图像几何失真　radar image geometric distortion
雷达微波中继装置　radar microwave link
雷达显示管　tonotron
雷达显示器【电子】　radar scope, radar indicator
雷达显示〔图〕　radar displays
雷达相干测量法　radar interferometry
雷达信标【航海】　radar beacon
雷达信号摄谱仪　radar signal spectrograph（RASAPH）
雷达研究发展中心(英国)　Radar Research and Development Establishment（RRDE）
雷达遥感【地理】　radar remote sensing
雷达引航图【航海】　radar navigation chart
雷达应答器【航海】　radar transponder
雷达荧光屏　radar screen
雷达雨量积分器【大气】　radar rainfall integrator
雷达雨量计　radar raingauge
雷达雨量计比对　radar-rain gauge comparison
雷达噪声　radar noise
雷达站【电子】　radar station
雷达站　radar site（或 station）
雷达〔照射〕体积　radar volume
雷达遮挡　radar obscuring
雷达转换开关　radar duplexer
雷达组网【大气】　radar network composite
雷电保护区,防雷区　lightning protection zone（LPZ）
雷电保护系统,防雷装置　lightning protection system（LPS）

雷电冲击　lightning impulse
雷电等位连接　lightning equipotential bonding
雷电感应　lightning induction
雷电干扰　lightning interference
雷电仪【大气】　ceraunometer, ceraunograph
雷电浪涌　lightning surge
雷电灾害损失　lightning disaster loss
雷击　thunderstroke, lightning stroke
雷击点　lightning striking point
雷击电磁脉冲　lightning electromagnetic impulse
雷击火【林】　lightning fire
雷击距　lightning striking distance
雷击跳闸　lightning outage
雷击损害概率　probability of lightning strike damage
雷诺平均　Reynolds averaging
雷诺数【大气】　Reynolds number (Re)
雷诺数相似性　Reynolds number similarity
雷诺条件　Reynolds condition
雷诺相似　Reynolds analogy
雷诺效应　Reynolds effect
雷诺应力【大气】　Reynolds stress
雷雨表　brontometer
雷雨锋　thundery front
雷雨计　brontograph
雷雨天空　thundery sky
雷雨云　thundercloud
雷雨云降水【大气】　thundery precipitation
雷雨云系【大气】　thundery cloud system
雷雨云砧　thunderhead
雷阵雨【大气】　thunder shower
镭　radium
泪珠状气球　teardrop balloon
类比气候模式　analog climate model
类比推理【计】　analogical inference
类比学习【计】　analogical learning

类地行星【天】　terrestrial planet
类木行星【天】　Jovian planet
类氢原子【物】　hydrogen-like atom
类型图　type map
累积带　accumulation zone
累积计算器　totalizer, totalizator
累积降水　accumulated precipitation
累积冷却〔量〕　accumulated cooling
累积浓度　cumulative concentration
累积频率　cumulative frequency
累积频率曲线　ogive curve
累积频率图　ogive
累积谱　cumulative spectrum
累积区域（冰川的）　accumulation area
累积误差【数】　cumulative error
累积雨量器　totalizer for rain, storage gauge
累积直方图　cumulative histogram
累计分布函数　cumulative distribution function (CDF)
累计雨量器　accumulation raingauge, pluviometer-association, totalizing rain gauge
累计雨量器【大气】　accumulative raingauge
累计值曲线　mass curve
累加法　adding method
累加器　accumulator
累加装置　sigma meter
棱镜【物】　prism
冷槽【大气】　cold trough
冷池【大气】　cold drop, cold pool
冷带　cold belt
冷岛　cold island
冷稻瘟病　rice blast induced by coldness
冷等离子体　cold plasma
冷等离子体流　cold plasma flow
冷低压【大气】　cold low
冷冻干燥【农】　freeze drying
冷堆　cold dome
冷风　cold wind

冷锋波动　cold-front wave
冷锋【大气】　cold front
冷锋雷暴　cold front thunderstorm
冷锋面　kataphalanx, cataphalanx
冷锋前飑线　pre-cold frontal squall line
冷锋型锢囚　cold-front type of occlusion
冷锋雨　cold front rain
冷锋云系【大气】　cold front cloud system
冷高压【大气】　cold high
冷锢囚　cold occlusion
冷性锢囚锋【大气】　cold occluded front
冷害　cool injury
冷害【大气】　cool damage
冷湖　cold lake
冷季　cold season
冷浸　cold soak
冷空气　cold air
冷空气爆发　cold-air outbreak
冷〔空气〕池　cold-air drop, cold pool
冷空气堆　pool of cold air
冷空气侵入　cold-air injection
冷空气输送带　cold conveyor belt
冷凝镜湿度表　chilled-mirror hygrometer
冷凝物　condensate
冷平流【大气】　cold advection
冷期　cold period
冷启动【计】　cold start
冷气汇　cold air sink
冷气团【大气】　cold air mass
冷区　cold sector
冷圈学【地理】　cryology
冷却　cooling
冷却表　frigorimeter
冷却度日【大气】　cooling degree-day
冷却杆（露点仪用）　cooling rod
冷却过程　cooling process
冷却计　frigorigraph
冷却剂　coolant
冷却率　cooling power

冷却率测定表　coolometer
冷却率风速表　cooling-power anemometer
冷却温度　cooling temperature
冷沙漠　cold desert, arctic desert
冷舌【大气】　cold tongue
冷事件　cold event
冷水壁　cold wall
冷温〔量〕　chilling
冷涡【大气】　cold vortex
冷雾　cold fog
冷下沉气流　cold downdraft
冷箱　cold box
冷楔　cold wedge
冷心低压　cold-core low
冷心反气旋　cold-core anticyclone
冷心高压　cold-core high
冷心〔海洋〕涡流　cold-core rings
冷心气旋　cold-core cyclone
冷心天气系统　cold core synoptic system
冷性反气旋【大气】　cold anticyclone
冷〔性〕锢囚　cold type occlusion
冷性锢囚低压　cold-occlusion depression
冷性气旋【大气】　cold cyclone
冷洋流　cold ocean current
冷源　cold source
冷云【大气】　cold cloud
冷云顶　cold top
厘巴（＝10 mb＝10 hPa）　centibar
厘（词头，10^{-2}）　centi-
厘米　centimeter(cm)
厘米·克·秒制　centimeter-gram-second system(cgs system)
离岸潮　rip tide
离岸风　wind form coast
离岸风【大气】　off-shore wind
离岸流　rip current
离极气流　ab-polar current
离解【物】　dissociation
离解复合〔过程〕【物】　dissociative recombination
离解光致电离【物】　dissociative photoi-

onization
离解热【物】 heat of dissociation
离群值【数】 outlier
离散【数】 dispersion
离散变换 discrete transform
离散分布【数】 discrete distribution
离散傅里叶变换【数】 discrete Fourier transform
离散化【数】【大气】 discretization
离散极盖区极光 discrete polar cap aurora
离散集【数】 discrete set
离散介质 dispersion medium
离散流体〔模型〕【力】 discrete fluid
离散谱【大气】 discrete spectrum
离散数据【地信】 discrete data
离散随机过程 discrete stochastic process
离散网格 discrete mesh
离散纵标法 discrete ordinates method
离体激波 detached shock wave
离线分析【物】 off-line analysis
离心不稳定度 centrifugal instability
离心机 centrifuge
离心加速度 centrifugal acceleration
离心力【物】【大气】 centrifugal force
离心率【数】 eccentricity
离心伸长 centrifugal stretching
离心势 centrifugal potential
离心水分当量 centrifuge moisture equivalent
离心效应 centrifugal effect
离子 ion
离子捕获〔学〕说 ion-capture theory
离子捕集器 ion trap
离子产生率 rate of ion production
离子传导 ionic conduction
离子电流 ion current
离子对 ion pair
离子反应【化】 ionic reaction
离子分离 ion detachment
离子-分子反应【化】 ion-molecule reaction
离子附着 ion attachment
离子活度 ionic activity
离子计数器【大气】 ion counter
离子交换【地质】 ion exchange
离子-离子复合 ion-ion recombination
离子流风速表 ionflo
离子密度 ion density
离子内动量输送 inter-ion momentum transfer
离子能量平衡 ion energy balance
离子能量损失速率 ion energy loss rate
离子浓度 ion concentration
离子平均寿命 ion mean life
离子迁移 ion migration
离子迁移率 ion mobility
离子迁移率【化】 ionic mobility
离子色谱法【化】 ion chromatography
离子寿命 ion life
离子束【物】 ion beam
离子酸度 ionic acidity
离子团 cluster ion
离子拖曳 ion drag
离子雾 ion atmosphere
离子原子互换 ion-atom interchange
离子云 ion cloud
离子柱 ion column
犁铧 plowshares
黎曼几何〔学〕【数】 Riemannian geometry
黎曼-希尔伯特问题 Riemann-Hilbert problem
黎明 daybreak, dawn
李雅普诺夫函数 Lyapunov function, Liapunov function
李雅普诺夫维数 Lyapunov dimension
李雅普诺夫稳定性【数】 Liapunov stability
李雅普诺夫稳定性【物】 Lyapunov stability
李雅普诺夫指数 Liapunov index

李雅普诺夫指数【数】 Liapunov exponent
李雅普诺夫指数谱 Liapunov exponent spectrum
里卡蒂微分方程【数】 Riccati differential equation
里奇数(弗罗德数的倒数) Reech number
里氏〔地震等级〕表 Richter scale
里氏震级【地】 Richter magnitude
里查森数【大气】 Richardson number
里查森网络 Richardson grid
理查德温度计 Richard thermograph
理查森外推〔法〕【数】 Richardson extrapolation
K 理论 K theory
理论气候学 climatonomy
理论气象学【大气】 theoretical meteorology
理论气象研究所(丹麦) Institute for Theoretical Meteorology (ITM)
理论【物】 theory
理想地平 ideal horizon
理想反射壁 perfect reflection wall
理想观测 perfect observation
理想黑体 perfect black body
理想流体 perfect fluid, ideal fluid【大气】
理想漫射辐射体 perfectly diffuse radiator
理想气候【大气】 ideal climate
理想气体 ideal gas, perfect gas
理想气体定律 ideal-gas laws
理想舒适区 desirable comfort zone
理想液体 ideal liquid
理想预报 perfect forecast
锂电池【化】 lithium cell
力 force
力的合成 resultant of forces
力的合成【物】 composition of forces
力管 solenoid
力管场 solenoid field, field of solenoid
力管〔的〕 solenoidal
力管项 solenoid term
力管效应 solenoid effect
力管指数 solenoidal index
力能学【大气】 energetics
力线 line of force
力学【物】 mechanics
历 calendar
历表【天】 ephemeris
历年【天】 calendar year
历时曲线【水利】 duration curve
历时统计 duration statistics
历史地理学【地理】 historical geography
历史地图集【地理】 historical atlas
历史回报 hindcast
历史气候【大气】 historical climate
历史气候记录 historical climatic record
历史气候序列 historical climate series
历史气候学 historical climatology
历史气候资料 historical climatic data
历史全球变化计划(IGBP) Past Global Changes(PAGES, IGBP)
历史时期 historical time
历史顺序原则 principle of historical sequence
〔历史天气〕个例推理 case-based reasoning
历史序列 historical sequence
厉风 severe gale
立春 Spring Beginning
立春(节气)【农】【大气】 Beginning of Spring
立冬(节气)【农】【大气】 Beginning of Winter
立法 legislation
立方 cube
立方〔晶〕系 cubic system
立秋(节气)【农】【大气】 Beginning of Autumn
立式 formulation

立体冰晶　spatial crystal
立体弧度　sr(steradian)
立体几何〔学〕【数】　solid geometry
立体角【数】　solid angle
立体摄影测量【测】　stereophotogrammetry
立体图　stereogram
立体图像　stereo image, stereographic image
立体像对【测】　stereopair
立体照片对　stereoscopic pair of photographs
立体枝状冰晶　spatial dendrite
立夏(节气)【农】【大气】　Beginning of Summer
立柱式贮存雨量器　standpipe storage precipitation gauge
利曼海流　Liman Current
利文斯通球　Livingstone sphere
利文斯通蒸发表　Livingstone atmometer
隶属度　membership grade
隶属关系【数】　membership
隶属函数　membership function
栗原网格　Kurihara grid
砾石洲　gravel bar
砾〔质沙〕漠　reg
粒径分布　particle size distribution
粒雪　eternal snow
粒子电荷　particle charge
粒子辐射　particle radiation
粒子监测仪　particle monitor
粒子摄影机　particle camera
粒子速度　particle velocity
粒子探测器　particle detector
粒子通量　particle flux
粒子吸收　particle absorption
连带勒让德函数【数】　associated Legendre function
连带系数　association coefficient
连接程序　linker
连接带　connecting band

连锁现象　interlocking phenomenon
连通〔磁〕力线　connected field line
连网【计】　networking
连续　continuation
连续波【物】　continuous wave
连续波发射器【电子】　continuous wave transmitter
连续波雷达【大气】　continuous-wave radar(CW radar)
连续电流【大气】　continuing current
连续方程【大气】　continuity equation
连续分布【数】　continuous distribution
连续观测　day-night observation
连续函数【数】　continuous function
连续记录　continuous record
连续监测　continuous monitoring
连续介质【物】　continuous medium
连续雷电　continuous thunder and lightning
连续〔频〕谱　continuous spectrum
连续谱【化】　continuum
连续区吸收　continuum absorption
连续曲线　continuous curve
连续吸收　continuous absorption
连续先导(闪电)　continuous leader
连续性　continuity
连续性大雨　continuous heavy rain
连续〔性〕方程【物】　continuity equation
连续性降水【大气】　continuous precipitation
连续性图　continuity chart
连续性中雨　continuous moderate rain
连续雨　steady rain, continuous rain
帘状极光　curtain aurora
联邦科学技术委员会(美国)　Federal Committee of Science and Technology (FCST)
联邦气象服务委员会(美国)　Federal Committee for Meteorological Services (FCMS)
联邦气象研究中心(澳大利亚)

Commonwealth Meteorology Research Centre(CMRC)
联合北极天气站　Joint Arctic Weather Stations(JAWS)
联合国　United Nations（UN）
联合国环境计划署　United Nations Environment Programme（UNEP）
联合国教科文组织　United Nations Educational, Scientific and Cultural Organization（UNESCO）
联合国开发计划署　United Nations Development Programme（UNDP）
联合国可持续发展委员会　United Nations Commission on Sustainable Development（UNCSD）
联合国粮〔食及〕农〔业〕组织　Food and Agricultural Organization of the United Nations（FAO）
联合国气候变化框架公约　United Nations Framework Convention on Climate Change（UNFCCC）
联合国千年发展目标　United Nations Millennium Development Goals
联合国沙漠化问题会议　United Nations Conference on Desertification（UNCD）
联合国社会发展研究所　United Nations Research Institute for Social Development（UNRISD）
联合国世界粮食会议　United Nations World Food Conference（UNWFC）
联合国原子辐射效应科学委员会　United Nations Scientific Committee on the Effects of Atomic Radiation（UNSCEAR）
联合计划办公室(全球气候观测系统)　Joint Planning Office(GCOS)
联合计划专家　Joint Planning Staff（JPS）
联合科学委员会(WMO/ICSU)　Joint Scientific Committee（JSC, WMO/ICSU）
联合气候研究基金　Joint Climate Research Fund（JCRF）
联合气象卫星咨询委员会(美国)　Joint Meteorological Satellite Advisory Committee(JMASC)
联合数值天气预报中心　Joint Numerical Weather Prediction Unit(JNWPU)
联合特征函数　joint characteristic function
联合研究中心(欧洲共同体委员会)　Joint Research Centre(CEC)
联合张弛　simultaneous relaxation
联合组织委员会(WMO/ICSU)　Joint Organizing Committee(JOC, WMO/ICSU)
联机【计】　on-line
联机分析处理【计】　on-line analytical processing
联立方程　simultaneous equation
链长【化】　chain length
链式法则【数】　chain rule
链式反应【物】　chain reaction
链烷烃　paraffins
链〔状〕闪〔电〕　chain lightning
良好能见度　good visibility
良好天气进场着陆　clear-weather approach
凉波　cool wave
凉季【大气】　cool season
两分〔法〕　dichotomy
两分类预报　binary prediction
两年〔的〕　biennial
亮带　bright band
亮带回波　bright band echo
亮度　brightness
亮度【物】　luminosity, luminance
亮度对比【大气】　contrast of luminance
亮度对比　brightness contrast
亮度分布　brightness distribution
亮度分量　brightness component
亮〔度〕温〔度〕　luminance temperature

亮度温度【大气】 brightness temperature
亮日照 bright sunshine
亮日照时数 bright sunshine duration
亮网络区(色球层) bright network
凉季【大气】 cool season
量 magnitude
量【数】 quantity
量杯 graduated glass cylinder
量纲 dimension
量纲【物】 dimensions
量纲参数 dimensional parameter
量纲方程 dimensional equation
量纲分析【物】【大气】 dimensional analysis
量规函数 calibrating function
量化信号【大气】 quantized signal
量级关系 ordering relationship
量距 span
量器 gauge, gage
量热器【物】 calorimeter
量热湿度表 thermal hygrometer
量热学【物】 calorimetry
〔量〕雪秤 density-of-snow gauge
量雪尺 snow level, insert for snow
〔量〕雪尺【大气】 snow scale
量雪尺【大气】 snow depth scale
量雪器 snow gauge
量雪箱 snow bin
量云器 nephometer
量子【物】 quantum
量子电子学【电子】 quantum electronics
量子化【物】 quantization
量子计算机【计】 quantum computer
量子可预测性分析 quantum predictability analysis
量子理论【物】 quantum theory
量子力学【物】 quantum mechanics
量子数【物】 quantum number
量子效率 quantum efficiency
列表【数】 tabulation
列车效应 train effect
列夫斯达图 Refsdal diagram
列联表【数】 contingency table
列纳效应 Lenard effect
列氏温标 Reaumur temperature scale
列氏温度表 Reaumur thermometer
列线 alignment
列线图 nomogram【大气】, nomograph, nomographic chart
列线图解法 nomography
烈飑 heavy squall
烈风【大气】 strong gale
裂变 fission
裂变产物化学【化】 fission product chemistry
裂变化学【化】 fission chemistry
裂变径迹定年法 fission track dating
裂变气体〔产物〕【化】 fission gas
裂变室 fission chamber
裂变物 fission product
裂变阈【物】 fission threshold
裂冰〔作用〕 calving
裂缝 fissure
裂流 rip
裂片结构 lobe structure
裂隙 crack
裂云〔天空〕(云量 6～9) broken
邻接矩阵【数】 adjacent matrix
邻近点检验 buddy check
林地沼泽【地理】 forest mire
林格曼数 Ringelmann number
林格曼图 Ringelmann chart
林格曼阴影 Ringelmann shades
林冠层【林】 forest canopy, canopy【大气】
林火〔天气〕预报【大气】 forest-fire 〔weather〕forecast
林吉曼小图 micro-Ringelmann chart
林克-福斯纳日射表 Linke-Feussner actinometer
林克浑浊因子 Linke turbidity factor

林克〔天空蓝度〕标　Linke-scale，Linke blue sky scale
林尼金层　Lineykin layer
林区【地理】　forest region
林学【林】　forest science，forestry
林业工程学【林】　forest engineering
林业经济学【林】　forest economics
林缘　forest border
临边变暗【大气】　limb darkening
临边发射辐射测量法　limb emission radiometry
临边发射压力调制辐射仪　limb emission pressure modulated radio-meter
临边反演【大气】　limb retrieval
临边辐射　limb radiance
临边辐射反演辐射仪　limb radiance inversion radiometer（LRIR）
临边辐射红外辐射仪　limb radiance infrared radiometer（LRIR）
临边辐射平流层探测器　limb radiance stratospheric sounder（LRSS）
临边昏暗〔现象〕　darkening towards the limb
临边扫描　limb scanning
临边扫描法【大气】　limb scanning method
临边扫描压力调制辐射仪　limb-scanning pressure modulated radio-meter
临边增亮【大气】　limb brightening
临界波长　critical wavelength
临界层　critical layer
临界潮〔水〕位　critical tidal level
临界成功指数　critical success index（CSI）
临界点【物】【大气】　critical point
临界风速　critical wind speed
临界感受器　critical receptor
临界高度　critical height
临界高度交互作用　critical level interaction
临界光长【大气】　critical day-length

临界角【物】　critical angle
临界雷诺数　critical Reynolds number
临界里查森数【大气】　critical Richardson number
临界流〔动〕　critical flow
临界流量　critical discharge
临界频率　critical frequency
临界破碎点　critical bursting point
临界区　critical region
临界山脉高度　critical mountain height
临界生长期　critical period of growth
临界水滴半径　critical drop radius
临界水深　critical depth
临界水深控制点　critical depth control
临界速度【物】　critical velocity
临界速率　critical speed
临界梯度　critical gradient
临界纬度　critical latitude
临界温度【物】　critical temperature
临界污染物　criteria pollutants
临界压力　critical pressure
临界液水含量　critical liquid water content
临界〔预报〕逃逸高度　critical level of escape
临界直减率　critical lapse-rate
临界值　critical value
临界状态　critical state
临界阻尼【物】　critical damping
临近预报　nowcasting
临近预报【大气】　nowcast
临时干旱影响区　emergency drought impact areas
临时业务试验机构　Interim Operational Test Facility（IOTF）
临阈信号　threshold signal
淋积矿床【地质】　leaching deposit
淋洗作用【土壤】　leaching
磷光　phosphorescence
磷酸二氢钾【电子】　potassium dihydrogen phosphate（KDP）

凛寒　keen
凛冽寒风　frigid blasts
灵敏度比　sensitivity ratio
灵敏度分析【数】　sensitivity analysis
灵敏度时间控制　sensitivity time control
灵敏度【物】　sensitivity
灵敏监视雷达　sensitive surveillance radar
岭回归　ridge regression
菱形截断　rhomboidal truncation
零层　zero layer, null layer
零差　homodyne
零磁倾角线　aclinic line
零度等温线　zero isotherm
零风速层顶【大气】　velopause
零级风【大气】　calm
零级云幂　ceiling zero
零阶闭合　zero-order closure
零阶不连续性　discontinuity of zero order
零零天气　zero-zero weather
零零条件　zero-zero
零厘米温度表【大气】　surface geothermometer
零平面位移　zero-plane displacement
零散阵雨　scattered shower
零时区　zero zone
零水平　zero-level
零维　zero-dimension
零维系统【物】　zero-dimensional system
零位电压【电子】　null voltage
零温度层　zero-temperature level
零下温度　subzero temperature
零线【物】　zero line
零压气球　zero-pressure balloon
零值磁偏线　agonic line
领海【航海】　territorial water, territorial sea
领航图　pilot chart
领空【地理】　territorial sky
领土【地理】　territory
领域知识　domain knowledge
领域专家【计】　domain expert
刘易斯数　Lewis number
浏览器【计】　browser
流　current
流变定律　flow law
流冰【海洋】　drift ice
流冰带【大气】　ice belt
流程图　flow chart
流程图【计】　flow diagram, flowchart
流出急流　exit jet
流出平流　advection of effluent
流出通量　outflux
流〔动〕　flow
流动窗方法　moving-window method
流动观测　moving observation
流动平衡　flow equilibrium
流动气象站　mobile weather station
流动人口【地理】　floating population
流动沙丘【地理】　wandering dune, moving dune
流动性　mobility
流管【物】　stream tube
流函数　stream function, current function
流溅　swash
流量表　flowmeter
流量测定　stream gauging
流量对照分析　double mass analysis
流量计　stream gauge
流量曲线　stream hydrograph
流量系数【水利】　discharge coefficient
流量演算　hydrological routing
流玫瑰图　current rose
流密度　current density
流明　lumen(lm)
流入边界　inflow boundary
流沙　shifting sand, saltation, quick sand
流水　ice drift
流速　flow velocity
流速管　Venturi tube
流速廓线仪　current profiler

流速流向曲线　current curve
流速-面积法　velocity-area method
〔流〕速〔水〕头　velocity head
流体　fluid
流体动力不稳定〔性〕　hydrodynamic instability
流体动力粗糙面　hydrodynamically rough surface
流体动力稳定〔性〕　hydrodynamic stability
流体动力学　hydrodynamics, fluid dynamics
流体动力学方程　hydrodynamic(al) equation, fluid dynamics equation
流体动力学基本方程　fundamental equations of hydrodynamics
流体动力学扩散　hydrodynamic(al) dispersion
流体动力噪声【海洋】　hydrodynamic noise
流体动能学　hydrokinetics
流体动压力　hydrodynamic pressure
流体静力波　hydrostatic wave
流体静力不稳定度【大气】　hydrostatic instability
流体静力层　hydrostatic layer
流体静力调整　hydrostatic adjustment
流体静力方程【大气】　hydrostatic equation
流体静力假设　hydrostatic assumption
流体静力近似【大气】　hydrostatic approximation
流体静力亏损　hydrostatic defect
流体静力模式　hydrostatic model
流体静力平衡　hydrostatic equilibrium, hydrostatic balance
流体静力稳定〔性〕　hydrostatic stability
流体静力学【物】　hydrostatics
流体静力延迟　hydrostatic delay
流体静压〔强〕【物】　hydrostatic pressure

流体静压位势　potential due to hydrostatic pressure
流体块　fluid parcel
流体力学　hydromechanics, fluid mechanics
流体力学体积【化】　hydrodynamic volume
流网　flow net
流线　flow line
流线【大气】　streamline
流线场　streamline field
流线分析　streamline analysis
流线流　streamline flow
流线式涡度　streamwise vorticity
流线图【大气】　streamline chart
流线型　streamline pattern
流线型风标　streamline vane
流泄风【大气】　drainage wind
流星　meteor【天】, shooting star
流星尘埃　meteoric dust
〔流星〕电离余迹　ionized trail
流星轨迹　meteor trajectory
流星列　meteor trains
〔流星〕气冠　gas cap
流星群　meteor stream（或 swarm）
流星烧蚀　meteoritic ablation
流星瞬现尾迹　meteor wake
流星体　meteoroid【天】, meteoritic material
流星尾迹　meteor trail
流星雨　meteor shower, meteoric shower
流形　manifold
流型　flow pattern
流域　basin, watershed, drainage area, drainage basin, river basin
流域出口　basin outlet
流域规划　river basin planning
流域回灌　basin recharge
流域平均降水量　mean basin precipitation
流域水量平衡　basin accounting

流域响应　basin response
流域形态【地理】　watershed morphology
流域预报　river forecast
流域滞时　basin lag
硫尘　sulphur dust
硫代甲醛　thioformaldehyde
硫化【化】　sulfurization
硫化氢　hydrogen sulfide
硫化羰　carbonyl sulfide
硫化雨　sulphur rain
硫酸（H_2SO_4）　sulphuric acid
硫酸铵　ammonium sulfate
硫酸气溶胶　sulfaric acid aerosol
硫酸雾　sulphur mist
硫酸盐　sulfate
硫酸盐气溶胶　sulfate aerosol
硫循环　sulphur cycle, sulfur cycle
硫氧化物（SO_x）　sulphur oxides
榴弹　grenade
榴弹测风法　grenade method
榴弹探测〔法〕　grenade sounding
柳井波　Yanai wave
六方晶系　hexagonal system
六氟化硫（SF_6）　sulfur hexafluoride
六角板状（冰晶）　hexagonal platelet
六角柱体（冰晶）　hexagonal column
龙格-库塔法【数】　Runge-Kutta method
龙骨　keel
龙卷　spout
龙卷【大气】　tornado
龙卷爆发　tornado outbreak
龙卷躲避所　tornado cave
龙卷风暴　tornadic storm
龙卷钩　tornado hook
龙卷回波【大气】　tornado echo
龙卷监测　tornado watch
龙卷雷暴　tornadic thunderstorm
龙卷漏斗柱　trunk, funnel（或 funnel column）
龙卷路径　tornado track
龙卷气旋　tornado cyclone

龙卷区　tornado belt
龙卷日珥　tornado prominence
龙卷通道　tornado alley
龙卷涡旋　tornado vortex
龙卷涡旋特征　tornado vortex signature
龙卷涡旋信号　tornadic vortex signature（TVS）
龙卷行迹　tornado swatch
龙卷预报　tornado forecast
龙卷预警　tornado warning
龙卷灾害　tornado damage
龙卷族　family of tornadoes
隆冬　midwinter
隆起　heave
垄作　ridge cultivating
漏斗云　pendant cloud, funnel cloud【大气】
漏斗云〔柱〕　funnel column
漏斗状对流层顶　tropopause funnel
漏警概率【电子】　alarm dismissal probability
漏水含水层　leaky aquifer
漏隙层积云　stratocumulus perlucidus（Sc pl）
漏隙高积云　altocumulus perlucidus
漏隙（云）　perlucidus（pe）
卤化碳　halocarbons
卤水　saltwater, brine
卤素【化】　halogen
陆半球　continental hemisphere
陆标　landmark, ground control point
陆标【航海】　terrestrial object
陆冰【地理】　land ice
陆成风　terrestrial wind
陆地测站月平均与总量报告　CLIMAT（report of monthly means and tatals from a land station）
陆地观测飞机　observation landplane
陆地环境【地质】　land environment
陆地生态系统　terrestrial ecosystem

陆地卫星【大气】 Land satellite
　(＝Landsat)
陆地卫星【地理】 Landsat
陆地效应【航海】 land effect
陆地侦察机　patrol landplane
陆风【大气】 land breeze
陆风风速　land breeze speed
陆风锋　land breeze front
陆风厚度　land breeze depth
陆风气候学　land breeze climatology
陆风温度　land breeze temperature
陆高海深曲线　hypsographic curve
陆架冰　shelf ice
陆龙卷　landspout
陆面参数化　land surface parameterization
陆面过程　land surface process
陆面蒸发　evaporation from land
陆面蒸发【地理】 land evaporation
陆棚【大气】 continental shelf
陆气交互作用　land-atmosphere interaction
陆圈　continental sphere
陆生植物【植】 terrestrial plant
陆雾　land fog
陆相【地质】 continental facies
陆旋风　terrestrial whirl wind
陆用气压表　land-barometer
陆源冰【海洋】 land-origin ice
陆照云光　land sky
录发（单角散射仪） Loofah
录音通信　voicespondence
路灯晕　streetlamp halo
路径　track
路径【物】 path
路径法　path-method
路径辐射亮度　path radiance
路线考察　moving exploration through line
路由器【计】 router
滤波【数】【大气】 filtering
滤波方程　filter equationm, filtered equations
滤波分析　filter analysis
滤波函数　filtering function, filter function
滤波近似法　filtering approximation
滤波器　filter
滤波楔形光谱仪　filter wedge spectrometer（FWS）
滤光光度表　filter photometer
滤光片　light filter
露【大气】 dew
露持续时间记录器　dew duration recorder
露滴　dew-drop
露点　fog point
露点【物】 dew-point
露点表　dew point meter
露点测定器　dew-point apparatus
露点锋【大气】 dew point front
露点公式　dewpoint formula
露点记录仪　dew-point recorder
露点湿度表【大气】 dew-point hygrometer
露点探空仪　dew-point radiosonde
露点图　depegram
露点位温　potential dew-point temperature
露点〔温度〕【大气】 dew-point〔temperature〕
露点温度　dew point temperature
露点线　dew-point line
露冠　dew-cap
露虹【大气】 dewbow
露量表【大气】 drosometer
露量计　drosograph
露量器【大气】 dewgauge
露面宝光　Heiligenschein, hellinis halo, shine, Cellini's halo
露水板　dew plate
鲁棒性　robustness
旅游点【地理】 tourist point
旅游业【地理】 tourism

旅游资源【地理】 tourist resources
率 quotiety
绿边 green rim
绿冰山 green iceberg
绿带城市【地理】 green belt city
绿带(无霜带) green belt
绿弓形 green segment
绿化【林】 greening
绿雷暴 green thunderstorm
绿日 green sun
绿色革命【地理】 green revolution
绿闪〔光〕 green flash
绿射线 green ray
绿天 green sky
绿雪 green snow
绿月 green moon
绿洲 oasis
绿洲耕作【地理】 oasis cultivation
绿洲效应【大气】 oasis effect
氯〔的〕氧化物(ClO_x) chlorine oxides
氯丁橡胶气球 polychloroprene balloon, neoprene balloon
氯度【大气】 chlorosity
氯仿 chloroform
氯氟碳化物【大气】,氯氟烃【大气】 chlorofluorocarbons (CFCs)
氯含量【大气】 chlorinity
氯化铵 ammonium chloride
氯化锂 lithium chloride
氯化氢 hydrogen chloride
氯化物 chlorine compounds
氯化物自动分析仪 automatic chloride analyzer
孪生低气压 dumbbell depression
孪生涡旋 vortex twinning
乱短波 chop
乱反射 irregular reflection
乱卷云 cirrus intortus
乱(云) intortus(in)

略图 schematic diagram
伦敦雾 London fog
伦琴射线 Rontgen ray
轮宽指数 ring-width index
轮形气压表 wheel-barometer
论证【数】 justification
罗别茨双金属天空辐射计 Robitzsch bimetallic pyranograph
罗别茨直接辐射计 Robitzsch actinograph
罗宾常数【数】 Robin constant
罗宾气球 Robin balloon
罗宾球 Robin sphere
罗经参考频率 gyro reference frequency
罗兰〔导航仪〕 long range navigation (LORAN)
罗兰C探空仪 Loran-C sonde
罗盘 compass
罗盘经纬仪【测】 compass theodolite
罗式温〔度表〕罩 Rosemount temperature housing
罗斯贝变形半径 Rossby radius of deformation
罗斯贝变形内半径 internal Rossby radius of deformation
罗斯贝波【大气】 Rossby wave
罗斯贝波方程 Rossby wave equation
罗斯贝波辐射 Rossby-wave radiation
罗斯贝波共振互作用 resonance interaction of Rossby waves
罗斯贝参数【大气】 Rossby parameter
罗斯贝公式 Rossby formula
罗斯贝-豪威茨波 Rossby-Haurwitz wave
罗斯贝廓线 Rossby profile
罗斯贝临界速度 Rossby critical velocity
罗斯贝流型 Rossby manifold
罗斯贝能量循环 Rossby energy cycle
罗斯贝平面 Rossby plane

罗斯贝数【大气】 Rossby number
罗斯贝数相似 Rossby number similarity
罗斯贝图 Rossby diagram
罗斯贝项 Rossby term
罗斯贝延迟〔量〕 Rossby retardation
罗斯贝域【大气】 Rossby regime
罗斯贝重力波 Rossby gravity wave
罗斯贝驻波 stationary Rossby wave
罗坦系统 long range and tactical navigation system(LORTAN)
罗维兹弧 vertical arcs of the 22° parhelia
罗伍渗压计 Rowe osmometer
罗伊库计数器 Royco counter
逻辑电路 logic circuit
逻辑斯谛分布【数】 logistic distribution
逻辑斯谛回归【数】 logistical regression
逻辑值 logical value
螺线层 spiral layer
螺线平衡 spiral equilibrium
螺线【数】 spiral
螺旋波 spiral wave
螺旋带 spiral band
螺旋度【物】 helicity
螺旋桨 propeller
〔螺〕旋桨〔式〕风标 propeller vane
螺旋桨式风速表 helicoidal anemometer
螺旋桨〔式〕风速表【大气】 propeller anemometer
螺旋扫描 spiral scanning
螺〔旋〕线【数】 helix
螺旋形扫瞄 helical scanning
螺旋形天线 helical antenna
螺旋雨带 spiral rain-band
螺旋雨带回波【大气】 spiral rain-band echo
螺旋云带【大气】 spiral cloud band

螺旋〔状〕结构 spiral structure
螺旋状卷云 spiral cirrus
螺旋状云型 spiral pattern
裸冰【大气】 bare ice
裸地【大气】 bare soil
洛伦茨-米理论 Lorenz-Mie theory
洛伦茨吸引子【物】 Lorenz attractor
洛伦茨转换 Lorenz conversion
洛伦兹变换 Lorentz transformation
洛伦兹加宽 Lorentz broadening
洛伦兹廓线 Lorentz profile
洛伦兹力 Lorentz force
洛伦兹-洛伦茨公式【物】 Lorenz Lorenz formule
洛伦兹能量循环 Lorentz energy cycle
洛伦兹线 Lorentz line
洛伦兹线型 Lorentzian lineshape
洛杉矶光化学烟雾〔事件〕 Los Angeles (photochemical) smog
洛施密特常量【物】 Loschmidt number
洛维茨〔晕〕弧 arcs of Lowitz
络合物【化】 complex
落差 head
落潮 ebb
落潮【航海】 ebb tide
落潮流【航海】 ebb stream, ebb current
落潮流落潮间隙 ebb interval
落潮强度 ebb strength
落尘图 fallout plot
落球法 falling-sphere method
落球〔法〕 falling sphere
落叶阔叶林【林】 deciduous broadleaved forest
落叶林【地理】 deciduous forest
落叶雪林气候 deciduous snow forest climate

M

马达通风测温防辐射罩　motor aspirated temperature shield
马登-朱利安振荡　Madden-Julian oscillation
马尔可夫过程　Markovian
马尔可夫过程【数】　Markov process
马尔可夫链　Markov chain
马尔可夫偏差　Markov deviation
马尔维纳斯海流　Malvinas Current
马尔文日照计　Marvin sunshine recorder
马格努斯公式　Magnus formulas
马格努斯力　Magnus force
马格努斯效应　Magnus effect
马古列斯方程　Margules's equation
马古列斯公式【大气】　Margules' formula
马古列斯模式　Margules's model
马哈兰诺距离　Mahalanobian distance
马赫数【大气】　Mach number
马赫原理【天】　Mach principle
马略特定律　Mariotte's law
马萨诸塞州理工学院（美国）　Massachusetts Institute of Technology(MIT)
马斯登方块　Marsden square
马斯登图　Marsden chart
马斯京根〔方〕法　Muskingum method
马斯克林高压　Mascarene high
马尾藻海〔域〕【海洋】　Sargasso Sea
马纬度【大气】　horse latitudes
马纬度高压　horse latitude high
马歇尔-帕尔默关系　Marshall-Palmer relation
马歇尔-帕尔默雷达雨量函数　Marshall-Palmer radar rainfall function
马歇尔-帕尔默雨滴谱　raindrop size distribution of Marshall and Palmer
马歇尔-帕尔默〔雨滴〕谱　Marshall and Palmer distribution
码分多址（一种数字通信技术）　code division multiple access(CDMA)
霾【大气】　haze
霾层　haze layer
霾层顶　haze horizon
霾滴　haze droplet
霾面　haze level
霾气溶胶　haze aerosol
霾线　haze line
迈尔斯检定法　Myers rating
迈克耳孙干涉仪【物】　Michelson interferometer
迈克耳孙直接辐射计　Michelson actinograph
迈内尔带（羟基发射带）　Meinel bands
迈则松图　Matheson diagram
麦基效应　Macky effect
麦克劳林公式【数】　Maclaurin formula
麦克劳林级数　Maclaurin series
麦克斯速度分布【物】　Maxwell velocity distribution
麦克斯韦-玻耳兹曼分布　Maxwell-Boltzmann distribution
麦克斯韦-玻耳兹曼碰撞方程　Maxwell-Boltzmann collision equation
麦克斯韦定律　Maxwell's law
麦克斯韦方程　Maxwell's equation
麦克斯韦方程组【物】　Maxwell equations
麦克斯韦速率分布【物】　Maxwell speed distribution
麦肯西高压　McKenzie high
脉冲　impulse
脉冲【电子】　pulse

脉冲编码　pulse coding
脉冲长度调制　pulse length modulation
脉冲长度(线距)　pulse length
脉冲持续时间　pulse duration
脉冲重复频率【电子】　pulse repetition (或 recurrence) frequency(PRF)
脉冲重复时间　pulse repetition time (PRT)
脉冲传播　pulse propagation
脉冲电离层探测器　pulse ionosound
脉冲电离层探测装置　pulse ionosonde
脉冲对处理　pulse-pair processing
脉冲对处理器　pulse pair processor (PPP)
脉冲多普勒雷达【电子】　pulse Doppler radar
脉冲幅度编码调制　pulse-amplitude-code modulation
脉冲光　pulsed light
脉冲光云高指示器　pulsed-light cloud-height indicator
脉冲积分　pulse integration
脉冲积分器　pulse integrator
脉冲激光器【物】　pulsed laser
脉冲计数式探空仪　pulse-counting sonde
脉冲加宽　pulse broadening
脉冲间歇周期　interpulse period (IPP)
脉冲宽度【电子】　pulse width
脉冲雷达【电子】【大气】　pulse radar, pulsed radar
脉冲频度调制　pulse number modulation
脉冲频率　pulse frequency
脉冲时间调制探空仪　pulse-time-modulated radiosonde
脉冲双基地散射　bistatic scattering of pulse
脉冲探测仪　pulse sounder
脉冲体积　pulse volume
脉冲调频　pulse frequency modulation
脉冲调相　pulse phase modulation

脉冲调制【电子】　pulse modulation
脉冲氙灯【物】　pulse xenon lamp
脉冲响应　impulse response
脉冲响应函数　impulse response function
脉冲星【物】　pulsar
脉冲云幂仪　pulse ceilometer
脉冲直接辐射表　pulse actinometer
脉冲周期　pulse period
脉动　pulsation
脉动带　belt of fluctuation
脉动极光　pulsating aurora
脉动气压　fluctuating pressure
脉动湿度　fluctuating humidity
脉动速度　fluctuation velocity, fluctuating velocity
脉动温度　fluctuating temperature
脉动状极光辉　pulsating auroral glow
脉幅调制【电子】　pulse-amplitude modulation
脉积压缩　pulse compression
脉宽调制【电子】　pulse-width modulation, pulse duration modulation(PDM)
脉码调制　pulse coded modulation
脉码调制【电子】　pulse code modulation (PCM)
脉时调制【电子】　pulse time modulation
脉位调制　pulse-position modulation
满潮时距　lunitidal interval
满宁方程　Manning equation
满足序列　complacent series
幔　veil
慢波　slow wave
慢波簇　slow manifold
慢调整　slow adjustment
慢离子　slow ion
慢模【物】　slow mode
慢模方程　slow mode equation
慢模约束初始化　slow mode constrained initialization
慢尾　slow tail

慢行冷锋【大气】 slowly moving cold front
慢性污染 chronic pollution
慢中子【物】 slow neutron
漫反射【大气】 diffuse reflection
漫灌 flood irrigation
漫灌【农】 flooding irrigation
漫散射 diffuse scattering
漫射 diffuse
漫射表 diffusometer
漫射场 diffuse field
漫〔射〕反射体 deffuse reflector
漫射辐射【大气】 diffuse radiation
漫射光 diffused light, deffuse light
漫射光照 diffuse illumination
漫射近似 diffusion approximation
漫射率因子 diffusivity factor
漫射强度 diffuse intensity
漫射入射强度 diffuse incident intensity
漫射太阳辐射【大气】 diffuse solar radiation
漫射体【物】 diffuser
漫射天光 diffuse skylight
芒克边界层(海洋) Munk boundary layer
芒塞尔色图 Munsell color chart
芒塞尔色系【测】 Munsell color system
芒塞尔土色卡【土壤】 Munsell soil charts
芒种(节气)【农】【大气】 Grain in Ear
忙音 busy tone
盲目飞行条件 blind (flying) condition
盲目预报 blind prognose
盲区 shadow, blind area
盲速 blind speed
毛发湿度表 Saussure's hygrometer
毛发湿度表【大气】 hair hygrometer
毛发湿度计【大气】 hair hygrograph
毛管持水量 capillary moisture capacity
毛管收集器 capillary collector
毛卷层云【大气】 cirrostratus filosus (Cs fib)
毛卷层云 cirrostratus fibratus
毛卷云【大气】 cirrus filosus (Ci fib)
毛卷云 cirrus fibratus
毛毛雨【大气】 drizzle
毛毛雨(英拼中文) mao mao yuh
毛毛雨滴 drizzle drop
毛毛雨雾 drisk
毛毛雨型 drizzle type
毛〔丝〕状〔云〕 filosus
毛〔细〕管 capillary
毛〔细〕管传导性 capillary conductivity, unsaturated hydraulic conductivity
毛〔细〕管订正 capillarity correction
毛〔细〕管静电计 capillary electrometer
毛〔细〕管孔隙 capillary interstice
毛〔细〕管力 capillary forces
毛〔细〕管涟波 capillary ripple, capillary wave
毛细管气相色谱仪【化】 capillary gas chromatograph
毛〔细〕管上升〔高度〕 capillary rise
毛〔细〕管水〔分〕 capillary water
毛〔细〕管水扩散〔作用〕 capillary diffusion, capillary movement
毛〔细〕管水头 capillary head
毛〔细〕管位势 capillary potential
毛〔细〕管吸力 capillary suction
毛〔细〕管现象 capillarity【大气】, capillary phenomenon
毛〔细〕管压力 capillary pressure
毛〔细〕管滞后 capillary hysteresis
毛细下降 capillary depression
毛细现象【物】 capillarity
毛细作用 capillary action
毛雨雾 drizzling fog
毛(云状) fibratus (fib)
毛重 gross weight
锚冰【大气】 anchor ice
锚槽【大气】 anchored trough
锚定平台 anchored platform
锚定气球 anchor balloon
锚定站 anchored station

茂草　tall grass
冒纳罗亚观象台(夏威夷)　Mauna Loa Observatory
帽〔状〕云　cloud cap, cap cloud
贸易风【大气】　trade winds
梅勒-山田参数化　Mellor-Yamada parameterization
梅默里周期　Memery period
梅森湿度表　Mayson's hygrometer
梅雨　mold rain
梅雨【大气】　Meiyu, plum rain
梅雨锋【大气】　Meiyu front
梅雨量　rainfall amount during Meiyu period
梅雨期【大气】　Meiyu period
煤成气【地质】　coal gas
煤地球化学【地质】　coal geochemistry
煤化作用【地质】　coalification
煤烟沉积　soot deposit
煤烟沉降　sootfall
煤烟含量　soot content
煤烟浓度　soot concentration
酶　enzyme
每日　diurnal
每日天气图　daily weather chart (=map)
每日预报　daily forecast
美国标准大气　United States Standard Atmosphere (USSA)
美国大气科学研究所　Institute for Atmospheric Sciences (IAS)
〔美国〕大气研究大学联合会　University Cooperation for Atmospheric Research (UCAR)
美国地球物理学联合会　American Geophysical Union (AGU)
美国电气和电子工程师协会　Institute of Electrical and Electronic Engineers (IEEE)
美国范登堡空军基地(加利福尼亚州)　Vandenberg Air Force Base, California (VAFB)

美国高空气象研究所　American Institute of Aerological Research (AIAR)
美国国际民航组织补充标准大气　U. S. Extension to the Standard ICAO Atmosphere
〔美国〕国家标准局　National Bureau of Standards (UBS)
〔美国〕国家潮汐基准面时期　National Tidal Datum Epoch
〔美国〕国家大气研究学会　National Institute of Atmospheric Research (NIAR)
美国国家地球卫星局　National Earth Satellite Service (NESS)
〔美国〕国家海洋大气局　National Oceanic and Atmospheric Administration (NOAA)
〔美国〕国家海洋大气局数据浮标处　NOAA Data Buoy Office (NDBO)
〔美国〕国家海洋大气局数据浮标中心　NOAA Data Buoy Center (NDBC)
〔美国〕国家海洋大气局有线气象通信系统　NOAA weather wire system (NWWS)
〔美国〕国家海洋卫星系统　National Ocean Satellite System (NOSS)
美国国家航空和航天局　National Aeronautics and Space Administration (NASA)
美国国家航空和航天局标准大气　NASA standard atmosphere
美国国家航空和航天理事会　National Aeronautics and Space Council (NASC)
〔美国〕国家航空天气咨询处　National Aviation Weather Advisory (Unit) (NAWAU)
〔美国〕国家航空咨询委员会　National Advisory Committee for Aeronautics (NACA)
美国国家航空咨询委员会标准大气　NACA Standard Atmosphere
〔美国〕国家环境空气质量标准　National

Ambient Air Quality Standards (NAAQS)

美国国家环境卫星局　National Environmental Satellite Service（NESS, NOAA）

〔美国〕国家环境卫星中心　National Environmental Satellite Center（NESC）

〔美国〕国家环境卫星、资料和信息局　National Environmental Satellite, Data and Information Service（NESDIS）

美国国家环境预报中心　National Centers for Environmental Prediction, USA（NCEP）

美国国家环境预报中心/美国国家大气研究中心再分析资料　NCEP/NCAR reanalysis

〔美国〕国家飓风和实验气象学研究所　National Hurricane and Experimental Meteorology Laboratory（NHEML）

〔美国〕国家飓风实验室　National Hurricane Research Laboratory（NHRL）

〔美国〕国家飓风中心（NOAA）　National Hurricane Center（NHC, NOAA）

〔美国〕国家科学基金会　National Science Foundation（NSF）

〔美国〕国家科学院　National Academy of Sciences（NAS）

美国国家科学院大气科学委员会　NAS Committee on Atmospheric Sciences（NASCAS）

美国国家科学院海洋委员会　NAS Committee on Oceanography（NASCO）

〔美国〕国家气候中心　National Climatic Center（NCC）

〔美国〕国家气象局　National Weather Service（NWS, USA）

〔美国〕国家晴空湍流研究委员会　National Committee for Clear Air Turbulence（NCCAT）

〔美国〕国家天气记录中心　National Weather Record Center（NWRC）

〔美国〕国家研究理事会　National Research Council（NRC）

〔美国〕国家业务气象卫星系统　National Operational Meteorological Satellite System（NOMSS）

美国海岸导航指南　United States Coast Pilot

〔美国〕海军天气研究中心　Naval Weather Research Facility（NWRF, US Navy）

美国航空〔天气〕代码　U. S. airways code

美国环境保护局　Environmental Protection Agency（EPA）

美国火箭学会　American Rocket Society（ARS）

美国极地研究会　American Polar Society（APS）

美国科学发展促进会　American Association for the Advancement of Science（AAAS）

〔美国〕肯尼迪空间中心　Kennedy Space Center（KSC）

〔美国〕空军地球物理实验室　Air Force Geophysics Laboratory（AFGL）

美国空军环境技术应用中心　United States Air Force Environmental Technical Application Center（USAFETAC）

〔美国〕空军剑桥研究室　Air Force Cambrige Research Laboratories（AFCRL, USAF）

〔美国〕空军全球天气中心　Air Force Global Weather Central（AFGWC）

美国空军〔研究发展司令部的〕标准大气　ARDC model atmosphere

美国兰利研究中心　Langley Research Center（LRC）

〔美国〕联邦航空局　Federal Aviation Administration（FAA）

美国联合台风预警中心（美国空海军）　Joint Typhoon Warning Center

（JTWC, USAF and US Navy）
美国刘易斯研究中心　Lewis Research Center (LeRC)
美国陆海军地面气象装备（高空探测用的地面设备）　Army-Navy ground meteorological device (AN/GMD)
美国洛克希德导弹和空间公司　Lockheed Missile and Space Company (LMSC)
美国气象局　United States Weather Bureau (USWB)
〔美国〕气象局气象站　Weather Service Meteorological Observatory (WSMO)
〔美国〕气象局预报台　Weather Service Forecast Office (WSFO)
美国气象学会　American Meteorological Society (AMS)
美国太阳能学会　Solar Energy Society of America (SESA)
美国天气中队　Weather Squadron, U.S.A.
〔美国〕西部试验场　Western Test Range (WTR, VAFB)
美国研究发展局　Research and Development Board of America (RDB)
〔美国〕业务和服务自动化系统　automation of field operation and services (AFOS)
美国原子能委员会　Atomic Energy Commission (AEC)
美中学术联络委员会（美国）　Committee on Scholarly Communication with the People's Republic of China (CSCPRC)
闷热　sultriness
闷热天气　muggy weather, oppressive weather
萌发【植】　germination
朦胧现象　obscuring phenomenon
猛烈雷暴　severe thunderstorm, rattler
蒙德极小期【天】　Maunder minimum
蒙哥马利函数　Montgomery function
蒙哥马利流函数　Montgomery stream function
蒙哥马利势　Montgomery potential
蒙古低压【大气】　Mongolian low
蒙古反气旋　Mongolia anticyclone
蒙古气旋【大气】　Mongolia cyclone
蒙气差　refraction
蒙特卡罗方法【数】【大气】　Monte Carlo method
蒙特卡罗模式　Monte Carlo model
蒙特利尔议定书　Montreal Protocol
濛雨天气【大气】　Crachin
孟加拉湾风暴【大气】　storm of Bay of Bengal
孟加拉湾水　Bay of Bengal Water
弥散长度　diffuse length
弥散极光亮斑　diffuse auroral patch
弥散剂　dispersing medium
弥散系数【土壤】　dispersion coefficient
弥散相态　dispersed phase
迷向【数】　isotropic
迷向随机函数　isotropic random function
迷向直线【数】　isotropic line
醚【化】　ether
米　metre, meter
米·吨·秒单位制　mts system
米尔恩问题　Milne problem
米尔河文化　Mill Creek culture
米尔黑德传真图像复制设备　Muirhead Facsimile Copier Equipment (MUFAX)
米后向散射　Mie backscattering
米后向散射系数　Mie backscattering efficiency
米兰科维奇更新世气候变化　Milankovitch Pleistocene climate variation
米兰科维奇假说　Milankovitch hypothesis
米兰科维奇理论　Milankovitch theory
米兰科维奇太阳辐射曲线　Milankovitch solar radiation curve

米兰科维奇循环　Milankovitch cycle
米兰科维奇振荡　Milankovitch oscillation
米勒气候分类法　Miller's climatic classification
米理论　Mie theory
米粒组织　granules
米粒组织【天】　granulation
米·千克·秒制　mks system
米散射【大气】　Mie scattering
米散射理论　Mie scattering theory
米消光　Mie extinction
米雪　granular snow
米雪【大气】　snow grains
米制　metric system
秘鲁〔冷〕海流　Peru Current
秘鲁/智利海流　Peru/Chile Current
密度【物】　density
密度差异混合　diapycnal mixing
密度订正　density correction
密度分布　density distribution
密度高度　density altitude
密度函数　density function
密度计　densometer
密度流【大气】　density current
密度逆增　density inversion
密度梯度　density gradient
密度通道　density channel
密度位温　density potential temperature
密度温度　density temperature
密度演化【天】　density evolution
密度跃层【大气】　pycnocline
密度涨落【物】　density fluctuation
密封水银气压表　sealed mercury barometer
密集厚冰　consolidated ice
密卷云【大气】　cirrus densus（Ci dens）
密卷云　cirrus spissatus（Ci spi）
密勒辐射图　möller chart, Mløler chart
密码　cypher, cipher
密码【电子】　cipher code, cipher

密码分析【电子】　cryptanalysis
密码学【电子】　cryptography
密实度　compactness
密钥【电子】　cipher key
密云　spissatus（spi）
密云圈　overcast circle
密云隙　breaks in overcast
幂窗　power window
幂函数【数】　power function
幂级数【数】　power series
幂律廓线　power-law profile
绵雨　long-lasting precipitation
棉〔花〕带气候　cotton-belt climate
棉兰老海流　Mindanao Current
棉兰老涡流　Mindanao Eddy
棉球云　cotton ball cloud
棉区百叶箱（广泛用于二级气象站）　cotton-region shelter
冕洞　coronal hole
冕〔状〕闪〔电〕　crown flash
面发光率　luminous emittance
面积-高程曲线　area-elevation curve
面积平均　area average
面积守恒　conservation of area
面积速度　areal velocity
面积指数　area index
面降水【量】【大气】　areal precipitation
面蚀【土壤】　surface erosion
面源　plane source, area source
面阵【电子】　area array
描述函数【电子】　describing function
描述数据　descriptive data
〔描述性〕概念模式　descriptive model
描述性气候学　descriptive climatology
描述性气象学　descriptive meteorology
秒　second
秒差距　parallax second
秒差距【天】　parsec
民用日【天】　civil day
民用时　civil time
民用曙暮光　civil twilight

敏感度【计】 susceptibility
敏感性检验 sensitivity test
敏感性实验 sensitivity experiment
敏感性研究 sensitivity study
敏感序列 sensitive series
明渠流 open channel flow
明线电阻温度表 open-wire resistance thermometer
明语气象报告【大气】 plain-language report
冥外行星【天】 trans-Plutonian planet
冥王星【天】 Pluto
命令【计】 command
模【数】 module
模糊变量 fuzzy variable
模糊测度 fuzzy measure
模糊分类 fuzzy classification
模糊集【计】 fuzzy set
模糊理论 fuzzy theory
模糊逻辑 fuzzy logic
模糊认识 fuzzy knowledge
模糊数学【计】 fuzzy mathematics
模糊贴近度 fuzzy neartude
模糊微分方程 fuzzy differential equation
模糊信息 fuzzy information
模糊〔性〕 fuzziness
模糊〔性〕【数】 fuzzy
模糊遮光法 unsharp masking
模块式自动气象观测系统 modular automated weather system（MAWS）
模拟 modeling，analog(ue)
模拟【物】 simulation
模拟磁强计 analog magnetometer
模拟法 analogue method
模拟计算机 analog computer
模拟降水 simulated rainfall
模拟降雨装置【土壤】 rain simulator
模拟龙卷 laboratory tornado
模拟判据 modeling cirteria
模拟气候 simulated climate

模拟实验 model experiment
模拟试验【航海】 simulation test
模〔拟〕数〔字〕通用天气图〔传真〕记录器，模〔拟〕数〔字〕通用天气图传真收片机 A/D universal weather graphic recorder
模〔拟〕数〔字〕转换 analog-to-digital conversion
模〔拟〕数〔字〕转换器 analog digital converter，A/D converter
模拟信号带宽压缩调制解调器 analog bandwidth compression modem
模式大气 model atmosphere
模式对流层 model troposphere
模式分辨率【大气】 model resolution
模式概念【计】 schema concept
模式流 modular flow
模式偏差 model biases
模式平流层 model stratosphere
模式缺陷 model deficiency
模式输出统计 model output statistics （MOS）
模式输出统计预报【大气】 model output statistic prediction（MOS prediction）
模式水 mode water
模式体系 model hlerarchy
模〔式〕【物】 mode
模式校正 model calibration
模数【数】 modulus
模数转换【计】 analog-to-digital convert
模数转换器【计】 analog-to-digital converter
模态【大气】 mode
模型 model
膜片 diaphragm
膜片压力表 diaphragm manometer
摩擦 friction
摩擦层 friction layer【大气】，frictional layer，layer of frictional influence
摩擦二次流 frictional secondary flow

摩擦风【大气】 antitriptic wind
摩擦辐合【大气】 frictional convergence
摩擦辐散【大气】 frictional divergence
摩擦副层 frictional sublayer
摩擦高度 friction height
摩擦耗散 frictional dissipation
摩擦力 frictional force
摩擦力矩 frictional torque
摩擦起电 triboelectrification
摩擦水头 friction head
摩擦速度【大气】 friction velocity
摩擦损失 friction loss
摩擦系数【化】 frictional coefficient
摩擦效应 frictional effect
摩擦曳力【大气】 frictional drag
摩擦应力 frictional stress
摩擦〔影响〕深度 depth of frictional influence
摩擦阻抗 frictional resistance
摩尔【化】 mole
摩尔比法【化】 mole ratio method
摩尔比热容 molar specific heat capacity
摩尔分数 mole fraction
摩尔内能【化】 molar internal energy
摩尔气体常数【化】 molar gas constant, universal gas constant【大气】
摩尔热容【物】 molar heat capacity
摩尔数 mole number
摩尔体积【物】 molar volume
摩尔质量平均【化】 molar mass average
摩勒图 Moeller chart
磨菇云 mushroom cloud
末次冰期 last glacial
末次冰期冰盛期 Last Glacial Maximum (LGM)
末次间冰期 last interglacial
末期 telophase
莫尔-戈辛斯基日射总量表 Moll-Gorczynski solarimeter
莫尔恰诺夫盘 Moltchanov board
莫尔热电堆 Moll thermopile

莫尔斯电码 signal needle code, Morse code
莫莱特小波 Morlet wavelet
莫勒日照计 Maurer sunshine chronograph
莫宁-奥布霍夫长度 Monin-Obukhov length
莫宁-奥布霍夫尺度长度 Monin-Obukhov scaling length
莫宁-奥布霍夫方程 Monin-Obukhov equation
莫宁-奥布霍夫相似理论 Monin-Obukhov similarity theory
莫桑比克〔暖〕海流 Mozambique Current
漠盖层【地理】 desert pavement
墨卡托拼图 Mercator mosaic
墨卡托投影【测】 Mercator projection
墨西哥高压 Mexican high
〔墨西哥〕湾流【大气】 Gulf Stream
墨西哥帽小波 Mexican hat wavelet
母板【计】 masterboard
母函数【数】 generating function
母体物质 parent material
母涡旋 maternal vortex
母云 parent cloud, mother-cloud
木材工艺学【林】 wood technology
木星【天】 Jupiter
木质部【植物】 xylem
目标 destination
目标〔代〕码【计】 target code
目标观测 targeted observations, adaptive observations
目标函数【电子】 target function
目标函数【数】 objective function
目标空间扫描器 object space scanner
目标区【大气】 target area
目标区-对照区交叉 target-control cross-over
目标推理〔法〕 goal-directed reasoning
目标位置显示 target position indication

目标位置显示器　target position indicator
目标信号　target signal
目测　eye observation
目测【大气】　visual observation
目测光度表　visual photometer
目的港【航海】　port of destination
目录【地信】　directory
目录和出版委员会(WMO)　Commission for Bibliography and Publications (CBP,WMO)
目视飞行　VFR flight, contact fight, visual flight
目视飞行规则【电子】　visual flight rules(VFR)
目视飞行规则天气　VFR weather
目视飞行规则最低条件　VFR terminal minimums
目视飞行气象条件　visual meteorological condition
目视飞行天气　contact weather
目视间或仪表飞行　on and off instruments
目视进场高度　visual contact height, approach-light contact height
牧业区【地理】　pastoral region
暮春　late spring
暮光　night twilight
穆斯堡尔效应【物】　Mossbauer effect

N

纳米晶体【物】　nano-crystal
纳维-斯托克斯方程【数】　Navier-Stokes equation
钠(Na)　sodium
钠层【地】　sodium layer
钠钙玻璃罩　sodalime glass dome
钠谱　sodium spectrum
钠云　sodium cloud
钠蒸气尾迹测风法　sodium vapor tail wind technique
氖　neon
奈弗防护罩　Nipher shield
奈奎斯特波数　Nyquist wave number
奈奎斯特采样频率【计】　Nyquist sampling frequency
奈奎斯特间距　Nyquist interval
奈奎斯特判据【电子】　Nyquist criterion
奈奎斯特频率　Nyquist frequency
奈奎斯特速度　Nyquist velocity
奈奎斯特约束　Nyquist constraint
耐寒性　hardiness
耐旱性　drought resistance
耐涝作物　waterlogging tolerant crop
耐热性【农】　heat tolerance
耐热植物　heat-resisting plant
耐湿植物　damp tolerant plant
耐盐作物　salt tolerant crop
耐渍性　waterlogging tolerance
南　south (S)
南半球　Southern Hemisphere
南半球冬季　Southern winter
南半球副热带无风带　calms of Capricorn
南半球交换中心　Southern Hemisphere Exchange Centre (SHEC)
南半球夏季　Southern summer
南北摆动　latitudinal fluctuation
南北半球质量交换　interhemispheric mass exchange
南赤道海流　South Equatorial Current (SEC)
南赤道逆流　South Equatorial Counter-

current(SECC)
南赤道漂流　South Equatorial Drift Current
南大风　souther
南大西洋辐合带　South Atlantic convergence zone
南大西洋高压　South Atlantic high
南大西洋海流　South Atlantic Current
南大西洋中央水　South Atlantic Central Water
南大洋　Southern Ocean
南东南　south southeast（SSE）
南方涛动【大气】　Southern Oscillation（SO）
南方涛动指数　Southern Oscillation Index（SOI）
南焚风　south foehn
南海　South China Sea
南海【海洋】　Nanhai Sea
南海低压【大气】　South China Sea depression
南海反气旋　South China Sea anticyclone
南海高压　South China Sea high
南海季风试验　South China Sea Monsoon Experiment（SCSMEX）
南海暖流【海洋】　Nanhai Warm Current
南海沿岸流【海洋】　Nanhai Coastal Current
南寒带　south frigid zone
南回归线　south tropic
南回归线【地理】　Tropic of Capricorn
南极　austral pole，South Pole，Antarctic Pole
南极表层水　Antarctic Surface Water
南极冰原　Antarctic Ice Sheet
南极臭氧洞【大气】　Antarctic ozone hole
南极臭氧分布　Antarctic ozone distribution

南极底层水　Antarctic Bottom Water（AABW）
南极反气旋　antarctic anticyclone
南极锋【大气】　antarctic front
南极辐合区　Antarctic Convergence Zone
南极辐散〔带〕　antarctic divergence
南极高压　antarctic high
南极光　southern lights
南极光【地】【大气】　aurora australis
南极海蒸气雾　antarctic sea smoke
南极极锋　Antarctic Polar Front
南极极涡　Antarctic Polar Vortex
南极界【生】　Antarctic realm
南极平流层涡旋　antarctic stratospheric vortex
南极气候【大气】　antarctic climate
南极气团　antarctic air（mass）
南极区　Antarctic Zone
南极圈　Antarctic Circle
南极绕极水　Antarctic Circumpolar Water
南极绕极洋流　Antarctic Circumpolar Current（ACC）
南极涛动　AAO（Antarctic Oscillation）
南极西半球冰原　West Antarctic Ice Sheet
南极研究科学委员会（ICSU）　Scientific Committee on Antarctic Research（SCAR，ICSU）
南极中层水　Antarctic Intermediate Water
南美大草原　pampas
南热带无风带　south tropical calm zone
南森瓶　Nansen bottle
南太平洋辐合带　South Pacific convergence zone（SPCZ）
南太平洋海流　South Pacific Current
南纬　southern latitude
南温带　south temperate zone
南西南　south southwest（SSW）

南亚高压【大气】 South Asia high
南印度洋海流 South Indian Ocean Current
南支急流【大气】 southern branch jet stream
南爪哇海流 South Java Current
难近冰极 pole of inaccessibility
挠曲波 flexural wave
内边界 inner boundary
内边界层 internal boundary layer(IBL)
内变量 internal variable
内禀电阻率【物】 intrinsic resistivity
内禀方程【物】 intrinsic equation
内禀渗透率 intrinsic permeability
内波【航海】【大气】 internal wave
内波实验 Internal Wave Experiment (IWEX)
内部防雷系统 internal lightning protection system
内参数 internal parameter
内〔层〕眼壁 inner eyewall
内插多项式 interpolation polynomial
内插权重 interpolation weight
内插误差 error of interpolation
内潮 internal tide
内尺度(微尺度) inner scale (microscale)
内存保护【计】 memory protection
内存储器 internal memory
内弹道学 interior ballistics
内迭代【数】 inner iteration
内辐射带 inner radiation belt
内海【航海】 internal sea
内积【数】【大气】 inner product
内角【数】 interior angle
内界面 internal boundary
内界面层 interfacial layer
内聚性 cohesion
内涝 inland inundation
内联网【计】 intranet
内联网安全【计】 intranet security

内流 inflow
内流水系 blind drainage
内流轴 axis of inflow
内陆 inland
内陆冰 inland ice
内陆海风 inland sea breeze
内陆湖 endorheic lake, closed lake
内陆率 inlandity
内陆盆地【地】 interior basin
内陆气候 inland climate
内陆沙漠 inland desert
内陆雾 inland fog
内模 internal mode
内摩擦【力】【大气】 internal friction 〔force〕
内摩擦 inner friction
内能 internal energy
内潜热 internal latent heat
内腔式气体激光器 intracavity gas laser
内区 inner region
内水【航海】 internal waters
内稳定性参数 internal stability parameter
内向通量 inward flux
内行星【天】 inferior planet
内晕珥 parry arc
内在一致性 internal consistency
能带【物】 energy band
能级【物】 energy level
能级图【物】 energy level diagram
能见度【大气】 visibility
能见度表 visibility recorder
能见度表【大气】 visiometer, visibility meter
能见度测定表【大气】 nephelometer
能见度极好 ceiling and visibility unlimited
能见度降低 visibility reduction
能见度目标物 visibility object
能见度目标物【大气】 visibility marker
能见极限 limit of visibility

能见圆锥　cone of visibility
能量【物】　energy
能量不等式【数】　energy inequality
能量传递　energy transfer
能量串级　cascade of energy
能量串级【大气】　energy cascade
能量递降　degradation of energy
能量法　energy method
能量方程　energy equation
能量分辨率【物】　energy resolution
能量分配　partition of energy
能量锋　energy front
能量锋区　energy frontal zone
能量耗散　energy dissipation
能量耗散率　energy dissipating rate
能量和拟能守恒模式　energy and enstrophy conserving model
能量恒等式【数】　energy identity
能量积聚　energy storage
能量交换　energy exchange
能量密度　energy density
能量密度谱【大气】　energy density spectrum
能量模　energy norm
能量频散　energy dispersion
能量平衡　energy balance
能量平衡方程　energy balance equation
能量平衡模式【大气】　energy balance model（EBM）
能量平衡气候模式　energy balance climate models
能量平衡气候学　energy balance climatology
能〔量〕谱密度　energy spectral density
能量球面　energy sphere
能量收支　energy budget
能量守恒【大气】　conservation of energy
能量守恒【物】　energy conservation
能量守恒定律【物】　law of conservation of energy
能量守恒模式　energy-conserving model
能量守恒原理　principle of conservation of energy
能〔量〕输送　energy transport
能量损失　energy loss
能量通量　energy flux
能量图　energy diagram
能量吸收　energy absorption
能量消散　dissipation of energy
能量学【大气】　energetics
能量学方法　energetics method
能量循环　energy cycle
能量转换　energy transformation, energy conversion
能量转换方程　energy conversion equation
能流　energy flow
能流密度【物】　energy flux density
能坡线　energy grade line
能谱【物】【大气】　energy spectrum
能斯特方程【化】　Nernst equation
能听度　audibility
能听极限　limit of audibility
能听区　zone of audibility
能源气象学【大气】　energy source meteorology
能源危机　energy crisis
能源〔子〕系统　power subsystem
尼尔森雨量器　Nilsson rain-gauge
尼龙测湿元件　nylon unit
泥流【土壤】　mud flow
泥流〔作用〕　sludging
泥盆纪【地质】　Devonian period
泥盆系【地质】　Devonian System
泥球　mud ball
泥沙产量　sediment yield
泥沙取样器　sediment sampler
泥石流　mud-rock flow, earth flow
泥石流【地质】　debris flow
泥石流侵蚀【地质】　debris flow erosion
泥凇　mud rime

泥炭沼泽　muskeg
泥雨　mud rain
霓　secondary bow
霓【大气】　secondary rainbow
拟合　fitting
拟合法　fitting method
拟能串级　enstrophy cascade
拟牛顿法【数】　quasi-Newton method
拟线性偏微分方程【数】　quasi-linear partial differential equation
逆变换　inverse transform
逆潮　reversed tide
逆尺度方向　upscale direction
逆尺度转换　upscale transfer
逆风　dead wind, opposing wind, anti-wind, adverse wind
逆风【大气】　head wind
逆风航程　upwind range
逆风顺风比　upwind-downwind ratio
逆辐射　counter radiation
逆环流圈　reverse cell
逆类比　antilog
逆流　countercurrent
逆命题【数】　converse proposition
逆切变　upshear
逆湿【大气】　moisture inversion
逆梯度　upgradient
逆梯度传送　upgradient transport
逆温【大气】　temperature inversion
逆温层　temperature inversion layer
逆温层【大气】　inversion layer
逆温层底　inversion base
逆温层顶　inversion lid
逆温层高度　inversion height
逆温冻结　inversion freezing
逆温风　inversion wind
逆温霾　inversion haze
逆温破坏　inversion break-up
逆温雾　inversion fog
逆温下气层　sub-inversion layer
逆温云　inversion cloud

逆转　inversion, backing
逆转风【大气】　backing wind
年　year
年报　annual report
年变〔化〕　annual variation
年波　annual wave
年差【航海】　magnetic annual change
年超过数系列　annual exceedance series
年代学　chronology
年调节　annual storage
年际变化　interannual change
年际变率【大气】　interannual variability
年际可预报性　interannual predictability
年际气候变率　interannual climate variability
年际气压差　year-to-year pressure difference
年鉴　year-book
年较差【大气】　annual range
年径流〔量〕　annual runoff
年距平　annual anomaly
年跨度　annual time span
年轮　tree ring, growth ring
年轮【植】　annual ring
年轮分析　tree-ring analysis
年轮记录　tree-ring record
年轮密度测定　ring densitometry
年轮气候学　tree-ring climatology
年轮气候学国际研究计划　International Project in Dendroclimatology (IPID)
年轮生态学　dendroecology
年轮学　dendrochronology
年内生长纹印　intra-annual growth band
年平均【大气】　annual mean
年平均海平面　mean annual sea level (MASL)
年平均降水量　mean annual precipitation
年平均温度　mean annual temperature
年气候　year-climate

年温度差　year temperature difference
年温度跃变〔事件〕　T-year event
年系列　annual series
年循环　annual cycle
n年一遇事件　n-year event
年总量【大气】　annual amount
年最大〔洪峰〕流量　annual flood, annual flood series
年洪峰系列　annual maximum series
年最小〔洪水〕系列　annual minimum series
黏度【物】　viscosity
黏土【土壤】　clay soil
黏性对流副区　viscous-convective subrange
黏〔性〕力【物】　viscous force
黏性流体【物】　viscous fluid
黏性体　viscous body
黏性系数　sticking coefficient
黏性系数【大气】　coefficient of viscosity
黏性阻尼　viscous damping
黏应力　viscous stress
黏滞副层　viscous sublayer
黏滞耗散【大气】　viscous dissipation
黏滞〔性〕系数　viscosity coefficient
黏滞〔性〕阻力　viscosity resistance
黏滞曳力　viscous drag
鸟瞰图【测】　bird's eye view map
鸟瞰图【地理】　airview map
鸟群飞散〔回波〕　bird burst
尿素撒播　urea seeding
宁静大气　zero-activity atmosphere
宁静日珥【天】　quiescent prominence
宁静水平　quiet level
宁静太阳热辐射　basic thermal radiation
宁静太阳射电　quiet solar radio
宁静太阳【天】　quiet sun
凝固【大气】　solidification
凝固点　freezing point
凝固点【物】　solidifying point

凝固热　heat of solidification, heat of freezing
凝华【大气】　deposition
凝华核　sublimation nucleus, deposition nucleus【大气】
凝胶【物】　gel
凝结参数化　condensation parameterization
凝结初相　initial condensation phase
凝结【大气】　condensation
凝结高度【大气】　condensation level (CL)
凝结过程【大气】　condensation process
凝结函数　condensation function
凝结核【大气】　condensation nucleus
凝结核计数器　condensation nuclei counter
凝结阶段　condensation stage
凝结热　heat of condensation
凝结水　sweat
凝结尾迹　contrail, condensation trail, vapor scarf
〔凝结〕尾迹生成图　contrail-formation graph
凝结温度　condensation temperature
凝结限度　condensation limit
凝结效率【大气】　condensation efficiency
凝结性　condensability
凝结压　condensation pressure
牛顿参考系　Newtonian reference frame
牛顿插值公式【数】　Newton interpolation formula
牛顿剪切黏度【化】　Newtonian shear viscosity
牛顿冷却　Newtonian cooling
牛顿冷却定律　Newton's law of cooling
牛〔顿〕(力的单位)　newton (N)
牛顿力学　Newtonian mechanics
牛顿流动【化】　Newtonian flow
牛顿流体　Newtonian fluid

牛顿摩擦定律　Newtonian friction law
牛顿黏度【化】　Newtonian viscosity
牛顿声速　Newtonian speed of sound
牛顿应力公式　Newton's formula for the stress
牛顿运动定律　Newton's laws of motion
牛顿坐标　Newtonian coordinates
扭矩【物】　torsional moment
扭曲项　twisting term
扭转【物】　torsion
扭转湿度表　torsion hygrometer
纽曼气压表　Newman barometer
农家肥【土壤】　farmyard manure
农民年　farmer's year
农事季节　farming season
农田防护林【农】　field safeguarding forest
农田生态系统【农】　field ecosystem
农田小气候　crop microclimate
农田小气候【大气】　microclimate in the fields
农学　agronomy
农谚【大气】　farmer's proverb
农业地形气候学　agrotopoclimatology
农业干旱　agricultural drought
农业界限温度【大气】　agricultural threshold temperature
农业气候　agroclimate
农业气候调查　agroclimatic investigation
农业气候分类【大气】　agroclimatic classification
农业气候分析【大气】　agroclimatic analysis
农业气候鉴定【大气】　agroclimatic evaluation
农业气候评价【大气】　agroclimatic evaluation
农业气候区划【大气】　agroclimatic division, agroclimatic demarcation
农业气候区域　agroclimatic region

〔农业〕气候生产潜力【大气】　agroclimatic potential productivity
农业气候手册　agroclimatic handbook
农业气候图集【大气】　agroclimatic atlas
农业气候相似【大气】　agroclimatic analogy
农业气候相似原理【农】　principle of agroclimatic analogy
农业气候学【大气】　agricultural climatology, agroclimatology
农业气候指标【大气】　agroclimatic index
农业气候志【大气】　agroclimatography
农业气候资源【大气】　agroclimatic resources
农业气象产量预报　agrometeorological yield forecast
农业气象电码　agrometeorological code
农业气象观测【大气】　agrometeorological observation
农业气象模式【大气】　agrometeorological model
农业气象条件　agrometeorological condition
农业气象信息【大气】　agrometeorological information
农业气象学【大气】　agricultural meteorology, agrometeorology
农业气象学委员会（WMO）　Commission for Agricultural Meteorology (CAgM, WMO)
农业气象旬报　ten-days agrometeorological bulletin
农业气象预报【大气】　agrometeorological forecast
农业气象灾害【大气】　agrometeorological hazard
农〔业〕气〔象〕站　agro-met station
农业气象站　agrometeorological station, agricultural meteorological station

农业气象指标【大气】 agrometeorological index
农业气象综合测录装置 agricultural meteorological recorder
农业生物气象学规划 agricultural biometeorology programme
农业天气预报 agricultural weather forecast
农业小气候【大气】 agricultural microclimate
农业小气候观测 agricultural microclimate observation
农作季节 agricultural seasons
浓〔白〕霜 depth hoar
浓度 concentration
浓度方差 concentration variance
浓度梯度 concentration gradient
浓积云【大气】 cumulus congestus (Cu con)
浓密高云 dense upper cloud
浓密(厚)高层云 altostratus densus
浓密(云) densus
浓湿雾 garua, camanchaca
浓缩盆地 concentration basin
浓雾 thick fog, gross fog, dense fog
浓(云) congestus (con)
努塞特数【力】 Nusselt number
怒涛 mountainous sea
怒涛【海洋】 precipitous sea
暖 warm
暖半年 warm half year
暖侧急流 warmward jet
暖池 warm pool, warm drop, warm-air drop
暖低压【大气】 warm low
暖冬害【农】 injury by warm winter
暖风 warm wind
暖锋波动 warm-front wave
暖锋【大气】 warm front
暖锋降水 warm front precipitation
暖锋面 warm front surface, anaphalanx

暖锋型 warm front type
暖锋型锢囚 warm front type occlusion
暖锋雨 warm front rain
暖锋云系【大气】 warm front cloud system
暖高压【大气】 warm high
暖〔海〕流 warm current
暖脊【大气】 warm ridge
暖季 warm season
暖空气 warm air
暖空气楔 warm air wedge
暖浪 warm wave
暖平流 warm air advection
暖平流【大气】 warm advection
暖气团【大气】 warm air mass
暖区【大气】 warm sector
暖舌【大气】 warm tongue
暖舌引导 warm-tongue steering
暖事件 warm event
暖输送带 warm conveyor belt
暖水层 warm water sphere
暖水团 warm water mass
暖水雾 warm water fog
暖涡 warm eddy
暖涡【大气】 warm vortex
暖雾【大气】 warm fog
暖心低压 warm-core low
暖心反气旋 warm-core anticyclone
暖心高压 warm-core high
暖心气旋 warm-core cyclone
暖心扰动 warm-core disturbance
暖心涡旋 warm-core rings
暖心系统 warm-core system
暖型锢囚 warm-type occlusion
暖性反气旋【大气】 warm anticyclone
暖性锢囚 warm occlusion
暖性锢囚锋【大气】 warm occluded front
暖性气旋【大气】 warm cyclone
暖夜 warm night
暖雨【大气】 warm rain

暖雨过程　warm rain process
暖云【大气】　warm cloud
挪威海深层水　Norwegian Sea Deep Water
挪威极地研究所　Norwegian Polar Institute(NPI)
挪威〔暖〕海流　Norwegian Current
挪威气旋模式　Norwegian cyclone model
挪威学派　Norwegian school
诺阿卫星【大气】　NOAA satellite
诺埃麦尔百叶箱　Neumayer screen
诺霍夫图　Neuhoff diagram
诺伦贝格光阵式探测器　Knollenberg optical array probe
诺伦贝格探测器　Knollenberg probe
诺曼德定理　Normand's theorem
诺伊曼边界条件　Neumann boundary condition
诺伊曼函数【数】　Neumann function

O

欧几里得几何〔学〕【数】　Euclidean geometry
欧几里得距离　Euclidean distance
欧几里得空间【数】　Euclidean space
欧拉变换　Eulerian change
欧拉测流法　Eulerian current measurement
欧拉方程　Eulerian equations
欧拉风　Eulerian wind
欧拉后差格式　Euler backward scheme
欧拉-拉格朗日微分方程【数】　Euler-Lagrange differential equation
欧拉平均公式　Eulerian mean formulation
欧拉时间导数　Eulerian time derivative
欧拉数【数】　Euler number
欧拉相关　Eulerian correlation
欧拉运动　Eulerian motion
欧拉坐标【大气】　Eulerian coordinates
欧盟　European Union (EU)
欧姆电流　ohmic current
欧〔姆〕(电阻单位)　ohm(Ω)
欧文计尘器　Owen's dust recorder
欧亚大陆　Eurasian continent
欧亚型　Eurasian pattern
欧洲地球物理学会　European Geophysical Society (EGS)
欧洲季风　European monsoon
欧洲经济委员会(联合国)　Economic Commission for Europe(ECE，UN)
欧洲空间管理局　European Space Agency (ESA)
欧洲空间控制中心　European Space Operations Centre (ESOC)
欧洲空间研究组织　European Space Research Organization (ESRO)
欧洲平时　European mean time
欧洲气象卫星　European Meteorological Satellite (EuMetSat)
欧洲气象卫星资料管理部　Meteosat Data Management Department (MDMD)
欧洲时间　European time
欧洲同步气象卫星　Meteosat
欧洲通信卫星组织　European Telecommunication satellites Organisation (EOTELSAT)
欧洲遥感卫星【大气】　European Remote Sensing Satellites (ERSS)
欧洲中期天气预报中心　European Center for Medium-range Weather Forecasts (ECMWF)
欧洲综合观测系统　European Union Composite Observing system (EUCOS)
偶氮化物【化】　azo compound

偶函数【数】 even function
偶极层【物】 dipole layer
偶极反气旋【大气】 dipole anticyclone
偶极环流型 bipolar circulation pattern
偶极矩【物】 dipole moment
偶极流型 dipole flow pattern
偶极子【力】 dipole
偶极子发射 dipole emission
偶极子流型 modon flow pattern
偶极子天线 dipole antenna
偶然误差【数】 accidental error

耦合 coupling
耦合差分方程 couped difference equation
耦合常微分方程 couped ordinary differential equation
耦合大气—水文系统 couped atmospheric-hydrologcal system
耦合方程 coupling equation
耦合模式 coupling model, coupled model
耦合系数 coupling coefficient
耦合系统 coupled system

P

爬升指示器 rate-of-climb indicator
帕尔默干旱强度 Palmer drought severity
帕尔默干旱强度指数 Palmer Drought Severity Index (PDSI)
帕雷托最优性【数】 Pareto optimality
帕卢奇图 Paluch diagram
帕塞瓦尔等式【数】 Parseval equality
帕塞瓦尔定理 Parseval's theorem
帕斯卡定律【物】 Pascal law
帕斯奎尔-吉福德稳定度分类 Pasquill-Gifford stability category
帕斯奎尔稳定度分类 Pasquill stability classes
帕特森〔水银〕气压表 Patterson barometer
拍频 beat frequency
拍频模 beat mode
拍频〔振荡〕 beating
拍频振荡器【电子】【大气】 beat frequency oscillator
排队论【数】 queueing theory
排队时间【数】 queueing time
排放 emission
排放标准 emission standard
（碳）排放交易 emission trading (ET)

排放贸易方案 Emissions Trading Scheme (ETS)
排放率【大气】 emission rate
排放清单 emission inventory
排放污染【航空】 exhaust pollution
排放物 effluent
排放限度 emission limit
排放烟羽 effluent plume
排放因子 emission factor
排涝设计标准 design criteria for surface drainage
排列【数】 permutation
排气 air exhaust
排气阀【航海】 exhaust valve
排气口【航海】 exhaust port
排气尾迹 exhaust trail, engine-exhaust trail
排气温度【航海】 exhaust temperature
排气烟羽【电力】 exhaust air plume
排水 drainage
排水井【电力】 drainage well
排污标准 effluent standards
排污河槽 effluent channel
排污极限 effluent discharge limits
排烟速度 flue gas velocity

排中律【数】 law of excluded middle
蒎烯($C_{10}H_{16}$) pinene
盘式〔测雪〕硬度器 disintegration
盘式流量计 disk meter
盘旋飞行路径 purl
判别分析 discriminatory analysis
判别分析【数】【大气】 discriminate analysis
判别式【数】 discriminant
判定函数 decision function
判据【物】 criterion
滂沱大雨 soaking downpour
庞加莱波 Poincare wave
庞加莱公式【数】 Poincare formula
庞加莱截面【物】 Poincare section
旁瓣【电子】 side lobe
旁瓣回波 side lobe echo
旁路【物】 by-pass
旁扫描声呐 side-scanning sonar
抛弃式深度温度计 Expendable Bathythermograph(XBT)
抛物环面天线【电子】 parabolic torus antenna
抛物面 paraboloid
抛物面反射器【电子】 paraboloidal reflector
抛物面镜【物】 paraboloidal mirror
抛物面天线【电子】 parabolic antenna
抛物线窗 parabolic window
抛物线方程 parabolic equation
抛物线方程法 parabolic equation method
抛物型 parabolic type
抛物柱面镜【物】 parabolic mirror
咆哮西风 shrieking sixties, roaring forties
咆哮西风带【大气】 brave west wind
炮兵标定气象条件 artillery meteorological condition
炮兵气象标准 artillery meteorological standard
炮兵气象勤务 artillery meteorological service
炮射探测系统 gunprobe
跑道 runway
跑道观测 runway observation
跑道〔海拔〕高度 runway elevation
跑道结冰 ice on runway
跑道积冰【大气】 icing on runway
跑道积水 runway puddle
跑道能见度 runway visibility
跑道能见度【大气】 runway visual range (RVR)
跑道视程【大气】 runway visual range
跑道温度 runway temperature
泡沫雪 foam crust
胚胎 embryo
胚胎帷幕 embryo curtain
佩尔捷效应【物】 Peltier effect
佩尔捷效应元件 Peltier element
佩弗渗压计 Peffer osmometer
佩克莱特数 Peclet number
佩斯日照计 Pers sunshine recorder
配料法(基于构成要素的预报方法) ingredients based forecasting methodology
配体【化】 ligand
配位化学【化】 coordination chemistry
配位键【化】 coordination bond
配置 allocation
喷粉机【农】 duster
喷气推进实验室(美国) Jet Propulsion Laboratory (JPL)
喷泉【地质】 fountain, fount
喷水冰层 ice blister, flooding ice
喷水撒播 water spray seeding
喷焰 flare, bright eruption
盆景【林】 bonsai
彭曼-蒙蒂思法 Penman-Monteith method
棚冰 tented ice
膨涨气球 extensible balloon
膨胀 expansion, dilatation
膨胀表 dilatometer

膨胀波　expansion wave
膨胀场　dilatation field
膨胀冷却　expansional cooling
膨胀率　expansibility
膨胀系数　coefficient of expansion, coefficient of thermal expansion
膨胀轴　dilatation axis
碰并【大气】　coagulation
碰并系数　collection efficiency【大气】, coagulation coefficient
〔碰冻〕增长率　rate of accretion
碰撞【大气】　collision
碰撞冰　impact ice
碰〔撞〕并〔合〕过程　collision-coalescence process
碰撞弛豫【物】　collision relaxation
碰撞猝灭【化】　collisional quenching
碰撞电离【物】　ionization by collision
碰撞动力学　collision dynamics
碰撞过程【物】　collision process
碰撞截面【物】　collision cross-section
碰撞理论　collision theory
碰撞率　collision efficiency
碰撞频率【物】　collision frequency
碰撞〔谱线〕增宽【物】【大气】　collision broadening
碰撞时间【物】　collision time
碰撞式测雹器　hail impactor
碰撞系数【大气】　collision efficiency
碰撞〔增长〕　agglomeration
碰撞张弛时间　collision relaxation time
批处理【计】　batch processing
批量(资料处理用语)　batch
霹雳　thunder clap, thunderbolt
皮尔逊分布　Pearson distribution
皮尔逊-莫斯科维茨海浪谱　Pierson-Moskowitz spectrum
皮尔逊型分布【数】　Pearson type distribution
皮托管　Pitot tube, Pitot-static tube, Pitot probe

皮托管风速表　Pitot tube anemometer
皮叶克尼斯环流定理　Bjerknes theorem of circulation
皮叶克尼斯环流定理【大气】　Bjerknes circulation theorem
疲劳断裂【海洋】　fatigue break
匹配　match
匹配滤波器　matched filter
片冰　sheet ice, frazil ice, lolly ice
片流　sheet flow, laminated current, laminar flow
片流下层　laminar sub-layer
片蚀　sheet wash
片蚀【土壤】　sheet erosion
片式矢〔量〕　laminar vector, lamellar vector
片式涡流　laminar vortex
片霜　sheet frost
片状晶体　tabular crystal
片状闪电【大气】　sheet lightning
片状烟流　laminar plume
偏差　deviation
偏导数【数】　partial derviative
偏度【数】　skewness
偏度系数　coefficient of skewness
偏航　crabbing
偏航角　crab angle, angle of yaw
偏航轴　yaw axis
偏回归【农】　partial regression
偏角　angle of declination
偏近点角　eccentric anomaly
偏离航路　off-airways
偏流订正角　drift-correction angle
偏流计　drift meter
偏敏感性　partial sensitivity
偏食【天】　partial eclipse
偏微分方程【数】　partial differential equation
偏微分方程组【数】　system of partial differential equations
偏微分系数　partial differential coeffi-

cient
偏相关　partial correlation
偏相关系数　coefficient of partial correlation
偏相关系数【化】　partial correlation coefficient
偏向角【物】　angle of deviation
偏向力　deviating force
偏斜　skew
偏斜曲线　skew curve
偏心轨道地球物理观测卫星　Eccentric-orbiting Geophysical Observatory（EGO）
偏心平面位置显示器　off-center PPI scope
偏移速度【地】　migration velocity
偏移速度分析【地】　migration velocity analysis
偏倚【数】　bias
偏振等倾线　polarization isocline
偏振电场　polarization electric field
偏振度【物】【大气】　degree of polarization
偏振辐射仪　polarizing radiometer
偏振光　polarized light
偏〔振〕光镜【物】　polariscope
偏振光雷达　polarization lidar
偏振光谱学【物】　polarization spectroscory
偏振计【物】　polarimeter
偏振雷达　polarimetric radar
偏振滤波　polarization filtering
偏振面　polarized plane
偏振仪　polariscope
偏振张量　polarizability tensor
偏正角　eccentric angle
偏轴角【物】　off-axis angle
偏转　deflection
偏转风速表　deflection anemometer
偏转锋　deflection front
偏转角　angle of deflection
偏转力　deflecting force

漂浮观测站　drift station
漂浮瓶　drift bottle
漂浮期　drift epoch
漂浮探空仪　floating radiosonde
漂浮式蒸发皿【大气】　floating pan
漂航【航海】　drifting
漂流　wind drift, drift current, wind-driven current
漂流测示器　drift sight
漂流浮标　drifting buoy
漂流杆　current float
漂流卡　drift card
漂流瓶　bottle post
漂流物　drifter
漂流鱼船【航海】　drift fishing boat
漂落冰针　poudrin
漂移　shift
漂移【物】　drift
漂移气球　floating balloon
飘尘【大气】　floating dust
氕【化】【大气】　protium
拼条图　strip chart
拼图　mosaic
拼图【大气】　picture mosaic
贫营养化【土壤】　oilgotrophication
频带　frequency band
〔频〕带宽〔度〕　band width
频带展宽　bandspread
频道【大气】　channel
频分多址【电子】　frequency division multiple access（FDMA）
频率　frequency
频率比法　frequency-ratio method
频率表式风速表　frequency-meter anemometer
频率-波数滤波　frequency-wavenumber filtering
频率波数偏移【地】　frequency-wavenumber migration
频率测探法【地】　frequency sounding method

频率范围　frequency range
频率方程　frequency equation
频率-方向谱　frequency-direction spectrum
频率分布　frequency-distribution
频率分集　frequency diversity
频率分裂　frequency splitting
频率分析　frequency analysis
频率函数　frequency function
频率合成器　frequency synthesizer
频率计　frequency meter
频率捷变　frequency agility
频率滤波　frequency filtering
频率曲线　frequency-curve
频率-时间分析　frequency-time analysis
频率-时间谱密度函数　frequency-time spectral density function
频率响应【大气】　frequency response
频率响应函数　frequency response function
频率展宽　frequency spread
频谱【大气】　frequency spectrum
频散波　dispersive wave
频散关系【大气】　dispersion relationship
频散图　dispersion diagram
频数多边形　frequency polygon
频数分布组合　grouping of frequency distributions
频数直方图【电子】　frequency distogram
频移【物】　frequency shift
频域【计】　frequency domain
频域平均　frequency-domain averaging
平安时期　halcyon days, alcyone days
平板层　slab layer
平板对称　slab symmetry
平板风速表　plate anemometer
平板厚度　slab thickness
平板近似　slab approximation
平板模式　slab model

平板扫描仪【计】　flat-bed scanner
平板式运动　slab like motion
平板（圆锥）辐射仪　flat plate (cone) radiometer
平板状冰山　barrier berg
平潮　stand
平赤道【天】　mean equator
平春分点【天】　mean equinox
平顶冰山　tabular iceberg, barrier iceberg
平珥　horizontal tangent arc
平方反比定律【农】　inverse-square law
平方根吸收　square root absorption
平恒星日　mean sidereal day, average sidereal day
平恒星时【天】　mean sidereal time
平衡　equilibrium, balance
平衡表　balance meter
平衡潮　equilibrium tide
平衡对流　equilibrant convection
平衡阀【航海】　blanced valve
平衡方程【大气】　balance equation
平衡风　balanced wind
平衡高度　balance height, equilibrium altitude
平衡过程　equilibration
平衡环　gimbal
平衡降深　equilibrium drawdown
平衡介质　stationary medium
平衡理论　equilibrium theory
平衡力　equilibrant
平衡力矩　equilibrant moment
平衡流　balanced flow
平衡面　balance level
平衡年　balance year
平衡抛物面　equilibrium paraboloid
平衡器　balancer
平衡球体　equilibrium spheroid
平衡区　equilibrium range
平衡时间　equilibration time
平衡太阳潮　equilibrium solar tide

平衡态　equilibrium state
平衡态气候　equilibrium climate
平衡态气候响应　equilibrium climate response
平衡条件【力】　equilibrium condition
平衡陀螺仪【航海】　balanced gyroscope
平衡温度　equilibrium temperature
平衡系统　equilibrium system
平衡线　equilibrium line
平衡线高度　equilibrium-line altitude
平衡斜压模式　balanced baroclinic model
平滑【数】　smoothing
平滑参数　smoothing parameter
平滑过程　smoothing process
平滑面　smooth surface
平滑算子　smoothing operator
平滑系数　smoothing coefficient
平缓冬季　coreless winter
平缓扰动法　smooth perturbation method(SPM)
平均　average
平均半径　mean radius
平均变率　average variability
平均表层温度　mean skin temperature
平均差　mean difference
平均场　average field
平均场方程　equation for average field
平均潮位　mean tide level
平〔均〕赤道日　mean equatorial day
平均大潮低潮〔面〕　mean low water springs
平均大潮高潮〔面〕　mean high water springs
平均大气原子序数　average atmospheric atom number
平均等温线　mean isotherm
平均等值线　normal isopleth
平均低潮　mean low water
平均低潮间隙【航海】　mean low water interval
平均低低潮〔面〕　mean lower low water

平均〔地〕磁扰〔动〕程度　average geomagnetic level
平均动能　mean kinetic energy
平均多普勒速度　mean Doppler velocity
平均分子量　mean molecular weight
平均风速【大气】　average wind velocity
平均风向量　mean wind vector
平均辐射温度　mean radiant temperature
平均高潮　mean high water
平均高潮间隙【航海】　mean high water interval
平均功率　average power
平均估计值直方图　mean estimate histogram
平均故障间隔　mean time before failure（MTBF）
平均故障时间　mean time to failure
平均海平面　mean sea level（MSL）
平均海平面趋势　mean sea level trends
平均海平面温度订正　reduction of temperature to mean sea-level
平均海平面资料常设局（英国）　Permanent Service for Mean Sea Level（PSMSL）
平均核　averaging kernel
平均环境风　mean environmental wind
平均环流　mean circulation
平均积分器　average integrator
平均经向环流　mean meridional circulation
平均经向流函数　mean meridional stream function
平均距平　mean anomaly, average departure
平均空隙流速（水文）　average interstitial velocity
平均敏感度　mean sensitivity
平均偏差【物】　mean deviation
平均年温度较差【大气】　mean annual range of temperature

平均谱　average spectrum
平均气候特征　mean climate characteristics
平均〔气〕流　mean flow
平均强度　average intensity
平均日地距离　mean distance between earth and sun
平均上升气流　mean ascending current
平均视差【天】　mean parallax
平均寿命　mean life, average life
平均寿命【物】　mean lifetime
平均水位　mean water level
平均速度　average velocity
平均速度【物】　mean velocity
平均算符　averaging operator
平均损失　mean loss
平〔均〕太阴日　mean lunar day
平均体积半径　mean-volume radius
平均体积直径　mean-volume diameter
平均通量　mean flux
平均图　mean map, mean chart
平均纬度【航海】　mean latitude
平均温度　temperature average, mean temperature
平均温度表　chronothermometer
平均温度年较差　mean annual range of temperature
平均误差　average error
平均误差【数】　mean error
平均下沉气流　mean descending current
平均夏令时　mean summer time（M. S. T）
平均夏雨等值线　isothermobrose
平均相对变率　relative mean variability
平均效益　mean utility
平均虚温　mean virtual temperature
平均压力计【航海】　mean pressure meter
平均曳引系数　average drag coefficient
平均有效直径　mean effective diameter
平均预报误差　mean forecast error
平均云高　mean cloud height
平均云量　mean cloudiness
平均运动【天】　mean motion
平均阵风速　mean gust speed
平均值　mean value, average value
平均滞后　average lag
平均逐日变化　mean day-to-day variation
平均自由程【物】　mean free path
平均最高高水位　mean higher high water
平流【大气】　advection
平流瓣　advection lobe
平流边界层　advective boundary layer
平流变化　advective change
平流层【大气】　stratosphere
平流层爆发性增温　sudden stratospheric warming
平流层爆发〔性〕增温【大气】　stratospheric sudden warming
平流层补偿　stratospheric compensation
平流层沉降　delayed (= stratospheric) fallout
平流层臭氧　stratospheric ozone
平流层顶【大气】　stratopause
平流层对流层雷达　ST radar
平流层〔放射性物〕沉降　stratospheric fallout
平流层辐射　stratosphere radiation
平流层光化学　stratospheric photochemistry
平流层和中间层探测器　stratospheric and mesospheric sounder (SAMS)
平流层化学　stratospheric chemistry
平流层环流指数　stratospheric circulation index
平流层极〔地〕涡〔旋〕　stratospheric polar vortex
平流层急流　stratosphere jet flow
平流层季风　stratospheric monsoon
平流层静稳气溶胶　unperturbed stratospheric aerosol
平流层临边红外监测　limb infrared

monitoring of the stratosphere(LIMS)
平流层零层　stratonull layer
平流层硫酸盐层　stratospheric sulfate layer
平流层能量学　energetics of stratosphere
平流层逆温　stratospheric inversion
平流层耦合作用　stratospheric coupling
【大气】, stratosphere coupling
平流层气球　stratostat
平流层气溶胶测量　stratospheric aerosol measurement(SAM)
平流层气溶胶【大气】　stratospheric aerosol
平流层气溶胶层　stratospheric aerosol layer
平流层气溶胶和气体试验　stratospheric aerosol and gas experiment (SAGE)
平流层上部　upper stratosphere
平流层探测装置　stratospheric sounding unit (SSU)
平流层污染【大气】　stratospheric pollution
平流层下部　lower stratosphere
平流层引导【大气】　stratospheric steering
平流层引导学说　theory of stratospheric steering
平流层云　stratospheric cloud
平流层增温　stratospheric warming
平流层振荡　stratospheric oscillation
平流层中部　middle stratosphere
平流尺度　advection scale
平流传播　advection propagation
平流方程【大气】　advection equation
平流辐射雾　advective-radiation fog
平流辐射雾【大气】　advection-radiation fog
平流过程　advection process
平流急流　advection jet
平流加速度　advective acceleration
平流假说　advective hypothesis

平流扩散方程　advection-diffusion equation
平流雷暴　advective thunderstorm
平流流出〔量〕　advection effluent
平流模式　advective model
平流模式　advection model
平流逆温　advection inversion
平流气压倾向　advective pressure tendency
平流区　advective region
平流时间尺度　advective time scale
平流霜【大气】　advection frost
平流速度　advection velocity
平流通量　advective flux
平流性雷暴【大气】　advective thunderstorm
平流雾【大气】　advection fog
平流项　advection term
平流效应　advection effect
平流形式　advective form
平流重力气流　advective-gravity flow
β平面【大气】　β-plane
平面波【大气】　plane wave
平面波脉冲　plane wave pulse
平面大气波　plane atmospheric wave
平面反照率　plane albedo
平面几何〔学〕【数】　plane geometry
f 平面近似　f-plane approximation
平面拷贝传真扫描器　flat copy facsimile scanner
平面流　plane flow
平面偏振【物】　plane polarization
平面平行大气　plane-parallel atmosphere
平面平行介质　plane parallel medium
平面平行近似　plane-parallel approximation
平面平行运动【物】　plane-parallel motion
平面切变显示器【大气】　plane shear indicator (PSI)
平面图【航海】　plane chart

平面位置显示器　PPI scope
平面位置显示器【大气】　plane position indicator(PPI)
平面枝状晶体　plane-dendritic crystal
平谱【天】　flat spectrum
平谱源【天】　fait-spectrum source
平时【天】　mean time
平水年【大气】　normal flow year
平朔望月　mean synodic lunar month
平台定位和资料中继　platform location and data relay
平太阳年　mean solar year
平太阳日　average solar day
平太阳日【天】　mean solar day
平太阳时　mean solar hour
平太阳时【天】【大气】　mean solar time
平太阳【天】　mean sun
平坦风　frank
平坦型最高(气温)　flat maximum
平稳高斯过程　stationary Gaussian process
平稳高斯时间序列　stationary Gaussian time series
平稳轨道　stationary orbit
平稳过程谱分解【数】　spectral decomposition of stationary process
平稳过程【数】　stationary process
平稳函数【数】　stationary function
平稳期　stationary phase
平稳时间序列【数】　stationary time series
平稳实随机函数　stationary real random function
平稳随机过程　stationary stochastic process, stationary random process
平稳随机函数　stationary random function
平稳〔信〕源　stationary source
平行层理【地质】　parallel bedding
平行光束辐射　parallel-beam radiation
平行光束近似　parallel-beam approximation
平行六面体　parallelepiped
平行〔谱〕带　parallel band
平行四边形【数】　parallelogram
平行通量　parallel flux
平行位移　parallel shift
平行线　parallel
平行移动【数】　parallel translation
平行褶皱【地质】　parallel fold
平移【物】　translation
平移波　wave of translation, translatory wave
平移场　translatory field
平移流　translation flow
平移能　translational energy
平移速度　translatory velocity
平移探空仪　transosonde
平移运动　translational motion
平原　plain
平正午【天】　mean noon
T 评分　Threat score
评估　assessment
评述　review
屏蔽层　shielding layer, screening layer
屏蔽效能　shielding effectiveness
屏蔽效应　shielding effect
屏障风　barrier wind
瓶采样　flask sampling
瓶颈效应【化】　bottleneck effect
瓶式温度计　bottle thermometer
坡度　slope
坡度流　slope flow, slope current
坡风【大气】　slope wind
坡风环流　slope wind circulation
坡面渗流　effluent seepage
破冰船【航海】　icebreaker
破坏风速【大气】　breaking wind speed
破坏性　destructiveness
破碎　sharttering
破云器　cloudbuster
剖面【大气】　cross-section

剖面分析　cross-sectional analysis
剖面检验　cross-section test
剖面图【大气】　cross-section diagram
剖面图【测】　profiles
葡萄糖【生化】　glucose
蒲福风级【大气】　Beaufort (wind) scale
蒲福风力　Beaufort force
蒲福数　Beaufort number
蒲福天气符号　Beaufort weather notation, Beaufort notation
普莱斯流速计　Price meter
普朗克常量【物】　Planck constant
普朗克定律　Planck's law
普朗克辐射定律　Planck's radiation law
普朗克辐射体　Planckian radiator
普朗克函数　Planck function
普朗克时间【天】　Planck time
普朗特混合长　Prandtl mixing length
普朗特混合长理论【大气】　Prandtl mixing-length theory
普朗特数【大气】　Prandtl number
普适常量【物】　universal constant
普适平衡假说　universal equilibrium hypothesis
普适气体常数【物】【大气】　universal gas constant
普通层云　stratus communis, common stratus
普通地理图【地理】　general geographic map
普通地理学【地理】　general geography
普通地图【地理】　general map
普通地图学【地理】　general cartography
普通地质学【地质】　physical geology
〔普通〕电报　telegraph exchange
普通辐射站　ordinary radiation station
普通卷层云　cirrostratus communis
普通流体力学　elementary fluid mechanics
普通平均　ordinary average

普通气象学　general meteorology
普通气象学(军用)　aerography
普通天文学【天】　general astronomy
普通自然地理学【地理】　general physical geography
普约绝对日射表　Pouillet pyrheliometer
谱【物】　spectrum
谱斑　plage
谱斑【天】　flocculus
谱变换模式　spectral transform model
谱表达式　spectral representation
谱参数　spectrum parameter
谱测度【数】　spectral measure
谱处理　spectral processing
谱传递函数　spectral transfer function
谱窗　spectrum window
谱窗【数】　spectral window
谱带　spectrum banding
〔谱〕带〔间〕隙　band gap
谱的拐点　spectrum inflection
谱的横坐标　spectrum abscissa
谱的极大值　spectrum maximum
谱的极小值　spectrum minimum
谱的可靠性　spectrum reliability
谱的纵坐标　spectrum ordinate
谱方法【数】　spectral method
谱放大率　spectrum amplification
谱分辨率【数】　spectral resolution
谱分布　spectrum distribution
谱分布函数　spectrum distribution function
谱分布函数【数】　spectral distribution function
谱分解【数】　spectral decomposition
谱分离　spectral separation
谱分析　spectrum analysis
谱分析【数】　spectral analysis
谱峰频率　spectrum peak frequency
谱辐射亮度　spectral radiance
谱辐照度　spectral irradiance
谱光电导　spectral photoconductivity

谱函数　spectrum function, spectral function
谱积分【数】　spectral integral
谱集　spectral set
谱计算　spectrum calculation
谱加宽　spectrum broadening, spectral broadening
谱间隔　spectrum interval, spectral interval
谱〔间〕隙　spectral gap, spectrum gap
谱间隙　gap in spectrum
谱矩　spectrum moment
谱空间　spectral space
谱宽　spectral width, spectrum width
谱扩散率　spectral diffusivity
谱滤波函数　spectral filter function
谱密度　spectrum density, spectral density
谱密度函数【数】　spectral density function
谱密度张量　spectral density tensor
谱模式　spectrum model
谱模式【大气】　spectral model
谱能　spectrum energy
谱频率　spectral frequency
谱平均　average over the spectrum
谱区　spectrum area
谱曲率　spectrum curvature
谱术语　spectrum terminology

谱数值分析　spectral numerical analysis
谱数值预报　spectral numerical prediction
谱水平　spectrum level
谱特性　spectrum characteristic
谱图　spectrum pattern
谱吸收　spectral absorption
〔谱〕线参数　line parameter
谱线干扰　spectral line interference
〔谱〕线加宽　line broadening
谱线宽度【化】　line width
谱线轮廓【化】　line profile
谱线强度　spectral line intensity
〔谱〕线吸收　line absorption
谱线因子【化】　line factor
谱相似〔性〕　spectral similarity
谱响应特征〔性〕　spectral-response characteristics
谱斜率　spectrum slope
谱型　spectral pattern
谱映射　spectral mapping
谱展开　spectral evolution
谱张量　spectrum tensor
谱指数　spectrum index
谱坐标　spectrum coordinate
瀑布【地理】　waterfall
瀑布效应　waterfall effect
瀑布云　cloud from waterfall

Q

栖留河　perched stream
栖留水　perched groundwater
期望　expectation, expectance
期望损失　expected loss
期望效益　expected utility
期望值　expected value
齐岸水位　bankfull stage
齐次方程　homogeneous equation

齐次微分方程【数】　homogeneous differential equation
齐次线性方程组【数】　system of homogeneous linear equations
齐纳二极管【电子】　Zener diode
齐性　homogeneity
奇点【数】　singular point, singularity
奇怪吸引子【数】【大气】　strange

attractor
奇异波　rogue wave
奇异对应点　singular corresponding point
奇异光学现象　rare optical phenomenon
奇异积分方程【数】　singular integral equation
奇异平流　singular advection
奇异谱分析　singular spectrun analysis (SSA)
奇异扰动【物】　singular perturbation
奇异矢量　singular vectors
奇异向量超增长　super-growth of singular vector
奇异向量分解　strange vector decomposition (SVD)
奇异性【物】　singularity
奇异值【数】　singular value
歧点【数】　bifurcation point
歧义函数　ambiguity function
歧义图　ambiguity diagram
棋盘式图　checkerboard diagram
旗云【大气】　banner cloud
旗〔状〕云　cloud banner, banner cloud
起转时间【大气】　spinup time
气波　air wave
气藏化合物　reservoir compounds
气动力　aerodynamic force
气动系数　aerodynamic coefficient
气动曳力　aerodynamic drag
气动阻力　aerodynamic resistance
气固反应　gas-solid reactions
气固色谱法【化】　gas solid chromatography
气候【大气】　climate
气候背景　climatic scenario
气候背景场　climatological background field
气候变化　climate change, climatic change
气候变化【大气】　climatic change
气候变化的天文学理论　astronomical theory of climate change
气候变化探测　climate change detection, climatic change detection (CCD)
气候变化展望　climate change projection
气候变率　climate variability, climatic variability
气候变率【大气】　climatic variability
气候变率与可预测性〔研究计划〕　Climate Variability and Predictability (CLIVAR)
气候变迁【大气】　climatic variation
气候标准平均值【大气】　climatological standard normals
气候病　meteoropathy
气候病理学　climatopathology【大气】, climatic pathology
气候波动　climatic vacillation
气候不连续　climatic discontinuity
气候不适应〔症〕　declimatization
气候不稳定性　climatic instability
气候长期调查测绘和预测计划（重建距今 18000 年的海面温度规划）　CLIMAP (Climate: Long-range Investigation Mapping and Prediction)
气候持续性　cliper
气候持续性【大气】　climatic persistence
气候重建【大气】　climatic reconstructlon
气候带【大气】　climatic zone, climatic belt
气候的自然变率　natural variability of climate
气候地带性　climatic zonation
气候对比　climatic contrast
气候恶化　climatic degeneration
气候恶化【大气】　climatic deterioration
气候反常　climate abnormality
气候反馈机制【大气】　climatic feedback

mechanism
气候反馈作用 climatic feedback interaction
气候非传递性 climatic intransitivity
气候非周期变化【大气】 climatic nonperiodic variation
气候分界【大气】 climatic divide, climate divide
气候分类【大气】 climatic classification
气候分区 climatological division
气候分析【大气】 climatic analysis
气候分析中心(NOAA) Climate Analysis Centre (CAC, NOAA)
气候风险分析【大气】 climatic risk analysis
气候锋【大气】 climatological front
气候辅助站 climatological substation
气候副区 climatic subdivision
气候改良 climatic amelioration, climate melioration
气候概率【大气】 climatic probability
气候概述 climatological summary
气候工作者 climatologist
气候共存态 climatic coexistance
气候观测【大气】 climatological observation
气候环境 climatic environment
气候极值 climatic extreme
气候极值检验 climatological limit check
气候记录 climatic record
气候监测 climate monitoring
气候监测【大气】 climatic monitoring
气候景观 climatic landscape
气候距平 climate anomaly
气候决定论 climatic determinism
气候可传递性 climatic transitivity
气候控制〔方案〕 climate control
气候控制室 biotron
气候控制〔因子〕 climatic control
气候疗法 climatic treatment

气候敏感性 climate sensitivity
气候敏感性【大气】 climatic sensitivity
气候敏感性实验【大气】 climate sensitivity experiment
气候模拟 climate simulation
气候模拟【大气】 climatic simulation
气候模式 climate model
气候年 climatic year, climatological year
气候漂移 climate drift
气候平均值 climatological normals
气候评价【大气】 climatic assessment
气候潜势 climatic potential
气候区 climatic province
气候区【地理】【大气】 climatic region
气候区划【大气】 climate regionalization
气候区划【地理】 climatic regionalization
气候趋势【大气】 climatic trend
气候生产力指数 climatic productivity index
气候生成 climatogenesis
气候生理学 climatic physiology
气候时间序列 climatic time series
气候实验室 climatizer
气候适宜〔期〕 climatic optimum
气候适应 acclimation
气候适应【大气】 climatic adaptation, acclimatization
气候适应过程 climatization
气候适应能力 weather resistance
气候舒适〔度〕 climatic comfort, climate comfort
气候数值模拟 climatic numerical modelling
气候特征 climatic characteristics
气候条件 climatological condition, climatic condition
气候统计【大气】 climatic statistics
气候统计学【大气】 climatological

statistics
气候透明系数　climatological transparent coefficient
气候突变　climate accident
气候突变【大气】　abrupt change of climate
气候图　climogram, climatic diagram, climograph, climatological chart, climatic chart, climagram
气候图【大气】　climatic map
气候图表　climatological diagram, climagraph
气候图集　climatological atlas
气候图集【大气】　climatic atlas
气候土壤型　climatic soil type
气候湍流　climatic turbulence
气候稳定性　constancy of climate
气候问题　climatic problem
气候系统【大气】　climatic system, climate system
气候现象　climatic phenomenon
气候相似　climate analogs
气候效应　climatic effect
气候心理学　climatic psychology
气候信号　climate signal
气候信息处理与分析系统　Climate Information Processing and Analysis System,（CIPAS）
气候形成分类法【大气】　genetic classification of climates
气候形成因子【大气】　factors for climatic formation
气候型　climatological pattern
气候型【大气】　climatic type
气候性土壤形成　climatic soil formation
气候学【大气】　climatology
气候〔学方法〕预报　climatological forecast
气候学和气象学应用委员会（WMO）　Commission for Climatology and Applications of Meteorology（CCAM, WMO）
气候学委员会（WMO）　Commission for Climatology（CCl, WMO）

气候雪线　climate snow line, climatic snow line
气候驯化【大气】　climatic domestication
气候旋回　climatic cycle
气候演变【大气】　climatic revolution
气候演替顶极　climatic climax
气候遥相关　climatic teleconnection
气候要求　climatic demand
气候要素【大气】　climatic element
气候异常【大气】　climatic anomaly
气候因子　climatic factor
气候应力荷载【大气】　climate stress load
气候应用及资料计划咨询委员会　Advisory Committee of Climate Application and Data Project（ACCADP）
气候应用计划　Climate Application Programme（CAP）
气候影响【大气】　climatic impact
气候影响评价计划　Climate Impact Assessment Programme（CIAP）
气候〔影响性〕反应　meteorotropic effect
气候影响研究计划　Climate Impact-Study Programme（CIP）
气候友好技术　climate-friendly technique
气候与人类健康　climate and human health（CHH）
气候预报　climate forecast
气候预测　climate prediction
气候预测【大气】　climatic prediction
气候约束〔因子〕　climatic constraint
气候月报　CLIMAT broadcast
气候韵律　climatic rhythm
气候灾害　climatic disaster
气候灾害【大气】　climate damage
气候栽培界限　climatic cultivation limit
气候噪声　climatic noise
气候噪声【大气】　climate noise
气候展望　climatic forecast, climate projection
气候站　climatological station

气候站海拔〔高度〕 climatological station elevation
气候站气压 climatological station pressure
气候站网 climatological station network, climatological network
气候障壁 climatic barrier
气候诊断【大气】 climatic diagnosis
气候振荡 climatic cycle
气候振荡【大气】 climatic oscillation
气候振动 climate fluctuations, climatic fluctuations
气候振动【大气】 climatic fluctuation
气候振幅 climatic amplitude
气候值 climatic value
气候植物区系 climatic plant formation
气候指示物 climatic indicator
气候指数 climatic index
气候志【大气】 climatography
气候治疗学 climatotherapy
气候周期性 climatic periodicity
气候周期性【大气】 climate periodicity
气候周期性变化【大气】 climatic periodic variation
气候转换 climatic transition
气候状态 climate state
气候状态矢量 climatic state vector, climate state vector (CSV)
气候资料 climatological data
气候资料【大气】 climatic data
气候资料计划 Climate Data Programme (CDP)
气候资料库 climatological data bank
气候资源 climatic resources
气候资源【大气】 climate resources
气候总体 climatic ensemble
气化 gasification
气辉【大气】 airglow
气辉层 glow-emitting layer
气辉辐射 airglow radiation
气辉光度计 airglow photometer
气辉光致离解 airglow photodissociation
气辉谱线强度 airglow spectral line intensity
气辉强度 airglow intensity
气辉闪〔光〕 elve
气辉消失 airglow quenching
气辉形成 airglow production
气汇 air sink
气界地质学 atmospheric geology
气界 air-sphere
气阱【大气】 air trap
气孔阻力 stomatal resistance
气块 parcel, air parcel【大气】, air package
气块法 parcel method
气块轨迹 air parcel trajectory
气块理论 parcel theory
气粒 air particle
气量表 gasometer
气流 air stream, stream, air flow, air current
气流表 air-meter
气流堆积〔现象〕 banking of the current
气流分离 separation of flow, flow separation
气流分支 flow splitting
气流综合测量仪 air flow multimeter
气煤【地质】 gas coal
气凝胶 aerogel
气泡 bubble, air bubble
气泡冰 bubbly ice
气泡对流 bubble convection
气泡高压 bubble high
气泡核 bubble nucleus
气泡理论 bubble theory
气泡破裂 bubble bursting
气泡水位计 bubble gauge
气瀑 air cascade, air cataract
气球 balloon, aerostat
气球测风【大气】 pibal
气球测风站 pilot balloon station

气球测量　balloon measurement
气球吊篮　balloon basket
气球发射卫星　satellite launched from a balloon（Saloon）
气球高空风计算尺　pilot balloon slide-rule
气球观测　balloon observation
气球〔护〕罩　balloon shroud
气球火箭　rockoon
气球火箭探测　rockoon sounding
气球净举力　free lift of balloon
气球平衡高度　static ceiling
气球气象仪　aerostat meteorograph
气球全升力　total lift of a balloon
气球上升速率　ascension rate of balloon
气球升限　ballonet ceiling
气球施放　pilot balloon ascent
气球探测　balloon sounding
气球天文学　balloon astronomy
气球卫星　balloon satellite
气球系留台　balloon bed
〔气〕球罩　shroud
气球罩　balloon cover
气圈　aerosphere, gas-sphere, air-sphere
气泉　air fountain
气溶胶【大气】　aerosol
气溶胶测量仪　aerosoloscope
气溶胶层　aerosol layer
气溶胶成分　aerosol composition
〔气溶胶〕粗粒　coarse particle mode
气溶胶电　aerosol electricity
气溶胶分布　aerosol distribution
气溶胶分析仪　aerosol analyzer
气溶胶负载　aerosol loading
气溶胶光度计　aerosol photometer
气溶胶光学厚度　aerosol optical thickness, aerosol optical depth
气溶胶化学【大气】　aerosol chemistry
气溶胶检测仪　aerosol detector
气溶胶结构　aerosol structure
气溶胶粒子　aerosol particle

气溶胶粒子谱【大气】　aerosol particle size distribution
气溶胶连续检测仪　continuous aerosol detector
气溶胶谱　aerosol size distribution
气溶胶气候效应　aerosol climatic effect（ACE）
气溶胶气候学　aerosol climatology
气溶胶散射相函数　aerosol scattering phase function
气溶胶探空仪　aerosolsonde
气溶胶特征试验　aerosol characterization experiment（ACE）
气溶胶吸湿性　aerosol wettability
气溶胶治疗〔法〕　aerosol therapy
气生植物【植】　aerophyte, aerial plant
气室　ballonet
气霜　frost in the air
气态　gas phase, gaseous state
气态示踪物法　gaseous tracer technique
气态与液态之比　vapour to liquid ratio
气体　gas
气体比重计　aerometer
气体常数【大气】　gas constant
气体地球化学测量【地质】　geochemical gas survey
气体电极【化】　gas electrode
气体定律　gas law
气体动力〔学〕粗糙度【大气】　aerodynamic roughness
气体动力粗糙度长度　aerodynamic roughness length
气体动力平滑度　aerodynamic smoothness
气体动力学　aerodynamics
气体动力学粗糙面　aerodynamically rough surface
气体动力学光滑面　aerodynamically smooth surface
气体放电　gaseous（electrical）discharge
气体分容量　fractional volume abundance

气体分析【化】 gasometry
气体分子运动论 kinetic theory of gases
气体过滤相关波谱学 gas filter correlation spectroscopy
气体激光器【物】 gas laser
气体扩散分离〔法〕【化】 gaseous diffusional separation
气体力学 pneumatology
气体粒子分布因子 gas-particle distribution factor
气体论 gas theory
气体密度天平【化】 gas density balance
气体输送 gas transfer
气体温度表【大气】 gas thermometer
气体污染 gaseous pollution
气体污染物 gaseous pollutant
气体吸收损耗 gaseous absorption loss
气体星云 gaseous nebula
气田【地质】 gas field
气团【大气】 air mass
气团保守性【大气】 conservative property of air mass
气团变性 modification of air mass, transformation of air mass, air-mass modification
气团变性【大气】 air-mass transformation
气团变性试验（日本） Air-mass Transformation Experiment（AMTEX, Japan）
气团分类【大气】 air-mass classification
气团分析【大气】 air-mass analysis
气团降水 air-mass precipitation
气团雷暴 air-mass thunderstorm
气团类型图 air-mass-type diagram
气团频率 air-mass frequency
气团气候学 air-mass climatology
气团气象学 air-mass meteorology
气团迁移 air-mass transport
气团侵入 invasion of air

气团识别 identification of air mass, air-mass identification
气团属性【大气】 air-mass property
气团特性 air-mass characteristic
气团图 air-mass chart
气团温度 air-mass temperature
气团雾【大气】 air-mass fog
气团源地【大气】 air-mass source
气团源〔地〕区〔域〕 air-mass source region
气团阵雨 air-mass shower
气味阈值 threshold of odor
气温【大气】 air temperature
气温距平 temperature departure
气温露点差 dew-point spread, dew-point deficit（=depression）
气温日较差 daily temperature range
气温异常 thermic anomaly
气温直减率【大气】 temperature lapse rate
气雾 aerial fog
气相 gaseous phase
气相动力学 gas-phase kinetics
气相聚合【化】 gaseous polymerization, gas-phase polymerization
气相色谱法【化】 gas chromatography
气相色谱仪 gas chromatograph
气相色谱仪-质谱仪系统 gas chromatography-mass spectrometer system
气象报告【大气】 meteorological report
气象报告范围 meteorological reporting area
气象〔标准〕视距 meteorological range
气象表 meteorological table
气象病【大气】 meteorotropic disease, meteoropathy
气象病理学 meteoropathology
气象参数 meteorological parameter
气象测量 meteorological survey
气象潮 wind setup, weather tide
气象潮【海洋】【大气】 meteorological

tide
气象潮分　meteorological constituent
气象车　meteorological car
气象赤道【大气】　meteorological equator
气象传送　meteorological transmission
气象档案　meteorological archive
气象导航【大气】　meteorological navigation
气象电报　meteorological message, weather message
气象电传通信网　meteorological teleprinter network
气象电〔传通〕信网　meteorological telecommunication network
气象电离层和气候卫星探测系统　Constellation Observing System for Meteorology, Ionosphere and Climate (COSMIC)
气象电码【大气】　meteorological code
气象定线【航海】　weather routing
气象动力学　meteorological dynamics
气象飞机　meteorological airplane
气象飞机【大气】　meteorological aircraft
气象分析　meteorological analysis
气象风洞【大气】　meteorological wind tunnel (MWT)
气象服务　meteorological service
气象符号　meteorological symbol
气象辅助设备　meteorological aid
气象干旱　meteorological drought
气象干旱指数　meteorological drought index
气象工作者　meteorologist
气象观测【大气】　meteorological observation
气象观测船　meteorological ship
气象观测平台【大气】　meteorological platform
气象观测塔　weather instrumentation tower
气象观测员【大气】　meteorological observer
气象观测站　meteorological observing station
气象官能症　meteorotropism, meteorological disease, meteorotropic disease
气象光学　meteorological optics
气象光学视距　meteorological optical range
气象广播　meteorological broadcast, meteorology broadcast
气象航线【大气】　meteorological shipping route
气象回波【航海】　meteorology echo
气象火箭【大气】　meteorological rocket
气象火箭探空仪　meteorological rocket sonde
气象火箭网　meteorological rocket network (MRN)
气象火箭站　meteorological rocket facility
气象基准面　meteorological datum plane
气象计　meteorograph
气象记录　meteorological record
气象监视台　meteorological watch office
气象减灾　meteorological hazard mitigation
气象经济决策　meteorological economic decision
气象静力学　meteorological statics
气象局　weather bureau, Meteorological Service
气象局(埃及)　Meteorological Authority
气象局(英国)　Meteorological Office
气象科技文献资料　records and documents of meteorological science and technology
气象雷达【大气】　meteorological radar
气象雷达方程【大气】　meteorological radar equation

气象雷达站　meteorological radar station
气象领域　meteorological realm
气象模拟　meteorological modeling
气象模式　meteorological model
气象能见度　meteorological visibility
气象年鉴　meteorological yearbook
气象勤务组　air weather group
气象情报区　flight information region
气象区　meteorological region
气象区〔域〕　meteorological district
气象热带　meteorological tropics
气象热力学　meteorological thermodynamics
气象日志　meteorological log
气象哨　meteorological post, weather shack
气象设备　meteorological equipment
气象设备发展实验室(美国)　Equipment Development Laboratory (EDL, NWS)
气象生理反应　meteoropathic reaction
气象声学　meteorological acoustics
气象史　meteorology history
气象数据的二进制通用表示格式　Binary Universal Form for Representation of meteorological data (BUFR)
气象数据收集与处理装置　meteorological data collecting and processing unit
气象损失　meteorological loss
气象塔　weather tower, meteorological tower
气象台　observatory, meteorological observatory【大气】
气象台站网　meteorological network, network of meteorological station【大气】
气象探测火箭　meteorological sounding rocket
气象〔探测〕气球　meteorological balloon
气象厅(日本)　Meteorological Agency

气象通报　Meteorological Bulletin
气象通信　meteorological communication, meteorological telecommunication
气象通信枢纽　meteorological telecommunication hub
气象通信中心　Meteorological Communications Center (M. C. C.)
气象统计学　meteorological statistics
气象图　meteorological map
气象网　network (of stations)
气象卫星　weather satellite, metsat
气象卫星【大气】　meteorological satellite
气象卫星传真收片机　weather satellite facsimile receiver
气象卫星地面站【大气】　meteorological satellite ground station
气象卫星接收天线　receiving antenna of meteorological satellite
气象卫星实验室　meteorological satellite laboratory (MSL)
气象卫星研究合作研究所(美国)　Cooperative Institute for Meteorological Satellite Studies (CIMSS)
气象卫星资料记录器　weather satellite recorder
气象稳定性　meteorological stability
气象无线电传播联合委员会(英国)　Joint Meteorological Radio Propagation Committee (JMRPC)
气象武器　meteorological weapon
气象现象　meteorological phenomenon
气象限制条件　limiting meteorological condition
气象信息　meteorological information
气象信息和辐射剂量获取系统　meteorological information and dose acquisition system (MIDAS)
气象信息网及显示系统　weather information network and display system (WINDS)

气象信息综合分析处理系统　Meteorological Information Comprehensive Analysis and Processing System（MICAPS）
气象学【大气】　meteorology
气象学报　Acta Meteorologica Sinica（AMS）
气象学和气候学特殊应用委员会（WMO）　Commission for Special Applications of Meteorology and Climatology（CoSAMC,WMO）
气象学会　meteorological society
气象学院　Meteorological Institute
气象研究飞行　meteorological research flight
气象研究公司（美国）　Meteorology Research, Inc.（MRI）
气象要素【大气】　meteorological element
气象要素等值线　isometeoric line
气象要素序列　meteorological element series
气象仪器【大气】　meteorological instrument
气象因子　meteorological factor
气象〔有线〕电报　meteorological telegraph
气象预报　meteorological forecast
气象预警　meteorological warning
气象预警电报　meteorological warning message
气象云室　meteorological cloud chamber
气象运动学　meteorological kinematics
气象灾害　meteorological disaster
气象噪声【大气】　meteorological noise
气象炸弹　meteorological bomb
气象战　meteorological warfare
气象站　meteorological station
气象侦察飞行　meteorological reconnaissance flight
气象侦察艇　weather patrol ship

气象中心　meteorological center
气象咨询报　meteorological advisory
气象资料处理系统　meteorological data handling system（MDHS）
气象〔资料〕档案存取系统　meteorological archive and retrieval system（MARS）
气象资料中心　meteorological data center
气象总台　main meteorological office（MMO）
气象组织　meteorological organization
气旋【大气】　cyclone
气旋波　cyclone wave
气旋波【大气】　cyclonic wave
气旋波动理论　wave theory of cyclones
气旋尺度　cyclonic scale
气旋的辐散理论【大气】　divergence theory of cyclones
气旋的阻碍学说　barrier theory of cyclones
气旋地转流　cyclo-geostrophic current
气旋对流理论　convection theory of cyclones
气旋风　cyclonic wind
气旋壕　moat
气旋环流　circulation in cyclone
气旋降水　cyclonic precipitation, cyclone precipitation
气旋路径　cyclone track, cyclone path
气旋模式　cyclone model
气旋期　cyclonic phase
气旋扰动　cyclonic disturbance
气旋生成【大气】　cyclogenesis
气旋生成的波动理论【大气】　wave theory of cyclogenesis
气旋生成的对流〔学〕说　convective theory of cyclogenesis
气旋生成的辐散理论　divergence theory of cyclogenesis
气旋式旋转　cyclonic rotation
气旋消散【大气】　cyclolysis

气旋〔性〕风暴　cyclonic storm
气旋性环流【大气】　cyclonic circulation
气旋〔性〕雷暴　cyclonic thunderstorm
气旋性切变【大气】　cyclonic shear
气旋性曲率【大气】　cyclonic curvature
气旋性温带风暴　cyclonic extratropical storm（CYCLES）
气旋性涡度【大气】　cyclonic vorticity
气旋〔性〕雨　cyclonic rain
气旋学说　theory of cyclone
气旋预警　cyclone warning
气旋中心　cyclonic centre
气旋周环　pericyclonic ring
气旋轴倾角　inclination of the axis of a cyclone
气旋族　family of cyclones
气旋族【大气】　cyclone family
气压　air pressure
气压【大气】　atmospheric pressure
气压鼻　pressure nose
气压变化　pressure change, barometric change
气压变化图　pressure-change chart
气压变量表　pressure variometer
气压变量【大气】　pressure variation
气压变量计　pressure variograph
气压表【大气】　barometer
气压表订正　barometer reduction, barometer correction
气压表读数　barometer reading
气压表高度【大气】　barometer level
气压表空盒　pressure capsule
气压表零点高度　elevation of the zero point of barometer
气压表器差订正表　barometric correction table
气压表水银槽　barometer cistern
气压表纬度订正〔值〕　latitude barometer correction
气压表匣　barometer box（或 case）
气压〔表值〕　barometric pressure

气压波　pressure wave【大气】, barometric wave
气压槽　pressure trough
气压测定法　barometry
气压测高表【大气】　pressure altimeter
气压测高常数　barometric constant, barometer constant
气压测高法　barometric hypsometry, barometric altimetry
气压测高公式　barometric equation, barometer formula
气压测量膜盒　aneroid chamber
气压层　barosphere
气压场　field of pressure
气压场【大气】　pressure field
气压传感器　pressure transducer, barosensor, barotron, baroreceptor, baroceptor
气压传感器阵列　pressure-sensor array
气压垂直坐标　pressure vertical coordinates
气压电阻　baroresistor
气压订正【大气】　barometric correction
气压订正表　barometric reduction table
气压方程　pressure equation
气压分布　pressure distribution
气压〔分布〕型〔式〕　pressure pattern
气压分析　baric analysis
气压风暴表　barocyclometer
气压风暴计　barocyclonometer
气压高度　pressure height, pressure altitude, barometric height
气压高度表　barometric altimeter
气压高度测量　barometric leveling
气压功率谱　air pressure power spectrum
气压计　atmospheric gauge, weather gauge, barometer
气压计【大气】　barograph
气压计海拔〔高度〕　barometer elevation
气压计误差　barometric errors

气压记录器　pressure recorder
气压检定箱　altichamber
气压降落　pressure drop
气压距平　pressure anomaly
气压开关　barometric switch，baroswitch【大气】
气压廓线　pressure profile
气压力矩　pressure torque
气压膜盒　barocell
气压平均法　pressure averaging method
气压谱　pressure spectrum
气压起伏　barometric fluctuation
气压起伏谱　pressure fluctuation spectrum
气压(铅直)速度　pressure velocity
气压倾向　barometric tendency，barometric characteristic
气压倾向【大气】　pressure tendency
气压倾向方程　pressure-tendency equation
气压倾向〔分布〕图　pressure-tendency chart
气压倾向特征　characteristic of the pressure tendency
气压丘　pressure dome
气压区　baric area
气压扰动　barometric disturbance，barometric depression
气压日际变化　interdiurnal pressure variation
气压日际等变线　isomentabole
气压升降率　barometric rate
气压梯度　barometric gradient
气压梯度【大气】　pressure gradient
气压梯度力【大气】　pressure-gradient force
气压图　pressure chart
气压微变化　microvariation of pressure
气压微扰　barometric ripple
气压温度计【大气】　barothermograph
气压温度湿度表　barothermohygrometer
气压温度湿度风速计　barothermohygroanemograph
〔气〕压温〔度〕湿〔度〕计【大气】　barothermohygrograph
气压系统【大气】　pressure system
气压相关　pressure correlation
气压效应　barometric effect
气压形势　baric topography
气压型飞行　pressure pattern flight
气压曳力　pressure drag
气压涌升　pressure jump
气压涌升传感器　pressure jump sensor
气压涌升探测器　pressure jump detector
气压月际变化　inter-monthly pressure variation
气压涌升　pressure surge
气压涌升线【大气】　pressure surge line
气压直减率　baric lapse rate
气压中心　pressure center
气压自动记录仪　barometrograph
气压自记曲线　barograph trace，barogram
气压最低值　barometric minimum
气压最高值　barometric maximum
气压坐标　pressure coordinates
气压坐标系　p system，pressure system，pressure coordinate system【大气】
气液色谱法【化】　gas liquid chromatography
(气)柱　column
气柱　atmospheric column，air column
〔气〕柱丰度【大气】　column abundance
气柱平均温度　mean temperature of air column
气柱重量图　weight chart
汽车废气　automobile exhaust，auto-exhaust
汽车废气排放　automobile emission
汽车尾气催化剂【化】　auto-exhaust catalyst
汽车污染　auto pollution

汽化【物】 vaporization
汽化器积冰 carburetor icing
汽化潜热 latent heat of vaporization
汽化热【物】 heat of vaporization
起步 initial step, first time step
起潮力 tide-producing force
起潮势 tide-producing potential
起电 electrization
起电〔作用〕 electrification
起飞预报 takeoff forecast
起伏场 fluctuating field
起偏器【物】 polarizer
起始 initial
起始读数 initial reading
起始时间 origin time
起转【航空】 spin-up
起转过程 spin-up process
起转时间 spin-up time
起转问题 spin-up problem
器示压力 gauge pressure
憩潮 slack water
恰当微分方程【数】 exact differential equation
千比特 kilobit
千岛海流 Kurile Current
千分率 per mille
千赫 kilohertz(kHz)
千焦耳 kilojoule
千卡 kilogram calorie, large calorie, kilocalorie
千克 kilogram(kg)
千粒重 thousand-grain weight
千米 kilometer(km)
千摩尔 kilomole
千年发展目标 Millennium Development Goals
千瓦-小时 kilowatt-hour
千周 kilocycle
迁移扩散 relocation diffusion
迁移气旋 migrating cyclone
牵引容量 hauling capacity

铅垂度 verticality
铅垂线 vertical line
前导黑子 preceding sunspot, leading sunspot
前辐照【农】 preirradiation
前估 prefigurance
前寒武纪【地质】 Precambrian
前进波【大气】 progressive wave
前进轨道 prograde orbit
前进式问题 marching problem, marching problem
前进运动 progressive motion
前馈控制【计】 feedforward control
前期降水指数 antecedent precipitation index
前期土壤湿度 antecedent soil moisture
前倾槽【大气】 forward-tilting trough
前生命化学【生化】 prebiotic chemistry
前向比辐射强度 forward specific intensity
前向积分 forward integration
前向积分格式 forward integration scheme
前向内插 forward interpolation
前向能见度 forward visibility
前向散射【大气】 forward scattering
前向散射计 forward scatter meter (FSM)
前向散射粒谱仪 forward scattering spectrometer
前向散射粒谱仪探头 forward scattering spectrometer probe(FSSP)
前向散射能见度仪 forward scatter visibility meter
前向散射原理 forward scattering theorem
前向通量密度 forward flux density
前向悬垂回波 forward overhang
前兆【地】 precursor
前兆天空 fore sky
前兆物 predecessor

前兆信号　precursor signal
前置放大器【电子】　preamplifier
钱塘〔江〕涌潮【海洋】　Qiantang River tidal bore
钳〔状〕闪〔电〕　pinched lightning
潜流　underrun
潜流【海洋】　undercurrent
潜能　latent energy
潜热【大气】　latent heat
潜热通量　latent heat flux
潜势法　potential method
潜势区　potential box
潜水蒸发【农】　evaporation from phreatic water
潜在不稳定【大气】　latent instability
潜在不稳定指数　latent instability index
潜在浮力　potential buoyancy
潜在结晶中心　potential centre of crystallization
潜在经济效益　potential economic effectiveness
潜在径流　potential runoff
潜在可预报性　potential predictability
潜在率　potential rate
潜在失水　potential loss
潜在蒸发【大气】　potential evaporation
潜在蒸发率　potential evaporation rate
潜在蒸散【大气】　potential evapotranspiration
潜折射本领　potential refractivity
浅边界层　shallow boundary layer
浅层单体环流　shallow cellular circulation
浅低压　shallow low【大气】, flat low
浅对流【大气】　shallow convection
浅对流参数化【大气】　shallow convection parameterization
浅对流系统　shallow convective system
〔浅海〕测深锤　coasting lead
浅水波【大气】　shallow water wave
浅水方程　shallow water equation
浅水近似　shallow water approximation
浅水理论　shallow water theory
浅水模式　shallow water model
浅水作用　shallow water effect
浅滩【测】　shoal
浅雾　shallow fog
浅云　shallow cloud
浅云街　shallow cloud street
欠定问题　underdetermined problem
欠定组【数】　underdetermined system
欠缺年轮　missing ring
欠阻尼【物】　underdamping
欠阻尼系统　underdamped system
嵌套区域模式　nested regional model
〔嵌〕套网格模式　nested grid model (NGM)
〔嵌〕套网格热带气旋模式　nested tropical cyclone model
〔嵌〕套网格问题　nesting problem
歉收年　lean year
腔体绝对日射表　cavity pyrheliometer
强飑　thick squall
强侧风　extreme crosswind
强度　intensity
强度〔等〕级　intensity level
强度调制　intensity modulation
强度调制显示器(雷达)　intensity-modulated indicator
强度量　intensive quantity
强度起伏　intensity fluctuation
强度-时间公式　intensity-duration formula
强度性质　intensive property
强对流　severe convection
强对流风暴　severe convective storm
强风　snorter, line blow, fresh gale
强风【大气】　strong breeze
强风暴　severe storm
强风暴观测　severestorm observation
强风暴观测卫星　Severe Storm Observation Satellite(SSOS)

强风暴和中尺度试验　severe environmental storms and mesoscale experiment
强风仪【大气】　strong wind anemograph
强辐射夜　radiation night
强环境风暴和中尺度试验　Severe Environmental Storm and Mesoscale Experiment(SESAME)
强回声〔区〕　retrogression-reflection
强雷暴监视　severe thunderstorm watch
强模型【物】　string model
强逆风　muzzler
强耦合系统【数】　strongly coupled system
强平稳过程【数】　strongly stationary process, strictly stationary process
强迫　forcing
强迫波　forced wave
强迫对流【大气】　forced convection
强迫风环流　forced wind circulation
强迫函数　forcing function
强迫项　forcing term
强迫云　forced cloud
强迫振荡【大气】　forced oscillation
强迫振动　forced vibration
强迫重建法　force-restore method
强起伏　strong fluctuation
强起伏区　strong fluctuation region
强切变　severely shear
强热带风暴【大气】　severe tropical storm
强撒播　heavy seeding
强沙尘暴【大气】　severe sand and dust storm
强霜冻侵蚀　cryoplanation
强台风【大气】　severe typhoon
强天气 T 指标　SWEAT index (severe weather threat index)
强天气综合预报工具　Severe Weather Integrated Forecasting Tools (SWIFT)
强纬向环流　high zonal circulation
强下沉气流　strong downdraft
强限制　strong constraint
强线飑　severe line squall
强线极限　strong-line limit
强线近似【大气】　strong-line approximation
强泻气流　downrush
强演变　strong evolution
羟基　hydroxyl
羟基发射(光谱)　hydroxyl emission
羟基离子　hydroxyl ion
羟自由基【化】　hydroxy radical
跷跷板结构　seesaw structure
乔木【林】　high forest
乔木【植】　arbor
乔唐日照计【大气】　Jordan sunshine recorder
乔治指数　George's index
桥臂压降式湿度表　pneumatic bridge hygrometer
切比雪夫多项式【数】　Chebyshev polynomial
切变　shear, shearing deformation
切变波　shear wave
切变波【大气】　shearing wave
切变不稳定　shear instability
切变不稳定【大气】　shearing instability
切变层【大气】　shear layer
切变产生〔量〕　shear production
切变脊　shear ridge
切变流　shear flow
切变能量产生　shear energy production
切变矢量　shear vector
切变速矢端迹图　shear hodograph
切变涡度【大气】　shearing vorticity
切变线【大气】　shear line
切变项　shear term
切变重力波　shear-gravity wave
切变轴　shear axis
切断低压【大气】　cut-off low
切断高压【大气】　cut-off high

切断过程　cutting-off process, breaking-off process
切断装置　cutting device
切线　tangent
切线性方程　tangent linear equation
切线性近似　tangent linear approximation
切线性模式　tangent linear model
切向风【大气】　tangential wind
切向加速度　tangential acceleration
切向速度　tangential velocity
切向应力【物】　tangential stresses
切应变　shear strain
切应力　shearing stress
切趾法　apodization
侵(腐)蚀性水　aggressive water
侵入　invasion, intrusion, incursion
侵入〔流〕　intrusions
侵蚀基准面(水文)　base level of erosion
侵蚀区　eroded field
侵蚀〔作用〕【大气】　erosion
侵占〔性〕生物气象学指数　aggressive biometeorological index
亲潮　Oyashio
亲和标记【生化】　affinity labeling
亲和冰　friendly ice
亲湿性核　hygroscopic wettable nuclei
亲水胶体【化】　hydrophilic colloid
亲水性材料　hydrophilic material
青藏低槽　Qinghai-Xizang trough
青藏反气旋　Tibetan anticyclone
青藏高压　Tibetan high
青藏高压【大气】　Qinghai-Xizang high
青藏高原　Tibetan Plateau
青藏高原季风　Qinghai-Xizang Plateau monsoon
青藏高原气象科学试验　Qinghai-Xizang Plateau Meteorological Science Experiment(QXPMEX)
青藏高原气象学　Qinghai-Xizang Plateau meteorology

青霾　blue haze
轻雹　slight hail
轻便风杯风速表　portable cup anemometer
轻便风速表　pocket anemometer
轻便湿度表　pocky hygrometer
轻便通风干湿表　pocket aspiration psychrometer
轻飑　light squall
轻冻　light freeze
轻度干旱　feebly arid
轻风【大气】　light breeze
轻旱　partial drought
轻浪　slight sea
轻离子　small ion, light ion
轻离子结合　small ion combination
轻霾　slight haze
轻黏土【土壤】　light clay soil
轻霜　light frost
轻微地震声　brontides
轻雾　thin fog, light fog
轻雾【大气】　mist
轻子　lepton
氢氟碳化物　HFCs (hydrofluorocarbons)
氢过氧化物【化】　hydroperoxide
氢检测装置　hydrogen detection system
氢离子　hydrogen ion
氢氯氟碳化物　hydrochlorofluorocarbons (HCFCs)
氢〔气〕　hydrogen(H)
氢气发生器　hydrogen generator
氢〔气谱〕线　hydrogen line
氢气球　hydrogen(-filled) balloon
氢温标　hydrogen scale of temperature
氢温〔度〕表　hydrogen thermometer
氢原子【物】　hydrogen atom
倾角　inclination, dip angle
倾角【数】　angle of inclination
倾盆大雨　downpour
倾向方程【大气】　tendency equation

倾向相关	tendency correlation
倾向性评分	bias score
倾斜	dip
倾斜不稳定性	slantwise instability
倾斜对流	slantwise convection
倾斜计	clinometer
倾斜角	canting angle
倾斜能见度	approach visibility
倾斜能见度【航空】【大气】	slant visibility
倾斜上击流	tilting updraft
倾斜项	tipping term, tilting term
倾斜效应	tilting effect
清除	scavenging
清除系数	scavenging coefficient
清洁发展机制	clean development mechanism(CDM)
清劲风【大气】	fresh breeze
清净空气	scavenging air
清明【农】(节气)【大气】	Fresh Green
情景	scenarios
晴	fairness
晴空	serene, clear air
晴空辐射	clear radiance
晴空化〔过程〕	clearing
晴空回波【大气】	clear air echo
晴空降水	clear sky precipitation
晴空(气柱)辐射	clear-column radiance
晴空热泡	blue thermal
晴空视线	clear line of sight (CLOS)
晴空湍流【大气】	clear-air turbulence (CAT)
晴天	clear sky【大气】, fair, fair weather, fine day, settled weather, sky clear, clear day
晴天电场【大气】	fair-weather electric field
晴天电流【大气】	fair-weather current
晴天电位梯度	fair-weather potential gradient
晴天电学	fair-weather electricity
晴天积云	fair-weather cumulus
晴天卷云	fair-weather cirrus
晴天效应	fine weather effect
晴夜	clear night
晴雨计	weather glass
穹窿	vault
穹面〔式〕	dome
琼斯韦普海浪谱	JONSWAP spectrum
丘陵【地理】	hill
秋	fall
秋冰	autumn ice
秋播期	fall seeding time
秋分	fall equinox
秋分【农】(节气)【大气】	Autumnal Equinox
秋分潮	autumn equinox tide
秋分点	First Point of Libra
秋分期	autumn equinoctial period
秋〔季〕	autumn
秋季生长期	fall growth period
秋季停止生长期	resting stage in autumn
秋老虎(中文)	chou lao hu
秋霜冻	fall frost
秋汛洪水	fall flood
秋雨	autumn rain
求和约定	summation notation
求积公式	quadrature formula
求积谱	quad-spectrum, quadrature spectrum【大气】
求积谱密度	quadrature spectral density
求积仪【测】	platometer, planimeter
求平均值	averaging
球	sphere
球部	bulb
球测云底〔高度〕	balloon ceiling
球对称	spherical symmetry
球根状云	bulbous cloud
球函数	spherical function
球壳【力】	spherical shell
球壳装置	spherical shell apparatus

球面波　spherical wave
球面调和函数【数】　spherical harmonics
球面度　steradian
球面反照率　spherical albedo
球面镜【物】　spherical mirror
球面曲率　spherical curvature
球面入射波　spherical incident wave
球面三角学【数】　spherical trigonometry
球面投影【测】　stereographic projection
球面投影【物】　spherical projection
球面透镜【物】　spherical lens
球面网格　spherical net
球〔面〕谐波　spherical harmonic wave
球〔面〕谐〔波〕分析　spherical harmonic analysis
球面坐标系　spherical coordinate system
球伞(供火箭探测用)　ballute
球陀螺　spherical top
球校准　sphere calibration
球形地面辐射表　spherical pyrgeometer
球形气球　spheric aerostat
球形全辐射表　spherical pyrradiometer
球型振荡　spheroidal oscillation
球载传感器　balloon-borne sensor
球载反射器　balloon-borne reflector
球载光〔雷〕达　balloon-borne laser radar
球载探测器　balloon-borne detector
球载湍流探头　balloon-borne turbulence probe
球状全天空辐射计　ball pyranometer
球〔状〕闪〔电〕　globe lightning
球状闪电【大气】　ball lightning
球状体　spheroid
球状投影　globular projection
区　zone
M 区　M-region
F 区　F region
E 区　E-region
D 区　D-region
区划图　regionalization map
区间估计【化】　interval estimation

区域标准气压表【大气】　regional standard barometer
区域标准器　regional standard
区域尺度模式　regional scale model
区域地质学　regional geology
区域电传通信枢纽(WWW)　Regional Telecommunication Hub(RTH, WWW)
区域分析　regional analysis
区域分析和预报系统　regional analysis and forecasting system(RAFS)
区域风　zone wind
区域辐射中心　regional radiation center (RRC)
区域观测和预报业务计划　Program of Regional Observation and Forecast Service(PROFS)
区域广播　regional broadcast
区域回归随机试验　regional regression randomized experiment
区域基本天气网　regional basic synoptic network
区域集合预报　regional ensemble forecast
区域监测　areal monitoring
区域降水　area precipitation, areal precipitation
区域模拟　regional modeling
区域模式【大气】　regional model
区域农业气候区划　regional agroclimatic delimitation
区域平均　domain average
区域平均雨量【大气】　area mean rainfall
区域谱模式　regional spectral model
区域气候【大气】　regional climate
区域气候模拟　regional climate modeling
区域气候学　regional climatology
区域气象电信网　Regional Meteorological Telecommunication Network(RMTN)

区域气象副中心　regional meteorological subcentre
区域气象中心　regional meteorological office
区域气象中心（世界气象组织）　Regional Meteorological Centre（RMC）
区域衰减因子　areal reduction factor
区域协会　Regional Association（RA）
区域性降水　regional precipitation
区域性空气污染　regional air pollution
区域性空气污染研究　Regional Air Pollution Study（RAPS）
区域性污染　regional pollution
区域预报　district forecast，area forecast
区域预报【大气】　regional forecast
区域预报中心　Area Forecast Centre（AFC）
区站号　station index number【大气】，station identifier，station designator，index number of station
曲管地温表　angle stem earth thermometer
曲管地温表【大气】　angle geothermometer
曲管温度表　bent stem thermometer，angle thermometer
曲流　meander
曲率【数】　curvature
曲率半径【数】　radius of curvature
曲率涡度【大气】　curvature vorticity
曲率效应　curvature effect
〔曲〕面积分【数】　surface integral
曲线　curve
N 曲线　N-curve
S 曲线〔法〕　S-curve〔method〕
曲线回归　curvilinear regression
曲线拟合【数】【大气】　curve fitting
曲线运动【物】　curvilinear motion
曲线坐标【数】【大气】　curvilinear coordinates
曲轴箱换气　crankcase ventilation
驱动函数　driving function

驱动机制　driving mechanism
驱动力【物】　driving force
驱动器　driver
屈光度　diopter
屈曲　buckling
趋肤效应【物】　skin effect
趋光性【农】　phototaxis
趋势　tendency
趋势法　trend method，tendency method
趋势分析　trend analysis
趋势廓线　tendency profile
趋势曲线　trend curve
趋势时距　tendency interval
趋势图　trend chart，tendency chart
趋势线　trend line
渠化　in pool
取土管　soil tube
〔取〕土钻　soil auger
取向　orientation
取向无序【化】　orientational disorder
取向涨落【物】　orientation fluctuation
取消　cancel
取雪管　snow tube，snow sampler
取雪装置　snow-sampling equipment
取样器　sampler
去季节化　deseasonalizing
去倾　detrending，detrend
去稳作用　destablization
圈状日珥　loop prominence
权重分布【数】　Weight distribution
权〔重〕函数【物】　weight function
权重因子　weight factor
全波理论【地】　full-wave theory
全粗糙流　fully-rough flow
全导数　total derivative
全对称【物】　compete symmetry
全反射【物】　total reflection
全分辨率【大气】　full resolution
全氟碳化物　perfluorocarbons
全辐射【大气】　total radiation
全辐射表　pyrradiometer
全概率　total probability
全光反射装置　holophote

全国自然灾害预警系统　nationwide natural disaster warning system（NADWARN）
全景【物】　panoramic view
全景摄影【测】　panoramic photography
全景摄影机【测】　panoramic camera, panorama camera
全景〔圆盘〕图像　full-disk image
全景圆盘〔云〕图　full-disk display, full-disc picture
全景圆盘云图【大气】　full-disc cloud picture
全量程气压计　panoramic barograph
全量程湿度计　panoramic hygrograph
全量程温度计　panoramic thermograph
全球变化　global change
全球变化的分析、研究和培训系统（IGBP 计划之一）　Global Change System for Analysis, Research and Training（START）
全球变化和地球生态系统（IGBP）　Global Change and Terrestrial Ecosystem（GCTE, IGBP）
全球变化研究计划（IGBP）　Global Change Research Program（GCRP, IGBP）
全球尺度　global scale
全球臭氧观测系统（WMO）　GO_3OS（Global Ozone Observing System）
全球大气研究计划【大气】　Global Atmospheric Research Program（GARP）
全球大气研究计划大西洋热带试验　GARP Atlantic Tropical Experiment（GATE）
全球大气研究计划第一期全球试验　First GARP Global Experiment（FGGE）
全球大气研究计划美国委员会　U. S. Committee for the GARP（USC-GARP）
全球电〔传通〕信系统【大气】　global telecommunication system（GTS）

全球电路　global circuit
全球导航卫星系统【地信】　Global Navigation Satellite System（GNSS）
全球定位系统【大气】　global positioning system（GPS）
全球动能　global kinetic energy
全球放射性沉降　world-wide fallout
全球分布　global distribution
全球分析　global analysis
全球风系【地】　global wind system
全球覆盖　global area coverage（GAC）
全球覆盖【航海】　global-mode coverage
全球观测系统　Global Observing System（GOS, WWW）
全球观测系统研究与可预报性试验　The Observing System Research and Predictability Experiment（THORPEX）
全球观测自动中继系统　Automatic Relay Global Observation System（ARGOS）
全球海平面观测系统　Global Sea Level Observing System（GLOSS）
全球海洋观测系统　Global Ocean Observing System（GOOS）
全球海洋观测系统联合科技委员会　Joint GOOS Scientific and Technical Committee（J-GOOS）
全球海洋通量联合研究　Joint Global Ocean Flux Study（with IGBP）
全球化　globalization
全球环境　global environment
全球环境变化的人文因素　Human Dimensions of Global Environmental Change（HDGEC）
全球环境变化与人类安全　Global Environmental Change and Human Security（GECHS）
全球环境变化与人类健康　Global Environmental Change and Human Health（GECHH）
全球环境监测系统　global environmental

monitoring system(GEMS)
全球环流　global circulation
全球环流模式　global circulation model
全球基本气候资料集　global baseline datasets
全球交互式大集合预报系统　THORPEX Interactive Grand Global Ensemble (TIGGE)
全球交互式预报系统　Global Interactive Forecast System (GIFS)
全球联合海洋台站网　Integrated Global Ocean Station System(IGOSS)
全球陆地观测系统　Global Terrestrial Observation System(GTOS)
全球陆地计划(IGBP 内)　Global Land Project (GLP)
全球模式　global model
全球内能　global internal energy
全球能量　global energy
全球能量与水循环试验　Global Energy and Water Cycle Experiment (GEWEX)
全球平均能量平衡　global mean energy balance
全球平均温度　global mean temperature
全球谱模式　global spectral model
全球气候　world climate, global climate【大气】
全球气候观测系统　Global Climate Observing System(GCOS)
全球气候系统【大气】　global climate system
全球气象资料　global meteorological data
全球/区域同化和预报增强系统(模式)(是中国自主研发的第一代数值预报模式)　Global/Regional Assimilation and Prediction Enhanced System (GRAPES)
全球输送　global transport
全球水平探测技术　global horizontal sounding technique(GHOST)
全球水系统计划(IGBP 内)　Global Water System Project (GWSP)
全球水循环　global water cycle
全球碳计划　Global Carbon Project (GCP)
全球天气试验　global weather experiment (GWE)
全球天气侦察　global weather reconnaissance
全球位能　global potential energy
全球温度　holospheric temperature
全球性大气污染　global air pollution
全球性海面升降【地理】　eustatic movement
全球性海面升降【海洋】　eustasy
全球性联系　global association
全球性气候变化　global climatologic change
全球预报　global prediction
全球增温　global warming
全球增温率　global warming rate
全球增温潜势　global warming potential (GWP)
全球资料处理系统　global data processing system (GDPS, WWW)
全球资料同化系统　Global Data Assimilation System (GDAS)
全球自然灾害预警系统　world-wide natural disaster warning system
全球总冰量　global ice volume
全球综合地球观测系统　Global Earth Observing System of Systems (GEOSS)
全日潮【航海】　diurnal tide
全日射　total solar irradiance
全日蚀　total solar eclipse
全色片【测】　panchromatic film
全色摄影　panchromatic photography
全升力　total lift
全食【天】　totality, total eclipse
全收敛【数】　complete convergence
全天光度计【大气】　all-sky photometer
全天候〔的〕　all-weather

全天候飞行　all-weather flight
全天候风向风速计　all-weather wind vane and anemometer
全天候机场【大气】　all-weather airport
全天候降落　all-weather landing
全天候自动着陆【电子】　all-weather automatic landing
全天景照相机　all-sky camera
全通滤波〔器〕　universal filter
全通滤波器　all-pass filter
全同粒子【物】　identical particles
全同模式孪生子积分实验　identical twin integrations
全透水井　fully penetrating well
全微分【数】　total differential
全涡度　total vorticity
全息方法　holographic method
全息术　holographic technique
全息术【物】　holography
全息图【物】　hologram
全息照相【物】　holograph
全息照相机　holoscope
全相关因子　total correlation factor
全向辐射仪　omnidirectional radiometer
全向气流表　omnidirectional air meter
全向天线【电子】　omnidirectional antenna
全新世测绘合作计划　COHMAP（Cooperative Holocene Mapping Project）
全新世【地质】　Holocene Epoch
全新世气候　Holocene climate
全新统【地质】　Holocene Series
全压力　total pressure
全总量指数　total-totals index
泉水引流　spring tapping
炔【化】　alkyne
缺水年　dry year
缺氧　anoxia
缺氧层　hypoxic layer
缺雨性　anhyetism
确定性〔的〕　deterministic
确定性【数】　determinacy
确定性水文学　deterministic hydrology
确定性系统【数】　deterministic system
确定性因子　certainty factor
确定性预报　deterministic prediction
确定性预报【大气】　deterministic forecast
确定预报　determinate forecast
确认【计】　acknowledgement
群波【航海】　group of waves
群落交错区【动物】　ecotone
群落生境【地理】　biotope
群谱　group spectrum
群速度【物】【大气】　group velocity
群体改良【农】　population improvement
群体密度　population density

R

燃烧尘　combustion dust
燃烧核【大气】　combustion nucleus, combustion nuclei
燃烧热【化】　heat of combustion
燃烧习性　fire behavior
染料激光器【物】　dye laser
壤砂土【土壤】　loamy sand
壤土【农】　loam soil

扰动　stirring, fluctuation
扰动边界层　disturbed boundary layer
扰动长度尺度　stirring length scale
扰动尺寸【化】　perturbed dimension
扰动法　method of perturbations
扰动方程【大气】　perturbation equation
扰动高层大气　disturbed upper atmosphere

扰动环 disturbance ring
扰动极区电流 disturbance polar current
扰动集合预报 perturbed ensemble forecast
扰动解 perturbation solution
扰动力 exciting force
扰动量 perturbation quantity, disturbed quantity
扰动模式 perturbed model
扰动气压梯度力 perturbation pressure gradient force(PPGF)
扰〔动〕天〔气〕电学 disturbed weather electricity
扰动土样【土木工程】 disturbed soil sample
扰动线 disturbance line
扰动选择 disturbance selection of perturbation
扰动运动 perturbation motion
扰动中心 center of disturbance
(扰流)抖振马赫数 buffeting Mach number
绕机气流 air around airplane
绕极环流【大气】 circumpolar circulation
绕极气旋 circumpolar cyclone
绕极涡旋【大气】 circumpolar vortex
绕极西风带【大气】 circumpolar westerlies
绕极旋风 circumpolar whirl
绕南极洋流【大气】antarctic circumpolar current (ACC)
热 heat
热斑【天】 host spot
热爆发 heat burst
热波导 thermoduct
热补偿净辐射表 heat-compensated net radiometer
热层【大气】 thermosphere
热层顶【大气】 thermopause
热层环流 thermospheric circulation

热层下部 lower thermosphere
热成低压【大气】 thermal low
热成风【大气】 thermal wind
热成风方程【大气】 thermal-wind equation, equation of thermal wind
热成〔风〕急流 thermal jet
热成风引导 thermal wind steering
热成高压【大气】 thermal high
热〔成〕积云 pyrocumulus
热成积云 heat cumulus
热成畸变 thermal distortion
热赤道【大气】 thermal equator, heat equator
热储量 heat storage
热传导 thermal conduction, heat conduction
热〔传〕导系数 coefficient of thermal conduction, coefficient of heat conduction, thermal conductivity
热刺激 thermal stimulus
热粗糙度【大气】 thermal roughness
热带 tropics, tropical belt, torrid zone
热带【地理】 tropical zone
热带飑 breather
热带飑线 tropical squall line
热带波 tropical wave
热带波扰动 tropical wave disturbance
热带槽 intertropical trough
热带草原气候 tropical savanna climate
热带常湿气候亚类【大气】 tropical rainy climate
热带潮湿气候 tropical wet climate
热带大陆气团 tropical continental air (mass), continental tropical air (mass)
热带低压【大气】 tropical depression
热带定高气球系统 tropical constant-level balloon system (TCLBS)
热带东风带【大气】 tropical easterlies
热带东风急流【大气】 tropical easterlies jet

热带冬干气候亚类【大气】 tropical winter dry climate, tropical steppe climate
热带对流层高空槽 tropical upper-tropospheric trough (TUTT)
热带多雨气候【大气】 tropical rainy climate
热带反气旋 tropical anticyclone
热带风暴 tropical revolving strom
热带风暴【大气】 tropical storm
热带风观测船 tropical wind observation ship (TWOS)
热带风能转换和参考层试验 Tropical Wind Energy Conversion and Reference Level Experiment (TWERLE)
热带锋 tropical front, intertropical front
热带辐合带 tropical convergence
热带辐合带【大气】 intertropical convergence zone (ITCZ)
热带干湿气候 tropical wet and dry climate
热带高空气团 tropical superior air mass
热带高压【大气】 tropical high
热带海洋地区低层分析 analysis of tropical oceanic low-levels (ATOLL)
热带海洋地区低层分析图 ATOLL Chart
热带海洋气团 maritime tropical air (mass)
热带海洋气团【大气】 tropical marine air mass
热带海洋与全球大气试验计划 Tropical Ocean and Global Atmosphere (TOGA)
热带环境 tropical environment
热带环流【大气】 tropical circulation
热带汇流区(带) intertropical confluence zone(或 band)
热带季风【大气】 tropical monsoon
热带季风气候 tropical monsoon climate
热带季风雨气候亚类【大气】 tropical monsoon rain climate
热带降雨测量〔卫星〕 tropical rainfall measuring mission (TRMM)
热带〔经向〕环流 tropical cell
热带局地强风暴 tropical severe local storm
热带美洲雨季 invierno
热带气候【大气】,热带气候类【大气】 tropical climate
热带气候学【大气】 tropical climatology
热带气团【大气】 tropical air mass
热带气团雾【大气】 tropical air fog
热带气象区域中心(美国) Regional Center for Tropical Meteorology (RCTM)
热带气象学 equatorial meteorology
热带气象学【大气】 tropical meteorology
热带气象研究计划 Project on Research in Tropical Meteorology (PRTM)
热带气旋【大气】 tropical cyclone
热带气旋计划 tropical cyclone program (TCP)
热带气旋类低涡【大气】 TCLV (tropical cyclone-like vortices)
热带气旋眼 eye of the tropical cyclone
热带扰动【大气】 tropical disturbance
热带试验 tropical experiment (TROPEX)
热带试验局(WMO/ICSU) Tropical Experiment Board (TEB, WMO/ICSU)
热带试验委员会(WMO/ICSU) Tropical Experiment Council (TEC, WMO/ICSU)
热带无风带 tropical calm zone
热带稀树草原 savanna
热带〔稀树〕草原气候【大气】 tropical steppe climate
热带系统 tropical system
热带雨林 hylea, hylaea, hileia, tropical rainforest【大气】

热带雨林气候【大气】 tropical rainforest climate
热带云区 intertropical cloud zone
热带云团【大气】 tropical cloud cluster
热带作物 tropical crop
热单位 thermal unit
热单位理论 heat-unit theory
热当量 heat equivalent, equivalent of heat
热导率 heat conductivity, thermometric conductivity
热导式热量计【化】 heat conduction calorimeter
热岛【大气】 heat island
热岛〔效应〕 thermal island〔effect〕
热岛效应【大气】 heat island effect
热〔的〕 hot
热等离子体 hot plasma
热低压 thermal depression, heat low
热低压【大气】 thermal low
热〔地〕带 hot belt
热点【地】 hot spot
热点效应 hot spot effect
热电堆 thermopile
热电偶 thermo-element
热电偶【电力】 thermocouple
热电偶风速表 thermocouple anemometer
热电偶露点湿度表 thermocouple dew point hygrometer
热电偶式干湿表 thermocouple psychrometer
热电偶套管 thermowell
热电偶温度表 thermoelectric thermometer, thermocouple thermometer
热电式浑浊度表 pyroelectric turbidity meter
热电式热线风速表 thermoelectric hot-wire anemometer
热电探测器 pyroelectric detector
热电效应【电子】 thermoelectric effect
热电直接辐射表 thermoelectric actinometer
热电直接辐射计 thermoelectric actinograph
热定标 thermal calibration
热对比 thermal contrast
热对流 heat convection
热对流【大气】 thermal convection
热发光定年法 thermoluminescence dating
热发射 thermal emission
热分解 thermal decomposition
热分析 thermal analysis
热风 hot wind
热辐射 radiation of heat
热辐射带 thermal band
热辐射计 bolograph
热辐射强度 caloradiance
热〔辐射〕扫描 thermal scanning
热辐射【物】 thermal radiation, heat radiation
热辐射仪【大气】 bolometer
热辐射仪自记曲线【大气】 bologram
热负荷 thermal load
热感探测仪 thermosonde
热感探空仪 thermasoned
热高压【大气】 thermal high
热功当量【物】【大气】 mechanical equivalent of heat
热功当量定律【化】 law of mechanical equivalent of heat
热惯性 thermal-inertia
热害 hoting damage
热害【大气】 hot damage
热害【地理】 heat damage
热函数 heat function
热焓 thermal enthalpy
热红外 thermal infrared
热湖【地理】 warm lake
热化学【化】 thermochemistry
热化学循环【化】 thermochemical cycle
热环流 thermal circulation
热汇【大气】 heat sink

热机　heat engine
热积云　thermic cumulus
热激发【物】　thermal excitation
热季【大气】　hot season
热降解　thermal degradation
热解〔作用〕　pyrolysis
〔热〕喀斯特地貌　thermokarst topography
热〔空气〕气球　hot-air balloon
热库【物】　heat reservoir
热扩散【物】　thermal diffusion
热扩散率　thermal diffusivity
热扩散因子　thermal diffusion factor
热浪　hot wave
热浪【大气】　heat wave
热浪健康预警系统　Heat Health Warning System（HHWS）
热雷暴【大气】　heat thunderstorm
热〔雷〕暴　heat storm
热雷阵雨　thermal thunder shower
热离子　thermion
热离子学　thermionics
热力边界层　thermal boundary layer
热力波　thermal wave
热力不稳定　thermal instability
热力层结　thermodynamic stratification
热力层结【大气】　thermal stratification
热力产生率　thermal production rate
热力潮　thermal tide
热力沉淀仪　thermal precipitator
热〔力〕粗糙度　thermal roughness
热力分层介质　thermally stratified medium
热力风化　insolation weathering
热力环流　thermodynamic circulation
热力间接环流　thermally indirect circulation
热力罗斯贝数　thermal Rossby number
热力内边界层　thermal internal boundary layer
热〔力〕泡　thermal

热力强迫作用　thermal forcing
热力势　thermodynamic potential
热力图　thermodynamic diagram, thermodynamic chart
热力学【物】　thermodynamics
热力学冰球温度【大气】　thermodynamic ice-bulb temperature
热力学参数　thermodynamic parameter
热力学第二定律　second law of thermodynamics
热力学第零定律【化】　zeroth law of thermodynamics
热力学第三定律【物】　third law of thermodynamics
热力学第一定律【物】　first law of thermodynamics
热力学定律　law of thermodynamics
热力学方程【大气】　thermodynamic equation
热力学概率【物】　thermodynamic probability
热力学过程【物】　thermodynamic process
热力〔学〕函数　thermodynamical function
热力学函数【物】　thermodynamic function
热力学混合层增长方法　thermodynamic method for mixed layer growth
热力学露点温度　thermodynamic dew-point temperature
热力学能　thermodynamic energy
热力学能量方程　thermodynamic energy equation
热力学平衡【物】　thermodynamic equilibrium
热力学湿球温度【大气】　thermodynamic wet-bulb temperature
热力学势【物】　thermodynamical potential
热力学霜点温度【大气】　thermodynamic

frost-point temperature
热力学速度极限　thermodynamic speed limit
热力学图【大气】　thermodynamic diagram
热力学温标【物】　thermodynamic scale of temperature, absolute temperature scale【大气】
热力学效率　thermodynamic efficiency
热力学性质　thermodynamic property
热力学状态函数　thermodynamic function of state
热〔力烟〕羽　thermal plume
热力张弛　thermal relaxation
热力直接环流　thermally direct circulation
热力滞后系数　thermal time constant
热力中性　thermoneutrality, thermal neutrality
〔热力〕中性带　thermoneutral zone
热粒子　thermal particle
热量垂直输送　vertical heat transport
热〔量〕单位　heat unit
热量的向上输送　upward heat transport
热量平衡　thermal balance
热量平衡【大气】　heat balance
热量收支【大气】　heat budget
热量收支测量仪器　heat budget instrument (HBI)
热量输入　heat input
热量输送【大气】　heat transfer
热量条件预报　heat condition forecasting
热量资源　thermal resources
热量资源【大气】　heat resource
热流　thermal current
热流板　fluxplate
热流测量板　heat flow measuring plate
热流换能器　heat-flow transducer
热流量方程【大气】　heat flow equation
热流仪　heat flow meter

热漫散射【物】　thermal diffuse scattering
热眠　heat dormancy
热敏电阻　thermal resistance
热敏电阻【电子】　thermistor
热敏电阻风速表【大气】　thermistor anemometer
热敏电阻热辐射仪　thermistor bolometer
热敏电阻温度表　thermistor thermometer
热膜风速表　hot-film anemometer
热膜探头（测风用）　hot-film probe
热那亚气旋　Genoa cyclone
热能　thermal energy, heat energy
热能方程　thermal equation of energy
热逆温　thermal inversion
热偶温度计　thermo-junction thermometer
热泡【大气】　thermal
热泡共振波　resonance waves over thermals
热泡核　thermal core
热膨胀【物】　thermal expansion
热平衡【物】　thermal equilibrium
热坡度　thermal slope
热坡度风　thermal slope wind
热启动【计】　warm start
热气旋生成　thermocyclogenesis
热迁移速度　thermophoretic velocity
热丘效应　thermal mountain effect
热日　thermal day, hot day
热容量　thermal capacity
热容〔量〕【物】　heat capacity
热容〔量〕测绘辐射仪　heat capacity mapping radiometer (HCMR)
热容〔量〕测绘卫星　Heat Capacity Mapping Mission (HCMM)
热容〔量〕方法　heat capacity method
热闪　summer lightning, heat lightning【大气】
热〔射〕线　heat ray
热渗透　thermo-osmosis
热实验室【化】　hot laboratory
热势　thermal potential

热试验【化】　hot test, hot run
热适应　thermal comfort
热室【化】　hot cell
热释电效应【物】　pyroelectric effect
热水平衡【地理】　heat-water balance
热水污染　hot water pollution
热损失估测仪　eupatheoscope
热塔　hot tower
热梯度　thermal gradient
热梯度扩散云室　thermal-gradient diffusion chamber
热通量　thermal flux, heat flux【大气】
热通量矢量　heat flux vector
热通量收支　heat flux budget
热湍流　thermal turbulence
热涡度【大气】　thermal vorticity
热涡度平流　thermal vorticity advection
热涡旋　thermal vortice
热污染　heat pollution【大气】, thermal pollution, calefaction
热下坡风　thermal downslope wind
热线风速表　thermo-anemometer
热线风速表【大气】　hot-wire anemometer
热线含水量仪　hot-wire liquid water content meter
热象仪【电子】　thermal imager
热效率【航空】　thermal efficiency
热效应　thermal effect
热效应【化】　heat effect
热型气候　thermal climate
热循环　heat cycle
热焰【地】　hot plume
热液过程【地质】　hydrothermal process
热引导【大气】　thermal steering
热应力　heat stress
热应力指数　index of thermal stress, heat stress index
热源【大气】　heat source
热云　hot cloud
热运动【物】　thermal motion
热晕　thermal blooming

热胀系数　coefficient of thermal expansion
热值　calorific value
热指数　heat index
热致发光【物】　thermoluminescence
热中性区　zone of thermal neutrality
热状况　thermal regime
人布植物【植】　androchore
人工冰核　artificial ice nucleus
人工尘埃　man-made dust
人工成核作用【大气】　artificial nucleation
人工催化剂【大气】　cloud seeding agent
人工岛【航海】　artificial island
人工地平　artificial horizon
人工干预　manual intervention
人工更新【林】　artificial reforestation
人工观测　manual observation
人工光源　artificial light source
人工核　artificial nucleus
人工湖【地理】　man-made lake
人工回灌　artificial recharge
人工降水【大气】　artificial precipitation
人工降雨【农】　rain-making
人工降雨【土壤】　rainfall simulation
人工接地体　manual earthing electrode
人工控制　artificial control
人工雷电抑制【大气】　lightning suppression
人工林【林】　forest plantation
人工气候室　climatic chamber, air conditioning room
人工气候室【大气】　phytotron
人工气候箱　climatic box
人工热流器　meteotron
人工神经网络【电子】　artificial neural net, artificial neural network(ANN)
人工数字化雷达　manually digitized radar (MDR)
人工通风　artificial ventilation
人工消散　artificial dissipation

人工小气候【大气】 artificial microclimate
人工引发闪电 artificially initiated lightning
人工影响电离层 ionospheric modification
人工影响降水 rain stimulation by artificial means, precipitation stimulation
人工影响气候 climatic modification, weather modification
人工影响气候【大气】 climate modification
人工源 man-made source
人工造雪机 snowmaker
人工增益调整 manual gain control
人工增雨 artificial precipitation stimulation
人工放射性【物】 artificial radioactivity
人工照明 artificial lighting, artificial illumination
人工智能【计】 artificial intelligence (AI)
人机对话【计】 man-machine dialogue, human-computer dialogue
人机对话数据存取系统 Man-computer Interactive Data Access System (McIDAS)
人机交互【计】 man-machine interaction
人机结合 man-machine mix【大气】, human-machine mix
人机结合天气预报【大气】 man-machine weather forecast
人机界面 human-machine interface
人均GDP per capita GDP
人均历史累计排放 historical cumulative emission per capita
人均排放〔量〕 emission per capita
人口地理学【地理】 population geography
人类地理学【地理】 anthropogeography
人类对气候影响 human influence on climate
人类活动引起的 anthropogenic
人类活动造成气候变化【大气】 anthropogenic climate change
人类气候 human climate
人类气候学 anthropoclimatology
人类圈 anthroposphere
人类生态学 anthropecology
人类生态学【地理】 human ecology
人类〔生物〕气候学 human climatology
人类生物气候学【大气】 human bioclimatology
人类生物气象学【大气】 human biometeorology
人类生物学 anthropobiology
人类世 anthropocene
人类影响气候 man's impact on climate
人体干燥率 anthropoclimatic drying power
人体舒适 human comfort
人为边界条件 artificial boundary condition
人为变旱 exsiccation
人为传播的 anthropochorous
人为排放 anthropogenic emissions
人为气候变化 man induced climatic change, anthropogenic climate change
人为气候突变 anthropogenic climate catastrophe
人为热 anthropogenic heat
人为污染 artificial contaminant
人为污染物清单 inventories of anthropogenic pollutants
人为误差 personal error
人〔为误〕差 personal equation
人为因素 anthropogenic factor
人文地理学【地理】 human geography
人文地图【测】 human map
人仪差【测】 personal and instrument equation
人与生物圈计划 Man and Biosphere

Program(MAB)
人员落水【航海】 man overboard
人造边界〔条件〕 artificial boundary
人造放射性元素【化】 artificial radio element
人造环境 microenvironment
人造钠云 artificial sodium cloud
人造气候 artificial climate
人造闪电 laboratory lightning
人造卫星 artificial satellite
人造雨 artificial rain
人造云 cloudier, rayon cloud, artificial cloud
人造〔站〕资料【大气】 bogus data
人致气候变化【大气】 anthropogenic climate change
〔认定〕实况站 ground truth site
〔认定〕实况资料 ground truth data
认证【物】 identification (ID)
认知科学【计】 cognitive science
认知任务分析 cognitive task analysis (CTA)
韧致辐射【物】 Bremsstrahlung
韧致辐射效应 Bremsstrahlung effect
日 day
日本地球资源卫星 Japanese Earth Resources Satellite (JERS-1)
日本电气公司 Nippon Electric Company (NEC)
日本海本征水 Japan Sea Proper Water
日本海流【大气】 Japan Current
日本海深层水 Japan Sea Deep Water
日本海中层水 Japan Sea Middle Water
日本湖沼学会 Japanese Society of Limnology (JSL)
日本南极研究探险 Japanese Antarctic Research Expedition (JARE)
日本农业气象学会 Society of Agricultural Meteorology of Japan (SAMJ)
日本〔暖〕海流(黑潮) Japan Current
日本气象厅 Japan Meteorological Agency (JMA)
日本气象学会 Japanese Meteorological Society (JMS)
日变程 daily course
日变化【大气】 diurnal variation, daily variation
日标【航海】 day mark
日差 diurnal inequality
日差【航海】 daily rate, chronometer rate
日成风 heliotropic wind
日承 kerns arc, circumhorizontal arc【大气】
日出 sunrise
日出时刻 sunrise time
日储水量 daily storage
日磁情指数 daily character figure
日-地关系 sun-earth relationship
日地关系【天】 solar-terrestrial relationship
日〔地〕距 solar distance
日地空间【地】 solar-terrestrial space
日地扰动 solar-terrestrial disturbance
日地物理〔学〕 heliogeophysics
日地物理学【地】 solar-terrestrial physics
日地物理学协会间委员会 Inter-Union Commission on Solar-Terrestrial Physics (IUCSTP)
日地研究委员会(美国) Committee on Solar-Terrestrial Research (CSTR)
日调节【水利】 daily regulation
日·度 day degree
日珥 prominence, solar prominence, Lowitz arc
日珥【天】 solar prominence
日光 sunbeam
日光层 heliosphere
日光温室 heliogreenhouse
日光仪 heliograph
日晷 solar spicule
日晷【天】 sundial

日晷仪　dial
日华　solar crown
日华【大气】　solar corona
日环流　diurnal circulation
日环食【天】　annular solar eclipse
日极值　daily extreme
日际　interdiurnal
日际变化　interdiurnal variation, day-to-day change
日际变化平均值　mean interdiurnal variability
日际变率　interdiurnal variability
日降雨量【铁道】　daily precipitation
日较差　diurnal range【大气】, daily range
日界线【航海】　date line, calendar line
日径图　sunpath diagram
日喀则群【地质】　Xigaze Group
日量　daily amount
日落　sunset
日落时刻　sunset time
日落效应　sunset effect
日冕　sun's corona
日冕瞬变【大气】　coronal transient
日冕物质抛射【天】【大气】　coronal mass ejection（CME）
日冕仪【天】　coronagraph
日面【天】　solar disk
日面边缘　solar limb
日面图【天】　heliographic chart
日面坐标【天】　heliographic coordinate
日平均【大气】　daily mean
日平均温度　mean daily temperature
日平均温度【大气】　daily mean temperature
日曝　solarization
日气压波　diurnal pressure wave
日射　insolation
日射表　insolameter, actinometer【大气】
日射病　ictus of sun, heliosis, sunstroke

日射测定表【大气】　actinometer
日射测定计【大气】　actinography
日射测定学【大气】　actinometry
日射大气光程　solar air mass
日射订正　solar radiation correction
日射光谱计　spectroheliograph
日射化学强度等值线　isactine
日射温度表　solar thermometer, solar-radiation thermometer
日射压　solar radiation pressure
日射自记曲线【大气】　actinogram
日射总量表　solarimeter
日射总量计　solarigraph
日食【天】　solar eclipse
日食天气　eclipse weather
日下点　subsolar point
日下晕　subsun
日心距离【天】　heliocentric distance
日心体系【天】　heliocentric system
日心引力常数【天】　heliocentric gravitational constant
日心坐标【天】　heliocentric coordinate
日循环　diurnal cycle
日耀极光　sunlit aurora
日夜闪电光学测量　night/day optical survey of lightning
日运动【天】　daily motion
日晕【大气】　solar halo
日晕（轮）　solar aureole
日载【大气】　circumzenithal arc
日照　sunlight
日照【大气】　sunshine
日照百分率【大气】　percentage of sunshine
日照标准【建筑】　daylight standard
日照计【大气】　sunshine recorder
日照记录　sunshine record
日照记录纸　sunshine card
日照开关　sunshine switch
日照亮斑【大气】　sun glint
日照时间　sunshine time, insolation duration【大气】, duration of sunshine

日照时间记录器　sunshine duration recorder
日照时数【大气】　sunshine duration
日照时数【农】　sunshine-hour
日照时数计算器　sunshine integrator
日照仪　sunshine autograph
日照纸　heliogramma
日振幅　diurnal amplitude, daily amplitude
日震学【天】　helioseismology
日致气压变化　solar barometric variation
日柱【大气】　sun pillar
日灼　sunscald
日灼谱　sunburning spectrum
日灼伤　sunstroke
日最低温度【大气】　daily minimum temperature
日最低温度平均值　mean daily minimum temperature
日最高温度【大气】　daily maximum temperature
日最高温度平均值　mean daily maximum temperature
冗余度　redundancy, redundance
荣格〔尺度〕谱　Junge size distribution
荣格律　Junge law
荣格模式　Junge model
荣格〔气溶胶〕层　Junge (aerosol) layer
荣格气溶胶〔尺度〕谱　Junge aerosol size distribution
容错计算机【计】　fault-tolerant computer
容积【物】　volume
容积式雪取样器　volumetric snow sampler
容积式蒸散计　volumetric lysimeter
容量订正　capacity correction
容许浓度　allowable concentration
容许浓度上限　admissible concentration limit
容许容量　acceptance capacity

容许误差　allowable error
容许误差【数】　tolerance error, admissible error
容许跃迁【物】　allowed transition
溶冰　rotten ice
溶度参数【化】　solubility parameter
溶剂【化】　solvent
溶剂合物【化】　solvate
溶剂效应【化】　solvent effect
溶胶-凝胶过程【化】　sol-gel process
溶〔解〕度　solubility
溶解热　heat of dissolution
溶解热【物】　heat of solution
溶〔解〕跃面　lysocline
溶解〔作用〕　dissolution
溶蚀【土壤】　corrosion
溶液【化】　solution
溶液特性　colligative property
溶质　solute
溶质效应　solute effect
熔化潜热　latent heat of fusion
熔化热【物】　heat of fusion
熔岩湖【地质】　lava lake
熔岩流【地质】　lava flow
融冰　spring sludge, ice-out
融冻泥流作用　congeliturbation
融化　melting
融化层　melting layer, melting level
融化带　melting band
融〔化〕点【大气】　melting point
融化指数【地理】　thawing index
融解热　heat of melting
融凝冰柱　penitent ice, nieve penitente
融凝雪　penitent snow
融土【地理】　melted soil
融雪风　snow-eater
融雪洪水　rush of snowmelt flood
融雪径流　snowmelt runoff
融雪量　thermal quality of snow
融雪侵蚀【土壤】　snowmelt erosion
融雪终止　ending of snowmelt

如果-则定则　　IF-THEN rule
铷钟【天】　rubidium clock
儒略历【天】　Julian calendar
儒略日　Julian day
蠕变【地】　creep
蠕变仪【地】　creepmeter
蠕动流　creeping flow
乳白〔的〕浑浊度　opalescent turbidity
乳白光　opalescence
乳白浑浊度　opalescence turbidity
乳白天空　whiteout, milky weather
乳色冰　milky ice
入海冰川　tidal glacier
入口区　entrance region
入流　influx
入流角　inflow angle
入流-蓄水-出流曲线　inflow-storage-discharge curve
入梅　beginning of the Meiyu period
入梅【大气】　onset of Meiyu
入侵空气　invading air
入射　incidence
入射辐射【大气】　incoming radiation
入射角【物】　incident angle, angle of incidence
入射罗斯贝波　incident Rossby wave
入射太阳辐射　incoming solar radiation
入射太阳通量　incident solar flux
入射线【物】　incident ray
入渗【大气】　infiltration
入渗量【大气】　infiltration capacity
入渗率【农】　infiltration rate
软雹　soft hail, graupel【大气】
软磁盘【计】　flexible disk
软风　slight air, light air【大气】
软辐射　soft radiation
软件　software
软件包【计】　software package
软件工程【计】　software engineering
软件开发周期【计】　software development cycle
软件库【计】　software library
软件配置【计】　software configuration
软件主体【计】　software agent
软决策【数】　soft decision
软科学　soft science
软粒子　soft particle
软盘【计】　floppy disk
软盘驱动器【计】　floppy disk drive
软水　soft water
软雪　ripe snow
软着陆　soft-land
锐度　acuity, sharpness
锐角三角形【数】　acute triangle
瑞典气象与水文研究所　Swedish Meteorological and Hydrological Institute (SMHI)
瑞利-贝纳不稳定性【物】　Rayleigh-Benard instability
瑞利-贝纳对流　Rayleigh-Benard convection
瑞利-贝纳对流旋转　Rayleigh-Benard convection rotation
瑞利波　Rayleigh wave
瑞利不稳定　Rayleigh instability
瑞利参数　Rayleigh parameter
瑞利大气　Rayleigh atmosphere
瑞利单体　Rayleigh cell
瑞利-德拜散射　Rayleigh-Debye scattering
瑞利分布　Rayleigh distribution
瑞利公式　Rayleigh's formula
瑞利光学厚度　Rayleigh optical depth
瑞利假设　Rayleigh hypothesis
瑞利-金斯定律　Rayleigh-Jeans law
瑞利-金斯公式【物】　Rayleigh-Jeans formula
瑞利摩擦　Rayleigh friction
瑞利平滑判据　Rayleigh smoothness criterion
瑞利散射【物】【大气】　Rayleigh scattering

瑞利散射定律　Rayleigh's scattering law
瑞利数【大气】　Rayleigh number
瑞利相函数　Rayleigh phase function
瑞士气象研究所　Swiss Meteorological Institute (SMI)
闰秒【天】　leap second
闰年　annus abundans, bissextile
闰年【天】　leap year
闰日　leap-day, intercalary day, adding day
若尔当代数【数】　Jordan algebra
弱场动力模式　weak-field dynamo mode
弱单体【大气】　weak cell
弱颠簸　cobblestone turbulence
弱非线性　weakly nonlinear
弱非线性理论　weakly nonlinear theory

弱锋　weak front
弱函数　minorant
弱回波穹窿【大气】　weak echo vault
弱回波区　weak echo region (WER)
弱季风　weak monsoon
弱解　weak solution
弱起伏　weak fluctuation
弱透水层【地质】　aquitard
弱纬向环流　low zonal circulation
弱稳定层　slightly stable layer
弱吸收极限　weak absorption limit
弱限制　weak constraint
弱线极限　weak-line limit
弱线近似【大气】　weak-line approximation
弱演变　weak evolution

S

萨顿扩散公式　Sutton's diffusion formula
萨赫勒干旱　sahelian drought
萨瓦纳气候【地理】　savanna climate
塞曼效应【物】　Zeeman effect
赛福尔-辛普森飓风等级　Saffir-Simpson hurricane scale
三胞模式　tricellular model
三杯风速表　three-cup anemometer
三边测量　trilateration
三波共振子　resonant triad
三次方程【数】　cubic equation
三次谐波【物】　third harmonic
三点测距　trilateration range and range rate (TRRR)
三叠纪【地质】　Triassic Period
三叠系【地质】　Triassic System
三分量风速仪　UVW anemometer, three coponent anemometer
三分量螺旋桨式风速表　three-component propeller anemometer
三分之二律　two-thirds law
三峰分布　trimodal distribution
三氟化硼检测器　boron trifluoride detector
三级环流【大气】　tertiary circulation
三级气候站　third-order climatological station
三极管【物】　triode
三角测量气球　triangulation balloon
三角插值【数】　trigonometric interpolation
三角对线矩阵　tridiagonal matrix
三角函数　triangular function
三角函数【数】　trigonometric function
三角级数【数】　trigonometric series
三角截断　triangular truncation
三角剖分【数】　triangulation
三角形窗　triangular window
三角洲　delta region

三阶闭合　third-order closure
三聚体【化】　trimer
三棱石　dreikanter
三面角反射器　trihedral reflector
三模态声〔雷〕达接收机　trimodal acoustic radar receiver
三圈〔经向〕环流【大气】　three-cell 〔meridional〕circulation
三日潮汐　terdiurnal tide
三十天预报　thirty-day forecast
三时次法　three-time level scheme
三体反应　three-body reaction
三体碰撞【天】　triple collision
三体散射长钉〔状回波〕　three-body scatter spike（TBSS）
三体相互作用　three-body interaction
三维变分分析　three-dimensional variational analysis
三维变分同化　three-dimensional variational assimilation
三维对流模式　three-dimensional convection mode
三维多普勒分析　triple-Doppler analysis
三维风观测　three dimensional wind observation
三维傅里叶转换　three-dimensional Fourier transform
三维结构【大气】　three-dimensional structure
三维模式　three-dimensional model
三维湍流　three dimensional turbulence
三维遥相关　three-dimensional teleconnection（teleconnexion）
三相　triple state
三相点【物】　triple point
三相自记器　triple register
三叶风向标　trivane
三原色【物】　three primary colors
三重积分【数】　triple integral
三重数积　triple scalar product
三重相关　triple correlation

三轴加速度仪　triaxial accelerometer
三轴稳定　three-axis stabilized
散布　spread
散布图　scattergram, scatter diagram【大气】
散度【数】【大气】　divergence
散度测定　divergence measurement
散度场　divergence field
散度定理【大气】　divergence theorem
散度方程【大气】　divergence equation
散度风　divergent wind
散度涡度比　divergence-vorticity ratio
散度阻尼　divergence damping
散见 E 层【地】　sporadic E
散焦【物】　defocusing
散粒噪声【物】　shot noise
散列　hashing
散流〔的〕　divective
散能分光计　energy dispersive spectrometer
散射【电子】　scatter
散射比　scattering ratio
散射传播　scatter propagation
散射辐射【大气】　scattered radiation
散射功率　scattered power
散射光【物】　scattered light
散射光光度计　scattered light photometer
散射函数　scattering function
散射计【电子】　scatterometer
散射角【物】　scattering angle
散射角平均余弦　mean cosine of scattering angle
散射截面【大气】　scattering cross-section
散射矩阵【物】　scattering matrix
散射面积比　scattering area ratio
散射面积系数　scattering area coefficient
散射能力　scattering power
散射通信【电子】　scatter communication

散射系数【大气】 scattering coefficient
散射系数仪 scattering coefficient meter
散射线 scattered ray
散射相函数 scattering phase function
散射效率 scattering efficiency
散射效率因子 scattering efficiency factor
NASA 散射仪 NASA Scatterometer
散射元薄层 slab of scatterer
散射振幅【物】 scattering amplitude
散雾器 fogbroom
桑给巴尔海流 Zanzibar Current
桑思韦特气候分类【大气】 Thornthwaite's climatic classification
桑斯威特湿度指数 Thornthwaite moisture index
桑斯威特蒸散计 Thornthwaite evapotranspirometer
桑塔格绝对日射表 Sonntag pyrheliometer
扫雷区【测绘】 mine-sweeping area
扫描【电子】 sweep
扫描 scanning, scan
扫描半径 scan radius
扫描波谱仪 scanning spectrometer
扫描成像分光光度计 scanning imaging spectrophotometer (SIS)
扫描点 flying-spot
扫描辐射仪【大气】 scanning radiometer (SR)
扫描辐射仪资料处理器 scanning radiometer data manipulator (SRDM)
扫描光〔雷〕达 scanning lidar
扫描激光多普勒速度计系统 scanning laser Doppler velocimeter system (SLDVS)
扫描镜 scan mirror
扫描探测器 scanning detector
扫描线 scan line
扫描仪【计】 scanner
扫描圆 scan circle
扫频干涉仪【电子】 swept frequency interferometer
色标查算表 color look-up table (CLUT)
色〔层〕谱法【物】 chromatography
色〔层〕谱图【物】 chromatogram
色〔层〕谱仪【物】 chromatograph
色调带 tone bar
色调显示 tone pattern
色球【天】 chromosphere
〔色球〕活动网络 active network
色散本领【物】 dispersion power
色散方程 dispersion equation
色散关系 dispersion relation
色散介质【物】 dispersive medium
色散曲线 dispersion curve
色散性 dispersivity
色闪烁 chromatic scintillation
色温【物】【大气】 color temperature
色相 hue
铯钟【天】 caesium clock
森林保护【农】 forest conservation
森林采运 logging
森林草原【地理】 forest steppe
森林草原气候 forest-steppe climate
森林风 forest wind
森林覆盖 canopy
森林冠层蒸腾 transpiring canopy
森林火险天气等级【大气】 weather grade of forest-fire
森林火险天气等级 forest fire-danger weather rating
森林火灾气象学 forest-fire meteorology
森林火险气象指数 forest fire-danger weather index
〔森〕林火〔灾〕烟云 forest-fire smoke
森林火灾云 forest-fire cloud
森林界限温度【大气】 forest limit temperature
森林内部 forest interior
森林气候【大气】 forest climate
森林气象学【大气】 forest meteorology
森林上限 upper forest limit

森林生态学【林】 forest ecology
森林生物学【林】 forest biology
森林水文学【地理】 forest hydrology
森林土壤【地理】 forest soil
森林线 forest line, tree line, isohyle, timber line
森林小气候【大气】 forest microclimate
森林烟 forest smoke
森林资源【林】 forest resources
杀霜【大气】 dark frost
沙霭 bai, sand mist
沙壁【大气】 wall of sand, sand wall
沙波 sand wave
沙尘暴 sand-dust storm, sand and dust storm【大气】
沙〔尘〕暴 sandstorm, duststorm
沙尘暴天气【大气】 sand and dust storm weather
沙尘天气【大气】 sand and dust weather
沙伐尔脱偏振仪 Savart polariscope
沙卷 sand devil
沙卷风 desert devil
沙龙卷 sand tornado
沙霾【大气】 sand haze
沙漠【地理】 sandy desert
沙漠草原 desert savanna, desert steppe
沙漠带 desert belt, desert zone
沙漠风 desert wind
沙漠风飑 desert wind squall
沙漠历史地理【地理】 desert historical geography
沙漠气候亚类【大气】 desert climate
沙漠气象学 desert meteorology
沙漠区 xerochore
沙丘【地理】 sand dune
沙雾 sand fog
沙性雪 sand snow
沙旋 sand whirl
沙照云光 sand blink
沙柱 sand pillar
砂壤土【土壤】 sandy loam soil

砂土【土壤】 sand soil
筛选 screening
筛选步骤 screening procedure
筛选多重线性回归 screening multiple linear regression
山本辐射图 Yamamoto radiation chart
山崩 landcreep
山崩风 landslip wind
山波云 hill wave cloud
山地标准时间(美国) Mountain Standard Time
山地波状云 mountain-wave cloud
山地电码 MONT code (mountain code 的缩写)
山地观测【大气】 mountain observation
山地空气 mountain air
山地平原风系 mountain-plains wind systems
山地气候【大气】 mountain climate (亦称 highland climate)
山地气候学【大气】 mountain climatology
山地气象学【大气】 mountain meteorology
山地天气 mountain weather
山地形态学 orography
山地障碍 mountain barrier
山地沼泽 upland mire
山地沼泽【地理】 mountain mire
山顶云 mountain cap cloud
山风 down valley wind, mountain wind
山风【大气】 mountain breeze
山峰【地理】 mountain peak
山谷风 mountain and valley breeze, mountain-valley wind
山谷风【大气】 mountain-valley breeze
山谷风系 mountain-valley wind systems
山谷逆温 valley inversion
山谷泄流 valley outflow jet
山洪 mountain torrent
山洪【力】 torrent
山间地区 intermountain region

山间盆地【地理】	intermountain basin
山口风	mountain-gap wind
山链【地质】	mountain chain
山麓冰川	piedmont glacier
山麓平原【地理】	piedmont plain
山脉【地质】	mountain range
山脉力矩	mountain torque
山脉坡度	mountain slope
山帽云【大气】	cap cloud
山区农业【农】	mountain region farming
山体效应【地理】	highland uplift effect
山雾	mountain fog【大气】, hill fog
山系【地质】	mountain system
山形【地理】	mountain forms
山岳背风波	mountain lee wave
闪电	bolt
闪电【大气】	lightning
闪电测绘传感器	Lightning Mapper Sensor (LMS)
闪电成像传感器	Lightning Imaging Sensor
闪电猝灭	lightning death
闪电电流	lightning current
闪电电流计	fulchronograph
闪电定位	location of lightning
闪电定位方法	lightning location method
闪电定位数据	lightning location data
闪电定位系统	lightning location system
闪电方向探测器	lightning direction finder
闪电放电	lightning discharge
闪电分支	branching of lightning
闪电光谱	lightning spectrum
闪电过电压	lightning overvoltage
闪电回波【大气】	lightning echo
闪电计数器	lightning-flash counter, lightning counter
闪电记录器	lightning recorder
闪电警报仪	lightning warning set (LWS)
闪电扩散	lightning diffusion
闪电熔岩	fulgurite
闪电事件	flash event
闪电探测和测距	lightning detection and ranging (LDAR)
闪电探测和定位系统	lightning detection and positioning system
闪电探测系统	lightning detection system, lightning mapping system
闪电探测〔站〕网	lightning detection network
闪电特性仪	klydonograph
闪电通道【大气】	lightning channel
闪电危害	lightning damage
闪光	glare, flash
闪〔光〕斑	glitter
闪光灯抽运激光器	flashlamp-pumped laser (FPL)
闪光光解系统【化】	flash photolysis systems
闪光率	flash rate
闪光谱【天】	flash spectrum
闪光视〔力减〕弱	glare asthenopia
闪光氙灯	xenon spark lamp
闪晃	shimmer
闪击	stroke
闪击间通道	interstroke channel
闪击间歇变化	interstroke change
闪击密度	stroke density
闪烁	flashing, twinkling, blink
闪烁【天】	scintillation
闪烁饱和	saturation of scintillation
闪烁计数器	scintillometer
闪烁计数器【物】	scintillation counter
闪烁晶体【物】	scintillation crystal
闪烁景	laurence
闪速存储器【计】	flash memory
扇形【数】	sector
扇形扩散	fanning
扇形滤波	fan filtering
扇形扫描	sector scan, fan scanning

扇形扫描显示器　sector scan indicator
伤害【农】　injury
商【数】　quotient
熵【物】【大气】　entropy
熵串级　entropy cascade
熵函数　entropy function
熵判据　entropy criterion
熵平衡方程　entropy balance equation
熵增量　entropy production
上包络（磁扰）　upper envelope
上标　superscript
上冰风　on-ice wind
上冲顶温度　overshooting top temperature
上冲云顶　overshooting top, overshooting cloud top
上传波　upgoing wave
上大气层　metasphere
上大气层顶　metapause
上对流层—下平流层　upper troposphere and lower stratosphere（UTLS）（200 hPa 至 30 hPa 区间）
上珥　supralateral tangent arcs, upper arcs
上风向　windward, upwind
上风效应　upwind effect
上轨道　northbound
上滑　overrunning
上滑锋　anafront
上滑锋【大气】　anabatic front
上滑冷锋　overrunning cold front
上滑面　upslide surface
上滑云【大气】　upglide cloud
上滑运动　upglide motion
上环天顶弧〔圈〕　upper circumzenithal arc
上亮带　upper bright-band
上流离子　upward streaming ions, upstreaming ion
上流离子流　upward ion current
上坡风　upslope wind
上坡风【大气】　anabatic wind
上坡雾【大气】　upslope fog

上切弧　upper tangential arc
上升　rise, ascent
上升电子　upgoing electron
上升空气　ascending air
上升力　lifting force
上升流【航海】　upwelling
上升气流　upward current, ascending current, anaflow
上升气流【大气】　upward flow
上升气流幕（上升气流薄层）　updraft curtains
上升日珥　ascending prominence
上升曲线　ascent curve
上升时间　rise time
上升速度　ascending velocity
上升运动　ascending motion
上下文内关键字【计】　keyword in context
上下弦　half moon
上弦【天】　first quarter
上现蜃景　superior mirage【大气】, upper mirage, looming
上新世【地质】　Pliocene Epoch
上新统【地质】　Pliocene Series
上行辐射　upwelling radiation
上曳气流　updraft
上曳气流【大气】　updraught
上曳气流速度　updraft velocity
上游效应　downstream development（对下游发展的）, downstream effect（指上游长波系统的变化逐渐影响到下游环流形势）
上游激波跃迁　upstream shock transition
上游解冻　upbank thaw
上游前差　forward-upstream differencing
上游切变瓣　upstream shear lobe
上游〔水面〕　headwaters
少水年【大气】　low flow year
少云　scattered sky, part cloudy
少云【大气】　partly cloudy
少云天空　sky slightly clouded

哨声【地】 whistler
蛇行路径 zigzag track
舍入误差【数】【大气】 round-off error, rounding error
设备 equipment, facility
设计暴雨【大气】 design torrential rain, design storm
设计暴雨雨型【水利】 design storm pattern
设计波〔浪〕 design wave
设计风速 designing wind speed
设计洪水【水利】 design flood
设计洪水频率【水利】 design flood frequency
设计洪水位【水利】 design flood level
设计枯水年【水利】 design low flow year
设计年径流量【水利】 design annual runoff
设计水位 design water level
设计因子 design factor
设想的全球气候 hypothetical global climate
社会地理学 social geography
社会环境 social environment
社会经济因素 socio-economic factor
社会生存排放 subsistence emission
社区空气传播的废物 community airborne waste
射程风 range wind
射程畸变 range distortion
射程误差 range error
射程延迟 range delay
射出长波辐射【大气】 outgoing longwave radiation (OLR)
射电【天】 radio
射电暴 radio burst
射电窗 radio window
射电极光 radio aurora
射电摄像仪 radiocinemograph
射电天体测量学【天】 radio astrometry
射电天文学【天】 radio astronomy
射电望远境【天】 radio telescope
射电新星【物】 radio nova
射电星云【物】 radio nebula
射电源【天】 radio source
射流测风仪 fluidic wind sensor
射流式双轴测风传感器 two-axis fluidic wind sensor
射流速度 efflux velocity
射频【铁道】 radio frequency (RF)
射频波段 radio frequency band
射频〔波〕谱【物】 radio-frequency spectrum
射频放大器【电子】 radio frequency amplifier
射频信号 RF signal
射束方程 ray equation
射束宽度(波束宽度) beam width
射线【数】 ray
α 射线 alpha-ray, α-ray
γ 射线 gamma-ray, γ-ray
X 射线 X-ray
α 射线电离 alpha ionization
射线光学 ray optics
X 射线晶体学【化】 X-ray crystallography
X 射线摄谱仪【物】 X-ray Spectrograph
射线弯曲 bending of ray, ray bending
X 射线吸收光谱法【化】 X-ray absorption spectrometry
射线雪量计 radioisotope snow-gauge
γ 射线雪量器 gamma ray snow gauge
X 射线荧光分析【物】 X-ray fluorescence analysis
涉氢作业 operation engaged in hydrogen
摄动【天】 disturbance, perturbation【大气】
摄动体【天】 disturbing body
摄谱仪【化】 spectrograph
摄氏度(温度单位) centigrade (℃)

摄氏温标【大气】 Celsius thermometric scale, centigrade temperature scale
摄氏温度表 Celsius' thermometer, centigrade thermometer
摄影测量学【测】 photogrammetry
摄影经纬仪 photogrammeter, kinetheodolite
摄影气压计 photographic barograph
摄影日照计 photographic sunshine recorder
摄影显微镜 photomicroscope
摄云机 cloud camera
摄云法 cloud photography
摄云仪 nephograph
伸长 elongation
伸长蜃景 towering
伸悬回波 sloping overhang echo
伸展变量 stretched variable
伸展变形 stretching deformation
伸展磁场 stretched-out field
伸展水平坐标 stretched horizontal coordinates
伸展项 stretching term
砷化镓【物】 gallium arsenide (GaAs)
深层渗漏 deep seepage
深层渗漏【水利】 deep percolation
深层水【海洋】 deep water
深度标志器 depth marker
深度 depth
深度线 depth line
深度优先搜索【自动】 depth-first search
深对流【大气】 deep convection
深对流模式 deep convection model
深根吸水植物 phreatophyte
深海 deep-sea
深海测深锤 deep-sea lead
深海沉积物 deep sea sediment
深海沉积物探针 deep ocean sediment probe (DOSP)
深海环境 abyssal environment
深海环流 deep ocean
深海考察器 deep research vehicle (DRV), deep ocean survey vehicle (DOSV)
深海流【海洋】 deep current
深海平原 abyssal plain
深海区 abyssal zone
深海系泊浮标 deep ocean moored buoy
深海系泊仪器站 deep-moored instrument station
深海压力计 deep ocean pressure gage
深海岩心 deep-sea core
深厚冰封 deep freeze
深厚东风带 deep easterlies
深厚信风 deep trades
深厚云带 deep cloud band
深井【石油】 deep well
深空探测器 deep space probe
深空探测设备 deep space instrumentation facility (DSIF)
深水波【大气】 deep-water wave
深水海洋波 deep-water ocean wave
深水声道 deep sound channel
深水温度仪【大气】 bathythermograph (BT)
神经网络【电子】 neural net, neural network
神经网络模型【计】 neural network model
神经元【计】 neuron
神经元网络【计】 neuron network
神经专家系统【计】 neural expert system
甚大〔尺度〕涡旋模拟 very large eddy simulation (VLES)
甚低频 very low frequency (VLF)
甚短期〔天气〕预报【大气】 very short-range〔weather〕forecast
甚高分辨率辐射仪【大气】 very high resolution radiometer (VHRR)
甚高频 very high frequency (VHF)
甚高频测向器 very high frequency

direction finder (VHFDF)
甚高频雷达【大气】 very high frequency radar, VHF radar
甚高频源 VHF source
甚小〔孔径〕地球站【电子】 Very Small aperture terminal (VSAT)
渗流【力】 seepage
渗流泉 seepage spring
渗流水 vadose water
渗流速度 seepage velocity
渗漏 influent seepage
渗漏计 drainage gauge
渗漏率【农】 percolation rate
渗漏式蒸散计 drainage evapotranspirometer
渗漏水【农】 percolation water
渗漏系数 leakage coefficient
渗润 infiltration
渗润量 infiltration capacity
渗润指数 infiltration index
渗水区 vadose zone
渗透【物】 osmosis
渗透深度【农】 penetration depth
渗透水 osmotic water
渗透吸力【农】 osmotic suction
渗透性 permeability
渗透〔性〕系数 permeability coefficient
渗透压【化】 osmotic pressure
渗透仪 infiltrometer
渗透作用【大气】 percolation
渗压表 osmometer
蜃景【大气】 mirage
升度 ascendent
升华【大气】 sublimation
升华曲线 sublimation curve
升华热【物】 heat of sublimation
升交点【天】【大气】 ascending node
升交〔点〕经度 ascending node longitude
升交〔点〕时刻 ascending node time
升交经度 longitude of ascending node
升腾〔学〕说 ascension (hydrothermal) theory
升压中心 center of rises, pressure-rise center
生成胞 generating cell
生成检验〔法〕 generate and test
生成区 generating area
生成热【化】 heat of formation
生存空间【地理】 living space
生化需氧量【大气】 biochemical oxygen demand(BOD)
生化作用 biochemical action
生活气候学 domestic climatology
生机温度 vital temperature
生境【地理】 habitat
生理病害【农】 physiological disease
生理成熟【农】 physiological maturity
生理干旱 physiological drought
生理气候学 physiological climatology
生理生存排放 survival emission
生命周期 life cycle
生态保护 ecological protection
生态地理学【地理】 ecogeography
生态工程【生】 ecological engineering
生态过程 ecological process
生态环境【大气】 ecologyical environment
生态活动【自然】 ecoactivity
生态农业 eco-agriculture
生态农业【农】 ecological agriculture
生态平衡 ecological equilibrium
生态平衡【地理】 ecological balance
生态气候【动物】 ecoclimate
生态气候适应 ecoclimatic adaptation
生态气候学 ecological climatology
生态气候学【大气】 ecoclimatology
生态气候预报 ecoclimatic forecast
生态气候预测 ecoclimate forecasting
生态区【地理】 ecotope
生态区域【地理】 ecoregion
生态圈【农】 ecosphere
生态容量【水产】 ecological capacity

生态设计【农】 ecological design
生态梯度【植物】 ecocline
生态条件 ecological condition
生态土壤学 ecopedology
生态危机【农】 ecological crisis
生态系统【地理】【大气】 ecosystem
生态系统【植物】 ecological system
生态系统多样性【生】 ecosystem diversity
生态系统平衡 ecosystem balance
生态小区【地理】 ecodistrict
生态效益【地理】 ecological benefit
生态型 ecotype
生态需水 ecological water requirement
生态学【大气】 ecology
生态演替【生】 ecological succession
生态遗传学【生】 ecogenetics
生态因子【植物】 ecological factor
生态阈值【生】 ecological threshold
生态最适度【生】 ecological optimum
生态灾难【自然】 ecocatastrophe
生土【土壤】 raw soil
生物冰核 biogenic ice nucleus
生物测定 bioassay
生物测试仪器 bioinstrumentation
生物处理【土壤】 biological treatment
生物大分子【生化】 biomacromolecule
生物带 life zone
生物地理学【植】 biogeography
生物地球化学 biogeochemistry
生物地球化学循环【大气】 biogeochemical cycle
生物电【电子】 bioelectricity
生物电池【电子】 bio-battery
生物电磁学 bioelectromagnetics
生物电子学【电子】 bioelectronics
生物多样性【植】 biodiversity
生物发光【物】 bioluminescence
生物反馈 biological feedback
生物反馈【电子】 biofeedback
生物分解 biological decomposition

生物风化【土壤】 biological weathering
生物工程〔学〕 bioengineering
生物固氮 biological nitrogen fixation (BNF)
生物光学 biooptics
生物痕量气体 biogenic trace gases
生物化学【生化】 biochemistry
生物活力温度界限 biokinetic temperature limit
生物降解 biodegradation, biological degradation
生物节律【植】 biological rhythm
生物介质 biological medium
生物聚合体【生化】 biopolymer
生物量【植】 biomass
生物流变学【力】 rheololgy
生物气候 bioclimate
生物气候的 bioclimatic
生物气候分区 bioclimate zonation
生物气候律 bioclimatic law
生物气候图 bioclimatograph
生物气候学 bioclimatics
生物气候学【大气】 bioclimatology
生物气象时间尺度模式 biometeorological time scale model
生物气象学【大气】 biometeorology
生物气象指数【大气】 biometeorological index
生物圈 vivosphere
生物圈【植】【大气】 biosphere
生物圈保护区 biosphere reserves
生物圈-大气交互作用 biosphere-atmosphere interaction
生物圈-反照率反馈 biosphere-albedo feedback
生物群 biota
生物群落 biocenosis, biocommunty
生物群系【植】 biome
生物生态学 bioecology
生物声学 bioacoustics
生物适应性 biocompatibility

生物天文学【天】 bioastronomy
生物统计学 biometry
生物温度 biotemperature
生物温室气体排放 biogenic greenhouse gas emission
生物无机化学【生化】 bioinorganic chemistry
生物物候学 biophenology
生物物理化学【生化】 biophysical chemistry
生物物理模式 biophysical model
生物物理〔学〕【物】 biophysics
生物雾 biofog
生物学定年法 biological dating method
生物学零度【大气】 biological zero point
生物〔学〕去污〔染〕 biological depollution
生物学最低温度 biological minimum temperature
生物循环 biocycle
生物有机化学【生化】 bioorganic chemistry
生物质燃烧【大气】 biomass burning
生物钟【植】 biological clock, bioclock
生物资源保护【地理】 living resources conservation
生雾温差 mist interval
生长【植】 growth
生长单位 growth unit
生长度-时 growing degree-hour (GDH)
生长季〔节〕 season of growth
生长率〔农〕 growth rate
生长期 duration of growing period, growing season
生长期【农】 growing period
生长期有效积温 growing degree-day
生长曲线 curve of growth
生长〔时〕期 period of growth
生长习性【农】 growth habit
生长因子【生化】 growth factor (GF)
生长指数 growth index
生长锥 increment cone
生殖增长 reproductive growth
声 sound
声爆 sonic boom
声波 acoustic wave (AW), sonic wave
声波【物】 sound wave
声波测温表 acoustic thermometer
声波导 acoustic waveguide
声波反射云 acoustic cloud
声〔波〕起伏 acoustic fluctuation
声场 acoustic field
声成像【物】 acoustic imaging
声传播系数 acoustic propagation constant
声传感 acoustic sensing
声导【物】 acoustic conductance
声导航系统 acoustic navigation system
声导率 acoustic conductivity
声导纳 acoustic mobility
声导纳【物】 acoustic admittance
声发 sound fixing and ranging (SOFAR)
声发技术 SOFAR technology
声发射 acoustic emission
声发射监测 acoustic emission monitoring
声发通道 SOFAR (sound fixing and ranging) channel
声反馈 acoustic feedback
声反射 acoustic reflection
声反射仿型 acoustic reflection profiling
声反射率 acoustical reflectivity
声共振【物】 acoustic resonance
声后向散射 acoustic backscattering
声环境 acoustic environment
声换能器 acoustic transducer
声回波团块结构 sound lump echoes
声混响 acoustic reverberation
声级 acoustical level
声级【物】 sound level
声级仪 sound level meter

声级仪(噪声仪)　acoustimeter
声截止频率　acoustic cutoff frequency
声抗【物】　acoustic reactance
声〔雷〕达　sound radar
声〔雷〕达【大气】　acoustic radar, sodar
声榴弹　sonic grenade
声脉冲　acoustic pulse
声呐　ASDIC (= sonar, Antisubmarine Detection Investigation Committee 的缩写)
声呐【航海】　sound navigation and ranging (sonar)
声呐扫海【测】　sonar sweeping
声呐图像【测】　sonar image
声呐遥感【地理】　sonar remote sensing
声能　acoustic energy, sonic energy
声能级　sound power level
声能密度　sound energy density
声能通量　sound energy flux
声能学　sonics
声频　sound frequency, audio frequency
声频参考信号　high-reference signal
声频调制探空仪　audio-modulated radiosonde
声频发生器　acoustic frequency generator
声频散　acoustic dispersion
声剖面图　acoustical profile
声谱　acoustical spectrum, acoustic spectrum
声强　acoustic intensity, sound intensity
声强计　acoustometer
声全息摄影扫描法　acoustical holography scanning technique
声容　acoustic capacitance
声散射　acoustic scattering
声闪烁【大气】　acoustical scintillation
声束　sound beam
声衰减常数　acoustical attenuation constant
声顺　acoustic compliance

声速　acoustic velocity, speed of sound
声速【物】　sound velocity
声探测器　acoustic sounder
声特征　acoustic signature
声特征阻抗　acoustic characteristic impedance
声通道　sound channel
声透镜　acoustic lens
声透射层　acoustically transparent layer
声位相常数　acoustical phase constant
声污染　acoustic pollution, sound pollution
声雾　acoustic fog
声吸收　acoustic absorption, sound absorption
声线　sound ray
声学测量　acoustical measurement
声学层析成像术　acoustic tomography
声学多普勒流体廓线仪　acoustic Doppler current profiler (ADCP)
声学多普勒探测器　acoustic Doppler sounder
声学多普勒系统　acoustic Doppler system
声〔学〕风速表　sonic anemometer
声〔学〕风速温度表　sonic anemometer-thermometer (SAT)
声学干涉仪测温　acoustic interferometer temperature sensing
声〔学〕海流计　acoustic ocean current meter
声〔学基〕阵　acoustic array
声学探测【大气】　acoustic sounding
声学探测和测距　acoustic detection and ranging (ACDAR)
声学-微波雷达　acoustic-microwave radar
声学温度表　sonic thermometer, acoustic thermometer【大气】
声学雨量器【大气】　acoustic raingauge
声压　acoustic pressure

声压等级仪　sound pressure level meter
声压积分仪　sound pressure integrator
声压级　sound pressure level
声音传播　sound propagation
声音全息记录器　holophone
声应答器　acoustic transponder
声影　sound shadow
声指令系统　acoustical command system
声质量　acoustic inertance, acoustic mass
声重力波【大气】　acoustic-gravity wave
声阻【物】　acoustic resistance
声阻抗【物】　acoustic impedance
圣爱尔摩火　St. Elmo's fire
圣马丁热期　Saint Martin's summer
盛夏　midsummer
盛行风【大气】　dominant wind, predominant wind
盛行风【大气】　prevailing wind
盛行风向　prevailing wind direction
盛行能见度　prevailing visibility
盛行西风　prevailing westerly winds
盛行西风带　prevailing westerlies
剩磁　remanent magnetism
剩余【数】　residue
剩余大气　residual atmosphere
剩余方差　residual variance
剩余非地转运动　residual ageostrophic motion
剩余环流　residual circulation
剩余经向环流　residual meridional circulation
剩余流函数　residual stream function
剩余平均环流　residual mean circulation
剩余平均经向环流　residual mean meridional circulation
剩余损耗【物】　residual loss
剩余洼地蓄水量　residual depression storage
剩余物　carry-over
失速马赫数　stalling Mach number
失稳过程　runaway process
失效【计】　failure
失效数据【计】　failure data
失真【电子】　distortion
失重　zero gravity
施蒂威图　Stuve diagram
施放前检定　pre-flight calibration
施密特数　Schmidt number（Sc）
施密特线【物】　Schmidt lines
施特藩问题【数】　stefan problem
施瓦氏解【物】　Schwarzchild solution
施瓦氏坐标【物】　Schwarzschild coordinates
施主杂质　donor impurity
湿不稳定的　wet unstable
湿不稳定度　moist instability
湿不稳定能量　moist-labile energy
湿沉降　wet precipitation
湿沉降【大气】　wet deposition
湿地　everglade, swamp
湿地【地理】　wetland
湿度【大气】　humidity
湿度表　humidometer
湿度表【大气】　hygrometer
湿度查算表　hygrometric tables, psychrometric table
湿度场【大气】　humidity field
湿度传感器　hygrotransducer
湿度当量　moisture equivalent
湿度反演【大气】　humidity retrieval
湿度方差　humidity variance
湿度公式　hygrometric formula, hygrometric equation
湿度混合比　humidity mixing ratio
湿度计【大气】　hygrograph
湿度计算尺　humidity slide-rule
湿度计算器　psychrometric calculator
湿度计算图　psychrometric chart
湿度廓线【大气】　moisture profile
湿度片　humidity strip
湿度起伏　humidity fluctuation
湿度器　hygroscope

湿度区　humidity province
湿〔度〕舌　moisture tongue
湿度调节仪　humidostat
湿度系数　psychrometric coefficient, humidity coefficient
湿度效应　humidity effect
湿度学　hygrology
湿度要素　humidity element
湿度仪　hygronom
湿度直减率　humidity lapse rate
湿度指示器　moisture indicator, atmidoscope
湿度指数　humidity index
湿度自记计　hygroautometer
湿度自记曲线　hygrogram
湿对流【大气】　moist convection
湿对流调整　moist convective adjustment
湿锋　wetting front
湿过程　moist process
湿害　excess moisture injury, wet injury
湿害【大气】　wet damage
湿焓　moist enthalpy, wet enthalpy
湿化　moistering
湿季　wet season
湿界　wetted perimeter
湿静力能　moist static energy, wet static energy
湿静力总能量　wet static total energy
湿绝热　condensation adiabat
湿绝热变化　wet adiabatic change
湿绝热〔的〕　wet adiabatic
湿绝热过程　saturation-adiabatic process
湿绝热过程【大气】　moist adiabatic process
湿绝热温度差　wet adiabatic temperature difference
湿绝热线【大气】　moist adiabat, saturation-adiabatics, wet adiabat
湿绝热直减率　saturation-adiabatic lapse rate, wet-adiabatic lapse rate
湿绝热直减率【大气】　moist adiabatic lapse rate
湿空气　moist air, wet air
湿空气冰面饱和水汽压【大气】　saturation vapour pressure of moist air with respect to ice
湿空气冰面相对湿度　relative humidity of moist air with respect to ice
湿空气发生器　humidity atmosphere producer
湿空气密度　density of moist air
湿空气水面饱和水汽压【大气】　saturation vapour pressure of moist air with respect to water
湿空气水面相对湿度　relative humidity of moist air with respect to water
湿力管　wet solenoid
湿霾　damp haze
湿敏电容　Humicap
湿敏电阻　humistor, hygristor
湿模式　moist model
湿能管　wet energy solenoid
湿气溶胶【大气】　aqueous aerosol
湿倾斜对流　moist slantwise convection
湿球　wet bulb
湿球冷却温度表　wet katathermometer
湿球位温　potential wet-bulb temperature
湿球位温【大气】　wet-bulb potential temperature
湿球温差　wet-bulb depression
湿球温度【大气】　wet-bulb temperature
湿球温度表【大气】　wet-bulb thermometer
湿球温度差值　wet bulb temperature difference
湿热的　muggy
湿热的　muggy
湿热多雨气候　tropical rain climate
湿热气候【大气】　warm-wet climate
湿热天气　sticky weather
湿润　humid

湿润带　humid zone
湿润低温气候　humid microthermal climate
湿润度【大气】　moisture index
湿润年【大气】　high flow year
湿润气候　moist climate, wet climate
湿润气候【大气】　humid cliamte
湿润区　humid region
湿润土地　spring land
湿润温和气候　humid temperate climate
湿润系数　moisture coefficient
湿润因子　moisture factor
湿润指数　index of wetness
湿润指数【农】　moist index
湿舌　moist tongue
湿舌【大气】　wet tongue
湿生长　wet growth
湿生植物【地理】　hygrophyte
湿态　hygrometric state
湿微暴流　wet microburst
湿位涡　moist potential vorticity
湿温气候　humid mehumidizer sothermal climate
湿雾【大气】　wet fog
湿线订正　wet-line correction
湿相当位温　wet-equivalent potential temperature
湿雪　wet snow【大气】, water snow, cooking snow
湿雪花　moist snow-flake
湿延迟　wet delay
湿有效能量　moist available energy(MAE)
湿有效位能　moist available potential energy
湿指数　wet index
十分位值　decile
十进二进制转换【数】　decimal-binary conversion
十进制　decimal system
十进制数字【计】　decimal digit
十六烷（$C_{16}H_{34}$）　cetane

十年　decade
十天　dekad
十位势米　dagpm
十亿分率（10^{-9}）　ppb（parts per billion）
十亿位　gigabit
十亿位每秒【计】　gigabits per second
十亿字节每秒【计】　gigabytes per second
"十字形"降落伞　"cross"parachute
十字晕　sun cross（＝cross）
石冰　stone ice
石膏【地质】　gypsum
石膏测湿元件(测土壤含水量)　gypsum unit
石化工厂　petrochemical plant
石化〔作用〕【地质】　lithification
石灰石涤气　limestone scrubbing
〔石灰〕熟化　slake
石灰性土【土壤】　calcareous soil
石炭二叠纪　Permo-Carboniferous era
石炭纪【地质】　Carboniferous Period
石英【物】　quartz
石英玻璃绝对日射表　quartz-glass pyrheliometer
石英玻璃温度表　quartz-glass thermometer
石英温度计　quartz thermometer
石油输出国组织　OPEC(Organization of the Petrolem Exporting Countries, Vienna, Austria)
石陨星　asiderite
时变边界条件　time-dependent boundary condition
时变传递函数　time-varying transfer function
时变粗糙面　time-varying rough surface
时变辐射单位强度　time-varying specific intensity
时变功率谱　time-varying power spectrum
时变空间谱密度　time-varying spatial

spectral density
时变流　time-dependent flow
时变滤波〔器〕　time-variable filter
时变谱　time-varying spectrum
时变谱密度　time-varying spectral density
时变散射截面　time-varying scattering cross section
时变随机介质　time-varying random medium
时变系统【数】　time-varying system
时标装置　time-marking device
时差定位【电子】　time-of-arrival location
时差【天】　equation of time
时程图　time-distance graph
时次　time level
时次观测　record observation
时段叠加法　superposed-epoch method
时分多址【电子】　time division multiple access(TDMA)
时高显示器　THI(time-height indicator)
时基【电子】　time base
时际变率　interhourly variability
时间变化　temporal variation
时间标准　Time Standard
时间标准差　temporal standard deviation
时间表　schedule
时〔间〕步〔长〕　time step
时间参数　time parameter
时间常量【物】　time constant
时间尺度【数】　time scale, temporal scale
时间窗　time window
时间代表性　time representativeness
时间导数　time derivative
时间等温线　chronoisotherm
时间分辨率　time-resolution
时间分辨率【大气】　temporal resolution

时间分布　temporal distribution
时间-高度剖面　time-height section
时间-高度显示器　time-height indicator
时间格式　time scheme
时间归一化法　time normalization
时间函数　time function
时间积分　time integration
时间记录器　chronograph
时间间隔　time interval
时间滤波　time filtering
时间滤波器　time filter
时间频谱　temporal frequency spectrum
时间平滑【大气】　time smoothing
时间平均　time average, time averaging
时间平均观测〔值〕　time-averaged measurement
时间平均〔气〕流　time-mean flow
时间平均图　time mean chart
时间剖面　time section
时间剖面图【大气】　time cross-section
时间起伏　temporal fluctuation
时间前移法　forward-marching techniques
时间推移　time lapse
时间相干性【物】　temporal coherence
时间相关　time correlation
时间相关函数　temporal correlation function
时间相关散射截面　time correlated scattering cross section
时间相关散射振幅　time-correlation scattering amplitude
时间响应　time response
时间向前差　forward time difference
时间序列【数】　time series
时间序列分析【数】　time series analysis
时间一致性检验　time consistency check
时间因子　temporal factor
时间中央差　centered time difference
时角　hour angle
时距探空仪　time-interval radiosonde

时空【物】 spacetime
时空变率 spatial-temporal variability
时空尺度 time and space scale
时空分辨率 spatial and time resolution
时空分布 spatial and temporal distribution
时空均匀性【物】 uniformity of space-time
时空取样 space-time sampling
时空相关 space-time correlation
时空转换 time-to-space conversion
时空坐标【物】 spacetime coordinates
时-面-深曲线 time-area-depth curve
时期 period
时区 time zone
时区号【航海】 zone description(ZD)
时-深-面值 DDA value
时现风 sporadic wind
时延【天】 time delay
时域【数】 time domain
时域反射法(测土壤湿度) time domain reflectometry(TDR)
时-域平均 time-domain averaging
时展资料【大气】 stretched data
时滞【数】 time lag
识别模式 recognition mode
识别信号【航海】 identity signal
实〔变〕元 actual argument
实部【数】 real part
实测天体物理学【天】 observational astrophysics
实地观测 in situ observation
实地监测 in situ monitoring
实根【数】 real root
实际大气 real atmosphere
实际飞行天气 actual flying weather
实际观测时间【大气】 actual time of observation
实际海拔〔高度〕 actual elevation
实际航速【航海】 speed over ground
实际气温 real air temperature
实际气压 actual pressure
实际温度等值线 thermo-isohyp
实际蒸发 actual evaporation
实时【计】 real time
实时测量 real-time measurement
实时处理【计】 real-time processing
实时传送 real-time transmission
实时传送系统 real-time transmission system(RTTS)
实时分析 real-time analysis
实时分析仪 real time analyzer
实时过程 real-time process
实时计算中心 Real Time Computing Center(RTCC)
实时气象资料 real-time meteorological data
实时视频【计】 real-time video
实时系统【计】 real-time system
实时显示【大气】 real time display
实时信号处理【电子】 real time signal processing
实时预报 real-time prediction
实时资料【大气】 real time data
实数【数】 real number
实线 solid line
实验 experiment
实验池 laboratory tank
实验气象学 experimental meteorology
实验气象学研究所(NOAA) Experimental Meteorology Laboratory(EML,NOAA)
实验室 laboratory(=lab)
实用分析 practical analysis
实用盐度单位 practical salinity units(PSU)
实在概率 probability of reality
实在能量 realized energy
实质导数 substantial derivative, individual derivative, material derivative
实质面 material surface
实质曲线 material curve

实质体积　material volume
实质坐标　material coordinates
拾波器　pickup
拾波线圈【物】　pick-up coil
食（日，月）　eclipse
食风　eclipse wind
食既【天】　second contact
食年　eclipse year
食欲的季节反应　seasonal effect on appetite
蚀损〔作用〕　erosion
史前考古学　prehistory
史前期　prehistoric period
矢积【物】　vector product
矢量编辑技术　vector editing technology
矢量场　vector field
矢量方程　vector equation
矢量分析　vector analysis
矢量风标　vector vane
矢量函数　vector function
矢量积分系统　vectorial integration system
矢量图　vector-diagram
矢量旋度　curl of vector
矢势【物】　vector potential
使用效率　end-use efficiency
始生代　Eozoic Era
始新世【地质】　Eocene Epoch
始新统【地质】　Eocene Series
驶线　steering line
示波器【物】　oscilloscope, oscillograph
A 示波器　A scope
B 示波器　B scope
示差气温表　differential air thermometer
示差温度表　differential thermometer
示差直接辐射表　differential actinometer
示意图　diagrammatic sketch
示踪分析　tracer analysis
示踪技术　tracer technique
示踪扩散实验【大气】　tracer diffusion experiment
示踪物　trace substance, tracer

示踪云　cloud tracer
世【地质】【大气】　Epoch
世界动力会议　World Power Conference（WPC）
世界辐射测量基准　world radiometric reference（WRR）
世界辐射中心　World Radiation Center（WRC）
世界海洋环流试验　World Ocean Circulation Experiment（WOCE）
世界海洋组织　World Oceanic Organization（WOO）
世界环境与发展委员会　World Commission on Environment and Development（WCED）
世界气候　world climate【大气】
世界气候会议　World Climate Conference（WCC）
世界气候计划　World Climate Programme（WCP）
世界气候研究计划　World Climate Research Program（WCRP）
世界气候资料信息检索服务　World Climatic Data Information Referral Service（INFOCLIMA）
世界气象日　World Meteorological Day
世界气象时段　World Meteorological Intervals（WMI）
世界气象学　world meteorology
世界气象中心（WMO）　World Meteorological Centre（WMC，WMO）
世界气象组织　World Meteorological Organization（WMO）
世界气象组织成员　WMO Member
世界气象组织代表大会　World Meteorological Congress
世界气象组织第二（亚洲）区域协会　Regional Association for Asia（RA Ⅱ）
世界气象组织第六（欧洲）区域协会　Regional Association for Europe（RA Ⅵ）
世界气象组织第三（南美）区域协会

Regional Association for South America (RA Ⅲ)
世界气象组织第四(北、中美)区域协会 Regional Association for North and Central America (RA Ⅳ)
世界气象组织第五(西南太平洋)区域协会 Regional Association for the South-West Pacific (RA Ⅴ)
世界气象组织第一(非洲)区域协会 Regional Association for Africa (RA Ⅰ)
世界气象组织全球综合观测系统 WMO Integrated Global Obseying System (WIGOS)
世界时 Z time, Zebra time (Z)
世界时【天】 universal time (UT)
世界〔台站〕网 reseau mondial
世界天气 world weather
世界天气监视网 World Weather Watch (WWW)
世界天气研究计划 World Weather Research Programme (WWRP)
世界卫生组织 World Health Organization (WHO)
世界资料中心(ICSU) World Data Centre (WDC, ICSU)
世界自然基金会 World Wide Fund for Nature (WWF)
示踪扩散实验【大气】 trace diffusion experiment
事后分析 post analysis
事件记录器 event recorder
事前分析 preanalysis
势【物】 potential
势函数【物】 potential function
势垒高度【电子】 barrier height
势流理论 potential flow theory
势能【物】【大气】 potential energy
势能图 potential energy diagram
势涡 irrotational vortex
试探 trial
试探规则 rule of thumb
试探解【数】 trial solution
试验场 test range
试验流域 experimental basin
试验设计 experiment design
试验数据 test data
试验预报 experimental forecast
试验预报中心 experimental forecast center
试验中心 test center
试液【化】 test solution
试纸【化】 test paper
视冰点 apparent freezing point
视差【天】 parallax
视场 field of view, visual field
视场光阑【物】 field stop
视场角 angular field-of-view
视程 visual range
视程表 videometer
视程公式 visual range formula
视程计 videograph
视程障碍 obstruction to vision
视程指数 visibility index
视地平线【测】 apparent horizon
视地下水流速 apparent groundwater velocity
视点【计】 eyepoint
视风【航海】 apparent wind
视高度【航海】 apparent altitude
视角 angle of view, visual angle
视角订正 viewing angle correction
视界圆锥 cone of vision
视距尺方法 stadia technique
视觉感阈【大气】 contrast threshold, threshold contrast
视亮度 apparent luminance, apparent brightness
视频 video frequency, vision frequency, video
视频积分处理器 video integrator and processor (VIP)
视频扫描器 video scanner

视频数据系统　videotex
视频信号　video signal
视频增益　video gain
视区内有降水　precipitation within sight
视热源　apparent heat source
视示力【大气】　apparent force
视示热　apparent heat
视水汽汇　apparent moist sink
视水汽源　apparent vapour source
视速度　apparent velocity
视太阳【航海】　apparent sun
视太阳日　apparent solar day
视太阳时　apparent solar time
视听距离　visual-aural range
视外区视像　parafoveal vision
视位置【航海】　apparent position
视涡〔度〕源　apparent vorticity source
视午　apparent noon
视线　line of sight
视线传播　line-of-sight propagation
视线范围　line-of-signt range
视像管【电子】　vidicon
视像映射　video mapping
视星等【天】　apparent magnitude
视应力　apparent stress
视直径　apparent diameter
视重力　apparent gravity
适暗视觉【物】　scotopic vision
适定问题【数】　well-posed problem
适定性　well-posedness
适耕性【土壤】　workability of soil
适旱植物　xerophobous plant
适航天气　weather above minimum
适航性　airworthiness
适配器【计】　adapter
适宜温度　favourable temperature
适应光照度　adaptation illuminance
适应过程　adjustment process, adjustment procedure, adaptation process
适应亮度　brightness level, adaptation level, adaptation brightness
适应发光率　adaptation luminance
适应回归　adaptive regression
适应性　adaptability
适应性对策　adaptation strategies
室内气候　room climate
室内气候【大气】　indoor climate
室内气候学【大气】　cryptoclimatology
室内气流速度　indoor air velocity
室内温度　indoor temperature
室内小气候　house microclimate, kryptoclimate【大气】, cryptoclimate【大气】
室内小气候学　cryptoclimatology, kryptoclimatology
释用　interpretation
嗜寒性的　cryophilic
收发〔合一〕两用机　single-unit transceiver
收发器【计】　transceiver
收发转换开关　TR-tube (transmit-receive tube)
〔收〕获月　harvest moon
收集　collection
收敛半径【数】　radius of convergence
收敛级数【数】　convergent series
收敛数值方案　convergent numerical scheme
收缩【数】　retraction
收缩系数　contraction coefficient
收缩轴　axis of contraction
手持风速表【大气】　hand anemometer
手控　manual control (M/C)
手选取样　grab sampling
手摇干湿表【大气】　sling psychrometer, whirling psychrometer
手摇温度表　whirling thermometer, sling thermometer
守恒变量图　conserved variable diagram
守恒参量图　conserved parameter diagram, conserved variable diagram
守恒差分格式　conservative difference scheme

守恒方程【数】 conservation equation
守恒格式【大气】 conservation scheme
守恒律【物】 conservation law
守恒散射 conservative scattering
首次阵风 first gust
寿命【化工】 lifetime
受旱地区 drought-striken region
受激分子 excited molecule
受激辐射【物】 stimulated radiation
受激离子 excited ion
受激原子 excited atom
受控访问系统【计】 controlled access system
受控系统【计】 controlled system
受摄体【天】 disturbed body
受体【生化】 receptor
受压磁层 compressed magnetosphere
受雨器 interceptometer
受雨桶 receiving bucket
受灾面积 damage area
授权【计】 grant
书夹涡旋 book-end vortices
枢轴高压 pivoting high
梳齿式探空仪 comb-type radiosonde
梳状测云器 comb nephoscope, cloud rake
舒曼共振 Schumann resonance
舒曼-龙格吸收带(氧的) Schumann-Runge band
舒曼-卢德兰极限 Schumann-Ludlam limit
舒适(环境,气候) amenity
舒适标准 comfort standard, comfort zone
舒适度曲线 comfort curve
舒适度图 fitness figure, comfort chart
舒适气流【大气】 comfort current
舒适区 comfort zone
舒适温度【大气】 comfort temperature
舒适指数【大气】 comfort index
疏密波 wave of condensation and rarefaction
疏散槽 diverging trough
疏水【电力】 drain
疏水〔的〕 hydrophobic
疏水核 hydrophobic nuclei
疏水胶体【化】 hydrophobic colloid
疏松雪 loose snow
输出 export
输出【物】 output
输出信号 output signal
输进 feeding
输入 input
输入区 intromission zone
输入-输出控制系统 input-output control system (IOCS)
输沙量-流量关系〔曲线〕 silt-discharge rating, sediment discharge rating
输送 transport, transfer, conveyance
输送带 conveyer(或 conveyor)belt
输送过程 transport processes
输送机制 transport mechanism
输运截面 transport cross section
输运理论 transport theory
熟土【土壤】 mellow soil
鼠标〔器〕【计】 mouse
曙光 sunrise colours
曙暮光 tenebrescence, owl-light
曙暮光【大气】 twilight
曙暮光弧 twilight arch, crepuscular arch, bright segment, arch twilight
曙暮光谱 twilight spectrum
曙暮光时订正 twilight correction
曙暮光现象 twilight phenomena
曙暮辉 twilight glow, crepuscular ray, twilight flash(或 glow)
曙暮紫光 purple twilight
束缚电荷【物】 bound charge
束缚电子【物】 bound electron
束轴 beam axis
束状波 beam wave
树表示【数】 tree representation

树木景带 dendrochore
树木年轮测定器 dendrograph
树木年轮气候学【大气】 dendroclimatology
树木年轮气候志【大气】 dendroclimatography
树木气候 tree climate
树算法【数】 tree algorithm
树图【数】 tree graph
树线移动 tree-line movement
树脂【化】 resin
数据编辑【地信】 data editing
数据编码【地信】 data encoding
数据表达【地信】 data presentaeion
数据表示【地信】 data representation
数据采集【地信】 data capture
数据采集系统【计】 data acquisition system
数据重组【计】 data reconstitution
数据窗 data window
数据〔磁〕带 data tape
数据〔打印〕格式 data layout
数据调制解调器 data modem
数据发布【地信】 data dissemination
数据分发【地信】 data distribution
数据分类系统 data classification system
数据分析系统【计】 data analysis system
数据更新【地信】 data updating
数据共享【计】 data sharing
数据管理【地信】 data management
数据恢复【计】 data restoration
数据集 data set
数据集成【地信】 data integration
数据集目录【地信】 dataset directory, dataset catalog
数据集文档【地信】 data set documentation
数据兼容性【地信】 data compatibility
数据交换系统 data exchange system
数据结构【计】 data structure

数据库 data bank
数据库管理系统 Database Management System（DBMS）
数据库【计】 data base
数据库控制系统 Database Control System（DBCS）
数据类型【计】 data type
数据流【计】 data flow
数据目录【计】 data directory
数据确认【计】 data validation
数据输出 data output
数据输入 data input
数据探空仪 datasonde
数据通信系统 data communication system
数据完整性【地信】 data completemess
数据网 data network
数据文件【地信】 data file
数据显示 data display
数据有效性【地信】 data validity
数据域【地信】 data field
数据源 data source
数据真实性【地信】 data reality
数据帧【计】 data frame
数据质量【地信】 data quality
数据属性【计】 data attribute
数理气候【大气】 mathematical climate
数理气象学 mathematical meteorology
数理统计〔学〕【数】 mathematical statistics
数理统计预报方法 forecasting by mathematical statistics
数量分布函数【化】 number distribution function
数量化理论 quantification theory
数量级【数】 order of magnitude
数码 code figure
数密度 number density
数模 digifax
数-模转换 digital-to-analog conversion
数模转换器【计】 digital-to-analog

converter
数浓度　number concentration
数学模拟　mathematical simulation
数学模型【数】　mathematical model
数学期望【数】　mathematical expectation
数学物理学　mathematical physics
数学限制条件　mathematical constraint
数值不稳定　numerical instability
数值的逐步逼近法　numreical step by step method
数值方案　numerical scheme
数值方法【数】　numerical method
数值分散　numerical dispersion
数值分析【数】　numerical analysis
数值积分【数】【大气】　numerical integration
数值计算【数】　numerical computation, numerical calculation
数值解【数】　numerical solution
数值滤波　mathematical filter
数值模拟【大气】　numerical simulation
数值模拟〔方法〕　numerical modeling
数值模拟实验　numerical simulation experiment
数值模式　numerical model
数值气候分类【大气】　numerical climatic classification
数值求积　numerical quadrature
数值实验【大气】　numerical experiment
数值天气预报【大气】　numerical weather prediction (NWP)
数值天气中心　Numerical Weather Center
数值稳定性【数】　numerical stability
数值预报　numerical forecasting, numerical forecast
数〔字〕　digit
数字并行输入　digital parallel input
数字处理　digital processing
数字传输　digital transmission
数字传真记录器　digital facsimile recorder

数字传真接口　digital fax interface (DFI)
数字地球【计】　digital Earth
数字地图【地理】　digital map
数字地图学【测】　digital cartography
数字地形模型【地理】　digital terrain model
数字电路【计】　digital circuit
数字电压表【电子】　digital voltmeter
数字读出机　digitiser, digitizer
数字高程矩阵【地信】　digital elevation matrix (DEM)
数字高程模型【地信】　digital elevation model (DEM)
数字化　digitizing, digitization
数字化地图【地信】　digitized map
数字〔化〕雷达　digital radar
数字〔化〕雷达回波〔信号〕处理器　digital radar echo signal processor (DIREP)
数字化取样　digital sampling
数字化误差　error of digitization, digitizing error
数字化云图【大气】　digitized cloud map
数字化资料　digital data
数字-话音计算机　digital-to-voice computer
数字-话音转换器　digital-to-voice converter
数字计算机【计】　digital computer
数字经纬仪　digital theodolite
数字雷达试验　digitized radar experiment
数字滤波　digital filtering
数字滤波器【农】　digital filter
数字模拟【地理】　digital simulation
数字扫描器　digital scanner
数字扫描转换器　digital scan converter
数字深度指示器　digital depth indicator
数字式日照〔时间〕记录器　digital sunshine duration recorder
数字式天气图记录器　digital weather-chart recorder

数字式温度表　digital thermometer
数字式温度探测器　digital temperature sensor
数字式烟雾监测仪　digital smoke monitor
数字视频积分处理器　digital video integrator and processor（DVIP）
数字水银气压表　digital mercury barometer
数字通信【电子】　digital communication
数字图像【计】　digital image
数字图像处理【测】　digital image-processing
数字图像镶嵌【地理】　digital image mosaic
数字无线系统【航海】　digital radio system
数字显示　digital display
数字信号【电子】　digital signal
数字相关技术　digital correlation technology
数字压缩调制解调器　digital compression modem
数字音频广播　digital audio broadcast（DAB）
数字影碟【计】　digital video disc
数字有线系统【航海】　digital line system
数字预报　digital forecast
数字照相机【计】　digital camera
数字正摄影像【测】　digital orthoimage
数字正摄影像图【测】　digital orthophoto to map
数字贮存　digital storage
数字资料处理机　digital data processor
数字资料获取系统　digital data acquisition system
刷形放电　brush discharge
衰变常数　decay constant
衰变率【物】　decay rate
衰减　attenuation, decay

衰减长度　attenuation length
衰减常数　attenuation constant
衰减截面【大气】　attenuation cross-section
衰减距离　decay distance
衰减模式　decaying mode
衰减区　decay area
衰减系数【大气】　attenuation coefficient
衰减因子　attenuation factor
衰退　fadeout
双半球反射比　bihemispherical reflectance
双边带　both sideband
双边谱　two-sided spectrum
双标准纬线圆锥投影　two satndard parallel conic projection
双波　double wave
双波长　dual-wavelength
双波长多普勒雷达　dual wavelength Doppler radar
双波长雷达【大气】　dual wavelength radar
双波束分光计　double-beam spectrophotometer
双层【物】　double layer
双潮　double tide
双程衰减【大气】　two-way attenuation
双低潮【海洋】　double ebb
双电子复合　dielectronic recombination
双端系统　double-ended system
双对流层顶　double tropopause
双多普勒分析【大气】　dual-Doppler analysis
双分子还原【化】　bimolecular reduction
双峰分布　bimodal distribution
双峰谱【大气】　bimodal spectrum
双锋　double front
双杆气压计　dual traverse barograph
双工【电子】　duplex
双管温度表　sheathed thermometer, double tube thermometer

双光子过程　two-photon process
双光子吸收【物】　two-photon absorption
双滑线电桥　double-slidewire bridge
双基地〔的〕　bistatic
双基地多普勒声学探测仪　bistatic Doppler acoustic sounder
双基地激光雷达【大气】　bistatic lidar
双基地截面　bistatic cross section
双基地雷达【电子】　bistatic radar
双基地雷达方程　bistatic radar equation
双基地雷达截面　bistatic radar cross section
双基地设置　bistatic arrangement
双基地声〔雷〕达　bistatic acoustic radar
双基地声学测风系统　bistatic acoustic wind monitor system
双基地声学探测器　bistatic acoustic sounder
双极磁区　bipolar magnetic region
双极〔黑子〕群　bipolar group
双极化测量　bipolarized measurement
双极晶体管【电子】　bipolar transistor
双极扩散　ambipolar diffusion
双极型(闪电)　bipolar pattern
双寄存器　double register
双尖点　double cusp, dual cusp
双阶谱　bispectrum
双阶谱分析　bispectrum analysis
双阶位相　biphase
双阶相干法　bicoherence
双阶相关　bicorrelation
双金属片　bimetallic strip
双金属片温度表　bimetallic thermometer, bimetal thermometer
双金属温度计【大气】　bimetallic thermograph
双金属元件　bimetal element
双金属直接辐射表　bimetallic actinometer
双金属直接辐射计　bimetallic actinograph
双经纬仪法　double-theodolite technique
双经纬仪观测【大气】　double-theodolite observation
双经纬仪技术　double-theodolite technique
双精度　double precision
双开尔文波　Double-Kelvin wave
双扩散对流　double diffusive convection
双累积曲线　double-mass curve
双链 RNA【生化】　double stranded RNA, dsRNA
双链 DNA【生化】　double stranded DNA, dsDNA
双流不稳定性【物】　two-stream instability
双流〔法〕　two-stream
双流近似　two-stream approximation
双偏振雷达【大气】　dual polarization radar
双频互相干函数　two-frequency mutual coherence function
双频雷达　dual-frequency radar
双频锁相接收机　dual frequency phase locked receiver
双频相关函数　two-frequency correlation function
双气旋【大气】　binary cyclones
双曲点【数】　hyperbolic point
双曲脐型突变　hyperbolic umbilic catastrophe
双曲脐〔型突变〕【物】　hyperbolic umbilic
双曲系　hyperbolic system
双曲线【数】　hyperbola
双曲线导航系统【航海】　hyperbolic navigation system
双曲线吸引子　hyperbolic attractor
双曲型　hyperbolic type
双热带气旋　tropical cyclone twins
双三次样条函数　bicubic spline functions

双散射　double scattering
双施放　twin flights
双水流绝对日射表　twin water-flow pyrheliometer
双水位测站　twin-gauge station
双随机矩阵　doubly stochastic matrix
双台风【大气】　binary typhoons
双探空仪　twin radiosonde
双探空仪探空　twin soundings, dual sounding
双通道辐射仪放大器　twin channel radiometer amplifier
双通道雷达【大气】　dual-channel radar
双通量理论　two-flux theory
双筒望远镜　binoculars
双稳性　bistabilism
双涡〔旋〕雷暴　double vortex thunderstorm
双线静电表　bifilar electrometer
双线性内插【计】　bilinear interpolation
双相关　double correlation
双向二极管【电子】　bidirectional diode
双向法　two-way approach
双向反射比　bi-directional reflectance
双向反射比分布函数　BRDF (bidirectional reflectance distribution function)
双向反射〔比〕因子【大气】　bidirectional reflectance factor
双向反射函数　bidirectional reflection function (BDRF)
双向风标　bivane
双向风向标　bidirectional wind vane
双向滑动　bi-directional slip
双向嵌套　two-way nesting
双向通信【电子】　both-way communication
双向影响　two-way interaction
双循环消去法　bicyclic elimination
双压法　two-pressure principle
双眼墙结构　double eye structure
双雨季〔的〕　birainy

双雨季气候　birainy climate
双元〔铷蒸气〕磁强计　double-cell magnetometer, dual-cell magnetometer
双原子分子【化】　diatomic molecule
双折射　double refraction
双折射【物】　birefringence
双正交小波变换【计】　bi-orthogonal wavelet transformation
双重突变　dual catastrophe
霜【大气】　frost
霜霭　frost mist
霜点【大气】　frost point
霜点法　frost-point technique
霜点湿度表　frost-point hygrometer
霜点温度表　frost-point thermometer
霜冻【大气】　frost injury
霜冻天气　frost weather
霜度　degrees of frost
霜害　frost hazard, frost damage, frostbite
霜花　frost flower
霜检测器　frost detector
霜降【农】【大气】　First Frost
霜坑　frost hollow
霜轮　frost ring
霜片　frost flakes
霜期　frost season
霜期【大气】　frost period
霜区　frost zone
霜日【大气】　frost day
霜凇　hard rime
霜〔透〕深〔度〕　frost depth
霜洼　frost pocket(或 hollow)
霜雾　frost fog
霜线　frost line
霜眼　frost hole
霜羽　frost feathers
霜柱　ice pillar
水　water
水半球　oceanic hemisphere
水包合物　clathrate hydrate

水波　water wave
水波折射　refraction of water waves
水槽　water reservoir
水槽理论　canal theory
水产资源【航海】　fishery resources
水成景观【地理】　aqual landscape
水当量　water equivalent
水导〔式〕雨量计【大气】　electric conductivity raingauge
水的冻融循环　freeze-thaw cycle of water
水的蒸发　evaporation of water
水滴　water drop
水滴集电器　water dropper, water dropping collector, drop collector
水滴破碎理论　breaking-drop theory
水底地形　bottom topography
水电　hydroelectricity
水发射率　water emissivity
水分　moisture
水分保持　water conservation
水分补偿　water compensation
水分差额【大气】　water budget
水分调整　moisture adjustment
水分管理　water management
水分距平指数　moisture anomaly index
水分〔快速〕测定仪　moisture teller
水分亏缺【植物】　water deficit
水分利用系数　water-use coefficient
水分连续方程　moisture-continuity equation
水分平衡　hydrologic balance（或 budget）
水分收支　water budget【大气】hydrologic accounting
水分体积百分率　moisture volume percentage
水分循环　hydrological cycle
水〔分〕循环【大气】　water cycle
水分循环系数　water circulation coefficient

水分再分布　moisture redistribution
水分直减率　hydrolapse
水分重量百分率　moisture weight percentage
水分状况　water regime
水分子　water molecule
水改道　diversion of water
水工建筑物　hydraulic structures
水光学　hydrooptics
水耗　water loss
水合【化】　hydration
水化学【地理】　hydrochemistry
水化学相　hydrochemical facies
水击　water hammer
水结构　water structure
水解【化】　hydrolysis
水解氮【土壤】　hydrolysable nitrogen
水库　reservoir
〔水库〕放水　drawoff
水库业务运行　reservoir operation
水雷〔危险〕区【测】　mine〔dangerous〕area
水力半径　hydraulic radius
水力流动　hydraulic flow
水力模拟　hydraulic analog
水力平均深度　hydraulic mean depth
水力坡线　hydraulic grade line
水力土壤蒸发器　hydraulic soil evaporimeter
水力相似〔性〕　hydraulic similarity
水力学【地理】　hydraulics
水力学法流量演算　hydraulic routing
水力阻抗性　hydraulic resistivity
水利经济学【地理】　hydroeconomics
水流记录器　water-flow recorder
水流扩展　water spreading
水流式绝对日射表　water-flow pyrheliometer
水龙卷　waterspout【大气】, dragon
水面饱和混合比【大气】　saturation mixing ratio with respect to water

水面浮标　surface float
水面过饱和　supersaturation with respect to water
水面温度　surface temperature of water
水面蒸发　water surface evaporation
水面蒸发【农】　evaporation from water surface
水沫起电　spray electrification
〔水内〕声速计　acoustic velocimeter
水能【地理】　hydropower
水凝胶【化】　hydrogel
水凝结指数　water condensation index
水凝物　hydrometeor
水凝物负荷　hydrometeor charge
水凝物曳力　hydrometeor drag
水平波　horizon wave
水平波长　horizontal wavelength
水平波数　horizontal wavenumber
水平尺度　horizontal scale
水平地带【地理】　horizontal zone
水平动量方程　horizontal momentum equation
水平对流卷涡　horizontal convective rolls
水平范围　horizontal extent
水平分辨率　horizontal resolution
水平风切变【大气】　horizontal wind shear
水平风矢量【大气】　horizontal wind vector
水平辐合　horizontal convergence
水平辐散【大气】　horizontal divergence
水平横向波　horizontal transverse wave
水平衡　water balance
水平衡【地理】　water budget
水平虹　horizontal rainbow
水平混合　horizontal mixing
水平集总采水器　integration water sampler
水平简化　horizontal simplification
水平降水【地理】　horizontal precipitation
水平结构方程　horizontal structure equation
水平距离　horizon distance
水平卷涡对流　horizontal roll convection
水平卷涡涡列　horizontal roll vortices
水平均一　horizontal homogeneity
水平扩散　horizontal diffusion
水平扩散系数　horizontal diffusion coefficient
水平面　horizontal plane
水平能见度【大气】　horizontal visibility
水平耦合　horizontal coupling
水平偏振　horizontal polarization
水平漂浮气球　horizontally floating balloon
水平平流　horizontal advection
水平气压力　horizontal pressure force
水平气压梯度力　horizontal pressure gradient force
水平切变　horizontal shear
水平区内蒸发面指数　index of evaporating surface of horizontal area
水平取向粒子　horizontal orientation particle
水平扫描　searchlighting
水平散度【大气】　horizontal divergence
水平时间剖面　horizontal time section
水平输送　horizontal transfer
水平探测技术　horizontal sounding technique
水平探测气球　horizontal sounding balloon
水平退耦　horizontal decoupling
水平网格点交错　staggering of horizontal grid
水平支【天】　horizontal branch
水-气界面　air-water interface
水-气界面散射　air-water interface scattering

水汽　water vapor【大气】, aqueous vapour, vapour(＝vapor)
水汽暴　steam devil
水汽不足〔量〕　moisture deficit
水汽测量仪　water vapour meter
水汽猝发　moisture burst
水汽带【大气】　water vapor bands
水汽订正　water-vapor correction
水汽反馈　water-vapor feedback
水汽反演【大气】　water-vapor retrieval
水汽方程　moisture equation
水汽辐合　water vapour convergence, moisture convergence
水汽含量　moisture content【大气】, water-vapor content
水汽集中池　moisture pooling
水汽可用指数　moisture available index
水汽廓线【大气】　water-vapor profile
水汽连续带　water vapor continuum
水汽密度　vapour density, water-vapour density
水汽摩尔分数　mole fraction of water vapor
水汽浓度　vapour concentration
水汽圈　psychrosphere, water atmosphere
水汽收支　moisture budget
水汽〔守恒〕方程【大气】　moist〔conservation〕equation
水汽输送【大气】　transfer of water vapor
水汽通道资料　moisture channel data
水汽通量　moisture flux, water-vapour flux, vapour flux
水汽图像　water vapour imagery
水汽〔推导〕风　water vapour winds
水汽尾迹　vapour trail
水汽-温室效应　water vapour-greenhouse effect
水汽吸收　water-vapour absorption
水汽吸收分光仪　water-vapour spectroscope

水汽线　vapor line
水汽学　atmology
水汽压　vapor pressure
水汽压【大气】　water-vapor pressure
水汽压亏值　vapor pressure deficit
水汽压曲线　vapor pressure curve
水汽压湿度计　vapor pressure hygrograph
水汽压温度表　vapor pressure thermometer
水汽应力　moisture stress
水汽有效率(饱和率)　moisture availability
水汽源　vapor source
水汽张力　vapor tension, moisture tension
水圈【地理】【大气】　hydrosphere
水热爆炸【地】　hydrothermal explosion
水热对流系统【地】　hydrothermal convection system
水热反应【地质】　hydrothermal reaction
水热活动【地】　hydrothermal activity
水热流体【地质】　hydrothermal fluid
水热喷发【地】　hydrothermal eruption
水热区【地质】　hydrothermal area
水热蚀变【地】　hydrothermal alteration
水热田【地】　hydrothermal field
水热系数　water-hot coefficient
水热系统【地质】　hydrothermal system
水热循环【地】　hydrothermal circulation
水热资源【地质】　hydrothermal resources
水溶胶【化】　hydrosol
水深图　bathymetric chart
水生生物学【海洋】　hydrobiology
水生形态〔的〕　hydromorphic
水生植物　hydrophytic plant
水生植物【林】　aquatic plant
水生植物【植】　hydrophyte
水声学　hydroacoustics

水蚀【土壤】 water erosion
水蚀程度【林】 degree of water erosion
水势【植物】 water potential
水体【地理】 water body
水听器校准器 hydrophone calibrator
水头 water head
水土保持 water and soil conservation
水土保持【土壤】 soil and water conservation
水土保持学【林】 soil and water conservation
水土流失 water and soil erosion
水土评价工具 Soil and Water Assessment Tool (SWAT)
水团【地理】 water mass
水团变性 water mass transformation
水团构成 water mass formation
水位 water level, river stage
水位【水力】 stage
水位标 staff gauge
水位表 water meter
水位尺 river gauge
水位关系 stage relation, gauge relation
水位计 water-stage recorder, water-level recorder
水位流量关系 stage discharge relation
水位流量关系绳套曲线 loop rating
水位-流速积分法 depth-velocity integration method
水温 water temperature
水温表 water-thermometer
水文测定法 hydrometric method
〔水文〕测量点 gauging site
〔水文〕测量截面 gauging section
水文测量学 hydrometry
水文地理学【地理】 hydrogeography
水文地球化学 hydrogeochemistry
水文地质边界 hydrogeological boundary
水文地质分区【地质】 hydrogeological division
水文地质条件【地质】 hydrogeological condition
水文地质学【地质】 hydrogeology
水文干旱 hydrological drought
水文年 hydrologic year
水文年度 water year
水文气候学 hydroclimatology
水文气象学【大气】 hydrometeorology
水文气象学家 hydrometeorologist
水文气象学委员会(WMO) Commission for Hydrological Meteorology (CHM, WMO)
水文气象研究中心(前苏联) Hydrometeorological Research Center (HRC)
水文气象预报 hydrometeorological forecast
水〔文〕情〔况〕 hydrologic regime
水文图【大气】 hydrograph
水文现象分带性 zonality of hydrological phenomena
水文学 hydrography
水文学【地理】【大气】 hydrology
水文〔学〕的 hydrographic
水文〔学〕方程 hydrologic equation
水文学委员会(WMO) Commission for Hydrology (CHY, WMO)
水文〔学〕性质 hydrologic properties
水文循环【大气】 hydrologic cycle
水文要素 hydrological element
水文雨量分析计划 hydrologic rainfall analysis project
水文预报 hydrological forecast, hydrological forecasting
水文预警 hydrological warning
水文站 hydrometric station
水文站网 hydrological network
水文指数 hydrologic index
水污染【地理】 water pollution
水污染监测 water pollution monitoring
水污染控制 water pollution control
水物质运动学 hygrokinematics
水雾 water fog

水系　river system, water system, hydrographical network
水下冰丘　bummock
水下浮标　subsurface float
水下光度表　hydrophotometer
水下声散射　underwater acoustic scattering
水星【天】　Mercury
水型　water type
水循环【大气】　water cycle
水循环系数【大气】　water circulation coefficient
水银　mercury(Hg), quicksilver
〔水银〕槽　cistern
水银槽气压表　cistern barometer
水银气压表【大气】　mercury barometer
水银气压表　mercurial barometer
水银气压表内管　barometric tube
〔水银气压表〕皮囊　wash-leather bag
〔水银〕气压柱　barometric column
水银温度表　mercury thermometer
水银温度表【大气】　mercury thermometer
水银柱　barometer column, mercury column
水应力作用　water stress effect
水盈值　water surplus
水俣病【地理】　Minamata disease
水域　hydrological basin
水源地　head-water point
水跃　hydraulic jump
水云【大气】　water cloud
水灾【地理】　flood damage, flood catastrophe
水闸　sluice, flood gate
水照云光　water sky
水蒸气【大气】　water vapor
水质　quality of water
水质【地理】　water quality
水质分析　water quality analysis
水质评价　water quality assessment
水质物　water substance
水资源【大气】　water resources
水资源保护【地理】　protection of water resources
顺潮　direct tide
顺磁性【物】　paramagnetism
顺风　downwind, following wind, fair wind
顺风【大气】　tail wind
顺风【航海】　favourable wind
顺切变　downshear
顺日向　deasil
顺时针　clock-wise
顺梯度　downgradient
顺梯度扩散　down-gradient diffusion
顺梯度输送　downgradient transfer, down-gradient transport
顺梯度通量　downgradient flux
顺序处理【计】　sequential processing
顺序存取【计】　sequential access
顺序式自动降水取样器　automatic sequential precipitation sampler
顺序统计学　order statistics
顺序张弛　sequential relaxation
顺转　veering
顺转风【大气】　veering wind
瞬变　transient variation
瞬变波【大气】　transient wave
瞬变〔的〕　transient
瞬变气候响应　transient climate response
瞬变数据　transient data
瞬变温跃层　transient thermocline
瞬变涡动【大气】　transient eddy
瞬变行星波　transient planetary wave
瞬变振荡　transient fluctuation
瞬间的　instantaneous
瞬时采样　instantaneous sampling
瞬时单位过程图　instantaneous unit hydrograph
瞬时风速　instantaneous wind speed

瞬时感〔觉〕 instantaneous sensation
瞬时功率 instantaneous power
瞬时功率谱密度 instantaneous power spectral density
瞬时锢囚 instant occlusion
瞬时几何视场 instantaneous geometric field of view (IGFOV)
瞬时谱 instantaneous spectrum
瞬时视场【电子】 instantaneous field of view (IFOV)
瞬时速度【物】 instantaneous velocity
瞬态分析【计】 transient analysis
瞬态问题 transient problem
朔【天】 new moon
朔望潮 syzygical tide
朔望月 synodic month
丝状冰 satin ice
斯波勒极小【天】 Sporrer minimum
斯蒂芬-玻耳兹曼常数 Stefan-Boltzmann constant
〔斯蒂文森〕百叶箱 Stevenson screen
斯科勒参数（大气重力波波动方程） Scorer parameter
斯梅尔马蹄变换 Smale horseshoe transform
斯梅尔马蹄【物】 Smale horseshoe
斯密森绝对日射表标尺 Smithsonian pyrheliometric scale
斯密森水柱式绝对日射表 Smithsonian water-flow pyrheliometer
斯密森银盘绝对日射表 Smithsonian silver disk pyrheliometer
斯涅耳定律【物】 Snell law
斯匹次卑尔根海流 Spitsbergen Current
斯普朗气压计 Sprung barograph
斯塔克效应【物】 Stark effect
斯坦福大学微波实验室（美国） Stanford University Microwave Laboratory (SU-ML)
斯坦通数 Stanton number
斯特藩-玻尔兹曼定律【物】 Stefan-Boltzmann law
斯特劳哈尔数 Strouhal number
斯特里特-费尔普斯方程 Streeter-Phelps equation
斯托克斯波 Stokesian wave, Stokes wave
斯托克斯参数 Stokes' parameter
斯托克斯定理 Stokes's theorem
斯托克斯定律 Stokes' law
斯托克斯公式【数】 Stokes formula
斯托克斯矩阵 Stokes' matrix
斯托克斯流函数 Stokes's stream function
斯托克斯漂移 Stokes's drift
斯托克斯漂移速度 Stokes's drift velocity
斯托克斯数 Stokes number
斯托克斯线【物】 Stokes line
斯托克斯荧光 Stokes fluorescence
斯瓦罗浮子 Swallow float
斯维尔德鲁普关系 Sverdrup relation
锶云 strontium cloud
死冰川 dead glacier
死火山【地质】 extinct volcano
死库容【水利】 dead reservoir capacity
死水 dead water, unfree water
死水位【水利】 dead water level
四边形元 quadrilateral element
四分差 quartile deviation
四分位差 semi-interquartile range
四分位数 quartile
四分位数间距 inter-quartile range
四阶矩 fourth order moment
四流理论 four flux theory
四氯化碳（CCl_4） carbon tetrachloride
四能级系统【物】 four-level system
四维变分同化 four-dimensional variational assimilation
四维谱密度 four dimensional spectral density
四维时空【物】 four dimensional space-time

四维资料同化【大气】 four-dimensional data assimilation
四维最优插值 four-dimensional optimal interpolation（4D－OI）
四用自记计 quadruple recorder
似然【数】 likelihood
似然函数【数】 likelihood function
伺服电〔动〕机【电子】 servo-motor
伺服放大器【电子】 servo amplifier
伺服环〔路〕 servo loop
伺服气压表 servo barometer
伺服系统【电子】 servo system
松冰团 brine slush
松弛法【数】 relaxation method
松弛算法【计】 relaxed algorithm
松雪 wild snow
松野格式 Matsuno scheme
凇碎过程 rime-splinter process
搜救卫星系统【航海】 search and rescue satellite system
搜救业务【航海】 search and rescue service
搜索空间 search space
搜索雷达 search radar
搜索算法【计】 search algorithm
苏格兰气象学会（英国） Scottish Meteorological Society（SMS）
苏格兰雾 Scotch mist
苏黎世数【天】 Zurich number
速度【物】 velocity
速度-波数滤波 velocity-wavenumber filtering
速度场 velocity field
速度方位处理 velocity-azimuth processing（VAP）
速度方位距离显示【大气】 velocity azimuth range display（VARD）
速度方位显示【大气】 velocity azimuth display（VAD）
速度方位显示风廓线 VAD wind profile
速度分布 velocity distribution
速度分解 resolution of velocity
速度辐散 velocity divergence
速度高度比 velocity-height ratio
速度环流 velocity circulation
速度混淆 velocity aliasing
速度夹卷 velocity entrainment
速度亏损定理 velocity-defect law
速度廓线 velocity profile
速度滤波 velocity filtering
速度模糊【大气】 velocity ambiguity
速度-频率滤波 velocity-frequency filtering
速度谱 spectrum of velocity, velocity spectrum
速度起伏 velocity fluctuation
速度切变 velocity shear
速度曲线 velocity curve
速度势 velocity potential
速度梯度 velocity gradient
速度折错 velocity folding
速度指示器 velocity indicator
速度指示相干积分器 velocity indication coherent integrator（VICI）
速率【物】 speed
速率常数【化】 rate constant
速率方程【化】 rate equation
速率系数 rate coefficient
速调管【电子】 klystron
速移极光 rapidly moving aurora
速阅资料 "Quick Look" data
宿落风 solore
〔宿〕主机【计】 host machine
宿主系统【计】 host system
塑封【电子】 plastic package
酸霭 acid mist
酸雹【大气】 acid hail
酸沉降【大气】 acid deposition
酸度 acidity【大气】, degree of acidity
酸度常数【化】 acidity constant
酸度廓线 acidity profile
酸酐【化】 anhydride

酸化【化】 acidification
酸露【大气】 acid dew
酸露点 acid dew point
酸霾 acid haze
酸霜【大气】 acid frost
酸污染 acid pollution
酸雾【大气】 acid fog
酸性降水 acid precipitation
酸性土【土壤】 acid soil
酸雪【大气】 acid snow
酸烟雾 acid fume
酸雨【大气】 acid rain
算法【数】 algorithm
算法语言【计】 algorithmic language
算符缩并【物】 contraction of operators
算术平均【数】 arithmetic mean
算子 operator
D-算子 D-operator
〔随〕地形坐标 terrain-following coordinate
随机爱尔沙色带模式【大气】 random Elsasser band model
随机变量 stochastic variable
随机变量【数】 random variable
随机并合 stochastic coalescence
随机并合方程 stochastic coalescence equation
随机播云试验 randomized seeding trial
随机初始扰动 stochastic initial perturbation
随机存储器【计】 random access memory (RAM)
随机存取【计】 random access
随机的 stochastic
随机〔的〕【数】 random
随机动力模式 stochastic dynamic model
随机动力预报 stochastic dynamic prediction
随机方程 stochastic equation
随机分析【数】 stochastic analysis
随机观测误差 random observing error
随机过程【数】 stochastic process
随机过程【自动】 random process
随机海〔况〕 random sea
随机函数【数】 random function
随机化 randomization
随机化检验【数】 randomized test
随机化试验 randomized experiment
随机混合 random mixing
随机交叉试验 randomized cross experiment
随机介质 random medium
随机决策 random decision
随机粒子模拟 stochastic particle modeling
随机模式 random model, stochastic model
随机起伏 random fluctuation
随机强迫 random forcing, stochastic forcing
随机取样 stochastic sampling
随机取样【地质】 random sampling
随机扰动 stochastic perturbation
随机事件 stochastic events
随机数【数】 random number
随机水文学 stochastic hydrology
随机通道测量系统 random access measurement system (RAMS)
随机微分方程【数】 stochastic differential equation
随机误差【数】 random error
随机响应 stochastic response
随机信号 random signal
随机性 randomness
随机样本【数】 random sample
随机游动【数】 random walk
随机预报【大气】 random forecast
随机噪声 stochastic noise
随机噪声【电工】 random noise
随机振幅 random amplitude, stochastic amplitude
随时〔间的〕变化 time-variation

岁差【天】 precession
碎冰【航海】 brash ice
碎冰片 ice splinter
碎波 surf
碎波带【海洋】 surf zone
碎层云【大气】 Fractostratus (Fs)
碎浮冰〔堆〕 brash
碎高层云 altostratus fractus
碎核 fragmentation nuclei
碎积云【大气】 Fractocumulus (Fc)
碎浪 breaker
碎浪深度 breaker depth
碎裂 splintering
碎乱天空 amorphous sky
碎片 debris, patch
碎片(云) pannus(pan)
碎片状云 shred cloud
碎石流 angular drift
碎雾 fog patch
碎啸冰 growler
碎雨云【大气】 Fractonimbus (Fn)
碎云 fractus(fra)
碎云量 fractional cloud amount
碎云幂 ragged ceiling
隧道效应【化】 tunnelling effect

损耗波 evanescent wave (EW)
损耗层 evanescent level
损耗函数 damage function
损失函数 loss function
羧基化【化】 carboxylation
缩并〔运算〕【数】 contraction
缩微胶卷 microfilm
缩微胶片 microfiche
缩微摄影 microfilming
缩微摄影【测】 microphotography
缩微摄影机 microfilmer
索贺(太阳和日球层观测站, 1995年发射)【天】 Solar and Heliospheric Observatory (SOHO)
索洛蒙纸 Solomon's paper
〔索马里〕大涡流 Great Whirl
索马里海流 Somali Current
索马里急流 Somali jet
索诺拉雷暴天气 Sonora weather
索引号码 index number
索状云 rope cloud
锁定 lock-in
锁模【物】 mode-locking
锁相放大器 lock-in amplifier
锁相【物】 phase-locking

T

塔层 tower layer
塔层微气象学 tower micrometeorology
塔成雾 tower-produced fog
塔台能见度 tower visibility
塔用传感器 tower sensor
塔状积云 towering cumulus
塔状云 turreted cloud
台地【地理】 tableland
台风【大气】 typhoon
台风飑 typhoon squall
台风潮 typhoon tide

台风发展 typhoon development
台风跟踪试验 typhoon tracking experiment (TTE)
台风警报【大气】 typhoon warning
台风监测 typhoon watch
台风路径 typhoon track
台风生成 typhoon genesis
台风委员会 Typhoon Committee (TC)
台风眼 eye of typhoon
台风眼【大气】 typhoon eye
台风业务试验 Typhoon Operational

Experiment（TOPEX）
台风雨　typhoon rain
台风预警　typhoon warning
台风〔云〕墙　typhoon bar
台风再生　regeneration of typhoon
台风中心　typhoon center
台风转向　recurvature of typhoon
台湾低压　Taiwan low
台湾地区中尺度试验　Taiwan Area Mesoscale Experiment（TAMEX）
台湾暖流　Taiwan Warm Current
台站比法　station-ratio method
台站海拔　station elevation
〔台〕站号　station index, station index number
台站年〔分析〕法　station-year method
台站气压表　station barometer
台站区号　block number
苔原【地理】　tundra
苔原气候【大气】　tundra climate
苔原区　bryochore
抬升　lifting
抬升对流　elevated convection
抬升凝结高度【大气】　lifting condensation level（LCL）
抬升凝结高度区　lifting condensation level zone
抬升雾　lifting fog
抬升指数　lifting index（LI）【大气】, lifted index
苔原气候亚类【大气】　tundra climate
太古代　Archeozoic Era
太古宇【地质】　Archean Eonothem
太古宙【地质】　Archean Eon
〔太空人〕轻便维生系统　portable life-support system（PLSS）
太平洋【海洋】　Pacific Ocean
太平洋板块【海洋】　Pacific Plate
太平洋-北美型　PNA（Pacific-North American pattern）, Pacific and North American（PNA）pattern

太平洋标准时间（美国）　Pacific Standard Time
太平洋赤道潜流【海洋】　Pacific Equatorial Undercurrent
太平洋副北极海流　Pacific Subarctic Current
太平洋高压【大气】　Pacific high
太平洋海洋环境实验室　Pacific Marine Environmental Laboratory（PMEL）
太平洋区域　Pacific region
太平洋深层水　Pacific Deep Water
太平洋十年振荡　Pacific Decadal Oscillation（PDO）
太平洋时间（美国）　Pacific Time
太平洋西部型　West Pacific pattern（WP）
太平洋型岸线【海洋】　Pacific-type coastline
太平洋型大陆边缘【海洋】　Pacific-type continental margin
太平洋洋中槽　Mid-Pacific trough
太阳【天】　sun
太阳半经【天】　solar radius
太阳变异度　solar variability
太阳标定　sun pointing
太阳波束　solar beam
太阳常数【大气】　solar constant
太阳潮【地】　solar tide
太阳赤纬　solar declination
太阳传感器　sun sensor
太阳磁周【天】　solar magnetic cycle
太阳大气　solar atmosphere
〔太阳〕大气潮　solar atmospheric tide
太阳地文说　solar-topographic theory
太阳电池帆板　solar paddle（或 panel）
太阳电子事件【地】　solar electron event
太阳短波辐射　solar short-wave radiation
太阳风【天】【大气】　solar wind
太阳风边界　solar wind boundary
太阳风顶【天】　heliopause

太阳辐射【大气】 solar radiation
太阳辐射表 solar radiometer
太阳辐射观测 solar radiation observation
太阳〔辐射〕加热率 solar heating rate
太阳辐射平衡 balance of solar radiation
太阳辐射衰减 attenuation of solar radiation
太阳辐射误差 solar radiation error
太阳辐照度【天】 solar irradiance
太阳高度 solar altitude, sun's altitude
太阳高度查算图 solar altitude diagram
太阳高能粒子 solar high energy particle
太阳光斑 solar facula
太阳光度【天】 solar luminosity
太阳光度计【大气】 heliograph
太阳光度计 sunphotometer
太阳光谱 solar spectrum
太阳光学望远镜 solar optical telescope
太阳黑子【天】 sunspot
太阳黑子极大 sunspot maximum
太阳〔黑子〕极大 solar maximum
太阳黑子极小 sunspot minimum
太阳〔黑子〕极小 solar minimum
〔太阳〕黑子群 spot group, group of sunspots
太阳黑子群【天】 sunspot group
太阳黑子数 sunspot number
太阳黑子相对数 sunspot relative number, relative sunspot number
太阳黑子周期 sunspot cycle【大气】, sunspot period, sunspot periodicity
太阳红外线 solar infrared
太阳后向散射紫外分光计 solar backscatter ultraviolet spectrometer (SBUVS)
太阳后向散射紫外辐射仪 solar back scatter ultraviolet radiometer (SBUV)
太阳后向散射紫外线/臭氧总量测绘系统 solar-backscattered ultraviolet/total ozone mapper system (SBUV/TOMS)
太阳活动【大气】 solar activity
太阳活动低潮 solar ebb
太阳活动区 solar active region, active solar region
太阳活动周【天】 solar cycle
太阳活动周期【大气】 solar cycle
太阳粒子发射 solar particle emission
太阳粒子事件 solar particle event
太阳盲区 solar blind region
太阳米粒 solar granule
太阳能 solar energy
太阳〔能〕电池 solar cell
太阳能技术 heliotechnics
太阳能利用装置 helioplant
太阳能区划【大气】 solar energy demarcation
太阳能资源【大气】 solar energy resources
太阳年 solar year
太阳气候【大气】 solar climate
太阳-气象关系 solar-meteorological relationship
太阳扰动 solar disturbance
太阳热 solar heat
太阳日【天】 solar day
太阳日变化 solar daily variation
太阳色球〔层〕 solar chromosphere
太阳射电 solar radio radiation
太阳射电〔发射〕 solar radio emission
太阳射电望远镜 solar radio telescope
太阳摄谱仪 solar spectrograph
太阳十字〔晕〕 solar cross
太阳时 solar time
太阳视运动 apparent motion of the sun
太阳天顶角 solar zenith angle
太阳-天气相关 sun-weather correlation
太阳天气效应 sun-weather effect
太阳通量密度 solar flux density
太阳同步轨道 sun-synchronous orbit
太阳同步卫星 sun-synchronous satellite

太阳同步气象卫星【大气】 solar synchronous meteorological satellite
太阳外差辐射仪 solar heterodyne radiometer
太阳〔望远〕镜 helioscope
太阳望远镜【天】 solar telescope
太阳微粒 solar corpuscle
太阳微粒发射【地】 solar corpuscular emission
太阳微粒辐射 solar corpuscular radiation
太阳微粒流 solar particle stream
太阳微粒射线 solar corpuscular ray
太阳微粒〔学〕说 solar corpuscular theory
太阳温度 solar temperature
太阳物理学【天】 solar physics
太阳系【天】 solar system
太阳信号 solar signal
太阳仰角 solar elevation
太阳耀斑【大气】 solar flare
〔太阳〕耀斑活动 solar flare activity
太阳耀斑扰动 solar flare disturbance
太阳直接辐射测量学【大气】 pyrheliometry
太阳指向器 solar pointer
太阳质子监测仪 solar proton monitor (SPM)
太阳质子事件【天】 solar proton event
太阳周边辐射 circumsolar radiation
太阳周年视运动 solar annual 〔apparent〕 motion
太阳紫外辐射 solar ultraviolet radiation
太阳紫外辐射监测仪 monitor of ultraviolet solar radiation (MUSR)
太阳紫外能监测仪 monitor of ultraviolet solar energy (MUSE)
太阴潮【大气】 lunar tide
太阴距离 lunar distance
太阴年 lunar year
太阴日 lunar day
太阴月 lunation (= lunar month)
态【物】 state
态空间【物】 state space
肽【生化】 peptide
钛酸铈湿度传感器 cerium titanate humidity transducer
泰加林【地理】 taiga
泰加林气候 taiga climate
泰勒定理 Taylor's theorem
泰勒多项式 Taylor polynomial
泰勒公式【数】 Taylor formula
泰勒级数 Taylor series
泰勒〔级数〕展开 Taylor 〔series〕 expansion
泰勒假设 Taylor's hypothesis
泰勒扩散定理 Taylor's diffusion theorem
泰勒-普罗德曼定理 Taylor-Proudman theorem
泰勒数 Taylor number
泰勒图 Taylor diagram
泰勒微尺度 Taylor microscale
泰勒涡 Taylor vortex
泰勒效应 Taylor effect
泰勒柱 Taylor column
泰罗斯大气(晴空)辐射模块 TIROS atmospheric radiance module (TARM)
泰罗斯平流层探测资料映射模块 TIROS Stratospheric Mapper (TSM)
泰罗斯卫星【大气】 TIROS (television and infrared observation satellite)
泰罗斯-N卫星【大气】 TIROS-N
泰罗斯-N卫星垂直探测器 TIROS-N operational vertical sounder
泰罗斯卫星业务系统 TIROS Operational System (TOS)
泰罗斯信息处理机 TIROS information processor (TIP)
泰罗斯业务垂直探测器 TIROS Operational Vertical Sounder (TOVS)
泰罗斯业务卫星 TIROS Operational Satellite (TOS)

贪青　remaining green
探测　detection, probe
探测概率,识别率　probability of detection（POD）
探测率【电子】　detectivity
探测气球　balloon-sonde
探测器【电子】　detector
探测与归因　detection and attribution
探空【大气】　sounding
探空报告　radiosonde report
探空火箭【大气】　sounding rocket
探空气球比较观测　comparative rabal
探空气球【地】　sounding balloon
探空气象仪　sounding meteorograph
探空设备系列　flight train
探空业务　radiosounding operation
探空气球【大气】　sounding balloon, radiosonde balloon
探空仪　sounder
探空仪【大气】　sonde
探空仪编码器　radiosonde encipher
探空仪参考信号　radiosonde reference signal
探空仪地面设备　radiosonde ground equipment
探空仪调制器　radiosonde modulator
探空仪发报机　radiosonde transmitter
探空仪防辐射罩　radiation shield for radiosonde
探空仪辐射误差　radiation error of radiosonde
探空仪观测　dropsonde observation
探空〔仪〕观测〔记录〕　radiosonde observation
探空仪记录器　radiosonde recorder
探空仪气球　radiosonde balloon
探空〔仪〕台站记录器　radiosonde station recorder
探空仪退绕器　radiosonde unwinder
探空仪误差　sonde error
探空仪校准　radiosonde calibration

探空仪转换器　radiosonde commutator
探空〔仪〕资料数字转换器　radiosonde data digitizer
探空原始记录　raw radio sounding record
探空站网　radiosonde network
探险者（常用于空间探测卫星）　Explorer
碳【大气】　carbon
碳池【大气】　carbon pool
碳定年法【大气】　carbon-dating
碳封存　carbon sequestration
碳黑催化　carbon-black seeding
碳汇【大气】　carbon sink
碳键机制　carbon bond mechanism
碳交易　carbon trading
碳库【大气】　carbon pool
碳膜湿度〔表〕元件　carbon-film hygrometer element
碳排放【量】　carbon emission
（碳）排放交易　emission trading（ET）
碳氢化合物【化】　hydrocarbon
碳湿敏电阻　carbon hygristor
碳收支　carbon budget
碳水化合物【化】　carbohydrate
碳同化【大气】　carbon assimilation
碳酸盐　carbonate
碳酸盐分析　carbonate analysis
碳同位素　carbon isotope
碳信用,碳权　carbon credit
碳循环【大气】　carbon cycle
碳营养【土壤】　carbon nutrition
碳源【大气】　carbon source
碳足迹　carbon footprint
汤姆孙散射【物】　Thomson scattering
汤森放电【电子】　Townsend discharge
汤森支架　Townsend support
羰基化【化】　carbonylation
羰基化合物　carbonyl compounds
淌凌　ice run
〔逃逸的〕临界高度　critical level

逸逸高度　level of escape
逸逸深度【物】　escape depth
逸逸速度　escape speed
逸逸速度【物】　velocity of escape
逸逸圆锥（外逸层中）　cone of escape
陶瓷半导体（元件）　ceramic semiconductor
陶瓷相对湿度传感器　ceramic relative humidity sensor
陶瓷蒸发计　clay atmometer
陶普生单位【大气】　Dobson unit（DU）
陶普生分光光度计【大气】　Dobson spectrophotometer
套网格【大气】　nested grid
套种　relay-interplanting
特大暴雨　extraordinary storm, excessive storm
特大洪水　enormous flood
特大雨量　excessive rainfall
特低频【电子】　ultra-low frequency（ULF）
特定降水密度　specific precipitation density
特高频【电子】　UHF（ultra-high frequency）
特佳能见度　exceptional visibility
特解【数】　particular solution
特纳角　Turner angle
特强沙尘暴【大气】　extreme severe sand and dust storm
特殊观测　special observation
特殊函数【数】　special function
特殊机场报告　special aerodrome report
特殊世界时段　Special World Intervals（SWI）
特殊天气报告　special weather report
特殊天气综述　special weather statements
特殊(性)点　significant point
特殊云　significant cloud
特藤斯公式　Tetens's formula

特性　characteristics, character
特性层【大气】　significant level
特性点　characteristic point
特性法　method of characteristics
特性曲线　response
特性曲线分析　response analysis
特选船舶站　selected ship station
特选气压面　selected pressure surface
特征　specificity, signature, feature
特征长度　characteristic length
特征方程　characteristic equation
特征分析技术　signature analysis techniques
特征根　characteristic root
特征函数【数】　characteristic function
特征矩阵【数】　characteristic matrix
特征频率　characteristic frequency
特征曲线【数】　characteristic curve
特征湿度　characteristic humidity
特征时间尺度　characteristic time scale
特征矢量　latent vector, characteristic vector
特征速度　characteristic velocity
特征通过时间　characteristic passage time
特征温度　characteristic temperature
特征线　characteristic line
特征向量【数】　characteristic vector
特征值　latent value, characteristic value
特征值问题　characteristic-value problem
特征坐标　characteristic coordinate
特种传感器微波成像仪　special sensor microwave imager（SSM/I）
特种传感器微波温度[仪]　special sensor microwave temperature（SSM/T）
藤田尺度　Fujita scale
藤田-皮尔逊尺度　Fujita-Pearson scale（FPP）
藤原效应　Fujiwara effect
梯度【数】　gradient
梯度风【大气】　gradient wind
梯度风锋　gradient wind front

梯度风高度　gradient wind level, gradient level
梯度风公式　gradient wind equation
梯度风速　gradient velocity
梯度观测　gradient observation
梯度里查森数　gradient Richardson number
梯度流　gradient flow, gradient current
梯度输送理论　gradient transport theory
梯度算子　gradient operator
梯级水电站　cascade hydroelectric station, step hydroelectric station
梯级先导(闪电)　stepped leader
梯形【数】　trapezoid
梯形窗　trapezoidal window
梯形法　trapezoidal scheme
梯状云　echelon cloud
体波　body waves
体积　bulk
体积百万分率　ppmv (parts per million by volume)
体积氯度【大气】　chlorosity
〔体积〕摩尔浓度　molarity (M/V)
〔体积〕摩尔浓度　molarity (M/V)
体积目标【大气】　volume target
体积平均【大气】　volume average
体积散射　volume scattering
体积扫描　volume scan
体积输送　volume transport
体积元【数】　volume element
体积中数直径　volume median diameter
体角散射系数　volume angular scattering coefficient
体膨胀率【物】　volume expansivity
体散射函数【大气】　volume scattering function
体散射系数　volume scattering coefficient
体温　somatic temperature
体温表　rafraichometer
体温过低　hypothermia
体温过高　hyperthermia
体吸收系数　volume absorption coefficient
体系　regime
体消光系数　volume extinction coefficient
体效应　bulk effect
体总散射系数　volume total scattering coefficient
替代能源　alternative energy
天　heaven
天波【电子】　sky wave
天波改正量【航海】　sky wave correction
天波雷达　sky wave radar
天波延迟【航海】　sky wave delay
天赤道【天】【大气】　celestial equator
天底〔点〕　nadir
天底角　nadir angle
天底太阳后向散射计　nadir solar backscatter meter
天电　strays, static, spherics, sferics, atmospheric interference
天电【大气】　atmospherics
天电定位　sferics fix
天电方位编码　(SFAZI)spherics azimuth code
天电干扰高度　noise level of atmospherics
天电观测　sferics observation
天电观测网　sferics network
天电记录器　sferics recorder
天电接收机　sferics receiver
天电强度计　atmoradiograph
天电强度仪　radiomaximograph
天电突增【天】　sudden enhancement of atmospherics
天电仪　keraunograph, ceraunograph
天电源　source of an atmospheric, sferics source
天电源地　foyer
天电源强度　activity of a foyer of atmospherics

天顶【天】 zenith
天顶期雨 zenithal rains
天顶角 zenith angle
天顶距【天】 zenith distance
天干 Heavenly Stems
天干【天】 celestial stem
天光 skylight
天基观测【大气】 space-based observation
天极【天】【大气】 celestial pole
天空 sky
天空背景 sky background
天空背景噪声 sky background noise
天空不明 obscured sky cover
天空反射镜 sky mirror
天空辐射【大气】 sky radiation
天空辐射表【大气】 sky radiometer diffusometer
天空辐射测量〔法〕 pyranometry
天空辐射计 pyranograph
天空辐射自记曲线 pyranogram
天空蓝度【大气】 blue of the sky
天空蓝度测定法 cyanometry
天空蓝度测定仪【大气】 cyanometer
天空亮度【大气】 sky luminance, sky brightness
天空亮度温度 sky brightness temperature
天空漫射辐射【大气】 diffuse sky radiation
天空偏振 sky polarization
天空形状 shape of the sky
天空遮蔽总量 total sky cover
天空状况 state of the sky
天空状况【大气】 sky condition
天落水 meteoric water
天气【大气】 weather
天气〔安全期〕窗 weather window
天气报告 synoptic report【大气】, weather report
天气背景 synoptic background

天气表格 synopsis
天气波 weather wave
天气参数 weather parameter
天气层 weathersphere
天气尺度【大气】 synoptic scale
天气尺度系统【大气】 synoptic scale system
天气电码 synoptic weather code, weather code
天气导航【大气】 weather routing
天气电码【大气】 synoptic code
天气分类 weather typing
天气分析 weather analysis
天气分析【大气】 synoptic analysis
天气服务【大气】 weather service
天气服务雷达 weather service radar（WSR）
天气符号 weather sign
天气符号【大气】 weather symbol
天气更新模式 synoptic update model
天气公报 forecast bulletin
天气观测 weather observation
天气观测【大气】 synoptic observation
天气观测时间【大气】 synoptic hour
天气广播 weather broadcast
天气过程【大气】 synoptic process
天气过程总降水量 total storm precipitation
天气好转报〔告〕 improvement report
天气合同 weather derivatives
天气回波【大气】 weather echo
天气汇报 debriefing
天气记录中心 weather record center
天气监测 weather monitoring
天气监视 weather watch
天气监视雷达 weather surveillance radar（WSR）
天气讲解（起飞前） briefing
天气警报【大气】 weather warning
天气警示报 weather advisory
天气控制 weather control

天气雷达【大气】 weather radar
天气雷达观测 weather radar observation
天气雷达信号积分器 weather radar signal integrator
天气雷达资料处理分析器 weather radar data processor and analyzer
天气历史顺序 sequence of weather
天气模拟 weather simulation
天气气候学【大气】 synoptic climatology
天气侵蚀 weather erosion
天气区界线 weather divide
天气趋势 weather trend
天气日记 weather diary
天气实况演变图【大气】 meteorogram
天气实体 weather regime
天气事件 weather event
天气释用 weather interpretation
天气图 weather map, weather chart, synoptic map
天气图【大气】 synoptic chart
天气图传真【大气】 weather facsimile (WEFAX)
天气图传真收片机 facsimile weather chart recorder
天气图的 synoptic
天气图分析 weather chart analysis
天气图记录器 weather chart recorder
天气图类型 weather-map type
天气图相关 map correlation
天气系统 weather system
天气系统【大气】 synoptic system
天气现象【大气】 weather phenomenon
天气形势 weather situation
天气形势【大气】 synoptic situation
天气形势预报 weather-prognostics
天气型【大气】 synoptic type, weather type
天气型更换 change of type
天气序列 weather sequence
天气学【大气】 synoptic meteorology
天气学模型 synoptic model

天气〔学〕体系 synoptic regime
天气学委员会(WMO) Commission for Synoptic Meteorology (CSM, WMO)
天气循环 weather cycle
天气严重指数 weather severity indicator
天气研究和预报模式 Weather Research and Forecast (WRF)
天气谚语 weather lore, weather proverb
天气要素 weather element
天气业务条例 weather service bill
天气异常 weather anomaly
天气有线广播 weather wire service
天气预报 weather prediction, synoptic forecasting, synoptic forecast
天气预报【大气】 weather forecast
天气预报模式 weather forecast model
天气预报员 weather forecaster, weatherman
天气预警 weather warning
天气预警公报 weather warning bulletin
天气预警系统 weather warning system
天气噪声 weather noise
天气展望 weather prospect
天气展望【大气】 weather outlook
天气站 synoptic station, weather station
天气侦察飞机 weather plane
天气侦察〔飞行〕 weather reconnaissance〔flight〕
天气重现 weather recurrence
天气转坏报〔告〕 deterioration report
天气状况 weather condition
天气资料【大气】 synoptic data
天气资料处理 weather data processing
天穹形状【大气】 apparent form of the sky
天球【天】【大气】 celestial sphere
天球地平圈 celestial horizon, astronomical horizon
天球坐标系【天】 celestial coordinate-system

天然放射性【物】 natural radioactivity
天然放射性元素【化】 natural radioelement
天然径流 natural flow
天然剩磁【地】 natural remanent magnetization (NRM)
天扫风 sky sweeper
天山准静止锋【大气】 Tianshan quasi-stationary front
天体 celestial body, heavenly body
天体测量学【天】 astrometry
天体地质学【地质】 astrogeology
天体光谱学【物】 astrospectroscopy
天体生物学【天】 astrobiology
天体视宁度【天】 astronomical seeing
天体图 celestial chart
天体物理学等离〔子〕体【物】 astrophysical plasma
天体物理学【天】 astrophysics
天体演化学【天】 cosmogony
天体子午圈 celestial meridian
天体坐标 celestial coordinate
天王星【天】 Uranus
天文常数 astronomical constant
天文船位【航海】 astronomical fix, celestial fix
天文大地网【测】 astro-geodetic network
天文单位【天】 astronomical unit (AU)
天文导航【天】 astronavigation, celestial navigation
天文地平 astronomical horizon
天文定位【航海】 celestial fixing
天文定位系统【测】 astronomical positioning system
天文光学【物】 astronomical optics
天文经度【航海】 astronomical longitude
天文年历 almanac
天文-气候指标 astro-climatic index
天文气象学 astrometeorology

天文日 astronomical day
天文三角形【航海】 astronomical triangle
天文闪烁 astronomical scintillation
天文曙暮光 astronomical twilight
天文望远镜【物】 astronomical telescope
天文纬度【航海】 astronomical latitude
天文学【天】 astronomy
天文学定年法 astronomical dating
天文学与地球物理学联合会 Federation of Astronomical and Geophysical Services
天文因子 astronomical factor
天文钟【航海】 chronometer
天文坐标【航海】 astronomical coordinate
天线 antenna
天线方向图【电子】 antenna pattern
天线极限 antenna limit
天线馈电 antenna feed
天线〔收发〕转换开关 duplexer
天线温度 antenna temperature
天线增益 aerial gain, antenna gain
天线罩【电子】 radome
天线阵【电子】 antenna array
天象仪【天】 planetarium
天轴【航海】 celestial axis
田间持水量 field water-holding capacity
田间持水量【大气】 field capacity
田间管理【农】 field management
田间技术【农】 field technique
田间需水量 farm water requirement
田间需水量【农】 field water requirement
田间最大持水量 field maximum moisture capacity
田园城市【地理】 garden city
填充塔 packed tower
填塞 filling, filling up
填塞气旋 filling cyclone
填图 map spotting, map plotting,

plotting, spotting, chart plotting
填图板　plotting board
填图符号【大气】　plotting symbol
填图格式　plotting model, station model
填图员　chart plotter
填注【水利】　depression storage
填隙的云凝结核　interstitial CCN
CFL条件【大气】　CFL condition
条带记录器　strip chart recorder
条件保守性　conditional conservatism
条件对称不稳定〔性〕　conditional symmetric instability(CSI)
条件分布【数】　conditional distribution
条件概率【数】　conditional probability
条件均值【数】　conditional mean
条件气候学　conditional climatology
条件取样　conditional sampling
条件线性回归　conditional linear regression
条件〔性〕不稳定【大气】　conditional instability
条件〔性〕不稳定大气　conditionally unstable atmosphere
条件〔性〕不稳定的　conditional unstable
条件〔性〕平衡　conditional equilibrium
条件〔性〕稳定〔度〕　conditional stability
条件语句　conditional statement
条码扫描器【计】　bar code scanner
条码阅读器【计】　bar code reader
条纹　streak
条状闪电【大气】　streak lightning
调幅【电子】　amplitude modulation (AM)
调幅器【电子】　amplitude modulator
调幅指示器　amplitude-modulated indicator
调和　harmonic
调和乘积谱　harmonic product spectrum
调和分析【数】　harmonic analysis
调和分析器　harmonic analyser
调和函数【数】【大气】　harmonic function
调和解【数】　harmonic solution

调和平均【数】　harmonic mean
调和预测　harmonic prediction
调节　accommodation
调节器　regulator
调节系数　accommodation coefficient
调偏指示器　deflection-modulated indicator, amplitude-modulated indicator
调频　frequency modulation (FM)
调频连续波雷达　frequency-modulated continuous-wave radar(＝FM-CW radar)
调试故障　debugging failure
调相　phase time modulation
调相【物】　phase modulation
调优法　evolutionary method
调优功率谱　evolutionary power spectrum
调优功率谱密度　evolutionary power spectral density
调优谱　evolutionary spectrum
调整列线图　alignment chart
调整时间　adjustment time
调制　modulation
调制不稳定〔性〕　modulational instability
调制传递函数　modulation transfer function (MTF)
调制定理　modulation theorem
调制定理　modulation theorem
调制解调装置　modem
调制器　modulator
调准装置　regulating device
跳点格式【大气】　stagger scheme
跳点网格【大气】　staggered grid
跳跃　jump
跳跃矩阵　transilient matrix
跳跃模式　jump model
贴地雾　ground-hugging fog
萜【化】　terpenes
萜类化合物【林】　terpenoids
铁磁性【地质】　ferromagnetism
听度表　audibility meter

听阈　auditory threshold
停表风速表　anemometer with stop-watch
停滞冰川　stagnant glacier
停滞空气　stagnant air
停滞气团　stagnated air mass
停滞压　stagnation pressure
通带　pass-band
通道【大气】　channel
通道急流　channel jet
Q通道(正交通道)　Q channel (quadrature channel)
通风　ventilation
通风电容器　aspiration condenser
通风电容仪(测大气电导率用)　aspirated electrical capacitor
通风干湿表　ventilated psychrometer
通风干湿表【大气】　aspirated psychrometer
通风混合层空气　venting mixed layer air
通风量　mass rate of air flow
通风气象计　aspirated meteorograph, aspiration meteorograph【大气】
通风器　aspirator
通风石英晶体温度表　aspirated quartz-crystal thermometer
通风温度表　aspirated thermometer, ventilated thermometer
通风系数　ventilation coefficient
通航水域【航海】　navigable waters
通解【数】　general solution
通量【大气】　flux
通量比方法　flux-ratio method
E-P通量【大气】　Eliassen-Palm flux
通量调整　flux adjustment
通量订正　flux correction
通量订正输送　flux-corrected transport
通量发射率　flux emissivity
通量聚集　flux aggregation
通量廓线关系　flux-profile relationships
通量里查森数【大气】　flux Richardson number
通量密度　flux density
通量密度低限　flux density threshold
通量散度　flux divergency
通量透射比　flux transmittance
通量透射率　flux transmissivity
通量形式　flux form
通量仪　fluxatron
通气层　zone of aeration
通气孔　vent hole
通信【计】　communication
通信接口【计】　communication interface
通信软件【计】　communication software
通信通道　communication channel
通信网络　communication network
通信卫星　communication satellite (ComSat), telesat
通信站　communication station
通信中心　communication centre
通用法则　generic rule
通用分组无线电业务系统　general packet radio service (GPRS)
通用函数【数】　universal function
通用〔气象〕广播　General Broadcast
通用透射函数　universal transmission function
通用信息处理器　versatile information processor
通用雨量计　universal rain gauge
通知　advice
同步　synchror
同步【电子】　synchronization
同步变化　synchronous change
同步并行算法【计】　synchronized parallel algorithm
同步传输【计】　synchronous transmission
同步单光子计数　synchronous single-photoelectron counting
同步观测　simultaneous observation
同步轨道卫星　synchronous orbit

satellite (SOS)
同步脉冲选通光子计数　synchronous pulse-gated photon counting
同步气象卫星　synchronous meteorological satellite (SMS)
同步数据缓冲器　synchronizer/data buffer (S/DB)
同步探测　synchronous detection
同步条件　in-step condition
同步通信　synchronous communication
同步卫星导航系统【电子】　navigation system of synchronous satellite
同步遥相关　synchronous teleconnection（或 telecon-nexion）
同调【数】　homology
同调代数【数】　homological algebra
同调湍流　homologous turbulence
同方差性　homoscedasticity
同化〔作用〕　assimilation
同时出现线　isopipteses
同时气压　simultaneous pressure
同时温度　simultaneous temperature
同时物理反演法　simultaneous physical retrieval method
同时相关　contemporary correlation, simulation correlation
同时性　isochronism
同时性【物】　simultaneity
同位角【数】　corresponding angles
同位素【物】　isotope
同位素比率　isotope ratio
同位素测温法【地质】　isotopic thermometry
同位素地球化学【化】　isotope geochemistry
同位素地热温标【地质】　isotopic geothermometer
同位素地质年代学【化】　isotope geochronology
同位素地质学【化】　isotope geology
同位素分析　isotopic analysis
同位素丰度　isotope abundance
同位素丰度【物】　isotopic abundance
同位素古气候学　isotopic paleoclimatology
同位素交换反应【地质】　isotopic exchange reaction
同位素交换平衡【地质】　isotopic exchange equilibrium
同位素年代测定【化】　isotope dating
同位素平衡【地质】　isotopic equilibrium
同位素示踪物　isotopic tracer
同位素水文学【化】　inotope hydrology
同温层【大气】　stratosphere
同系光线【物】　homologous ray
同相　in phase
同相和 90°相差的通道　I and Q Channels(in-phase and quadrature channels)
同向线　isogonic line
同心光束【物】　concentric beam
同心双眼　two concentric eyes
同心透镜【物】　concentric lens
同心眼壁【大气】　concentric eyewalls
同心眼壁变化周期　concentric eyewall cycle
同心圆【数】　concentric circles
同形滤波〔器〕　homomorphic filter
同质结激光器【电子】　homojunction laser
同质结太阳电池【电子】　homojunction solar cell
同轴电缆【电子】　coaxial cable
同轴天线【电子】　coaxial antenna
酮　ketone
酮体【生化】　ketone body
统计参数　statistical parameter
统计插值法【大气】　statistical interpolation method
统计带模式【大气】　statistical band model
统计的　statistical

统计动力模式【大气】 statistical-dynamic model (SDM)
统计动力预报【大气】 statistical-dynamic prediction
统计独立〔性〕 statistical independence
统计反演法 statistical inversion method, statistical inversion technique
统计方法 statistical method
统计分析【数】 statistical analysis
统计功率 statistical power
统计假设【数】 statistical hypothesis
统计检验 statistical test
统计均匀的粗糙面 statistically homogeneous rough surface
统计控制法 statistical regulation
统计量【数】 statistic
统计模式【大气】 statistical model
统计内插 statistical interpolation
统计气候 statistical climate
统计气候学【大气】 statistical climatology
统计权重 statistical weight
统计热力学 statistical thermodynamics
统计释用 statistical interpretation
统计-数值预报 statistical-numerical prediction
统计天气预报 statistical weather forecast
统计同化法 statistical assimilation method
统计图 cartogram
统计物理〔学〕【物】 statistical physics
统计显著性检验 statistical significance test
统计相关〔性〕 statistical dependence
统计信息 statistical information
统计〔学〕【数】 statistics
统计预报【大气】 statistical forecast
统计资料 statistical data
统一场论【物】 unified field theory
投弃式温深仪【大气】 expendable bathythermograph (XBT)

投影【物】 projcetion
投掷焰弹 dropping flare
投掷焰弹系统 droppable pyrotechnic flare system
透辐射热性 diathermance, diathermancy
透光 translucidus (tra)
透光层 photic zone
透光层积云【大气】 stratocumulus translucidus (Sc tra)
透光层云 stratus translucidus (St tra)
透光带 euphotic zone
透光度监测仪 transmittance monitor
透光高层云【大气】 altostratus translucidus (As tra)
透光高积云【大气】 altocumulus translucidus (Ac tra)
透光率 light transmissivity
透光云量 transparent sky cover
透过率函数指数和拟合 exponential sum fitting of transmission function (ESFT)
透镜 lens
透镜组【物】 combination of lenses
透明冰 transparent ice
透明层 clear layer
透明〔的〕 transparent
透明度 transparency
透明度表 potometer
透明度系数 coefficient of transparency
透明系数 transparent coefficient
透明云 transparent cloud
透气率【土壤】 air permeability
透射【物】 transmission
透射比【物】 transmittance
透射表 transmittance meter, transmissormeter
透射测量技术 transmissormetry
透射函数 transmission function
透射矩阵 transmission matrix
透射距离 transmission range

透射率【物】 transmissivity
透射密度 transmission density
透射曲线 transmission curve
透射损耗 transmission loss
透射系数 coefficient of transmission
透射系数【物】 transmission coefficient
透视投影 perspective projection
透视网格 perspective grid
透雨 soaking rain
凸透镜【物】 convex lens
秃积雨云【大气】 cumulonimbus calvus (Cb calv)
秃状(云) calvus (calv)
突变【物】 catastrophe
突变报〔告〕 sudden change report
突变几何形 catastrophe geometry
突发电离层骚扰【地】 sudden ionospheric disturbance (SID)
突发雷暴 pop-up thunderstorm
突发阵风 sharp-edged gust
突然增温 sudden warming
图 figure, map, chart, graph
F图 Feynmann diagram
图标资料 mapped data
图集 atlas
图〔解〕 diagram
图解【物】 graphical solution
图解法【物】 graphic method, diagram method
图解静力学 graphical statics
图里潘辐射计 Tulipan radiometer
图例【测】 legend, map legend
图名【地信】 map name, map title
图片传真〔机〕 photofax
图森特公式 Toussaint's formula
图示 graphic representation
图示气候学 cartographical climatology
图示湿度表 hygrodeik
图式地面 patterned ground
图韦尔-汶布斯指数 Teweles-Wobus index
图文电视【电子】 teletext

图像 picture
图像【物】 imagery
图像比例尺 image scale factor
图像变换【物】 image transform
图像处理【大气】 image processing
图像处理系统 image processing system
图像存储系统【地信】 image storage system
图像导航 image navigation
图像对准 image alignment
图像分辨率 resolution of imagery
图像分辨率【大气】 image resolution
图像分割【计】 image segmentation
图像分析照相机系统 image dissector camera system (IDCS)
图像复合【地理】 image complex
图像复原【计】 image restoration
图像合成【农】 image composition
图像互相干函数 image mutual coherence function
图像几何测定【地理】 image geometry
图像劣化【物】 image degradation
图像模拟【地理】 image simulation
图像判读【农】 image interpretation
图像匹配【地信】 image matching
图像配准 image registration
图像平滑【地信】 image smoothing
图像平面扫描器 image plane scanner
图像清晰度【物】 image definition
图像识别【地理】 imagery recognition
图像释用统计 mainterpretation statistics (MIS), map interpretation statistics (MIS)
图像数字化【地理】 image digitizing
图像数字化【物】 image digitization
图像数字仪【天】 photo-digitizing system
图像投影变换【地理】 image projection transformation
图像网格定位 image gridding
图像压缩【计】 image compression
图像增强 image sharpening

图像增强【物】 image enhancement
图像增强器 image intensifier
图像重建【物】 image reconstruction
图像资料处理系统 image data processing system(IDAPS)
图形识别 pattern recognition
图形识别技术【大气】 pattern recognition technique
图型相关 pattern correlation
途中飞行时间 time en route
土坝 earth hummock, earth mound
土地处理系统【农】 land treatment system
土地分类【地理】 land classification
土地覆盖分类【农】 land cover classification
土地改良【地理】 land improvement
土地功能【地理】 land function
土地管理【地理】 land management
土地规划【农】 plan of land utilization
土地利用【地理】 land use
土地利用变化 land use change
土地利用分类【地理】 land use classification
土地利用类型【地理】 land use type
土地利用率【农】 land utilization rate
土地利用制度【地理】 land use system
土地评价【地理】 land evaluation
土地生产力【农】 productivity of land
土地生产率【地理】 land productivity
土地生产能力【地理】 land capability
土地特性【地理】 land characteristics
土地退化 land degradation
土地信息系统【地理】 land information system
土地要素【地理】 land element
土地质量【地理】 land quality
土地资源【农】 land resources
土力学【地质】 soil mechanics
土内水流 interflow
土壤【土壤】 soil

土壤表面温度【大气】 temperature of the soil surface
土壤不透水层 watertight stratum
土壤测湿法 soil hygrometry
土壤尘埃 soil dust
土壤传导率 soil conductivity
土壤带 soil zone
土壤单位体积干重 bulk density of soil
土壤导热率 thermal conductivity of soil
土壤导热系数 coefficient of soil thermal conductivity
土壤导温系数 coefficient of soil thermometric conductivity
土壤地理学【地理】 pedogeography
土壤地理〔学〕【土壤】 soil geography
土壤地球化学【土壤】 soil geochemistry
土壤发射率 soil emissivity
土壤肥力监测【土壤】 soil fertility monitoring
土壤肥力【土壤】 soil fertility
土壤腐殖质化学【土壤】 soil humus chemistry
土壤干旱 soil drought
土壤耕性【土壤】 soil tilth
土壤耕作【土壤】 soil tillage
土壤含水量【大气】 soil water content
土壤化学【土壤】 soil chemistry
土〔壤化育〕层 soil horizon
土壤环境【土壤】 soil environment
土壤环境容量【土壤】 soil environment capacity
土壤环境质量【土壤】 soil environment quality
土壤环境质量评价【土壤】 soil environment quality assessment
土壤积热仪 thermo-integrator
土壤碱化作用【土壤】 soil alkalization
土壤胶体化学【土壤】 soil colloid chemistry
土壤结构【土壤】 soil structure
土壤净化【土壤】 soil purification

土壤绝对湿度　absolute moisture of the soil
土壤〔绝对〕湿度【大气】　〔absolute〕soil moisture
土壤空气　soil air, soil atmosphere
土壤空气交换【土壤】　soil air exchange
土壤空气扩散【土壤】　soil air diffusion
土壤空气容量【土壤】　soil air capacity
土壤空气状况　soil air regime
土壤空气组成【土壤】　soil air composition
土壤孔〔隙〕度【土壤】　soil porosity
土壤矿物化学【土壤】　soil mineral chemistry
土壤利用【土壤】　soil utilization
土壤零温度层　zero curtain
土壤毛〔细〕管上升水　capillary rise of soil moisture
土壤模式　soil model
土壤钠质化作用【土壤】　soil sodication
土壤气候【大气】　soil climate
土壤气候条件　edapho-climatic condition
土壤气候学　soil climatology
土壤气体　soil gas
土壤侵蚀〔学〕【土壤】　soil erosion
土壤区划【土壤】　soil regionalization
土壤圈【地理】　pedosphere
土壤热通量　soil heat flux
土壤热通量〔测量〕板　soil heat flux plate
土壤热状况【土壤】　soil heat regime, soil thermal exchange
土壤蠕动【土壤】　soil creep
土壤生产力【土壤】　soil productivity
土壤生产力分级【土壤】　soil productivity grading
土壤生态系统【土壤】　soil ecosystem
土壤生物地球化学【土壤】　soil bio-geochemistry
土壤生物化学【土壤】　soil biochemistry
土壤湿度【土壤】　soil moisture
土壤湿度计　soil hygrograph, tensiometer
土壤湿度监测仪　soil humidity monitor
土壤湿度廓线　soil moisture profile
土壤实际密度　actual density of soil
土壤水分　water in soil, hygroscopic moisture, soil water
土壤水分差值　soil moisture deficit
土壤水〔分〕含量　soil moisture content
土壤水分平衡【大气】　soil water balance
土壤水分张力　soil moisture tension
土壤水势【大气】　soil water potential
土壤水文学【地理】　pedohydrology
土壤水压　soil water pressure
土壤酸度　soil acidity
土壤酸化作用【土壤】　soil acidification
土壤通气〔性〕【土壤】　soil areation
土壤退化　soil degradation
土壤温度【大气】　soil temperature
土壤温度表　soil depth thermometer
土壤温度表【大气】　soil thermometer
土壤温度计　soil thermograph
土壤污染【土壤】　soil pollution
土壤污染化学【土壤】　soil pollution chemistry
土壤物理〔学〕【土壤】　soil physics
土壤相对含水量　relative water content of soil
土壤相对湿度　relative moisture of the soil
土壤相对湿度【大气】　relative soil moisture
土壤学【土壤】　soil science, pedology
土壤氧化还原体系【土壤】　soil redox system
土壤有机质【土壤】　soil organic matter
土壤有效水分　available soil moisture
土壤原有机质【土壤】　native soil organic matter

土壤蒸发【大气】 soil evaporation
土壤蒸发【农】 evaporation from soil
土壤蒸发表 soil evaporimeter
土壤蒸发皿 soil evaporation pan, soil pan
土壤-植物-大气系统 soil-plant-atmosphere continuum (SPAC)
土壤质地【土壤】 soil texture
土壤中可用水分 available moisture of the soil
土壤资源 soil resources
土壤自净化功能【土壤】 soil self-purification function
土壤总含水量 holard
土石流 soil flow
土星【天】 Saturn
土宜【土壤】 soil feasibility
土族【土壤】 soil family
吐水【植】 guttation
钍射气【大气】 thoron (Th)
湍降冰川 cascading glacier
湍流【物】 turbulence, turbulent flow
湍流埃克曼层 turbulent Ekman layer
湍流半经验理论 semi-empirical theory of turbulence
湍流闭合 turbulence closure
湍流闭合法 turbulence closure scheme
湍流边界层【大气】 turbulent boundary layer
湍流层【地】 turbosphere
湍流层顶【地】 turbopause
湍流长度尺度 turbulence length scales
湍流场 turbulent field
湍流尺度 turbulent scale, scale of turbulence
湍流尺度应力 turbulent scale stress
湍流传输 turbulent transfer
湍流传输系数 turbulent transfer coefficients
湍流动能 turbulent kinetic energy, turbulence kinetic energy

湍流〔度〕 turbulency
湍流度 turbulence severity, turbulivity
湍流对流 turbulent convection
湍流分量 turbulence component
湍流惯性次区 turbulent inertial sub-range
湍流过程 turbulent process
湍流混合 turbulent mixing
湍流夹卷 turbulent entrainment
湍流间歇性 turbulent intermittency
湍流交换【大气】 turbulent exchange
湍流结构 turbulent structure
湍流扩散【大气】 turbulent diffusion
湍流扩散率 turbulent diffusivity
湍流扩散系数 turbulent diffusion coefficient
湍流雷诺数 turbulent Reynolds number
湍流雷诺应力 turbulence Reynolds stress
湍流 K 理论【大气】 K theory of turbulence
湍流脉动 turbulent fluctuation
湍流媒质 turbulent medium
湍流摩擦速度 turbulent friction velocity
湍流内尺度 inner scale of turbulence
湍流能量 turbulent energy
湍流能量【大气】 turbulence energy
湍流能量耗散率 turbulent energy dissipation rate
湍流能通量 turbulent energy flux
湍流逆温 turbulent inversion
湍流逆温【大气】 turbulence inversion
湍流凝结高度【大气】 turbulence condensation level
湍流谱 spectrum of turbulence
湍流谱【大气】 turbulence spectrum
湍流强度 turbulent intensity, turbulence intensity, intensity of turbulence
湍流切应力 turbulence shear stress, turbulent shear stress
湍流区 turbulent zone

湍流热输送　turbulent heat transfer
湍流声学　aero-thermo-acoustics
湍流时间尺度　turbulence time scale, turbulent time scale
湍流输运　turbulent transport
湍流速度尺度　turbulence velocity scale
湍流速度矢量　turbulent velocity vector
湍流体　turbulence body
湍流跳跃理论　transilient turbulenle theory
湍流通量【大气】　turbulent flux
湍流统计理论　turbulent statistical theory
湍流统计描述　statistical description of turbulence
湍流微尺度　turbulent microscale
湍流尾流　turbulent wake
湍流涡旋　turbulent vortex
湍流相似理论　turbulent similarity theory, similarity theory of turbulence【大气】
湍流消散　turbulent dissipation
湍流效应　turbulence effect
湍流烟流　turbulent plume
湍流因子　turbulence factor
湍流云【大气】　turbulence cloud
湍流运动　turbulent motion
湍流阵风　turbulent gust
湍流指数廓线　exponential profile of turbulence
湍涡【大气】　eddy
湍振边界【航空】　surge line
湍振【力】　surge
推测场　guess field
推荐航线　recommended route
推理【计】　inference
推理〔方〕法　rational method
推理规则　rules of inference
推理机　inference engine
推理机【计】　inference machine
推力风速表　thrust-anemometer
推论　deduction

推算航行法　dead reckoning
退波流　undertow
退潮　falling tide
退化　degeneracy, degeneration
退化【动物】　retrogression
退化双曲〔型〕方程【数】　degenerate hyperbolic equation
退化椭圆〔型〕方程【数】　degenerate elliptic equation
退化旋转潮波　degenerate amphidrome
退化状态　degenerate state
退回　retreater
退极化【物】　depolarization
退极化效应　depolarization effect
退偏振比【大气】　depolarization ratio
退水段　falling limb
退水曲线　depletion curve
退水曲线【地理】　recession curve
退行【大气】　retrogression
退行速度　velocity of recession
托管地区【地理】　trust territory
托里拆利管　Torricelli's tube
托里拆利真空　Torricelli's vacuum
拖板计程器　chip log
拖尾气球　lizard-balloon
拖曳锋　trailing front
拖曳气球　balloon drag
拖曳系数【大气】　drag coefficient
脱机【计】　off-line
脱机处理【计】　off-line processing
脱机输出　output spooling
脱硫【化】　desulfurization
脱硫系统　scrubber
脱氢【化】　dehydrogenation
脱水〔作用〕　dehydration
脱盐【农】　salinization
脱盐〔作用〕　desalination
脱氧【化】　deoxygenation
脱氧核糖核酸【生化】　deoxyribonucleic acid（DNA）
陀螺差【航海】　gyrocompass error

陀罗方位【航海】 gyrocompass bearing
陀罗航向【航海】 gyrocompass course
陀螺参考系 gyro reference system
陀螺光学导航 gyro erected optical navigation(GEON)
陀螺罗经【航海】 gyrocompass
陀螺仪【物】 gyroscope
椭圆 ellipse
椭圆轨道 elliptical orbit
椭圆函数【数】 elliptic function
椭圆积分【数】 elliptic integral
椭圆偏振 elliptic polarization
椭圆退偏振比 elliptical depolarization ratio
椭圆余弦波【物】 cnoidal wave
椭圆坐标 elliptical coordinates
拓扑结构 topological structure
拓扑〔学〕【数】 topology

W

挖泥船【航海】 dredger
蛙风暴 frog storm
蛙跳格式 frog leap scheme
蛙跳格式【数】 leap-frog scheme
蛙跃差分 leapfrog differencing
蛙跃法 leapfrog method
蛙跃式积分格式 leapfrog integration scheme
瓦劳特日光温度表 Vallot heliothermometer
瓦〔特〕 watt（W）
外暴流 outburst
外边界 outer boundary
外参数 external parameter
外层 outer layer
外层【化】 outer sphere
外层空间 deep space
外层空间【地】 outer space
外〔层〕涡旋 outer vortex
外差 heterodyne
外差探测 heterodyne detection
外场观测【大气】 field observation
外场试验 field experiment
外尺度 outer scale
外大气层 outer atmosphere
外弹道学 exterior ballistics
外迭代【数】 out iteration
外毒素【生化】 ectotoxin, exotoxin
外反射【物】 external reflection
外功 external work
外积〔向量积〕 outer product
外加宽 foreign broadening
外角【数】 exterior angle
外界梯度 environmental gradient
外扩散系数 outdiffusion coefficient
外力 external force
外力作用 external forcing
外连网【计】 extranet
外流 outflow
外流边界 outflow boundary
外〔流〕急流 outflow jet
外流流域 exorheic〔basin〕, exorheic region
外流螺旋(云或气流) outflow spiral
外流轴 axis of outflow
外罗斯贝变形半径 external Rossby radius of deformation
外貌 aspect
外模态 external mode
外气层 region of escape
外切晕 circumscribed halo
外强迫【大气】 external forcing
外区 outer region
外水分循环 external water circulation

外推【数】 extrapolation
外推法【大气】 extrapolation method
外围设备【计】 peripheral device, peripherals, peripheral equipment
外行星 outer planet
外行星【天文】 superior planet
外压力 external pressure
外延附生【物】 epitaxy(=epitaxis)
外眼墙 outer eyewall
外逸层 exoatmosphere
外〔逸〕层【大气】 exosphere
外逸层顶 exopause
外逸辐射 escape radiation
外逸速度 escape velocity
外逸速率 escape rate
外源起电 exogenous electrification
外源影响 exogenic influences
外展 abduction
外周环流 peripheral circulation
弯曲比 swirl ratio
弯曲等压线 curved isobar
弯液面 meniscus
湾流涡旋 Gulf Stream Rings
湾流系统 Gulf Stream System
湾流延伸 Gulf Stream Extension
蜿蜒路径 meandering course
完全电离等离子体 fully ionized plasma
完全冻结 complete freeze-up
完全辐射 perfect radiation
完全辐射体 perfect radiator
完全混合层 well-mixed layer
完全气体 ideal gas
完全气体【物】 perfect gas
完全气体定律 perfect-gas law
完全守恒格式【数】 complete conservation scheme
完全湍流的 fully turbulent
完全预报 perfect prognostic
完全预报【大气】 perfect prediction (PP)
完全预报法 perfectprog method (PPM),
perfect prognosis method
完整性 integrity
完整约束 holonomic constraint
顽〔磁〕力 coercive force
烷【化】 alkane
晚冰期气候 late glacial stage climate
晚材 latewood
晚春 deep spring
晚霜 late frost
晚霜【农】 spring frost
晚霞 sunset glow
碗状尘暴 dust bowl
万维网【计】 world wide web (WWW)
万亿分率(10^{-12}) ppt (parts per trillion)
万有引力常数 universal gravitational constant
万有引力定律 universal gravitational law
王水【化】 aqua regia
网格【数】 grid
网格北 grid north
网格尺度 mesh scale
网格地图【地理】 grid map
〔网〕格点法 grid point method
网格点【数】 grid point
网格分辨率 grid resolution
网格航行 grid navigation
网格嵌套 grid telescoping
网格系 grid system
网格子午线 grid meridians
网关【计】 gateway
网际协议【计】 internet protocol
网络地址【计】 net address
网络分析 network analysis
网络管理【计】 network management
网络航向 grid heading
网络偏差 grid variation
网络蠕虫【计】 network worm
网络软件 network software
网络新闻【计】 netnews, network news

网民【计】 net citizen, netizen
网罩蒸发皿 screened pan
网状〔云〕 lacunosus (la)
网状层积云 stratocumulus lacunosus (Sc la)
网状高积云 altocumulus lacunosus
网状卷积云 cirrocumulus lacunosus
往返测距站 turn around ranging station (TARS)
往复潮流 reversing tidal current
往复〔潮〕流 reversing current, rectilinear current
望【天】 full moon
望远镜 telescope
危害分析【农】 hazard analysis
危角 coffin corner
危险半圆 dangerous half
危险半圆【航海】 dangerous semicircle
危险区【航空】 danger area
危险天气通报【大气】 hazardous weather message
危险天气预警 hazardous weather warning
危险天气警报【大气】 severe weather warning
危险线(指洪水水位) danger line
危险象限【航海】 dangerous quadrant
危险信号 danger signal
危险指数 hazard index
威德尔深层水 Weddell Deep Water
威德尔旋涡边界 Weddell Gyre Boundary
威尔克斯 λ 统计量【数】 Wilks λ-statistic
威尔逊云室【物】 Wilson cloud chamber
威斯康星－玉米(冰期) Wisconsin-Wiirm
微变化 microvariation
微表层 microlayer
微波【物】 microwave
微波超声学 microwave ultrasonics
微波传播 microwave propagation

微波大气探测辐射仪 microwave atmospheric sounding radiometer (MASR)
微波大气遥感探测器 microwave atmospheric remote sensor
微波辐射 microwave radiation
微波辐射测量 microwave radiometric measurement
微波辐射仪【大气】 microwave radiometer
微波干燥【农】 microwave drying
微波高度计 microwave altimeter
微波-红外探测 microwave-infrared sounding
微波后向散射 microwave backscattering
微波激射【物】 maser
微波激射器 microwave amplification by stimulated emission of radiation (MASER)
微波雷达 microwave radar
微波气象学 microwave meteorology
微波散射计 microwave scatterometer
微波扫描波谱仪 scanning microwave spectrometer (SCAMS)
微波探测 microwave probing
微波探测装置 microwave sounding unit (MSU)
微波图像【大气】 microwave image
微波温湿廓线〔探测〕仪 microwave temperature-humidity profiler
微波遥感【农】 microwave remote sensing
微波折射计 microwave refractometer
微尘学 koniology, coniology
微尺度〔天气〕系统 microscale weather system
微尺度湍流 microscale turbulence
微尺度效应 microscale effect
微带天线【电子】 microstrip antenna
微滴 droplet
微滴学说 drop theory
微电子学【物】 microelectronics
微动气压器 statoscope

微分【数】 differential
微分奥米加 Differential Omega
微分持水量 differential water capacity
微分反射率 differential reflectivity
微分方程【数】 differential equation
微分分析【大气】 differential analysis
微分分析仪 differential analyser
微分光学吸收 differential optical absorption
微分后向散射截面 differential back-scattering cross section
微分截面 differential cross section
微分迁移率分析仪 differential mobility analyzer
微分散射截面 differential scattering cross section
微分衰减 differential attenuation
微分算符 derivative operator
微分算子 differential operator
微分同胚 diffeomorphism
微分图 differential chart
微分吸收 differential absorption
微分吸收法【大气】 differential absorption technique
微分吸收激光雷达【大气】 differential absorption lidar（DIAL）
微分吸收光〔雷〕达温度表 differential absorption lidar thermometer
微分吸收和散射 differential absorption and scattering（DAS）
微分吸收技术 differential absorption technique
微分吸收截面 differential absorption cross section
微分吸收湿度计【大气】 differential absorption hygrometer
微分吸收应用 differential absorption application
微分系数 differential coefficient
微分相函数 differential phase function
微分相〔位漂〕移 differential phase shift

微分运动学 differential kinematics
微风【大气】 gentle breeze
微封闭技术 microencapsulation technique
微锋 microfront
微功率日照检测仪 micro-power sunshine detector
微观〔的〕 microscopic
微观地理学 microgeography
微观物理过程 microphysical process
微观物理特性 microphysical property
微观物理学 microphysics
微观物理学参数化 microphysics parameterization
微机数据采集加工系统【农】 micro-computer data acquisition and processing system
微机信息系统【农】 microcomputer information system
微积分〔学〕【数】 calculus
微结构【物】 microstructure
微浪 smooth sea
微粒 corpuscle, particulate
微粒辐射 corpuscular radiation
微粒流 corpuscular stream
微粒热源 corpuscular heat source
微粒说 emission theory
微粒说【物】 corpuscular theory
微粒污染物 particulate pollutant
微粒宇宙线 corpuscular cosmic ray
微量分析【化】 micro analysis
微量元素【农】 microelement
微脉动【地】 micropulsation
微米 micron
微黏度 microviscosity
微气象计 micrometeorograph
微气象学【大气】 micrometeorology
微〔气〕压计 microbarovariograph
微气压计 micro-pressure gauge
微〔气〕压记录图 microbarogram
微气压图 microbarm

微扰【物】 perturbation
微扰法 method of (small) perturbation
微扰技术 perturbation technique
微扰论【物】 perturbation theory
微商【数】 derivative, differential quotient
微生态气候 ecidioclimate
微生物学 microbiology
微涡 microvortex
微物理模式 microphysical model
微下击暴流【大气】 microburst
微下击暴流线 microburst line
微〔信息〕处理机 microprocessor
微星假说 planetesimal hypothesis
微型电流 microcircuit
微〔型计算〕机 microcomputer
微压表 micromanometer
微压计【大气】 microbarograph
微应力 microstress
微雨 dribble
微雨量计 trace recorder, ombrograph
微雨量器 micropluviometer, ombrometer
微阵雨 sprinkle
微振动 microtremor
微震 microseisms, tremor
微震【地】 microearthquake
微震仪 microseismograph
韦〔伯〕 weber (Wb)
韦伯-费希纳定律 Weber-Fechner law
韦伯数 Weber number
韦布尔测绘点 Weibull plotting position
韦布尔分布【数】 Weibull distribution
韦格纳-〔贝吉龙〕-芬德森理论 Wegener-〔Bergeron〕-Findeisen theory
韦格通风离子测定仪 Weger aspirator
围圩 polder
帷幕〔状〕极光 dramundan
维持 maintenance
维恩单波分配定律 Wien's distribution law
维恩定律 Wien's law

维恩辐射定律 Wien's law of radiation
维恩位移律【物】 Wien's displacement law
维尔德蒸发器 Wild's evaporimeter
维甘德视程计 Wigand visibility meter
维纳-霍普夫积分方程【数】 Wiener-Hopf integral equation
维生作用 vital role
伪彩色【电子】 pseudo color
伪彩色处理显示 pseudo-colour enhancement
伪彩色图像 pseudocolor image
伪彩色云图【大气】 pseudo-color cloud picture
伪彩色增强 pseudocolor enhancement
伪锋 false front, fictitious front
伪卷云【大气】 cirrus nothus (Ci not)
伪谱 pseudo-spectrum
伪谱近似 pseudospectral approximation
伪随机数【数】 pseudo-random number
伪相关 spurious correlation
伪噪声 pseudo noise
伪噪声码【电子】 pseudo noise code
伪重力波 spurious gravity-wave
伪装目标云 decay cloud
尾波 coda wave
尾低压 wake low
尾迹 trail
尾迹波【物】 trailing wave
尾流 wake stream, wake【大气】
尾流捕捉 wake capture
尾流低压【大气】 wake depression
尾流环流 circulation in wake
尾流湍流 wake turbulence
尾流效应 wake effect
尾流轴 wake axis
尾气 end gas
尾随台风 wake-following typhoon
尾涡 wake vortex
尾涡度 wake vorticity

尾翼截面（风向标的） airfoil section
尾云 tail cloud
尾状卷云 cirrus caudatus
尾追龙卷 chasing tornado
纬度 latitude
纬度-时间分布 latitude-time distribution
纬度-时间剖面 latitude-time section
纬度效应 latitude effect
纬圈平均动能 zonal kinetic energy
纬圈平均有效位能 zonal available potential energy
纬向变化 longitudinal variation
纬向波数 zonal wavenumber
纬向的 zonal
纬向〔的〕 longitudinal
纬向动量方程 zonal momentum equation
纬向非对称加热 longitudinally asymmetric heating
纬向分布 zonal distribution
纬向风【地】 zonal wind
纬向风廓线 zonal wind profile
纬向风速廓线 zonal wind-speed profile
纬向环流【大气】 zonal circulation
纬向能量 zonal energy
纬向平均 zonal mean
纬向平均模式 zonally averaged models
纬向平均运动 zonal mean motion
纬向气候温度 zonal climatologic temperature
纬向气流 zonal flow
纬向气压梯度 zonal pressure gradient
纬向西风带 zonal westerlies
纬向指数 zonal index
萎蔫【植物】 wilting
萎蔫点【植物】【大气】 wilting point
卫生气象学 heigyne meteorology, hygieno-meteorology
卫星【天】 satellite
卫星温度 satellite temperature
（SATEM）
卫星白天 satellite day (SD)
卫星测风 wind measurement from satellite
卫星测高 satellite altimetry
卫星城 satellite town
卫星大地测量学 satellite geodesy
卫星〔大气〕密度计 satellite density gauge
卫星导航系统 satellite navigation system
卫星导航仪 satellite navigator
卫星地面接收站 earth station
卫星对地观测委员会 Committee on Earth Observation Satellites (CEOS)
卫星多普勒定位 satellite Doppler positioning
〔卫星〕二次资料利用站 secondary data utilization station (SDUS)
卫星风 satellite winds
卫星风估计 satellite wind estimate
卫星辐射率 satellite radiance
卫星覆盖区【航海】 satellite coverage
卫星跟踪天线 satellite tracking antenna
卫星观测 satellite observation (SATOB)
卫星光〔雷〕达 satellite laser radar
卫星光〔雷〕达探测 satellite laser radar probing
卫星轨道 satellite orbit
卫星海洋学 satellite oceanography
卫星和中尺度气象研究计划 Satellite and Mesometeorological Research Project (SMRP)
卫星黑夜 satellite night (SN)
卫星红外辐射光谱仪 satellite infrared radiometer spectrometer (SIRS)
卫星红外光谱仪 satellite infrared spectrometer (SIRS)
卫星红外光谱仪电源 SIRS instrument power supply (SIPS)
卫星红外光谱仪控制组件 SIRS control unit module (SCUM)

卫星监测　satellite monitoring
卫星判释信息　satellite interpretation message（SIM）
卫星气候学【大气】　satellite climatology
卫星气象学【大气】　satellite meteorology
卫星区域站　satellite field service station（SFSS）
卫星闪电探测器　satellite lightning sensor
卫星摄动轨道【航海】　satellite disturbed orbit
卫星寿命　lifetime of satellite
卫星数字电视广播　digital video broadcast-satellite（DVB-S）
卫星探测【大气】　satellite sounding
卫星探测反演【大气】　inversion of satellite sounding
卫星探空（资料）　SATEM
卫星天顶角　satellite zenith angle
卫星天气报告　satellite weather bulletin
卫星天气信息系统　satellite weather information system（SWIS）
卫星通信【电子】　satellite communication
卫星图片接收器　satellite photo receiver
卫星图像热带气旋分类　tropical cyclone clssification system from satellite imagery
卫星温度反演　satellite temperature retrieval
〔卫星〕星历　satellite ephemeris
〔卫星〕星下点【大气】　sub-satellite point（SSP）
卫星遥感探测　satellite remote sensing observation
〔卫星〕一次资料利用站　primary data utilization station（PDUS）
卫星影像　satellite image
卫星影像导航　satellite image navigation
卫星云迹风【大气】　satellite derived wind
卫星云迹风　satellite cloud-tracked wind
卫星云图【大气】　satellite cloud picture
卫星云图传真收片机　satellite picture recorder
卫星云图导航　satellite picture navigation
卫星云图分析【大气】　satellite cloud picture analysis
卫星运载火箭　satellite launching vehicle（SLV）
卫星照片　satellite photograph
卫星照片判读　satellite photographic study
卫星中继的天气资料收集平台　satellite weather data collection platform
卫星转播资料　satellite relay of data
卫星姿态　attitude of satellite
卫星资料剖析　satellite perspective
未饱和　unsaturation
未饱和导水性　unsaturated hydraulic conductivity
未饱和空气　non-saturated air, unsaturated air
未饱和气流　unsaturated flow
未饱和区　unsaturated zone
未饱和下沉气流　unsaturated downdraft
未滤波模式　unfiltered model
未扰〔动〕大气　unperturbed atmosphere
未扰〔动〕模式　unperturbed model
未扰动太阳状况　undisturbed solar condition
未失真的　undistorted
未污染的大气　non-contaminated atmosphere
未知层　ignorosphere
位密度　potential density
位能【大气】　potential energy
位容　potential vloume
位矢　position vector
位势不稳定【大气】　potential instability
位势不稳定指数　potential instability index
位势场　geopotential field
位势〔等高〕图　geopotential topography
位势高度【大气】　geopotential height
位势厚度　geopotential thickness
位〔势〕流　potential flow

位势米【大气】 geopotential meter
位势拟能 geopotential enstrophy
位势趋势方程 geopotential tendency equation
位势梯度 potential gradient
E 位势涡度 Ertel potential vorticity
位势涡度【大气】 potential vorticity
位势涡度方程 potential vorticity equation
位势英尺 geopotential foot
位速率 bit rate
位温【大气】 potential temperature
位温高度图 thetagram
位温梯度 potential temperature gradient
位温坐标 potential temperature coordinate
位温坐标【大气】 θ-coordinate
位温坐标系(θ坐标系) theta coordinate
位涡【大气】 potential worticity
位涡度拟能 potential enstrophy
位涡度拟能守恒格式 potential enstrophy conserving scheme
位涡守恒【大气】 conservation of potential vorticity
位相谱【大气】 phase spectrum
位形【物】 configuration
位移 displacement, judder
位移电流 displacement current
位移高度 displacement height
位移厚度 displacement thickness
位移距离 displacement distance
位折射率 potential refractive index, potential index of refraction
位蒸散比 potential evapotranspiration ratio
位置 site
位置显示器 position indicator
位置线 line of position (L. O. P.)
畏寒植物 frigofuge
温标 temperature scale, scale of thermometer, thermometric scale
温差电效应【物】 thermoelectric effect

温差商数 temperature-difference quotient
温带 extratropical belt (=zone)
温带【大气】 temperate belt
温带【地理】 temperate zone
温带冰川 temperate glacier
温带低压 extra-tropical low
温带冬雨气候 temperate climate with winter rain
温带对流层顶 extratropical tropopause
温带多雨气候 temperate rainy climate
温带风暴 extra-tropical storm
温带环流 extra-tropical circulation
温带环流数值模式 extra-tropical circulation numerical model
温带局强风暴 extra-tropical severe local storm
温带气候【大气】 temperate climate
温带气旋【大气】 extratropical cyclone
温带西风带【大气】 temperate westerlies
温带西风带指数 temperate westerlies index
温带夏雨气候 temperate climate with summer rain
温带行星波 extratropical planetary wave
温带雨林 temperate rainforest
温度 temperature
温度变程 march of temperature
温度变化 temperature variation
温度表【大气】 thermometer
温度表订正〔值〕 thermometer correction
温度表防辐射罩 radiation shield for thermometer
温度表架 thermometer support
温度表球部 thermometer bulb
温度表柱 thermometer column
温度波 temperature wave
温度补偿 temperature compensation
温度槽 thermal trough

温度场【大气】 temperature field
温度垂直廓线辐射仪【大气】 vertical temperature profile radiometer (VTPR)
温度垂直梯度 vertical temperature gradient
温度大陆度 thermal continentality
温度带 temperature belt, temperature zone
温度〔的〕平流变化 advective change of temperature
温度订正 temperature correction【大气】, temperature reduction
温度动力变化 dynamic temperature change
温度对比 temperature contrast
温度对数气压斜交图表 skew T-log P diagram
温度-对数压力图【大气】 T-ln$p$ diagram
温度反演【大气】 temperature retrieval
温度方差 temperature variance
温度高度图 temperature altitude chart
温度基准探空仪 temperature reference sonde
温度极值 temperature extremes
温度脊 thermal ridge
温度计【大气】 thermograph
温度计订正表 thermograph correction card
温度记录器 thermometrograph
温度记录仪 temperature recorder
温度较差【大气】 temperature range
温度结构函数 temperature structure function
温度结构系数 temperature structure coefficient
温度距平 temperature anomaly
温度绝对年较差 absolute annual range of temperature
温度控制 temperature control
温度廓线【大气】 temperature profile
温度廓线记录器 temperature profile recorder (TPR)
温度露点差 temperature-dewpoint spread
〔温度〕露点差【大气】 depression of the dew point
温度脉动 temperature fluctuation
温度平流 thermal advection
温度平流【大气】 temperature advection
温度谱 spectrum of temperature
温度区域 temperature province
温度日际变化 interdiurnal temperature variation
温度日较差【大气】 daily range of temperature
温度随时间的变化 temperature history
温度特定死亡率比 temperature-specific mortality ratio
温度梯度【大气】 temperature gradient
温度微尺度 temperature microscale
温度系数 temperature coefficient
温度效率 temperature (或 thermal) efficiency
温度效应 temperature effect
温度盐度计 thermosalinograph
温度月际变化 inter-monthly temperature variation
温度跃层 epilimnion
温度直减率【大气】 temperature lapse rate, lapse rate of temperature
温〔度〕滞〔后〕 thermal hysteresis
温度自记曲线【大气】 thermogram
温高图 pastagram
温和期 miothermic period
温和气候 mild climate
温和天气 mild weather
温量指数 warmth index
温暖指数 index of warmth
温泉 hot spring
温熵图【大气】 tephigram
温熵图【化】 temperature-entropy chart

温深仪【大气】 bathythermograph (BT)
温深仪玻〔璃〕片 bathythermograph slide
温深仪〔读数〕网格 bathythermograph grid
温深仪复制记录 bathythermograph print
温湿表 hygrothermometer, thermohygrometer【大气】
温湿计【大气】 thermohygrograph
温湿度 humiture
温湿红外辐射仪 temperature humidity infrared radiometer (THIR)
温湿计 hygrothermograph, thermohygrograph
温湿图 hythergraph【大气】, temperature-moisture diagram
温湿仪【大气】 hygrothermoscope
温湿指数 temperature-moisture index, temperature humidity index(THI), moisture-temperature index
温湿自记曲线【大气】 thermohygrogram
温湿综合图 thermohyet diagram
温湿作用 hyther
温时曲线 time-temperature curve
温室 solar house, greenhouse
温室二氧化碳加浓【农】 greenhouse carbon dioxide enrichment
温室管理【农】 greenhouse management
温室加热【农】 greenhouse heating
温室气候 glass-house (=greenhouse) climate, greenhouse climate【大气】
温室气体【大气】 greenhouse gases (GHG)
温室气体排放 greenhouse gas emission
温室气体稳定【化】 greenhouse gas stabilization
温室强迫〔作用〕 greenhouse forcing
温室通风【农】 greenhouse ventilation
温室效应 hothouse effect

温室效应【大气】 greenhouse effect
温室〔效应〕增温 greenhouse warming
温室状态 greenhouse state
温效比 temperature-efficiency ratio (T-E radio), thermal-efficiency ratio
温效指数 temperature-efficiency index (T-E index)
温压表 thermobarometer
温压计 thermobarograph
温压曲线 temperature pressure curve
温盐关系 T-S relation
温盐环流【海洋】 thermohaline circulation (THC)
温盐曲线 temperature-salinity curve, T-S curve
温盐图 temperature-salinity diagram, T-S diagram
温跃层以下海层 subthermocline
温周期 thermoperiodicity, thermoperiod
温周期〔性〕【大气】 thermoperiodism
文恩图 Venn diagram
文件【计】 file
文件存储器【计】 file memory
文件存取【计】 file access
文件分配【计】 file allocation
文件名【计】 file name
文件目录【计】 file directory
文图里飞机温度表 Venturi aircraft thermometer
文图里效应 Venturi effect
文献记录时期 Documental period
文字说明 descriptive text
纹理 striation, texture
纹线 streakline
闻阈 threshold of audibility, threshold of hearing
吻合度 goodness of fit
稳定 stationary
稳定本〔机〕振〔荡〕器 stalo (stable local oscillator)
稳定本机振荡器 stable local oscillator

稳定边界层　stable boundary layer，（SBL）
稳定边界层强度　SBL strength
稳定波　stable wave
稳定层结　stable stratification
稳定产量　firm yield
稳定大气　stable atmosphere
稳定度　degree of stability
稳定度参数【大气】　stability parameter
稳定度长度　stability length
稳定度分类　stability categories
稳定度理论　stability theory
稳定〔度〕判据　stability criterion
稳定〔度〕条件　stability condition
稳定度图　stability chart
稳定度指数　stability index
稳定度指数【大气】　index of stability
稳定风　constant wind
稳定风暴　steady state storm
稳定风压【大气】　steady wind pressure
稳定河槽　regime channel
稳定节点　stable node
稳定空气【大气】　stable air
稳定控制　permanent control
稳定廓线　stable profile
稳定力　stabilizing force
稳定流形　stable manifold
稳定片〔流〕层　stable lamina
稳定平衡态　stable equilibrium
稳定气团【大气】　stable air mass
稳定天线平台　stable antenna platform
稳定系统　stabilization system
稳定型　stable type
稳定性【物】　stability
稳定性分析【物】　stability analysis
稳定因子　stabilizing factor
稳定运动　stable motion
稳定振荡　stable oscillation
稳定作用　stabilization
稳健回归　robustness regression
稳健控制　robust control
稳健性【数】　robustness
稳水器　still-well gauges
稳压电源【计】　constant voltage power supply
涡层　vortex sheet
涡动【大气】　eddy
涡动场　eddy field
涡动传导　eddy conduction
涡动传导率【大气】　eddy conductivity
涡动传导系数　eddy conduction coefficient
涡动传输系数　eddy transfer coefficients
涡动动量输送　eddy momentum transfer
涡动动量通量　eddy momentum flux
涡动动能【大气】　eddy kinetic energy
涡动方程　eddy equation
涡动交换系数　eddy exchange coefficient
涡动扩散【大气】　eddy diffusion
涡动扩散率　eddy diffusivity
涡动扩散系数　eddy diffusion coefficient
涡动连续　eddy continuum
涡动龙卷　eddy tornado
涡动摩擦　eddy friction
涡动能量　eddy energy
涡动拟能【大气】　enstrophy
涡动黏滞率【大气】　eddy viscosity
涡动平流【大气】　eddy advection
涡动切应力【大气】　eddy shearing stress
涡动热通量　eddy heat flux
涡动输送　eddy transport
涡动水汽通量　eddy moisture flux
涡动速度　eddy velocity
涡动通量【大气】　eddy flux
涡动通量参数化　eddy flux parameterization
涡动通量仪　evapotron
涡动系数　eddy coefficient
涡动相关【大气】　eddy correlation
涡动相关法　eddy correlation method
涡动相关蒸发传感器　eddy-correlation evaporation sensor

涡动协方差　eddy covariance
涡动应力　eddy stress
涡动应力张量　eddy-stress tensor
涡动有效位能　eddy available potential energy
涡动阻力　eddy resistance
涡度【大气】　vorticity
涡度瓣　vorticity lobe
涡度场　field of vorticity
涡度方程　equation of vorticity
涡度方程【大气】　vorticity equation
涡度路径　vorticity path
涡度拟能【大气】　enstrophy
涡度拟能模　enstrophy norm
涡度拟能守恒方案　enstrophy conserving scheme
涡度平流【大气】　vorticity advection
涡度强迫　vorticity forcing
涡度守恒　conservation of vorticity
涡度输送　vorticity transfer
涡度输送假设　vorticity transport hypothesis
涡度输送理论　vorticity transport theory
涡度中心　vorticity center
涡管　vortex tube
涡管露点湿度表　vortex dew-point hygrometer
涡管温度表　vortex thermometer
涡汇　eddy sink, vortex sink
涡卷　vortex roll
涡列　vortex train
涡流　eddy flow, vortex flow, eddy current
涡流累积〔法〕　eddy accumulation
涡流系数　eddy current coefficient
涡流效应　eddy current effect
涡偶　modon, vortex pair
涡谱　eddy spectrum
涡旋【大气】　eddy, vortex
涡旋崩溃　vortex breakdown
涡旋尺度　eddy size, vortex scale

涡旋传导系数　coefficient of eddy conduction, eddy conductivity
涡旋轨迹　vortex trail
涡旋合并　vortex merging
涡〔旋〕环　vortex-rings
涡旋结构　eddy structure
涡旋扩散系数　coefficient of eddy diffusion, eddy diffusivity
涡旋冷却管　vortex cooling tube
涡〔旋〕列　vortex streets
涡旋罗斯贝波　vortex Rossby waves
涡旋黏滞系数　coefficient of eddy viscosity
涡旋伸展　vortex stretching
涡旋式风速传感器　vortex speed sensor
涡旋收缩　vortex shrinking
涡〔旋〕丝　vortex filament
涡旋特征　vortex signature
涡〔旋〕线　vortex line
涡旋曳离　vortex shedding
涡旋雨　vortex rain
涡旋云街　vortex cloud street
涡旋云系【大气】　vortex cloud system
涡旋运动　eddy motion, vortex motion
涡旋状回波【大气】　whirling echo
涡源　eddy source, vortex source
沃尔夫数【天】　Wolf number
沃尔什变换【物】　Walsh transform
沃尔什序贯谱　Walsh sequential spectrum
沃尔泰拉积分方程【数】　Volterra integral equation
沃克环流　Walker circulation
沃克环流〔圈〕【大气】　Walker cell
沃克曼-雷诺效应　Workman-Reynolds effect
沃拉斯顿棱镜【物】　Wollaston prism
沃斯混浊因子　Volz turbidity factor
沃斯偏光器　Voss polariscope
沃土　loam
沃伊特廓线　Voigt profile

握索结【航海】 manrope knot
乌拉尔山阻塞高压【大气】 Ural blocking high
污染【农】 pollution, contamination
污染病 pollution disease
污染当量 pollutional equivalent
污染剂量【地理】 pollution dose
污染监测 pollutant monitoring
污染空气 polluted air
污染控制 pollution control
污染气象学【大气】 pollution meteorology
〔污染〕事件标准 episode criteria
污染水平【农】 pollution level
污染物 contaminant, pollutant
污染物排放 pollutant emission
污染物迁移 pollutant transport
污染〔物〕通量 pollution flux
污染物指数 pollutant index
污染物质降解【水利】 degradation of pollutant
污染系数【石油】 damage factor
污染预报 pollution prediction
污染预警系统 pollution warning system
污染源【地质】 pollution source
污染源排放物 source emission
污染源现场监测 in situ source monitoring
污染指数【地质】 pollution index
污水灌溉 wastewater irrigation
污水灌溉【土壤】 sewage irrigation
污水井 sinkhole
污水排放限制 effluent limitations
无冰区【航海】 ice free
无潮点 amphidromic point, amphidrome
无潮区 amphidromic region
无尘大气 dust-free atmosphere
无磁等离子体 unmagnetized plasma
无代表性的 non-representative
无定向〔的〕 astatic
无定向风 baffling wind
无定形雪 amorphous snow

无定形云 amorphous cloud
无定状高积云 altocumulus informis
无定状（云） informis
无冻期 freeze free period
无风层 calm layer
无风带 calm zone, calm belt【大气】, belt of calms
无风眼 calm central eye
无锋飑线 non-frontal squall line
无缝钢管 seamless steel pipe
无缝合成橡胶气球 synthetic seamless rubber balloon
无缝〔隙〕预报 seamless forecast
无辐散层 level of non-divergence
无辐散层【大气】 non-divergence level
无辐散风〔分量〕 non-divergent wind (component)
无辐散流 non-divergent flow
无辐散模式 nondivergent model
无辐散运动【大气】 nondivergent motion
无〔覆〕水冰 dried ice
无功功率【物】 reactive power
无规高频型天电 irregular high-frequency type atmospherics
无规相〔位〕近似【物】 random phase approximation
无滑动边界条件 no-slip boundary condition
无滑动条件 no-slip condition
无回波穹 echo-free vault
无机化学【化】 inorganic chemistry
无畸变滤波〔器〕 distortionless filter
无浪（风浪零级） calm sea
无量纲 dimensionless
无量纲参数 non-dimensional parameter
【大气】dimensionless parameter
无量纲方程 non-dimensional equation
【大气】dimensionless equation
无量纲化【物】 nondimensionalization
无量纲量 non-dimensional quantity

无量纲谱　non-dimensional spectrum
无量纲数　non-dimenstional number, dimensionless number
无量纲数群【化工】　dimensionless group
无量纲坐标　non-dimensional coordinate
无流的　arheic(＝areic, aretic, arhetic)
无碰撞等离子体激波　collissionless plasma shock
无偏估计【数】　unbiased estimate
无偏检验【数】　unbiased test
无倾角线【大气】　aclinic line
无穷〔的〕【数】　infinite
无穷小【数】　infinitesimal
无扰运动　undisturbed motion
无人飞行器　unmanned spacecraft
无人回收运载工具　unmanned recovery vehicle
无人科学卫星　unmanned scientific satellite
无人区　Empty Quarter
无散射大气【大气】　non-scattering atmosphere
无升力气球　non-lift balloon
无释方差　unexplained variance
无霜期【大气】　duration of frost-free period
无霜区　no-frost zone
无霜日【农】　frost free day, day without frost
无条件回归　unconditional regression
无限连分数　infinite continued fraction
无限远能见度　unrestricted visibility
无限云幂高　unlimited ceiling
无线电波　radio wave
无线电波传播　radio wave propagation
无线电波导　radio duct
无线电波的云层吸收　radio wave absorption by clouds
无线电波射线　radio wave ray
无线电波射线曲率　radio wave ray curvature
无线电波探空　radio-wave sounding
无线电测风　radio wind finding
无线电测风【大气】　radio wind observation
无线电测风靶　rawin target
无线电测风观测【大气】　radiowind observation
无线电测风气球　radio pilot〔balloon〕
无线电测风仪　rawin, radio wind finding
无线电测风站　radiowind station
无线电测距　radio-range finding
无线电测距接收机　radio-range receiver
无线电测距器　radio distance finder
无线电测向　radio direction finding (RDF)
无线电测向计　radio goniometer
无线电测向雷达　radio direction finding radar
无线电测向仪　radio goniograph
无线电传送　radio transmission
无线电传真变换器　radiofacsimile converter
无线电传真〔照片〕　radiophotograph
无线电导航测向器　radio range goniometer
无线电导航【船舶】　radio navigation
无线电导航设备　radio navigation aid
无线电地平　radio horizon
无线电定位器　radiolocator
无线电定向　radiobearing
无线电定向法　radio goniometry
无线电定向仪　radio direction finder
无线电洞　radio hole
无线电浮标【电子】　radio-beacon buoy
无线电高度表【航空】　radio altimeter
无线电跟踪测风气球　radio-tracked pilot balloon
无线电海洋学　radio oceanography
无线电〔航向〕信标　radio frequency

range
无线电经纬仪【大气】 radio theodolite
无线电经纬仪记录式接收机 radio theodolite recording receiver
无线电流星 radio meteor
无线电罗盘【电子】 radio compass
无线电能 radio energy
无线电气候学 radio climatology
无线电气象计 radiometeorograph
无线电气象学【大气】 radio meteorology
无线电气象学协会间委员会 Inter-Union Commission on Radio-Meteorology(IUCRM)
无线电气象业务【航海】 radio weather service
无线电哨声 radio whistler
无线电射束 radio beam
无线电蜃景 radio mirage
无线电-声学探测系统 radio acoustic sounding system（RASS）
无线电时号【航海】 radio time signal
无线电衰减 radio fadeout
无线电探空 radio-sondage
无线电探空【大气】 radiosounding
无线电探空测风观测 rawinsonde observation
无线电探空测风系统 radiosonde-radiowind system
无线电探空测风仪【大气】 rawinsonde
无线电探空测风站 radiosonde and radiowind station（RWS），rawinsonde station
无线电探空气球 rabal
无线电探空仪 radiometer sonde
无线电探空仪【大气】 radiosonde
〔无线电〕探空〔仪〕观测〔记录〕 RAOB（radiosonde observation）
无线电探空〔仪〕接收记录器 radiosonde receiver-recorder
〔无线电〕探空仪系统 radiosonde system
无线电探空站 radiosonde station

无线电天气预警系统 radio weather warning system
无线电通信中断 radio blackout
无线电信标【航海】 radio beacon
〔无线电〕信号强度 signal strength
无线电寻呼【电子】 radio paging
无线电掩星法 radio occultation technique
无线电云幂器 radio ceiling
无线电噪声【铁道】 radio noise
无线电窄束定向器 narrow beam radiogoniometer
无线电折射仪【大气】 radio-refractometer
无线电追踪 radio tracking
无线线路【航海】 radio link
无效 invain
无效温度 ineffective temperature
无形飑 white squall
无旋场 nonrotational field
无旋〔的〕 irrotational
无旋流 irrotational flow
无旋向量 irrotational vector
无旋运动【大气】 irrotational motion
无旋〔转〕风（分量） irrotational wind（component）
无压地下水 phreatic water
无烟区 smokeless zone
无液压力计 aneroid manometer
无意识气候影响 inadvertent climate modification
无意识天气影响 inadvertent weather modification
无意识云影响 inadvertent cloud modification
无因次变量 nondimensional variable
无因次参数 non-dimensional parameter
无因次方程 non-dimensional equation
无涌【海洋】 no swell
无雨区 rainless region
无源辐射计 passive radiometer

无源卫星　passive satellite
无云　cloudless
无云区　cloud-free area
无云视线　cloud-free line-of-sight
　（CFLOS）
无运动层　level of no motion
无运动面　surface of no motion
无蒸腾森林覆盖　nontranspiring canopy
无阻尼振荡　undamped oscillation
五边形【数】　pentagon
五次曲线【数】　quintics
五点法　five-point method
五分位数　quintile
五进和十进风制　five and ten system
五镜头空中照相机　five lens aerial camera
五日预报　five day forecast
五色晕　pleochroic halo
五通道扫描辐射计　five channel scanning radiometer
五氧化二氮(N_2O_5)　dinitrogen pentoxide
午后效应　afternoon effect
物候测定学　phenometry
物候倒置　reverse of phenological sequence
物候分区【大气】　phenological division
物候关系　phenological relation
物候观测【大气】　phenological observation
物候季　phenoseason
物候季节　phenological season
物候历　phenological calendar
物候模拟　phenological simulation
物候模拟模式　phenological simulation model
物候谱　phenological spectrum
物候谱【大气】　phenospectrum
物候期　phenological phase
物候期【大气】　phenophase
物候日【大气】　phenodate
物候事件　phenological event

物候图【大气】　phenogram
物候学【大气】　phenology
物候学定律【农】　phenology law
物候学图　phenological chart
物理初始化　physical initialization
物理存储器　physical memory
物理大地测量学【测】　physical geodesy
物理地球化学【地】　physical geochemistry
物理方程　physical equation
物理风化【土壤】　physical weathering
物理光学【物】　physical optics
物理海洋实时系统　Physical Oceanographic Real-Time System(PORTS)
物理海洋学【海洋】　physical oceanography
物理合理性检验　physical limit check
物理化学【化】　physical chemistry
物理化学特性　physical and chemical properties
物理机制　physical mechanism
物理流体力学　physical hydrodynamics
物理模〔态〕【大气】　physical mode
物理气候　physical climate
物理气候学【大气】　physical climatology
物理气象学【大气】　physical meteorology
物理统计预报　physico-statistic prediction
物理限制条件　physical constraint
物理〔学〕【物】　physics
物理预报　physical forecasting
物质边界　material boundary
物质薄层　substantial sheet
物质面　substantial surface
物种起源说【植】　theory of origin species
物种起源【植】　origin of species
误差【数】　error
误差传播【数】　error propagation
误差定律　law of error
误差方差　error variance

误差分布【数】 error distribution
误差分析【数】 errror anahysis
误差估计【数】 error estimate
误差函数 error function
误差混淆 error contamination
误差检测电码 error-detecting code
误差检验 error detection
误差校正【数】 error correction
误差协方差 error covariance
误差协方差矩阵 error covariance matrix
误差诊断 error diagnosis
误差正态定律 normal law of errors
误差正态〔分布〕律 normalization law of errors
误差正态曲线 normal curve of error
雾【大气】 fog
雾层顶 fog horizon
雾堤【大气】 fog bank
雾滴【大气】 fog-drop
雾滴收集器 fog water collector
雾风 fog wind
雾海 sea of fog
雾害 fog damage
雾号【航海】 fog signal

雾虹 fog bow, mistbow, mistbow, white rainbow
雾化器【航海】 atomizer
雾级标准 fog scale
雾降水 fog precipitation
雾晶 fog crystal
雾量表(计) fogmeter
雾量计 fog-gauge
雾区 fog-region
雾日 fog day
雾室 fog chamber
雾收集器 fog collector
雾凇 soft rime
雾凇【大气】 rime
雾凇冰 rime ice, kernel ice
雾凇传感器 rime sensor
雾凇探棒 rime rod
雾天 greasy weather
雾天能见度仪 fog visiometer
雾消(太阳加热消雾) burn-off
雾信号器 megafog
雾雨 fog rain
雾预警【航海】 fog warning
雾中航行【航海】 navigating in fog

X

夕静 evening calm
夕没【天】 heliacal setting
夕阳 sunset colours
西 west(W)
西岸气候 west coast climate
西澳大利亚海流 West Australia Current
西北 northwest(NW)
西北大风 northwester(=nor'wester)
西北大西洋中央水 Western North Atlantic Central Water(WNACW)
西北太平洋反气旋 Northwest Pacific anticyclone

西北太平洋高压 Northwest Pacific high
西北太平洋中央水 Western North Pacific Central Water(WNPCW)
西边界流 western boundary current
西伯利亚反气旋 Siberian anticyclone
西伯利亚高压【大气】 Siberian high
〔西〕东向输送 west-east transport
西尔〔降雨〕强度计 Sil intensity gauge
西非龙卷 West African tornado
西风爆发 westerly wind burst
西风波 westerly wave
西风槽 trough in westerlies

西风槽【大气】 westerly trough
西风带 westerly zone, westerly belt
西风〔带〕【大气】 westerlies
西风带分支 splitting of westerlies
西风多雨带 westerlies rain belt
西风急流【大气】 westerly jet
西风漂流【大气】 west wind drift
西风涡 westerly vortex
西格陵兰海流 West Greenland Current
西〔门子〕(电导单位，1 S＝1 A/V) S (siemens)
西南 southwest (SW)
西南大风 southwester (＝sou'wester)
西南〔低〕涡【大气】 Southwest China vortex
西南风带 southwesterlies
西南季风 southwest monsoon
西南季风海流 Southwest Monsoon Current
西南气流 southwesterly flow
西西北 west northwest (WNW)
西西南 west southwest (WSW)
吸附层【化】 adsorption layer
吸附等温线 adsorption isotherm
吸附剂 adsorbent
吸附膜【化】 adsorption film
吸附热【物】 heat of adsorption
吸附指示剂【化】 adsorption indicator
吸附滞后【化】 hysteresis of adsorption
吸附〔作用〕【大气】 adsorption
吸管 suction tube
吸管式风速表 suction anemometer
吸光度【化】 absorbance
吸能反应【生化】 endergonic reaction
吸气区 suction zone
吸热反应 endothermic reaction
吸声系数 sound absorption coefficient
吸湿〔的〕 hygroscopic(al)
吸湿率 hygroscopicity
吸湿水 hygroscopic water
吸湿系数【农】 hygroscopic coefficient
吸湿性核【大气】 hygroscopic nuclei
吸湿性粒子 hygroscopic particle
吸收 absorption
吸收本领 absorptive power
吸收比【物】【大气】 absorptance
吸收分光计 absorption spectrometer
吸收光度计 absorptiometer
吸收〔光〕谱 absorption spectrum
吸收〔光谱〕带【大气】 absorption band
吸收光谱电化学【化】 absorption spectroelectrochemistry
吸收光学厚度 absorption optical thickness
吸收过滤器 absorbent filter
吸收函数 absorbing function, absorption function
吸收剂 absorbing agent
吸收剂量 absorbed dose
吸收截面【大气】 absorption cross-section
吸收介质 absorbing medium
吸收率 absorptivity(α), specific absorption
吸收能力 absorbing power
吸收谱【大气】 absorption spectrum
吸收〔谱〕线【大气】 absorption line
吸收器 absorber
吸收溶液 absorbent solution
吸收〔式〕湿度表【大气】 absorption hygrometer
吸收损耗 absorption loss
吸收太阳辐射 absorbed solar radiation
吸收太阳通量 absorbed solar flux
吸收系数 absorption coefficient
吸收液体 absorption liquid
吸收因子 absorption factor
吸收质量 absorption mass
吸收指数 index of absorption
吸引 attraction
吸引子【数】【大气】 attractor
吸涨〔作用〕〔植〕 imbibition
汐 evening tide

希尔伯特空间【数】 Hilbert space
析离低压 breakaway depression
析因实验【化】 factorial experiment
矽肺 silicosis
悉克斯〔最高最低〕温度表 Six's thermometer
烯【化】 alkene
烯烃 olefine
稀薄气体动力学【力】 rarefied gas dynamics
稀薄因子 tenuity factor
稀释 dilution
稀释热【化】 heat of dilution
稀疏波【物】 rarefaction wave
稀疏矩阵算子核心排放（排放源处理程序） Sparse Martix Operator Kernel Emission (SMOKE)
稀土金属【化】 rare earth metal, rare metal
稀土元素【化】 rare earth element, rare element
稀有气体【化】 rare gase
喜氮植物 nitrophile
喜冬植物 chimonophilous plant
喜光树种【林】 intolerant tree species
喜光植物 light-loving plant
喜湿植物 hygrophilous plant
喜雪的 chionophile
喜雪植物 chionophilous plant
喜阳植物 heliophil(e)
喜氧细菌 aerobacteria
喜阴植物 ombrophyte
喜雨植物 ombrophile
系泊声浮标系统 moored acoustic buoy system (MABS)
系集离散 ensemble spread
系留气球 kytoon, kite balloon, kitoon, tethered balloon, captive balloon
系留气球大气廓线仪 tethered balloon profiler, tethersonde profiler
系留气球探测【大气】 captive balloon sounding
系留气球探空仪 tethersonde
系留探空仪 wiresonde
系数 coefficient
系数数据库 coefficient data base (CDB)
系统抽样【数】 systematic sampling
系统分析【数】 systematic analysis
系统复原 system reset
系统工程 system engineering
系统观测 systematic observations
系统函数 system function
系统兼容性【计】 system compatibility
系统科学【数】 systems science
系统理论【数】 systems theory
系统误差【数】 systematic error
系统植物学【植】 plant systematics
系综【物】 ensemble
系综理论 ensemble theory
细胞对流【大气】 cellular convection
细胞环流【大气】 cellular circulation
细胞生理学 cellular physiology
细胞状云 cell cloud
细胞状云【大气】 cellular pattern
〔细〕胞状云〔型〕 cellular cloud pattern
细度【物】 finesse
细颗粒物 PM-2.5
细粒密度【物】 fine-grained density
细粒子 fine particles
细毛毛雨 thin drizzle
细砂土【土壤】 fine sand soil
细调【物】 fine adjustment
细网格【大气】 fine mesh, fine-mesh grid
细网格模式 fine mesh model
细线 fine line
峡谷【地理】 canyon, gorge
峡谷风 canyon wind, channeling, ravine wind
峡谷风【大气】 gorge wind
峡湾 fiord, fjord
狭道风 gap wind

狭缝【物】 slit
狭缝函数 slit function
狭管近似 narrow channel approximation
狭管效应 canalization, funnelling
狭水道航行【航海】 navigating in narrow channel
霞【大气】 twilight colours
霞〔光〕 sunglow
下包络(磁扰) lower envelope
下标 subscript
下冰风 off-ice wind
下沉 subsidence
下沉空气 descending air
下沉面 surface of subsidence
下沉逆温【大气】 subsidence〔temperature〕inversion
下沉气流【大气】 downward flow, subsidence flow
下沉区 descending area
下沉无云区 descending cloudless region
下沉运动 sinking motion
下吹风 fall wind, katabatic wind
下垫面 substrate, underlying earth's surface, underlying surface
下垫面反照率【大气】 albedo of underlying surface
下垫面辐射平衡 radiation balance of underlying surface
下珥 lower arc
下轨道 southbound
下滑 downslide
下滑锋 katafront, catafront, katabatic front
下滑航线 glide path
下滑〔冷〕锋【大气】 katabatic cold front
下滑面 downslide surface
下滑运动 downglide motion
下环天顶弧(圈) lower circumzenithal arc

下幻日 under-parhelion
下幻月 under-moon
下击暴流【大气】 downburst
下击暴流带 downburst swath, burst swath
下击暴流团 downburst cluster
下击暴流族 family of downburst cluster
下降 descent
下降度 descendent
下降法 descent method
下降流【航海】 downwelling
下降气流 descending current
下落末速 terminal fall velocity
下落速度 fall speed, fall velocity
下偏珥 low parry arc
下坡风【大气】 downslope wind
下坡风暴 downslop windstorm, slope windstorm
下坡运动 downslope
下切弧(晕的) lower tangential arc
下切冷锋 undercutting cold front
下切晕弧 infralateral tangent arcs
下日 undersun
下渗井 infiltration well
下渗演算 infiltration routing
下投式测风探空仪 dropwindsonde
下投式导航测风探空仪 navaid dropwindsonde
下投式探空仪 parachute radiosonde
下投式探空仪【大气】 dropsonde (D/P)
下弦【天】 last quarter
下弦月 decrescent
下现蜃景【大气】 inferior mirage, lower mirage, sinking mirage
下泻气流 drainage flow
下曳气流【大气】 downdraft
下曳气流速度 downdraft velocity
下一代高空探测系统 next generation upper air system
下溢 underflow
下游 downstream, tailwater

下游【航海】 lower reach
下游切变瓣 downstream shear lobe
下载【计】 download
夏半年 summer half year
夏半球 summer hemisphere
夏材 summerwood
夏干区 summer dry region
夏干温暖气候 warm climate with dry summer
夏〔季〕 summer
夏季的 aestival (=estival)
夏季对流降水 summer convective precipitation
夏季风【大气】 summer monsoon
夏令 estival (=aestival)
夏令时【天】 daylight-saving time (=summer time)
夏眠 aestivation
夏日 summer day
夏雾 summer fog
夏至【大气】(节气) Summer Solstice
夏至点 summer solstice
仙波【大气】 angel echo
先导流光(闪电) pilot streamer
先导〔流光〕【大气】 leader〔streamer〕
先导闪击【大气】 leader stroke
先导〔作用〕 leader
先进的 sophisticated
先进的大气探测和成像辐射仪 Advanced Atmospheric Sounding and Imaging Radiometer (AASIR)
先进的光导摄像系统 Advanced Vidicon Camera System (AVCS)
先进的沿轨扫描辐射仪 Advanced Along-Track Scanning Radiometer (AATSR)
先进地球观测卫星【大气】 Advanced Earth Observing Satellite (ADEOS)
先进技术卫星 Advanced Technology Satellite (ATS)
先进甚高分辨率辐射仪【大气】 Advanced Very High Resolution Radiometer (AVHRR)
先进甚高分辨率辐射仪极地探索者〔计划〕 AVHRR Polar Pathfinder(APP)
先进泰罗斯-N卫星【大气】 Advanced TIROS-N (ATN)
先进泰罗斯业务垂直探测器 Advanced TIROS Operational Vertical Sounder (ATOVS)
先进天气交互处理系统 Advanced Weather Interactive Processing System (AWIPS)
先进微波探测装置【大气】 Advanced Microwave Sounding Unit (AMSU)
先进温湿探测器 Advanced Moisture and Temperature Sounder (AMTS)
先进星载热发射和反射辐射仪(EOS) Advanced Spaceborne Thermal Emission and Reflection radiometer, EOS (ASTER)
先进云风系统 advanced cloud wind system
先驱者(行星和行星际探测器) Pioneer
先验分布【数】 prior distribution
先验概率【数】 prior probability
先验估计【数】 a priori estimate
先验理由 a priori reason
先验信息 prior information
纤维光学【物】 fiber optics
纤维素纸(测湿用) cellulosed paper
纤维状冰 fibrous ice
纤状卷云 cirro-filum (或 thread)
氙〔气〕(Xe) xenon
衔接闪流 junction streamer
衔接条件 junction condition
嫌风植物 anemophobe
嫌雪植物 chionophobous plant
显生宙【地质】 Phanerozoic Eon
显示板【电子】 display panel
显示格式【计】 display format, display mode

显示积分法　explicit integration method
显示屏【电子】　display screen
显示〔器〕　display
显示器　indicator
B 显示器【大气】　B－scope，B－display
显示终端　video terminal
显示终端【计】　display terminal
显式并行性【计】　explicit parallelism
显式差分格式【数】【大气】　explicit difference scheme
显式分解方法　explicit splitting technique
显式平滑　explicit smoothing
显式时间差分　explicit time difference
显著地平　sensible horizon
显著性　significance
显著性差异　significant difference
显著性检验　test of significance
显著性检验【数】　significance test
显著性水平　level of significance
显著性水平【数】【大气】　significance level
现场定域监测　in situ localized monitoring
现存冰川　living glacier
现存雪线　existing snowline
现代气候　present climate
现蕾期　flower-bud appearing stage
现时监测　present monitoring
现时预报【大气】　nowcast
现象　phenomenon
现象律　phenomenological law
现在天气　current weather
现在天气【大气】　present weather
现在位置　present position
限幅【电子】　amplitude limiting
限幅雷达测高计　pulse-limited radar altimeter
限幅器【电子】　amplitude limiter
限角　limiting angle
限量雨量器　limit-gauge

限束雷达测高表　beam-limited radar altimeter
线　line
线飑　line-squall
线飑云　line-squall cloud
线动量　linear momentum
线端涡旋　line-end vortices
线对流【大气】　line convection
线积分　line integral
线间隔　line spacing
线宽【物】　line breadth
线缆【计】　cable
线雷暴　line thunderstorm
线粒体 RNA【生化】　mitochondrial RNA
线粒体 DNA【生化】　mitochondrial DNA
线流　filamental flow
线平均　line average
线强　line intensity，line strength
线速度　linear velocity
线退偏比　linear depolarization ratio
线涡　line vortex
线形分子　linear molecule
线形函数　line shape function
线性　linear
线性变换　linear transformation
线性标度　linear scale
线性波　linear wave
线性不稳定〔度〕　linear instability
线性垂直风切变　rectilinear wind shear
线性代数【数】　linear algebra
线性地转平衡　linear geostrophic balance
线性度　linearity
线性反演【大气】　linear inversion
线性方程　linear equation
线性风暴　line storm
线性规划　linear programming
线性函数　linear function
线性合成【化】　linear synthesis

线性化　linearization
线性化〔微分〕方程　linearized (differential) equation
线性回归　linear regression
线性计算不稳定　linear computational instability
线性计算不稳定【大气】　computational instability
线性滤波【数】　linear filtering
线性内插　linear interpolation
线性能量转换【农】　linear energy transfer
线性黏弹性【化】　linear viscoelasticity
线性偏微分方程【数】　linear partial differential equation
线性平流方程　linear advection equation
线性齐次方程　linear homogeneous equation
线性倾向去除　linear-trend removal
线性全球平衡　linear global balance
线性山岳波理论　linear mountain wave theory
线性熵标　linear entropy scale
线性湿度表　linear action hygrometer
线性水波理论　linear water wave theory
线性算子　linear operator
线性推移　linear transition
线性微分方程　linear differential equation
线性温度场　linear field of temperature
线性稳定度分析　linear stability analysis
线性吸收系数　linear absorption coefficient
线性系统　linear system
线性限制法　linear constrained method
线性相关　linear correlation
线性运动场　linear field of motion
线性正切方程　linear tangent equations
线性转置模式　linear inverse model (LIM)

线源　line source
线状风暴　derecho
线状回波　line echo
线状谱　line spectrum
线状云　line cloud
陷波【海洋】　trapped wave
陷阱【物】　trap
霰【大气】　graupel
相【大气】　phase
相变【大气】　phase change, change of phase, phase transformation
相变【物】【大气】　phase transition
相变点　transformation point
相变平衡条件【物】　equilibrium condition of phase transition
相参数　phase parameter
相长干涉【物】　constructive interference
相程　phase path
相当辐射大气　equivalent radiative atmosphere
〔相当〕黑体温度　TBB (black body temperature equivalent)
相当深度　equivalent depth
相当位温　potential equivalent temperature
相当温度【大气】　equivalent temperature
相当正压大气【大气】　equivalent barotropic atmosphere
相当正压高度　equivalent barotropic level
相当正压模式【大气】　equivalent barotropic model
相电压【物】　phase voltage
相对臭氧浓度　relative ozone concentration
相对磁导率【物】　relative permeability
相对等高线　relative contour
相对递减率　relative reduction
相对电容率【物】　relative permittivity
相对动量　relative momentum

相对发光效率　relative luminous efficiency
相对风　relative wind
相对辐散　relative divergence
相对高度　relative height
相对高度型　relative hypsography
相对海平面　relative sea level
相对角动量　relative angular momentum
相对均匀性　relative homogeneity
相对孔径【物】　relative aperture
相对流　relative current
相对论【物】　relativity
相对密度　relative density
相对能见度　relative visibility
相对浓度　relative concentration
相对频率　relative frequency
相对谱　relative spectrum
相对日照　relative sunshine
相对散射函数　relative scattering function
相对散射强度　relative scattering intensity
相对湿度【大气】　relative humidity
相对湿度探头　relative humidity probe
相对时代　relative age
相对速度【大气】　relative velocity
相对太阳黑子数　relative sunspot number
相对梯度　relative gradient
相对梯度流　relative gradient current
相对稳定〔度〕　relative stability
相对涡度【大气】　relative vorticity
相对误差【数】　relative error
相对相位变化　relative phase change
相对响应　relative response
相对形势　relative topography
相对雨量系数　relative pluviometric coefficient
相对运动【物】　relative motion
相对蒸发　relative evaporation
相对蒸发器　relative evaporation gauge
相对坐标系　relative coordinate system

相干　coherency
相干靶〔标〕　coherent target
相干波【物】　coherent wave
相干长度　coherence length
相干场　coherent field
相干存储滤波器【大气】　coherent memory filter(CMF)
相干带宽　coherence bandwidth
相干度【物】　degree of coherence
相干多普勒声〔雷〕达　coherent acoustic Doppler radar
相干辐射【物】　coherent radiation
相干光【物】　coherent light
相干光〔雷〕达　coherent light detection and ranging
相干光〔雷〕达【物】　coherent optical radar
相干光源【物】　coherent source
相干光自适应技术　coherent optical adaptive technique(COAT)
相干合成孔径成像雷达　coherent synthetic aperture imaging radar (CSAIR)
相干回波【大气】　coherent echo
相干积分　coherent integration
相干检测【数】　coherent detection
相干结构　coherent structure
相干雷达　【大气】coherent radar
相干谱　coherence spectrum
相干强度　coherent intensity
相干散射【物】　coherent scattering
相干时间　coherence time
相干视频信号【大气】　coherent-video signal
相干态【物】　coherent state
相干位相　coherent phase
相干系数　coefficient of coherency
相干性【大气】　coherence
相干元素　coherence element
相干振荡器　coherent oscillator, coho (coherent oscillator 的缩写)
相关【物】【大气】　correlation

相关比　correlation ratio
相关变换法　correlation-transform method
相关变数　correlation variable
相关表　correlation table
相关长度【化】　persistence length
相关处理机　correlation processor
相关概率表【大气】　contingency table
相关法　correlation method
相关分析【数】　correlation analysis
相关峰　correlation peak
相关光谱学　correlation spectroscopy
相关函数【数】　correlation function
相关〔矩〕阵　correlation matrix
相关距离　correlation distance
相关谱　correlation spectrum
相关三角形　correlation triangle
相关探测和测距　COrrelation Detection And Ranging (CODAR)
相关天气学　correlation synoptics
相关图　correlation diagram, correlogram
相关系数　coefficient of correlation
相关系数【数】【大气】　correlation coefficient
相关斜率法　correlation slope method
相关因子　correlation factor
相关预报【大气】　correlation forecasting
相函数【大气】　phase function
相合估计【数】　consistent estimate
相互作用　interaction, interplay
相畸变　phase distortion
相加性常数　additive constant
相角【物】　phase angle
相矩阵　phase matrix
相空间【物】　phase space
相控阵【电子】　phase array
相控阵天线　phased-array antenna
相邻视场法　adjacent field-of-view method

相律　phase rule
相平衡【化】　phase equilibrium
相平面　phase plane
相容数值格式　consistent numerical scheme
相容性【数】　consistency
相容性【物】　compatibility
相似〔定〕律　similarity law
相似方法　analog method
相似理论　similarity theory, theory of similarity
相似〔理论〕关系〔式〕　similarity relationship
相似模式　analog model
相似性　equality, resemblance
相似性【计】　similarity
相似性假设　similarity hypothesis
相似转换　similarity transformation
相速　phase speed
相速(度)【物】【大气】　phase velocity
相态不稳定　phase instability
相同气候　homoclime
相图　phase diagram, phase portrait
相〔位〕【物】　phase
相位编码【航海】　phase coding
相〔位〕差【物】　phase difference
相位常量【物】　phase constant
相位结构函数　phase structure function
相位匹配【物】　phase matching
相位频率谱　phase frequency spectrum
相位平均　phase averaging
相位屏理论　phase screen theory
相位谱　phase spectrum
相位相干　phase coherence
相位相关函数　phase correlation function
相位响应　phase response
相位延迟【物】　phase delay
相位涨落【物】　phase fluctuation
相位滞后　phase lag
相位滞后谱　phase-lag spectrum
相消干涉　destructive interference

相序【航海】 phase sequence
相移【电子】 phase shift
相移键控【电子】 phase-shift keying
香农理论【数】 Shannon theory
香农熵【电子】 Shannon entropy
箱室法 box method
箱〔室〕模式【大气】 box models
箱形风筝 box kite
响度 loudness
响应参数化 responsive parameterization
响应函数【物】 response function
响应距离 response distance, response length
响应率 responsivity
响应谱 response spectrum
响应时间 response time
响应速率 rate of response
向岸风 off-sea wind, wind toward coast, inshore wind
向岸风【大气】 on-shore wind
向岸流 inshore current
向东 eastward
向风面 luvside
向后差分【数】【大气】 backward difference
向极的 poleward
向量【数】 vector
Q 向量 Q vector
向量积【数】 cross product
向量计算机【计】 vector computer
向量微分算子 del-operator
向南 southward
向前差分【数】【大气】 forward difference
向前倾斜 forward tilting
向前向后格式 forward-backward scheme
向前-向后散射【化】 forward-backward scattering
向日面 sunward side

向上大气辐射 upward atmospheric radiation
向上地球辐射 upward terrestrial radiation
向上反射地球辐射 reflected terrestrial radiation
向上〔全〕辐射【大气】 upward 〔total〕 radiation
向上延拓 upward continuation
向上运动 upward motion
向外长波辐射【大气】 outgoing longwave radiation(OLR)
向外辐射 outgoing radiation
向晚层积云 stratocumulus vesperalis (Se ves)
向下大气辐射【大气】 downward atmospheric radiation
向下〔全〕辐射【大气】 downward 〔total〕 radiation
向下输送 downwash
向下延拓 downward continuation
向下运动 downward motion
向心加速度【物】【大气】 centripetal acceleration
向心力【物】 centripetal force
项圈云 collar cloud
象限【数】 quadrant
象限分析 quadrant analysis
象限角 quadrant angle
象限静电计【物】 quadrant electrometer
象限能见度 quadrant visibility
象牙针高程 elevation of ivory point
象牙针尖 ivory point
像【物】 image
像差【物】 aberration
像平面【物】 image plane
像〔强度〕级 image level
像散【物】 astigmatism
像散光束 astigmatic beam
像素【大气】 pixel
像素值 pixel value

像元【物】 picture element
像元【物】【大气】 image element, pixel
像元尺寸【地信】 cell size
像元分辨率【地信】 cell resolution
橡胶气球 rubber balloon
削台计划 Hurricane Modification Program (STORMFURY)
消雹 hail mitigation
消冰区 ablation area
消电离【电子】 deionization
消光 extinction
消光截面 extinction cross section
消光系数【大气】 extinction coefficient, coefficient of extinction
消光效率 extinction efficiency
消光佯谬 Extinction paradox
消耗 consumption
消耗性温深仪【大气】 expendable bathythermograph, XBT
消混淆 dealiasing
消减【地】 subduction
消暖雾系统 thermal fog dispersal system
消去法 elimination method
消融区 zone of wastage
消散波 decaying wave
消散常数 dissipation constant
消散函数 dissipation function
消散阶段 dissipative stage
消散期 decaying stage
消散尾迹 dissipation trail (=distrail)
消散中的锋 frontolytical front
消退 recession
消雾 fog clearing, fog dispersal, fido
消雾【大气】 fog dissipation
消雾作业 fog dispersal operation
消旋控制电子设备 despin control electronics (DSE)
消云 cloud dispersal
消云【大气】 cloud dissipation
消转 spin-down
消转过程 spin-down process

消转时间【大气】 spin-down time
消转效应 spin-down effect
硝化【化】 nitration
硝化作用 nitrification
硝酸 nitric acid
硝酸铵 ammonium nitrate
硝酸根 nitrate radical
硝酸根离子 nitrate ion
硝酸过氧化乙酰 peroxy acetyl nitrate (PAN)
硝酸氯($ClNO_3$, 常写为 $ClONO_2$) chlorine nitrate
硝酸三水合物 nitric acid trihydrate (NAT)
硝酸盐 nitrate
硝态氮【土壤】 nitrite-nitrogen
小冰雹 small hail
小冰期【大气】 Little Ice Age
小冰山 bergy bit, floeberg
小波变换【数】 wavelet transform
小波分析【数】 wavelet analysis
小参数法 small parameter method
小侧风 light crosswind
小潮【航海】 neap tide
小潮差 neap range
小潮升【航海】 neap rise
小尺度 microscale, misoscale
小尺度分析 microanalysis
小尺度扰动(雷达用语) blop
小尺度天气 microweather
小尺度天气图 microsynoptic map
小尺度湍流 microturbulence
小尺度系统【大气】 microscale system
小船预警 small-craft warning
小春雪 lambing storm, lamb-storm, lamb-shower
小岛国联盟 Alliance of Small Island States (AOSIS)
小段步长 fractional step
小风暴 whirlies
小浮冰 glacon

小股铅直气流　draft
小海风　minor sea breeze
小寒【农】【大气】　Lesser Cold
小滑雪　sluff
小急流　jetlet
小集水口雨量筒　small orifice raingauge
小角近似　small angle approximation
小角散射【物】　small angle scattering
小块冰　calf
小雷暴　slight thunderstorm
小笠原〔副热带〕高压　Ogasawara high
小满【农】【大气】　Lesser Fullness
小年　off-year
小片冰晶　platelet
小平面模式　facet model
小气候【大气】　microclimate
小气候测量　microclimatic measurement
小气候观测　microclimatic observation
小气候热岛　microclimatic heat island
小气候适宜期　Little Climatic Optimum
小气候学【大气】　microclimatology
小气候因素　microclimatic factor
小气旋　microcyclone, misocyclone
小曲折　sinuosity
小扰动　small perturbation
小扰动法【大气】　perturbation method
小生境【地理】　microhabitat
1小时降水量　hourly precipitation
小时-距离表　hourly distance scale
小暑【大气】　Slight Heat
小暑【农】【大气】　Lesser Heat
小水滴　water droplet
小穗【生】　spikelet
小天气学　microsynoptic meteorology
小涡闭合　small eddy closure
小涡理论　small eddy theory
小雾滴　fog droplet
小行星　asteroid, planetoid
小行星带　asteroid belt
小型超级单体　mini-supercell
小〔型〕环流　minor circulation

小型计算机　minicomputer
小型净辐射表　miniature net radiometer
小〔型〕龙卷气旋　mini-tornado cyclone
小型气象传真机　Minifax
小型热带气旋　midget tropical cyclone
小型台风　midget tropical storm, midget typhoon
小型探空仪　minisonde
小型〔卫星〕资料利用站　small-scale data utilization station（SDUS）
小型蒸发皿【大气】　evaporation pan
小型〔自动〕气候站　miniature climatological station（MINICLIM）
小旋风　cyclonette
小雪【农】【大气】　Light Snow
小雪丘　snow igloo
小雨【大气】　light rain
小圆〔航路〕　small circle
小月　moonlet
小云块　cloudlet
〔小〕晕　small halo
小阵雨　slight shower of rain
小振幅　small amplitude
小振幅理论　small amplitude theory
肖特基势垒【电子】　Schottky barrier
肖特基势垒二极管【电子】　Schottky barrier diode
肖沃特稳定〔度〕指数　Showalter stability index
效率　efficiency
效益　cost-effectiveness
效益函数　utility function
效应【数】　effect
β效应【大气】　β-effect, beta effect
楔线　wedge line
楔形等压线　wedge isobar
协变　covariation
协变量　covariant
协变性　covariability
协变张量【物】　covariant tensor
协调世界时【天】　coordinated universal

time (UTC)
协方差【数】【大气】 covariance
协方差函数 covariance function
协方差〔矩〕阵 covariance matrix
协谱【大气】 cospectrum
协同学【物】 synergetics
协同〔作用〕 synergism
协议【计】 protocol
挟沙能力 sediment-carrying capacity
斜槽 tilted trough
斜称张量【数】 skew-symmetric tensor
斜度 steepness
斜对称 skew-symmetry
斜对流可用位能 slantwise convection available potential energy
斜方晶体【地质】 orthorhombic system
斜交〔的〕 heterotropic
斜角等距圆柱投影 oblique equidistant cylindrical projection
斜角笛卡儿坐标 oblique Cartesian coordinates
斜角墨卡托投影 oblique Mercator projection
斜距 slant range
斜距分子吸收 slant path molecular absorption
斜率法 slope method
斜率函数【数】 slope function
斜能见度【大气】 oblique visibility, slant visibility
斜坡函数 ramp function
斜坡结构 ramp structure
斜射 oblique incidence
斜视程 oblique visual range
斜温〔性〕 thermoclinic
斜温层,温跃层【大气】 thermocline
斜温性 thermoclinicity
斜压 barocline
斜压边界层 baroclinic boundary layer
斜压波【大气】 baroclinic wave
斜压波活动 baroclinic wave activity

斜压不稳定【大气】 baroclinic instability
斜压大气【大气】 baroclinic atmosphere
斜压带 baroclinic zone
斜压的 baroclinic
斜压过程【大气】 baroclinic process
斜压环带 baroclinic annulus
斜压流体 baroclinic fluid
斜压模式【大气】 baroclinic model
斜压模〔态〕【大气】 baroclinic mode
斜压气流 baroclinic flow
斜压扰动【大气】 baroclinic disturbance
斜压适应 baroclinic adjustment
斜压数值模式 baroclinic numerical model
斜压条件 baroclinic condition
斜压涡 baroclinic eddy
斜压性【大气】 baroclinicity (= baroclinity, barocliny)
斜压叶 baroclinic leaf
斜压预报 baroclinic forecast
斜压运动 baroclinic motion
斜压转矩矢量 baroclinic torque vector
斜压状态 barocline state
斜压准地转流 baroclinic quasi-geostrophic flow
斜轴投影 oblique projection
斜坐标【数】 oblique coordinates
谐波 harmonics
谐波标度盘 harmonic dial
谐波级数 harmonic series
谐振荡 harmonic oscillation
谐振辐射 harmonic radiation
谐振子【物】 harmonic oscillator
泄地电流【电工】 earth-current
泄离 shedding
泄漏 leakage
泄漏电流 leakage current
泄水区 discharge area
泻流速度 effusion velocity
屑冰 acicular ice, frazil
谢齐方程（水文） Chezy equation

心射切面投影　gnomonic projection
芯件【计】　chipware
芯片【计】　chip
辛普森挡污器　Simpson dirt trap
新冰　young ice
新冰川期　Neoglacial
新冰川作用　neoglaciation
新陈代谢【生化】　metabolism
新陈代谢率　metabolic rate
新〔成〕冰　new ice
新积雪　fresh snow cover
新几内亚海岸流　New Guinea Coastal Current
新几内亚海岸潜流　New Guinea Coastal Undercurrent
新近纪【地质】　Neogene Period
新近系【地质】　Neogene System
新生代【地质】　Cenozoic Era
新生界【地质】　Cenozoic Erathem
新石器时代　Neolithic Age, New Stone Age
新西兰地球物理学会　New Zealand Geophysical Society（NZGS）
新西兰气象局　New Zealand Meteorological Service（NZMS）
新仙女木时期　Younger Dryas
新仙女木事件　Younger Dryas event
新鲜空气　fresh air
新鲜气团　fresh air mass
〔新〕鲜重〔量〕　fresh weight
新星【天】　nova
新雪　new snow
新一代〔多普勒〕气象雷达　Next Generation Weather Radar（NEXRAD）
新元古代【地质】　Neoproterozoic Era
新元古界【地质】　Neoproterozoic Erathem
新月形沙丘　barchan, barchane, barkhan
信标　beacon
信锤　messenger
信风　trade
信风【大气】　trade-winds
信风潮　trade surge
信风赤道槽　trade-wind equatorial trough
信风带【大气】　trade-wind belt, trade-wind zone, zone of trade wind
信风锋【大气】　trade-wind front
信风环流【大气】　trade-wind circulation
信风积云　trade cumulus, trade-wind cumulus
信风积云平衡　trade cumulus equilibrium
信风空气　trade air
信风逆温【大气】　trade inversion, trade-wind inversion
信风圈　trade-wind cell
信风沙漠　trade-wind desert
信干比【电子】　signal to jamming ratio
信号【电子】　signal
信号变率处理器　signal variability processor（SVP）
信号灯　call lamp
信号发生器【电子】　signal generator
信号强度　signal intensity
信号速度　signal velocity
信息　information
信息安全【地信】　information safety
信息采集【地信】　information collection
信息产业【计】　information industry
信息检索系统【地信】　information retrieval system
信息技术【地信】　information technology
信息科学【数】　information science
信息量　information content
信息论　information theory
信息内容　information contents
信息容量　data capacity
信息融合【地信】　information fusion
信息熵【数】　entropy of information
信息提取【地理】　information extraction

信息系统【地信】 information system
信息效率【化】 information efficiency
信息效益【化】 information profitability
信息学【数】 informatics
信息压缩【地理】 information compression
信杂比【电子】 signal to clutter ratio
信噪比【物】 signal-to-noise ratio (SNR)
信噪问题 signal-to-noise problem
星风【天】 stellar wind
星光闪烁 stellar scintillation
星基导航系统 satellite based navigational system
星际尘埃【天】【大气】 interstellar dust
星际气体 interplanetary gas
星际气体【天】 interstellar gas
星际湍流 interstellar turbulence
星位角【天】 parallactic angle
星系【天】 galaxy
星下点 subpoint
星下点轨迹【电子】 subtrack, subpoint track
星下点路径 sub-satellite point track
星云 nebula
星云假说 nebular hypothesis
星云〔学〕说 nebular theory
星载〔电离层〕测高仪 satellite-borne ionosonde
星载气象雷达 satellite-borne weather radar
星载闪电摄像仪 satellite-borne lightning mapper sensor
星载探测仪器的瞬时视场 footprint
星照计 starshine recorder
星状结晶 stellar crystal
星〔状〕闪〔电〕 stellar lightning
行波【电子】 travelling wave
行波管【电子】 travelling-wave tube (TWT)
行道树【林】 street tree

行洪河道 flood channel
行迹 swath
行迹宽度 swath width
行进波 marching wave
行近流速 approach velocity
行星【天】 planet
行星边界层【大气】 planetary boundary layer(PBL)
行星边界层厚度 depth of planetary bouandary layer
行星边界层急流 planetary boundary layer jet
行星波【大气】 planetary wave
行星波公式 planetary-wave formula
行星尺度【大气】 planetary scale
行星尺度系统【大气】 planetary-scale system
行星大地测量学【测】 planetary geodesy
行星大气【大气】 planetary atmosphere
行星大气实验研究所（英国伦敦大学） Laboratory for Planetary Atmosphere (LPA)
行星反照率【大气】 planetary albedo
行星地转（近似） planetary-geostrophic
行星风 planetary wind
行星风带【大气】 planetary wind belt
行星风系【大气】 planetary wind system
行星光行差【天】 planetary aberration
行星环流 planetary circulation
行星际尘埃 interplanetary dust
行星际磁场【地】【大气】 interplanetary magnetic field(IMF)
行星际空间【天】 interplanetary space
行星平衡温度 planetary equilibrium temperature
行星温度【大气】 planetary temperature
行星涡度【大气】 planetary vorticity
行星涡度梯度 planetary vorticity gradient
行星涡度效应【大气】 planetary vorticity effect

行星无线电掩星试验　planetary radio occultation experiment
行星运动　planetary motion
行星-重力波　planetary-gravity wave
行星驻波　stationary planetary wave
行政地理【地理】　administrative geography
行政界线【地理】　administrative boundary
形变【数】　deformation
形变运动　straining motion
形成层活动　cambial initial
形成函数【化】　formation function
V形等压线【大气】　V-shaped isobars
V形低压【大气】　V depression，V-shaped depression
形式解【数】　formal solution
形式曳力　form drag
〔形势〕预报【大气】　prognosis
〔形势〕预报图　prog chart
形态学　morphology
形状参数【数】　shape parameter
形状函数　shape function
形状算子【数】　shape operator
型　type
α型前导　type-α leader
β型前导　type-β leader
型式　pattern
88型天气监测多普勒雷达　WSR-88D（weather surveillance rader-1988 Doppler）
e型吸收　e-type absorption
A型显示　A-scope display
A〔型〕显〔示〕　A-display
R型显示器　R-scope
A型显示器【大气】　A-scope indicator
R-型因子分析　R-mode factor analysis
Ω型阻塞　omega block
性能指数　performance index
休梅期　break of the Meiyu period
休眠【植】　dormancy
休眠火山【地质】　dormant volcano
修平〔处理〕　hanning

修正【物】　correction
修正贝塞耳函数　modified Bessel function
修正γ分布　modified Gamma distribution
修正冯·卡曼谱　modified von Karman spectrum
修正折射率　modified refractive index
修正折射率差　modified refractivity
溴化合物　bromine compounds
溴甲烷　bromomethane
虚变元　dummy argument
虚部【数】　imaginary part
虚副层　virtual sublayer
虚功【物】　virtual work
虚功原理　principle of virtual work
虚〔假〕变量　dummy variable
虚警〔报〕率　false alarm ratio(FAR)
虚警概率【电子】　false alarm probability
虚拟冲击器　virtual impactor
虚拟存储〔器〕【计】　virtual memory
虚拟存取法　virtual access method
虚拟地图【地信】　virtual map
虚拟观测　bogus observation
虚拟机【计】　virtual machine
虚拟计算机【计】　virtual computer
虚拟涡旋　bogus vortex
虚拟现实【计】　virtual reality
虚黏滞率　fictitious viscosity
虚数【数】　imaginary number
虚太阳　fictitious sun
虚位温　virtual potential temperature
虚位移【物】　virtual displacement
虚位移原理　principle of virtual displacement
虚温【大气】　virtual temperature
虚线　dashed line
虚像【物】　virtual image
虚压　virtual pressure
虚应力　virtual stress

虚重力　virtual gravity
需光量　light requirement
需热常数　thermal constant
需热量　thermometric constant
需水关键期【大气】　critical period of 〔crop〕water requirement
需水量【植物】　water requirement, water need
序参数　order parameter
序贯分析【数】　sequential analysis
序号　serial number
序列【数】　sequence, series, succession
序列变率　intersequential variability
序列估计　sequence estimation
序列算法　sequential algorithm
序列同化　sequential assimilation
序列相关　serial correlation
序列相关系数　serial correlation coefficient
序数【数】　ordinal
畜牧气象学【大气】　animal husbandry meteorology
畜牧业　animal husbandry
絮状（云）　floccus（flo）
絮状层积云　stratocumulus floccus（Sc flo）
絮状高积云【大气】　altocumulus floccus（Ac flo）
絮状卷积云　cirrocumulus floccus
絮状卷云　cirrus floccus
蓄洪方程　storage equation
蓄水　pondage
蓄水函数　storage function
蓄水率　storage ratio, storativity
蓄水期　storage term
蓄水效应　storage effect
蓄水演算　storage routing
玄武岩【地质】　basalt
悬壁涡流　cliff-eddy
悬冰川【地理】　hanging glacier
悬垂回波　echo overhang, overhang echo

悬锤式水尺　wire-weight gauge
悬浮　suspension
悬浮尘　suspended dust
悬浮灰分　suspended ash
悬浮颗粒物　suspended particulate
悬浮粒子　suspended particle
悬浮态　suspended phase, suspending phase
悬浮体【化】　suspensoid, suspended matter
悬浮微粒　particulate loading
悬浮物【海洋】　seston
悬浮物散射光【大气】　airlight
悬浮相　suspended phase
悬浮载荷　suspended load
悬挂式探空仪　suspended sonde
悬球状积雨云　cumulonimbus mammatus
悬球状积云　mammato-cumulus
悬球状〔云〕　mammatus, mamma（mam）
悬球状云　mammato cloud, pocky cloud
悬着水　hanging water
旋杯式海流计　cup current meter
旋度【数】　rotation, curl
旋风　whirlwind（＝whirl）
旋风集尘器　cyclone collector
旋杆取样器　rotorod sampler
旋光物质【物】　optical active substance
旋光性【物】　optical rotation
旋衡的　cyclostrophic
旋衡风【大气】　cyclostrophic wind
旋衡辐合　cyclostrophic convergence
旋衡辐散　cyclostrophic divergence
旋衡函数　cyclostrophic function
旋衡流　cyclostrophic flow
旋衡输送　cyclostrophic transport
旋衡性平衡　cyclostrophic balance
旋桨式风标风速计　propeller-vane anemonmeter
旋桨式流速仪　propeller-type current

meter
旋球状层积云　stratocumulus mammatus
旋涡　whirlpool
旋翼式风速感应器　helicopter
旋轴点　spin axis point
旋转　rotational
旋转波【地】　rotational wave
旋转不稳定〔度〕　rotational instability
旋转场　rotational field
旋转潮流　rotary tidal current
旋转 EOF 分析　rotated EOF analysis
旋转风暴　revolving storm
旋转风〔分量〕　rotational wind〔component〕
旋转风速表　rotation anemometer
旋转风约束初始化　rotational wind constrained initialization
旋转关节　rotary joint, rotating joint
旋转光束〔式〕云幂仪　rotating-beam ceilometer（BBC）
旋转减弱时间【大气】　spindown time
旋转角【数】　angle of rotation
旋转镜　rotoscope
旋转雷诺数　rotating Reynolds number
旋转流　rotary current
旋转流体　revolving fluid
旋转流体动力〔学〕　rotating fluid dynamics
旋转马赫数　rotational Mach number
旋转模　rotational mode
旋转模型　rotational model
旋转谱　rotational spectrum
旋转扫描示差偏振仪　rotating scanning differential polarimeter
旋转适应过程　rotating adaptation process
旋转特征向量　rotated eigenvector
旋转相似性判断　rotational similarity criterion
旋转向量序列　rotary vector series
旋转运动　rotational motion
旋转轴【数】　axis of rotation

旋转坐标系　rotating coordinate frame
选单【计】　menu
选通风速表　switching anemometer
选择不稳定〔度〕　selective instability
选择调制辐射仪　selective chopper radiometer（SCR）
选择定则【物】　selection rule
选择假设　selection hypothesis
选择滤波　selective filtering
选择散射　selective scattering
选择稳定〔度〕　selective stability
选择吸附　selective adsorption
选择吸收【物】　selective absorption
选择性阻尼　selective damping
选择作用　selective action
选址　siting
薛定谔方程【数】　Schrodinger equation
学科　discipline
雪【大气】　snow
雪霭　snow mist
雪暴　blizzard
雪暴【大气】　snow storm
雪暴预警　blizzard warning
雪暴灾害　blizzard fatality
雪崩【大气】　snowquake, snow tremor, snow slide, avalanche
雪崩气浪　avalanche wind
雪飑　snow squall
雪冰　snow ice
雪沉降物　snowout
雪带　nival belt, snow belt
雪的软化（熟化）　ripening of snow
雪的形态变化　metamorphosis of snow
雪堤　snowbank
雪堆　snowpack, snow patch
雪幡【大气】　snow virga
雪反射率　snow reflectivity
雪反照率-温度反馈　snow albedo-temperature feedback
雪盖【大气】　snow cover
雪盖冰　snow covered ice

雪盖水当量 water equivalent of snow cover
雪冠 snow cap
雪害 snow injury
雪荷载【大气】 snow load
雪花【大气】 snowflake
雪花环 snow garland
雪花静电 snow static
雪迹 snow trail
雪浆 slush
雪晶【大气】 snow crystal
雪晶浓度计 snow crystal concentration meter
雪聚成球 balling
雪卷 snow roller
雪雷暴 snow thunderstorm
雪量【大气】 snowfall〔amount〕
雪量计 snow-cover meter
雪量累计器 snow totalizer
雪量器 nivometer
雪林气候 snow forest climate
雪盲〔症〕【大气】 snow blindness
雪密度【大气】 snow density
雪面波纹 skavler
雪〔面〕波〔纹〕【大气】 sastrugi
雪面壳 skare
雪泥 snow sludge, snow slush
雪泥冰 shuga ice, sludge
雪喷 snow geyser
雪片 flake
雪旗 snow banner
雪球 snowball
雪球假说【古生物】 Snowball Earth hypothesis
雪日【大气】 snow day
雪融 snow melt
雪山 snowberg
雪深【大气】 snow depth
雪深测定仪 snow depthmeter
雪深光学测量仪 optical snow depthmeter
雪生单体 snow generating cell

雪蚀 nivation, snow erosion
雪蚀【土壤】 snow erosion
雪水 snow-broth, snow water
雪水当量【大气】 snow-water equivalent, water equivalent of snow
雪丸 snow pellets, tapioca snow
雪纹 snow ripple
雪席 snow mat
雪线【大气】 snow line
雪效应 snow effect
雪压【大气】 snow pressure
雪烟 snow smoke
雪檐 cornice
雪移 snowcreep
雪映天 snow sky, snow blink
雪羽 snow plume
雪原 snowfield
雪原气候 snow climate
雪云 snow cloud
雪灾【大气】 snow damage
雪障 snow barrier, snowbreak
雪照云光 snow blink
雪枕 snow pillow
雪质 quality of snow
雪中白霜 sugar snow
雪灼 snow burn
雪阻 road damped by snow
血红蛋白【化】 hemoglobin
血红素蛋白【化】 hemoprotein
血雪【大气】 blood-snow
血雨【大气】 blood-rain
熏烟 smudging
熏烟法 fumigation
旬平均 ten-days average
寻常胞 ordinary cell
寻常波【物】 ordinary wave
寻常光【物】 ordinary light
寻常光线 ordinary ray
寻找时间 seektime
寻址 addressing

巡检用标准气压表　travelling standard barometer
巡洋舰【航海】　cruiser
循环　recursion, cycle
循环边界条件　cyclic boundary condition
循环过程　cycle process
循环卷积　circular convolution
循环码　cyclic code
循环相关　circular correlation
循环小数【数】　recurring decimal
循环校验　cyclic check
汛期　flood season
汛期【大气】　flood period
逊常　below normal

Y

压板风速表　pressure-plate anemometer
压板风速仪　swinging plate anemometer
压电晶体【物】　piezocrystal, piezoelectric crystal
压电陶瓷【电子】　piezoelectric ceramic
压电吸附式湿度表　piezoelectric sorption hygrometer
压电效应【物】　piezoelectric effect
压高公式　barometric formula
压高公式【大气】　barometric height formula
压高转换效应　inverted barometer effect
压管风速表　pressure tube anemometer
压管风速计　anemobiagraph
压力　pressure
压力表　air-gauge
压力冰　pressure ice
压力调制池　pressure modulated cell (PMC)
压力调制辐射仪　pressure modulated radiometer(PMR)
压力调制红外辐射仪　pressure modulated infrared radiometer(PMIR)
压力风速表　pressure anemometer
压力功　pressure work
压力换算　pressure scaling
压力〔谱线〕增宽【大气】　pressure broadening
压力器　pressure gauge
压力球风速计　pressure-sphere anemometer
压力〔水〕头　pressure head, hydraulic head
压力系数　pressure coefficient
压力验潮仪　pressure tide gauges
压力中心　center of pressure
压流　baric flow
压强计【物】　piezometer
压容管　isobaric isosteric solenoid
压熵图　piphigram
压缩　compression
压缩比【电子】　compression ratio
压缩波　compressional wave, longitudinal wave, compression wave, compressibility wave
压缩码资料格式　packed data format
压缩脉冲雷达高度计　compressed pulse radar altimeter
压缩评价　compression evaluation
压缩数字传输　compressed digital transmission
压缩增温　compression heating
压缩轴　contraction axis
压温表　barothermometer
压温湿探测　pressure-temperature humidity sounding

压温相关　pressure-temperature correlation
压性　piezotropy
压性〔的〕　piezotropic
压性方程　piezotropic equation, equation of piezotropy
压性系数　coefficient of piezotropy, piezocoefficient
压应力　pressure stress
压致〔谱线〕增宽【物】　pressure broadening
哑铃形分布　dumbbell distribution
雅各布光象　Jacob's ladder
雅可比迭代法　Jacobi iterative method
雅可比迭代〔法〕【数】　Jacobi iteration
雅可比积分【天】　Jacobi's integral
雅可比能量转换　Jacobian energy transformation
雅可比算子　Jacobian operator
雅可比行列式【数】　Jacobian
亚澳季风系统　Asian-Australian monsoon system
亚暴　substorm
亚北方〔气候〕期　subboreal climatic phase
亚北方晚期〔气候〕　Late Subboreal climatic phase
亚北方早期〔气候〕　Early Subboreal climatic phase
亚北方中期〔气候〕　Middle Subboreal climatic phase
亚北极水　Subarctic Water
亚北极中层水　Subarctic Intermediate Water
亚格鲁〔加热式〕温度表　Yaglou thermometer
亚毫米波【电子】　submillimerter
亚绝热状态　sub-adiabatic state
亚临界　sub-critical
亚临界分岔　subcritical bifurcation
亚临界流　subcritical flow
亚临界稳定度　subcritical stability
亚洛伦兹谱线　sub-Lorentzian spectral lines
亚马孙平原（巴西）　Amazon Basin
亚声速〔的〕　subsonic
亚声〔速〕运动　subsonic motion
亚松弛【数】　underrelaxation
亚速尔反气旋　Azores anticyclone
亚速尔高压【大气】　Azores high
亚速尔海流　Azores Current
亚太清洁发展与气候伙伴计划（2005年制订）　APP（Asia-Pacific Partnership of Clean Development and Climate）
亚稳　metastable
亚稳度　metastability
亚稳态【物】　metastable state
亚硝酸　nitrous acid
亚原子粒子　subatomic particle
亚洲和太平洋经济社会理事会（联合国）　Economic and Social Council of Asia and Pacific（ESCAP, UN）
亚洲和远东经济委员会（联合国）　Economic Commission for Asia and the Far East（ECAFE, UN）
亚洲及太平洋理事会　Asian and Pacific Council（ASPAC）
氩离子激光器【电子】　argon ion laser
氩〔离子〕激光器【物】　argon〔ion〕laser
氩〔气〕【大气】　argon
烟【大气】　smoke
烟尘层顶　smoke horizon
烟尘计　smokemeter
烟尘图　smoke chart
烟囱【电力】　stack
烟囱高度　stack height
烟囱排放　stack emission
烟囱排放物　stack effluent
烟囱式气流　chimney current
烟囱效应　stack effect
烟〔囱〕云　chimney cloud
烟道气〔体〕　flue gas

烟道气净化器　flue gas purifier
烟害　smoke injury
烟灰　smoke fly ash
烟灰云【大气】　ash cloud
烟迹　smoke trail
烟流【大气】　smoke plume
烟流〔上升〕模式　plume〔rise〕model
烟霾　smoke haze, smaze
烟幕【大气】　smoke screen
烟雾【大气】　smog, smoky fog, smoke fog
烟雾层　smog layer
烟雾层顶　smog horizon
烟雾锋　smog front
烟雾挥发度指数　smoke volatility index
烟雾浓度显示仪　smoke density indicator
烟雾气溶胶【大气】　smog aerosol
烟雾色调　smoke shade
烟雾指数　smog index
烟羽　plume
烟羽【大气】　smoke plume
烟羽高度【大气】　plume height
烟羽类型【大气】　plume type
烟羽密度　smoke plume density
烟羽抬升【大气】　plume rise
烟云【大气】　smoke cloud
烟云柱　cloud column
淹没　drown
延迟　delay
延迟时间【计】　delay time
延迟线【计】　delay line
延伸预报　extended-range forecast, extended-term forecast
延伸预报【大气】　extended forecast
延时图像传输【大气】　delay picture transmission（DPT）
延展荚状云　whaleback cloud
严冰　winter ice
严冬　severe winter, killing freeze
严寒　bitter cold, pinching frost, severe cold
严酷气候　Stern climate
严霜　severe frost, killing frost
严重污染　severe contamination
岩床【地质】　sill
岩浆水　Juvenile water, magmatic water
岩石圈【大气】　lithosphere
岩石地层学【地质】　lithostratigraphy
岩石学　lithology
芯样品　core sample
岩心分析　core analysis
炎热干旱区　hot arid zone
沿岸海流　longshore current, littoral current
沿岸急流　coastal jet
沿岸流　alongshore current
沿岸流【航海】　coastal current
沿岸泥沙流【海洋】　littoral drift, longshore drift
沿谷风　along-valley wind
沿谷风系　along-valley wind system
沿轨道扫描　along-track scanning
沿海地区冷海流　Maritime Province Cold Current
沿海海洋环境　coastal marine environment
沿海航行指南　Coast Pilot
沿坡风系　along-slope wind system
研究报告　research report
研究论文　research paper
研究性〔的〕消雾计划　fog investigation dispersal operation（FIDO）
盐度　salinity
盐度电桥　salinity bridge
盐度计　salinometer
盐度-温度和深度　salinity-temperature-depth（STD）
盐粉播撒【大气】　salt-seeding
盐风灾害　salty wind damage
盐核【大气】　salt nucleus
盐湖【地理】　salt lake
盐粒　salt particle

盐卤　bittern
盐霾　salt haze
盐漠【地理】　salt desert
盐侵　salt intrusion
盐侵【海洋】　saltwater intrusion
盐舌　salinity tongue
盐霜　salt efflorescence
盐酸　hydrochloric acid
盐酸气体　hydrochloric acid gas
盐线　salt line
盐跃层　halocline
盐泽　tidal marsh
盐指【海洋】　salt fingering
颜色反应　color reaction
衍射【大气】　diffraction
衍射波　diffracted wave
衍射峰值　diffraction peak
衍射光谱　diffraction spectrum
衍射角【物】　angle of diffraction, diffraction angle
衍射耦合谐振腔　diffraction coupled resonator (DCR)
衍射区　diffraction region, diffraction zone
衍射条纹　diffraction fringe
衍射图样【物】　diffraction pattern
衍射现象　diffraction phenomenon
衍射线　diffracted ray
衍射晕　diffraction halo
衍生云　genitus
衍生灾害　derivative disaster
掩【天】　occultation
掩日法　solar occultation
掩星　obscuration
掩星法【大气】　occultation method
眼壁　eye wall
眼壁柱　eye-wall chimney
眼涩度　eye irritation
眼窝正视　photopic vision
演变　development
演化【物】　evolution

演化植物学【植】　evolutionary botany
厌阳植物　heliophobe
厌氧条件　anaerobic condition
厌氧微生物　anaerobia
厌雨植物　ombrophobe
验潮计　marigraph
验潮器　automatic tide gauge
验潮图　marigram
验电器【物】　electroscope
验温器　thermoscope
验压器　baroscope
验证【物】　verification
验证统计　verification statistics
验证样本　verification sample
堰　weir
焰弹　pyrotechnic flare
焰弹发生器　pyrotechnic generator
焰状极光　flaming aurora
燕尾突变　swallow-tail catastrophe
扬沙【大气】　blowing sand
扬蚀【土壤】　raising erosion
阳极　anode
阳极电流【化】　anodic current
阳离子交换容量　cation exchange capacity
阳历【天】　solar calendar
阳坡　adret
阳坡【地理】　sunny slope
阳伞效应　umbrella effect
洋　ocean
洋底边界层　benthic boundary layer
洋壳【海洋】　oceanic crust
洋流【大气】　ocean current
洋流图【航海】　ocean current chart
洋盆【地理】　oceanic basin
洋中槽【大气】　tropical upper-tropospheric trough
洋中脊【海洋】　mid-ocean ridge
洋中隆【地质】　mid-ocean rise
仰角　elevation angle, vertical angle
仰角【数】　angle of elevation

仰角位置指示器　elevation position indicator（EPI）
氧饱和　oxygen saturation
氧18冰芯纪录　oxygen-18 ice core record
氧分子带　oxygen band
氧化反应　oxidizing reaction
氧化还原电池【电子】　redox cell
氧化还原指示剂【化】　oxidation-reduction indicator
氧化还原〔作用〕【化】　redox
氧化环境【地】　oxidizing environment
氧化剂【大气】　oxidant
氧化剂【化】　oxidizing agent
氧化腈【化】　nitrile oxide
氧化铝湿度表　alumin(i)um oxide hygrometer
氧化铝〔湿度〕传感器　alumin(i)um oxide sensor
氧化铝湿度元件　alumin(i)um oxide humidity element
氧化镁法　magnesium oxide method, magnesia method
氧化塘【农】　lagoon
氧化物【化】　oxide
氧化物比色分析仪　colorimetric oxidant analyzer
氧化物催化剂【化】　oxide catalyst
氧化亚氮　nitrous oxide(N_2O)
氧化〔作用〕　oxidation
氧亏欠　oxygen deficiency
氧〔气〕　oxygen(O)
氧气分压　oxygen partial pressure
氧气罐　hydrogen pot
氧气含量　oxygen content
氧气绿线　green oxygen line
氧同位素比　oxygen isotope ratio
氧同位素纪录　oxygen isotope record
氧吸收带　oxygen absorption band
氧吸收谱　oxygen absorption spectrum
氧效应【农】　oxygen effect
氧循环　oxygen cycle

样本【数】　sample
样本百分位数【数】　sample percentile
样本分位数【数】　sample quantile
样本函数【数】　sample function
样本均值【数】　sample mean
样本空间【数】　sample space
样本量【数】　sample size
样本偏差　sample deviation
样本〔容〕量　sample capacity
样本四分位数【数】　sample quartile
样本似然函数　sample likelihood function
样本中位数【数】　sample median
样本资料　data sample
样条插值　spline interpolation
样条〔函数〕　spline function
样条〔函数〕【数】　spline
样条函数展开　spline-function expansion
样条拟合【数】　spline fitting
幺矩阵【数】　unit matrix
遥测　remote measurement
遥测【电子】　telemetry
遥测表　telemeter
遥测方式　remote mode
遥测干湿表　telepsychrometer, remote reading psychrometer
遥测光度表　telephotometer
遥测记录　tele-recording
遥测接收机　telemetric receiver
遥测气象计　telemeteorograph
遥测气象仪器学　telemeteorography
遥测气象仪器制造学　telemeteorometry
遥测气象站　remote meteorological station
遥测台　observatory-remote
遥〔测〕探〔空〕　remote sounding
遥测图像　remote picture
遥测温度表【大气】　telethermometer
遥测温度计　remote thermometer, telemetering thermometer, distance

遥测温度器 telethermoscope
遥测雪深计 snow-depth telemeter
遥测雨量计 telemetering precipitation gauge, distance rainfall recorder
遥测雨量计【大气】 telemetering pluviograph
遥测自动气象站 telemetering automatic weather station
遥感〔测量〕【地信】 remote sensing
遥感技术 remote sensing technique
遥感平台 remote sensing platform
遥感信息 remote sensing information
遥感仪器 remote sensing instrument
遥感资料 remotely sensed data
遥控 distance control, telecontrol
遥控机械装置 teleoperator
遥控空气取样器 remote controlled air sampler
遥控装置 remote control equipment
遥相关 teleconnexion
遥相关【大气】 teleconnection
遥相关型 teleconnection pattern
遥响应 remote response
要冲城市【地理】 gateway city
要素 element
〔要素间〕一致性检验 internal consistency check
耀斑辐射电离层效应 flare-radiation ionospheric effect
耀斑回波 flare echo
业务模式时限 operational model timing
业务使用 operational use
业务数值模式 operational numerical model
业务水文学咨询委员会（WMO） Advisory Committee for Operational Hydrology（ACOH,WMO）
业务线扫描系统 Operational Linescan System（OLS）
业务预报【大气】 operational forecast,

operational prediction
业余气象站 amateur weather station
业余预报 amateur forecast
叶绿素【植】 chlorophyll
叶绿体 RNA【生化】 chloroplast RNA
叶绿体 DNA【生化】 chloroplast DNA
叶面积指数【农】 leaf area index（LAI）
叶面湿润 leaf wetness
叶面湿润持续时间 leaf wetness duration
叶面湿润度 surface wetness
叶面湿润期 surface wetness duration
叶温【大气】 leaf temperature
页面存取时间【计】 page access time
页状冰 shale ice
曳光弹 trailing flare
曳力 drag
曳力定律 drag law
曳式锋【大气】 trailing front
曳引度 tractionability
曳引力 tractive force, traction
夜风 night wind
夜光云【大气】 luminous night cloud, noctilucent cloud
夜辉光谱 night air-glow spectrum
夜间 nighttime
夜间边界层 nocturnal boundary layer
夜间低空急流 low-level nocturnal jet
夜间电离层 nighttime ionosphere
夜间对流 nocturnal convection
夜间飞行 night flight
夜间辐射【大气】 nocturnal radiation
夜间急流【大气】 nocturnal jet
夜间雷暴 nocturnal thunderstorm
夜间冷却 nocturnal cooling
夜间能见度 night visibility
夜间能见距离 night visual range
夜间逆温 nocturnal inversion
夜间气辉 nightglow
夜间气象能见度 meteorological visibility at night

夜间视距　penetration range
夜间下泄风　nocturnal drainage wind
夜间最低气温　nocturnal minimum temperature
夜露　night-dew
夜气辉　permanent aurora
夜天发光　night-sky luminescence
夜天辐射　night sky radiation
夜天光【大气】　night-sky light, light of night sky
液化气体　gasol
液化热【化】　heat of liquefaction
液化天然气【地质】　liquified natural gas
液化〔作用〕　liquefaction
液晶　liquid crystal
液晶显示【计】　liquid crystal display（LCD）
液泡追踪物　liquid-bubble tracer
液态　liquid state
液态含水量　liquid water content(LWC)
J-W 液态含水量仪　J-W meter
液态空气　liquid air
液态水　liquid water
液态水光程　liquid water path(LWP)
液态水含量　liquid water-holding capacity, liquid-water content (LWC)
液态水混合比　liquid water mixing ratio
液态水静力能　liquid water static energy
液态水位温　liquid water potential temperature
液态水载荷　liquid water loading
液态云滴取样器　cloud water droplet sampler
液体　liquid
液体色谱仪　liquid chromatograph
液体温度表　liquid thermometer
液〔体〕限〔度〕　liquid limit
液相　liquid phase
液相反应【化】　liquid phase reaction
液相色谱法　liquid chromatography

液压梯度　hydraulic gradient
液压系统【航海】　hydraulic system
液柱气压表　liquid-column barometer
一般飞行报告　general notice
一般年【大气】　normal flow year
一般农业气象站　ordinary agricultural meteorological station
一般气候站【大气】　ordinary climatological station
一般天气预报　general forecast
一般照明【航海】　general lighting
一次测风雷达　primary wind-finding radar
一次空间平均　single space mean
一次雷达【电子】　primary radar
一次雷达技术　primary radar technique
一次散射　single scattering
一次散射【大气】　primary scattering
一冬冰【大气】　first year ice
一级标准【化】　primary standard
一级标准绝对日射表　primary standard pyrheliometer
一级标准气压表　primary standard barometer
一级反应【化】　first order reaction
一级环流　primary circulation
一级基准仪器　absolute instrument
一级绝对腔体辐射仪　primary absolute cavity radiometer（PACRAD）
一级气候站　first order climatological station
一级气象站　first order station
一级生态系统　first order ecosystem
一级图谱【化】　first order spectrum
一级相变【化】　first order phase transition
一级资料使用站　primary data user station（PDUS）
一阶半闭合　one-and-a-half order closure
一阶闭合　first order closure
一阶闭合模式　first order closure model

一阶不连续性　discontinuity of first order
一阶段模式　one-tier model
一阶多次散射近似　first order multiple scattering approximation
一阶矩　first moment
一阶平滑　first order smoothing
一阶平滑近似　first order smoothing approximation
一阶扰动解　first order perturbation solution
一阶线性微分方程【数】　linear differential equation of first order
一阶自相关　first order autocorrelation
一年冰【大气】　first year ice
一年两次的　biannual
一年生植物【植】　annual plant
一维模式　one-dimensional model
一维云探测器　one-dimensional cloud probe
一氧化氮【大气】　nitric oxide（NO）
一氧化硫(SO)　sulphur monoxide
一氧化氯二聚物(Cl_2O_2)　chlorine monoxide dimmer
一氧化氯基（ClO）　chlorine monoxide radical
一氧化碳【大气】　carbon monoxide
一元二次方程【数】　quadratic equation with one unknown
一元时间序列　univariate time series
一元最优内插　univariate optimum interpolation
一致空间【数】　uniform space
一致收敛性【数】　uniform convergence
一致性检验【计】　consistency check
一致性约束【计】　consistency constraint
伊迪波　Eady wave
伊尔明格海流　Iriminger Current
伊格尼尔定律　Egnell's law
伊莱亚森-帕尔姆通量　E-P flux（Eliassen-Palm flux）
伊藤方程　Ito equation
伊藤过程【数】　Ito process
医疗气候学【大气】　medical climatology
医疗气象学【大气】　medical meteorology
医学地理学【地理】　medical geography
遗传算法【大气】　GA（genetic aglorithm）
仪表船　instrumentation ship
仪表飞行　instrument flight
仪表飞行规则　instrument flight rules（IFR）
仪表飞行气象条件　instrument meteorological condition（IMC）
仪表飞行天气　instrument weather
仪表飞行最低着陆气象〔条件〕　IFR terminal minimums
仪表着陆系统　instrument landing system
仪器　apparatus
M 仪器　M meter
仪器暴露　instrument exposure
仪器标定　calibration of instrument, calibration tank
仪器补偿　compensation of instrument
仪器测定　instrumentation
仪器订正　instrument correction
仪器分析【化】　instrumental analysis
仪器和观测方法委员会（WMO）　Commission for Instruments and Methods of Observation（CIMO,WMO）
仪器记录时期(气候)　instrumental period
仪器露置　exposure of instruments
仪器漂移　instrumental drift
仪器误差　instrumental error
仪器响应　instrumental response
仪器响应函数　instrument response function
仪器载荷　instrument payload
仪器中子活化分析【化】　instrumental neutron activation analysis
移出【计】　shift-out

移动　migration
移动船舶　mobile ship
移动船舶站【大气】　mobile ship station
移动〔的〕　migratory
移动订正　removal correction
移〔动方〕向　direction of movement
移动风浪区　moving fetch
移动〔观测〕平台　movable platform
移动接收机　moving receiver
移动目标显示雷达　radar indicating moving target
移动目标指示　moving-target indication（MTI）
移动平均模型　moving average（MA）model
移动式遥测站　mobile telemetering station
移动〔污染〕源　mobile source
移动性反气旋　travelling anticyclone, migratory anticyclone
移动性高压　migratory high
移动重力波　moving gravity wave
移频键控　frequency shift keying
移入【计】　shift-in
移行〔罗斯贝比〕波　propagating（Rossby）wave
乙腈　acetonitrile
乙醛　acetaldehyde
乙炔　acetylene
乙酸　acetic acid
乙烷　ethane
乙烯　ethene, ethylene
以色列气象学会　Isreal Meteorological Society(IMS)
以太波　ether wave
以太风【物】　ether wind
以太网【计】　Ethernet
钇铝石榴石激光器【电子】　Yttrium Aluminum Garnet（YAG）Laser
蚁酸　formic acid
异步传送模式【计】　asynchronous transfer mode
异步通信　asynchronous communication
异常　abnormality
异常传播　abnormal propagation, anomalous propagation
异常的　anomalous
异常电离　anomalous ionization
异常回波　angel
异常回波【大气】　angel echo
异常　anomaly
异常能听区　zone of abnormal audibility
异常台风移动　unusual typhoon movement
异常太阳活动〔性〕　unusual solar activity
异常梯度风　anomalous gradient wind
异常天气　unusual weather, extraordinary weather, abnormal weather
异常天气型　unusual synoptic pattern
异常云线　anomalous（cloud）line
异常折射　abnormal refraction
异常直减率　abnormal lapse rate
异丁烯醛　methacrolein
异化　alienation
异密层　schlieren
异戊二烯　isoprene
异相〔位〕【物】　out-of-phase
异质成核　heterogeneous nucleation
异质结【电子】　heterojunction
异质结构【电子】　heterostructure
异质结太阳电池【电子】　heterojunction solar cell
抑制　suppress
抑制〔作用〕　inhibition
易变卷云　inconstant cirrus
逸散层底【地】　exobase
溢出　overflow
溢洪道　spillway
溢洪道容量　spillway capacity
溢洪道设计洪水　spillway design flood
翼架温度表　strut thermometer
翼式风力发电机　wing aerogenerator

因次分析【大气】 dimensional analysis
因式分解 factoring
因数 factor
因数分解【数】 factorization
因素效应【化】 factorial effect
因特网【计】 Internet
因特网电话【计】 Internet phone
因子分析【大气】 factor analysis
阴沉 heavy overcast
阴沉〔的〕天气 dull weather
阴沉天空 ugly sky, ugly weather
阴电 resinous electricity
阴极 cathode
阴极电流【化】 cathodic current
阴极射线 cathode ray
阴极射线测向器 cathode-ray direction-finder, cathode-ray direction finding
阴极射线定向器 cathode-ray radiogoniometer
阴极射线管 cathode-ray tube (CRT)
阴极射线示波器 cathode-ray oscilloscope
阴历 lunar calendar
阴坡【地理】 ubac
阴日 overcast day
阴天 overcast sky, thick weather, heavy weather
阴天【大气】 overcast
阴温 shade temperature
阴影区 shaded area
阴影效应【物】 shadow effect
阴影有效比 shading effect ratio
音频 voice-frequency
殷墟 Yin Dynasty ruin
银光风暴 silver storm
银光霜 silver frost
银河【天】 Milky Way
银河光 galactic light
银河系【天】 Galactic System, Galaxy
银河系辐射 galactic radiation
银河宇宙射线 galactic cosmic ray

银盘绝对日射表 silver-disk pyrheliometer
霪雨 excessive rain
引潮力【地】 tide-generating force
引潮位【地】 tide-generating potential
引潮〔重力〕位势【海洋】 tidal〔gravitational〕potential
引导 steering
引导层 steering surface
引导法则 steering law
引导风 steering wind
引导概念 steering concept
引导高度【大气】 steering level
引导气流 steering flow
引导气流【大气】 steering current
引导中心等离子体 guiding centre plasma
引导着陆〔系统〕 ground-controlled approach
引发闪电 sympathetic discharge
引理 lemma
引力常量【物】 gravitational constant
引力潮【地】 gravitational tide
引力模式【地理】 gravity model
引力势【物】 gravitational potential
引力势能 gravitational potential energy
引力中心 barycenter, center of attraction
引水渠 approach channel
引曳速度 drawing velocity
隐蔽作用 sequestration
隐藏层 hidden layer
隐藏性雷暴 embeded thunderstorm
隐藏周期性 hidden periodicity
隐槽 masked trough
隐锋 masked front
隐函数【数】 implicit function
隐模式 implicit model
隐嵌积云 embeded cumulus
隐生宙【地质】 Cryptozoic Eon
隐式并行性【计】 implicit parallelism

隐式差分方程【数】 implicit difference equation
隐式差分格式 implicit difference scheme
隐式法【化】 implicit method
隐式积分方案 implicit integration scheme
隐式平滑 implicit smoothing
隐式时间差分 implicit time difference
隐式时间格式 implicit time scheme
印第安夏【大气】 Indian summer
印度低压【大气】 Indian low
印度国家卫星系统 Indian National Satellite System(INSAT)
印度季风【大气】 Indian monsoon
印度空间研究组织 Indian Space Research Organization(ISRO)
印度尼西亚〔上层海水〕输送流 Indonesian Throughflow
印度气象局 Indian Meteorological Department(IMD)
印度洋【海洋】 Indian Ocean
印度洋板块【海洋】 Indian Plate
印度洋赤道急流 Indian equatorial jet
印度洋季风 Indian Ocean monsoon
印度洋平均大潮低潮面 Indian spring low water
印度洋深水 Indian Deep Water
印度洋试验 Indian Ocean Experiment (INOEX)
印度洋偶极子 Indian Ocean dipole
印痕 impression
印缅低槽 Indian and Burma trough
印刷电路板 printed(circuit) board
应变【物】 strain
应变数 dependent variable
应变张量 strain tensor
应答器【电子】 transponder, transmitter-responder
应急减洪 emergency flood mitigation
应急解题〔法〕 jury problem

应力【物】 stress
应力效应 stress effect
应力型风速表 strain-gage anemometer
应力张量【物】 stress tensor
应用地理学【地理】 applied geography
应用光学【物】 applied optics
应用技术卫星【大气】 Application Technology Satellite(ATS)
应用气候学【大气】 applied climatology
应用气象学【大气】 applied meteorology
应用软件【计】 application software
应用水文学 applied hydrology
应用物理〔学〕【物】 applied physics
英格兰雷暴 hurly-burly
英国皇家气象学会 Royal Meteorological Society(RMS)
〔英国〕皇家学会 Royal Society(RS)
英国南极测量局 British Antarctic Survey(BAS)
英国南极研究委员会 British National Committee on Antarctic Research (BNCAR)
英国气象局 United Kingdom Meteorological Office(UKMO)
英里 mile
英联邦 British Commonwealth of Nations(BCN)
英制单位区域预报国际电码 ARFOT
英制热〔量〕单位 British thermal unit
迎风 aweather, upstream
迎风差分【数】 upwind difference
迎风差格式【大气】 upstream scheme
迎风的 weather-side
迎风面 windward side
荧光 fluorescence, fluorescent light
荧光猝灭法【化】 fluorescence quenching method
荧光分光光度计 fluorescence spectrophotometer
荧光分析 fluorescence analysis
荧光光谱【化】 fluorescence spectrum

荧光粒子自动计数器　automatic fluorescent particle counter
荧光散射　fluorescent scattering
荧光色谱　fluorescence chromatogram
荧光示踪物　fluorescent tracer
盈水河　gaining stream
影区【地】　shadow zone
影响　impact
影响半径　radius of influence
影响场　influence field
影响函数【数】　influence function
影响区　area of influence, region of influence
影像质量【地理】　image quality
映射【数】　mapping
映射函数【数】　mapping function
硬磁盘【计】　hard disk
硬度　hardness
硬度测量计　penetrometer
硬辐射　hard radiation
硬件【计】　hardware
硬件模型　hardware model
硬科学　hard science
硬盘　hardpan
硬盘【计】　rigid disk
硬雪壳　marble crust
永磁体【物】　permanent magnet
永冻　permafrost
永冻地　perennially frozen ground（＝permofrost）
永冻面　pergelisol table, permafrost table
永冻气候　climate of eternal frost, eternal frost climate
永冻气候亚类【大气】　perpetual frost climate
永冻上层　supragelisol
永冻土　pergelisol【大气】, ever frozen soil, neve frozen soil, permafrost【大气】
永冻土活动层　suprapermafrost layer
永冻作用　pergelation

永久磁化　permanent magnetization
永久积雪　firn snow, neve penitent
永久积雪【大气】　firn
永久积雪作用　firnification
永久记录簿　permanent register
永久镜冰　firnspiegel
永久性低压【大气】　permanent depression
永久性反气旋【大气】　permanent anticyclone
永久性高压【大气】　permanent high
永久〔性海〕流　permanent current
永久性温跃层【海洋】　permanent thermocline
永久性自动中尺度观测网　stationary automated mesonet (SAM)
永久雪限　firn limit
永久雪线　neve line
永久雪线【大气】　firn line
永枯点　permanent wilting point
涌潮　eager (＝ragre), tidal bore
涌浪　ocean swell, swell
涌〔浪等〕级　swell scale
涌流　surge current
涌升流　upwelling current
涌升水域　upwelling pocket
用户导向的　user-oriented
用户电报【电子】　telex
用户端信息　user-end information
用户密钥【计】　user key
用户区【计】　user area
用户需求　user requirement
用户注册【计】　user log-in, user log-on
用户注销【计】　user log-off
用于数据交换的字符表示格式　character representation form for deta exchange (CREX)
优生学【生】　eugenics
优势半径　predominant radius
优势反应率　over-all rate
优势木【林】　dominant tree

优势云覆盖区　preferred cloud coverage
优势周期　preferred period
优先权【数】　priority
幽〔灵〕闪〔电〕　sprite
尤因-唐〔全球气候变化〕理论　Ewing-Donn theory
邮件发送清单【计】　maillist, mailing list
邮件分发器【计】　mail exploder
油藏【地质】　oil pool
油膜　oil slick
油酸【生化】　oileic acid
油田【地质】　oil field
油页岩【地质】　oil shale
油脂状冰　grease ice, ice fat
游标【物】　vernier
游标尺　vernier scale
游览地　tourist site
游离酸性　free acidity
有害因子　adverse factor
有机地球化学【地】　organic geochemistry
有机肥料【土壤】　organic fertilizer
有机过氧化物　organic peroxides
有机化学【化】　organic chemistry
有机气溶胶　organic aerosol
有机酸　organic acids
有机碳　organic carbon
有机污染物　organic pollutant
有机硝酸盐　organic nitrates
有界导数法　bounded-derivative method
有界弱回波区　bounded weak echo region（BWER）
有理函数【数】　rational function
有理数【数】　rational number
有理指数定律【化】　law of rational indices
有利天气　favourable weather
有偏估计【数】　biased estimate
有人测站　attended station
有色雨　colored rain
有无降雨传感器　on/off rain sensor

有雾〔的〕　foggy
有限差比　finite difference ratio
有限差分　finite difference, finite differencing
有限差分法　method of finite difference, finite difference method
有限差分方程　finite difference equation
有限差分近似　finite difference approximation
有限差〔分〕商　finite difference quotient
有限带宽白噪声　bandwidth limited white noise
有限傅里叶级数　finite Fourier series
有限厚度　finite depth
有限频宽功率损失　finite bandwidth power loss
有限区模式【大气】　limited-area model（LAM）
有限区细网格模式【大气】　limited area fine-mesh model（LFM）
有限区〔域〕嵌套网格模式　limited-area nested grid model（LNGM）
有限〔区域〕细网格模式　limited fine-mesh model
有限区〔域〕预报　limited-area prediction
有限区〔域〕预报模式　limited-area forecast model
有限体积方程　finite volume equation
有限维向量空间【数】　finite dimensional vector space
有限元　finite element
有限元法【数】　finite element method
有限元分析【数】　finite element analysis
有限元模式　finite element model
有限源法　finite source method
有限振幅【大气】　finite amplitude
有限振幅不稳定度　finite amplitude instability
有限振幅对流　finite amplitude convection
有限振幅扰动　finite amplitude

disturbance
有效半径 effective radius
有效边界层 effective boundary layer
有效波 significant wave
有效波【地】 effective wave
有效波长 effective wavelength
有效波高 significant wave height
有效长度 effective length
有效场 effective field
有效尺度参数 effective size parameter
有效粗糙〔度〕长度 effective roughness length
有效大气透射【地】 effective atmospheric transmission
有效带宽 effective band width
有效地球半径【大气】 effective earth radius
有效地球辐射 effective terrestrial radiation
有效风能【大气】 available wind energy
有效风速 available wind velocity
有效风速【大气】 effective wind speed
有效风向 effective wind direction
有效辐射层 effective radiation layer
有效辐射【大气】 effective radiation
有效高度 virtual height
有效各向同性辐射功率 effective isotropic radiated power(EIRP)
有效功率 effective power
有效荷载 effective load
有效洪水预警 effective flood warning
有效积温 total effective temperature
有效积温【大气】 effective accumulated temperature
有效降水【大气】 effective precipitation
有效降雨强度 effective rainfall intensity
有效降雨损失量 effective abstractions
有效焦距 effective focal length (EFL)
有效接收孔径 effective receiver aperture
有效接收面积 effective receiver area

有效截面 effective section
有效可降水分 effective precipitable water
有效孔径 effective aperture
有效孔隙度 effective porosity
有效库容 available storage capacity
有效扩散系数 effective diffusion coefficient
有效亮度温度 effective brightness temperature
有效量程 effective span
有效流量 effective discharges
有效流域面积 active basin area
有效脉冲长度 effective pulse length
有效面积 effective area
有效能见度 potential visibility
有效能见度【大气】 effective visibility
有效能量 available energy
有效融雪量 effective snowmelt
有效射线 effective ray
有效渗透率 effective permeability
有效生长能量 effective growth energy
有效声中心 effective acoustic center
有效数字【数】 significant digit, significant figure
有效衰减系数 effective attenuation factor
有效水分【大气】 available water, effective moisture
有效水资源利用 effective use of water resources
有效太阳辐射【大气】 available solar radiation
有效通量 effective flux
有效统计量 efficient statistic
有效湍流通量 effective turbulent flux
有效位能【大气】 available potential energy（APE）
有效位温 effective potential temperature
有效温度【大气】 effective temperature
有效温度指数 effective temperature index
有效向外辐射 effective outgoing radiation

有效性【数】 validity
有效压头【航海】 effective head
有效烟囱高度【大气】 effective stack height
有效样本数 effective sample size
有效夜间辐射 effective nocturnal radiation
有效液水含量 effective liquid water content
有效应力【力】 effective stress
有效雨量【农】 effective rainfall
有效雨量过程线 effective rainfall hyetograph
有效张量 effective tensor
有效阵风速度 effective gust velocity
有效蒸发势〔能〕 EAPE (evaporative available potential energy)
有效蒸散【大气】 actual(或 effective) evapotranspiration
有效值 effective value
有效字符 valid character
有心力 central force
有序人类活动 orderly human activities
有源腔体辐射仪 active cavity radiometer(ACR)
有源散射气溶胶粒谱仪 active scattering aerosol spectrometer (ASAS)
有源散射气溶胶粒谱仪探测器 active scattering aerosol spectrometer probe (ASASP)
有源元件 active element
有源元件【电子】 active component
有组织大涡旋 organized large eddies
有组织对流 organized convection
右手式旋转 right-handed rotation
右手〔旋转〕坐标 right-handed coordinates
右手直角坐标系 right-handed rectangular coordinates
右移强风暴 severe right moving storm
诱导 induction

诱导反应【化】 induced reaction
诱导回灌 induced recharge
诱导期【化】 induction period
诱导试验 induction experiment
诱导效应【化】 inductive effect
诱发裂变【化】 induced fission
诱生槽 impressed trough
余变量 covariable, covariate
余赤纬 codeclination
余函数 complementary function
余辉 afterglow
余数【数】 remainder
余维〔数〕【物】 codimension
余纬 colatitude
余弦窗 cosine window
余弦矩形窗 cosine-tapered rectangular window
余震【地】 aftershock
鱼类资源【海洋】 fish resources
鱼鳞天【大气】 mackerel sky
鱼鳞状卷积云 cirrocumulus "mackerel"
渔礁【航海】 fish reef
渔业海洋学【海洋】 fisheries oceanography
渔业气象学 fisheries meteorology
隅角频率法 corner-frequency method
隅角效应 corner effect
逾量衰减 excess attenuation
与非门【计】 NAND gate
宇【地质】 eonothem
宇称守恒【物】 parity conservation
宇航员 astronaut
宇宙 cosmos
宇宙【天文】 universe
宇宙背景辐射【地】 cosmic background radiation
宇宙常数 cosmical constant
宇宙尘【大气】 cosmic dust
宇宙尘成分分析 composition analysis of cosmic dust
宇宙成因的放射性核素 cosmogenic

radionuclides (CRNs)
宇宙丰度【物】 cosmic abundance, universal abundance
宇宙辐射 cosmic radiation
宇宙航行 astronautics
宇宙空间 aerospace, astrospace
宇宙年代学【地质】 cosmochronology
宇宙气象学 cosmical meteorology
宇宙〔射电〕噪声 jansky noise
宇宙射电噪声【地】 cosmic radio noise
宇宙〔射〕线 cosmic ray
宇宙速度 cosmic velocity
宇宙探测器 space probe
宇宙卫星 Kosmos, Cosmos
宇宙线【大气】 cosmic ray
宇宙线产生的放射性同位素 cosmogenic radioisotopes
宇宙线簇射 cosmic-ray shower
宇宙线电离 cosmic-ray ionization
宇宙学【天】 cosmology
宇宙〔演化〕论【物】 cosmism
宇宙演化【物】 universe evolution
宇宙医学 space medicine
宇宙噪声【物】 cosmic noise
羽角【海洋】 feather angle
羽状晶体 feathery crystal
羽状移动【海洋】 feathering
雨【大气】 rain
雨胞 rain cell
雨飑 rain squall
雨层云【大气】 nimbostratus (Ns)
雨成湖 pluvail lake
雨代法 gradex method
雨带【大气】 rain belt, rain band
雨滴【大气】 rain drop
雨滴大小【土壤】 raindrop size
雨滴动能【土壤】 raindrop kinetic energy
雨滴冻结 raindrop freezing
雨滴记录器 raindrop recorder
雨滴破碎 drop breakup
雨滴谱 raindrop spectrum

雨滴谱【大气】 raindrop size distribution
雨滴谱仪 rain droplet collector, raindrop spectrograph
雨滴谱仪【大气】 disdrometer, raindrop disdrometer
雨滴侵蚀【土壤】 raindrop erosion, rain erosion
雨滴选集器 raindrop sorting collector
雨幡 precipitation trails, trails of rain, trails of precipitation
雨幡【大气】 rain virga
雨幡回波 streamer echoes
雨幡回波【大气】 elevated echo
雨迹 trace of rainfall
雨季【大气】 rainy season
雨季湖 hamun
雨夹雪 mixed rain and snow, rain and snow mixed, rain and snow
雨夹雪【大气】 sleet
雨量【大气】 rainfall 〔amount〕
雨量报 rainfall message
雨量比 pluviometric quotient
雨量变率 rainfall variability
雨量表 hyetometer
雨量测定法【大气】 pluviometry
雨量场 rain field
雨量-持续时间-频数曲线 depth-duration-frequency curve
雨量-持续时间曲线 depth-duration curve
雨量赤道 hyetal equator
雨量分布【农】 rainfall distribution
雨量分布学 hyetography
雨量计 hyetograph
雨量计【大气】 recording rain gauge, pluviograph
雨量计曲线 hyetographic curve
雨量-面积-持续时间分析 depth-area-duration analysis
雨量-面积公式 depth-area formula
雨量逆减 rainfall inversion, inversion of precipitation, inversion of rainfall

雨量器　pluviometer, udometer, pluvioscope
雨量器【大气】　rain gauge
雨量器防护罩　rain gauge shield
雨量器风档　rain gauge wind shield
雨〔量〕强〔度〕　raininess
雨量强度　rainfall density
雨量损失　rainfall loss
雨量桶　vessel of the rain gauge
雨量筒　non-recording rain gauge, precipitation gauge
雨量筒防风圈　Alter shield
雨量图【大气】　rainfall chart
雨量系数　hyetal coefficient, pluviometric coefficient
雨量型　rainfall regime
雨量站　precipitation station, rainfall station
雨量指数　pluvail index
雨林　raintrees
雨林【植物】　rain forest
雨林气候　rainforest climate
雨期　pluvial, wet spell, rain spell
雨强分布型　rainfall intensity pattern
雨强计　rain-rate gauge, rate-of-rainfall recorder, rate-of-rainfall gauge
雨强计【大气】　rainfall intensity recorder
雨强重现周期　rainfall intensity return period
雨区　hyetal region, rain area
雨区移置【水利】　storm transposition
雨日　day of rain, wet day
雨日【大气】　rain day
雨日【农】　rainy day
雨蚀〔作用〕　rainwash
雨蚀〔作用〕【农】【大气】　rainfall erosion
雨衰减　rain attenuation
雨水(节气)【大气】　rainwater
雨水冻结壳　rain crust
雨水含盐度　rain salinity
雨水清洁作用　cleaning effect by rain
雨水渗透【建筑】　rain penetration
雨凇　glazed frost, verglas, silver thaw, glaze ice

雨凇【大气】　glaze
雨凇暴　glaze storm
雨温比　pluviothermic ratio
雨洗【大气】　rain-out, scavenging by precipitation, rain washout
雨型　rainfall pattern
雨养农田　rainfed cropland
雨养农业【农】　rainfed farming
雨影　precipitation shadow
雨影【大气】　rain shadow
雨影效应　rainshadow effect
雨云　nimbus (Nb), rain cloud
雨云极化　rain cloud polarization
雨云气象卫星　Nimbus meteorological satellite
雨云卫星【大气】　NIMBUS
雨云-E卫星微波波谱仪　NIMBUS E microwave spectrometer (NEMS)
雨云卫星资料处理中心　Nimbus Data Handling Facility (NDHF)
语境分析【计】　context analysis
语言分析　speech analysis
语义网络【计】　semantic network
语音合成【计】　speech synthesis
语音识别【计】　speech recognition
玉米热单位　corn heat unit
玉石【地质】　jade
育苗室【大气】　phytotron
预白化　prewhitening
预饱和器　gross saturator
预报　prediction
预报差异　forecast difference
预报方程　predictive equation, prognostic equation
预报〔方法〕　forecasting
预报分析循环　forecast-analysis cycle
预报集成　forecast ensemble
预报技术　forecasting technique
预报检验【大气】　forecast verification
预报检验评分　forecast verification score
预报技术　forecast skill

预报鉴定　forecast identification
预报距平　forecast anomaly
预报量【大气】　predictand
预报敏感性　forecast sensitivity
预报判据　forecasting criterion
预报偏低(少)的　under predicated
预报评分【大气】　forecast score
预报评估　forecast evaluation
预报期　forecasting period
预报区【大气】　forecast area, forecast district, forecast zone
预报时段　forecast period
预报时效　period of validity
预报示范项目　Forecast Demonstration Project (FDP)
预报术语　forecast terminology
预报提前时间　forecast lead time
预报图【大气】　forecast chart, prognostic chart
预报误差　forecast error, forecasting error
预报因子【大气】　predictor
预报员　forecaster
预报云〔方程组〕　prognostic clouds
预报中心　forecasting center
预报责任区【大气】　responsible forecasting area
预报准确率【大气】　forecast accuracy
预测　foreshadow
预处理【电子】　pretreatment, preprocessing
预防　precaution
预活化　preactivation
预计到达时间　estimated time of arrival (ETA)
预警　early warning, warning
预警阶段　warning stage
预警雷达【电子】　early warning radar
预警系统　early warning system
预警〔信度〕水平　alarm level
预冷水槽　pre-cooling water reservoir
预离解　predissociation
预热时间　warm up time
预设置值　prelims

预选容许误差　pre-selected tolerance
预研究　pilot study
预应力【地质】　prestress
预兆　symbol
预兆〔性〕天空　emissary sky
域　domain
域名【计】　domain name(DN)
阈温　threshold temperature
阈限度　threshold limit value
阈值感应元件　threshold sensor
阈值　threshold
阈值【数】　threshold value
阈值条件【物】　threshold condition
阈值自回归　threshold autoregression
遇险【航海】　distress
遇险报警【航海】　distress alerting
元磁暴　elementary storm
元反应【化】　elementary reaction
元古代　Proterozoic Era
元古宇【地质】　Proterozoic Eonothem
元古宙【地质】　Proterozoic Eon
元量涡旋　elemental vortex
元数据【计】　metadata
元数据实体　metadata entity
元数据元素　metadata element
元素地球化学【地】　geochemistry of element
元素丰度【化】　abundance of element
元素起源【地质】　origin of element
元素周期表【化】　periodic table of elements
元素周期律【化】　periodic law of elements
园林工程【林】　landscape engineering
园林学【林】　landscape architecture
原地　in situ
原地测量【地】　in situ measurement
原电池　galvanic cell
原聚体【生化】　protomer
原理　principle
原力矩　raw moment
原林【林】　log
原〔起〕始气体　precursor gas

原生空气　primary air
原生水　connate water, eternal water
原生污染物【大气】　primary pollutant
原生灾害　primary disaster
原始地球　primitive earth
原始方程【大气】　primitive equation
原始方程模式【大气】　primitive equation model
原始林【林】　virgin forest
原始农业【农】　primitive farming
原始谱密度估计量　raw spectral density estimate
原始生物化学【生化】　protobiochemistry
原始数据　initial data
原始数据【计】　raw data
原始土壤【土壤】　primitive soil
原始资料　primary data, original data, crude data
原水【土木】　raw water
原型【计】　prototype
原永冻层　passive permafrost
原子　atom
原子〔爆炸〕烟云　atomic cloud
原子弹云　atomic bomb cloud
原子光谱【物】　atomic spectrum
原子轨道【化】　atomic orbital
原子核【物】　atomic nucleus
原子核物理〔学〕【物】　atomic nuclear physics
原子结构【物】　atomic structure
原子模型【化】　atomic model
原子能【物】　atomic energy
原子时【航海】　atomic time(AT)
原子物理〔学〕【物】　atomic physics
原子吸收谱测量　atomic absorption spectrometry
原子序数【化】　atomic number
原子荧光光谱法【化】　atomic fluorescence spectrometry
原子质量单位【物】　atomic mass unit
原子钟【天】　atomic clock
圆　circle
圆变量　circular variable
圆对称　circular symmetry
圆函数　circle function
圆孔衍射【物】　circular hole diffraction
圆偏振【物】　circular polarization
圆频率【物】　circular frequency
圆退偏〔振〕比　circular depolarization ratio (CDR)
圆涡旋　circular vortex
圆形温度计　circular thermograph
圆振荡　circular oscillation
圆周【数】　circumference
圆周角【数】　angle in a circular segment
圆周速度　circumferential velocity
圆周运动　circular motion
圆柱度【航海】　cylindricity
圆柱函数【数】　cylindrical function
圆柱投影　cylindrical projection
圆柱形净辐射表　cylindrical net radiometer
〔圆〕柱坐标　circular cylindrical coordinates, cylindrical coordinates
圆锥辐射表　cone radiometer
圆锥角　cone angle
圆锥截面　conic section
圆锥体　cone
圆锥投影　conical projection
圆锥形扩散　coning
圆锥〔形〕扫描　conical scanning
源程序【计】　source program
源地　source region, source, origin
源地强度　source strength
源函数　source function
源函数系数　source function coefficient
源汇流　source-sink flow
源项　source terms
远岸流　off-shore current
远场　far field
远程访问服务器【计】　remote access server
远程访问【计】　remote access
远程教学【计】　remote instruction
远程教育【计】　distance education
远程雷电状况　spheric state

远程数据库访问【计】 remote database access
远程终端【计】 remote terminal
远地点潮差 apogean range
远地点【天】【大气】 apogee
远红外【化】 far infrared
远红外辐射 far infrared radiation
远红外区 far-infrared region
远幻日【大气】 paranthelion
远幻月【大气】 parantiselene
远近线 line of apsides
远距离传感仪 remote sensor
远距离分辨测湿法 remote range-resolved hygrometry
远距离外差探测 remote heterodyne detection
远距离温度诊断 remote temperature diagnostic
远距离作用 remote forcing
远日点【天】【大气】 aphelion
远闪 distant flash
远心点【天】 apocenter
远洋渔业【航海】 distant fishery
远月潮 apogean tide
远月点 apocynthion
远紫外辐射 far ultraviolet radiation
约分【数】 reduction of a fraction
约翰逊-威廉姆斯液态含水量仪 Johnson-Williams liquid water meter (=J-W meter)
约翰逊-威廉姆斯液态水探测器 Johnson-Williams liquid water probe
约化波数 reduced wave number
约化公式 reduced form
约化核【数】 reduced kernel
约化入射强度 reduced incident intensity
约化摄动 reductive perturbation
约化网格 reduced grid
约化质量 reduced mass
约化重力 reduced gravity
约束方程 equation of constrain
约束条件【数】 constraint condition
约数 rounding
月 month
月报 monthly bulletin, monthly record
月潮间隙 local establishment, lunitidal interval
月-地关系 moon-earth relation
月光 moonshine
月虹 lunar rainbow, moonbow
月华 lunar corona
月华【大气】 lunar aureole
月际变化 inter-monthly variation
月际变率【大气】 inter-monthly variability
月尖型突变 cusp catastrophe
月径幻觉 moon illusion
月绝对最低温度 absolute monthly minimum temperature
月绝对最高温度 absolute monthly maximum temperature
月可预报性 monthly predictability
月亮潮【大气】 lunar tide
月平均【大气】 monthly mean
月平均等值线 isomenal
月平均日最低温度 mean daily minimum temperature for a month
月平均日最高温度 mean daily maximum temperature for a month
月平均温度【农】 monthly mean temperature
月平均最低温度 mean monthly minimum temperature
月平均最高温度 mean monthly maximum temperature
月气候概述 monthly climatological summary
月球【天】 moon
月球探测器 lunar probe, moonshot
月球正面【天】 near side of the moon
月食【天】 lunar eclipse
月雾色 moonmist
月象 phase of moon
月晕【大气】 lunar halo

月震【地质】 moonquake
月致气压变化 lunar barometric variation
月柱 moon pillar
月总和 monthly sum
月总量 monthly amount
月最低温度 monthly minimum temperature
月最高温度 monthly maximum temperature
跃步法 step-over method
跃阶 step
跃迁态理论 transition state theory
越程 skip distance
越赤道海流 cross-equatorial current
越赤道气流【大气】 cross-equatorial flow
越冬 overwinter, overyearing
越冬防治【农】 overwintering control
越冬作物【农】 overwintering crop
云【大气】 cloud
云层 cloud layer, cloud veil
云〔层〕分析【大气】 nephanalysis
云层分析图 neph chart
云层间〔飞行〕 between layers
云层间目视飞行规则 VFR between layers
云簇 cloud groups
〔云〕催化剂【大气】 seeding agent
云带【大气】 cloud band, cloud zone
云单体轮廓线 cell outline
云导风 cloud winds
云的分类 cloud classification
〔云的〕附加特征 supplementary feature 〔of clouds〕
云的阶段 cloud stage
云〔的〕频率 cloud frequency
云的人工影响【大气】 cloud modification
云〔的〕日志 cloud diary
云的识别 cloud identification
云的微〔观〕结构 cloud microstructure
云的形成 cloud formation
云堤【大气】 bank of clouds, cloud bank
云滴 cloud drop
云滴【大气】 cloud droplet
云滴采样器【大气】 cloud-particle sampler
云滴凝结器【大气】 nepheloscope
云滴谱【大气】 cloud droplet-size distribution
云滴谱探测仪 watersonde
云滴谱仪【大气】 cloud droplet collector
云滴取样器 cloud-drop sampler
云底【大气】 cloud base
云底高度 cloud base height, height of cloud base
云底高度测录器 cloud-base recorder
云底高度探测气球 cloud-base height balloon
云底亮度图 sky map
云底云顶雷达指示器 radar cloud-base and cloud-top indicator
云地〔间〕放电【大气】 cloud-to-ground discharge, ground discharge
云地闪电 cloud-to-ground flash
云顶【大气】 cloud top
云顶高度【大气】 cloud top height (CTH)
云顶高度估计系统 cloud top height estimation system(CTHES)
云顶夹卷不稳定〔性〕 cloud-top entrainment instability
云顶目视飞行规则 visual flight rule on top
云顶温度【大气】 cloud top temperature
云动力学【大气】 cloud dynamics
云幡 streamer
云反馈【大气】 cloud feedback
云反照率【大气】 cloud albedo
云〔际〕放电【大气】 cloud discharge

云分类【大气】 cloud classification
云风估计系统 cloud wind estimation system
云风矢 cloud-motion wind vector
云符号 cloud symbol
云-辐射交互作用 cloud-radiation interaction
云辐射强迫 cloud radiative forcing
云幞 scarf
云覆盖区 cloud covered area
云覆盖区【大气】 cloud coverage
云干扰 cloud contamination
云高 height of cloud
云高【大气】 cloud height
云高测量法 cloud-height measurement method
云高度 cloud level
云高度表 nephohypsometer
云高遥测仪 cloud telemeter
云高指示器 cloud height indicator
云观测 cloud observation
云冠 cloud crest, sansan
云光学路径 cloud optical depth
云海 cloud deck, cloud sea
云含水量【大气】 water content of cloud
云厚度【大气】 vertical extent of a cloud
云虹 cloudbow(亦称 fogbow, mistbow, white rainbow)
云辉光 cloud luminance
云回波【大气】 cloud echo
云计算 cloud computing
云际放电【大气】 intercloud discharge, cloud-to-cloud discharge
云际闪电 intercloud lightning
云际退耦〔作用〕 cloud decoupling
云街【大气】 cloud street
云结构【大气】 cloud structure
云块 cloud mass
云类【大气】 cloud variety
云粒 cloud particle

云〔粒子〕核 cloud nucleus(cloud nuclei)
云粒子成像器 cloud particle imager
云粒子印模仪 cloud particle replicator
云量 cloud cover, cloud fraction, cloudiness, cover of cloud, sky cover, amount of cloud
云量【大气】 cloud amount
云林 cloud forest
云幔 cloud canopy, velium
云帽 cloud cap
云密度 cloud density
云幂 ceiling
云幂【大气】 cloud ceiling
云幂灯【大气】 ceiling projector, cloud searchlight, search light
云幂灯【大气】 ceiling light
云幂法 ceilometry
云幂分类 ceiling classification
云幂高度 ceiling height
云幂高度指示器 ceiling height indicator
云幂计 ceilograph
云幂警告器 ceiling alarm
云幂气球【大气】 ceiling balloon
云幂器 cloud height meter
云幂仪【大气】 ceilometer, cloud ceilometer
〔云幂仪〕光脉冲发射器 light-pulse projector
〔云幂仪〕光脉冲接收器 light-pulse receiver
云模式【大气】 cloud model
云幕【大气】 cloud ceiling
云幕灯 projector
云内放电【大气】 intracloud discharge
云内闪电 intracloud flash, intracloud lightning
云凝结核【大气】 cloud condensation nuclei (CCN)
云凝结核计数器 cloud condensation nucleus counter
云片 cloud-sheet

云起电　cloud electrification
云墙　wall cloud
云区　cloud field, cloud sector
云闪　cloud flash
云上〔飞行〕　on top
云室　cloud simulator
云室【大气】　cloud chamber
云属【大气】　cloud genera
云衰减【大气】　cloud attenuation
云素　cloud element
云速　speed of cloud movement
云图　cloud atlas, cloud chart, cloud picture, cloud map
云图动画〔显示〕【大气】　cloud image animation
云团【大气】　cloud cluster
云外飞行条件　clear-of clouds condition
云微物理学【大气】　cloud microphysics
云物理学【大气】　cloud physics
云吸收　cloud absorption
云系　nephsystem
云系【大气】　cloud system
云系分界线　nephcurve
云隙　rift in clouds
云隙辉　sun drawing water
云下层　subcloud
云下气层　subcloud layer, sublayer
云下清除　below-cloud scavenging
云线【大气】　cloud line
云相图　cloud-phase chart
云向　cloud direction
云消　clearance, clearing
云型　cloud pattern, cloud type
云学　nephology
云影　cloud shadow
云运动矢量【大气】　cloud motion vector
云中飞行　cloud flying
云中光学密度　optical density of a cloud
云中含水量　water content of clouds
云中能见度　visibility in cloud
云〔中气〕流　cloud current
云中闪电　cloud lightning discharge
云种　species of clouds

云种【大气】　cloud species
云周围环境体系　cloud-environment system
云属　genera of cloud
云属【大气】　cloud genera（genus 的复数）
云状【大气】　cloud form
云总数目　cloud population
云总体　cloud ensemble
云族　family of clouds
云族【大气】　cloud étage
匀速运动【物】　uniform motion
陨石　aerolite, meteorite
陨石尘埃　meteoritic dust
陨石学【地质】　meteoritics
陨石雨【地质】　meteorite shower
陨星【天】　meteorite
运筹学【数】　operations research
运动　motion
运动场　motion field
运动方程【大气】　equation of motion
运动锋　kinematic front
运动后寒冷　after-exercise chill
运动粒子　moving particle
运动黏度【物】, 运动黏滞性【大气】　kinematic viscosity
运动黏滞系数　coefficient of kinematic viscosity
运动随机介质　moving random medium
运动外延　kinematic extrapolation
运动学【物】　kinematics
运动学边界条件【大气】　kinematic boundary condition
运动学参考系【天】　kinematical reference system
运动学赤道　kinematic equator
运动学方程【物】　kinematical equation
运动学分析　kinematical analysis（＝kinematic analysis）
运动学通量　kinematic flux
运动〔学〕相似性　kinematic similarity
运输量　traffic
运算地址　arithmetic address

运算控制键　operation control key (OCK)
运算指令　operation order
运行方式【计】　run mode
运行时间【计】　running time
运载工具　vehicle
运载火箭　launch vehicle
运载气球　carrier balloon
运载气球系统　carrier balloon system
晕【大气】　halo
22°晕切弧　tangent arc to 22° halo
46°晕切弧　tangent arc to 46° halo

Z

杂项数据(资料)　miscellaneous data
灾变论【地理】　catastrophe theory
灾变说　catastrophism
灾变性气候变化　catastrophic climatic change
灾变性事件　catastrophic event
灾害　damage
灾害地理学【地理】　hazard geography
灾害天气短时临近预报系统【大气】　severe weather automatic nowcast system (SWAN)
灾害性天气【大气】　severe weather
灾害性天气征兆指数【大气】　severe weather threat index
灾害预报【医】　damage forecasting
栽培年　grower's year
载波　carrier wave
载波电话【电子】　carrier telephone
载频　carrier frequency
载人轨道空间站　manned orbital space station
载人空间飞行　manned space flight
载人气球　manned balloon
载体【化】　carrier
载体效应【化】　support effect
再冻雪壳　sun crust
再冻〔作用〕　regelation
再分析〔资料〕　reanalysis
再辐射　reradiation
再活化　reactivation
再生　regeneration
再生期　representative period
再循环水　recycled water
再预报　reforecast
再造林　reforestation
暂时变化　transitory variation
暂态运动【物】　transient motion
暂用标准干湿表　interim reference psychrimeter (IRP)
暂用标准日照计　interim reference sunshine recorder
早材　earlywood
早材【植物】　springwood
早潮　morning tide
早春　early spring
早期沉降　early fallout
早期风暴　early storm
早霜　early frost
造林　afforestation, forestation
造陆运动【地质】　epeirogeny
造陆作用【地质】　epeirogenesis
造山循环　orogenic cycle
造山运动【地质】　orogeny
造山作用【地质】　orogenesis
噪声【物】　noise
Q 噪声　Q noise
噪声本底　noise background
噪声比　noise ratio
噪声等效辐射率　noise equivalent radiance (NER)

噪声等效辐射差 noise equivalent radiance difference
噪声等效温度差 noise equivalent temperature difference
噪声功率 noise power
噪声过滤 noise filtering
噪声级 noise level
噪声计 psophometer
噪声控制平滑器 noise control smoother
噪声谱 noise spectrum
噪声温度 noise temperature
噪声污染 noise pollution
噪声系数【电子】 noise factor
噪声抑制 noise suppression
噪声阈值 noise threshold
噪声指数 noise index, noise figure
增补 alimentation
增长波 growing wave
增长模 growing mode
增广矩阵【数】 augmented matrix
空中承载分数(CO_2 增加的) airborne fraction
增湿器 humidifier
增量【数】 increment
增强 reinforcement
增强模式【地信】 enhanced mode
增强曲线 enhancement curve
增强图像 enhanced imagery
增强网络【天】 enhanced network
增强温室效应 enhanced greenhouse effect
增强显示红外卫星云图 enhanced infrared satellite cloud picture
增强显示卫星云图 enhanced satellite cloud picture
增强 V 型 enhanced "V"
增强云图 enhanced image, enhanced picture
增强云图【大气】 enhanced cloud picture
增湿剂 humidizer
增湿作用【农】 humidification
增水计划 precipitation enhancement program(PEP)
增温期气候变化 anathermal climatic change
增温式干湿表 elevated temperature psychrometer
增压舱 normal-air cabin
增益 gain
增益公式 formula for gain
增益函数 magnification function, gain function
增益因子 gain factor
憎湿性核 insoluble wettable nuclei
闸 gating
栅栏 Wild fence
栅屏湍流 grid turbulence
栅状测云器 grid nephoscope
窄带辐射 narrowband radiation
窄带记录器 narrow-sector recorders (NSR)
窄浮冰带 ice strip
窄角(12°)照相机 narrow angle camera
窄视场扫描系统 narrow-field scanning system
窄束发射机 narrow beam transmitter
窄束方程 narrow beam equation
窄束脉冲 narrow beam pulse
窄线回波 thin line echo
摘要 abstract
粘合【化】 adhesion
粘合剂【化】 adhesive
粘合〔作用〕 agglutination
斩波器 chopper
展开轴 axis of dilatation
展宽 VISSR stretched VISSR data
展宽网格 stretched grid
展宽网格资料 stretched gridded data
展宽〔云〕图 stretched image
展览温室【林】 conservatory
展望 outlook
占空因数【电子】 duty factor, duty cycle
占星术【天】 astrology

战斗机航空电码　COMBAR code（或 combat aircraft code）
战斗机气象报告　combat aircraft meteorological report(COMBAR)
战列舰【航海】　battle ship
站　station
站点　siter
〔站〕点降水　point precipitation
站圈【大气】　station circle
站网密度　network density
站址【大气】　station location
张弛逼近　nudging
张弛法【大气】　relaxation method
张弛涡动累积　relaxed eddy accumulation
张弛振荡器　relaxation oscillator
张角　opening angle
张力【物】　tension
张力饱合区　tension saturated zone
张力系数　coefficient of tension
张量【物】　tensor
张量代数【数】　tensor algebra
张量积【数】　tensor product
章动【天】　nutation
章动传感器　nutation sensor
章动角【物】　angle of nutation
涨潮【大气】　rising tide, flood tide
涨潮流【航海】　flood stream, flood current
涨落潮　ebb and flow
涨水段　rising limb
长草日数　grass-growing days
帐帷状极光　streamers
障碍急流　barrier jet
障碍气流征兆　obstacle flow signature
障碍物　obstruction
障碍物回波反射　back echo reflection
障板　baffle
朝霞　sunrise glow
沼气　marsh gas
沼泽　morass, marsh, bog
兆达因　megadyne
兆赫（百万周/秒）　megahertz
兆周　megacycle
照常排放〔情景〕【大气】　business as usual
照度计【物】　illuminometer
照度余弦定律　cosine law of illumination
照射体积　target volume, pulse volum
照相测光【天】　photographic photometry
照相机　camera
照相流星　photographic meteor
照相投影　photographic projection
照准仪　alidade
罩　mantle
遮蔽系数　sheltering coefficient
遮风板　wind shield
遮光率　shade ratio
遮光罩辐射计　shadow radiometer
遮光罩天空辐射计　shadow band pyranometer
遮阳系数　shading coefficient
折点　break point
折叠　folding
折叠频率　folding frequency
折叠速度　folding velocity
折反射望远镜【天】　catadioptric telescope
e折减时间【大气】　e-folding time
折射定律【物】　refraction law
折射计【物】　refractometer
折射角　angle of refraction
折射率　index of refraction
折射率【物】　refraction index, refractive index
折射率结构常数　refractive index structure constant
折射率起伏　refractive index fluctuation
折射率因子　refractive index factor
折射霾　refraction haze
折射模数　refraction modulus, refractive modulus
折射本领　refractivity
折射图　refraction diagram
折射望远镜【天】　refracting telescope,

refractor
折射误差　refraction error
折射系数　refraction coefficient
折射线　refracted ray
折射修正　refraction correction
折射预测　refraction prediction
寒冷单位　chill unit
寒冷时数　chill hour, chilling hour
褶皱【地质】　fold
褶皱带【地质】　fold belt
褶皱型突变　fold catastrophe
针冰　needle ice
针叶林【林】　coniferous forest
针叶树　conifer
真北　true north
真赤道【天】　true equator
真春分点【天】　true equinox
真地平　real horizon
真地平【航海】　true horizon
真地平圈【航海】　celestial horizon
真点播电视【计】　real video on demand
真冻结温度　true freezing point
真风【航海】【大气】　true wind
真风向　true wind direction
真高度　true altitude
真近点角【天】　true anomaly
真空【物】　vacuum
真空电子器件【电子】　vacuum electron device
真空订正　vacuum correction
真空度【电子】　degree of vacuum
真空管静电表　vacuum-tube electrometer
真空速〔度〕　true airspeed
真空微电子学【电子】　vacuum microeletronics
真力　true force
真平均　true mean
真平均温度　true mean temperature
真潜热　true latent heat
真潜〔在〕不稳定〔度〕　real latent instability
真太阳　true solar
真太阳日　true solar day
真太阳时　true solar time, true solar hour
真午　true noon
真误差　true error
真值【物】　true value
真值函数　truth-function
真重力　true gravity
砧云丘　anvil dome
砧状（云的）　incus(inc)
砧状积雨云　cumulonimbus incus
砧状云　anvil cloud
诊断　diagnosis(=diagnose)
诊断方程【大气】　diagnostic equation
诊断分析【大气】　diagnostic analysis
诊断模式【大气】　diagnostic model
诊断天气分析　diagnostic weather analysis
诊断研究　diagnostic study
阵侧风　turbulent crosswind
阵尘　shower of ash
阵发（风的）　rush
阵风　blast, windblast
阵风【大气】　gust
阵风〔持续〕时间【大气】　gust duration
阵风分量　gust component, gustiness-component
阵风锋【大气】　gust front
阵风负荷　gust load
阵风频数　gust frequency
阵风频数时段　gust frequency interval
阵风矢量　gust vector
阵风衰减时间　gust decay time
阵风探测器　gust probe
阵风探空仪　gustsonde
阵风梯度距离　gust-gradient distance
阵风系数　gust factor, gustiness factor
阵风性　gustiness
阵风阵息　gust and lulls
阵风振幅【大气】　gust amplitude
阵风最大递减　maximum gust lapse

阵风最大递减时段　maximum gust lapse interval
阵风最大递减时间　maximum gust lapse time
阵风最大风速　gust peak speed
阵列【计】　array
阵天线【电子】　array antenna
阵性降水【大气】　showery precipitation
阵性雨夹雪　rain and snow shower
阵雪　snow flurry, snow shower
阵雪【大气】　showery snow
阵雨　rain shower, shower
阵雨【大气】　showery rain
阵雨带　shower street
阵雨公式　shower formula
阵雨云　shower cloud
振荡　vacillation
振荡【物】【大气】　oscillation
振荡波　wave of oscillation, oscillatory wave
振荡方程　oscillation equation
振荡级数　oscillation series
振荡(气压表的)　pumping (of barometer)
振荡现象　vacillation phenomena
振荡周期　oscillation period, vacillation cycle
振动【物】　vibration
振动带　vibration band
振动能　vibration energy
振动谱　vibration spectra
振动数　number of vibration
振动体　oscillating body
振动周期　period of oscillation
振〔动〕子　oscillator
振幅　amplitude
振幅结构函数　amplitude structure function
振幅谱　amplitude spectrum
振幅相关函数　amplitude correlation function
振幅响应　amplitude response

振幅-周期图　amplitude-period graph
振铃【电子】　ringing
振膜式传感器　vibrating diaphragm transducer
振转带　vibration-rotation band
振转谱　vibration-rotation spectra
震旦纪【地质】　Sinian Period
震旦系【地质】　Sinian System
震级【地】　earthquake magnitude
震源【地】　seismic source, hypocenter
震中　epicenter
震中【地】　epifocus
震中分布【地】　epicenter distribution
震中烈度【地】　epicenter intensity
征兆　forerunner
征兆日　key day, control day
蒸发【大气】　evaporation
蒸发表　evaporometer, atmometer, atmidometer
蒸发侧定法　evaporimetry, atmometry
蒸发调节范围　zone of evaporative regulation
蒸发公式　evaporation formula
蒸发观测　evaporation observation
蒸发过程　evaporation process
蒸发计【大气】　evaporograph
蒸发可能率　evaporation opportunity
蒸发力【大气】　potential evaporation
蒸发量　evaporation capacity
蒸发量曲线　evaporogram
蒸发率　evaporativity, evaporation rate
蒸发率测定法　atmidometry
蒸发皿【大气】　evaporating pan
蒸发皿　evaporating dish
蒸发皿系数　pan coefficient
蒸发皿蒸发　pan evaporation
蒸发能力　evaporation power, evaporative power, evaporative capacity
蒸发器　evaporation gauge, evaporation hook gauge
蒸发曲线　evaporation curve

蒸发热【物】 heat of evaporation
蒸发霜 evaporation frost
蒸发通量 evaporation flux
蒸发尾迹 evaporation trail
蒸发系数 evaporation coefficient
蒸发仪 evaporimeter
蒸发抑制剂 evaporation suppressor
蒸发蒸腾比 ratio of evaporation to transpiration
蒸发指数【农】 evaporation index
蒸馏 distillation
蒸馏法【航海】 distillation method
蒸汽 steam
蒸汽暴 stream devil
蒸汽轻雾 steam mist, sea mist
蒸汽雾 steam fog, water smoke
蒸汽雾【大气】 evaporation fog
蒸散 flyoff
蒸散【大气】 evapotranspiration
蒸散表 evapotranspirometer
蒸散量测定装置 lysimeter
蒸散势【大气】 potential evapotranspiration
蒸腾【大气】 transpiration
蒸腾表 phytometer
蒸腾率 water-use ratio, transpiration ratio
蒸腾速率 transpiration rate
蒸腾系数 transpiration coefficient
蒸腾效率 transpiration efficiency
整合资料阅读机 integrated data viewer (IDV)
整流【物】 rectification
整流器【物】 rectifier
整数【数】 integer
整体边界层【大气】 bulk boundary layer
整体参数化 bulk parameterization
整体传输法 bulk transfer method
整体法 bulk method
整体公式 bulk formula
整体混合层模式 bulk mixed layer model
整体空气动力学方法 bulk aerodynamic method
整体里查森数【大气】 bulk Richardson number
整体流 bulk flow
整体模数 bulk modulus
整体拟合 global fitting
整体平均【大气】 bulk average
整体气体动力学曳力公式 bulk aerodynamic drag formulation
整体热通量 bulk heat flux
整体输送 bulk transport
整体输送定律 bulk transfer law
整体输送系数 bulk transfer coefficient
整体水 bulk water
整体水参数化 bulkwater parameterization
整体水〔云〕模式 bulkwater (cloud) model
整体速度 bulk velocity
整体湍流尺度 bulk turbulence scale
整体稳定边界层增长 bulk stable boundary layer growth
整体相 bulk phase
整体行星边界层 bulk planetary boundary layer
整体性质 bulk property
整体曳力方案 bulk-drag scheme
整体运动 bulk motion
正 positive
正半定矩阵【数】 positive semi-definite matrix
正变压〔的〕 anallobaric, anabaric
正变压线【大气】 anallobar
正变压中心 isallobaric high
正变压中心【大气】 anallobaric center
正侧风 direct crosswind
正常大气 normal atmosphere
正常风 normal wind
正常态【物】 normal state
正常星系【天】 normal galaxy

正常漩涡星系【天】 normal spiral galaxy
正常因子 normal factor
正常值 normal value
正常重力 normal gravity
正常重力位【地】 normal gravity potential
正等温涡度平流 positive isothermal vorticity advection
正地闪 positive ground flash
正电荷 positive charge
正电子【物】 positron
正定〔的〕 positive definite
正定矩阵【数】 positive difinite matrix
正定平流方案 positive definite advection scheme
正反馈 positive feedback
正方波 square wave
〔正〕方形网格【数】 square net，square mesh
正放电 positive discharge
正负电子对撞机【物】 electron-positron collider
正割投影 secant projection
正规测站 authorized station
正规方程【数】 normal equation
正规函数【数】 normal function
正规模〔态〕初值化【大气】 normal mode initialization（NMI）
正规模〔态〕【大气】 normal mode
正环流【大气】 direct circulation
正极 positive pole
正交 orthogonal，normal
正交变换【数】 orthogonal transformation
正交多项式【数】 orthogonal polynomials
正交函数【大气】 orthogonal function
正交化【化】 orthogonalization
正交积分 quadrature
正交模方程 normal mode equation
正交模解 normal mode solution
正交偏振 cross polarization
正交偏振信号 cross-polarized signal
正交曲线坐标 rectangular curvilinear coordinate
正交曲线坐标【数】 orthogonal curvilinear coordinates
正交试验【农】 orthogonal experiment
正交天线 orthogonal antennas
正〔交〕投影【数】 orthogonal projection
正交涡度 crosswise vorticity
正交系数【农】 orthogonal coefficient
正交线 orthogonal lines
正交性 orthogonality
正交坐标【数】 orthogonal coordinates
正静力稳定 positive static stability
正距平中心 pleion
正离子 positive ion（或 cation）
正粒子 positive particle
正切弧 tangent arc
正区 positive area
正入射【物】 normal incidence
正射投影【测】 orthographic projection
正射像片【测】 orthophoto
正射影像地图【测】 orthophotomap
正态分布【数】 normal distribution
正态扩散 normal dispersion
正态频率 normal frequency
正态曲线【农】 normal curve
正态总体 normal population
正温 thermotropic
正温〔大气〕模式【大气】 thermotropic model
正涡度平流【大气】 positive vorticity advection（PVA）
正午 midday
正弦波 sinusoidal wave
正弦窗 sine window
正弦电流表 sine galvanometer
正弦函数 sine function
正弦函数展开 sine-function expansion
正弦曲线【数】 sine curve
正弦式 sinusoid

正弦型　sinusoidal pattern
正相关　positive correlation
正向反应【化】　forward reaction
正向链接　forward chaining
正斜槽　positive tilt
正形图　orthomorphic map
正形圆锥投影　conformal conic projection
正压波【大气】　barotropic wave
正压波活动　barotropic wave activity
正压不稳定【大气】　barotropic instability
正压大气【大气】　barotropic atmosphere
正压〔的〕　barotropic
正压方程　barotropic equation, equation of barotropy
正压过程　barotropic process
正压流体　barotropic fluid
正压罗斯贝波　barotropic Rossby wave
正压模式【大气】　barotropic model
正压模〔态〕【大气】　barotropic mode
正压区　barotropic zone
正压扰动　barotropic disturbance
正压数值模式　barotropic numerical model
正压条件　barotropic condition
正压涡度方程【大气】　barotropic vorticity equation
正压涡流　barotropic eddy
正压系数　coefficient of barotropy
正压〔性〕【大气】　barotropy
正压压力函数　barotropic pressure function
正压预报　barotropic forecast
正压原始方程　barotropic primitive equation
正压转矩矢量　barotropic torque vector
正压状态　barotropic state
正异常【地】　positive anomaly
正云地闪电　positive cloud-to-ground lightning
正则边界【数】　regular boundary
正则方程【物】　canonical equation
正则平流　regular advection
正则摄动　regular perturbation
正则坐标【物】　canonical coordinates
正轴　positive axis
正轴投影　normal projection
政府间海洋委员会　Intergovernmental Oceanographic Commission(IOC)
政府间海运咨询组织　Inter-Governmental Maritime Consultative Organization(IMCO)
政府间气候变化专门委员会【大气】　Intergovernmental Panel on Climate Change(IPCC)
政府间气候变化专门委员会
　第二次评估报告　IPCC SAR;
　第三次评估报告　IPCC TAR;
　第四次评估报告　IPCC AR4
帧每秒【计】　frame per second
帧同步器　frame synchronizer
支持程序【计】　support program
支持向量机　support vector machines (SVM)
支流【地理】　tributary
芝加哥学派　Chicago school
枝状　dendrite
枝状冰晶【大气】　dendritic crystal
枝状晶体　dendritic crystal
枝状雪花晶体　dendric snow crystals
枝状雪晶　dendritic snow crystal
知识产业【计】　knowledge industry
知识工程【计】　knowledge engineering
知识获取【计】　knowledge acquisition
知识库【计】　knowledge base
知识库系统【计】　knowledge base system
知识提取【计】　knowledge extraction
脂蛋白【生化】　lipoprotein
脂质【生化】　lipid
脂状冰　lard ice
直布罗陀外流水　Gibraltar outflow water
直串梯级先导　dart-stepped leader
直窜先导　dart leader

直达声波　direct sound〔wave〕
直读式地面站【大气】　direct read-out ground station
直读式红外辐射计　direct read-out infrared radiometer（DRIR）
直读式图像分析仪　direct read-out image dissector（DRID）
直读式温度表【大气】　direct-reading thermometer
直读式仪器　direct-reading instrument
直方图【大气】　histogram
直根【植】　taproot
直管地温表【大气】　tube-typed geothermometer
直击强风　straight blow
直积【数】　direct product
直减率　lapse rate
直减率线　lapse line
直角【数】　right angle
直角笛卡儿式坐标　rectangular Cartesian coordinate
直角三角形【数】　right triangle
直角坐标【数】　rectangular coordinate
直接传送系统　direct transmission system
直接存取【计】　direct access
直接读出　direct read-out
直接法【物】　direct method
直接分段法　direct segment method
直接辐射【大气】　direct radiation
直接辐射表【大气】　pyrheliometer
直接辐射计　actinograph
直接辐射强迫　direct radiative forcing
直接辐射自记曲线　actinogram
直接〔观测〕方式　direct mode
直〔接广〕播　direct broadcast
直接环流〔图〕【大气】　direct circulation
直接〔环流〕胞　direct cell
直接径流　direct runoff, direct flow
直接气溶胶效应　direct aerosol effect
直接日射测量学【大气】　pyrheliometry
直接数值模拟　direct numerical simulation
直接太阳辐射　direct solar radiation
直接探测　direct detection
直接吸收【物】　direct absorption
直径　diameter
直流电　direct current
直切珥　vertical tangent arc
直射　perpendicular incidence
直升机尾流　helicopter wake
直视测云器　direct vision nephoscope
直视棱镜　direct vision prism
直纹〔曲〕面【数】　ruled surface
直线风　straight wind
直行风　straight-line wind
直展云　vertical-development cloud, heap cloud
直展云【大气】　cloud with vertical development
直照　direct lighting
植被　vegetation, plant cover
植被型　vegetation type
植被指数【大气】　vegetation index
植物保护【农】　plant protection
植物病害【农】　plant disease
植物病理学【农】　plant pathology
植物带　botanical zone
植物地理学【植】　phytogeography, plant geography
〔植物〕风灼伤　windburn
植〔物〕冠〔盖〕　plant canopy
植物化学【植】　phytochemistry
植物胚胎学【植】　plant embryology
植物气候　plant climate
植物气候区　plant climate zone
植物气候学【大气】　phytoclimatology
植物气候学【植】　plant climatology
植物器官学【植】　plant organography
植物群落【地理】　plant community
植物生长季　vegetation season
植物生长期　vegetative period
植物生理学【植】　plant physiology
植物生态学【植】　plant ecology,

phytoecology
植物生物气象学 phytological biometeorology
植物生物学【植】 plant biology
植物温度 plant temperature
植物物候学 plant phenology
植物〔小〕气候【大气】 phytoclimate
植物学【植】 plant science, botany
植物夜光 sylvanshine
植物园【植】 botanical garden
植物志 flora
只读存储器【计】 read only memory
只读碟【计】 compact disc-read only memory (CD-ROM)
纸示湿度表 paper hygrometer
指标 index arm (of sextant), index
指标差 index error
指标订正 index correction
指导〔性〕终点预报 guidance terminal forecast (GTF)
指极星 pointer
指令【计】 instruction
指令地址寄存器【计】 instruction address register
指令和资料接收站 command and data acquisition station (CDA), read-out station
指令钟子系统 command clock subsystem
指示灯 indicator lamp
指示高度 indicated altitude (IA)
指示函数【数】 indicator function
指示空速 indicated air speed (IAS)
指示码 symbolic figure, indicator figure
指示气流速度 indicated air speed (IAS), indicated airspeed
指示式测风仪 windicator
指示式风速仪 wind minder anemometer, wind minder vane
指示式湿度表 indicating hygrometer
指示线图 indicator diagram
指示组 symbolic figure group, key group
指数 exponent
U 指数 U figure, U index
指数逼近【数】 exponential approximation
指数窗 exponential window
指数大气 exponential atmosphere
K 指数【地】 K index
K_p 指数（地磁活动） K_p
C 指数（地磁活动） C index
Ap 指数（地磁扰动） Ap index
指数分布 exponential distribution
指数函数【数】 exponential function
指数核近似 exponential kernel approximation
指数积分 exponential integral
指数律【数】 exponential law
指数平滑 exponential smoothing
指数谱 exponential spectrum
指数前的系数 preexponential factor
指数趋势 index trend
指数循环【大气】 index cycle
指纹法 fingerprint method
指向性【电子】 directivity
指向性函数 directivity function
指向性图案 directivity pattern, beam pattern
指向性因数 directivity factor
指向性增益 directional gain
指向性指数 directivity index
指印现象 fingering
指针式气压表 pointer barometer
指状筏冰 finger rafting
至点潮 solstitial tide
致洪降水 flood-leading rainfall
志留纪【地质】 Silurian Period
志留-泥盆纪冰期 Siluro-Devonian ice age
志留系【地质】 Silurian System
志愿观测者 voluntary observer
志愿协作计划 Voluntary Co-operation Programme (VCP)
制表机 tabulating machine

制导激光束【物】 guiding laser beam
制动风杯风速表 bridled-cup anemometer
制动风速表 bridled anemometer
制冷机 refrigerator
制冷剂【航海】 refrigerant
制图分级【测】 cartographic hierarchy
质点 material point
质点【物】 particle
α质点 alpha-particle
质量 quality
质量比例 mass proportion
质量标准【农】 quality standard
质量调节系数 mass accommodation coefficient
质量辐合 mass convergence
质量辐散 mass divergence
质量管理【电子】 quality management
质量过剩【物】 mass excess
质量控制【电子】 quality control
质量亏损【物】 mass defect
质量流调整 mass flow adjustment
质量浓度 mass concentration
质量平衡【海洋】 mass balance
质量评价【农】 quality evaluation
质量迁移法 mass-transfer method
质量散射系数 mass scattering coefficient
质量收支【海洋】 mass budget
质量守恒【大气】 conservation of mass, mass conservation
质量输送 mass transport
质量通量 mass flux
质量吸收系数 mass absorption coefficient
质量消光截面 mass extinction cross section
质量消光系数 mass extinction coefficient
质能约束 mass/energy constraint
质谱仪 mass spectrometer
质心 center of mass
质心坐标 center-of-mass coordinate
质子【物】 proton
质子层【地】 protonosphere
质子雨 proton precipitation

质子-质子反应【天】 proton-proton reaction
致电离辐射 ionizing radiation
致密 compaction
致密化【力】 densification
致密射电源 compact radio source
致偏磁场 deflecting magnetic field
致死剂量 lethal dose
致死临界温度 zero point
致死浓度 lethal concentration
致死湿度 fatal humidity
致死温度 killing temperature, fatal temperature, deadly temperature
致死温度【农】 thermal death point
致死紫外辐射 lethal ultraviolet radiation
致雨系统 rain-producing system
秩相关 rank correlation
智利海流 Chile Current
智能卡【计】 IC card
智能库 working memory
滞差 lag error
滞弹性【物】 anelasticity
滞弹性方程 anelastic equation
滞弹性近似 anelastic approximate
滞点 stagnation point
滞洪区 retarding basin
滞后 lag
滞后交叉相关 lag cross correlation
滞后时间 lag time
滞后系数【大气】 lag coefficient
滞后现象 hysteresis
滞后效应 carry-over effect
滞后效应【海洋】 lag effect
滞留期 detention period, residence time
滞留区 stagnation area
滞育【农】 diapause
置信点 fiducial point
〔置〕信度【大气】 degree of confidence
置信区间 fiducial interval, confidence band

置信区间【数】 confidence interval
置信区域【数】 confidence region
置信上限【数】 confidence upper limit
置信水平【数】【大气】 confidence level
置信系数【数】 confidence coefficient
置信限　fiducial limits
置信限【数】 confidence limit
中白垩世　middle Cretaceous
中部标准时间(美国)　Central Standard Time
中侧弧　mesolateral arc
中层大气【大气】 middle atmosphere
中层大气物理学【大气】 middle atmospheric physics
中层大气研究计划　Middle Atmospheric Programme(MAP)
中层大气研究计划指导委员会(WMO/ICSU)　Middle Atmosphere Programme Steering Committee (MAPSC, WMO/ICSU)
中层气旋　mid-level cyclone
中层水　intermediate water
中层云　middle-level cloud
中常能见度　moderate visibility
中尺度背风坡低压区　mesoscale lee low pressure area
中尺度背风坡涡旋　mesoscale lee vortex
中尺度【大气】 mesoscale
α中尺度【大气】 meso-α scale
β中尺度【大气】 meso-β scale
γ中尺度【大气】 meso-γ scale
中尺度大气运动　mesoscale atmospheric motion
中尺度单体对流　mesoscale cellular convection (MCC)
中〔尺度〕低压　mesolow
中尺度低压【大气】 mesoscale low
中尺度低压区　mesoscale low pressure area
中尺度对流风暴复合体　mesoscale convective storm complex
中尺度对流复合体【大气】 mesoscale convective complex (MCC)
中尺度对流系统【大气】 mesoscale convective system (MCS)
中尺度对流云带　mesoscale convective 〔cloud〕band
中尺度对流云团　mesoscale convective 〔cloud〕cluster
中〔尺度〕反气旋　mesoanticyclone
中〔尺度〕分析　mesoanalysis
中〔尺度〕高压　mesohigh
中尺度高压区　mesoscale high pressure area
中尺度观测网站　mesonet station
中尺度急流　mesojet
中尺度夹卷不稳定度　mesoscale entrainment instability
中尺度降水区　mesoscale precipitation area
中尺度降水中心　mesoscale precipitation core
中尺度结构　mesoscale structure
中尺度空中逆温　mesoscale elevated inversion
中尺度模式【大气】 mesoscale model
中尺度气候学　mesoclimatology
中尺度气象模拟　mesoscale meteorological modeling
中尺度气象学【大气】 mesometeorology
中尺度气象学　mesoscale meteorology, mesosynoptic meteorology
中尺度气象研究合作研究所(美国)　Cooperative Institute for Mesoscale Meteorological Studies (CIMMS)
中〔尺度〕气旋　mesocyclone
中〔尺度〕气旋回波特征　mesocyclone signature
中尺度气压系统　mesoscale pressure system
中尺度扰动　mesoscale disturbance
中尺度天气系统　mesoscale weather sys-

中尺度天气学　mososcale synoptics
中尺度图　mesochart
中尺度网　mesonet
中尺度涡街　mesoscale vortex street
中尺度涡流　mesoscale eddy
中尺度涡旋　mode eddies（亦称 mesoscale eddies）
中尺度无组织对流　mesoscale unorganized convection（MUC）
中尺度系统【大气】　mesoscale system
中尺度雨带　mesoscale rainband
中尺度运动　mesoscale motion
中尺度组织　mesoscale organization
中等热带风暴　moderate tropical storm
中等视场辐射仪　medium field-of-view radiometer
中〔度〕侧风　moderate crosswind
中度〔的〕　moderate
中度斜压性　moderate baroclinity
中断　break
中断面法（推求断面流量）　midsection method
中断矢量　interrupt vector
中对流层　middle troposphere
中对流层锋　midtropospheric front
中对流层逆温　midtropospheric inversion
中二叠世　middle Permian
中非线性　moderately nonlinear
中分辨红外辐射仪　medium resolution infrared radiometer（MRIR）
中分辨〔率〕成像光谱辐射仪（EOS）　MODerate-resolution Imaging Spectroradiometer（MODIS）
中国海岸流　China Coastal Current
中国南极考察队　Chinese National Antarctic Research Expedition（CHINARE）
中国气象局　China Meteorological Administration（CMA）
中国气象学会【大气】　Chinese Meteorological Society（CMS）
中国新一代天气雷达　China new generation weather radar（CINRAD）
中国异常台风科学试验（1993—1994）　China Abnormal Typhoon Scientific Experiment（CATEX）
中和【化】　neutralization
中和温度　neutral temperature
中红外　middle infrared
中继站【计】　relay station
中继〔转发〕发射机　retransmitter
中间层臭氧　mesospheric ozone
中间层【大气】　mesosphere
中间层【电子】　intermediate layer
中间层顶【大气】　mesopause
中间层环流【大气】　mesospheric circulation
中间层急流　mesospheric jet
中间层-平流层-对流层雷达　mesosphere-stratosphere-troposphere（MST）radar
中间层纬向平均环流　zonal mean mesospheric circulation
中间层云　mesospheric cloud
中间层最高温度点　mesopeak
中间场　intermediate field
中间尺度　intermediate scale，medium-scale
中间尺度反气旋　intermediate anticyclone，intermediate cyclone
中角（78°）照相机　medium angle camera
中浪　moderate sea
中离子　intermediate ion
中能质子和电子检测器　medium energy proton and electron detector（MEPED）
中泥盆世　middle Devonian
中黏土【土壤】　medium clay soil
中频　medium frequency（MF），intermediate frequency（IF）
中频放大器【电子】　intermediate

frequency amplifier
中频信号　IF signal
中期水文预报　medium-term hydrological forecast
中期〔天气〕预报【大气】　medium-range〔weather〕forecast
中期预报　mid-range forecast
中期振荡　medium-range oscillation
中气候【大气】　mesoclimate
中气旋龙卷　mesocyclone tornado
中日本海冷海流　Mid-Japan Sea Cold Current
中生代【地质】　Mesozoic Era
中生界【地质】　Mesozoic Erathem
中生植物【植】　mesophyte
中石器时代　Mesolithic age
中世纪初寒冷期　Early Medieval Cool Period
中世纪极大值　Medieval Maximum
中世纪暖期(气候)　Medieval Warm Epoch（MWE）
中世纪气候适宜期　Medieval Climate Optimum
中世纪〔太阳黑子〕极大期　Grand Maximum
中世纪温暖期　Medieval Warm Period, Medieval Mild Phase
中暑　heat exhaustion, heat-stroke
中水年【大气】　normal flow year
中太平洋飓风中心(美国)　Central Pacific Hurricane Center（CPHC）
中微子【物】　neutrino
中位半径　median radius
中位数【数】　median
中位体积半径　median volume radius, median volume diameter
中位直径　median diameter
中纬度　middle latitude
中纬〔度〕低压　mid-latitude depression
中纬度反气旋　mid-latitude anticyclone
中纬〔度〕高压带　middle-latitude high-pressure belt
中纬〔度〕平静型　middle-latitude equable regime
中纬〔度〕气团　middle-latitude air mass
中纬度气旋　mid-latitude cyclone
中纬〔度〕西风〔带〕　middle-latitude westerlies
中纬〔度〕系统　middle-latitude system
中纬强迫　mid-latitude forcing
中温　mesotherm
中温〔的〕　mesothermal
中温气候【大气】　mesothermal climate
中温植物〔的〕　mesothermic
中温植物型　mesothermal type
中涡旋　mideddy, mesovortex
中午　noon
中雾　moderate fog
中小气候情况　climatomesochore
中心差分【数】　central difference, centered difference
中心等离子体片　central plasma sheet
中心低压　central depression
中心点　center
中心核　central core
中心极限　central limit
中心极限定理【数】　central limit theorem
中心矩　central moment
中心气压　central pressure
中心无风区　central calm
中心眼　central eye
中心有限差分　centered finite difference
中心子午线中天　central meridian passage
中新世【地质】　Miocene Epoch
中新统【地质】　Miocene Series
中型〔卫星〕资料利用站　medium-scale data utilization station
中性波　neutral wave
中性层　neutrosphere
中性层顶　neutropause
中性大气　neutral atmosphere

中性点　neutral point
中性电离漂浮　neutral ionization drift
中性风　neutral wind
中性浮力点　point of neutral buoyancy
中性浮力高度　level of neutral buoyancy
中性锢囚　neutral occlusion
中性锢囚锋【大气】　neutral occluded front
中性扩散系数　neutral diffusion coefficient
中性粒子　neutral particle
中性面　neutral surfaces
中性模　neutral mode
中性能量平衡　neutral energy balance
中性片　neutral sheet
中性平衡　neutral equilibrium, indifferent equilibrium
中性平面　neutral plane
中性气体平衡　neutral gas equilibrium
中性气旋　neutral cyclone
中性条件　neutrality condition
中性土【土壤】　neutral soil
中性拖曳不稳定性　neutral drag instability
中性稳定【大气】　neutral stability
中性稳定度　indifferent stability
中〔性〕线【物】　neutral line
中性线　Busch lemniscate
中性振荡　neutral oscillation
中央处理器【计】　central processing unit (CPU)
中央数据分发中心　central data distribution facility (CDDF)
中央水　central water
中雨【大气】　moderate rain
中元古代冰期　mid-Proterozoic ice age
中元古代【地质】　Mesoproterozoic Era
中元古界【地质】　Mesoproterozoic Erathem
中云　medium cloud, mediocris (med), medium-level cloud
中云【大气】　middle cloud
中展积云【大气】　cumulus mediocris (Cu med)
中值　mid-value
中值定理【数】　mean value theorem
中志留世　middle Silurian
中侏罗世　middle Jurassic
中子【物】　neutron
中子测井【地】　neutron logging
中子含水量测定器　neutron moisture gauge
中子活化分析【电子】　neutron activation analysis
中子激活　neutron activation
中子〔热能〕慢化土壤湿度探头　neutron thermalization soil probe
中子散射法　neutron-scattering method
中子输送理论　neutron transport theory
中子〔土壤〕含水量仪　neutron moisture meter
中子土壤水分测定仪　neutron soil moisture meter
终白霜　last hoarfrost
终点　terminal point
终读数　final reading
终端　terminal
终端设备　terminal equipment, terminal device
终降雪　last snowfall
终日　last date
终杀霜　last killing frost
终霜【大气】　latest frost
终霜【农】　last frost
终雪　last snow
终止区域　terminal area
钟形孤波　bell soliton
钟形圆锥〔法〕　bell taper
钟制风速表　clockwork anemometer
种肥【土壤】　seed fertilizer
种群【地理】　population
种群动态【海洋】　population dynamics
种群生态【海洋】　population ecology

种植密度【农】 planting density
种植期 planting season
仲春 depth of spring
众数半径 mode radius
重冰 heavy ice
重大误差检验 gross error check
重离子含量 heavy ion content
重离子【物】 heavy ion
重力 gravitation, gravity force
重力【物】 gravitg
重力波 gravitational wave
重力波【大气】 gravity wave
重力波破碎 gravity wave breaking
重力波拖曳【大气】 gravity wave drag
重力波拖曳参数化 gravity wave drag parameterization
重力波曳力 gravity wave drag
重力不稳定 gravitational instability
重力测量【地】 gravity measurement
重力测量学【地】 gravimetry
重力场 gravitational field
重力场【地】 gravity field
重力等位面【地】 equipotential surface of gravity
重力订正 gravity correction
重力对流 gravitational convection
重力分离 gravitational separation
重力风 gravitational wind, gravity wind
重力加速度 acceleration of gravity, gravitational acceleration
重力加速度【地】 gravity acceleration
重力流【大气】 gravity current, gravity flow
重力落差 gravitational head
重力模 gravitational mode
重力内波【大气】 internal gravity wave
重力平衡 gravitational equilibrium
重力水 gravitational water, gravity water
重力梯度测量【地】 gravity gradient survey
重力梯度带【地】 gravity gradient zone
重力梯度稳定 gravity gradient stabilization
重力梯度仪【地】 gravity gradiometer
重力外波【大气】 external gravity wave
重力位【地】 gravity potential
重力位面 level surface
〔重力〕位势 geopotential
重力位势差距平 anomaly of geopotential difference
〔重力〕位势限制初始化 geopotential constrained initialization
重力稳定度 gravitational stability
重力压〔力〕 gravitational pressure
重力仪【地】 gravimeter
重力雨量器 gravimetric raingauge
重量【物】 weight
重量测湿法 gravimetric hygrometry
重量法【化】 gravimetric method
重量〔分析〕法【化】 gravimetric analysis
〔重量〕摩尔浓度 molality（M/W）
重量因子【化】 gravimetric factor
重黏土【土壤】 heavy clay soil
重氢含量 denterium content
重水【化】 heavy water
重雾 heavy fog
重心【物】 center of gravity
重心高度【航海】 height of centre of gravity
重要气象信息【大气】 significant meteorological information
重要天气【大气】 significant weather
重要天气报告 significant weather report
重要天气图【大气】 significant weather chart
周每秒（相当于赫兹） cycle per second （cps）
周年风【大气】 anniversary wind
周年光行差【航海】 annual aberration
周年视差【天】 annual parallax
周期【大气】 period

周期边界条件　periodic boundary condition
周期变化　periodic variation
周期风　periodic wind
周期轨道【天】　periodic orbit
周期函数【数】　periodic function
周期化【数】　periodization
周期加倍　period doubling
周期谱　period spectrum
周期图【大气】　periodogram
周期图法　periodogram method
周期图分析【数】　periodogram analysis
周期图矢量　periodogram vector
周期吸引子　periodic attractor
周期信号　periodic signal
周期〔性海〕流　periodic current
周期性【物】　periodicity
周期运动　periodic motion
周期振动　periodic oscillation
周日变风　diurnal wind
周日波　diurnal wave
周日加热（海洋中）　diurnal heating
周日冷却（海洋中）　diurnal cooling
周日太阳潮　diurnal solar tide
轴　axis（axes）
轴对称　axial symmetry
轴对称流　axisymmetric flow
轴对称湍流　axisymmetric trubulence
轴对称涡旋　axisymmetric vortex
轴晶【化】　axialite
轴流风速表　axial-flow anemometer
轴向力　axial force
轴向散射粒谱仪　axially scattering spectrometer
宙【地质】　eon
绉纹卷云　cirro-ripples
昼侧磁层　dayside magnetosphere
昼侧极光　dayside aurora
昼侧极光卵　dayside auroral oval
昼长【大气】　daylength
昼风　day breeze

昼间飞行【航空】　day（＝light）flight
昼夜　diel, around-the-clock
昼夜垂直移动【海洋】　diurnal vertical migration
皱波【物】　ripple wave
皱点　fold point
骤风　blash
侏罗纪【地质】　Jurassic Period
侏罗系【地质】　Jurassic System
珠母云【大气】　mother of pearl cloud, nacreous cloud
珠状热敏电阻　bead thermistor
珠〔状〕闪〔电〕【大气】　pearl lightning, beaded lightning
蛛网状闪电　spider lightning
烛光　candle
烛〔状〕冰　candle ice
逐步　stepwise
逐步回归【数】　stepwise regression
逐步回归分析　stepwise regression analysis
逐步近似〔法〕　step-by-step（或 successive）approximation
逐步判别法　stepwise discriminatory analysis
逐步序列检验　stepwise sequential test
逐层剥皮法　onion peeling method
逐次超松弛【数】　successive over-relaxation
逐次迭代　successive iteration
逐次订正　successive correction
逐次订正〔分析〕法【大气】　successive correction analysis
逐次消元法　successive elimination
逐段逼近法　piecewise approximation method
逐段回归方法　stage wise regression procedure
逐级散射　successive order of scattering
〔逐日〕磁变示数　character figure
逐时观测　hourly observation

逐线法　line-by-line method
逐线积分〔法〕　line-by-line integration
逐项滤波　successive filtering
主波　major wave
主〔波〕瓣(射束的)　major lobe, main lobe
主槽　major trough
主成分分析【数】　principal component analysis(PCA)
主成分【化】　major constituent
主程序　master routine, master program
主尺度　dominant scale
主从计算机【计】　master/slave computer
主存储器　main memory
主大西洋气候期　Main Atlantic Climate Phase
主导风向【大气】　predominant wind direction
主低压　primary depression, primary low
主点　principle point
主动技术　active technique
主动监测技术　active monitoring technique
主动监测系统　active monitoring system
主动上滑锋　active anafront
主动实验　active experiment
主动卫星　active satellite
主动系统　active system
主动下滑锋　active katafront
主动遥感技术【大气】　active remote sensing technique
主动遥感系统　active remote sensing system
主动引导　active steering
主动制导　active guidance
主锋　primary front
主锋【大气】　principal front
主干电信网　main telecommunications network(MTN)
主干线　main trunk circuit
主根【植】　axial root
主观分析　subjective analysis

主观估计　subjective assessment
主虹　primary bow, primary rainbow
主机脉冲　mainframe pulse
主级环流【大气】　primary circulation
主极小　primary minimun
主脊　major ridge
主计算机　mainframe computer
主截面　principle section
主流　main current, mother current
主〔螺旋〕带　principal band
主脉冲　main pulse
主面　principle plane
主模式　host model
主气旋　primary cyclone
主奇异向量　leading singular vector
主闪击　main stroke
主特征向量　leading eigenvector
主退水曲线　master recession curve
主〔温〕跃层【海洋】　main thermocline
主向风(即东、南、西、北风)　cardinal wind
主效应【化】　main effect
主〔要〕环流　major circulation
主要降水中心　main precipitation core
主要权重　principal weight
主站【计】　master station
主帧　major frame
主振荡型　principal oscillation pattern
主轴【数】　principal axis
住宅小气候　apartment microclimate
驻波　immobile wave
驻波【物】【大气】　standing wave
驻涡　standing eddies, stationary eddy
驻云【大气】　standing cloud
柱电阻　columnar resistance
柱丰度　column abundance
柱面极坐标　cylindrical polar coordinates
柱面坐标【数】【大气】　cylindrical coordinates
柱模式【大气】　column model

柱数密度　column number density
柱状晶体　columnar crystal
柱状涡旋　columnar vortex
专家系统【计】【大气】　expert system
专利　patent
专门技术　expertise
专题【地信】　thematic
专题丛书,论丛　monograph series
专题地图【测】　thematic map
专题地图学【地理】　thematic cartography
专题影像【地信】　thematic image
专题制图仪【地理】　thematic mapper
专业农业气象站　agricultural meteorological station for special purpose
专业气象服务　specialized meteorological service
专用地图【地理】　applied map
专属经济区【航海】　exclusive economic zone
转报设备　message switching facility
转杯风速表　cup anemometre【大气】, Robinson's anemometer
转杯液水含量仪　spinning bowl LWC meter
转臂风速计　pivot arm anemometer
转动动能　rotational energy
转动惯性　rotational inertia
转动能级　rotational energy level
转动谱带【大气】　rotation band
转风点　amphidromos
转化云　mutatus
转换　conversion, transition
转换率　turnover rate
转换器　converter
转换器式探空仪　commutator radiosonde
转换时间　turnover time
转换温度　reversal temperature
转镜式 Q 开关　rotating mirror Q-switching
转矩　rotation moment, moment of rotation

转矩【物】　torque
转盘　rotating annulus (或 dishpan)
转盘实验　dishpan experiment, dish experiment
转盘实验【大气】　rotating dishpan experiment
转速测量〔法〕　tachometry
转向　recurvature
转向点　turning point
转向区　zone of recurvature
转向纬度　recurvature latitude, turning latitude
转移函数　transmissibility function
转折点不稳定度　inflection point instability
转振光谱　rotation-vibration spectrum
转子　rotor
转子气流　rotor streaming, rotor flow
转子效应　rotor effect
装配模块　load module
装置　installation
状态变化　change of state
状态变量　state variable
状态参量　state parameter
状态方程【大气】　equation of state
状态方程的非线性关系　nonlinearity of the equation of state
状态曲线　state curve
撞冻效率　accretion efficiency
撞冻〔增长〕【大气】　accretion
撞击采样器　impactor, impactometer, impinger
撞击采样仪　impactometer
追肥【土壤】　top dressing
追算技术　hindcasting technique
锥表船用雨量器　conical marine raingauge
锥头温度表　conical-head thermometer
锥形冰雹【大气】　conical hail
锥形射束　conical beam
锥形针　conical point

准保守性的　quasi-conservative
准不连续性　quasi-discontinuity
准常定低压　quasi-permanent low
准单调格式　quasi-monotonic scheme
准等压过程　quasi-isobaric process
准地转　quasi-geostrophic, quasigeostrophic
准地转定标　quasi-geostrophic scaling
准地转Ω方程　quasi-geostrophic omega equation
准地转方程　quasi-geostrophic equations
准地转近似　quasi-geostrophic approximation, pseudo-geostrophic approximation
准地转理论　quasi-geostrophic theory
准地转流　quasi-geostrophic current, quasi-geostrophic flow
准地转滤波　quasi-geostrophic filtering
准地转模式【大气】　quasi-geostrophic model
准地转平衡　quasi-geostrophic equilibrium
准地转湍流　quasi-geostrophic turbulence
准地转位涡度　quasi-geostrophic potential vorticity
准地转位涡度方程　quasi-geostrophic potential vorticity equation
准地转系统　quasi-geostrophic system
准地转运动【大气】　quasi-geostrophic motion
准反射　quasi-reflection
准非可递系统　almost intransitive system
准分子激光器【电子】　excimer laser
准分子【物】　quasi-molecule
准灰体大气　quasi-gray atmosphere
准活动中心　quasi-center of action
准极地太阳同步轨道　quasi-polar sun-synchronous orbit
准静力过程　quasi-static process
准静力近似　quasi-hydrostatic approximation
准静力系统　quasi-hydrostatic system
准静止　quasi-stationary
准静止锋【大气】　quasi-stationary front
准静止扰动　quasi-stationary perturbation
准均匀的　quasi-homogeneous
准拉格朗日〔嵌〕套网格模式　quasi-Lagrangian nested grid model（QNGM）
准拉格朗日坐标　quasi-Lagrangian coordinates
准两年振荡【大气】　quasi-biennial oscillation（QBO）
准两年周期　quasi-biennial period
准两周振荡　quasi-two weeks oscillation
准平衡【物】　quasi-equilibrium
准平衡态　almost-equilibrium state
准平原【地理】　peneplain
准确度【物】【大气】　accuracy
准确度级　order of accuracy
准三年振荡　quasi-triennial oscillation
准双周振荡　quasi-biweekly oscillation（QBWO）
准水平输送　quasi-horizontal transport
准水平运动　quasi-horizontal motion
准稳定　quasi-steady
准稳定时间序列　quasi-stationary time series
准稳定态　quasi steady state
准无辐散〔的〕【大气】　quasi-nondivergent, quasi-nondivergence
准无辐散气流　quasi-nondivergent flow
准线　directrix
准线性波列　quasi-linear wave train
准线状对流系统　quasi-linear convective system
AIC准则【数】　Akaike information criterion（AIC）
准正压流体　quasi-barotropic fluid
准直波束　collimated beam
准直强度　collimated intensity
准直望远镜　auto-collimator

准直仪　collimator
准周期　almost-period
准周期性【大气】　quasi-periodic
桌布云　tablecloth
桌状冰山　table iceberg
浊度【物】　turbidity
浊度测定法　nephelometry
浊度计【机械】　turbidimeter
浊度系数　turbidity coefficient
着陆〔天气〕预报【大气】　landing 〔weather〕forecast
仔冰　calved ice
咨询　consultation
咨询委员会　advisory committee
咨询性预报　advisory forecast
咨询〔预报〕区　advisory area
姿态测量传感器　attitude measurement sensor
姿态角【航海】　attitude angle
姿态精度　attitude precision
姿态控制　attitude control
姿态确定　attitude determination
姿态水平传感器　attitude horizon sensor
姿态误差　attitude error
资料　data（datum 的复数形式）
资料冲击　data impact
资料处理　data handling
资料处理【计】　data processing
资料处理中心　Data Processing Centre
资料传送　data transmission
资料分析加工中心　Data Processing Analysis Facility
资料分析中心　data analysis center
资料覆盖范围　data coverage
资料格式　data format
资料管理系统　data management system（DMS）
资料管理中心　Data Management Center（DMC）
资料记录〔输出〕器　datalogger

资料监测　data monitoring
资料监测系统　data monitoring system
资料检验　data checking
资料接收中心　data acquisition facility
资料利用站　data utilization station
资料量　data volume
资料判读　data interpretation
资料融合　data fusion
资料收集　data acquisition
资料收集和定位系统　data collection and location system（DCLS）
资料收集和观测平台定位系统　data collection and platform location system（DCPLS）
资料收集平台【大气】　data collection platform（DCP）
资料收集系统【大气】　data collection system
资料收集站网　data gathering network
资料剔除　data rejection
资料提取　data extraction
资料同化【大气】　data assimilation
资料同化模式　data assimilation model
资料同化频率　data assimilation frequency
资料同化系统　data assimilation system
资料同化周期　data assimilation cycle
资料稀少地区　data-void area
资料稀少区域　sparse data area
资料系统试验　data system test
资料选择　data selection
资料压缩　data compression
资料要求　data requirement
资料自动编辑转接系统　automatic data editing and switching system（ADESS）
资源共享　resource-sharing
资源信息系统　resource information system
子波【物】　wavelet
子程序【计】　subprogram
子弹花瓣状〔冰晶〕　bullet rosettes
子〔环流〕胞　daughter cell

子集【数】 subset
子矩阵【数】 submatrix
子流形【数】 submanifold
子模型【数】 submodel
子区域 subrange
子体云 daughter cloud
子图像 subimage
子网度【计】 degree of subnet
子午面 meridian plane
子午圈 meridian circle
子午线【天】 meridian
子系统【数】 subsystem
子夜 midnight
紫光 purple light
紫辉 purple glow
紫〔辉光〕斑 purple spot
紫外差分吸收光〔雷〕达 ultraviolet differential absorption lidar(UV-DIAL)
紫外分光辐照计 UV spectroirradiometer(UVS)
紫外分光光度表 spectrophotometric dosimeter
紫外辐射 ultraviolet radiation
紫外辐射表 ultraviolet radiation meter
紫外辐射后向散射法【大气】 backscatter ultraviolet technique
紫外辐射仪 UV radiometer
紫外光 ultraviolet light
紫外光度测定法 ultraviolet photometry
紫外光谱 ultraviolet spectrum
紫外光谱仪 ultraviolet spectrometer
紫外和红外湿度计 UV and IR hygrometers
紫外激光器 ultraviolet laser
紫外检测仪 UV detector
紫外湿度表 UV hygrometer
紫外吸收监测仪 ultraviolet absorbance monitor
紫外线 ultraviolet (UV)
紫外线【物】 ultraviolet ray
紫外线表 UV dosimeter

紫外〔线〕遥感 ultraviolet remote sensing
紫外线指数 ultraviolet index (UVI)
紫外灾难 ultraviolet catastrophe
自伴算子【数】 self-adjoint operator
自变量 argument
自变量【数】 independent variable
自标定 self-calibrating
自〔乘〕谱 autospectrum
自动报警装置 auto-alarm
自动编码 autocode
自动标准地磁观测台 automatic standard magnetic observatory
自动标准地磁遥测台 automatic standard magnetic observatory-remote
自动测试设备【电子】 automatic test equipment
自动测向器(测天电方向用) automatic direction-finder
自动测站 automatic station
自动程序控制 automatic program control
自动程序设计【计】 automatic programming
自动传递 automatic transmission
〔自动〕滴谱仪 drop-size spectrometer
自动电导率分析仪 automatic conductivity analyzer
自动对流 autoconvection
自动对流不稳定度 autoconvective instability
自动对流梯度 autoconvection gradient
自动对流直减率 autoconvective lapse rate
自动多频电离层记录器 automatic multifrequency ionospheric recorder
自动发射 automotive emission
自动分析 automated analysis
自动分析仪 auto-analyzer
自动分页【计】 automatic paging
自动浮粒分析仪 automatic suspended solids analyzer

自动跟踪　autotrack, automatic tracking
自动跟踪控制　automatic following control
〔自动〕海洋气象浮标站　automatic meteorological oceanographic buoy
自动恒温箱　thermautostat
自动监测仪　automonitor
自动检测　auto monitoring
自动检验【计】　automatic check
自动降水取样器　automatic precipitation sampler
自动降水收集器　automatic precipitation collector
自动控制　automatically control
自动控制扫描器　auto command scanner
自动灵敏度控制　automatic sensibility control
自动逻辑推理【计】　automated logic inference
自动频率控制【电子】　automatic frequency control
自动气候测录装置　automatic climatological recording equipment (ACRE)
自动气象观测站　automatic meteorological observing station
自动气象站　automatic weather station (AWS), robot weather station
自动气象站【大气】　automatic meteorological station
自动日照计　automatic sunshine recorder
自动洒水灭火系统　automatic sprinkler system
自动数据交换系统　automatic data exchange system
自动搜索　automated search
自动台站管理　automatic station keeping
自动天气观测系统　automatic weather observing system (AWOS)
自动天气资料加工和通信控制系统　automatic weather-data processing and communication control system (APCS)
自动天文导航　automatic celestial navigation
自动同步〔系统〕　selsyn
自动图像传输【大气】　automatic picture transmission (APT)
自动图像传输信号模拟器　APT signal simulator
自动无线电雨量计　automatic radio rain-gauge
自动误差校正装置　automatic error request equipment
自动校平装置【航海】　autolevelling assembly
自动校正　autocorrection
自动验证系统【计】　automated verification system
自动遥测气象观测系统　remote automatic meteorological observing system (RAMOS)
自动音量控制【电子】　automatic volume control
自动远距离跟踪　automatic range tracking (A. R. T.)
自动云量记录器　automatic cloud amount recorder
自动增益控制【电子】　automatic gain control
自动蒸发皿　automatic evaporation pan (AUTOVAP)
自动蒸发站　automatic evaporation station
自〔动〕正压大气　autobarotropic atmosphere
自〔动〕正压的　autobarotropic
自〔动〕正压状态　autobarotropy
自动转化(小水滴成对并合)　auto-conversion
自动装置　robot device, automat(on)
自动资料处理【大气】　automatic data processing

自发电风速表 self-generating anemometer
自发冻结 spontaneous freezing
自发对流 spontaneous convection
自发发射【物】 spontaneous emission
自发核化 spontaneous nucleation
自发凝华 spontaneous sublimation
自发凝结 spontaneous condensation
自发性振荡 self-exited oscillation
自反函数【数】 self-reciprocal function
自回归法 autoregression method
自回归过程 autoregressive process
自回归滑动平均 auto-regressive moving average (ARMA)
自回归滑动平均模型【数】【大气】 autoregressive moving-average model
自回归积分滑动平均 auto-regressive integrated moving average (ARIMA)
自回归级数 autoregressive series
自回归模式 autoregressive model
自回归模型【数】【大气】 autoregression model
自计风标 registering weather vane, registering wind
自计雨量计【水利】 rainfall recorder
自记笔 recording pen, recording pointer
自记测风器【大气】 wind-recorder
自记底层海流计 direct-recording bottom current meter
自记电位计 recording potentiometer
自记反照率表 recording albedometer
自记风速计 self-registering anemometer
〔自记〕风速计 recording anemometer
自记干湿计 recording psychrometer
自记记录【大气】 autographic records
自记经纬仪 recording theodolite
〔自记〕冷却计 recording frigorimeter
自记能见度仪 recording visibility meter
自记气压计 self-registering barometer, self-recording barometer
〔自记〕气压计 recording barometer

自记器 automatic recorder, self-recorder
自记设备 recording equipment
自记湿度计 self-recording hygrometer
自记微雨量计 recording ombrometer
自记温度计 self-registering thermometer
自记无线电气象仪 automatic radiometeorograph
自记仪器 autographic instrument, recording instrument
自记雨量计 recording pluviometer, udomograph, self-recording rain gauge, registering rain gauge
自记云幂仪 recording ceilometer
自记蒸发皿 recording pan
自记钟 recording clock
自加宽 self-broadening
自净【化】 self-cleaning
自净过程 self-cleaning process
自流地下水 artesian groundwater
自流井 artesian well
自流水〔含水层〕 artesian aquifer
自流水盆地 artesian basin
自碰并 self-collection
自洽性【物】 self-consistency
自清能力 self-scavenging ability
自然雹胚 natural hail embryo
自然保护区 natural reserve
自然本底 natural background
自然边界条件 natural boundary condition
自然地理学【地理】 physical geography
自然地貌【航海】 natural feature
自然对流 natural convection
自然对数(以 e 为底) Napierian logarithm, natural logarithm
自然法则 natural law
自然高分子【化】 natural polymer
自然光【物】 natural light
自然光照 natural illumination
自然虹吸式雨量计 natural siphon rain-gauge
自然环境 physical environment

自然环境【地理】 natural environment
〔自然〕火险(因天气而发生的火灾危险) fire hazard
自然季节【农】 natural season
自然季节现象 natural seasonal phenomenon
自然加宽 natural broadening
自然景观【地理】 natural landscape
自然科学 natural science
自然科学家 natural scientist
自然控制 natural control
自然历【地理】 natural calendar
自然硫循环 natural sulfur cycle
自然频率 natural frequency
自然释放【地质】 natural release
自然数【数】 natural number
自然天气季节【大气】 natural synoptic season
自然天气区【大气】 natural synoptic region
自然天气周期【大气】 natural synoptic period
自然通风干湿表 naturally ventilated psychrometer
自然通风干燥【农】 natural-draftdrying
自然线宽【电子】 natural linewidth
自然选择【农】 natural selection
自然源 natural source
自然灾害 natural disaster, natural calamity
自然灾害【地理】 natural hazard
自然正交函数 natural orthogonal function
自然资源【地理】 natural resources
自然资源保护【地理】 conservation of natural resources
自然综合体【地理】 natural complex
自然坐标〔系〕【大气】 natural coordinates
自然植被 natural vegetation
自热 self-heating

自热式露点传感器 self-heating dew-point sensor
自〔身〕加强 self-development
自生混沌 self-generated chaos
自〔生〕起电 autogenous electrification
〔自〕适应观测 adaptive observations
〔自〕适应观测网站 adaptive observational network
自适应光学 adaptive optics
自适应控制 adaptive control
自适应雷达【电子】 adaptive radar
自适应滤波 adaptive filtering
自适应神经网络 adaptive neural network
〔自〕适应网格 adaptive grid
自适应遥测【电子】 adaptive telemetry
自相干【物】 self coherence
自相关 self correlation
自相关 autocorrelation【大气】
自相关函数【数】 autocorrelation function
自相关谱 autocorrelation spectrum
自相关图 autocorrelogram
自相关系数 autocorrelation coefficient
自协方差 auto-covariance
自协方差谱 autocovariance spectrum
自旋【物】 spin
自旋标记【生化】 spin labeling
自旋量子数【物】 spin quantum number
自旋扫描摄云机 spin scan cloud camera (SSCC)
自旋体【物】 spin wave
自旋稳定 spin stabilization
自旋线圈磁强计 spinning coil magnetometer
自旋轴 spin axis
自旋-转动带 spin-rotation band
自由表面近似【物】 free surface approximation
自由波【大气】 free wave
自由〔波〕模 free mode
自由程【物】 free path

自由磁流振荡　free hydromagnetic oscillation
自由大气【大气】　free atmosphere
自由大气中 CO_2 浓度富集　Free-Air CO_2 Enrichmeht（FACE）
自由低频变异度　free low frequency variability
自由电子【物】　free electron
自由度【物】【大气】　degree of freedom
自由对流【大气】　free convection
自由对流标定速度　free convection scaling velocity
自由对流尺度　free convection scaling
自由对流高度【大气】　level of free convection（LFC）
自由对流高度【大气】　free convection level（FCL）
自由风　free wind
自由浮动式系留浮标　free floating moored buoy
自由高空气球　free aerostat
自由〔光〕谱区　free spectral range
自由含水层　water table aquifer
自由滑动边界条件　free-slip boundary condition
自由基【化】　free radical
自由空间【物】　free space
自由空气　free air
自由空气异常【地】　free air anomaly
自由流线　free streamline
自由罗斯贝波　free Rossby wave
自由落体加速度　free fall acceleration
自由面　free surface
自由面条件　free-surface condition
自由膨胀　free expansion
自由气流马赫数　free-stream Mach number
自由气球　free balloon
自由水　free water
自由水含量　free-water content
自由水舌　free nappe

自由湍流　free turbulence
自由下沉高度　level of free sink（LFS）
自由振荡　free oscillation
自由周期　free period
自愿援助计划（WWW）　Voluntary Assistance Programme（VAP，WWW）
自振铷蒸气磁强计　self-oscillating Rb-vapor magnetometer
自治系统【计】　autonomous system
自组织　self-organization
渍害【大气】　wet damage
字符串【计】　character string
字符每秒【计】　characters per second
字节【计】　byte
字码　code word
字母代码　alphabetic code
字母数字资料　alphanumeric data
字母显示器　alphascope
字体【计】　character style
宗谷暖流　Soya Warm water
宗谷暖洋流　Soya Warm Current
综合垂直剖面图　composite vertical cross-section
综合反射率〔因子〕　composite reflectivity
综合分析　integrated analysis
综合观测系统　composite observational system
综合监测网　integrated monitoring network
综合考察【地理】　integrated survey
综合评估　integrated assessment
综合气候学　synthetic climatology，complex climatology
综合水文图　synthetic hydrograph
综合水样〔品〕　composite water sample
综合探测系统　integrated sounding system（ISS）
综合天气观测　synoptic weather observation
综合〔天气〕预报图　composite forecast chart

综合图　composite map
综合〔形势〕预报图　composite prognostic chart
综合业务数字网【计】　integrated service digital network (ISDN)
综合治理　comprehensive treatment
综合自然地理学【地理】　integrated physical geography
综合自然区划【地理】　integrated physicogeographical regionalization
棕霾　brown haze
棕色尘雾　brown fume
棕雪　brown snow
棕云　brown cloud
鬃积雨云【大气】　cumulonimbus capillatus, Cb cap
鬃状（云）　capillatus (cap)
总残留量　total residue
总尺寸　overall dimension
总初级生产〔能〕力　gross primary production
总电导率　total conductivity
总发射〔功〕率　total emissive power
总反应【化】　overall reaction
总方差　total variance
总辐射【大气】　global radiation
总辐射表【大气】　pyranometer
总辐射和太阳漫射辐射积分器　global and diffuse solar radiation integrator
总辐射平均强度表　lucimeter
总辐射平均强度表指数　heliometric index
总光合强度　gross photosynthetic intensity
总含水量　total water content
总含水量测量仪【大气】　total water content instrument (TWCI)
总和原则　summation principle
总活性氮　total reactive nitrogen, NO$_y$
总角动量【大气】　absolute angular momentum
总截留损失〔量〕　gross interception loss
总截面【物】　total cross section

总露点　total dew point
总能检测器　total energy detector (TED)
总能量　total energy
总能量方程　total energy equation
总能头线　total head line
总强度　total intensity
总日射自记曲线　solarigram
总散射截面　total scattering cross section
总散射系数　total scattering coefficient
总势能　total potential energy
总衰减系数　total attenuation coefficient
总水分混合比　total-water mixing ratio
总水量　water inventory
总水头　total head
总酸度　total acidity
总太阳能监测仪　total solar energy monitor (TSEM)
总碳〔量〕【大气】　total carbon
总梯度流　total gradient current
总温度　total temperature
总吸收　integrated absorption
总线【计】　bus
总线网【计】　bus network
总悬浮粒子【数】　total suspended particulates (TSP)
总有机物　total organic matter (TOM)
总云量【大气】　total cloud cover
总蒸发　total evaporation
总直接辐射　total direct radiation
纵波【大气】　longitudinal wave
纵断面　longitudinal section
纵横比【物】　aspect ratio
纵向风　longitudinal wind
纵〔向〕滚涡　longitudinal rolls, longitudinal roll vortices
纵向速度谱　spectrum of longitudinal velocity
纵振动　longitudinal vibration
纵坐标　ordinate

走时　travel time
走时曲线　time curve, time front
足迹模拟　footprint modeling
阻碍学说　barrier theory
阻挡层　barrier layer, detaining layer
阻抗匹配【物】　impedance matching
阻力　resistance
阻力板　drag plate
阻力加速〔度〕　drag acceleration
阻力理论　drag theory
阻尼【物】　damping
阻尼比　damping ratio
阻尼波　damped wave
阻尼程序　damping procedure
阻尼风速表　drag anemometer
阻尼辐射【天】　damping radiation
阻尼格式　damping scheme
阻尼〔固有〕频率　damped〔natural〕frequency
阻尼碰撞【物】　damped collision
阻尼球风速表　drag sphere anemometer
阻尼系数【电工】　damping coefficient
阻尼项　damping term
阻尼因子　damping factor
阻尼振荡　damped oscillation
阻尼振动【物】　damped vibration
阻尼振子【物】　damped oscillator
阻尼作用【物】　damping action
阻塞〔高压〕　block
阻塞高压【大气】　blocking high（或anticyclone）
阻塞脊　blocking ridge
阻塞形势　blocking pattern
阻塞形势【大气】　blocking situation
阻塞作用　blocking action
阻止作用　interception
组成　composition
组合带　combination band
组合〔谱〕线【物】　combination line
组合原理　combination principle
组间方差　interclass variance

组距　class interval
组网〔天气〕雷达　network radar
组元　constituent
组装　package
钻井记录〔曲线〕　driller's log
钻井平台【航海】　drilling platform
钻石尘　diamond dust
最大冰川期　glacial maximum
最大不模糊距离【大气】　maximum unambiguous range
最大不模糊速度　maximum unambiguous velocity
最大测量深度【航海】　maximum measuring depth
最大持水量　maximum water-holding capacity
最大冻土深度【大气】　maximum depth of frozen ground
最大防护距离【大气】　maximum shelter distance
最大风高度型　maximum-wind topography
最大风速【大气】　maximum wind speed
最大风速半径　radius of maximum wind
最大风速层【大气】　maximum wind level
最大风速和风切变图　maximum-wind and shear chart
最大风压【大气】　maximum wind pressure
最大高度【航海】　maximum height
最大公因数【数】　greatest common divisor
最大回波高度　maximum echo height
最大降水带　belt of maximum precipitation
最大降水量【大气】　maximum precipitation
最大可能洪水　maximum probable flood, probable maximum flood
最大可能降水　maximum possible precipitation
最大可能日照时数　maximum possible

sunshine duration
最大扩散流〔动〕 maximum diffusion flow
最大能见度 maximum visibility
最大起升高度【航海】 maximum height of lift
最大熵方法 maximum entropy method (MEM)
最大熵谱 maximum entropy spectrum
最大熵谱分析 maximum entropy spectrum analysis
最大设计平均风速【大气】 maximum design wind speed
最大深-面-时资料 maximum depth-area-duration data
最大输送速度 maximum transport velocity
最大水汽压〔力〕 maximum vapour pressure
最大水汽张力 maximum water vapour tension
最大瞬时风速【大气】 maximum instantaneous wind speed
最大似然法 maximum likelihood method
最大似然估计【数】 maximum likelihood estimate
最大似然滤波 maximum likelihood filter
最大似然判据 maximum likelihood criterion
最大污染水平（饮用水标准） maximum contaminant level(MCL)
最大西风〔带〕 maximum zonal westerlies
最大吸湿水【农】 maximum hygroscopicity
最大吸湿量【土壤】 maximum hygroscopicity
最大协同作用 potential synergism
最大雪深 maximum depth of snow cover
最大允许浓度 maximum acceptable concentration
最大涨潮流速 flood strength

最大阵风 peak gust
最大阵风速 maximum gust speed
最大正变压 isallobaric maximum
最大值原理【数】 maximum principle
最大总降水量 maximum total precipitation content
最低本站气压 minimum station pressure
最低草温 grass minimum〔temperature〕
最低草温表 grass minimum thermometer
最低飞行高度 minimum flight altitude
最低光量 light minimum
最低光限 limen
最低能级 minimum energy state
最低能见度条件 marginal visibility condition
最低气象条件 weather minimum
最低气象条件【大气】 meteorological minimum
最低天文潮位 lowest astronomical tide
最低尾迹条件 mintra
最低温度 minimum temperature, lowest temperature
最低温度表【大气】 minimum thermometer
最低温〔度〕预报 minimum temperature prediction
最短航线 direct route
最高本站气压 maximum station pressure
最高辐射温度表 maximum radiation thermometer
最高气压 maximum pressure
最高日射温度 solar maximum temperature
最高生长温度【农】 maximum growth temperature
最高水位水尺 crest gauge
最高天文潮 highest astronomical tide
最高温度【大气】 maximum temperature, highest temperature
最高温度表【大气】 maximum thermometer

最高最低温度表　maximum and minimum thermometer
最佳逼近【数】　best approximation
最佳超松弛参数　optimum overrelaxation parameter
最佳初值化　optimum initialization
最佳飞行　optimum flight
最佳分辨率　optimum resolution(OR)
最佳观测网　optimal network
最佳观测系统　optimum observing system
最佳航速【航海】　optimum speed
最佳航线　optimum track route, optimum track line
最佳航线【大气】　optimum route
最佳路径　best track
最佳滤波　optimum filter
最佳能见度　excellent visibility
最佳拟合【数】　best fit
最佳气候　optimum climate
最佳湿度　optimum humidity
最佳数　Best number
最佳线性不变估计【数】　best linear invariant estimate
最可几大小　most frequent size
最冷期等水温线　isocryme
最冷月份平均温度等值线　isoryme
最平缓罗斯贝波模　gravest Rossby wave mode
最适温度【大气】　optimum temperature
最适温度【农】　optimal temperature
最适需水　optimum water need
最小　minimum
最小二乘逼近【数】　least squares approximation
最小二乘法　least square method
最小二乘法【数】　method of least squares
最小二乘法拟合【化】　least square fitting
最小二乘方　least square
最小二乘方过程　least-square procedure
最小二乘方极小化　least square minimization
最小二乘估计【数】　least squares estimate
最小二乘解【数】　least squares solution
最小飞行航程　aerologation
最小分辨角【物】　angle of minimum resolution
最小风时【海洋】　minimum duration
最小负变压　isallobaric minimum
最小可探测信号　minimum detectable signal(MDS)
最小年流量　minimum annual flow
最小偏向　minimum deviation
最小偏向角　angle of minimum deviation
最小显著差数【农】　least significant difference
最小相对湿度　minimum relative humidity
最小信息法　minimum information method
最小值　minimum value
最优策略　optimum strategy
最优插值法【大气】　optimum interpolation method
最优出水量　optimal yield
最优估计【化】　optimal estimate
最优化【数】　optimization
最优解【物】　optimum solution
最优决策　optimum decision
最优控制【航海】　optimum control, optimal control
最优内插〔法〕【大气】　optimum interpolation
最优内插质量控制　optimum interpolation quality control
最优气候值,最优气候均态　optimal climate normals
最优扰动　optimal perturbation
最优值【化】　optimal value
最终地磁参考场　Definitive Geomagnetic

Reference Field (DGRF)
最终预报误差　final prediction error (FPE)
最终增温　final warming
左手〔旋〕坐标　left-handed coordinates
左手坐标系　left-handed coordinate system
作物病害环境监测器　crop disease environment monitor(CDEM)
作物历　crop calendar
作物气候　crop climate
作物气候界限　climatic limit of crops
作物气候生态型　crop climatic ecotype
作物气候适应性　crop climatic adaptation
作物气象【大气】　meteorology of crops
作物生理〔学〕【农】　plantphysiology
作物水分指数　crop moisture index
作物天气　crop-weather
作物-天气分析统计模式　statistically based crop-weather analysis model
作物-天气模式　crop-weather model
作物系数　crop coefficient
作物形态〔学〕【农】　plant morphology
作物需水量【大气】　crop water requirement
〔作物〕需水临界期【大气】　critical period of〔crop〕water requirement

作物遗传〔学〕【农】　plant genetics
作物育种〔学〕【农】　plant breeding
作物预测　crop forecast
作业风险　operational risk
作业中心　operation center
作用量谱密度　action spectral density
作战气象条件　operational meteorological condition
坐标　coordinate
坐标变换【数】　coordinate transformation
坐标差　difference coordinate
z 坐标【大气】　z-coordinate
P 坐标【大气】　P-coordinate
坐标平面　coordinate plane
坐标曲面　coordinate surface
坐标曲线【数】　coordinate curves
坐标系　system of coordinates
坐标系【数】　coordinate system
σ 坐标系【大气】　sigma-coordinate system
P-6 坐标系　P-6 coordinate system
θ 坐标系　theta（或 θ）coordinate system
坐标线　coordinate line
坐标原点　origin of coordinates
坐标轴【数】　axis of coordinates, coordinate axis

附录1

风力等级划分标准(蒲福风级表)

风力等级	中文名称	英文名称	风速 (m/s)	风速 (km/h)	陆 地 现 象	海面状态
0	静风	calm	0～0.2	小于1	静,烟直上	平静如镜
1	软风	light air	0.3～1.5	1～5	烟能够表示风向,但风向标不能转动	微浪
2	轻风	slight breeze	1.6～3.3	6～11	人面感觉有风,树叶有微响,风向标能够转动	小浪
3	微风	gentle breeze	3.4～5.4	12～19	树叶及微枝摆动不息,旗帜展开	小浪
4	和风	moderate breeze	5.5～7.9	20～28	能吹起地面灰尘和纸张,树的小枝微动	轻浪
5	清劲风	fresh breeze	8.0～10.7	29～38	有叶的小树枝摇摆,内陆水面有小波	中浪
6	强风	strong breeze	10.8～13.8	39～49	大树枝摆动,电线呼呼有声,举伞困难	大浪
7	疾风	near gale	13.9～17.1	50～61	全树摇动,迎风步行感觉不便	巨浪
8	大风	gale	17.2～20.7	62～74	微枝折毁,人向前行感觉阻力甚大	猛浪
9	烈风	strong gale	20.8～24.4	75～88	建筑物有损坏(烟囱顶部及屋顶瓦片移动)	狂涛
10	狂风	storm	24.5～28.4	89～102	陆上少见,可使树木拔起、建筑物损坏严重	狂涛
11	暴风	violent storm	28.5～32.6	103～117	陆上很少,有则必有重大损毁	非凡现象
12	飓风	hurricane	32.7～36.9	118～133	陆上绝少,其摧毁力极大	非凡现象
13	飓风		37.0～41.4	134～149	陆上绝少,其摧毁力极大	非凡现象
14	飓风		41.5～46.1	150～166	陆上绝少,其摧毁力极大	非凡现象
15	飓风		46.2～50.9	167～183	陆上绝少,其摧毁力极大	非凡现象
16	飓风		51.0～56.0	184～201	陆上绝少,其摧毁力极大	非凡现象
17	飓风		56.1～61.2	202～220	陆上绝少,其摧毁力极大	非凡现象

附录 2

云的分类

云族	云属		云类		
	中文名	简写	中文名	简写	拉丁文名
低云	积云	Cu	淡积云 碎积云 浓积云	Cu hum Fc Cu cong	Cumulus humilis Fractocumulus Cumulus congestus
	积雨云	Cb	秃积雨云 鬃积雨云	Cb calv Cb cap	Cumulonimbus calvus Cumulonimbus capillatus
	层积云	Sc	透光层积云 蔽光层积云 积云性层积云 堡状层积云 荚状层积云	Sc tra Sc op Sc cug Sc cast Sc lent	Stratocumulus translucidus Stratocumulus opacus Stratocumulus cumulogenitus Stratocumulus castellanus Stratocumulus lenticularis
	层云	St	层云 碎层云	St Fs	Stratus Fractostratus
	雨层云	Ns	雨层云 碎雨云	Ns Fn	Nimbostratus Fractonimbus
中云	高层云	As	透光高层云 蔽光高层云	As tra As op	Altostratus translucidus Altostratus opacus
	高积云	Ac	透光高积云 蔽光高积云 荚状高积云 积云性高积云 絮状高积云 堡状高积云	Ac tra Ac op Ac lent Ac cug Ac flo Ac cas	Altocumulus translucidus Altocumulus opacus Altocumulus lenticularis Altocumulus cumulogenitus Altocumulus floccus Altocumulus castellanus
高云	卷云	Ci	毛卷云 密卷云 伪卷云 钩卷云	Ci fil Ci dens Ci not Ci unc	Cirrus filosus Cirrus densus Cirrus nothus Cirrus uncinus
	卷层云	Cs	毛卷层云 薄幕卷层云	Cs fil Cs nebu	Cirrostratus filosus Cirrostratus nebulosus
	卷积云	Cc	卷积云	Cc	Cirrocumulus

注：表中拉丁文名及简写采用全国科学技术名词审定委员会公布的大气科学名词。

附录3

二十四节气英文译名

中文名	英文名
二十四节气	twenty-four solar terms
立春	Beginning of Spring, Spring Beginning
雨水	Rain Water
惊蛰	Awakening from Hibernation
春分	Vernal Equinox, Spring Equinox
清明	Fresh Green
谷雨	Grain Rain
立夏	Beginning of Summer
小满	Lesser Fullness
芒种	Grain in Ear
夏至	Summer Solstice
小暑	Lesser Heat
大暑	Greater Heat
立秋	Beginning of Autumn
处暑	End of Heat
白露	White Dew
秋分	Autumnal Equinox
寒露	Cold Dew
霜降	First Frost
立冬	Beginning of Winter
小雪	Light Snow
大雪	Heavy Snow
冬至	Winter Solstice
小寒	Lesser Cold
大寒	Greater Cold

注:表中英文名采用全国科学技术名词审定委员会公布的大气科学、天文学和农学名词。

附录 4.1

世界气象组织/亚太经社理事会(WMO/ESCAP)台风委员会关于西北太平洋和南海热带气旋的命名方案*

1 目标

(1)台风委员会命名表将用于国际媒体以及向国际航空和航海发布的公报中。也用于台风委员会成员向国际社会发布的公报中。

(2)供各成员用当地语言发布热带气旋警报时使用(如果希望使用命名)。这将有助于人们对逐渐接近的热带气旋提高警觉,增加警报的效用。各成员仍将继续使用热带气旋编号。

2 命名方法

(1)在西北太平洋和南海地区采用一套热带气旋命名表。

(2)邀请台风委员会所有成员以及该区域 WMO 的有关成员贡献热带气旋名字。

(3)每个有关的成员贡献等量的热带气旋名字;命名表按顺序命名,循环使用;命名表共有五列,每列分两组,每组里的名字按每个成员的字母顺序依次排列。

3 名字的选择

(1)命名原则:每个名字不超过 9 个字母,容易发音,在各成员语言中没有不好的意义,不会给各成员带来任何困难,不是商业机构的名字。

(2)选取的名字应得到全体成员的认可(一票否决)。

4 命名表

(1)台风委员会通过的西北太平洋和南海热带气旋命名表共有 140 个名字,分别来自柬埔寨、中国、朝鲜、中国香港、日本、老挝、中国澳门、马来西亚、密克罗尼西亚联邦、菲律宾、韩国、泰国、美国和越南(各贡献 10 个)。

(2)各成员可以根据发音或意义将命名表翻译成当地语言。

5 命名的业务程序

(1)区域专业气象中心——东京台风中心负责按照台风委员会确定的命名表在给达到热带风暴及其以上强度的热带气旋编号的同时命名,按热带气旋命名、编号(加括号)的次序排列。国际民航组织(ICAO)东京热带气旋咨询中心以及中国和日

---

* 1998 年 12 月 1—7 日在菲律宾马尼拉召开的台风委员会第 31 届会议上通过。原 140 个热带气旋命名表因改动较大,未列;现附录 4.2 所列 140 个命名为 2012 年最新通过的命名表。

本全球海上遇险安全系统(GMDSS)Ⅺ海区气象广播发布的公报也采用相同的命名和编号。

(2)鼓励各成员尽可能多地交换观测资料,确保区域专业气象中心——东京台风中心能得到最好的资料和信息以完成任务。

(3)热带气旋名字按预先确定的次序依次命名。热带气旋在其整个生命史中保持名字不变。为避免混乱,对通过国际日期变更线进入西北太平洋的热带气旋,东京台风中心只给编号不给新命名,即:维持原有命名不变。负责给北太平洋中部热带气旋命名的美国中太平洋飓风中心也同意对从西向东越过国际日期变更线的热带气旋维持东京台风中心的命名。

(4)台风委员会所有成员在向国际社会(包括媒体、航空、航海)发布警报公报时都将使用东京台风中心分配的命名和编号。

(5)对造成特别严重灾害的热带气旋,台风委员会成员可以申请将该热带气旋使用的名字从命名表中删去(永久命名),也可以因为其他原因申请删除名字。每年的台风委员会届会将审议台风命名表。

6 执行计划

热带气旋命名表及其相关的业务程序从2000年1月1日开始执行。

附录 4.2

西北太平洋和南海热带气旋命名表*

（自 2012 年 3 月 1 日起执行）

第 1 列		第 2 列		第 3 列		第 4 列		第 5 列		名字来源
英文名	中文名	英文名	中文名	英文名	中文名	英文名	中文名	英文名	中文名	
Damrey	达维	Kong-rey	康妮	Nakri	娜基莉	Krovanh	科罗旺	Sarika	莎莉嘉	柬埔寨
Haikui	海葵	Yutu	玉兔	Fengshen	风神	Dujuan	杜鹃	Haima	海马	中国
Kirogi	鸿雁	Toraji	桃芝	Kalmaegi	海鸥	Mujigae	彩虹	Meari	米雷	朝鲜
Kai-tak	启德	Man-yi	万宜	Fung-wong	凤凰	Choi-wan	彩云	Ma-on	马鞍	中国香港
Tembin	天秤	Usagi	天兔	Kammuri	北冕	Koppu	巨爵	Tokage	蝎虎	日本
Bolaven	布拉万	Pabuk	帕布	Phanfone	巴蓬	Champi	蔷琵	Nock-ten	洛坦	老挝
Sanba	三巴	Wutip	蝴蝶	Vongfong	黄蜂	In-Fa	烟花	Muifa	梅花	中国澳门
Jelawat	杰拉华	Sepat	圣帕	Nuri	鹦鹉	Melor	茉莉	Merbok	苗柏	马来西亚
Ewiniar	艾云尼	Fitow	菲特	Sinlaku	森拉克	Nepartak	尼伯特	Nanmadol	南玛都	密克罗尼西亚
Maliksi	马力斯	Danas	丹娜丝	Hagupit	黑格比	Lupit	卢碧	Talas	塔拉斯	菲律宾
Gaemi	格美	Nari	百合	Jangmi	蔷薇	Mirinae	银河	Noru	奥鹿	韩国
Prapiroon	派比安	Wipha	韦帕	Mekkhala	米克拉	Nida	妮妲	Kulap	玫瑰	泰国
Maria	玛利亚	Francisco	范斯高	Higos	海高斯	Omais	奥麦斯	Roke	洛克	美国
Son-Tinh	山神	Lekima	利奇马	Bavi	巴威	Conson	康森	Sonca	桑卡	越南
Bopha	宝霞	Krosa	罗莎	Maysak	美莎克	Chanthu	灿都	Nesat	纳沙	柬埔寨
Wukong	悟空	Haiyan	海燕	Haishen	海神	Dianmu	电母	Haitang	海棠	中国
Sonamu	清松	Podul	杨柳	Noul	红霞	Mindulle	蒲公英	Nalgae	尼格	朝鲜
Shanshan	珊珊	Lingling	玲玲	Dolphin	白海豚	Lionrock	狮子山	Banyan	榕树	中国香港
Yagi	摩羯	Kajiki	剑鱼	Kujira	鲸鱼	Kompasu	圆规	Washi	天鹰	日本
Leepi	丽琵	Faxai	法茜	Chan-hom	灿鸿	Namtheun	南川	Pakhar	帕卡	老挝
Bebinca	贝碧嘉	Peipah	琵琶	Linfa	莲花	Malou	玛瑙	Sanvu	珊瑚	中国澳门
Rumbia	温比亚	Tapah	塔巴	Nangka	浪卡	Meranti	莫兰蒂	Mawar	玛娃	马来西亚
Soulik	苏力	Mitag	米娜	Soudelor	苏迪罗	Rai	雷伊	Guchol	古超	密克罗尼西亚
Cimaron	西马仑	Hagibis	海贝思	Molave	莫拉菲	Malakas	马勒卡	Talim	泰利	菲律宾
Jebi	飞燕	Neoguri	浣熊	Goni	天鹅	Megi	鲇鱼	Doksuri	杜苏芮	韩国
Mangkhut	山竹	Rammasun	威马逊	Atsani	艾莎尼	Chaba	暹芭	Khanun	卡努	泰国
Utor	尤特	Matmo	麦德姆	Etau	艾涛	Aere	艾利	Vicente	韦森特	美国
Trami	潭美	Halong	夏浪	Vamco	环高	Songda	桑达	Saola	苏拉	越南

* 根据 2012 年 2 月 6—11 日在浙江杭州举行的 ESCAP/WMO 台风委员会第 44 届会议的决定，"RAI"取代"凡亚比"（FANAPI）成为台风命名表中的新成员，经与香港天文台、澳门地球物理暨气象局和我国台湾地区气象部门协商，一致同意"RAI"的中文译名为"雷伊"。

附录 5

热带气旋等级*

(2006 年 6 月 15 日实施)

1 范围

本标准规定了我国预报责任区内热带气旋的等级及其划分原则。

本标准适用于我国预报责任区内热带气旋的业务和科学研究。有关热带气旋的业务规定可参照本标准执行。

2 术语和定义

下列术语和定义适用于本标准。

2.1 热带气旋 tropical cyclone

生成于热带或副热带洋面上,具有有组织的对流和确定的气旋性环流的非锋面性涡旋的统称,包括热带低压、热带风暴、强热带风暴、台风、强台风和超强台风。

2.2 风力等级 wind scale

根据风对地面(或海面)物体影响程度而定出的等级,用来估计风速的大小。

注:常用的风力等级系英国人蒲福(Beaufort)于 1805 年拟定,故又称"蒲福风力等级(Beaufort scale)",自 0~12 共分 13 个等级。自 1946 年以来,风力等级又作了扩充,增加到 18 个等级(0~17 级)。

2.3 海平面气压 sea-level pressure

由本站气压推算到平均海平面高度上的气压值。

2.4 平均风速 mean wind speed

在给定的某一时间内风速的平均值。

注:平均风速是风速的一种统计量。在观测规范中,以正点前 2 min 至正点内的平均风速作为该正点的风速。

2.5 热带气旋强度 tropical cyclone intensity

热带气旋底层(近地面或海面,下同)中心附近的最大平均风速或最低海平面气压。

2.6 预报责任区 responsible forecasting area

---

* 根据中国气象局关于实施《热带气旋等级》国家标准(GBT 19201—2006)的通知。

各级气象台站按服务责任或行政责任区划规定而制作、发布热带气旋预报和警报的区域。

注：我国预报责任区指 105°E～180°E、赤道以北的区域。

2.7 最大风力 maximum wind

在给定的某一时段内或某一期间内热带气旋底层中心附近所出现的平均风速的最大值。

注：最大风力通常以风级表示。

3 缩略语

下列缩略语适用于本标准。

STS 强热带风暴(severe tropical storm)。

STY 强台风(severe typhoon)。

Super TY 超强台风(super typhoon)。

TC 热带气旋(tropical cyclone)。

TD 热带低压(tropical depression)。

TS 热带风暴(tropical storm)。

TY 台风(typhoon)。

4 热带气旋的等级

4.1 热带气旋等级划分的原则

热带气旋等级的划分以其底层中心附近最大平均风速为标准。

4.2 热带气旋等级划分

热带气旋分为热带低压、热带风暴、强热带风暴、台风、强台风和超强台风 6 个等级。

热带气旋等级划分表

热带气旋等级	底层中心附近最大平均风速(m/s)	底层中心附近最大风力(级)
热带低压(TD)	10.8～17.1	6～7
热带风暴(TS)	17.2～24.4	8～9
强热带风暴(STS)	24.5～32.6	10～11
台风(TY)	32.7～41.4	12～13
强台风(STY)	41.5～50.9	14～15
超强台风(Super TY)	$\geqslant 51.0$	16 或以上

附录 6

常见的与大气科学有关的英文版 SCI 期刊

序号	期刊全称(缩写)	中译刊名	主办者或出版单位	影响因子*
1	Acta Meteorologica Sinia	气象学报	中国气象学会	0.874
2	Advances in Atmospheric Sciences (Adv Atmos Sci)	大气科学进展	IAMAS 中国委员会及中科院大气物理所	0.691
3	Advances in Space Research (Adv Space Res)	空间研究进展	Elsevier(爱思维尔出版公司)	(0.548) 1.079
4	Agricultural and Forest Meteorology (Agr Forest Meteorol)	农业和林业气象学	Elsevier	(2.811) 3.197
5	Annales Geophysicae (Ann Geophys-Germany)	地球物理学纪录	EGU(欧洲地球物理学协会)	(1.610) 1.648
6	Atmospheric Chemistry and Physics (Atmos Chem Phys)	大气化学和物理	EGU	(2.670) 4.881
7	Atmospheric Environment (Atmos Environ)	大气环境	Elsevier	(2.562) 3.139
8	Atmosphere-Ocean (Atmos Ocean)	大气和海洋	加拿大气象和海洋学会	(1.021) 1.000
9	Atmospheric Research (Atmos Res)	大气研究	Elsevier	(0.863) 1.811
10	Australian Meteorological Magazine (Aust Meteorol Mag)	澳大利亚气象杂志	澳大利亚气象和海洋学会	(0.529) 1.143

* 表中影响因子(impact factor)带括号者为 2004 年值,未加括号者为 2009 年值,取自 Journal Citation Report (JCR),可作为期刊影响力的一个参考指标;近几年来,许多期刊的影响因子又有提升,读者可自行查阅 JCR 等资料。

续表

序号	期刊全称(缩写)	中译刊名	主办者或出版单位	影响因子
11	Bulletin of the American Meteorological Society (B Am Meteorol Soc)	美国气象学会通报	AMS(美国气象学会)	(2.605) 3.123
12	Boundary-Layer Meteorology (Bound-Lay Meteorol)	边界层气象学	Springer(施普林格出版社)	(1.988) 2.127
13	Chinese Science Bulletin (Chinese Sci Bull)	中国科学通报	中国科学出版社	(0.683) 0.898
14	Climate Dynamics (Clim Dynam)	气候动力学	Springer	(3.497) 3.917
15	Climate of the Past (Clim Past)	历史气候	EGU	3.826
16	Climate Research (Climate Res)	气候研究	Inter-Research(国际研究出版社,德国)	(1.575) 2.25
17	Climatic Change (Climatic Change)	气候变化	Springer	(2.035) 3.635
18	Dynamics of Atmospheres and Oceans (Dynam Atmos Oceans)	大气和海洋动力学	Elsevier	(1.116) 1.788
19	Earth-Science Reviews (Earth-Sci Rev)	地球科学评论	Elsevier	(4.543) 6.942
20	Geophysical Research letters (Geophy Res Lett)	地球物理研究通信	AGU(美国地球物理协会)	(2.378) 3.204
21	Global Biogeochemical Cycles(Global Biogeochem Cy)	全球生物地球化学循环	AGU	(2.864) 4.294
22	IEEE Transactions on Geosciences and Remote Sensing (T-GRS) (IEEE Trans Geosci RemSens)	电气和电子工程师协会地球科学和遥感学报	电气和电子工程师协会(IEEE)	(2.185) 2.34

续表

序号	期刊全称(缩写)	中译刊名	主办者或出版单位	影响因子
23	International Journal of Biometeorology (Int J Biometeorol)	国际生物气象杂志	John Wiley and Sons（约翰·威利父子出版社）	(1.275) 1.84
24	International Journal of Climatology (Int J Climatol)	国际气候学杂志	John Wiley and Sons	(1.658) 2.347
25	Journal of Aerosol Science (J Aerosol Sci)	气溶胶科学杂志	Elsevier	(1.861) 2.529
26	Journal of Applied Meteorology and Climatology (J Appl Meteorol Climatol)	应用气象和气候学杂志（原应用气象杂志）	AMS	(1.472) 1.894
27	Journal of the Air and Waste Management Association (J Air Waste Manage)	空气和废物管理协会会刊	国际空气和废物管理协会	(1.357) 1.67
28	Journal of Atmospheric Chemistry (J Atmos Chem)	大气化学杂志	Springer	(2.046) 1.427
29	Journal of Atmospheric and Oceanic Technology (J Atmos Ocean Tech)	大气和海洋技术杂志	AMS	(1.700) 1.588
30	Journal of Atmospheric and Solar-Terrestrial Physics (J Atmos Sol-Terr Phy)	大气和日地物理学杂志	Elsevier	(1.517) 1.643
31	Journal of Climate (J Climate)	气候杂志	AMS	(3.500) 3.363
32	Journal of the Atmospheric Sciences (J Atmos Sci)	大气科学杂志	AMS	(2.954) 2.911
33	Journal of Hydrometeorology (J Hydrometeorol)	水文气象杂志	AMS	(1.896) 2.739

续表

序号	期刊全称(缩写)	中译刊名	主办者或出版单位	影响因子
34	Journal of the Meteorological Society of Japan (J Meteorol Soc Jpn)	日本气象学会集志	日本气象学会	(1.286) 1.104
35	Journal of Geophysical Research (J Geophys Res)	地球物理研究杂志	AGU	(2.839) 3.082
36	Journal of Physical Oceanography (J Phys Ocean)	物理海洋学杂志	AMS	(1.893)
37	Meteorological Applications (Meteorol Appl)	气象应用	英国皇家气象学会	(0.506) 1.467
38	Meteorology and Atmospheric Physics (Meteorol Atmos Phys)	气象学和大气物理	Springer	(1.097) 0.872
39	Monthly Weather Review (Mon Wea Rev)	每月天气评论	AMS	(1.859) 2.238
40	Natural Hazards (Nat Hazards)	自然灾害	Springer	(0.362) 1.217
41	Nature(Nature)	自然	英国自然出版集团	(32.182) 34.48
42	Palaeogeography, Palaeoclimatology, Palaeoecology (Palaeogeography, Palaeoclimatology, Palaeoecology)	古地理,古气候,古生态	Elsevier	(1.766) 2.646
43	Physics and Chemistry of the Earth, Part B-Hydrology, Oceans and Atmosphere (Phys Chem Earth)	地球的物理学和化学 B 辑:水文,海洋和大气	Elsevier	(0.577) 0.975

续表

序号	期刊全称(缩写)	中译刊名	主办者或出版单位	影响因子
44	Quarterly Journal of the Royal Meteorological Society (Q J Roy Meteor Soc)	皇家气象学会季刊	英国皇家气象学会	(1.844) 2.522
45	Quaternary Science Reviews (Quaternary Sci Rev)	第四纪科学评论	Elsevier	(3.323) 4.245
46	Radio Science (Radio Sci)	无线电科学	AGU	(1.007) 1.012
47	Remote Sensing Environment (Rem Sen Environ)	环境遥感	Elsevier	(1.992) 3.943
48	Science (Science)	科学	AAAS(美国科学促进会)	(31.853) 29.747
49	Science in China Series A—Mathmatics (Sci China Ser A)	中国科学 A 辑：数学	中国科学出版社	(0.34) 0.584
50	Science in China Series B—Chemistry (Sci China Ser B)	中国科学 B 辑：化学	中国科学出版社	(0.84) 0.830
51	Science in China Series C—Life Sciences (Sci China Ser C)	中国科学 C 辑：生命科学	中国科学出版社	(0.396) 0.691
52	Science in China Series D—Earth Sciences (Sci China Ser D)	中国科学 D 辑：地球科学	中国科学出版社	(0.909) 0.880
53	Science in China Series E—Technological Sciences (Sci China Ser E)	中国科学 E 辑：技术科学	中国科学出版社	(0.376) 0.682
54	Science in China Series F—Information Sciences (Sci China Ser F)	中国科学 F 辑：信息科学	中国科学出版社	0.387

续表

序号	期刊全称(缩写)	中译刊名	主办者或出版单位	影响因子
55	Science in China Series G—Physics, Mechanics and Astronomy (Sci China ser G)	中国科学 G 辑：物理学、力学、天文学	中国科学出版社	1.040
56	Space Weather (Space Wea)	空间天气	AGU	1.845
57	Surveys in Geophysics (Surv Geophys)	地球物理学探索	Springer	(1.405) 3.179
58	Tellus Series A—Dynamic Meteorology and Oceanography	大地 A 辑：动力气象学和海洋学	Blackwell 出版社	(1.603) 2.214
59	Tellus Series B—Chemical and Physical Meteorology	大地 B 辑：化学和物理气象学	Blackwell 出版社	(1.854) 4.278
60	Terrestrial Atmospheric and Oceanic Sciences (Terr Atmos Ocean Sci)	地球大气和海洋科学	台湾地球科学联合会	0.643
61	Theoretical and Applied Climatolopy (Theor Appl Climatol)	理论和应用气候学	Springer	(0.964) 1.776
62	Water, Air, & Soil Pollution (Water Air Soil Poll)	水,空气和土壤污染	Springer	(1.058) 1.676
63	Weather and Forecasting (Wea Forecast)	天气和预报	AMS	(0.852) 1.663

附录 7

部分与大气科学有关的单位和机构的英文译名

北京大学物理学院大气与海洋科学系　Department of Atmospheric and Oceanic Sciences, School of Physics, Peking University (Beijing 100871)
北京师范大学地理学与遥感科学学院　School of Geography, Beijing Normal University (Beijing 100875)
北京师范大学全球变化与地球系统科学研究院　College of Global Change and Earth System Science, Beijing Normal University (Beijing 100875)
成都信息工程学院　Chengdu University of Information Technology (Chengdu 610225)
国家海洋局海洋环境预报中心　National Marine Environmental Forecasting Center, State Ocean Administration (Beijing 100081)
国家气候中心　National Climate Center (Beijing 100081)
国家气象信息中心　National Meteorological Information Center (Beijing 100081)
国家气象中心　National Meteorological Center (Beijing 100081)
国家卫星气象中心　National Satellite Meteorological Center (Beijing 100081)
国家遥感中心　National Remote Sensing Center of China (Beijing 100036)
国家自然科学基金委员会　National Natural Science Foundation of China (Beijing 100085)
华东师范大学资源与环境科学学院　School of Resources and Environment Science, East China Normal University (Shanghai 100062)
兰州大学大气科学学院　College of Atmospheric Sciences, Lanzhou University (Lanzhou 730000)
南京大学大气科学学院　School of Atmospheric Sciences, Nanjing University (Nanjing 210093)
南京信息工程大学　Nanjing University of Information Science and Technology (Nanjing 210044)
气象出版社　China Meteorological Press (Beijing 100081)
清华大学地球系统科学研究中心　The Center for Earth System Science, Tsinghua University (Beijing 100084)
清华大学全球变化研究院　College for Global Change Studies, Tsinghua University (Beijing 100084)
沈阳农业大学　Shenyang Agricultural University (Shenyang 110161)
云南大学资源环境与地球科学学院大气科学系　Department of Atmospheric Sciences, School of Resources, Environment and Earth Sciences, Yunnan University (Kunming 650091)
浙江大学环境与资源学院　College of Environmental and Resource Sciences, Zhejiang University (Hangzhou 310027)

中国海洋大学海洋环境学院海洋气象学系　Department of Oceanic and Meteorological Sciences, College of Physical and Environmental Oceanography, Ocean University of China (Qingdao 266003)

中国环境科学研究院　Chinese Research Academy of Environmental Sciences (Beijing 100012)

中国科学技术大学地球和空间科学学院　School of Earth and Space Sciences, University of Science and Technology of China (Hefei 230026)

中国科学院　Chinese Academy of Sciences (CAS) (Beijing 100864)

中国科学院安徽光学和精密机械研究所　Anhui Institute of Optics and Fine Mechanics, CAS (Hefei 230031)

中国科学院成都山地灾害与环境研究所　Chengdu Institute of Mountain Hazards and Environment, CAS (Chengdu 610041)

中国科学院大气物理研究所　Institute of Atmospheric Physics, CAS (Beijing 100029)

中国科学院大学　University of CAS (Beijing 100049)

中国科学院地理科学与资源研究所　Institute of Geographic Sciences and Natural Resources Research, CAS (Beijing 100101)

中国科学院地球环境研究所　Institute of Earth Environment, CAS (Xi'an 710075)

中国科学院地质与地球物理研究所　Institute of Geology and Geophysics, CAS (Beijing 100029)

中国科学院东北地理与农业生态研究所　Northeast Institute of Geography and Agricultural Ecology, CAS (Changchun 130012)

中国科学院寒区旱区环境与工程研究所　Cold and Arid Regions Environmental and Engineering Research Institute, CAS (Lanzhou 730000)

中国科学院空间科学与应用研究中心　Center for Space Science and Applied Research, CAS (Beijing 100080)

中国科学院南海海洋研究所　South China Sea Institute of Oceanology, CAS (Guangzhou 510301)

中国科学院南京地理与湖泊研究所　Nanjing Institute of Geography and Limnology, CAS (Nanjing 210008)

中国科学院青藏高原研究所　Institute of Tibetan Plateau Research, CAS (Beijing 100085)

中国科学院生态环境研究中心　Research Center for Eco-Environment Sciences, CAS (Beijing 100085)

中国科学院遥感应用研究所　Institute of Remote Sensing Application, CAS (Beijing 100101)

中国科学院植物研究所　Institute of Botany, CAS (Beijing 100093)

中国林业科学研究院　Chinese Academy of Forestry (Beijing 100091)

中国农业大学资源与环境学院　College of Resources and Environment, China Agricultural University (Beijing 100094)

中国农业科学院农业环境与可持续发展研究所　Institute of Environment and

Sustainable Development in Agriculture, The Chinese Academy of Agricultural Sciences (Beijing 100081)
中国气象报社　China Meteorological News Press (Beijing 100081)
中国气象局　China Meteorological Administration (CMA) (Beijing 100081)
　　×××省(区)气象局　×××　Provincial (Autonomous Regional) Meteorological Service
　　×××直辖市气象局　×××　Municipal Meteorological Service
　　省(区,市)气象科学研究所　Research Institute of Meteorological Science
　　地(市)气象局　Meteorological Office
　　县气象站　Weather Station
中国气象局北京城市气象研究所　Institute of Urban Meteorology, CMA (Beijing 100089)
中国气象局成都高原气象研究所　Institute of Plateau Meteorology, CMA (Chengdu 610071)
中国气象局广州热带海洋气象研究所　Guangzhou Institute of Tropical and Marine Meteorology, CMA (Guangzhou 510080)
中国气象局公共气象服务中心　CMA Public Meteorological Service Center (PMSC) (Beijing 100081)
中国气象局兰州干旱气象研究所　Institute of Arid Meteorology, CMA (Lanzhou 730020)
中国气象局气象干部培训学院　CMA Training Center (Beijing 100081)
中国气象局气象探测中心　CMA Meteorological Observation Center (MOC) (Beijing 100081)
中国气象局上海台风研究所　Shanghai Typhoon Institute, CMA (Shanghai 200030)
中国气象局沈阳大气环境研究所　Institute of Atmospheric Environment, CMA (Shengyang 110016)
中国气象局乌鲁木齐沙漠气象研究所　Institute of Desert Meteorology, CMA (Urumqi 830002)
中国气象局武汉暴雨研究所　Institute of Heavy Rain, CMA (Wuhan 430074)
中国气象科学研究院　Chinese Academy of Meteorological Sciences (Beijing 100081)
中国气象学会　The Chinese Meteorological Society (Beijing 100081)
中国人民解放军理工大学气象海洋学院　Institute of Meteorology, PLA University of Science and Technology (Nanjing 211101)
中华人民共和国科学技术部　Ministry of Science and Technology of the People's Republic of China (Beijing 100862)
中山大学环境科学与工程学院大气科学系　Department of Atmospheric Sciences, School of Environmental Science and Engineering, Sun Yat-Sen University (Guangzhou 510275)

附录8 部分与大气科学有关的国内地名（特殊拼法）的英文拼写

阿尔泰山〔新〕*	Altay Mountains
阿克苏市〔新〕	Aksu City
阿拉善盟〔蒙〕	Alxa League
阿勒泰地区〔新〕	Altay Prefecture
艾比湖〔新〕	Ebinur Lake
艾丁湖〔新〕	Aydingkol Lake
安多〔藏〕	Amdo
巴丹吉林沙漠〔蒙〕	Badain Jaran Desert
巴颜喀拉山〔青〕	Bayan Har Mountains
巴音布鲁克〔新〕	Bayanbulak
巴音郭勒〔蒙〕	Bayan Gol
班戈〔藏〕	Baingoin
博斯腾湖〔新〕	Bosten Lake
布尔津河〔新〕	Burqin River
茶卡盐湖〔青〕	Caka Salt Lake
察尔汗盐湖〔青〕	Qarhan Salt Lake
察哈尔右翼后旗〔蒙〕	Qahar Right Wing Rear Banner
柴达木盆地〔青〕	Qaidam Basin
长江	Changjiang River（或 Yangtze River）
昌都地区〔藏〕	Qamdo Prefecture
大兴安岭〔蒙,黑〕	Da Hinggan Mountains
当雄〔藏〕	Damxung
定日〔藏〕	Tingri
东海	East China Sea（或 Donghai Sea）
二连浩特〔蒙〕	Erenhot
额济纳旗〔蒙〕	Ejin Banner
额尔齐斯河〔新〕	Ertix River
鄂伦春自治旗	Oroqen Autonomous Banner
鄂尔多斯〔蒙〕	Ordos
改则〔藏〕	Gêrzê
甘孜〔川〕	Garzê
格尔木〔青〕	Golmud
哈尔滨〔黑〕	Harbin
海拉尔〔蒙〕	Hailar
和田〔新〕	Hotan
红其拉甫〔新〕	Kunjirap
呼和浩特〔蒙〕	Hohhot
呼伦贝尔盟〔蒙〕	Hulun Buir League
黄海	Huanghai Sea（或 Yellow Sea）
黄河	Huanghe River（或 Yellow River）
江孜〔藏〕	Gyangzê
喀喇昆仑山〔新〕	Karakorum Mountains
可可西里山〔青,藏〕	Hoh Xil Range
克拉玛依市〔新〕	Karamay City
科尔沁左翼中旗〔蒙〕	Horqin Left Wing Middle Banner
库车〔新〕	Kuqa
库尔勒〔新〕	Kolra
拉萨〔藏〕	Lhasa
林芝〔藏〕	Nyingchi
罗布泊〔新〕	Lop Lake
马尔康县〔川〕	Barkam County
玛多〔青〕	Madoi
玛纳斯河〔新〕	Manas River
毛乌素沙地〔蒙〕	Mau Us Desert
墨脱〔藏〕	Mêdog
墨竹工卡〔藏〕	Maizhokunggar
那曲〔藏〕	Nagqu

* 新为新疆,蒙为内蒙古,藏为西藏,余类推;表示此地名的所在区域

南海	South China Sea (或 Nanhai Sea)	翁牛特旗〔蒙〕	Ongniud Banner
		乌兰察布盟〔蒙〕	Ulanqab League
南海诸岛	South China Sea Islands	乌兰浩特〔蒙〕	Ulanhot
		乌鲁木齐〔新〕	Urümqi
念青唐古拉山〔藏〕	Nyainqêntanglha Range	希夏邦马峰〔藏〕	Mount Xixabangma
		喜马拉雅山〔藏〕	Himalayas
诺木洪〔青〕	Nomhon	锡林郭勒盟〔蒙〕	Xilin Gol League
青藏高原	Qinghai-Xizang Plateau (或 Qinghai-Tibetan Plateau 或 Tibetan Plateau)	锡林浩特〔蒙〕	Xilinhot
		小兴安岭〔黑〕	Xiao Hinggan Mountains
		雅鲁藏布江〔藏〕	Yarlung Zangbo River
日喀则〔藏〕	Xigazê		
赛里木湖〔新〕	Sayram Lake	叶尔羌河〔新〕	Yarkant River
山西省	Shanxi Province	伊犁河〔新〕	Ili River
陕西省	Shaanxi Province	玉门〔藏〕	Yumai
塔克拉玛干沙漠〔新〕	Taklimakan Desert	则克台〔新〕	Zekti
塔里木河〔新〕	Tarim River	泽当〔藏〕	Zêtang
塔里木盆地〔新〕	Tarim Basin	哲里木盟〔蒙〕	Jirem League
腾格里沙漠〔蒙〕	Tengger Desert	珠穆朗玛峰〔藏〕	Mount Qomolangma (或 Mt. Qomolangma)
吐鲁番盆地〔新〕	Turpan Depression		
土门〔藏〕	Tumain	准噶尔盆地〔新〕	Junggar Basin

附录9 美国各州的名称和首府以及它们的规定中译名

序号	州名 英文	州名 中文	英文缩写	首府 英文	首府 中文
1	Alabama	亚拉巴马	AL	Mountgomery	蒙哥马利
2	Alaska	阿拉斯加	AK	Juneau	朱诺
3	Arizona	亚利桑那	AZ	Phoenix	菲尼克斯
4	Arkansas	阿肯色	AR	Little Rock	小石城
5	California	加利福尼亚	CA	Sacramento	萨克拉门托
6	Colorado	科罗拉多	CO	Denver	丹佛
7	Connecticut	康涅狄格	CT	Hartford	哈特福德
8	Delaware	特拉华	DE	Dover	多佛
9	Florida	佛罗里达	FL	Tallahassee	塔拉哈西
10	Georgia	佐治亚	GA	Atlanta	亚特兰大
11	Hawaii	夏威夷	HI	Honolulu	火奴鲁鲁(檀香山)
12	Idaho	爱达荷	ID	Boise	博伊西
13	Illinois	伊利诺伊	IL	Springfield	斯普林菲尔德
14	Indiana	印第安纳	IN	Indianapolis	印第安纳波利斯
15	Iowa	艾奥瓦	IA	Des Moines	得梅因
16	Kansas	堪萨斯	KS	Topeka	托皮卡
17	Kentucky	肯塔基	KY	Frankfort	法兰克福
18	Louisiana	路易斯安那	LA	Baton Rouge	巴吞鲁日
19	Maine	缅因	ME	Augusta	奥古斯塔
20	Maryland	马里兰	MD	Annapolis	安纳波利斯
21	Massachusetts	马萨诸塞	MA	Boston	波士顿
22	Michigan	密歇根	MI	Lansing	兰辛
23	Minnesota	明尼苏达	MN	St. Paul	圣保罗
24	Mississippi	密西西比	MS	Jackson	杰克逊
25	Missouri	密苏里	MO	Jefferson City	杰斐逊城

续表

序号	州　名		英文缩写	首　府	
	英　文	中　文		英　文	中　文
26	Montana	蒙大拿	MT	Helena	海伦娜
27	Nebraska	内布拉斯加	NE	Lincoln	林肯
28	Nevada	内华达	NV	Carson City	卡森城
29	New Hampshire	新罕布什尔	NH	Concord	康科德
30	New Jesey	新泽西	NJ	Trenton	特伦顿
31	New Mexico	新墨西哥	NM	Santa Fe	圣菲
32	New York	纽约	NY	Albany	奥尔巴尼
33	North Carolina	北卡罗来纳	NC	Raleigh	罗利
34	North Dakota	北达科他	ND	Bismarck	俾斯麦
35	Ohio	俄亥俄	OH	Columbus	哥伦布
36	Oklahoma	俄克拉何马	OK	Oklahoma City	俄克拉何马城
37	Oregon	俄勒冈	OR	Salem	塞勒姆
38	Pennsylvania	宾夕法尼亚	PA	Harrisburg	哈里斯堡
39	Rhode Island	罗得岛	RI	Providence	普罗维登斯
40	South Carolina	南卡罗来纳	SC	Columbia	哥伦比亚
41	South Dakota	南达科他	SD	Pierre	皮尔
42	Tennessee	田纳西	TN	Nashville	纳什维尔
43	Texas	得克萨斯	TX	Austin	奥斯汀
44	Utah	犹他	UT	Salt Lake City	盐湖城
45	Vermont	佛蒙特	VT	Montpelier	蒙彼利埃
46	Virginia	弗吉尼亚	VA	Richmond	里士满
47	Washington	华盛顿	WA	Olympia	奥林匹亚
48	West Virginia	西弗吉尼亚	WV	Charleston	查尔斯顿
49	Wisconsin	威斯康星	WI	Madison	麦迪逊
50	Wyoming	怀俄明	WY	Cheyenne	夏延

主要参考文献

本书编写组. 1987. 英汉大气科学词汇. 北京:气象出版社.
国际气象辞典编译组. 1994. 中英法俄西国际气象词典. 北京:气象出版社.
全国科学技术名词审定委员会. 2009. 大气科学名词(第三版). 北京:科学出版社.
世界气象组织常用缩略词词典编译组. 2000. 世界气象组织常用缩略语词典. 北京:气象出版社.
Glickman T S et al. 2000. *Glossary of Meteorology* (Second Edition). Boston: American Meteorological Society.
Holton J R et al. 2002. *Encyclopedia of Atmospheric Sciences*. Boston, London, New York: Academic Press.